CHILTON'S GUIDE TO

FUEL INJECTION and FEEDBACK CARBURETORS

Managing Editor John H. Weise, S.A.E. □ **Assistant Managing Editor** David H. Lee, A.S.E., S.A.E.
Technical Editor Tony Molla, S.A.E.

Service Editors Nick D'Andrea, Robert McAnally,
Michael D. Powers, Jack T. Kaufmann, Lawrence C. Braun
Editorial Consultants Edward K. Shea, S.A.E., Stan Stephenson

Production Manager John Cantwell
Manager Editing & Design Dean F. Morgantini
Art & Production Coordinator Robin S. Miller
Mechanical Paste-up Supervisor Margaret A. Stoner
Mechanical Artists Bill Gaskins, Cynthia Fiore

National Sales Manager Albert M. Kushnerick □ **Assistant** Jacquelyn T. Powers
Regional Managers Joseph Andrews, Jr., James O. Callahan, David Flaherty

OFFICERS
President Lawrence A. Fornasieri
Vice President & General Manager John P. Kushnerick

CHILTON BOOK COMPANY Chilton Way, Radnor, Pa. 19089

 ISBN 0-8019-7488-7
Library of Congress Catalog Card. No. 83-45323
7890 4321098

HOW TO USE THIS MANUAL

For ease of use, this manual is divided into sections:

The **CONTENTS**, inside the front cover, summarize the subjects covered in each section.

To quickly locate the proper service section, use the chart on the following pages. It references applicable **CAR MODELS** and **SERVICE SECTIONS** for major engine performance control systems.

It is recommended that the service technician be familiar with the applicable **GENERAL INFORMATION** and **SERVICE PRECAUTIONS** (if any) before testing or servicing the system.

Major service sections are grouped by individual vehicle or component manufacturers. Each manufacturer's sub-section contains:

- ☐ **GENERAL INFORMATION** Information pertaining to the operation of the system, individual components and the overall logic by which components work together.
- ☐ **SERVICE PRECAUTIONS (if any)** Precautions of which the service technician should be aware to prevent injury or damage to the vehicle or components.
- ☐ **TESTS, ADJUSTMENTS, AND COMPONENT R & R** Performance tests and specifications, adjustments and component replacement procedures.
- ☐ **FAULT DIAGNOSIS** Complete troubleshooting for the entire system.

SAFETY NOTICE

Proper service and repair procedures are vital to the safe, reliable operation of all motor vehicles, as well as the personal safety of those performing repairs. This manual outlines procedures for servicing and repairing vehicles using safe effective methods. The procedures contain many NOTES, CAUTIONS and WARNINGS which should be followed along with standard safety procedures to eliminate the possibility of personal injury or improper service which could damage the vehicle or compromise its safety.

It is important to note that repair procedures and techniques, tools and parts for servicing motor vehicles, as well as the skill and experience of the individual performing the work vary widely. It is not possible to anticipate all of the conceivable ways or conditions under which vehicles may be serviced, or to provide cautions as to all of th epossible hazards that may result. Standard and accepted safety precautions and equipment should be used when handling toxic or flammable fluids, and safety goggles or other protection should be used during cutting, grinding chiseling, prying, or any other process that can cause material removal or projectiles.

Some procedures require the use of tools specially designed for a specific purpose. Before substituting another tool or procedure, you must be completely satisfied that neither your personal safety, nor the performance of the vehicle will be endangered.

PART NUMBERS

Part numbers listed in this reference are not recommendations by Chilton for any product by brand name. They are references that can be used with interchange manuals and aftermarket supplier catalogs to locate each brand supplier's discrete part number.

Although information in this manual is based on industry sources and is as complete as possible at the time of publication, the possibility exists that some car manufacturers made later changes which could not be included here. While striving for total accuracy, Chilton Book Company cannot assume responsibility for any errors, changes, or omissions that may occur in the compilation of this data.

CONTENTS

SECTION 1 ELECTRONICS & AUTOMOTIVE COMPUTERS

SECTION 2 DIAGNOSTIC EQUIPMENT & SPECIAL TOOLS

SECTION 3 ON-BOARD DIAGNOSTIC SYSTEMS

SECTION 4 FEEDBACK CARBURETORS

SECTION 5 FUEL INJECTION SYSTEMS

SECTION 6 GLOSSARY

COMPUTERIZED ENGINE CONTROL APPLICATIONS

Manufacturer	Year	Engine	SECTION 4 Feedback Carburetor	SECTION 5 Fuel Injection System
American Motors	1980	4 cyl	R-E2SE	—
		6 cyl	C-BBD	—
	1981	4 cyl	R-E2SE	—
		6 cyl	C-BBD	—
	1982	4 cyl	R-E2SE①	—
		6 cyl	C-BBD	—
	1983–85	4 cyl	E2SE	—
		6 cyl	BBD	—
Audi	1978–84	all	—	CIS
	1984–85	all	—	CIS
BMW	1978–82	all	—	AFC③
	1983–85	all	—	AFC③
Buick	See General Motors			
Cadillac	See General Motors			
Chevrolet	See General Motors			
Chrysler	1979–80	4 cyl	H-5220	—
		6 cyl	C-BBD	—
		8 cyl	C-TQ	—
	1981–82	4 cyl	H-6520②	—
		6 cyl	H-6145	—
		8 cyl	C-BBD	—
		8 cyl	—	EFI

AFC—Air Flow Controlled
CIS—Constant Injection System
TBI—Throttle Body Injection
C—Carter carburetor
R—Rochester carburetor
H—Holley carburetor
M—Motorcraft carburetor
EFI—Electronic Fuel Injection
CFI—Central Fuel Injection
① Carter BBD used on engine with CEC system
② Mikuni carburetor on some models
③ CIS on 320i models
④ Holley 6500 on California models
⑤ EFI on Lincoln and Mark IV. Carter thermo quad (TQ) on 4 bbl models
⑥ Digital Fuel Injection on Cadillac models
⑦ Bosch K-Jetronic (KE-Jetronic on 1984 and later models)
⑧ AFC on California; TBI on Federal
⑨ MFI, SFI fuel injection systems on Buick models
⑩ E2SE, E4SE, M2SE or M4SE on carbureted engines
⑪ TBI or MFI on fuel injected engines

Manufacturer	Year	Engine	SECTION 4 Feedback Carburetor	SECTION 5 Fuel Injection System
Chrysler	1983–85	1.6L	H-6520	—
		1.7L	H-6520	—
		2.2L	—	AFC
		2.6L	—	TBI
		3.7L	H-6145	—
		5.2L	C-TQ	—
Datsun	See Nissan			
Fiat	1980–85	1.7L	—	AFC
		2.0L	—	AFC
Ford	1978–79	2.3L	H-5200④	—
		2.8L	M-2700VV	—
		5.0L	M-2700VV	—
		5.8L	H-2150	—
	1980	4 cyl	H-5200	—
		6 cyl	C-YFA	—
		8 cyl	C-BBD⑤	—
		8 cyl	M-7200VV	—
	1981	4 cyl	H-6500	—
		6 cyl	H-1946	—
		8 cyl	—	CFI
		8 cyl	M-7200VV	—
	1982	4 cyl	H-6500	—
		6 cyl	C-YFA	—
		8 cyl	H-4180C	—
		8 cyl	—	CFI

AFC—Air Flow Controlled
CIS—Constant Injection System
TBI—Throttle Body Injection
C—Carter carburetor
R—Rochester carburetor
H—Holley carburetor
M—Motorcraft carburetor
EFI—Electronic Fuel Injection
CFI—Central Fuel Injection
① Carter BBD used on engine with CEC system
② Mikuni carburetor on some models
③ CIS on 320i models
④ Holley 6500 on California models
⑤ EFI on Lincoln and Mark IV. Carter thermo quad (TQ) on 4 bbl models
⑥ Digital Fuel Injection on Cadillac models
⑦ Bosch K-Jetronic (KE-Jetronic on 1984 and later models)
⑧ AFC on California; TBI on Federal
⑨ MFI, SFI fuel injection systems on Buick models
⑩ E2SE, E4SE, M2SE or M4SE on carbureted engines
⑪ TBI or MFI on fuel injected engines

Continued

COMPUTERIZED ENGINE CONTROL APPLICATIONS

Manufacturer	Year	Engine	SECTION 4 Feedback Carburetor	SECTION 5 Fuel Injection System
Ford	1983–85	1.6L	—	EFI
		2.3L	C-YFA	—
		2.3L HSC	H-6149	—
		5.0L	—	CFI
		5.0L	H-4180C	—
General Motors	1978–79	4 cyl	R-E2SE	—
		6 cyl	R-E2SE	—
		8 cyl	R-E2SE, E4SE	—
		8 cyl ⑩	—	AFC
	1981	all	R-E2SE, E4SE, R-M4ME, M2ME	⑥
	1982	all (carb)	R-E2SE, E4SE, R-M4ME, M2ME	—
		all (fuel inj)	—	TBI ⑥
	1983	all (carb)	R-E2SE, E4SE, R-M2ME, M4ME	—
		4 cyl (fuel inj)	—	MFI
		8 cyl (fuel inj)	—	TBI
	1984–85	4 cyl	—	MFI
		6 cyl	—	⑨
		8 cyl	⑩	⑪

AFC—Air Flow Controlled
CIS—Constant Injection System
TBI—Throttle Body Injection
C—Carter carburetor
R—Rochester carburetor
H—Holley carburetor
M—Motorcraft carburetor
EFI—Electronic Fuel Injection
CFI—Central Fuel Injection
① Carter BBD used on engine with CEC system
② Mikuni carburetor on some models
③ CIS on 320i models
④ Holley 6500 on California models
⑤ EFI on Lincoln and Mark IV. Carter thermo quad (TQ) on 4 bbl models
⑥ Digital Fuel Injection on Cadillac models
⑦ Bosch K-Jetronic (KE-Jetronic on 1984 and later models)
⑧ AFC on California; TBI on Federal
⑨ MFI, SFI fuel injection systems on Buick models
⑩ E2SE, E4SE, M2SE or M4SE on carbureted engines
⑪ TBI or MFI on fuel injected engines

Manufacturer	Year	Engine	SECTION 4 Feedback Carburetor	SECTION 5 Fuel Injection System
Isuzu	1983–85	Impulse	—	AFC
Mazda	1984–85	13B	—	AFC
Mercedes-Benz	1978–85	all	—	CIS⑦
Mitsubishi	1983–85	1.8L	—	TBI
		2.6L	—	TBI
Nissan (Datsun)	1982–85	2.8L	—	AFC
		3.0L	—	AFC
Oldsmobile	See General Motors			
Peugeot	1980–85	1.9L	—	CIS
Pontiac	See General Motors			
Porsche	1979	all	—	CIS
	1980–85	924, 944	—	AFC
	1980–85	911, 928⑮	—	CIS
Renault	1981–85	18i, Fuego	—	AFC
	1983–85	Alliance, Encore	—	⑧

AFC—Air Flow Controlled
CIS—Constant Injection System
TBI—Throttle Body Injection
C—Carter carburetor
R—Rochester carburetor
H—Holley carburetor
M—Motorcraft carburetor
EFI—Electronic Fuel Injection
CFI—Central Fuel Injection
① Carter BBD used on engine with CEC system
② Mikuni carburetor on some models
③ CIS on 320i models
④ Holley 6500 on California models
⑤ EFI on Lincoln and Mark IV. Carter thermo quad (TQ) on 4 bbl models
⑥ Digital Fuel Injection on Cadillac models
⑦ Bosch K-Jetronic (KE-Jetronic on 1984 and later models)
⑧ AFC on California; TBI on Federal
⑨ MFI, SFI fuel injection systems on Buick models
⑩ E2SE, E4SE, M2SE or M4SE on carbureted engines
⑪ TBI or MFI on fuel injected engines

Continued

COMPUTERIZED ENGINE CONTROL APPLICATIONS

Manufacturer	Year	Engine	SECTION 4 Feedback Carburetor	SECTION 5 Fuel Injection System
Saab	1979–80	99,900	—	CIS
	1981–85	900, Turbo	—	CIS
Subaru	1983–85	1.8L Turbo	—	AFC
Toyota	1980–85	all (fuel inj)	—	AFC
Triumph	1980–81	TR7, TR8	—	AFC
Volkswagen	1979	Type 1, 2	—	AFC
	1980–85	Vanagon	—	AFC
	1980–85	all 4 cyl exc. Vanagon	—	CIS
Volvo	1978–85	all (fuel inj)	—	CIS

NOTE: Always check the underhood emission sticker to determine exactly which engine control system is used on the vehicle

AFC—Air Flow Controlled
CIS—Constant Injection System
TBI—Throttle Body Injection
C—Carter carburetor
R—Rochester carburetor
H—Holley carburetor
M—Motorcraft carburetor
EFI—Electronic Fuel Injection
CFI—Central Fuel Injection

① Carter BBD used on engine with CEC system
② Mikuni carburetor on some models
③ CIS on 320i models
④ Holley 6500 on California models
⑤ EFI on Lincoln and Mark IV. Carter thermo quad (TQ) on 4 bbl models
⑥ Digital Fuel Injection on Cadillac models
⑦ Bosch K-Jetronic (KE-Jetronic on 1984 and later models)
⑧ AFC on California; TBI on Federal
⑨ MFI, SFI fuel injection systems on Buick models
⑩ E2SE, E4SE, M2SE or M4SE on carbureted engines
⑪ TBI or MFI on fuel injected engines

Electronics and Automotive Computers

SECTION 1

INDEX

SECTION 1 ELECTRONICS & AUTOMOTIVE COMPUTERS

All matter is made up of tiny particles called molecules. Each molecule is made up of two or more atoms. Atoms may be divided into even smaller particles called protons, neutrons and electrons. These particles are the same in all matter and differences in materials (hard or soft, conductive or nonconductive) occur only because of the number and arrangement of these particles. Protons and neutrons form the nucleus of the atom, while electrons orbit around the nucleus much the same way as the planets of the solar system orbit around the sun.

The proton is a small positive natural charge of electricity, while the neutron has no electrical charge. The electron carries a negative charge equal to the positive charge of the proton. Every electrically neutral atom contains the same number of protons and electrons, the exact number of which determines the element. The only difference between a conductor and an insulator is that a conductor possesses free electrons in large quantities, while an insulator has only a few. A material must have very few free electrons to be a good insulator, and vice-versa. When we speak of electricity, we're talking about these electrons.

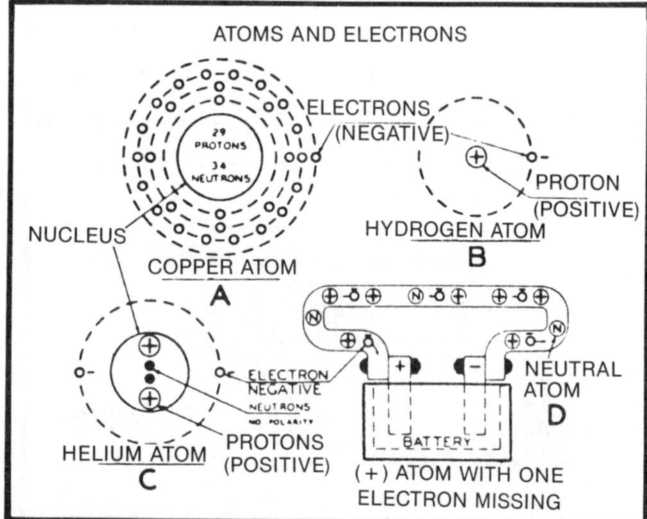

Typical atoms of copper (A), hydrogen (B) and helium (C), showing electron flow through a battery (D)

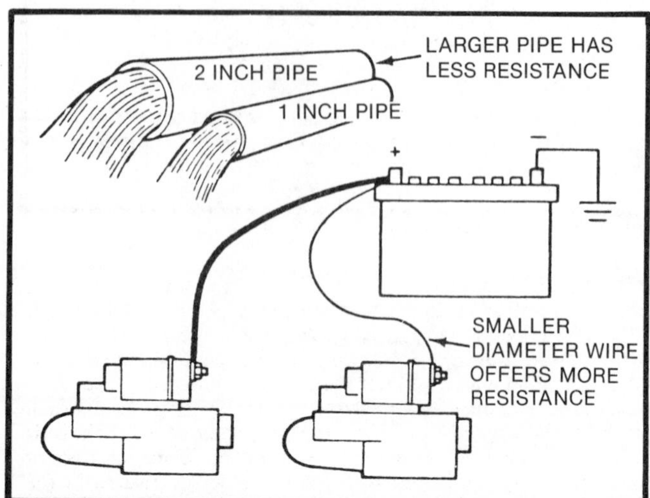

Electircal resistance can be compared to water flow through a pipe. The smaller the wire (pipe), the more resistance to the flow of electrons (water)

In a conductor, the movement of the free electrons is hindered by collisions with the adjoining atoms of the material (matter). This hindrance to movement is called RESISTANCE and it varies with different materials and temperatures. As temperature increases, the movement of the free electrons increases, causing more frequent collisions and therefore increasing resistance to the movement of the electrons. The number of collisions (resistance) also increases with the number of electrons flowing (current). Current is defined as the movement of electrons through a conductor such as a wire. In a conductor (such as copper) electrons can be caused to leave their atoms and move to other atoms. This flow is continuous in that every time an atom gives up an electron, it collects another one to take its place. This movement of electrons is called electric current and is measured in amperes. When 6.28 billion, billion electrons pass a certain point in the circuit in one second, the amount of current flow is called one ampere.

The force or pressure which causes electrons to flow in any conductor (such as a wire) is called voltage. It is measured in volts and is similar to the pressure that causes water to flow in a pipe. Voltage is the difference in electrical pressure measured between two different points in a circuit. In a 12 volt system, for example, the force measured between the two battery posts is 12 volts. Two important concepts are voltage potential and polarity. Voltage potential is the amount of voltage or electrical pressure at a certain point in the circuit with respect to another point. For example, if the voltage potential at one post of the 12 volt battery is zero, the voltage potential at the other post is 12 volts with respect to the first post. One post of the battery is said to be positive (+); the other post is negative (−) and the conventional direction of current flow is from positive to negative in an electrical circuit. It should be noted that the electron flow in the wire is opposite the current flow. In other words, when the circuit is energized, the current flows from positive to negative, but the electrons actually flow from negative to positive. The voltage or pressure needed to produce a current flow in a circuit must be greater than the resistance present in the circuit. In other words, if the voltage drop across the resistance is greater than or equal to the voltage input, the voltage potential will be zero—no voltage will flow through the circuit. Resistance to the flow of electrons is measured in ohms. One volt will cause one ampere to flow through a resistance of one ohm.

Magnetism and Electromagnets

Electricity and magnetism are very closely associated because when electric current passes through a wire, a magnetic field is created around the wire. When a wire, carrying electric current, is wound into a coil, a magnetic field with North and South poles is created

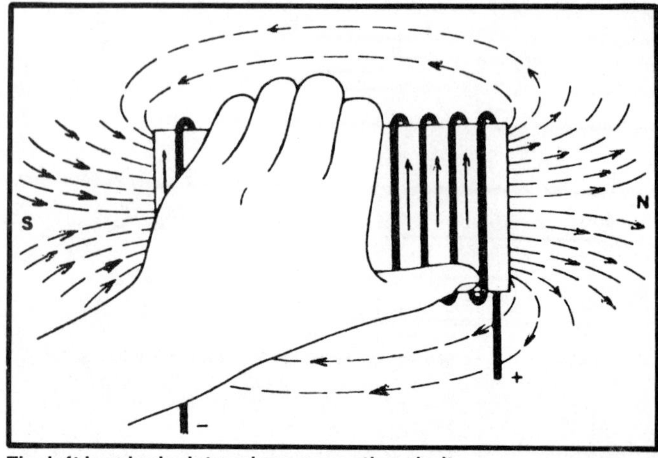

The left hand rule determines magnetic polarity.

just like in a bar magnet. If an iron core is placed within the coil, the magnetic field becomes stronger because iron conducts magnetic lines much easier than air. This arrangement is called an electromagnet.

The direction of current flow is determined by the direction of the magnetic lines of force and the direction of motion of the magnetic field with respect to the conductor. The direction of current flow can be determined by using the "right hand rule". Grasp the conductor with the right hand with the fingers on the leading side of the conductor and pointed in the direction of the magnetic lines of force. The thumb will then point in the direction of current flow.

UNITS OF ELECTRICAL MEASUREMENT

There are three fundamental characteristics of a direct-current electrical circuit: volts, amperes and ohms.

VOLTAGE is the difference of potential between the positive and negative terminals of a battery or generator. Voltage is the pressure or electromotive force required to produce a current of one ampere through a resistance of one ohm.

AMPERE is the unit of measurement of current in an electrical circuit. One ampere is the quantity of current that will flow through a resistance of one ohm at a pressure of one volt.

OHM is the unit of measurement of resistance. One ohm is the resistance of a conductor through which a current of one ampere will flow at a pressure of one volt.

Ohms Law

Ohms law is a statement of the relationship between the three fundamental characteristics of an electrical circuit. These rules apply to direct current only.

$$\text{AMPERES} = \frac{\text{VOLTS}}{\text{OHMS}} \quad \text{or} \quad \frac{E}{R}\, I = \frac{3}{M}$$

$$\text{OHMS} = \frac{\text{VOLTS}}{\text{AMPERES}} \quad \text{or} \quad R = \frac{E}{I}$$

$$\text{VOLTS} = \text{AMPERES} \times \text{OHMS} \quad \text{or} \quad E = I \times R$$

Ohms law provides a means to make an accurate circuit analysis without actually seeing the circuit. If, for example, one wanted to check the condition of the rotor winding in an alternator whose specifications indicate that the field (rotor) current draw is normally 2.5 amperes at 12 volts, simply connect the rotor to a 12 volt battery and measure the current with an ammeter. If it measures about 2.5 amperes, the rotor winding can be assumed good.

An ohmmeter can be used to test components that have been removed from the vehicle in much the same manner as an ammeter. Since the voltage and the current of the rotor windings used as an earlier example are known, the resistance can be calculated using Ohms law. The formula would be:

$R = \frac{E}{I}$ Where: $E = 12$ volts
$I = 2.5$ amperes

If the rotor resistance measures about 4.8 ohms when checked with an ohmmeter, the winding can be assumed good. By plugging in different specifications, additional circuit information can be determined such as current draw, etc.

Electrical Circuits

An electrical circuit must start from a source of electrical supply and return to that source through a continuous path. There are two basic types of circuit; series and parallel. In a series circuit, all of the elements are connected in chain fashion with the same amount of current passing through each element or load. No matter where an ammeter is connected in a series circuit, it will always read the samė. The most important fact to remember about a series circuit is that the sum of the voltages across each element equals the source voltage. The total resistance of a series circuit is equal to the sum of the individual resistances within each element of the circuit. Using ohms law, one can determine the voltage drop across each element in the circuit. If the total resistance and source voltage is known, the amount of current can be calculated. Once the amount of current (amperes) is known, values can be substituted in the Ohms law formula to calculate the voltage drop across each individual element in the series circuit. The individual voltage drops must add up to the same value as the source voltage.

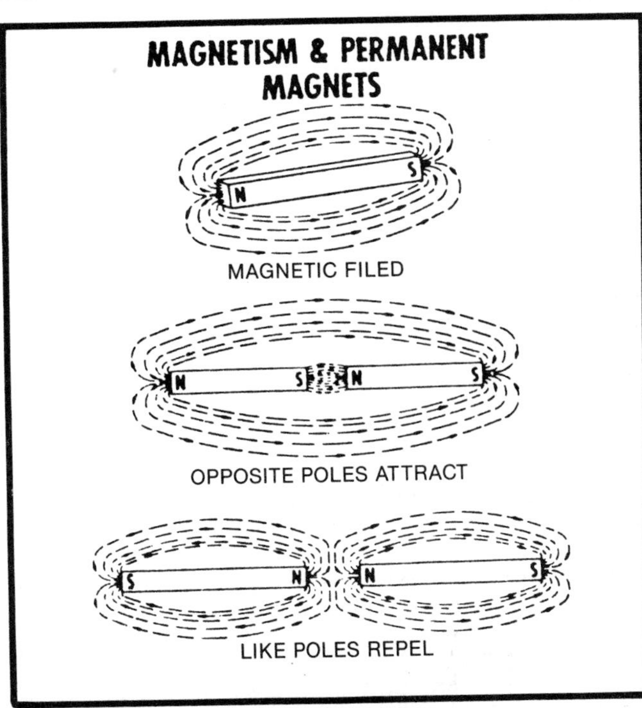

Magnetic field surrounding a bar magnet

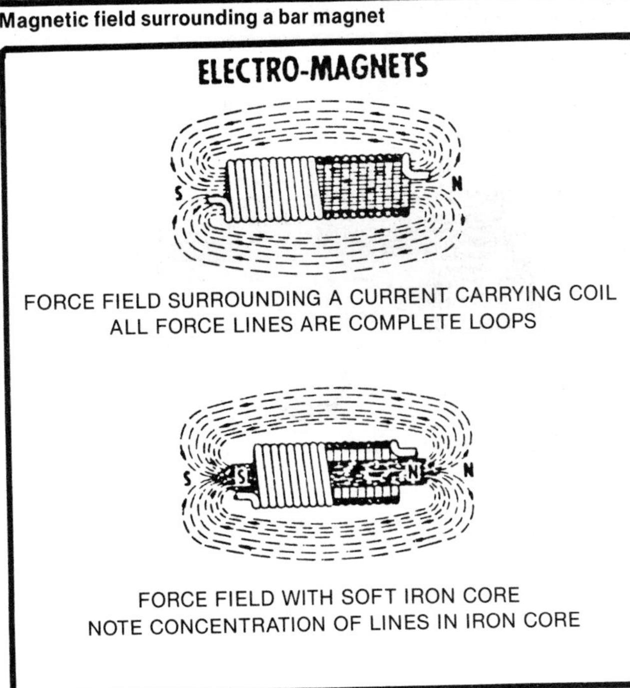

Magnetic field surrounding an electromagnet

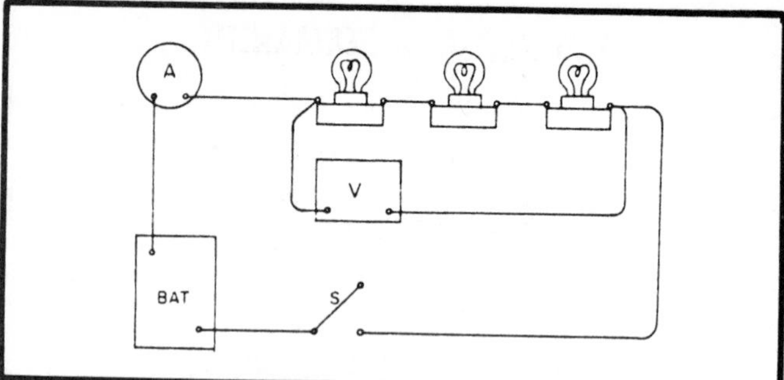

Example of a series circuit

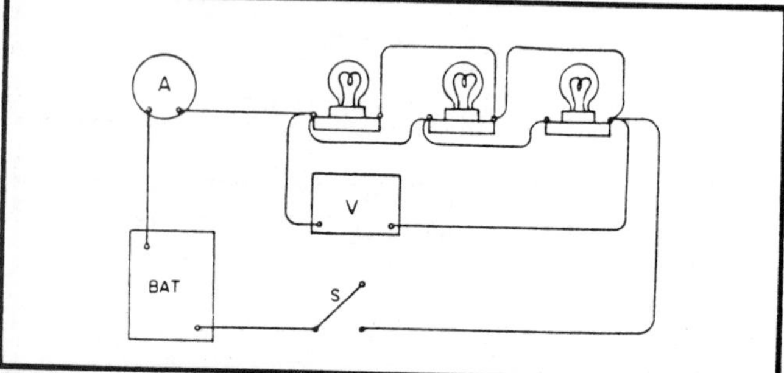

Example of a parallel circuit

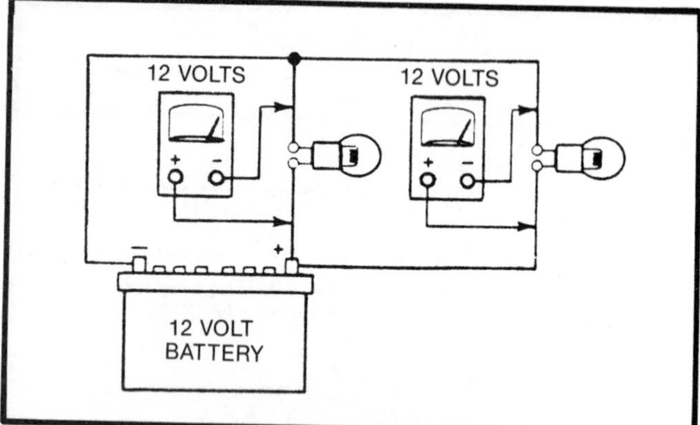

Voltage drop in a parallel circuit. Voltage drop across each lamp is 12 volts

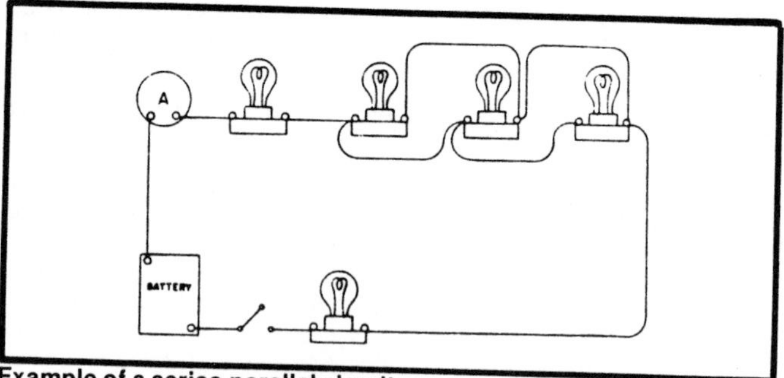

Example of a series-parallel circuit

By measuring the voltage drops, you are in effect measuring the resistance of each element within the circuit. The greater the voltage drop, the greater the resistance. Voltage drop measurements are a common way of checking circuit resistances in automotive electrical systems. When part of a circuit develops excessive resistance (due to a bad connection) the element will show a higher than normal voltage drop. Normally, automotive wiring is selected to limit voltage drops to a few tenths of a volt.

A parallel circuit, unlike a series circuit, contains two or more branches, each branch a separate path independent of the others. The total current draw from the voltage source is the sum of all the currents drawn by each branch. Each branch of a parallel circuit can be analyzed separately. The individual branches can be either simple circuits, series circuits or combinations of series-parallel circuits. Ohms law applies to parallel circuits just as it applies to series circuits, by considering each branch independently of the others. The most important thing to remember is that the voltage across each branch is the same as the source voltage. The current in any branch is that voltage divided by the resistance of the branch.

A practical method of determining the resistance of a parallel circuit is to divide the product of the two resistances by the sum of two resistances at a time. Amperes through a parallel circuit is the sum of the amperes through the separate branches. Voltage across a parallel circuit is the same as the voltage across each branch.

BASIC SOLID STATE

The term "solid state" refers to devices utilizing transistors, diodes and other components which are made from materials known as semiconductors. A semiconductor is a material that is neither a good insulator or a good conductor; principally silicon and germanium. The semiconductor material is specially treated to give it certain qualities that enhance its function, therefore becoming either p-type (positive) or n-type (negative) material. Most modern

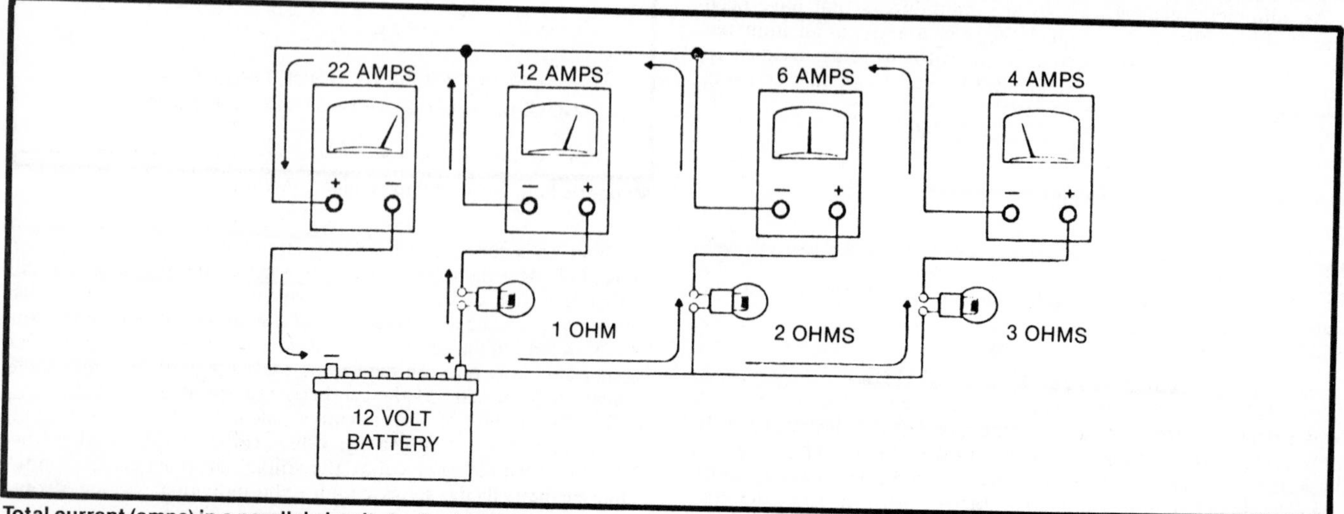

Total current (amps) in a parallel circuit. 4 + 6 + 12 = 22 amps

semiconductors are constructed of silicon and can be made to change their characteristics depending on whether its function calls for an insulator or conductor.

The simplest semiconductor function is that of the diode or rectifier (the two terms mean the same thing). A diode will pass current in one direction only, like a one-way valve, because it has low resistance in one direction and high resistance on the other. Whether the diode conducts or not depends on the polarity of the voltage applied to it. A diode has two electrodes, an anode and a cathode. When the anode receives positive (+) voltage and the cathode receives negative (−) voltage, current can flow easily through the diode. When the voltage is reversed, the diode becomes non-conducting and only allows a very slight amount of current to flow in the circuit. Because the semiconductor is not a perfect insulator, a small amount of reverse current leakage will occur, but the amount is usually too small to consider.

Like any other electrical device, diodes have certain ratings that must be observed and should not be exceeded. The forward current rating indicates how much current can safely pass through the diode without causing damage or destroying it. Forward current rating is usually given in either amperes or milliamperes. The voltage drop across a diode remains constant regardless of the current flowing through it. Small diodes designed to carry low amounts of current need no special provision for dissipating the heat generated in any electrical device, but large current carrying diodes are usually mounted on heat sinks to keep the internal temperature from rising to the point where the silicon will melt and destroy the diode. When diodes are operated in a high ambient temperature environment, they must be de-rated to prevent failure.

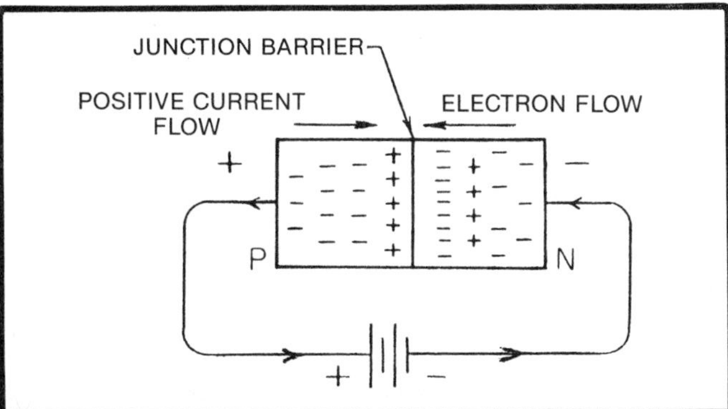

Diode with foreward bias

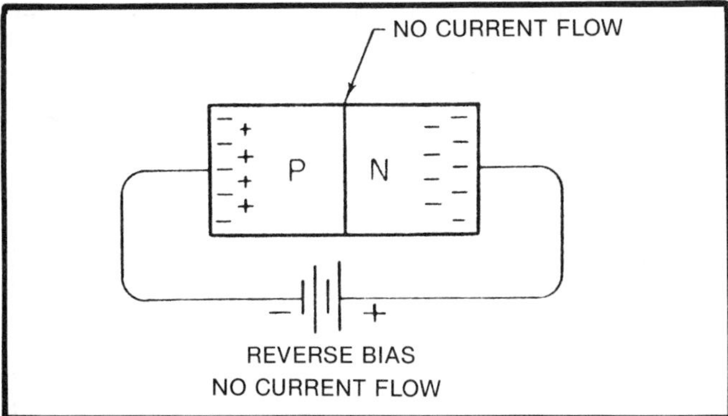

Diode with reverse bias

Another diode specification is its peak inverse voltage rating. This value is the maximum amount of voltage the diode can safely handle when operating in the blocking mode. This value can be anywhere from 50–1000 volts, depending on the diode, and if exceeded can damage the diode just as too much forward current will. Most semiconductor failures are caused by excessive voltage or internal heat.

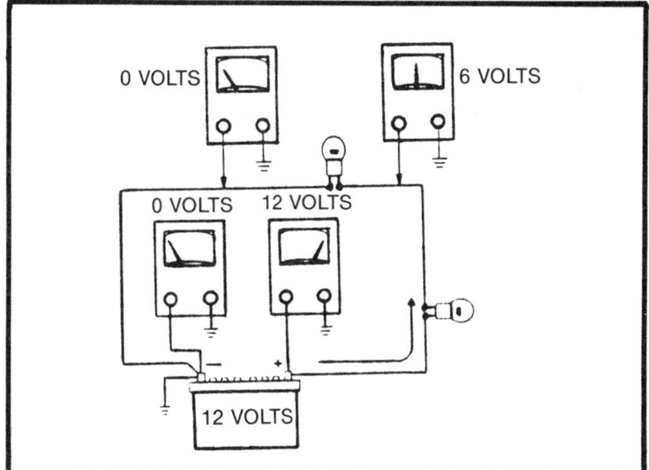

Voltage drop in a series circuit

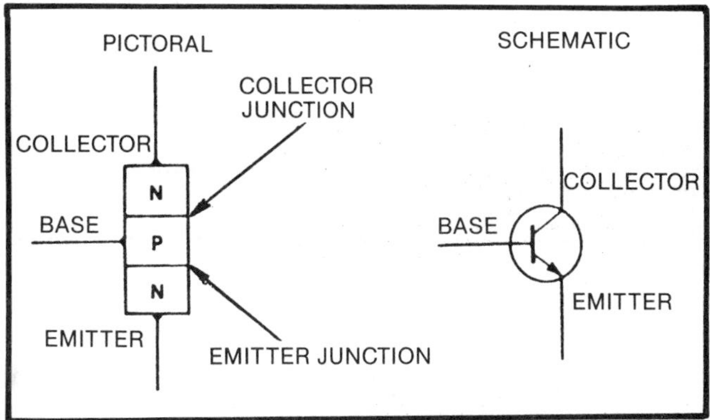

NPN transistor showing both pictoral and schematic illustrations

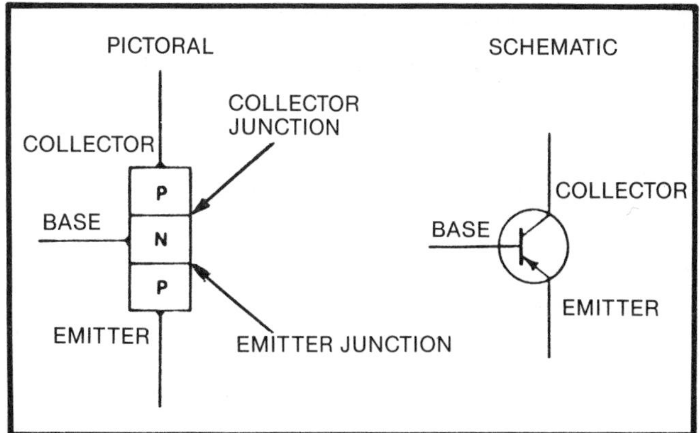

PNP transistor showing both pictoral and schematic illustrations

One can test a diode with a small battery and a lamp with the same voltage rating. With this arrangement one can find a bad diode and determine the polarity of a good one. A diode can fail and cause either a short or open circuit, but in either case it fails to function as a diode. Testing is simply a matter of connecting the test bulb first in one direction and then the other and making sure that current flows in one direction only. If the diode is shorted, the test bulb will remain on no matter how the light is connected.

ZENER DIODES

Normally, exceeding the reverse voltage rating of a conventional diode will cause it to fail; the Zener diode is an exception to this rule. Because of this characteristic, it can perform the important function of regulation and become the essential ingredient in the solid state voltage regulator. A Zener diode behaves like any other silicon diode in the forward direction and, up to a point, in the reverse direction also. As the reverse voltage increases, very little reverse current flows since this is normally the non-conducting direction; but when the reverse voltage reaches a certain point, the reverse current suddenly begins to increase. In a conventional diode, this is known as breakdown and the heat caused by the resulting high leakage current would melt the semiconductor junction. In a Zener diode, due to its particular junction construction, the resulting heat is spread out and damage does not occur. Instead the reverse current (called Zener current) begins to increase very rapidly with only a very slight increase in the reverse (or Zener) voltage. This characteristic permits the Zener diode to function as a voltage regulator. The specific reverse voltage at which the Zener diode becomes conductive is controlled by the diode manufacturing process. Almost any Zener voltage rating can be obtained, from a few volts up to several hundred volts.

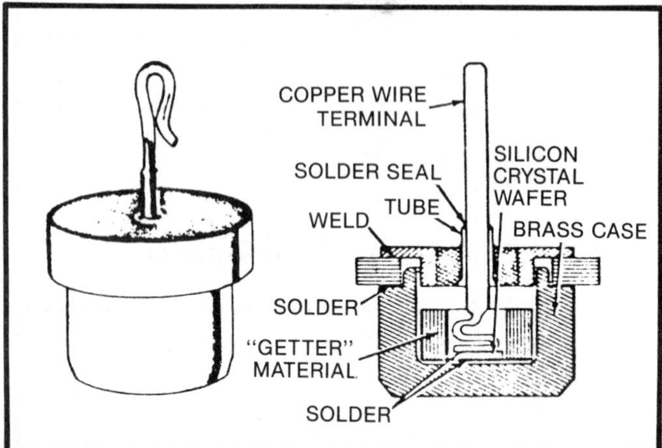

Cross section of typical alternator diode

TRANSISTORS

The transistor is an electrical device used to control voltage within a circuit. A transistor can be considered a "controllable diode" in that, in addition to passing or blocking current, the transistor can control the amount of current passing through it. Simple transistors are composed of three pieces of semiconductor material, P and N type, joined together and enclosed in a container. If two sections of P material and one section of N material are used, it is known as the PNP transistor; if the reverse is true, then it is known as an NPN transistor. The two types cannot be interchanged. Most modern transistors are made from silicon (earlier transistors were made from germanium) and contain three elements; the emitter, the collector and the base. In addition to passing or blocking current, the transistor can control the amount of current passing through it and because of this can function as an amplifier or a switch.

The collector and emitter form the main current-carrying circuit

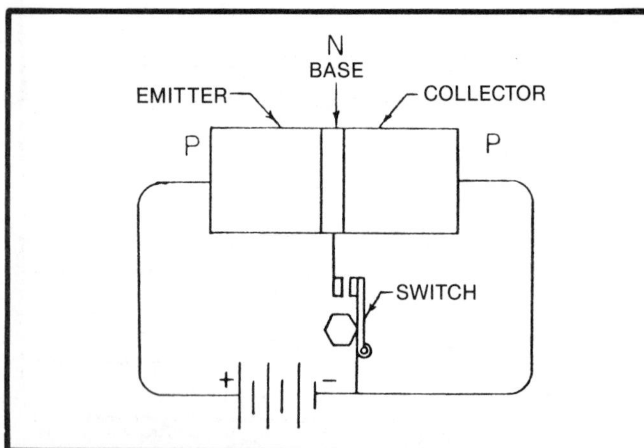

PNP transistor with base switch open (no current flow)

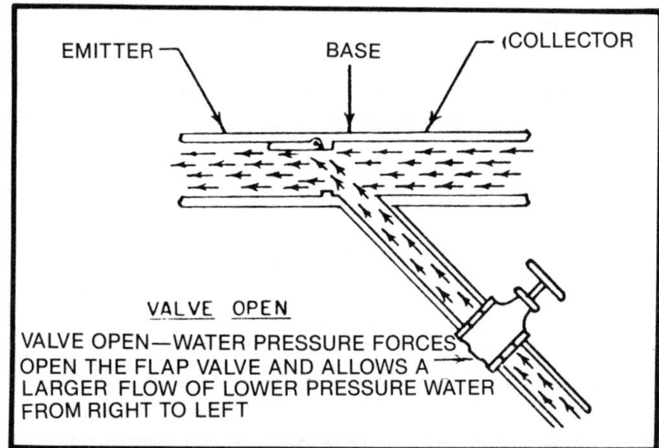

Hydraulic analogy to transistor function with the base circuit energized

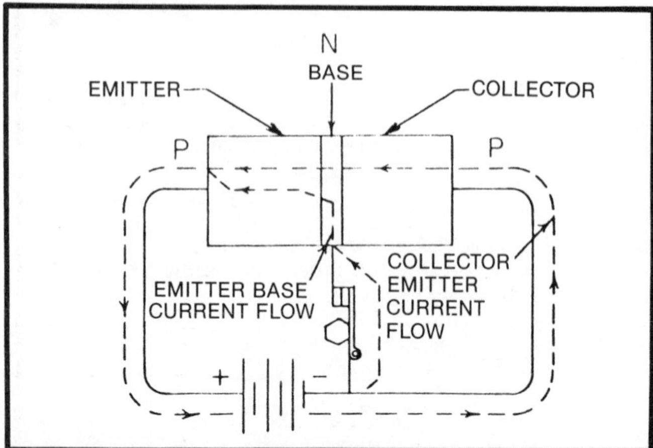

PNP transistor with base switch closed (base emitter and collector emitter current flow)

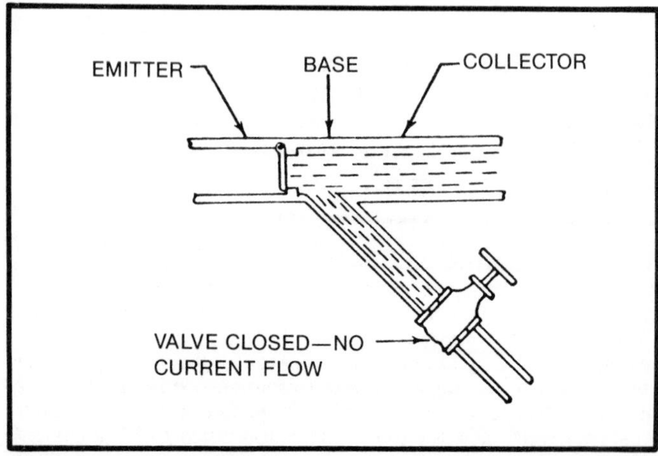

Hydraulic analogy to transistor function with the base circuit shut off

of the transistor. The amount of current that flows through the collector-emitter junction is controlled by the amount of current in the base circuit. Only a small amount of base-emitter current is necessary to control a large amount of collector-emitter current (the amplifier effect). In automotive applications, however, the transistor is used primarily as a switch.

When no current flows in the base-emitter junction, the collector-emitter circuit has a high resistance, like to open contacts of a relay. Almost no current flows through the circuit and transistor is considered OFF. By bypassing a small amount of current into the base circuit, the resistance is low, allowing current to flow through the circuit and turning the resistor ON. This condition is known as "saturation" and is reached when the base current reaches the maximum value designed into the transistor that allows current to flow. Depending on various factors, the transistor can turn on and off (go from cutoff to saturation) in less than one millionth of a second.

Much of what was said about ratings for diodes applies to transistors, since they are constructed of the same materials. When transistors are required to handle relatively high currents, such as in voltage regulators or ignition systems, they are generally mounted on heat sinks in the same manner as diodes. They can be damaged or destroyed in the same manner if their voltage ratings are exceeded. A transistor can be checked for proper operation by measuring the resistance with an ohmmeter between the base-emitter terminals and then between the base-collector terminals. The forward resistance should be small, while the reverse resistance should be large. Compare the readings with those from a known good transistor. As a final check, measure the forward and reverse resistance between the collector and emitter terminals.

MICROPROCESSORS, COMPUTERS AND LOGIC SYSTEMS

Mechanical or electromechanical control devices lack the precision necessary to meet the requirements of modern Federal emission control standards, and the ability to respond to a variety of input conditions common to normal engine operation. To meet these requirements, manufacturers have gone to solid state logic systems and microprocessors to control the basic functions of fuel mixture, spark timing and emission control.

One of the more vital roles of solid state systems is their ability to perform logic functions and make decisions. Logic designers use a shorthand notation to indicate whether a voltage is present in a circuit (the number 1) or not present (the number 0), and their systems are designed to respond in different ways depending on the output signal (or the lack of it) from various control devices.

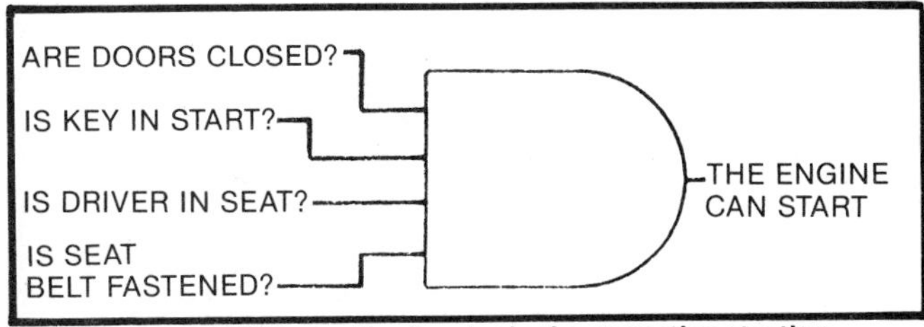

Multiple input AND operation in a typical automotive starting circuit

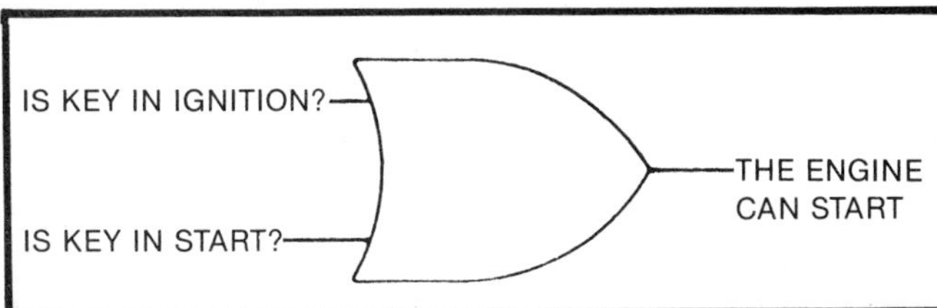

Typical two-input OR circuit operation

There are three basic logic functions or "gates" used to construct a control system: the AND gate, the OR gate or the NOT gate. Stated simply, the AND gate works when voltage is present in two or more circuits which then energize a third (A and B energize C). The OR gate works when voltage is present at either circuit A or circuit B which then energizes circuit C. The NOT function is performed by a solid state device called an "inverter" which reverses the input from a circuit so that, if voltage is going in, no voltage comes out and vice versa. With these three basic building blocks, a logic designer can create complex systems easily. In actual use, a logic or decision making system may employ many logic gates and receive inputs from a number of sources (sensors), but for the most part, all utilize the basic logic gates discussed above.

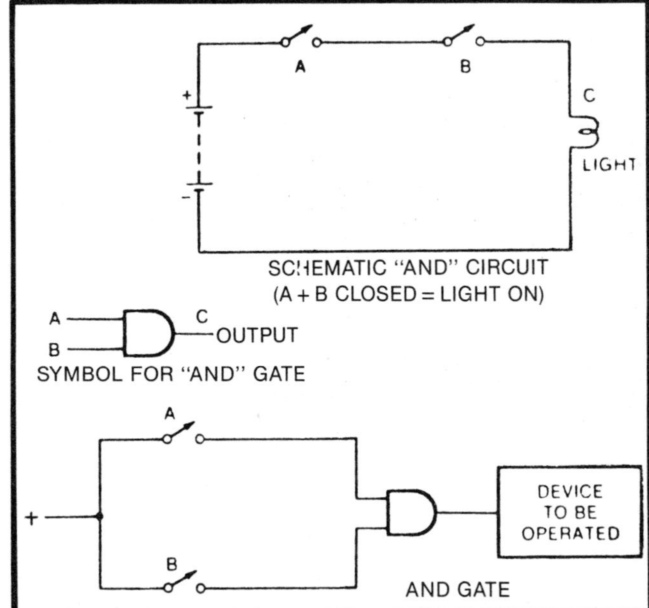

Schematic of an AND circuit and how it responds to various input signals

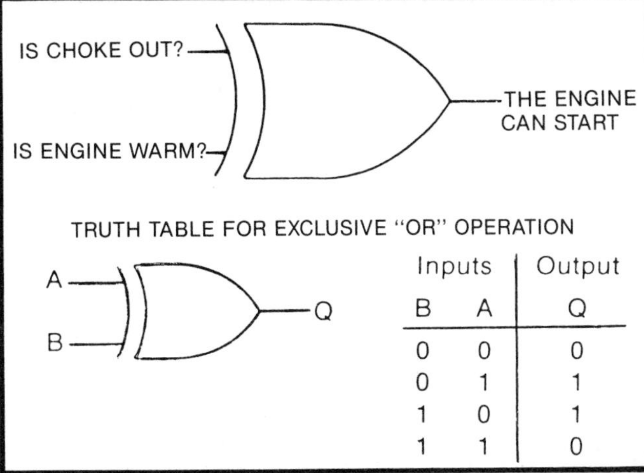

Inputs		Output
B	A	Q
0	0	0
0	1	1
1	0	1
1	1	0

Exclusive OR (EOR) circuit operation

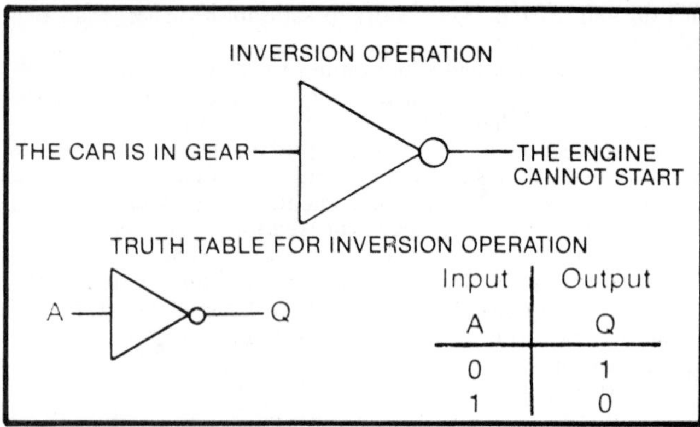

Inversion operation characteristics

NOTE: There is one more basic logic gate, called the Exclusive OR Gate (EOR) that is commonly used where arithmetical calculations (addition and subtraction) must be performed.

Stripped to its bare essentials, a decision making system is made up of three subsystems:

a. Input devices (sensors)
b. Logic circuits (computer control unit)
c. Output devices (actuators or controls)

The input devices are usually nothing more than switches or sensors that provide a voltage signal to the control unit logic circuits that is read as a 1 or 0 (on or off) by the logic circuits. The output devices are anything from a warning light to solenoid operated valves, motors, etc. In most cases, the logic circuits themselves lack sufficient output power to operate these devices directly. Instead, they operate some intermediate device such as a relay or power transistor which in turn operates the appropriate device or control. Many problems diagnosed as computer failures are really the result of a malfunctioning intermediate device like a relay and this must be kept in mind whenever troubleshooting any computer based control system.

The logic systems discussed above are called "hardware" systems, because they consist only of the physical electronic components (gates, resistors, transistors, etc.). Hardware systems do not contain a program and are designed to perform specific or "dedicated" functions which cannot readily be changed. For many simple automotive control requirements, such dedicated logic systems are perfectly adequate. When more complex logic functions are required, or where it may be desirable to alter these functions (e.g. from one model car to another) a true computer system is used. A computer can be programmed through its software to perform many different functions and, if that program is stored on a separate chip called a ROM (Read Only Memory), it can be easily changed simply by plugging in a different ROM with the desired program. Most on-board automotive computers are designed with this capability. The on-board computer method of engine control offers the manufacturer a flexible method of responding to data from a variety of input devices and of controlling an equally large variety of output controls. The computer response can be changed quickly and easily by simply modifying its software program.

Microprocessors

The microprocessor is the heart of the microcomputer. It is the thinking part of the computer system through which all the data from the various sensors passes. Within the microprocessor, data is acted upon, compared, manipulated or stored for future use. A microprocessor is not necessarily a microcomputer, but the differences between the two are becoming very minor. Originally, a mircoprocessor was a major part of a microcomputer, but nowadays microprocessors are being called "single-chip microcomputers". They contain all the essential elements to make them behave as a computer, including the most important ingredient—the program.

All computers require a program. In a general purpose computer, the program can be easily changed to allow different tasks to be performed. In a "dedicated" computer, such as most on-board automotive computers, the program isn't quite so easily altered.

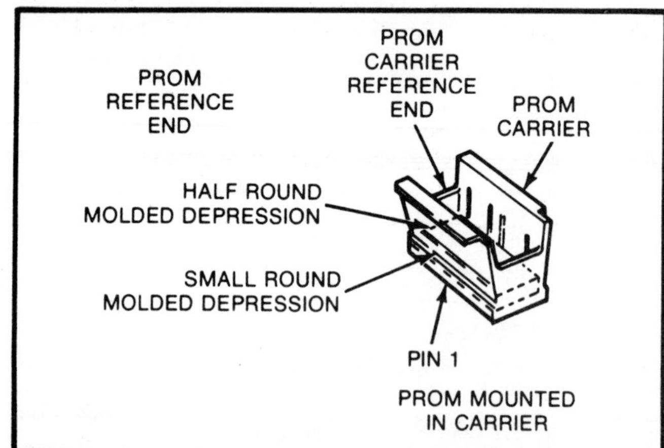

Typical PROM showing carrier and reference markings for installation

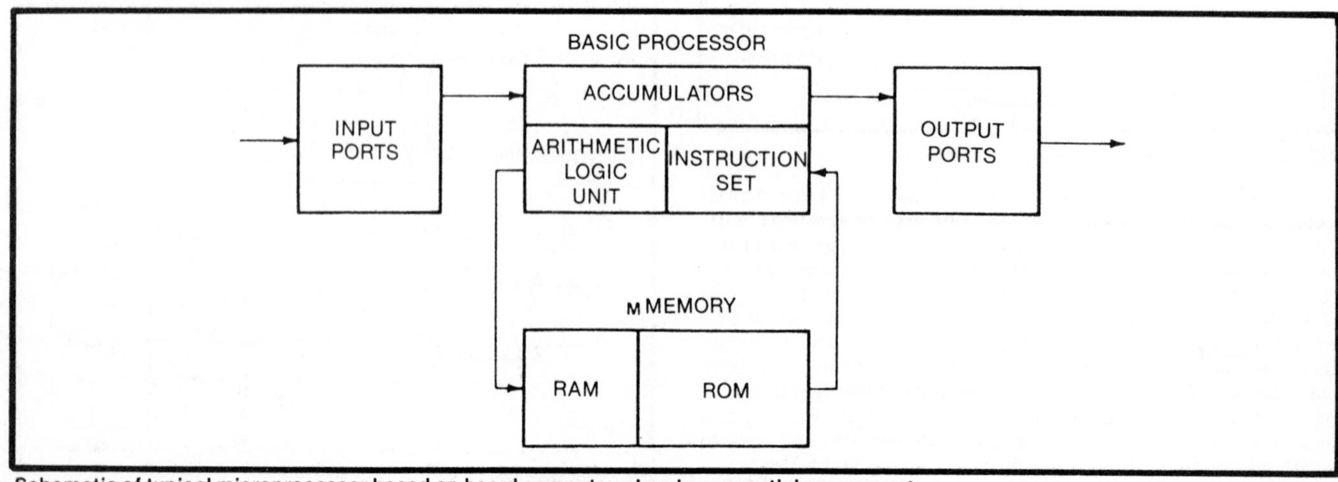

Schematic of typical microprocessor based on-board computer showing essential components

These automotive computers are designed to perform one or several specific tasks, such as maintaining an engine's air/fuel ratio at a specific, predetermined level. A program is what makes a computer smart; without a program a computer can do absolutely nothing. The term "software" refers to the program that makes the hardware do what you want it to do. The software program is simply a listing in sequential order of the steps or commands necessary to make a computer perform the desired task.

Before the computer can do anything at all, the program must be fed into it by one of several possible methods. A computer can never be "smarter" than the person programming it, but it is a lot faster.

Although it cannot perform any calculation or operation that the programmer himself cannot perform, its processing time is measured in millionths of a second.

Because a computer is limited to performing only those operations (instructions) programmed into its memory, the program must be broken down into a large number of very simple steps. Two different programmers can come up with two different programs, since there is usually more than one way to perform any task or solve a problem. In any computer, however, there is only so much memory space available, so an overly long or inefficient program may not fit into the memory. In addition to performing arithmetic functions (such as with a trip computer), a computer can also store data, look up data in a table and perform the logic functions previously discussed. A Random Access Memory (RAM) allows the computer to store bits of data temporarily while waiting to be acted upon by the program. It may also be used to store output data that is to be sent to an output device. Whatever data is stored in a RAM is lost when power is removed from the system by turning off the ignition key, for example.

Computers have another type of memory called a Read Only Memory (ROM) which is permanent. This memory is not lost when the power is removed from the system.

Most programs for automotive computers are stored on a ROM memory chip. Data is usually in the form of a look-up table that saves computing time and program steps. For example, a computer designed to control the amount of distributor advance can have this information stored in a table. The information that determines distributor advance (engine rpm, manifold vacuum and temperature) is coded to produce the correct amount of distributor advance over a wide range of engine operating conditions. Instead of the computer computing the required advance, it simply looks it up in a pre-programmed table.

However, not all engine control functions can be handled in this manner; some must be computed.

There are several ways of programming a ROM, but once programmed, the ROM cannot be changed. If the ROM is made on the same chip that contains the microprocessor, the whole computer must be altered if a program change is needed. For this reason, a ROM is usually placed on a separate chip. Another type of memory is the Programmable Read Only Memory (PROM) that has the program "burned in" with the appropriate programming machine.

Like the ROM, once a PROM has been programmed, it cannot be changed. The advantage of the PROM is that it can be produced in small quantities economically, since it is manufactured with a blank memory. Program changes for various vehicles can be made readily. There is still another type of memory called an EPROM (Erasable PROM) which can be erased and programmed many times, but they are used only in research and development work, not on production vehicles.

The primary function of the on-board computer is to achieve the necessary emission control while maintaining the maximum fuel economy. It should be remembered that all on-board computers perform all the functions described; some perform only one function, such as spark control or idle speed control. Others handle practically all engine functions. As computers increase in complexity and memory capacity, more engine and non-engine (e.g. suspension control, climate control, etc.) functions will be controlled by them.

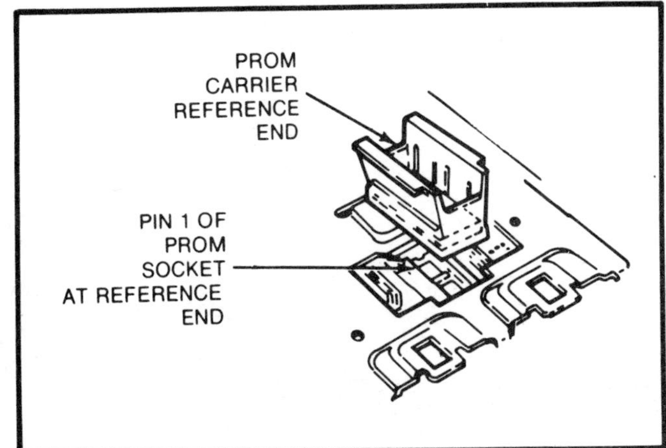

Installation of PROM unit in GM on-board computer

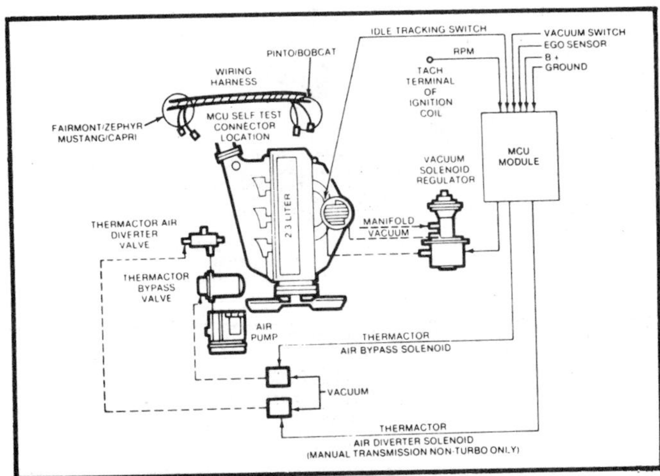

Feedback carburetor system components

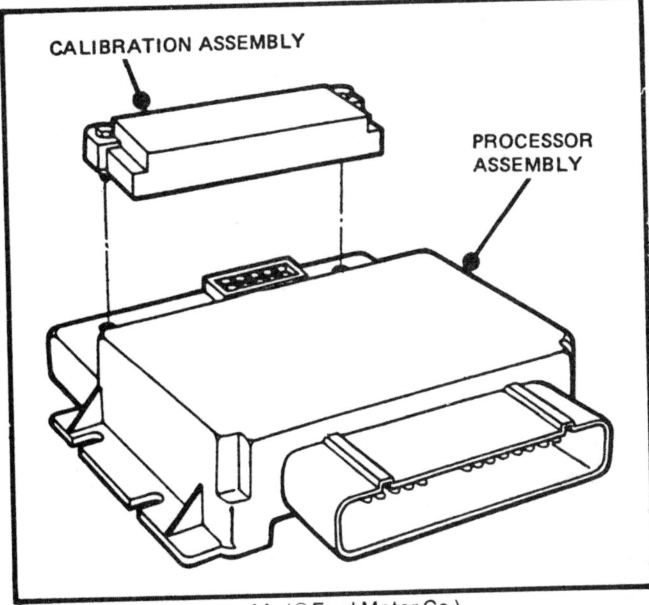

Electronic control assembly (© Ford Motor Co.)

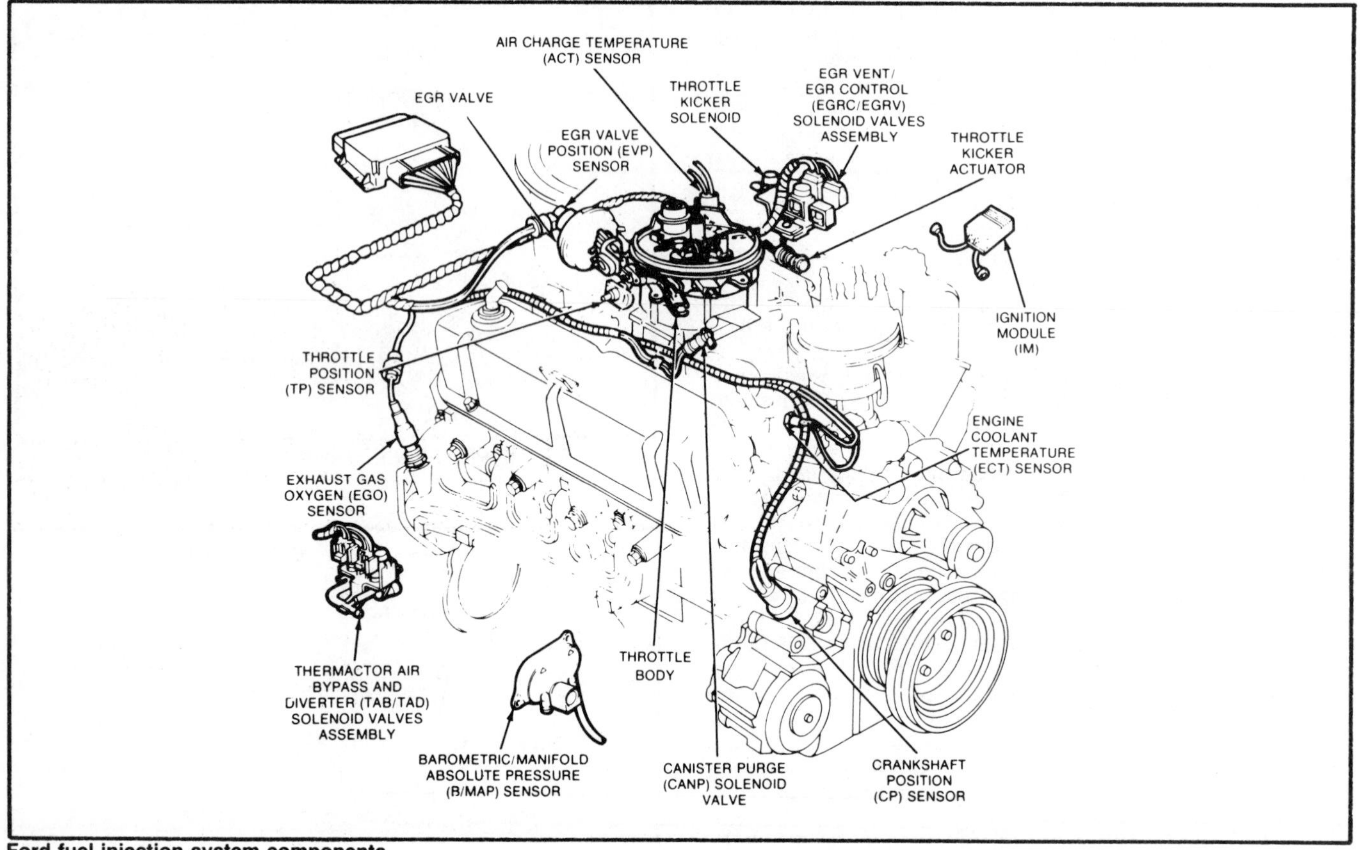

Ford fuel injection system components

Diagnostic Equipment and Special Tools

SECTION 2

INDEX

SPECIAL TOOLS AND DIAGNOSTIC EQUIPMENT

General Information

At the rate which both import and domestic manufacturers are incorporating electronic engine control systems into their production lines, it won't be long before every new vehicle is equipped with one or more on-board computers. These computers and their electronic components (with no moving parts) should theoretically last the life of the car, provided nothing external happens to damage the circuits or memory chips. While it is true that electronic components should last longer than a similar, mechanical system, it is also true that any computer-based system is extremely sensitive to electrical voltages and cannot tolerate careless or haphazard testing or service procedures. An inexperienced individual can literally do major electronic circuit damage looking for a minor problem by using the wrong kind of test equipment or connecting test leads or connectors with the ignition switch ON in any computerized control system. When selecting test equipment, make sure the manufacturers instructions state that the tester is compatible with whatever manufacturer or type of electronic control system is being serviced. Read all instructions and system operation information carefully and double check all test points before installing probes or making any equipment connections.

The following outlines basic diagnosis techniques for dealing with most computerized engine control systems. Along with a general explanation of the various types of test equipment available to aid in servicing modern electronic automotive systems, basic electrical repair techniques for wiring harnesses and weatherproof connectors is given. Read this basic information before attempting any repairs or testing on any computerized system to provide the background of information necessary to avoid the most common and obvious mistakes that can cost both time and money. Although the actual test and replacement procedures are simple, a few different rules apply when dealing with microprocessor-based, on-board computer control systems. Read all service precautions carefully before attempting anything. Likewise, the individual system sections for electronic engine controls, fuel injection and feedback carburetors (both import and domestic) should be read from the beginnning to the end before any repairs or diagnosis is attempted. Although the component replacement and testing procedures are basically simple in themselves, the systems are not, and unless one has a thorough understanding of all particular components and their function within a particular system (fuel injection, for example), the logical test sequence that must be followed to isolate the problem becomes impossible. Minor component malfunctions can make a big difference, so it is important to know how different components interact and how each component affects the operation of the overall computer control system to find the ultimate cause of a problem without replacing electronic components unnecessarily.

GENERAL SAFETY PRECAUTIONS

CAUTION

Whenever working on or around any computer-based microprocessor control system such as if found on most electronic fuel injection, feedback carburetor or emission control systems, always observe these general precautions to prevent the possibility of personal injury or damage to electronic components. Additional precautions (not covered here) may be found in the individual system sections.

- Never install or remove battery cables with the key ON or the engine running. Jumper cables should be connected and disconnected with the key OFF to avoid possible power surges that can damage electronic control units. Engines equipped with computer controlled systems should avoid both giving and getting jump starts due to the possibility of serious damage to computer components from arcing in the engine compartment when connections are made with the ignition ON.
- Always remove the battery cables before charging the battery in the car. Never use a high-output charger on an installed battery or attempt to use any type of "hot shot" (24 volt) starting aid.
- Never remove or attach wiring harness connectors with the ignition switch ON, especially to the electronic control unit.
- When checking compression on engines with AFC injection systems, unplug the cable from the battery to the main power or fuel pump relays before cranking the engine to prevent fuel delivery and possible engine starting during tests. Always look for uniform compression readings between cylinders, rather than specific values.
- Always depressurize fuel injection systems before attempting to disconnect any fuel lines. Although only fuel injection-equipped vehicles use a pressurized fuel system, it's a good idea to exercise caution whenever disconnecting any fuel line or hose during any test procedures. Take adequate precautions to avoid a fire hazard.
- Always use clean rags and tools when working on any open fuel system and take care to prevent any dirt from entering the fuel lines. Wipe all components clean before installation and prepare a clean work area for disassembly and inspection of components. Use lint-free towels to wipe components and avoid using any caustic cleaning solvents.
- Do not drop any components (especially the on-board computer unit) during service procedures and never apply 12 volts directly to any component (like a fuel injector) unless instructed specifically to do so. Some component electrical windings are designed to safely handle only 4 or 5 volts in operation and can be destroyed in approximately 1½ seconds if 12 volts are applied directly to the connector.
- Remove the electronic control unit(s) if the vehicle is to be placed in an environment where temperatures exceed approximately 176 degrees F (80 degrees C), such as a paint spray booth or when arc or gas welding near the control unit location in the car. Control units can be damaged by heat and sudden impact (like a fall from about waist height onto a concrete floor).

UNDERHOOD EMISSION STICKER

The Vehicle Emission Control Information Label is located somewhere in the engine compartment (fan shroud, valve cover, radiator support, hood underside, etc.) of every vehicle produced for sale in the USA or Canada. The label contains important emission specifications and setting procedures, as well as a vacuum hose schematic with emission components identified. This label is permanently attached and cannot be removed without destroying. The specifications shown on the label are correct for the vehicle the label is mounted on. If any difference exists between the specifications shown on the label and those shown in any service manual, those shown on the label should be used.

NOTE: When servicing the engine or emission control systems, the vehicle emission control information label should be checked for up-to-date information, vacuum schematics, etc. Some manufacturers also list the type of engine control system installed.

An additional information label may also be used for any applicable engine calibration numbers. There are several styles and types of calibration labels, but they all carry basically the same information. Finally, a color coded schematic of the vacuum hoses is included on most emission labels. Although the color coding on the schematic represents the actual color of the vacuum hose tracer, there may be instances where an individual hose color may be different.

The emission control decal contains information on the engine normal and fast idle speeds, enriched idle speed, exhaust emission information (model year standards the engine meets), initial timing, spark plug type and gap, vacuum hose routing, and special

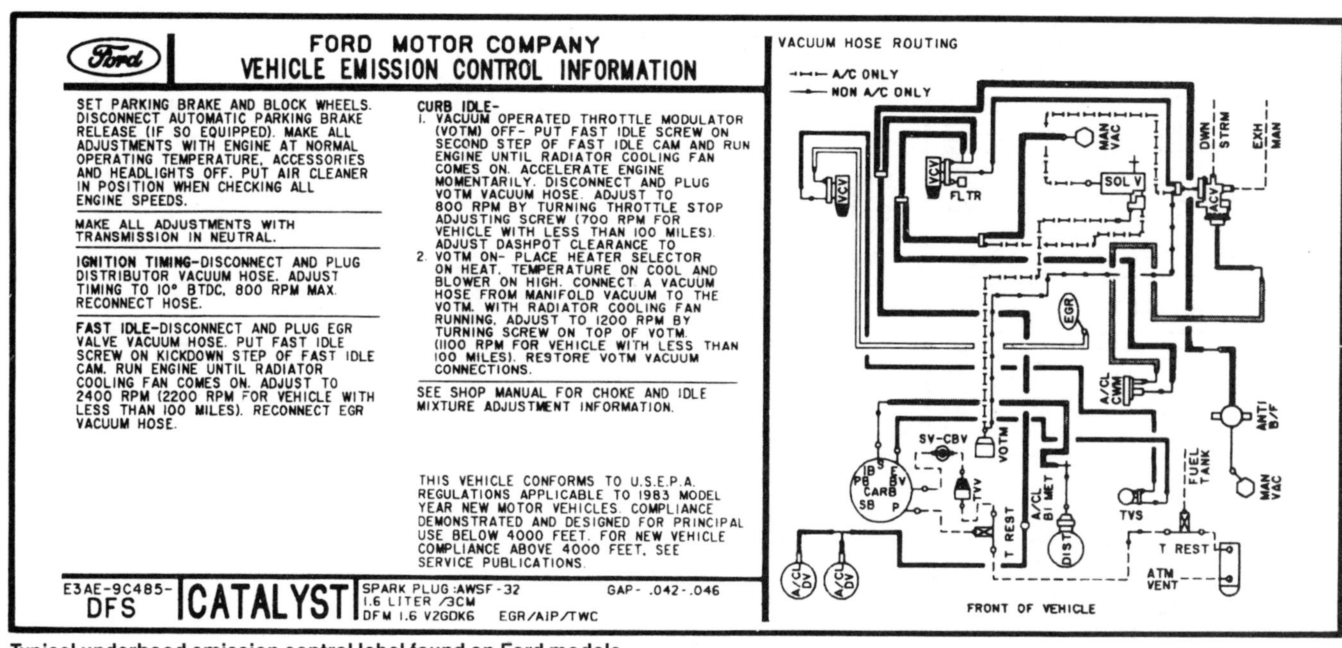
FORD MOTOR COMPANY
VEHICLE EMISSION CONTROL INFORMATION

SET PARKING BRAKE AND BLOCK WHEELS. DISCONNECT AUTOMATIC PARKING BRAKE RELEASE (IF SO EQUIPPED). MAKE ALL ADJUSTMENTS WITH ENGINE AT NORMAL OPERATING TEMPERATURE, ACCESSORIES AND HEADLIGHTS OFF. PUT AIR CLEANER IN POSITION WHEN CHECKING ALL ENGINE SPEEDS.

MAKE ALL ADJUSTMENTS WITH TRANSMISSION IN NEUTRAL.

IGNITION TIMING-DISCONNECT AND PLUG DISTRIBUTOR VACUUM HOSE. ADJUST TIMING TO 10° BTDC, 800 RPM MAX. RECONNECT HOSE.

FAST IDLE-DISCONNECT AND PLUG EGR VALVE VACUUM HOSE. PUT FAST IDLE SCREW ON KICKDOWN STEP OF FAST IDLE CAM. RUN ENGINE UNTIL RADIATOR COOLING FAN COMES ON. ADJUST TO 2400 RPM (2200 RPM FOR VEHICLE WITH LESS THAN 100 MILES). RECONNECT EGR VACUUM HOSE.

CURB IDLE-
1. VACUUM OPERATED THROTTLE MODULATOR (VOTM) OFF- PUT FAST IDLE SCREW ON SECOND STEP OF FAST IDLE CAM AND RUN ENGINE UNTIL RADIATOR COOLING FAN COMES ON. ACCELERATE ENGINE MOMENTARILY. DISCONNECT AND PLUG VOTM VACUUM HOSE. ADJUST TO 800 RPM BY TURNING THROTTLE STOP ADJUSTING SCREW (700 RPM FOR VEHICLE WITH LESS THAN 100 MILES). ADJUST DASHPOT CLEARANCE TO
2. VOTM ON- PLACE HEATER SELECTOR ON HEAT, TEMPERATURE ON COOL AND BLOWER ON HIGH. CONNECT A VACUUM HOSE FROM MANIFOLD VACUUM TO THE VOTM. WITH RADIATOR COOLING FAN RUNNING, ADJUST TO 1200 RPM BY TURNING SCREW ON TOP OF VOTM. (1100 RPM FOR VEHICLE WITH LESS THAN 100 MILES). RESTORE VOTM VACUUM CONNECTIONS.

SEE SHOP MANUAL FOR CHOKE AND IDLE MIXTURE ADJUSTMENT INFORMATION.

THIS VEHICLE CONFORMS TO U.S.E.P.A. REGULATIONS APPLICABLE TO 1983 MODEL YEAR NEW MOTOR VEHICLES. COMPLIANCE DEMONSTRATED AND DESIGNED FOR PRINCIPAL USE BELOW 4000 FEET. FOR NEW VEHICLE COMPLIANCE ABOVE 4000 FEET, SEE SERVICE PUBLICATIONS.

E3AE-9C485-DFS CATALYST SPARK PLUG :AWSF-32 GAP- .042-.046
1.6 LITER /3CM
DFM 1.6 V2GDK6 EGR/AIP/TWC

Typical underhood emission control label found on Ford models.

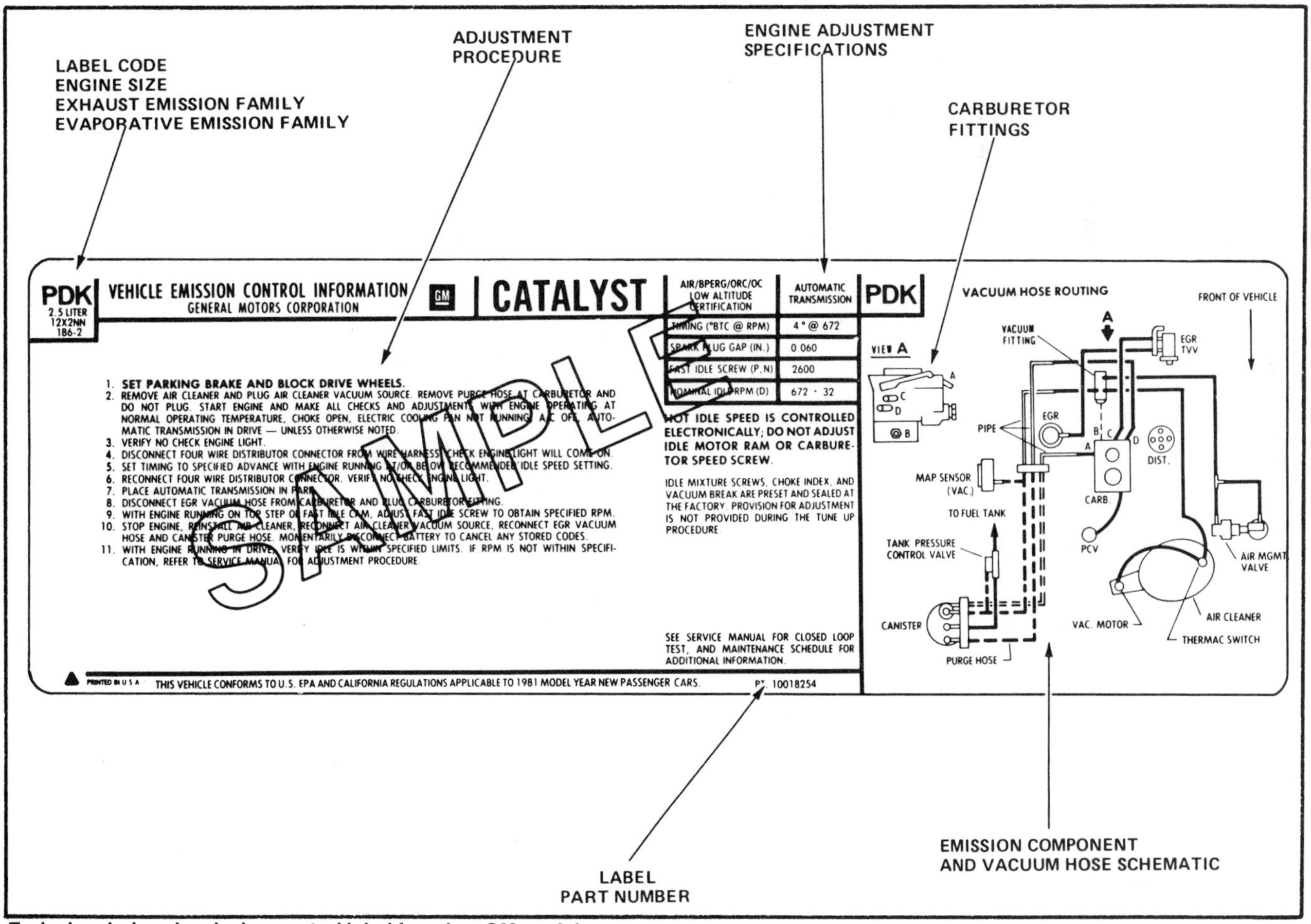

LABEL CODE
ENGINE SIZE
EXHAUST EMISSION FAMILY
EVAPORATIVE EMISSION FAMILY

ADJUSTMENT PROCEDURE

ENGINE ADJUSTMENT SPECIFICATIONS

CARBURETOR FITTINGS

PDK
2.5 LITER
12X2NN
1B6-2

VEHICLE EMISSION CONTROL INFORMATION
GENERAL MOTORS CORPORATION

GM CATALYST

AIR/BPERG/ORC/OC LOW ALTITUDE CERTIFICATION	AUTOMATIC TRANSMISSION
TIMING (°BTC @ RPM)	4° @ 672
SPARK PLUG GAP (IN.)	0.060
FAST IDLE SCREW (P,N)	2600
NOMINAL IDLE RPM (D)	672 · 32

PDK

1. SET PARKING BRAKE AND BLOCK DRIVE WHEELS.
2. REMOVE AIR CLEANER AND PLUG AIR CLEANER VACUUM SOURCE. REMOVE PURGE HOSE AT CARBURETOR AND DO NOT PLUG. START ENGINE AND MAKE ALL CHECKS AND ADJUSTMENTS WITH ENGINE OPERATING AT NORMAL OPERATING TEMPERATURE, CHOKE OPEN, ELECTRIC COOLING FAN NOT RUNNING, A/C OFF, AUTOMATIC TRANSMISSION IN DRIVE — UNLESS OTHERWISE NOTED.
3. VERIFY NO CHECK ENGINE LIGHT.
4. DISCONNECT FOUR WIRE DISTRIBUTOR CONNECTOR FROM WIRE HARNESS. CHECK ENGINE LIGHT WILL COME ON.
5. SET TIMING TO SPECIFIED ADVANCE WITH ENGINE RUNNING AT/OR BELOW RECOMMENDED IDLE SPEED SETTING.
6. RECONNECT FOUR WIRE DISTRIBUTOR CONNECTOR. VERIFY NO CHECK ENGINE LIGHT.
7. PLACE AUTOMATIC TRANSMISSION IN PARK.
8. DISCONNECT EGR VACUUM HOSE FROM CARBURETOR AND PLUG CARBURETOR FITTING.
9. WITH ENGINE RUNNING ON TOP STEP OF FAST IDLE CAM, ADJUST FAST IDLE SCREW TO OBTAIN SPECIFIED RPM.
10. STOP ENGINE, REINSTALL AIR CLEANER, RECONNECT AIR CLEANER VACUUM SOURCE, RECONNECT EGR VACUUM HOSE AND CANISTER PURGE HOSE. MOMENTARILY DISCONNECT BATTERY TO CANCEL ANY STORED CODES.
11. WITH ENGINE RUNNING IN DRIVE, VERIFY IDLE IS WITHIN SPECIFIED LIMITS. IF RPM IS NOT WITHIN SPECIFICATION, REFER TO SERVICE MANUAL FOR ADJUSTMENT PROCEDURE.

SAMPLE

HOT IDLE SPEED IS CONTROLLED ELECTRONICALLY; DO NOT ADJUST IDLE MOTOR RAM OR CARBURETOR SPEED SCREW.

IDLE MIXTURE SCREWS, CHOKE INDEX, AND VACUUM BREAK ARE PRESET AND SEALED AT THE FACTORY. PROVISION FOR ADJUSTMENT IS NOT PROVIDED DURING THE TUNE UP PROCEDURE.

SEE SERVICE MANUAL FOR CLOSED LOOP TEST, AND MAINTENANCE SCHEDULE FOR ADDITIONAL INFORMATION.

PRINTED IN U.S.A. THIS VEHICLE CONFORMS TO U.S. EPA AND CALIFORNIA REGULATIONS APPLICABLE TO 1981 MODEL YEAR NEW PASSENGER CARS. PT. 10018254

VACUUM HOSE ROUTING
FRONT OF VEHICLE
VIEW A
VACUUM FITTING
EGR TVV
PIPE
EGR
DIST.
MAP SENSOR (VAC.)
CARB.
TO FUEL TANK
TANK PRESSURE CONTROL VALVE
PCV
AIR MGMT VALVE
CANISTER
VAC. MOTOR
AIR CLEANER
THERMAC SWITCH
PURGE HOSE

EMISSION COMPONENT AND VACUUM HOSE SCHEMATIC

LABEL PART NUMBER

Typical underhood emission control label found on GM models.

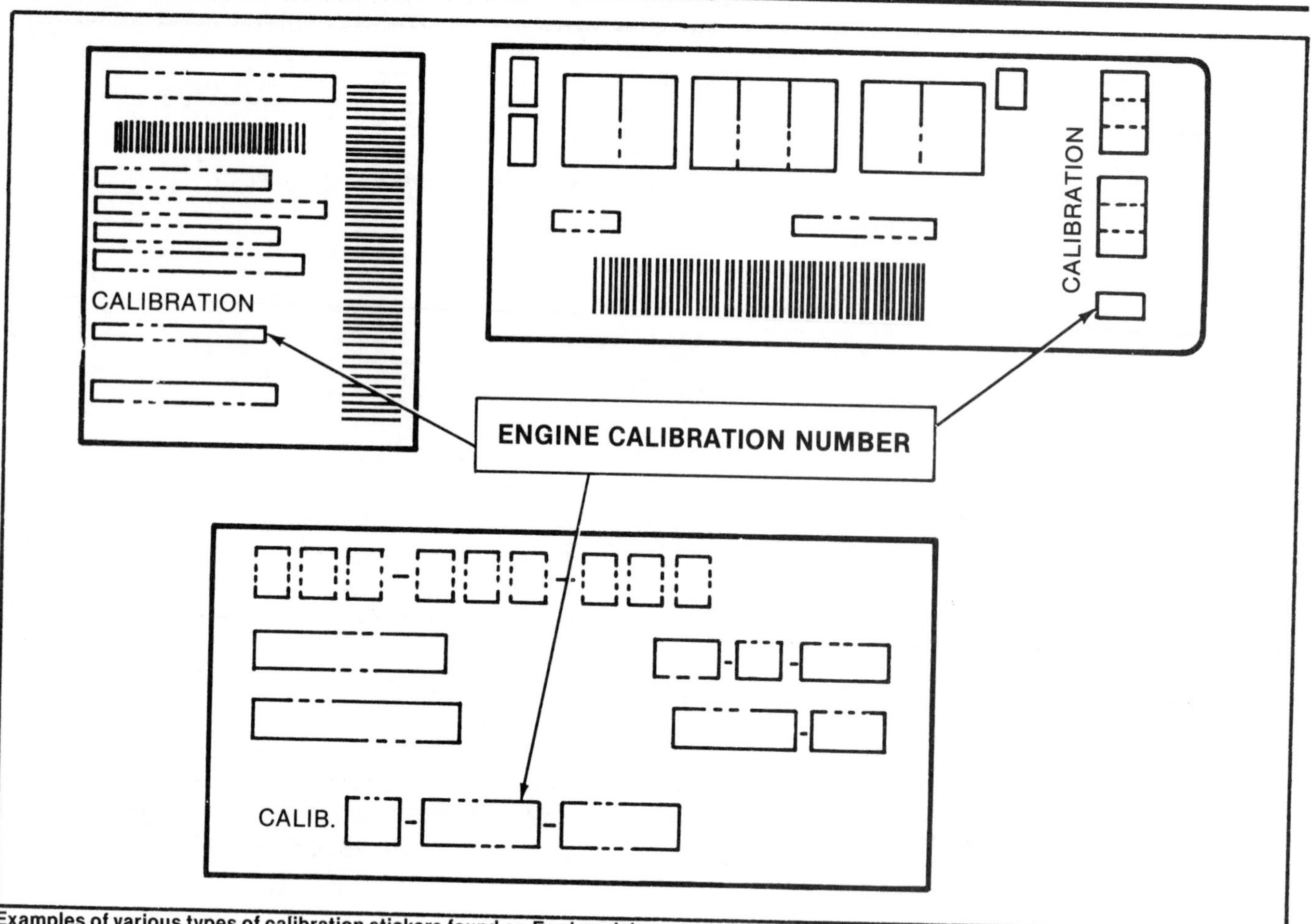

Examples of various types of calibration stickers found on Ford models.

instructions on carburetor and ignition timing adjustment procedures. Some manufacturers use more than one label or decal per car, but all contain important information tailored to that specific vehicle and engine. Any late changes made during the model year will be printed on the emission control label and any required changes made to the vehicle (such as adjustments for high altitude operation) should be marked on the decal for future reference.

WIRING DIAGRAMS

Wiring diagrams are used to show, on paper, how the electrical system of a car is constructed. These diagrams use color coding and symbols that identify different circuits. If an electrical system is at fault, an electrical diagram will show:

- Where the circuit receives battery voltage
- What switches control current flow
- What devices use current flow to do a job
- Where the circuit is grounded.

When testing the problem system, look for this pattern of source, flow controls, and work done by the devices. Similar patterns can be seen in fuel and air delivery systems.

VACUUM DIAGRAMS

Vacuum-controlled systems are used to regulate an engine's operation for better emission control. Vacuum systems also are used to control air conditioners, headlamp doors, power brakes, and other devices. Several of these systems are often used on the same engine. They have become quite complex, and are often connected to each other to change their own operation, as well as engine operation. The engine compartment of many late-model cars is crisscrossed with vacuum lines and hoses, each of which does a specific job. When disconnecting these lines to test or service engine parts, be sure to reinstall them correctly. To help properly route and connect vacuum lines, manufacturers provide vacuum diagrams for each engine and vehicle combination.

NOTE: Vacuum hoses, like electrical wires, can be color coded for easier identification. The vacuum diagram can be printed in color or the color name can be printed near the line in the drawing.

A vacuum diagram is especially needed with late-model engines. The routing and connection pattern of vacuum lines can vary a great deal on a given engine during a single model year. One diagram may apply to an engine sold in California but a different diagram is used when the same engine is sold in a high-altitude area. The problem is compounded when carmakers make running changes in emission systems during a model year. This can result in two identical engines in the same geographical region having slightly different devices and systems. In this case, trying to route and connect vacuum lines properly without the help of a factory vacuum diagram can be next to impossible.

After establishing what area is to be tested, you can study that area in detail. If it is a vacuum system, a vacuum diagram will show:

- The source of vacuum
- Where specific hoses should be connected
- What switches and solenoids control vacuum flow
- What devices are affected by the presence or absence of vacuum.

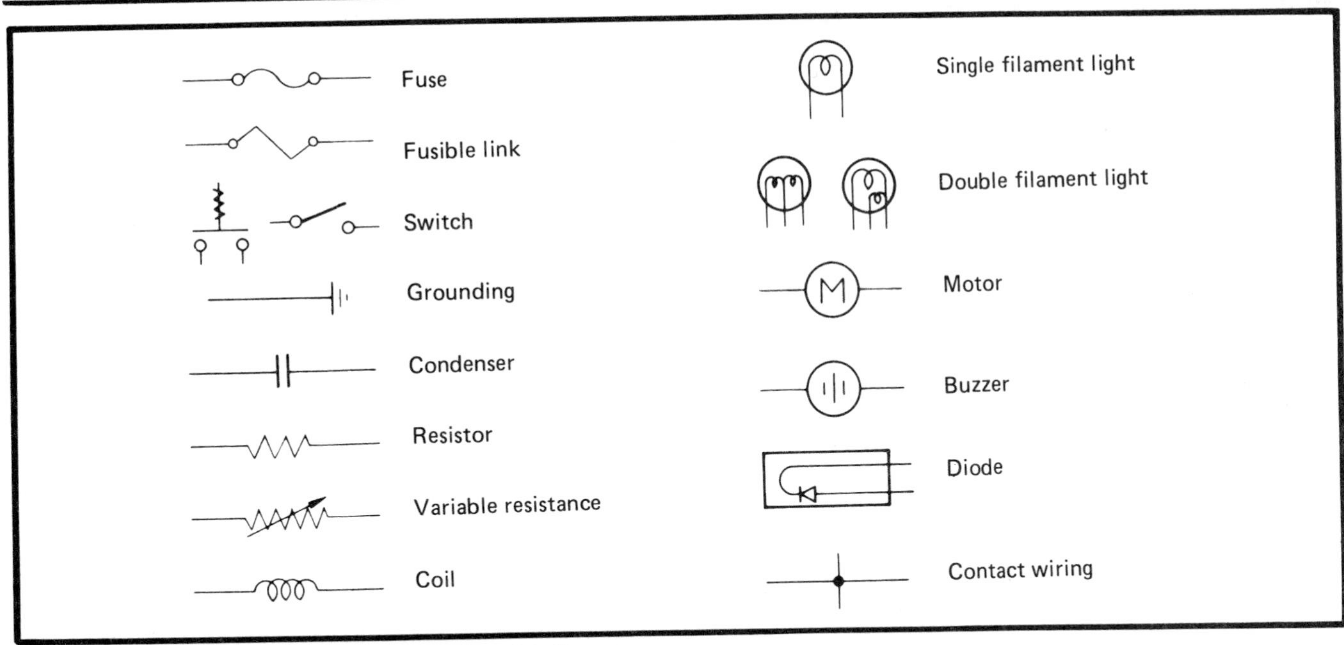

Examples of various electrical symbols found on wiring diagrams

Troubleshooting

BASIC DIAGNOSIS TECHNIQUES

When diagnosing engine performance because of a specific condition, such as poor idling or stalling, it used to be possible to fix most common problems by adjusting timing or idle mixture. However, on today's vehicles, there are many emission-related engine controls that could be responsible for poor performance. The sensitivity of electronic control systems makes organized troubleshooting a must; since almost any component malfunction will affect performance, it's important that you approach the problem in a logical, organized manner. There are some basic troubleshooting techniques that are standard for diagnosing any problem:

1. **Establish when the problem occurs.** Does the problem appear only under certain conditions such as cold start, hot idle or hard acceleration? Were there any noises, odors, or other unusual symptoms? Make notes on any symptoms found, including warning lights and trouble codes, if applicable.
2. **Isolate the problem area.** To do this, make some simple tests and visual under-hood observations; then eliminate the systems that are working properly. For example, a rough idle could be caused by an electrical or a vacuum problem. Check for the obvious problems such as broken wires or split or disconnected vacuum hoses. Always check the obvious before assuming something complicated is the cause. Most of the time, it's NOT the computer.
3. **Test for problems systematically** to determine the cause once the problem area is isolated. Are all the components functioning properly? Is there power going to electrical switches and motors? Is there vacuum at vacuum switches and/or actuators? Is there a mechanical problem such as bent linkage or loose mounting screws? Doing careful, systematic checks will often turn up most causes on the first inspection without wasting time checking components that have little or no relationship to the problem. A no-start or rough running condition on a fuel injected engine (for example) could be caused by a loose air cleaner cover, clogged filter, or a blown fuse. Always start with the easy checks first, before going ahead with any diagnostic routines.
4. **Test all repairs after the work is done.** Make sure that the problem is fixed. Some causes can be traced to more than one component, so a careful verification of repair work is important to pick up additional malfunctions that may cause a problem to reappear or a different problem to arise. A blown fuse, for example, is a simple problem that may require more than another fuse to repair. If you don't look for a problem that caused a fuse to blow, a shorted wire may go undetected and cause another problem.

DIAGNOSIS CHARTS

The diagnostic tree charts are designed to help solve problems on fuel injected vehicles by leading the user through closely defined conditions and tests so that only the most likely components, vacuum and electrical circuits are checked for proper operation when troubleshooting a particular malfunction. By using the trouble trees to eliminate those systems and components which normally

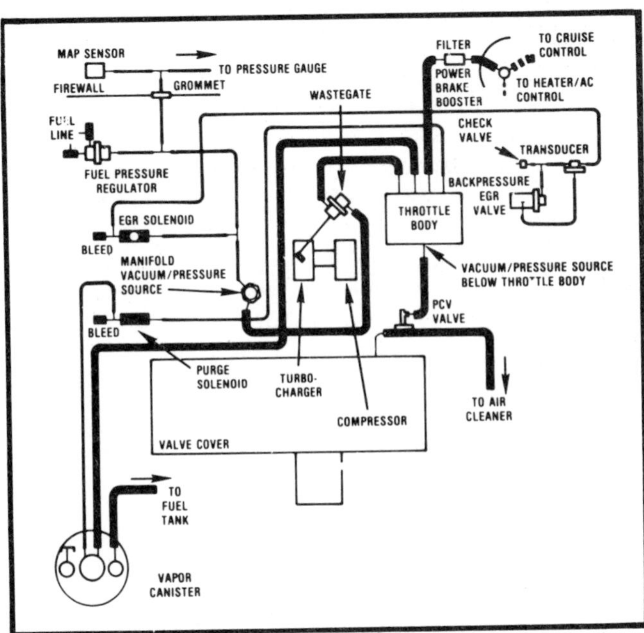

Typical vacuum schematic for a tubrocharged four cylinder engine.

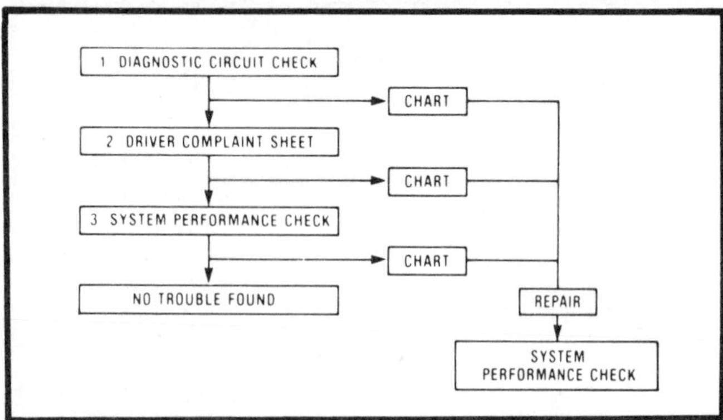

Basic trouble diagnosis procedure

will not cause the condition described, a problem can be isolated within one or more systems or circuits without wasting time on unnecessary testing. Experience has shown that most problems tend to be the result of a fairly simple and obvious cause, such as loose or corroded connectors or air leaks in the intake system; making a careful inspection of components during testing is essential to quick and accurate troubleshooting. Frequent references to the various system testers for fuel injection, feedback carburetors and computerized engine controls will be found in the text and in the diagnosis charts. These devices or their compatible equivalents are necessary to perform some of the more complicated test procedures listed, but many of the electronic fuel injection components can be functionally tested with the quick checks outlined in the "On-Car Service" procedures. Aftermarket electronic system testers are available from a variety of sources, as well as from the manufacturer, but care should be taken that the test equipment being used is designed to diagnose the particular engine control system accurately without damaging the electronic control unit (ECU) or components being tested. You should understand the basic theory of electricity, and know the meaning of voltage, amps, and ohms. You should understand what happens in a circuit with an open or a shorted wire. You should be able to read and understand a wiring diagram. You should know how to use a test light, how to connect and use a tachometer, and how to use jumper wires to by-pass components to test circuits. These techniques are all covered later in this section.

You should also be familiar with the digital volt-ohm meter, a particularly essential tool. You should be able to measure voltage, resistance, and current, and should be familiar with the controls of the meter and how to use it correctly. Without this basic knowledge, the diagnosis trees are impossible to use.

NOTE: Some test procedures incorporate other engine systems in diagnosis routines, since many conditions are the result of something other than the fuel, air, or ignition systems alone. Electrical and vacuum circuits are included wherever possible to help identify small changes incorporated into later models that may affect the diagnosis results. Circuit numbers are OEM designations and should only be used for reference purposes.

Because of the specific nature of the conditions listed in the individual diagnosis procedures, it's important to understand exactly what the definitions of various engine operating conditions are:

• STALLS—engine stops running at idle or when driving. Determine if the stalling condition is only present when the engine is either hot or cold, or if it happens consistently regardless of operating temperature.
• LOADS UP—engine misses due to excessively rich mixture. This usually occurs during cold engine operation and is characterized by black smoke from the tailpipe.
• ROUGH IDLE—engine runs unevenly at idle. This condition can range from a slight stumble or miss up to a severe shake.
• TIP IN STUMBLE—a delay or hesitation in engine response when accelerating from idle with the car at a standstill. Some slight hesitation conditions are considered normal when they only occur during cold operation and gradually vanish as the engine warms up.
• MISFIRE—rough engine operation due to a lack of combustion in one or more cylinders. Fouled spark plugs or loose ignition wires are the most common cause.
• HESITATION—a delay in engine response when accelerating from cruise or steady throttle operation at road speed. Not to be confused with the tip in stumble described above.
• SAG—engine responds initially, then flattens out or slows down before recovering. Severe sags can cause the engine to stall.
• SURGE—engine power variation under steady throttle or cruise. Engine will speed up or slow down with no change in the throttle position. Can happen at a variety of speeds.
• SLUGGISH—engine delivers limited power under load or at high speeds. Engine loses speed going up hills, doesn't accelerate as fast as normal, or has less top speed than was noted previously.
• CUTS OUT—temporary complete loss of power at sharp, irregular intervals. May occur repeatedly or intermittently, but is usually worse under heavy acceleration.
• POOR FUEL ECONOMY—significantly lower gas mileage than is considered normal for the model and drivetrain in question. Always perform a careful mileage test under a variety of road conditions to determine the severity of the problem before attempting corrective measures. Fuel economy is influenced more by external conditions, such as driving habits and terrain, than by a minor malfunction in the fuel delivery system (carburetor or injector) that doesn't cause another problem (like rough operation).

BATTERY TESTING

A variety of battery testing methods and instruments are available, from simple hydrometer checks to specialized load testers. Each method has its features and limitations. Most manufacturers specify a particular type of battery test. Since these vary, check the particular manufacturer's requirements. On models with computerized engine controls, all tests should be done with the battery removed.

Battery Load Test

This is frequently called a high-rate discharge test, since it approximates the current drawn by the starter. These testers may have either a fixed load or an adjustable load controlled by a *carbon pile*. The latter provides greater flexibility in testing because the test load can be adjusted to match the rating of the battery. The basic test procedure is to draw current out of the battery equal to three times the battery's ampere-hour rating (e.g. If amp-hour rating = 60, load current equals 3 x 60 or 180 amperes). While this load is being applied, the voltage of the battery is measured and then compared to specified load voltage.

If this voltage is above the specified minimum, the battery is considered to be in a serviceable state. However, if the voltage is below the minimum, the battery may be either defective or merely in a discharged state. The specific gravity should be checked. If it is below a certain level, typically 1.225, it is necessary to recharge the battery and then retest. A discharged battery should not be replaced on the basis of a load test alone. However, if the battery is sufficiently charged and still fails the load test, it may be considered defective and should be replaced.

Battery Cell Test

The battery cell test measures individual cell voltages which are then compared for variations. If the cell-to-cell variation exceeds a certain amount, for example, 0.05 volt, the battery is considered defective. The test is made with only a small load on the battery, such as that of the headlights. Cell voltages must be measured through the electrolyte between two adjacent cells. Special, immersion-type test prods connected to a suitable voltmeter are required for this. The cell prods are transferred to each cell pair, in turn, noting the voltage readings of each.

BATTERY CHARGING

If a battery is not at full charge, it should be recharged before being put back into service. This is especially important for computer-equipped and electronic ignition models that require a fully charged battery for proper operation. If a battery installed in an operating vehicle is found to be partly discharged, it too should be recharged. The cause for its discharged condition should also be investigated and corrected. It may be a malfunctioning charging system. A number of battery chargers are available and each type is designed for a particular charging requirement.

Constant Current Chargers

The constant current method of charging is one of the oldest methods in use. It is principally used where a number of batteries must be charged at the same time. With this method, all the batteries being charged are connected in series, using short connector straps between each. The string of batteries is then connected to the charger which is adjusted to produce a fixed charging current, typically about 6 or 7 amperes, and all batteries receive the same charge rate.

NOTE: The batteries under charge must be frequently checked for specific gravity and removed when they reach full charge. Since the charge rate remains constant and does not taper off as for other types of chargers, there is danger of overcharging.

Fast Chargers

The fast charger was developed to enable a severely discharged battery to be quickly brought up to a serviceable state of charge. It is based on the principle that a discharged, but otherwise healthy, battery can readily absorb a moderately high charge rate for a short period. Because of their high output, fast chargers must be used with care. Not all batteries can or should be charged at the maximum rate available from the charger. Charging at too high a rate will simply overheat the battery and cause excessive gassing.

CAUTION

A fast charge should never be used to bring a battery up to full charge. This should be accomplished by slow charging. Some chargers automatically switch from fast to slow charge at the end of the preset fast-charge period. Never use a fast charger on an installed batter or damage to the on-board computer can result.

Slow Chargers

A slow charger has a much lower output current than a fast charger. Generally, they do not produce more than about ten amperes. Because their mode of operation is known as the *constant potential* method of charging, they can bring a battery to full charge. The constant potential method simulates the charging system of a vehicle by holding the voltage at a fixed level. As the battery comes up on charge, its counter voltage increases, causing the charge rate to gradually decrease. This is why slow chargers are said to give a tapering charge; when the battery charge is low, the charge current is high, but it tapers to a low value as the battery reaches full charge.

Although the charge rate of a slow charger does taper off, it never drops to zero. It is possible to overcharge a battery with a slow charger if it is left on too long. About 15 hours is the recommended maximum charging period with a slow charger. Many batteries reach full charge in much less time. To prevent accidental overcharging, the automatic battery charger was developed. Basically, this is a slow charger containing an electronic circuit that automatically switches off the charging current when battery voltage reaches a predetermined level. It works on the principle that the voltage of a semi-charged battery will remain low during charge but will become high when full charge is reached. This increase in voltage is sensed by a zener-controlled circuit that electronically turns the charger off. When battery voltage drops below a certain level, due to the normal self-discharge of the battery, this circuit automatically turns the charger on. Because of this feature, an automatic charger can be left connected to a battery for long periods of time without danger of overcharging.

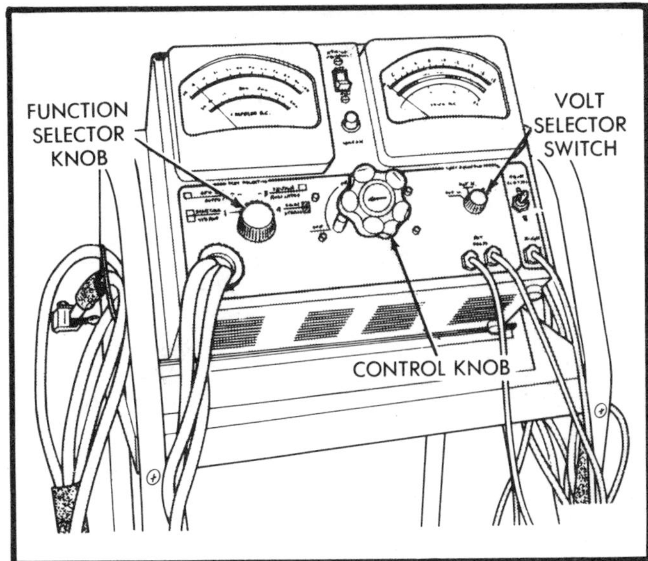

Typical battery load tester with separate voltmeter and ammeter

Trickle Chargers

Trickle chargers are effectively used for keeping batteries freshly charged and ready for use. Avoid continuous overcharge for long periods of time. Whenever such chargers are used which have an output of less than one ampere, it is very important that such "trickle" chargers be used in strict accordance with directions. Continuous overcharging for an indefinite time, even though at a very low rate, can be very destructive to the grids of the positive plates, causing them to disintegrate.

NOTE: Pinpointing trouble in the ignition, electrical, or fuel system can sometimes only be detected by the use of special test equipment. In fact, test equipment and the wiring diagrams are your best tools. The following describes commonly used test equipment and explains how to put it to best use in diagnosis.

Electrical Test Equipment

JUMPER WIRES

Jumper wires are simple, yet extremely valuable, pieces of test equipment. Jumper wires are merely short pieces of 16, 18 or 20 gauge wire that are used to bypass sections of an electrical circuit. The simplest type of jumper is merely a length of multistrand wire with an alligator clip at each end. Jumper wires are usually fabricated from standard automotive wire and whatever type of connector (alligator clip, spade connector or pin connector) that is required for the particular system being tested. The well-equipped tool box will have several different styles of jumper wires in several different lengths. Some jumper wires are made with three or more terminals coming from a common splice for special-purpose testing. In cramped, hard-to-reach areas it is advisable to have insulated boots over the jumper wire terminals in order to prevent accidental grounding, sparks, and possible fire, especially when testing fuel system components.

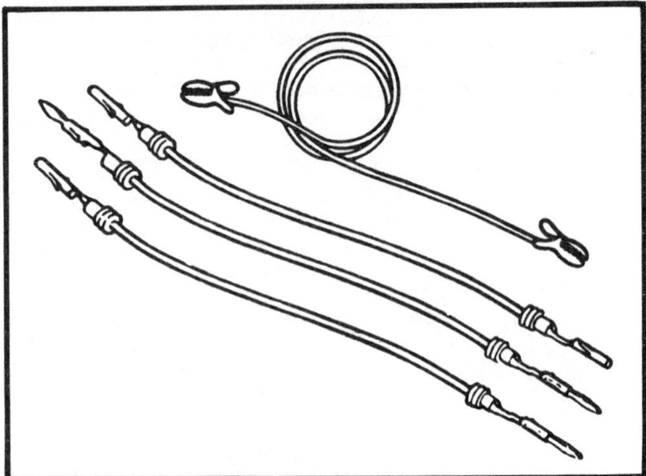

Typical jumper wires with various terminal ends

CAUTION

On-board computers are especially sensitive to arcing anywhere in the system with the ignition switch ON. Circuit damage is possible during testing.

Jumper wires are used primarily to locate open electrical circuits, on either the ground (−) side of the circuit or on the hot (+) side. If an electrical component fails to operate, connect the jumper wire between the component and a good ground. If the component operates only with the jumper installed, the ground circuit is open. If the ground circuit is good, but the component does not operate, the circuit between the power feed and component is open. You can sometimes connect the jumper wire directly from the battery to the hot terminal of the component, but first make sure the component uses 12 volts in operation. Some electrical components, such as fuel injectors, are designed to operate on about 4 volts and running 12 volts directly to the injector terminals can burn out the windings. When in doubt, check the voltage input to the component and measure how much voltage is being applied normally. See the information under "voltmeter." By moving the jumper wire successively back from the lamp toward the power source, you can isolate the area of the circuit where the open circuit is located. When the component stops functioning, or the power is cut off, the open is in the segment of wire between the jumper and the point previously tested.

CAUTION

Never use jumpers made from wire that is of lighter gauge than used in the circuit under test. If the jumper wire is of too small gauge, it may overheat and possibly melt. Never use jumpers to bypass high-resistance loads (such as motors) in a circuit. Bypassing resistances, in effect, creates a short circuit which may, in turn, cause damage and fire. Never use a jumper for anything other than temporary bypassing of components in a circuit.

12-VOLT TEST LIGHT

The 12 volt (unpowered) test light is used to check circuits and components while electrical current is flowing through them. It is used for voltage and ground tests. Twelve volt test lights come in different styles but all have three main parts—a ground clip, a sharp probe, and a light. The most commonly used test lights have pick-type probes. To use a test light, connect the ground clip to a good ground and probe the wires or connectors wherever necessary with the pick. The pick should be sharp so that it can penetrate wire insulation to make contact with the wire without making a large hole in the insulation. The wrap-around light is handy in hard-to-reach areas or where it is difficult to support a wire to push a probe pick into it. To use the wrap around light, hook the wire to probed with the hook and pull the trigger. A small pick will be forced through the wire insulation into the wire core. After testing, the wire can be resealed with a dab of silicone or electrical tape.

CAUTION

Do not use a test light to probe electronic ignition spark plug or secondary coil wires. Never use a pick-type test light to pierce wiring on computer controlled systems unless specifically instructed to do so.

Like the jumper wire, a 12-volt test light is used to isolate opens (breaks or shorts) in circuits. But, whereas the jumper wire is used to bypass the open to operate the load, the 12-volt test light is used to locate the presence or absence of voltage in a circuit. If the test light glows, you know that there is power up to that point; if the

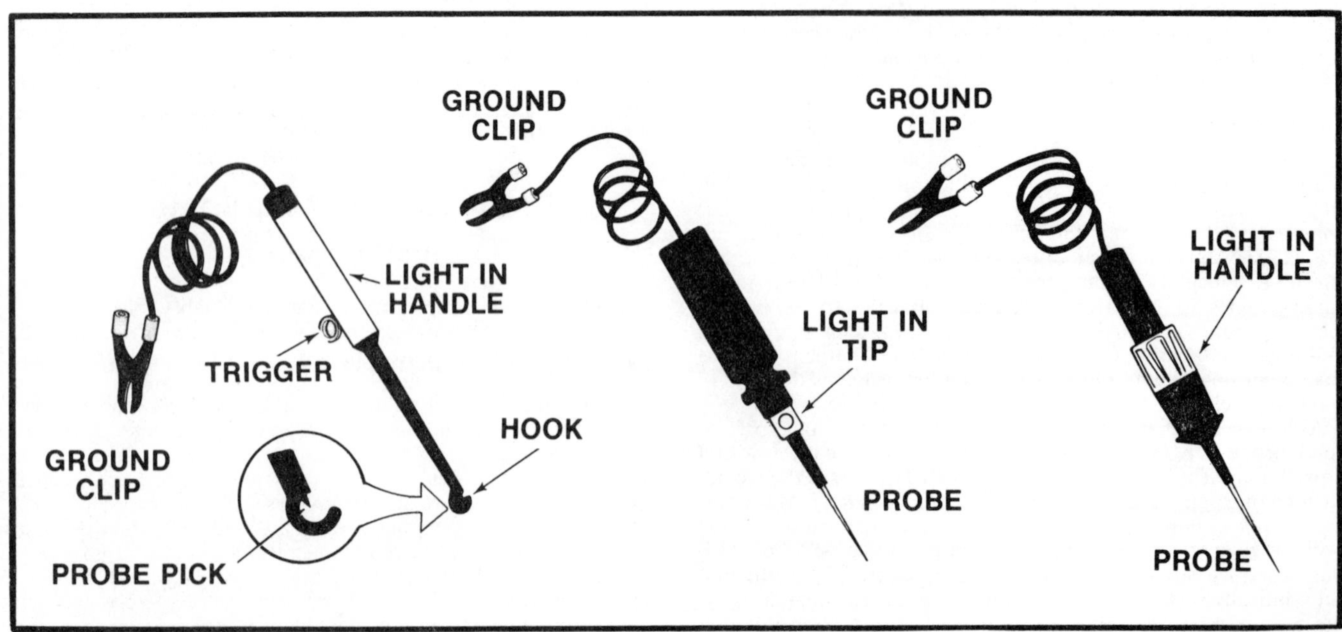

Typical 12 volt test lights

test light does not glow when its probe is inserted into the wire or connector, you know that there is an open circuit (no power). Move the test light in successive steps back along the wire harness, toward the power source, until the light in the handle does glow. When it does glow, the open circuit is between the probe and point previously probed.

NOTE: The test light does not detect that 12 volts (or any particular amount of voltage) is present in a circuit; it only detects that some voltage is present. It is advisable before using the test light to touch its terminals across the battery posts to make sure the light is operating properly. The light should glow brightly.

SELF-POWERED TEST LIGHT (CONTINUITY TESTER)

The typical continuity tester is powered by a 1.5 volt penlight battery. One type of self-powered test light is similar in design to the 12–volt test light; this type has both the battery and the light in the handle and pick-type probe tip. The second type has the light toward the open tip, so that the light illuminates the contact point. The self-powered test light is a dual-purpose piece of test equipment. It can be used to test for either open or short circuits when power is disconnected from the circuit (continuity test). A powered test light should not be used on any computer controlled system or component unless specifically instructed to do so. Many engine sensors (like the oxygen sensor) can be destroyed by even this small amount of voltage applied directly to the terminals.

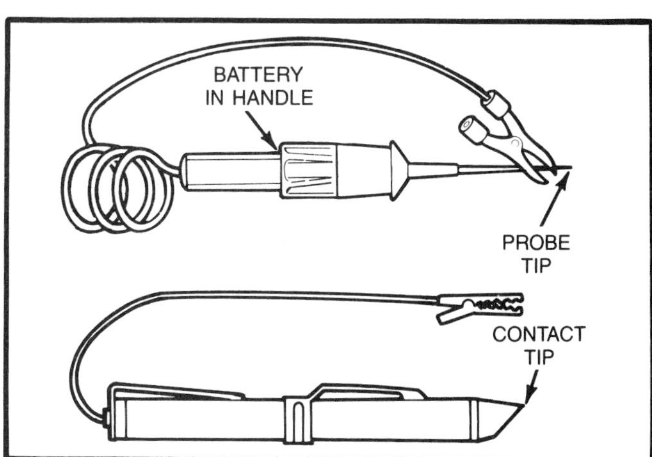

NOTE: The following procedures are meant to be general in nature and apply to most electrical systems. See the individual system sections for specific circuit testing procedures.

Open Circuit Testing

To use the self-powered test light to check for open circuits, first isolate the circuit from the vehicle's 12 volt power source by disconnecting the battery positive (+) terminal or the wiring harness connector. Connect the test light ground clip to a good ground and probe sections of the circuit sequentially with the test light (start from either end of the circuit). If the light is out, the open wire or connection is between the probe and the circuit ground. If the light is on, the open is between the probe and end of the circuit toward the power source.

Short Circuit Testing

By isolating the circuit both from power and from ground, and using a self-powered test light, you can check for shorts to ground in the circuit. Isolate the circuit from power and ground. Connect the test light ground clip to a good ground and probe any easy-to-reach test point in the circuit. If the light comes on, there is a short somewhere in the circuit. To isolate the short, probe a test point at either end of the isolated circuit (the light should be on). Leave the test light probe connected and open connectors, switches, remove parts, etc., sequentially, until the light goes out. When the light goes out, the short is between the last circuit component opened and the previous circuit opened.

NOTE: The 1.5 volt battery in the test light does not provide much current. A weak battery may not provide enough power to illuminate the test light even when a complete circuit is made (especially if there are high resistances in the circuit). Always make sure that the test battery is strong. To check the battery, briefly touch the ground clip to the probe; if the light glows brightly the battery is strong enough for testing. Never use a self-powered test light to perform checks for opens or shorts when power is applied to the electrical system under test. The 12-volt vehicle power will burn out the 1.5 volt light bulb in the test light.

VOLTMETER

A voltmeter is used to measure electrical voltage at any point in a circuit, or to measure the voltage drop across any part of a circuit. It can also be used to check continuity in a wire or circuit by indicating current flow from one end to the other. Analog voltmeters usually have various scales on the meter dial, while digital models only have a four or more digit readout. Both will have a selector switch to allow the selection of different voltage ranges. The voltmeter has a positive and a negative lead; to avoid damage to the meter, always connect the negative lead to the negative (−) side of circuit (to ground or nearest the ground side of the circuit) and always connect the positive lead to the positive (+) side of the circuit (to the power source or the nearest power source). Note that the negative voltmeter lead will always be black and that the positive voltmeter will always be some color other than black (usually red). Depending on how the voltmeter is connected into the electrical circuit, it has several uses.

NOTE: Some engine control systems (like the GM CCC or C4) require the use of a voltmeter with a high impedence rating of 20 megohms or higher. A standard, 9-volt powered analog voltmeter can damage computer components if used improperly.

A voltmeter can be connected either in parallel or in series with a circuit and it has a very high resistance to current flow. When connected in parallel, only a small amount of current will flow through the voltmeter current path; the rest will flow through the normal circuit current path and the circuit will work normally. When the

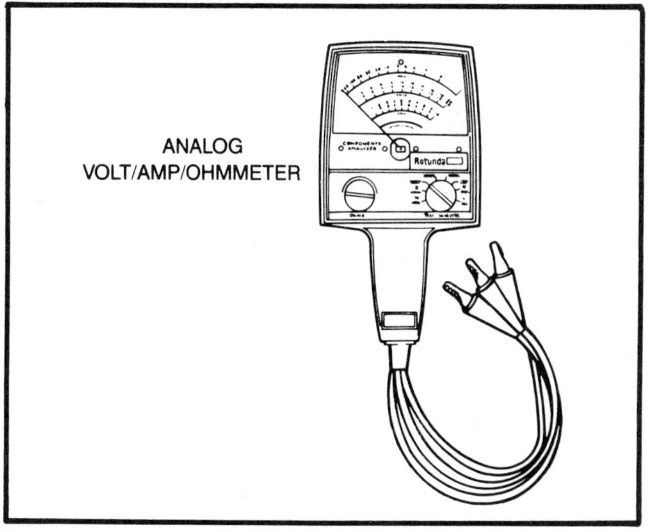

Typical analog-type voltmeter

voltmeter is connected in series with a circuit, only a small amount of current can pass through the voltmeter and flow through the circuit. The circuit will not work properly, but the voltmeter reading will show if the circuit is complete and being energized through the wiring harness and/or electronic control unit.

Available Voltage Measurement

Set the voltmeter selector switch to the 20V position and connect the meter negative lead to the negative post of the battery. Connect the positive meter lead to the positive post of the battery and turn the ignition switch ON to provide a load. Read the voltage level measured on the meter or digital display. A well-charged battery should register over 12 volts. If the meter reads below 11.5 volts, the battery power may be insufficient to operate the electrical system properly. This test determines the amount of voltage available from the battery and should be the first step in any electrical trouble diagnosis procedure. Many electrical problems, especially on computer controlled systems, can be caused by a low state of charge in the battery. Excessive corrosion at the battery cable terminals at the battery, ground or starter can cause a poor contact that will prevent proper charging and full battery current flow.

Normal battery voltage is 12 volts when fully charged. When the battery is supplying current to one or more circuits it is said to be "under load". When everything is off the electrical system is under a "no-load" condition. A fully charged battery may show about 12.5 volts at no load; will drop to 12 volts under medium load; and will drop even lower under heavy load. If the battery is partially discharged the voltage decrease under heavy load may be excessive, even though the battery shows 12 volts or more at no load. When allowed to discharge further, the battery's available voltage under load will decrease more severely. For this reason, it is important that the battery be fully charged during all testing procedures to avoid errors in diagnosis and incorrect test results. See the general battery charging precautions outlined earlier.

Voltage Drop

When current flows through a resistance, the voltage beyond the resistance is reduced (the larger the current, the greater the reduction in voltage). When no current is flowing, there is no voltage drop because there is no current flow. All points in the circuit which are connected to the power source are at the same voltage as the power source. The total voltage drop always equals the total source voltage. In a long circuit with many connectors, a series of small, unwanted voltage drops due to corrosion at the connectors can add up to a total loss of voltage which impairs the operation of the normal loads in the circuit.

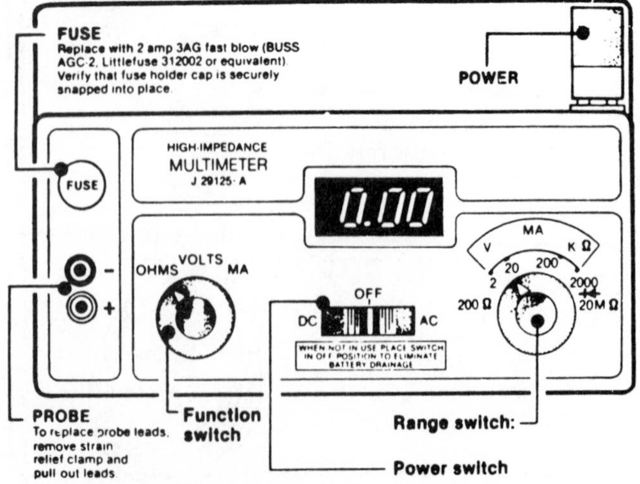

A high impedence digital multimeter (volt-ohmmeter) is necessary for testing computerized engine control systems

INDIRECT COMPUTATION OF VOLTAGE DROPS

1. Set the voltmeter selector switch to the 20 volt position.
2. Connect the meter negative lead to a good ground.
3. Probe all resistances in the circuit with the positive meter lead.
4. Operate the circuit in all modes and observe the voltage readings. Record the voltage values for later reference.

DIRECT MEASUREMENT OF VOLTAGE DROPS

1. Set the voltmeter switch to the 20 volt position.
2. Connect the voltmeter negative lead to the ground side of the resistance load to be measured.
3. Connect the positive lead to the positive side of the resistance or load to be measured.
4. Read the voltage drop directly on the 20 volt scale and record the results for later reference.

Too high a voltage indicates too high a resistance. If, for example, a blower motor runs too slowly, you can determine if there is too high a resistance in the resistor pack. By taking voltage drop readings in all parts of the circuit, you can isolate the problem. Too low a voltage drop indicates too low a resistance. If, for example, a blower motor runs too fast in the MED and/or LOW position, the problem can be isolated in the resistor pack by taking voltage drop readings in all parts of the circuit to locate a possibly shorted resistor. The maximum allowable voltage drop under load is critical, especially if there is more than one high resistance problem in a circuit because all voltage drops are cumulative. A small drop is normal due to the resistance of the conductors.

High Resistance Testing

1. Set the voltmeter selector switch to the 4 volt position.
2. Connect the voltmeter positive lead to the positive post of the battery.
3. Turn on the headlights and heater blower to provide a load for proper circuit operation.
4. Probe various points in the circuit with the negative voltmeter lead.
5. Read the voltage drop on the 4 volt scale. Some average maximum allowable voltage drops are:

FUSE PANEL..0.7 volt
IGNITION SWITCH ....0.5 volt
HEADLIGHT SWITCH..............................0.7 volt
IGNITION COIL (+)................................0.5 volt
ANY OTHER LOAD.................................1.3 volt

NOTE: Voltage drops are all measured while a load is operating; without current flow, there will be no voltage drop.

OHMMETER

The ohmmeter is designed to read resistance (ohms) in a circuit or component. Although there are several different styles of ohmmeters, all will have a selector switch or a number of buttons which permits the measurement of different ranges of resistance (usually the selector switch allows the multiplication of the meter reading by 10, 100, 1000, and 10,000). A calibration knob allows the analog type meter to be set at zero for accurate measurement. Digital type ohmmeters are usually self-calibrating. Since all ohmmeters are powered by an internal battery (usually 9 volts), the ohmmeter can be used as a self-powered test light or continuity tester. When the ohmmeter is connected, current from the ohmmeter flows through the circuit or component being tested. Since the ohmmeter's internal resistance and voltage are known values, the amount of current flow through the meter depends on the resistance of the circuit or component being tested.

The ohmmeter can be used to perform continuity test for open or short circuits (either by observation of the meter needle or as a self-powered test light), and to read actual resistance in a circuit. It should be noted that the ohmmeter is used to check the resistance of a component or wire while there is no voltage applied to the circuit (wire harness disconnected). Current flow from an outside voltage source (such as the vehicle battery) can damage the ohmmeter, so the circuit or component should be isolated from the vehicle 12 volt electrical system before any testing is done. Since the ohmmeter uses its own voltage source, either test lead can be connected to any test point.

NOTE: When checking diodes or other solid state components, the ohmmeter leads can only be connected one way in order to measure current flow in a single direction. Make sure the positive (+) and negative (−) terminal connections are as described in the test procedures to verify the one-way diode operation.

In using the ohmmeter for making continuity checks, do not be too concerned with the actual resistance readings. Zero resistance, or any resistance readings, indicate continuity in the circuit. Infinite resistance indicates an open in the circuit. A high resistance reading where there should be none indicates a problem in the circuit. Checks for short circuits are made in the same manner as checks for open circuits except that the circuit must be isolated from both power and normal ground. Infinite resistance indicates no continuity to ground, while zero resistance indicates a dead short to ground. This test is useful for testing temperature sensors.

Measuring Resistance

The batteries in an analog ohmmeter will weaken with age and temperature, so the ohmmeter must be calibrated or "zeroed", before taking measurements, to insure accurate readings. To zero the meter, place the selector switch in its lowest range and touch the two ohmmeter leads together. Turn the calibration knob until the meter needle is exactly on zero.

NOTE: All analog (needle) type ohmmeters must be zeroed before use, but some digital ohmmeter models are automatically calibrated when the switch is turned on. Self-calibrating digital ohmmeters do not have an adjusting knob, but it's a good idea to check for a zero readout before use by touching the leads together. Some computer controlled systems require the use of a digital ohmmeter with at least 10 megohms impedance for testing. Before any test procedures are attempted, make sure the ohmmeter used is compatible with the electrical system or damage to the on-board computer could result.

To measure the resistance of a circuit, first isolate the circuit from the vehicle power source by disconnecting the battery cables or the harness connector. Make sure the key is OFF when disconnecting any components or the battery. Where necessary, also isolate at least one side of the circuit to be checked to avoid reading parallel resistances. Parallel circuit resistances will always give a lower reading than the actual resistance of either of the branches. When measuring the resistance of parallel circuits, the total resistance will always be lower than the smallest resistance in the circuit. Connect the meter leads to both sides of the circuit (wire or component) and read the actual measured ohms on the meter scale. Make sure the selector switch is set to the proper ohm scale for the circuit being tested to avoid misreading the ohmmeter test value. Specific component tests are outlined in the individual system sections.

CAUTION

Never use an ohmmeter for testing with power applied to the circuit. Like the self-powered test light, the ohmmeter is designed to operate on its own power supply. The normal 12 volt automotive electrical system current could damage the meter.

AMMETERS

An ammeter measures the amount of current flowing through a circuit in units called amperes or amps. Amperes are units of electron flow which indicate how fast the electrons are flowing through the circuit. Since Ohms Law dictates that current flow in a circuit is equal to the circuit voltage divided by the total circuit resistance, increasing voltage also increases the current level (amps). Likewise, any decrease in resistance will increase the amount of amps in a circuit. At normal operating voltage, most circuits have a characteristic amount of amperes, called "current

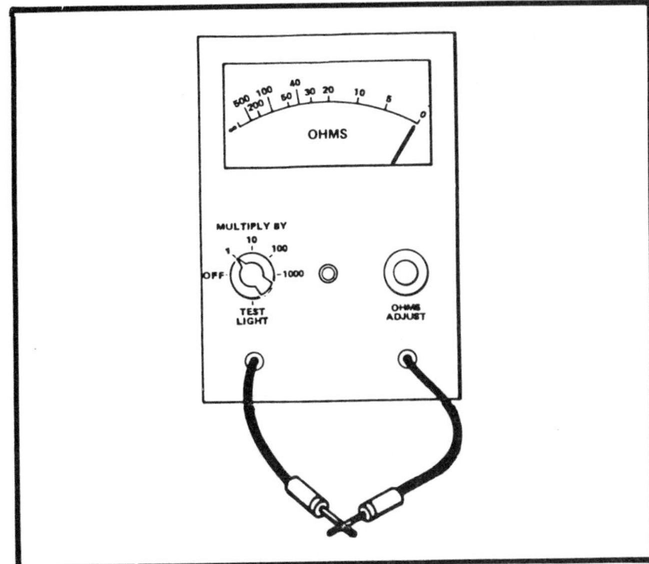

Analog ohmmeters must be calibrated before use by touching the probes together and turning the adjustment knob

draw" which can be measured using an ammeter. By referring to a specified current draw rating, measuring the amperes, and comparing the two values, one can determine what is happening within the circuit to aid in diagnosis. An open circuit, for example, will not allow any current to flow so the ammeter reading will be zero. More current flows through a heavily loaded circuit or when the charging system is operating.

An ammeter is always connected in series with the circuit being tested. All of the current that normally flows through the circuit must also flow through the ammeter; if there is any other path for the current to follow, the ammeter reading will not be accurate. The ammeter itself has very little resistance to current flow and therefore will not affect the circuit, but it will measure current draw only when the circuit is closed and electricity is flowing. Excessive current draw can blow fuses and drain the battery, while

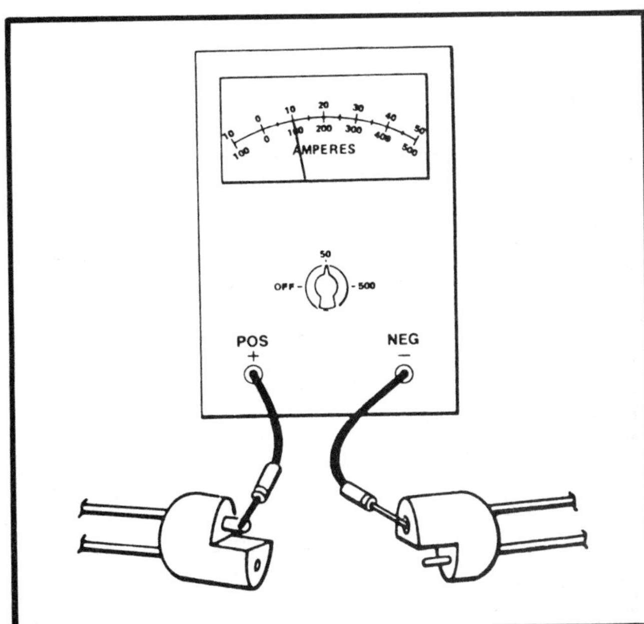

An ammeter must be connected in series with the circuit being tested

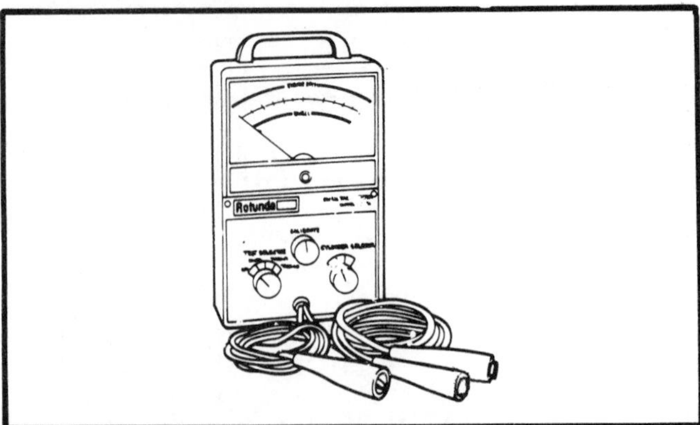
Analog tach-dwell meter with insulated spring clip test leads

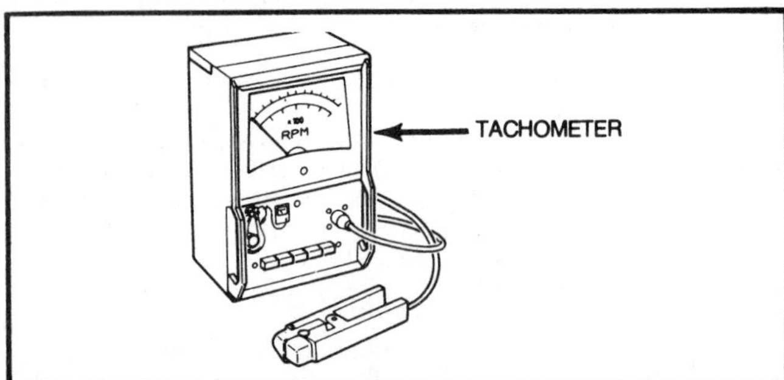

Analog tachometer with inductive pick-up test lead

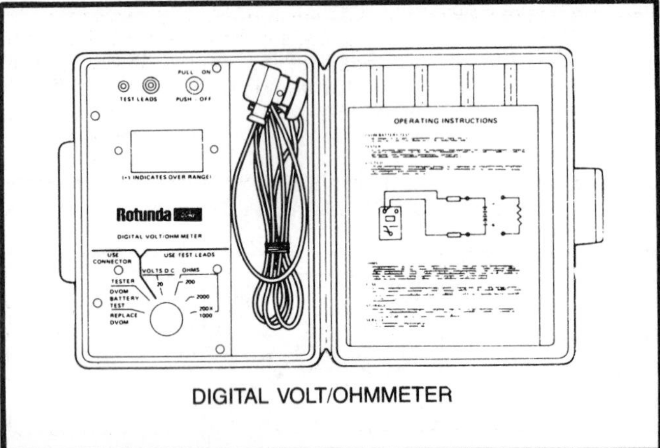

Digital Volt-ohmmeter

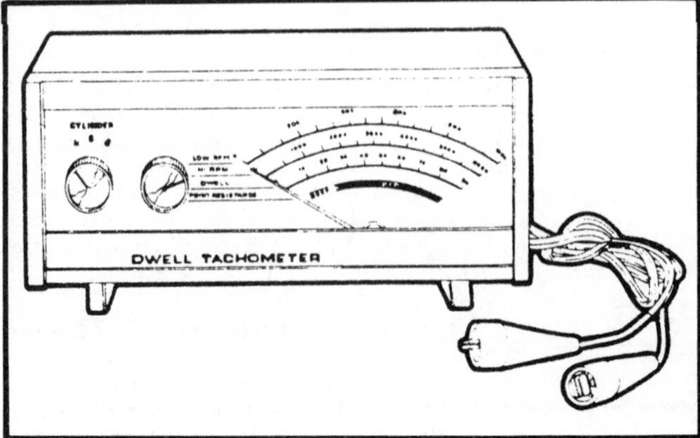

Tach-dwell combination meter

a reduced current draw can cause motors to run slowly, lights to dim and other components to not operate properly. The ammeter can help diagnose these conditions by locating the cause of the high or low reading.

TACHOMETER

The function of the tachometer is to measure engine speed, and it is primarily used for adjusting idle rpm and setting engine speed for other test purposes, such as analyzing the charging system or ignition system. Tachometers are basic instruments for all tuneup work. A wide variety of automotive tachometers are available, from single-range "idle" tachometers to multi-range "universal" instruments. The most common test tachs are dual-range, primary operated units. This means the instrument is connected to the ignition primary circuit, usually at the coil positive (+) terminal. Some have a cylinder selector switch to match the instrument to the number of cylinders in the engine being tested. Others dispense with this switch and provide a separate scale range for each number of cylinders.

NOTE: Some tachometers are designed to use a crankshaft harmonic balancer pickup for the rpm signal.

The "universal" or "secondary" type tachometer is connected to a spark plug wire. This eliminates the need for a cylinder selector switch, although these instruments usually incorporate a 2-cycle/4-cycle switch. This allows them to function accurately on either four-cycle engines (standard automotive), two-cycle engines (outboards) or four-cycle engines with two-cycle or magnetotype ignition systems which produce a spark for every revolution of the crankshaft (outboards, some motorcycles, other small engines). Secondary tachs are useful for servicing electronic ignition systems, but not all primary tachs operate properly on electronic ignition systems.

Single-range tachometers usually measure up to 2000 or 2500 rpm and are used for idle adjustments and other low-speed testing. Dual-range tachs usually measure from 0–1000 rpm and 0–5000 rpm, although ranges up to 10,000 rpm are also available. The higher ranges are primarily for road testing or dynamometer use. The low range is principally used for accurate idle and timing-speed adjustments, since it is commonly scaled in 20-rpm increments. To provide even closer low-speed readings, some low-range tachometers are expanded. That is, the low end of the range is eliminated, thus stretching out or magnifying the range from about 400 rpm to 1000 rpm. The higher speed ranges are primarily for other purposes, such as checking alternator output at certain speeds or noting the setting of the fast-idle adjustment. Some of the various uses of a tachometer include idle speed resets and adjustments, checking ignition timing advance, ignition system tests, mixture adjustments (if applicable) and emissions testing.

MULTIMETERS

Different combinations of test meters can be built into a single unit designed for specific tests. Some of the more common combination test devices are known as Volt-Amp testers, Tach-Dwell meters, or Digital Multimeters. The Volt-Amp tester is used for charging system, starting system or battery tests and consists of a voltmeter, an ammeter and a variable resistance carbon pile. The voltmeter will usually have at least two ranges for use with 6, 12 and 24 volt systems. The ammeter also has more than one range for testing various levels of battery loads and starter current draw and the carbon pile can be adjusted to offer different amount of resistance. The Volt-Amp tester has heavy leads to carry large amounts of current and many later models have an inductive ammeter pickup that clamps around the wire to simplify test connections. On some models, the ammeter also has a zero-center scale to allow testing of charging and starting systems without switching leads or polarity. A digital multimeter is a voltmeter, ammeter and ohmmeter combined in an instrument which gives a digital readout. These are often used when testing solid state circuits because of their high input impedance (usually 10 megohms or more).

The tach-dwell meter combines a tachometer and a dwell (cam angle) meter and is a specialized kind of voltmeter. The tachometer scale is marked to show engine speed in rpm and the dwell scale is marked to show degrees of distributor shaft rotation. In most electronic ignition systems, dwell is determined by the control unit, but the dwell meter can also be used to check the duty cycle (operation) of some electronic engine control systems. Some tach-dwell meters are powered by an internal battery, while others take their power from the car battery in use. The battery powered testers usually required calibration much like an ohmmeter before testing.

TIMING LIGHT

A timing light is basically a hand-held stroboscope that, when connected to the No. 1 spark plug wire, will flash on and off every time the plug fires. This flashing strobe light will "freeze" the timing marks to allow a check of the ignition timing during service. The timing light may be powered by the car battery, or it can be designed to run off of house current (120 volts), but most of them are designed to operate on a 12 volt current. Some timing lights have an adapter to allow connection to the No. 1 spark plug wire and others use an inductive pickup that simply clamps around the plug wire, but under no circumstances should a spark plug wire be punctured with a sharp probe to make a connection. Simple timing lights do nothing more than flash when the plug fires, but there are other types called adjustable timing lights that contain circuitry which delays the flashing of the light. This delay feature is adjustable and the amount of deadly in degrees is measured by a meter built into the timing light. The adjustable timing light can be used to test advance mechanisms and timing control systems.

OSCILLOSCOPE

An oscilloscope is a sophisticated electronic test device that shows the changing voltage levels in a electrical circuit over a period of time. The scope has a display screen much like the picture tube of a television set and displays a line of light called a trace which indicates the voltage levels. The screen is usually marked with voltage and time scales. The ignition waveform pattern on the oscilloscope screen indicates the point of ignition, spark duration, voltage level, coil/condenser operation, dwell angle and the condition of the spark plugs and wires.

The oscilloscope patterns formed by electronic systems vary slightly from one manufacturer to another, but generally the firing section of a trace can be interpreted the same way for all models because the same basic thing is happening to all systems. The length of the dwell section, for example, is not important to an electronic ignition system, although some systems are designed to lengthen the dwell at higher engine rpm. The dwell reading that indicates the duty cycle of a fuel mixture control device, on the other hand, is very important and in fact may go a long way toward indicating a problem or verifying that everything is functioning normally.

NOTE: Follow the manufacturer's instructions for all oscilloscope connections and test procedures. Make sure the scope being used is compatible with the system being tested before any diagnosis is attempted.

SOLDERING GUN

Soldering is a quick, efficient method of joining metals permanently. Everyone who has the occasion to make electrical repairs should know how to solder. Electrical connections that are soldered are far less likely to come apart and will conduct electricity far better than connections that are only "pig-tailed" together. The most popular (and preferred) method of soldering is with an electric soldering gun. Soldering irons are available in many sizes and wattage ratings. Irons with high wattage ratings deliver higher temperatures and recover lost heat faster. A small soldering iron rated for no more than 50 watts is recommended for home use, especially on electrical projects where excess heat can damage the components being soldered.

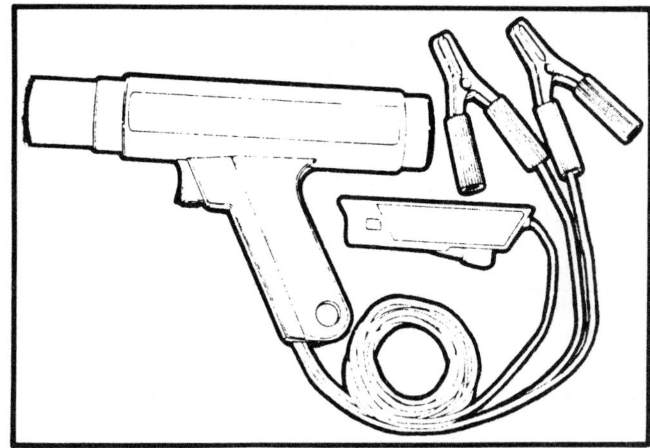

Typical timing light with inductive pick-up

There are three ingredients necessary for successful soldering—proper flux, good solder and sufficient heat. A soldering flux is necessary to clean the metal of tarnish, prepare it for soldering and to enable the solder to spread into tiny crevices. When soldering electrical work, always use a resin flux or resin core solder, which is non-corrosive and will not attract moisture once the job is finished. Other types of flux (acid-core) will leave a residue that will attract moisture, causing the wires to corrode. Tin is a unique metal with a low melting point. In a molten state, it dissolves and alloys easily with many metals. Solder is made by mixing tin (which is very expensive) with lead (which is very inexpensive). The most common proportions are 40/60, 50/50 and 60/40, the percentage of tin always being listed first. Low-priced solders often contain less tin, making them very difficult for a beginner to use because more heat is required to melt the solder. A common solder is 40/60 which is well suited for all-around general use, but 60/40 melts easier, has more tin for a better joint and is preferred for electrical work.

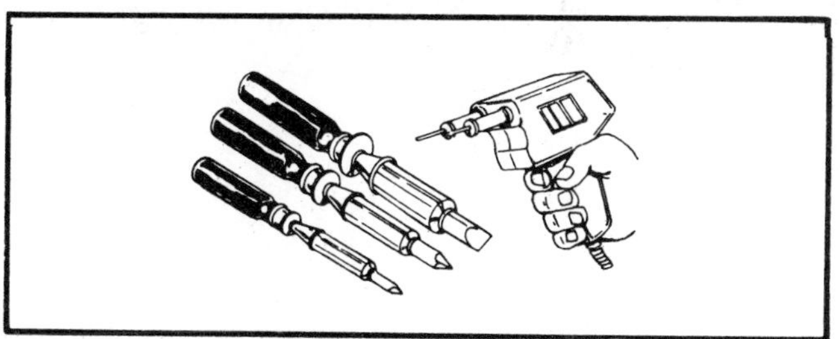

Various types of soldering guns

Soldering Techniques

Successful soldering requires that the metals to be joined be heated to a temperature that will melt the solder, usually somewhere around 360–460°F., depending on the tin content of the solder. Contrary to popular belief, the purpose of the soldering iron is not to melt the solder itself, but to heat the parts being soldered to a temperature high enough to melt solder when it is touched to the work. Melting flux-cored solder on the soldering iron will usually destroy the effectiveness of the flux.

NOTE: Soldering tips are made of copper for good heat conductance, but must be "tinned" regularly for quick transference of heat to the project and to prevent the solder from sticking to the iron. To "tin" the iron, simply heat it and touch flux-cored solder to the tip; the solder will flow over the tip. Wipe the excess off with a rag. Be careful; soldering iron will be hot.

Tinning the soldering iron before use

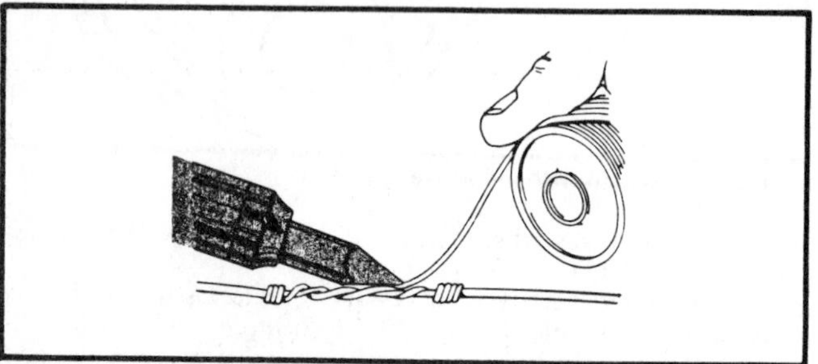
Proper soldering method. Allow the soldering iron to heat the wire first, then apply solder as shown

After some use, the tip may become pitted. If so, simply dress the tip smooth with a smooth file and "tin" the tip again. An old saying holds that "metals well-cleaned are half soldered." Flux-cored solder will remove oxides, but rust, bits of insulation and oil or grease must be removed with a wire brush or emery cloth. For maximum strength in soldered parts, the joint must start off clean and tight. Weak joints will result in gaps too wide for the solder to bridge.

If a separate soldering flux is used, it should be brushed or swabbed on only those areas that are to be soldered. Most solders contain a core of flux and separate fluxing is unnecessary. Hold the work to be soldered firmly. It is best to solder on a wooden board, because a metal vise will only rob the piece to be soldered of heat and make it difficult to melt solder. Hold the soldering tip with the broadest face against the work to be soldered. Apply solder under the tip close to the work; using enough solder to give a heavy film between the iron and piece being soldered, while moving slowly and making sure the solder melts properly. Keep the work level or the solder will run to the lowest part, and favor the thicker parts, because these require more heat to melt the solder. If the soldering tip overheats (the solder coating on the face of the tip burns up), it should be retinned. Once the soldering is completed, let the soldered joint stand until cool. Tape and seal all soldered wire splices after the repair has cooled.

WIRE HARNESS REPAIR PROCEDURES

Condition	Location	Correction
Non-continuity	Using the electric wiring diagram and the wiring harness diagram as a guideline, check the continuity of the circuit in question by using a tester, and check for breaks, loose connector couplings, or loose terminal crimp contacts. Crimp by using pliers Solder	**Breaks**—Reconnect the point of the break by using solder. If the wire is too short and the connection is impossible, extend it by using a wire of the same or larger size. Solder Be careful concerning the size of wire used for the extension **Loose couplings**—Hold the connector securely, and insert it until there is a definite joining of the coupling. If the connector is equipped with a locking mechanism, insert the connector until it is locked securely. **Loose terminal crimp contacts**—Remove approximately 2 in. (5mm) of the insulation covering from the end of the wire, crimp the terminal contact by using a pair of pliers, and then, in addition, complete the repair by soldering.
Short-circuit	Using the electric wiring diagram and the wiring harness diagram as a guideline, check the entire circuit for pinched wires.	Remove the pinched portion, and then repair any breaks in the insulation covering with tape. Repair breaks of the wire by soldering.
Loose terminal	Pull the wiring lightly from the connector. A special terminal removal tool may be necessary for complete removal.	Raise the terminal catch pin, and then insert it until a definite clicking sound is heard. Catch pin

Note: There is the chance of short circuits being caused by insulation damage at soldered points. To avoid this possibility, wrap all splices with electrical tape and use a layer of silicone to seal the connection against moisture. Incorrect repairs can cause malfunctions by creating excessive resistance in a circuit.

WIRE HARNESS AND CONNECTORS

Repairs and Replacement

The on-board computer (ECM) wire harness electrically connects the control unit to the various solenoids, switches, and sensors in the engine compartment. Most connectors in the engine compartment are protected against moisture and dirt which could create oxidation and deposits on the terminals. This protection is important because of the very low voltage and current levels used by the computer and engine sensors. All connectors have a lock which secures the male and female terminals together; a secondary lock holds the seal and terminal into the connector. Both terminal locks must be released when disconnecting ECM connectors.

These special connectors are weather-proof and all repairs require the use of a special terminal and the tool required to service it. This tool is used to remove the pin and sleeve terminals. If removal is attempted with an ordinary pick, there is a good chance that the terminal will be bent or deformed. Unlike standard blade type terminals, these terminals cannot be straightened once they are bent. Make certain that the connectors are properly seated and all of the sealing rings in place when connecting leads. On some models, a hinge-type flap provides a backup, or secondary locking feature for the terminals. Most secondary locks are used to improve the connector reliability by retaining the terminals if the small terminal lock tangs are not positioned properly.

Molded-on connectors require complete replacement of the connection. This means splicing a new connector assembly into the harness. All splices in on-board computer systems should be soldered to insure proper contact. Use care when probing the connections or replacing terminals in them; it is possible to short between opposite terminals. If this happens to the wrong terminal pair, it is possible to damage certain components. Always use jumper wires between connectors for circuit checking. NEVER probe through the weather-proof seals.

Open circuits are often difficult to locate by sight because corrosion or terminal misalignment are hidden by the connectors. Merely wiggling a connector on a sensor or in the wiring harness may correct the open circuit condition. This should always be considered when an open circuit or failed sensor is indicated. Intermittent problems may also be caused by oxidized or loose connections. When using a circuit tester for diagnosis, always probe connectors from the wire side. Be careful not to damage sealed connectors with test probes.

All wiring harnesses should be replaced with identical parts, using the same gauge wire and connectors. When signal wires are spliced into a harness, use wire with high temperature insulation only. With the low current and voltage levels found in the system, it is important that the best possible connection at all wire splices be made by soldering the splices together. It is seldom necessary to replace a complete harness. If replacement is necessary, pay close attention to insure proper harness routing. Secure the harness with suitable plastic wire clamps to prevent vibrations from causing the harness to wear in spots or contact hot engine components.

NOTE: Weather-proof connections cannot be replaced with standard connections. Instructions are provided with replacement connector and terminal packages. Some wire harnesses have mounting indicators (usually pieces of colored tape) to mark where the harness is to be secured.

Replacement of Open or Shorted Wires

In making wiring repairs, it's important that you always replace broken or shorted wires with the same gauge wire. The heavier the wire, the smaller the gauge number. Wires are usually color-coded to aid in identification and, whenever possible, the same color coded wire should be used for replacement. A wire stripping and crimping tool is necessary to install solderless terminal connectors properly. Test all crimps by pulling on the wires; it should not be possible to pull the wires out of a good crimp.

NOTE: Some late model on-board computer control systems require that all wiring be soldered to insure proper contact.

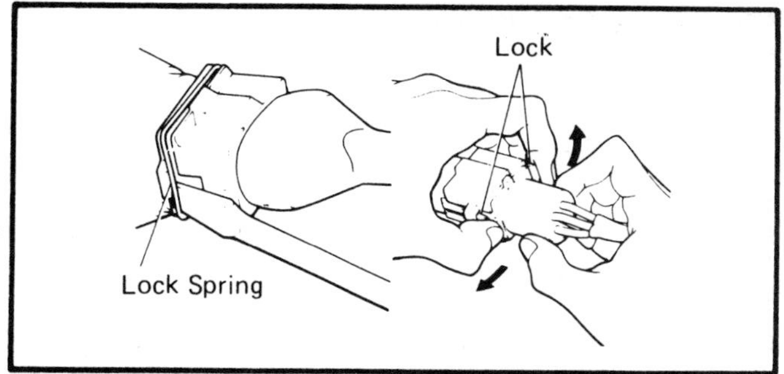

Some electrical connectors use a lock spring instead of the molded locking tabs

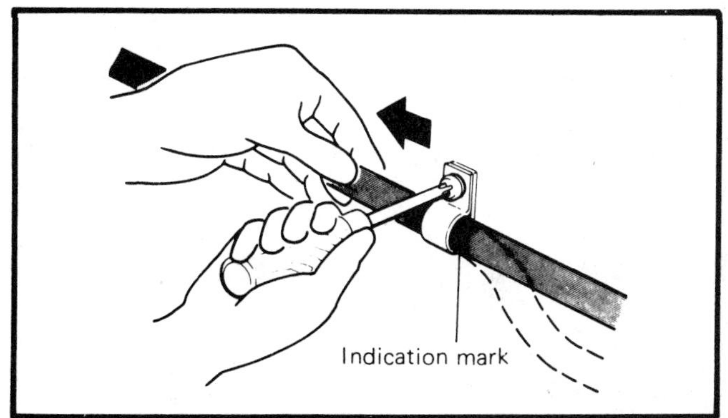

Secure the wiring harness at the indication marks, if used, to prevent vibrations from causing wear and a possible short

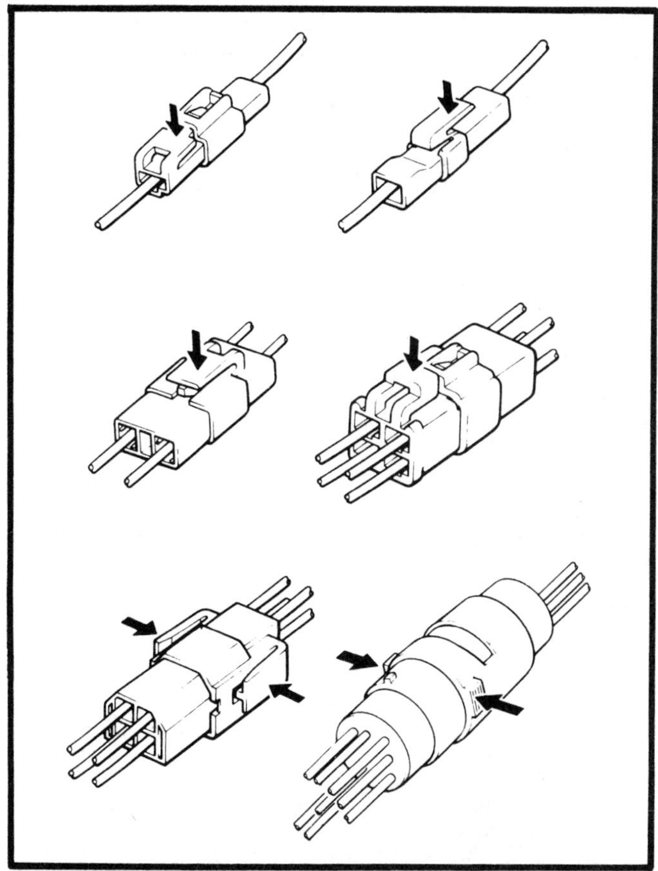

Various types of locking harness connectors. Depress the locks at the arrows to separate the connector

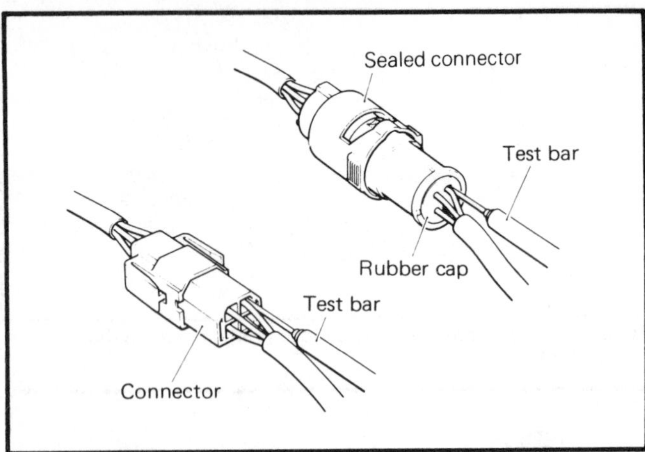

Probe all connectors from the wire side when testing

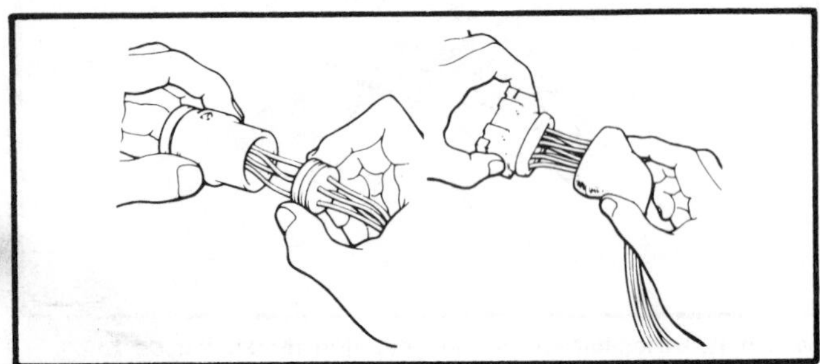
Slide back the weatherproof seals or boots on sealed terminals for testing, if necessary

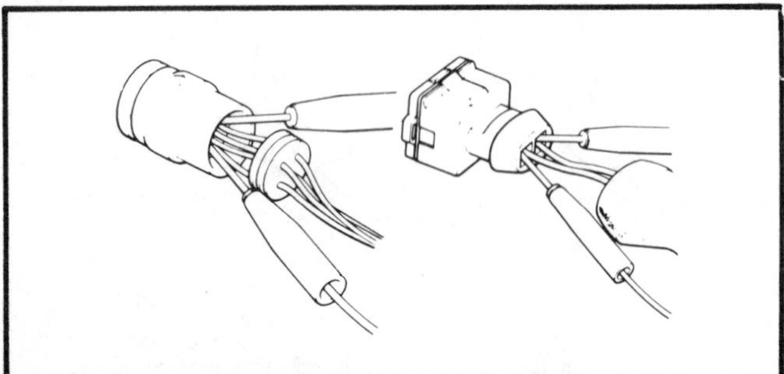
Correct method of testing weatherproof connectors—do not pierce connector seals with test probes

Wires which are open, exposed or otherwise damaged are repaired by simple splicing. Where possible, if the wiring harness is accessible and the damaged place in the wire can be located, it is best to open the harness and check for all possible damage. In an inaccessible harness, the wire must be bypassed with a new insert, usually taped to the outside of the old harness.

Fusible Links

When replacing fusible links, be sure to use fusible link wire, NOT ordinary automotive wire. Make sure the fusible wire segment is of the same gauge and construction as the one being replaced and double the stripped end when crimping the terminal connector for a good contact. The melted (open) fusible link segment of the wiring harness should be cut off as close to the harness as possible, then a new segment spliced in as described. In the case of a damaged fusible link that feeds two harness wires, the harness connections should be replaced with two fusible link wires so that each circuit will have its own separate protection.

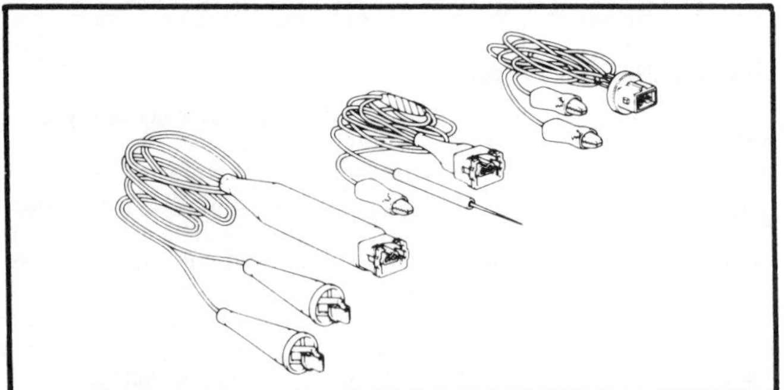
Special purpose testing connectors for use on fuel injected engines

NOTE: Most of the problems caused in the wiring harness are usually due to bad ground connections. Always check all vehicle ground connections for corrosion or looseness before performing any power feed checks to eliminate the chance of a bad ground affecting the circuit. Electrical wiring connectors can be classified according to the type of terminals (such as pin terminals or flat terminals), the number of poles (terminals), whether they are male or female, whether they have a locking device or not, etc.

Repairing Hard Shell Connectors

Unlike molded connectors, the terminal contacts in hard shell connectors can be replaced. Weatherproof hard-shell connectors (like solid state ignition module connectors) with the leads molded into the shell have non-replaceable terminal ends. Replacement usually involves the use of a special terminal removal tool that depress the locking tangs (barbs) on the connector terminal and allow the connector to be removed from the rear of the shell. The connector shell should be replaced if it shows any evidence of burning, melting, cracks, or breaks. Replace individual terminals that are burnt, corroded, distorted or loose.

The insulation crimp must be tight to prevent the insulation from sliding back on the wire when the wire is pulled. The insulation must be visibly compressed under the crimp tabs, and the ends of the crimp should be turned in for a firm grip on the insulation. The wire crimp must be made with all wire strands inside the crimp. The terminal must be fully compressed on the wire strands with the ends of the crimp tabs turned in to make a firm grip on the wire. Check all connections with an ohmmeter to insure a good contact. There should be no measurable resistance between the wire and the terminal when connected.

Diagnostic Connectors

Some cars have diagnostic connectors mounted on the firewall or fender panel. Each terminal is connected in parallel with a voltage test point in the electrical system. Many test equipment companies make special testers that plug into the connector. These have the advantage of fast hookup and test time, but in most cases the same tests can be done with a suitable voltmeter. By connecting the voltmeter negative (−) lead to the diagnostic ground terminal and the positive (+) lead to the other terminals in turn, one can make seven or eight voltage tests from a single point on the car. Jumper wires may be necessary to connect some voltmeters to the connector.

On GM models, later model diagnostic connector has a variety of information available. Called Serial Data, there are several tools on the market for reading this information. These tools do not make the use of diagnostic charts unnecessary. They do not tell exactly where a problem is in a given circuit. However, with an understanding of what each position on the equipment measures, and knowledge of the circuit involved, the tools can be very useful in getting information which would be more time consuming to get with other equipment. In some cases, it will provide information that is either extremely difficult or impossible to get with other equipment. When a chart calls for a sensor reading, the diagnostic

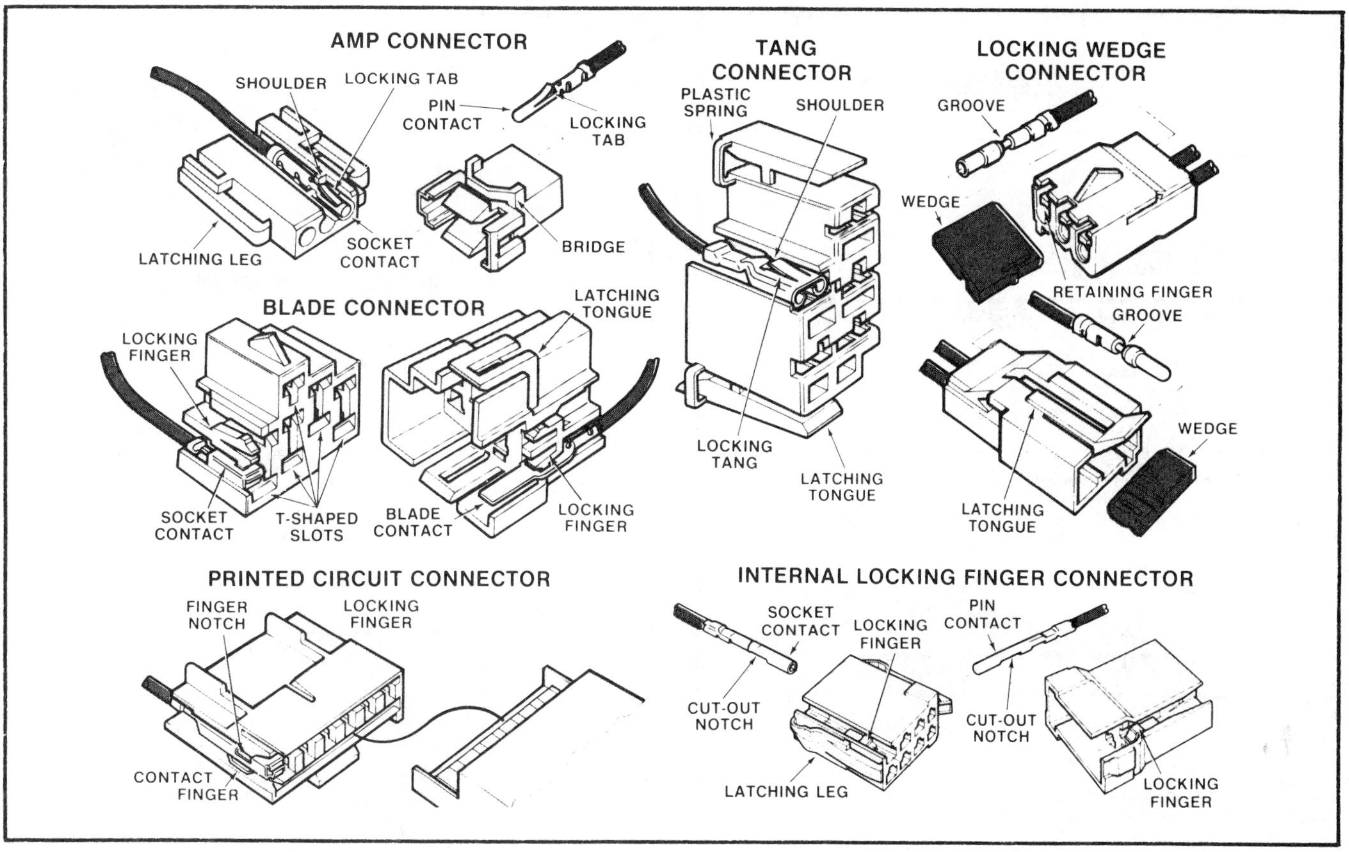

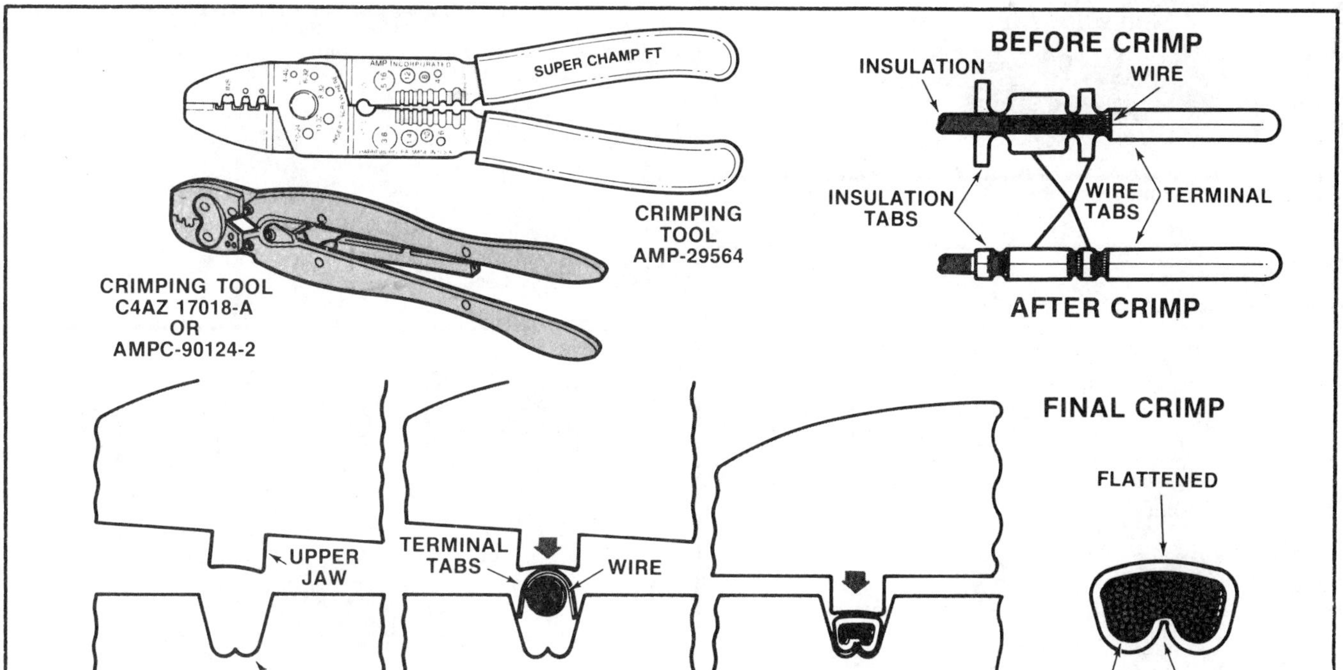

Correct method of crimping terminals with special tool

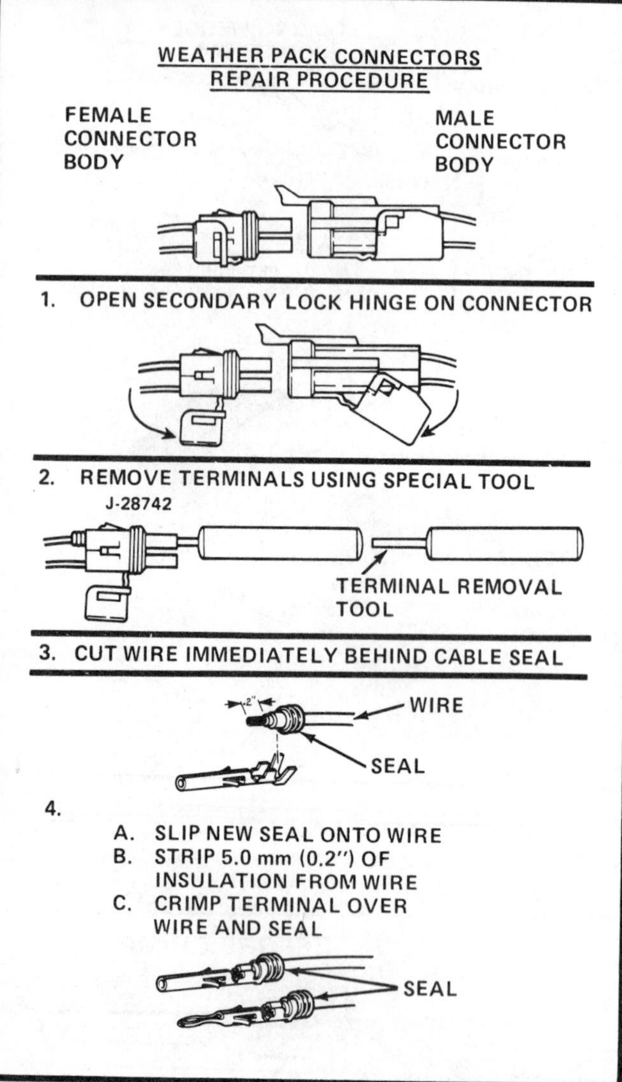

Repairing GM Weatherpak connectors. Note special terminal removal tools

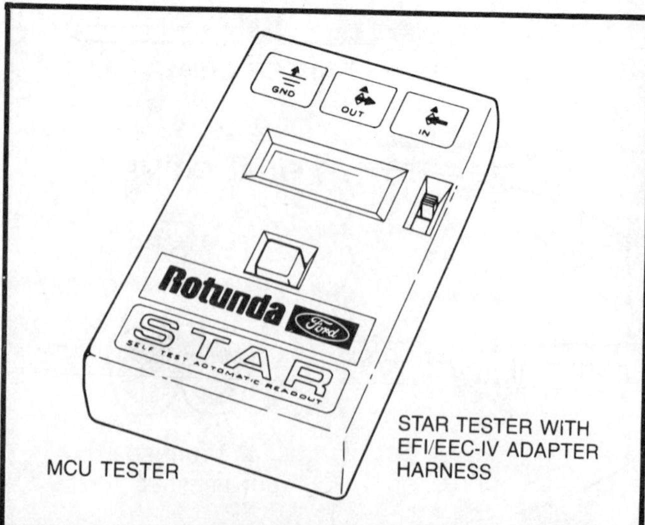

Self-Test and Automatic Readout (STAR) tester used for obtaining trouble codes from Ford MCU and EEC IV systems

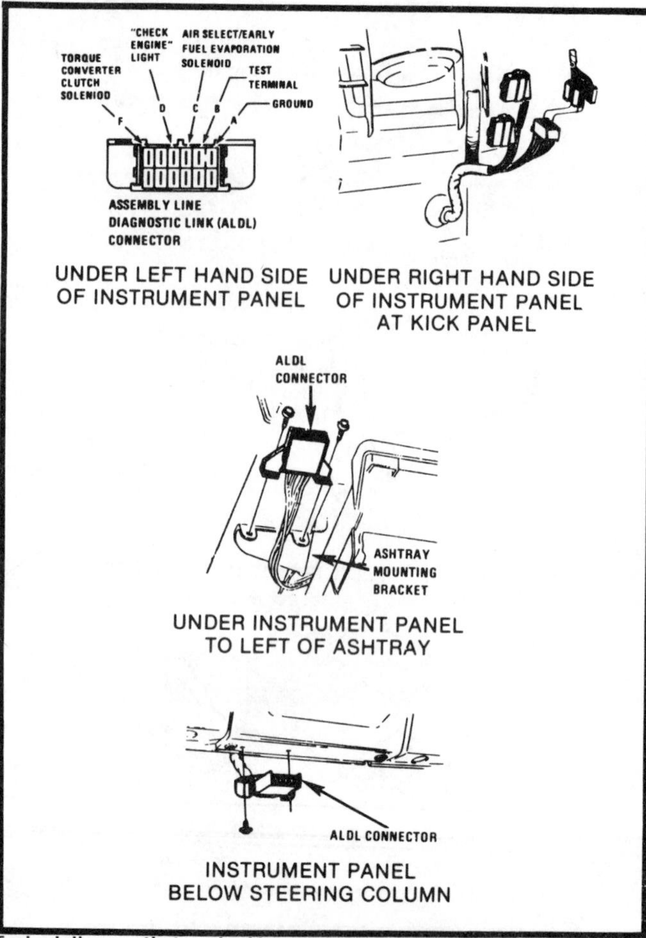

Typical diagnostic terminal locations on GM models. The diagnosis terminals are usually mounted under the dash or in the engine compartment

connector tool can be used to read the following voltage signals directly:

- Park/Neutral
- Throttle Position Sensor
- Manifold Absolute Pressure Sensor
- Coolant Temperature Sensor
- Vehicle Speed Sensor

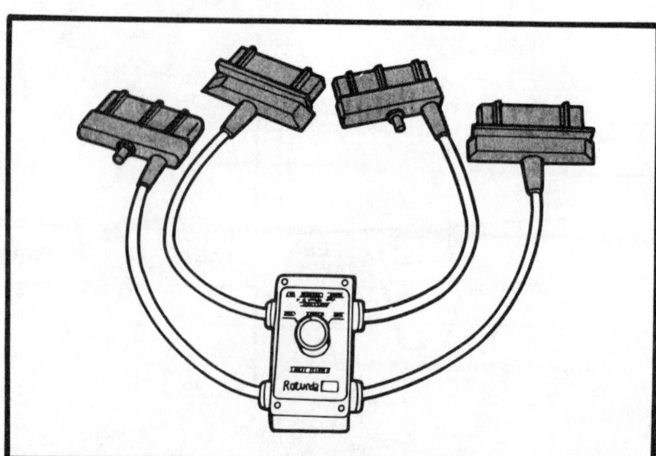

Typical adapter wiring harness for connecting tester to diagnostic terminal

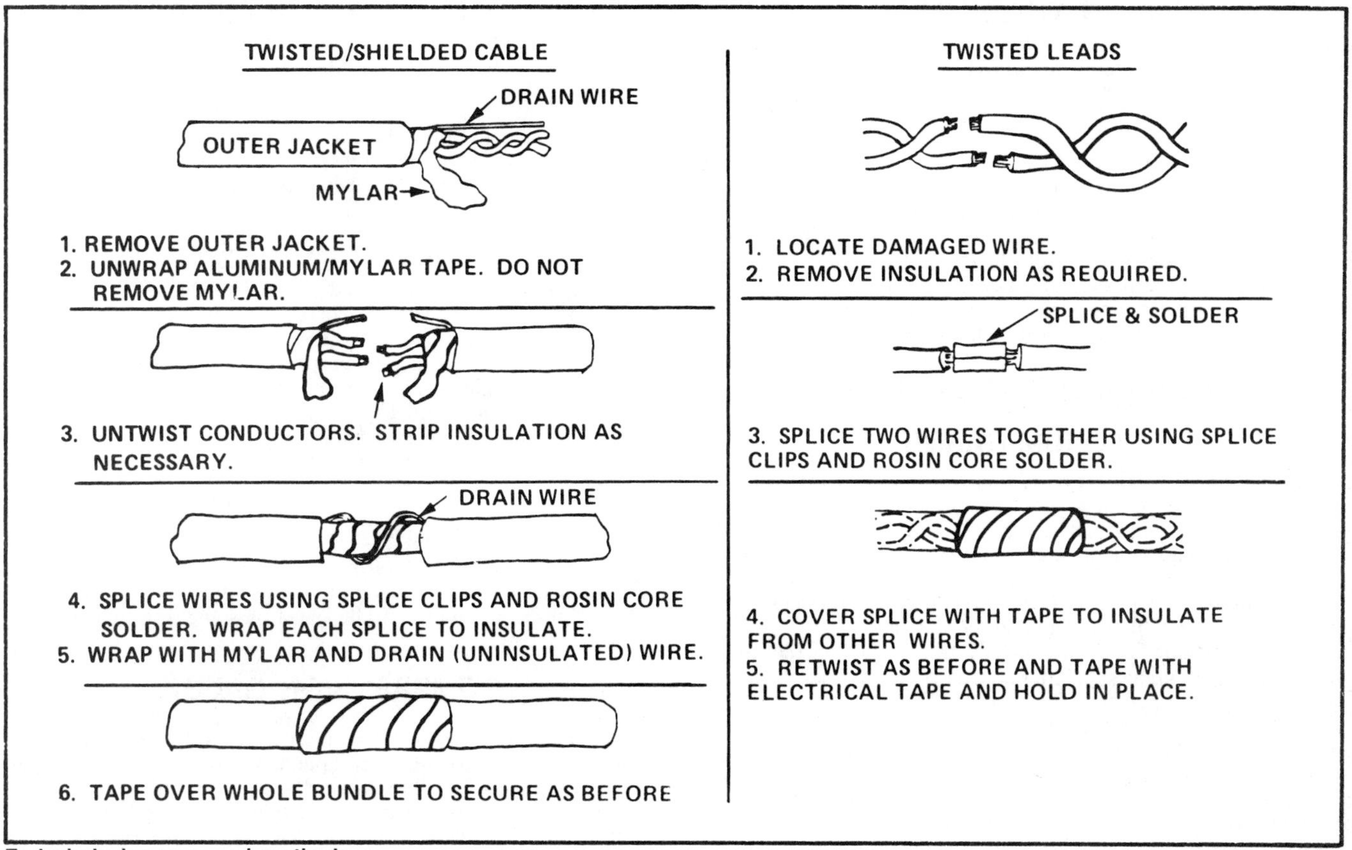

TWISTED/SHIELDED CABLE

DRAIN WIRE
OUTER JACKET
MYLAR

1. REMOVE OUTER JACKET.
2. UNWRAP ALUMINUM/MYLAR TAPE. DO NOT REMOVE MYLAR.

3. UNTWIST CONDUCTORS. STRIP INSULATION AS NECESSARY.

DRAIN WIRE

4. SPLICE WIRES USING SPLICE CLIPS AND ROSIN CORE SOLDER. WRAP EACH SPLICE TO INSULATE.
5. WRAP WITH MYLAR AND DRAIN (UNINSULATED) WIRE.

6. TAPE OVER WHOLE BUNDLE TO SECURE AS BEFORE

TWISTED LEADS

1. LOCATE DAMAGED WIRE.
2. REMOVE INSULATION AS REQUIRED.

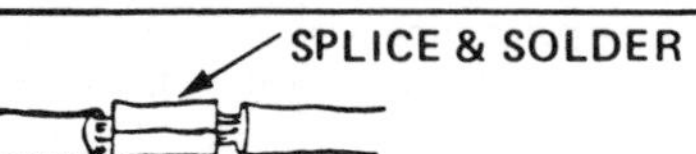

3. SPLICE TWO WIRES TOGETHER USING SPLICE CLIPS AND ROSIN CORE SOLDER.

4. COVER SPLICE WITH TAPE TO INSULATE FROM OTHER WIRES.
5. RETWIST AS BEFORE AND TAPE WITH ELECTRICAL TAPE AND HOLD IN PLACE.

Typical wire harness repair methods

When the diagnosis tool is plugged in on GM models, it takes out the timer that keeps the fuel injection system in open loop for a certain period of time. Therefore, it will go closed loop as soon as the car is started, if all other closed loop conditions are met. This means that if, for example, the air management operation were checked with the diagnosis tool plugged in, the air managment system would not function normally.

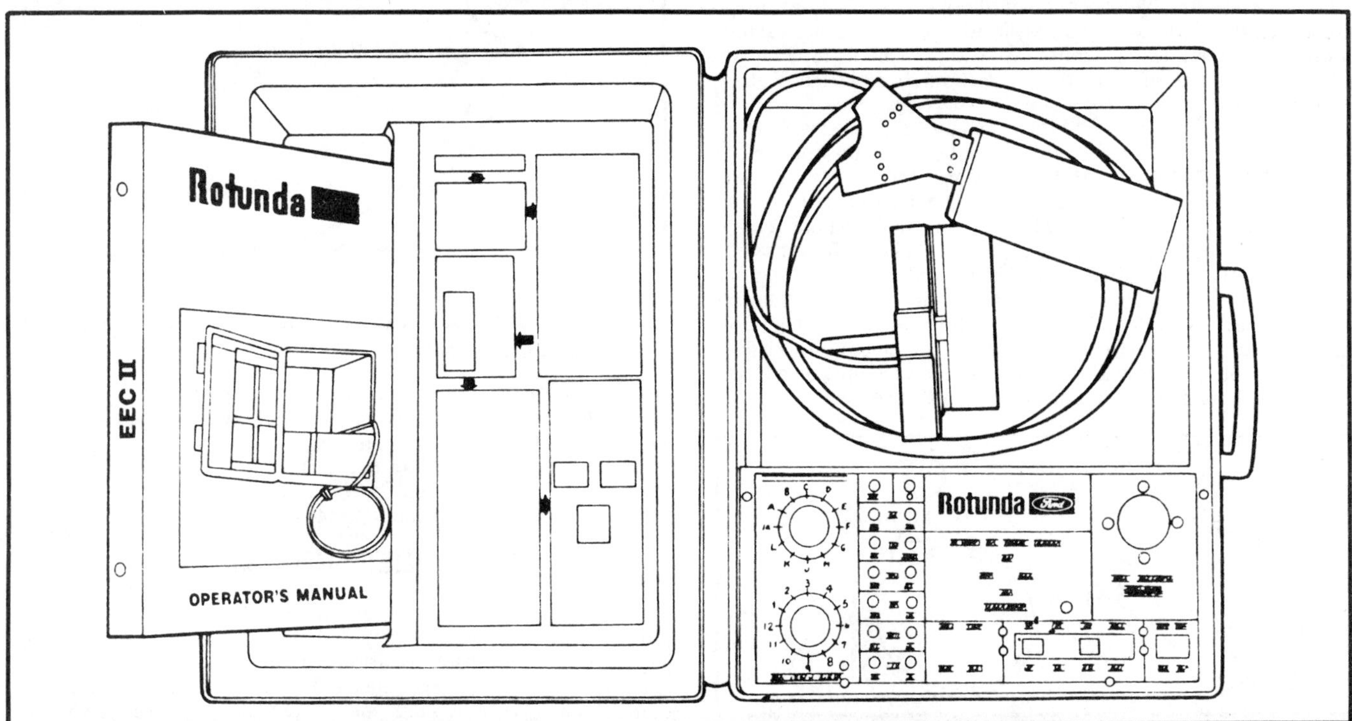

More elaborate electronic engine control testers are necessary for testing Ford's EEC II and EEC III systems

CAUTION

Some manufacturers engine control systems can be damaged by using the underhood diagnostic connector for testing

The diagnosis tester is helpful in cases of intermittent operation. The tool can be plugged in and observed while driving the car under the condition where the light comes ON momentarily, or the engine driveability is poor momentarily. If the problem seems to be related to certain areas that can be checked on the diagnosis tool, then those are the positions that should be checked while driving the car. If there does not seem to be any correlation between the problem and any specific circuit, the diagnosis tool can be checked on each position, watching for a period of time to see if there is any change in the readings that indicates intermittent operation.

Mechanical Test Equipment

VACUUM GAUGE

How much vacuum an engine produces depends on efficiency; a badly worn engine cannot produce as much vacuum as one in good condition. Piston rings, valves, carburetion, ignition timing and exhaust all have a predictable effect on engine vacuum, making it possible to diagnose engine condition and performance by measuring the amount of vacuum. There are two types of vacuum produced by an engine-manifold and ported vacuum.

Manifold vacuum is drawn directly from a tap on the intake manifold and is greatest at idle and decreases as the throttle is opened. At wide-open throttle, there is very little vacuum in the intake manifold. Ported vacuum is drawn from an opening in the carburetor just above the throttle valve. When the throttle valve is closed during idle or deceleration, ported vacuum is low or non-existent. As the throttle is opened, it uncovers the port and the vacuum level becomes essentially the same as manifold vacuum. Ported vacuum is used to operate vacuum switches and diaphragms; it is not used to diagnose engine problems.

Many late model engines have two other vacuum sources at the carburetor, used to control EGR and other emission control devices. EGR ported vacuum is taken from a tap in the carburetor barrel slightly above the ported vacuum tap. The EGR ported vacuum tap often has two openings in the barrel so that vacuum is applied to the EGR control valve in two stages. Another vacuum source on late model carburetors is venturi vacuum, usually used with a vacuum amplifier to control EGR during and slightly above cruising speed.

Most vacuum readings are taken at engine idle speed with a vacuum gauge that measures the difference between atmospheric and intake manifold pressure. Most gauges are graduated in inches of mercury (in. Hg), although a device called a manometer reads vacuum in inches of water (in. H2O). The normal vacuum reading usually varies between 18 and 22 in. Hg at sea level. To test engine vacuum, the vacuum gauge must be connected to a source of manifold vacuum. Many engines have a plug in the intake manifold which can be removed and replaced with an adapter fitting. Connect the vacuum gauge to the fitting with a suitable rubber hose or, if no manifold plug is available, connect the vacuum gauge to any device using manifold vacuum, such as EGR valves, etc.

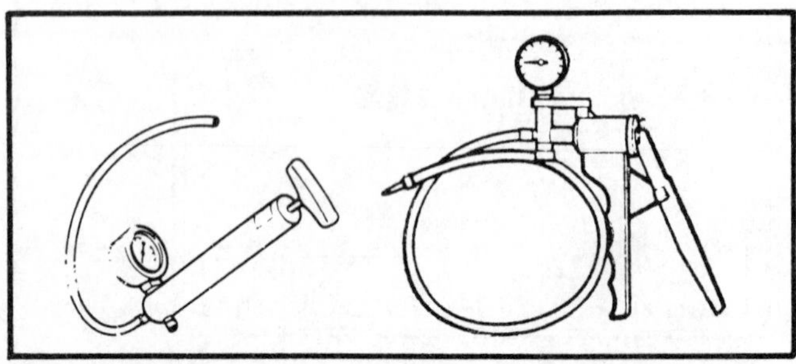

Typical hand vacuum pumps

HAND VACUUM PUMP

Small, hand-held vacuum pumps come in a variety of designs. Most have a built-in vacuum gauge and allow the component to be tested without removing it from the vehicle. Operate the pump lever or plunger to apply the correct amount of vacuum required for the test specified in the diagnosis routines. The level of vacuum in inches of Mercury (in. Hg) is indicated on the pump gauge. For some testing, an additional vacuum gauge may be necessary.

Testing Vacuum Components

Intake manifold vacuum is used to operate various systems and devices on late model cars. To correctly diagnose and solve problems in vacuum control systems, a vacuum source is necessary for testing. In some cases, vacuum can be taken from the intake manifold when the engine is running, but vacuum is normally provided by a hand vacuum pump. These hand vacuum pumps have a built-in vacuum gauge that allow testing while the device is still attached to the car. For some tests, an additional vacuum gauge may be necessary.

MANOMETER

A manometer measures vacuum in inches of water and is required for synchronizing minimum idle speed on some throttle body fuel injection systems.

COMPRESSION GAUGE

A compression gauge is designed to measure the amount of air pressure in psi that a cylinder is capable of producing. Some gauges have a hose that screws into the spark plug hole while others have a tapered rubber tip which is held in the spark plug hole. Engine compression depends on the sealing ability of the rings, valves, head gasket and spark plug gaskets. If any of these parts are not sealing well, compression will be lost and the power output of the engine will be reduced. The compression in each cylinder should be measured and the variation between cylinders noted. The engine should be cranked through five or six compression strokes while it is warm, with all the spark plugs removed.

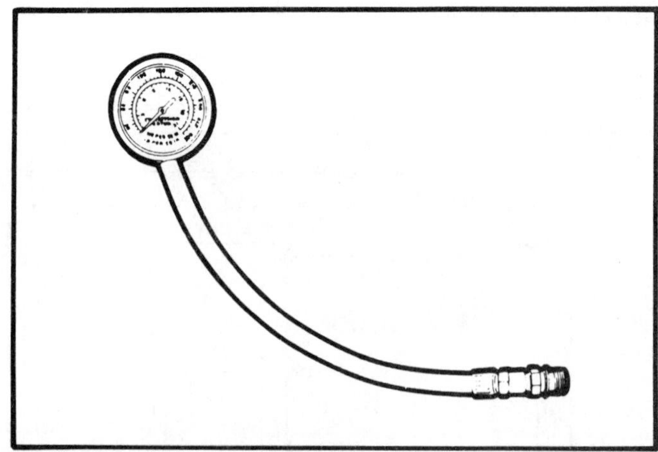

Typical compression tester

FUEL PRESSURE GAUGE

A fuel pressure gauge is required to test the operation of any fuel injection system. Some systems also need a three way valve to check the fuel pressure in various modes of operation, or special adapters for making fuel connections. Always observe the cautions outlined in the individual fuel system sections when working around any pressurized fuel system.

USING A VACUUM GAUGE

White needle = steady needle ***Dark needle = drifting needle***

The vacuum gauge is one of the most useful and easy-to-use diagnostic tools. It is inexpensive, easy to hook up, and provides valuable information about the condition of your engine.

Indication: Normal engine in good condition

Gauge reading: Steady, from 17–22 in./Hg.

Indication: Sticking valve or ignition miss

Gauge reading: Needle fluctuates from 15–20 in./Hg. at idle

Indication: Late ignition or valve timing, low compression, stuck throttle valve, leaking carburetor or manifold gasket.

Gauge reading: Low (15–20 in./Hg.) but steady

Indication: Improper carburetor adjustment, or minor intake leak at carburetor or manifold

Gauge reading: Drifting needle

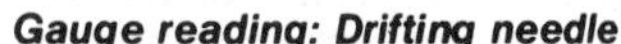

Indication: Weak valve springs, worn valve stem guides, or leaky cylinder head gasket (vibrating excessively at all speeds).

Gauge reading: Needle fluctuates as engine speed increases

Indication: Burnt valve or improper valve clearance. The needle will drop when the defective valve operates.

Gauge reading: Steady needle, but drops regularly

Indication: Choked muffler or obstruction in system. Speed up the engine. Choked muffler will exhibit a slow drop of vacuum to zero.

Gauge reading: Gradual drop in reading at idle

Indication: Worn valve guides

Gauge reading: Needle vibrates excessively at idle, but steadies as engine speed increases

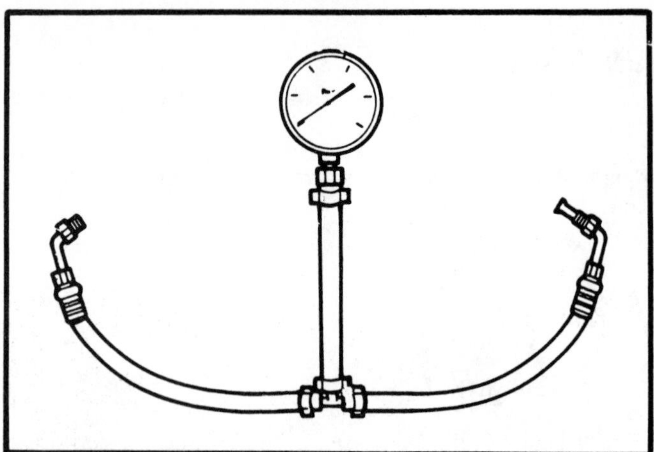

Fuel pressure gauge with tee adapter

Repairing Fuel Lines

CAUTION

Fuel supply lines on vehicles equipped with fuel injected engines will remain pressurized for long periods of time after engine shutdown. The pressure must be relieved before servicing the fuel system.

Vehicles equipped with nylon fuel tubes and push connect fittings have three types of service that can be performed to the fuel lines; replacing nylon tubing (splicing nylon to nylon), replacing push connector retainer clip, and replacing damaged push connect tube end. These nylon lines replace the conventional steel tubing. The individual tubes are taped together by the manufacturer and are supplied as an assembly. The plastic fuel tube assembly is secured to the body rails with nylon wrap-around clips and push-in pins. To make hand insertion of the barbed connectors into the nylon easier, the tube end must be soaked in a cup of boiling water for one minute immediately before pushing the barbs into the nylon. Damaged push connectors must be discarded and replaced with new push connectors. If only the retaining clip is damaged, replace the clip.

CAUTION

The plastic fuel lines can be damaged by torches, welding sparks, grinding and other operations which involve heat and high temperatures. If any repair or service operation will be used which involves heat and high temperatures locate all fuel system components, especially the plastic fuel lines to be certain they will not be damaged.

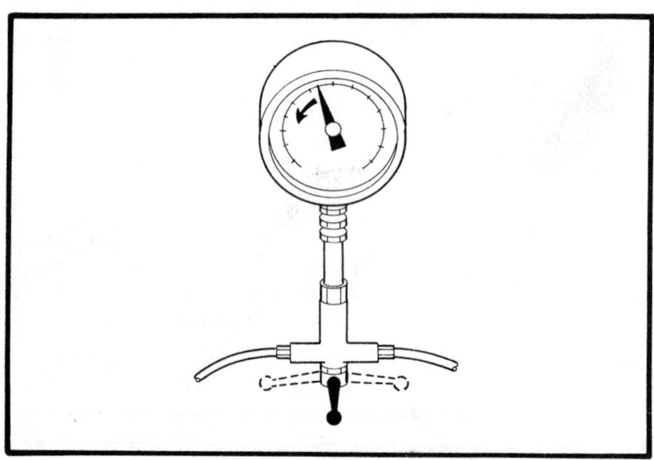

Fuel pressure gauge with three-way valve adapter. Some fuel systems use test points with quick disconnect fittings

On-Board Diagnostic Systems

SECTION 3

INDEX

NOTE: Please refer to the Application Chart at the front of this Manual.

SELF-DIAGNOSING ELECTRONIC ENGINE CONTROL

General Information

Some on-board computer systems are equipped with a self-diagnosis capability to allow retrieval of stored trouble codes from the ECU memory. The number of codes stored and the meaning of the code numbers varies from one manufacturer to another. By activating the diagnostic mode and counting the number of flashes on the CHECK ENGINE or ECU lights, it is possible to ask the computer where the problem is (which circuit) and narrow down the number of pin connectors tested when diagnosing an AFC fuel injection problem. It should be noted that the trouble codes are only an indication of a circuit malfunction, not a component analysis. A trouble code that indicates a specific component has failed may be caused by a loose connection, rather than some mechanical problem. Use the trouble codes as a guide for further diagnosis or you may replace expensive electronic devices unnecessarily.

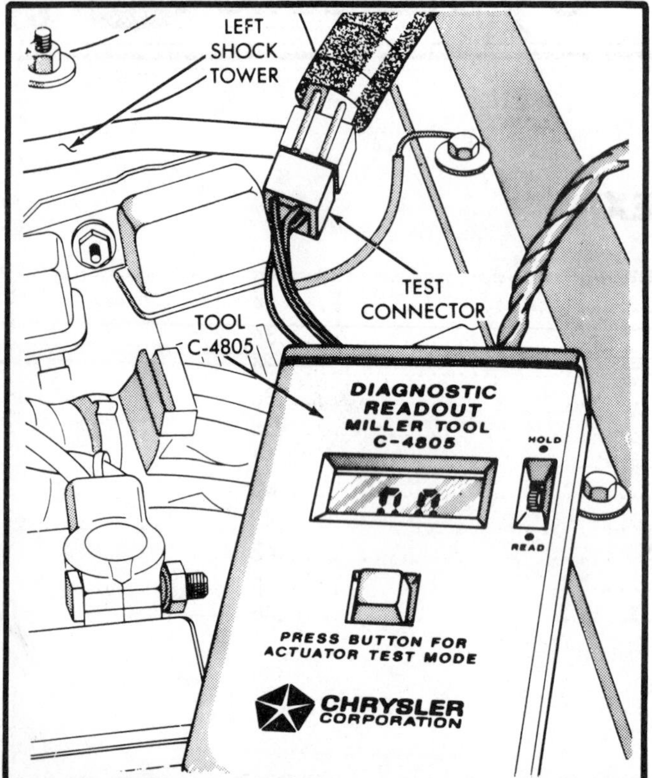

Diagnostic read out tool used to obtain trouble codes from Chrysler ESA system

NOTE: Remember to clear the trouble codes as outlined in the various sections whenever all diagnosis procedures are complete, or the same trouble code may remain stored in the computer memory even though repairs have been made.

On-Board Diagnosis Procedures

The following information describes how to activate the self-diagnosis mode on several of the most popular engine control systems. Do not attempt to enter the diagnostic mode on any electronic computer based system unless all indicated special tools are available. Follow the instructions carefully to avoid misdiagnosis or possible circuit damage from improper connections or procedures. It is very important to cancel the diagnosis mode when all service codes have been recorded in order to restore the computer to its normal operating mode. Some systems may be self-canceling, but most require specific steps to be taken or the computer may be damaged or fail to operate properly when the ignition is switched ON.

Chrysler ESA System

The electronic Spark Advance (ESA) on-board computer is programmed to monitor several different component systems simultaneously. If a problem is detected in a monitored circuit often enough to indicate a malfunction, a fault code is stored in the computer memory for eventual display on a diagnostic readout tool (Part No. C-4805 or equivalent). If the problem is repaired or disappears, the computer cancels the fault code after 30 ignition ON/OFF cycles. If a fault code appears on the diagnostic readout tool, perform a careful visual check of all wiring and vacuum connections. Many problems are the result of loose, disconnected, or cracked vacuum hoses or wiring connectors.

The readout tool can be used to test the engine control system in three different modes; the diagnostic test mode, the circuit actuation test (ATM) mode, and the switch test mode. The diagnostic test mode is used to retrieve fault codes that are stored in the computer memory. The circuit actuation test mode is used to test specific component circuits and systems. The switch test mode is used to test switch circuits and operation. When a fault code appears on the display screen, it indicates that the spark control computer has detected an abnormal signal in the system. Fault codes indicate a problem in a circuit, but do not necessarily identify the failed component.

CHRYSLER ESA SYSTEM TROUBLE CODES

Code Number	Circuit
00	Diagnostic readout tool connected properly
11	02 solenoid control circuit
13	Canister purge solenoid circuit ①
14	Battery has been disconnected
16	Radiator fan control relay ②
17	Electronic throttle control vacuum solenoid ①
18	Vacuum operated secondary control solenoid system ①

CHRYSLER ESA SYSTEM TROUBLE CODES

Code Number	Circuit
21	Distributor pick-up system
22	02 system full rich or full lean
24	Computer (ECU) failure
25	Radiator fan coolant sensor ①
26	Engine temperature sensor
28	Odometer sensor (mileage counter) ③
31	Engine has not been started since battery disconnection
32	Computer (ECU) failure
33	Computer (ECU) failure
55	End of message (trouble codes)
88	Start of message (trouble codes)

① 2.2L engines only
② 2.2L engines only. Disregard if air conditioned
③ Manual transmission only

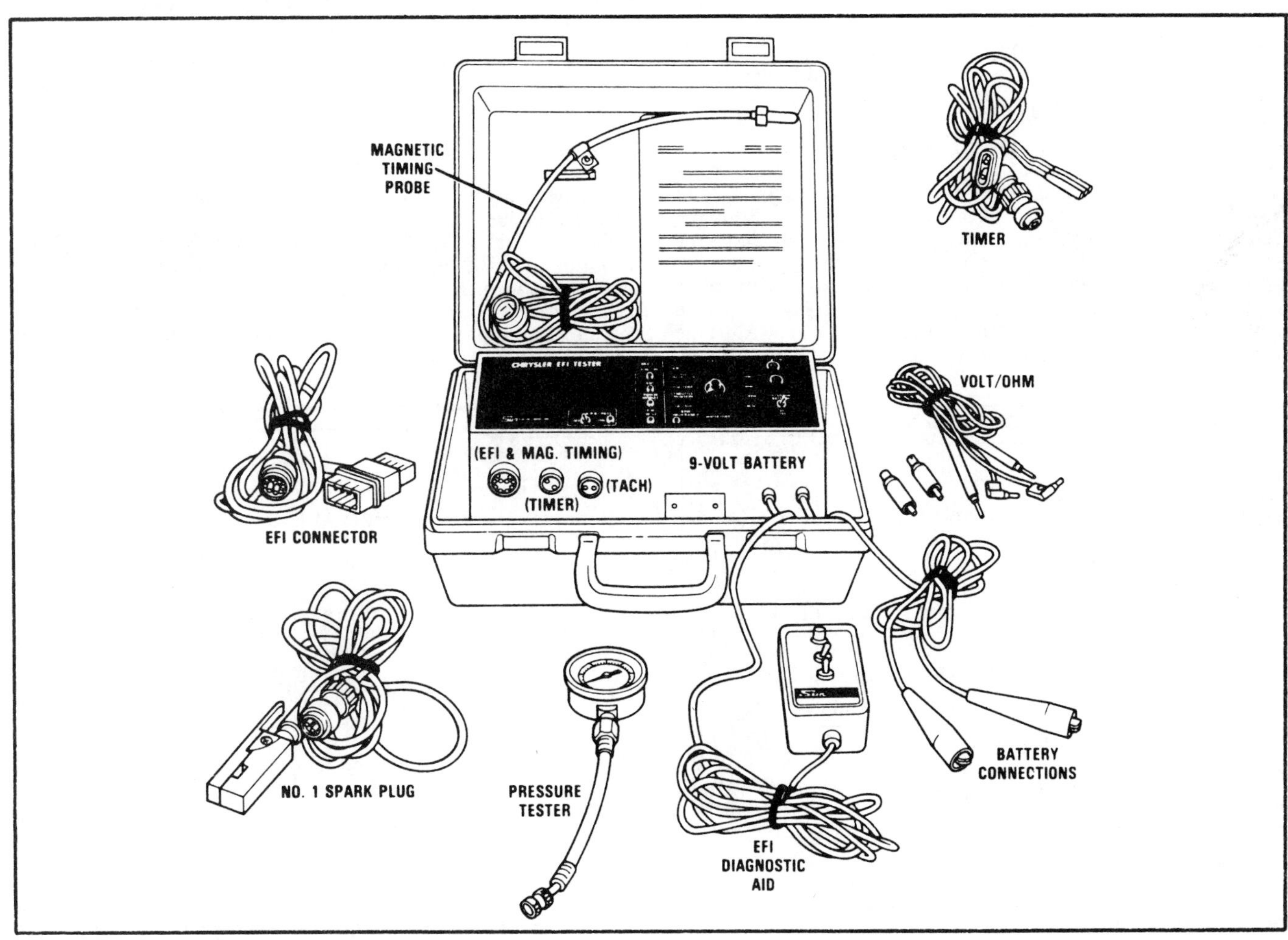

Tester used to diagnose Chrysler EFI system

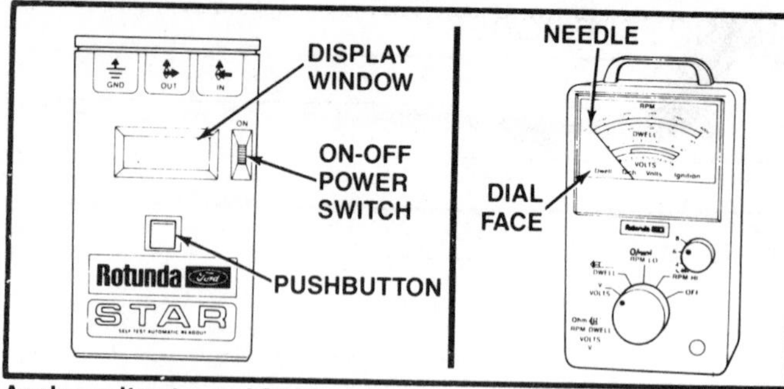

Analog voltmeter and STAR tester

Activating Circuit Actuation Test (ATM) Mode

Place the system into the diagnostic test mode as previously described and wait for code 55 to appear on the diagnostic readout tool display screen. Press the ATM button down until the desired test code appears. The computer will continue to turn the selected circuit on and off for as long as five minutes or until the ATM button is pressed again or the ignition switch is turned OFF. If the ATM button is not pressed a second time, the computer will continue cycling the selected circuit for five minutes and then shut the system off.

TROUBLE CODES

91 Oxygen sensor feedback solenoid activated
92 Shift indicator light activated (manual transmission only)
93 Canister purge solenoid activated
96 Fan relay activated
97 Electronic throttle control solenoid activated
98 Vacuum operated secondary control solenoid activated

Activating Switch Test Mode

Place the system into the diagnostic test mode as previously described and wait for code 55 to appear on the diagnostic readout tool display screen. Check that both the air conditioning and heated rear window switches are in the OFF position. Press the ATM button and immediately move the read/hold switch to the read position; wait for code 00 to appear on the display screen. Turn the air conditioning switch to the ON position. If the computer is receiving information (switch input) the display will change to 88 when the switch is turned ON and switch back to 00 when the switch is turned OFF. Repeat the test for the heated rear window switch. Code 00 indicates the switch is OFF and code 88 indicates that the switch is ON.

Ford EEC IV System

The 2.3L system is similar to the 1.6L, with the addition of a "keep alive" memory in the ECA that retains any intermittent trouble codes stored within the last 20 engine starts. With this system, the memory is not erased when the ignition is switched OFF. A self-diagnosis capability is built into the EEC IV system to aid in troubleshotting. The primary tool necessary to read the trouble codes stored in the system is an analog voltmeter or special Self Test Automatic Readout (STAR) tester (Motorcraft No. 007-0M004, or equivalent). While the self-test is not conclusive by itself, when activated it checks the EFC IV system by testing its memory integrity and processing capability. The self-test also verifies that all sensors and actuators are connected and working properly.

When a service code is displayed on an analog voltmeter, each code number is represented by pulses or sweeps of the meter needle. A code 3, for example, will be read as three needle pulses followed by a six-second delay. If a two digit code is stored, there will be a two second delay between the pulses for each digit of the number. Code 23, for example, will be displayed as two needle pulses, a two second pause, then three more pulses followed by a four second pause. All testing is complete when the codes have been repeated once. The pulse format is ½ second ON-time for each digit, 2 seconds OFF-time between digits, 4 seconds OFF-time between codes and either 6 seconds (1.6L) or or 10 seconds (2.3L) OFF-time before and after the half-second separator pulse.

NOTE: If using the STAR tester, or equivalent, consult the manufacturers instructions included with the unit for correct hookup and trouble code interpretation.

In addition to the service codes, two other types of coded information are outputted during the self-test; engine identification and fast codes. Engine ID codes are one digit numbers equal to one-half the number of engine cylinders (e.g. 4 cylinder is code 2, 8 cylinder is code 4, etc.). Fast codes are simply the service codes transmitted at 100 times the normal rate in a short burst of information. Some meters may detect these codes and register a slight meter deflection just before the trouble codes are flashed. Both the ID and fast codes serve no purpose in the field and this meter deflection should be ignored.

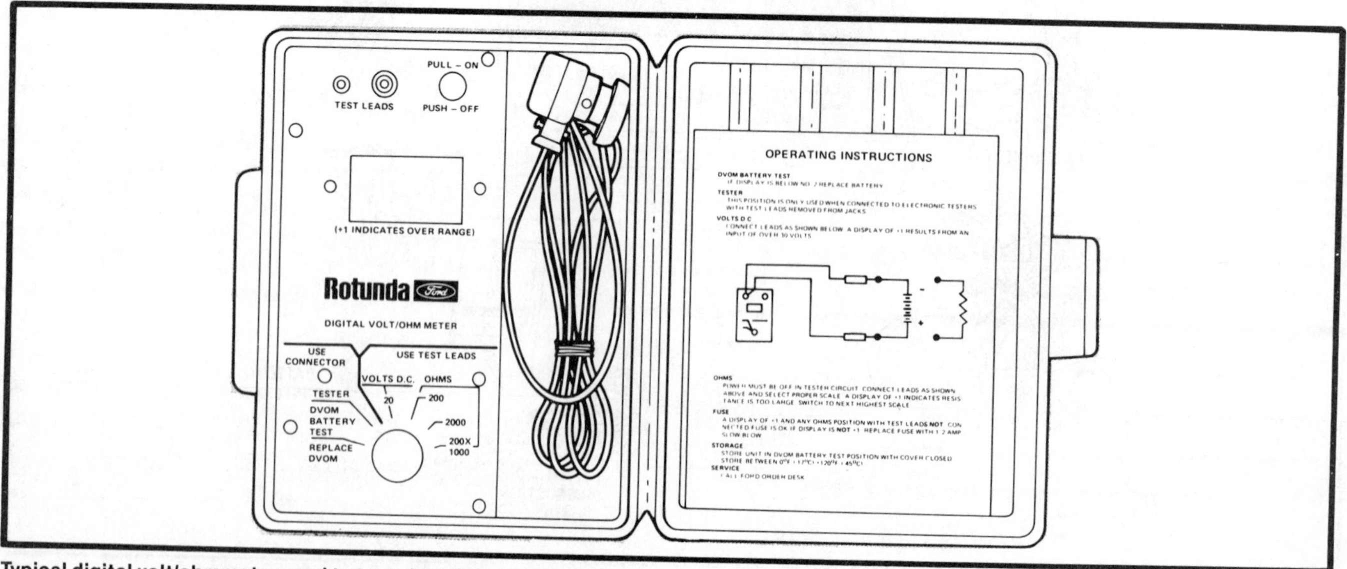

Typical digital volt/ohmmeter used to test electronic engine control systems

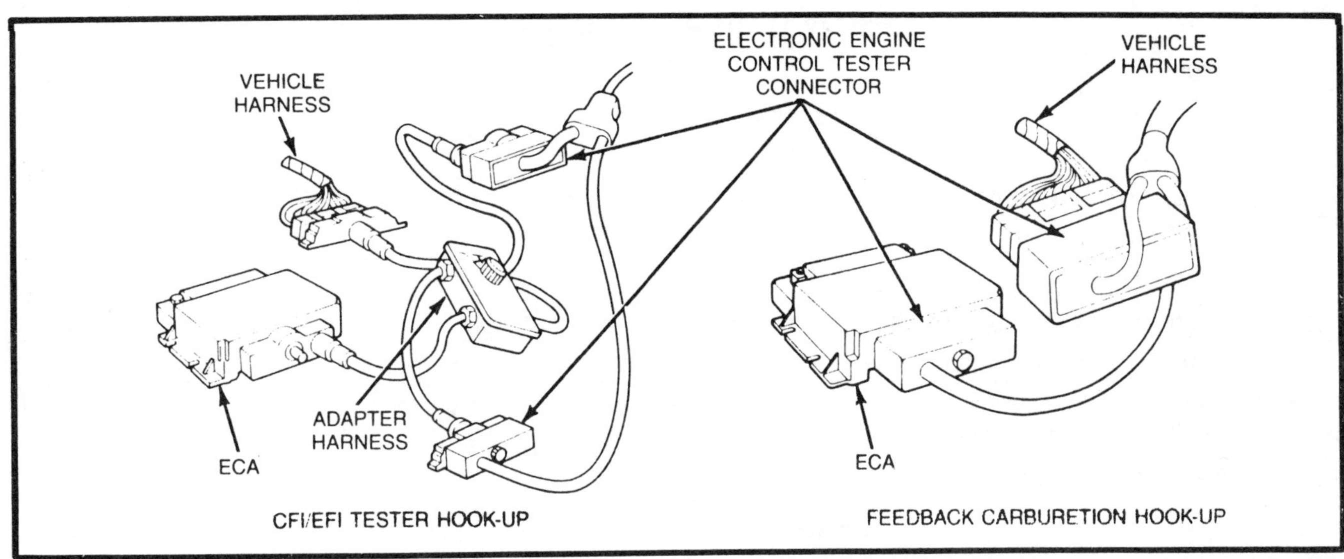

Ford EEC tester connector harness showing computer hookup

Activating Self-Test Mode on EEC IV

Turn the ignition key OFF. On the 2.3L engine, connect a jumper wire from the self-test input (STI) to pin 2 (signal return) on the self-test connector. On the 1.6L engine, connect a jumper wire from pin 5 self-test input to pin 2 (signal return) on the self-test connector. Set the analog voltmeter on a DC voltage range to read from 0–15 volts, then connect the voltmeter from the battery positive (+) terminal to pin 4 self-test output in the self-test connector. Turn the ignition switch ON (engine off) and read the trouble codes on the meter needle as previously described. A code 11 means that the EEC IV system is operating properly and no faults are detected by the computer.

NOTE: This test will only detect "hard" failures that are present when the self-test is activated. For intermittent problems, remove the voltmeter clip from the self-test trigger terminal and wiggle the wiring harness. With the voltmeter still attached to the self-test output, watch for a needle deflection that signals an intermittent condition has occurred. The meter will deflect each time the fault is induced and a trouble code will be stored. Reconnect the self-test trigger terminal to the voltmeter to retrieve the code.

EEC IV TROUBLE CODES (2.3L)

Code	Diagnosis
11	Normal operation (no codes stored)
12	Incorrect high idle rpm value
13	Incorrect curb idle rpm value
14	Erratic Profile Ignition Pickup (PIP) signal
15	Read Only Memory (ROM) failure
21	Incorrect engine coolant temperature (ECT) sensor signal
22	Incorrect barometric pressure (BAP) sensor signal
23	Incorrect throttle position sensor (TPS) signal
24	Incorrect vane air temperature (VAT) sensor signal
26	Incorrect vane air flow (VAF) sensor signal
41	System always lean
42	System always rich
51	Engine coolant temperature (ECT) sensor signal too high
53	Throttle position sensor (TPS) signal too high
54	Vane air temperature (VAT) sensor signal too high
56	Vane air flow (VAF) sensor signal too high
61	Engine coolant temperature (ECT) signal too low

EEC IV TROUBLE CODES (2.3L)

Code	Diagnosis
63	Throttle position sensor (TPS) signal too low
64	Vane air temperature (VAT) signal too low
66	Vane air flow (VAF) sensor signal too low
67	A/C compressor clutch ON
73	No vane air temperature (VAT) signal change when engine speed is increased
76	No vane air flow (VAF) signal change when engine speed is increased
77	Engine speed not increased to check VAT and VAF signal change

NOTE: Incorrect sensor signals could be out of range or not being received by the control unit. Perform wiring harness and sensor checks to determine the cause, or check for additional codes to indicate high or low reading

EEC IV TROUBLE CODES (1.6L)

Code	Diagnosis
11	Normal operation (no codes stored)
12	Incorrect high idle rpm value
13	Incorrect curb idle rpm value
15	Read Only Memory (ROM) failure
21	Incorrect engine coolant temperature (ECT) sensor signal
23	Incorrect throttle position sensor (TPS) signal
24	Incorrect vane air temperature (VAT) sensor signal
26	Incorrect vane air flow (VAF) sensor signal
41	System always lean
42	System always rich
67	Neutral/Drive switch in Neutral

NOTE: Incorrect rpm values could be high or low and an incorrect sensor signal could be caused by a defective sensor or a wiring harness problem. Use the trouble codes to isolate the circuit, then continue diagnosis to determine the exact cause of the problem

General Motors C4 and CCC Systems

When an electrical or electronic malfunction is detected, the CHECK ENGINE light will illuminate on the dash. When any input circuit (such as the engine temperature sensor) is supplying unreasonable information, the computer will substitute a fixed value from its programmed memory so the vehicle can be driven. If such a substitution occurs, the CHECK ENGINE light will come on and a numerical trouble code will be stored in the on-board computer memory to indicate that a malfunction has occurred. If the problem is intermittent, the light will go out but the trouble code will remain stored until the battery is disconnected or the system fuse is removed from the fuse panel. To eliminate the trouble codes stored for an occasional stray voltage or other non-malfunction reason, the computer is programmed to erase the trouble code memory after a certain number of ignition switch ON/OFF cycles.

The CHECK ENGINE light on the instrument panel is used as a warning lamp to tell the driver that a problem has occurred in the electronic engine control system. When the self-diagnosis mode is activated by grounding the test terminal of the diagnostic connector, the check engine light will flash stored trouble codes to help isolate system problems. The electronic control module (ECM) has a memory that knows what certain engine sensors should be under certain conditions. If a sensor reading is not what the ECM thinks it should be, the control unit will illuminate the check engine light and store a trouble code in its memory. The trouble code indicates what circuit the problem is in, each circuit consisting of a sensor, the wiring harness and connectors to it and the ECM.

NOTE: Some models have a "Service Engine Soon" light instead of a "Check Engine" display

Activating Diagnosis Mode

To retrieve any stored trouble codes, an under dash diagnosis connector is provided. On the C4 system, the test lead can be identified by a green plastic connector with an integral clip. The test lead is usually located under the instrument panel at the extreme right (passenger) side, usually above the right hand kick panel. Grounding the test terminal while the ignition is ON will cause the system to display any stored trouble codes by flashing the CHECK ENGINE light in two-digit code sequences. For example, a code 12 is displayed as one flash followed by a pause and two more flashes. After a longer pause, the pattern will repeat itself two more times and the cycle will continue to repeat itself unless the engine is started, the test lead is disconnected or the ignition power is interrupted by turning the switch OFF or disconnecting the battery.

The Assembly Line Communications Link (ALCL) is a diagnostic connector located in the passenger compartment, usually under the left side of the instrument panel. It has terminals which are used in the assembly plant to check that the engine is operating properly before shipment. One of the terminals is the diagnostic test terminal and another is the ground. By connecting the two terminals together with a jumper wire, the diagnostic mode is activated and the control unit will begin to flash trouble codes using the check engine light. When the test terminal is grounded with the key ON and the engine stopped, the ECM will display code 12 until the test terminal is disconnected. Each trouble code will be flashed three times, then code 12 will display again. The ECM will also energize all controlled relays and solenoids when in the diagnostic mode to check function.

When the test terminal is grounded with the engine running, it will cause the ECM to enter the Field Service Mode. In this mode, the SERVICE ENGINE SOON light will indicate whether the system is in Open or Closed Loop operation. In open loop, the light will flash $2^1/_2$ times per second; in closed loop, the light will flash once per second. In closed loop, the light will stay out most of the time if the system is too lean and will stay on most of the time if the system is too rich.

NOTE: The vehicle may be driven in the Field Service mode and system evaluated at any steady road speed. This mode is useful in diagnosing driveability problems where the system is rich or lean too long.

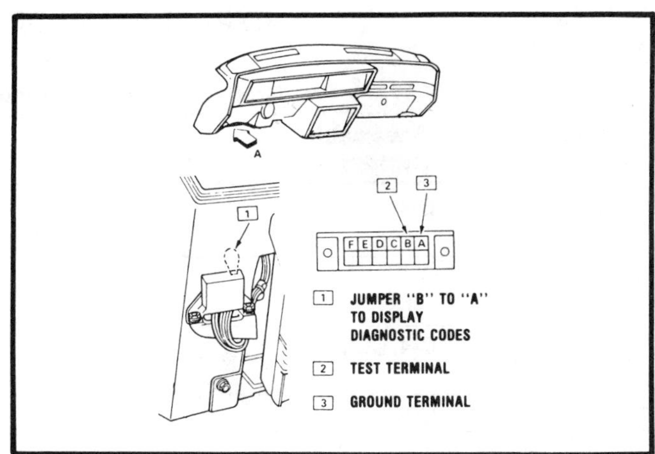

Typical ALCL connector

Trouble codes should be cleared after service is completed. To clear the trouble code memory, disconnect the battery for at least 10 seconds. This may be accomplished by disconnecting the ECM harness from the positive battery pigtail or by removing the ECM fuse. The vehicle should be driven after the ECM memory is cleared to allow the system to readjust itself. The vehicle should be driven at part throttle under moderate acceleration with the engine at normal operating temperature. A change in performance should be noted initially, but normal performance should return quickly.

CAUTION

The ignition switch must be OFF when disconnecting or reconnecting power to the ECM.

EXPLANATION OF TROUBLE CODES GM C-4 AND CCC SYSTEMS

(Ground test lead or terminal AFTER engine is running.)

Trouble Code	Applicable System	Notes	Possible Problem Area
12	C-4, CCC		No tachometer or reference signal to computer (ECM). This code will only be present while a fault exists, and will not be stored if the problem is intermittent.
13	C-4, CCC		Oxygen sensor circuit. The engine must run for about five minutes (eighteen on C-4 equipped 231 cu in. V6) at part throttle (and under road load—CCC equipped cars) before this code will show.
13 & 14 (at same time)	C-4	Except Cadillac and 171 cu in. V6	See code 43.
13 & 43 (at same time)	C-4	Cadillac and 171 cu in. V6	See code 43.
14	C-4, CCC		Shorted coolant sensor circuit. The engine has to run 2 minutes before this code will show.
15	C-4, CCC		Open coolant sensor circuit. The engine has to operate for about five minutes (18 minutes for C-4 equipped 231 cu in. V6) at part throttle (some models) before this code will show.
21	C-4		Shorted wide open throttle switch and/or open closed-throttle switch circuit (when used).
	C-4, CCC		Throttle position sensor circuit. The engine must be run up to 10 seconds (25 seconds—CCC System) below 800 rpm before this code will show.

EXPLANATION OF TROUBLE CODES GM C-4 AND CCC SYSTEMS

(Ground test lead or terminal AFTER engine is running.)

Trouble Code	Applicable System	Notes	Possible Problem Area
21 & 22 (at same time)	C-4		Grounded wide open throttle switch circuit (231 cu in. V6, 151 cu in. 4 cylinder).
22	C-4		Grounded closed throttle or wide open throttle switch circuit (231 cu in. V6, 151 cu in. 4 cylinder).
23	C-4, CCC		Open or grounded carburetor mixture control (M/C) solenoid circuit.
24	CCC		Vehicle speed sensor (VSS) circuit. The car must operate up to five minutes at road speed before this code will show.
32	C-4, CCC		Barometric pressure sensor (BARO) circuit output low.
32 & 55 (at same time)	C-4		Grounded +8V terminal or V(REF) terminal for barometric pressure sensor (BARO), or faulty ECM computer.
34	C-4	Except 1980 260 cu in. Cutlass	Manifold absolute pressure (MAP) sensor output high (after ten seconds and below 800 rpm).
34	CCC	Including 1980 260 cu in. Cutlass	Manifold absolute pressure (MAP) sensor circuit or vacuum sensor circuit. The engine must run up to five minutes below 800 RPM before this code will set.
35	CCC		Idle speed control (ISC) switch circuit shorted (over ½ throttle for over two seconds).
41	CCC		No distributor reference pulses to the ECM at specified engine vacuum. This code will store in memory.
42	CCC		Electronic spark timing (EST) bypass circuit grounded.
43	C-4		Throttle position sensor adjustment (on some models, engine must run at part throttle up to ten seconds before this code will set).
44	C-4, CCC		Lean oxygen sensor indication. The engine must run up to five minutes in closed loop (oxygen sensor adjusting carburetor mixture), at part throttle and under road load (drive car) before this code will set.
44 & 55 (at same time)	C-4, CCC		Faulty oxygen sensor circuit.
45	C-4, CCC	Restricted air cleaner can cause code 45	Rich oxygen sensor system indication. The engine must run up to five minutes in closed loop (oxygen sensor adjusting carburetor mixture), at part throttle under road load before this code will set.
51	C-4, CCC		Faulty calibration unit (PROM) or improper PROM installation in electronic control module (ECM). It takes up to thirty seconds for this code to set.
52 & 53	C-4		"Check Engine" light off: Intermittent ECM computer problem. "Check Engine" light on: Faulty ECM computer (replace).
52	C-4, CCC		Faulty ECM computer.
53	CCC	Including 1980 260 cu in. Cutlass	Faulty ECM computer.
54	C-4, CCC		Faulty mixture control solenoid circuit and/or faulty ECM computer.

EXPLANATION OF TROUBLE CODES GM C-4 AND CCC SYSTEMS

(Ground test lead or terminal AFTER engine is running.)

Trouble Code	Applicable System	Notes	Possible Problem Area
55	C-4	Except 1980 260 cu. in. Cutlass	Faulty oxygen sensor, open manifold absolute pressure sensor or faulty ECM computer (231 cu in. V6). Faulty throttle position sensor or ECM computer (except 231 cu. in. V6). Faulty ECM computer (151 cu in. 4 cylinder)
55	CCC	Including 1980 260 cu in. Cutlass	Grounded +8 volt supply (terminal 19 of ECM computer connector), grounded 5 volt reference (terminal 21 of ECM computer connector), faulty oxygen sensor circuit or faulty ECM computer.

GM PORT INJECTION TROUBLE CODES

Trouble Code	Circuit
12	Normal operation
13	Oxygen sensor
14	Coolant sensor (low voltage)
15	Coolant sensor (high voltage)
21	Throttle position sensor (high voltage)
22	Throttle position sensor (low voltage)
24	Speed sensor
32	EGR vacuum control
33	Mass air flow sensor
34	Mass air flow sensor
42	Electronic spark timing
43	Electronic spark control
44	Lean exhaust
45	Rich exhaust
51	PROM failure
52	CALPAK
55	ECM failure

Isuzu I-Tec System

The self-diagnosis system is designed to monitor the input and output signals of the sensors and actuators and to store any malfunctions in its memory as a trouble code. When the electronic control unit detects a problem, it will activate a CHECK ENGINE light on the dash board. To activate the trouble code readout, locate the diagnosis connector near the control unit and connect the two leads with the ignition ON. The trouble codes stored in the memory will be displayed as flashes of the CHECK ENGINE light, the flashes corresponding to the first and second digit of a two digit number. The Isuzu I-TEC system is capable of storing three trouble codes which are displayed in numerical sequence no matter what order the faults occur in. Each trouble code will be displayed three times, then the next code is displayed. The control unit will display all trouble codes stored in its memory as long as the diagnostic lead is connected with the key ON. A code 12 indicates that the I-TEC system is functioning normally and that no further testing is necessary. After service, clear the trouble codes by disconnecting the No. 4 fuse from the fuse holder. All codes stored will be automatically cleared whenever the main harness connector is disconnected from the control unit.

NOTE: For further trouble diagnosis procedures, see the Isuzu I-TEC section.

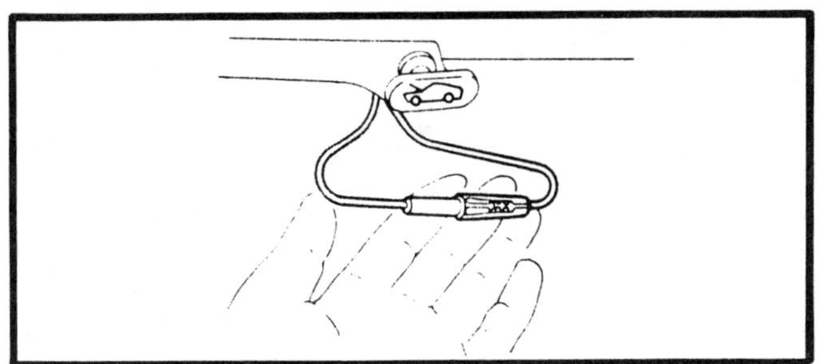

Isuzu trouble code connector

ISUZU I-TEC SYSTEM TROUBLE CODE CHART

Trouble Code	ECU Circuit	Possible Cause
12	Normal operation	No testing required
13	Oxygen sensor	Open or short circuit, failed sensor
44	Oxygen sensor	Low voltage signal
45	Oxygen sensor	High voltage signal
14	Coolant temperature sensor	Shorted with ground (no signal)
15	Coolant temperature sensor	Incorrect signal
16	Coolant temperture sensor	Excessive signal (harness open)
21	Throttle valve switch	Idle and WOT contacts closed at the same time
43	Throttle valve switch ①	Idle contact shorted
65	Throttle valve switch	Full throttle contact shorted
22	Starter signal	No signal
41	Crank angle sensor	No signal or wrong signal
61	Air flow sensor	Weak signal (harness shorted or open hot wire)
62	Air flow sensor	Excessive signal (open cold wire)
63	Speed sensor ①	No signal
66	Detonation sensor	Harness open or shorted to ground
51, 52, 55	ECU malfunction	Incorrect injection pulse or fixed timing

Mazda EGI System

The Mazda System Checker 83 (49 G040 920) is necessary to troubleshoot the EGI electronic control system with the test connector located next to the ECU harness connectors at the control unit.

Mazda EGI tester used to read trouble codes

With the tester, the on-board diagnosis system will read out trouble codes (1 through 6) to indicate problems in different circuits within the fuel injection system. Follow the manufacturer's instructions included with the tester for all wiring and sensor checks using the special tester.

CAUTION

Do not attempt to disconnect or reconnect the control unit main harness connector with the ignition switch ON, or the ECU can be damaged or destroyed.

MAZDA TROUBLE CODES

Code Number	Circuit
1	Engine speed (rpm) signal
2	Air flow meter
3	Coolant temperature sensor
4	Oxygen sensor
5	Throttle position sensor
6	Atmospheric pressure sensor

Mitsubishi ECI System

The Mitsubishi self-diagnosis system monitors the various input signals from the engine sensors and enters a trouble code in the on-board computer memory if a problem is detected. There are nine monitored items, including the "normal operation" code which can be read by using a special ECI tester (MD9984 06 or equivalent) and adapter. The adapter connects the ECI tester to the diagnosis connector located on the right cowl, next to the control unit. Because the computer memory draws its power directly from the battery, the trouble codes are not erased when the ignition is switched OFF. The memory can only be cleared (trouble codes erased) if a battery cable is disconnected or the main ECU wiring harness connector is disconnected from the computer module. The trouble codes will not be erased if the battery cable or harness connector is reconnected within 10 seconds.

CAUTION

Before any ECU harness connectors are removed or installed, make sure the ignition is switched OFF or the control unit may be damaged. Make sure the connector is seated properly and the lock is in its correct position.

If two or more trouble codes are stored in the memory, the computer will read out the codes in order beginning with the lowest number. The needle of the ECI tester will swing back and forth between 0 and 12 volts to indicate the trouble code stored. There is no memory for code No. 1 (oxygen sensor) once the ignition is switched OFF, so it is necessary to perform this diagnosis with the engine running. The oxygen sensor should be allowed to warm up for testing (engine at normal operating temperature) and the trouble code should be read before the ignition is switched OFF. All other codes will be read out with the engine ON or OFF. If there are no trouble codes stored in the computer (system is operating normally), the ECI tester will indicate a constant 12 volts on the meter. Consult the instructions supplied with the test equipment to insure proper connections for diagnosis and testing of all components.

If there is a problem stored, the meter needle will swing back and forth every 0.4 seconds. Trouble codes are read by counting the pulses, with a two second pause between different codes. If the battery voltage is low, the self-diagnosis system will not operate properly, so the battery condition and state of charge should be checked before attempting any self-diagnosis inspection procedures. After completing service procedures, the computer trouble code memory should be erased; by disconnecting the battery cable or main harness connectors to the control unit for at least 10 seconds. See the Mitsubishi ECI section for details on further trouble diagnosis procedures.

MITSUBISHI ECI TROUBLE CODES

Trouble Code	ECU Circuit	Possible Cause
1	Oxygen sensor	Open circuit in wire harness, faulty oxygen sensor or connector
2	Ignition signal	Open or shorted wire harness, faulty igniter
3	Air flow sensor	Open or shorted wire harness, loose connector, defective air flow sensor
4	Boost pressure sensor	Defective boost sensor, open or shorted wire harness or connector
5	Throttle position sensor	Sensor contacts shorted, open or shorted wire harness or connector
6	ISC motor position sensor	Defective throttle sensor open or shorted wire harness or connector, defective ISC servo
7	Coolant temperature sensor	Defective sensor, open or shorted wire harness or connector
8	Speed sensor	Malfunction in speed-sensor circuit, open or shorted wire harness or connector

Nissan E.C.C.S. System

The self-diagnostic system determines the malfunctions of signal systems such as sensors, actuators and wire harness connectors based on the status of the input signals received by the E.C.C.S. control unit. Malfunction codes are displayed by two LED's (red and green) mounted on the side of the control unit. The self-diagnosis results are retained in the memory clip of the ECU and displayed only when the diagnosis mode selector (located on the left side of the ECU) is turned fully clockwise. The self-diagnosis system on the E.C.C.S. control unit is capable of displaying malfunctions being checked, as well as trouble codes stored in the memory. In this manner, an intermittent malfunction can be detected during service procedures.

CAUTION

Turn the diagnostic mode selector carefully with a small screwdriver. Do not press hard to turn or the selector may be damaged.

Activating Diagnosis Mode

Service codes are displayed as flashes of both the red and green LED. The red LED blinks first, followed by the green LED, and the two together indicate a code number. The red LED is the tenth digit, and the green LED is the unit digit. For example; when the red light blinks three times and the green light blinks twice, the code displayed is 32. All malfunctions are classified by code numbers. When all service procedures are complete, erase the memory by disconnecting the battery cable or the ECU harness connector. Removing the power to the control unit automatically erases all trouble codes from the memory. Never erase the stored memory before performing self diagnosis tests.

NISSAN E.C.C.S. TROUBLE CODES

Code Number	ECU Circuit	Test Point Pin Numbers	Normal Test Results
11	Crank Angle Sensor	Check harness for open circuit	Continuity
12	Air Flow Meter	Ground terminal 26① connect VOM @ 26–31 Apply 12v @ E–D② connect VOM @ B–D VOM @ 12-GND③ VOM @ C–F②	IGN ON—1.5–1.7 volts 1.5–1.7 volts Continuity Continuity
13	Cylinder Head Temperature Sensor	VOM @ 23–26①	Above 68 deg F–2.9 kΩ Below 68 deg F–2.1 kΩ
14	Speed Sensor	VOM @ 29–GND④	Continuity
21	Ignition Signal	VOM @ 3–GND VOM @ 5–GND Check power transistor terminals to base plate	Continuity Continuity Continuity
22	Fuel Pump	VOM @ 108–GND⑤ Pump connectors Pump relay: VOM @ 1–2 VOM @ 3–4 12v @ 1–2, VOM @ 3–4	IGN ON–12 volts Continuity Continuity [if] Continuity
23	Throttle Valve Switch	VOM @ 18–25⑥ VOM @ 18–GND VOM @ 25–GND	Continuity [if] [if]
24	Neutral/Park Switch	VOM @ Switch terminals	Neutral–0Ω Drive–∞Ω
31	Air Conditioner	VOM @ 22–GND①	IGN ON–12 volts
32	Start Signal	VOM @ 9–GND③	12 volts with starter S terminal disconnected
34	Detonation Sensor	Disconnect sensor and check timing with engine running	Timing should retard 5 degrees above 2000 rpm
41	Fuel Temperature Sensor	VOM @ 15–GND③ VOM @ Sensor terminals	Above 68 deg. F–2.9 kΩ Below 68 deg.F–2.1 kΩ Resistance (ohms) should decrease as temperature rises
44	Normal Operation—no further testing required		

NOTE: Make sure test equipment will not damage the control unit before testing

VOM—Volt/ohm meter ∞—Infinite resistance ③20-pin harness connector
GND—Ground ①16-pin harness connector ④16-pin connector at ECU
Ω—Ohms (kΩ = kilo-ohms) ②6-pin air flow meter connection ⑤Throttle valve switch connector

Renault TBI System

On 1983 and later Renault models, the on-board self-diagnosis system will illuminate a test bulb if a malfunction exists. When the trouble code terminal at the diagnostic connector in the engine compartment is connected to a test bulb, the system will flash a trouble code if a malfunction has been detected.

CAUTION

Be extremely careful when making test connections. Never apply more than 12 volts to any point or component in the TBI system.

The self-diagnosis feature of the electronic control unit (ECU) provides support for diagnosing system problems by recording six possible failures should they be encountered during normal engine operation. Additional tests should allow specific tracing of a failure to a single circuit or component. Multiple failures of different circuits or components must be diagnosed separately. It is possible that the test procedures can cause false failure codes to be set.

NOTE: In the following procedures, no specialized service equipment is necessary. It is necessary to have available a volt/ohmmeter (with at least 10 megohms impedence), a 12 volt test light and an assortment of jumper wires and probes.

Trouble Code Test Lamp

If the ECU is functional, service diagnosis codes can be obtained by connecting a No. 158 test bulb to pins D2-2 and D2-4 of the large diagnostic connector. With the test bulb installed, push the wide open throttle (WOT) switch lever on the throttle body and, with the idle speed control (ISC) motor plunger also closed, have an assistant turn the ignition switch ON while observing the test bulb. If the ECU is functioning normally, the test bulb should light for a moment and then go out. This will always occur regardless of the failure condition and serves as an indication that the ECU is functional.

After the initial illumination, the ECU will cycle through and flash a single digit code if any system malfunctions have been detected by the ECU during normal engine operation. The ECU is capable of storing various trouble códes in its memory. The initial trouble detected will be flashed first and then a short pause will separate the second trouble code stored. There will be a somewhat longer pause between the second code and the repeat cycle of the first code again to provide distinction between codes like 3-6 and 6-3. Although the two codes indicate the same two failures, the last code stored indicates the most recent failure.

If further testing fails to indicate the cause of the trouble code, an intermittent problem exists and marginal components should be suspected. The most common cause of an intermittent problem is corrosion or loose connections. If the trouble code is erased and quickly returns with no other symptoms, the ECU should be suspected. If the ECU is determined to be malfunctioning, it must be replaced. No repairs should be attempted. It is important to note that the trouble memory is erased if the ECU power is interrupted by disconnecting the wire harness from the ECU, disconnecting either battery terminal, or allowing the engine to remain unstarted in excess of five days. It is equally important to erase the trouble memory when a defective component is replaced.

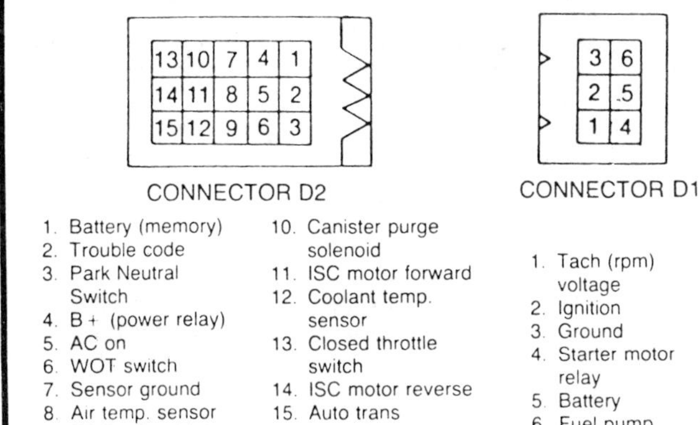

Diagnostic connector terminals on Renault TBI system

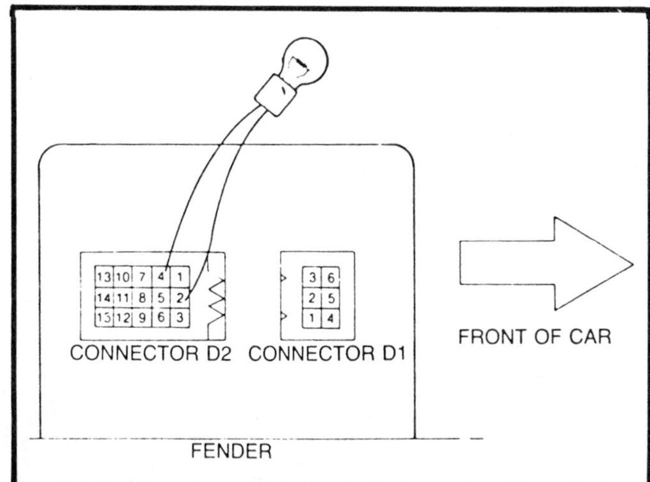

Trouble code test lamp connections

TBI TROUBLE DIAGNOSIS

Condition or Trouble Code	Possible Cause	Correction
CODE 1 (poor low air temp. engine performance).	Manifold air/fuel temperature (MAT) sensor resistance is not less than 1000 ohms (HOT) or more than 100 kohms (VERY COLD)	Replace MAT sensor if not within specifications. Refer to MAT sensor test procedure.
CODE 2 (poor warm temp. engine performance-engine lacks power).	Coolant temperature sensor resistance is less than 300 ohms or more than 300 kohms (10 kohms at room temp.).	Replace coolant temperature sensor. Test MAT sensor. Refer to coolant temp. sensor test and MAT sensor test procedures.

TBI TROUBLE DIAGNOSIS

Condition or Trouble Code	Possible Cause	Correction
CODE 3 (poor fuel economy, hard cold engine starting, stalling, and rough idle).	Defective wide open throttle (WOT) switch or closed (idle) throttle switch or both, and/or associated wire harness.	Test WOT switch operation and associated circuit. Refer to WOT switch test procedure. Test closed throttle switch operation and associated circuit. Refer to closed throttle switch test procedure.
CODE 4 (poor engine acceleration, sluggish performance, poor fuel economy).	Simultaneous closed throttle switch and manifold absolute pressure (MAP) sensor failure.	Test closed throttle switch and repair/ replace as necessary. Refer to closed throttle switch test procedure. Test MAP sensor and associated hoses and wire harness. Repair or replace as necessary. Refer to MAP sensor test procedure.
CODE 5 (poor acceleration, sluggish performance).	Simultaneous WOT switch and manifold absolute pressure (MAP) sensor failure.	Test WOT switch and repair or replace as necessary. Refer to WOT switch test procedure. Test MAP sensor and associated hoses and wire harness. Repair or replace as necessary. Refer to MAP sensor test procedure.
CODE 6 (poor fuel economy, bad driveability, poor idle, black smoke from tailpipe).	Inoperative oxygen sensor.	Test oxygen sensor operation and replace if necessary. Test the fuel system for correct pressure. Test the EGR solenoid control. Test canister purge. Test secondary ignition circuit. Test PCV circuit. Refer to individual component test procedure.
No test bulb flash.	No battery voltage at ECU (J1-A with key on). No ground at ECU (J1-F).	Repair or replace wire harness, connectors or relays.
	Simultaneous WOT and CTS switch contact (Ground at both D2 Pin 6 and D2 Pin 13).	Repair or replace WOT switch, CTS switch, harness or connectors.
	No battery voltage at test bulb (D2 Pin 4).	Repair wire harness or connector.
	Defective test bulb.	Replace test bulb.
	Battery voltage low (less than 11.5V).	Charge or replace battery, repair vehicle wire harness.

Toyota EFI System

The Toyota electronic control unit uses a dash-mounted CHECK ENGINE light that illuminates when the control unit detects a malfunction. The memory will store the trouble codes until the system is cleared by removing the EFI fuse with the ignition OFF. To activate the trouble code readout and obtain the diagnostic codes stored in the memory, first check that the battery voltage is at least 11 volts, the throttle valve is fully closed, transmission is in neutral, the engine is at normal operating temperature and all accessories are turned OFF.

Activating Diagnosis Mode

Turn the ignition switch ON, but do not start the engine. Locate the Check Engine Connector under the hood near the ignition coil and use a short jumper wire to connect the terminals together. Read the diagnostic code as indicated by the number of flashes of the CHECK ENGINE light. If normal system operation is occurring, the light will blink once every three seconds to indicate a code 1 (no malfunctions). The light will blink once every three seconds to indicate a trouble code stored in the memory, with three second pauses between each code number. The diagnostic code series will be repeated as long as the CHECK ENGINE terminals are connected together. After all trouble codes are recorded, remove the jumper wire and replace the rubber cap on the connector. Cancel the trouble codes in the memory by removing the STOP fuse for about 30 seconds. If the diagnostic codes are not erased, they will be reported as new problems the next time the diagnosis mode is activated. Verify all repairs by clearing the memory, road testing the car, then entering the diagnosis mode again to check that a code 1 (no malfunctions) is stored. For more information on trouble diagnosis procedures, see the Toyota Fuel Injection section.

Feedback Carburetors

SECTION 4

INDEX

NOTE: Please refer to the Application Chart at the front of this Manual.

FEEDBACK CARBURETORS

General Information

The need for better fuel economy combined with increasingly strict emission control regulations dictate a more exact control of the engine air/fuel mixture. A number of computer controlled carburetor systems have been developed in response to these needs. They are described in the following sections.

Essentially, all of these systems operate on the same principle. By controlling the air/fuel mixture exactly, more complete combustion can occur in the engine, and more thorough oxidation and reduction of the exhaust gases can be achieved in the catalytic converter.

The systems under discussion use variations of the same basic components: an oxygen sensor, to monitor exhaust gas oxygen content; a variable-mixture carburetor; a three-way catalytic converter capable of oxidizing HC and CO and reducing NOx; an air injection system to supply the converter with additional oxygen for the oxidation reaction; and a computer to monitor the process and adjust it according to continually changing engine and environmental conditions.

These systems operate in the following manner: The oxygen sensor, installed in the exhaust manifold upstream of the catalytic converter, reads the oxygen content of the exhaust gases. It generates an electrical signal and sends it to the computer. The computer then decides how to adjust the mixture to keep it at the correct air/fuel ratio. For example, if the mixture is too lean, the computer signals the carburetor that more fuel is needed. The computer signal activates a mixture control device on the carburetor, which enrichens the mixture accordingly. The monitoring process is a continual one, so that fine mixture adjustments are going on at all times.

The object of all of the systems is to maintain the optimum air/fuel mixture, which is chemically correct for theoretically complete combustion. The stoichiometric ratio is 14.7:1 (air to fuel). At that point, the catalytic converter's efficiency is greatest in oxidizing and reducing HC, CO, and NOx into carbon dioxide (CO_2), water (H_2O), and free oxygen and nitrogen (O_2 and N_2 respectively).

Most of the systems have two modes of operation: closed loop and loop. Closed loop operation occurs when the converter and oxygen sensor have warmed to efficient levels. All sensors become interdependent in this mode: the oxygen sensor sends signals to the computer, which signals the carburetor, which adjusts the mixture, which changes the oxygen sensor's readings, which go back to the computer, and so on. A continual and ongoing feedback of information and adjustment is achieved. Open loop operation generally takes place when the engine is still cold. In this mode, the computer simply provides a predetermined and invariable signal; the signal may affect only the carburetor air/fuel ratio, although in most installations it also provides a fixed spark advance signal to the electronic ignition module as well. In open loop operation, no signals are accepted by the computer.

Specific components used in these systems vary a great deal. The air injection system may be either an air pump or a pulse air design. Air injection may be into the exhaust manifold or directly into the catalytic converter. Mixture control on the carburetor can be a direct linkage to the metering valve or rods, or may be a vacuum control.

The Chrysler and Ford designs, and most 1980 and later G.M. C-4 and CCC systems, include electronic ignition as part of the system. Spark advance is controlled by the computer; the distributor pick-up and electronic ignition module are incorporated as signal devices for the computer, which decides when to order ignition firing as a function of interpretation of all sensor inputs.

One thing all of the systems have in common is a certain amount of nonadjustability. Mixture is always nonadjustable in these installations. If a mixture screw is provided on the carburetor, it is concealed under a staked-in plug, or locked in place. In many cases, ignition timing is also non-adjustable; idle speed, choke setting and fast idle rpm may also be either fixed during manufacture or under the control of the computer, and therefore not adjustable.

Unfortunately, most of the systems are so sophisticated and complicated that they do not respond favorably to conventional troubleshooting techniques. Problems with these devices almost invariably require the use of special test equipment and careful diagnosis procedures.

CAUTION

The use of improper test equipment can damage some computerized systems. See the "Tools" section for details.

AMERICAN MOTORS FEEDBACK SYSTEMS

General Information

American Motors introduced feedback systems on all cars (except Eagle) in 1980. Two different, but similar, systems are used. The four cylinder engine uses the G.M. C-4 feedback system, which is covered earlier in this section. Component usage is identical to that of the G.M. 151 cu. in four cylinder engine, including an oxygen sensor, a vacuum switch (which is closed at idle and partial throttle positions), a wide open throttle switch, a coolant temperature sensor (set to open at 150°F), an Electronic Control Module (ECM) equipped with modular Programmable Read Only Memory (PROM), and a mixture control solenoid installed in the air horn on the E2SE carburetor. A "Check Engine" light is included on the instrument panel as a service and diagnostic indicator.

The six cylinder engine is equipped with a Computerized Emission Control (CEC) System. 1980 CEC components include an oxygen sensor; two vacuum switches (one ported and one manifold) to detect three operating conditions: idle, partial throttle, and wide open throttle; a coolant temperature switch: a Micro Computer Unit (MCU), the control unit for the system which monitors all data and sends an output signal to the carburetor; and a stepper motor installed in the main body of the BBD carburetor, which varies the position of the two metering pins controlling the size of the air bleed orifices in the carburetor. The MCU also interprets signals from the distributor (rpm voltage) to monitor engine rpm.

On 1981 and later models with CEC, the number of sensors has been increased. Three vacuum operated electric switches, two mechanically operated electric switches, one engine coolant switch and an air temperature operated switch are used to detect and send engine opeating data to the MCU concerning the following engine operating conditions; cold engine start-up and operation; wide open throttle; idle (closed throttle); and partial and deep throttle.

Both AMC systems are conventional in operation. As in other feedback systems, two modes of operation are possible: open loop and closed loop. Open loop operation occurs during engine starting, cold engine operation, cold oxygen sensor operation, engine idling, wide open throttle operation, and low battery voltage operation. In open loop, a fixed air/fuel mixture signal is provided by the ECM or MCU to the carburetor, and oxygen sensor data is ignored. Closed loop operation occurs at all other times, and in this mode all signals are used by the control unit to determine the optimum air/fuel mixture.

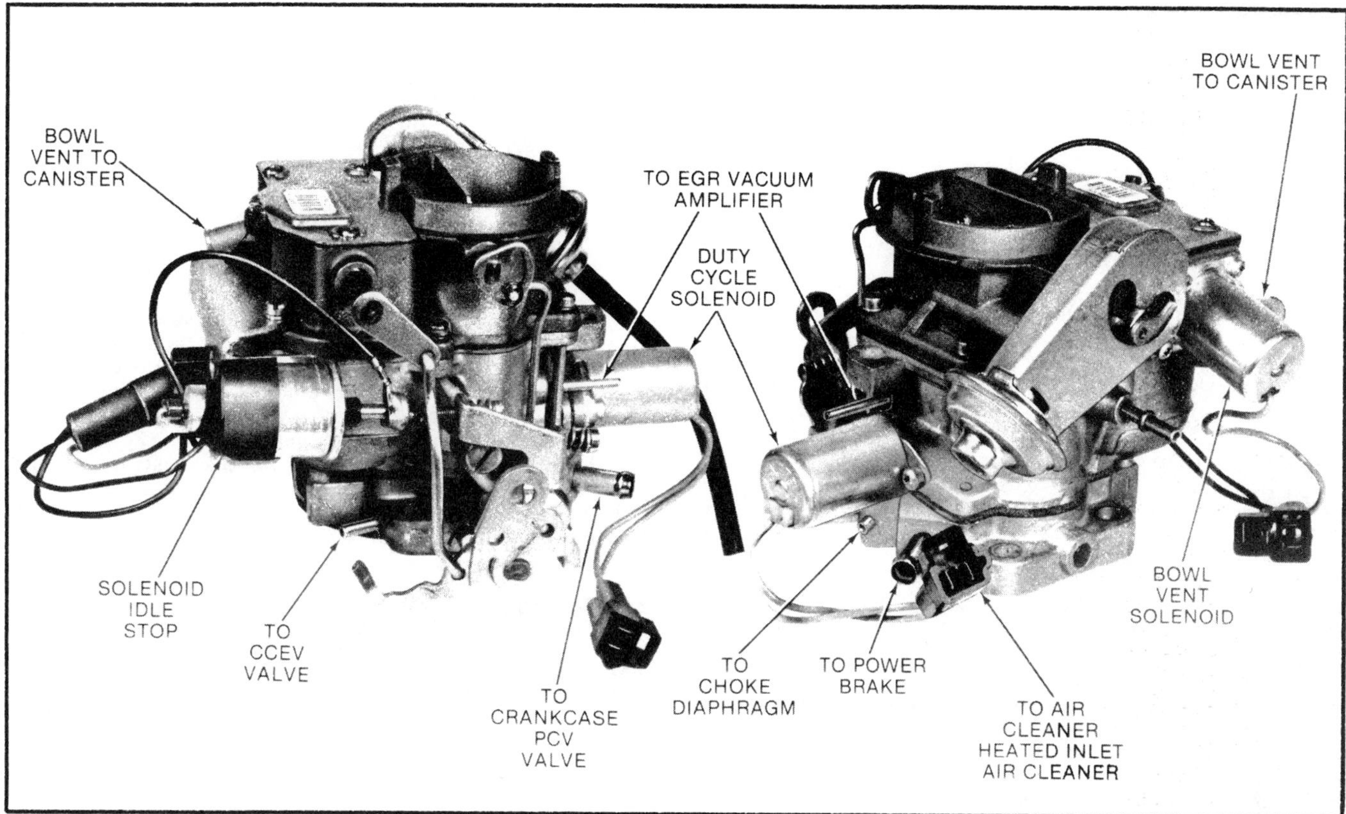

Carter BBD feedback carburetor assembly—typical

CARTER BBD FEEDBACK CARBURETOR

General Information

The Carter model BBD dual venturi carburetor utilizes three basic fuel metering systems. The idle System provides a mixture for idle and low speed performance; the Accelerator Pump System provides additional fuel during acceleration; the Main Metering System, provides an economical mixture for normal cruising conditions. In addition to these three basic systems, there is a fuel inlet system that constantly supplies the fuel to the basic metering systems, and a choke system (with electric assist) which temporarily enriches the mixture to aid in starting and running a cold engine.

The function of the O^2 Feedback Solenoid is to provide limited regulation of the fuel-air ratio of a feedback carburetor in response to electrical signals from the on-board computer. It performs this task by metering air-flow and operates in parallel with a conventional fixed main metering jet. The O^2 feedback solenoid includes, in addition to a protective case, such attaching surfaces and passageways as are required for the directing and containing of airflow when the solenoid assembly is correctly installed in a feedback carburetor designed for its use. If problems are experienced with the electronic feedback carburetor system, the system and its related components must be checked before carburetor work is begun. Close attention should be given to vacuum hose condition, hose connections, and any area where leaks could develop. In addition, the related wiring and its connectors must be checked for excessive resistance. In normal service, the mixture should not require adjustment. The idle set rpm can be checked without removal of the tamper resistant plug.

NOTE: Tampering with the carburetor is a violation of Federal Law. Adjustment of the carburetor idle air-fuel mixture can only be done under certain circumstances.

Adjustment should only be considered if an idle defect exists after normal diagnosis has revealed no other faulty condition, such as, incorrect idle speed, faulty hose or wire connection, etc. Also, it is important to make sure the on-board computer systems are operating properly. Adjustment of the carburetor idle air-fuel mixture should also be performed after a major carburetor overhaul. Upon completion of the carburetor adjustment, it is important to reinstall the plug.

Fuel Inlet System

All fuel enters the fuel bowl through the fuel inlet fitting in the bowl. The fuel inlet needle seats directly in the fuel inlet seat. The fuel inlet needle is controlled by dual floats and are hinged by a float fulcrum pin. The fuel inlet system must constantly maintain the specified level of fuel as the basic fuel metering systems are calibrated to deliver the proper mixture only when the fuel is at this level. When the fuel level in the bowl drops, the float also drops permitting additional fuel to flow past the fuel inlet needle into the bowl. The float chamber is vented to the vapor canister.

Idle System

Fuel used during curb idle and low speed operation flows through the main metering jet into the main well. Fuel continues into an idle fuel pick up tube where the fuel is mixed with air which enters

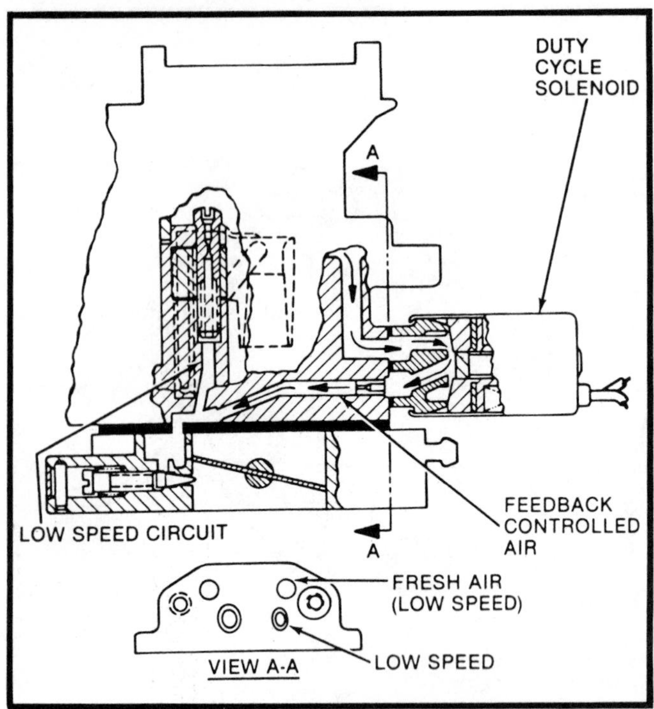

Low speed fuel control circuit on Carter BBD feedback carburetor

through idle by pass air bleeds located in the venturi cluster screws. At curb idle the fuel and air mixture flows down the idle channel and is further mixed or broken up by air entering the idle channel through the transfer slot which is above the throttle valve at curb idle. The idle system is equipped with a restrictor in the idle channel, located between the transfer slot and the idle port, which limits the maximum attainable idle mixture.

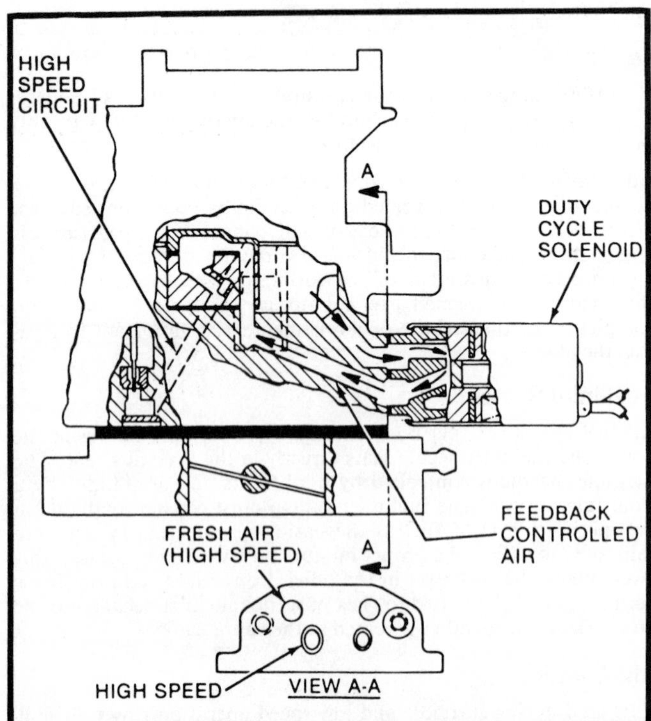

High speed fuel control circuit on Carter BBD feedback carburetor

During low speed operation the throttle valve moves exposing the transfer slot and fuel begins to flow through the transfer slot as well as the idle port. As the throttle valves are opened further and engine speed increases the air flow through the carburetor also increases. This increased air flow creates a low pressure area in the venturi and the main metering system begins to discharge fuel.

Accelerator Pump System

When the throttle valves are opened suddenly the air flow through the carburetor responds almost immmediately. However, there is a brief time interval or lag before the fuel can overcome its inertia and maintain the desired fuel-air ratio. The piston type accelerating pump system mechanically supplies the fuel necessary to overcome this deficiency for a short period of time. Fuel enters the pump cylinder from the fuel bowl through a port in the bottom of the pump well below the normal position of the pump piston. When the engine is turned off, fuel vapors in the pump cylinder are vented through the area between the pump piston and pump cup.

As the throttle lever is moved, the pump link is operated through a system of levers. A pump drive spring pushes the pump piston down. Fuel is forced through a passage around the pump discharge check ball and out the pump discharge jets which are located in the venturi cluster.

Main Metering System

As the engine approaches cruising speed the increased air flow through the venturi creates low pressure area in the venturi of the carburetor. Near atmospheric pressure present in the fuel bowl causes the fuel to flow to the lower pressure area created by the venturi and magnified by the booster venturi. Fuel flows through the main jet into the main well. The air enters through the main well air bleeds. The mixture of fuel and air being lighter than raw fuel, responds faster to changes in venturi vacuum and is also more readily vaporized when discharged into the venturi.

Exhaust Gas Recirculation System (EGR)

All BBD carburetors have an additional vacuum port for the exhaust gas recirculation system. The EGR system is controlled by the venturi vacuum control system. This system utilizes a venturi port in the side of the carburetor above the throttle valve.

SERVICE PROCEDURES

Oxygen Sensor Replacement

1. Disconnect the two wire plug.
2. Remove the sensor from the exhaust manifold on the four cylinder, or the exhaust pipe on the six.
3. Clean the threads in the manifold or pipe.
4. Coat the threads of the replacement sensor with an electrically-conductive antiseize compound. Do not use a conventional antiseize compound, which may electrically insulate the sensor.
5. Install the sensor. Installation torque is 25 ft. lbs. for the four cylinder, 31 ft. lbs. for the six cylinder.
6. Connect the sensor lead. Do not push the rubber boot into the sensor body more than 1/2 inch above the base.
7. If the sensor's pigtail is broken, replace the sensor. The wires cannot be spliced or soldered.

Vacuum Switch Replacement

The vacuum switches are mounted in a bracket bolted to the left inner fender panel in the engine compartment. They are not replaceable individually; the complete unit must be replaced.

1. Tag all the vacuum hoses, then disconnect them from the switches. Disconnect the electrical plugs. The four cylinder has two plugs and the six has one.
2. Remove the switch and bracket assembly from the fender panel.
3. Installation is the reverse.

Stepper Motor Replacement

The BBD stepper motor is installed in the side of the main body of the carburetor.

1. Remove the air cleaner case.
2. Disconnect the electrical plug.
3. Remove the retaining screw and remove the motor from the side of the carburetor. Be careful not to drop the metering pins or the spring when removing the motor.
4. Installation is the reverse.

CARBURETOR

Removal and Installation

1. Remove the air cleaner.
2. Disconnect the fuel and vacuum lines. It might be a good idea to tag them to avoid confusion when the time comes to put them back.
3. Disconnect the choke rod.
4. Disconnect the accelerator linkage.
5. Disconnect the automatic transmission linkage.
6. Unbolt and remove the carburetor.
7. Remove the base gasket.
8. Before installation, make sure that the carburetor and manifold sealing surfaces are clean.
9. Install a new carburetor base gasket.
10. Install the carburetor and start the fuel and vacuum lines.
11. Bolt down the carburetor evenly.
12. Tighten the fuel and vacuum lines.
13. Connect the accelerator and automatic transmission linkage. If the transmission linkage was disturbed, it will have to be adjusted.
14. Connect the choke rod.
15. Install the air cleaner. Adjust the idle speed and mixture. Depending on the vintage, it may not be necessary (or possible) to adjust the idle mixture.

Disassembly

1. Place carburetor on repair stand to protect throttle valves from damage and to provide a stable base for working.
2. Remove retaining clip from accelerator pump arm link and remove link.
3. Remove step-up piston cover plate and gasket from top of air horn.
4. Remove the screws and locks from the accelerator pump arm and the vacuum piston rod lifter. Then slide the pump lever out of the air horn. The vacuum piston and step-up rods can now be lifted straight up and out of the air horn as an assembly.
5. Remove vacuum hose from between carburetor main body and choke vacuum diaphragm at main body tap.
6. Remove choke diaphragm, linkage and bracket assembly and place to one side to be cleaned as a separate item.
7. Remove retaining clip from fast idle cam link and remove link from choke shaft lever.
8. Remove fast idle cam retaining screw and remove fast idle cam and linkage.

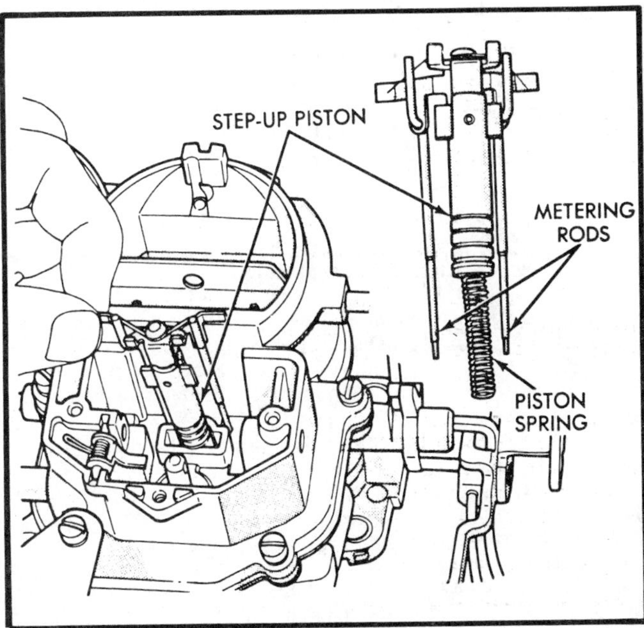

Step-up piston and metering rods showing location of assembly in the float bowl

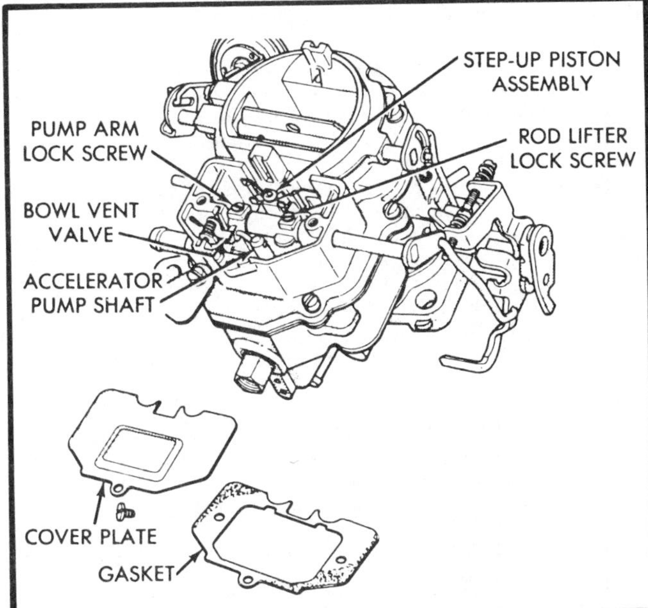

Step-up piston assembly and related components

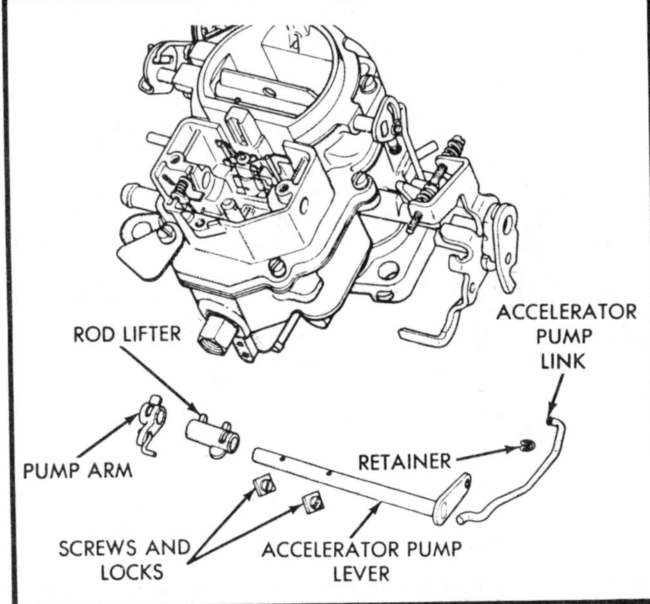

Exploded view of accelerator pump lever

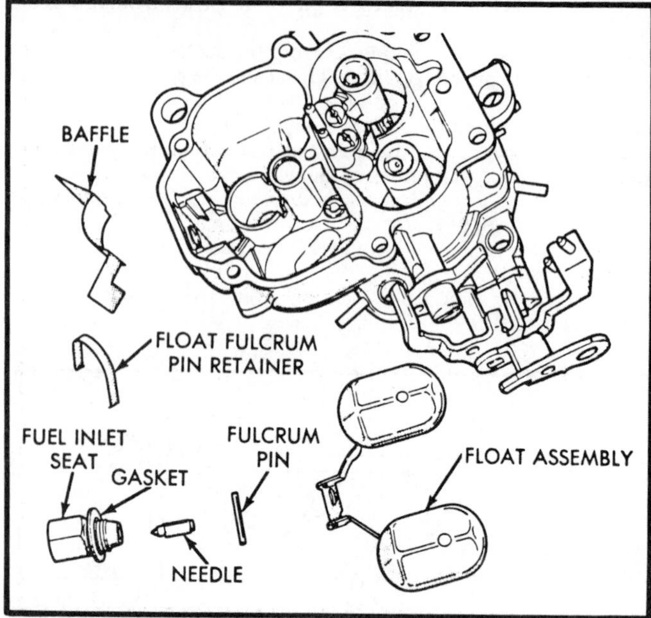

Exploded view of Carter BBD float assembly

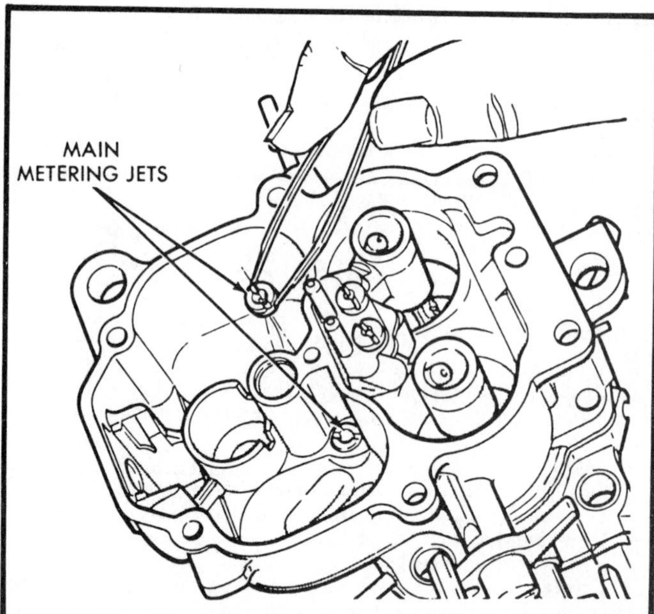

Remove and install main metering jets carefully from the float bowl

9. Remove air horn retaining screws and lift air horn straight up and away from main body. Discard gasket.

10. Invert the air horn and compress accelerator pump drive spring and remove "S" link from pump shaft. Pump assembly can now be removed.

11. Remove fuel inlet needle valve, seat, and gasket from main body.

12. Lift out float fulcrum pin retainer and baffle. Then lift out floats and fulcrum pin.

13. Remove main metering jets.

14. Remove venturi cluster screws, then lift venturi cluster and gaskets up and away from main body. Discard gaskets. Do not remove the idle orifice tubes or main vent tubes from the cluster. They can be cleaned in a solvent and dried with compressed air.

15. Invert carburetor and drop out acccelerator pump discharge and intake check balls.

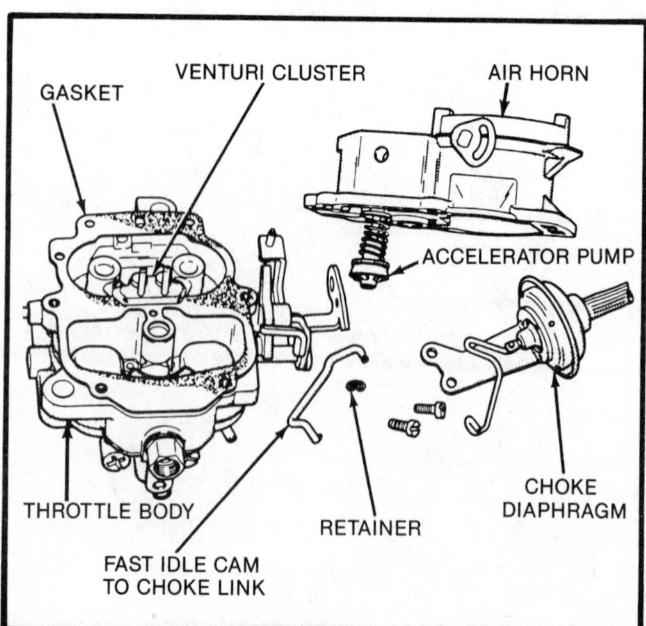

Air horn and throttle body components

16. Using procedure outlined in "Propane Assisted Idle Set Procedure" in the adjustment area at the end of this section. Remove idle mixture screws from throttle body.

17. Remove screws that attach throttle body to main body. Separate bodies.

18. Test freeness of choke mechanism in air horn. The choke shaft must float free to operate correctly. If choke shaft sticks, or appears to be gummed from deposits in air horn, a thorough cleaning will be required.

19. The carburetor now has been disassembled into three main units: the air horn, main body, and throttle body. Components are disassembled as far as necessary for cleaning and inspection.

Cleaning Carburetor Parts

Efficient carburetion depends greatly on careful cleaning and inspection during overhaul, since dirt, gum, water, or varnish in or on the carburetor parts are often responsible for poor performance.

Overhaul the carburetor in a clean, dustfree area. Carefully disassemble the carburetor, referring often to the exploded views. Keep all similar and lookalike parts segregated during disassembly and cleaning to avoid accidental interchange during assembly. Make a note of all jet sizes.

When the carburetor is disassembled, wash all parts (except diaphragms, electric choke units, pump plunger, and any other plastic, leather, fiber, or rubber parts) in clean carburetor solvent. Do not leave parts in the solvent any longer than is necessary to sufficiently loosen the deposits. Excessive cleaning may remove the special finish from the float bowl and choke valve bodies, leaving these parts unfit for service. Rinse all parts in clean solvent and blow them dry with compressed air or allow them to air dry. Wipe clean all cork, plastic, leather, and fiber parts with a clean, lint-free cloth.

Blow out all passages and jets with compressed air and be sure that there are no restrictions or blockages. Never use wire or similar tools to clean jets, fuel passages, or air bleeds. Clean all jets and valves separately to avoid accidental interchange.

Inspection and Reassembly

THROTTLE BODY

Check throttle shaft for excessive wear in throttle body. If wear is extreme, it is recommended that throttle body assembly be replaced rather than installing a new shaft in old body. Install idle mixture screws in body. The tapered portion must be straight and smooth. If tapered portion is grooved or ridged, a new idle mixture screw should be installed to insure having correct idle mixture control.

MAIN BODY

1. Invert main body and place insulator in position, then place throttle body on main body and align. Install screws and tighten securely.
2. Install accelerator pump discharge check ball (5/32 inch diameter) in discharge passage. Drop accelerator pump intake check ball (3/16 inch diameter) into bottom of the pump cylinder.
3. To check the accelerator pump system; fuel inlet and discharge check balls, proceed as follows:
4. Pour clean gasoline into carburetor bowl, approximately 1/2 inch deep. Remove pump plunger from container of mineral spirits and slide down into pump cylinder. Raise plunger and press lightly on plunger shaft to expel air from pump passage.
5. Using a small clean brass rod, hold discharge check ball down firmly on its seat. Again raise plunger and press downward. No fuel should be emitted from either intake or discharge passage. If any fuel does emit from either passage, it indicates the presence of dirt or a damaged check ball or seat. Clean passage again and repeat test. If leakage is still evident, stake check ball seats (place a piece of drill rod on top of check ball and lightly tap drill rod with a hammer to form a new seat), remove and discard old balls, and install new check balls. The fuel inlet check ball is located at bottom of the plunger well. Remove fuel from bowl.
6. Install discharge check ball. Install new gaskets on venturi cluster, then install in position in main body. Install cluster screws and tighten securely.
7. Install main metering jets.

FLOAT

The carburetors are equipped with a synthetic rubber tipped fuel inlet needle. The needle tip is viton, which is not affected by gasoline and is stable over a wide range of temperatures. The tip is flexible enough to make a good seal on the needle seat, and to give increased resistance to flooding.

NOTE: The use of the synthetic rubber tipped inlet needle requires that care be used in adjusting the float setting. Care should be taken to perform this accurately in order to secure the best performance and fuel economy.

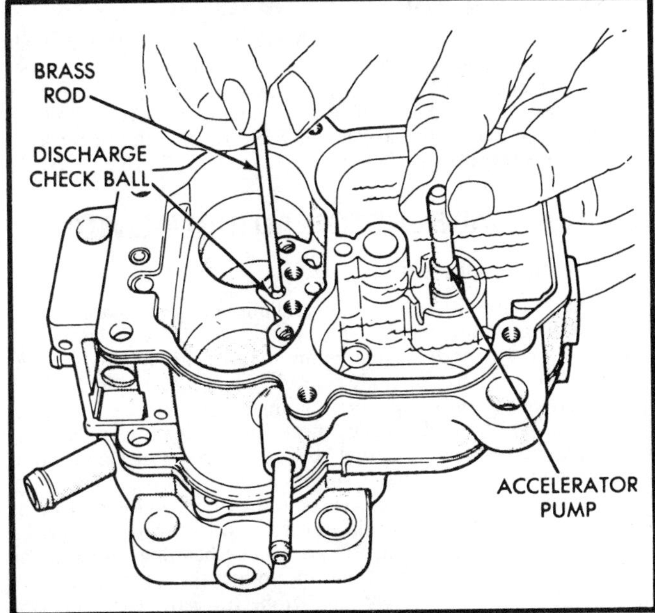

Hold down the check ball with a small brass rod when testing accelerator pump intake and discharge

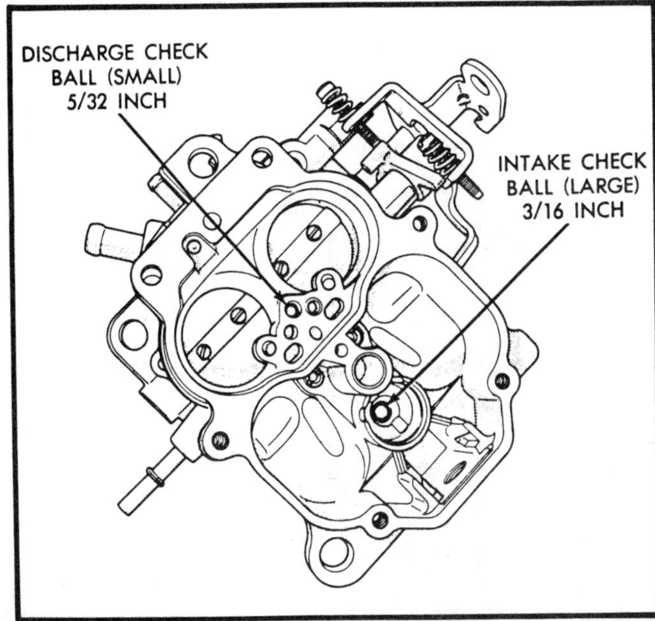

Location of accelerator pump intake and discharge check balls

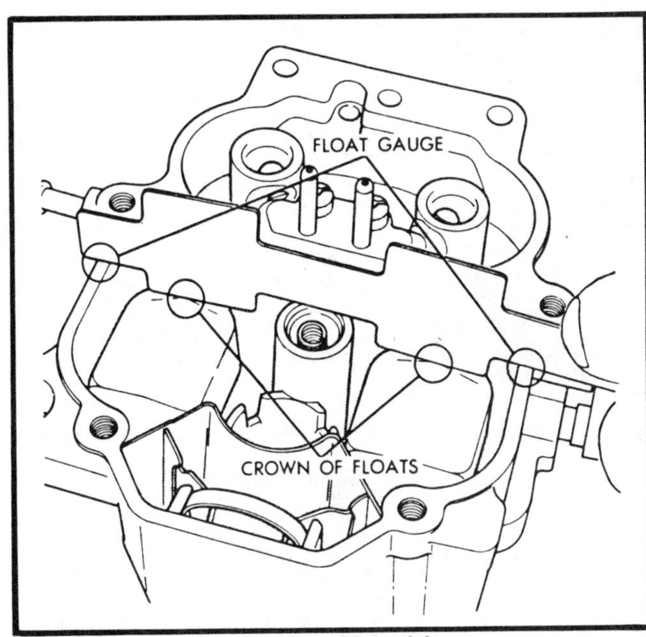

Checking float setting on Carter BBD models

When replacing the needle and seat assembly (inlet fitting), the float level must also be checked, as dimensional difference between the old and new assemblies may change the float level. Refer to carburetor adjustments for adjusting procedure.

1. Install floats and fulcrum pin.
2. Install fulcrum pin retainer and baffle.
3. Install fuel inlet needle valve, seat and gasket in main body.

AIR HORN

1. Place the accelerator pump drive spring on pump plunger shaft then insert shaft into air horn. Compress spring far enough to insert "S" link.
2. Install the pump lever, vacuum piston rod lifter and accelerator pump arm in the air horn.
3. Drop intake check ball into pump bore. Install baffle into main body. Place step-up piston spring in piston vacuum bore.

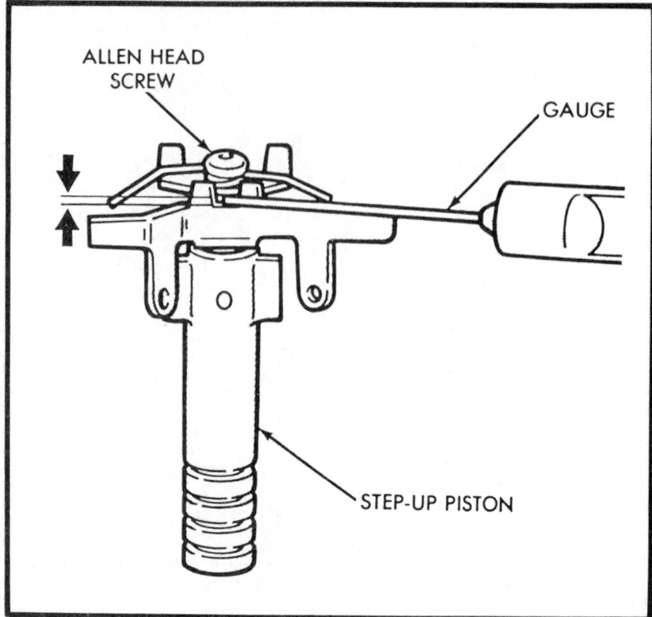

Adjusting or qualifying the step-up piston—see the text for details

Position a new gasket on the main body and install air horn.

4. Install air horn retaining screws and tighten alternately, a little at a time, to compress the gasket securely.

5. To qualify the step-up piston, adjust gap by turning the allen head calibration screw on top of the piston. See the adjustment section for proper measurement. Record number of turns and direction to obtain this dimension for this must be reset to its original position after the vacuum step-up piston adjustment has been made.

6. Carefully position vacuum piston metering rod assembly into bore in air horn making sure metering rods are in main metering jets. Then place the two lifting tangs of the plastic rod lifter under piston yoke. Slide shaft of the accelerator pump lever through rod lifter and pump arm. Install two locks and adjusting screws, but do not tighten until after adjustment is made.

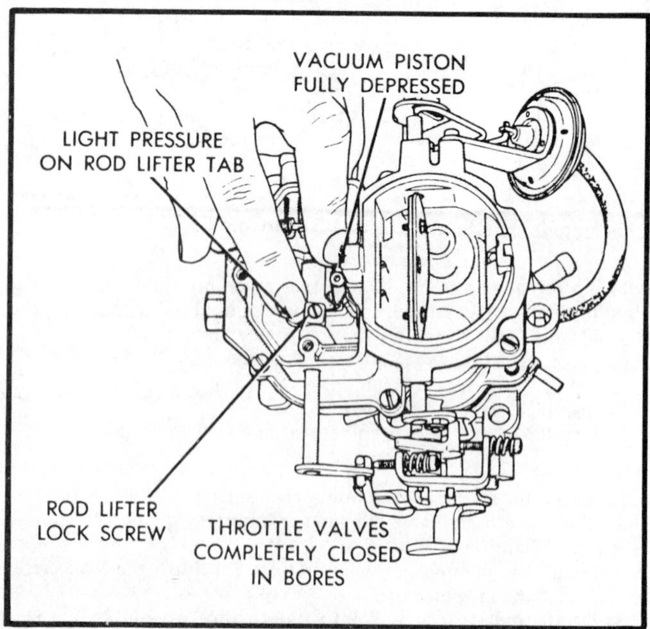

Adjusting the step-up piston assembly on Carter BBD models

7. Install fast idle cam and linkage. Tighten retaining screw securely.

8. Connect accelerator pump linkage to pump lever and throttle lever. Install retaining clip.

CHOKE VACUUM DIAPHRAGM

Inspect the diaphragm vacuum fitting to insure that the passage is not plugged with foreign material. Leak check the diaphragm to determine if it has internal leaks. To do this, first depress the diaphragm stem, then place a finger over the vacuum fitting to seal the opening. Release the diaphragm stem. If the stem moves more than 1/16 inch in 10 seconds, the leakage is excessive and the assembly must be replaced.

Install the diaphragm assembly on the air horn as follows:

1. Engage choke link in slot in choke lever.
2. Install diaphragm assembly, secure with attaching screws.
3. Inspect rubber hose for cracks before placing it on correct carburetor fitting. Do not connect vacuum hose to diaphragm fitting until after vacuum kick adjustment has been made.
4. Loosen choke valve attaching screws slightly. Hold valve closed, with fingers pressing on high side of valve. Tap valve lightly with a screw driver to seat in air horn. Tighten attaching screws securely and stake by squeezing with pliers.

CHOKE VACUUM KICK

The choke diaphragm adjustment controls the fuel delivery while the engine is running. It positions the choke valve within the air horn by action of the linkage between the choke shaft and the diaphragm. The diaphragm must be energized to measure the vacuum kick adjustment. Vacuum can be supplied by an auxiliary vacuum source.

CHOKE UNLOADER

The choke unloader is a mechanical device to partially open the choke valve at wide open throttle. It is used to eliminate choke enrichment during cranking of an engine. Engines which have been flooded or stalled by excessive choke enrichment can be cleared by use of the unloader. Refer to carburetor adjustments for adjusting procedure.

FAST IDLE SPEED

Fast idle engine speed is used to overcome cold engine friction, stalls after cold starts and stalls due to mixture icing. Make adjustment to the specifications. Refer to carburetor adjustment section for procedure.

SOLENOID IDLE STOP

Solenoid idle stops are used on vehicles equipped with either air conditioning or a heated backlite. The SIS is energized when either of those two accessories are turned on. The SIS will not change the throttle position, however, when rough engine idle occurs and the throttle is opened by the driver, the SIS will maintain a position that increases the idle speed.

BBD CARBURETOR ADJUSTMENTS

Float Setting Adjustment

1. Install floats with fulcrum pin and pin retainer in main body.
2. Install needle, seat and gasket in body and tighten securely.
3. Invert main body (catch pump intake check ball) so that weight of floats only, is forcing needle against seat. Hold finger against retainer to fully seat fulcrum pin.
4. Using a straight-edge scale, check float setting. The measurement from the surface of the fuel bowl to the crown of each float at center should be as indicated in specifications.
5. If an adjustment is necessary, hold the floats on the bottom of the bowl and bend the float lip toward or away from the needle. Recheck the setting again then repeat the lip bending operation as required. When bending the float lip, do not allow the lip to push against the needle as the synthetic rubber tip can be compressed sufficiently to cause a false setting which will affect correct level of fuel in the bowl.

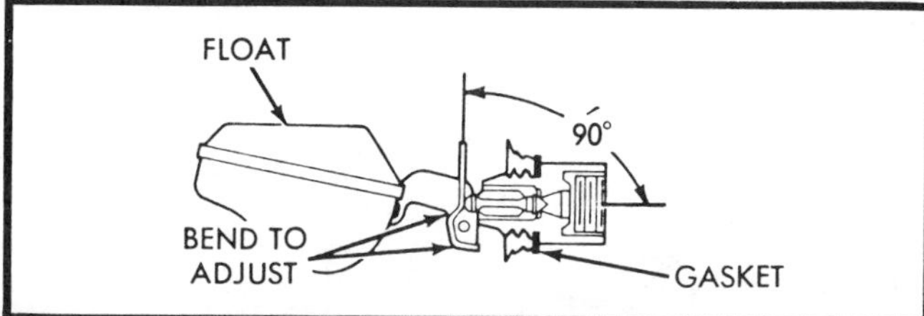

Bend the float lever as indicated to adjust the fuel level

Qualifying Step-Up Piston Adjustment

If the step-up piston assembly is removed or the mechanical rod lifter adjustment is disturbed, the step-up piston must be readjusted or "Qualified". Qualifying places the step-up piston in a "mean" or central position.

1. Remove step-up piston cover plate and gasket.
2. Remove rod lifter lock screw and remove step-up piston assembly.
3. Adjust gap in the step-up piston by turning the allen head calibration screw on top of the piston. See specifications for proper measurement. Record number of turns and direction to obtain this dimension, for this must be reset to its original position after the vacuum step-up piston adjustment has been made.
4. Install step-up piston assembly and rod lifter lock screw.
5. Proceed to Vacuum Step-Up Piston Adjustment.

Vacuum Step-Up Piston Adjustment

1. Back off idle speed screw until throttle valves are completely closed. Count number of turns so that screw can be returned to its original position.
2. Fully depress step-up piston while holding moderate pressure on the rod lifter tab. While in this position, tighten rod lifter lock screw.
3. Release piston and rod lifter, then return idle speed screw to its original position.
4. Reset the allen head calibration screw on top of the step-up piston to its original position as recorded under Qualifying Step-Up Piston Adjustment.

NOTE: If this adjustment is changed the Step-Up Piston must be Re-Qualified.

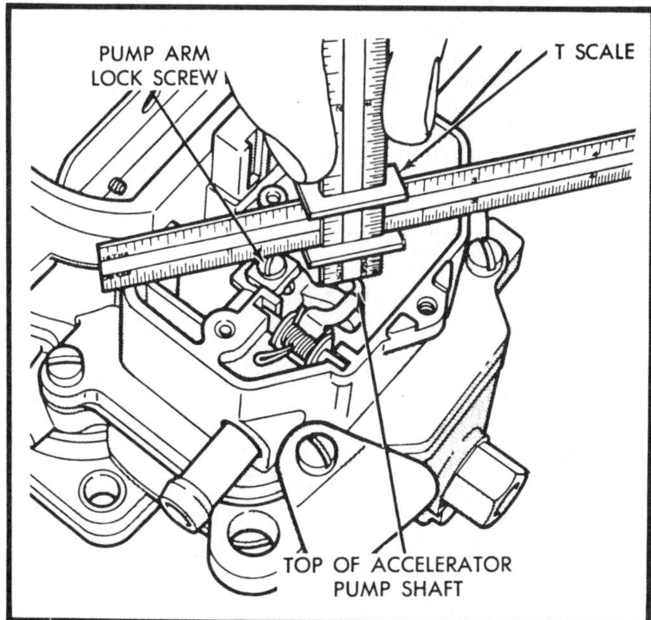

Measuring the accelerator pump stroke on Carter BBD models

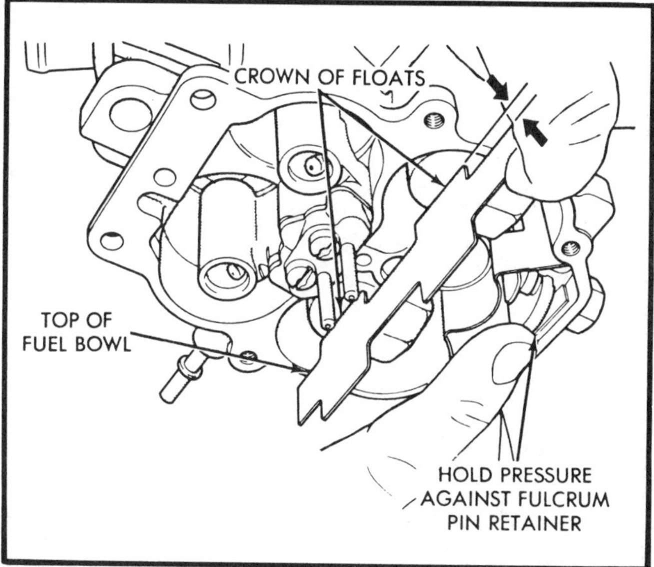

Adjusting the float level on Carter BBD models

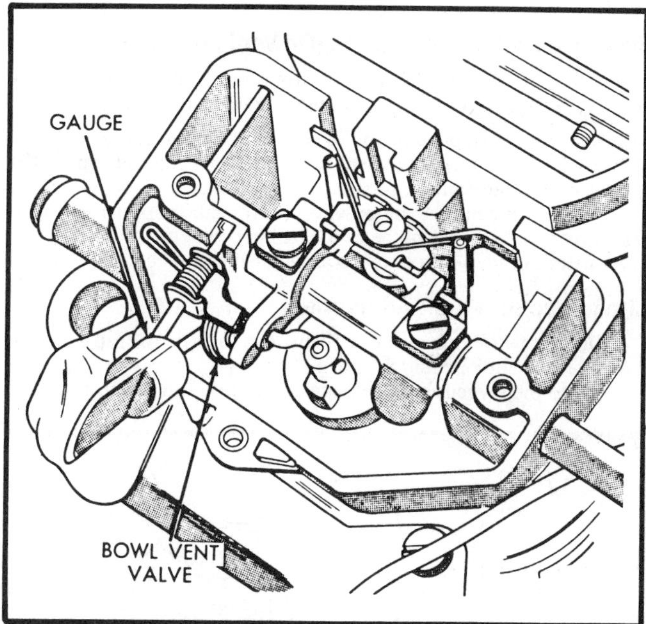

Insert the gauge between the top of the bowl vent and seat when making bowl vent valve adjustment

Accelerator Pump Stroke Measurement (at Idle)

To establish an approximate curb idle on newly assembled carburetors, back off idle speed screw to completely close the throttle valve (fast idle cam must be in open choke position). Then turn the idle speed screw clockwise until it just contacts stop. Then continue to turn two complete turns.

1. Be sure accelerator pump "S" link is in the outer hole of the pump arm if arm has two holes.
2. Measure the distance between surface of air horn and top of accelerator pump shaft this measurement must be as shown in specifications.
3. To adjust pump travel, loosen pump arm adjusting lock screw and rotate sleeve until proper measurement is obtained, then tighten lock screw.

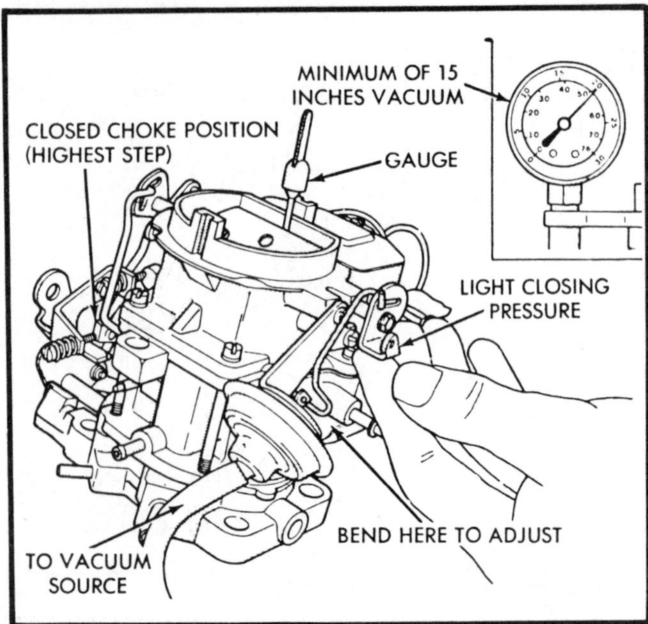

Choke vacuum kick adjustment on Carter BBD models

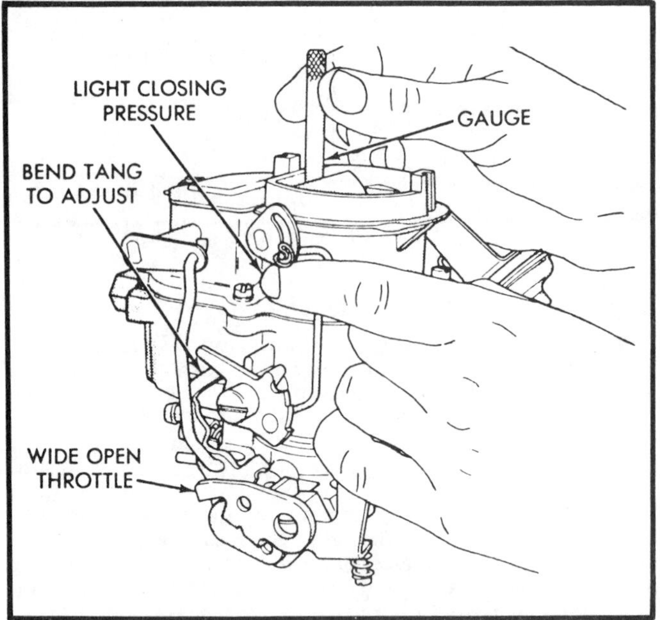

Choke unloader adjustment—Carter BBD models

Bowl Vent Valve Adjustment (On Vehicle at Curb Idle)

NOTE: The accelerator pump stroke adjustment and the idle speed must be properly set before making this adjustment.

1. Remove the step-up piston cover plate and gasket.
2. Measure by inserting specified gauge between the top of the bowl vent and the seat.
3. Adjust by bending the bowl vent lever tab. Support the bowl vent lever assembly before bending the bowl vent tab.
4. Install the step-up piston cover plate and gasket.

Choke Vacuum Kick Adjustment

1. Open throttle, close choke then close throttle to trap fast idle cam at closed choke position.
2. Disconnect vacuum hose from carburetor and connect to hose from auxiliary vacuum source with small length of tube. Apply a vacuum of 15 or more inches of mercury.
3. Apply sufficient closing force on choke lever to completely compress spring in diaphragm stem without distorting linkage. Note: solid rod stem of diaphragm extends to an internal stop as spring compresses.
4. Measure by inserting specified gauge between top of choke valve and air horn wall at throttle lever side.
5. Adjust by changing diaphragm link length (open or close U-bend).
6. Check for free movement between open and adjusted positions. Correct any misalignment or interference by rebending link and readjust.
7. Reinstall vacuum hose on correct carburetor fitting.

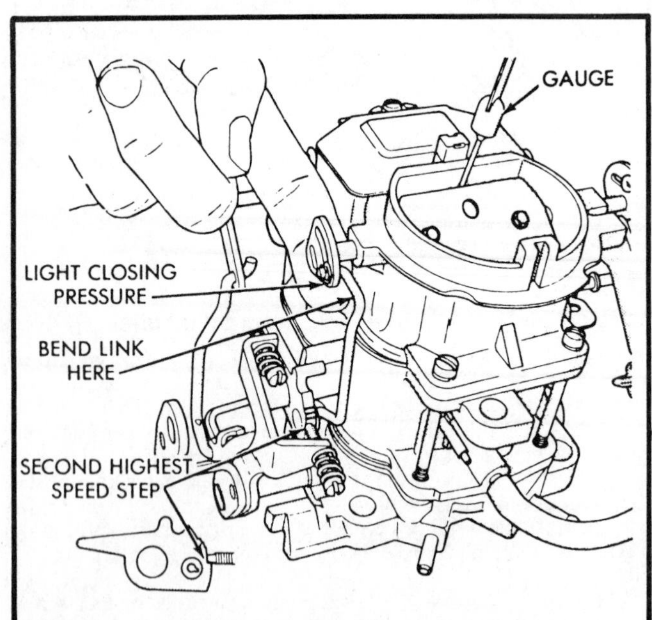

Fast idle cam position adjustment—insert gauge as shown

Fast Idle Cam Position Adjustment

1. With fast idle speed adjusting screw contacting second highest speed step on fast idle cam, move choke valve towards closed position with light pressure on choke shaft lever.
2. Measure by inserting specified gauge between top of choke valve and air horn wall at throttle lever side.
3. Adjust by bending fast idle connector rod at angle until correct valve opening has been obtained.

Choke Unloader Adjustment

1. Hold throttle valves in wide open position.
2. Lightly press finger against control lever to move choke valve toward closed position.
3. Measure by inserting specified gauge between top of choke valve and air horn wall at throttle lever side.
4. Adjust by bending tang on throttle lever.

Solenoid Idle Stop Adjustment

Before checking or adjusting any idle speed, check ignition timing and adjust if necessary. Disconnect and plug the vacuum hose at the EGR valve. On vehicles equipped with a carburetor ground switch, connect a jumper wire between the switch and a good ground. On models not equipped with a Spark Control Computer (SCC), disconnect and plug the vacuum hoses from the carburetor at the heated air temperature sensor and at the OSAC valve and remove the air cleaner. The air cleaner may be propped up on SCC equipped vehicles but not removed. Disconnect and plug the 3/16 inch diameter control hose at the canister. Remove the PCV valve from the cylinder head cover and allow the valve to draw underhood air. Connect a tachometer to the engine.

1. On Canadian vehicles equipped with a 318, 5.2L engine, proceed to Idle Set RPM Adjustment.
2. Turn on air conditioning and set blower on low. Disconnect the air conditioning compressor clutch wire.
3. On non-air conditioned vehicles, connect a jumper wire between the battery positive post and the solenoid idle stop lead wire.

CAUTION

Use care in jumping to the proper wire on the solenoid. Applying battery voltage to other than the correct wire will damage the wiring harness.

4. Open the throttle slightly and allow the solenoid plunger to extend.
5. Adjust engine rpm to the correct specification by turning the screw on the throttle lever.
6. Turn off air conditioning and reconnect clutch wire or disconnect solenoid jumper wire. Do not remove carburetor ground switch wire.
7. Proceed to Idle Set RPM Adjustment.

Idle Set RPM Adjustment

NOTE: This adjustment is to be performed only after the Solenoid Idle Stop Adjustment Procedure.

1. Disconnect the engine harness lead from the O^2 sensor and ground the engine harness lead.

CAUTION

Care should be exercised so that no pulling force is put on the wire attached to the O^2 sensor. Use care in working around the sensor as the exhaust manifold is extremely hot.

2. Remove and plug the vacuum hose at the vacuum transducer on the SCC. Connect an auxiliary vacuum supply to the vacuum transducer and set at 16 inches of vacuum.
3. Allow the engine to run for two minutes to allow the effect of disconnecting the O^2 sensor to take place.
4. If the idle set rpm is not correct, turn the screw on the solenoid to obtain the correct rpm.
5. Proceed to Fast Idle Speed Adjustment Procedure.

Fast Idle Speed Adjustment

NOTE: This adjustment is to be performed only after the Solenoid Idle Stop and Idle Set RPM Adjustments have been completed.

1. Open the throttle slightly and place the fast idle adjusting screw on the second highest step of the fast idle cam.
2. With the choke fully open, adjust the fast idle speed screw to obtain the correct fast idle rpm.
3. Return to idle then reposition the adjusting screw on the second highest step of the fast idle cam to verify speed. Readjust if necessary.
4. Return to idle and turn off engine. Unplug and reconnect vacuum hoses at the EGR valve and canister. Remove the tachometer. Remove all ground and jumper wires. Reinstall the air cleaner and, if disconnected, unplug and reconnect vacuum hoses from the carburetor to the heated air temperature sensor, and OSAC valve on the air cleaner. If disconnected, reconnect vacuum line on SCC. Reconnect O^2 sensor wire if disconnected.

NOTE: Idle speed with engine in normal operating condition (everything connected) may vary from set speeds. Do not readjust.

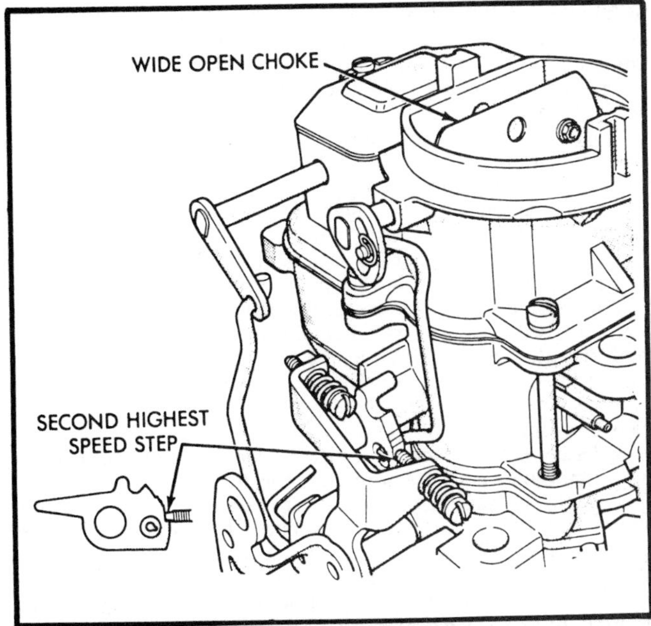

Fast idle speed adjustment—Carter BBD models

Concealment Plug Removal

1. Remove carburetor from engine.
2. Remove throttle body from carburetor.
3. Place the throttle body in a vise with the concealment plugs facing up and the gasket surfaces protected from the vise jaws.
4. Drill a 5/64 inch pilot hole at a 45° angle towards concealment plugs as shown in illustration.
5. Redrill hole to 1/8 inch.
6. Install a blunt punch into the hole and drive out the plug. Repeat procedure on the opposite side.
7. Reinstall the carburetor on the engine. The carburetor does not have to be removed to install new concealment plugs.
8. Proceed to propane assisted idle set procedure.

Propane Assisted Idle Set Procedure

NOTE: Tampering with the carburetor is a violation of Federal law. Adjustment of the carburetor idle air fuel mixture can only be done under certain circumstances, as explained below. Upon completion of the carburetor adjustment, the concealment plug must be replaced.

This procedure should only be used if an idle defect still exists after normal diagnosis has revealed no other faulty conditions such

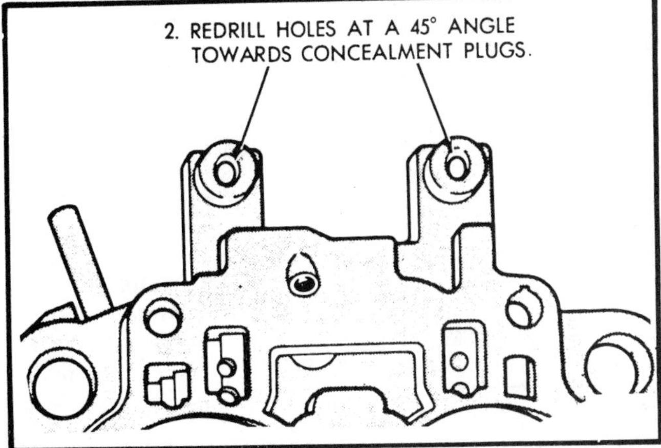

Removing plugs to adjust idle mixture—see text for details

as incorrect basic timing, incorrect idle speed, faulty hose or wire connections, etc. Also, it is important to make sure the on-board computer systems are operating properly. Adjustment of the carburetor air fuel mixture should be performed after a major carburetor overhaul.

1. Remove the concealment plug. Set the parking brake and place the transmission in neutral. Turn all lights and accessories off. Connect a tachometer to the engine. Start the engine and allow it to warm up on the second highest step of the fast idle cam until normal operating is reached. Return engine to idle.

2. Disconnect and plug the vacuum hose at the EGR valve. On vehicles equipped with a carburetor ground switch, connect a jumper wire between the switch and a good ground. On vehicles not equipped with a Spark Control Computer (SCC), disconnect and plug the vacuum hoses at the heated air temperature sensor and at the OSAC valve. Remove the air cleaner except on vehicles equipped with SCC where the air cleaner cannot be removed but may be propped up.

3. Disconnect the hose from the heated air door sensor at the carburetor and install the propane supply hose in its place.

4. With the propane bottle upright and in a safe location, remove the PCV valve from the cylinder head cover and allow the valve to draw underhood air. Disconnect and plug the 3/16 inch diameter control hose from the canister.

5. Disconnect engine harness lead from the O^2 sensor and ground the engine harness lead.

CAUTION

Care should be exercised so that no pulling force is put on the wire attached to the O^2 sensor. Use care in working around the sensor as the exhaust manifold is extremely hot.

6. Remove and plug the vacuum line at the vacuum transducer at the SCC. Connect an auxiliary vacuum supply to the vacuum transducer and apply 16 inches of vacuum.

7. Allow the engine to run for two minutes to allow the effect of disconnecting the O^2 sensor to take place.

8. Open the propane main valve. Slowly open the propane metering valve until the maximum engine rpm is reached. When too much propane is added, engine rpm will decrease. "Fine tune" the metering valve to obtain the highest engine rpm.

9. With the propane still flowing, adjust the idle speed screw on the solenoid to obtain the correct propane rpm. Again, "fine tune" the metering valve to obtain the highest engine rpm. If there has been a change in the maximum rpm, readjust the idle speed screw to the specified propane rpm.

10. Turn off the propane main valve and allow the engine speed to stabilize. Slowly adjust the mixture screws by equal amounts, pausing between adjustments to allow engine speed to stabilize, to obtain the smoothest idle at the correct idle set rpm.

11. Turn on the propane main valve and "Fine tune" the metering valve to obtain the highest engine rpm. If the maximum engine speed is more than 25 rpm different than the specified propane rpm, repeat Steps 8–12.

12. Turn off propane main and metering valves. Remove the propane supply hose and reinstall the heated air sensor hose. Reinstall new concealment plug. If installed, remove O^2 sensor ground wire and reconnect O^2 sensor. If disconnected, reconnect vacuum line on SCC. Perform the Solenoid Idle Stop, Idle Set RPM, and Fast Idle Speed Adjustment Procedures.

Float Adjustment - On Vehicle

1. Remove air cleaner assembly and air cleaner gasket.
2. Remove air cleaner mounting bolt assembly.
3. Remove choke assembly.
4. Disconnect vacuum kick diaphragm hose.
5. Remove retaining clip from accelerator pump arm link and remove link.
6. Remove fast idle cam retaining clip and remove link.
7. Remove step-up piston cover plate and gasket from top of air horn.
8. Remove metering rod lifter lock screw.
9. Lift step-up piston and metering rod assembly straight up and out of the air horn.
10. Remove air horn retaining screws.
11. Lift air horn straight up and away from main body.
12. Remove float baffle.
13. Overfill the float bowl by depressing the float manually or from an external supply to within ¼ to ⅜ inch from the top of the fuel bowl. This will provide adequate inlet needle seating force.
14. Using two wrenches, back off flare nut and tighten inlet fitting to 200 inch pounds (23 N m).
15. Firmly seat the float pin retainer by hand while measuring height. The measurement from the surface of the fuel bowl to the crown of each float at center should be as indicated in specifications. If adjustment is necessary, hold the floats on the bottom of the bowl and bend the float lip toward or away from the needle.

NOTE: When bending the float lip, do not allow the lip to push against the needle as the synthetic rubber tip can be compressed sufficiently to cause a false setting which will affect correct level of fuel in the bowl.

16. Install float baffle.
17. Install air horn on main body. Be sure the leading edge of the accelerator pump cup is not damaged as it enters the pump bore. Alternately tighten attaching screws to compress gasket evenly.
18. Alternately tighten attaching screws to compress gasket evenly.
19. Install step-up piston and metering rod assembly in air horn.
20. Install metering rod lifter lock screw.
21. Back off idle speed screw until throttle valves are completely closed. Count the number of turns so that screw can be returned to its original position.
22. Fully depress step-up piston while holding light pressure on metering rod lifter tab and tighten lock screw.
23. Install step-up piston cover plate and gasket.
24. Return idle speed screw to its original position.
25. Install fast idle cam link and retaining clip.
26. Install accelerator pump arm link and retaining clip.
27. Connect vacuum kick hose.
28. Install choke assembly.
29. Install air cleaner mounting bolt assembly.
30. Install air cleaner gasket and air cleaner.
31. Check idle speed with tachometer. If adjustment is necessary, refer to Emission Control Information Label in engine compartment for proper specifications.

CARTER BBD SPECIFICATIONS
Chrysler Products

Year	Model ②	Float Level (in.)	Accelerator Pump Travel (in.)	Bowl Vent (in.)	Choke Unloader (in.)	Choke Vacuum Kick	Fast Idle Cam Position	Fast Idle Speed (rpm)	Automatic Choke Adjustment
'78	8136S	¼	0.500①	0.080	0.280	0.110	0.070	1500	Fixed
	8137S	¼	0.500①	0.080	0.280	0.100	0.070	1600	Fixed

CARTER BBD SPECIFICATIONS
Chrysler Products

Year	Model ②	Float Level (in.)	Accelerator Pump Travel (in.)	Bowl Vent (in.)	Choke Unloader (in.)	Choke Vacuum Kick	Fast Idle Cam Position	Fast Idle Speed (rpm)	Automatic Choke Adjustment
'78	8177S	¼	0.500①	0.080	0.280	0.100	0.070	1600	Fixed
	8175S	¼	0.500①	0.080	0.280	0.160	0.070	1400	Fixed
	8143S	¼	0.500①	0.080	0.280	0.150	0.070	1500	Fixed
'79	8198S	¼	0.500①	0.080	0.280	0.100	0.070	1600	Fixed
	8199S	¼	0.500①	0.080	0.280	0.100	0.070	1600	Fixed
'80	8233S	¼	0.500①	0.080	0.280	0.130	0.070	1500	Fixed
	8235S	¼	0.500①	0.080	0.280	0.130	0.070	1700	Fixed
	8237S	¼	0.500①	0.080	0.280	0.110	0.070	1500	Fixed
	8239S	¼	0.500①	0.080	0.280	0.110	0.070	1500	Fixed
	8286S	¼	0.500①	0.080	0.280	0.100	0.070	1400	Fixed
'81-'82	8290S	¼	0.500①	0.080	0.280	0.100	0.070	1600	Fixed
	8291S	¼	0.500①	0.080	0.280	0.130	0.070	1400	Fixed
	8292S	¼	0.500①	0.080	0.280	0.130	0.070	1600③	Fixed
'83	8290S	¼	0.470①	0.080	0.280	0.100	0.070	1600	Fixed
	8291S	¼	0.470①	0.080	0.280	0.130	0.070	1400	Fixed
	8369S	¼	0.500①	0.080	0.280	0.130	0.070	1500	Fixed
'84	8385S	¼	0.470①	0.080	0.280	0.130	0.070	1400	Fixed
	8369S	¼	0.500	0.080	0.280	0.130	0.070	1500	Fixed

CARTER BBD SPECIFICATIONS
American Motors

Year	Model ①	Float Level (in.)	Accelerator Pump Travel (in.)	Choke Unloader (in.)	Choke Vacuum Kick	Fast Idle Cam Position	Fast Idle Speed (rpm)	Automatic Choke Adjustment
'78	8128	¼	0.496	0.280	0.150	0.110	1600	Index
	8129	¼	0.520	0.280	0.128	0.095	1500	1 Rich
'79	8185	¼	0.470	0.280	0.140	0.110	1600	1 Rich
	8186	¼	0.520	0.280	0.150	0.110	1500	1 Rich
	8187	¼	0.470	0.280	0.140	0.110	1600	1 Rich
	8221	¼	0.530	0.280	0.150	0.110	1600	1 Rich
'80	8216	¼	0.520	0.280	0.140	0.090	1850	2 Rich
	8246	¼	0.520	0.280	0.140	0.095	1850	2 Rich
	8247	¼	0.520	0.280	0.150	0.095	1700	1 Rich
	8248	¼	0.520	0.280	0.150	0.095	1700	1 Rich
	8253	¼	0.470	0.280	0.128	0.095	1850	2 Rich
	8256	¼	0.470	0.280	0.128	0.093	1850	2 Rich
	8278	¼	0.542	0.280	0.140	0.093	1850	Index

CARTER BBD SPECIFICATIONS
American Motors

Year	Model ①	Float Level (in.)	Accelerator Pump Travel (in.)	Choke Unloader (in.)	Choke Vacuum Kick	Fast Idle Cam Position	Fast Idle Speed (rpm)	Automatic Choke Adjustment
'81	8310	¼	0.525	0.280	0.140	0.095	1850	Index
	8302	¼	0.500	0.280	0.128	0.095	1850	1 Rich
	8303	¼	0.500	0.280	0.128	0.090	1700	1 Rich
	8306	¼	0.500	0.280	0.128	0.090	1700	1 Rich
	8307	¼	0.500	0.280	0.128	0.095	1850	1 Rich
	8308	¼	0.500	0.280	0.128	0.095	1850	2 Rich
	8309	¼	0.520	0.280	0.128	0.093	1700	2 Rich
'82	8338	¼	0.520	0.280	0.140	0.095	1850	1 Rich
	8339	¼	0.520	0.280	0.140	0.095	1850	1 Rich
'83	8360	¼	0.520	0.280	0.140	0.095	1850	Fixed
	8364	¼	0.520	0.280	0.140	0.095	1700	Fixed
	8367	¼	0.520	0.280	0.140	0.095	1700	Fixed
	8362	¼	0.520	0.280	0.140	0.095	1850	Fixed
'84	8383	¼	0.520	0.280	0.140	0.095	1850	½–1½ Rich
	8384	¼	0.520	0.280	0.140	0.095	1700	½–1½ Rich

① Model numbers located on the tag or casting

CHRYSLER ELECTRONIC FEEDBACK CARBURETOR

The Chrysler Electronic Feedback Carburetor (EFC) system was introduced in mid 1979 on Volarés and Aspens sold in California with the six cylinder engine. The system is a conventional one, incorporating an oxygen sensor, a three-way catalytic converter, an oxidizing catalytic converter, a feedback carburetor, a solenoid-operated vacuum regulator valve, and a Combustion Computer. Also incorporated into the system are Chrysler's Electronic Spark Control, and a mileage counter which illuminates a light on the instrument panel at 15,000 mile intervals, signaling the need for oxygen sensor replacement.

In Chrysler's system, "Combustion Computer" is a collective term for the Feedback Carburetor Controller and the Electronic Spark Control computer, which are housed together in a case located on the air cleaner. The feedback carburetor controller is the information processing component of the system, monitoring oxygen sensor voltage (low voltage/lean mixture, high voltage/rich mixture), engine coolant temperature, manifold vacuum, engine speed, and engine operating mode (starting or running). The controller examines the incoming information and then sends a signal to the solenoid-operated vacuum regulator valve (also located in the Combustion Computer housing), which then sends the proper rich or lean signal to the carburetor.

The 1 bbl Holley R-8286A carburetor (and 1 bbl Holley 6145 on later models) is equipped with two diaphragms, controlling the idle system and the main metering system. The diaphragms move tapered rods, which vary the size of the orifices in the idle system air bleed and the main metering system fuel flow. A "lean" command from the controller to the vacuum regulator results in increased vacuum to both diaphragms, which simultaneously raise both the idle air bleed rod (increasing idle air bleed) and the main metering rod (reducing fuel flow). A "rich" command reduces vacuum level, causing the spring-loaded rods to move in the other direction, enrichening the mixture.

Both closed loop and open loop operation are possible in the EFC system. Open loop operation occurs under any one of the following conditions: coolant temperature under 150°F; oxygen sensor temperature under 660°F; low manifold vacuum (less than 4.5 in. Hg. engine cold, or less than 3.0 in Hg. engine hot): oxygen sensor failure; or hot engine starting. Closed loop operation begins when engine temperature reaches 150°F.

Air injection is supplied by an air pump. At cold engine temperature, air is injected into the exhaust manifold upstream of both catalytic converters. At operating temperature, an air switching valve diverts air from the exhaust to an injection point downstream from the three-way catalyst, but upstream of the conventional oxidizing catalyst.

In 1980, the system was modified slightly and used on all California models, and on all 318 4-bbl. V8's nationwide. The 1980 and later system is used with Electronic Spark Advance (ESA), not Electronic Spark Control (ESC)—see the previous description in this section and the Electronic Ignition Systems section in this book for details. Differences lie in the deletion of some components within the combustion computer. The start timer, vacuum transducer count-up clock and memory throttle transducer, and ambient air temperature sensor are not used.

The feedback system for the six cylinder engines is essentially unchanged. The four and eight cylinder systems differ from the six mainly in the method used to control the carburetor mixture. Instead of having vacuum-controlled diaphragms to raise or lower the mixture rods, the carburetors are equipped with an electric solenoid valve, which is part of the carburetor. These later carburetors include the Carter BBD and Carter Thermo-Quad.

Other differences between the systems are minor. On the four cylinder, the ignition sensor is the Hall Effect distributor, but it functions in the same manner as the six cylinder pick-up coil. The eight cylinder uses two pick-up coils (a Start pick-up and a Run pick-up); troubleshooting is included in the "Lean Burn/Electronic Spark Control" section. The four and six cylinder engines use a 150°F coolant switch; the eight cylinder uses a 150°F switch with Combustion Computer 4145003, and a 98°F switch with Computer 4145088. The eight cylinder uses a 150°F switch with Combustion Computer 4145003, and a 98°F switch with Computer 4145088. The eight cylinder engine has a detonation sensor (see the "Lean Burn/Electronic Spark Control" section), and the six and eight cylinder engines have a charge temperature switch to monitor intake charge temperature. Below approximately 60°F, the switch prevents EGR timer function and EGR valve operation; additionally, on eight cylinder engines, air injection is routed upstream of the exhaust manifolds.

Finally, the replacement interval for the oxygen sensor has been doubled, from 15,000 to 30,000 miles. Replacement procedures and odometer resetting are the same as for the 1979 six cylinder system.

Note that two completely different troubleshooting procedures have been included here. Use the "1979" procedure only for 1979 Aspens and Volarés with the six cylinder engine. Use the "1980 and Later" procedure as applicable.

Troubleshooting

1979 MODELS

Troubleshooting requires the use of a few special tools. A 0-5 in. Hg. vacuum gauge accurate within 1/2 in. Hg.; a 0-30 in. Hg. vacuum gauge; a hand vacuum pump with vacuum gauge; two short lengths of 3/16 in. I.D. vacuum hose; two 3/16 in. vacuum tees; and a jumper wire approximately five feet long.

Before performing any tests, check all vacuum hoses for leaks, breaks, kinks, or improper connections, all electrical connections for soundness, and all wires for fraying or breaks. Check for leakage at both the intake and exhaust manifolds.

1. Warm the engine to normal operating temperature. Install a tee into the control vacuum hose which runs to the carburetor. Install the 0-5 in. vacuum gauge on the tee. Start the engine and allow it to idle. The vacuum gauge should read 2.5 in. for approximately 100 seconds, then fall to zero, then gradually rise to between 1.0 and 4.0 in. The reading may oscillate slightly.
2. If the vacuum reading is incorrect, increase the engine speed to 2000 rpm. If vacuum reads between 1.0 and 4.0 in., return the engine to idle. If the reading is now correct, the system was not warmed up, originally, but is OK.
3. If the gauge is correct at 2000 rpm but not at idle, the carburetor must be replaced.
4. If the vacuum is either above 4.0 in. or below 1.0 in., follow the correct troubleshooting procedure given next. Note that in most cases of system malfunction, control vacuum will be either 0 in. or 5.0 in.

Control Vacuum Above 4.0 In. Hg.

Start the engine, apply the parking brake, place the transmission in Neutral, and place the throttle on the next to lowest step of the fast idle cam.

1. Remove the PCV hose from the PCV valve. Cover the end of the hose with your thumb. Gradually uncover the end of the hose until the engine runs rough. If control vacuum gets lower as the hose is uncovered, the carburetor must be replaced; however, complete Step 2 before replacing it. If control vacuum remains high, continue with the tests.
2. Before replacing the carburetor, examine the heat shield. Interference may exist between the heat shield and the mechanical power enrichment valve lever. If so, the carburetor will be running rich. Correct the problem and repeat Step 1.

NOTE: A new heat shield is used starting in 1979 which has clearance for the enrichment lever. Earlier heat shields should not be used unless modified for clearance.

3. Disconnect the electrical connector at the solenoid regulator valve. Control vacuum should drop to zero. If not, replace the solenoid regulator valve.
4. Disconnect the oxygen sensor wire. Use the jumper wire to connect the *harness* lead to the negative battery terminal.

CAUTION

Do not connect the oxygen sensor wire to ground or to the battery.

Control vacuum should drop to zero in approximately 15 seconds. If not, replace the Combustion Computer. If it does, replace the oxygen sensor. Before replacing either part, check the Computer to sensor wire for continuity.

Control Vacuum Below 1.0 In. Hg.

1. Start the engine and allow it to idle in Neutral. Disconnect the vacuum hose at the computer transducer and connect the hose to the 0-30 in. Hg. vacuum gauge. The gauge should show manifold vacuum (above 12 in.). If not, trace the hose to its source and then connect it properly to a source of manifold vacuum.

NOTE: The following Steps should be made with the engine warm, parking brake applied, transmission in Neutral, and throttle placed on the next to lowest step of the fast idle cam.

2. Remove the air cleaner cover. Gradually close the choke plate until the engine begins to run roughly. If control vacuum increases to 5.0 in. as the choke is closed, go to Step 3. If control vacuum remains low, go to Step 4.
3. Disconnect the air injection hose from its connection to a metal tube at the rear of the cylinder head. Plug the tube. If control vacuum remains below 1.0 in., replace the carburetor. If control vacuum returns to the proper level, reconnect the air injection hose and disconnect the 3/16 in. vacuum hose from the air switching valve. If control vacuum remains below 1.0 in., replace the air switching valve. If control vacuum rises to the proper level, check all hoses for proper connections, then, if correct, replace the coolant vacuum switch.
4. Check that the bottom nipple of the solenoid regulator valve is connected to manifold vacuum. Disconnect the solenoid regulator electrical connector. Use the jumper wire to connect one terminal of the solenoid regulator lead to the positive battery terminal. Connect the other terminal of the solenoid regulator lead to ground. Control vacuum should rise above 5.0 in. If not, replace the solenoid regulator. If so, go to the next step.
5. Disconnect the 5 terminal connector at the computer. The terminals are numbered 1 to 5, starting at the rounded end. Connect a jumper wire from terminal 2 in the harness to a ground. Control vacuum should rise to 5 in. If not, trace the voltage to the battery to discover where it is being lost. If so, go to the next step.

NOTE: Wiring harness problems are usually in the connectors. Check them for looseness or corrosion.

6. Disconnect the oxygen sensor wire. Use a jumper wire to connect the *harness* lead to the positive battery terminal.

CAUTION

Do not connect the oxygen sensor wire to the battery or to a ground.

Control vacuum should rise to 5 in. in approximately 15 seconds. If not, replace the computer. If so, replace the oxygen sensor.

Ignition Timing

1. Ground the carburetor switch with a jumper wire.
2. Connect a timing light to the engine.

3. Start the engine. Wait one minute.
4. With the engine running at a speed not greater than the specified curb idle rpm (see the emission control sticker in the engine compartment), adjust the timing to specification.
5. Remove the ground wire after adjustment.

Curb Idle Adjustment

Adjust the curb idle only after ignition timing has been checked and set to specification.

1. Start the engine and run in Neutral on the second step of the fast idle cam until the engine is fully warmed up and the radiator becomes hot. This may take 5 to 10 minutes.
2. Disconnect and plug the EGR hose at the EGR valve.
3. Ground the carburetor switch with a jumper wire.
4. Adjust the idle rpm in Neutral to the curb idle rpm figure given on the emission control sticker in the engine compartment.
5. Reconnect the EGR hose and remove the jumper wire.

Oxygen Sensor Replacement

1. Disconnect the negative battery cable. Remove the air cleaner.
2. Disconnect the sensor electrical lead. Unscrew the sensor using Chrysler special tool C-4589.
3. Installation is the reverse. Before installation, coat the threads of the sensor with a nickel base anti-seize compound. Do not use other type compounds since they may electrically insulate the sensor. Torque the sensor to 35 ft. lbs.

Mileage Counter Reset

The mileage counter will illuminate every 15,000 miles, signaling the need for oxygen sensor replacement. After replacing the oxygen sensor, reset the counter as follows:

1. Locate the mileage counter. It is spliced into the speedometer cable, covered by a rubber boot.
2. Slide the rubber boot up the speedometer cable to expose the top of the mileage counter. Turn the reset screw on top of the counter to reset. Replace the boot.

1980 AND LATER MODELS

ESA System Tests

1. Connect a timing light to the engine.
2. Disconnect and plug the vacuum hose at the vacuum transducer. Connect a vacuum pump to the transducer fitting and apply 14–16 in. Hg. of vacuum.
3. With the engine at normal operating temperature, raise the speed to 2000 rpm. Wait one minute, then check the timing advance. Specifications are as follows (timing in addition to basic advance):

1980 4 cyl. M/T: 20°–28°
1980 4 cyl. A/T: 31°–39°
1981 1.7 A/T: 31°–39°
1981 1.7 M/T Fed.: 34°–42°
39°–47°
1981 2.2 M/T Fed.: 24°–32°
29°–37°
1981 2.2 M/T Cal.: 19°–27°
1981 2.2 A/T: 21°–29°
1980 6 cyl.: 10°–18°
1980 8 cyl.: 15°–23°
1981 6 cyl.: 16°–24°
1981 8 cyl.: 30°–38°
1982 1.7 A/T Fed.: 51°–59°
1982 1.7 M/T: 46°–54°
1982 2.2 A/T: 41°–49°
1982 2.2 M/T: 43°–51°
1982 6 cyl.: 16°–24°
1982 8 cyl.: EFI 19°–27°
1982 8 cyl.: 30°–38°
1983 1.7 M/T: 26°–34°
1983 2.2 Can.: 20°–28°
1983 2.2 Fed.: 24°–32°
1983 2.2 Hi. Alt.: 30°–38°
1983 2.2 Cal. M/T: 30°–38°
1983 2.2 Cal. A/T: 27°–35°
1983 225 Fed.: 10°–18°
1983 225 Calif.: 4°–12°
1983 318-2 USA: 26°–34°
1983 318-2 Can.: 21°–29°
1983 318-4 Can.: 21°–29°
1983 318-4 Fed.: 16°–22°
1983 318 EFI: 16°–24°

Air Switching System Tests

1. Remove the vacuum hose from the air switching valve; connect a vacuum gauge to the hose.
2. Start the engine. With the engine cold, engine vacuum should be present on the gauge until the engine coolant temperature is as follows:

1980 four cylinder M/T: 98°F
1980 four cylinder A/T: 135°F
1981–83 four cylinder 2.2L
California: 125°F
1981–83 four cylinder 2.2L 49 states:
150°F
6 cylinder all: 150°F
8 cylinder with computer 4145003: 150°F
8 cylinder with computer 4145088: 98°F

On the 8 cylinder models, the charge temperature switch must be open and fuel mixture temperature above 60°F.

3. When the indicated temperatures are reached, vacuum should drop to zero. If no vacuum is present on the gauge before the temperature is reached:

On the four cylinder, check the vacuum supply and the Coolant Controlled Engine Vacuum Switch (CCEVS); on the six and eight cylinder, check the vacuum supply, air switching solenoid, coolant switch (and charge temperature switch on the eight), and the wiring and connections to the computer. If all these systems are OK, it is possible that the computer is faulty, preventing air switching.

4. With the engine warm on the four cylinder, no vacuum should be present; if there is vacuum, check the CCEVS.
5. With the engine warm on the six and eight: on the 1980 six, vacuum should be present for 100 seconds; on the 1981 six for 65 seconds; on the 1980 eight with 4145003 for 25 seconds; on the 1980 eight with 4145088 for 90 seconds; on the 1981 Cal. eight for 20 seconds; on the 1981 Fed. 4 bbl. eight for 30 seconds; on the 1981 Fed. 2 bbl. eight for 90 seconds after the engine starts. After the period indicated, vacuum should drop to zero. If there is no vacuum, check as follows:

Connect a voltmeter to the light green wire on the air switching solenoid. On the eight cylinder, also disconnect the coolant switch and charge temperature switch. Start the engine; voltage should be less than one volt. Allow the warm-up schedule to finish (time as specified at the beginning of this step). The solenoid should deenergize and the voltmeter should then read charging system voltage. If not, replace the solenoid and repeat the test. If the voltmeter indicates charging system voltage before the warm-up schedule finishes, replace the computer.

6. Test the air switching valve: remove the air supply hose from the valve. Remove the vacuum hose from the valve and connect a vacuum pump to the fitting. Start the engine; air should be discharged from the side port. When vacuum is applied, discharge from the side port should cease, and air should then be discharged from the bottom port.

EFC Tests

Check all vacuum hose connections and the spark advance schedule before performing these tests. Check the resistance of all related wiring, and examine all electrical connections for soundness. On the four cylinder, connect a vacuum pump to the vacuum transducer and apply 10 (16–1981) in. Hg. of vacuum. On all engines, start the engine and allow it to reach normal operating temperature.

NOTE: After a hot restart, run the engine at 1200–2000 rpm for at least two minutes before continuing. DO NOT GROUND THE CARBURETOR SWITCH.

1. On the four and eight and 1981 six cylinder engines, disconnect the electrical connector from the regulator solenoid. Engine speed should increase at least 50 rpm. (If not, on the four cylinder only disconnect the four-way tee from the air cleaner temperature sensor and repeat the test. If no response, replace the computer.) Connect the regulator solenoid; engine speed should return to 1200–

2000 rpm. Disconnect the six (twelve-1981 reardrive) pin connector from the computer, and connect a ground to the #15 harness connector pin. Engine speed should drop 50 rpm. If not, check for carburetor air leaks, and service the carburetor as necessary.

On 1980 six cylinders, tee a 0–5 in. Hg. vacuum gauge into the vacuum regulator supply line to the carburetor. Disconnect the regulator wiring. With the engine idling and no voltage to the regulator, vacuum should be zero, and engine speed should increase by at least 50 rpm. Using a jumper wire, apply battery voltage to one terminal of the regulator, and ground the other terminal, vacuum should rise to 5 in. Hg., and engine speed should drop at least 50 rpm. If not, replace the regulator and repeat the test; if still faulty, replace the computer.

2. With the engine cold, check the coolant switch. It should have continuity to ground on the four cylinder, or have a resistance of less than 10 ohms on the six and eight. With the engine warm (above 150°F) the switch should be open.

3. With the engine hot, disconnect the coolant temperature switch. *Do not ground the carburetor switch.* Maintain an engine speed of 1200–2000 rpm (use a tachometer). Disconnect the oxygen sensor electrical lead at the sensor and connect a jumper wire to the harness end of the connector. Ground the other end of the jumper wire. The engine speed should increase (at least 50 rpm) for 15 seconds, then return to 1200–2000 rpm. (If not, on the four cylinder *only*, disconnect the four-way tee from the air cleaner temperature sensor, allowing the engine to draw in air. Repeat the test; if no response, replace the computer.) Next, connect the end of the jumper wire to the positive battery terminal; engine speed should drop. If the computer fails these tests, replace it. Reconnect the wires.

4. To test the oxygen sensor, run the engine at 1200–2000 rpm. Connect a voltmeter to the solenoid output wire which runs to the carburetor (18 DGN). Hold the choke plate closed. Over the next ten seconds, voltage should drop to 3 V. or less. If not, disconnect the air cleaner temperature sensor four-way tee and repeat the test. If no response, replace the computer.

Disconnect the PCV hose and/or the canister purge hose. Over the next ten seconds, voltage should be over 9 V. Voltage should then drop slightly, and remain there until the vacuum hoses are reconnected.

If the oxygen sensor fails these tests, replace it. Reconnect all wires.

Holley 6145 Feedback Carburetor

GENERAL INFORMATION

The Holley Model 6145 carburetor is a single venturi concentric downdraft electronic feedback carburetor designed for use on 225/3.7 engines. The fuel bowl completely surrounds the venturi. Dual nitrophyl floats control a fuel level which permits high angularity operation to meet the most severe driving conditions. The closed-cell nitrophyl material also eliminates the possibility of malfunction due to a punctured metal float.

Principal sub-assemblies include a bowl cover, carburetor body and throttle body. A thick gasket between the throttle body and main body retards heat transfer to the fuel to resist fuel percolation in warm weather. To correctly identify the carburetor model, always check the part number stamped on the main body or attached tag. The carburetor includes four basic fuel metering systems. The idle system provides a mixture for smooth idle and a transfer system for low speed operation. The main metering system provides an economical mixture for normal cruising conditions. A fuel regulator solenoid responsive to the oxygen sensor. The accelerator system provides additional fuel during acceleration. The power enrichment system provides a richer mixture when high power output is desired.

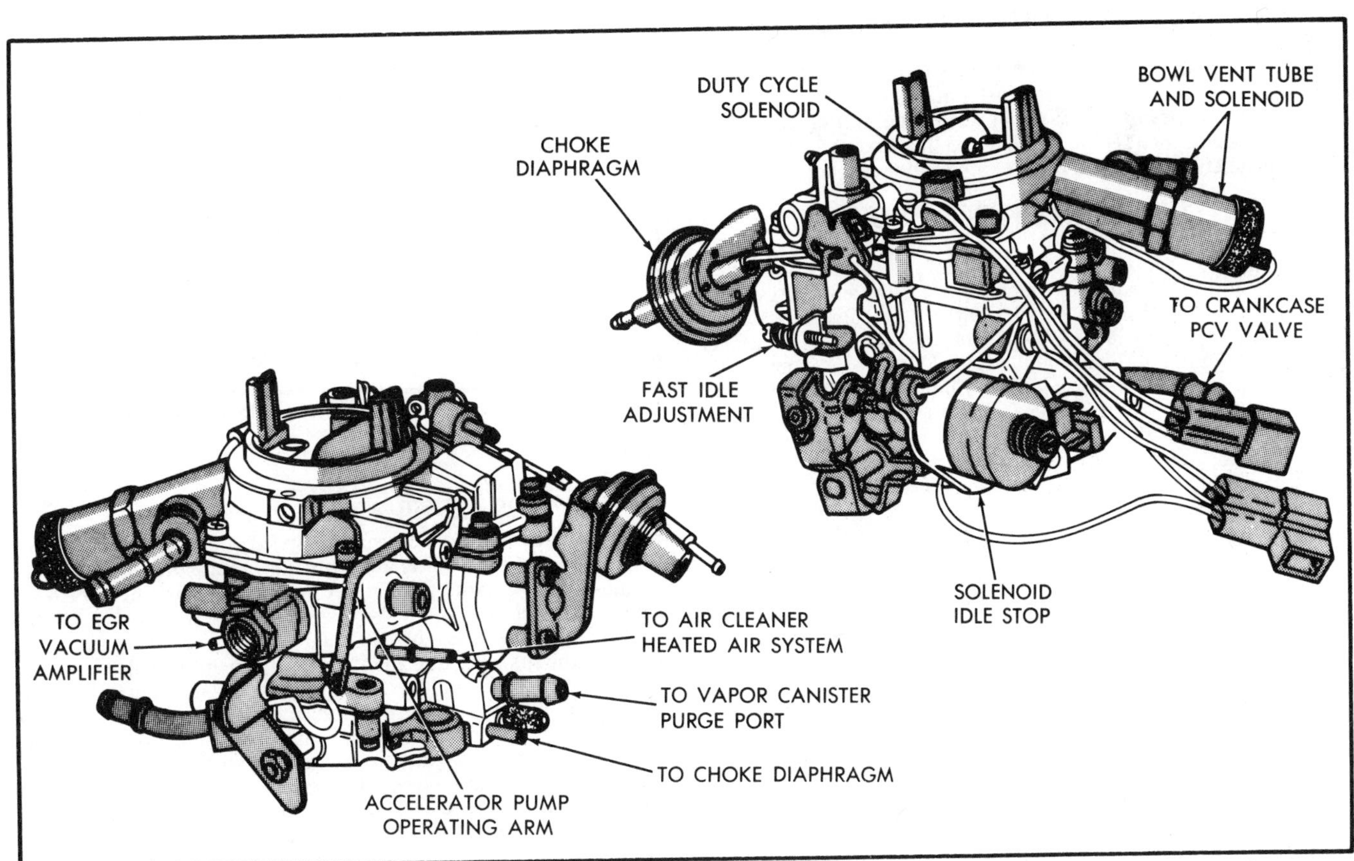

Holley 6145 feedback carburetor

In addition to these four basic systems, there is a fuel inlet system that constantly supplies the fuel to the basic metering systems. A choke system temporarily enriches the mixture to aid in starting and running a cold engine. If problems are experienced with the Electronic Feedback Carburetor System, the system and its related components must be checked before carburetor work is begun. Close attention should be given to vacuum hose condition, hose connections, and any area where leaks could develop. In addition, the related wiring and its connectors must be checked for excessive resistance. Refer to group 25 Emission Control Systems for system operation and service diagnosis. In normal service, the mixture should not require adjustment. The idle set rpm can be checked without removal of the tamper resistant plug.

NOTE: Tampering with the carburetor is a violation of Federal Law. Adjustment of the carburetor idle air-fuel mixture can only be done under certain circumstances.

Adjustment should only be considered if an idle defect still exists after normal diagnosis has revealed no other faulty condition, such as, incorrect idle speed, faulty hose or wire connection, etc. Also, it is important to make sure the Combustion Computer systems are operating properly. Adjustment of the carburetor idle air-fuel mixture should also be performed after a major carburetor overhaul. Upon completion of the carburetor adjustment, it is important to reinstall the plug. The proper procedure is outlined later in this group.

Fuel Inlet System

All fuel enters the fuel bowl through the fuel inlet fitting in the carburetor body. The "Viton" tipped fuel inlet needle seats directly in the fuel inlet fitting. The needle is retained by a cap that permits the fuel to flow out holes in the side of the cap. The design of the fuel bowl eliminates the necessity of a fuel baffle. The fuel inlet needle is controlled by a dual lung nitrophyl float (a closed cellular buoyant material which cannot collapse or leak) and a stainless steel float lever which is hinged by a stainless steel float shaft.

The fuel inlet system must constantly maintain the specified level of fuel as the basic fuel metering systems are calibrated to deliver the proper mixture only when the fuel is at this level. When the fuel level in the bowl drops the float also drops permitting additional fuel to flow past the fuel inlet needle into the bowl. All carburetors with external bowl vents are vented to the vapor canister.

O_2 FEEDBACK SOLENOID

The function of the O_2 Feedback Solenoid is to provide limited regulation of the fuel-air ratio of a feedback carburetor in response to electrical signals from the Spark Control Computer. It performs this task by metering main system fuel, and operates in parallel with a conventional fixed main metering jet. In the power-off mode the solenoid valve spring pushes upward through the main system fuel valve. At this point, the solenoid controlled main metering orifice is fully uncovered, so the richest condition exists within the carburetor for any given airflow. In the power-on mode the field windings are energized and a magnetic flux path is established from the armature, through the armature bushing, the solenoid case and the pole piece, returning once again to the armature. This attraction between the pole piece and the armature assembly moves the valve operating push rod and main system valve downward against the valve spring. This movement continues until the main system valve bottoms on the main system valve seat and stops before the armature contacts the pole piece. In this position, the solenoid controlled main metering orifice is fully sealed. These conditions remain unchanged until the electrical signal from the Spark Control Computer to the solenoid is switched off. This solenoid position offers the leanest condition within the carburetor for any given airflow.

Main system fuel may be regulated between the richest and leanest limits by controlling the amount of time that the solenoid is in the power-on position. Under normal operating conditions a voltage signal (12 volts nominally) is applied to the field windings at a frequency of 10 Hz. By controlling the duration of this voltage signal the ratio of power-on time to total time, referred to as duty cycle is established. The O_2 feedback solenoid includes, in addition to a protective case, such attaching surfaces and passageways as are required for the directing and containing of main system fuel when the solenoid assembly is correctly installed in a feedback carburetor designed for its use.

Idle System

Fuel used during curb idle and low speed operation flows through the main metering jet into the main well. An angular connecting idle well intersects the main well. An idle tube is installed in the idle well. Fuel travels up the idle well and mixes with air which enters through the idle air bleed located in the bowl cover. At curb idle the fuel and air mixture flows down the idle channel and is further mixed or broken up by air entering the idle channel through the transfer slot which is above the throttle plate at curb idle. The idle system is equipped with a restrictor in the idle channel, located between the transfer slot and the idle port, which limits the maximum attainable idle mixture. During low speed operation the throttle plate moves exposing the transfer slot and fuel begins to flow through the transfer slot as well as the idle port. As the throttle plates are opened further and engine speed increases, the air flow through the carburetor also increases. This increased air flow creates a vacuum or depression in the venturi and the main metering system begins to discharge fuel.

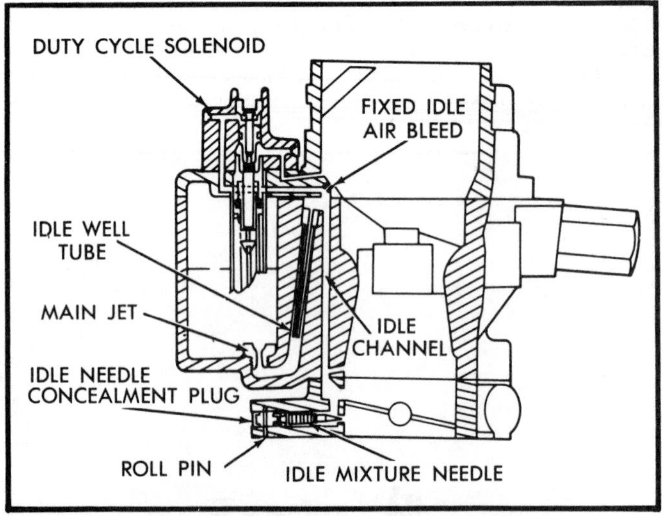

Idle fuel control circuit on Holley 6145 feedback carburetor

Main Metering System

As the engine approaches cruising speed, the increased air flow through the venturi creates a low pressure area in the venturi of the carburetor. Near-atmospheric pressure present in the fuel bowl causes the fuel to flow to the lower pressure area created by the venturi and magnified by the dual booster venturi. Fuel flows through the main jet into the main well. Then, air enters through high speed air bleed and mixes in the main well through holes in the main well tube. The mixture of fuel and air being lighter than raw fuel responds faster to changes in venturi vacuum and is also more readily vaporized when discharged into the venturi.

The main discharge nozzle passage is a part of the dual booster venturi which is an integral part of the main body casting. The main metering system is calibrated to deliver a lean mixture for best overall economy. When additional power is required a vacuum operated power system enriches the fuel-air mixture.

Power Enrichment System (Modulated Power Valve)

During high speed (or low manifold vacuum) the carburetor must provide a mixture richer than is needed when the engine is running at cruising speed under no great power requirements. Added fuel for power operation is supplied by a vacuum modulated power enrichment system. A vacuum passage in the throttle body transmits manifold vacuum to the vacuum piston chamber in the bowl cover. Under light throttle and light load conditions, there is sufficient vacuum acting on the vacuum piston to overcome the piston spring tension. When the throttle valves are opened to a 55° angle, vacuum that is acting on the piston is bled to atmosphere and manifold vacuum is closed off, which in turn insures proper mixture for this throttle opening. The vent port is right in line with the throttle shaft. The throttle shaft has a small hole drilled through it. When the throttle valve is opened to 55° the hole in the throttle shaft will line up with the port in the base of the carburetor and vent the piston vacuum chamber to atmosphere allowing the spring loaded piston to open the power valve. As engine power demands are reduced, and the throttle valve begins to close, manifold vacuum increases. The increased vacuum acts on the vacuum piston, overcoming the tension of the piston spring. This closes the power valve and shuts off the added supply of fuel which is no longer required.

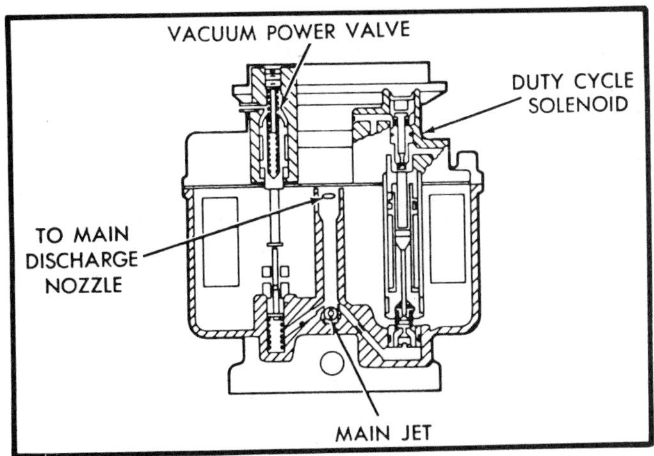

Main metering system on Holley 6145 models

Accelerating Pump System

When the throttle plates are opened suddenly, the air flow through the carburetor responds almost immediately. However, there is a brief time interval or lag before the fuel can overcome its inertia and maintain the desired fuel-air ratio. The piston type accelerating pump system mechanically supplies the fuel necessary to overcome this deficiency for a short period of time. Fuel enters the pump cylinder from the fuel bowl through the pump cup with the fuel level well above the normal position of the pump piston.

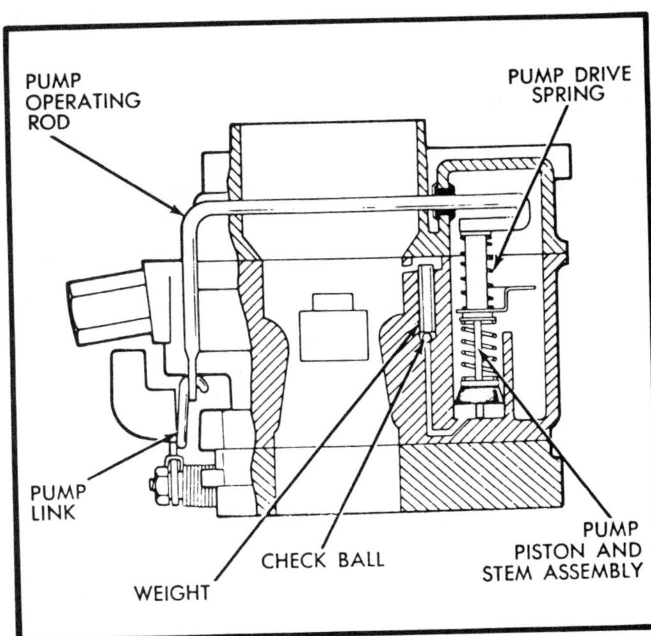

Accelerator pump system on Holley 6145 models

As the throttle lever is moved, the pump link operating through a system of levers pushing the pump piston down seating the pump cup against the face of the stem. Fuel is forced through a passage around the pump discharge needle valve and out the pump discharge jet which is drilled in the main body. When the pump is not in operation, vapors or bubbles forming in the pump cylinder can escape around the pump stem through the inlet of the floating piston cup.

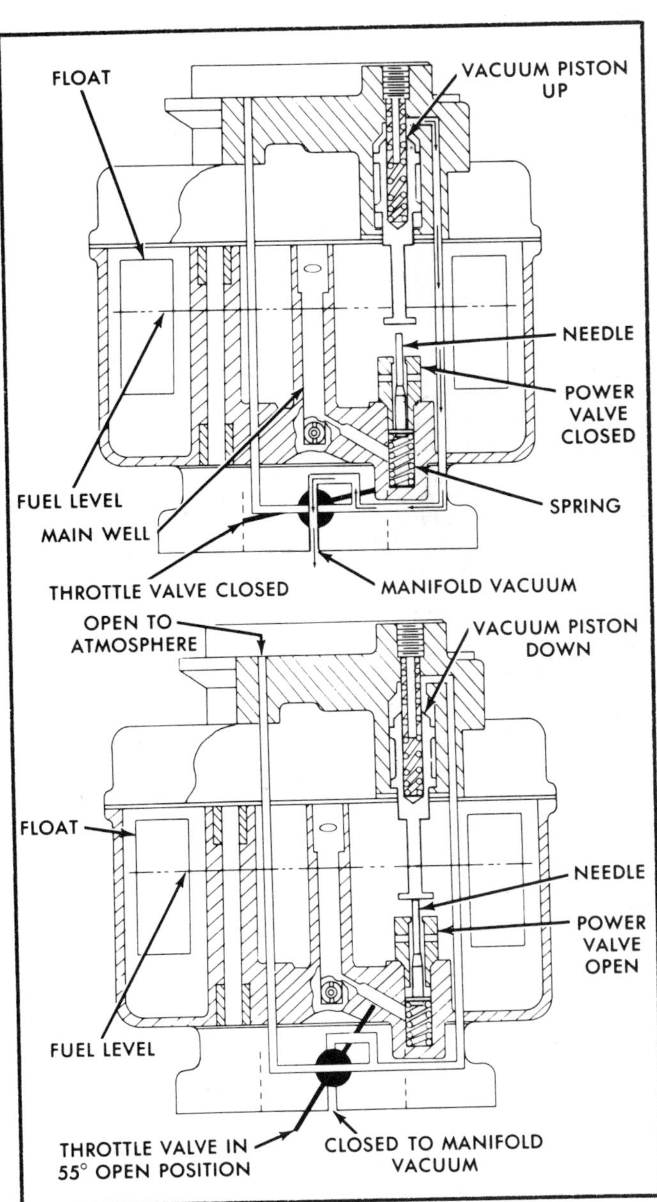

Power enrichment system with modulated power valve

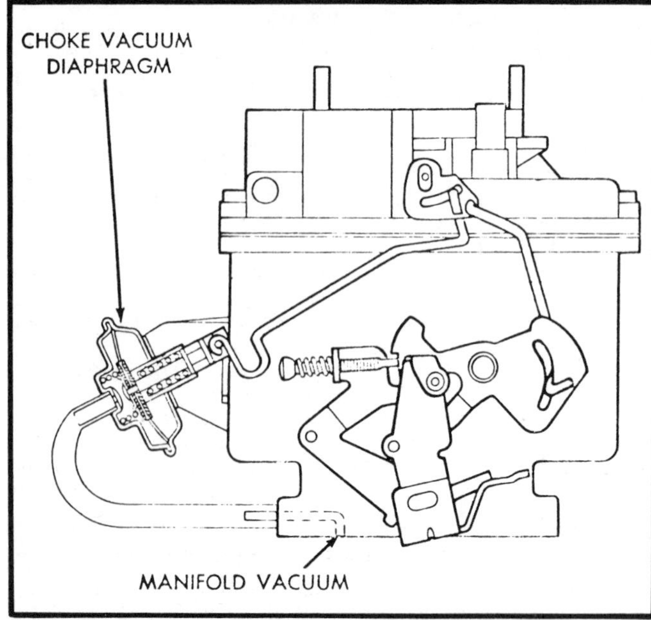

Automatic choke system showing vacuum connection

Automatic Choke System

The automatic choke provides richer fuel-air mixture required for starting and operating a cold engine. A bi-metal spring inside the choke housing, which is installed in a well in the intake manifold, pushes the choke valve toward the closed position. When the engine starts, manifold vacuum is applied to the choke diaphragm through a rubber hose from the throttle body. The adjustment of the choke valve opening when the engine starts by the choke diaphragm, is called vacuum kick. Manifold vacuum alone is not strong enough to provide the proper degree of choke opening during the entire choking period. The impact of in rushing air past the offset choke valve provides the additional opening force. As the engine warms up, manifold heat transmitted to the choke housing relaxes the bi-metal spring until it eventually permits the choke to open fully. An electric heater assists engine heat to open the choke rapidly in summer temperatures.

Exhaust Gas Recirculation (EGR System)

The venturi vacuum control system utilizes a vacuum tap at the throat of the carburetor venturi to provide a control signal for the Exhaust Gas Recirculation (EGR) system.

Carburetor Service Procedures

A thorough road test and check of minor carburetor adjustments should precede major carburetor service. Specifications for some adjustments are listed on the Vehicle Emission Control Information label found in each engine compartment.

Many performance complaints are directed at the carburetor. Some of these are a result of loose, misadjusted or malfunctioning engine or electrical components. Others develop when vacuum hoses become disconnected or are improperly routed. The proper approach to analyzing carburetor complaints should include a routine check of such areas.

1. Inspect all vacuum hoses and actuators for leaks. Refer to the vacuum hose routing diagram label located under the hood in the engine compartment for proper hose routing.
2. Tighten intake manifold bolts and carburetor mounting bolts to specifications.
3. Perform cylinder compression test.
4. Clean or replace spark plugs as necessary.
5. Test resistance of spark plug cables. Refer to "Ignition System Secondary Circuit Inspection," Engine Electrical Section.
6. Inspect ignition primary wire and vacuum advance operation. Test coil output voltage, primary and secondary resistance. Replace parts as necessary. Refer to "Ignition System" and make necessary adjustment.
7. Reset ignition timing.
8. Check carburetor idle mixture and speed adjustment. Adjust throttle stop screw to specifications.
9. Test fuel pump for pressure and vacuum.
10. Inspect manifold heat control valve in exhaust manifold for proper operation.
11. Remove carburetor air filter element and blow out dirt gently with an air hose. Install a new recommended filter element if necessary.
12. Inspect crankcase ventilation system.
13. Road test vehicle as a final test.

Cleaning Carburetor Parts

There are many commercial carburetor cleaning solvents available which can be used with good results. The choke diaphragm, choke heater and some plastic parts of the carburetor can be damaged by solvents. Avoid placing these parts in any liquid. Clean the external surfaces of these parts with a clean cloth or a soft brush. Shake dirt or other foreign material from the stem (plunger) side of the diaphragm. Compressed air can be used to remove loose dirt but should not be connected to the vacuum diaphragm fitting.

NOTE: If the commercial solvent or cleaner recommends the use of water as a rinse, HOT water will produce better results. After rinsing, all trace of water must be blown from the passages with air pressure. Never clean jets with a wire, drill, or other mechanical means, because the orifices may become damaged, causing improper performance. When checking parts removed from the carburetor, it is recommended that new parts be installed whenever the old parts are questionable.

Carburetor Removal

CAUTION

Do not attempt to remove the carburetor from the engine of a vehicle that has just been road tested. Allow the engine to cool sufficiently to prevent accidental fuel ignition or personal injury.

1. Disconnect battery ground cable.
2. Remove air cleaner.
3. Remove fuel tank pressure vacuum filler cap. The fuel tank should be under a small pressure.
4. Place a container under fuel inlet fitting to catch any fuel that may be trapped in fuel line.
5. Disconnect fuel inlet line using two wrenches to avoid twisting the line.
6. Disconnect throttle linkage, choke linkage and all vacuum hoses.
7. Remove carburetor mounting bolts or nuts and carefully remove carburetor from engine compartment. Hold carburetor level to avoid spilling fuel from fuel bowl.

Carburetor Installation

1. Inspect the mating surfaces of carburetor and intake manifold. Be sure both surfaces are clean and free of nicks, burrs or other damage. Place a new flange gasket on manifold surface.

NOTE: Some flange gaskets can be installed up-side down or backwards. To prevent this, match holes in the flange gasket to holes on bottom of carburetor, then place gasket properly on intake manifold surface.

2. Carefully place carburetor on manifold without trapping choke rod under carburetor linkage.
3. Install carburetor mounting bolts or nuts and tighten alternately, a little at a time, to compress flange gasket evenly. The nuts

or bolts must be drawn down tightly to prevent vacuum leakage between carburetor and intake manifold.

4. Connect throttle and choke linkage and fuel inlet line.

5. Check carefully for worn or loose vacuum hose connections. Refer to the vacuum hose routing diagram label located under the hood in the engine compartment.

6. Check to be sure the choke plate opens and closes fully when operated.

7. Check to see that full throttle travel is obtained.

8. Install air cleaner. The air cleaner should be cleaned or replaced at this time to insure proper carburetor performance.

9. Connect battery cable.

10. Check carburetor idle mixture adjustment. Refer to Emission Control Label in engine compartment.

CAUTION

The practice of priming an engine by pouring gasoline into the carburetor air horn for starting after servicing the fuel system, should be strictly avoided. Cranking the engine, and then priming by depressing the accelerator pedal several times should be adequate.

Diagnosing carburetor complaints may require that the engine be started and run with the air cleaner removed. While running the engine in this mode it is possible that the engine could backfire. A backfiring situation is likely to occur if the carburetor is malfunctioning, but removal of the air cleaner alone can lean the air fuel ratio in the carburetor to the point of producing an engine backfire. The battery cable should be removed from the negative terminal of the battery before any fuel system component is removed. This precaution will prevent the possibility of ignition of fuel during servicing.

Disassembly

1. Place carburetor assembly on repair stand to prevent damage to throttle valves and to provide a suitable base for working.

2. Remove wire retainer and bowl vent assembly.

3. Remove Solenoid Idle Stop (SIS).

4. Remove the fast idle cam retaining clip, fast idle cam and link. Disconnect link.

5. Remove choke vacuum diaphragm, link and bracket assembly. Disengage link from slot in choke lever. Place the diaphragm to one side and clean as special items.

6. Remove nut and washer from throttle shaft. Remove throttle lever and link. Note the hole position of lever.

7. Remove screws and duty cycle solenoid.

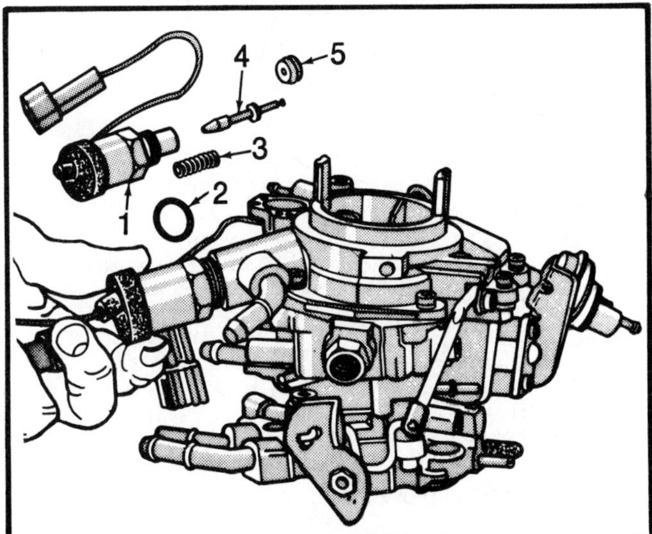

Exploded view of bowl vent solenoid assembly

8. Remove the bowl cover screws. Separate the bowl cover from the carburetor body by tapping with a plastic hammer or handle of a screwdriver. Do not pry cover off with screwdriver blade. Lift bowl cover straight up until vacuum piston stem, accelerator pump, and main well tube are clear of the main body.

9. Remove the bowl cover gasket. If any gasket material is remaining on either surface, remove with a suitable cleaner. Do not use a metal scraper such as a carbon scraper or screwdriver on either the bowl cover surface or carburetor body surface. A nylon or hard plastic material such as a delrin body moulding remover, may be used as a suitable scraper.

10. Remove the accelerating pump operating rod retainer screw and retainer.

11. Remove accelerator pump assembly retaining screw and pump assembly.

12. Rotate the pump operating rod and remove from bowl cover.

13. Remove the pump operating rod grommet.

14. The power piston assembly retaining ring is staked in position and care must be taken at removal. Remove staking with a suitable sharp tool then remove the vacuum piston from the bowl cover by depressing the piston and allowing it to snap up against the retaining ring.

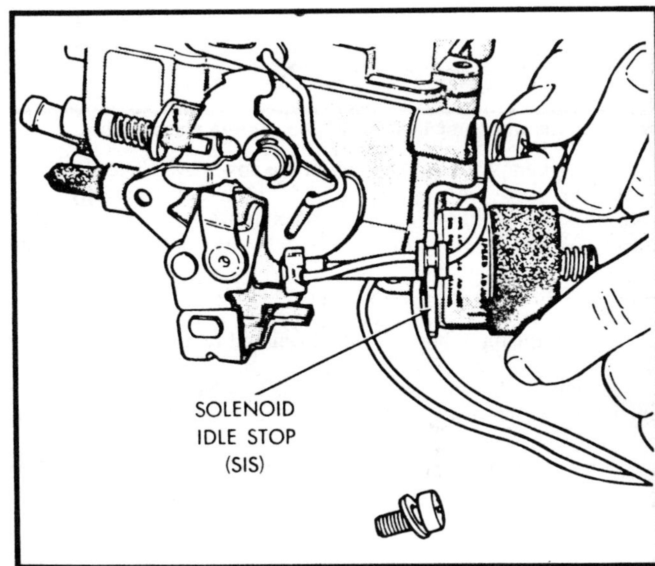

Remove the solenoid idle stop with its bracket

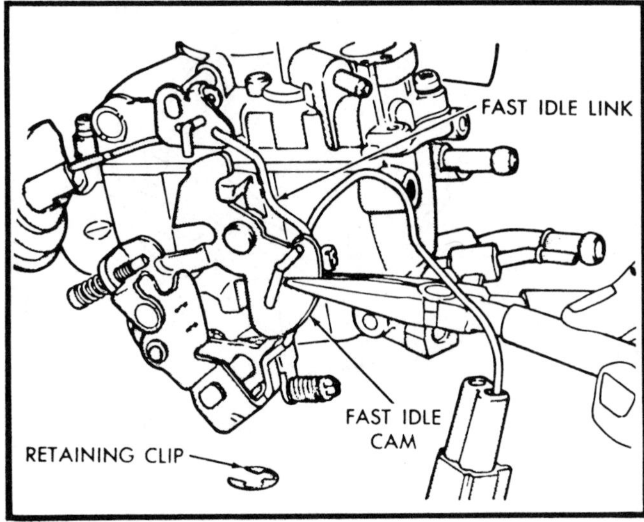

Removing fast idle cam and link—typical

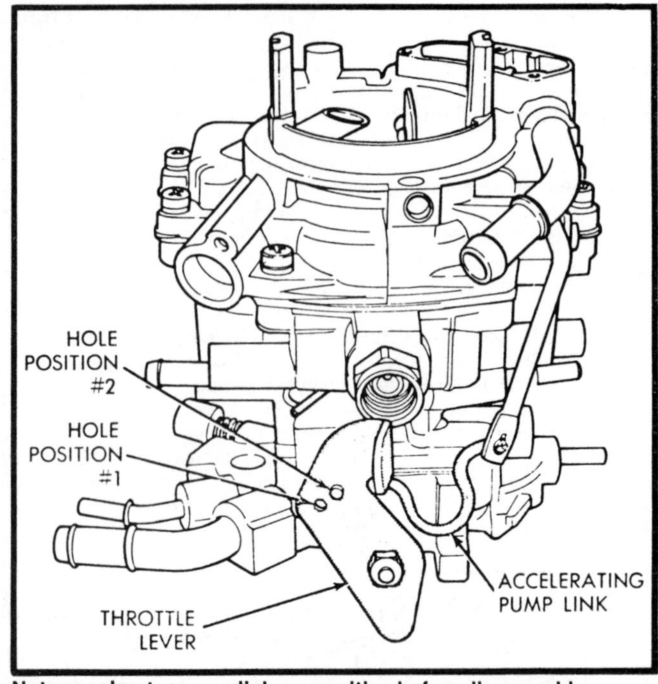

Note accelerator pump linkage position before disassembly

15. This completes disassembly of the bowl cover. The main well tube cannot be removed and must be blown out carefully from both inside and outside of the cover. Remove the fuel inlet fitting valve assembly from the main body. Separate the gaskets from the parts.

16. Remove the spring float shaft retailer, float shaft and float assembly.

17. Turn the main body upside down and remove the pump discharge check ball and weight.

18. Remove the main jet with Special Tool Number C-3748. A $^3/_8$ inch or wider screwdriver can also be used, but be sure it has a good square blade.

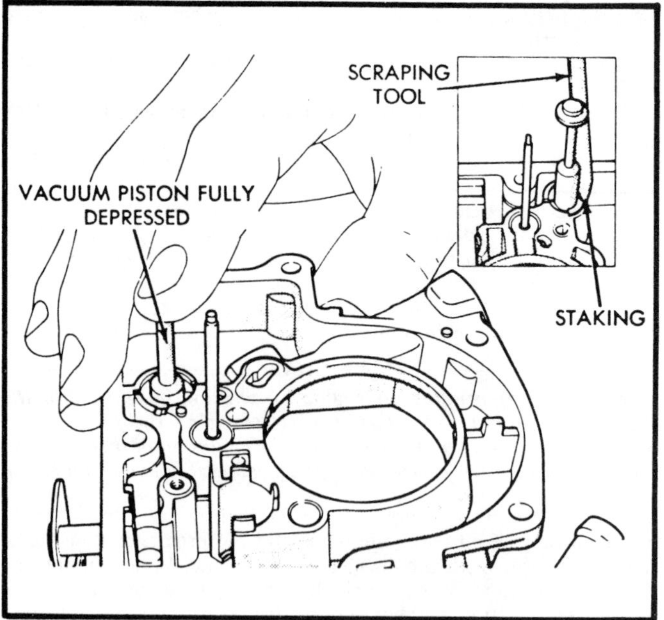

Exercise caution when removing vacuum piston—see text for details

19. Carefully depress the power valve needle with a $^3/_8$ inch wide screwdriver until screwdriver blade is squarely seated in slot on top of the valve. Remove valve. The power valve assembly consists of seat needle and spring. All of these components of the service power valve should be used if replacement is required.

20. This completes disassembly of the main body. Remove the three main body-to-throttle body screws. Separate the throttle body from the main body and remove the gasket.

Inspection and Reassembly

THROTTLE BODY

1. Check throttle shaft for excessive wear in body. If wear is

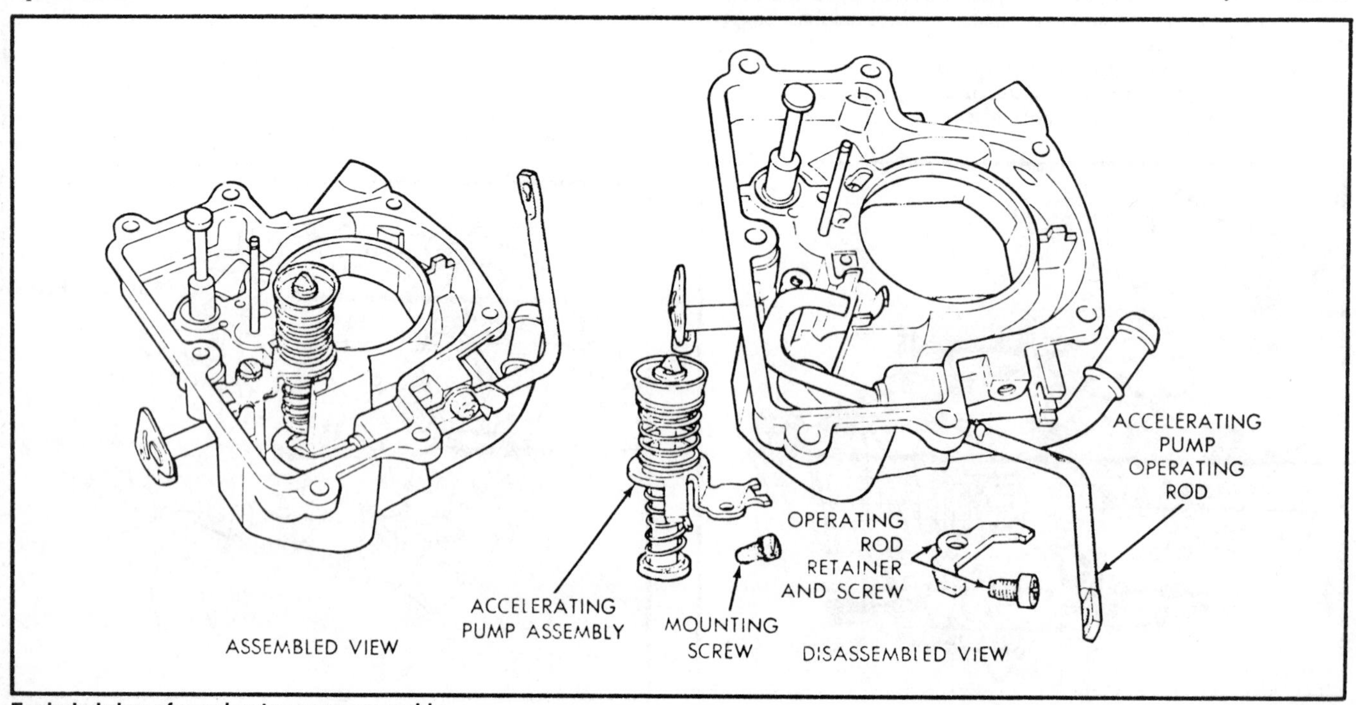

Exploded view of accelerator pump assembly

extreme, it is recommended that carburetor assembly be replaced rather than installing a new shaft in old body.

2. Install idle mixture screw and spring in body. (The tapered portion of the screw must be straight and smooth. If tapered portion is grooved or ridged, a new idle mixture screw should be installed to insure having correct idle mixture control).

3. Using a new gasket, install throttle body to main body, torque screws to 30 inch pounds (3 Nm).

BOWL COVER

1. Before installing the vacuum piston assembly in bowl cover, be sure and remove all staking from the retainer cavity. Install the spring and piston in the vacuum cylinder, seat the retainer and stake lightly with a suitable tool.

2. Test the accelerator pump discharge check ball and seat prior to assembly by coating the pump piston with oil, or filling the fuel bowl with clean fuel. Install accelerating pump discharge check ball and weight. Hold the discharge check ball and weight down with a small brass rod and operate the pump plunger by hand. If the check ball and seat are leaking, no resistance will be experienced when operating the plunger. If the valve is leaking, remove weight and stake the ball using a suitable drift punch. Exercise care when staking the ball to avoid damaging the bore containing the pump weight. After staking the old ball, remove and replace with new ball from tune-up kit. Install weight and re-check for leaks. If no leaks, remove check ball and weight from main body and install accelerator pump, pump operating rod, and rod retainer in bowl cover.

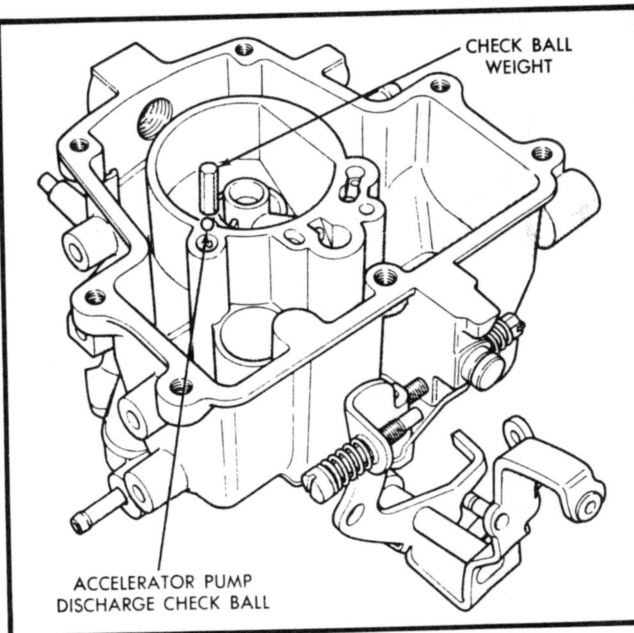

Accelerator pump discharge check ball and weight installation

MAIN BODY

1. Install power valve assembly in bottom of fuel bowl and tighten securely. Be sure needle valve operates freely.

2. Install main metering jet with special tool, C-3748, or equivalent.

3. Install float shaft and position the assembly in the float shaft cradle. Insert retaining spring. Check float alignment to prevent binding against bowl casting.

NOTE: A nitrophyl float can be checked for fuel absorption by lightly squeezing between fingers. If wetness appears on surface or float feels heavy (check with known good float), replace the float assembly.

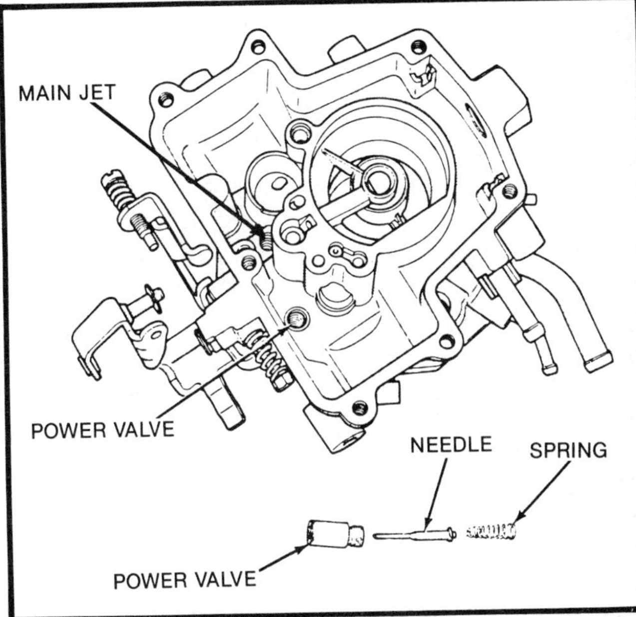

Location of main jet and power valve in float bowl

4. Place a new gasket on fuel inlet fitting and install assembly into main body, tighten securely. When replacing the needle and seat assembly (inlet fitting), the float level must also be checked, as dimensional difference between the old and new assemblies may change the float level. If necessary, adjust as follows:

5. Place bowl cover gasket on top of the fuel bowl. Hold gasket in place and invert the bowl (refer to carburetor adjustments). Place a straight edge across the gasket surface. The portion of the float lungs farthest from the fuel inlet should just touch the

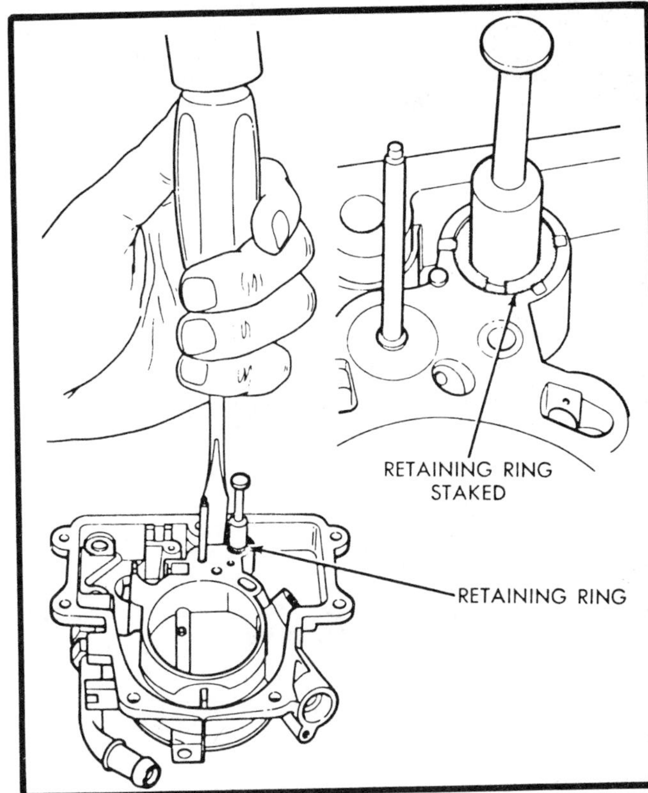

Correct method of staking the vacuum piston retaining washer

straight edge. If adjustment is necessary, bend the float tang to obtain this adjustment.

6. Insert check ball and weight into accelerator pump discharge well.
7. Install bowl cover gasket on air horn.
8. Carefully position bowl cover on the fuel bowl. Be sure the leading edges of the accelerator pump cup are not damaged as it enters the pump bore. Be careful not to damage the main well tube.
9. Install seven bowl cover screws and tighten alternately, a little at a time, to compress the gasket evenly (30 in lb).
10. Position a new O^2 feedback solenoid gasket on the air horn. Install new "O" Ring seal on duty cycle solenoid. Lubricate lightly with petroleum jelly and carefully install solenoid into carburetor. Install and tighten mounting screws securely. Route wire and tighten screw.
11. Install fast idle cam and link.
12. Install choke vacuum diaphragm.
13. Install Solenoid Idle Stop (SIS).
14. Install bowl vent assembly.

CHOKE UNLOADER

The choke unloader is a mechanical device designed to partially open the choke valve at wide open throttle. It is used to eliminate choke enrichment during cranking of an engine. Engines which have flooded or stalled by excessive choke enrichment can be cleared by use of the unloader. Refer to carburetor adjustments for adjusting procedure.

CHOKE VACUUM DIAPHRAGM

Inspect the diaphragm vacuum fitting to insure that the passage is not plugged with foreign material. Leak check the diaphragm to determine if it has internal leaks. To do this, first depress the diaphragm stem, then place a finger over the vacuum, fitting to seal the opening. Release the diaphragm stem. If the stem moves more than 1/16 inch in 10 seconds, the leakage is excessive and the assembly must be replaced.

CHOKE VACUUM KICK

The choke diaphragm adjustment controls the fuel delivery while the engine is running. It positions the choke valve within the air horn by action of the linkage between the choke shaft and the diaphragm. The diaphragm must be energized to measure the vacuum kick adjustment. Vacuum can be supplied by an auxiliary vacuum source.

FAST IDLE SPEED

Fast idle engine speed is used to overcome cold engine friction, stalls after cold starts and stalls due to carburetor icing. Adjustment must be made to specifications. If adjustment is necessary with less than 300 miles/500 km on the odometer, the specifications must be reduced 75 RPM. Refer to carburetor adjustments for procedure.

Float Setting Adjustment

1. Install float shaft and position the assembly in the float shaft cradle.
2. Install retaining spring and place bowl cover gasket on top of the fuel bowl.
3. Hold gasket in place and invert the bowl.
4. Place a straight edge across the gasket surface. The portion of the float lungs, farthest from the fuel inlet, should just touch the straight edge. If adjustment is necessary, bend the float tang.

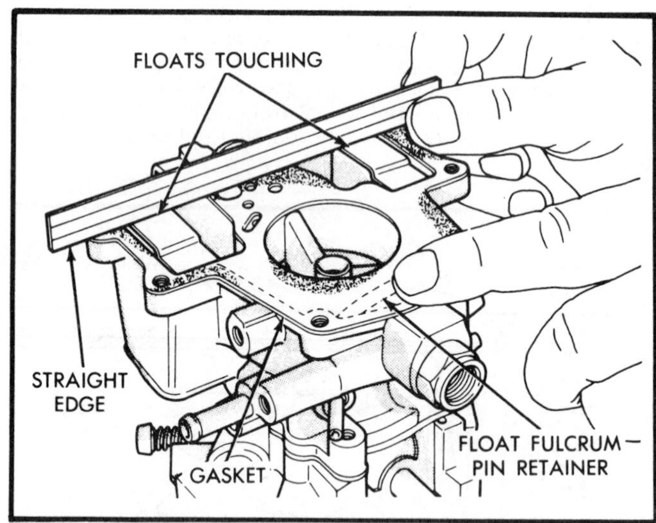

Adjusting float level on Holley 6145 feedback carburetor

MODEL 6145
Chrysler Corporation

Year	Carb. Part No. ①	Float Level (in.)	Accelerator Pump Adjustment (in.)	Bowl Vent Clearance (in.)	Fast Idle (rpm)	Choke Unloader Clearance (in.)	Vacuum Kick (in.)	Fast Idle Cam Position (in.)	Choke
'81	R-9129A	②	1.615③	④	2000	.250	.150	.090	Fixed
'82	R-9936A	②	1.616③	④	1950	.250	.150	.090	Fixed
	R-9695A	②	1.615③	④	1950⑤	.250	.150	.090	Fixed
'83	R-40042A	②	1.615③	④	2000	.250	.150	.090	Fixed

① Located on a tag attached to the carburetor
② Flush with the top of the main body casting to .050″ above
③ Position #2
④ Not Adjustable
⑤ Cordoba and Mirada—2000 rpm

Choke Vacuum Kick Adjustment

1. Open throttle, close choke then close throttle to trap fast idle cam at closed choke position.
2. Disconnect vacuum hose from carburetor and connect to hose of auxiliary vacuum source with small length of tube. Apply a vacuum of 15 or more inches of mercury.
3. Apply sufficient closing force on choke lever to completely compress spring in diaphragm stem without distorting linkage. Cylindrical stem of diaphragm extends to a stop as spring compresses.
4. Measure by inserting specified gauge between top of choke valve and air horn wall at throttle lever side.
5. Insert a 5/64 inch allen wrench into the choke diaphragm and turn to adjust the choke vacuum kick.

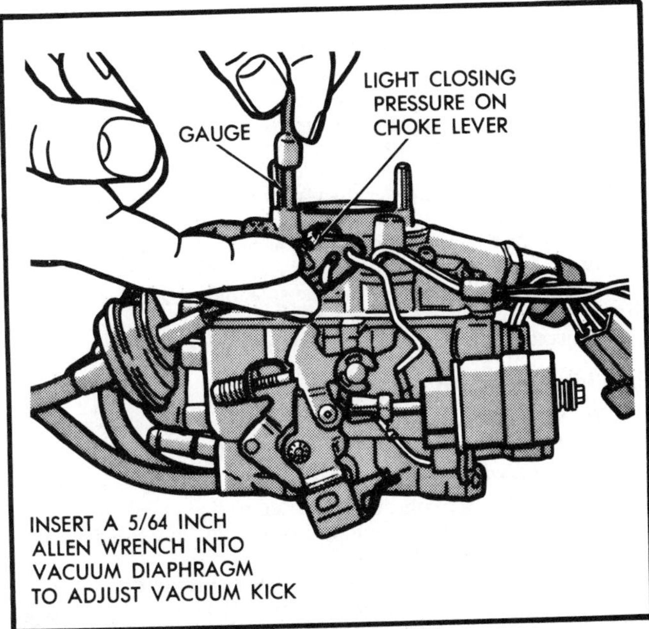

Choke vacuum kick adjustment—Holley 6145

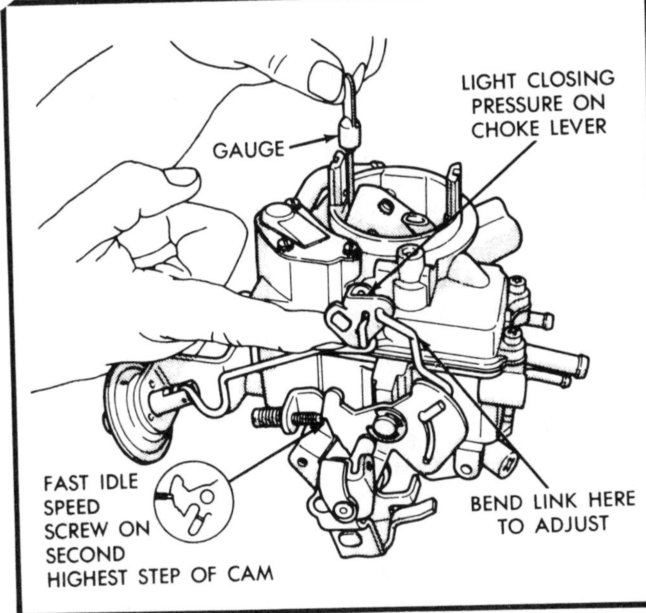

Fast idle cam position adjustment—Holley 6145

6. Check for free movement between open and adjusted positions. Correct any misalignment or interference by rebending link and readjust.
7. Replace vacuum hose on correct carburetor fitting.

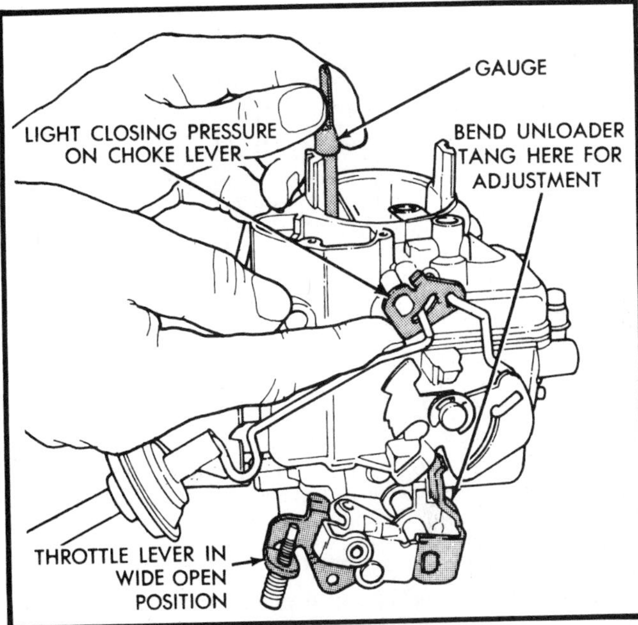

Choke unloader adjustment—Holley 6145

Fast Idle Cam Position Adjustment

1. With fast idle speed adjusting screw contacting second highest speed step on fast idle cam, move choke valve towards closed position with light pressure on choke shaft lever.
2. Measure by inserting specified gauge between top of choke valve and air horn wall at throttle lever side.
3. Adjust by bending fast idle connector rod at angle until correct valve opening has been obtained.

Choke Unloader Adjustment

1. Hold throttle valves in wide open position.
2. Lightly press finger against control lever to move choke valve toward closed position.
3. Measure by inserting specified gauge between top of choke valve and air horn wall at throttle lever side.
4. Adjust by bending tang on throttle lever.

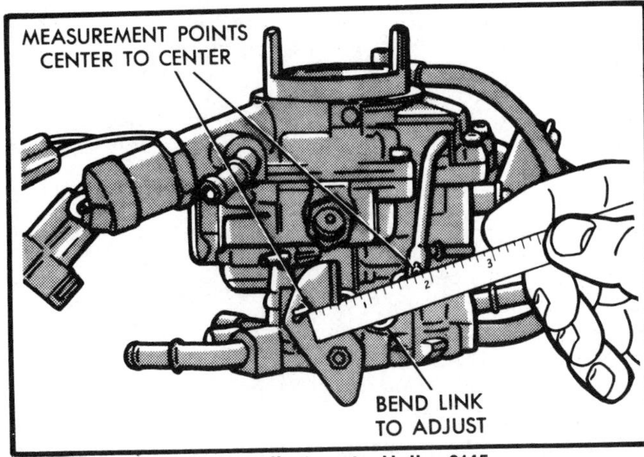

Accelerator pump stroke adjustment—Holley 6145

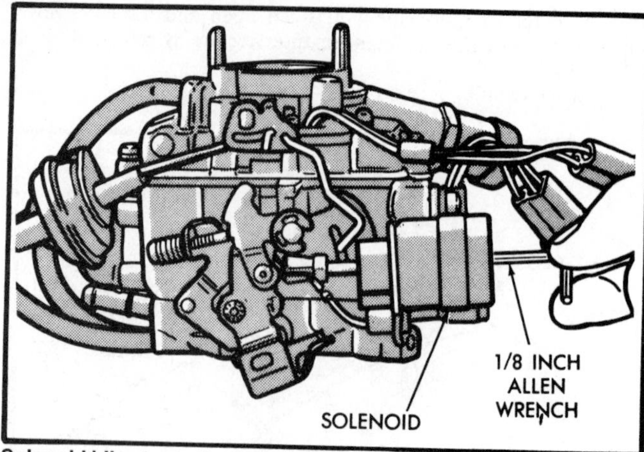

Solenoid idle stop adjustment—Holley 6145

Accelerator Pump Piston Stroke Adjustment

Position the throttle in the curb idle position with the accelerator pump operating link in the proper slot in the throttle lever. Measure the pump operating link as shown in above illustration. This measurement must be to specifications.

Solenoid Idle Stop Adjustment

Before checking or adjusting any idle speed, check ignition timing and adjust if necessary. Disconnect and plug the vacuum hose at the EGR valve. Also, connect a jumper wire between the carburetor switch and a good ground. Disconnect and plug the ³/₁₆ inch diameter control hose at the canister. The air cleaner cannot be removed but may be propped up to provide access to the carburetor. Remove the PCV valve from the cylinder head cover and allow the valve to draw underhood air. Connect a tachometer to the engine.

1. Turn on air conditioning and set the blower on low.
2. Disconnect the air conditioning clutch wire.
3. On non-airconditioned models, connect a jumper wire between the battery positive post and the SIS lead wire.

CAUTION

Use care in jumping to the proper wire on the solenoid. Applying battery voltage to other than the correct wire will damage the wiring harness.

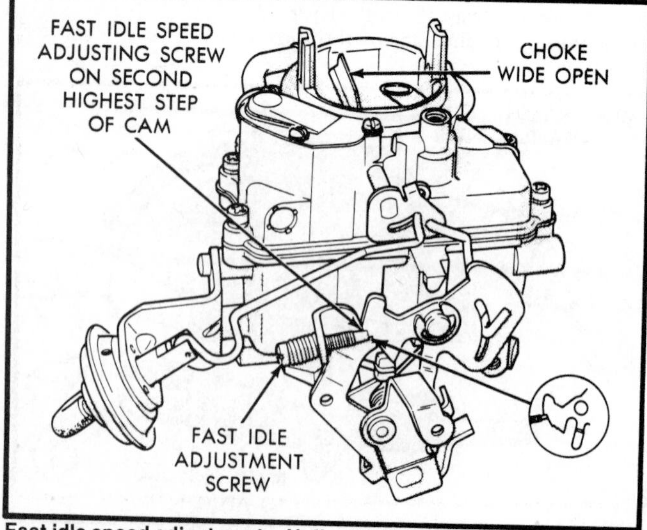

Fast idle speed adjustment—Holley 6145

4. Open throttle slightly to allow the solenoid plunger to extend.
5. Remove the adjusting screw and spring from the solenoid.
6. Insert a ¹/₈ inch allen wrench into the solenoid and adjust to the correct engine speed.
7. Turn off air conditioning and replace clutch wire or remove jumper wire.
8. Replace solenoid screw and spring.
9. Proceed to Idle Set RPM Adjustment.

Idle Set RPM Adjustment

NOTE: This adjustment is to be performed only after the Solenoid Idle Stop Adjustment Procedure.

1. Remove the exhaust manifold heat shield for access to the O^2 sensor.
2. Disconnect the engine harness lead from the O^2 sensor and ground the engine harness lead.

CAUTION

Care should be exercised so that no pulling force is put on the wire attached to the O^2 sensor. Use care in working around the sensor as the exhaust manifold is extremely hot.

Remove and plug the vacuum line at the vacuum transducer on the Spark Control Computer (SCC). Connect an auxiliary vacuum supply to the vacuum transducer and set at 16 inches of vacuum.

3. Allow the engine to run for two minutes to allow the effect of disconnecting the O^2 sensor to take place.
4. Adjust the idle set screw on the solenoid to obtain the correct idle set rpm.
5. Proceed to the Fast Idle Speed Adjustment.

Fast Idle Speed Adjustment

NOTE: This adjustment is to be performed only after the Solenoid Idle Stop and Idle Set RPM Adjustment Procedures.

1. Open throttle slightly and place the fast idle speed screw on the second highest step of the fast idle cam.
2. With the choke fully open, adjust the fast idle speed screw to obtain the correct fast idle rpm.
3. Return to idle then reposition the adjusting screw on the second highest step of the fast idle cam to verify fast idle speed. Readjust if necessary.
4. Return to idle and turn off the engine. Unplug and reconnect the vacuum hoses at the EGR valve and canister. Reconnect the vacuum line at the SCC. Install exhaust manifold heat shield. Remove the tachometer, reinstall the PCV valve, reconnect the O^2 sensor, and remove the ground wire.

NOTE: The idle speed with the engine in normal operating condition (everything connected) may vary from set speeds. DO NOT READJUST.

Concealment Plug Removal

1. Remove carburetor from engine.
2. Remove throttle body from the carburetor.
3. Clamp the throttle body in a vise with the gasket surfaces protected from the vise jaws.
4. Drill a ⁵/₆₄ inch hole in the casting surrounding the idle mixture screw, then redrill the hole to ¹/₈ inch.
5. Insert a blunt punch into the hole and drive out the plug.
6. Reassemble the carburetor and reinstall it on the engine.
7. Proceed to Propane Assisted Idle Set Procedure.

NOTE: Tampering with the carburetor is a violation of Federal law. Adjustment of the carburetor idle air fuel mixture can only be done under certain circumstances, as explained below. Upon completion of the carburetor adjustment, it is important to restore the plug.

This procedure should only be used if an idle defect still exists after normal diagnosis has revealed no other faulty condition such as incorrect basic timing, incorrect idle speed, faulty wire or hose connections, etc. Also, it is important to make sure that the combustion computer systems are operating properly. Adjustment of the carburetor idle air fuel mixture should be performed after a major carburetor overhaul.

1. Remove concealment plug. Set the parking brake and place transmission in neutral. Turn all lights and accessories off. Connect a tachometer to the engine. Remove exhaust manifold heat shield for access to the O^2 sensor. Start engine and allow it to warm up on the second highest step of the fast idle cam until normal operating temperature is reached. Return engine to idle.

2. Disconnect and plug the vacuum hose at the EGR valve. Connect a jumper wire between the carburetor switch and a good ground. Do not remove the air cleaner. The air cleaner may be propped up to provide access to the carburetor.

3. Disconnect vacuum supply hose from the choke diaphragm at the tee and install propane supply hose in its place. Other connections at the tee must remain in place.

4. With the propane bottle upright and in a safe location, remove the PCV valve from the valve cover and allow the valve to draw underhood air. Disconnect and plug the 3/16 inch diameter control hose from the canister.

5. Disconnect the engine harness lead from the O^2 sensor and ground the engine harness lead.

CAUTION

Care should be exercised so no pulling force is put on the wire attached to the O^2 sensor. Use care in working around the sensor as the exhaust manifold is extremely hot.

6. Remove and plug the vacuum line at the vacuum transducer on SCC. Connect an auxiliary vacuum supply to vacuum transducer and set at 16 inches of vacuum. Allow engine to run for two minutes to allow the effect of disconnecting the O^2 sensor to take place.

7. Open propane main valve. Slowly open propane metering valve until maximum engine rpm is reached. When too much propane is added, engine rpm will decrease. "Fine tune" the metering valve to obtain the highest engine rpm.

8. With the propane still flowing, adjust the idle speed screw on solenoid to obtain the specified propane rpm. Again, "fine tune" the metering valve to obtain the highest engine rpm. If there has been a change in the maximum rpm, readjust the idle speed screw to the specified propane rpm.

9. Turn off propane main valve and allow engine to run for one minute to stabilize. Slowly adjust the mixture screw, pausing between adjustments to allow the engine to stabilize, to achieve the smoothest idle at the specified idle set rpm.

10. Turn on propane main valve. "Fine tune" metering valve to obtain the highest engine rpm. If the maximum engine speed is more than 25 rpm different than the specified propane rpm, repeat Steps 7–9.

11. Turn off propane main valve and metering valve.

12. Remove the propane supply hose and reinstall the vacuum supply hose from the choke. Remove O^2 sensor ground wire and reconnect O^2 sensor. Reconnect vacuum line on SCC.

13. Install concealment plug and perform Solenoid Idle Stop, Idle Set RPM, and Fast Idle Speed Adjustments.

Float Setting (On Vehicle)

1. Remove air cleaner assembly and air cleaner gasket.
2. Remove bowl vent hose.
3. Disconnect choke assembly.
4. Remove fast idle cam retaining clip, cam and link.
5. Remove vacuum choke diaphragm (vacuum kick), link, and hose.
6. Remove nut and washer from throttle shaft. Remove throttle lever and link. Note the hole position of lever.
7. Remove seven bowl cover screws and lift bowl cover straight up until vacuum piston stem, accelerator pump, and main well tube are clear of the main body.
8. Depress float manually to allow residual line pressure to overfill the bowl to within 1/8–1/4 inch below the top of the fuel bowl. If line pressure is not enough to fill the bowl, use an external supply.
9. Using two wrenches, back off flare nut and tighten inlet fitting to recommended torque (170 inch-pounds) (19 N m).
10. Place bowl cover gasket on top of fuel bowl.
11. Firmly seat the float pin retainer by hand while measuring float height with a straight edge across the gasket surface. The portion of the float lungs farthest from the fuel inlet should just touch the straight edge. If adjustment is necessary, bend the float tang.
12. Remove bowl cover gasket. Drain fuel from accelerator

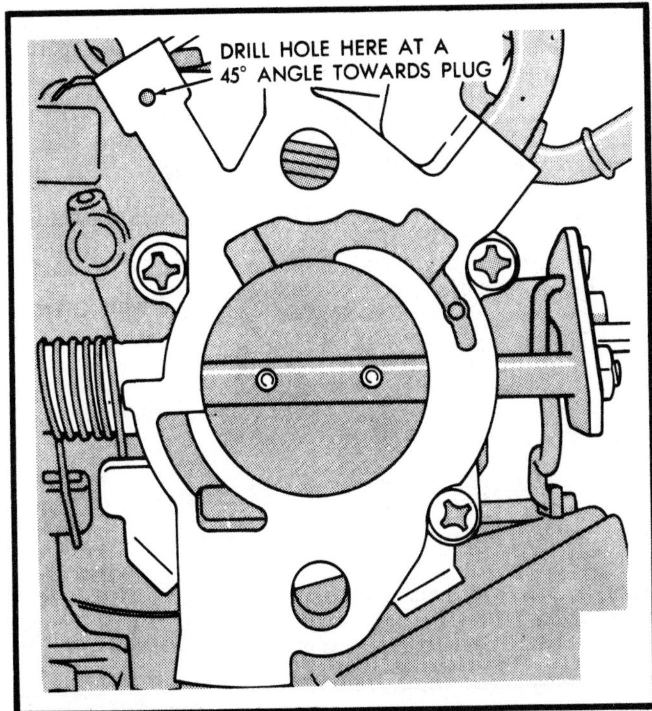

Removing idle mixture plugs from Holley 6145

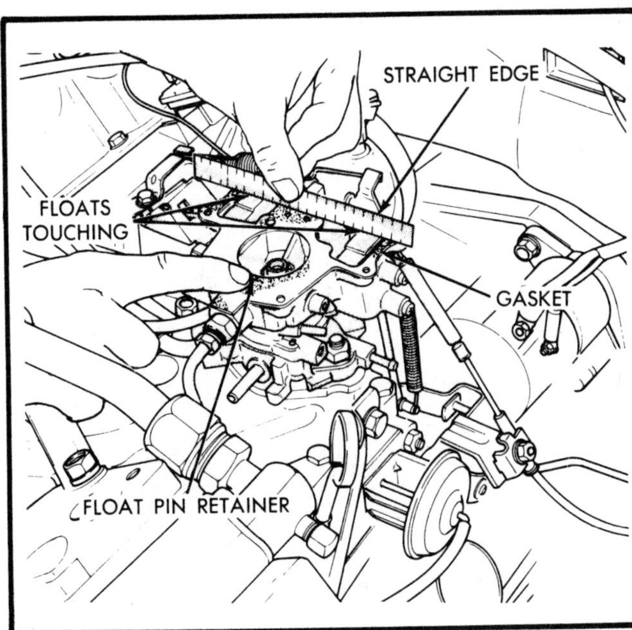

Adjusting float level on the car. Make sure the float pin retainer is properly seated

pump well with syringe type tool. This is done to prevent the discharge check ball and weight from leaving its position during assembly of the bowl cover.

13. Position bowl cover gasket on bowl cover. Carefully install bowl cover on main body. Be sure the leading edge of the accelerator pump cup is not damaged as it enters the pump bore and that the weight and discharge check ball remains in place. Be careful not to damage the main well tube.

14. Install seven bowl cover screws and tighten alternately to compress the gasket evenly.

15. Install pump rocker arm and linkage assembly.

16. Install vacuum kick diaphragm, link and hose.

17. Install fast idle cam retaining clip, cam and link.

18. Install bowl vent hose.

19. Connect choke assembly.

20. Install air cleaner and gasket.

21. Check idle speed with tachometer. If adjustment is necessary, use the "Propane Procedure."

Holley 6520 Carburetor

GENERAL INFORMATION

Holley model 6520 is a staged dual venturi carburetors. The primary bore or venturi is smaller than the secondary bore. The secondary stage is operated by a vacuum controlled diaphragm. The primary stage includes a curb idle and transfer system, diaphragm type accelerator pump system, main metering system, and power enrichment system. On model 6520 carburetors there is also an O_2 Feedback Solenoid that is responsive to the oxygen sensor. The secondary stage includes a main metering system and power system. Both the primary and secondary venturi draw fuel from a common fuel bowl. The electric automatic choke has a bimetal two stage heating element. In normal service the mixture should not require adjustment. The idle RPM can be checked without removing the tamper resistant plug.

Fuel Inlet System

Fuel under pressure from the fuel pump enters the fuel bowl through the fuel inlet line. The fuel inlet needle is controlled by a dual lung monocellular nitrophyl float (a closed cell buoyant material which cannot collapse or leak). A fuel inlet needle clip hooks over the front lever to insure the needle is pulled off the seat when the float drops. When the fuel level in the bowl drops the float drops permitting additional fuel to flow past the fuel inlet needle into the bowl. These carburetors are vented to the vapor canister.

Idle System

Fuel used during curb idle and low speed operation flows from the fuel through the primary main jet into the main well. From the main well it flows into the idle well and through the idle tube. Air enters through the primary idle air bleed. This air-fuel mixture

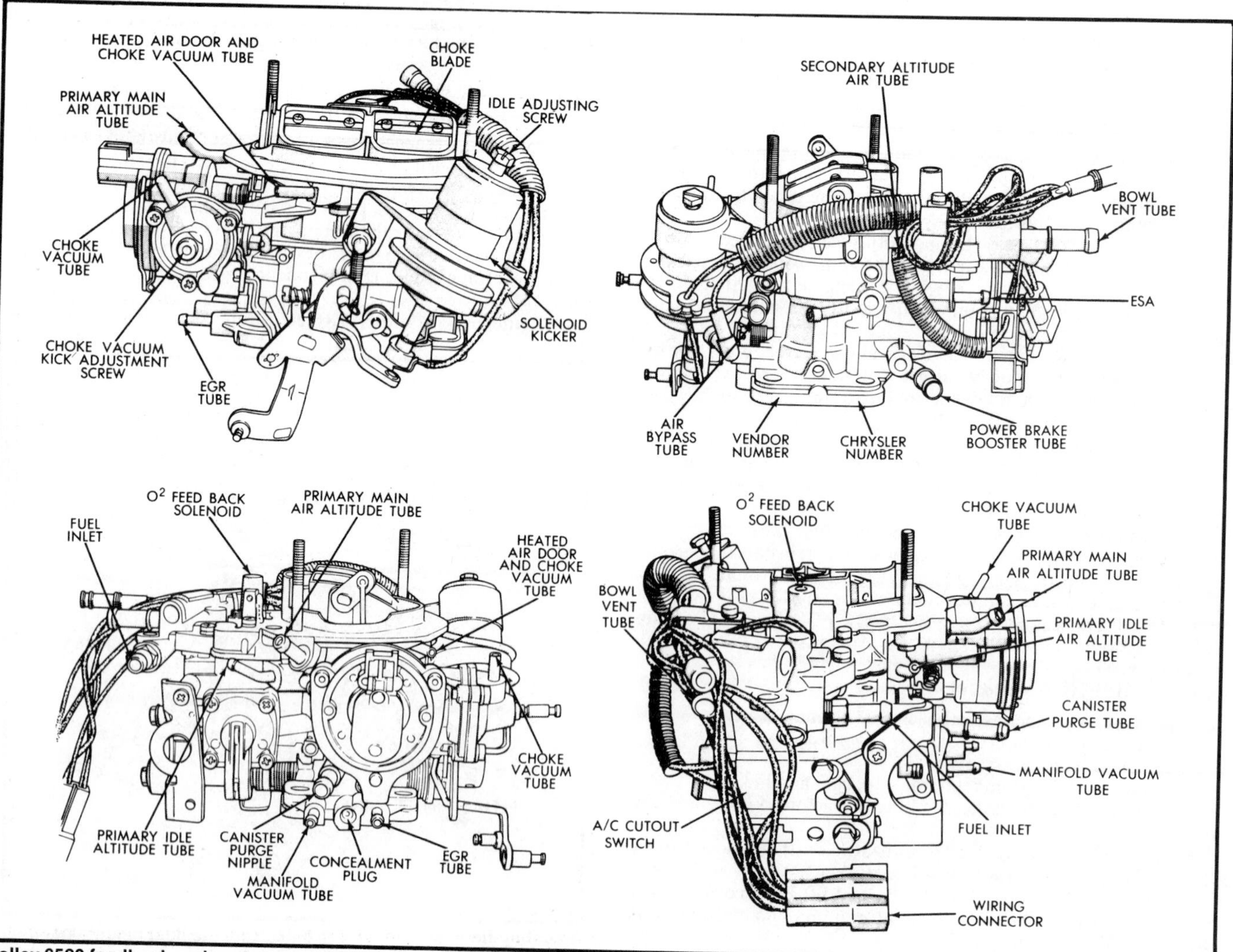

Holley 6520 feedback carburetor assembly

moves down the idle passages and past the idle transfer slot. This slot serves as an additional air bleed during curb idle operation. The air-fuel mixture then moves past the idle mixture adjusting screw tip which controls the amount of the mixture at curb idle. At speeds above idle the throttle valve exposes the transfer slot to vacuum and the air-fuel mixture is drawn out of the transfer slot as well as the idle passage.

Main Metering System 6520 Models

As the throttle valves continue opening, the airflow through the carburetor increases and creates a low pressure area in the venturi. This low pressure causes fuel to flow from the fuel bowl through the main jets and into the main wells. In addition to the main jet, fuel also enters the main well through another parallel circuit. Fuel flow in this circuit is regulated by the O_2 feedback solenoid. Air fom the main air bleed mixes with the fuel through holes in the sides of the main well tube. The mixture is then drawn from the main well tube and discharged through the venturi nozzle. By controlling the amount of fuel released, the solenoid regulates the total air-fuel mixture. The solenoid acts in direct response to the engine demand. As airflow through the carburetor increases, the amount of air-fuel mixture discharged also increases and the idle system tapers off.

ACCELERATOR PUMP SYSTEM

A diaphragm type accelerator pump system is located in the side of the carburetor body. When the throttle valves are opened quickly, air flow through the carburetor responds rapidly. Since fuel is heavier than air, there is a very brief time lag before fuel flow can sustain the proper air-fuel ratio. During this lag the accelerating pump system mechanically supplies the required additional fuel, until the proper air-fuel ratio can be maintained by the other metering systems.

When the throttle valves are closed, the diaphragm returns against its cover. Fuel is drawn through the inlet, past the inlet ball check valve and into the pump chamber. A discharge check ball prevents air from being drawn into the pump chamber. The moment the throttle valves are opened, the diaphragm rod is pushed inward, the diaphragm forces fuel from the pump chamber into the discharge passage. The intake check ball prevents fuel from returning to the fuel bowl. Fuel under pressure unseats the discharge check ball and is forced through the pump discharge valve where it sprays into the primary venturi through the pump discharge nozzle. Excess fuel and pump chamber vapors are discharged back into the fuel bowl through a restriction.

Manifold vacuum is applied to the power valve diaphragm from a passage in the base of the carburetor body. The passage connects to the air horn and the top of the power valve diaphragm. During idle and normal driving conditions, manifold vacuum is high enough to overcome the power valve spring tension, and the valve remains closed. When manifold vacuum drops below a predetermined level the diaphragm spring overcomes the vacuum and the diaphragm stem depresses the power valve stem. Fuel then flows through the power valve, power valve restriction, and into the main well. As engine load requirements decrease, manifold vacuum increases and overcomes the diaphragm spring. The spring in the power valve assembly then closes the valve.

SECONDARY PROGRESSION SYSTEM

Secondary metering circuit becomes operational when primary throttle opening exceeds 45°

SECONDARY POWER ENRICHMENT SYSTEM

The secondary system is also provided with an air velocity operated power system for full power operation. As the secondary throttle valve approaches wide open position, air velocity through the secondary venturi creates a low pressure at the discharge port. Fuel flows from the bowl through a restricted vertical channel. As this occurs, air enters through a calibrated air bleed and mixes with the fuel. This mixture is discharged through the discharge passage.

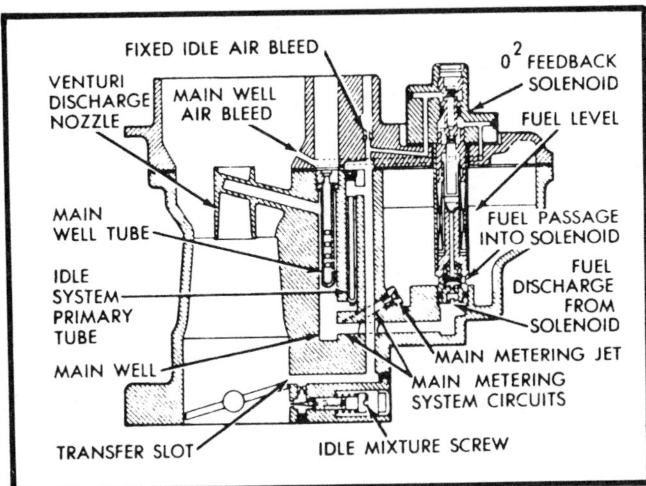

Main metering system on Holley 6520 feedback carburetor

Altitude Compensation System

Some models are equipped with an altitude compensation system. Less dense air at higher altitudes is compensated for by the use of variable air bleeds which act in parallel with the normal fixed air bleeds in the carburetor idle, primary, and secondary circuits. These air bleeds consist of tapered needles which are moved in orifices by an aneroid bellows. The bellows is contained in an altitude compensation module which is mounted in the engine compartment. Outside air passes through an air filter, through the compensation module, and into the carburetor. The system is calibrated so that there is no air flow until atmospheric pressure is below a preset level. Air flow gradually increases as atmospheric pressure decreases.

Solenoid Kicker System

The solenoid kicker is the combination of a position-holding solenoid, and a vacuum diaphragm assembly. When the engine is running, the solenoid is extended to hold the throttle at curb idle.

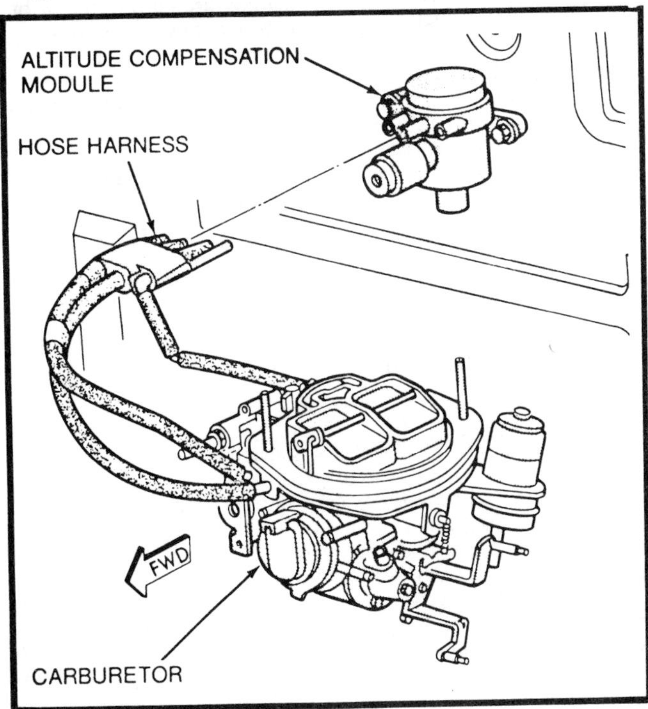

Typical altitude compensation system

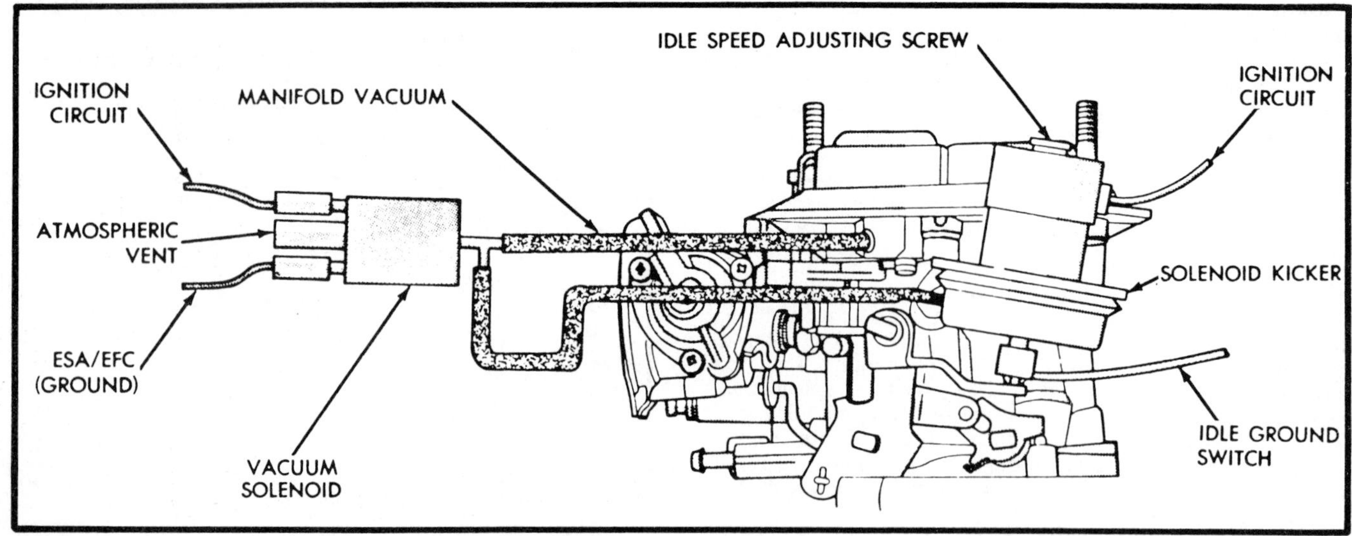

Solenoid kicker system showing vacuum and electrical connections

When the engine is off, the solenoid retracts, allowing the throttle to close to prevent after-running (dieseling). Curb idle is adjusted with a screw on top of the solenoid. The diaphragm assembly extends when manifold vacuum is applied to open the throttle a fixed amount above the idle position. For any assembly, the stroke is fixed, but different calibrations may have different strokes. Manifold vacuum is supplied to the kicker by a three-port solenoid valve which receives its control signal from the ESA/EFC module. Control inputs which may trigger the kicker include: air conditioning "ON", electrically heated rear window defroster (EBL) "ON", and high engine coolant temperature at start up, which actuation ceases after a specified time period.

Electronic Feedback Control (EFC) System

The EFC system is essentially an emissions control system which utilizes an electronic signal, generated by an exhaust gas oxygen sensor to precisely control the air-fuel mixture ratio in the carburetor. This in turn allows the engine to produce exhaust gases of the proper composition to permit the use of a three-way catalyst. The three-way catalyst is designed to convert the three pollutants (1) hydrocarbons (HC), (2) carbon monoxide (CO), and (3) oxides of Nitrogen (NOx) into harmless substances.

There are two operating modes in the EFC system:

Open Loop—air fuel ratio is controlled by information programmed into the computer at manufacture.

Closed Loop—air fuel ratio is varied by the computer based on information supplied by the oxygen sensor.

When the engine is cold, the system will be operating in the open loop mode. During that time, the air fuel ratio will be fixed at a richer level. This will allow proper engine warm up. Also, during this period, air injection (from the air injection pump) will be injected upstream in the exhaust manifold.

OXYGEN (O_2) SOLENOID

The function of the O_2 Feedback Solenoid is to provide limited regulation of the fuel-air ratio of a feedback carburetor in response to electrical signals from the Spark Control Computer. It performs this task by metering main system fuel, and operates in parallel with a conventional fixed main metering jet. In the power-off mode the solenoid valve spring pushes upward through the main system fuel valve. At this point, the solenoid controlled main metering orifice is fully uncovered, so the richest condition exists within the carburetor for any given airflow. In the power-on mode the field windings are energized and a magnetic flux path is established from the armature, through the armature bushing, the solenoid case and the pole piece, returning once again to the armature. This attraction between the pole piece and the armature assembly moves the valve operating push rod and main system valve downward against the valve spring. This movement continues until the main system valve bottoms on the main sytem valve seat and stops before the armature contacts the pole piece. In this position, the solenoid controlled main metering orifice is fully sealed. These conditions remain unchanged until the electrical signal from the Spark Control Computer to the solenoid is switched off. This solenoid position offers the leanest condition within the carburetor for any given airflow.

Main system fuel may be regulated between the richest and leanest limits by controlling the amount of time that the solenoid is in the power-on position. Under normal operating conditions a voltage signal (12 volts nominally) is applied to the field windings at a frequency of 10Hz. By controlling the duration of this voltage signal to the ratio of power-on time to total time, referred to as duty cycle is established.

The O_2 feedback solenoid includes attaching surfaces and passageways as are required for the directing and containing of main system fuel when the solenoid assembly is correctly installed in a feedback carburetor designed for its use.

OXYGEN SENSOR

The oxygen sensor is a device which produces electrical voltage. The sensor is mounted in the exhaust manifold and must be heated by the exhaust gas before producing a voltage. When there is a large amount of oxygen present (lean mixture) the sensor produces a low voltage. When there is a lesser amount present, it produces a higher voltage. By monitering the oxygen content and converting it to electrical voltage, the sensor acts as a rich-lean switch. The voltage is transmitted to the Spark Control Computer. The computer sends a signal to the Oxygen Feedback Solenoid mounted on the carburetor to change the air-fuel ratio back to stoichiometric.

Vacuum Operated Secondary System

The VOS system uses a vacuum diaphragm to operate the secondary throttle blade. At lower speeds and engine loads the throttle blade remains closed. When engine speed increases to a point where more breathing capacity is needed, the vacuum controlled secondary throttle blade begins to open. Venturi vacuum from the primary and secondary venturi act on the vacuum diaphragm. At high engine speeds when engine requirements approach the capacity of the primary bore, the increased primary venturi moves the diaphragm which compresses the diaphragm spring. This action moves the diaphragm link and lever which causes the secondary throttle blade to open. The position of the secondary throttle blade depends on the strength of the vacuum signal. This is determined by the air flow through the primary and secondary bores. As the air flow increases, a greater secondary throttle blade opening will result and the secondary bore will supply a greater portion of the engine's requirements. As higher speeds are reached, the secondary throttle blade will be wide open. The restriction in the vacuum passage to the secondary diaphragm limits the rate at which the secondary blade will open. This allows the vacuum to build up slowly at the diaphragm, resulting in an controlled secondary throttle blade opening rate. The secondary throttle blade begins to open where vacuum is created at the primary and secondary vacuum pick up holes. This vacuum assists in the operation of the secondary diaphragm.

When engine speed is reduced, venturi vacuum in the throttle bores becomes weaker, reducing the vacuum level in the secondary diaphragm chamber. As vacuum drops, the load from the diaphragm spring will start closing the secondary blade. The diaphragm spring is assisted by the design of the secondary blade. The blade is slightly offset. When the blades are closing, the combined force of manifold vacuum and the air stream has greater effect on the larger, upstream areas of the blade, forcing the blades to a closed position. The vacuum control solenoid bleeds off vacuum from the diaphragm during warm up and low air flow conditions. Secondary throttle blade operation can only occur if certain conditions occur simultaneously. Coolant temperature must be above 140°F, 30 seconds must have elapsed since start up, engine speed must be above 1800 rpm and manifold vacuum must be below 8 inches Hg.

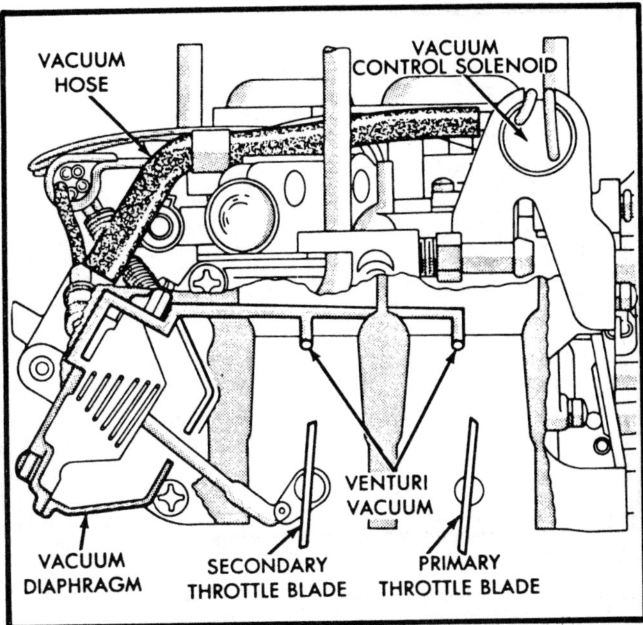

Vacuum secondary system on Holley 6520 feedback carburetor

Electric Choke Assembly

An electric heater and switch assembly is sealed within the choke housing. Electrical current is supplied through the oil pressure switch. A minimum of 2.7 kPa (4PSI) oil pressure is necessary to close the contacts in the oil pressure switch and feed current to the automatic choke system. Electrically must be present when the engine is running to open the choke and keep it open.

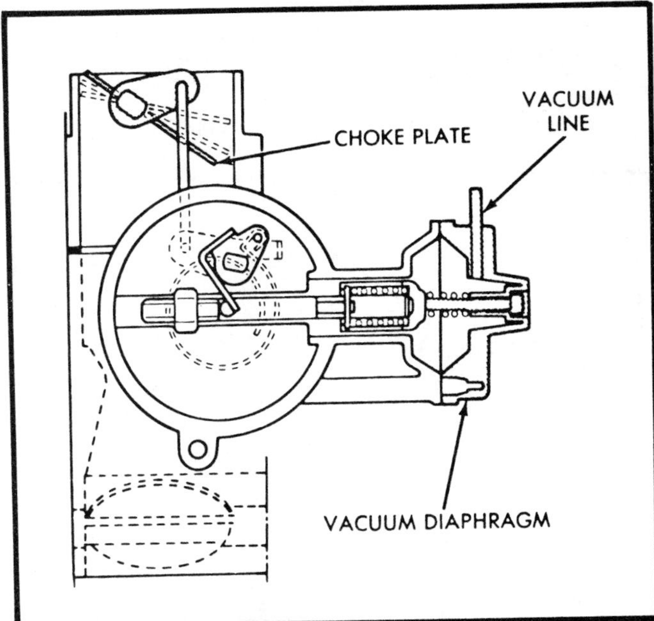

Typical electric choke system

NOTE: Use care when removing the choke thermostat not to lose the plastic bushing located between the thermostat loop and pin. The loop and bushing must be placed over the pin during assembly. Choke assembly removal is not necessary during normal service. The choke assembly must never be immersed in fluid as damage to the internal switch and heater assembly will result.

The heater can be tested with a direct B+ connection. The choke valve should reach the open position within five minutes.

CAUTION

Operation of any type, including idling should be avoided if there is any loss of choke power. Under this condition, any loss of power to the choke will cause the choke to remain fully on during the operation of the vehicle. This will cause a very rich mixture to burn and result in abnormally high exhaust system temperatures, which may cause damage to the catalyst or other underbody parts of the car. It is advised that the electric choke power not be disconnected to trouble shoot cold start problems.

CARBURETOR OVERHAUL

Disassembly

1. Disconnect and remove the choke operating rod and choke rod seal.

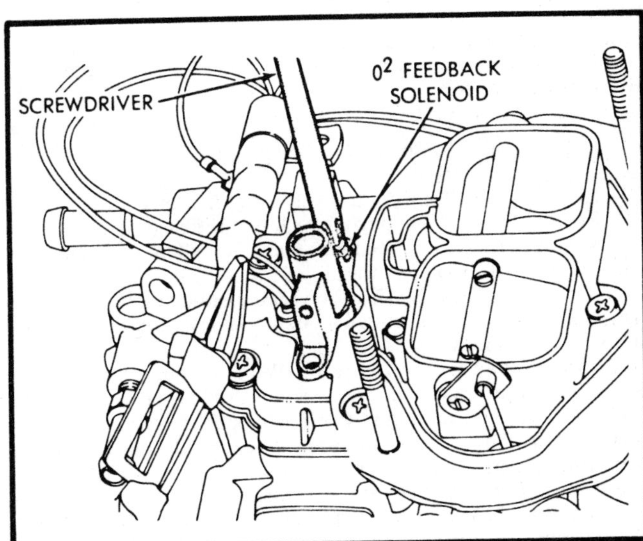

Removing feedback solenoid from the air horn

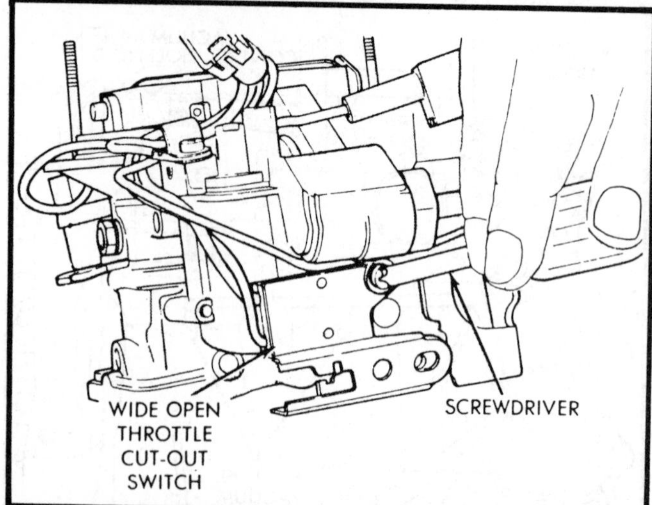

Mark the location of the mounting screws before removing the WOT cut-out switch

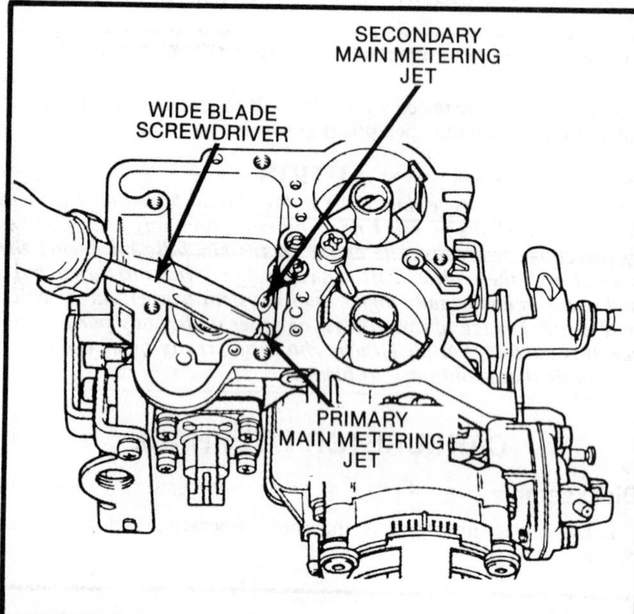

Location of primary and secondary main metering jets—note sizes before removal

2. Remove the two solenoid retaining screws and remove the solenoid from the carburetor.
3. Remove the vacuum control valve and filter assembly.
4. Remove the wiring clip from the top of the carburetor.
5. Remove the two 02 feedback solenoid retaining screws and gently lift the solenoid from the air horn.
6. Remove the clip securing the vacuum control rod to the link and remove the vacuum control mounting screws. Remove the vacuum control diaphragm.
7. Remove the wide open throttle (WOT) cutout switch mounting screws after marking the location for proper reassembly. Remove the harness mounting screws and open the retaining clip.
8. Remove the air horn mounting screws and separate the air horn assembly from the carburetor body.
9. Remove the float level pin, float and float inlet needle.
10. Remove the fuel inlet seal and gasket.
11. Remove the main secondary metering jet after noting its size to make sure it is installed in its proper position on reassembly.

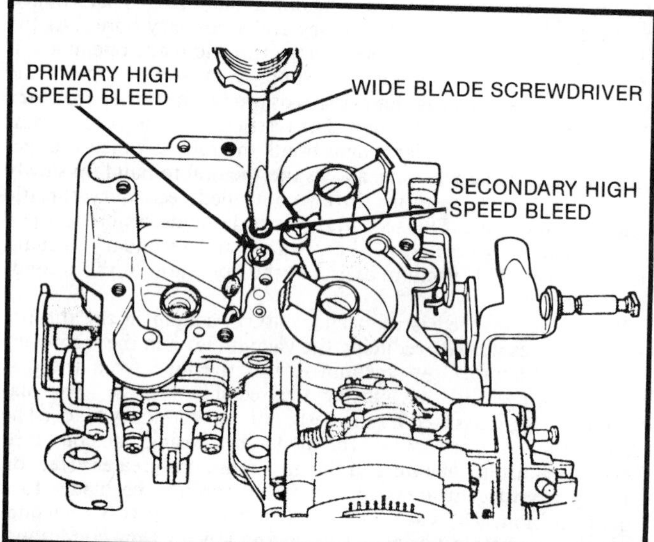

Location of primary and secondary high speed bleed—note sizes before removal

12. Remove the primary metering jet, again after noting its size to insure proper reassembly.
13. Remove the secondary high speed bleed and secondary main well tube after noting sizes for reassembly.
14. Remove the primary high speed bleed and primary main well tube after noting sizes for reassembly.
15. Remove the discharge nozzle screw, discharge nozzle and gasket.
16. Invert the carburetor body and drop out accelerator pump discharge weight ball and check ball.

NOTE: Both balls are the same size.

17. Remove the accelerator pump cover screws. Remove the cover, accelerator pump diaphragm and spring.
18. Remove the choke diaphragm cover screws and remove the cover and spring.
19. Rotate the choke shaft and lever assembly counterclockwise.

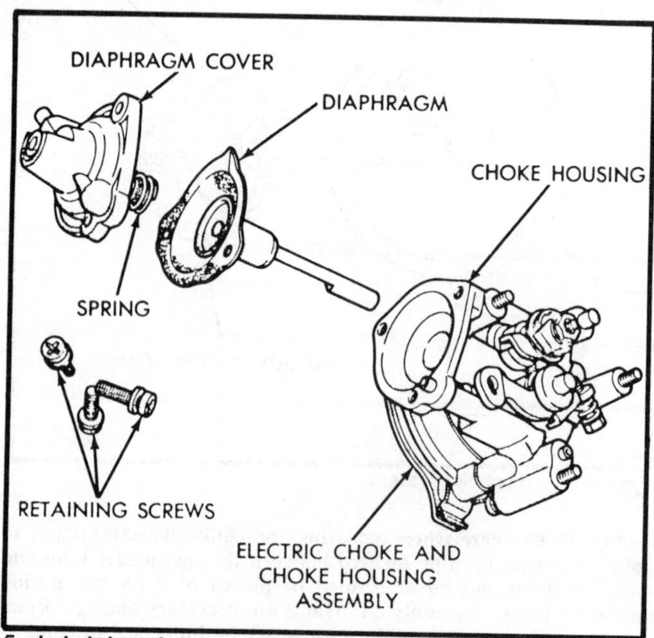

Exploded view of choke diaphragm assembly on Holley 6520 models

Rotate the choke diaphragm assembly clockwise and remove it from the housing. Remove the end of the lower screw from the housing.

NOTE: If the choke assembly is to be replaced, the diaphragm cover must also be replaced.

20. See the concealment plug removal procedure for removal of the idle mixture screw from the carburetor body.

Assembly

1. If the entire choke assembly was removed, position it on the carburetor and install the three retaining screws. If the choke diaphragm was serviced rotate the choke shaft counter-clockwise. Insert the diaphragm with a clockwise motion. Position spring and cover over diaphragm and install the retaining screws. **Make certain the fast idle link has been installed properly.**
2. Install accelerator pump spring, diaphragm cover and screws.
3. Install accelerator pump discharge check ball in discharge passage. Check accelerator and seat prior to assembly by filling the fuel bowl with clean fuel. Hold discharge check ball down with a small brass rod and operate pump plunger by hand. If the check ball and seat are leaking, no resistance will be experienced when operating plunger. If valve is leaking, stake ball using a suitable drift punch.

CAUTION

Exercise care when staking ball to avoid damaging bore containing the pump weight. After staking old ball, remove and replace with new ball from tune-up kit. Install weight ball and recheck for leaks.

4. Install new gaskets, discharge nozzle and screw.
5. Install primary main well tube and high speed bleed (primary).
6. Install secondary main well tube and secondary high speed bleed.
7. Install primary main metering jet (on this carburetor the primary main metering jet will have a smaller number stamped on it than the secondary main metering jet).
8. Install secondary main metering jet (on this carburetor the secondary main metering jet will have a larger number stamped on it than the primary main metering jet).
9. Install needle and set assembly.
10. Invert air horn and insert gauge or drill .480 in. (12.2 mm) between air horn and float.
11. Using a small screwdriver bend tang to adjust dry float level.
12. Position depth gauge and check float drop.
13. Using a small screwdriver bend tang to adjust float drop to 1 7/8 in. (48 mm).
14. Position new gasket on air horn; install choke rod seal and choke operating rod.
15. Carefully locate air horn to carburetor. Install new choke operating rod retainers on choke shaft lever and fast idle cam pickup lever and connect choke operating rod.
16. Install and tighten the 5 airhorn to carburetor body screws evenly in stages to 30 inch lbs. (3 N m).
17. Install and tighten idle stop solenoid screws. Reinstall anti-rattle spring.
18. Install wide open throttle cut-out switch, move switch so that air conditioning clutch circuit is open in throttle position 10° before to wide open.
19. Position a new O_2 feedback solenoid gasket on the air horn. Install new "O" ring seal on O_2 feedback solenoid. Lubricate lightly with petroleum jelly and carefully install solenoid into carburetor. Install and tighten mounting screws securely. Route wiring through clamp.
20. Install vacuum solenoid.
21. Install vacuum control valve.
22. Install and tighten harness mounting screw.

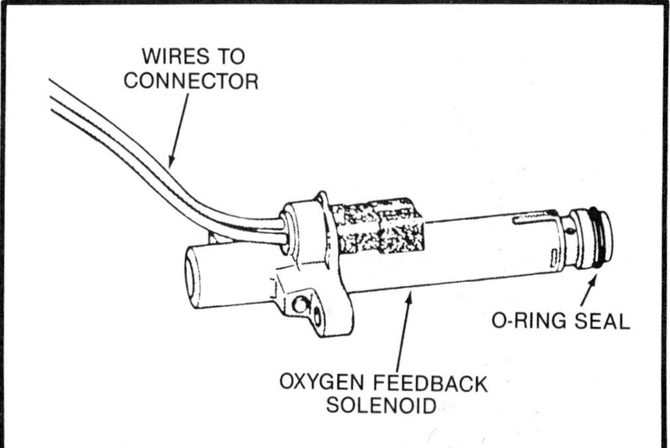

Typical oxgen feedback solenoid

MODEL 6520
Chrysler Corporation

Year	Carb. Part No. ①	Accelerator Pump	Dry Float Level (in.)	Float Drop (in.)	Vacuum Kick (in.)	Fast Idle RPM
'81	R-9052A	#2 hole	.480	1.875	.070	1400②
	R-9053A	#2 hole	.480	1.875	.070	1400②
	R-9054A	#2 hole	.480	1.875	.040	1400②
	R-9055A	#2 hole	.480	1.875	.040	1400②
	R-9060A	#2 hole	.480	1.875	.030	1100②
	R-9061A	#2 hole	.480	1.875	.030	1100②
	R-9602A	#2 hole	.480	1.875	.035	1500②
	R-9603A	#2 hole	.480	1.875	.035	1500②
	R-9125A	#2 hole	.480	1.875	.030	1200②
	R-9126A	#2 hole	.480	1.875	.030	1200②

MODEL 6520
Chrysler Corporation

Year	Carb. Part No. ①	Accelerator Pump	Dry Float Level (in.)	Float Drop (in.)	Vacuum Kick (in.)	Fast Idle RPM
'81	R-9604A	#2 hole	.480	1.875	.035	1600②
	R-9605A	#2 hole	.480	1.875	.035	1600②
'82	R-9822A	#2 hole	.480	1.875	.080	1400
	R-9823A	#2 hole	.480	1.875	.080	1400
	R-9824A	#2 hole	.480	1.875	.065	1400
	R-9503A	#3 hole	.480	1.875	.085	1300
	R-9504A	#3 hole	.480	1.875	.085	1300
	R-9505A	#3 hole	.480	1.875	.100	1600
	R-9506A	#3 hole	.480	1.875	.100	1600
	R-9750A	#3 hole	.480	1.875	.085	1300
	R-9751A	#3 hole	.480	1.875	.085	1300
	R-9509A	#3 hole	.480	1.875	.085	1600
	R-9510A	#3 hole	.480	1.875	.085	1600
	R-9752A	#3 hole	.480	1.875	.100	1600
	R-9753A	#3 hole	.480	1.875	.100	1600
	R-9507A	#3 hole	.480	1.875	.085	1300
	R-9508A	#3 hole	.480	1.875	.085	1300
'83	R-40003A	#3 hole	.480	1.875	.070	1400
	R-40004A	#3 hole	.480	1.875	.080	1500
	R-40005A	#3 hole	.480	1.875	.080	1350
	R-40006A	#3 hole	.480	1.875	.080	1275
	R-40007A	#3 hole	.480	1.875	.070	1400
	R-40008A	#3 hole	.480	1.875	.070	1600
	R-40010A	#3 hole	.480	1.875	.080	1500
	R-40012A	#3 hole	.480	1.875	.070	1600
	R-40014A	#3 hole	.480	1.875	.080	1275
	R-40080A	#2 hole	.480	1.875	.045	1400
	R-40081A	#3 hole	.480	1.875	.045	1400
'84	R-400641A	#3 hole	.480	1.875	.080	1500
	R-400651A	#3 hole	.480	1.875	.080	1600
	R-400811A	#2 hole	.480	1.875	.080	1500
	R-400821A	#2 hole	.480	1.875	.080	1600
	R-40071A	#3 hole	.480	1.875	.080	1500
	R-40122A	#2 hole	.480	1.875	.080	1500

① Located on tag attached to the carburetor
② With radiator fan running

CARBURETOR ADJUSTMENTS

Float Setting Adjustment

Invert the air horn and, with the gasket removed, insert a .480 inch (12.2mm) gauge between the air horn and the float. Bend the tang with a small screwdriver to obtain the proper level.

Float Drop Adjustment

Position the depth gauge and check the float drop. Bend the tang with a small screwdriver to adjust the float drop to the correct level.

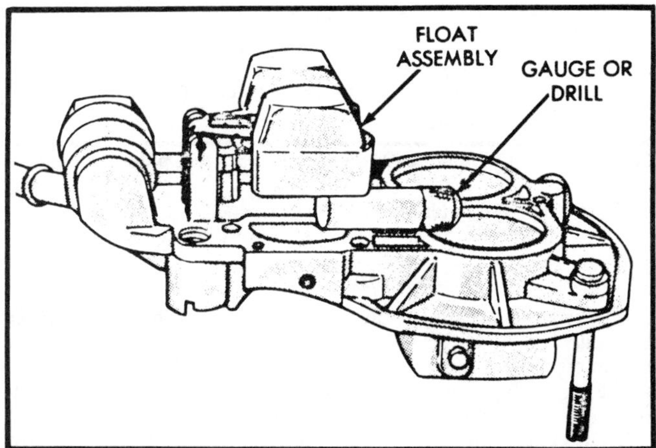

Dry float level adjustment—Holley 6520 feedback carburetor

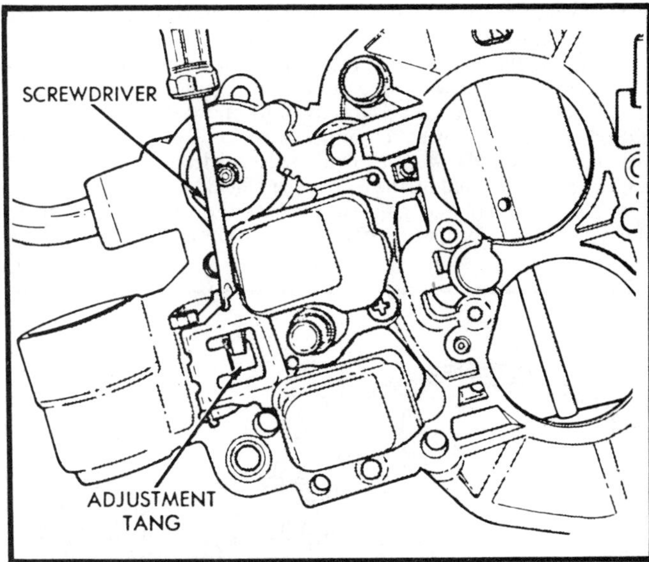

Bend the tang to adjust the float level

Fast Idle Adjustment

1. Unplug the connector at the radiator fan and install a jumper wire so that the fan will run continuously.
2. Remove the PCV valve from the moulded rubber connector, allow the PCV valve to draw underhood air.
3. Disconnect and plug the vacuum connector at the CVSCC.
4. Install a tachometer.
5. Ground the carburetor switch with a jumper wire.
6. Disconnect O_2 system test connector located on left fender shield by shock tower.
7. Start the engine and run until normal operating temperature is obtained.

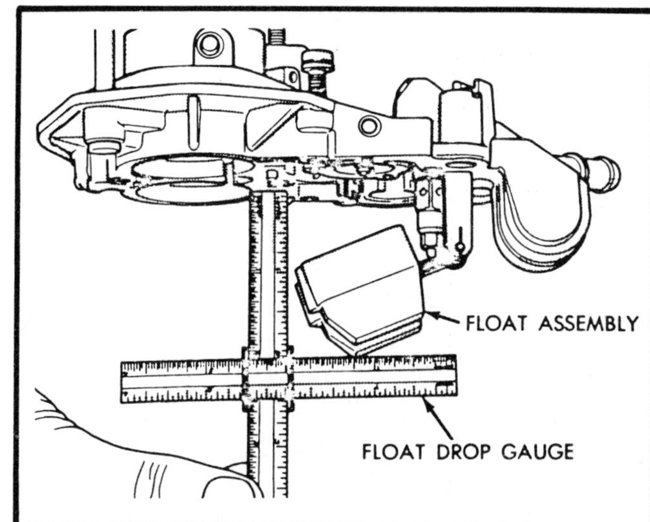

Measuring float drop on Holley 6520 models

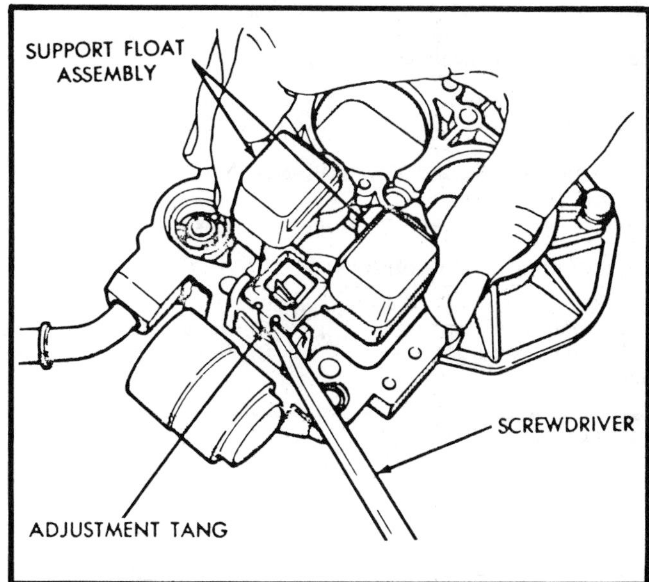

Adjusting float drop on Holley 6520 models

8. Open throttle slightly and place adjustment screw on slowest speed step of fast idle cam.
9. With the choke valve fully open, adjust the fast idle speed to the above specifications. Return to idle then replace adjusting screw on the slowest speed step of fast idle speed cam to verify fast idle speed, readjust if necessary.
10. Turn off engine, remove jumper wire and reconnect the radiator fan. Unplug and reconnect the vacuum connector. Reinstall the PCV valve and remove tachometer. Reconnect O_2 system test connector.

Choke Vacuum Kick Adjustment

1. Open throttle, close choke, then close throttle to trap fast idle system at closed choke condition.
2. Disconnect vacuum hose from carburetor and replace with hose from auxiliary vacuum source. Apply a vacuum of 15 in. Hg.
3. Apply sufficient closing force to position choke valve at smallest opening possible without distorting linkage system. An internal spring will compress to a stop inside choke system.
4. Measure by inserting drill or gauge in center of area between

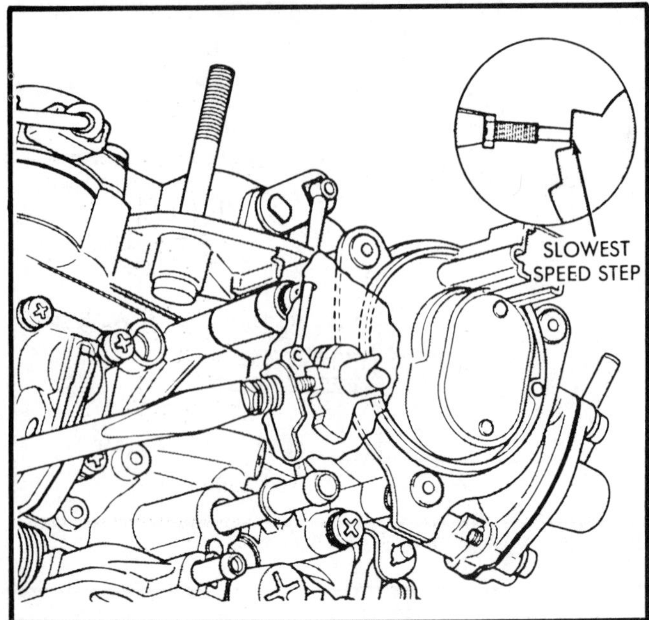

Fast idle speed adjustment—Holley 6520

top of choke valve and air horn wall at primary throttle end of carburetor. See Specifications above for drill or gauge size to be used.

5. Adjust by rotating allen head screw in center of diaphragm housing.

6. Replace vacuum hose on carburetor fitting.

Concealment Plug Removal

1. Remove air cleaner cross over assembly.
2. Remove canister purge and diverter valve vacuum hoses.
3. Center punch at a point 1/4 inch from end of mixture screw housing.

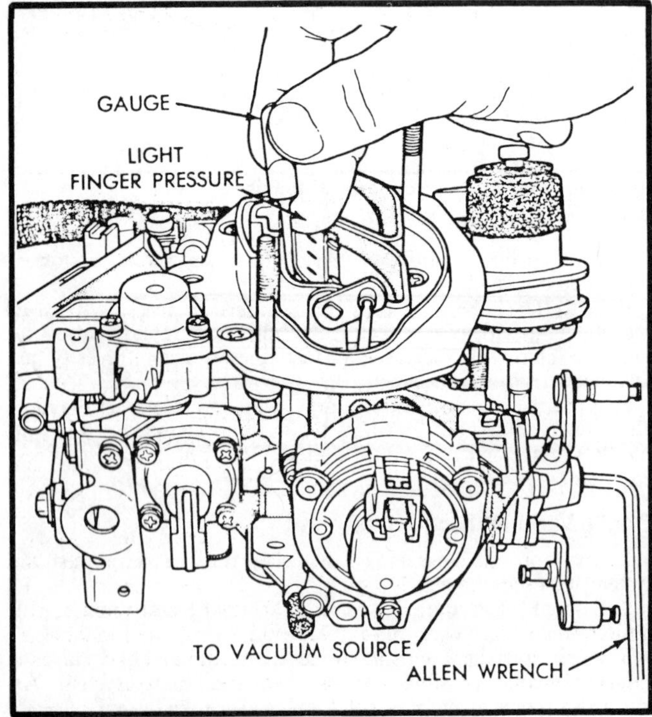

Choke vacuum kick adjustment

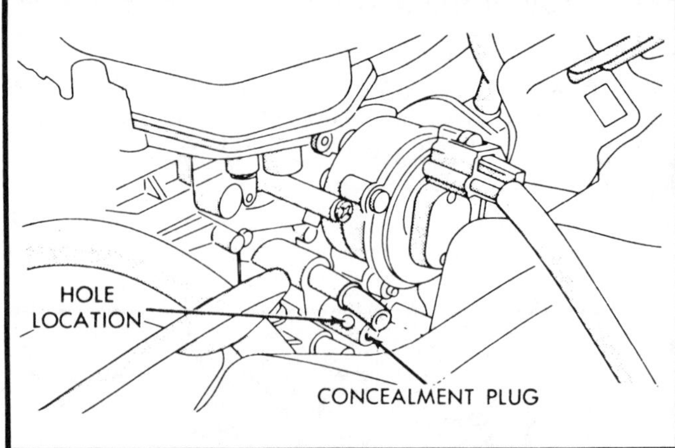

Drill a hole as indicated to remove the idle mixture plug

4. Drill through outer housing with 3/16 inch drill bit.
5. Pry out concealment plug and save for reinstallation.

Propane Idle Set Procedure

1. Remove concealment plug. Set the parking brake and place transaxle in neutral. Turn off all lights and accessories. Connect a tachometer to the engine. Start the engine and allow it to warm up on the second highest step on the fast idle cam until normal operating temperature is reached. Return engine to idle.

2. Disconnect and plug the vacuum connector at CVSCC. Disconnect the vacuum hose to the heated air door sensor at the three way connector, and in its place, install the supply hose from the propane bottle. Make sure that both valves are fully closed and that the bottle is upright and in a safe location.

3. Unplug the connector at the radiator fan and install a jumper wire so that the fan will run continuously. Remove the PCV valve from the molded rubber connector and allow the valve to draw underhood air. Connect a jumper wire between the carburetor switch and ground. On 6520 equipped vehicles disconnect the O_2 test connector located on the left fender shield.

4. Open the propane main valve. With the air cleaner in place, slowly open the propane metering valve until maximum engine

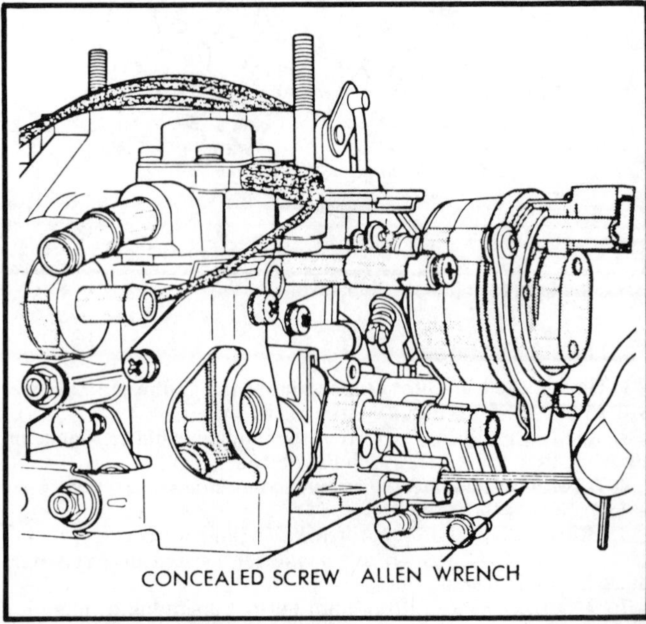

Adjusting idle mixture—Holley 6520

RPM is reached. When too much propane is added, engine speed will decrease. "Fine Tune" the metering valve for the highest engine RPM.

5. With the propane still flowing, adjust the idle speed screw on top of the solenoid to get the specified RPM. "Fine Tune" the propane metering valve to get the highest engine RPM. If there has been a change in the maximum RPM, readjust the idle speed to the specified propane RPM.

6. Turn off the propane main valve and allow engine speed to stabilize. With air cleaner in place, slowly adjust the mixture screw to achieve the specified idle set RPM. Pause between adjustments to allow engine speed to stabilize.

7. Turn on the main propane valve. "Fine Tune" the metering valve to get the highest engine RPM. If the maximum engine speed is more than 25 RPM different than the specified propane RPM repeat Steps 4–7.

8. Turn off both valves on propane bottle. Remove propane supply hose and reinstall the vacuum hose.

9. Reinstall the concealment plug. The plug can be installed with the carburetor on the engine.

10. Perform the fast idle speed adjustment procedure starting at Step 7.

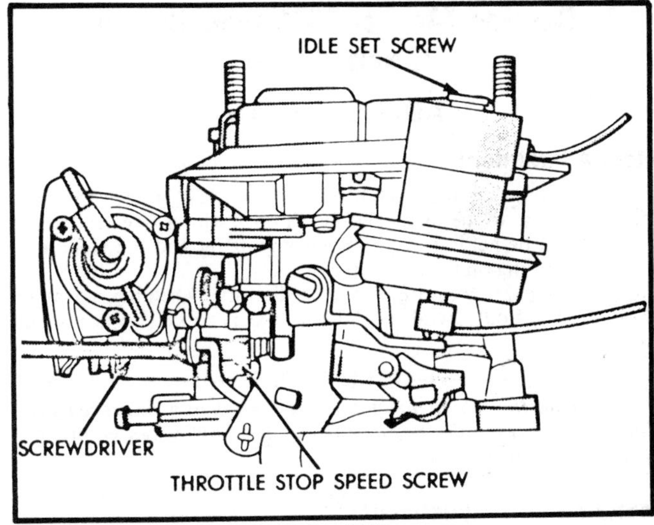

Anti-dieseling adjustment—Holley 6520

FORD FEEDBACK CARBURETOR ELECTRONIC ENGINE CONTROL

General Information

This system, first used on 1978 Pinto and Bobcat models sold in California with the 2.3 liter four cylinder engine, actually consists of three subsystems: a two part catalytic converter, a Thermactor (air pump) system, and an electronically controlled feedback carburetor.

The converter consists of two catalytic converters in one shell. The front section is designed to control all three engine emissions (NOx, HC, and CO). The rear section acts only on HC and CO. There is a space between the two sections which serves as a mixing chamber. Air is pumped into this area by the Thermactor system to assist in the oxidation of HC and CO.

The Thermactor system is the same as that found on conventional Ford models, with the addition of a second air control valve and a second exhaust check valve.

An electronically controlled feedback carburetor is used to precisely calibrate fuel metering. The air/fuel ratio is externally controlled and variable. It is adjusted according to conditions by the Electronic Control Unit ECU), 1978–79, or the Microprocessor Control Unit (MCU), 1980 and later. There are two modes of operation: closed loop control and open loop control. Under closed loop operation, each component in the chain is sensitive to the signals sent by the other components. This means that the carburetor mixture is being controlled by the vacuum regulator/solenoid, which is adjusted by the control unit, which is receiving signals from the oxygen sensor in the exhaust manifold, which is measuring a mixture determined by the carburetor, and so on. In this case, the feedback loop is complete. Under open loop operation, the carburetor air/fuel mixture is controlled directly by the control unit according to a predetermined setting. Open loop operation takes place when the coolant temperature is below 125°F, or when the throttle is closed, during idle or deceleration.

The control unit receives signals from the exhaust gas oxygen sensor, the throttle angle vacuum switch, and the cold temperature vacuum switch, analyzes them, and sends out commands to the vacuum solenoid/regulator, which in turn adjusts, by means of vacuum, the height of the carburetor fuel metering rod. In this way, the fuel mixture is adjusted according to conditions. The control unit also varies the transition time from rich to lean (and vice versa) according to engine rpm. The rpm signal is taken from the coil connector TACH terminal.

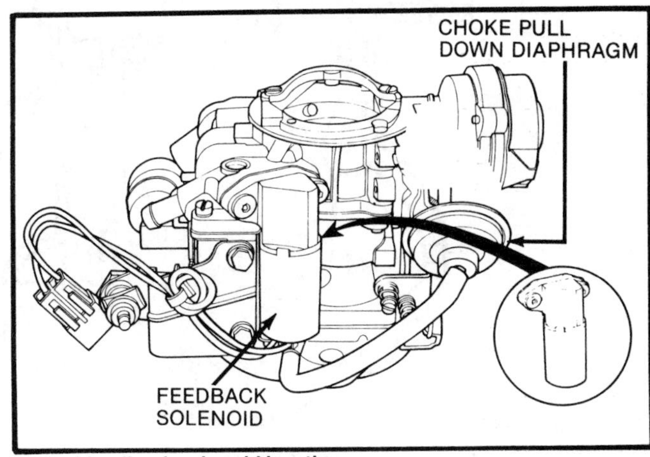

Typical feedback solenoid location

There are two differences between the ECU, used in 1978 and 1979, and the MCU, used in 1980 and later models. The MCU is programmable, enabling it to be used with many different engine calibrations. Additionally, the MCU controls the Thermactor solenoid values, thus directing the air flow to the exhaust manifold, the catalytic converter mixing chamber, or the atmosphere when air flow is not needed or wanted.

Because of the complicated nature of the Ford sytem, special diagnostic tools are necessary for troubleshooting and repair. No attempt at testing or repair should be made unless both the Feedback Control Tester (Ford part No. T78L-50-FBC-1) and a digital volt/ohmmeter (Ford part No. T78L-50-DVOM) are available. A tachometer, vacuum gauge, hand vacuum pump and gauge, and a special throttle rpm tool are also required for diagnosis. No trouble-shooting procedures will be given here, since they are supplied with the testing equipment.

Carter Model YFA Feedback Carburetor

ADJUSTMENTS

Float Adjustment

1. Invert the air horn assembly and check the clearance from the top of the float to the surface of the air horn with a T-scale. The air horn should be held at eye level when gauging and the float arm should be resting on the needle pin.
2. Do not exert pressure on the needle valve when measuring or adjusting the float. Bend the float arm as necessary to adjust the float level.

CAUTION

Do not bend the tab at the end of the float arm as it prevents the float from striking the bottom of the fuel bowl when empty and keeps the needle in place.

Metering Rod Adjustment

1. Remove the air horn. Back out the idle speed adjusting screw until the throttle plate is seated fully in its bore.
2. Press down on the upper end of the diaphragm shaft until the diaphragm bottoms in the vacuum chamber.
3. The metering rod should contact the bottom of the metering rod well. The lifter link at the outer end nearest the springs and at the supporting link should be bottomed.
4. On models not equipped with an adjusting screw, adjust by bending the lip of the metering rod is attached.
5. On models with an adjusting screw, turn the screw until the metering rod just bottoms in the body casting. For final adjustment, turn the screw one additional turn clockwise.

Fast Idle Cam Adjustment

1. Put the fast idle screw on the second highest step of the fast idle cam against the shoulder of the high step.
2. Adjust by bending the choke plate connecting rod to obtain the specified clearance between the lower edge of the choke plate and the air horn wall.

Choke Unloader Adjustment

1. With the throttle valve held wide open and the choke valve held in the closed position, bend the unloader tang on the throttle lever to obtain the specified clearance between the lower edge of the choke valve and their air horn wall.

Automatic Choke Adjustment

1. Loosen the choke cover retaining screws.

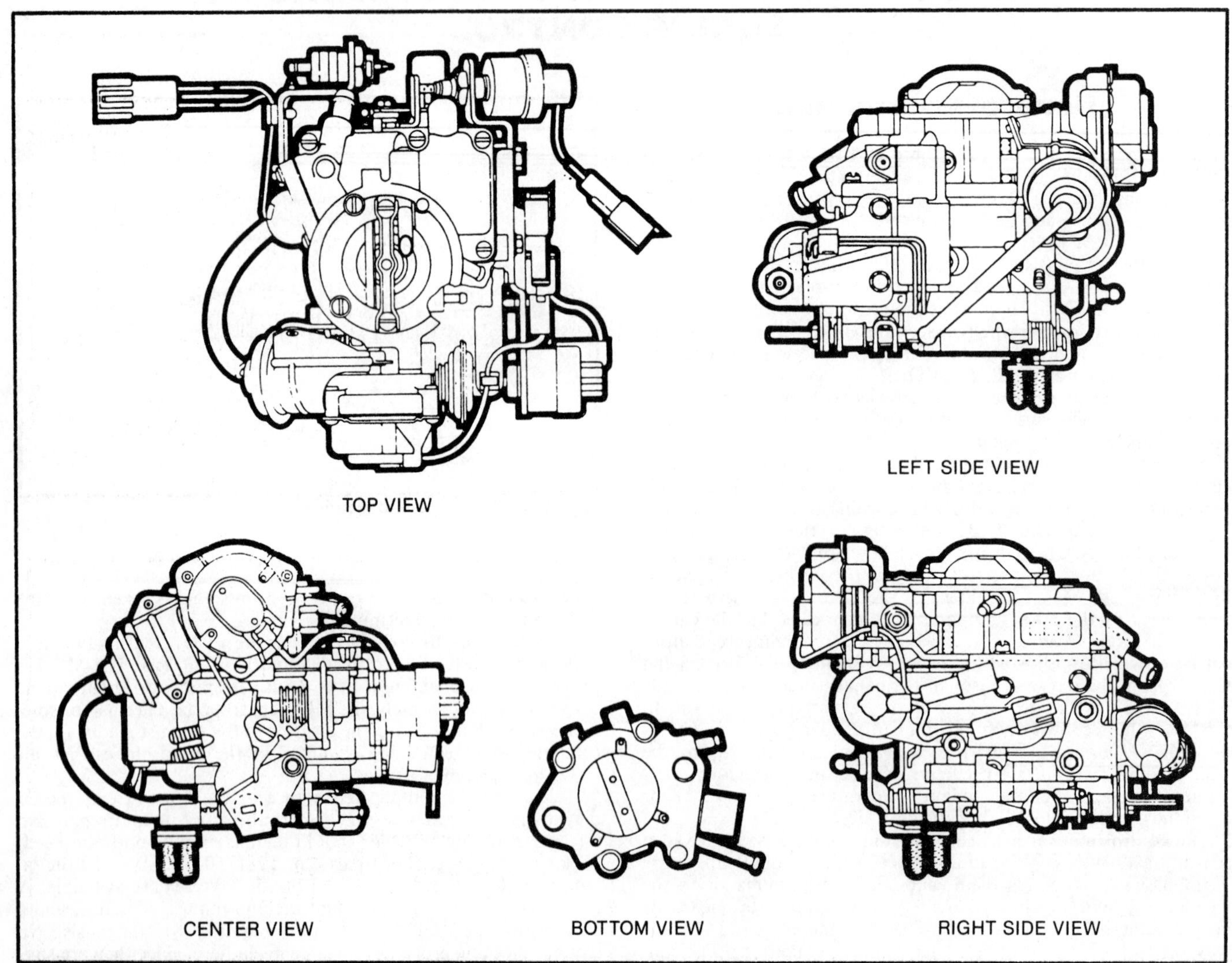

Carter YFA feedback carburetor assembly—1 bbl shown

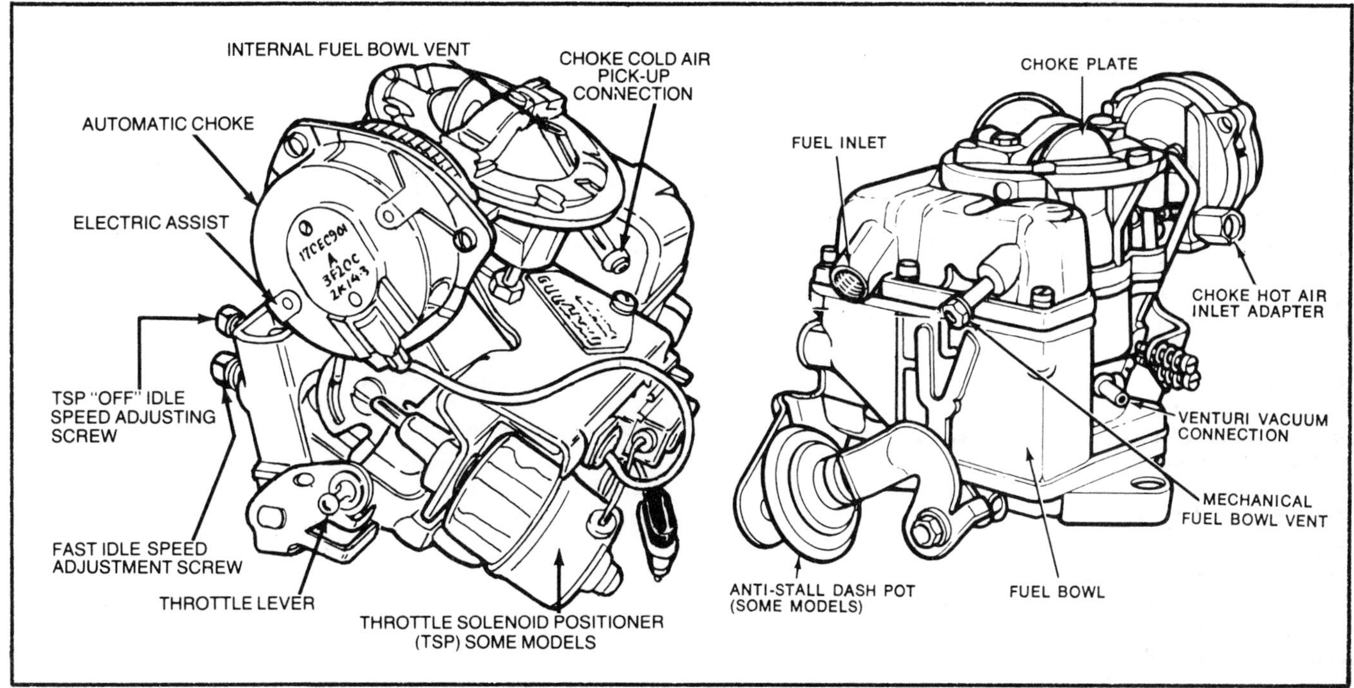

Component locations on Carter YFA feedback carburetor

2. Turn the choke cover so that the index mark on the cover lines up with the specified mark on the choke housing.

Choke Plate Pulldown Adjustment—1983-84

PISTON TYPE CHOKE

NOTE: This adjustment requires that the thermostatic spring housing and gasket (choke cap) are removed. Refer to the "Choke Cap" removal procedure below.

1. Remove the air cleaner assembly, then the choke cap.
2. Bend a 0.026 in. diameter wire gauge at a 90 degree angle approximately 1/8 in. from one end. Insert the bent end of the gauge between the choke piston slot and the right hand slot in the choke housing. Rotate the choke piston lever counterclockwise until the gauge is shut in the piston slot.
3. Apply light pressure on the choke piston lever to hold the gauge in place. Then measure the clearance between the lower edge of the choke plate and the carburetor bore using a drill with the diameter equal to the specified pulldown clearance.

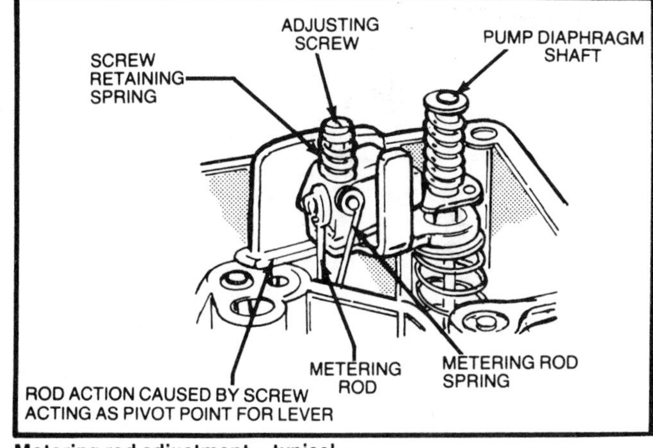

Metering rod adjustment—typical

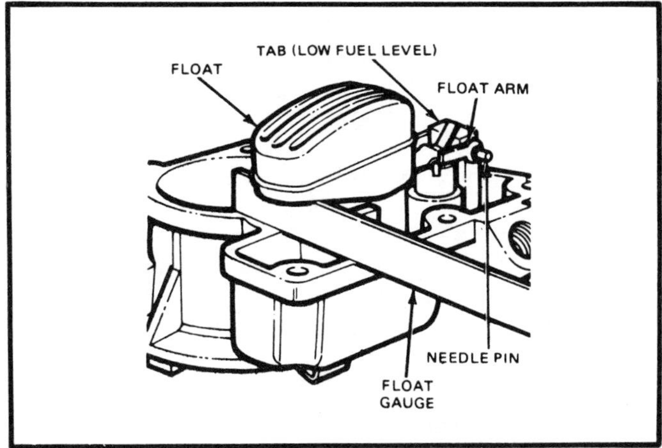

Float level adjustment—typical

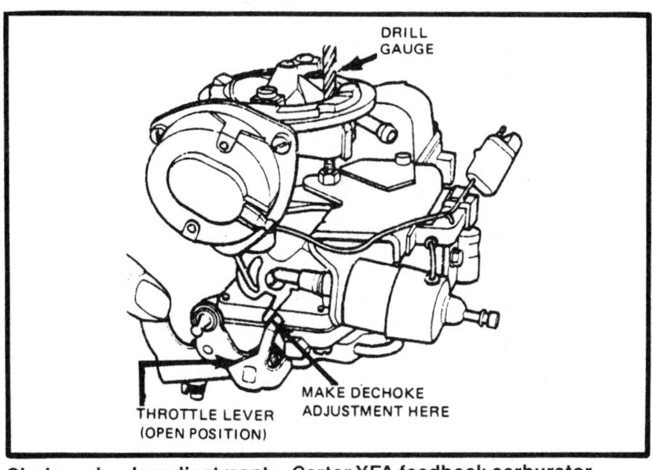

Choke unloader adjustment—Carter YFA feedback carburetor

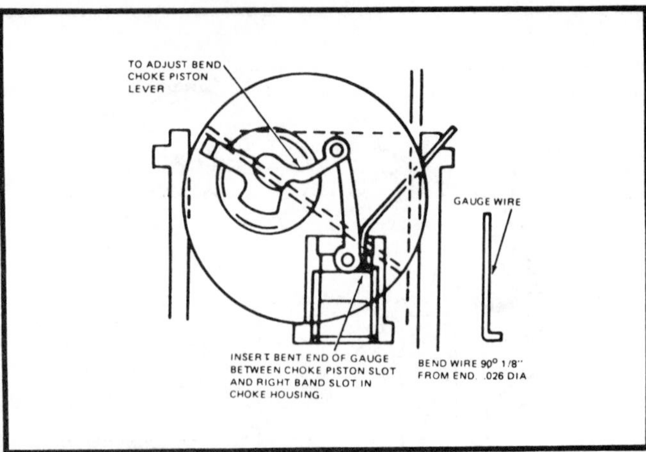

Piston type choke plate pulldown adjustment

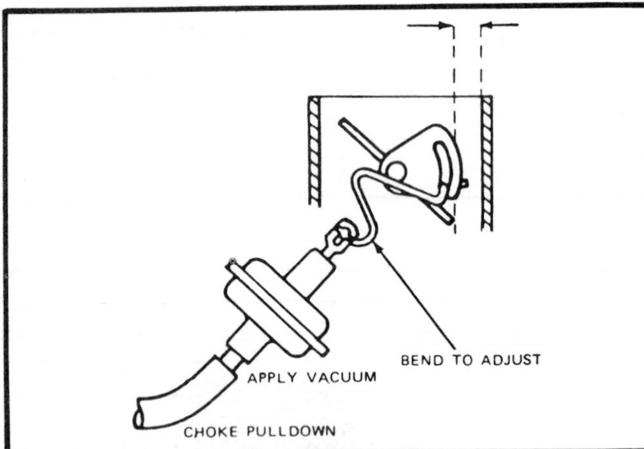

Diaphragm type choke plate pulldown adjustment

4. Bend the choke piston lever to obtain the proper clearance.
5. Install the choke cap.

DIAPHRAGM TYPE CHOKE

1. Activate the pulldown motor by applying an external vacuum source.
2. Close the choke plate as far as possible without forcing it.
3. Using a drill of the specified size, measure the clearance between the lower edge of the choke plate and the air horn wall.
4. If adjustment is necessary, bend the choke diaphragm link as required.

Choke Cap Removal

1983–84

NOTE: The automatic choke has two rivets and a screw, retaining the choke cap in place. There is a locking and indexing plate to prevent misadjustment.

1. Remove the air cleaner assembly from the carburetor
2. Check choke cap retaining ring rivets to determine if mandrel is well below the rivet head. If mandrel appears to be at or within the rivet head thickness, drive it down or out with a 1/16 inch diameter punch.
3. Use a 1/8 inch diameter of No. 32 drill (.128 inch diameter) for drilling the rivet heads. Drill into the rivet head until the rivet head comes loose from the rivet body.
4. After the rivet head is removed, drive the remaining portion of the rivet out of the hole with a 1/8 inch diameter punch.

NOTE: This procedure must be followed to retain the hole size.

5. Repeat Steps 1–4 for the remaining rivet.
6. Remove the screw in the conventional manner.

Choke Cap Installation

1. Install choke cap gasket.
2. Install the locking and indexing plate.
3. Install the notched gasket.
4. Install choke cap, making certain that bimetal loop is positioned around choke lever tang.

CARTER YF, YFA SPECIFICATIONS
American Motors

Year	Model ①	Float Level (in.)	Fast Idle Cam (in.)	Unloader (in.)	Choke
'78-'79	7201	0.476	0.195	0.275	Index
	7228	0.476	0.195	0.275	1 Rich
	7229	0.476	0.195	0.275	1 Rich
	7235	0.476	0.195	0.275	Index
	7267	0.476	0.195	0.275	1 Rich
	7232	0.476	0.201	0.275	2 Rich
	7233	0.476	0.201	0.275	1 Rich
'83–'84	7700	0.600	0.175	0.370	Fixed
	7701	0.600	0.175	0.370	Fixed
	7702	0.600	0.175	0.370	Fixed
	7703	0.600	0.175	0.370	Fixed

① Model numbers located on the tag or casting

CARTER YF, YFA, YFA-FB SPECIFICATIONS
Ford Motor Co.

Year	Model ①	Float Level (in.)	Fast Idle Cam (in.)	Choke Plate Pulldown (in.)	Unloader (in.)	Dechoke (in.)	Choke
'78	D7BE-AA,AB,BA	25/32	0.140	—	0.250	—	Index
	D7BE-FA,HB, GB,GC	25/32	0.140	—	0.250	—	2 Rich
	D7BE-NA,DA	25/32	0.140	—	0.250	—	1 Rich
'79	D9BE-RA D9DE-CB,DB, AA,BA,CA,EA	25/32	0.140	—	0.250	—	1 Rich
'80	DEDE-GA,HA EODE-JA,NA, LA,MA	25/32	0.140	—	0.250	—	2 Rich
'83	E3ZE-LA	0.650	0.140	0.260	—	0.220	—
	E3ZE-MA	0.650	0.140	0.260	—	0.220	—
	E3ZE-TB	0.650	0.140	0.240	—	0.220	—
	E3ZE-UA	0.650	0.140	0.240	—	0.220	—
	E3ZE-VA	0.650	0.140	0.260	—	0.220	—
	E3ZE-YA	0.650	0.140	0.260	—	0.220	—
	E3ZE-NB	0.650	0.160	0.260	—	0.220	—
	E3ZE-PB	0.650	0.160	0.260	—	0.220	—
	E3ZE-ASA	0.650	0.160	0.260	—	0.220	—
	E3ZE-APA	0.650	0.140	0.240	—	0.220	—
	E3ZE-ARA	0.650	0.140	0.240	—	0.220	—
	E3ZE-ADA	0.650	0.140	0.260	—	0.220	—
	E3ZE-AEA	0.650	0.140	0.260	—	0.220	—
	E3ZE-ACA	0.650	0.140	0.260	—	0.220	—
	E3ZE-ATA	0.650	0.160	0.260	—	0.220	—
	E3ZE-ABA	0.650	0.140	0.260	—	0.220	—
	E3ZE-UB	0.650	0.140	0.240	—	0.220	—
	E3ZE-TC	0.650	0.140	0.240	—	0.220	—
'84	E4ZE-HC,DB	0.650	0.140	0.260	—	0.270	—
	E4ZE-MA,NA	0.650	0.140	0.240	—	0.270	—
	E4ZE-PA,RA	0.650	0.140	0.260	—	0.270	—

① Model number located on the tag or casting

5. While holding cap in place, actuate choke plate to make certain bimetal loop is properly engaged with lever tang. Set retaining clamp over choke cap and orient clamp to match holes in casting (holes are not equally spaced). Make sure retaining clamp is not upside down.
6. Place rivet in rivet gun and trigger lightly to retain rivet (1/8 inch diameter 1/2 inch long 1/4 inch diameter head).
7. Press rivet fully into casting after passing through retaining clamp and pop rivet (mandrel breaks off).
8. Repeat this step for the remaining rivet.
9. Install screw in conventional manner. Tighten to (17–20 lb-in.)

Choke Plate Clearance (Dechoke) Adjustment

1. Remove the air cleaner assembly.
2. Hold the throttle plate fully open and close the choke plate as far as possible without forcing it. Use a drill of the proper diameter to check the clearance between the choke plate and air horn.
3. If the clearance is not within specification, adjust by bending the arm on the choke lever of the throttle lever. Bending the arm downward will decrease the clearance, and bending it upward will increase the clearance. **Always recheck the clearance after making any adjustment.**

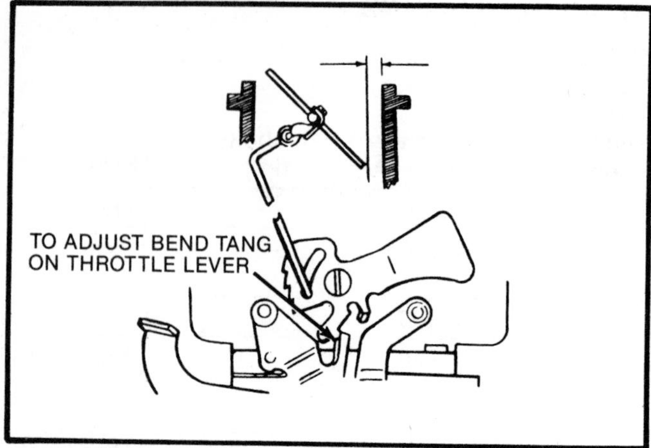

Dechoke adjustment—Carter YFA

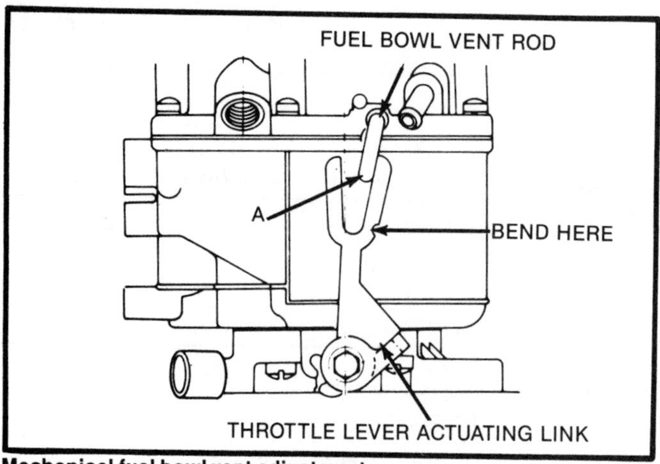

Mechanical fuel bowl vent adjustment

Mechanical Fuel Bowl Vent Adjustment

1. Start the engine and wait until it has reached normal operating temperature before proceeding.
2. Check engine idle rpm and set to specificatons.
3. Check DC motor operation by opening throttle off idle. The DC motor should extend. Release the throttle, and the DC motor should retract when in contact with the throttle lever.
4. Disconnect the idle speed motor in the idle position.
5. Turn engine Off.
6. Open the throttle lever so that the throttle lever actuating lever does not touch the fuel bowl vent rod.
7. Close the throttle lever to the idle set position and measure the travel of the fuel bowl vent rod at point A. The distance measured represents the travel of the vent rod from where there is no contact with the actuating lever to where the actuating lever moves the vent rod to the idle set position. The travel of the vent rod at point A should be 0.100–0.150 in. (2.54–3.81 mm).
8. If adjustment is required, bend the throttle actuating lever at notch shown.
9. Reconnect the idle speed control motor.

Motorcraft 2700 VV Carburetor

GENERAL INFORMATION

In exterior appearance, the variable venturi carburetor is similar to conventional carburetors and, like a conventional carburetor, it uses a normal float and fuel bowl system. However, the similarity ends there. In place of a normal choke plate and fixed area venturis, the 2700 VV carburetor has a pair of small oblong castings in the top of the upper carburetor body where you would normally expect to see the choke plate. These castings slide back and forth across the top of the carburetor in response to fuel-air demands. Their movement is controlled by a spring-loaded diaphragm valve regulated by a vacuum signal taken below the venturis in the throttle bores. As the throttle is opened, the strength of the vacuum signal increases, opening the venturis and allowing more air to enter the carburetor.

Fuel is admitted into the venturi area by means of tapered metering rods that fit into the main jets. These rods are attached to the venturis, and, as the venturis open or close in response to air demand, the fuel needed to maintain the proper mixture increases

MOTORCRAFT MODEL 2700 VV SPECIFICATIONS
Ford Products

Year	Model	Float Level (in.)	Float Drop (in.)	Fast Idle Cam Setting (notches)	Cold Enrichment Metering Rod (in.)	Control Vacuum (in. H_2O)	Venturi Valve Limiter (in.)	Choke Cap Setting (notches)	Control Vacuum Regulator Setting (in.)
'78	Pinto, Bobcat	1³⁄₆₄	1¹⁵⁄₃₂	4 Rich/2nd step	.125	5.0	¹³⁄₃₂	Index	—
	All other	1³⁄₆₄	1¹⁵⁄₃₂	1 Rich/3rd step	.125	5.0	⁶¹⁄₆₄	Index	—
'79	D9ZE-LB	1³⁄₆₄	1¹⁵⁄₃₂	1 Rich/2nd step	.125	①	②	Index	.230
	D84E-KA	1³⁄₆₄	1¹⁵⁄₃₂	1 Rich/3rd step	.125	5.5	⁶¹⁄₆₄	Index	—
'80	All	1³⁄₆₄	1¹⁵⁄₃₂	1 Rich/4th step	.125	③	④	⑤	.075
'81	EIAE-AAA	1.010-1.070	1.430-1.490	1 Rich/4th step	⑥	③	④	Index	—
	D9AE-AZA	1.015-1.065	1.435-1.485	1 Rich/4th step	.125	③	④	Index	—

① Venturi Air Bypass 6.8-7.3
Venturi Valve Diaphragm 4.6-5.1
② Limiter Setting .38-.42
Limiter Stop Setting .73-.77
③ See text
④ Opening gap: 0.99-1.01
Closing gap: 0.94-0.98
⑤ See underhood decal
⑥ 0°F—0.490 @ starting position
75°F—0.475 @ starting position

or decreases as the metering rods slide in the jets. In comparison to a conventional carburetor with fixed venturis and a variable air supply, this system provides much more precise control of the fuel-air supply during all modes of operation. Because of the variable venturi principle, there are fewer fuel metering systems and fuel passages. The only auxiliary fuel metering systems required are an idle trim, accelerator pump (similar to a conventional carburetor), starting enrichment, and cold running enrichment.

ADJUSTMENTS

NOTE: Adjustment, assembly and disassembly of this carburetor require special tools for some of the operations. Some 1978-81 models equipped with the 2700 VV carburetor experienced engine stalling, stumbling and poor performance. According to a Ford Motor Company service bulletin No. 81-9-10 issued in May of 1981, this condition may be caused by fluids trapped in the venturi valve diaphragm cavity which eventually deteriorate the diaphragm, resulting in a leak.

A quick check to verify if the above condition exists is to visually observe the venturi valve action while running the engine. Throttle movement from idle to just above idle should show a corresponding movement of the venturi valves. If no venturi valve action is observed and the valves are not sticking, there may be a leak in the diaphragm (stalls may be encountered while performing this check).

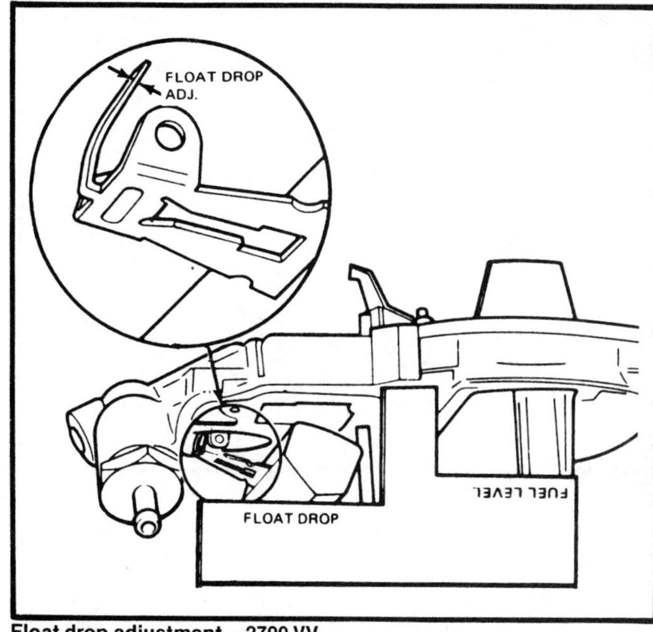

Float drop adjustment—2700 VV

Float Level Adjustment

1. Remove and invert the upper part of the carburetor, with the gasket in place.
2. Measure the vertical distance between the carburetor body, outside the gasket, and the bottom of the float.
3. To adjust, bend the float operating lever that contacts the needle valve. Make sure that the float remains parallel to the gasket surface.

Float Drop Adjustment

1. Remove and hold upright the upper part of the carburetor.
2. Measure the vertical distance between the carburetor body, outside the gasket, and the bottom of the float.
3. Adjust by bending the stop tab on the float lever that contacts the hinge pin.

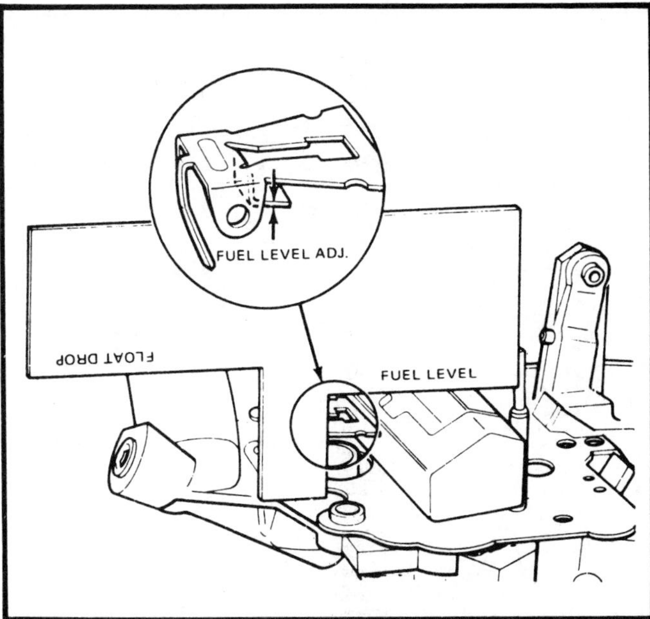

Float level adjustment—Motorcraft 2700 VV feedback carburetor

Fast Idle Speed Adjustment

1. With the engine warmed up and idling, place the fast idle lever on the step of the fast idle cam specified on the engine compartment sticker or in the specifications chart. Disconnect and plug the EGR vacuum line.
2. Make sure the high speed cam positioner lever is disengaged.
3. Turn the fast idle speed screw to adjust to the specified speed.

Fast Idle Cam Adjustment

You will need a special tool for this job: Ford calls it a stator cap (#T77L-9848-A). It fits over the choke thermostatic lever when the choke cap is removed.

1. Remove the choke coil cap. On 1980 California model and all 1981 and later models, the choke cap is riveted in place. The top

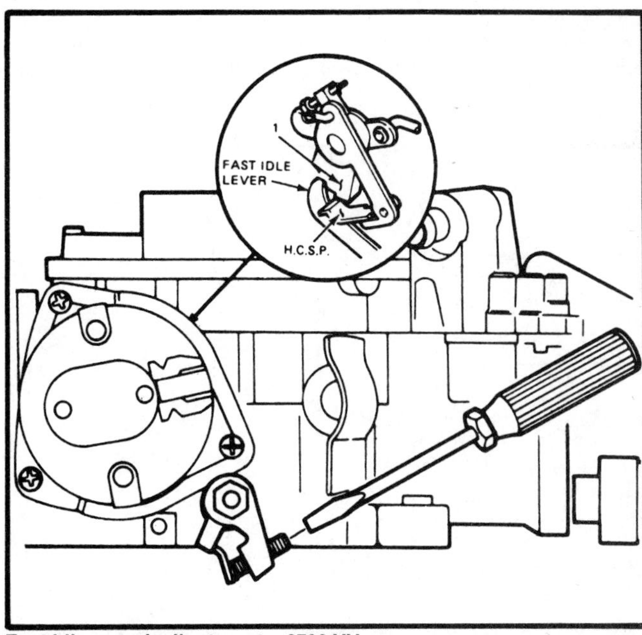

Fast idle speed adjustment—2700 VV

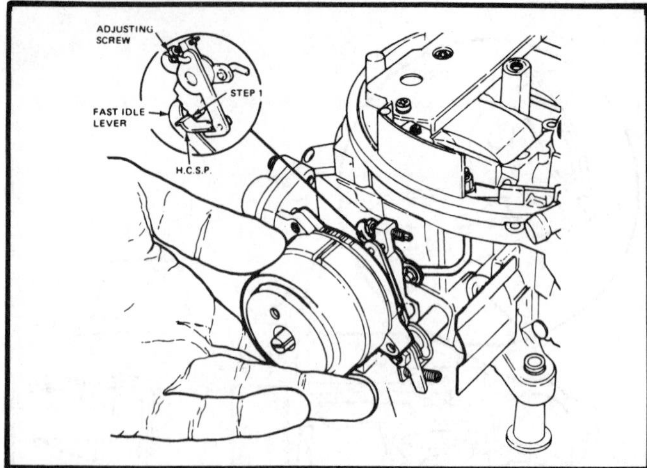

Fast idle cam adjustment—2700 VV

rivets will have to be drilled out: the bottom rivets will have to be driven out from the rear. New rivets must be used upon installation.

2. Place the fast idle lever in the corner of the specified step of the fast idle cam (the highest step is first) with the high speed cam positioner retracted.

3. If the adjustment is being made with the carburetor removed, hold the throttle lightly closed with a rubber band.

4. Turn the stator cap clockwise until the lever contacts the fast idle cam adjusting screw.

5. Turn the fast idle cam adjusting screw until the index mark on the cap lines up with the specified mark on the casting.

6. Remove the stator cap. Install the choke coil cap and set to the specified housing mark.

Cold Enrichment Metering Rod Adjustment

A dial indicator and the stator cap are required for this adjustment.

1, Remove the choke coil cap. See Step 1 of the "Fast Idle Cam Adjustment."

2. Attach a weight to the choke coil mechanism to seat the cold enrichment rod.

3. Install and zero a dial indicator with the tip on top of the enrichment rod. Raise and release the weight to verify zero on the dial indicator.

4. With the stator cap at the index position, the dial indicator should read the specified dimension. Turn the adjusting nut to correct.

5. Install the choke cap at the correct setting.

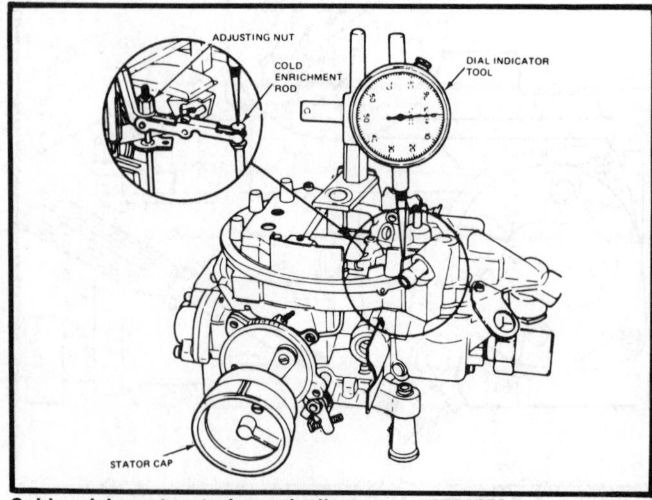

Cold enrichment metering rod adjustment—2700 VV

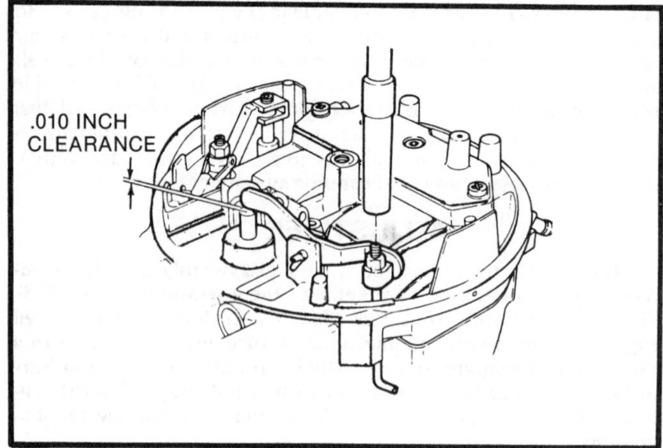

Internal vent adjustment—2700 VV

Internal Vent Adjustment

1978 ONLY

This adjustment is required whenever the idle speed adjustment is changed.

1. Make sure the idle speed is correct.

2. Place of 0.010 in. feeler gauge between the accelerator pump stem and the operating link.

3. Turn the nylon adjusting nut until there is a slight drag on the gauge.

Venturi Valve Limiter Adjustment

1. Remove the carburetor. Take off the venturi valve cover and the two rollers.

2. Use a center punch to loosen the expansion plug at the rear of the carburetor main body on the throttle side. Remove it.

3. Use an Allen wrench to remove the venturi valve wide open screw.

4. Hold the throttle wide open.

5. Apply a light closing pressure on the venturi valve and check the gap between the valve and the air horn wall. To adjust, move the venturi valve to the wide open position and insert an Allen

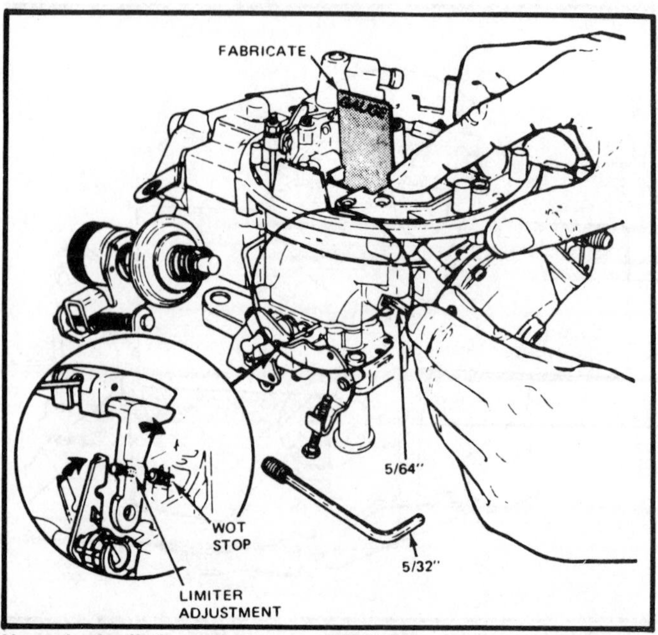

Venturi valve limiter adjustment—2700 VV

wrench into the stop screw hole. Turn clockwise to increase the gap. Remove the wrench and check the gap again.

6. Replace the wide open stop screw and turn it clockwise until it contacts the valve.

7. Push the venturi valve wide open and check the gap. Turn the stop screw to bring the gap to specifications.

8. Reassemble the carburetor with a new expansion plug.

Control Vacuum Regulator Adjustment

There are two systems used. The earlier system's C.V. R. rod threads directly through the arm. The revised system, introduced in late 1977, has a 3/8 in. nylon hex adjusting nut on the C.V. R. rod and a flange on the rod.

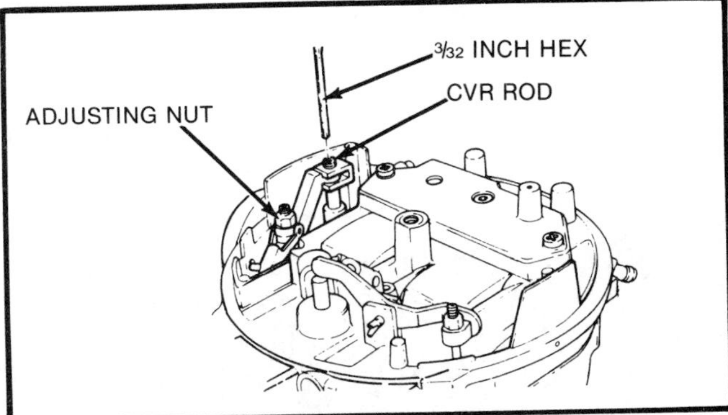

Control vacuum regulator adjustment—2700 VV

EARLY SYSTEM

1. Make sure that the cold enrichment metering rod adjustment is correct.

2. Rotate the choke coil cap half a turn clockwise from the index mark. Work the throttle to set the fast idle cam.

3, Press down lightly on the regulator rod. If there is no down travel, turn the adjusting screw counter-clockwise until some travel is felt.

4. Turn the regulator rod clockwise with an Allen wrench until the adusting nut just begins to rise.

5. Press lightly on the regulator rod. If there is any down travel, turn the adjusting screw clockwise in 1/4 turn increments until it is eliminated.

6. Return the choke coil cap to the specified setting.

REVISED SYSTEM

The cold enrichment metering rod adjustment must be checked and set before making this adjustment.

1, After adjusting the cold enrichment metering rod, leave the dial indicator in place but remove the stator cap. Do not rezero the dial indicator.

2. Press down on the C.V. R. rod until it bottoms on its seat. Measure this amount of travel with the dial indicator.

3. If the adjustment is incorrect, hold the 3/8 in. C.V. R. adjusting nut with a box wrench to prevent it from turning. Use a 3/32 in. Allen wrench to turn the C.V. R. rod; turning counter-clockwise will increase the travel, and vice versa.

High Speed Cam Positioner Adjustment

1979 ONLY

1. Place the high speed cam positioner in the corner of the specified cam step, counting the highest step as the first.

2. Place the fast idle lever in the corner of the positioner.

3. Hold the throttle firmly closed.

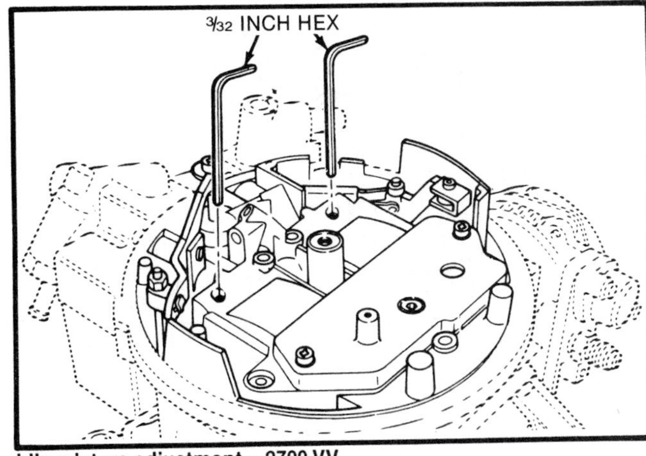

Idle mixture adjustment—2700 VV

4. Remove the diaphragm cover. Adjust the diaphragm assembly clockwise until it lightly bottoms. Turn it counterclockwise 1/2 to 1 1/2 turns until the vacuum port and diaphragm hole line up.

5. Replace the cover.

Motorcraft 7200 VV Carburetor

The Motorcraft model 7200 variable venturi (VV) carburetor shares most of its design features with the model 2700 VV. The major difference between the two is that the 7200 is designed to work with Ford's EEC (electronic engine control) feedback system. The feedback system precisely controls the air/fuel ratio by varying signals to the feedback control monitor located on the carburetor, which opens or closes the metering valve in response. This expands or reduces the amount of control vacuum above the fuel bowl, leaning or richening the mixture accordingly.

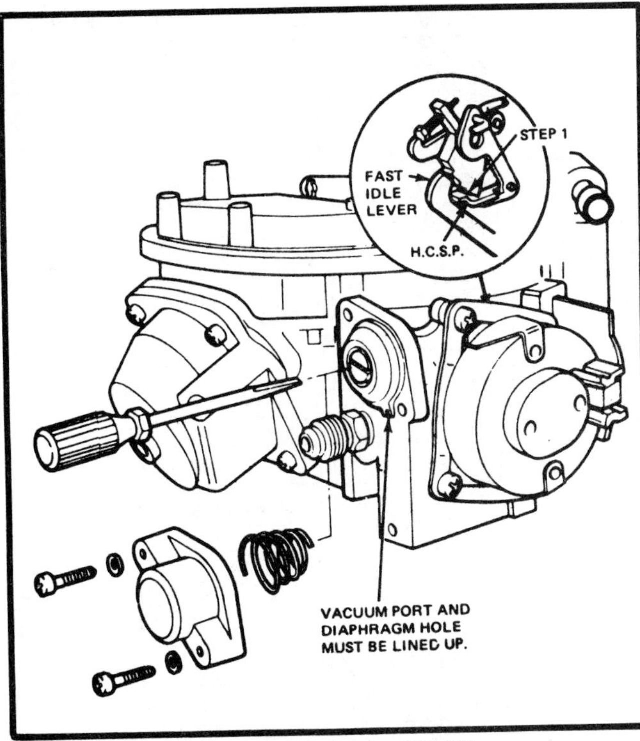

High speed cam positioner adjustment—2700 VV

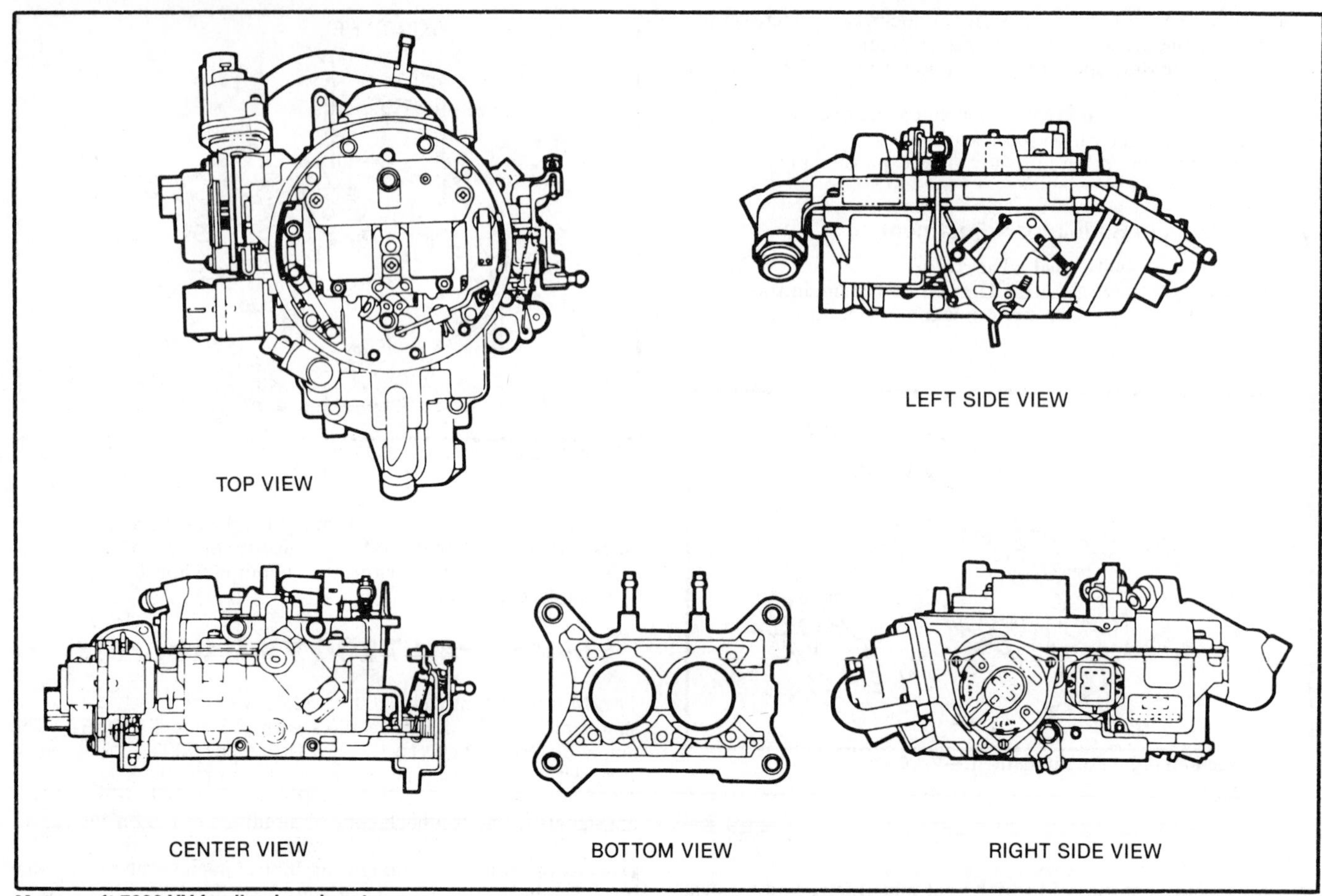

Motorcraft 7200 VV feedback carburetor

MOTORCRAFT MODEL 7200 VV SPECIFICATIONS

Year	Model	Float Level (in.)	Float Drop (in.)	Fast Idle Cam Setting (notches)	Cold Enrichment Metering Rod (in.)	Control Vacuum (in. H_2O)	Venturi Valve Limiter (in.)	Choke Cap Setting (notches)
'79	D9AE-ACA	1 3/64	1 15/32	1 Rich/3rd step	.125	7.5	.73-.77①	Index
	D9ME-AA	1 3/64	1 15/32	1 Rich/3rd step	.125	7.5	.73-.77①	Index
'80	All	1 3/64	1 15/32	1 Rich/3rd step	.125	②	③	④
'81	D9AE-AZA	1.015-1.065	1.435-1.485	1 Rich/3rd step	.125	②	⑤	Index
	EIAE-LA	1.015-1.065	1.435-1.485	0.360/2nd step	⑦	②	⑥	1 Rich
	EIAE-SA	1.015-1.065	1.435-1.485	0.360/2nd step	⑦	②	⑥	1 Rich
	EIAE-KA	1.010-1.070	1.430-1.490	0.360/2nd step	⑩	②	⑭	Index
	EIDE-AA	1.010-1.070	1.430-1.490	0.360/2nd step	⑩	②	⑭	Index
	EIVE-AA	1.015-1.065	1.435-1.485	0.360/2nd step	⑦	②	③	Index
'82	E2AE-LB	1.010-1.070	1.430-1.490	0.360/2nd step	⑧	②	⑨	Index
	E2DE-NA	1.010-1.070	1.430-1.490	0.360/2nd step	⑧	②	⑨	Index
	E2AE-LC	1.010-1.070	1.430-1.490	0.360/2nd step	⑧	②	⑨	Index
	E25E-FA	1.010-1.070	1.430-1.490	0.360/2nd step	⑧	②	⑨	Index
	E25E-GB	1.010-1.070	1.430-1.490	0.360/2nd step	⑧	②	⑨	Index
	E2SE-GA	1.010-1.070	1.430-1.490	0.360/2nd step	⑧	②	⑨	Index

MOTORCRAFT MODEL 7200 VV SPECIFICATIONS

Year	Model	Float Level (in.)	Float Drop (in.)	Fast Idle Cam Setting (notches)	Cold Enrichment Metering Rod (in.)	Control Vacuum (in. H_2O)	Venturi Valve Limiter (in.)	Choke Cap Setting (notches)
'82	E2AE-RA	1.010-1.070	1.430-1.490	0.360/2nd step	⑩	②	⑨	Index
	E1AE-ACA	1.010-1.070	1.430-1.490	0.360/2nd step	⑩	②	⑨	Index
	E2SE-DB	1.010-1.070	1.430-1.490	0.360/2nd step	⑪	②	⑨	Index
	E2SE-DA	1.010-1.070	1.430-1.490	0.360/2nd step	⑪	②	⑨	Index
	E1AE-SA	1.010-1.070	1.430-1.490	0.360/2nd step	⑫	②	⑬	1 Rich
	E2AE-MA	1.010-1.070	1.430-1.490	0.360/2nd step	⑫	②	⑬	1 Rich
	E2AE-MB	1.010-1.070	1.430-1.490	0.360/2nd step	⑫	②	⑬	1 Rich
	E2AE-TA	1.010-1.070	1.430-1.490	0.360/2nd step	⑫	②	⑬	Index
	E2AE-TB	1.010-1.070	1.430-1.490	0.360/2nd step	⑫	②	⑬	Index
	E25E-AC	1.010-1.070	1.430-1.490	0.360/2nd step	⑪	②	⑨	Index
	E1AE-AGA	1.010-1.070	1.430-1.490	0.360/2nd step	⑫	②	⑨	Index
	E2AE-NA	1.010-1.070	1.430-1.490	0.360/2nd step	⑫	②	⑨	Index
'83	E2AE-NA	1.010-1.070	1.430-1.490	0.360/2nd step	⑫	②	⑨	Index
	E2AE-AJA	1.010-1.070	1.430-1.490	0.360/2nd step	⑫	②	⑨	Index
	E2AE-APA	1.010-1.070	1.430-1.490	0.360/2nd step	⑫	②	⑨	Index
'84	E2AE-AJA	1.010-1.070	1.430-1.490	0.360/2nd step	⑫	②	⑨	Index
	E2AE-APA	1.010-1.070	1.430-1.490	0.360/2nd step	⑫	②	⑨	Index

① Limiter Stop Setting: .99-1.01
② See text
③ Opening gap: 0.99-1.01
Closing gap: 0.39-0.41
④ See underhood decal
⑤ Maximum opening: .99/1.01
Wide open on throttle: .94/.98
⑥ Maximum opening: .99/1.01
Wide open on throttle: .74/.76
⑦ 0°F—0.490 @ starting position
75°F—0.475 @ starting position
⑧ 0°F—0.525 @ starting position
75°F—0.445 @ starting position
⑨ Maximum opening: .99/1.01
Wide open on throttle: .39/.41
⑩ 0°F—0.490 @ starting position
75°F—0.445 @ starting position
⑪ 0°F—0.525 @ starting position
75°F—0.475 @ starting position
⑫ 0°F—0.490 @ starting position
75°F—0.460 @ starting position
⑬ Maximum opening: .99/1.01
Wide open on throttle: .74/.76
⑭ Maximum opening: .99/1.01
Wide open on throttle: .48/.52

ADJUSTMENTS

Float Level, Float Drop, Fast Idle Speed Adjustments

These adjustments are performed in the same manner as for the 2700VV. See that section for procedures.

Fast Idle Cam Adjustment

This procedure is the same as for the 2700 VV. Use the procedure in that section. The 7200 VV used on California models has a choke cover held on with rivets. The carburetor must be removed to remove the rivets. With the carburetor removed, the top two rivets can be drilled out with a 1/8 in. drill bit. Drill only through the rivet head. The bottom rivet is located in a blind hole and must be removed by lightly tapping the backside of the retainer ring with a punch. The cover must be installed with replacement rivets, Ford part no. 388575, or the equivalent.

Cold Enrichment Metering Rod Adjustment

This adjustment is made in the same manner as for the 2700 VV. See the paragraph under the Fast Idle Cam Adjustment above concerning the riveted choke used on California models.

Internal Vent, Venturi Valve Limiter Adjustments

These adjustments are the same as for the 2700 VV. See that section for details.

Control Vacuum Regulator Adjustment

Use the Revised System procedure in the 2700 VV section. Note that the control vacuum is not adjustable on any 7200 carburetor; only the regulator is adjustable.

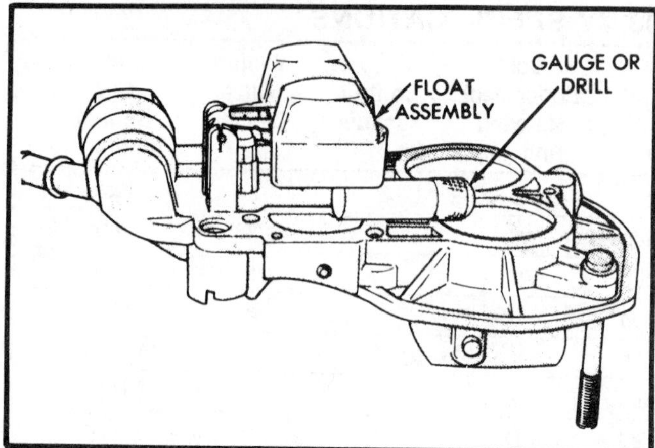

Float level adjustment—typical

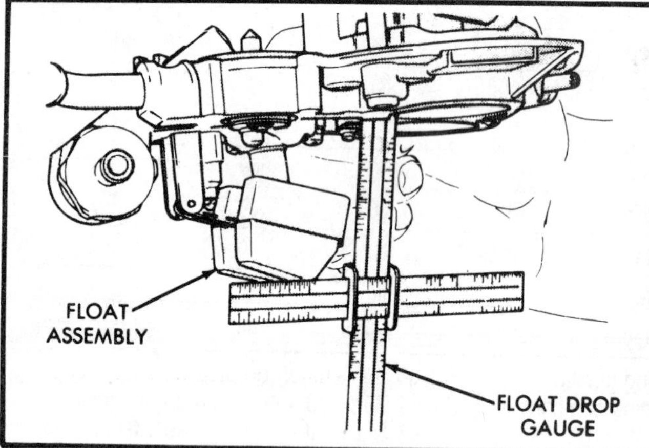

Measuring float drop—typical

High Speed Cam Positioner, Idle Mixture Adjustments

Procedures are the same as for the 2700 VV. See that section for details. Like the 2700 VV, the 7200 idle trim is preset at the factory and non-adjustable.

Holley 6520 Carburetor

ADJUSTMENTS

The model 6520 has the electronic feedback system. On the 6520 always check the condition of hoses and related wiring before making carburetor adjustments.

Float Setting and Float Drop Adjustment

1. Remove and invert the air horn.
2. Insert a 0.480 inch gauge between the air horn and float.
3. If necessary, bend the tang on the float arm to adjust.
4. Turn the air horn right side up and allow the float to hang freely. Measure the float drop from the bottom of the air horn to the bottom of the float. It should be exactly 1$^7/_8$ in. Correct by bending the float tang.

Vacuum Kick Adjustment

1. Open the throttle, close the choke, then close the throttle to trap the fast idle system at the closed choke position.
2. Disconnect the vacuum hose to the carburetor and connect it to an auxiliary vacuum source.
3. Apply at least 15 inches Hg. vacuum to the unit.

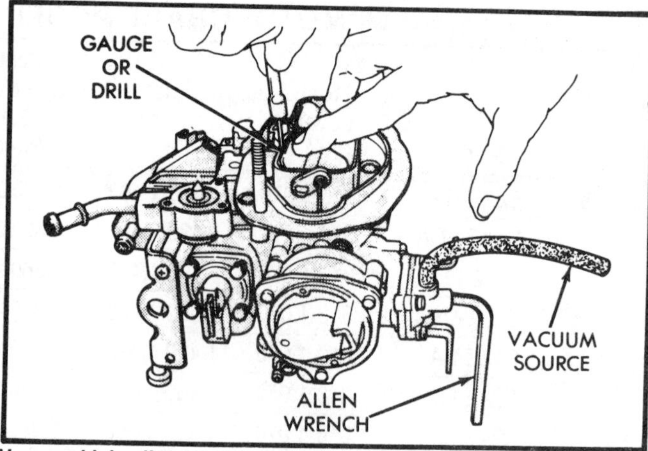

Vacuum kick adjustment—typical

4. Apply sufficient force to close the choke valve without distorting the linkage.
5. Insert a gauge (see Specification Chart) between the top of the choke plate and the air horn wall.
6. Adjust by rotating the Allen screw in the center diaphragm housing.
7. Replace the vacuum hose.

Throttle Position Transducer Adjustment

1978 ONLY

1. Disconnect the wire from the transducer.
2. Loosen the locknut.
3. Place an $^{11}/_{16}$ in. gauge between the outer portion of the transducer and the mounting bracket.
4. To adjust the gap, turn the transducer.
5. Tighten the locknut.

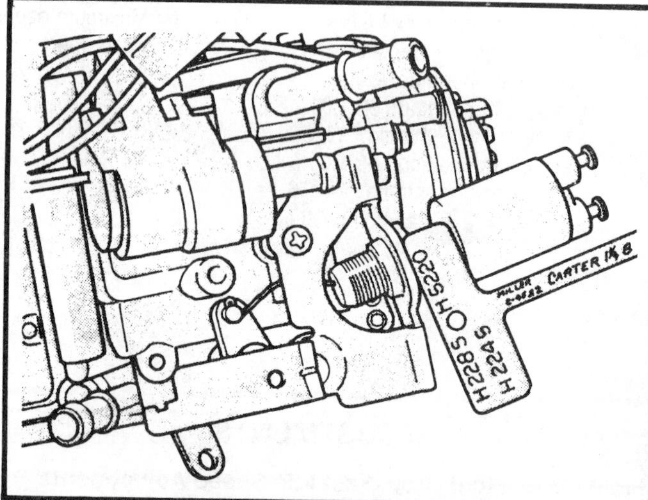

Throttle position transducer adjustment—typical

Fast Idle Speed Adjustment

1. Remove the air cleaner, disconnect and plug the EGR line, but do not disconnect the spark control computer vacuum line. Use a jumper wire to ground the idle stop switch (through 1979). Turn the air conditioning off.
2. Disconnect the radiator fan electrical connector and use a jumper wire to complete the circuit at the fan. Do not short to ground, as this will damage the system.
3. With the parking brake set and the transmission in Neutral (engine still off), open the throttle and place the fast idle screw on the slowest step of the cam.

4. Start the engine and check the idle speed. If it continues to rise slowly, the idle stop switch is not grounded properly.

5. Adjust the fast idle with the screw, moving the screw off the cam each time to adjust. Allow the screw to fall back against the cam and the speed to stabilize between each adjustment.

Holley 6500 Carburetor

This is a Holley-Weber Unit used on 1978 and later Pinto and Bobcat California models with the 2.3L engine. It is also used on all 1981–82 models with the 2.3L engine equipped with the Feedback Electronic Engine Control System. With the exception of an externally variable fuel metering system in place of the fuel enrichment valve, it is identical to the model Motorcraft 5200.

ADJUSTMENTS

Fast Idle Cam

1978–79 MODELS

1. Insert a 5/32 in. drill between the lower edge of the choke plate and the air horn wall.

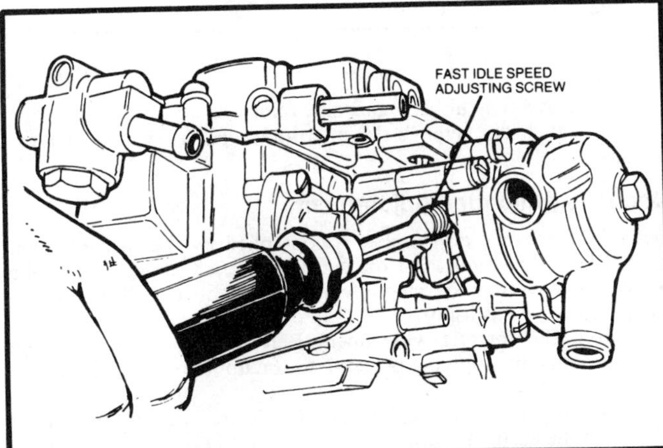

Fast idle speed adjustment—water heated choke assembly shown

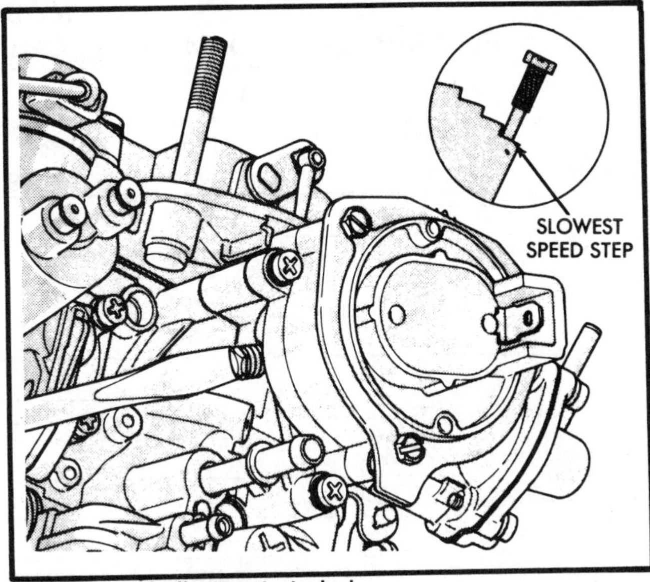

Fast idle speed adjustment—typical

2. With the fast idle screw held on the second step of the fast idle cam, measure the clearance between the tang of the choke lever and the arm of the fast idle cam.

3. Bend the choke lever tang to adjust it if it is not up to specification.

1980–82

1. Place the fast idle screw on the second step of the fast idle cam against the shoulder of the top step.

2. Apply light pressure (downward) on the choke lever tang, and, using the proper size drill, measure the clearance between the lower edge of the choke plate and the air horn wall.

3. Bend the choke lever tang down to increase clearance and up to decrease the clearance.

NOTE: On 1982 and later models Ford recommends that if an adjustment is necessary, the choke lever should be replaced since the lever tang is hardened. This may also be necessary for 1980-81 models.

MODEL 6500
Ford Bobcat, Pinto, Mustang, Capri, Fairmont, Zephyr, Granada, Cougar

Year	Carb. Iden.	Dry Float Level (in.)	Pump Hole Setting	Choke Plate Pulldown (in.)	Fast Idle Cam Linkage (in.)	Dechoke (in.)	Choke Setting
'78-'79	D9EE-AFC	0.455	2	0.236	0.118	0.236	2 Rich
	D9EE-AJC, D9EE-AKC	0.460	2	0.236	0.118	0.236	1 Rich
	D9EE-AGC	0.460	2	0.236	0.118	0.236	2 Rich
'80	EOEE-NA, VA	0.460	2	0.236	0.118	0.393	①
	EOEE-NC, NV	0.460	2	0.236	0.118	0.157	①
	EOEE-ND, VD	0.460	2	0.236	0.118	0.393	①
	EOZE-AFA, SA	0.460	2	0.236	0.118	0.393	①
	EOZE-AFC, SC	0.460	—	0.236	0.118	0.393	①

MODEL 6500
Ford Bobcat, Pinto, Mustang, Capri, Fairmont, Zephyr, Granada, Cougar

Year	Carb. Iden.	Dry Float Level (in.)	Pump Hole Setting	Choke Plate Pulldown (in.)	Fast Idle Cam Linkage (in.)	Dechoke (in.)	Choke Setting
'81	EIZE-RA	0.460	3	0.240	0.120	0.400	—
	EIZE-SA	0.460	3	0.240	0.120	0.400	—
	EIDE-DA	0.460	3	0.240	0.120	0.400	—
	EIDE-EA	0.460	3	0.240	0.120	0.400	—
'82	E2ZE-ARA	.41–.51	2	0.275	0.118	0.393	—
	E2ZE-APA	.41–.51	2	0.275	0.118	0.393	—
	E2ZE-VA	.41–.51	3	0.275	0.118	0.393	—
	E2ZE-ADA	.41–.51	3	0.275	0.118	0.393	—
	E2ZE-ACA	.41–.51	3	0.275	0.118	0.393	—
	E2ZE-UA	.41–.51	3	0.275	0.118	0.393	—

① See underhood decal

Choke Plate Pulldown

THROUGH 1980

1. Remove the choke thermostatic spring cover.
2. Pull the water cover and the thermostatic spring cover assembly or the electric choke assist assembly out of the way.
3. Set the fast idle cam on the second step.
4. Push the diaphragm stem against its stop and insert the specified gauge between the lower edge of the choke valve and the air horn wall.
5. Apply sufficient pressure to the upper edge of the choke valve to take up any slack in the choke linkage.
6. Turn the adjusting screw in or out to adjust the choke plate-to-air horn clearance.

1981–82

NOTE: The following procedure requires the removal of the carburetor and also the choke cap which is retained by two rivets.

1. Remove the carburetor from the engine.
2. Remove the choke cap as follows:
   a. Check the rivets to determine if mandrel is well below the rivit head. If mandrel is within the rivit head thickness, drive it down or out with a $^1/_{16}$ inch diameter tip punch.
   b. With a $^1/_8$ inch diameter drill, drill into the rivet head until the rivet head comes loose from the rivet body. Use light pressure on the drill bit or the rivet will just spin in the hole.
   c. After drilling off the rivet head, drive the remaining rivet out of the hole with a $^1/_8$ inch diameter punch.
   d. Repeat steps (a thru c) to remove the remaining rivet.
3. Remove the plastic dust cover.
4. Place the fast idle adjusting screw on the high step of the fast idle cam.
5. Attach a rubber band to remove the slack from the choke linkage. Push the diaphragm stem back against the stop screw.
6. Using the specified diameter drill check the clearance between the lower edge of the choke plate and the air horn wall.
7. If adjustment is necessary, obtain a replacement kit containing a new choke pulldown diaphragm cover, adjusting screw and cup plug.
8. After installing the adjusting screw in the cover, adjust the pulldown by turning the screw clockwise to decrease and counterclockwise to increase the setting.
9. After making the adjustment, install a new plug in the choke pulldown adjustment access opening.
10. Remove the rubber band and reinstall the choke cap using rivets ($^1/_8$ inch diameter × $^1/_2$ inch long with a $^1/_4$ inch diameter head).

Dechoke (Unloader) Adjustment

Dechoke clearance adjustment is controlled by the fast idle cam adjustment. The figures in the specification chart refer to choke plate clearance between the plate and the air horn wall. Clearance can be measured as follows:

1. Hold the throttle wide open. Remove any slack from the choke linkage by applying pressure to the upper edge of the choke valve.
2. Measure the distance between the lower edge of the choke plate and the air horn wall.
3. Adjust by bending the tab on the fast idle lever where it touches the cam.

Fast Idle Speed

Set the fast idle speed with the fast idle screw positioned on the second step of the fast idle cam and with the engine at operating temperature.

Remove the EGR line at the valve and plug it. If the car is equipped with a spark delay valve, remove the valve and route the distributor advance vacuum signal directly to the distributor advance diaphragm. On all manual transmission models, remove and plug the vacuum line to the distributor. If the distributor also

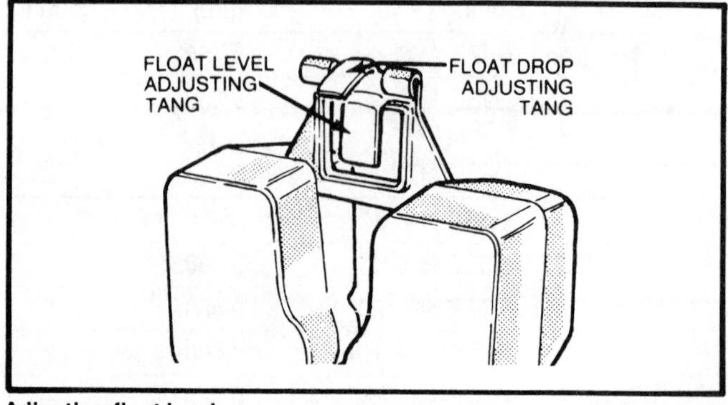

Adjusting float level

has a retard diaphragm, leave the hose connected to it alone. If the engine has a deceleration valve, remove this hose at the carburetor and plug it. Finally, if the car has air conditioning it must be off before adjusting the fast idle.

Float Level Adjustment

With the bowl cover held upside down and the float tang resting lightly on the spring loaded fuel inlet needle, measure the clearance between the edge of the float and the bowl cover. To adjust the level, bend the float tang up or down as required. Adjust both floats equally.

Secondary Throttle Stop Screw

1. Turn the secondary throttle stop screw counterclockwise until the secondary throttle plate seats in its bore.
2. Turn the screw clockwise until it touches the tab on the secondary throttle lever.
3. Add 1/4 turn clockwise for four-cylinder engines.

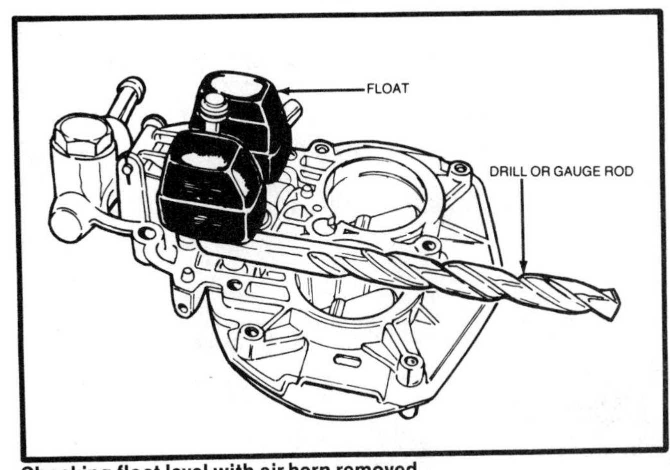

Checking float level with air horn removed

GENERAL MOTORS ELECTRONIC FUEL CONTROL SYSTEM

General Information

Similar in both principle and hardware to the Ford Feedback Carburetor System, the G.M. Electronic Fuel Control System is used in 1978 and 1979 on the Pontiac 151 cubic inch (2.5 liter) four cylinder engine installed in Sunbirds, Starfires, and Monzas sold in California. It is designed to closely regulate the air/fuel ratio through electronic monitoring. A two part catalytic converter oxidizes all three pollutants, but does not have either the mixing chamber or air injection used in the Ford system. Other components in this system include an oxygen sensor which monitors the oxygen content in the exhaust, an Electronic Control Unit (ECU) which receives signals from the oxygen sensor, engine temperature switch, and vacuum input switch and sends a control signal to the vacuum modulator, the vacuum modulator which adjusts the carburetor and/fuel mixture, and a carburetor equipped with feedback diaphragms.

The ECU monitors the voltage output of the oxygen sensor. Lean mixtures reduce voltage, rich mixtures increase voltage. Adjustments of the signal sent to the vacuum modulator are made by the ECU according to the oxygen sensor output. Unlike the Ford system, there is no open loop or closed loop operation. The oxygen sensor input is a constant function. However, the ECU may limit the amount of leanness applied by the vacuum modulator according to signals received from two sources. If the temperature switch indicates that the engine is cold, the ECU allows a slightly richer mixture. If the vacuum input switch indicates low vacuum (heavy engine load), the ECU reduces the rate at which the system goes lean.

BASIC TROUBLESHOOTING

1. Before any tests are made, check the vacuum hoses for leaks, breaks, kinks, or improper connections. Inspect the wiring for breaks, shorts, or fraying. Be sure the electrical connector at the ECU is tight. Disconnect the wire from the vacuum switch (3B), and connect a test light between it and the positive battery terminal. Run the engine at 1500 rpm, with the transmission in Neutral. The test light should go on and off as the vacuum hose is removed and replaced at the switch. If not, replace the switch.
2. Turn the ignition switch to run (engine off).
3. The vacuum modulator should emit a steady clicking sound. If so, go to Step 4. If not, ground one end of a jumper wire to a ground, the other to the brown wire in the modulator connector (5B).
   a. If the modulator clicks once, check the ECU connector for tightness. If it's ok, remove the ECU connector and touch 5A with the jumper wire. If there's no click, there is an open in the brown wire. If it clicks once, replace the ECU.
   b. If the modulator does not click when the brown wire is grounded, connect a test light to a ground and the pink wire at the modulator (1D). If the test light goes on, replace the modulator. If not, there is an open in the pink wire.
4. If a clicking sound is heard, use a T-fitting to attach a vacuum gauge between the center port of the vacuum modulator and the carburetor. Start the engine, allow it to reach operating temperature, and let the engine idle. Automatic transmission should be in Drive (front wheels blocked, parking brake on), manual in Neutral.
   a. If the gauge reads above 7 in. Hg., replace the vacuum modulator.
   b. If the gauge reads 2-4 in. Hg., shift the transmission to Neutral or Park, with an automatic, and increase the engine speed to 3500 rpm. If the reading is still 2-4 in., the system is ok; either the ignition or fuel supply is faulty. If it reads below 2-4 in., the carburetor is faulty.
   c. If the guage reads below 2-4 in., shift to Neutral or Park, with automatic transmission, and increase the engine speed to 3500 rpm. If it now reads 2-4 in., adjust the idle speed to read 2-4 in. at normal idle, according to the emission sticker in the engine compartment. If the reading is still below that figure, go to the next step.
5. Remove the oxygen sensor wire (4B).
   a. If the vacuum gauge reads above 1 in. Hg., disconnect the modulator connector. If the vacuum falls below 1 in., replace the ECU. If it stays above 1 in., replace the vacuum modulator.
   b. If it reads below 1 in., connect a jumper wire from the positive battery terminal to the oxygen sensor terminal (4B). Go to the next step.
6. If the gauge reads below 4 in., go to Step 7. If it reads 4-7 in., leave the jumper connected, and disconnect the vacuum hose at the center port of the modulator. Set the fast idle screw on the high step of the cam and not the engine rpm. Reconnect the hose and note the rpm.
   a. If the engine speed drops 50 rpm or more when the hose is reconnected, replace the oxygen sensor.
   b. If it drops less than 50 rpm the problem is in the carburetor.
7. If after Step 5b the gauge still reads below 4 in., remove the jumper wire and reconnect the oxygen sensor wire. Ground the jumper wire and connect the other end to the brown wire at the modulator connector (5B).
   a. If the gauge reads 4-7 in., go to the next step.

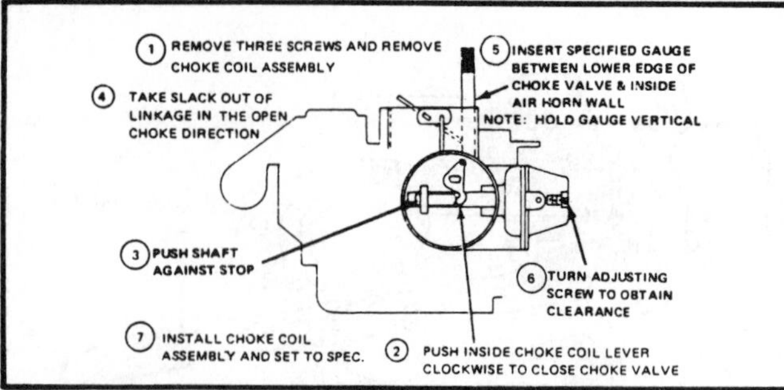

Vacuum break adjustment—1979 and earlier models

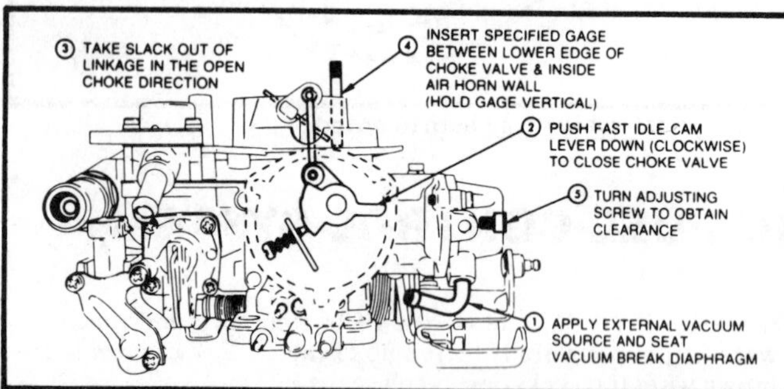

Vacuum break adjustment—1980 and later Holley 6510C feedback carburetor

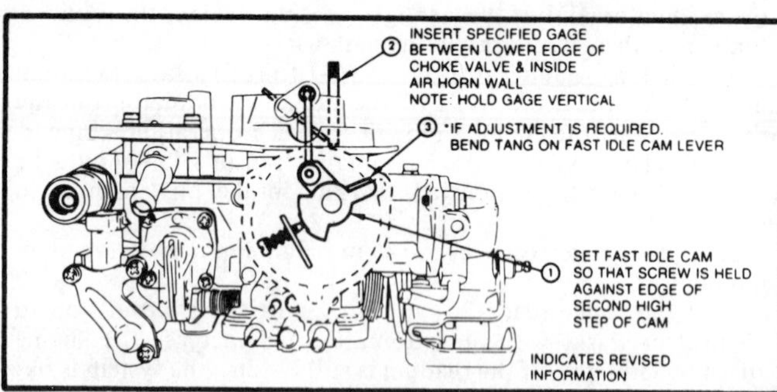

Fast idle cam adjustment—Holley 6510C models

b. If the gauge reads below 4 in., remove and plug the modulator vacuum hose from the carburetor. If the reading is still below 4 in., replace the modulator. If it is above 4 in., the problem is in the carburetor.

8. If the gauge reads 4–7 in. in Step 7a. remove the jumper wire at the modulator, and ground the engine temperature switch wire (2B).

a. If the vacuum gauge reads 4–7 in., replace the temperature switch.

b. If it reads under 4 in., go to the ECU Wiring Harness Continuity Check (diagram). If after the check the reading is still below 4 in., repair the ECU harness. If not, replace the ECU.

Holley 6510-C Carburetor

ADJUSTMENTS

NOTE: The 6510-C is used on subcompact GM cars with the 4-151 engine through 1979. In 1980 and later, it is used only on the Chevette and T-1000.

Vacuum Break Adjustment

THROUGH 1979

1. Remove the choke coil assembly.

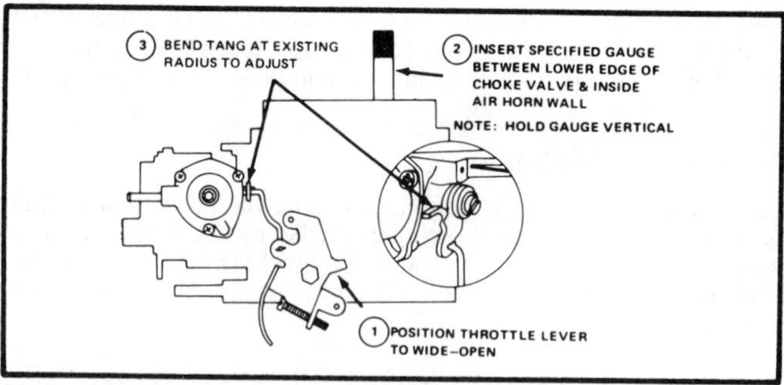

Choke unloader adjustment—typical

2. Push the choke coil lever clockwise to close the choke valve.
3. Push the choke shaft against its stop.
4. Take the slack out of the linkage, in the open direction.
5. Insert the specified gauge between the lower edge of the choke plate and the air horn wall. Turn the adjusting screw on the diaphragm housing to adjust.

1980 AND LATER

1. Attach a hand vacuum pump to the vacuum break diaphragm. Apply vacuum until the diaphragm is seated.
2. Push the fast idle cam lever down to close the choke plate.
3. Take the slack out of the linkage in the open choke position.
4. Insert the specified gauge between the lower edge of the choke plate and the air horn wall.
5. If the clearance is incorrect, turn the screw in the end of the diaphragm to adjust.

Fast Idle Cam Adjustment

1. Set the fast idle cam so that the screw is on the second highest step of the fast idle cam.
2. Insert the specified gauge between the lower edge of the choke valve and the air horn wall.
3. Bend the tang on the arm to adjust.

Unloader Adjustment

1. Place the throttle in the wide open position.
2. Insert a 0.350 inch gauge between the lower edge of the choke valve and the air horn wall.
3. Bend the tang on the choke arm to adjust.

Choke Cap Setting

1978-79 ONLY

1. Loosen the retaining screws.
2. Make sure that the choke coil lever is located inside the coil tang.
3. Turn the cap to the specified setting.
4. Tighten the retaining screws.

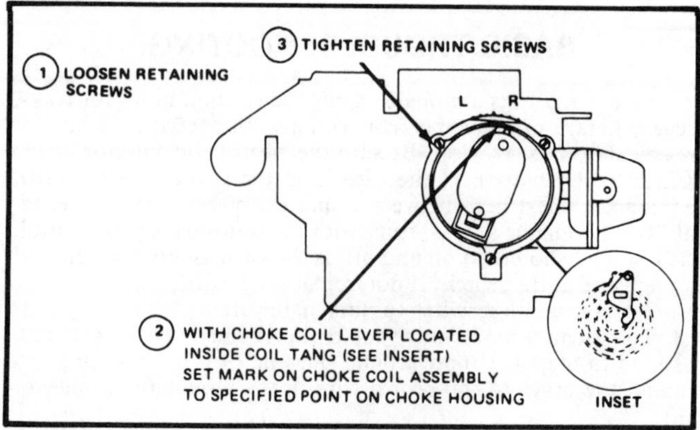

Choke cap adjustment—1979 and earlier models

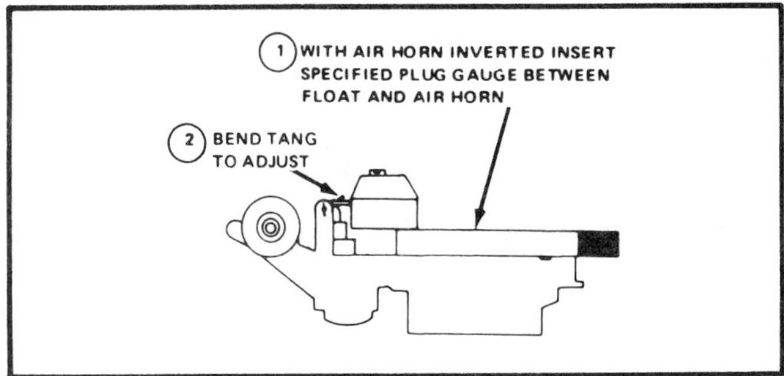

Float level adjustment—typical

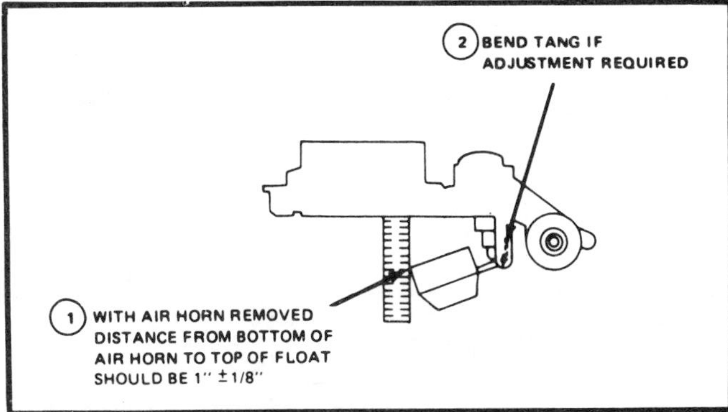

Float drop adjustment—1979 and earlier models

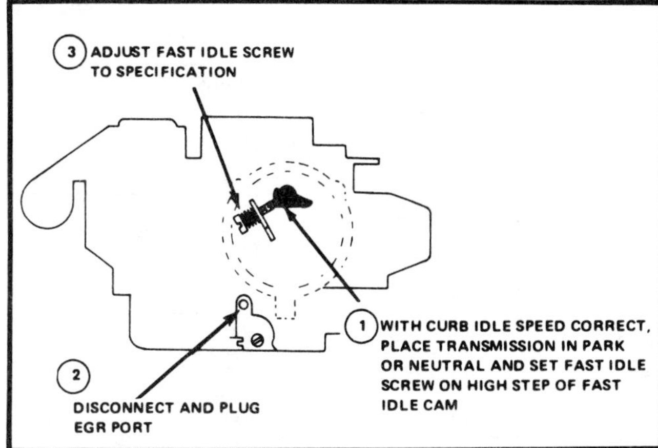

Fast idle speed adjustment—typical

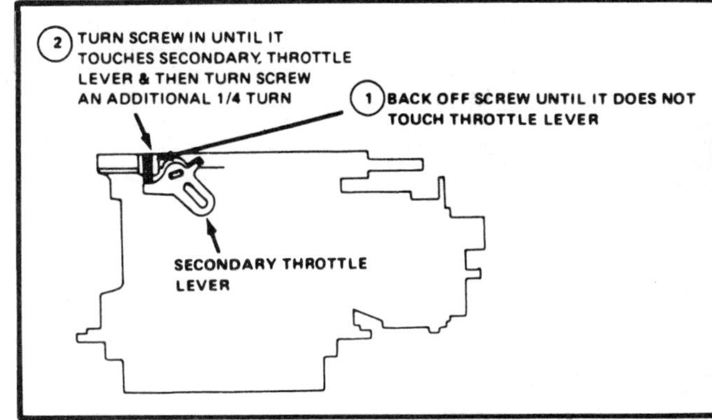

Secondary throttle stop screw adjustment—typical

Fast Idle Adjustment

1. With the curb idle speed correct, place the fast idle screw on the highest cam step and adjust to the specified rpm.

NOTE: The EGR line must be disconnected and plugged.

Float Level Adjustment

1. Remove and invert the air horn.
2. Place the specified gauge between the air horn and the float.
3. If necessary, bend the float arm tang to adjust.

Float Drop Setting

THROUGH 1979 ONLY

1. Hold the air horn right side up. The distance between the bottom of the air horn and the top of the float should be 1 inch.
2. If necessary, bend the tang on the side of the float arm support, to adjust.

Secondary Throttle Stop Screw Adjustment

1. Back off the screw until it does not touch the lever.
2. Turn the screw in until it touches the lever, then turn it an additional 1/4 turn.

MODEL 6510-C
General Motors Corporation

Year	Part Number	Vacuum Break Adjustment (in.)	Fast Idle Cam Adjustment (in.)	Unloader Adjustment (in.)	Fast Idle Adjustment (rpm)	Float Level Adjustment (in.)	Choke Setting
'78	10001056, 10001058	.325	.150	.350	2400	.520	1 Rich
'79	10008489, 10008490	.250	.150	.350	2400	.520	1 Rich
	10008491, 10008492	.250	.150	.350	2200	.520	2 Rich
	10009973, 10009974	.275	.150	.350	2400	.520	2 Rich

MODEL 6510-C
General Motors Corporation

Year	Part Number	Vacuum Break Adjustment (in.)	Fast Idle Cam Adjustment (in.)	Unloader Adjustment (in.)	Fast Idle Adjustment (rpm)	Float Level Adjustment (in.)	Choke Setting
'80	All w/manual	.275	.130	.350	2600	.500	Fixed
	All w/automatic	.300	.130	.350	2500	.500	Fixed
'81	14004768	.300	.130	.350	①	.500	Fixed
	14004769	.300	.130	.350	①	.500	Fixed
	14004770	.300	.130	.350	①	.500	Fixed
	14004771	.300	.130	.350	①	.500	Fixed
	14004777	.300	.130	.350	①	.500	Fixed
'82	14032364	.270	.080	.350	①	.500	Fixed
	14032365	.270	.080	.350	①	.500	Fixed
	14032366	.270	.080	.350	①	.500	Fixed
	14032367	.270	.080	.350	①	.500	Fixed
	14032368	.270	.080	.350	①	.500	Fixed
	14032369	.270	.080	.350	①	.500	Fixed
	14032370	.270	.080	.350	①	.500	Fixed
	14032371	.270	.080	.350	①	.500	Fixed
	14033392	.270	.080	.350	①	.500	Fixed
	14033393	.270	.080	.350	①	.500	Fixed
	14047072	.270	.080	.350	①	.500	Fixed
'83	14048827	.270	.080	.350	①	.500	Fixed
	14048828	.300	.080	.350	①	.500	Fixed
	14048829	.270	.080	.350	①	.500	Fixed
'84	14068690	.270	.080	.350	①	.500	Fixed
	14068691	.270	.080	.350	①	.500	Fixed
	14068692	.300	.080	.350	①	.500	Fixed

① See underhood decal

GENERAL MOTORS COMPUTER CONTROLLED CATALYTIC CONVERTER (C-4) SYSTEM, AND COMPUTER COMMAND CONTROL (CCC) SYSTEM

General Information

The GM designed Computer Controlled Catalytic Converter System (C-4) System), introduced in 1979 and used on GM cars through 1980, is a revised version of the 1978-79 Electronic Fuel Control System (although parts are not interchangeable between the systems). The C-4 System primarily maintains the ideal air/fuel ratio at which the catalytic converter is most effective. Some versions of the system also control ignition timing of the distributor.

The Computer Command Control System (CCC System), introduced on some 1980 California models and used on all 1981 and later carbureted car lines, is an expansion of the C-4 System. The CCC System monitors up to fifteen engine/vehicle operating conditions which it uses to control up to nine engine and emission control systems. In addition to maintaining the ideal air/fuel ratio for the catalytic converter and adjusting ignition timing, the CCC System also controls the Air Management System so that the catalytic converter can operate at the highest efficiency possible. The system also controls the lockup on the transmission torque converter clutch (certain automatic transmission models only), adjusts idle speed over a wide range of conditions, purges the evaporative emissions charcoal canister, controls the EGR valve operation and operates the early fuel evaporative (EFE) system. Not all engines use all of the above sub-systems.

There are two operation modes for both the C–4 System and the CCC System: closed loop and open loop fuel control. Closed loop fuel control means the oxygen sensor is controlling the carburetor's air/fuel mixture ratio. Under open loop fuel control operating conditions (wide open trottle, engine and/or oxygen sensor cold), the oxygen sensor has no effect on the air/fuel mixture.

NOTE: On some engines, the oxygen sensor will cool off while the engine is idling, putting the system into open loop operation. To restore closed loop operation, run the engine at part throttle and accelerate from idle to part throttle a few times.

COMPUTER CONTROLLED CATALYTIC CONVERTER (C-4) SYSTEM OPERATION

Major components of the system include an Electronic Control Module (ECM), an oxygen sensor, and electronically controlled variable-mixture carburetor, and a three-way oxidation-reduction catalytic converter.

The oxygen sensor generates a voltage which varies with exhaust gas oxygen content. Lean mixtures (more oxygen) reduce voltage; rich mixtures (less oxygen) increase voltage. Voltage output is sent to the ECM.

An engine temperature sensor installed in the engine coolant outlet monitors coolant temperatures. Vacuum control switches and throttle position sensors also monitor engine conditions and supply signals to the ECM.

The Electronic Control Module (ECM) monitors the voltage input the oxygen sensor along with information from other input signals. It processes these signals and generates a control signal sent to the carburetor. The control signal cycles between ON (lean command) and OFF (rich command). The amount of ON and OFF time is a function of the input voltage sent to the ECM by the oxygen sensor. The ECM has a calibration unit called a PROM (Programable Read Only Memory) which contains the specific instructions for a given engine application. In other words, the PROM unit is specifically programed or "tailor made" for the system in which it is installed. The PROM assembly is a replacable component which plugs into a socket on the ECM and requires a special tool for removal and installation.

NOTE: To prevent ECM damage on 1980 and later Cadillac C-4 and CCC equipped vehicles, the power supply feeding the ECM must not be interrupted with the ignition switch in the run, start or ACC position. The ignition switch must be placed in the *off* position when performing the following service operations:

1. Disconnecting or connecting either battery cable.

2. Removing or replacing the fuse providing continuous battery power to the ECM.

3. Disconnecting or connecting any ECM connectors.

4. Disconnecting (unless the engine is running) or connecting jumper cables (the jumper polarity must be correct, as even a momentary reversal of cable polarity may cause ECM damage).

5. Disconnecting or connecting a battery charger (it is recommended that the battery be charged with both battery cables disconnected using the battery side terminal adapter, AC Delco ST-1201 or equivalent).

On some 231 cu in. V6 engines, the ECM controls the Electronic Spark Timing System (EST), AIR control system, and on the Turbo-charged 231 cu in. C–4 System it controls the early fuel evaporative system (EFE) and the EGR valve control (on some models). On some 350 V8 engines, the ECM controls the electronic module retard (EMR) system, which retards the engine timing 10 degrees during certain engine operations to reduce the exhaust emissions.

NOTE: Electronic Spark Timing (EST) allows continuous spark timing adjustments to be made by the ECM. Engines with EST can easily be identified by the absence of vacuum and mechanical spark advance mechanisms on the distributor. Engines with EMR systems may be recognized by the presence of five connectors, instead of the HEI module's usual four.

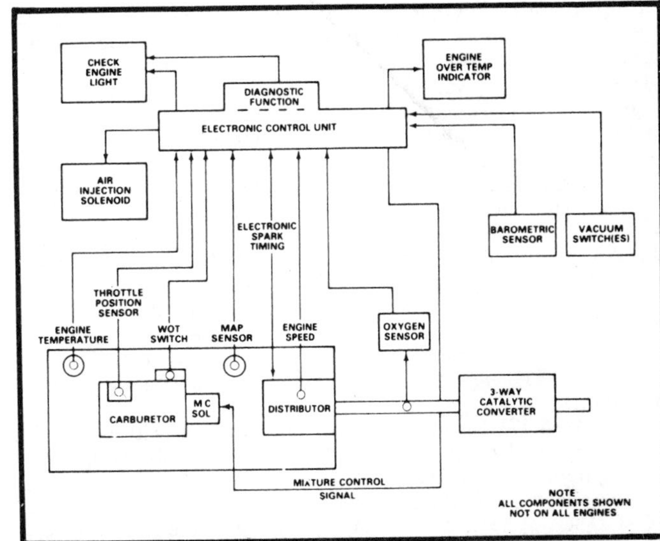

Schematic of GM Computer Controlled Catalytic Converter (C4) system

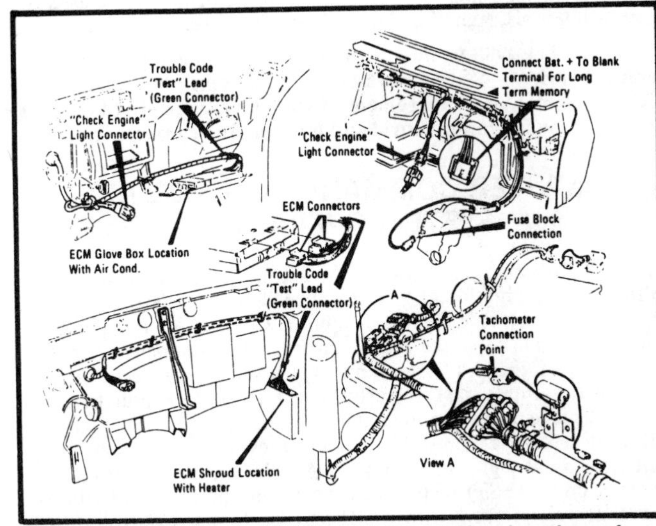

Typical 1979 C4 system wiring harness layout. Note locations of connectors, test joints and diagnostic terminal

To maintain good idle and driveability under all conditions, other input signals are used to modify the ECM output signal. Besides the sensors and switches already mentioned, these input signals include the manifold absolute pressure (MAP) or vacuum sensors and the barometric pressure (BARO) sensor. The MAP or vacuum sensors sense changes in manifold vacuum, while the BARO sensor senses changes in barometric presure. One important function of the BARO sensor is the maintenance of good engine performance at various altitudes. These sensors act as throttle position sensors on some engines. See the following paragraph for description.

A Rochester Dualjet carburetor is used with the C–4 System. It may be an E2SE, E2ME, E4MC or E4ME model, depending on engine application. An electronically operated mixture control solenoid is installed in the carburetor float bowl. The solenoid controls the air/fuel mixture metered to the idle and main metering systems. Air metering to the idle system is controlled by an idle air bleed valve. It follows the movement of the mixture solenoid to control the amount of air bled into the idle system, enriching or leaning out the mixture as appropriate. Air/fuel mixture enrichment occurs when the fuel valve is open and the air bleed is closed.

All cycling of this system, which occurs ten times per second, is controlled by the ECM. A throttle position switch informs the ECM of open or closed throttle operation. A number of different switches are used, varying with application. The four cylinder engine (151 cu. in.) uses two vacuum switches to sense open throttle and closed throttle operation. The V6 engines (except the 231 cu. in. turbo V6) use two pressure sensors–MAP (Manifold Absolute Pressure) and BARO (Barometric Pressure)–as well as a throttle-actuated wide open throttle switch mounted in a bracket on the side of the float bowl. The 231 cu in. turbo V6, and V8 engines, use a throttle position sensor mounted in the carburetor bowl cover under the accelerator pump arm. When the ECM receives a signal from the throttle switch, indicating a change of position, it immediately searches its memory for the last set of operating conditions that resulted in an ideal air/fuel ratio, and shifts to that set of conditions. The memory is continually updated during normal operation.

Some 1980 173 cu in. V6 engines are equipped with a Pulsair control solenoid which is operated by the ECM. Likewise, many C-4 equipped engines with AIR systems (Air Injection Reaction systems) have an AIR system diverter solenoid controlled by the ECM. These systems are similar in function to the AIR Management system used in the CCC System. See below for information. Most C-4 Systems include a maintenance reminder flag connected to the odometer which becomes visible in the instrument cluster at regular intervals, signaling the need for oxygen sensor replacement.

NOTE: The 1980 Cutlass with 260 cu in. V8 engine is equipped with a hybrid C-4 System which includes some functions of the CCC System (Air Management, EGR valve control, Idle speed control, canister purge control and transmission converter clutch).

COMPUTER COMMAND CONTROL (CCC) SYSTEM OPERATION

The CCC has many components in common with the C-4 system (although they should not be interchanged between systems. These include the Electronic Control Module (ECM), which is capable of monitoring and adjusting more sensors and components than the ECM used on the C-4 System, an oxygen sensor, an electronically controlled variable-mixture carburetor, a three way catalytic converter, throttle position and coolant sensors, a barometric pressure (BARO) sensor, a manifold absolute pressure (MAP) sensor, a "check engine" light on the instrument cluster, and an Electronic Spark Timing (EST) distributor, which on some engines (turbocharged) is equipped with an Electronic Spark Control (ESC) which retards ignition spark under some conditions (detonation, etc.).

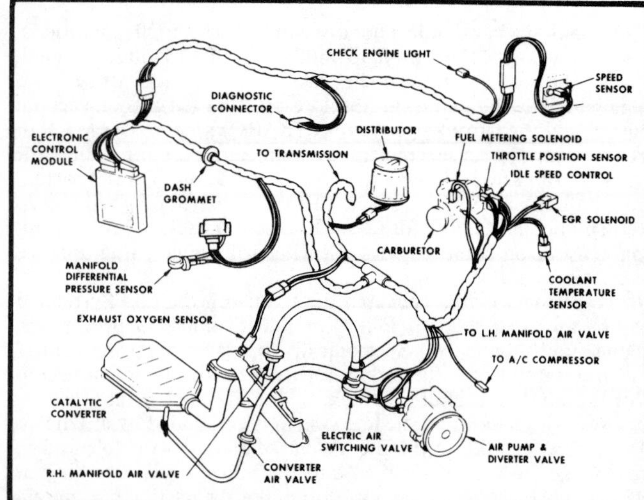

Components of GM Computer Command Control (CCC) system

Components used almost exclusively by the CCC System include the Air Injection Reaction (AIR) Management System, charcoal canister purge solenoid, EGR valve control, vehicle speed sensor located in the instrument cluster), transmission torque converter clutch solenoid (automatic transmission models only), idle speed control, and early fuel evaporative (EFE) system.

See the operation descriptions under C-4 System for those components (except the ECM) the CCC System shares with the C-4 System.

The CCC System ECM, in addition to monitoring sensors and sending a control signal to the carburetor, also control the following components or sub-systems: charcoal canister purge, AIR Management System, idle speed control, automatic transmission converter lockup, distributor ignition timing, EGR valve control, EFE control, and the air conditioner compressor clutch operation. The CCC ECM is equipped with a PROM assembly similar to the one used in the C-4 ECM. See above for description.

The AIR Management System is an emission control which provides additional oxygen either to the catalyst or the cylinder head ports (in some cases exhaust manifold). An AIR Management System, composed of an air switching valve and/or an air control valve, controls the air pump flow and is itself controlled by the ECM. A complete description of the AIR system is given towards the front of this unit repair section. The major difference between the CCC AIR System and the systems used on other cars is that the flow of air from the air pump is controlled electrically by the ECM, rather than by vacuum signal.

The charcoal canister purge control is an electrically operated solenoid valve controlled by the ECM. When energized, the purge control solenoid blocks vacuum from reaching the canister purge valve. When the ECM de-energizes the purge control solenoid, vacuum is allowed to reach the canister and operate the purge valve. This releases the fuel vapors collected in the canister into the induction system.

The EGR valve control solenoid is activated by the ECM in similar fashion to the canister purge solenoid. When the engine is cold, the ECM energizes the solenoid, which blocks the vacuum signal to the EGR valve. When the engine is warm, the ECM de-energizes the solenoid and the vacuum signal is allowed to reach and activate the EGR valve.

The Transmission Converter Clutch (TCC) lock is controlled by the ECM through an electrical solenoid in the automatic transmission. When the vehicle speed sensor in the instrument panel signals the ECM that the vehicle has reached the correct speed, the ECM energizes the solenoid which allows the torque converter to mechanically couple the engine to the transmission. When the brake pedal is pushed or during deceleration, passing, etc., the ECM returns the transmission to fluid drive.

The idle speed control adjusts the idle speed to load conditions, and will lower the idle speed under no-load or low-load conditions to conserve gasoline.

The Early Fuel Evaporative (EFE) system is used on some engines to provide rapid heat to the engine induction system to promote smooth start-up and operation. There are two types of system: vacuum servo and electrically heated. They use different means to achieve the same end, which is to pre-heat the incoming air/fuel mixture. They are controlled by the ECM.

BASIC TROUBLESHOOTING

NOTE: The following explains how to activate the Trouble Code signal light in the instrument cluster and gives an explanation of what each code means. This is not a full C-4 or CCC System troubleshooting and isolation procedure.

Before suspecting the C-4 or CCC System or any of its components as faulty, check the ignition system including distributor, timing, spark plugs and wires. Check the engine compression, air cleaner, and emission control components not controlled by the ECM. Also check the intake manifold, vacuum hoses and hose connectors for leaks and the carburetor bolts for tightness.

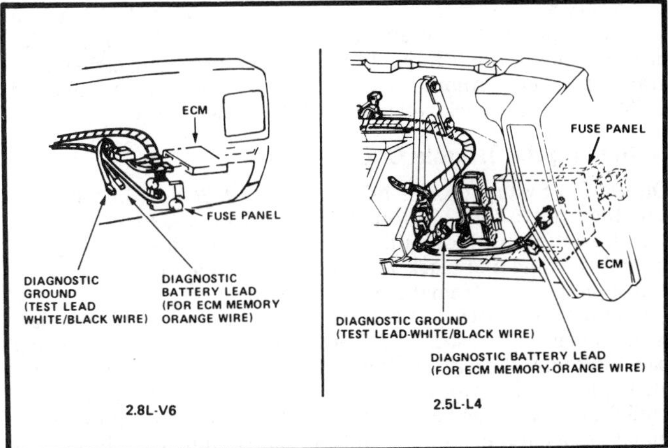

Typical diagnostic test lead locations—1980 models

The following symptoms could indicate a possible problem with the C-4 or CCC System.

1. Detonation
2. Stalls or rough idle-cold
3. Stalls or rough idle-hot
4. Missing
5. Hesitation
6. Surges
7. Poor gasoline mileage
8. Sluggish or spongy performance
9. Hard starting-cold
10. Objectionable exhaust odors (that "rotten egg" smell)
12. Cuts out
13. Improper idle speed (CCC System and C-4 equipped 1980 Cutlass with 260 cu. in. engine only)

As a bulb and system check, the "Check Engine" light will come on when the ignition switch is turned to the ON position but the engine is not started.

The "Check Engine" light will also produce the trouble code or codes by a series of flashes which translate as follows. When the diagnostic test lead (C-4) or terminal (CCC) under the dash is grounded, with the ignition in the ON position and the engine not running, the "Check Engine" light will flash once, pause, then flash twice in rapid succession. This is a code 12, which indicates that the diagnostic system is working. After a longer pause, the code 12 will repeat itself two more times. The cycle will then repeat itself until the engine is started or the ignition is turned off.

NOTE: The C-4 equipped 1980 Cutlass with 260 cu in. V8 engine has a test terminal similar to the kind used on the CCC System.

When the engine is started, the "Check Engine" light will remain on for a few seconds, then turn off. If the "Check Engine" light remains on, the self-diagnostic system has detected a problem. If the test lead (C-4) or test terminal (CCC) is then grounded, the trouble code will flash three times. If more than one problem is found, each trouble code will flash three times. Trouble codes will flash in numerical order (lowest code number to highest). The trouble codes series will repeat as long as the test lead or terminal is grouned.

A trouble code indicates a problem with a given circuit. For example, trouble code 14 indicates a problem in the cooling sensor circuit. This includes the coolant sensor, its electrical harness, and the Electronic Control Module (ECM).

Since the self-diagnostic system cannot diagnose every possible fault in the system, the absence of a trouble code does not mean the system is trouble-free. To determine problems within the system which do not activate a trouble code, a system performance check must be made. This job should be left to a qualified technician.

In the case of an intermittant fault in the system, the "Check Engine" light will go out when the fault goes away, but the trouble code will remain in the memory of the ECM. Therefore, if a trouble code can be obtained even though the "Check Engine" light is not on, the trouble code must be evaluated. It must be determined if the fault is intermittant or if the engine must be at certain operating conditions (under load, etc.) before the "Check Engine" light will come on. Some trouble codes will not be recorded in the ECM until the engine has been operated at part throttle for about 5 to 18 minutes.

On the C-4 System, the ECM erases all trouble codes every time the ignition is turned off. In the case of intermittent faults, a long term memory is desirable. This can be produced by connecting the orange connector/lead from terminal "S" of the ECM directly to the battery (or to a "hot" fuse panel terminal). This terminal must

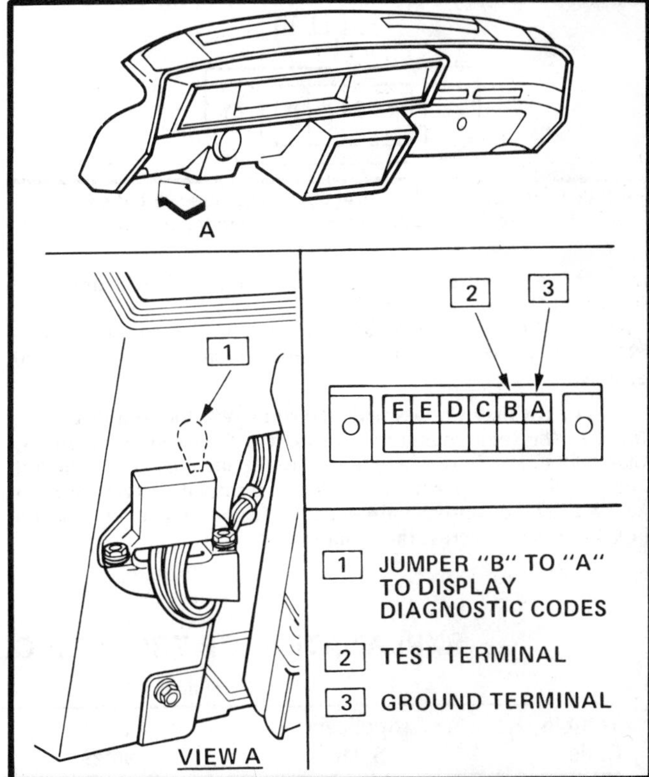

Typical 1981 and later CCC test terminal location. Ground the test terminal to activate the trouble code readout

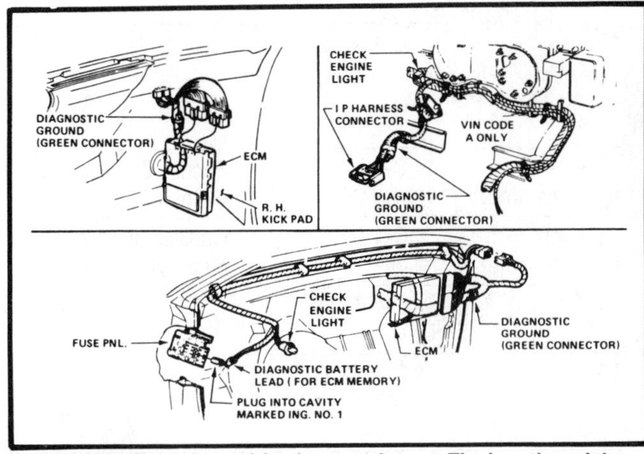

Typical 1980 C4 system wiring harness layout. The location of the diagnostic test terminal depends on the position of the on-board computer (ECM) and the body style

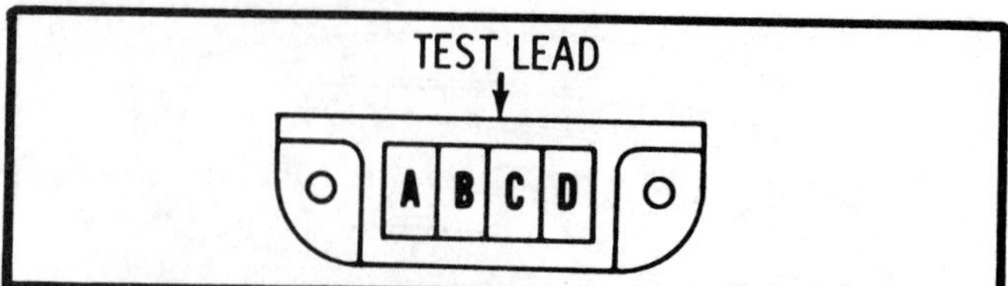

The 1980 Oldsmobile 260 cu. in. (4.3L) V8 equipped with the C4 system uses a test terminal similar to CCC-equipped models

be disconnected after diagnosis is complete or it will drain the battery.

On the CCC System, a trouble code will be stored until terminal "R" of the ECM has been disconnected from the battery for 10 seconds.

NOTE: On 1980 Cutlass with 260 cu in. V8, the trouble code is stored in the same manner as on the CCC System. In addition, some 1980 Buicks have a long term constant memory similar to that used on the CCC System. In which case terminal S (terminal R on the 3.8 Liter V6) must be disconnected in the same manner as on the CCC System to erase the memory.

An easy way to erase the computer memory on the CCC System is to disconnect the battery terminals from the battery. If this method is used, don't forget to reset clocks and electronic preprogramable radios. Another method is to remove the fuse marked ECM in the fuse panel. Not all models have such a fuse.

Activating the Trouble Code

On the C-4 System (except 1980 Cutlass with 260 cu. in V8), activate the trouble code by grounding the trouble code test lead. Use the illustrations to locate the test lead under the instrument panel (usually a white and black wire or a wire with a green connector). Run a jumper wire from the lead to ground.

On the CCC System and the C-4 System used on the 1980 Cutlass with 260 cu. in V8, locate the test terminal under the instrument panel. Ground the test lead. On many systems, the test lead is situated side by side with a ground terminal. In addition, on some models, the partition between the test terminal and the ground terminal has a cut out section so that a spade terminal can be used to connect the two terminals.

NOTE: Ground the test lead or terminal according to the instructions given in "Basic Troubleshooting", above.

EXPLANATION OF TROUBLE CODES GM C-4 AND CCC SYSTEMS

(Ground test lead or terminal AFTER engine is running.)

Trouble Code	Applicable System	Notes	Possible Problem Area
12	C-4, CCC		No tachometer or reference signal to computer (ECM). This code will only be present while a fault exsists, and will not be stored if the problem is intermittent.
13	C-4, CCC		Oxygen sensor circuit. The engine must run for about five minutes (eighteen on C-4 equipped 231 cu in. V6) at part throttle (and under road load—CCC equipped cars) before this code will show.
13 & 14 (at same time)	C-4	Except Cadillac and 171 cu in. V6	See code 43.
13 & 43 (at same time)	C-4	Cadillac and 171 cu in. V6	See code 43.
14	C-4, CCC		Shorted coolant sensor circuit. The engine has to run 2 minutes before this code will show.
15	C-4, CCC		Open coolant sensor circuit. The engine has to operate for about five minutes (18 minutes for C-4 equipped 231 cu in. V6) at part throttle (some models) before this code will show.
21	C-4		Shorted wide open throttle switch and/or open closed-throttle switch circuit (when used).
	C-4, CCC		Throttle position sensor circuit. The engine must be run up to 10 seconds (25 seconds—CCC System) below 800 rpm before this code will show.
21 & 22 (at same time)	C-4		Grounded wide open throttle switch circuit (231 cu in. V6, 151 cu in. 4 cylinder).
22	C-4		Grounded closed throttle or wide open throttle switch circuit (231 cu in. V6, 151 cu in. 4 cylinder).
23	C-4, CCC		Open or grounded carburetor mixture control (M/C) solenoid circuit.
24	CCC		Vehicle speed sensor (VSS) circuit. The car must operate up to five minutes at road speed before this code will show.

EXPLANATION OF TROUBLE CODES GM C-4 AND CCC SYSTEMS

(Ground test lead or terminal AFTER engine is running.)

Trouble Code	Applicable System	Notes	Possible Problem Area
32	C-4, CCC		Barometric pressure sensor (BARO) circuit output low.
32 & 55 (at same time)	C-4		Grounded +8V terminal or V(REF) terminal for barometric pressure sensor (BARO), or faulty ECM computer.
34	C-4	Except 1980 260 cu in. Cutlass	Manifold absolute pressure (MAP) sensor output high (after ten seconds and below 800 rpm).
34	CCC	Including 1980 260 cu in. Cutlass	Manifold absolute pressure (MAP) sensor circuit or vacuum sensor circuit. The engine must run up to five minutes below 800 RPM before this code will set.
35	CCC		Idle speed control (ISC) switch circuit shorted (over ½ throttle for over two seconds).
41	CCC		No distributor reference pulses to the ECM at specified engine vacuum. This code will store in memory.
42	CCC		Electronic spark timing (EST) bypass circuit grounded.
43	C-4		Throttle position sensor adjustment (on some models, engine must run at part throttle up to ten seconds before this code will set).
44	C-4, CCC		Lean oxygen sensor indication. The engine must run up to five minutes in closed loop (oxygen sensor adjusting carburetor mixture), at part throttle and under road load (drive car) before this code will set.

ROCHESTER FEEDBACK CARBURETORS

Model Identification

General Motors Rochester carburetors are identified by their model number. The first number indicates the number of barrels, while one of the last letters indicates the type of choke used. These are V for the manifold mounted choke coil, C for the choke coil mounted on the carburetor, and E for electric choke, also mounted on the carburetor. Model numbers ending in A indicate an altitude-compensating carburetor.

ANGLE DEGREE TOOL

An angle degree tool is recommended by Rochester Products Division, for use to confirm adjustments to the choke valve and related linkages on their late model two and four barrel carburetors, in place of the plug type gauges.

Decimal and degree conversion charts are provided for use by technicians who have access to an angle gauge and not plug gauges. It must be remembered that the relationship between the decimal and the angle readings are not exact, due to manufacturers tolerances.

To use the angle gauge, rotate the degree scale until zero (0) is opposite the pointer. With the choke valve completely closed, place the gauge magnet squarely on top of the choke valve and rotate the bubble until it is centered. Make the necessary adjustments to have the choke valve at the specified degree angle opening as read from the degree angle tool.

NOTE: The carburetor may be off the engine for adjustments. Be sure the carburetor is held firmly during the use of the angle gauge.

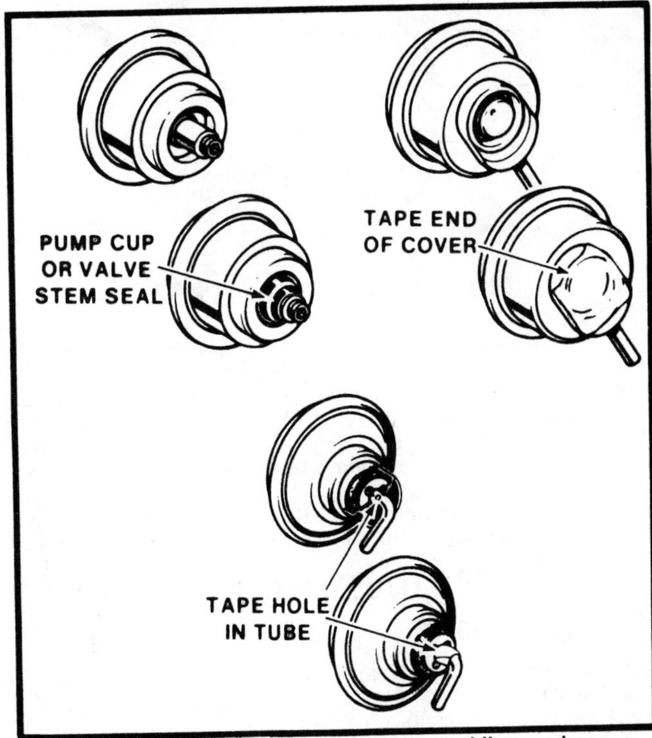

Plugging air bleed holes in vacuum break assemblies used on Rochester E2SE feedback carburetors

ANGLE DEGREE TO DECIMAL CONVERSION
Model M2MC, M2ME and M4MC Carburetor

Angle Degrees	Decimal Equiv. Top of Valve	Angle Degrees	Decimal Equiv. Top of Valve
5	.023	33	.203
6	.028	34	.211
7	.033	35	.220
8	.038	36	.227
9	.043	37	.234
10	.049	38	.243
11	.054	39	.251
12	.060	40	.260
13	.066	41	.269
14	.071	42	.277
15	.077	43	.287
16	.083	44	.295
17	.090	45	.304
18	.096	46	.314

ANGLE DEGREE TO DECIMAL CONVERSION
Model M2MC, M2ME and M4MC Carburetor

Angle Degrees	Decimal Equiv. Top of Valve	Angle Degrees	Decimal Equiv. Top of Valve
19	.103	47	.322
20	.110	48	.332
21	.117	49	.341
22	.123	50	.350
23	.129	51	.360
24	.136	52	.370
25	.142	53	.379
26	.149	54	.388
27	.157	55	.400
28	.164	56	.408
29	.171	57	.418
30	.179	58	.428
31	.187	59	.439
32	.195	60	.449

Models 2SE and E2SE

The Rochester 2SE and E2SE Varajet II carburetors are two barrel, two stage down-draft units. Most carburetor components are aluminum, although a zinc choke housing is used on four cylinder engines installed in 1980 models. The E2SE is used both in conventional installations and in the Computer Controlled Catalytic Converter System. In that installation the E2SE is equipped with an electrically operated mixture control solenoid, controlled by the Electronic Control Module. The 2SE and E2SE are also used on the AMC four cylinder in 1980–83.

ADJUSTMENTS

Float Adjustment

1. Remove the air horn from the throttle body
2. Use your fingers to hold the retainer in place, and to push the float down into light contact with the needle.
3. Measure the distance from the toe of the float (furthest from the hinge) to the top of the carburetor (gasket removed).

ANGLE DEGREE TO DECIMAL CONVERSION
Model 4MV Carburetor

Angle Degrees	Decimal Equiv. Top of Valve	Angle Degrees	Decimal Equiv. Top of Valve
5	.019	33	.158
6	.022	34	.164
7	.026	35	.171
8	.030	36	.178
9	.034	37	.184
10	.038	38	.190
11	.042	39	.197
12	.047	40	.204
13	.051	41	.211
14	.056	42	.217
15	.060	43	.225
16	.065	44	.231
17	.070	45	.239
18	.075	46	.246
19	.080	47	.253
20	.085	48	.260
21	.090	49	.268
22	.095	50	.275
23	.101	51	.283
24	.106	52	.291
25	.112	53	.299
26	.117	54	.306
27	.123	55	.314
28	.128	56	.322
29	.134	57	.329
30	.140	58	.337
31	.146	59	.345
32	.152	60	.353

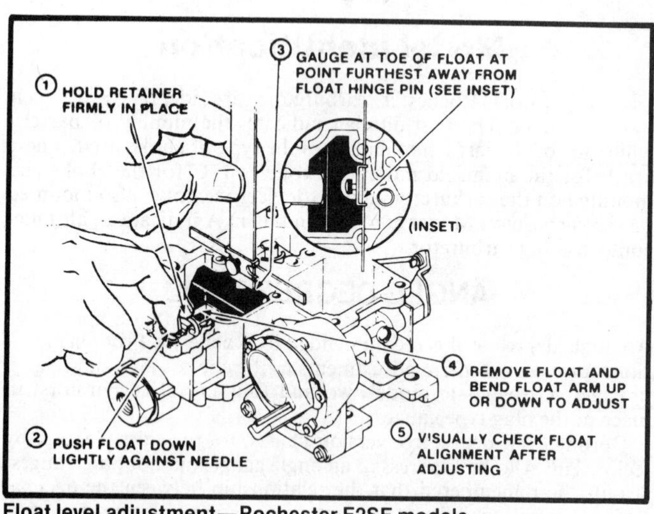

③ GAUGE AT TOE OF FLOAT AT POINT FURTHEST AWAY FROM FLOAT HINGE PIN (SEE INSET)
① HOLD RETAINER FIRMLY IN PLACE
(INSET)
④ REMOVE FLOAT AND BEND FLOAT ARM UP OR DOWN TO ADJUST
⑤ VISUALLY CHECK FLOAT ALIGNMENT AFTER ADJUSTING
② PUSH FLOAT DOWN LIGHTLY AGAINST NEEDLE

Float level adjustment—Rochester E2SE models

4. To adjust, remove the float and gently bend the arm to specification. After adjustment, check the float alignment in the chamber.

NOTE: Some models have a float stabilizer spring. If used, remove the spring with float. Use care when removing.

Pump Adjusltment

1. With the throttle closed and the fast idle screw off the steps of the fast idle cam, measure the distance from the air horn casting to the top of the pump stem.

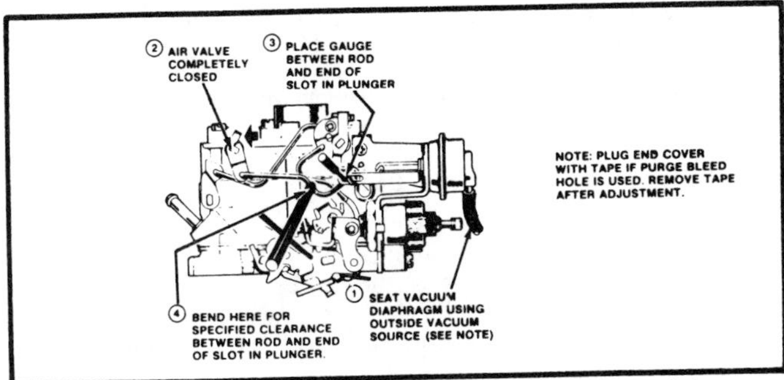

Air valve rod adjustment—1980 GM and 1980-82 AMC models

2. To adjust, remove the retaining screw and washer and remove the pump lever. Bend the end of the lever to correct the stem height. Do not twist the lever or bend it sideways.

3. Install the lever, washer and screw and check the adjustment. When correct, open and close the throttle a few times to check the linkage movement and alignment.

NOTE: No pump adjustment is required on 1981 and later models.

Fast Idle Adjustment

1. Set the ignition timing and curb idle speed, and disconnect and plug hoses as directed on the emission control decal.

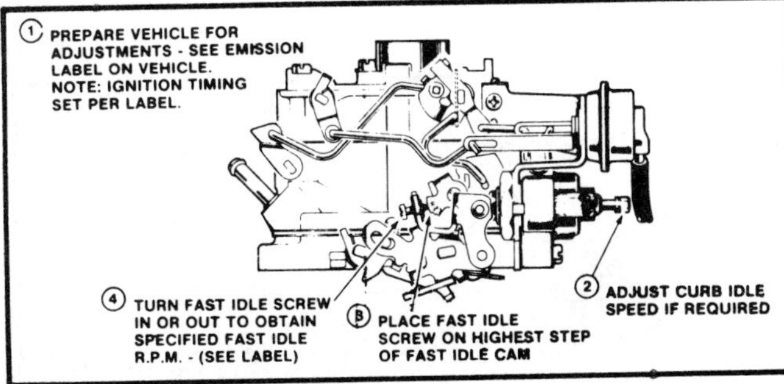

Fast idle adjustment—typical

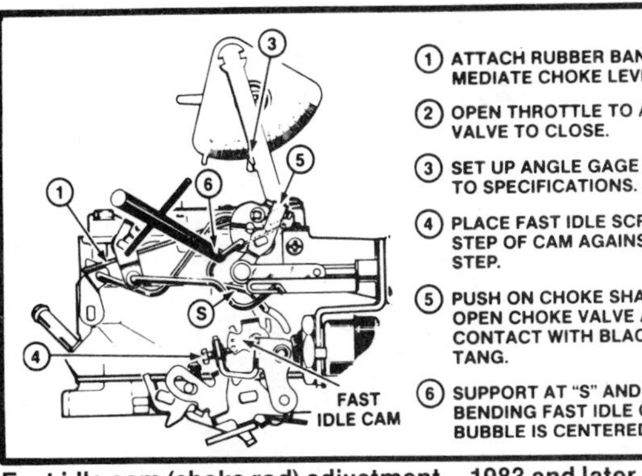

Fast idle cam (choke rod) adjustment—1983 and later models

2. Place the fast idle screw on the highest step of the dam.

3. Start the engine and adjust the engine speed to specification with the fast idle screw.

Choke Coil Lever Adjustment

1. Remove the three retaining screws and remove the choke cover and coil. On models with a riveted choke cover, drill out the three rivets and remove the cover and choke coil.

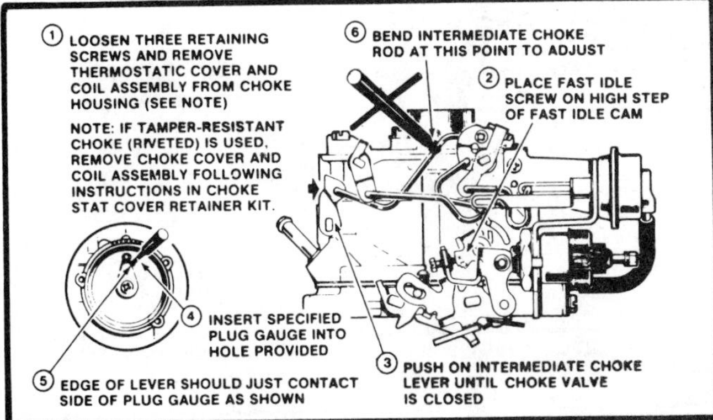

Choke coil level adjustment on E2SE carburetor—typical

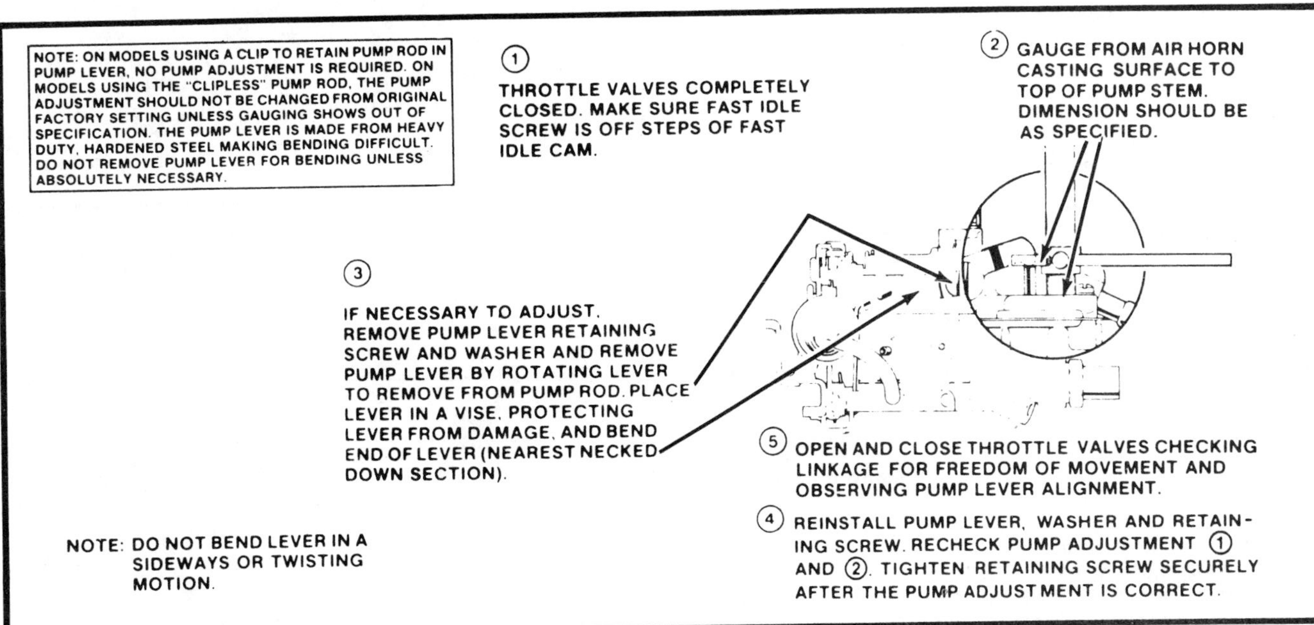

Accelerator pump adjustment—typical

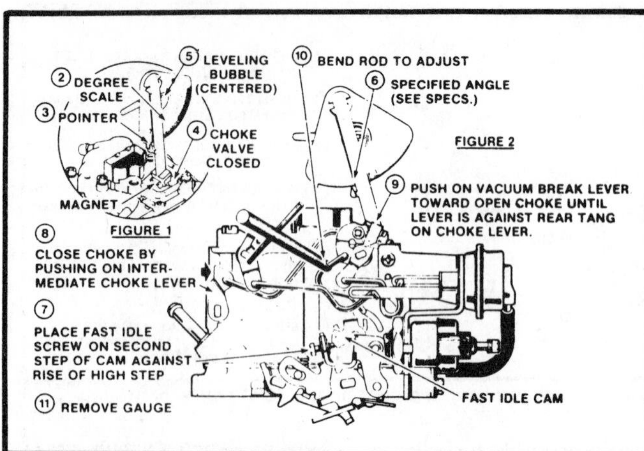

Fast idle cam adjustment—1982 and earlier models

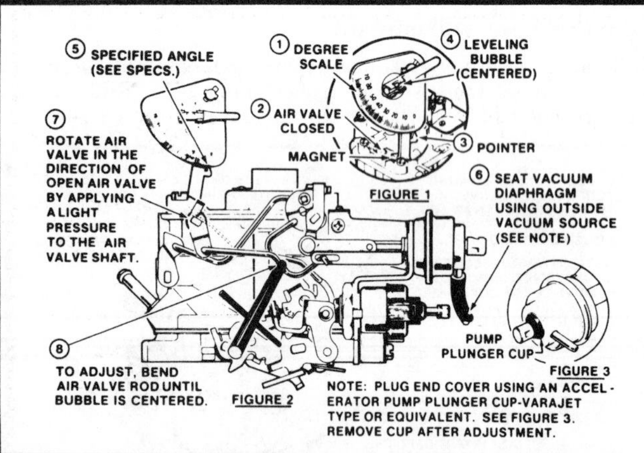

Rochester E2SE air valve adjustment—1981-82 4 cyl. except GM J-body

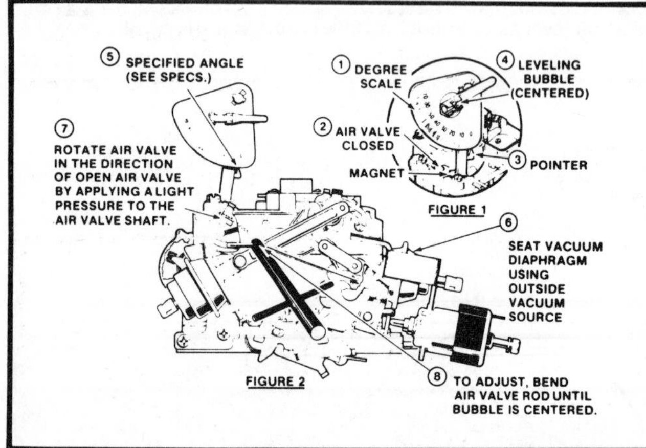

Rochester E2SE air valve adjustment—1981-82 V6 engine

NOTE: A choke stat cover retainer kit is required for reassembly.

2. Place the fast idle screw on the high step of the dam.
3. Close the choke by pushing in on the intermediate choke lever. On front wheel drive models, the intermediate choke lever is behind the choke vacuum diaphragm.
4. Insert a drill or gauge of the specified size into the hole in the choke housing. The choke lever in the housing should be up against the side of the gauge.
5. If the lever does not just touch the gauge, bend the intermediate choke rod to adjust.

Fast Idle Cam (Choke Rod) Adjustment

1980–82 MODELS

NOTE: A special angle gauge should be used.

1. Adjust the choke coil lever and fast idle first.
2. Rotate the degree scale until it is zeroed.
3. Close the choke and install the degree scale onto the choke plate. Center the leveling bubble.
4. Rotate the scale so that the specified degree is opposite the scale pointer.
5. Place the fast idle screw on the second step of the cam (against the high step). Close the choke by pushing in the intermediate lever.
6. Push on the vacuum break lever in the direction of opening choke until the lever is against the rear tang on the choke lever.
7. Bend the fast idle cam rod at the U to adjust angle to specifications.

1983–84 MODELS

Refer to the illustration for the adjustment procedure on these models.

Air Valve Rod Adjustment

1980 MODELS

1. Seat the vacuum diaphragm with an outside vacuum source. Tape over the purge bleed hole if present.
2. Close the air valve.

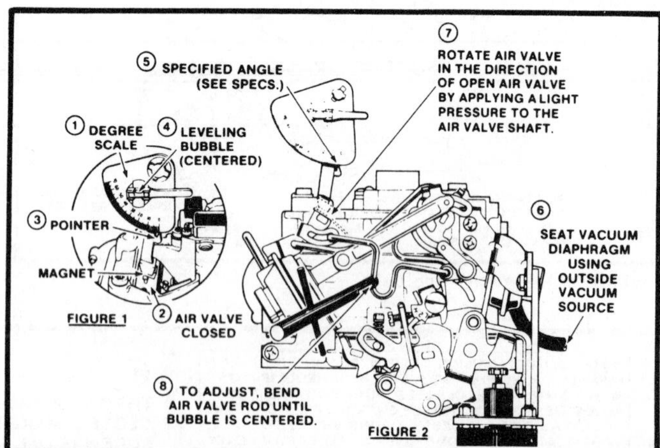

Rochester E2SE air valve adjustment—1982 GM J-body models

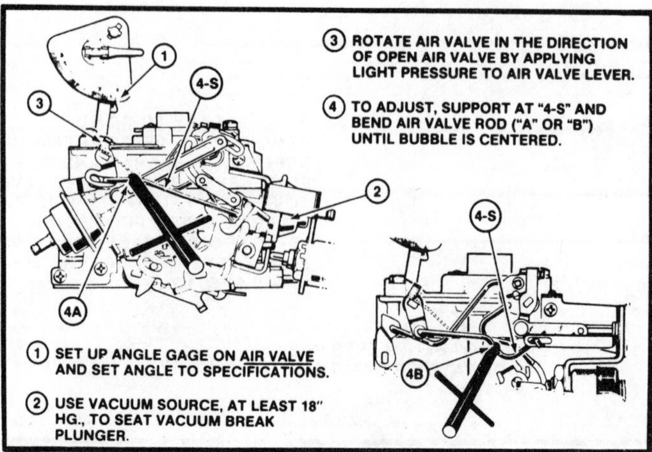

Rochester E2SE air valve adjustment—1983 and later models

3. Insert the specified gauge between the rod and the end of the slot in the plunger on fours, or between the rod and the end of the slot in the air valve on V6s.
4. Bend the rod to adjust the clearance.

1981-82 MODELS

1. Align the zero degree mark with the pointer on an angle gauge.
2. Close the air valve and place a magnet on top of it.
3. Rotate the bubble until it is centered.
4. Rotate the degree scale until the specified degree mark is aligned with the pointer.
5. Seat the vacuum diaphragm using an external vacuum source.
6. On four cylinder models plug the end cover. Unplug after adjustment.
7. Apply light pressure to the air valve shaft in the direction to open the air valve until all the slack is removed between the air link and plunger slot.
8. Bend the air valve link until the bubble is center.

1983-84 MODELS

Refer to the illustration for the adjustment procedure on these models.

Primary Side Vacuum Break Adjustment

1980 GM MODELS
1980-83 AMERICAN MOTORS

1. Follow Steps 1-4 of the "Fast Idle Cam Adjustment".
2. Seat the choke vacuum diaphragm with an outside vacuum source.
3. Push in on the intermediate choke lever to close the choke valve, and hold closed during adjustment.
4. Adjust by bending the vacuum break rod until the bubble is centered.

1981-82 GM MODELS

NOTE: Prior to adjustment, remove the vacuum break from the carburetor. Place the bracket in a vise and using the proper safety precautions, grind off the adjustment screw cap then reinstall the vacuum break.

1. Rotate the degree scale on the measuring gauge until the zero is opposite the pointer.
2. Seat the choke vacuum diaphragm by applying an external vacuum source of over 5" vacuum to the vacuum break.

NOTE: If the air valve rod is restricting the vacuum diaphragm from seating it may be necessary to bend the air valve rod slightly to gain clearance. Make an air valve rod adjustment after the vacuum break adjustment.

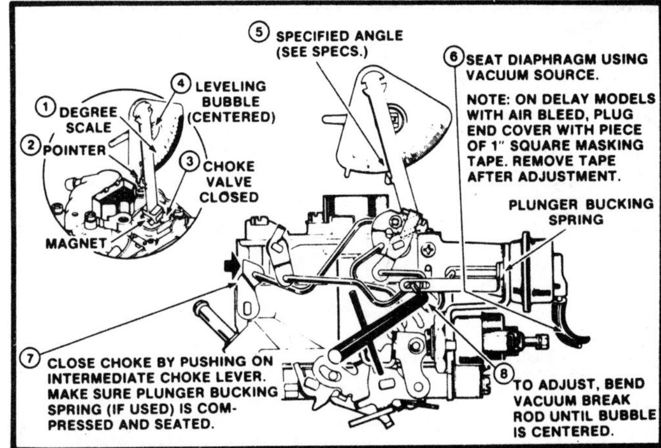

Primary vacuum break adjustment—1980 GM and 1980 and later AMC models with 4 cyl. engine

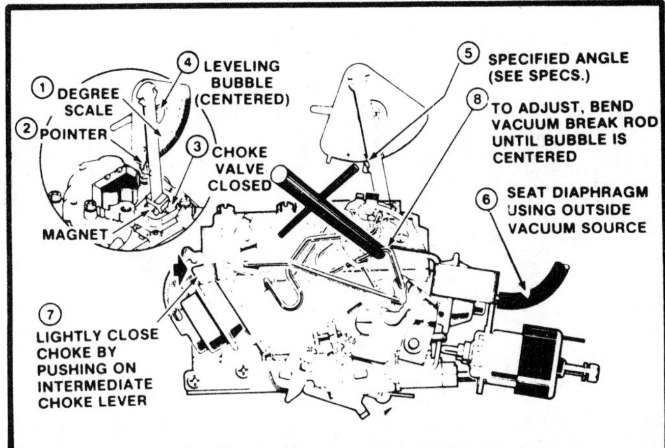

Primary vacuum break adjustment—1980 models

3. Read the angle gauge while lightly pushing on the intermediate choke lever so that the choke valve is toward the close position.
4. Use a 1/8 in. hex wrench and turn the screw in the rear cover until the bubble is centered. Apply a silicone sealant over the screw head to seal the setting.

1983-84 GM MODELS

Refer to the illustration for the adjustment procedure on these models.

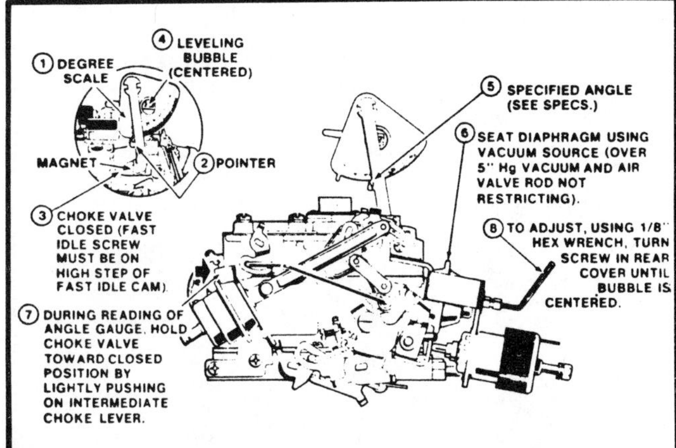

Primary vacuum break adjustment—1981-82 GM A and X-body models with V6

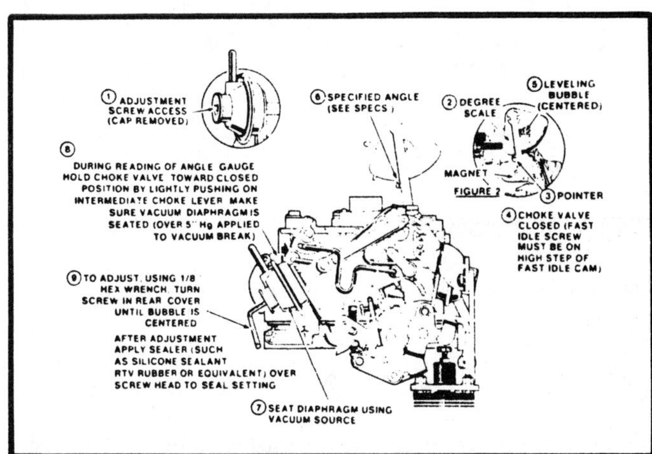

Secondary vacuum break adjustment—1982 GM J-body

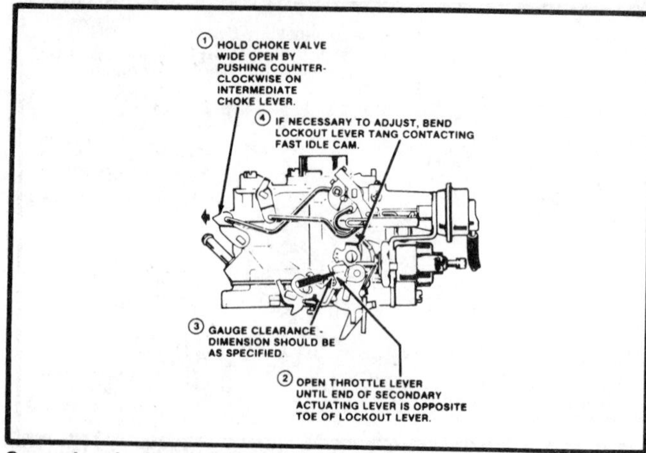

Secondary lockout adjustment—typical

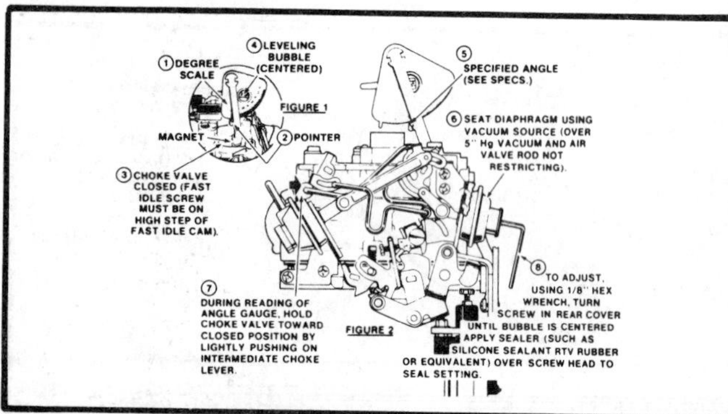

Rochester E2SE primary vacuum break adjustment—1982 GM J-body models with 4 cyl. engine

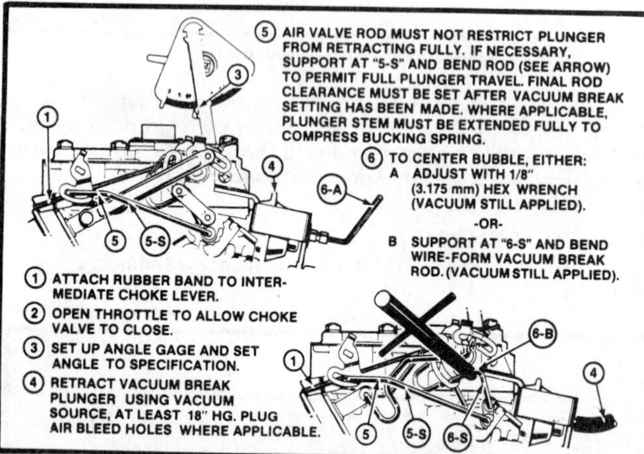

Rochester E2SE primary vacuum break adjustment—1983 and later models

Electric Choke Setting

This procedure is only for those carburetors with choke covers retained by screws. Riveted choke covers are preset and nonadjustable.

1. Loosen the three retaining screws.
2. Place the fast idle screw on the high step of the cam.
3. Rotate the choke cover to align the cover mark with the specified housing mark.

NOTE: The specification "index" which appears in the specification table refers to the mark between "1 notch lean" and "1 notch rich".

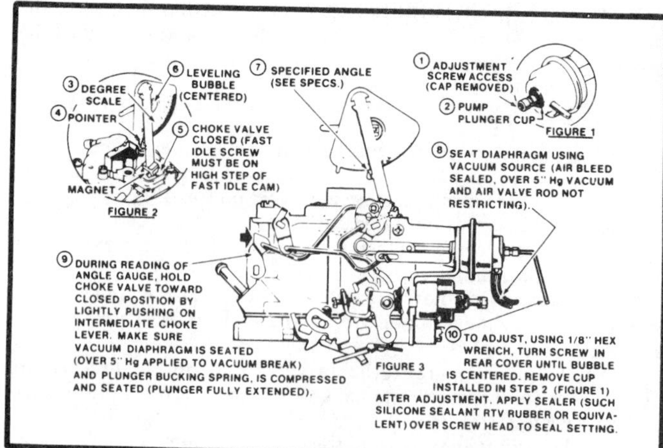

Rochester E2SE primary vacuum break adjustment—1981-82 GM A and X-body models with 4 cyl. engine

Secondary Vacuum Break Adjustment

1980 MODELS

This procedure is for V6 installations in front wheel drive models only.

1. Follow Steps 1–4 of the "Fast Idle Cam Adjustment".
2. Seat the choke vacuum diaphragm with an outside vacluum source.
3. Push in on the intermediate choke lever to close the choke valve, and hold closed during adjustment. Make sure the plunger spring is compressed and seated, if present.
4. Bend the vacuum break rod at the U next to the diaphragm until the bubble is centered.

1981–82 GM MODELS

NOTE: Prior to adjustment, remove the vacuum break from the carburetor. Place the bracket in the vise and using the proper safety precautions, grind off the adjustment screw cap then reinstall the vacuum break. Plug the end cover using an accelerator pump plunger cup or equivalent. Remove the cup after the adjustment (A and X series only).

1. Rotate the degree scale on the measuring gauge until the zero is opposite the pointer.
2. Seat the choke vacuum diaphragm by applying an external vacuum source of over 5 in. vacuum to the vacuum break.

NOTE: If the air valve rod is restricting the vacuum diaphragm from seating it may be necessary to bend the air valve rod lightly to gain clearance. make an air valve rod adjustment after the vacuum break adjustment.

3. Read the angle gauge while lightly pushing on the intermediate choke lever so that the choke valve is toward the close position.
4. Use a $^1/_8$ in. hex wrench and turn the screw in the rear cover until the bubble is centered. Apply a silicone sealant over the screw head to seal the setting.

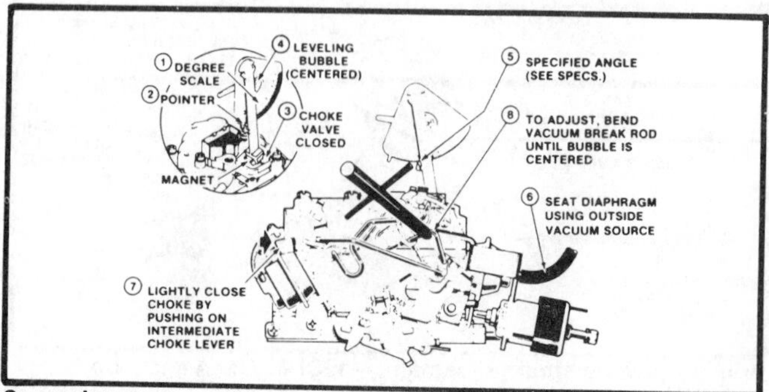

Secondary vacuum break adjustment—1980 models

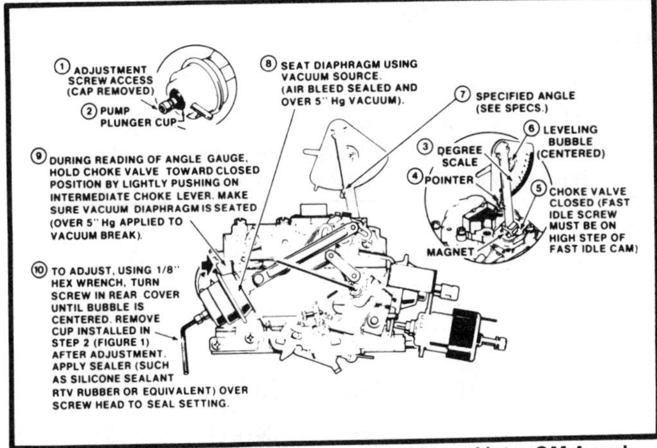

Secondary vacuum break adjustment—1981 and later GM A and X-body models

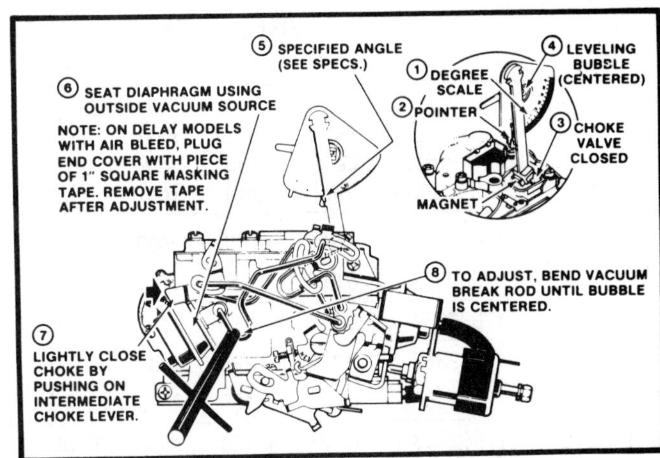

Rochester E2SE choke unloader adjustment—typical

1983–84 GM MODELS

Refer to the illustration for the adjustment procedure on these models.

Choke Unloader Adjustment

THROUGH 1982

1. Follow Steps 1-4 of the "Fast Idle Cam Adjustment".
2. Install the choke cover and coil, if removed, aligning the marks on the housing and cover as specified.
3. Hold the primary throttle wide open.
4. If the engine is warm, close the choke valve by pushing in on the intermediate choke lever.
5. Bend the unloader tang until the bubble is centered.

1983–84 MODELS

Refer to the illustration for the adjustment procedure on these models.

① ATTACH RUBBER BAND TO INTERMEDIATE CHOKE LEVER.
② OPEN THROTTLE TO ALLOW CHOKE VALVE TO CLOSE.
③ SET UP ANGLE GAGE AND SET ANGLE TO SPECIFICATIONS.
④ HOLD THROTTLE LEVER IN WIDE OPEN POSITION.
⑤ PUSH ON CHOKE SHAFT LEVER TO OPEN CHOKE VALVE AND TO MAKE CONTACT WITH BLACK CLOSING TANG.
⑥ ADJUST BY BENDING TANG UNTIL BUBBLE IS CENTERED.

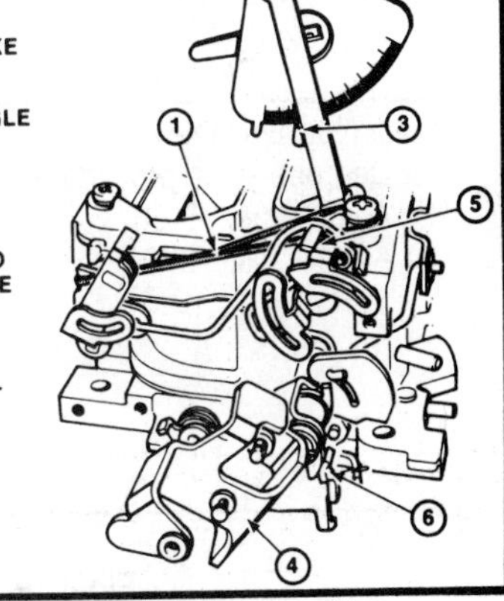

Rochester E2SE choke unloader adjustment—1983 and later models

2SE, E2SE CARBURETOR SPECIFICATIONS
American Motors

Year	Carburetor Identification	Float Level (in.)	Pump Rod (in.)	Fast Idle (rpm)	Choke Coil Lever (in.)	Fast Idle Cam (deg./in.)	Air Valve Rod (in.)	Primary Vacuum Break (deg./in.)	Choke Setting (notches)	Choke Unloader (deg./in.)	Secondary Lockout (in.)
'80	17080681	3/16	17/32	2400	.142	18/0.096	.018	20/.110	Fixed	32/.195	N.A.
	17080683	3/16	1/2	2400	.142	18/0.096	.018	20/.110	Fixed	32/.195	N.A.
	17080686	3/16	1/2	2600	.142	18/0.096	.018	20/.110	Fixed	32/.195	N.A.
	17080688	3/16	1/2	2600	.142	18/0.096	.018	20/.110	Fixed	32/.195	N.A.
'81	17081790	0.256	0.128	2600	0.085	25/0.142	.011	19/.103	Fixed	32/.195	0.065
	17081791	0.256	0.128	2400	0.085	25/0.142	.011	19/.103	Fixed	32/.195	0.065
	17081792	0.256	0.128	2400	0.085	25/0.142	.011	19/.103	Fixed	32/.195	0.065
	17081794	0.256	0.128	2600	0.085	25/0.142	.011	19/.103	Fixed	32/.195	0.065
	17081795	0.256	0.128	2600	0.085	25/0.142	.011	19/.103	Fixed	32/.195	0.065
	17081796	0.208	0.128	2400	0.065	25/0.142	.011	19/.103	Fixed	32/.195	0.065
	17081797	0.208	0.128	2600	0.085	25/0.142	.011	19/.103	Fixed	32/.195	0.085
	17081793	0.256	0.128	2400	0.085	25/0.142	.011	19/.103	Fixed	32/.195	0.065

2SE, E2SE CARBURETOR SPECIFICATIONS
American Motors

Year	Carburetor Identification	Float Level (in.)	Pump Rod (in.)	Fast Idle (rpm)	Choke Coil Lever (in.)	Fast Idle Cam (deg./in.)	Air Valve Rod (in.)	Primary Vacuum Break (deg./in.)	Choke Setting (notches)	Choke Unloader (deg./in.)	Secondary Lockout (in.)
'82	17082385	0.256	0.128	2400	0.085	18/.096	2①	21/.117	Fixed	34/.211	0.065
	17082383	0.256	0.128	2400	0.085	18/.096	2①	21/.117	Fixed	34/.211	0.065
	17082380	0.216	0.128	2400	0.085	18/.096	2①	21/.117	Fixed	34/.211	0.065
	17082386	0.125	0.128	2400	0.065	18/.096	2①	19/.103	Fixed	34/.211	0.065
	17082387	0.125	0.128	2600	0.085	18/.096	2①	19/.103	Fixed	34/.211	0.065
	17082388	0.125	0.128	2500	0.085	18/.096	2①	19/.103	Fixed	34/.211	0.065
	17082389	0.125	0.128	2500	0.085	18/.096	2①	19/.103	Fixed	34/.211	0.065
'83–'84	1982380	0.216②	0.128	2500③	0.085	18/.096	2①	21/.117	Fixed	34/.211	0.065
	1983384	0.138	0.128	2700	0.085	18/.096	2①	19/.103	Fixed	34/.211	0.065
	1983385	0.138	0.128	2500	0.085	18/.096	②①	19/.103	Fixed	34/.211	0.065

N.A.: Not Available
① Degrees—see procedure
② Auto. trans.—.138
③ Auto. trans.—2700

2SE, E2SE CARBURETOR SPECIFICATIONS
General Motors—U.S.A.

Year	Carburetor Identification	Float Level (in.)	Pump Rod (in.)	Fast Idle (rpm)	Choke Coil Lever (in.)	Fast Idle Cam (deg./in.)	Air Valve Rod (in.)	Primary Vacuum Break (deg./in.)	Choke Setting (notches)	Secondary Vacuum Break (deg./in.)	Choke Unloader (deg./in.)	Secondary Lockout (in.)
'79	17059674	13/64	1/2	2400	.120	18/0.096	.025	19/.103	2 Rich	—	32/.195	.030
	17059675	13/64	17/32	2200	.120	18/0.096	.025	21/.117	1 Rich	—	32/.195	.030
	17059676	13/64	1/2	2400	.120	18/0.096	.025	19/.103	2 Rich	—	32/.195	.030
	17059677	13/64	17/32	2200	.120	18/0.096	.025	21/.117	1 Rich	—	32/.195	.030
'80	17059614	3/16	1/2	2600	.085	18/.096	.025	17/.090	Fixed	—	36/.227	.120
	17059615	3/16	5/32	2600	.085	18/.096	.025	19/.103	Fixed	—	36/.227	.120
	17059616	3/16	1/2	2600	.085	18/.096	.025	17/.090	Fixed	—	36/.227	.120
	17059617	3/16	5/32	2600	.085	18/.096	.025	19/.103	Fixed	—	36/.227	.120
	17059618	3/16	1/2	2600	.085	18/.096	.025	17/.090	Fixed	—	36/.227	.120
	17059619	3/16	5/32	2600	.085	18/.096	.025	19/.103	Fixed	—	36/.227	.120
	17059620	3/16	1/2	2600	.085	18/.096	.025	17/.090	Fixed	—	36/.227	.120
	17059621	3/16	5/32	2600	.085	18/.096	.025	19/.103	Fixed	—	36/.227	.120
	17059650	3/16	3/32	2600	.085	27/.157	.025	30/.179	Fixed	38/.243	30/.179	.120
	17059651	3/16	3/32	1900	.085	27/.157	.025	22/.123	Fixed	23/.120	30/.179	.120
	17059652	3/16	3/32	2000	.085	27/.157	.025	30/.179	Fixed	38/.243	30/.179	.120
	17059653	3/16	3/32	1900	.085	27/.157	.025	22/.123	Fixed	23/.120	30/.179	.120
	17059714	11/16	5/32	2600	.085	18/.096	.025	23/.129	Fixed	—	32/.195	.120
	17059715	11/16	3/32	2200	.085	18/.096	.025	25/.142	Fixed	—	32/.195	.120

2SE, E2SE CARBURETOR SPECIFICATIONS
General Motors—U.S.A.

Year	Carburetor Identification	Float Level (in.)	Pump Rod (in.)	Fast Idle (rpm)	Choke Coil Lever (in.)	Fast Idle Cam (deg./in.)	Air Valve Rod (in.)	Primary Vacuum Break (deg./in.)	Choke Setting (notches)	Secondary Vacuum Break (deg./in.)	Choke Unloader (deg./in.)	Secondary Lockout (in.)
'80	17059716	11/16	5/32	2600	.085	18/.096	.025	23/.129	Fixed	—	32/.195	.120
	17059717	11/16	3/32	2200	.085	18/.096	.025	25/.142	Fixed	—	32/.195	.120
	17059760	1/8	5/64	2000	.085	17.5/.093	.025	20/.110	Fixed	33/.203	35/.220	.120
	17059762	1/8	5/64	2000	.085	17.5/.093	.025	20/.110	Fixed	33/.203	35/.220	.120
	17059763	1/8	5/64	2000	.085	17.5/.093	.025	20/.110	Fixed	33/.203	35/.220	.120
	17059774	5/32	1/2	①	.085	18/0.096	.018	19/.103	Fixed	—	32/.195	.012
	17059775	5/32	17/32	①	.085	18/0.096	.018	21/.117	Fixed	—	32/.195	.012
	17059776	5/32	1/2	①	.085	18/0.096	.018	19/.103	Fixed	—	32/.195	.012
	17059777	5/32	17/32	①	.085	18/0.096	.018	21/.117	Fixed	—	32/.195	.012
	17080674	3/16	1/2	①	.085	18/0.096	.018	19/.103	Fixed	—	32/.195	.012
	17080675	3/16	1/2	①	.085	18/0.096	.018	21/.117	Fixed	—	32/.195	.012
	17080676	3/16	1/2	①	.085	18/0.096	.018	19/.103	Fixed	—	32/.195	.012
	17080677	3/16	1/2	①	.085	18/0.096	.018	21/.117	Fixed	—	32/.195	.012
'81	17081650	1/4	Fixed	2600	.085	17/.090	1②	25/.142	Fixed	34/.211	35/.220	.012
	17081651	1/4	Fixed	2400	.085	17/.090	1②	29/.171	Fixed	35/.220	35/.220	.012
	17081652	1/4	Fixed	2600	.085	17/.090	1②	25/.142	Fixed	34/.211	35/.220	.012
	17081653	1/4	Fixed	2600	.085	17/.090	1②	29/.171	Fixed	35/.220	35/.220	.012
	17081670	5/32	Fixed	2600	.085	18/.096	1②	19/.103	Fixed	—	32/.195	.012
	17081671	5/32	Fixed	2600	.085	33.5/.207	1②	21/.117	Fixed	—	32/.195	.012
	17081672	5/32	Fixed	2600	.085	18/.096	1②	19/.103	Fixed	—	32/.195	.012
	17081673	5/32	Fixed	2600	.085	33.4/.207	1②	21/.117	Fixed	—	32/.195	.012
	17081740	1/4	Fixed	2400	.085	17/.090	1②	25/.142	Fixed	35/.220	35/.220	.012
	17081742	1/4	Fixed	2400	.085	17/.090	1②	25/.142	Fixed	35/.220	35/.220	.012
'82	17081600	5/16	Fixed	①	③	24/.136	1②	20/.110	Fixed	27/.157	35/.220	③
	17081601	5/16	Fixed	①	③	24/1.36	1②	20/.110	Fixed	27/.157	35/.220	③
	17081607	5/16	Fixed	①	③	24/.136	1②	20/.110	Fixed	27/.157	35/.220	③
	17081700	5/16	Fixed	①	③	24/.136	1②	20/.110	Fixed	27/.157	35/.220	③
	17081701	5/16	Fixed	①	③	24/.136	1②	20/.110	Fixed	27/.157	35/.220	③
	17082196	5/16	Fixed	①	.085	18/.096	1②	21/.117	Fixed	19/.103	27/157	③
	17082316	1/4	Fixed	2600	.085	17/.090	1②	30/.179	Fixed	34/.211	45/.304	③
	17082317	1/4	Fixed	260	.085	17/.090	1②	30/.179	Fixed	35/.220	45/.304	③
	17082320	1/4	Fixed	2800	.085	25/.142	1②	30/.179	Fixed	35/.220	45/.304	③
	17082321	1/4	Fixed	2600	.085	25/.142	1②	30/.179	Fixed	35/.220	45/.304	③
	17082390	13/32	Fixed	2500	.085	17/.090	1②	26/.149	Fixed	34/.211	35/.220	.011–.040
	17082391	13/32	Fixed	2600	.085	25/.142	1②	29/.171	Fixed	35/.220	35/.220	.011–.040
	17082490	13/32	Fixed	2500	.085	17/.090	1②	26/.149	Fixed	34/.211	35/.220	.011–.040
	17082491	13/32	Fixed	2600	.085	25/.142	1②	29/.171	Fixed	35/.220	35/.220	.011–.040
	17082640	1/4	Fixed	2600	.085	17/.090	1②	30/.179	Fixed	34/.211	45/.304	③

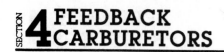

2SE, E2SE CARBURETOR SPECIFICATIONS
General Motors—U.S.A.

Year	Carburetor Identification	Float Level (in.)	Pump Rod (in.)	Fast Idle (rpm)	Choke Coil Lever (in.)	Fast Idle Cam (deg./in.)	Air Valve Rod (in.)	Primary Vacuum Break (deg./in.)	Choke Setting (notches)	Secondary Vacuum Break (deg./in.)	Choke Unloader (deg./in.)	Secondary Lockout (in.)
'82	17082641	¼	Fixed	2400	.085	17/.090	1②	30/.179	Fixed	35/.220	45/.304	③
	17082642	¼	Fixed	2800	.085	25/.142	1②	30/.179	Fixed	35/.220	45/.304	③
'83	17083356	13/32	Fixed	①	.085	22/.123	1②	25/.142	Fixed	35/.220	30/.179	.025
	17083357	13/32	Fixed	①	.085	22/.123	1②	25/.142	Fixed	35/.220	30/.179	.025
	17083358	13/32	Fixed	①	.085	22/.123	1②	25/.142	Fixed	35/.220	30/.179	.025
	17083359	13/32	Fixed	①	.085	22/.123	1②	25/.142	Fixed	35/.220	30/.179	.025
	17083368	13/32	Fixed	①	.085	22/.123	1②	25/.142	Fixed	35/.220	30/.179	.025
	17083369	13/32	Fixed	①	.085	22/.123	1②	25/.142	Fixed	35/.220	30/.179	.025
	17083370	13/32	Fixed	①	.085	22/.123	1②	25/.142	Fixed	35/.220	30/.179	.025
	17083391	13/32	Fixed	①	.085	28/.164	1②	30/.179	Fixed	35/.220	38/.243	.025
	17083392	13/32	Fixed	①	.085	28/.164	1②	30/.179	Fixed	35/.220	38/.243	.025
	17083393	13/32	Fixed	①	.085	28/.164	1②	30/.179	Fixed	35/.220	38/.243	.025
	17083394	13/32	Fixed	①	.085	28/.164	1②	30/.179	Fixed	35/.220	38/.243	.025
	17083395	13/32	Fixed	①	.085	28/.164	1②	30/.179	Fixed	35/.220	38/.243	.025
	17083396	13/32	Fixed	①	.085	28/.164	1②	30/.179	Fixed	35/.220	38/.243	.025
	17083397	13/32	Fixed	①	.085	28/.164	1②	30/.179	Fixed	35/.220	38/.243	.025
	17083450	¼	Fixed	①	.085	28/.164	1②	27/.157	Fixed	35/.220	45/.304	.025
	17083451	¼	Fixed	①	.085	28/.164	1②	27/.157	Fixed	35/.220	45/.304	.025
	17083452	¼	Fixed	①	.085	28/.164	1②	27/.157	Fixed	35/.220	45/.304	.025
	17083453	¼	Fixed	①	.085	28/.164	1②	27/.157	Fixed	35/.220	45/.304	.025
	17083454	¼	Fixed	①	.085	28/.164	1②	27/.157	Fixed	35/.220	45/.304	.025
	17083455	¼	Fixed	①	.085	28/.164	1②	27/.157	Fixed	35/.220	45/.304	.025
	17083456	¼	Fixed	①	.085	28/.164	1②	27/.157	Fixed	35/.220	45/.304	.025
	17083630	¼	Fixed	①	.085	28/.164	1②	27/.157	Fixed	35/.220	45/.304	.025
	17083631	¼	Fixed	①	.085	28/.164	1②	27/.157	Fixed	35/.220	45/.304	.025
	17083632	¼	Fixed	①	.085	28/.164	1②	27/.157	Fixed	35/.220	45/.304	.025
	17083633	¼	Fixed	①	.085	28/.164	1②	27/.157	Fixed	35/.220	45/.304	.025
	17083634	¼	Fixed	①	.085	28/.164	1②	27/.157	Fixed	35/.220	45/.304	.025
	17083635	¼	Fixed	①	.085	28/.164	1②	27/.157	Fixed	35/.220	45/.304	.025
	17083636	¼	Fixed	①	.085	28/.164	1②	27/.157	Fixed	35/.220	45/.304	.025
'84	17072683	9/32	Fixed	①	.085	28/.164	1②	25/.142	Fixed	35/.220	45/.304	.025
	17074812	9/32	Fixed	①	.085	28/.164	1②	25/.142	Fixed	35/.220	45/.304	.025
	17084356	9/32	Fixed	①	.085	22/.123	1②	25/.142	Fixed	30/.179	30/.179	.025
	17084357	9/32	Fixed	①	.085	22/.123	1②	25/.142	Fixed	30/.179	30/.179	.025
	17084358	9/32	Fixed	①	.085	22/.123	1②	25/.142	Fixed	30/.179	30/.179	.025
	17084359	9/32	Fixed	①	.085	22/.123	1②	25/.142	Fixed	30/.179	30/.179	.025
	17084368	⅛	Fixed	①	.085	22/.123	1②	25/.142	Fixed	30/.179	30/.179	.025
	17084370	⅛	Fixed	①	.085	22/.123	1②	25/.142	Fixed	30/.179	30/.179	.025

2SE, E2SE CARBURETOR SPECIFICATIONS
General Motors—U.S.A.

Year	Carburetor Identification	Float Level (in.)	Pump Rod (in.)	Fast Idle (rpm)	Choke Coil Lever (in.)	Fast Idle Cam (deg./in.)	Air Valve Rod (in.)	Primary Vacuum Break (deg./in.)	Choke Setting (notches)	Choke Unloader (deg./in.)	Secondary Lockout (in.)	
'84	17084430	11/32	Fixed	①	.085	15/.077	1②	26/.149	Fixed	30/.179	30/.179	.025
	17084431	11/32	Fixed	①	.085	15/.077	1②	26/.149	Fixed	38/.243	42/.277	.025
	17084434	11/32	Fixed	①	.085	15/.077	1②	26/.149	Fixed	38/.243	42/.277	.025
	17084435	11/32	Fixed	①	.085	15/.077	1②	26/.149	Fixed	38/.243	42/.277	.025
	17084452	5/32	Fixed	①	.085	28/.164	1②	25/.142	Fixed	38/.243	42/.277	.025
	17084453	5/32	Fixed	①	.085	28/.164	1②	25/.142	Fixed	35/.220	45/.304	.025
	17084455	5/32	Fixed	①	.085	28/.164	1②	25/.142	Fixed	35/.220	45/.304	.025
	17084456	5/32	Fixed	①	.085	28/.164	1②	25/.142	Fixed	35/.220	45/.304	.025
	17084458	5/32	Fixed	①	.085	28/.164	1②	25/.142	Fixed	35/.220	45/.304	.025
	17084532	5/32	Fixed	①	.085	28/.164	1②	25/.142	Fixed	35/.220	45/.304	.025
	17084534	5/32	Fixed	①	.085	28/.164	1②	25/.142	Fixed	35/.220	45/.304	.025
	17084535	5/32	Fixed	①	.085	28/.164	1②	25/.142	Fixed	35/.220	45/.304	.025
	17084537	5/32	Fixed	①	.085	28/.164	1②	25/.142	Fixed	35/.220	45/.304	.025
	17084538	5/32	Fixed	①	.085	28/.164	1②	25/.142	Fixed	35/.220	45/.304	.025
	17084540	5/32	Fixed	①	.085	28/.164	1②	25/.142	Fixed	35/.220	45/.304	.025
	17084542	1/8	Fixed	①	.085	28/.164	1②	25/.142	Fixed	35/.220	45/.304	.025
	17084632	9/32	Fixed	①	.085	28/.164	1②	25/.142	Fixed	35/.220	45/.304	.025
	17084633	9/32	Fixed	①	.085	28/.164	1②	25/.142	Fixed	35/.220	45/.304	.025
	17084635	9/32	Fixed	①	.085	28/.164	1②	25/.142	Fixed	35/.220	45/.304	.025
	17084636	9/32	Fixed	①	.085	28/.164	1②	25/.142	Fixed	35/.220	45/.304	.025

① See underhood decal
② Measurement in degrees
③ Not available

2SE, E2SE CARBURETOR SPECIFICATIONS
General Motors—Canada

Year	Carburetor Identification	Float Level (In.)	Pump Rod (in.)	Fast Idle (rpm)	Choke Coil Lever (in.)	Fast Idle Cam (deg./in.)	Air Valve Rod (deg.)	Primary Vacuum Break (deg./in.)	Choke Setting (notches)	Secondary Vacuum Break (deg./in.)	Choke Unloader (deg./in.)	Secondary Lockout (in.)
'81	17059660	1/4	17/32	①	.085	24/.136	1	30/.179	Fixed	32/.195	30/.179	②
	17059662	1/4	17/32	①	.085	24/.136	1	30/.179	Fixed	37/.195	30/.179	②
	17059651	1/4	17/32	①	.085	24/.136	1	30/.179	Fixed	32/.195	30/.179	②
	17059666	1/4	17/32	①	.085	24/.136	1	26/.149	Fixed	32/.195	30/.179	②
	17059667	1/4	17/32	①	.085	24/.136	1	26/.149	Fixed	32/.195	30/.179	②
	17059622	5/32	17/32	①	.085	18/.096	1	17/.090	Fixed	—	36/.227	②
	17059623	5/32	17/32	①	.085	18/.096	1	19/.103	Fixed	—	36/.227	②
	17059624	5/32	17/32	①	.085	18/.096	1	17/.090	Fixed	—	36/.227	②

2SE, E2SE CARBURETOR SPECIFICATIONS
General Motors—Canada

Year	Carburetor Identification	Float Level (In.)	Pump Rod (in.)	Fast Idle (rpm)	Choke Coil Lever (in.)	Fast Idle Cam (deg./in.)	Air Valve Rod (deg.)	Primary Vacuum Break (deg./in.)	Choke Setting (notches)	Secondary Vacuum Break (deg./in.)	Choke Unloader (deg./in.)	Secondary Lockout (in.)
'82	17082440	¼	19/32	①	.085	24/.136	1	30/.179	Fixed	32/.195	45/.304	②
	17082441	¼	19/32	①	.085	24/.136	1	30/.179	Fixed	32/.195	45/.304	②
	17082443	¼	19/32	①	.085	24/.136	1	30/.179	Fixed	32/.195	45/.304	②
	17082460	¼	19/32	①	.085	18/.096	1	21/.117	Fixed	—	36/.227	②
	17082461	¼	19/32	①	.085	18/.096	1	21/.117	Fixed	—	36/.227	②
	17082462	¼	19/32	①	.085	18/.096	1	21/.117	Fixed	—	36/.227	②
	17082464	⅛	19/32	①	.085	18/.096	1	21/.117	Fixed	—	36/.227	②
	17082465	⅛	19/32	①	.085	18/.096	1	21/.117	Fixed	—	36/.227	②
	17082466	⅛	19/32	①	.085	18/.096	1	21/.117	Fixed	—	36/.227	②
	17082620	7/16	19/32	①	.085	24/.136	1	30/.179	Fixed	32/.195	45/.304	②
	17082621	7/16	19/32	①	.085	24/.136	1	30/.179	Fixed	32/.195	45/.304	②
	17082622	7/16	19/32	①	.085	24/.136	1	30/.179	Fixed	32/.195	45/.304	②
	17082623	7/16	19/32	①	.085	24/.136	1	30/.179	Fixed	32/.195	45/.304	②
'83	17083311	5/16	Fixed	①	.085	24/.136	1	18/.096	Fixed	20/.110	35/.220	.025
	17083401	5/16	Fixed	①	.085	24/.136	1	18/.096	Fixed	20/.110	35/.220	.025
	17083440	¼	19/32	①	.085	24/.136	1	28/.164	Fixed	32/.195	40/.260	.025
	17083441	¼	19/32	①	.085	24/.136	1	28/.164	Fixed	32/.195	40/.260	.025
	17083442	¼	19/32	①	.085	24/.136	1	28/.164	Fixed	32/.195	40/.260	.025
	17083443	¼	19/32	①	.085	24/.136	1	28/.164	Fixed	32/.195	40/.260	.025
	17083444	¼	19/32	①	.085	24/.136	1	28/.164	Fixed	32/.195	40/.260	.025
	17083445	¼	19/32	①	.085	24/.136	1	28/.164	Fixed	32/.195	40/.260	.025
	17083460	¼	19/32	①	.085	18/.096	1	19/.103	Fixed	—	36/.227	.025
	17083461	¼	19/32	①	.085	18/.096	1	18/.096	Fixed	—	36/.227	.025
	17083462	¼	19/32	①	.085	18/.096	1	19/.103	Fixed	—	36/.227	.025
	17083464	⅛	19/32	①	.085	18/.096	1	19/.103	Fixed	—	36/.227	.025
	17083465	⅛	19/32	①	.085	18/.096	1	20/.110	Fixed	—	36/.227	.025
	17083466	⅛	19/32	①	.085	18/.096	1	19/.103	Fixed	—	36/.227	.025
	17083620	7/16	19/32	①	.085	24/.136	1	28/.164	Fixed	32/.195	40/.260	.025
	17083621	7/16	19/32	①	.085	24/.136	1	28/.164	Fixed	32/.195	40/.260	.025
	17083622	7/16	19/32	①	.085	24/.136	1	28/.164	Fixed	34/.195	40/.260	.025
	17083623	7/16	19/32	①	.085	24/.136	1	28/.164	Fixed	32/.195	40/.260	.025
'84	17084312	5/16	Fixed	①	.085	24/.136	1	18/.096	Fixed	20/.110	35/.220	.025
	17084314	5/16	Fixed	①	.085	29/.171	1	16/.083	Fixed	20/.110	30/.179	.025
	17084480	¼	Fixed	①	.085	24/.136	1	28/.164	Fixed	32/.195	45/.304	.025
	17084481	¼	Fixed	①	.085	24/.136	1	28/.164	Fixed	32/.195	45/.304	.025
	17084482	¼	Fixed	①	.085	24/.136	1	28/.164	Fixed	32/.195	45/.304	.025
	17084483	¼	Fixed	①	.085	24/.136	1	28/.164	Fixed	32/.195	45/.304	.025
	17084484	¼	Fixed	①	.085	24/.136	1	28/.164	Fixed	32/.195	45/.304	.025

2SE, E2SE CARBURETOR SPECIFICATIONS
General Motors—Canada

Year	Carburetor Identification	Float Level (In.)	Pump Rod (in.)	Fast Idle (rpm)	Choke Coil Lever (in.)	Fast Idle Cam (deg./in.)	Air Valve Rod (deg.)	Primary Vacuum Break (deg./in.)	Choke Setting (notches)	Secondary Vacuum Break (deg./in.)	Choke Unloader (deg./in.)	Secondary Lockout (in.)
'84	17084485	1/4	Fixed	①	.085	24/.136	1	28/.164	Fixed	32/.195	45/.304	.025
	17084486	1/4	Fixed	①	.085	24/.136	1	28/.164	Fixed	32/.195	45/.304	.025
	17084487	1/4	Fixed	①	.085	24/.136	1	28/.164	Fixed	32/.195	45/.304	.025
	17084620	7/16	Fixed	①	.085	24/.136	1	26/.149	Fixed	32/.195	45/.304	.025
	17084621	7/16	Fixed	①	.085	24/.136	1	26/.149	Fixed	32/.195	45/.304	.025
	17084622	7/16	Fixed	①	.085	24/.136	1	26/.149	Fixed	32/.195	45/.304	.025
	17084623	7/16	Fixed	①	.085	24/.136	1	26/.149	Fixed	32/.195	45/.304	.025

① See underhood decal
② Not available

Secondary Lockout Adjustment

1. Pull the choke wide open by pushing out on the intermediate choke lever.
2. Open the throttle until the end of the secondary actuating lever is opposite the toe of the lockout lever.
3. Gauge clearance between the lockout lever and secondary lever should be as specified.
4. To adjust, bend the lockout lever where it contacts the fast idle cam.

Model E2ME Carburetor

The Dualjet E2ME Model 210 is a variation of the M2ME, modified for use with the Electronic Fuel Control System (also called the Computer Controlled Catalytic Converter, or C-4, System). An electrically operated mixture control solenoid is mounted in the float bowl. Mixture is thus controlled by the Electronic Control Module, in response to signals from the oxygen sensor mounted in the exhaust system upstream of the catalytic converter.

ADJUSTMENTS

Float Level Adjustment

See the illustration for float level adjustment for all carburetors.

③ GAUGE FROM TOP OF CASTING TO TOP OF FLOAT – GAUGING POINT 3/16" BACK FROM END OF FLOAT AT TOE (SEE INSET)
① HOLD RETAINER FIRMLY IN PLACE
(INSET)
TOE
② PUSH FLOAT DOWN LIGHTLY AGAINST NEEDLE
④ REMOVE FLOAT AND BEND FLOAT ARM UP OR DOWN TO ADJUST
GAUGING POINT (3/16" BACK FROM TOE)
⑤ VISUALLY CHECK FLOAT ALIGNMENT AFTER ADJUSTING

Typical float level adjustment on Rochester E2ME and M2ME models

The E2ME procedure is the same except for adjustment (step 4 in the figure). For the E2ME only, if the float level is too high, hold the retainer firmly in place and push down on the center of the float to adjust.

If the float level is too low on the E2ME, lift out the metering rods. Remove the solenoid connector screws. Turn the lean mixture

① ATTACH RUBBER BAND TO INTERMEDIATE CHOKE LEVER.
② OPEN THROTTLE TO ALLOW CHOKE VALVE TO CLOSE.
③ SET UP ANGLE GAGE AND SET ANGLE TO SPECIFICATION.
④ RETRACT VACUUM BREAK PLUNGER USING VACUUM SOURCE, AT LEAST 18" HG. PLUG AIR BLEED HOLES WHERE APPLICABLE.
WHERE APPLICABLE, PLUNGER STEM MUST BE EXTENDED FULLY TO COMPRESS PLUNGER BUCKING SPRING.
⑤ TO CENTER BUBBLE, EITHER:
A. ADJUST WITH 1/8" (3.175 mm) HEX WRENCH (VACUUM STILL APPLIED)
-OR
B. SUPPORT AT "5-S", BEND WIRE-FORM VACUUM BREAK ROD (VACUUM STILL APPLIED)

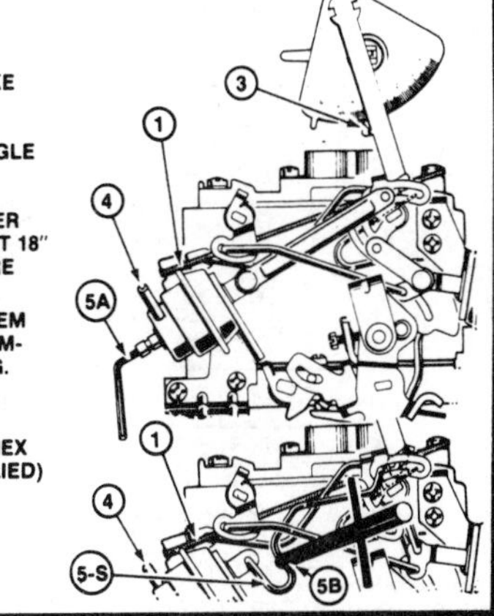

Secondary vacuum break adjustment—1983 and later models

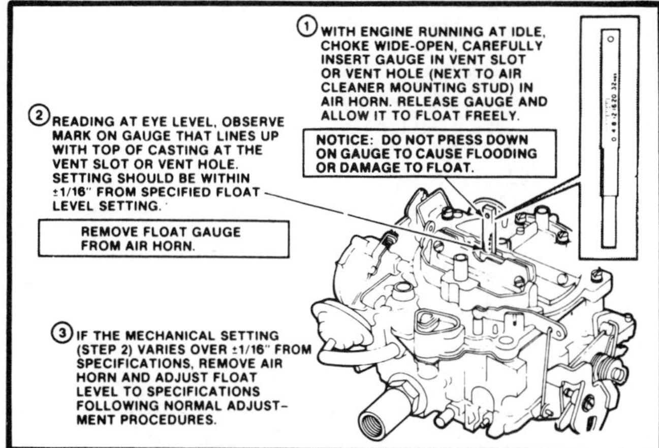

① WITH ENGINE RUNNING AT IDLE, CHOKE WIDE-OPEN, CAREFULLY INSERT GAUGE IN VENT SLOT OR VENT HOLE (NEXT TO AIR CLEANER MOUNTING STUD) IN AIR HORN. RELEASE GAUGE AND ALLOW IT TO FLOAT FREELY.
② READING AT EYE LEVEL, OBSERVE MARK ON GAUGE THAT LINES UP WITH TOP OF CASTING AT THE VENT SLOT OR VENT HOLE. SETTING SHOULD BE WITHIN ±1/16" FROM SPECIFIED FLOAT LEVEL SETTING.
NOTICE: DO NOT PRESS DOWN ON GAUGE TO CAUSE FLOODING OR DAMAGE TO FLOAT.
REMOVE FLOAT GAUGE FROM AIR HORN.
③ IF THE MECHANICAL SETTING (STEP 2) VARIES OVER ±1/16" FROM SPECIFICATIONS, REMOVE AIR HORN AND ADJUST FLOAT LEVEL TO SPECIFICATIONS FOLLOWING NORMAL ADJUSTMENT PROCEDURES.

Checking float level on the car with external float gauge

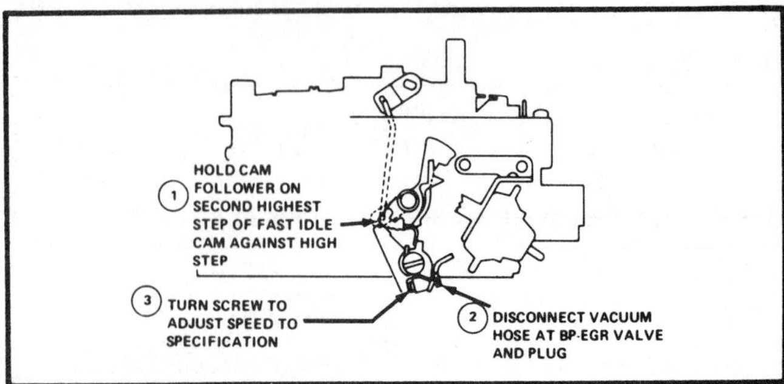

Fast idle speed adjustment—Rochester E2ME and M2ME models

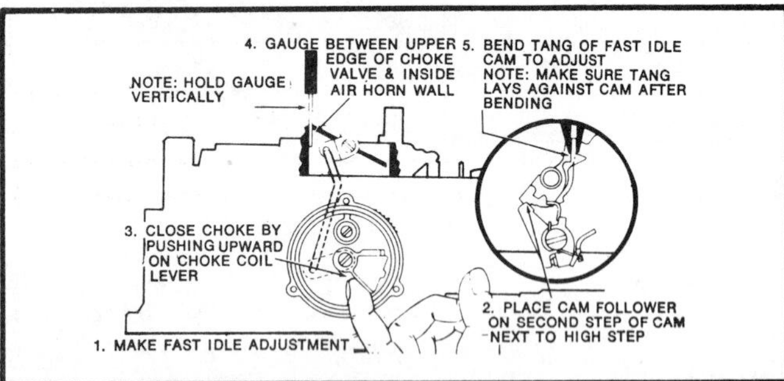

Fast idle cam (choke rod) adjustment—typical

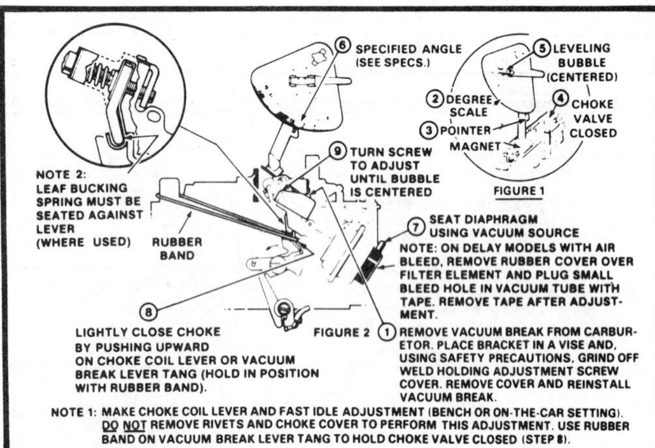

Front vacuum break adjustment—1981-82 models

solenoid screw in clockwise, counting the exact number of turns until the screw is lightly bottomed in the bowl. Then turn the screw out counterclockwise and remove it. Lift out the solenoid and connector. Remove the float and bend the arm up to adjust. Install the parts, installing the mixture solenoid screw in until it is lightly bottomed, the turning it out the exact number of turns counted earlier.

Fast Idle Speed

1. Place the fast idle lever on the high step of the fast idle cam.
2. Turn the fast idle screw out until the throttle valves are closed.
3. Turn the screw in to contact the lever, then turn it in the number of turns listed in the specifications. Check this preliminary setting against the sticker figure.

Fast Idle Cam (Choke Rod) Adjustment

1. Adjust the fast idle speed.
2. Place the cam follower lever on the second step of the fast idle cam, holding it firmly against the rise of the high step.
3. Close the choke valve by pushing upward on the choke coil lever inside the choke housing, or by pushing up on the vacuum break lever tang.
4. Gauge between the upper edge of the choke valve and the inside of the air horn wall.
5. Bend the tang on the fast idle cam to adjust.

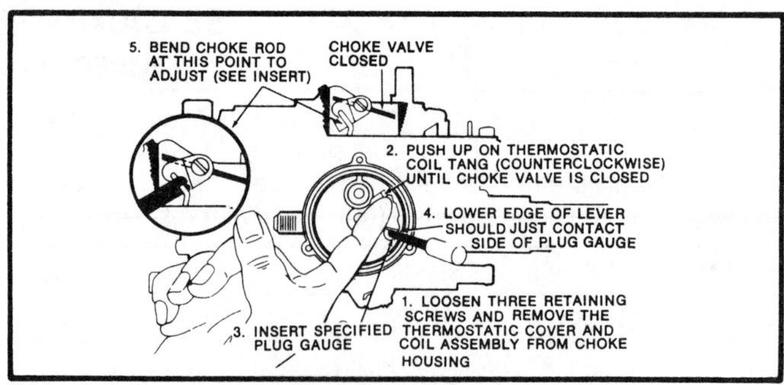

Typical choke coil lever adjustment

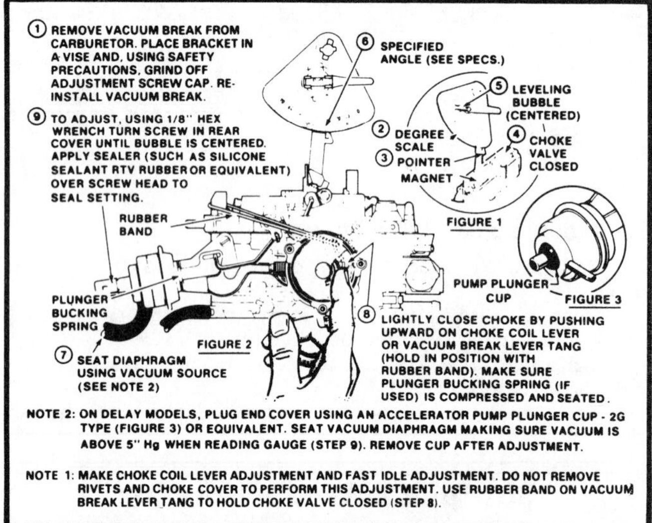

Rear vacuum break adjustment—1981-82 E2ME models

Choke Coil Lever Adjustment

1. Remove the choke cover and thermostatic coil from the choke housing. On models with a fixed choke cover, drill out the rivets and remove the cover. A stat cover kit will be required for assembly.
2. Push up on the coil tang (counterclockwise) until the choke valve is closed. The top of the choke rod should be at the bottom of the slot in the choke valve lever. Place the fast idle cam follower on the high step of the cam.
3. Insert a 0.120 in. plug gauge in the hole in the choke housing.
4. The lower edge of the choke coil lever should just contact the side of the plug gauge.
5. Bend the choke rod to adjust.

Front/Rear Vacuum Break Adjustment

1978-80 MODELS

1. Seat the front diaphragm, using an outside vacuum source. If there is an air bleed hole on the diaphragm, tape it over.
2. Remove the choke cover and coil. Rotate the inside coil lever counter-clockwise. On models with a fixed choke cover (riveted), push up on the vacuum break lever tang and hold it in position with a rubber band.
3. Check that the specified gap is present between the top of the choke valve and the air horn wall.
4. Turn the front vacuum break adjusting screw to adjust.

5. To adjust the rear vacuum break diaphragm, perform Steps 1-3 on the rear diaphragm, but make sure that the plunger bucking spring is compressed and seated in Step 2. Adjust by bending the link at the bend nearest the diaphragm.

1981-84 MODELS

On these models a choke valve measuring gauge J-26701 or equivalent is used to measure angle (degrees instead of inches). See illustration for procedure.

Unloader Adjustment

1. With the choke valve completely closed, hold the throttle valves wide open.
2. Measure between the upper edge of the choke valve and air horn wall.
3. Bend the tang on the fast idle lever to obtain the proper measurement.

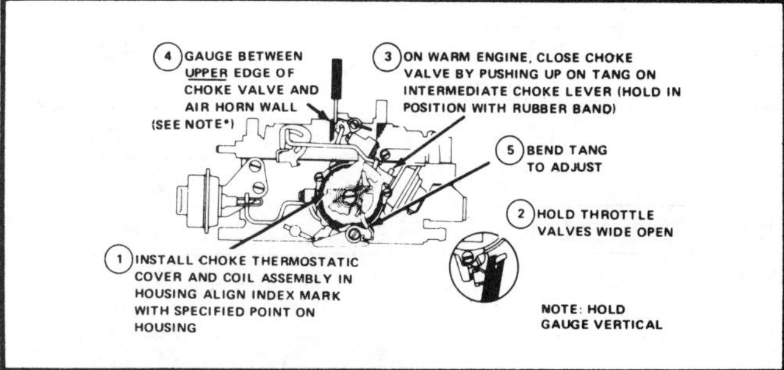

Typical choke unloader adjustment

Mixture Control Solenoid Plunger Adjustments

1. Remove air horn, mixture control solenoid plunger, air horn gasket and plastic filler block, using normal service procedures.
2. Check carburetor for cause of incorrect mixture:
   a. M/C solenoid bore or plunger worn or sticking.
   b. Metering rods for incorrect part number, sticking, or rods or springs not installed properly.
   c. Foreign material in jets.

3. Remove throttle side metering rod. Install mixture control solenoid gaging tool, J-33815-1, BT-8253-A or equivalent, over the throttle side metering jet rod guide, and temporarily reinstall the solenoid plunger into the solenoid body.
4. Holding the solenoid plunger in the DOWN position, use Tool J-28696-10, BT-7928, or equivalent, to turn lean mixture (solenoid) screw counterclockwise until the plunger breaks contact with the gaging tool. Turn slowly clockwise until the plunger makes contact with the gaging tool. The adjustment is correct when the solenoid plunger is contacting BOTH the SOLENOID STOP and the GAGING TOOL

NOTE: If the total difference in adjustment required less than 3/4 turn of the lean mixture (solenoid) screw, the original setting was within the manufacturer's specifications.

5. Remove solenoid plunger and gaging tool, and reinstall metering rod and plastic filler block.
6. Invert air horn, and remove rich mixture stop screw from bottom side of air horn, using Tool J-28696-4, BT7967 A or equivalent.
7. Remove lean mixture screw plug and the rich mixture stop screw plug from air horn, using a suitable sized punch.
8. Reinstall rich mixture stop screw in air horn and bottom lightly, then back screw out 1/4 turn.
9. Reinstall air horn gasket, mixture control solenoid plunger and air horn to carburetor.
10. Adjust M/C Solenoid Plunger travel.
   a. Insert float gage down "D" shaped vent hole. Press down on gage and release, observing that the gage moves freely and does not bind. With gage released, (plunger UP position), read

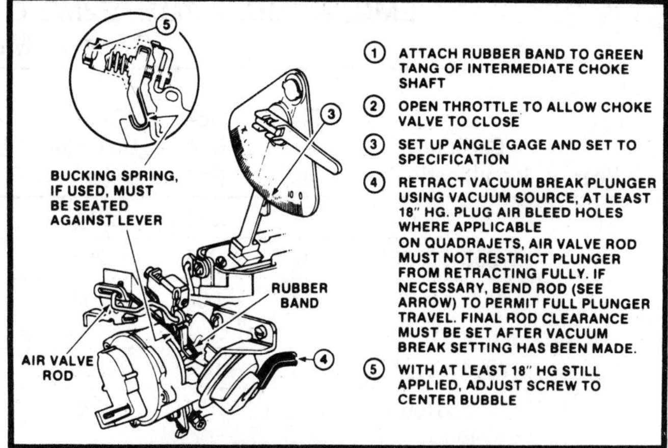

BUCKING SPRING, IF USED, MUST BE SEATED AGAINST LEVER
RUBBER BAND
AIR VALVE ROD
1 ATTACH RUBBER BAND TO GREEN TANG OF INTERMEDIATE CHOKE SHAFT
2 OPEN THROTTLE TO ALLOW CHOKE VALVE TO CLOSE
3 SET UP ANGLE GAGE AND SET TO SPECIFICATION
4 RETRACT VACUUM BREAK PLUNGER USING VACUUM SOURCE, AT LEAST 18" HG. PLUG AIR BLEED HOLES WHERE APPLICABLE
ON QUADRAJETS, AIR VALVE ROD MUST NOT RESTRICT PLUNGER FROM RETRACTING FULLY. IF NECESSARY, BEND ROD (SEE ARROW) TO PERMIT FULL PLUNGER TRAVEL. FINAL ROD CLEARANCE MUST BE SET AFTER VACUUM BREAK SETTING HAS BEEN MADE.
5 WITH AT LEAST 18" HG STILL APPLIED, ADJUST SCREW TO CENTER BUBBLE

Front vacuum break adjustment—1983 and later E2ME models

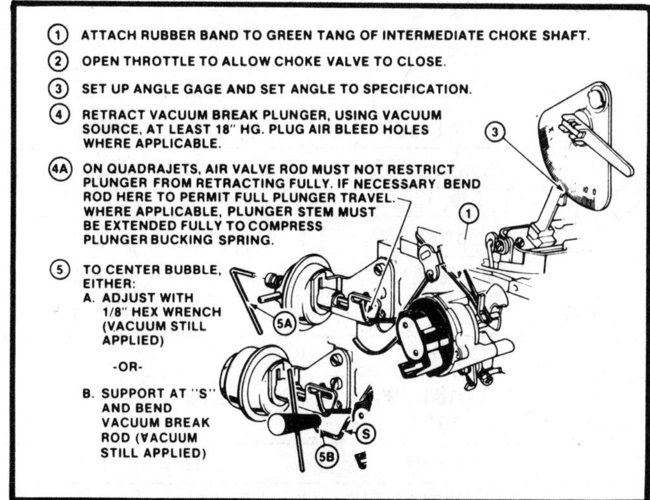

1 ATTACH RUBBER BAND TO GREEN TANG OF INTERMEDIATE CHOKE SHAFT.
2 OPEN THROTTLE TO ALLOW CHOKE VALVE TO CLOSE.
3 SET UP ANGLE GAGE AND SET ANGLE TO SPECIFICATION.
4 RETRACT VACUUM BREAK PLUNGER, USING VACUUM SOURCE, AT LEAST 18" HG. PLUG AIR BLEED HOLES WHERE APPLICABLE.
4A ON QUADRAJETS, AIR VALVE ROD MUST NOT RESTRICT PLUNGER FROM RETRACTING FULLY. IF NECESSARY BEND ROD HERE TO PERMIT FULL PLUNGER TRAVEL. WHERE APPLICABLE, PLUNGER STEM MUST BE EXTENDED FULLY TO COMPRESS PLUNGER BUCKING SPRING.
5 TO CENTER BUBBLE, EITHER:
A. ADJUST WITH 1/8" HEX WRENCH (VACUUM STILL APPLIED)
-OR-
B. SUPPORT AT "S" AND BEND VACUUM BREAK ROD (VACUUM STILL APPLIED)

Rear vacuum break adjustment—1983 and later E2ME models

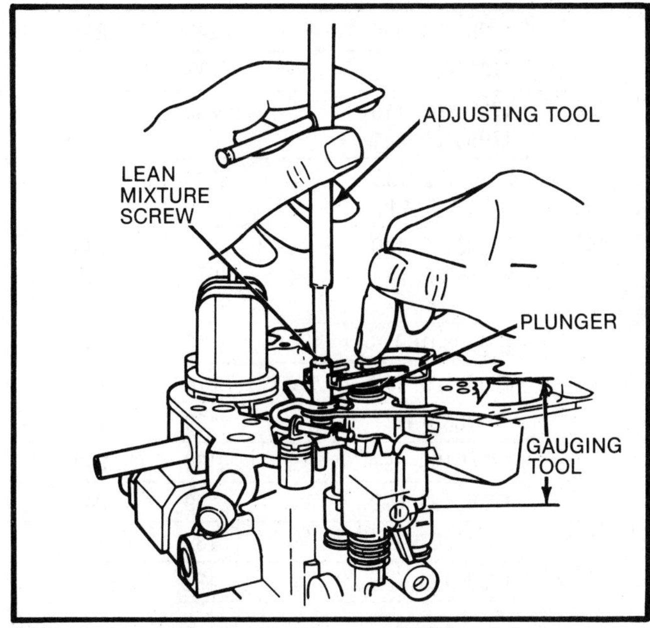

Adjusting lean mixture solenoid screw—E2ME models

2MC, M2MC, M2ME, E2ME CARBURETOR SPECIFICATIONS
General Motors—U.S.A.

Year	Carburetor Identification ①	Flat Level (in.)	Choke Rod (in.)	Choke Unloader (in.)	Vacuum Break Lean or Front (deg./in.)	Vacuum Break Rich or Rear (deg./in.)	Pump Rod (in.)	Choke Coil Lever (in.)	Automatic Choke (notches)
'78	All Chev.	1/4	.314	.314	.136	—	9/32②	.120	Index
	17058150, 152	3/8	.065	.203	.203	.133	1/4②	.120	2 Rich
	17058151, 156, 158	3/8	.065	.203	.229	.133	11/32③	.120	2 Rich
	17058154, 155	3/8	.071	.220	.157	.260	11/32③	.120	2 Rich
	17058160 Buick	11/32	.133	.220	.149	.227	1/4②	.120	2 Lean
	17058160 Pontiac	11/32	.126	.203	.142	.195	1/4②	.120	2 Rich
	17058192	1/4	.074	.350	.117	.103	9/32②	.120	1 Rich
	17058450	3/8	.065	.203	.146	.289	11/32③	.120	2 Rich
	17058496	1/4	.077	.243	.136	.211	3/8③	.120	1 Rich
'79	17059134, 135, 136, 137	11/32	.243	.243	.157	—	9/32②	.120	1 Lean
	17059150, 152	3/8	.065	.203	.203	.133	1/4②	.120	2 Rich
	17059151	3/8	.071	.220	.243	.142	11/32③	.120	2 Rich
	17059154	3/8	.071	.220	.157	.260	11/32③	.120	2 Rich
	17059160	11/32	.110	.195	.129	.187	1/4②	.120	2 Rich
	17059180, 190, 191	11/32	.039	.243	.103	.090	1/4②	.120	2 Rich
	17059193	5/16	.139	.243	.117	.187	Fixed	.120	Fixed
	17059194	5/16	.139	.243	.117	.179	Fixed	.120	Fixed
	17059196	5/16	.139	.243	.164	.136	Fixed	.120	Fixed
	17059430, 432	9/32	.243	.243	.157	—	1/4②	.120	1 Lean
	17059434, 436	13/32	.243	.243	.171	—	1/4②	.120	1 Lean
	17059492, 493	11/32	.139	.277	.129	.117	9/32②	.120	1 Rich
'80	17080108, 110 17080130, 131	3/8	.243	.243	.142	—	5/16②	.120	Fixed
	17080132, 133, 147, 148, 149	5/16	.243	.243	.142	—	5/16②	.120	Fixed
	17080138, 140	3/8	.243	.243	.142	—	5/16②	.120	Fixed
	17080150, 152, 153	3/8	.071	.220	.243	.157	11/32③	.120	Fixed
	17080160	5/16	.110	.243	.168	.207	1/4②	.120	Fixed
	17080190, 192	9/32	.074	.243	.123	.110	1/4②	.120	Fixed
	17080191	11/32	.139	.243	.096	.096	1/4②	.120	Fixed
	17080195, 197	9/32	.139	.243	.103	.071	1/4②	.120	Fixed
	17080490, 492	5/16	.139	.243	.117	.203	1/4②	.120	Fixed
	17080491	5/16	.139	.243	.117	.220	1/4②	.120	Fixed
	17080493, 495	5/16	.139	.243	.117	.179	3/8	.120	Fixed

2MC, M2MC, M2ME, E2ME CARBURETOR SPECIFICATIONS
General Motors—U.S.A.

Year	Carburetor Identification ①	Flat Level (in.)	Choke Rod (in.)	Choke Unloader (in.)	Vacuum Break Lean or Front (deg./in.)	Vacuum Break Rich or Rear (deg./in.)	Pump Rod (in.)	Choke Coil Lever (in.)	Automatic Choke (notches)
'80	17080494	5/16	.139	.243	.117	.179	1/4②	.120	Fixed
	17080496, 498	5/16	.139	.243	.117	.203	Fixed	.120	Fixed
'81	17080185, 187	9/32	.139	.243	19/.103	14/.071	1/4②	.120	Fixed
	17080191	11/32	.139	.243	18/.096	18/.096	1/4②	.120	Fixed
	17081130, 131, 132, 133	11/32	.110	.243	25/.142	—	Fixed	.120	Fixed
	17081138, 140	11/32	.110	.260	25/.142	—	Fixed	.120	Fixed
	17081150, 152	13/32	.071	.220	24/.136	36/.227	Fixed	.120	Fixed
	17081160	11/32	.074	.220	24/.136	37/.234	④	.120	Fixed
	17081196	5/16	.139	.243	28/.164	24/.136	④	.120	Fixed
	17081190, 193	5/16	.139	.243	21/.117	31/.187	Fixed	.120	Fixed
	17081191, 194	5/16	.139	.243	28/.164	24/.136	④	.120	Fixed
	17081198	3/8	.139	.243	28/.164	24/.136	④	.120	Fixed
	17081192, 197	5/16	.139	.243	21/.117	30/.179	④	.120	Fixed
	17081199	3/8	.096	.243	18/.096	24/.136	Fixed	.120	Fixed
'82	17082130, 132, 138, 140	3/8	.110	.164	27/.157	—	④	④	Fixed
	17082150	13/32	.071	.220	24/.136	38/.243⑤	④	④	Fixed
	17082182, 184	5/16	.096	.195	28/.164	24/.136	④	④	Fixed
	17082192, 194	5/16	.096	.195	28/.164	24/.136	④	④	Fixed
	17082196	5/16	.096	.157	21/.117	19/.103	④	④	Fixed
	17082497	5/16	.113	.195	28/.164	24/.136	④	.120	Fixed
'83	17082130, 132	3/8	.110	.243	27/.157	—	④	.120	Fixed
	17083190, 192	5/16	.096	.195	28/.164	24/.136	④	.120	Fixed
	17083193	5/16	.090	.157	23/.129	28/.164	④	.120	Fixed
	17083194	5/16	.090	.220	27/.157	25/.142	④	.120	Fixed
'84	17082130	3/8	.110	.243	27/.157	None	④	.120	Fixed
	17082132	3/8	.110	.243	27/.157	None	④	.120	Fixed
	17084191	5/16	.096	.195	28/.164	24/.136	④	.120	Fixed
	17084193	5/16	.090	.220	27/.157	25/.142	④	.120	Fixed
	17084194	5/16	.090	.220	27/.157	25/.142	④	.120	Fixed
	17084195	5/16	.090	.220	27/.157	25/.142	④	.120	Fixed

① The carburetor identification number is stamped on the float bowl, next to the fuel inlet nut.
② Inner hole
③ Outer hole
④ Not Adjustable
⑤ High altitude—0.206

2MC, M2MC, M2ME, E2ME CARBURETOR SPECIFICATIONS
General Motors—Canada

Year	Carburetor Identification ①	Float Level (in.)	Choke Rod (in.)	Choke Unloader (in.)	Vacuum Break Lean or Front (deg./in.)	Vacuum Break Rich or Rear (deg./in.)	Pump Rod (in.)	Choke Coil Lever (in.)	Automatic Choke (notches)
'81	17080191	11/32	.139	.243	18/.096	18/.096	1/4②	.120	Fixed
	17081492	9/32	.139	.243	17/.090	19/.103	1/4②	.120	Fixed
	17081493	9/32	.139	.243	17/.090	19/.103	1/4②	.120	Fixed
	17081170	13/32	.110	.243	25/.142	—	1/4②	.120	Fixed
	17081171	13/32	.110	.243	25/.142	—	1/4②	.120	Fixed
	17081174	9/32	.110	.243	25/.142	—	1/4②	.120	Fixed
	17081175	9/32	.110	.243	25/.142	—	1/4②	.120	Fixed
'82	17082174	9/32	.110	.243	25/.142	—	5/16②	.120	Fixed
	17082175	9/32	.110	.243	25/.142	—	5/16②	.120	Fixed
	17082492	9/32	.139	.243	17/.090	19/.103	1/4②	.120	Fixed
	17082172	9/32	.110	.243	25/.142	—	5/16②	.120	Fixed
	17082173	9/32	.110	.243	25/.142	—	5/16②	.120	Fixed
'83–'84	17083172	9/32	.139	.243	17/.090	19/.103	1/4②	.120	Fixed

① The carburetor identification number is stamped on the float bowl, next to the fuel inlet nut.
② Inner hole

at eye level and record the reading of the gage mark (in inches) that lines up with the top of air horn casting, (upper edge).

b. Lightly press down on gage until bottomed, (plunger DOWN position). Read and record (in inches) the reading of the gage mark that lines up with top of air horn casting.

c. Subtract gage UP position (Step 1) from gage DOWN position (Step 2), and record difference. This difference is total plunger travel. Insert external float gage in vent hole and, with Tool J-28696-10, BT-7928, or equivalent, adjust rich mixture stop screw to obtain 4/32 in. total plunger travel.

11. With solenoid plunger travel correctly set, install plugs (supplied in service kits) in the air horn, as follows:

a. Install plug, hollow end down, into the access hole to lean mixture (solenoid) screw, and use suitably sized punch to drive plug into the air horn until the top of plug is even with the lower. Plug must be installed to retain the screw setting and to prevent fuel vapor loss.

b. Install plug, with hollow end down, over the rich mixture stop screw access hole, and drive plug into place so that the top of the plug is 1/16 in. below the surface of the air horn casting.

NOTE: Plug must be installed to retain screw setting.

12. To check the M/C solenoid dwell, first disconnect vacuum line to the canister purge valve and plug it. Ground diagnostic "test" terminal and run engine until it is at normal operation temperature (upper radiator hose hot) and in closed loop.

13. Check M/C dwell at 3000 RPM. If within 10° – 50°, calibration is complete. If higher than 50°, check carburetor for cause of rich condition. If below 10°, look for cause of lean engine condition such as vacuum leaks. If none found, check for cause of lean carburetor.

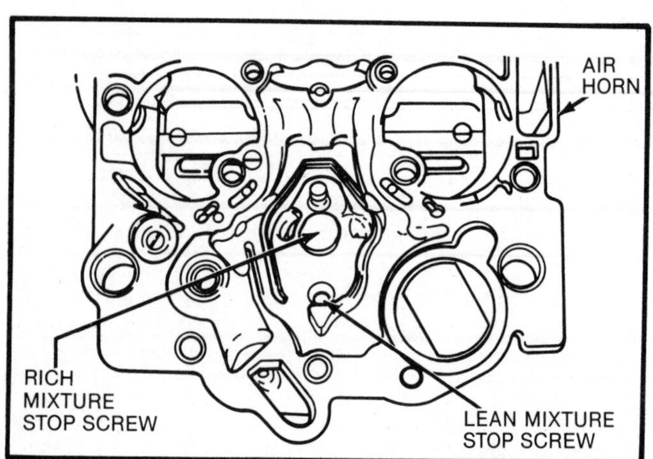

Location of rich and lean mixture stop screws—E2ME models

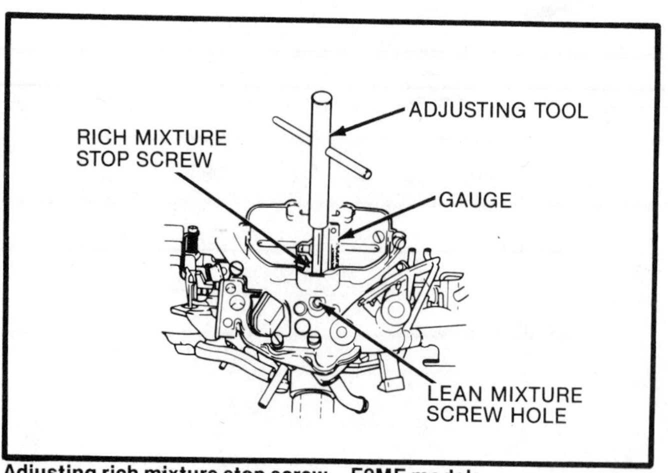

Adjusting rich mixture stop screw—E2ME models

Idle Mixture and Speed Adjustment

A cover is in place over the idle air bleed valve, and the access holes to the idle mixture needles are sealed with hardened plugs, to seal the factory settings, during original equipment production. These items are NOT to be removed unless required for cleaning, part replacement, improper dwell readings, or if the System Performance Check indicates the carburetor is the cause of the trouble.

Remove the idle air bleed cover. A missing cover indicates that the idle air bleed valve setting has been changed from its original factory setting. With cover removed, look for presence (or absence) of a letter identification on top of idle air bleed valve.

If no identifying letter appears on top of the valve, begin Procedure A, below. If the valve is identified with a letter, begin Procedure B.

PROCEDURE A (NO LETTER ON IDLE AIR BLEED VALVE)

Carburetors WITHOUT letter identification on the idle air bleed valve are adjusted by presetting the idle air bleed valve to a gage dimension if the valve was serviced prior to on-vehicle adjustment. Install idle air bleed valve gaging Tool J-33815-2, BT-8253-B, or equivalent, in throttle side "D" shaped vent hole in the air horn casting. The upper end of the tool should be positioned over the open cavity next to the idle air bleed valve.

While holding the gaging tool down lightly, so that the solenoid plunger is against the solenoid stop, adjust the idle air bleed valve so that the gaging tool will pivot over and just contact the top of the valve. The valve is now preset for on-vehicle adjustment. Remove gaging tool.

1. Disconnect the vacuum hose from the canister purge valve and plug it.
2. Start engine and allow it to reach normal operating temperature.
3. While idling in Drive (Neutral for manual transmission), use a screwdriver to slowly turn the valve counterclockwise or clockwise, until the dwell reading varies within the 25-35° range, attempting to be as close to 30° as possible.

NOTE: Perform this step carefully. The air bleed valve is very sensitive and should be turned in 1/8 turn increments only.

4. If, after performing Steps b and c above, the dwell reading does not vary and is not with the 25-35° range, it will be necessary to remove the plugs and to adjust the idle mixture needles.

Idle Mixture Needle Plug Removal

1. Remove the carburetor from the engine, following normal service procedures, to gain access to the plugs covering the idle mixture needles.
2. Invert carburetor and drain fuel into a suitable container.

CAUTION

Take precautions to avoid the risk of fire. Do not smoke.

3. Place carburetor on a suitable holding fixture, with manifold side up. Use care to avoid damaging linkage, tubes, and parts protruding from air horn.
4. Make two parallel cuts in the throttle body, one on each side of the locator points beneath the idle mixture needle plug (manifold side), with a hacksaw.

NOTE: The cuts should reach down to the steel plug, but should not extend more than 1/8 in. beyond the locator points. The distance between the saw cuts depends on the size of the punch to be used.

5. Place a flat punch near the ends of the saw marks in the throttle body. Hold the punch at a 45° angle and drive it into the throttle body until the casting breaks away, exposing the steel plug. The hardened plug will break, rather than remaining intact. It is not necessary to remove the plug in one piece, but remove the loose pieces.

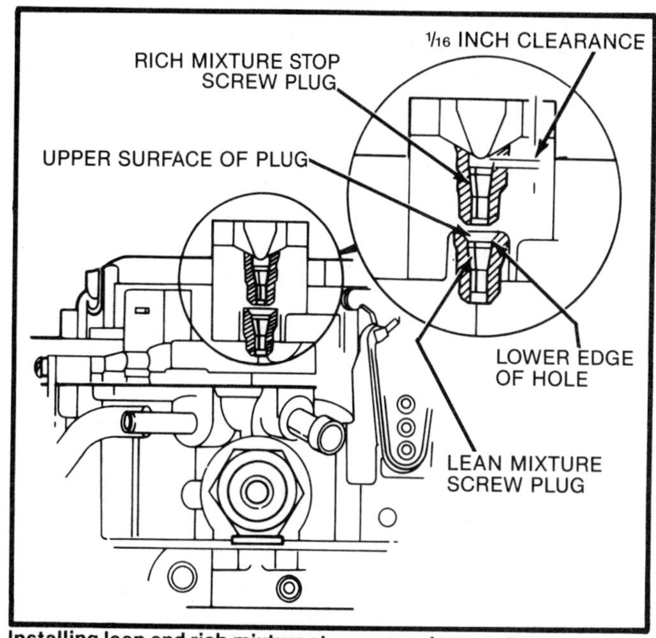

Installing lean and rich mixture stop screw plugs on E2ME models

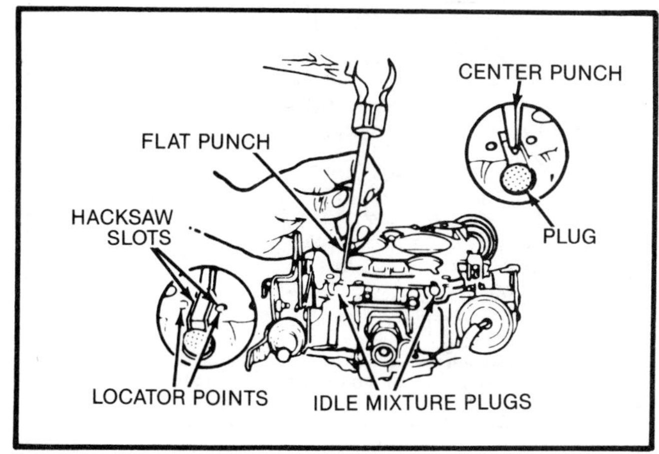

Removing idle mixture plugs—E2ME models

6. Repeat this procedure with the other mixture needle.

Setting the Idle Mixture Needles

1. Using Tool J-29030, BT-7610B, or equivalent, turn both idle mixture needles clockwise until they are lightly seated, then turn each mixture needle counterclockwise the number of turns specified in an applicable manufacturer's specifications.
2. Reinstall carburetor on engine using a new flange mounting gasket, but do not install air cleaner and gasket at this time.
3. Readjust the idle air bleed valve to finalize correct dwell reading.
4. Start engine and run until fully warm, and repeat the idle air bleed valve adjustment.
5. If unable to set dwell to 25–35°, and the dwell is below 25°, turn both mixture needles counterclockwise an additional turn. If dwell is above 35°, turn both mixture needles clockwise an additional turn. Readjust idle air bleed valve to obtain dwell limits.
6. After adjustments are complete, seal the idle mixture needle openings in the throttle body, using silicone sealant, RTV rubber, or equivalent. The sealer is required to discourage unnecessary adjustment of the setting, and to prevent fuel vapor loss in that area.

7. On vehicles without a carburetor-mounted Idle Speed Control or Idle Load Compensator, adjust curb idle speed if necessary.

8. Check, and only if necessary adjust, fast idle speed as described on Vehicle Emission Control Information label.

PROCEDURE B (LETTER APPEARS ON IDLE AIR BLEED VALVE)

Carburetors WITH letter identification on the idle air bleed valve are adjusted by installing air bleed valve gaging tool J-33815-2, BT-8253-B, or equivalent, in throttle side "D" shaped vent hole in the air horn casting. The upper end of the tool should be positioned over the open cavity next to the idle air bleed valve. While holding the gaging tool down lightly, so that the solenoid plunger is against the solenoid stop, adjust the idle air bleed valve so that the gaging tool will pivot over and just contact the top of the valve. The valve is now set properly. No further adjustment of the valve is necessary. Remove gaging tool.

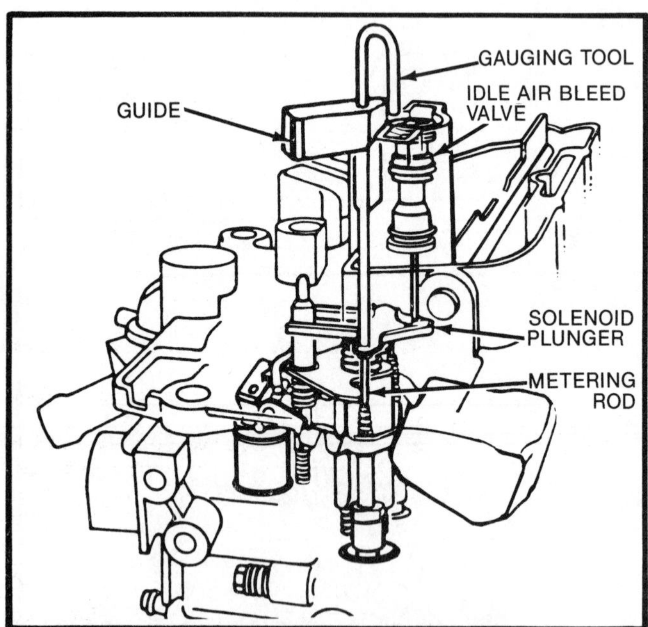

Correct installation of idle air bleed valve gauging tool

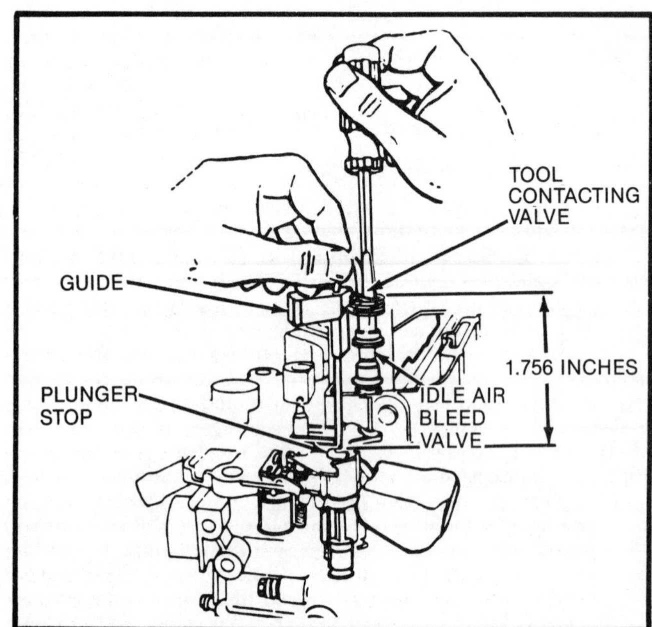

Positioning idle air bleed valve—E2ME models

Adjusting the Idle Mixture Needles

1. Remove idle mixture needle plugs, following instructions in PROCEDURE A.

2. Using Tool J-29030-B, BT-7610-B, or equivalent, turn each idle mixture needle clockwise until lightly seated, then turn each mixture needle counterclockwise 3 turns.

3. Reinstall carburetor on engine, using a new flange mounting gasket, but do not install air cleaner or gasket at this time.

4. Disconnect vacuum hose to canister purge valve and plug it.

5. Start engine and allow it to reach normal operating temperature.

6. While idling in Drive (Neutral for manual transmission), adjust both mixture needles equally, in 1/8 turn increments, until dwell reading varies within the 25-35° range, attempting to be as close to 30° as possible.

7. If reading is too low, turn mixture needles counterclockwise. If reading is too high, turn mixture needles clockwise. Allow time for dwell reading to stabilize after each adjustment.

NOTE: After adjustments are complete, seal the idle mixture needle openings in the throttle body, using silicone sealant, RTV rubber, or equivalent. The sealer is required to discourage unnecessary readjustment of the setting, and to prevent fuel vapor loss in that area.

8. On vehicles without a carburetor-mounted Idle Speed Control or Idle Load Compensator, adjust curb idle speed if necessary.

9. Check, and if necessary, adjust fast idle speed, as described on the Vehicle Emission Control Information label.

THROTTLE POSITION SENSOR (TPS)

The plug covering the TPS adjustment screw is used to provide a tamper-resistant design and retain the factory setting during vehicle operation. Do not remove the plug unless diagnosis indicates the TPS Sensor is not adjusted correctly, or it is necessary to replace the air horn assembly, float bowl, TPS, or TPS adjustment screw. This is a critical adjustment that must be performed accurately to ensure proper vehicle performance and control of exhaust emissions. Remove TPS plug if not already removed.

TPS Plug Removal

1. Using a 5/64 in. drill, drill a 1/16 in. to 1/8 in. deep hole in aluminum plug covering TPS adjustment screw. Use care in drilling to prevent damage to adjustment screw head.

2. Start a No. 8, 1/2 in. long self-tapping screw in drilled hole

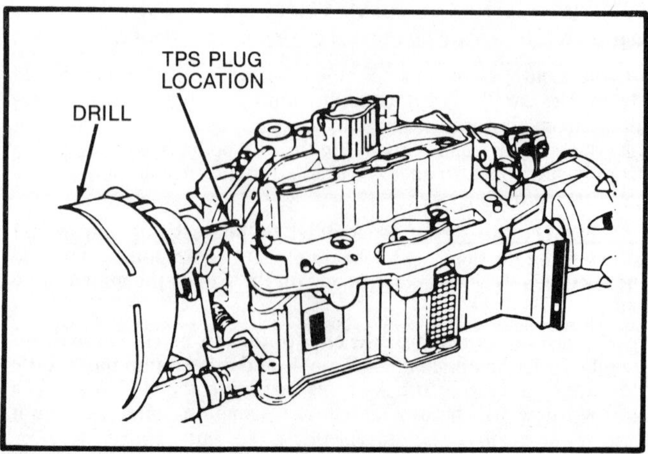

Removing throttle position sensor (TPS) adjustment plug

THROTTLE POSITION SWITCH ADJUSTMENT SPECIFICATIONS

Year	Engine	Carburetor	Throttle Position	Voltage
'80–'85	All	E2SE	Curb Idle	.26
'80–'84	8,Y,A,9	E2ME, E4ME	Curb Idle High Step of Fast Idle	.46 .77
	4,E	E2ME, E4ME	Curb Idle High Step of Fast Idle	.46 .97
	L,H	E2ME, E4ME	Curb Idle	.56
'85	All	E2ME, E2MC	Curb Idle Automatic 3-spd Man 4-spd Man	.46 .57 .60
	All	E4ME, E4MC	Curb Idle	.41

turning screw in only enough to ensure good thread engagement in hole.

3. Placing a wide-blade screwdriver between screw head and air horn casting, pry against screw head to remove plug. Discard plug.

TPS Adjustment

NOTE: Adjustment is required if voltage is different than specifications by more than ± .05 volts.

1. Using Tool J-28696 or equivalent, remove TPS adjustment screw.
2. Connect digital voltmeter J29125-A from TPS connector center "C" to bottom terminal "C" (Jumpers for access can be made using terminals).
3. With ignition ON, engine stopped, reinstall TPS adjustment screw, and with tool J-28696, BT-7967A or equivalent, quickly adjust screw to obtain specified TPS voltage with A/C off.
4. After adjustment, install a new plug (supplied in service kits) in air horn, driving plug in place until flush with raised pump lever boss on casting.

NOTE: Plug must be installed to retain the TPS adjustment screw setting. If plug is not available, remove screw and apply GM Threadlock® Adhesive or equivalent to screw threads, then repeat adjustment.

5. Clear trouble code memory after adjustment.

Removal and Installation

1. Disconnect the M/C Solenoid and TPS electrical connector.
2. Disconnect the idle speed control or idle speed solenoid electrical connector.
3. Loosen the attaching screws and remove idle speed control or, idle speed solenoid or idle load compensator.
4. Disconnect the upper choke lever from the end of choke shaft by removing retaining screw. Rotate the upper choke lever to remove choke rod from slot in lever.
5. Remove the choke rod from lower lever inside the float bowl casting by holding lower lever outward with small screwdriver and twisting rod counterclockwise.
6. Using tool J25322 or a 2.4 mm (3/32 in.) drift punch, drive roll pin (pump lever pivot pin) inward until end of pin is against air cleaner locating boss on air horn casting. Remove pump lever and lever from pump rod.
7. Remove front vacuum break hose from tube on float bowl.
8. Remove air horn-to-bowl screws; then remove the two countersunk attaching screws located next to the venturi. Remove the air horn from float bowl by lifting it straight up.
9. Remove Solenoid-metering rod plunger by lifting straight up.
10. Remove air horn gasket by lifting it from the dowel locating pins on float bowl. DISCARD GASKET.
11. Stake the TPS in bowl as follows:
    a. Lay a flat tool or metal piece across bowl casting to protect gasket sealing surface.
    b. Use a small screwdriver to depress TPS sensor lightly and hold against spring tension.
    c. Observing safety precautions, pry upward with a small chisel or equivalent to remove bowl staking, making sure prying force is exerted against the metal piece and not against the bowl casting.
    d. Push up from bottom on electrical connector and remove TPS and connector assembly from bowl.
12. Install the TPS and connector assembly in float bowl by aligning groove in electrical connector with slot in float bowl casting. Push down on connector and sensor assembly so that connector and wires are located below bowl casting surface. Be sure green TPS actuator plunger is in place in air horn.
13. Install the air horn, holding down on pump plunger assembly against return spring tension, and aligning pump plunger stem with hole in gasket, and aligning holes in gasket over TPS plunger, solenoid plunger return spring, metering rods, solenoid attaching screw and electrical connector. Position gasket over the two dowel locating pins on the float bowl.
14. Solenoid-metering rod plunger, holding down on air horn gasket and pump plunger assembly, and aligning slot in end of plunger with solenoid attaching screw. Remove the solenoid.
15. Carefully lower air horn assembly onto float bowl while positioning the TPS Adjustment Lever over the TPS sensor, and guiding pump plunger stem through seal in air horn casting. To ease installation, insert a thin screwdriver between air horn gasket and float bowl to raise the TPS Adjustment Lever positioning it over the TPS sensor.
16. Install the air horn screws and lockwashers, and two countersunk screws (located next to the carburetor venturi area). Tighten all screws evenly and securely, following air horn screw tightening sequence.
17. Install the front vacuum break and bracket assembly on the air horn, using two attaching screws. Tighten screws securely.
18. Reconnect the upper end of pump rod to pump lever. Place lever between raised bosses on air horn casting, making sure lever engages TPS actuator plunger and the pump plunger stem. Align hole in pump lever with holes in air horn casting bosses. Using a small drift or rod the diameter of the pump lever roll pin will aid alignment. Using diagonal (sidecutter) pliers on the end of the roll pin, pry the roll pin only enough to insert a thin bladed screwdriver between the end of the pump lever roll pin and the air cleaner locating boss on air horn casting. Use screwdriver to push pump lever roll pin back through the casting until end of pin is flush with casting bosses on the air horn.

NOTE: Use care installing the roll pin to prevent damage to the pump lever bearing surface and casting bosses.

19. Install the choke rod into lower choke lever inside bowl cavity. Install choke rod in slot in upper choke lever, and position lever on end of choke shaft, making sure flats on end of shaft align with flats in lever. Install attaching screw and tighten securely. When properly installed, the lever will point to the rear of the carburetor, and the number on the lever will face outward.
20. Install the idle speed control motor.
21. Install the M/C solenoid, TPS and idle speed control or idle speed solenoid electrical connectors.
22. Clear trouble code memory after replacement.
23. Check TPS voltage and refer to TPS adjustment procedure if required.

MIXTURE CONTROL SOLENOID

Checking Plunger Travel

Mixture control solenoid plunger travel should be checked before proceeding with any carburetor adjustments or disassembly. Using Float Gage J-9789-130, BT-7720, or equivalent, (used to check float level setting externally), insert gage in the vertical "D" shaped vent hole in the air horn casting, next to the Idle Air Bleed Cover.

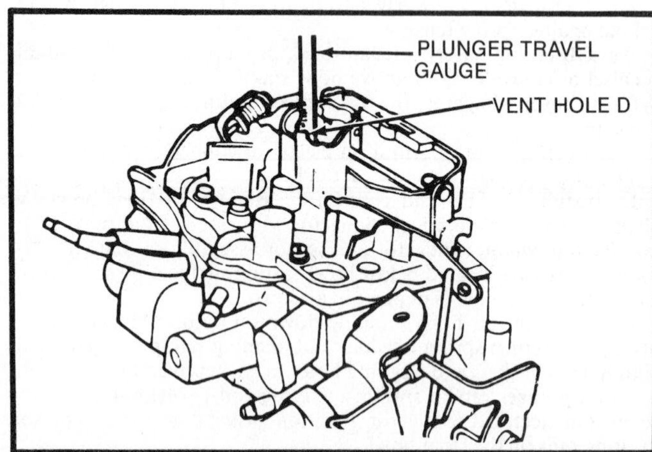

Checking plunger travel—E2ME models

NOTE: It may be necessary to file or grind material off the gage to allow it to enter the vent hole freely. Gage will be used to check total mixture control solenoid plunger travel.

1. Insert float gage down "D" shaped vent hole. Press down on gage and release, observing that the gage moves freely and does not bind. With gage released, (plunger UP position), read at eye level and record the reading of the gage mark (in inches) that lines up with the top of air horn casting, (upper edge).
2. Lightly press down on gage until bottomed, (plunger DOWN position). Read and record mm (inches) the reading of the gage mark that lines up with top of air horn casting.
3. Subtract gage UP position (Step 1) from gage DOWN position Step 2), and record difference. This difference is total plunger travel.
4. If total plunger travel (Step 3) is between 2.4 mm and 4.8 mm ($^{2}/_{32}$ in. and $^{6}/_{32}$ in.), proceed to Idle Air Bleed Valve Adjustment. If it is less than 2.4 mm ($^{2}/_{32}$ in.) or greater than 4.8mm ($^{6}/_{32}$ in.), adjust mixture control solenoid plunger travel indicated below.

Adjusting Plunger Travel

1. Remove air horn, mixture control solenoid plunger, air horn gasket and plastic filler block, using normal service procedures.
2. Remove throttle side metering rod. Install mixture control

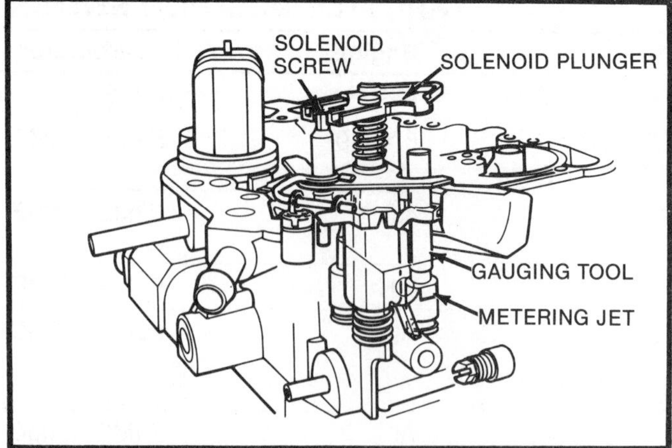

Correct installation of mixture control solenoid gauging tool

solenoid gaging tool, J-33815-1, BT-8253-A, or equivalent, over the throttle side metering jet rod guide, and temporarily reinstall the solenoid plunger into the solenoid body.
3. Holding the solenoid plunger in the DOWN position, use Tool J-28696-10, BT-7928, or equivalent, to turn lean mixture (solenoid) screw counterclockwise until the plunger breaks contact with the gaging tool. Turn slowly clockwise until the plunger just contacts the gaging tool. The adjustment is correct when the solenoid plunger is contacting both the solenoid stop and the gaging tool.

If the total difference in adjustment required less than $^{3}/_{4}$ turn of the lean mixture (solenoid) screw, the original setting was within the manufacturer's specifications.
4. Remove solenoid plunger and gaging tool, and reinstall metering rod and plastic filler block.
5. Invert air horn, and remove rich mixture stop screw from bottom side of air horn, using Tool J-28696-4, BT7967A or equivalent.
6. Remove lean mixture screw plug and the rich mixture stop screw plug from air horn, using a suitable sized punch.
7. Reinstall rich mixture stop screw in air horn and bottom lightly, then back screw out $^{1}/_{4}$ turn.
8. Reinstall air horn gasket, mixture control solenoid plunger and air horn to carburetor.
9. Insert external float gage in vent hole and, with Tool J-28696-10, BT-7928, or equivalent, adjust rich mixture stop screw to obtain $^{4}/_{32}$ in. total plunger travel.
10. With solenoid plunger travel correctly set, install plugs (supplied in service kits) in the air horn. Install plug, hollow end down, into the access hole to lean mixture (solenoid) screw, and use suitably sized punch to drive plug into the air horn until the top of the plug is even with the lower edge of hole chamfer. Plug must be installed to retain the screw setting and to prevent fuel vapor loss. Install plug, with hollow end down, over the rich mixture stop screw access hole, and drive plug into place so that the top of the plug is $^{1}/_{16}$" below the surface of the air horn casting.

NOTE: Plug must be installed to retain screw setting.

Removal and Installation

1. Disconnect the M/C solenoid and TPS electrical connector.
2. Disconnect the idle speed control electrical connector.
3. Loosen the attaching screws and remove idle speed control. Disconnect the upper choke lever from the end of the choke shaft by removing retaining screw. Rotate upper choke lever to remove choke rod from slot in lever. Disconnect the choke rod from lower lever inside the float bowl casting. Remove rod by holding lower lever outward with small screwdriver and twisting rod counterclockwise.
4. With a small drift punch of the correct size or J25322 tool drive roll pin (pump lever pivot pin) inward until end of pin is

against air cleaner locating boss on air horn casting. Remove pump lever and lever from pump rod.

5. Remove the front vacuum break hose from tube on float bowl.

6. Loosen the horn-to-bowl screws; then remove the two countersunk attaching screws located next to the venturi. Remove the air horn from float bowl by lifting it straight up.

7. Remove the solenoid-metering rod plunger by lifting straight up. Remove the air horn gasket by lifting it from the dowel locating pins on float bowl. Discard gasket.

8. Remove the plastic filler block over float bowl. Carefully lift each metering rod out of the guided metering jet, checking to be sure the return spring is removed with each metering rod.

9. Remove the lean mixture screw using tool J28696-10, BT-7928 or equivalent.

10. Remove the screw attaching solenoid connector to float bo̸wl and lift the mixture control solenoid and connector assembly from the float bowl.

NOTE: If a service replacement Mixture Control Solenoid package is installed, the solenoid and plunger MUST be installed as a matched set.

11. Install the mixture control solenoid carefully in the float chamber, aligning pin on end of solenoid with hole in raised boss at bottom of bowl. Align solenoid connector wires to fit in slot in bowl, or plastic insert if used.

12. Install the lean mixture (solenoid) screw through hole in solenoid bracket and tension spring in bowl, engaging first six screw threads to assure proper thread engagement.

13. Install mixture control solenoid gaging Tool, J-33815-1, BT-8253-A or equivalent, over the throttle side metering jet rod guide, and temporarily install solenoid plunger.

14. Holding the solenoid plunger against the Solenoid Stop, use Tool J-28696-10, BT-7938, or equivalent, to turn the lean mixture (solenoid) screw slowly clockwise, until the solenoid plunger just contacts the gaging tool. The adjustment is correct when the solenoid plunger is contacting BOTH the Solenoid Stop and the Gaging Tool. Remove solenoid plunger and gaging tool.

15. Align solenoid wires with notch in plastic insert (if used) and connector lugs with recesses in bowl. Install connector attaching screw. Tighten screw securely.

CAUTION

DO NOT overtighten the screw; it could cause damage to the connector.

16. Install the plastic filler clock over float valve, pressing downward until properly seated (flush with bowl casting surface).

17. Install metering rod and spring assembly through hole in plastic filler block and gently lower the metering rod into the guided metering jet, until large end of spring seats on the recess on end of jet guide.

NOTE: Do not force metering rod down in jet. Use extreme care when handling these critical parts to avoid damage to rod and spring. If service replacement metering rods, springs and jets are installed, they must be installed in matched sets.

18. Install the air horn, holding down on pump plunger assembly against return spring tension and aligning pump plunger stem with hole in gasket, and aligning holes in gasket over TPS plunger, solenoid plunger return spring, metering rods, solenoid attaching screw and electrical connector. Position gasket over the two dowel locating pins on the float bowl.

19. Install the solenoid-metering rod plunger, holding down on aligning slot in end of plunger with solenoid attaching screw. Be sure plunger arms engage top of each metering rod.

20. Carefully lower air horn assembly onto float bowl while positioning the TPS Adjustment Lever over the TPS sensor, and guiding pump plunger stem through seal in air horn casting. To ease installation, insert a thin screwdriver between air horn gasket and float bowl to raise the TPS Adjustment Lever positioning it over the TPS sensor.

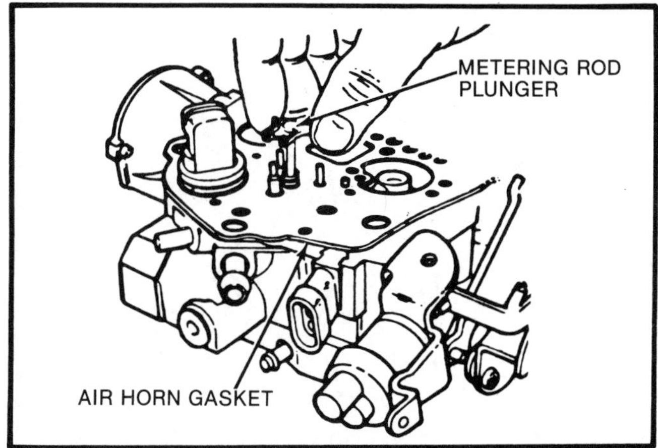

Location of solenoid metering plunger in throttle body—E2ME models

NOTE: Make sure that the tubes are positioned properly through the holes in the air horn gasket. Do not force the air horn assembly onto the bowl, but lower it lightly into place.

21. Install attaching screws and lockwashers, and two countersunk screws (located next to the carburetor venturi area). Tighten all screws evenly and securely, following air horn screw tightening sequence.

IDLE SPEED CONTROL (ISC)

On some carburetors, there is an idle speed control (ISC) to control idle engine speed with the ECM. It is mounted on the carburetor, and moves the throttle lever to open or close the throttle blades at idle. A switch in the housing closes when the throttle lever contacts the ISC plunger. When the switch is closed the ECM controls RPM of the engine. When it is open, the motor is unoperative. When the ignition is turned off, the motor extends slightly, therefore, if the ISC contacts did not close, the engine would have a high idle speed after repeated starts. When A/C is turned "on", the ECM, because of change on engine load, adjusts ISC to maintain idle speed.

On-Car Speed Adjustments

When a new Idle Speed Control (ISC) assembly is installed, a base (minimum) and extended (maximum) rpm speed check must be performed and adjustments made as required. When making a low and high speed adjustment, the low speed adjustment is always made first. DO NOT use the ISC plunger to adjust curb idle speed as the idle speed is controlled by the ECM. Before starting the engine, place transmission selector level in park or neutral, set parking brake, and block drive wheels. Prior to adjustment, check for identification letter on the ISC adjustment plunger Adjust plunger, if required, to Dimension "B".

CAUTION

Do not disconnect or connect ISC connector with ignition ON as damage to the ECM may occur.

1. Connect tachometer.

2. Connect a dwell meter to mixture control (M/C) solenoid dwell (green) lead. Set dwell meter on the six cylinder scale.

3. Turn A/C OFF.

4. Start engine and run until stabilized by entering "closed loop" (dwell meter needle starts to vary).

5. Turn ignition OFF.

6. Unplug connector from ISC motor.

7. Fully retract ISC plunger by applying 12 volts to terminal "C" of the ISC motor connection and ground lead to terminal "D" of the ISC motor connection. It may be necessary to install jumper

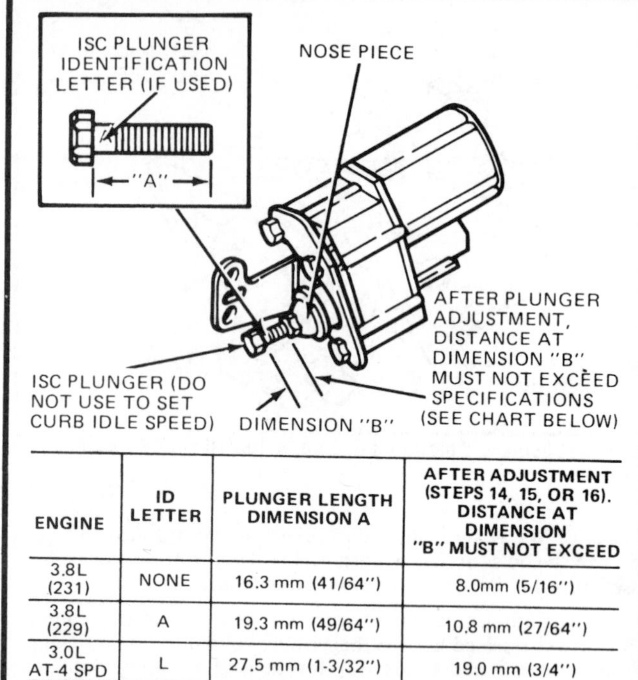

ENGINE	ID LETTER	PLUNGER LENGTH DIMENSION A	AFTER ADJUSTMENT (STEPS 14, 15, OR 16). DISTANCE AT DIMENSION "B" MUST NOT EXCEED
3.8L (231)	NONE	16.3 mm (41/64")	8.0mm (5/16")
3.8L (229)	A	19.3 mm (49/64")	10.8 mm (27/64")
3.0L AT-4 SPD	L	27.5 mm (1-3/32")	19.0 mm (3/4")
3.0L AT-3 SPD	J	30.0 mm (1-3/16")	21.5 mm (27/32")
4.1L	NONE	16.3 mm (41/64")	8.0 mm (5/16")

Idle speed control (ISC) plunger adjustment—typical

leads from the ISC motor in order to make proper connections (refer to Special Tools).

NOTE: Do not leave battery voltage applied to motor longer than necessary to retract ISC plunger. Prolonged contact will damage motor. Also, NEVER connect voltage source across terminals "A" and "B" as damage to the internal throttle contact switch will result.

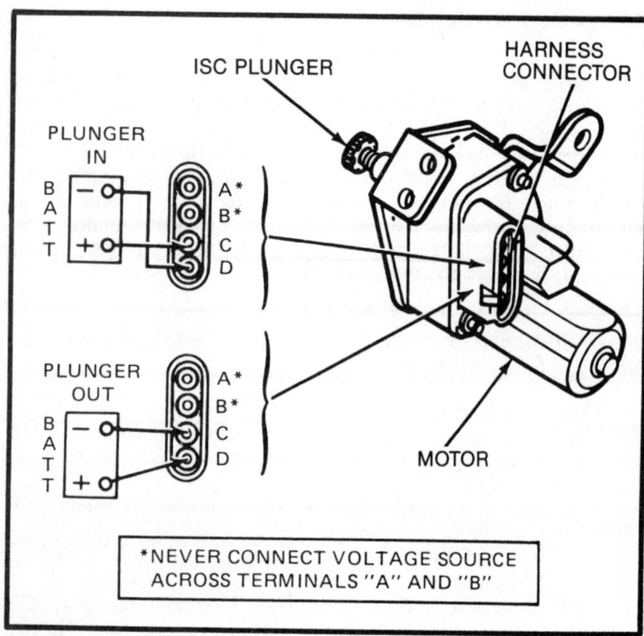

Do not use the ISC plunger to set curb idle speed

8. Start engine and wait until dwell meter needle starts to vary, indicating "closed loop" operation.
9. With parking brake applied and drive wheels blocked, place transmission in DRIVE.
10. With ISC plunger fully retracted, adjust carburetor base (slow) idle stop screw if required to obtain minimum RPM idle speed.
11. Place transmission in PARK and fully extend ISC plunger by apply 12 volts to terminal "D" of the ISC motor connection and ground lead to terminal "C" is the ISC motor connection. Apply voltage only long enough to extend plunger.

NOTE: When the plunger is in the fully extended position, the plunger must be in the contact with the throttle lever to prevent possible internal damage to unit.

12. Place transmission in DRIVE.
13. With ISC plunger fully extended, check maximum rpm. Adjust ISC plunger if required using tool J29607, BT-8022 or equivalent to obtain maximum rpm speed.
14. Recheck maximum rpm adjustment with voltage applied to ISC motor. Motor will ratchet at full extension with power applied. Readjust if necessary.
15. Place transmission in Park or Neutral and turn ignition OFF. Disconnect 12 volt DC power source, jumper leads, ground lead and tachometer.
16. Reconnect 4 terminal harness connector to ISC motor.
17. Remove blocks from drive wheels.

Rochester Quadrajet Models

The Rochester Quadrajet carburetor is a two stage, four-barrel downdraft carburetor. It has been built in many variations designated as 4MC, 4MV, M4MC, M4MCA, M4ME, M4MEA, E4MC, and E4ME. See the beginning of the Rochester section for an explanation of these designations. The primary side of the carburetor is equipped with two primary bores and a triple venturi with plain tube nozzles. During off idle and part throttle operation, the fuel is metered through tapered metering rods operating in specially designed jets positioned by a manifold vacuum responsive piston. The secondary side of the carburetor contains two secondary boxes. An air valve is used on the secondary side for metering control and supplements the primary bore. The secondary air valve operates tapered metering rods which regulate the fuel in constant proportion to the air being supplied.

Fast Idle Speed Adjustment

1. Position the fast idle lever on the high step of the fast idle cam.
2. Be sure that the choke is wide open and the engine warm. Plug the EGR vacuum hose. Disconnect the vacuum hose to the front vacuum break unit, if there are two.
3. Make a preliminary adjustment by turning the fast idle screw out until the throttle valves are closed, then screwing it in the specified number of turns after it contacts the lever (see the carburetor specifications).

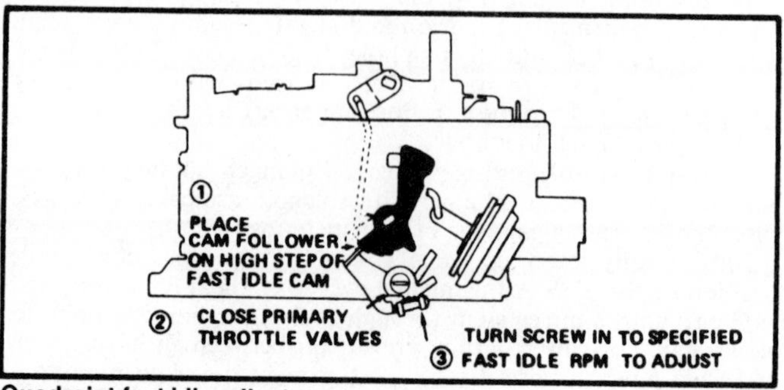

Quadrajet fast idle adjustment

4. Use the fast idle screw to adjust the fast idle to the speed, and under the conditions, specified on the engine compartment sticker or in the specifications chart.

Choke Rod (Fast Idle Cam) Adjustment

1. Adjust the fast idle and place the cam follower on the second step of the fast idle cam against the shoulder of the high step.
2. Close the choke valve by exerting counterclockwise pressure on the external choke lever. Remove the coil assembly from the choke housing and push upon the choke coil lever. On models with a fixed (riveted) choke cover, push up on the vacuum break lever tang and hold in position with a rubber band.
3. Insert a gauge of the proper size between the upper edge of the choke valve and the inside air horn wall.
4. To adjust, bend the tang on the fast idle cam. Be sure that the tang rests against the cam after bending.

Primary (Front) Vacuum Break Adjustment

1978–81 MODELS

1. Loosen the three retaining screws and remove the thermostatic cover and coil assembly from the choke housing through 1979.
2. Seat the front vacuum diaphragm using an outside vacuum source. If there is a diaphragm unit bleed hole, tape it over.
3. Push up on the inside choke coil lever until the tang on the vacuum break lever contacts the tang on the vacuum break plunger. On models with a fixed choke coil cover, push up on the vacuum break lever tang.
4. Place the proper size gauge between the upper edge of the choke valve and the inside of the air horn wall.
5. To adjust, turn the adjustment screw on the vacuum break plunger lever.
6. Install the vacuum hose to the vacuum break unit.

1982–84 MODELS

On these models a choke valve measuring gauge J-26701 or equivalent is used to measure angle (degrees instead of inches). See illustration for procedure.

Secondary (Rear) Vacuum Break Adjustment

1978–80 MODELS

1. Remove the thermostatic cover and coil assembly from the choke housing through 1979.

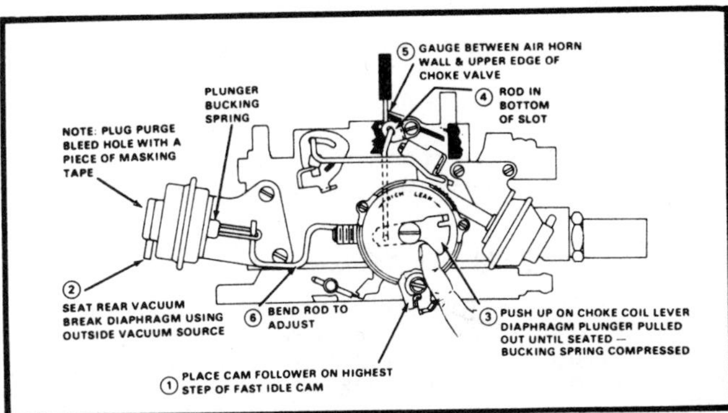

Quadrajet rear vacuum break adjustment (without adjusting screw)—1980 and earlier models

2. Tape over the bleed hole in the rear vacuum break diaphragm and seat the diaphragm using an outside vacuum source. Make sure the diaphram plunger bucking spring, if any, is compressed. On delay models (1980), plug the end cover with a pump plunger cup or equivalent and remove after adjustment.
3. Close the choke by pushing up on the choke coil lever inside the choke housing. On models with a fixed choke coil cover, push up on the vacuum break lever tang and use a rubber band to hold in place.
4. With the choke rod in the bottom of the slot in the choke lever, measure between the upper edge of the choke valve and the air horn wall with a wire type gauge.
5. To adjust, bend the vacuum break rod at the first bend near the diaphragm except on 1980 models with a screw at the rear of the diaphragm; on those models, turn the screw to adjust.
6. Remove the tape covering the bleed hole of the diaphragm and connect the vacuum hose.

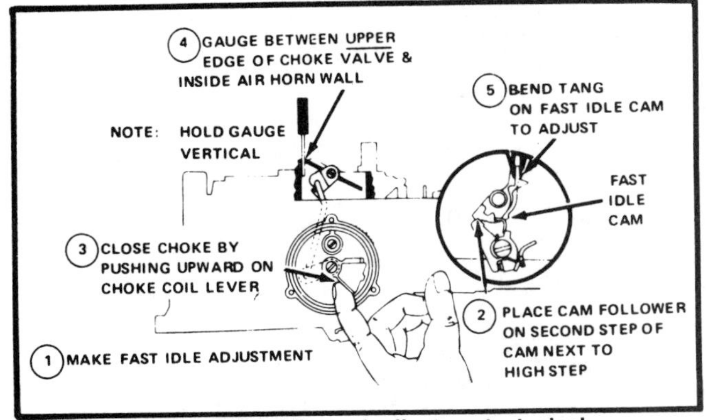

Quadrajet choke rod (fast idle cam) adjustment—typical

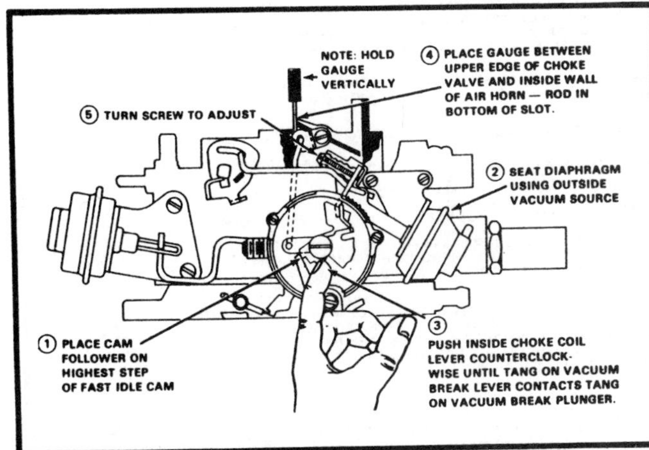

Quadrajet front vacuum break adjustment—1981 and earlier

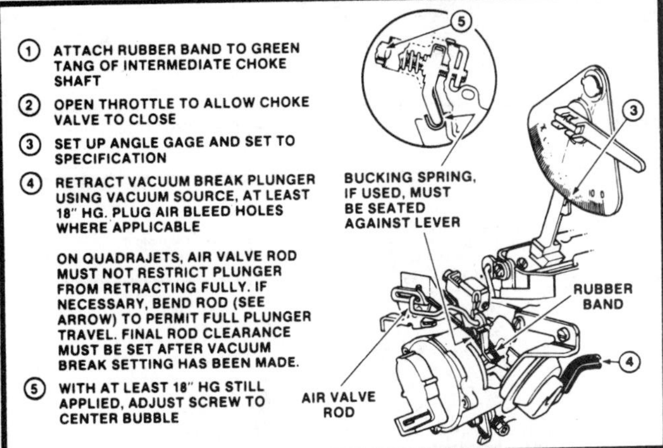

Quadrajet front vacuum break adjustment—1982 and later

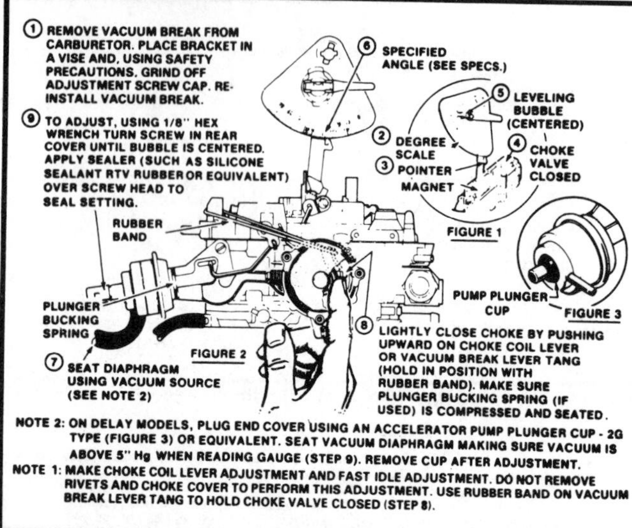

Quadrajet rear vacuum break adjustment—1981-82 models

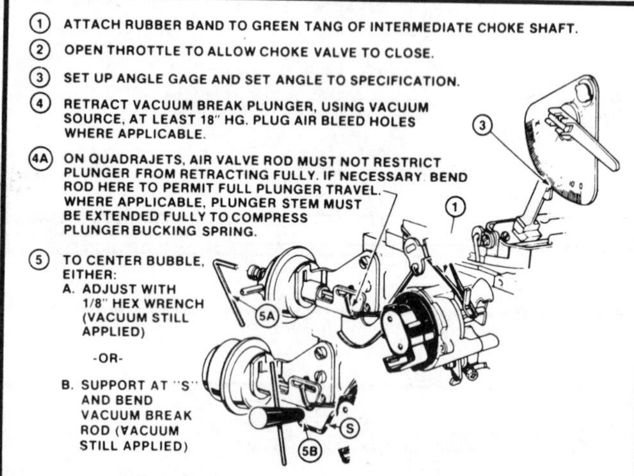

Quadrajet rear vacuum break adjustment—1983 and later models

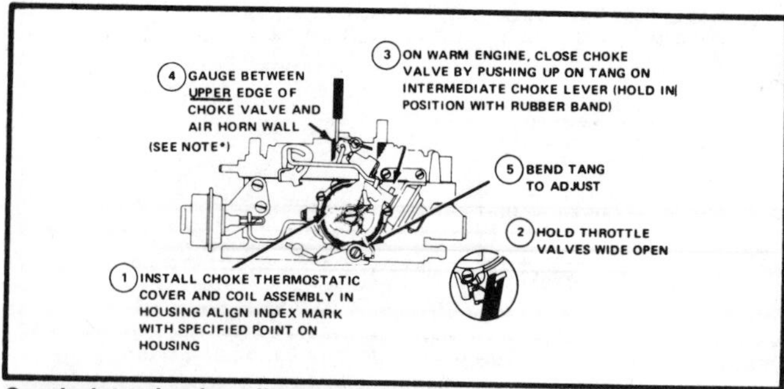

Quadrajet unloader adjustment—typical

1981-84 MODELS

On these models a choke valve measuring gauge J-26701 or equivalent is used to measure angle (degrees instead on inches). See illustration for procedure.

Choke Unloader Adjustment

1. Push up on the vacuum break lever to close the choke valve, and fully open the throttle valves.

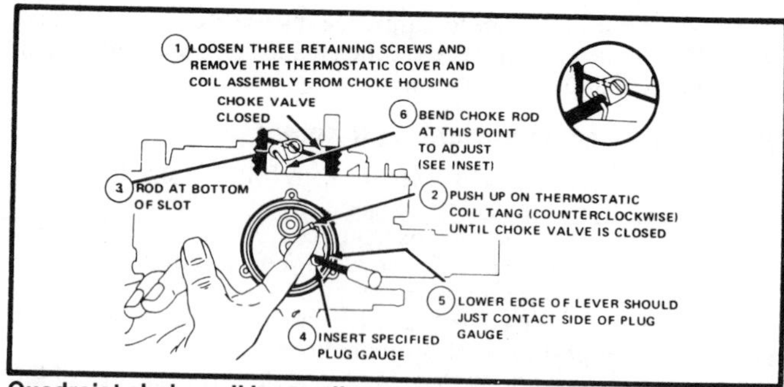

Quadrajet choke coil lever adjustment—typical

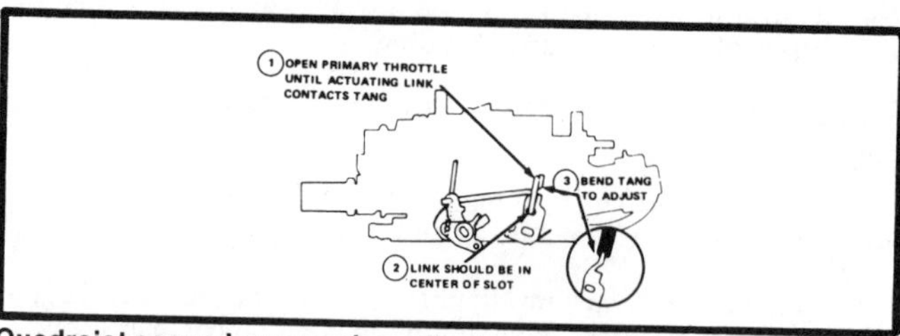

Quadrajet secondary opening adjustment with two point linkage

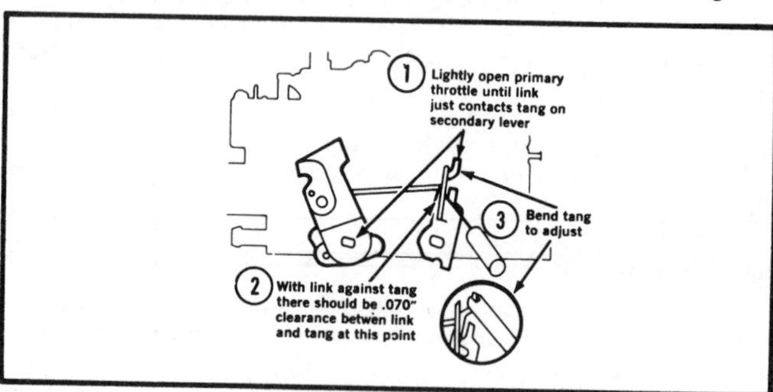

Secondary opening adjustment with three point linkage

2. Measure the distance from the upper edge of the choke valve to the air horn wall.
3. To adjust, bend the tang on the fast idle lever.

4MV Choke Coil Rod Adjustment

1. Close the choke valve by rotating the choke coil lever counterclockwise.
2. Disconnect the thermostatic coil rod from the upper lever.
3. Push down on the rod until it contacts the bracket of the coil.
4. The rod must fit in the notch of the upper lever.
5. If it does not, it must be bent on the curved portion just below the upper lever.

MC, ME Choke Coil Lever Adjustment

1. Remove the choke cover and thermostatic coil from the choke housing. On models with a fixed (riveted) choke cover, the rivets must be drilled out. A choke stat kit is necessary for assembly. Place the fast idle cam follower on the high step.
2. Push up on the coil tang (counter-clockwise) until the choke valve is closed. The top of the choke rod should be at the bottom of the slot in the choke valve lever.
3. Insert a 0.120 in. drill bit in the hole in the choke housing.
4. The lower edge of the choke coil lever should just contact the side of the plug gauge.
5. Bend the choke rod at the top angle to adjust.

Secondary Closing Adjustment

This adjustment assures proper closing of the secondary throttle plates.

1. Set the slow idle as per instructions in the appropriate car section. Make sure that the fast idle cam follower is not resting on the fast idle cam and the choke valve is wide open.

2. There should be 0.020 in. clearance between the secondary throttle actuating rod and the front of the slot on the secondary throttle lever with the closing tang on the throttle lever resting against the actuating lever.

3. Bend the secondary closing tang on the primary throttle actuating rod or lever to adjust.

Secondary Opening Adjustment

1. Open the primary throttle valves until the actuating link contacts the upper tang on the secondary lever.

2. With two point linkage, the bottom of the link should be in the center of the secondary lever slot.

3. With three point linkage, there should be 0.070 in. clearance between the link and the middle tang.

4. Bend the upper tang on the secondary lever to adjust as necessary.

Float Level Adjustment

With the air horn assembly removed, measure the distance from the air horn gasket surface (gasket removed) to the top of the float at the toe (3/16 in. back from the toe).

NOTE: Make sure the retaining pin is firmly held in place and that the tang of the float is lightly held against the needle and seat assembly.

Remove the float and bend the float arm to adjust except on carburetors used with the computer controlled systems (E4MC and E4ME). For those carburetors, if the float level is too high, hold the retainer firmly in place and push down on the center of the float to adjust. If the float level is too low on models with the computer controlled system, lift out the metering rods. Remove the solenoid connector screw. Turn the lean mixture solenoid screw in clockwise, counting and recording the exact number of turns until the screw is lightly bottomed in the bowl. Then turn the solenoid and connector. Remove the float and bend the arm up to adjust. Install the parts, turning the mixture solenoid screw in until it is lightly bottomed, then unscrewing it the exact number of turns counted earlier.

Air Valve Spring Adjustment

To adjust the air valve spring windup, loosen the Allen head lockscrew and turn the adjusting screw counter-clockwise to remove all spring tension. With the air valve closed, turn the adjusting screw clockwise the specified number of turns after the torsion spring contacts the pin on the shaft. Hold the adjusting screw in this position and tighten the lockscrew.

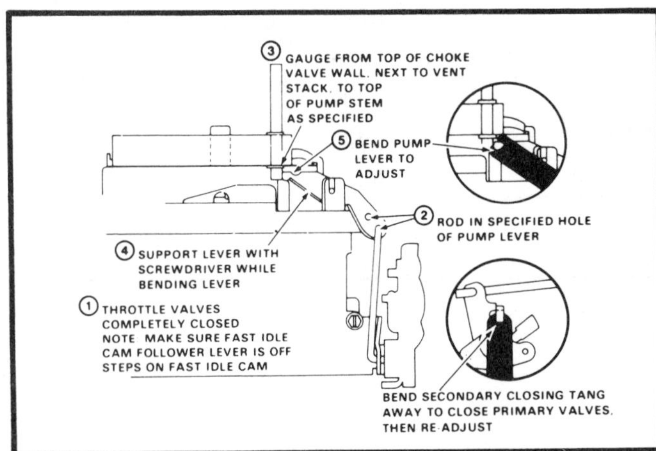

Quadrajet accelerator pump rod adjustment

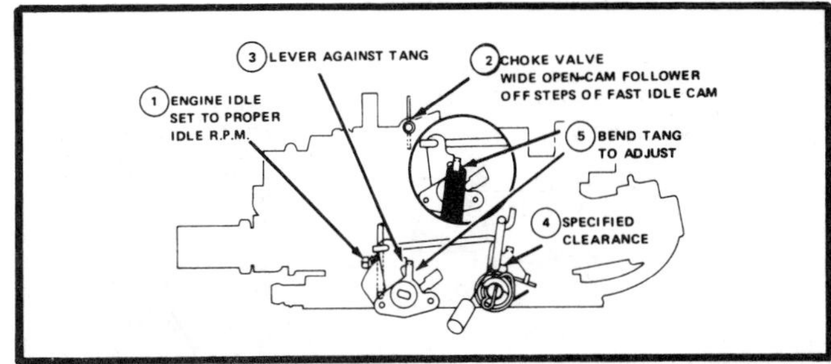

Quadrajet secondary closing adjustment

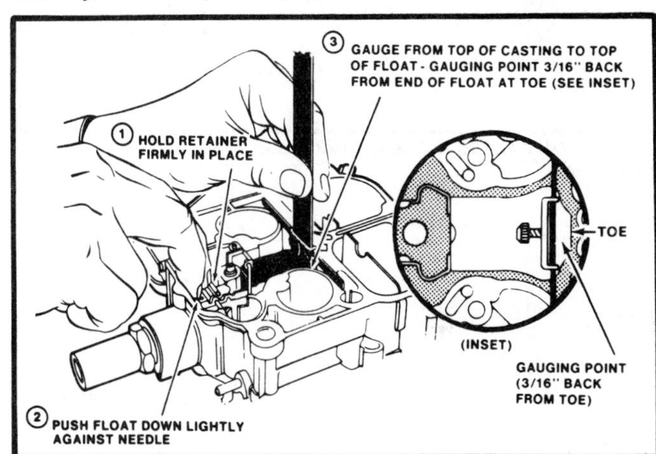

Quadrajet float level adjustment

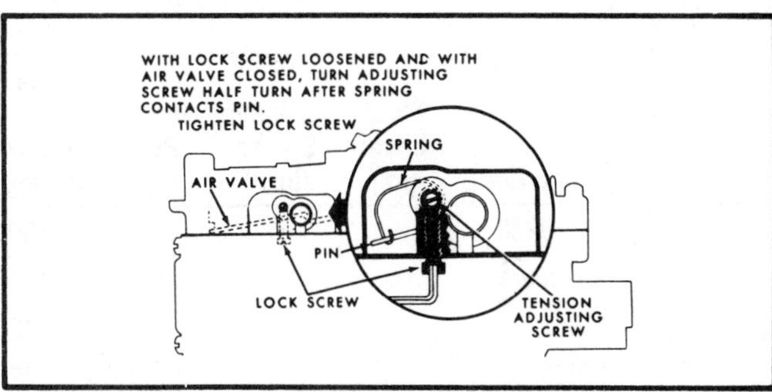

Quadrajet air valve spring setting—typical

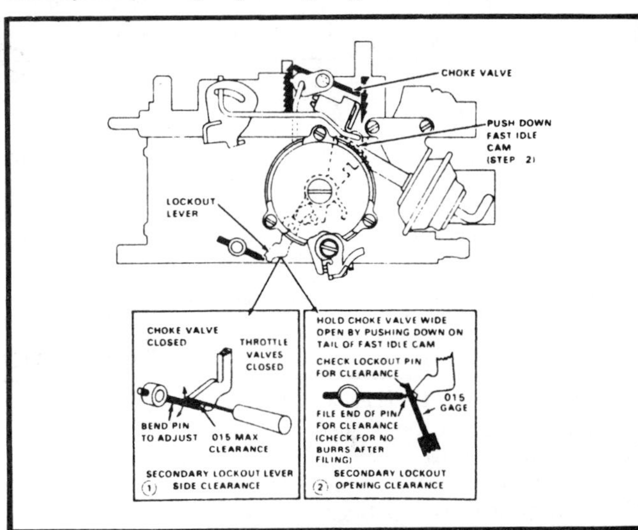

Secondary lockout adjustment—Rochester E4ME and E4MC carburetors

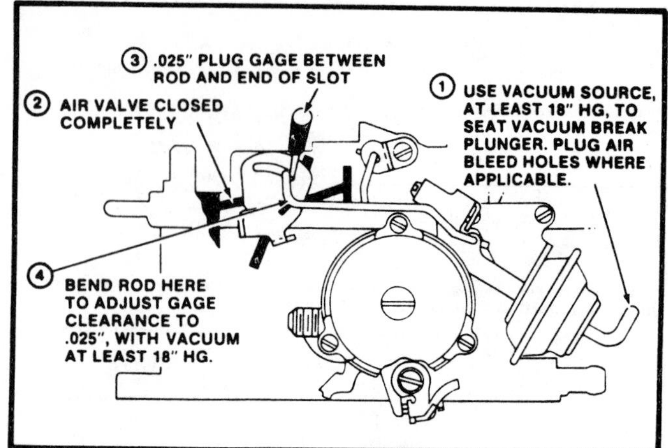

Front air valve rod adjustment—Rochester E4ME and E4MC carburetors

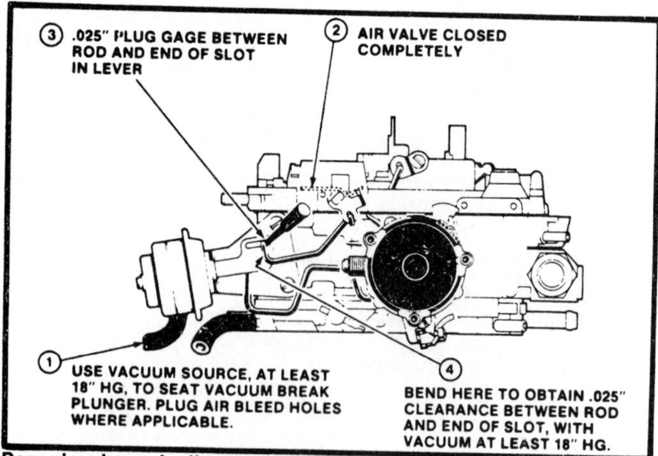

Rear air valve rod adjustment—Rochester E4ME and E4MC carburetors

QUADRAJET CARBURETOR SPECIFICATIONS
Cadillac

Year	Carburetor Identification ①	Float Level (in.)	Air Valve Spring (turn)	Pump Rod (in.)	Primary Vacuum Break (in./deg.)	Secondary Vacuum Break (in./deg.)	Secondary Opening (in.)	Choke Rod (in.)	Choke Unloader (in.)	Fast Idle Speed (rpm)
'78	17058230	13/32	1/2	3/8	0.150	0.165	③	0.080	0.230	1500
	All others	13/32	1/2	3/8	0.140	0.250	③	0.080	0.230	1400
'79	17059230	13/32	1/2	9/32 ②	0.142	0.234	0.015	0.083	0.142	1000
	17059232	13/32	1/2	9/32 ②	0.142	0.234	0.015	0.083	0.142	1500
	17059530	13/32	1/2	9/32 ②	0.149	0.164	0.015	0.083	0.142	1500
	17059532	13/32	1/2	9/32 ②	0.149	0.164	0.015	0.083	0.142	1500
'80	17080230	7/16	1/2	9/32 ②	0.149	0.136	③	0.083	0.220	1450
	17080530	17/32	1/2	Fixed	0.142	0.400	③	0.083	0.260	1350
'81	17081248	3/8	5/8	Fixed	0.164	0.136	③	0.139	0.243	④
	17081289	13/32	5/8	Fixed	0.164	0.136	③	0.139	0.243	④
'82	17082246	3/8	5/8	Fixed	0.149/26	0.149/26	③	0.139	0.195	④
	17082247	13/32	5/8	Fixed	0.164/28	0.136/24	③	0.139	0.243	④
'83	17082266	3/8	5/8	Fixed	0.149/26	0.149/26	③	0.071	0.195	④
	17082267	3/8	5/8	Fixed	0.149/26	0.149/26	③	0.071	0.195	④

① The carburetor identification number is stamped on the float bowl, near the secondary throttle lever.
② Inner hole
③ No measurement necessary on two point linkage; see text.
④ See underhood decal.

QUADRAJET CARBURETOR SPECIFICATIONS
Buick

Year	Carburetor Identification ①	Float Level (in.)	Air Valve Spring (turn)	Pump Rod (in.)	Primary Vacuum Break (in./deg.)	Secondary Vacuum Break (in./deg.)	Secondary Opening (in.)	Choke Rod (in.)	Choke Unloader (in.)	Fast Idle Speed ④ (rpm)
'78	17058240	7/32	3/4	9/32	0.117	0.117	②	0.074	0.243	⑤
	17058241	5/16	3/4	3/8	0.120	0.103	②	0.096	0.243	⑤
	17058250	13/32	1/2	9/32	0.129	0.183	②	0.096	0.220	⑤
	17058253	13/32	1/2	9/32	0.129	0.183	②	0.096	0.220	⑤
	17058254	15/32	1/2	9/32	0.136	—	②	0.103	0.220	⑤
	17058257	13/32	1/2	9/32	0.136	0.231	②	0.103	0.220	⑤
	17058258	13/32	1/2	9/32	0.136	0.231	②	0.103	0.220	⑤
	17058259	13/32	1/2	9/32	0.136	0.231	②	0.103	0.220	⑤
	17058582	15/32	7/8	9/32	0.179	—	②	0.314	0.277	⑤
	17058584	15/32	7/8	9/32	0.179	—	②	0.314	0.277	⑤
	17058282	15/32	7/8	9/32	0.157	—	②	0.314	0.277	⑤
	17058284	15/32	7/8	9/32	0.157	—	②	0.314	0.277	⑤
	17058228	15/32	1	9/32	0.179	—	②	0.314	0.277	⑤
	17058502	15/32	7/8	9/32	0.164	—	②	0.314	0.277	⑤
	17058504	15/32	7/8	9/32	0.164	—	②	0.314	0.277	⑤
	17058202	15/32	7/8	9/32	0.157	—	②	0.314	0.277	⑤
	17058204	15/32	7/8	9/32	0.157	—	②	0.314	0.277	⑤
	17058540	7/32	3/4	9/32	0.117	0.117	②	0.074	0.243	⑤
	17058550	13/32	1/2	9/32	0.136	0.231	②	0.103	0.220	⑤
	17058553	15/32	1/2	9/32	0.129	0.231	②	0.096	0.220	⑤
	17058559	15/32	1/2	9/32	0.136	—	②	0.096	0.231	⑤
'79	17059240	7/32	3/4	9/32	0.117	0.117	②	0.074	0.179	⑥
	17059243	7/32	3/4	9/32	0.117	0.117	②	0.074	0.179	⑥
	17059540	7/32	3/4	9/32	0.117	0.129	②	0.074	0.243	⑥
	17059543	7/32	3/4	9/32	0.117	0.129	②	0.074	0.243	⑥
	17059242	7/32	3/4	9/32	0.066	0. 066	②	0.074	0.179	⑥
	17059553	13/32	1/2	9/32	0.136	0.230	②	0.103	0.220	⑥
	17059555	13/32	1/2	9/32	0.149	0.230	②	0.103	0.220	⑥
	17059250	13/32	1/2	9/32	0.129	0.182	②	0.096	0.220	⑥
	17059253	13/32	1/2	9/32	0.129	0.182	②	0.096	0.220	⑥
	17059208	15/32	7/8	9/32	—	0.129	②	0.314	0.277	⑥
	17059209	15/32	7/8	9/32	—	0.129	②	0.314	0.277	⑥
	17059210	15/32	1	9/32	0.157	—	②	0.243	0.243	⑥
	17059211	15/32	1	9/32	0.157	—	②	0.243	0.243	⑥
	17059228	15/32	1	9/32	0.157	—	②	0.243	0.243	⑥
	17059241	5/16	3/4	3/8	0.120	0.113	②	0.096	0.243	⑥
	17059247	5/16	3/4	3/8	0.110	0.103	②	0.096	0.243	⑥
	17059272	15/32	5/8	3/8	0.136	0.195	②	0.074	0.220	⑥

QUADRAJET CARBURETOR SPECIFICATIONS
Buick

Year	Carburetor Identification①	Float Level (in.)	Air Valve Spring (turn)	Pump Rod (in.)	Primary Vacuum Break (in./deg.)	Secondary Vacuum Break (in./deg.)	Secondary Opening (in.)	Choke Rod (in.)	Choke Unloader (in.)	Fast Idle Speed④ (rpm)
'80	17080240	3/16	9/16	9/32③	0.083	0.083	②	0.074	0.179	⑥
	17080241	7/16	3/4	9/32③	0.129	0.114	②	0.096	0.243	⑥
	17080242	13/32	9/16	9/32③	0.077	0.096	②	0.074	0.220	⑥
	17080243	3/16	9/16	9/32③	0.083	0.083	②	0.074	0.179	⑥
	17080244	5/16	5/8	9/32③	0.096	0.071	②	0.139	0.243	⑥
	17080249	7/16	3/4	9/32③	0.129	0.114	②	0.096	0.243	⑥
	17080253	13/32	1/2	9/32③	0.149	0.211	②	0.090	0.220	⑥
	17080259	13/32	1/2	9/32③	0.149	0.211	②	0.090	0.220	⑥
	17080270	15/32	5/8	3/8⑦	0.149	0.211	②	0.074	0.220	⑥
	17080271	15/32	5/8	3/8⑦	0.142	0.211	②	0.110	0.203	⑥
	17080272	15/32	5/8	3/8⑦	0.129	0.175	②	0.074	0.203	⑥
	17080502	1/2	7/8	Fixed	0.136	0.179	②	0.110	0.243	⑥
	17080504	1/2	7/8	Fixed	0.136	0.179	②	0.110	0.243	⑥
	17080540	3/8	9/16	Fixed	0.103	0.129	②	0.074	0.243	⑥
	17080542	3/8	9/16	Fixed	0.103	0.066	②	0.074	0.243	⑥
	17080543	3/8	9/16	Fixed	0.103	0.129	②	0.074	0.243	⑥
	17080553	15/32	1/2	Fixed	0.142	0.220	②	0.090	0.220	⑥
	17080554	15/32	1/2	Fixed	0.142	0.211	②	0.090	0.220	⑥
81	17081202 204	11/32	7/8	Fixed	0.157⑧	—	②	0.110	0.243	⑩
	17081203 207	11/32	7/8	Fixed	0.157⑧	—	②	0.110	0.243	⑩
	17081216 218	11/32	7/8	Fixed	0.157⑧	—	②	0.110	0.243	⑩
	17081242	3/8	9/16	Fixed	0.090⑧	0.077⑨	②	0.139	0.243	⑩
	17081243	5/16	9/16	Fixed	0.103⑧	0.090⑨	②	0.139	0.243	⑩
	17081245	3/8	5/8	Fixed	0.164⑧	0.136⑨	②	0.139	0.243	⑩
	17081247	3/8	5/8	Fixed	0.164⑧	0.136⑨	②	0.139	0.243	⑩
	17081248 249	3/8	5/8	Fixed	0.164⑧	0.136⑨	②	0.139	0.243	⑩
	17081253 254	15/32	1/2	Fixed	0.142⑧	0.227⑨	②	0.071	0.220	⑩
	17081270	7/16	5/8	Fixed	0.136⑧	0.211⑨	②	0.074	0.220	⑩
	17081272	5/8	5/8	Fixed	0.136⑧	0.260⑨	②	0.074	0.220	⑩
	17081274	5/8	5/8	Fixed	0.136⑧	0.220⑨	②	0.083	0.220	⑩
	17081289	5/8	5/8	Fixed	0.164⑧	0.136⑨	②	0.139	0.243	⑩
'82	17082202	11/32	7/8	Fixed	0.110/20	—	②	0.110	0.243	⑤
	17082204	11/32	3/8	Fixed	0.110/20	—	②	0.110	0.243	⑤
	17082244	7/16	9/16	Fixed	0.117/21	0.083/16	②	0.139	0.195	⑤
	17082245	3/8	5/8	Fixed	0.149/26	0.149/26	②	0.139	0.195	⑤

QUADRAJET CARBURETOR SPECIFICATIONS
Buick

Year	Carburetor Identification ①	Float Level (in.)	Air Valve Spring (turn)	Pump Rod (in.)	Primary Vacuum Break (in./deg.)	Secondary Vacuum Break (in./deg.)	Secondary Opening (in.)	Choke Rod (in.)	Choke Unloader (in.)	Fast Idle Speed ④ (rpm)
'82	17082246	3/8	5/8	Fixed	0.149/26	0.149/26	②	0.139	0.195	⑤
	17082247	13/32	5/8	Fixed	0.164/28	0.136/24	②	0.139	0.243	⑤
	17082248	13/32	5/8	Fixed	0.164/28	0.136/24	②	0.139	0.243	⑤
	17082251	15/32	1/2	Fixed	0.142/25	0.304/45	②	0.071	0.220	⑤
	17082253	15/32	1/2	Fixed	0.142/25	0.227/36	②	0.071	0.220	⑤
	17082264	7/16	9/16	Fixed	0.117/20	0.083/16	②	0.139	0.195	⑤
	17082265	3/8	5/8	Fixed	0.149/26	0.149/26	②	0.139	0.195	⑤
	17082266	3/8	5/8	Fixed	0.149/26	0.149/26	②	0.139	0.195	⑤
	17082267	3/8	5/8	Fixed	0.164/28	0.136/24	②	0.139	0.243	⑤
	17082268	13/32	5/8	Fixed	0.164/28	0.136/24	②	0.139	0.243	⑤
'83	17082265	3/8	5/8	Fixed	0.149/26	0.149/26	②	0.139	0.195	⑪
	17082266	3/8	5/8	Fixed	0.149/26	0.149/26	②	0.139	0.195	⑪
	17082267	3/8	5/8	Fixed	0.149/26	0.149/26	②	0.096	0.195	⑪
	17082268	3/8	5/8	Fixed	0.149/26	0.149/26	②	0.096	0.195	⑪
	17083242	9/32	9/16	Fixed	0.110/20	—	②	0.139	0.243	⑪
	17083244	1/4	9/16	Fixed	0.117/21	0.083/16	②	0.139	0.195	⑪
	17083248	3/8	5/8	Fixed	0.149/26	0.149/26	②	0.139	0.195	⑪
	17083250	7/16	1/2	Fixed	0.157/27	0.271/42	②	0.071	0.220	⑪
	17083253	7/16	1/2	Fixed	0.151/27	0.269/41	②	0.071	0.220	⑪
	17083553	7/16	1/2	Fixed	0.157/27	0.269/41	②	0.071	0.220	⑪
'84	17084201	11/32	7/8	Fixed	0.157/27	—	②	0.110	0.243	⑪
	17084205	11/32	7/8	Fixed	0.157/27	—	②	0.243	0.243	⑪
	17084208	11/32	7/8	Fixed	0.157/27	—	②	0.110	0.243	⑪
	17084209	11/32	7/8	Fixed	0.157/27	—	②	0.243	0.243	⑪
	17084210	11/32	7/8	Fixed	0.157/27	—	②	0.110	0.243	⑪
	17084240	5/16	1	Fixed	0.136/24	—	②	—	0.195	⑪
	17084244	5/16	1	Fixed	0.136/24	—	②	—	0.195	⑪
	17084246	5/16	1	Fixed	0.123/22	0.136/24	②	—	0.195	⑪
	17084248	5/16	1	Fixed	0.136/24	—	②	—	0.195	⑪
	17084252	7/16	1/2	Fixed	0.157/27	0.269/41	②	—	0.220	⑪
	17084254	7/16	1/2	Fixed	0.157/27	0.269/41	②	—	0.220	⑪

① The carburetor identification number is stamped on the float bowl, near the secondary throttle lever.
② No measurement necessary on two point linkage; see text
③ Inner hole
④ On high step of cam, automatic in Park
⑤ 3 turns after contacting lever for preliminary setting
⑥ 2 turns after contacting lever for preliminary setting
⑦ Outer hole
⑧ Front
⑨ Rear
⑩ 4½ turns after contacting lever for preliminary setting
⑪ See underhood decal

QUADRAJET CARBURETOR SPECIFICATIONS
Chevrolet

Year	Carburetor Identification ①	Float Level (in.)	Air Valve Spring (turn)	Pump Rod (in.)	Primary Vacuum (deg./in.)	Secondary Vacuum (deg./in.)	Secondary Opening (in.)	Choke Rod (in.)	Choke Unloader (in.)	Fast Idle Speed ④ (rpm)
'78	17058202	15/32	7/8	9/32	0.179	—	⑤	0.314	0.277	⑥
	17058203	15/32	7/8	9/32	0.179	—	⑤	0.314	0.277	⑥
	17058204	15/32	7/8	9/32	0.179	—	⑤	0.314	0.277	⑥
	17058210	15/32	1/2	9/32	0.203	—	⑤	0.314	0.277	⑥
	17058211	15/32	1/2	9/32	0.203	—	⑤	0.314	0.277	⑥
	17058228	15/32	7/8	9/32	0.203	—	⑤	0.314	0.277	⑥
	17058502	15/32	7/8	9/32	0.187	—	⑤	0.314	0.277	⑥
	17058504	15/32	7/8	9/32	0.187	—	⑤	0.314	0.277	⑥
	17058582	15/32	7/8	9/32	0.203	—	⑤	0.314	0.277	⑥
	17058584	15/32	7/8	9/32	0.203	—	⑤	0.314	0.277	⑥
'79	17059203	15/32	7/8	1/4	0.157	—	⑤	0.243	0.243	⑦
	17059207	15/32	7/8	1/4	0.157	—	⑤	0.243	0.243	⑦
	17059216	15/32	7/8	1/4	0.157	—	⑤	0.243	0.243	⑦
	17059217	15/32	7/8	1/4	0.157	—	⑤	0.243	0.243	⑦
	17059218	15/32	7/8	1/4	0.164	—	⑤	0.243	0.243	⑦
	17059222	15/32	7/8	1/4	0.164	—	⑤	0.243	0.243	⑦
	17059502	15/32	7/8	1/4	0.164	—	⑤	0.243	0.243	⑦
	17059504	15/32	7/8	1/4	0.164	—	⑤	0.243	0.243	⑦
	17059582	15/32	7/8	11/32	0.203	—	⑤	0.243	0.314	⑦
	17059584	15/32	7/8	11/32	0.203	—	⑤	0.243	0.314	⑦
	17059210	15/32	1	9/32	0.157	—	⑤	0.243	0.243	⑦
	17059211	15/32	1	9/32	0.157	—	⑤	0.243	0.243	⑦
	17029228	15/32	1	9/32	0.157	—	⑤	0.243	0.243	⑦
'80	17080202	7/16	7/8	1/4⑧	0.157	—	⑤	0.110	0.243	⑩
	17080204	7/16	7/8	1/4⑧	0.157	—	⑤	0.110	0.243	⑩
	17080207	7/16	7/8	1/4⑧	0.157	—	⑤	0.110	0.243	⑩
	17080228	7/16	7/8	9/32⑧	0.179	—	⑤	0.110	0.243	⑩
	17080243	3/16	9/16	9/32⑧	0.016	0.083	⑤	0.074	0.179	⑩
	17080274	15/32	5/8	5/16⑨	0.110	0.164	⑤	0.083	0.203	⑩
	17080282	7/16	7/8	11/32⑨	0.142	—	⑤	0.110	0.243	⑩
	17080284	7/16	7/8	11/32⑨	0.142	—	⑤	0.110	0.243	⑩
	17080502	1/2	7/8	Fixed	0.136	0.179	⑤	0.110	0.243	⑩
	17080504	1/2	7/8	Fixed	0.136	0.179	⑤	0.110	0.243	⑩
	17080542	3/8	9/16	Fixed	0.103	0.066	⑤	0.074	0.243	⑩
	17080543	3/8	9/16	Fixed	0.103	0.129	⑤	0.074	0.243	⑩
'81	17081202	11/32	7/8	Fixed	0.149	—	⑤	0.110	0.243	⑪
	17081203	11/32	7/8	Fixed	0.149	—	⑤	0.110	0.243	⑪
	17081204	11/32	7/8	Fixed	0.149	—	⑤	0.110	0.243	⑪

QUADRAJET CARBURETOR SPECIFICATIONS
Chevrolet

Year	Carburetor Identification ①	Float Level (in.)	Air Valve Spring (turn)	Pump Rod (in.)	Primary Vacuum (deg./in.)	Secondary Vacuum (deg./in.)	Secondary Opening (in.)	Choke Rod (in.)	Choke Unloader (in.)	Fast Idle Speed ④ (rpm)
'81	17081207	11/32	7/8	Fixed	0.149	—	⑤	0.110	0.243	⑪
	17081216	11/32	7/8	Fixed	0.149	—	⑤	0.110	0.243	⑪
	17081217	11/32	7/8	Fixed	0.149	—	⑤	0.110	0.243	⑪
	17081218	11/32	7/8	Fixed	0.149	—	⑤	0.110	0.243	⑪
	17081242	5/16	9/16	Fixed	0.090	0.077	⑤	0.139	0.243	⑪
	17081243	1/4	9/16	Fixed	0.103	0.090	⑤	0.139	0.243	⑪
'82	17082202	11/32	7/8	Fixed	0.157	—	⑤	0.110	0.243	⑫
	17082204	11/32	7/8	Fixed	0.157	—	⑤	0.110	0.243	⑫
	17082203	11/32	7/8	Fixed	0.157	—	⑤	0.243	0.243	⑫
	17082207	11/32	7/8	Fixed	0.157	—	⑤	0.243	0.243	⑫
'83	17083202	11/32	7/8	Fixed	—	27/.157	⑤	0.110	0.243	⑬
	17083203	11/32	7/8	Fixed	—	27/.157	⑤	0.243	0.243	⑬
	17083204	11/32	7/8	Fixed	—	27/.157	⑤	0.110	0.243	⑬
	17083207	11/32	7/8	Fixed	—	27/157	⑤	0.243	0.243	⑬
	17083216	11/32	7/8	Fixed	—	27/.157	⑤	0.110	0.243	⑬
	17083218	11/32	7/8	Fixed	—	27/.157	⑤	0.110	0.243	⑬
	17083236	11/32	7/8	Fixed	—	27/.157	⑤	0.110	0.243	⑬
	17083506	7/16	7/8	Fixed	27/.157	36/.227	⑤	0.110	0.227	⑬
	17083508	7/16	7/8	Fixed	27/.157	36/.227	⑤	0.110	0.227	⑬
	17083524	7/16	7/8	Fixed	25/.142	36/.227	⑤	0.110	0.227	⑬
	17083526	7/16	7/8	Fixed	25/.142	36/.227	⑤	0.110	0.227	⑬
'84	17084201	11/32	7/8	Fixed	.157/27	—	⑤	0.110	0.243	⑬
	17084205	11/32	7/8	Fixed	.157/27	—	⑤	0.243	0.243	⑬
	17084208	11/32	7/8	Fixed	.157/27	—	⑤	0.110	0.243	⑬
	17084209	11/32	7/8	Fixed	.157/27	—	⑤	0.243	0.243	⑬
	17084210	11/32	7/8	Fixed	.157/27	—	⑤	0.110	0.243	⑬
	17084507	7/16	1	Fixed	.157/27	.227/36	⑤	0.110	0.227	⑬
	17084509	7/16	1	Fixed	.157/27	.227/36	⑤	0.110	0.227	⑬
	17084525	7/16	1	Fixed	.142/25	.227/36	⑤	0.110	0.227	⑬
	17084527	7/16	1	Fixed	.142/25	.227/36	⑤	0.110	0.227	⑬

① The carburetor identification number is stamped on the float bowl, near the secondary throttle lever.
② Without vacuum advance.
③ With automatic transmission; vacuum advance connected and EGR disconnected and the throttle positioned on the high step of cam.
④ With manual transmission; without vacuum advance and the throttle positioned on the high step of cam.
⑤ No measurement necessary on two point linkage; see text.
⑥ 3 turns after contacting lever for preliminary setting.
⑦ 2 turns after contacting lever for preliminary setting.
⑧ Inner hole
⑨ Outer hole
⑩ 4 turns after contacting lever for preliminary setting.
⑪ 4½ turns after contacting lever for preliminary setting
⑫ 3⅛ turns after contacting lever for preliminary setting
⑬ See underhood sticker

QUADRAJET CARBURETOR SPECIFICATIONS
Oldsmobile

Year	Carburetor Identification ①	Float Level (in.)	Air Valve Spring (turn)	Pump Rod (in.)	Primary Vacuum Break (in./deg.)	Secondary Vacuum Break (in./deg.)	Secondary Opening (in.)	Choke Rod (in.)	Choke Unloader (in.)	Fast Idle Speed ④ (rpm)
'78	17058202	15/32	7/8	9/32	0.157	—	④	0.314	0.277	⑤
	17058204	15/32	7/8	9/32	0.157	—	④	0.314	0.277	⑤
	17058250	13/32	1/2	9/32	0.129	0.183	④	0.096	0.220	⑤
	17058253	13/32	1/2	9/32	0.129	0.183	④	0.096	0.220	⑤
	17058257	13/32	1/2	9/32	0.136	0.230	④	0.103	0.220	⑤
	17058258	13/32	1/2	9/32	0.136	0.230	④	0.103	0.220	⑤
	17058259	13/32	1/2	9/32	0.136	0.183	④	0.103	0.220	⑤
	17058502	15/32	7/8	9/32	0.164	—	④	0.314	0.277	⑤
	17058504	15/32	7/8	9/32	0.164	—	④	0.314	0.277	⑤
	17058553	13/32	1/2	9/32	0.136	0.230	④	0.103	0.220	⑤
	17058555	13/32	1/2	9/32	0.136	0.230	④	0.103	0.220	⑤
	17058582	15/32	7/8	9/32	0.179	—	④	0.314	0.277	⑤
	17058584	15/32	7/8	9/32	0.179	—	④	0.314	0.277	⑤
'79	17059202	1/2	7/8	1/4	0.164	—	④	0.314	0.243	⑥
	17059207	15/32	7/8	1/4	0.157	—	④	0.243	0.243	⑥
	17059216	15/32	7/8	1/4	0.157	—	④	0.243	0.243	⑥
	17059217	15/32	7/8	1/4	0.157	—	④	0.243	0.243	⑥
	17059218	15/32	7/8	9/32	0.164	—	④	0.243	0.243	⑥
	17059222	15/32	7/8	9/32	0.164	—	④	0.243	0.243	⑥
	17059250	13/32	1/2	9/32	0.129	0.183	④	0.096	0.220	⑥
	17059251	13/32	1/2	9/32	0.129	0.183	④	0.096	0.220	⑥
	17059253	13/32	1/2	9/32	0.129	0.183	④	0.096	0.220	⑥
	17059256	13/32	1/2	9/32	0.136	0.195	④	0.103	0.220	⑥
	17059258	13/32	1/2	9/32	0.136	0.195	④	0.103	0.220	⑥
	17059502	15/32	7/8	1/4	0.164	—	④	0.243	0.243	⑥
	17059504	15/32	7/8	1/4	0.164	—	④	0.243	0.243	⑥
	17059553	13/32	1/2	9/32	0.136	0.230	④	0.103	0.220	⑥
	17059554	13/32	1/2	9/32	0.136	0.230	④	0.103	0.220	⑥
	17059582	15/32	7/8	11/32	0.203	—	④	0.243	0.314	⑥
	17059584	15/32	7/8	11/32	0.203	—	④	0.243	0.314	⑥
'80	17080202	7/16	7/8	1/4 ⑦	0.157	—	④	0.110	0.243	⑤
	17080204	7/16	7/8	1/4 ⑦	0.157	—	④	0.110	0.243	⑤
	17080250	13/32	1/2	9/32 ⑦	0.149	0.211	④	0.090	0.220	⑤
	17080251	13/32	1/2	9/32 ⑦	0.149	0.211	④	0.090	0.220	⑤
	17080252	13/32	1/2	9/32 ⑦	0.149	0.211	④	0.090	0.220	⑤
	17080253	13/32	1/2	9/32 ⑦	0.149	0.211	④	0.090	0.220	⑤
	17080259	13/32	1/2	9/32 ⑦	0.149	0.211	④	0.090	0.220	⑤
	17080260	13/32	1/2	9/32 ⑦	0.149	0.211	④	0.090	0.220	⑤

QUADRAJET CARBURETOR SPECIFICATIONS (Cont'd)
Oldsmobile

Year	Carburetor Identification ①	Float Level (in.)	Air Valve Spring (turn)	Pump Rod (in.)	Primary Vacuum Break (in./deg.)	Secondary Vacuum Break (in./deg.)	Secondary Opening (in.)	Choke Rod (in.)	Choke Unloader (in.)	Fast Idle Speed ④ (rpm)
'80	17080504	1/2	7/8	⑧	0.136	0.179	④	0.110	0.243	⑤
	17080553	15/32	1/2	⑧	0.142	0.220	④	0.090	0.220	⑤
	17080554	15/32	1/2	⑧	0.142	0.211	④	0.090	0.220	⑤
'81	17081250	13/32	1/2	9/32⑦	0.149⑨	0.211⑩	④	0.090	0.220	⑤
	17081253	15/32	1/2	⑧	0.142⑨	0.227⑩	④	0.071	0.220	⑤
	17081254	15/32	1/2	⑧	0.142⑨	0.227⑩	④	0.071	0.220	⑤
	17081248	3/8	—	⑧	0.164⑨	0.136⑩	④	0.139	0.243	⑤
	17081289	13/32	—	⑧	0.164⑨	0.136⑩	④	0.139	0.243	⑤
'82	17082202	11/32	7/8	Fixed	0.110/20	—	④	0.110	0.243	⑤
	17082204	11/32	3/8	Fixed	0.110/20	—	④	0.110	0.243	⑤
	17082244	7/16	9/16	Fixed	0.117/21	0.083/16	④	0.139	0.195	⑤
	17082245	3/8	5/8	Fixed	0.149/26	0.149/26	④	0.139	0.195	⑤
	17082246	3/8	5/8	Fixed	0.149/26	0.149/26	④	0.139	0.195	⑤
	17082247	13/32	5/8	Fixed	0.164/28	0.136/24	④	0.139	0.243	⑤
	17082248	13/32	5/8	Fixed	0.164/28	0.136/24	④	0.139	0.243	⑤
	17082251	15/32	1/2	Fixed	0.142/25	0.304/45	④	0.071	0.220	⑤
	17082253	15/32	1/2	Fixed	0.142/25	0.227/36	④	0.071	0.220	⑤
	17082264	7/16	9/16	Fixed	0.117/20	0.083/16	④	0.139	0.195	⑤
	17082265	3/8	5/8	Fixed	0.149/26	0.149/26	④	0.139	0.195	⑤
	17082266	3/8	5/8	Fixed	0.149/26	0.149/26	④	0.139	0.195	⑤
	17082267	3/8	5/8	Fixed	0.164/28	0.136/24	④	0.139	0.243	⑤
	17082268	13/32	5/8	Fixed	0.164/28	0.136/24	④	0.139	0.243	⑤
'83	17082265	3/8	5/8	Fixed	0.149/26	0.149/26	②	0.139	0.195	⑪
	17082266	3/8	5/8	Fixed	0.149/26	0.149/26	②	0.139	0.195	⑪
	17082267	3/8	5/8	Fixed	0.149/26	0.149/26	②	0.096	0.195	⑪
	17082268	3/8	5/8	Fixed	0.149/26	0.149/26	②	0.096	0.195	⑪
	17083242	9/32	9/16	Fixed	0.110/20	—	②	0.139	0.243	⑪
	17083244	1/4	9/16	Fixed	0.117/21	0.083/16	②	0.139	0.195	⑪
	17083248	3/8	5/8	Fixed	0.149/26	0.149/26	②	0.139	0.195	⑪
	17083250	7/16	1/2	Fixed	0.157/27	0.271/42	②	0.071	0.220	⑪
	17083253	7/16	1/2	Fixed	0.157/27	0.269/41	②	0.071	0.220	⑪
	17083553	7/16	1/2	Fixed	0.157/27	0.269/41	②	0.071	0.220	⑪
'84	17084201	11/32	7/8	Fixed	0.157/27	—	②	0.110	0.243	⑪
	17084205	11/32	7/8	Fixed	0.157/27	—	②	0.243	0.243	⑪
	17084208	11/32	7/8	Fixed	0.157/27	—	②	0.110	0.243	⑪
	17084209	11/32	7/8	Fixed	0.157/27	—	②	0.243	0.243	⑪
	17084210	11/32	7/8	Fixed	0.157/27	—	②	0.110	0.243	⑪
	17084240	5/16	1	Fixed	0.136/24	—	②	—	0.195	⑪

QUADRAJET CARBURETOR SPECIFICATIONS
Oldsmobile

Year	Carburetor Identification ①	Float Level (in.)	Air Valve Spring (turn)	Pump Rod (in.)	Primary Vacuum Break (in./deg.)	Secondary Vacuum Break (in./deg.)	Secondary Opening (in.)	Choke Rod (in.)	Choke Unloader (in.)	Fast Idle Speed ④ (rpm)
'84	17084244	5/16	1	Fixed	0.136/24	—	②	—	0.195	⑪
	17084246	5/16	1	Fixed	0.123/22	0.136/24	②	—	0.195	⑪
	17084248	5/16	1	Fixed	0.136/24	—	②	—	0.195	⑪
	17084252	7/16	1/2	Fixed	0.157/27	0.269/41	②	—	0.220	⑪
	17084254	7/16	1/2	Fixed	0.157/27	0.269/41	②	—	0.220	⑪

① The carburetor identification number is stamped on the float bowl, next to the secondary throttle lever.
② 1800 rpm on Omega and 400 cu. in. engines with the cam follower on the highest step of the fast idle cam; 900 rpm on all others with the fast idle cam follower on the lowest step of the fast idle cam.
④ No measurement necessary on two point linkage; see text.
⑤ 3 turns after contacting lever for preliminary setting.
⑥ 2 turns after contacting lever for preliminary setting.
⑦ Inner hole
⑧ Not Adjustable
⑨ Front
⑩ Rear
⑪ See underhood sticker

QUADRAJET CARBURETOR SPECIFICATIONS
Pontiac

Year	Carburetor Identification ①	Float Level (in.)	Air Valve Spring (turn)	Pump Rod (in.)	Primary Vacuum Break (in./deg.)	Secondary Vacuum Break (in./deg.)	Secondary Opening (in.)	Choke Rod (in.)	Choke Unloader (in.)	Fast Idle Speed ② (rpm)
'78	17058202	15/32	—	9/32	0.157	—	④	0.314	0.277	③
	17058204	15/32	—	9/32	0.157	—	④	0.314	0.277	③
	17058241	5/16	3/4	3/8	0.117	0.103	④	0.096	0.243	③
	17058250	13/32	1/2	9/32	0.119	0.167	④	0.088	0.203	③
	17058253	13/32	1/2	9/32	0.119	0.167	④	0.088	0.203	③
	17058258	13/32	1/2	9/32	0.126	0.212	④	0.092	0.203	③
	17058263	17/32	5/8	3/8	0.164	0.260	④	0.129	0.220	③
	17058264	17/32	1/2	3/8	0.149	0.260	④	0.129	0.220	③
	17058266	17/32	1/2	3/8	0.149	0.260	④	0.129	0.220	③
	17058272	15/32	5/8	3/8	0.126	0.195	④	0.071	0.222	③
	17058274	17/32	1/2	3/8	0.149	0.260	④	0.129	0.220	③
	17058276	17/32	1/2	3/8	0.149	0.260	④	0.129	0.220	③
	17058278	17/32	1/2	3/8	0.149	0.260	④	0.129	0.220	③
	17058502	15/32	—	9/32	0.164	—	④	0.314	0.277	③
	17058504	15/32	—	9/32	0.164	—	④	0.314	0.277	③
	17058553	13/32	1/2	9/32	0.126	0.212	④	0.092	0.203	③
	17058582	15/32	7/8	9/32	0.179	—	④	0.314	0.277	③
	17058584	15/32	7/8	9/32	0.179	—	④	0.314	0.277	③

QUADRAJET CARBURETOR SPECIFICATIONS
Pontiac

Year	Carburetor Identification ①	Float Level (in.)	Air Valve Spring (turn)	Pump Rod (in.)	Primary Vacuum Break (in./deg.)	Secondary Vacuum Break (in./deg.)	Secondary Opening (in.)	Choke Rod (in.)	Choke Unloader (in.)	Fast Idle Speed ② (rpm)
'79	17058263	17/32	5/8	3/8	0.164	0.243	④	0.129	0.220	⑤
	17059250,253	13/32	1/2	9/35	0.129	0.183	④	0.096	0.220	⑤
	17059241	5/16	3/4	3/8	0.120	0.113	④	0.096	0.243	⑤
	17059271	9/16	5/8	3/8	0.142	0.227	④	0.010	0.203	⑤
	17059272	15/32	5/8	3/8	0.136	0.195	④	0.074	0.220	⑤
	17059502,504	15/32	7/8	1/4	0.164	—	④	0.243	0.243	⑤
	17059553	13/32	1/2	9/32	0.136	0.230	④	0.103	0.220	⑤
	17059582,584	15/32	7/8	11/32	0.203	—	④	0.243	0.314	⑤
'80	17080249	7/16	3/4	9/32⑥	0.129	0.114	④	0.096	0.243	③
	17080270	15/32	5/8	3/8⑦	0.149	0.211	④	0.074	0.220	③
	17080272	15/32	5/8	3/8⑦	0.129	0.175	④	0.074	0.203	③
	17080274	15/32	5/8	5/16⑥	0.110	0.164	④	0.083	0.203	③
	17080502	1/2	7/8	⑧	0.136	0.179	④	0.110	0.243	③
	17080504	1/2	7/8	⑧	0.136	0.179	④	0.110	0.243	③
	17080553	15/32	1/2	⑧	0.142	0.220	④	0.090	0.220	③
'81	17081202,204	11/32	7/8	⑧	0.157⑩	—	④	0.110	0.243	⑨
	17081203,207	11/32	7/8	⑧	0.157⑩	—	④	0.110	0.243	⑨
	17081216, 17,218	11/32	7/8	⑧	0.157⑩	—	④	0.110	0.243	⑨
	17081242	3/8	9/16	⑧	0.090⑩	0.077⑪	④	0.139	0.243	⑨
	17081243	5/16	9/16	⑧	0.103⑩	0.090⑪	④	0.139	0.243	⑨
	17081245	3/8	5/8	⑧	0.164⑩	0.136⑪	④	0.139	0.243	⑨
	17081247	3/8	5/8	⑧	0.164⑩	0.136⑪	④	0.139	0.243	⑨
	17081248,249	3/8	5/8	⑧	0.164⑩	0.136⑪	④	0.139	0.243	⑨
	17081253,254	15/32	1/2	⑧	0.142⑩	0.227⑪	④	0.071	0.220	⑨
	17081270	7/16	5/8	⑧	0.136⑩	0.211⑪	④	0.074	0.220	⑨
	17081272	7/16	5/8	⑧	0.136⑩	0.260⑪	④	0.074	0.220	⑨
	17081274	7/16	5/8	⑧	0.136⑩	0.220⑪	④	0.083	0.220	⑨
	17081289	13/36	5/8	⑧	0.164⑩	0.136⑪	④	0.139	0.243	⑨
'82	17082202	11/32	7/8	Fixed	0.110/20⑭	—	④	0.110	0.243	⑫⑮
	17082204	11/32	3/8⑬	Fixed	0.110/20⑭	—	④	0.110	0.243	⑫⑮
	17082203	11/32	7/8	Fixed	0.157/27	—	④	0.243	0.243	⑮
	17082207	11/32	7/8	Fixed	0.157/27	—	④	0.243	0.243	⑮
	17082244	7/16	9/16	Fixed	0.117/21	0.083/16	④	0.139	0.195	⑫
	17082245	3/8	5/8	Fixed	0.149/26	0.149/26	④	0.139	0.195	⑫
	17082246	3/8	5/8	Fixed	0.149/26	0.149/26	④	0.139	0.195	⑫
	17082247	13/32	5/8	Fixed	0.164/28	0.136/24	④	0.139	0.243	⑫
	17082248	13/32	5/8	Fixed	0.164/28	0.136/24	④	0.139	0.243	⑫
	17082251	15/32	1/2	Fixed	0.142/25	0.304/45	④	0.071	0.220	⑫

QUADRAJET CARBURETOR SPECIFICATIONS
Pontiac

Year	Carburetor Identification ①	Float Level (in.)	Air Valve Spring (turn)	Pump Rod (in.)	Primary Vacuum Break (in./deg.)	Secondary Vacuum Break (in./deg.)	Secondary Opening (in.)	Choke Rod (in.)	Choke Unloader (in.)	Fast Idle Speed ② (rpm)
'83	17082253	15/32	1/2	Fixed	0.142/25	0.227/36	④	0.071	0.220	⑫
	17082264	7/16	9/16	Fixed	0.117/20	0.083/16	④	0.139	0.195	⑫
	17082265	3/8	5/8	Fixed	0.149/26	0.149/26	④	0.139	0.195	⑫
	17082266	3/8	5/8	Fixed	0.149/26	0.149/26	④	0.139	0.195	⑫
	17082267	3/8	5/8	Fixed	0.164/28	0.136/24	④	0.139	0.243	⑫
	17082268	13/32	5/8	Fixed	0.164/28	0.136/24	④	0.139	0.243	⑫
'83	17082265	3/8	5/8	Fixed	0.149/26	0.149/26	②	0.139	0.195	⑯
	17082266	3/8	5/8	Fixed	0.149/26	0.149.26	②	0.139	0.195	⑯
	17082267	3/8	5/8	Fixed	0.149/26	0.149/26	②	0.096	0.195	⑯
	17082268	3/8	5/8	Fixed	0.149/26	0.149/26	②	0.096	0.195	⑯
	17083242	9/32	9/16	Fixed	0.110/20	—	②	0.139	0.243	⑯
	17083244	1/4	9/16	Fixed	0.117/21	0.083/16	②	0.139	0.195	⑯
	17083248	3/8	5/8	Fixed	0.149/26	0.149/26	②	0.139	0.195	⑯
	17083250	7/16	1/2	Fixed	0.157/27	0.271/42	②	0.071	0.220	⑯
	17083253	7/16	1/2	Fixed	0.157/27	0.269/41	②	0.071	0.220	⑯
	17083553	7/16	1/2	Fixed	0.157/27	0.269/41	②	0.071	0.220	⑯
'84	17084201	11/32	7/8	Fixed	0.157/27	—	②	0.110	0.243	⑪
	17084205	11/32	7/8	Fixed	0.157/27	—	②	0.243	0.243	⑪
	17084208	11/32	7/8	Fixed	0.157/27	—	②	0.110	0.243	⑪
	17084209	11/32	7/8	Fixed	0.157/27	—	②	0.243	0.243	⑪
	17084210	11/32	7/8	Fixed	0.157/27	—	②	0.110	0.243	⑪
	17084240	5/16	1	Fixed	0.136/24	—	②	—	0.195	⑪
	17084244	5/16	1	Fixed	0.136/24	—	②	—	0.195	⑪
	170804246	5/16	1	Fixed	0.123/22	.136/24	②	—	0.195	⑪
	17084248	5/16	1	Fixed	0.136/24	—	②	—	0.195	⑪
	17084252	7/16	1/2	Fixed	0.157/27	.269/41	②	—	0.220	⑪
	17084254	7/16	1/2	Fixed	0.157/27	.269/41	②	—	0.220	⑪

① The carburetor identification number is stamped on the float bowl, near the secondary throttle lever.
② On highest step.
③ 1½ turns after contacting lever for preliminary setting
④ No measurement necessary on two point linkage; see text.
⑤ 2 turns after contacting lever for preliminary setting.
⑥ Inner hole
⑦ Outer hole
⑧ Not adjustable
⑨ 4½ turns after contacting lever for preliminary setting
⑩ Front
⑪ Rear
⑫ 3 turns after contacting lever for preliminary setting
⑬ Firebird—7/8
⑭ Firebird—0.157 in./27°
⑮ Firebird—3⅛ turns after contacting lever for preliminary setting
⑯ See underhood sticker

QUADRAJET CARBURETOR SPECIFICATIONS
All Canadian Models

Year	Carburetor Identi-fication①	Float Level (in.)	Air Valve Spring (turn)	Pump Rod (in.)	Primary Vacuum Break (deg./in.)	Secondary Vacuum Break (deg./in.)	Secondary Opening (in.)	Choke Rod (in.)	Choke Unloader (in.)	Fast Idle Speed② (rpm)
'81	17080201	15/32	7/8	9/32②	—	23/0.129	④	0.314	0.277	⑤
	17080205	15/32	7/8	9/32②	—	23/0.129	④	0.314	0.277	⑤
	17080206	15/32	7/8	9/32②	—	23/0.129	④	0.314	0.277	⑤
	17080290	15/32	7/8	9/32②	—	26/0.149	④	0.314	0.277	⑤
	17080291	15/32	7/8	9/32②	—	26/0.149	④	0.314	0.277	⑤
	17080292	15/32	7/8	9/32②	—	26/0.149	④	0.314	0.277	⑤
	17080213	3/8	1	9/32②	23/0.129	30/0.179	④	0.234	0.260	⑤
	17080215	3/8	1	9/32②	23/0.129	30/0.179	④	0.234	0.260	⑤
	17080298	3/8	1	9/32②	23/0.129	30/0.179	④	0.234	0.260	⑤
	17080507	3/8	1	9/32②	23/0.129	30/0.179	④	0.234	0.260	⑤
	17080513	3/8	1	9/32②	23/0.129	30/0.179	④	0.234	0.260	⑤
	17081250	13/32	1/2	9/32②	26/0.149	34/0.211	④	0.090	0.220	⑤
	17080260	13/32	1/2	9/32②	26/0.149	34/0.211	④	0.090	0.220	⑤
	17081276	15/32	5/8	5/16②	20/0.110	28/0.164	④	0.083	0.203	⑤
	17081286	13/32	1/2	9/32②	18/0.096	34/0.211	④	0.077	0.220	⑤
	17081287	13/32	1/2	9/32②	18/0.096	34/0.211	④	0.077	0.220	⑤
	17081282	3/8	5/8	9/32②	20/0.110	—	④	0.110	0.243	⑤
	17081283	3/8	7/8	9/32②	20/0.110	—	④	0.110	0.243	⑤
	17081284	1/2	7/8	9/32③	20/0.110	—	④	0.110	0.243	⑤
	17081285	1/2	7/8	9/32③	20/0.110	—	④	0.110	0.243	⑤
	17080243	3/16	9/16	9/32②	14.5/0.075	16/0.083	④	0.075	0.179	⑤
	17081295	13/32	9/16	9/32②	14.5/0.075	13/0.066	④	0.075	0.220	⑤
	17081294	5/16	5/8	9/32②	24.5/0.139	14/0.071	④	0.139	0.243	⑤
	17081290	13/32	7/8	9/32②	46/0.314	24/0.136	④	0.314	0.277	⑤
	17081291	13/32	7/8	9/32②	46/0.314	24/0.136	④	0.314	0.277	⑤
	17081292	13/32	7/8	9/32②	46/0.314	24/0.136	④	0.314	0.277	⑤
	17081506	13/32	7/8	9/32②	46/0.314	36/0.227	④	0.314	0.227	⑤
	17081508	13/32	7/8	9/32②	46/0.314	36/0.227	④	0.314	0.227	⑤
	17080202	7/16	7/8	1/4②	20/0.110	—	④	0.110	0.243	⑤
	17080204	7/16	7/8	1/4②	20/0.110	—	④	0.110	0.243	⑤
	17080207	7/16	7/8	1/4②	20/0.110	—	④	0.110	0.243	⑤
'82	17082280	3/8	7/8	9/32②	25/0.142	—	④	0.110	0.243	⑤
	17082281	3/8	7/8	9/32②	25/0.142	—	④	0.110	0.243	⑤
	17082282	3/8	7/8	9/32②	25/0.142	—	④	0.110	0.243	⑤
	17082283	3/8	7/8	9/32②	25/0.142	—	④	0.110	0.243	⑤
	17082286	13/32	1/2	9/32②	22/0.123	34/0.211	④	0.077	0.243	⑤
	17082287	13/32	1/2	9/32②	22/0.123	34/0.211	④	0.077	0.243	⑤
	17082288	3/8	7/8	9/32②	25/0.142	—	④	0.110	0.243	⑤

QUADRAJET CARBURETOR SPECIFICATIONS
All Canadian Models

Year	Carburetor Identification①	Float Level (in.)	Air Valve Spring (turn)	Pump Rod (in.)	Primary Vacuum Break (deg./in.)	Secondary Vacuum Break (deg./in.)	Secondary Opening (in.)	Choke Rod (in.)	Choke Unloader (in.)	Fast Idle Speed② (rpm)
'82	17082289	3/8	7/8	9/32②	25/0.142	—	④	0.110	0.243	⑤
	17082296	1/2	7/8	9/32②	25/0.142	—	④	0.110	0.243	⑤
	17082297	1/2	7/8	9/32②	25/0.142	—	④	0.110	0.243	⑤
'83	17080213	3/8	1	9/32	23/.129	30/.179	④	0.234	0.260	⑤
	17082213	9/32	1	9/32	23/.129	30/.179	④	0.234	0.260	⑤
	17082282	3/8	7/8	9/32	25/.142	—	④	0.110	0.243	⑤
	17082283	3/8	7/8	9/32	25/.142	—	④	0.110	0.243	⑤
	17082286	13/32	1/2	9/32	23/.129	34/.211	④	0.107	0.220	⑤
	17082287	13/32	1/2	9/32	23/.129	34/.211	④	0.107	0.220	⑤
	17082296	1/2	7/8	9/32	25/.142	—	④	0.110	0.243	⑤
	17082297	1/2	7/8	9/32	25/.142	—	④	0.110	0.243	⑤
	17083280	3/8	7/8	9/32	25/.142	—	④	0.110	0.243	⑤
	17083281	3/8	7/8	9/32	25/.142	—	④	0.110	0.243	⑤
	17083282	3/8	7/8	9/32	25/.142	—	④	0.110	0.243	⑤
	17083283	3/8	7/8	9/32	25/.142	—	④	0.110	0.243	⑤
	17083290	13/32	7/8	9/32	—	24/.136	④	0.314	0.251	⑤
	17083292	13/32	7/8	9/32	—	24/.136	④	0.314	0.251	⑤
	17083298	3/8	1	9/32	23/.129	30/.179	④	0.234	0.260	⑤
'84	17084280	3/8	7/8	9/32②	23/.129	—	④	0.110	0.243	⑤
	17084281	3/8	7/8	9/32②	23/.129	—	④	0.110	0.243	⑤
	17084282	3/8	7/8	9/32②	23/.129	—	④	0.110	0.243	⑤
	17084283	3/8	7/8	9/32②	23/.129	—	④	0.110	0.243	⑤
	17084284	3/8	7/8	9/32②	23/.129	—	④	0.110	0.243	⑤
	17084285	3/8	7/8	9/32②	23/.129	—	④	0.110	0.243	⑤

MITSUBISHI FEEDBACK CARBURETOR

General Information

The feedback carburetor system consists of various kinds of sensors and actuators and an on-board computer. If any feedback carburetor system components fail, interruption of the fuel supply or failure to supply the proper amount of fuel for engine operating conditions will result. This usually results in symptoms that include hard starting, unstable idle or poor driveability. If any of these symptoms are noted, first perform basic engine checks for ignition system operation or obvious problems such as loose connections or vacuum hoses, etc. The Mitsubishi FBC system can only be diagnosed with a special tester and adapter. Since instructions for connecting and using the diagnostic tester is included with the unit, they will not be covered here.

NOTE: The tester used is the same one for the Mitsubishi Electronically Controlled Injection (ECI).

ADJUSTMENTS

Float Level Adjustment

1. Invert the float chamber cover assembly without a gasket.
2. Position the universal float level gauge or suitable depth gauge and measure the distance from the bottom of the float to the surface of the float chamber cover. It should be 20mm (.787 in.). If the reading is not within 1 mm (.0394 in.), the shim under the needle seat must be changed. Adding or removing a shim will change the float level by three times the thickness of the shim.

3. To remove the float, slide the pin out and remove the float and needle.
4. Unscrew the retainer and remove the needle seat used with pliers.

NOTE: When removing the needle seat, clamp the chamferred portion with the jaws of the pliers. DO NOT clamp the machined surface of the needle seat that fits in the bore.

5. Check the filter for blockage or damage and replace if necessary.
6. Install a new O-ring into the needle seat.
7. Install the shim and filter into the needle seat.
8. Insert the needle seat assembly into the float chamber cover.

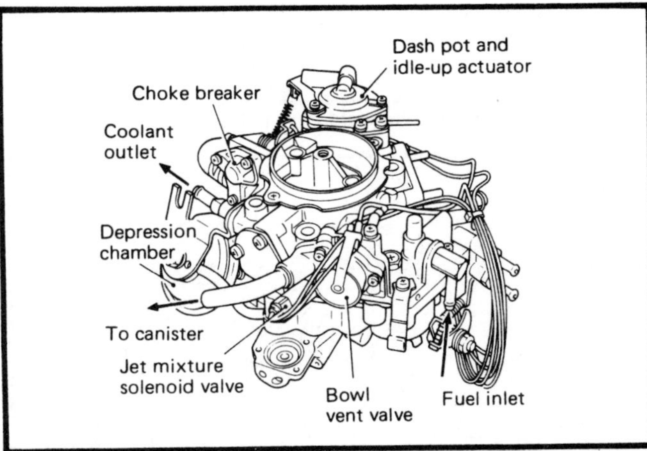

Mitsubishi feedback carburetor assembly showing component locations

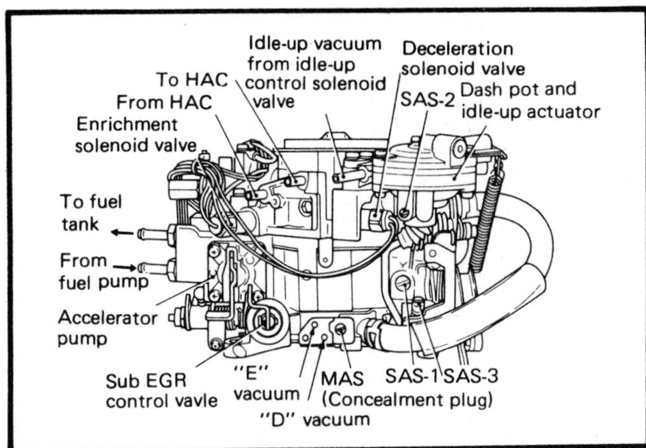

Mitsubishi feedback carburetor assembly showing component locations

9. Install the needle seal retainer and tighten the screw firmly.
10. Insert the needle into the seat.
11. Install the float and insert the retaining pin.
12. Check the distance from the bottom of the float to the surface of the float chamber cover and readjust if necessary.

Carburetor Removal

1. Disconnect the battery.
2. Drain the coolant down to intake manifold level or below.
3. Remove the air cleaner.
4. Place a suitable container under the fuel inlet fitting to catch any fuel that may be trapped in the line, then disconnect the fuel line from the carburetor inlet nipple.

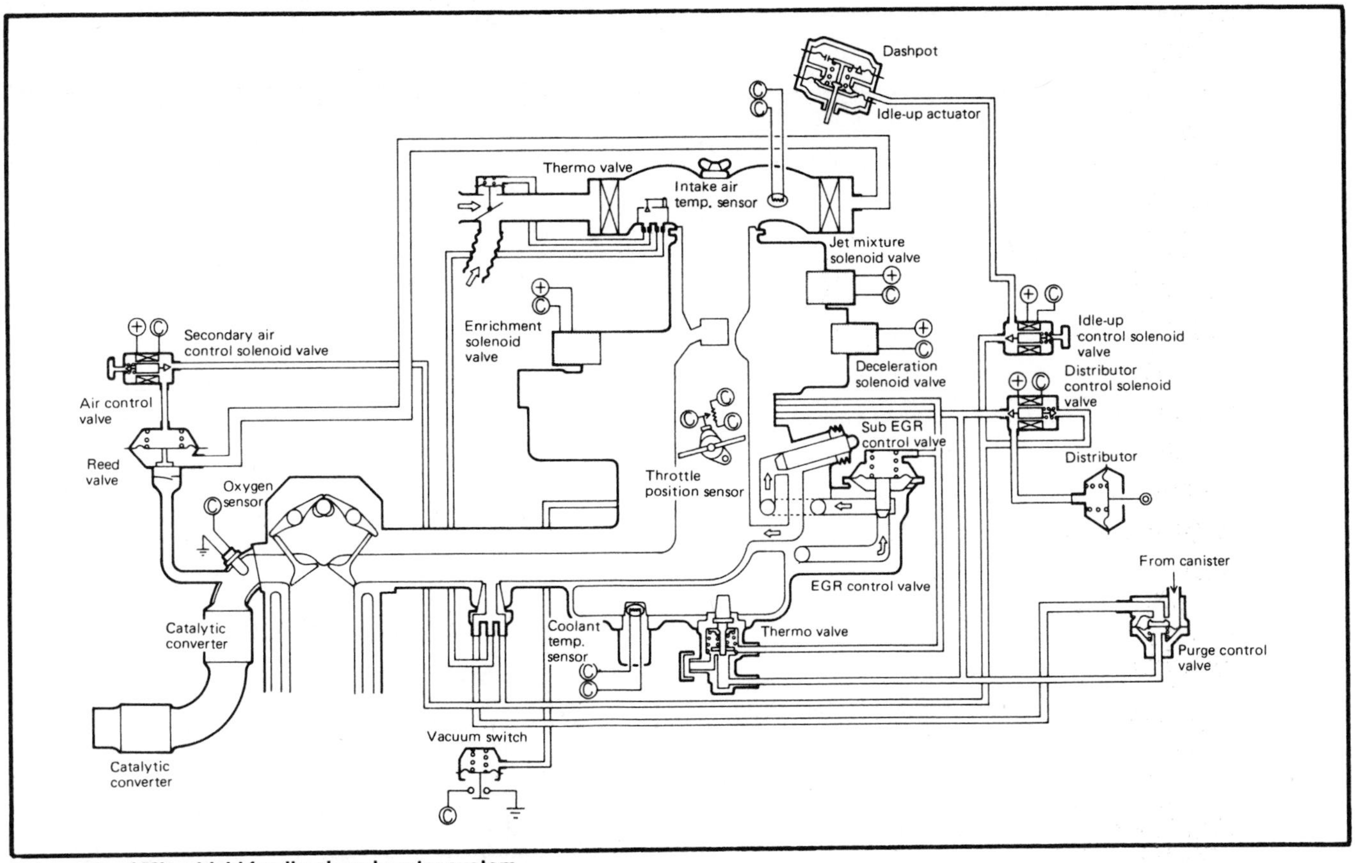

Schematic of Mitsubishi feedback carburetor system

Measuring float level with gauge (float chamber cover inverted)

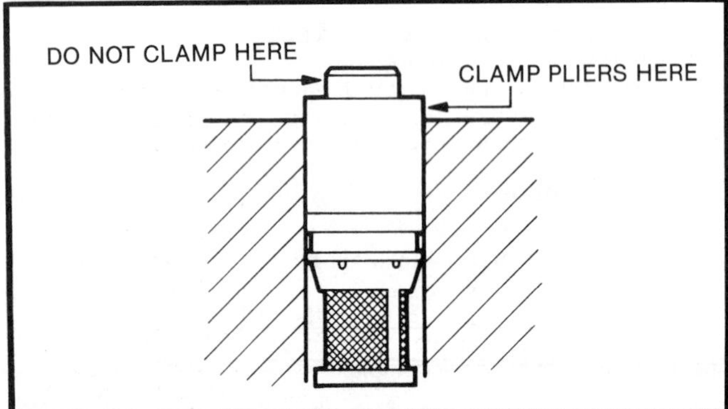

Be extremely careful where the pliers clamp the needle seat assembly during removal

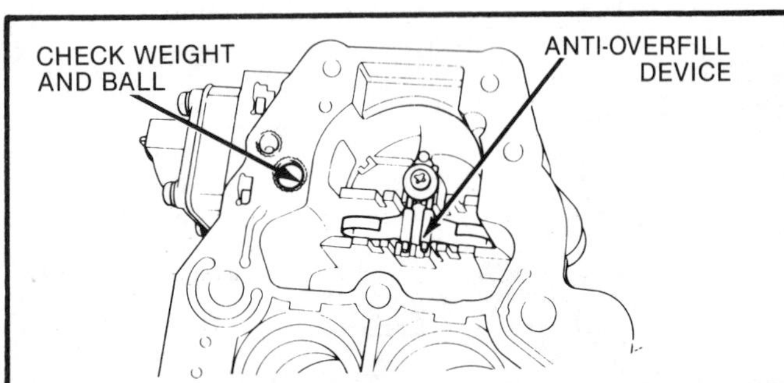

Location of check weight and ball of the anti-overfill device

5. Disconnect the vacuum hoses from the carburetor.
6. Disconnect the throttle cable from the carburetor.
7. Remove the carburetor mounting bolts and carefully remove the carburetor from the engine. Hold the carburetor level to avoid spilling fuel from the float bowl.

Disassembly

CAUTION

Do not remove the following parts from the feedback carburetor:
- *Choke valves*
- *Choke levers and related parts*
- *Round nut of accelerator pump link*
- *Adjusting screws except the idle speed adjusting screws, idle mixture adjusting screw and dashpot adjusting screw*
- *Throttle valves*

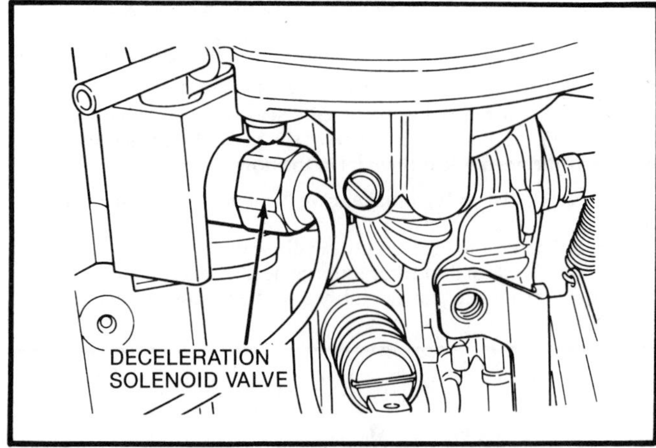

Location of deceleration solenoid valve in float chamber cover

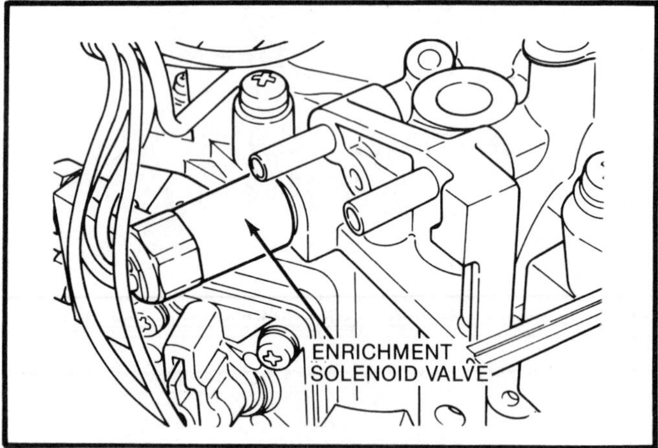

Location of enrichment solenoid valve in float chamber cover

1. Pull the water hose off the nipple of the throttle body and from the nipple of the wax element portion.
2. Grind down the head of the choke cover lock screws by using a hand grinder.
3. Remove the throttle return spring and the damper spring.
4. Remove the vacuum hose from the vacuum chamber and the throttle body.
5. Remove the accelerator pump rod from the throttle lever.
6. Remove the dashpot/idle-up actuator rod (manual transaxle), or idle-up actuator rod (automatic transaxle) from the free lever.

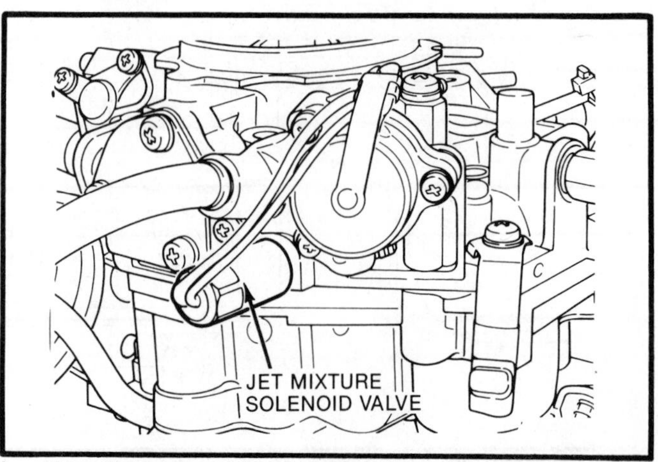

Location of jet mixture solenoid valve in float chamber cover

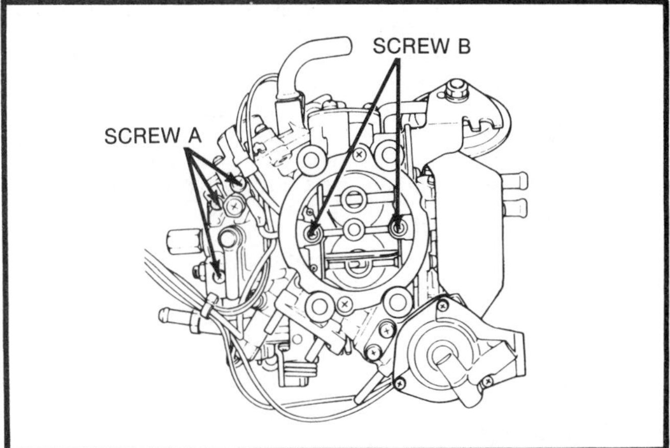

Remove the float chamber cover screws (B) to remove the throttle body, then float chamber cover screws (A) to remove the float chamber from the main body

7. Remove the actuator from the float chamber cover.
8. Remove the deceleration solenoid valve from the float chamber cover.
9. Remove the enrichment solenoid valve from the float chamber cover.
10. Remove the jet mixture solenoid valve from the float chamber cover.
11. Remove the depression chamber rod from the secondary throttle lever.
12. Remove the vacuum (depression) chamber. To remove the depression chamber, first remove the choke breaker cover, then remove the depression chamber attaching screws.
13. Remove the float chamber cover screws (B) and remove the throttle body.
14. Remove the remaining chamber cover screws (A) and remove the float chamber from the main body.
15. Remove the check weight and ball and the steel ball of the anti-overfill device.
16. Pull off the pin and remove the float.
17. Remove the needle valve retainer and then remove the needle valve assembley with pliers.
18. Remove the main jets from the jet blocks. When the main jet is to be removed, use a screwdriver with the proper blade for the slot in the jet.
19. Remove the pilot jet retainer and pull out the secondary pilot jet with pliers.
20. Remove the accelerator pump mounting screws and remove the pump cover-link assembly, diaphragm, spring, body and gasket from main body.
21. Remove the snap ring from the sub EGR control valve pin.
22. Remove the pin and then remove the link from the valve, then remove the little steel ball and spring from the sub EGR control valve.
23. Remove the sub EGR control valve from the throttle body.

Assembly

The carburetor is assembled in the reverse of the disassembly procedure, paying special attention to the following items:

a. Clean all parts before assembly
b. Check that no air or fuel passages are blocked
c. Check for smooth operation of all throttle and choke linkage
d. Make sure the sub EGR valve operates smoothly
e. When replacing a main or pilot jet, the old jet and the new jet must be the same size. The jets are selected after an exact flow measurement at the factory and must not be interchanged

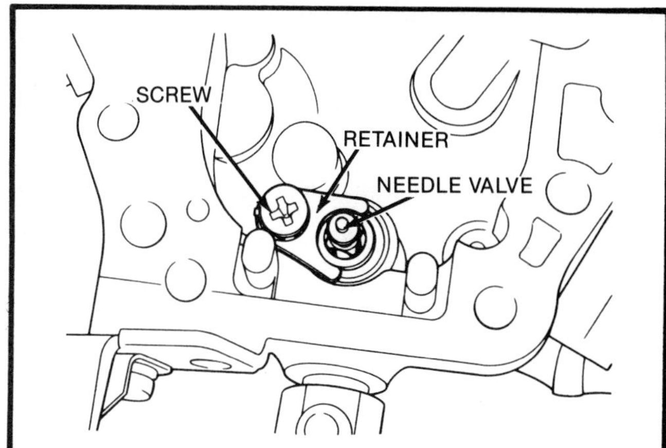

Location of needle valve assembly in float bowl

Installation

1. Inspect the mating surfaces of the carburetor and intake manifold and make sure both surfaces are clean, smooth and free from any gasket material. Look for nicks, burrs or other damage.
2. Place a new carburetor gasket on the intake manifold.
3. Carefully place the carburetor on the intake manifold.
4. Install the carburetor mounting bolts and tighten them alternately to compress the gasket evenly. The nuts must be drawn down tightly to prevent vacuum leakage between the carburetor and the intake manifold.
5. Connect the throttle cable, vacuum hoses and fuel hoses.
6. Check for worn or loose vacuum hose connections.
7. Check to be sure the choke plate open fully when operated.
8. Check to see that full throttle travel is obtained.
9. Install the air cleaner and replace the element, if necessary.
10. Reconnect the battery cable. Start the engine and reset the idle speed to the specifications listed on the underhood emission sticker.

CAUTION

Priming the carburetor by pouring gasoline into the air horn while cranking the engine should be strictly avoided. Depressing the accelerator pedal several times while cranking the engine should be adequate.

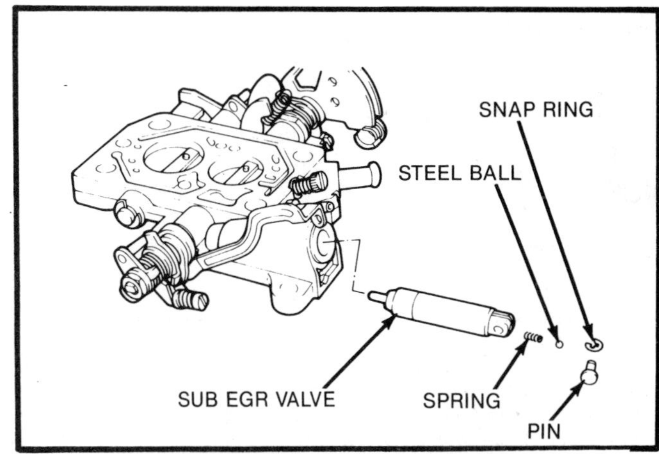

Sub EGR valve assembly showing internal components

Idle Speed and Mixture Adjustment

NOTE: The carburetor must be removed and held in a suitable fixture for this procedure. DO NOT clamp the carburetor body in a vise. Fabricate a holding plate from heavy brackets or sheet metal and bolt the carburetor mounting flange to it.

1. Remove the carburetor from the engine.
2. Mount the carburetor in a suitable holding fixture and remove the concealment plug.
3. Reinstall the carburetor on the engine.
4. Start the engine and allow it to reach normal operating temperature. Make sure all accessories are off and the transaxle is in neutral.
5. Turn the ignition OFF and disconnect the negative terminal from the battery for about three seconds, then reconnect it.
6. Disconnect the oxygen sensor.
7. Run the engine for more than five seconds at 2000-3000 rpm, then let the engine return to idle for two minutes.
8. Using a suitable meter, set the engine CO and idle speed to the specifications listed on the underhood emission control sticker by turning the idle speed and mixture adjusting screws until the correct values are obtained.
9. Reconnect the oxygen sensor.
10. Reset the engine speed to specifications by turning the idle speed adjusting screw, if necessary.
11. Install the concealment plug to seal the idle mixture screw.

Fuel Injection Systems

SECTION 5

INDEX

NOTE: Please refer to the Application Chart at the front of this Manual.

FUEL INJECTION

General Information

There are two basic types of fuel injection systems currently in production, Port and Throttle Body. Constant Injection Systems (CIS) are a mechanically controlled, Port type of fuel injection that uses linkage between an air flow sensor and fuel distributor to regulate the air/fuel mixture by moving a piston up and down. Air Flow Controlled (AFC) Port and Throttle Body (TBI) injection systems use an intake air flow (or air mass) sensor and an electronic control unit that monitors engine conditions and then regulates the fuel mixture by sending electrical impulses to the injector(s). By varying the length of time the solenoid-type fuel injector is open, more or less fuel is delivered. A fuel injection systems is called PORT type when the injectors are mounted in the cylinder head and spray the fuel charge directly behind the intake valve. Throttle Body Injection (TBI) uses one or two injectors mounted atop the intake manifold (much like a conventional carburetor) that spray the fuel charge down through a throttle body butterfly valve. The fuel charge is drawn into the intake manifold and distributed to the cylinders in the conventional manner.

Fuel injection combined with electronics and various engine sensors provides a precise fuel management system that meets all the demands for improved fuel economy, increased performance and lower emissions more precisely and reliably than is possible with a conventional carburetor. A fuel injected engine generally averages ten percent more power and fuel economy with lower emissions than a carbureted engine. Even "feedback" carburetors with computer controls cannot achieve the accurate fuel metering necessary to meet the lower emission levels required by upcoming Federal standards. Because of its precise control, fuel injection allows the engine to operate with a stoichiometric or optimum fuel ratio of 14.7 parts air to one part fuel throughout the entire engine rpm range, under all operating conditions. This 14.7:1 fuel mixture assures that all the carbon and hydrogen is burned in the combustion chamber during the power stroke. This produces the lowest combination of emissions from unburned hydrocarbons, carbon monoxide and oxides of nitrogen in the exhaust gases. By using an oxygen sensor to measure the oxygen content of the exhaust gases, the on-board computer (control unit) can constantly adjust and "fine tune" the fuel mixture in response to changing temperature, load and altitude conditions. In addition, this degree of fuel control allows the catalytic converter to function at peak efficiency for emission control.

It's important to identify all system components, how they work and their relationship to one another in operation before attempting any maintenance or repair procedures on any particular type of fuel injection. All fuel injection systems are delicate and vulnerable to damage from rust, dirt, water and careless handling. The shock of hitting a cement floor when dropped from about waist height can ruin a control unit, for example. Because of the close tolerances involved in fuel injector construction (25 millionths of an inch on some), any rust or dirt particles in the fuel lines can ruin machined surfaces or block the nozzle completely. Water in the fuel can do more damage than a well-placed grenade and some fuel additives or chemicals may damage fuel lines or components like the oxygen sensor. Many fuel injection components, while similar in both appearance and function, can vary from one manufacturer to another so it is very important to identify exactly which type of injection system is being used on a particular engine, and to become familiar with all related sensors and equipment. In addition, there are manufacturer's modifications within the major groups of CIS, AFC and TBI type systems that utilize different sensors and electronic control units that are similar in appearance but are not interchangeable with any other system. For example, the Bosch L-Jetronic AFC injection system has four modifications that are designated LH, LH II, LU and Motronic. Each individual system has some sensors, capabilities and characteristics unique to its own design. Although they may look pretty much the same to the casual observer, each PROM or memory unit is programmed differently for each type of engine it is supposed to regulate. For this reason, the careful recording of part numbers or engine serial numbers and the correct identification of vehicle make, model and year is vital to insure the correct replacement parts are obtained. Always double check all numbers before installing any new component.

COMPARING CARBURETOR AND FUEL INJECTION CIRCUITS

Carburetor Component	Fuel Injection Component
Accelerator pump	Wide open throttle (WOT) switch
Fast idle cam	Auxiliary air bypass device
Choke pull-off	Thermo-time switch
Float and needle valve	Pressure regulator
Float bowl	Fuel manifold
Power valve metering rods	Manifold pressure or airflow sensor
Metering jets	Fuel injectors
Choke assembly	Cold start injector
Carburetor linkage	Throttle valve cable
Idle speed adjusting screw	Bypass air adjusting screw
Fast idle solenoid	Idle air control valve

NOTE: These are general comparisons. Some fuel injection systems may not use all of the components listed, while others incorporate additional sensors, actuators and control capabilities such as ignition timing or idle speed control systems.

COMPARING CARBURETOR CIRCUITS TO FUEL INJECTION COMPONENTS

It makes fuel injection a little easier to understand by comparing the functions of major systems and components in the carburetor to those in a typical port fuel injection system. Fuel injection is basically just a more precise way of doing the same thing—delivering the correct air/fuel mixture to the combustion chambers at the correct time under all engine operating conditions.

One major difference is that the accelerator pedal actually feeds fuel to the intake manifold when depressed (via the accelerator pump linkage) on the carburetor, even if the engine is stopped and the ignition switch is turned OFF. On a fuel injection system, the accelerator pedal merely opens or closes the butterfly-type throttle plate, allowing more air to flow into the intake system. There is no direct, mechanical linkage connection that provides fuel for the engine. The amount of fuel delivered by the port fuel injection system is regulated either by movement of the fuel distributor plunger (CIS systems), or by the electronic control unit that sets the fuel mixture according to the output signals from various sensors (AFC system). It's impossible to flood a fuel injected engine by pumping the accelerator pedal with the ignition switch off since the pedal only opens and closes an air valve. No fuel can be delivered unless the control unit is energized and sends impulses to the injectors.

To assist cold starting, modern carburetors incorporate an automatic choke assembly which consists of mechanical linkage operating a butterfly valve that reduces the amount of incoming air, thereby increasing the amount of vacuum that pulls on the main fuel discharge nozzle in the carburetor. This increased vacuum draws a larger quantity of fuel from the float bowl, providing the richer fuel mixture necessary for cold starting. The butterfly valve (choke plate) movement is controlled either manually, or by a bimetal coil spring that responds to temperature and is usually heated either by hot water from the cooling system, hot air from the exhaust manifold, or by an electric element in the choke spring housing. Because it is a mechanical system, it is vulnerable to malfunctions due to loose or sticking linkage, broken springs, and incorrect adjustment, causing hard starting or flooding problems on cold engines. All fuel injection systems incorporate a cold start system that does essentially the same thing as the carburetor choke system, but in a much more precise way with different components. Some fuel injection systems use an extra fuel injector called a "cold start valve" that is mounted in the intake manifold downstream from the throttle plate. This solenoid-type injector provides an extra fuel charge when the ignition is switched ON with the engine cold and is regulated through a temperature sensitive switch designed to cut off the signal to the cold start injector after about 12 seconds. This temperature sensitive switch, called a "thermo-time switch," is also designed to lock out the cold start system operation when the engine is warm and the enriched fuel mixture is not necessary for starting. Some computer-controlled fuel injection systems are designed to enrich the fuel mixture by providing longer signal to the injectors (increasing the amount of time the nozzle is open) without the need for a separate cold start injector. The longer the solenoid-type injectors are open, the more fuel is delivered to the combustion chambers.

NOTE: Because some cold start injectors are designed to operate when the key is switched ON, it is possible to flood a cold engine by turning the key on and off a few times. Although the effect is the same as pumping the accelerator on a carbureted engine, the cause is totally different. Some manufacturers design the cold start valve to spray only when the starter is cranking to avoid this problem.

Safety Precautions

CAUTION

Whenever working on or around any fuel injection system, always observe these general precautions to prevent the possibility of personal injury or damage to fuel injection components.

- Never install or remove battery cables with the key ON or the engine running. Jumper cables should be connected with the key OFF to avoid power surges that can damage electronic control units. Engines equipped with computer controlled fuel injection systems should avoid both giving and getting jump starts due to the possibility of serious damage to components due to arcing in the engine compartment.
- Always remove the battery cables before charging the battery. Never use a high-output charger on an installed battery or attempt to use any type of "hot shot" (24 volt) starting aid.
- Never remove or attach wiring harness connectors with the ignition switch ON, especially to the electronic control unit.
- When checking compression on engines with AFC injection systems, unplug the cable from the battery to the relays.
- Always depressurize the fuel system before attempting to disconnect any fuel lines.
- Always use clean rags and tools when working on an open fuel injection system and take care to prevent any dirt from entering the system. Wipe all components clean before installation and prepare a clean work area for disassembly and inspection of components. Use lint-free cloths to wipe components and avoid using any caustic cleaning solvents.
- Do not drop any components during service procedures and never apply 12 volts directly to a fuel injector unless instructed specifically to do so. Some injectors windings are designed to safely handle only 4 or 5 volts and can be destroyed in seconds if 12 volts are applied directly to the connector.
- Remove the electronic control unit if the vehicle is to be placed in an environment where temperatures exceed approximately 176 degrees F (80 degrees C), such as a paint spray booth or when arc or gas welding near the control unit location in the car.

BOSCH CONTINUOUS INJECTION SYSTEM (CIS)

K and KE-Jetronic

GENERAL INFORMATION

The Bosch CIS fuel injection system differs from the electronic AFC system in that injection takes place continuously: it is controlled through variation of the fuel flow rate through the injectors, rather than by variation of the fuel injection duration as on the AFC system. Prior to 1983, CIS used no electronic computer, and is an electro-mechanical system that will provide suitable air/fuel mixtures under all driving conditions.

Air
Fuel
Electric fuel pump
Fuel accumulator
Air filter
Fuel filter
Air-flow sensor
Mixture control unit
Fuel distributor
Throttle valve
Mixture
Injection valves
Intake tubes
Combustion chamber

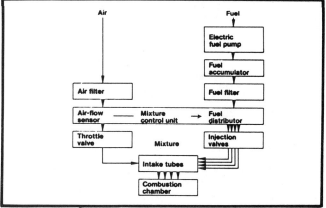

Basic operation schematic of K-Jetronic fuel injection system

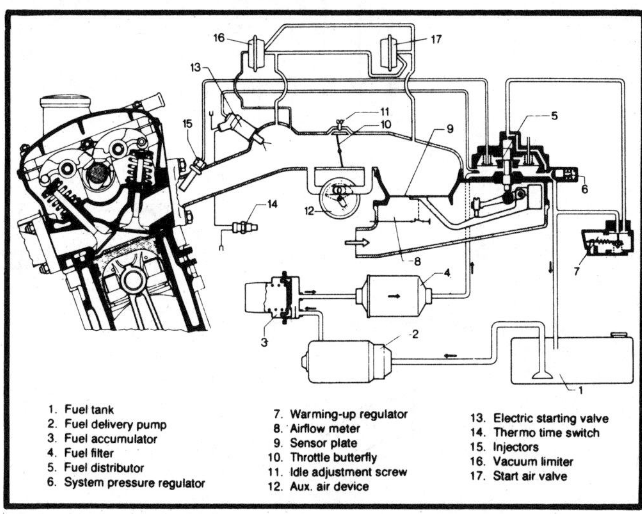

Typical K-Jetronic fuel injection system (without push valve)

SECTION 5 FUEL INJECTION SYSTEMS

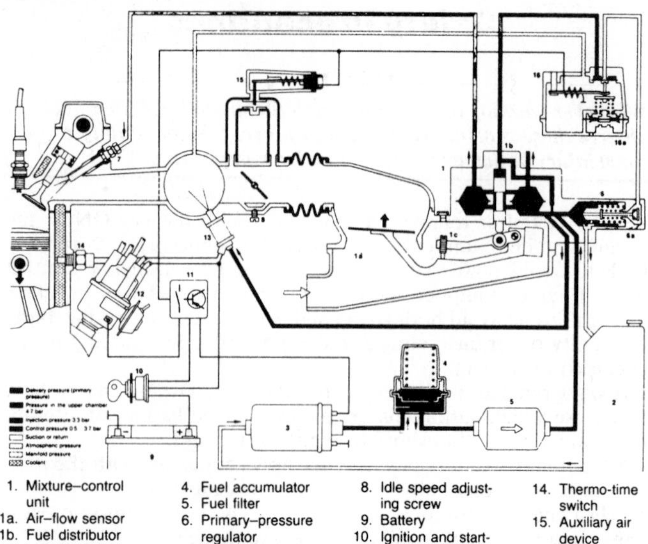

1. Mixture–control unit
1a. Air–flow sensor
1b. Fuel distributor
1c. Idle–mixture adjusting screw
2. Fuel tank
3. Electric fuel pump
4. Fuel accumulator
5. Fuel filter
6. Primary–pressure regulator
6a. Push–up valve
7. Fuel injection valve
8. Idle speed adjusting screw
9. Battery
10. Ignition and starting switch
11. Control relay
12. Ignition distributor
13. Start valve
14. Thermo-time switch
15. Auxiliary air device
16. Warm–up regulator
16a. Full–load diaphragm

Typical K-Jetronic fuel injection system (with push valve)

The complete CIS system consists of the following components: air/fuel control unit (housing both air flow sensor and fuel distributor), electric fuel pump(s) (and fuel pressure accumulator), fuel filter, control pressure regulator, fuel injectors, auxiliary air valve, cold start injector, engine sensors and various switches and relays.

The heart of the early (non-electronic) CIS system is the air/fuel control unit. It consists of an air flow sensor and a fuel distributor. Intake air flows past the air cleaner and through the air venturi raising or lowering the counterbalanced air flow sensor plate. The plate is connected to a pivoting lever which moves the control plunger in the fuel distributor in direct proportion to the intake air flow.

NOTE: The KE type CIS system uses a new electronic actuator which replaces several mixture control components, including the push valve, frequency valve and control pressure regulator.

The fuel distributor, which controls the amount of fuel to the injectors, consists of a line pressure regulator, a control plunger, and (4, 6, or 8) pressure regulator valves (one for each injector). The line pressure regulator maintains the fuel distributor inlet pressure at a constant psi., and will recirculate fuel to the tank if pressure exceeds this value. The control plunger, which is connected to the air flow sensor plate, controls the amount of fuel available to each of the pressure regulator valves. The presure regulator valves maintain a constant-fuel pressure differential between the inlet and outlet sides of the control plunger. This is independent of the

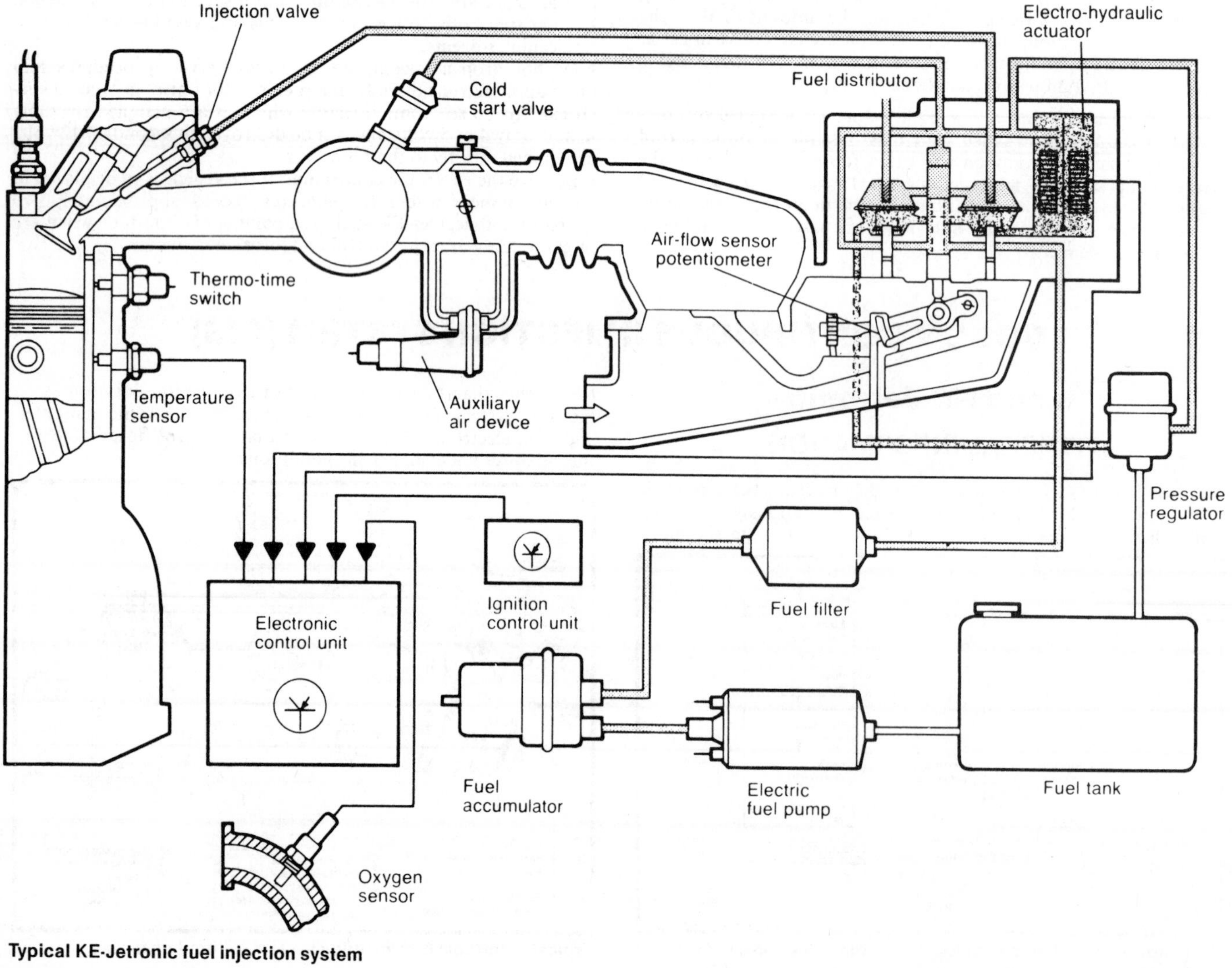

Injection valve
Electro-hydraulic actuator
Fuel distributor
Cold start valve
Air-flow sensor potentiometer
Thermo-time switch
Temperature sensor
Auxiliary air device
Pressure regulator
Ignition control unit
Electronic control unit
Fuel filter
Fuel accumulator
Electric fuel pump
Fuel tank
Oxygen sensor

Typical KE-Jetronic fuel injection system

amount of fuel passing through the valves, which varies according to plunger height.

The fuel distributor on the KE-Jetronic is different than the earlier K model. The system pressure regulator piston and pressure compensating valve are no longer installed and the adjusting screws and compression springs are moved to the lower chamber. An additional strainer has been installed in front of the differential pressure valve and a strainer with a permanent magnet is in the supply line to the electrohydraulic actuator to catch any rust particles in the fuel. The pressure on top of the control plunger is now equal to system pressure and there is a pressure measuring connection located in the lower chamber.

The main difference between the K and KE-Jetronic is the electrohydraulic actuator used for mixture correction. The actuator is flanged onto the fuel distributor and acts as a pressure regulator which operates as a plate valve. The place valve position can be varied, causing a differential pressure change in the actuator and lower chamber and thereby correcting the mixture. The signal for actuation comes from an electronic control unit which adjusts the mixture according to the engine operating conditions.

With the ignition ON, the control unit is connected to battery voltage. A voltage correction circuit prevents fluctuations when components are energized and controls the operating voltage to approximately 8 volts. The amount of cranking enrichment depends on coolant temperature. A timing element regulates the enrichment after one second to the warm-up plus the after-starting value. This value remains constant as long as the engine is cranked. The after-start enrichment establishes smooth running characteristics after starting: the amount of after-start enrichment also depends on coolant temperatures, as does warm-up enrichment. The lower the temperature, the higher the current rate at the actuator and the greater the fuel enrichment.

NOTE: The KE-Jetronic system also controls altitude enrichment, based on a signal from an altitude correction capsule.

The injectors themselves are spring loaded and calibrated to open at a preset pressure. They are not electrically operated as on the AFC fuel injection system.

The control pressure regulator, located on the intake manifold, acts to regulate the air/fuel mixture according to engine temperature. When the engine is cold, the control pressure regulator enriches the mixture. This is accomplished when a certain amount of fuel is bled off into a separate control pressure system. The control pressure regulator maintains this fuel at a set pressure. The regulator is connected to the upper side of the fuel distributor control plunger. When the engine temperature is below operating parameters, a bi-metal spring in the regulator senses this and reduces the fuel pressure on top of the plunger, allowing the plunger to rise further and channel more fuel to the regulator valves and injectors, thereby enriching the mixture. When the engine warms, the bi-metal spring in the regulator increases the pressure back to the preset value leaning the air/fuel mixture back to its normal operating ratio.

The auxiliary air valve provides extra air for the richer mixture during warm-up, thus raising the engine speed and improving cold driveability. The auxiliary air valve, which also has a temperature sensitive bi-metal spring, works directly with the control pressure regulator. At cold start-up, the valve is fully open. As the engine warms, an electric coil slowly closes the valve (4-8 minutes max.), blocking off the extra air and lowering the idle speed.

The cold start injector, located on the inlet duct, sprays extra fuel into the intake air stream during starter motor operation when the engine is cold.

The thermal time switch, located on the cylinder head, actuates the cold injector. The switch has a bi-metal spring which senses coolant temperature and an electric coil which limits the cold start injector spray to about 12 seconds, to prevent flooding the engine.

The fuel accumulator has a check valve which keeps residual fuel pressure from dropping below a minimum psi when the engine or fuel pump are shut off. The fuel system is always pressurized, preventing vapor lock in a hot start situation.

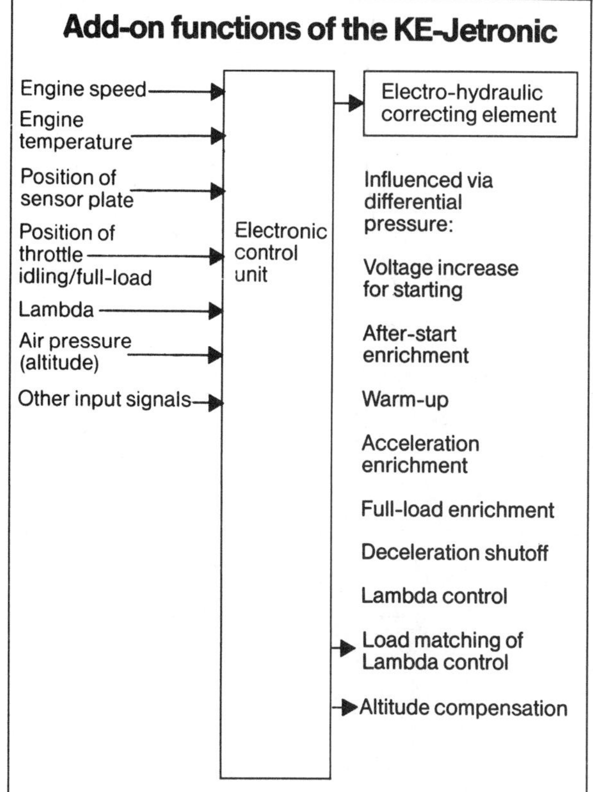

Schematic of KE-Jetronic operation

CAUTION

Because the fuel system is constantly under pressure, it is very important to follow the procedures outlined under "Relieving Fuel System Pressure" before attempting to disconnect any fuel lines.

Engine emissions are controlled by means of a catalytic converter and/or Lambda (Oxygen) Sensor system, which provides a signal to regulate the fuel mixture according to the measured oxygen content of the exhaust gases.

NOTE: On KE type CIS fuel injection systems the oxygen sensor is electrically heated for faster operation.

On-Car Service

Water in the fuel, and the resulting rust, are the number one enemies of any fuel injection system, especially CIS. Rust particles inflict more damage more often on K-Jetronic than any other Bosch injection system. Because normal operating pressure on K systems ranges from 65-85 psi, the fuel pump usually pushes any

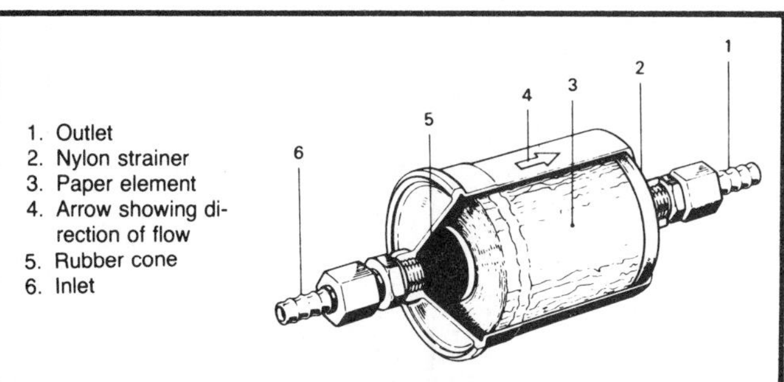

Cross section of typical fuel filter

TROUBLESHOOTING—K-JETRONIC

Engine does not start or starts poorly when **cold**	Engine does not start or starts poorly when **warm**	Irregular idle (engine shakes) during warm-up	Irregular idle (engine shakes) with engine **warm**	Engine does not draw fuel smoothly (backfires)	Engine misfires under full load	Insufficient power	Engine runs on (diesels)	Excessive fuel consumption	Flat spot during acceleration	Idle CO value too **high**	Idle CO value too **low**	Idle speed cannot be adjusted (too high)	Engine stalls immediately after starting	Cause(s)
►	►	►	►		►				►		►			Vacuum system leaking (see *Vacuum Leaks*)
►	►		►			►	►		►	►	►			Air flow sensor plate and/or control plunger not moving smoothly (see *Air Flow Sensor Movement*)
	►						►							Air flow sensor plate stop incorrectly set (see *Air Flow Sensor Position*)
►		►												Auxiliary air valve does not open (see *Auxiliary Air Valve*)
												►		Auxiliary air valve does not close (see *Auxiliary Air Valve*)
►	►				►								►	Electric fuel pump not operating (see *Fuel Pump*)
►														Defective cold start system (see *Cold Start System*)
		►	►				►	►		►				Leaking cold start valve (see *Cold Start System*)
►		►												Incorrect cold control pressure (see *Warm-up Regulator*)
	►		►	►	►	►			►				►	Warm control pressure too high (see *Warm-up Regulator*)
			►	►		►		►	►	►			►	Warm control pressure too low (see *Warm-up Regulator*)
					►	►			►				►	Incorrect system pressure (see *System Pressure*)
	►													Fuel system pressure leakage (see *Fuel Leaks*)
	►	►	►		►		►							Injection valve(s) leaking, opening pressure too low (see *Testing Injectors*)
		►	►			►			►					Unequal fuel delivery between cylinders (see *Comparative Test*)
	►	►	►	►			►	►	►	►	►			Basic idle and/or CO adjustment incorrect (see *Idle and CO Adjustment*)
						►								Throttle plate does not open completely

water right through the fuel filter. A small plastic filter in the center of the fuel distributor traps the water, which then rusts out the fuel distributor. Not only will the rust block the tiny metering slits in the fuel distributor, it can also flow into the injectors and warmup compensator. Finding rust in one part usually means there's rust in the other. When rust particles block a metering slit, the cylinder fed by that slit will either die completely or develop an intermittent misfire closely resembling the conditions created by a bad spark-plug. If the plug is known to be good, look for rust or water in the fuel. Disconnect the injector lines and spread fresh white paper towels on the workbench. Remove the entire fuel distributor and tap it gently against the bench. Shake and tap it again. If rust particles shake out of those injector line openings, replace the fuel distributor and flush the lines.

Water contamination can be prevented with regular fuel filter changes. Filters for these cars are expensive, but a rusty fuel distributor can't be fixed.

PRELIMINARY CHECKS

NOTE: Before doing any injection system testing, first check the three basics; ignition, compression and the fuel gauge.

Always make a complete ignition check first and then perform all the injection pressure tests in sequence.

The cause of many CIS complaints is a minor vacuum leak. Vacuum leaks fool the fuel metering system and affect the mixture. This gives poor gas economy and poor performance. Look for leaks at:

1. EGR valve
2. Intake manifold
3. Cold start injector
4. Air sensor boot
5. Brake booster vacuum lines
6. Air ducts

Also check vacuum limiter, evaporative canister, and A/C door actuator.

To quick test the system for leaks, disconnect the auxiliary air valve hose, block open the throttle, and apply pressure to the hose. Use a spray bottle of soapy water to hit all the fittings where leaks could occur.

RELIEVING FUEL SYSTEM PRESSURE

CAUTION

Fuel pressure must be relieved before attempting to disconnect any fuel lines.

1. Carefully loosen the fuel line on the control pressure regulator (large connection).
2. Wrap a clean rag around the connection while loosening to catch any fuel.

Alternate Method For Relieving Fuel Pressure

1. Disconnect the electrical plug from the cold start valve.
2. Using a jumper wire, apply 12 volts to the cold start valve terminal for 10 seconds.
3. Reconnect the cold start valve.

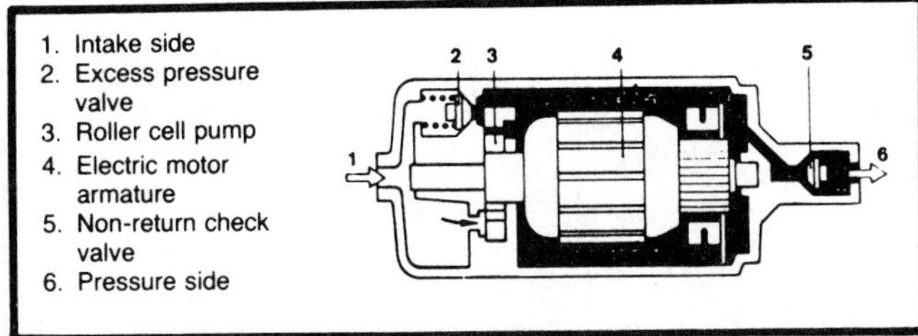

Typical roller cell pump

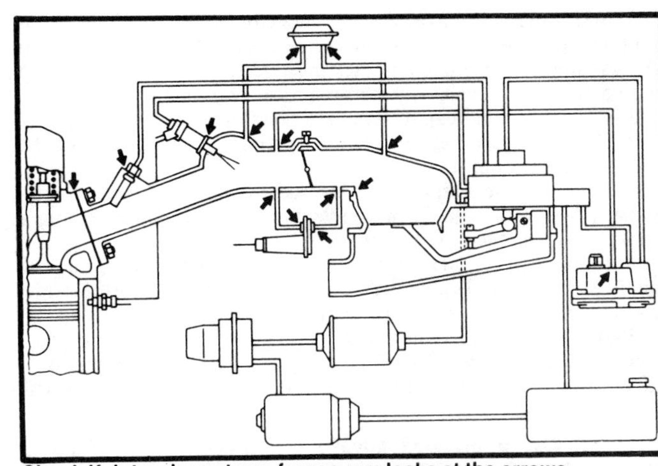

Check K-Jetronic systems for vacuum leaks at the arrows

FUEL PUMP TESTS

The electric fuel pump is a roller cell type with the electric motor permanently surrounded by fuel. An eccentrically mounted roller disc is fitted with rollers resting in notches around the circumference. These rollers are pressured against the thrust ring of the pump housing by centrifugal force. The fuel is carried in the cavities between the rollers and the pump delivers more than the maximum requirement to the engine so that fuel pressure is maintained under all operating conditions. During starting, the pump runs as long as the ignition key is turned. The pump continues to run once the engine starts, but has a safety device that will stop the pump if the ignition is turned on but the engine stops moving.

NOTE: Some models utilize a priming fuel pump, mounted in the tank, in addition to the main fuel pump.

Fuel Pump Power Check

Remove the round cover plate from the top of the fuel pump and measure the voltage between the positive and negative terminals when the pump is operating. The lowest permissible voltage is 11.5 V. Disconnect the terminals connecting the pump to the air flow sensor and the warm-up regulator if these are to be checked later. If the pump is dead, check the fuse, the ground, and try bridging the relay with a jumper wire. During winter months, water in the gas can cause the pump to freeze and then blow a fuse. K-Jetronic type pumps need 12 volts and a good ground. A relay located somewhere in the pump's power wire operates the pumps. And like L-Jetronic systems, the relay may be a hide-and-seek item. All K-Jetronic pumps are fused someplace in the car, usually right in the vehicle's fuse panel.

Fuel Pump Safety Circuit Check

The pump will only run if the starter motor is actuated or if the engine is running.

1. Remove the air filter.
2. Turn on the ignition and briefly depress the sensor plate.
3. Remove the coil wire from the distributor.
4. Connect a voltmeter to the positive fuel pump terminal and ground.
5. Actuate the starter. The voltmeter should indicate 11 volts.
6. If the fuel pump runs only when the sensor plate is depressed or only when the engine is cranked, replace the fuel pump relay. If the pump is already running when the ignition is turned ON, replace the safety switch.

Fuel Volume Test

The test point for fuel pump operation depends on the type of fuel distributor. Fuel distributors WITH push valves (most vehicles covered in this manual) have two fuel lines connecting with the warm-up regulator, the other fuel line connects to a "T" and returns to the gas tank. Determine whether the vehicle to be tested is fitted with or without a push valve before proceeding further.

1. Remove the gas cap to vent tank pressure.
2. Disconnect the return fuel line leading to the gas tank at the appropriate test point.
3. Hold the line coming from the fuel distributor in a large container (1500 cc or larger). Inflexible metal fuel lines may require a rubber hose to reach the container.
4. Remove the electric plug from the warm-up regulator and the auxiliary air valve. Bridge the electric safety circuit for 30 seconds. Delivery rate should be about one quart in 30 seconds.
5. If fuel quantity is not within specification, check for sufficient voltage supply to the fuel pump (minimum 11.5 volts) or a dirty fuel filter. If all the above are satisfactory, replace the electric fuel pump.

NOTE: Use a bridging adapter to energize the fuel pump through the relay connector on early models without push valves.

PRESSURE TESTS

Diagnose a K-Jetronic system in this order: check cold control pressure first, hot control pressure next, then primary pressure, and finally rest pressure. The gauge for testing K-Jetronic always tees into the line running between the fuel distributor and the warmup compensator. With the 3-way valve open, the gauge remains teed-in and therefore reads control pressure. Remember that control pressure is derived or fed from the primary circuit pressure so a pressure change in one circuit means a change in the other.

Closing the valve shuts off fuel flow to the warm-up compensator and forces the gauge to read only one circuit—the primary circuit.

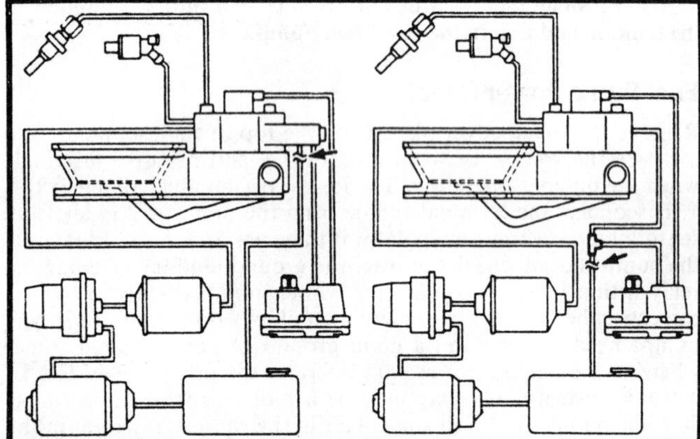

Fuel pressure test points with push valve (top) and without push valve (bottom). Most 1978 and later K-Jetronic systems have push valves.

When you shut the engine off and leave the 3-way valve open, the teed-in gauge should show the system's holding pressure. This residual pressure is called rest pressure. The valve functions are: *valve open—control pressure; valve closed—primary pressure; valve open and engine off—rest pressure.*

Control Pressure Test-Engine Cold

1. Connect a pressure gauge between the fuel distributor and the control pressure regulator.
2. Remove the electrical connector from the regulator.
3. Idle the engine for no more than one minute.
4. Read the control pressure from the chart for cold control. If control pressure is wrong, check the control pressure (warmup) regulator for proper operation.

CIS PRESSURE TEST SPECIFICATIONS

(All measurements in psi)

Manufacturer	Normal Fuel Pressure	Injector Opening Pressure	Rest Pressure	System Pressure
Audi (all models)	49-55	36-59	23-35	65-78 ①
BMW 320i	49-55	36-52	23-35	65-75
Mercedes-Benz (all models)	49-55	36-52	41	65-75
Peugeot 505	49-55	44-59	38	65-75
Porsche (all models)	38-55	44-59	14-25	65-75
Saab (all models)	49-55	44-59	14-22	65-75
VW (all models)	49-55	51-59	35	68-78
Volvo (all models)	49-55	37-51	25	65-75 ①

NOTE: Minimum-maximum test ranges are given. Exact pressures may vary.
①Turbo models: 75-84 psi

Control Pressure Test—Engine Warm

1. Carry out this test when poor performance has been experienced when the engine is warm. Connect gauge and 3-way valve.
2. Open the valve.
3. Connect the warm-up regulator plug.
4. Leave the ignition switched on until rest pressure is present.
5. Refer to the test values.
6. If the value should differ, replace the warm-up regulator.

NOTE: Perform primary and rest pressure tests before disconnecting the gauge and 3-way valve.

Primary Pressure Test

Close the 3-way valve. Switch on the ignition. Refer to the test values. If the line pressure should differ from the recommended values this may be due to:

- Insufficient pressure from fuel pump Blockage of the strainer in the tank
- Leakage in the fuel line

When you close the 3-way valve with the engine idling and the primary pressure reads low, there could be a fuel pump problem or a primary regulator problem. To isolate on from the other, locate

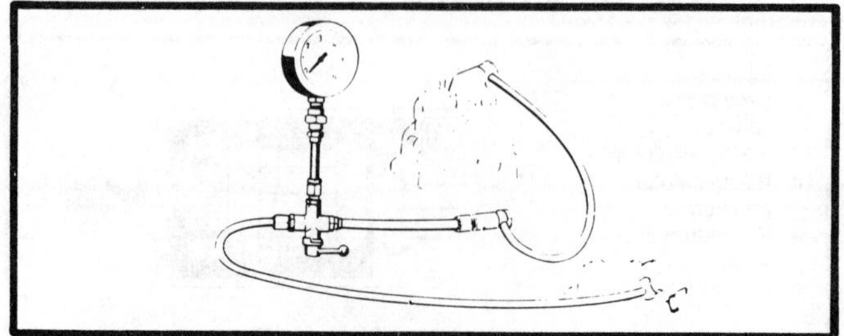

Typical fuel pressure test connection

the return line leading from the fuel distributor back to the gas tank. Now plug the return line securely at some convenient point. Have an assistant switch on the pump *just long enough* to get a pressure reading. If the pump is good, the pressure will jump almost instantly to 100 psi or higher. Should the primary pressure exceed 116 psi during this momentary test, the check valve in the intake side of the fuel pump will vent the pressure back into the gas tank.

If the fuel pump produces good pressure with the return line plugged, then the primary pressure regulator is causing a low primary pressure reading.

Rest Pressure Test

Good rest pressure enables a K-Jetronic-equipped engine to start easily when it's hot. If the K system loses rest pressure for any reason, the injector lines will vapor lock. Then the driver has to crank the engine until the pump can replenish the lines with liquid fuel.

If rest pressure drops off too quickly, start the engine and run it long enough to build the system pressure back up. Stop the engine, pull the electrical connector off the cold start injector and remove the injector from the intake manifold. If the cold start injector is dripping, replace it and re-check the rest pressure.

If the cold start valve is holding pressure, reinstall it and run the engine again to build up pressure. Stop the engine and close the 3-way valve. If the pressure reading remains steady now, the return side of the system—the warm-up compensator—is leaking. Replace the compensator.

If the rest pressure still drops rapidly after closing the 3-way valve, the leak is in the feed side of the system. This could mean a bad fuel pump, bad accumulator, or a bad primary pressure regulator.

The fuel pump's intake hose is flexible. Clamp this hose shut and recheck rest pressure. If the system holds pressure, then the check ball in the output side of the pump is leaking.

A sharp eye and ear will help isolate a bad accumulator. Watch the pressure gauge as a helper shuts off the engine. If the accumulator is good, the pressure will drop, stop for a second, and then the gauge needle will actually rise—pressure will increase for a moment—before the pressure stabilizes. If you don't see that slight upward needle movement, the accumulator is leaking rest pressure. Furthermore, a bad accumulator will cause a loud, tell-tale groaning noise at the fuel pump. If still in doubt about the accumulator, try blocking off its return line.

The same primary pressure regulator that can cause low primary pressure can also leak *rest* pressure. If you've eliminated the fuel pump and the accumulator, install a primary regulator repair kit.

Component Removal and Testing

LINE PRESSURE REGULATOR

The line pressure regulator ensures that the pressure in the circuit remains constant when the fuel pump is in operation and also controls the recirculation of fuel to the tank. When the fuel pump is switched off, the regulator will cause a rapid pressure drop to approximately 2.5 bar (kp/cm^2, 35 psi), i.e. the rest pressure, which is maintained by means of the O-ring seal and the quantity of fuel contained in the fuel accumulator. The purpose of the rest pressure is to prevent the fuel from vaporizing in the circuit when the engine is warm, which would otherwise make restarting difficult.

The line pressure regulator forms an integral unit with a shut-off valve to which the return fuel line from the control pressure regulator is connected. When the fuel pump is operating, the shut-off valve is actuated mechanically by the control pressure regulator, whereupon the return fuel from the control pressure regulator bypasses the shut-off valve to the return line.

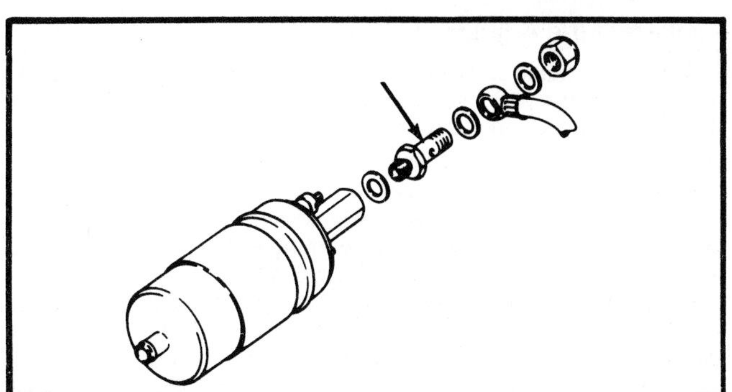

Fuel pressure regulator on fuel pump (arrow)

When the fuel pump stops running and the line pressure regulator valve is pressed into its seating, the shut-off valve is also pressed into its seating, preventing the fuel system from emptying through the control pressure return.

Removal and Installation

The pressure regulator is removed by loosening the screw plug on the fuel distributor and removing the copper gasket, shim, spring, control piston and O-ring. On 1979 and later systems, a push valve is attached to the large screw plug. Make sure all parts are kept clean and in order for assembly. Replace the O-ring and copper gasket, especially if any fuel leakage is noted.

CAUTION

Relieve fuel system pressure before removing the screw plug.

NOTE: The piston in front of the shim is matched to the housing and is not replaceable.

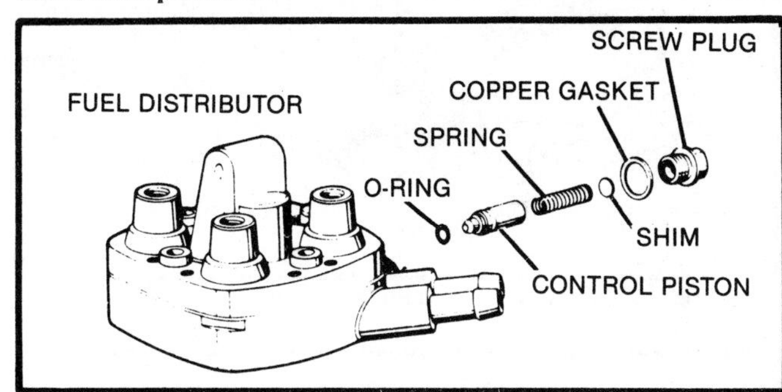

Fuel pressure regulator without push valve

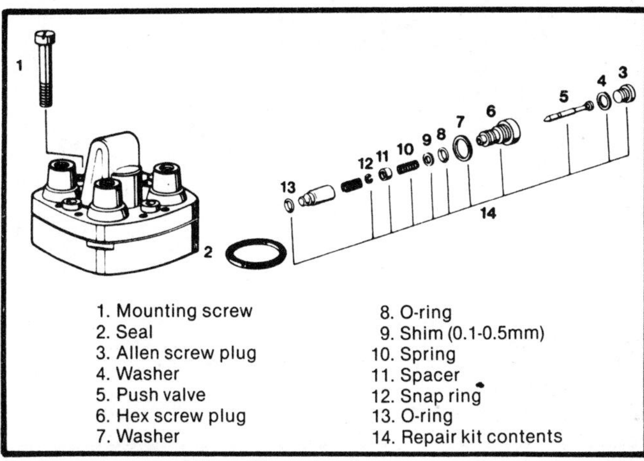

Fuel pressure regulator with push valve

Fuel Pressure Adjustment

NOTE: On some early models, the fuel pressure can be adjusted on the regulator with the adjusting screw. Adjust to 28 psi. If a slight turn of the screw shows no change of pressure, replace the regulator.

If the fuel distributor is fitted with a push valve, loosen the large screw plug with attached push valve assembly. Change the adjusting shim as required to raise or lower fuel system pressure. Each 0.1mm increase in shim thickness increases fuel pressure by 2.2 psi.

If the fuel distributor doesn't have a push valve, remove the screw plug and change the shim as required. Again, each 0.1mm increase is shim thickness will increase system pressure by 2.2 psi.

WARM-UP REGULATOR

When the engine is cold, the warm-up regulator reduces control pressure, causing the metering slits in the fuel distributor to open further. This enrichment process prevents combustion miss during the warm-up phase of engine operation and is continually reduced as temperature rises. The warm-up regulator is a spring controlled flat seat diaphragm-type valve with an electrically heated bi-metal spring. When cold, the bi-metal spring overcomes the valve spring pressure, moving the diaphragm and allowing more fuel to be diverted out of the control pressure circuit thereby lowering the control pressure. When the bi-metal spring is heated electrically or by engine temperature the valve spring pushes the diaphragm up, allowing less fuel to be diverted thereby raising the control pressure. When the bi-metal spring lifts fully off the valve spring the warm-up enrichment is completed and control pressure is maintained at normal level by the valve spring. The warm-up regulator should be checked when testing fuel pressure.

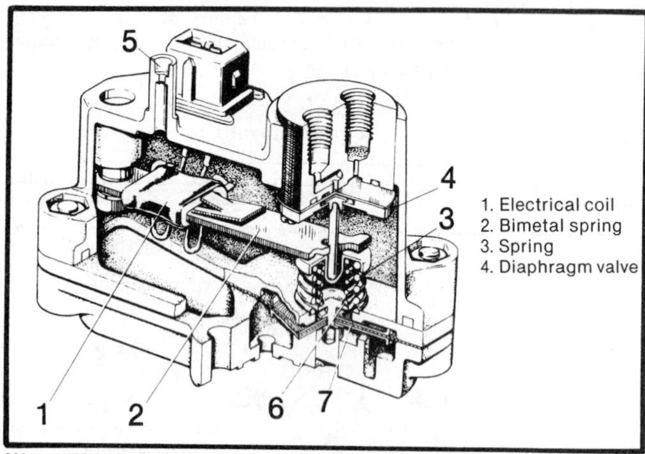

Warm-up regulator typical of California models

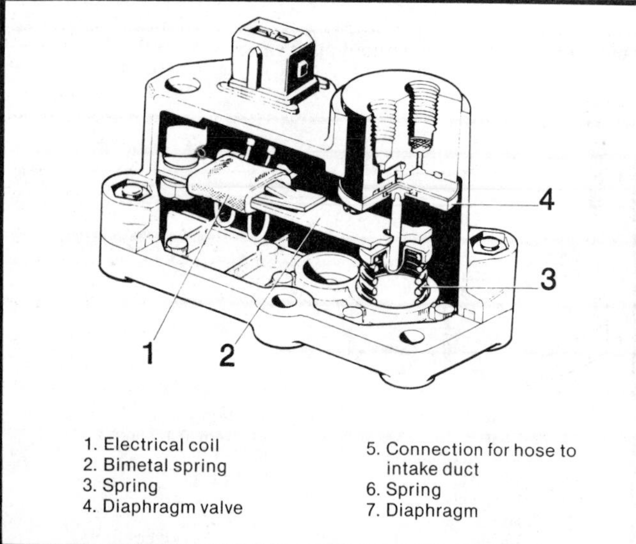

Warm-up regulator typical of Federal (49 States) models

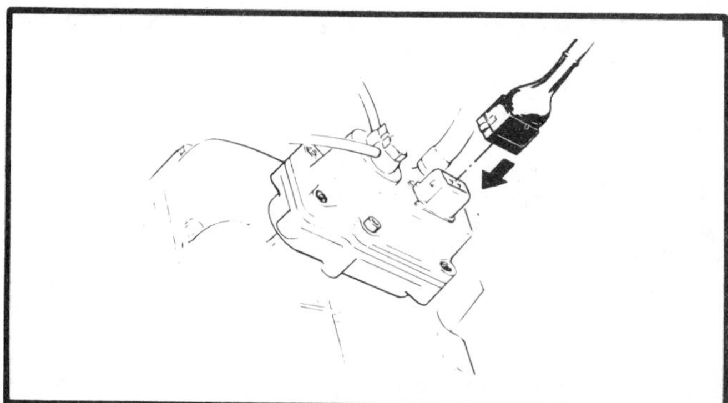

Control pressure regulator showing connector

Testing

1. Disconnect the terminal from the warm-up regulator and connect a voltmeter across the contacts in the connector. Ensure that the ignition is switched on and that the safety circuit connection at the air flow sensor has been bypassed. The lowest permissible voltage is 11.5V.
2. Check that there are no breaks in the heater coil of the regulator by connecting a test lamp in series with the coil. If the coil is found to be damaged, the warm-up regulator should be replaced.
3. Connect an ohmmeter across the terminals in the regulator socket. Resistance must be 16–22 ohms. If the resistance is not within the specifications, the regulator will require replacement.

NOTE: The system is under considerable constant pressure. The only practical test that should be attempted is one using an ohmmeter. Be sure the engine is at normal operating temperature. There should be no loose fuel fittings or other fire hazards when the electrical connections are disengaged.

Removal and Installation

1. Disconnect the negative battery cable.
2. Unplug the electrical connector.
3. Depressurize the fuel system as previously described.

NOTE: Wrap a cloth around the connection to catch any escaping fuel.

4. Remove both of the fuel lines.
5. Remove the vacuum hose.
6. Unscrew the four mounting bolts and remove the control pressure regulator.
7. Installation is in the reverse order of removal.

FUEL ACCUMULATOR

This device maintains the fuel system pressure at a constant level under all operating conditions. The pressure regulator, incorporated into the fuel distributor housing, keeps delivery pressure at approximately 5.0 bar (73 psi). Because the fuel pump delivers more fuel than the engine can use, a plunger shifts in the regulator to open a port which returns excess fuel to the tank. When the engine is switched off and the primary pressure drops, the pressure regulator closes the return port and prevents further pressure reduction in the system.

The fuel accumulator has a check valve which keeps residual fuel pressure from dropping below a pre-determined pressure when the engine or fuel pump are shut off.

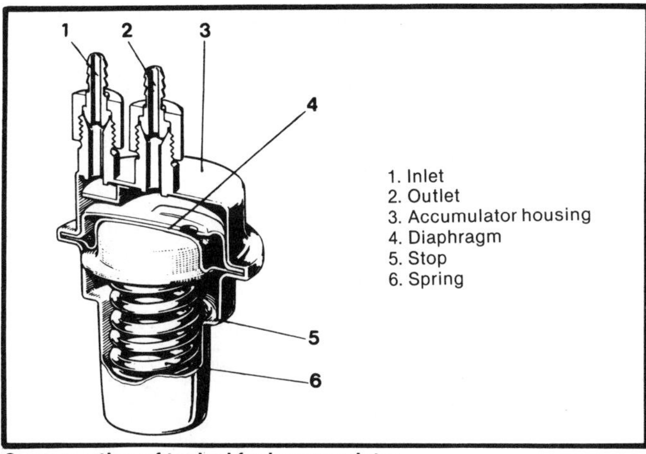

Cross section of typical fuel accumulator

Removal and Installation

The fuel accumulator is usually mounted near or on the fuel pump bracket somewhere near the fuel tank. It may be necessary to remove the fuel pump assembly to gain access to the fuel accumulator. Loosen the hose clamps and the retaining clamp and remove the accumulator. Reverse the procedure to install.

CAUTION

Relieve fuel system pressure before disconnecting any fuel lines.

Testing

If the pump fails a volume test, raise the car and trace the fuel feed line back to the pump. Look for crimped fuel lines from a carelessly placed jack. Plug the accumulator return hose and repeat the volume test. If this yields good fuel volume, the accumulator diaphragm is ruptured. Soak a new fuel filter with clean gas and install it. Remember to wet the filter beforehand. When 80 psi of incoming fuel hits a dry filter, it can tear off some of the element and carry the paper debris into the rest of the system.

FUEL DISTRIBUTOR

K-Jetronic

The fuel distributor meters the correct amount of fuel to the individual cylinders according to the position of the air flow sensor plate. A control plunger opens or closes metering slits in the barrel, allowing more or less fuel to pass into the system. Control pressure assures that the plunger follows the movement of the sensor plate immediately. Control pressure is tapped from primary fuel pressure through a restriction bore. It acts through a damping restriction on the control plunger, eliminating any oscillations of the sensor plate due to a pulsating air flow. 1978 and later models incorporate a "push valve" in the fuel distributor to maintain pressure when the engine is switched off. The push valve is a one-way device mounted in the primary pressure regulator which is held open in normal operation by the pressure regulator plunger. Differential pressure valves in the fuel distributor hold the drop in pressure at the metering slits at a constant value.

KE-Jetronic

The internal components and operation of the KE type fuel distributor is slightly different than the K model. With the engine running, a constant system pressure of approximately 79 psi (5.4 bar) is present at the fuel inlet. The plate valve is adjusted depending on current intensity and in this manner determines the flow rate, in combination with a fixed orifice (0.3mm in diameter) at the fuel distributor outlet. The pressure change in the lower chamber causes movement of the diaphragm and regulates the fuel volume flowing to the injectors.

During cold start and warm-up, current at the electrohydraulic actuator (EHA) is approximately 8–120 milliamps. The plate valve is positioned in the direction of the intake port and the differential pressure drop in the lower chamber is approximately 6–22 psi (0.4–1.5 bar). With increasing coolant temperature, the current at the actuator drops to approximately 8 milliamps and the differential pressure drops at the same rate, down to approximately 6 psi (0.4 bar). For acceleration enrichment, the current to the actuator is determined by the coolant temperature and the amount of sensor plate deflection. The plate valve is moved closer to the intake port and the differential pressure decreases by approximately 22 psi (1.5 bar).

NOTE: Acceleration enrichment is cancelled at approximately 176°F (80°C).

The airflow sensor position indicator operates with approximately 8 volts supplied constantly. During acceleration, a voltage signal is transmitted to the control unit, depending on the position of the airflow sensor plate. The control unit provides acceleration enrichment as an impulse which increases the instantaneous current value. During acceleration enrichment, Lambda (oxygen sensor) control is influenced by the control unit.

NOTE: With the accelerator pedal at idle, the microswitch on the side of the airflow sensor is closed and no enrichment is possible.

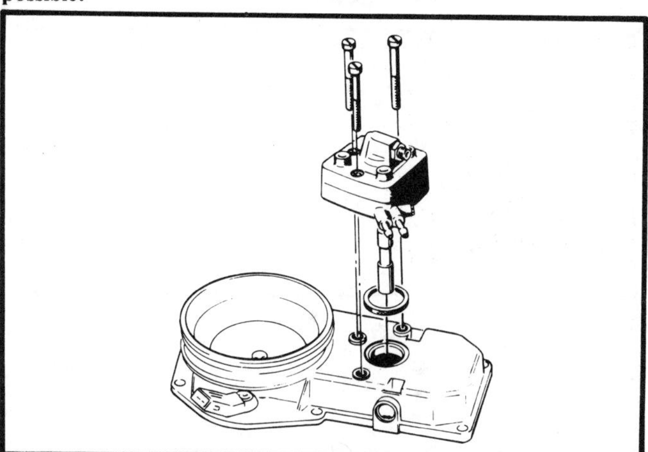

Fuel distributor mounting screws

The throttle valve switch receives a constant 8 volt signal from the control unit. With the throttle valve fully open (switch closed), approximately 8 mA of current flows to the electrohydraulic actuator, independent of engine speed. At full load enrichment, the plate valve moves in the direction of the intake port and the differential pressure in the lower chamber of the fuel distributor is approximately 6 psi (0.4 bar) below system pressure. Under deceleration, the circuit to the control unit is closed by the microswitch. The speed at which deceleration shutoff occurs depends on coolant temperature. The lower the temperature, the higher the speed at which restart of fuel injection begins. With the microswitch closed, the current at the actuator is approximately 45 mA and the plate valve moves away from the intake port. The pressure difference between the upper and lower chamber is cancelled and system pressure is present in the lower chamber. Operational signals from the control unit will change the direction of the current flow at the actuator plate valve; the plate valve then opens. When the lower chamber pressure changes, pressure and spring force push the diaphragm against the ports to the injectors and cut off the fuel supply.

Removal and Installation

1. Release the pressure in the system by loosening the fuel line on the control pressure regulator (large connector). Use a clean rag to catch the fuel that escapes.

2. Mark the fuel lines in the top of the distributor in order to put them back in their correct positions.

NOTE: Using different colored paints is usually a good marking device. When marking each line, be sure to mark the spot where it connects to the distributor.

3. Clean the fuel lines, then remove them from the distributor. Remove the little looped wire plug (the CO adjusting screw plug). Remove the two retaining screws in the top of the distributor.

CAUTION

When removing the fuel distributor be sure the control plunger does not fall out from underneath. Keep all parts clean.

4. If the control plunger has been removed, moisten it with gasoline before installing. The small shoulder on the plunger is inserted first.

NOTE: Always use new gaskets and O-ring when removing and installing fuel distributor. Lock all retaining screws with Loctite® or its equivalent.

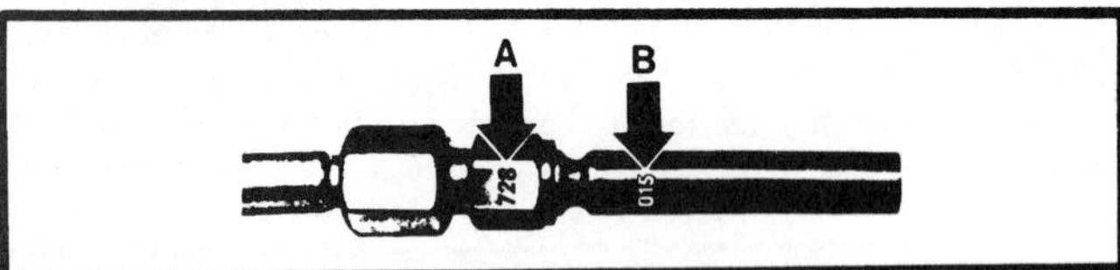

Identification numbers of fuel injector hexagon (A) and shaft (B)

FUEL INJECTORS

The fuel injection valves are mounted in the intake manifold at the cylinder head, and continuously inject atomized fuel upstream of the intake valves.

A spring loaded valve is contained in each injector, calibrated to open at a fuel start pressure of 47–54 psi. The valves also contain a small fuel filter.

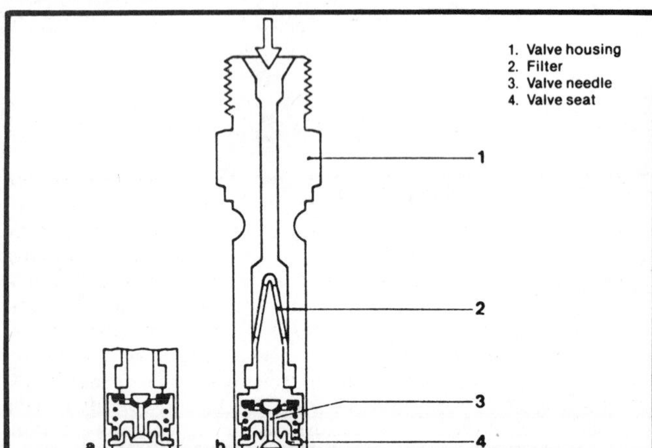

Cross section of typical K-Jetronic fuel injector showing closed (a) and open (b) positions

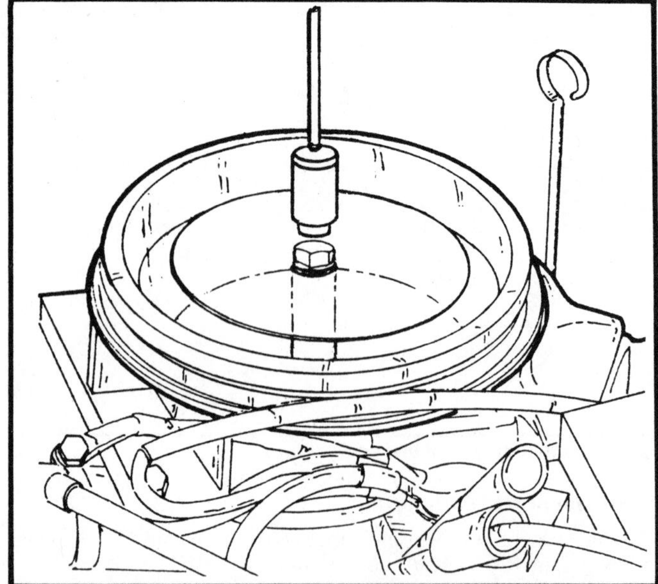
Lifting air flow sensor plate with magnet (updraft type)

The fuel injector (one per cylinder) delivers the fuel allocated by the fuel distributor into the intake tubes directly in front of the intake valves of the cylinders. The injectors are secured in a special holder in order to insulate them from engine heat. The insulation is necessary to prevent vapor bubbles from forming in the fuel injection lines which would lead to poor starting when the engine is hot. The fuel injectors have no metering function: they open when the pressure exceeds 47–54 psi and are fitted with a valve needle that "chatters" at high frequency to atomize the fuel. When the engine is switched off, the injection valve closes tightly, forming a seal that prevents fuel from dripping into the intake tubes.

Testing

NOTE: Leave the fuel lines attached to the injectors for testing, but be careful not to crimp any metal fuel lines.

Remove the rubber intake hose leading from the mixture control unit to the throttle unit. Expose the air flow sensor plate and bypass the safety circuit with a bridging adapter. Use a small magnet to lift the sensor plate during the test.

To check spray pattern, remove the injectors, one at a time, and hold them over a beaker. Switch the ignition key on and disconnect the electrical connector at the air-flow sensor to activate the fuel pump. Move the air flow sensor plate. The injector should provide a dose of uniformly atomized fuel at about a 15–52 degree wide angle.

To check injection quantity, connect the removed injectors via hoses to equal sized beakers. Switch on the ignition. Disconnect the electrical connector at the airflow sensor to activate the fuel pump. Run the pump for approximately 30 seconds to pressurize the system, then connect the air flow sensor to stop the fuel pump. Lift or

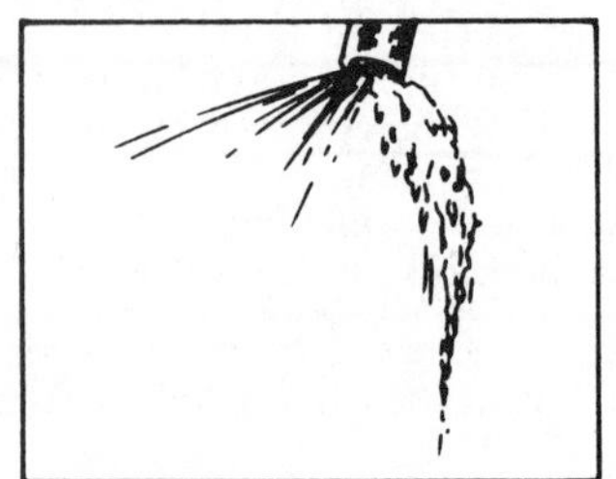
Uneven spray

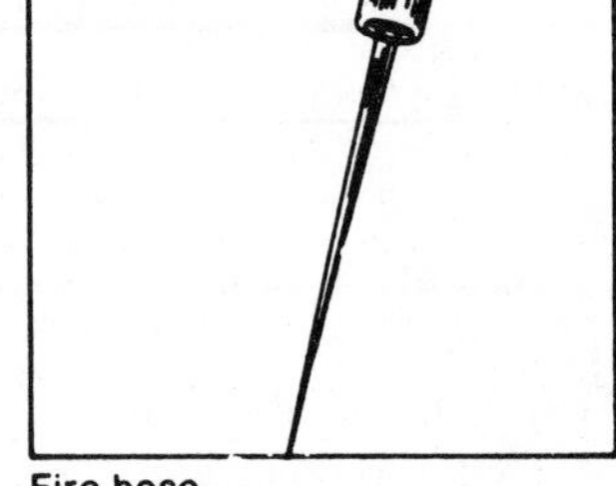
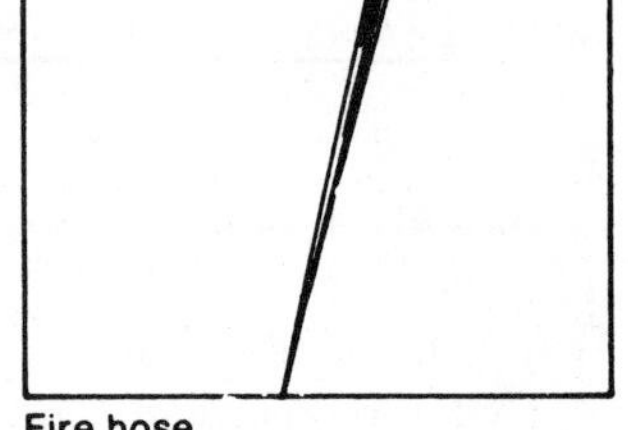
Fire hose

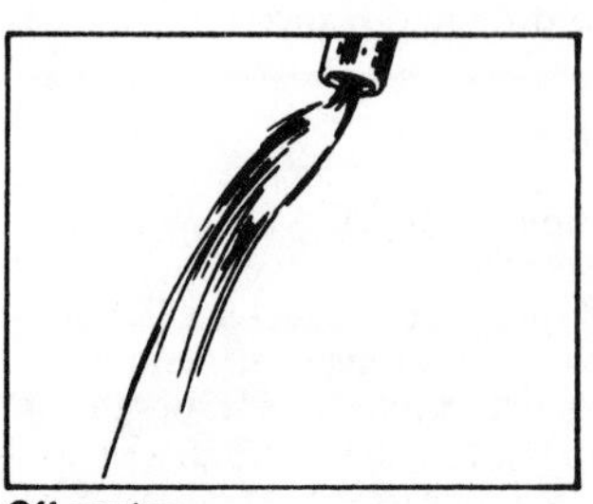
Off center

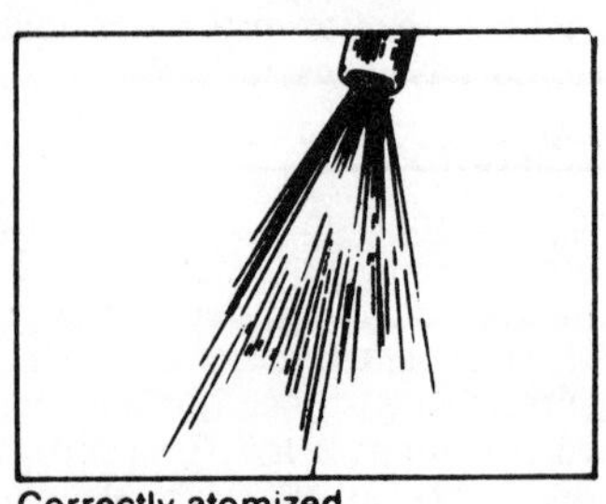
Correctly atomized

Fuel injector spray patterns

depress the airflow sensor plate halfway until one of the beakers fills up. Check the beakers. If injection quantity deviates more than 20% between injectors, isolate the problem by swapping the lowest and highest (in fuel quantity) injectors and repeating the test. If the same injector still injects less, clean or replace that injector and fuel supply line. If the other injector is now faulty, the fuel distributor is defective.

The check for injector leak-down (when closed) can now be conducted. Injector leakage (more than slight seepage) may be due to airflow sensor plate set to incorrect height, seizing of fuel distributor plunger, or internal leaks in the fuel distributor. Connect the airflow sensor connector to de-activate the fuel pump and switch off the ignition. Check for injector leakage at rest pressure. Depress the sensor plate to open the fuel distributor slots. Maximum permissible leakage is one drop per 15 seconds. If all injectors leak, the problem may be excessive rest pressure.

NOTE: The injectors can be replaced individually.

COLD START VALVE

The cold start valve is mounted near the throttle valve housing and is connected to the pressure line. The valve, which is operated by a solenoid coil, is actuated by a thermo-time switch which is controlled by the engine temperature.

During cold starting, part of the fuel mixture is lost due to condensation on the cold cylinder walls. To compensate for this loss, extra fuel is injected by means of a solenoid-operated cold start valve. Fuel is delivered downstream of the throttle valve. A thermo-time switch determines the injection period of the cold start valve according to either engine temperature or an electrically heated bimetal strip which de-energizes the cold start valve after approximately 8 seconds to prevent flooding. The cold start valve does not function when the engine is warm.

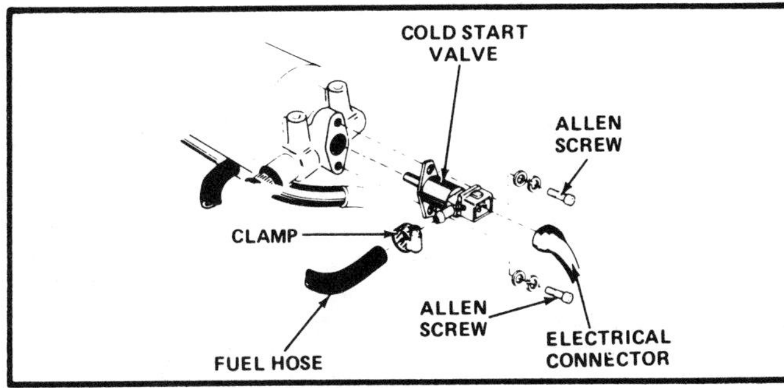

Typical cold start valve installation

Removal and Installation

The cold start valve is mounted in the intake manifold, downstream from the air flow sensor. The valve is usually retained with two small Allen screws and can be removed for inspection without disconnecting the fuel line. When removing the valve, take care not to damage the valve body which is usually made of plastic. Make sure the valve is clean and the sealing O-ring is seated properly and in good condition when installing. Do not overtighten the mounting screws.

CAUTION

Depressurize the fuel system before attempting to remove any fuel hoses. Use plenty of rags around fuel connections to absorb and deflect any fuel spray.

Testing

Remove the cold start injector from the intake manifold and hold over a beaker. With a cold engine (95°F or lower coolant temperature), the injector should spray during starter operation (max. 12 seconds). If not, check the voltage between the terminals of the

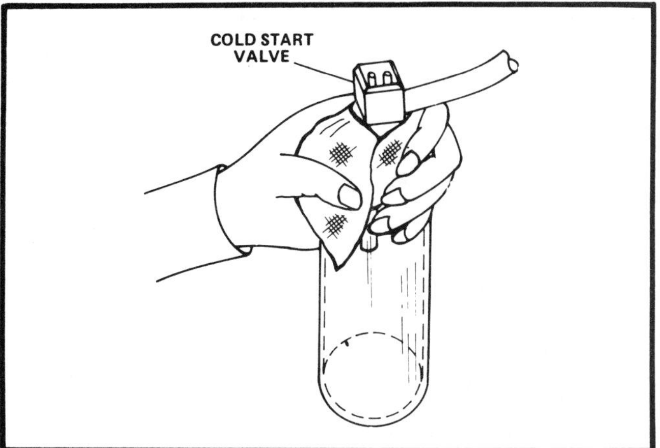

Testing the cold start valve

injector when the starter is on. Voltage indicates a bad cold start injector. No voltage indicates a faulty thermo-time switch or wiring. Ground one terminal and connect the second to the positive side of the coil. When you run the pump, 10–30 seconds, you should get a good, cone-shaped spray. Dry the injector, disconnect jumpers, and energize the pump again. There shouldn't be any fuel. If it drips, replace it. Check the thermo-time switch when it's below 95°F. Disconnect the cold start valve and hook up a test light across its connector. Ground the No. 1 coil terminal and run the starter. The light should glow for several seconds and then go out. If not, replace the thermo-time switch.

NOTE: Remove the fuel pump relay and attach a bridging adaptor to energize the fuel pump during testing. See the "Tools" section for specifics on adapter construction.

With the starter off, attach a test relay to operate fuel pump. Check for cold start valve leakage. Maximum allowable leakage is one drop per minute. Any excessive leakage is reason enough to replace the cold start valve. Don't forget to wipe the valve nozzle with a clean towel after every test and before installing.

THERMO-TIME SWITCH

The thermo time switch, located on the cylinder head, actuates the cold start valve. It has a bi-metal spring which senses coolant temperature and an electric coil which limits the cold start valve spray to 12 seconds. A thermo-time switch also measures water temperature and opens the cold start valve, located on the intake header, a varying amount of each time the engine is started, depending on the conditions. With a hot engine (coolant temperature over 95°F), the injector should not operate. If it does, the thermo-time switch is defective. Also, on a cold engine, the cold start valve should not inject fuel for more than 12 seconds (during starter cranking). If it does, the thermo-time switch is defective.

Testing

NOTE: To perform the following test properly, the engine must be cold with a coolant temperature below 95°F (35°C).

With the use of a test lamp, the switch can be tested at various temperatures for continuity. The operating time is eight seconds at −4°F (−20C) and declines to 0 seconds at +59°F (+15C.).

When the engine temperature is below approximately 113°F (+45°FC), current is allowed to flow for a certain period (depending on the temperature) while the starter motor is running.

Check that the switch closes when the engine is started by means of connecting a test lamp in series across the contacts of the cold start valve plug.

It is not possible to make a more accurate check of the cut-in time or temperature. If the condition of the switch is at all in doubt, it should be replaced.

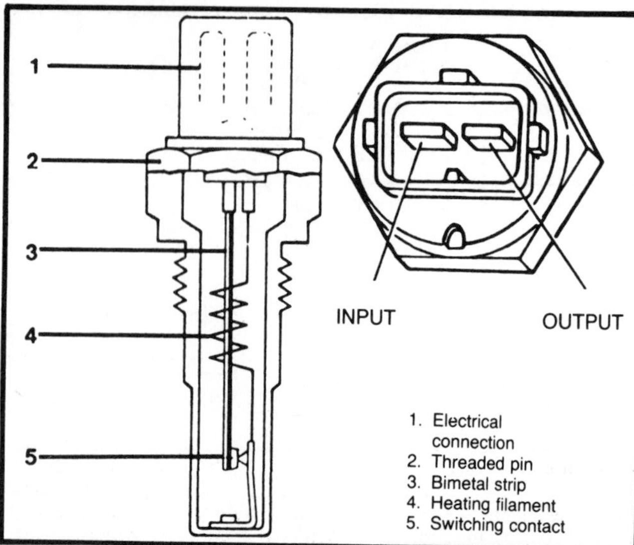

Typical thermo-time switch showing construction and terminal location

1. With the engine cold, remove the harness plug from the cold start valve and connect a test light across the harness plug connections.
2. Connect a jumper wire from coil terminal No. 1 to ground.
3. Operate the starter. If the test light does not light for about 8 seconds, replace the switch. Removal of the switch requires that the engine be cold and the cooling system drained.

Removal and Installation

1. Disconnect the negative battery cable. Drain the cooling system.
2. Locate the thermo-time switch on the left side of the engine block and disconnect the electrical connection.
3. Unscrew the switch and remove it.
4. Installation is in the reverse order of the removal. Refill the cooling system.

CAUTION

The engine must be cold when removing the switch.

AIR FLOW SENSOR

This device measures the amount of air drawn in by the engine. It operates according to the suspended body principle, using a counterbalanced sensor plate that is connected to the fuel distributor

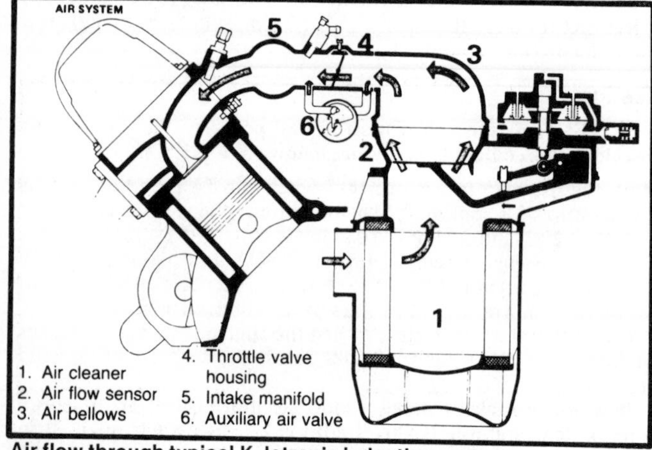

Air flow through typical K-Jetronic induction system

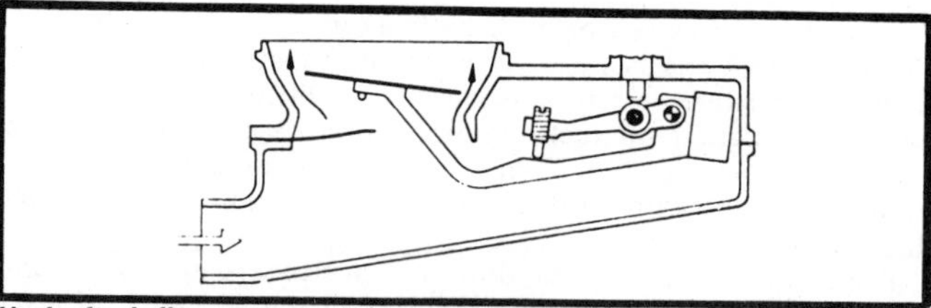
Updraft air flow sensor plate movement

control plunger by a lever system. A small leaf spring assures that the sensor plate assumes the correct zero position when the engine is stationary.

The air flow sensor consists of an air venturi tube in which an air flow sensor plate moves. The air flowing into the venturi from the air cleaner lifts the air flow sensor plate, allowing the air to flow through. The greater the amount of air, the higher the sensor plate will be raised.

The air flow sensor plate is fitted to a lever which is compensated by a counterweight. The lever acts on the control plunger in the fuel distributor which is pressed down by the control pressure, thus counteracting the lifting force of the air flow sensor plate.

The height to which the air flow sensor plate is raised is governed by the magnitude of the air flow.

The air/fuel mixture varies with the engine load. The inclination of the venturi walls therefore varies in stages in order to provide a correct air/fuel mixture at all loads. Thus, the mixture is enriched at full load and leaned at idle.

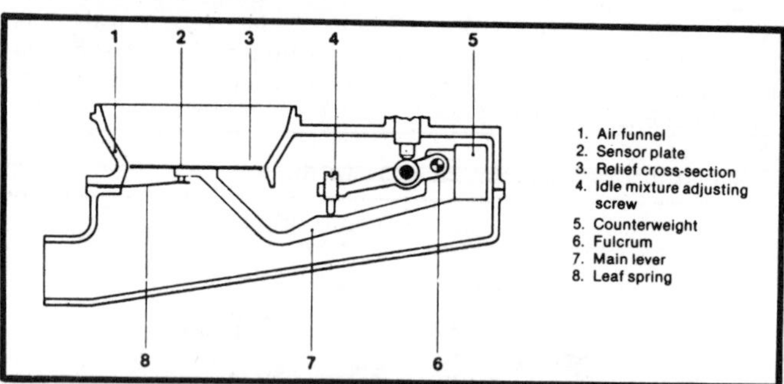

Updraft air flow sensor in zero position

The lever acts on the control plunger in the fuel distributor by means of an adjustable link with a needle bearing at the contact point. The basic fuel setting, and thus the CO setting, is adjusted by means of the adjustment screw on the link. This adjustment is made with a special tool and access to the screw can be gained through a hole in the air flow sensor between the air venturi and the fuel distributor. The CO adjustment is sealed on later models.

A rubber bellows connects the air flow sensor to the throttle valve housing.

NOTE: On KE systems, the sensor plate rest position is angled upward, not horizontal as is the K system.

Testing

The air flow sensor plate in the fuel distributor must operate smoothly in order to do a good job of measuring air. Remove the air boot and check the sensor plate movement. When released, the plate should fall freely with one or two bounces. If the plate sticks, loosen the mounting screws and retighten them uniformly. Clean the funnel and sensor plate too, as these can get dirty from PCV fumes.

Check for leakage in the inlet system between the air flow sensor and the engine Air leaking into the system may result in poor engine performance, owing to the fact that it bypasses the air flow sensor, causing a lean mixture.

Leakage can occur in the following places:

a. At the rubber bellows between the air flow sensor and the throttle valve housing.

b. At the gasket on the flange of the cold start valve.

c. At the gasket between the throttle valve housing and the inlet manifold.

d. At the gasket between the inlet manifold and the cylinder head.

e. At the hose connections on the throttle valve housing, auxiliary air valve or inlet manifold.

f. Via the crankcase ventilation hose from the oil filler cap, dip stick or valve cover gasket.

NOTE: Move the sensor plate gently with a small magnet.

The plate should not bind, and although the plate will offer some resistance when depressed (due to the control pressure), it should return to its rest position when released. Be careful not to scratch the plate or venturi.

To check the air flow sensor contact switch, depress or lift the sensor plate by hand. The fuel injectors should buzz, and the fuel pump should activate. If the pump operates, but the injectors do not buzz, check the fuel pressure. If the pump does not operate, check for a short in the air flow sensor connector.

Sensor Plate Position Adjustment

NOTE: The air flow sensor plate adjustment is critical. The distance between the sensor plate and the plate stop must be 0–0.2 in. The plate must also be centered in the venturi, and must not contact the venturi walls.

1. Remove the air cleaner assembly.
2. Using a 0.004 in. (0.10mm) feeler gauge, check the clearance around the sensor plate at four opposite points around the plate.
3. If necessary, loosen the bolt in the center of the sensor plate and center the plate. Torque the bolt to 3.6 ft. lbs. (4.9 Nm).

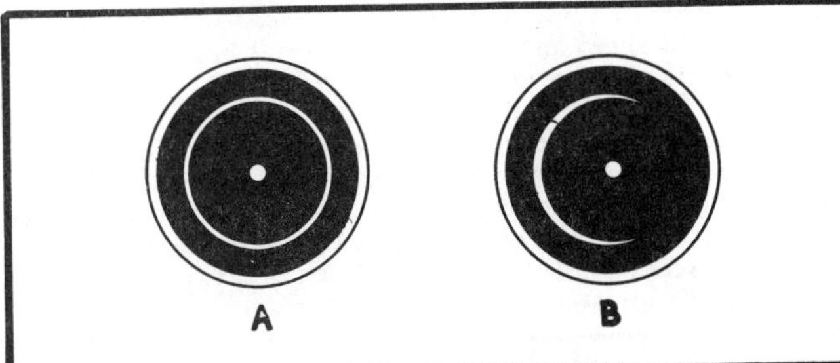

Sensor plate showing correct (A) and incorrect (B) centering in the venturi

CAUTION

Do not scratch the venturi or the plate.

Sensor Plate Height Adjustment

NOTE: The sensor plate height adjustment must be checked under fuel pressure.

1. Install a pressure gauge in the line between the fuel distributor and the control pressure regulator as previously described.
2. Remove the rubber elbow from the air flow sensor assembly.
3. Remove the fuel pump replay from the fuse panel and install a bridge adapter on pre-1979 models. Check that the fuel pressure is within specifications as previously described.
4. The sensor plate should be flush or 0.02 in. (0.5mm) below the beginning of the venturi taper. If necessary to adjust, remove the mixture control from the intermediate housing and bend the spring accordingly.

NOTE: With the sensor plate too high, the engine will run on and with the sensor plate too low, poor cold and warm engine start-up will result. If the sensor plate movement is erratic, the control piston can be sticking.

5. Recheck the pressure reading after any adjustments.
6. Remove the pressure gauge, reconnect the fuel lines and install the fuel pump relay, if removed.

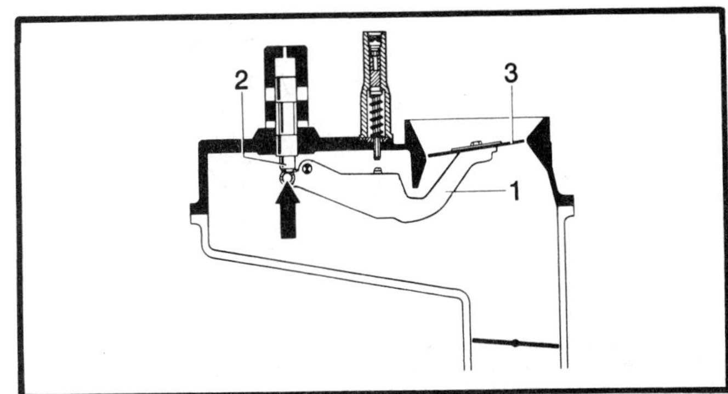

Air flow sensor used on KE-Jetronic systems showing counterbalanced sensor arm (1), fuel distributor plunger (2) and angled air sensor plate (3)

7. Reset the idle speed if necessary. See the individual car sections for details. Anytime an adjustment is made on the fuel system, or if a component is replaced, the idle CO should be reset with a CO meter.

Removal and Installation

1. Relieve the fuel system pressure as previously described.

NOTE: Wrap a cloth around the connection to catch any escaping fuel.

2. Thoroughly clean all fuel lines on the fuel distributor and then remove them.
3. Remove the rubber air intake duct.
4. Remove the air flow sensor/fuel distributor as a unit.
5. Remove the three retaining screws and remove the fuel distributor.

NOTE: When installing the air flow sensor, always replace the O-ring and gaskets. Use Loctite® on all retaining bolts.

6. Installation is in the reverse order of removal.

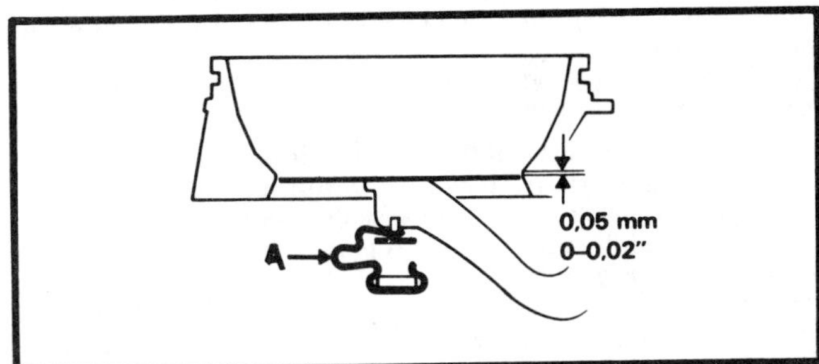

Correct sensor plate height in venturi. Adjust by bending spring. (A)

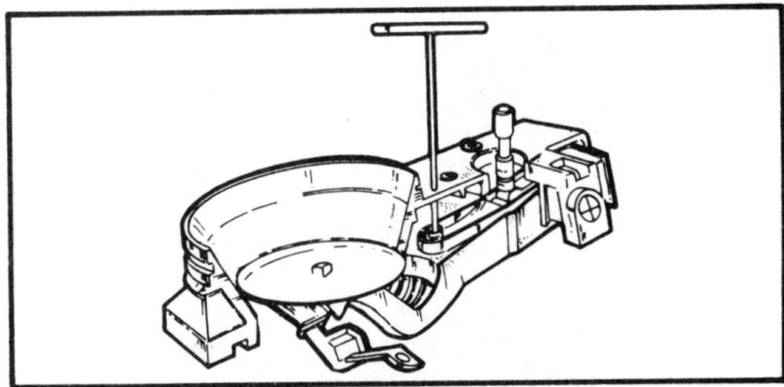

Cutaway of the air flow sensor assembly showing the direction of air flow and proper placement of the mixture (CO) adjusting tool. Never accelerate the engine with the tool in place

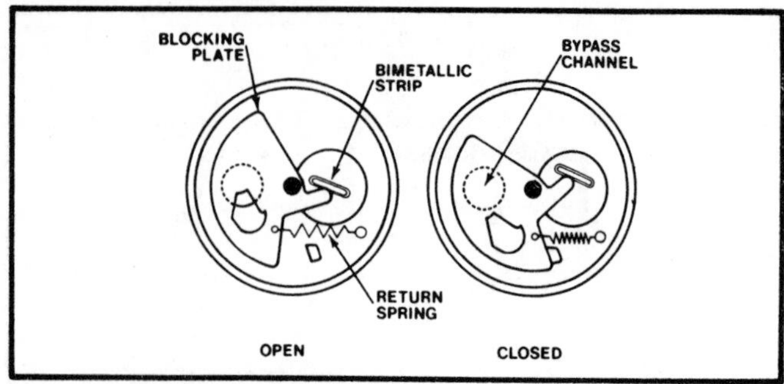

Internal components of the auxiliary air regulator showing cold (open) and warm (closed) positions

AUXILIARY AIR REGULATOR

The auxiliary air regulator allows more air/fuel mixture when the engine is cold in order to improve driveability and provide idle stabilization. The increased air volume is measured by the air flow sensor and fuel is metered accordingly. The auxiliary air regulator contains a specially shaped plate attached to a bi-metal spring. The plate changes position according to engine temperature, allowing the moist air to pass when the engine is cold. As the temperature rises, the bi-metal spring slowly closes the air passage. The bi-metal spring is also heated electrically, allowing the opening time to be limited according to engine type. The auxiliary air regulator does not function when the engine is warm.

Testing

NOTE: The engine must be cold to perform this test.

1. Disconnect the electrical terminal-plug and the two air hoses at the auxiliary air regulator.
2. Voltage must be present at the terminal plug with the ignition switch ON. Check the continuity of the heater coil by connecting a test light or ohmmeter to the terminals on the regulator.
3. Use a mirror to look through the bore of the regulator. If the air valve is not open, replace the auxiliary air regulator.
4. Connect the terminal plug and the two air hoses to the auxiliary air regulator.
5. Start the engine: the auxiliary air regulator bore should close within five minutes of engine operation by the cut-off valve.

NOTE: A quick check of the auxiliary air regulator can be made by unplugging the electrical connector with the engine cold. Start the engine and pinch the hose between the regulator and the intake manifold—the idle speed should drop. Reconnect the air regulator and allow the engine to warm up. With the engine at operating temperature, pinching the hose to the intake manifold should not affect the idle speed. If it does, replace the auxiliary air regulator.

Removal and Installation

1. Locate the auxiliary air regulator on the rear of the intake manifold. Disconnect the electrical connection and remove the air hoses.

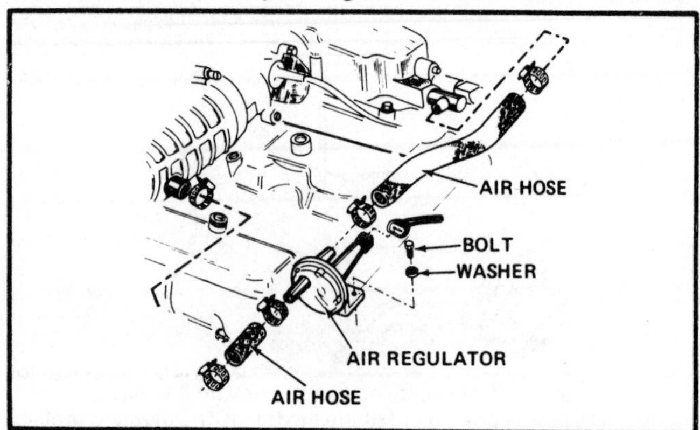

Typical auxiliary air regulator installation

2. Remove the mounting bolts and remove the regulator.
3. Installation is in the reverse order of removal. Make sure all hose and electrical connections are tight.

THROTTLE VALVE

The throttle valve housing is connected to the intake manifold and, in addition to the throttle valve, it contains the idling air passage and the idling adjustment screw, connections for the hoses to the auxiliary air valve, and the cold start valve and the vacuum outlet for ignition timing.

NOTE: Some later models with electronic engine controls do not use vacuum advance units on the distributor.

Adjustment

The stop screw is set by the factory and should not be moved. If for some reason it is moved, adjust as follows:

1. Turn the screw counterclockwise until a gap is visible between the stop and the screw.
2. Turn the screw until it just touches the stop.
3. Turn the screw clockwise an additional ½ turn.
4. Adjust the idle speed and CO and check the linkage for proper operation and free movement.

Idle Adjustments

The idle speed screw (called a bypass screw by some manufacturers) on the throttle body housing of a K-Jetronic system bleeds air into the manifold when you increase idle speed—or cuts it off to slow the engine down. When idle speed changes the idle mixture always changes to some degree, and vice-versa. Therefore, you must juggle the speed and mixture adjustments back and forth to get the specified idle speed within the right range of CO. Unlike a carburetor, changing the idle CO changes the mixture throughout the entire rpm range. Never rev the engine with the Allen wrench sitting in the mixture screw or it may damage the air sensor plate.

Engine oil temperature is the most critical factor in getting an accurate CO adjustment on any K-Jetronic system. Don't touch the mixture screw until the oil temperature is between 140–176°F (60–80°C). CO adjustment is sealed on late K and all KE systems.

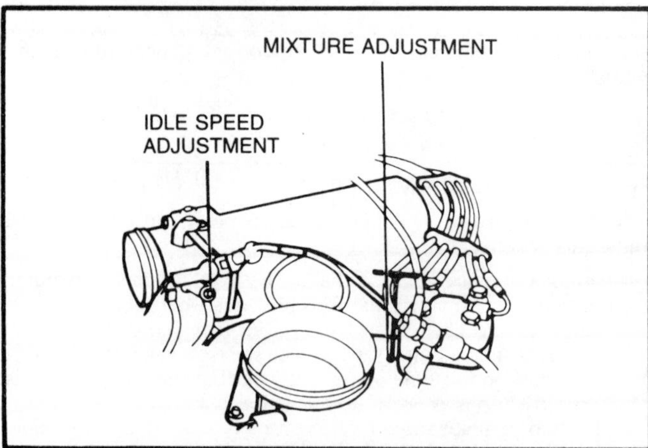

Typical adjustment points on K-Jetronic systems

Lambda (Closed Loop) System

K AND KE-JETRONIC

With the advent of the 3-way catalytic converter, the K-Jetronic system has undergone two major modifications; the addition of a Lambda or Oxygen Sensor system (K-Jetronic) and the integration of an electrohydraulic actuator (KE-Jetronic). Both systems use an electronic control unit (ECU) for precise mixture (read emissions) control. Various new components have been integrated, such as an oxygen sensor, an electronic control unit (ECU), and a frequency valve, which convert K-Jet into a closed loop fuel system. In order for the 3-way converter to work effectively, the air/fuel mixture must be kept within a very precise range. Between 15 and 14 parts of air to one part of fuel, there's an ideal ratio called "stoichiometric" where HC, CO, and NO_x emissions are all at a minimum. Bosch calls this ratio (or stoichiometric point) "Lambda". Combined with an oxygen (or Lambda) sensor, mounted in the exhaust manifold, the air/fuel ratio in a fuel injected system can be controlled within a tolerance of 0.02%.

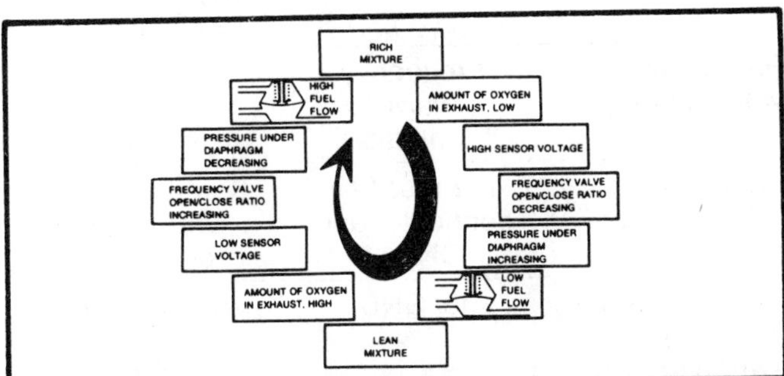

Operational schematic of K-Jetronic Lambda (Oxygen) Sensor System closed loop mode of operation

The oxygen sensor in the exhaust system monitors the oxygen content of the exhaust gases. It produces a small amount of voltage that varies depending on the amount of oxygen present in the exhaust. This voltage signal is sent to the ECU. The ECU, in turn, then signals the frequency valve to enrichen or lean the mixure. The voltage signal is approximately one volt. The frequency valve is located between the fuel distributor and the fuel return line. It does not change control pressure in the K-Jet system; it alters system pressure in the lower chamber of the fuel distributor for each cylinder's differential pressure valve.

NOTE: On the KE type CIS injection system, an electronic actuator replaces the frequency valve and conrol pressure regulator, as described under "Fuel Distributor" earlier. The control units are different for K and KE systems.

Operation

When the oxygen sensor signals a rich mixture to the ECU, it will close the frequency valve. This causes pressure in the lower chamber of the fuel distributor to increase and push the diaphragm up to reduce fuel quantity. If the air/fuel mixture is too lean, the frequency valve will be open and reduce pressure in the lower chamber. The diaphragm is then pushed down and the amount of fuel is increased. The valve opens and closes many times per second. The average pressure is determined by the ratio of valve openings and closings. A higher open-to-closed ratio would provide a richer mixture, a lower open-to-closed ratio would give a leaner mixture. This is called the duty cycle of the frequency valve.

In the KE-Jetronic system, the Lambda control is integrated into the control unit which monitors the input signals from the various engine sensors (including the oxygen sensor), amplifies these signals and then calculates the correct output signal for the electrohydraulic actuator on the fuel distributor to modify the mixture as necessary. There is not frequency valve or control pressure regulator on KE systems. In addition, the oxygen sensor has three wires; one of which supplies current to a heating element that warms the oxygen sensor up to operating temperature faster and doesn't allow it to cool off at idle. The Lambda control system on the KE-Jetronic system is designed to cut out (go to open loop mode) under the following operating conditions.

- Oxygen sensor not warmed up to operating temperature
- During deceleration (fuel shut-off)

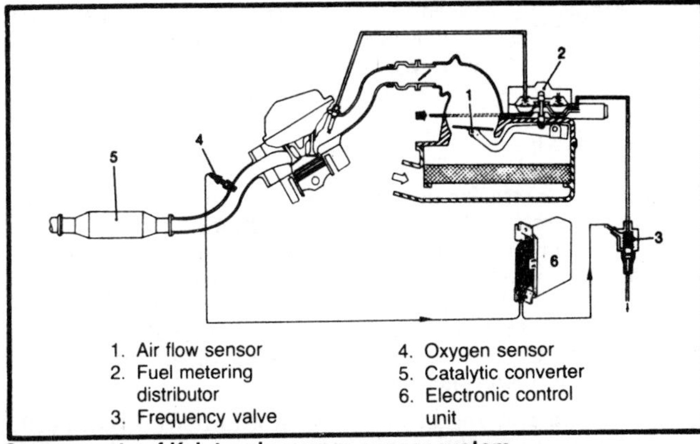

1. Air flow sensor
2. Fuel metering distributor
3. Frequency valve
4. Oxygen sensor
5. Catalytic converter
6. Electronic control unit

Components of K-Jetronic oxygen sensor system

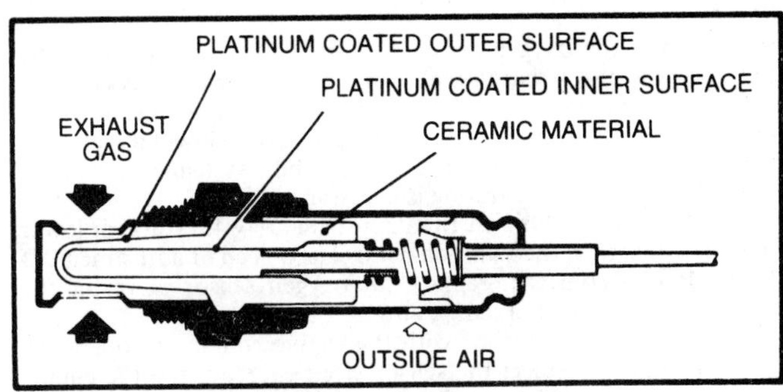

Cross section of a typical oxygen sensor

- At full load operation (wide open throttle)
- Whenever the engine coolant temperature is below 59°F (15°C) and until engine temperature reaches approximately 104°F (40°C)

The heating element of the oxygen sensor is usually energized from a terminal on the fuel pump relay and is heated as long as the fuel pump is running.

NOTE: Do not interchange different types of oxygen sensors (one, two, or three wire) when servicing. Do not let grease or oil contaminate the sensor tip while it's removed. Replace the oxygen sensor at 30,000 mile intervals.

Testing

Bosch makes a tester for the Lambda system that measures the open-to-closed ratio, or duty cycle. The Bosch KDJE 7453 tester (KDJ#-P600 for KE systems) reads out the duty cycle on a percentage meter. A reading of 60, for instance, would be a frequency valve pulse rate of 60%. You can also use a sensitive dwell meter that reads to at least 70° to measure duty cycle on the 4-cyl. scale. Manufacturers using K-Jet with Lambda provide a test socket, so that a duty cycle tester can easily be hooked up. Audi's test connection, for instance, is located behind the throttle valve housing. Saab and most other makers locate the test socket in the wiring harness right near the underhood relay box.

CAUTION

If connected improperly, an analog meter may damage the oxygen sensor. Read the manufacturer's instructions before any testing

The following is a general procedure for checking the duty cycle with a dwell meter on K-Jetronic systems.

1. Disconnect the thermo switch wiring. Shorting the switch will enrich the mixture to about 60% for a cold engine. The meter needle should stay steady, which indicates an "open loop." This means the system isn't being affected by the oxygen sensor. Reconnect the thermo switch.
2. Disconnect the oxygen sensor with a warm engine. The meter should register a 50% signal.

3. Reconnect the oxygen sensor. You should see a change from open to closed loop. The needle will stay steady in the middle at 50% for about a minute. When the oxygen sensor warms up, it should signal the frequency valve for closed loop operation. That will be shown by a swinging needle on the meter, when the system is working correctly.

The KE-Jetronic system requires the use of a special tester for diagnosis. Before tracing possible faults in the oxygen sensor regulating system itself, make sure that the symptoms are not caused by mechanical faults in the engine, ignition system or other components in the injection system. For example, an incorrectly adjusted exhaust valve may have a considerable effect on regulation of the system. Aside from the dwell meter test described earlier, extensive testing of the Oxygen Sensor System requires the use of a special Bosch tester or its equivalent equipment.

NOTE: Before testing Lambda system with tester KDJE-P600, or equivalent, unplug the oxygen sensor and run the engine at idle until it reaches normal operating temperature. Read and record the voltage.

1. Connect the oxygen sensor, run the engine at approximately 2000 rpm and check that the readout on the test meter stays around the recorded value (taken with the sensor disconnected) and that the reading never exceeds ± 0.8 volts above or below that value. If these test results are achieved, the Lambda system is operating properly and no further testing is necessary.
2. If the readout is higher than approximately 4.8 volts, or lower than 2.1 volts, then the Lambda control is in need of adjustment. If the readout is constant, replace the oxygen sensor or repair the open circuit between the sensor and the control unit.
3. Test the oxygen sensor with the engine running at operating temperature. Disconnect the sensor electrical lead, run the engine at approximately 2000 rpm, then check the voltage of the sensor to ground. If the voltage does not exceed 450 mV, replace the oxygen sensor. If equipped with a heating element, check the circuit by measuring the voltage at the connector with the fuel pump relay removed and the sockets bridged at the connector. It should read approximately 12 volts and have a current value of 0.5A or greater.

OXYGEN SENSOR

Removal and Installation

Before fitting the oxygen sensor, coat all threads and gaskets with an antiseize compound (e.g. Never Seize® or Molycote 1000®). Do not apply compound to the sensor body.

NOTE: The joint between the oxygen sensor and the exhaust manifold must be gas-tight. Check that the other joints between the cylinder head cover and the muffler are tight.

CAUTION

The oxygen sensor is highly sensitive to knocks and must be handled carefully. Torque the sensor to 15–30 ft. lbs.

ELECTRONIC CONTROL UNIT (ECU)

Unlike the early K-Jetronic's mechanical control, the later K and KE-Jetronic system control is accomplished by an electronic unit (ECU) that receives and processes signals from various engine sensors to regulate the fuel mixture and accommodate several other functions not present in the early K-Jet system. The control unit is connected to battery voltage and is provided with a voltage correction (safety) circuit to prevent harmful voltage surges. The control unit on the KE-Jetronic system measures:

- Engine speed (tachometer)
- Coolant temperature (sensor in cylinder head)
- Airflow sensor position (position sensor)
- Full load signal (throttle valve switch)
- Idle speed signal (microswitch)
- Oxygen (Lambda) sensor signal
- Signal from altitude compensator

K-JETRONIC LAMBDA SYSTEM TEST SPECIFICATIONS

Circuit Tested	Terminal Numbers	Normal Reading
Check for short circuit	2–4 ①	0 ohms-wiring OK (look for short if continuity exists)
Ground check	8-GND ②	key ON-12 volts key OFF-0 volts
Frequency valve windings	Frequency valve terminals ①	2–3 ohms
Frequency valve harness connector ②	harness (+) connector and GND ②	12 volts with key ON and ECU connected
Battery voltage from ignition switch to ECU	15–GND ②	key ON-12 volts
Relay operation (key ON)	Relay terminal 87-GND ②	12 volts-OK
	Relay terminal 87b-GND ②	12 volts-OK
	Relay terminal 30-GND ②	12 volts
	Relay terminal 86-GND ②	12 volts

NOTE: Turn ignition off between tests and when removing or installing wiring harness connector. Probe all terminals carefully from the rear; test probes can spread terminals and cause a poor connection. Make sure test equipment is compatible with Lambda system or components may be damaged.

GND-Ground
① Ohmmeter connections
② Voltmeter connections

The signals are converted into the appropriate current values and sent to the electrohydraulic actuator and idle speed air valve. The control unit provides an enrichment signal during cranking and establishes smooth running characteristics after starting.

NOTE: The amount of after-start enrichment depends on coolant temperature.

The engine speed is limited by the control unit, which actuates the EHA to shut off fuel to the injectors at a preset rpm. Fuel mixture is also adjusted according to signals from an altitude compensator.

Removal and Installation

Locate the control unit and remove any covers or fastening bolts. The ECU is mounted somewhere in the passenger compartment, usually under the dash or one of the front seats. Carefully pull the unit clear to expose the main connector. Release the clip and remove the connector. Remove the control unit. Reverse the procedure for installation. Make sure all connectors are clean.

NOTE: Some control units are mounted under the passenger seat. Move the seat all the way back to gain access.

MODULATING VALVE

Removal and Installation

NOTE: During removal and fitting of the modulating valve, prevent the rubber valve retainer from coming into contact with gasoline. The rubber is of a special grade to prevent vibrations from the valve being transmitted to the body. The rubber swells considerably if allowed to come into contact with gasoline.

1. Disconnect the electric cable.
2. Disconnect the small-bore line to the modulating valve. Grip the hexagonal nut closest to the hose (14 mm) and undo the valve nut (17 mm).
3. Disconnect the modulating valve return line from the warmup regulator, from the fuel distributor and the joint in the return line on the latter.
4. Remove and disconnect the valve and return lines.
5. Installation is the reverse of removal.

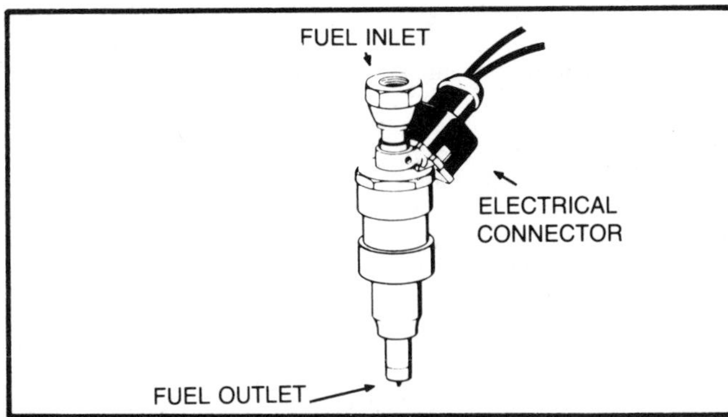

Typical modulating or frequency valve assembly

Constant Idle Speed (CIS) Control System

GENERAL INFORMATION

This electronic control system is designed to correct engine idle speed deviations very quickly; holding the idle speed constant even when loads such as power steering, air conditioning or automatic transmission are applied to the engine. Sensors on the engine read the engine speed (rpm), operating temperature and throttle position, then supply a voltage signal to a separate control unit which in turn adjusts the engine speed by feeding more or less air into the injection system by means of an idle air device. The control unit adjusts the idle speed according to pre-set (programmed) "ideal" rpm values contained in its memory.

The constant idle speed system is designed to be maintenance-free and will operate regardless of wear or changes in ignition timing. Under normal circumstances, no adjustment of the idle speed is necessary or possible. The idle air valve is located in a bypass hose which routes air around the throttle valve much the same way as the auxiliary air valve this system replaces. The idle speed control unit processes voltage signals from the various sensors on:

- Engine speed (tachometer)

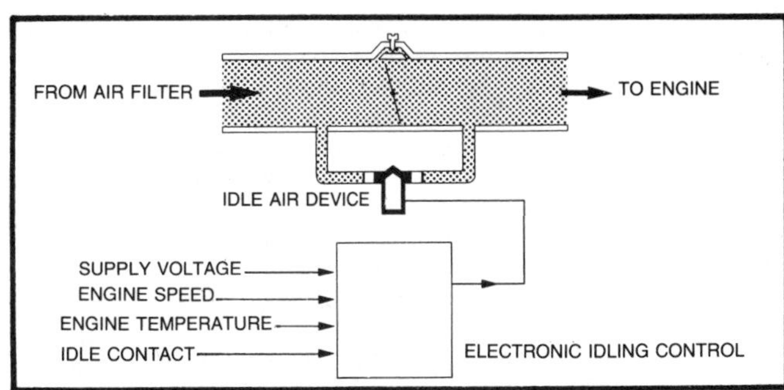

Schematic of Bosch Constant Idle Speed control system

- Coolant temperature (sensor in cylinder head)
- Throttle position (microswitch)
- Shift lever position (automatic transmission)
- A/C compressor cut-in signal

With the ignition OFF, the idle air valve is opened fully by a set of return springs. With the ignition ON, the idle air valve is controlled by a specific voltage from the control unit, providing an air valve opening dependent on coolant temperature. In this manner, the idle speed is constantly controlled between approximately 1000 rpm @ −4° F (−20°C) to roughly 750 rpm @ 68°F (20°C). The control unit also receives a voltage signal when the transmission shift lever is in PARK or NEUTRAL; this voltage signal usually drops when the transmission is engaged (DRIVE) to maintain the idle speed at the factory-established specification (usually around 650–700 rpm). The same type of voltage signal is produced by the air conditioner compressor, which also drops when the compressor engages, sending another signal to the idle air valve to once again maintain engine idle speed under load.

Testing

NOTE: Testing the constant idle system (CIS) requires the use of Bosch tester KDJE-P600, or equivalent, along with the proper test cable adapter.

1. Connect test cable to idle air valve and tester. The engine should be idling at normal operating temperature.
2. Press the IR 100% button and note the reading on the tester. It should be 27–29% @ 670–770 rpm. If the reading is correct, the CIS system is operating properly.
3. If the reading on the tester is higher or lower, adjust to the correct value or test the microswitch for proper operation.
4. If the readout is 0%, test for voltage on the plug for the idle air valve. There should be 12 volts from the feed side of the pin socket to ground. If not, check for an open circuit in the idle air valve harness and repair as necessary.
5. Check the resistance of the idle air valve. It should be approximately 12 ohms between terminals 2 and 3; and the same value between terminals 2 and 1. If not, replace the idle air valve assembly.

NOTE: If the voltage signal to the idle air valve is correct (12 volts), check the ground line to the control unit. Replace the control unit if no ground connection is present. Make sure the ignition switch is OFF before disconnecting or reconnecting any CIS system components.

BOSCH AIR FLOW CONTROLLED (AFC) FUEL INJECTION

General Information

The most common type of AFC fuel injection is the Bosch L-Jetronic system. The L stands for "Luftmengenmessung" which means "air flow management." The L-Jetronic AFC injection system is used, with various modifications, by both European and Asian manufacturers. Different versions of the basic L-Jetronic design include LH, LH II, LU, LU Digital and Motronic. Although similar in design and operation, each separate system uses slightly different components. LH-Jetronic, for example, measures the incoming air mass with a heated wire built into the air flow meter, instead of a flap valve.

Aside from different terminology, most Asian and European AFC systems are similar in both appearance and function. For this reason, it is important to accurately identify the particular type of AFC system being serviced. A small diagram of the fuel injection system is usually found under the hood on a sticker somewhere near the emission control label; it's very important to understand how the different components function and affect one another when troubleshooting any AFC fuel injection system.

AIR FLOW CONTROLLED (AFC) SYSTEMS

Type of System	Primary Control Measurement
L-Jetronic	Air flow
LH-Jetronic	Air mass
LH II-Jetronic	Air mass
LU-Jetronic	Air flow (Utilizes a hybrid ECU that converts analog signals to digital map)
LU-Digital Jetronic	Air flow (Digital ECU with inputs)
Motronic	Air flow with digital control of ignition and injection
K-Jetronic	Air flow Controlling fuel flow non-electronically
KE-Jetronic	Electronically controlled K-Jetronic
E.C.C.S. ①	Air mass
I-TEC ②	Air flow
EGI ③	Air flow

NOTE: Identification numbers are located on control unit housing. Compare replacement part numbers to original unit before installing.

ECU—Electronic Control Unit
① Nissan (Datsun) fuel injection system
② Isuzu fuel injection system
③ Mazda fuel injection system

L-JETRONIC

The Bosch L-Jetronic is an electronically controlled system that injects the fuel charge into the engine intake ports in intermittent pulses by means of electromagnetic, solenoid-type fuel injectors (one per cylinder). The quantity of fuel injected per pulse depends on the length of time the injector is open, which is determined by an electrical impulse signal from the electronic control unit (ECU) that reacts to inputs from various engine sensors. The main control sensor is the air flow meter, which measures the amount of air being inducted into the engine. Sensors for engine temperature, engine speed (rpm), intake air temperature, throttle position, exhaust gas oxygen content and barometric pressure also feed information into the control unit to help determine the fuel injection quantity (mixture) that will produce the best performance with

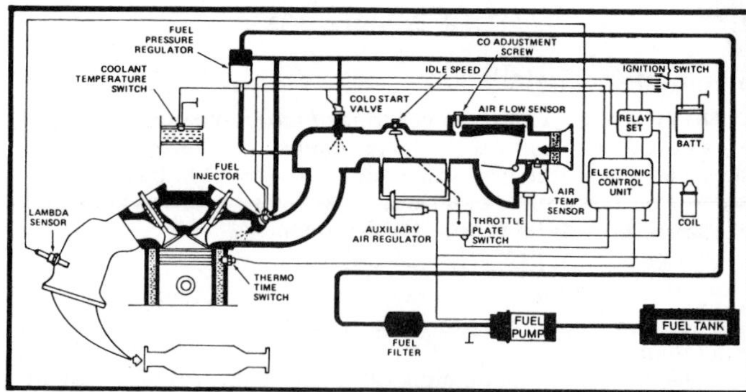

Typical European L-Jetronic injection system schematic

the least emissions. The L-Jetronic control unit is used in both analog and digital versions which are not interchangeable.

Fuel is supplied to the injectors under fairly constant pressure by an electric roller cell pump and a fuel pressure regulator that responds to manifold vacuum and returns excess gasoline to the tank. A set of points in the intake air sensor assures that the pump only gets current when the engine is running or being cranked. This eliminates flooding in the event that an injector springs a leak.

Three components that aren't connected to the computer assist in starting and warm-up. The cold start valve in the intake manifold injects extra fuel while the engine is being cranked. It's controlled by the thermotime switch, a thermostatic device that energizes the cold start valve for 3 to 10 seconds. Its bimetal strip is affected by both engine temperature and an electrical heating coil. The auxiliary air regulator bypasses the throttle plate to provide extra air during warm-up, which prevents stalling. A heating element inside causes it to close gradually.

Operation

The L-Jetronic system measures intake air and meters the proper fuel to obtain the correct air/fuel ratio under a wide range of driving conditions.

We can break the L-Jetronic into three basic systems: air intake; fuel supply and electronic control. Some components like the air flow meter are part of two systems. The air intake system consists of the air cleaner, air flow meter, throttle housing, connecting hoses, air valve (air regulator) and the manifold. The fuel system includes the tank, fuel pump, fuel damper, filter, fuel rail, cold start valve, fuel injectors, and connecting lines. Some European L-Jet systems don't use a fuel damper. Both Datsun/Nissan and Toyota employ dampers to reduce pulsation from pump output.

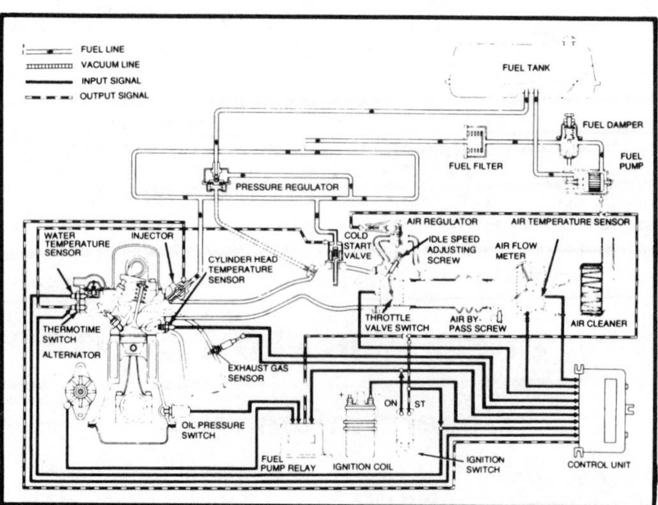

Typical Japanese L-Jetronic injection system schematic

TROUBLESHOOTING—L-JETRONIC

Engine cranks but does not start	Engine starts but then dies	Rough or unstable idle	Idle speed incorrect	CO value incorrect	Erratic running	Engine misses when driving	Fuel consumption too high	No maximum power	Cause	Correction
●	●	●	●		●	●	●	●	Defect in ignition system	Check battery, distributor, plugs, coil and timing
●	●	●	●		●	●	●	●	Mechanical defect in engine	Check compression, valve adj. and oil pressure
●	●	●	●	●	●	●		●	Leaks in air intake system (false air)	Check all hoses and connections; eliminate leaks
●	●	●						●	Blockage in fuel system	Check fuel tank, filter and lines for free flow
●									Relay defective; wire to injector open	Test relay; check wiring harness
●									Fuel pump not operating	Check pump fuse, pump relay and pump
●	●	●		●	●		●	●	Fuel system pressure incorrect	Check pressure regulator
●									Cold start valve not operating	Test for spray; check wiring and thermo-time switch
●	●	●	●	●	●		●		Cold start valve leaking	Check valve for leakage
●									Thermo-time switch defective	Test for resistance readings vs. temperature
●	●	●	●						Auxiliary air valve not operating correctly	Must be open with cold engine; closed with warm
●	●	●		●			●		Temperature sensor defective	Test for 2-3 kΩ at 68° F.
●	●	●	●	●	●		●	●	Air flow meter defective	Check pump contacts; test flap for free movement
		●	●					●	Throttle butterfly does not completely close or open	Readjust throttle stops
		●						●	Throttle valve switch defective	Check with ohmmeter and adjust
	●	●	●	●					Idle speed incorrectly adjusted	Adjust idle speed with bypass screw
		●		●	●	●		●	Defective injection valve	Check valves individually for spray
	●	●	●	●	●		●		CO concentration incorrectly set	Readjust CO with screw on air flow meter
●	●	●			●	●			Loose connection in wiring harness or system ground	Check and clean all connections
●	●	●				●		●	Control unit defective	Use known good unit to confirm defect

Dampers have what looks like an adjusting screw, but *don't attempt any adjustment*. The Datsun 200SX injection system doesn't use a cold start valve.

The electronic control system is made up of the electronic control unit (ECU) and several sensors. Each sensor signals measurements on engine condition to the ECU, which then computes fuel needs. Injectors are then opened and closed by ECU command to feed the needed amount of fuel. Injectors are connected in parallel. All supply fuel at the same time. They open twice for each rotation of the engine camshaft, injecting 1/2 the needed fuel each time.

The big sensor, and the one that gives the system its name, is the air flow sensor. This is located in the large airbox. A flap in the box opens and closes in response to air being drawn into the engine. The ECU is then signalled as to the amount of air being taken in. An air temperature sensor in the box sends that information to the ECU. Engine rpm is picked up from the negative side of the coil. A throttle valve switch tells the control unit how much the throttle is open. Engine temperature is transmitted from a coolant temperature sensor. Later model Datsuns/Nissans use a cylinder head temperature sensor instead of a coolant temperature sensor. A signal from the ignition switch tells the ECU when the engine is being started. An O_2 sensor has been added to feed exhaust gas oxygen content to the ECU.

LH AND LH II-JETRONIC

These systems are further developments of the L-Jetronic. LH is an abbreviation of "Luft-Hitzdracht" which means "hot air wire." Both the LH and LH II work on the same principle as the L system, but instead of an air flow sensor flap to measure intake air quantity, these systems use a heated platinum wire to measure the air mass. In this manner, altitude influences are eliminated from the injection systems input quantities. A very thin platinum wire is stretched across the air intake opening in the air flow meter and forms part of a bridge circuit. Air flowing over the wire draws heat from it, changing the wire's electrical resistance as the temperature changes. An electronic amplifier instantly responds to any such resistance change and regulates the current to the wire so as to maintain it at a virtually constant temperature. The current necessary to maintain the wire temperature is the measure of the air mass flowing into the induction system, and this signal is used by the ECU to determine the injector opening time to adjust the air/fuel mixture.

Operation

The LH and LH II-Jetronic fuel injection are electronic systems with the same basic components as the L-Jetronic. The systems consist of a control unit with various engine sensors that measure and monitor engine operating conditions and adjust the fuel mixture for optimum performance with minimum emissions, according to load, throttle position, temperature, etc. The control unit regulates the fuel quantity by varying the length of time the injectors are open and regulates idle speed by varying an air control valve opening. The system measures the air mass entering the engine by means of a heated platinum wire. Because dirt and impurities may accumulate on the wire surface and affect the voltage signal, the system is designed to clean the wire each time the engine is turned off to eliminate impurities and corrosion.

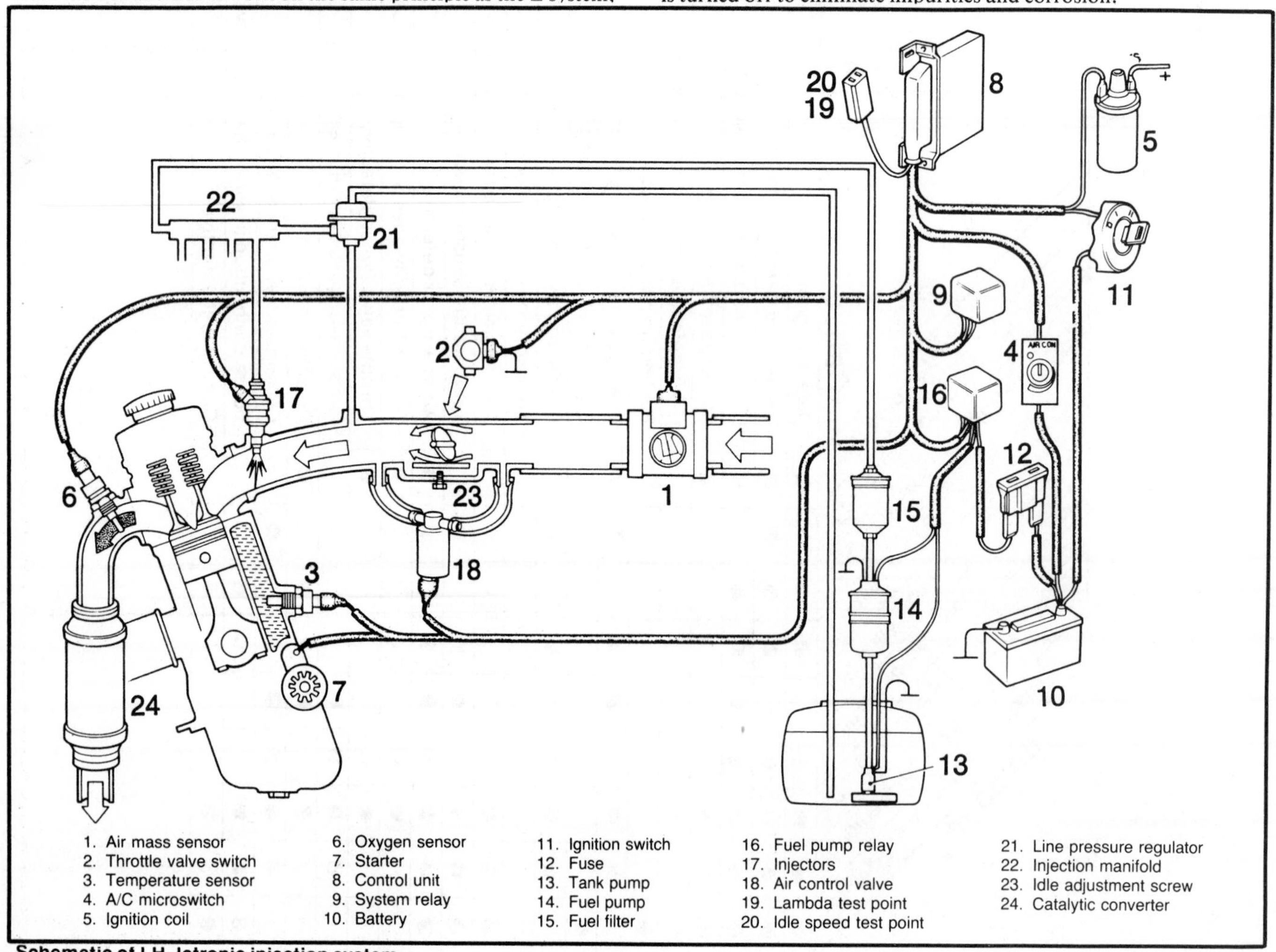

Schematic of LH-Jetronic injection system

The LH and LH II injection systems operate with a moderate fuel pressure which is held constant by a fuel pressure regulator. Fuel injection is by means of electrically controlled solenoid-type injectors which spray fuel directly behind the engine intake valves. The injection duration is usually measured in milliseconds and is controlled by the electronic control unit (ECU). The ECU receives signals from a set of engine sensors, the most important of which is the air mass meter that continuously measures the mass of air entering the intake manifold. When the engine stops, the wire is heated up to 1050-1920°F for less than one second to burn off any dirt, corrosion or impurities on the filament. If allowed to build up these impurities would cause false signals to be sent to the control unit and affect the air/fuel mixture. In addition to the air flow meter, the other sensors providing input to the ECU include:

- Throttle valve position sensor (switch)
- Vacuum switch (indicating part and full load idle conditions)
- Engine speed sensor (tachometer connection)
- Oxygen sensor (Lambda system)
- Coolant temperature sensor
- Cold starting control (enrichment program)

The electronic control unit receives and processes signals from the engine sensors and calculates the correct fuel mixture according to a preprogrammed memory unit contained within the ECU assembly. *The ECU and its connector should be handled with the utmost care during all testing and/or service procedures.*

NOTE: Each individual variant of the L-Jetronic injection system uses a different control unit. They are not interchangeable.

MOTRONIC

The Motronic fuel injection system combines the digital control of individual systems such as fuel injection and ignition into a single unit. The heart of the Motronic system is a microcomputer that is programmed according to dynomometer data on a specific engine's characteristics. In operation, various engine sensors deliver data on engine speed (rpm), crankshaft position and temperature (engine and ambient air). From this input, the control unit determines the ideal spark advance and fuel quantity, up to 400 times a second. In this manner, spark advance and fuel quantity is tailored exactly to the engine operating conditions such as idling, part load, full load, warmup, deceleration and transient modes. Optimal fuel injection and ignition settings improve the engine's overall performance while reducing fuel consumption and emissions. Motronic systems can be made to incorporate other features from modified L-Jetronic systems, such as the air mass sensor wire in the intake air meter and electronic control of automatic transmission. Its basic operation is very similar to the previously described L, LH and LH II injection systems, except for the automatic ignition timing control.

Service Precautions

- Make sure the ignition is switched OFF before removing or installing any component connectors, especially to the control unit.
- Do not replace any control unit without first checking all wiring and components, otherwise an existing fault could ruin a new ECU the same way the old one was damaged.
- Never check connector terminals from the front, especially the main harness connector to the ECU. Excessive force can damage connector terminals and cause other faults. The correct testing method is to remove the connector cover and check the terminals from the holes provided in the side of the connector. Terminal numbers are usually stamped on the side of the connector.
- When removing the control unit connector, release the lock tabs, then fold out the connector. Do not pull the ECU connector straight out. All electrical connectors should have some kind of lock device to keep the connectors from coming apart.
- Remove any electronic control unit before subjecting the car to any temperature in excess of 176°F (80°C), such as when baking paint in an oven.

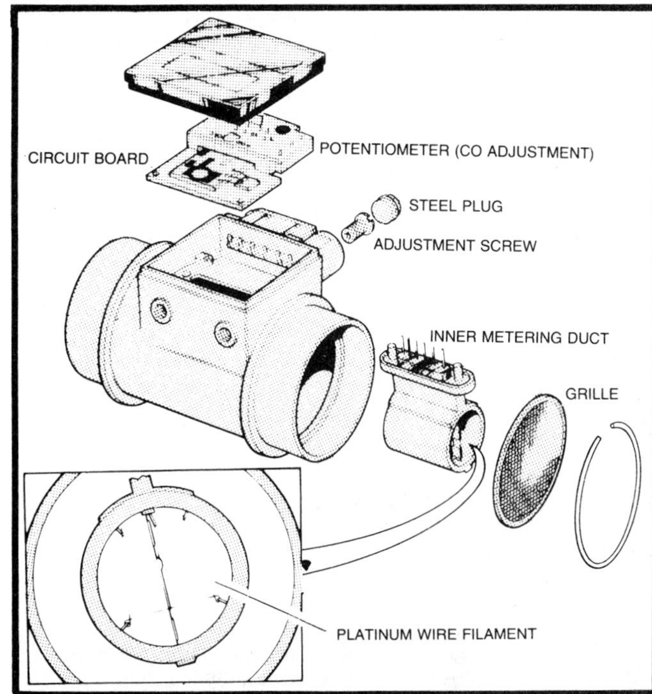

Exploded view of air mass sensor used on LH and LH II injection systems

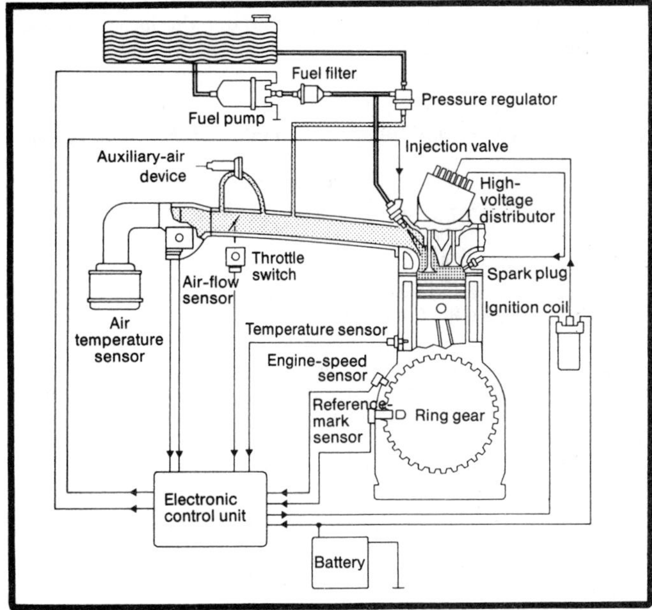

Schematic of Motronic fuel injection system

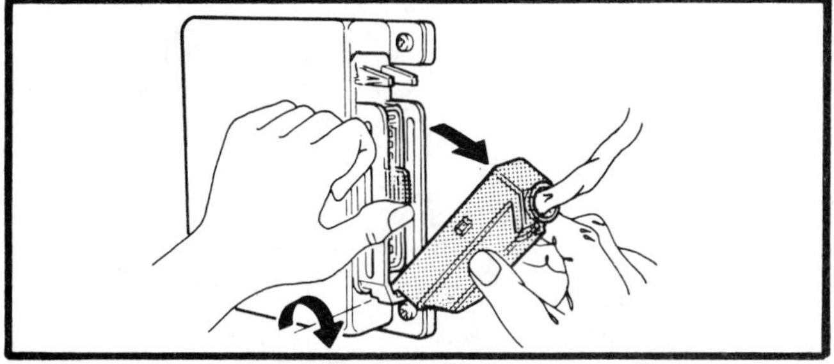

Release the locks, then fold out the ECU connector. Do not pull the connector straight out

• Cleanliness is extremely important when working on or around the fuel system. Do not allow dirt or grease to contaminate the fuel system. Always use new gaskets and seals during reassembly of components using them. Clean all fuel connections before removal, and prepare a clean work area.

• Disconnect the ignition system when performing compression tests. Any arcing to the injectors or injector wiring can damage the control unit. Arcing to the ignition coil low tension side can damage vehicles equipped with Hall Effect distributors.

• Never use a quick charger on the battery or attempt to use a "hot shot" 24 volt starting aid. Maximum charging current should not exceed 15 amps.

• Always depressurize the fuel system before attempting to disconnect any fuel lines or components.

• Harsh chemicals such as carburetor cleaner should not be used to find vacuum leaks. The oxygen sensor is easily contaminated and ruined by certain chemicals and fuel additives commonly used on carbureted engines.

On-Car Service

Water is one of the biggest killers of electronic fuel injection (EFI) system components, regardless of the type of system. The high pressures generated by the fuel pump can and will push water and fuel right through the fuel filter into the rest of the system.

Water problems can be minimized by changing filters regularly, using water absorbing fuel additives (especially in colder weather), and by installing a water filter.

Several companies manufacture water separation/filtration systems which electronically detect the presence of water in the fuel. They can be used in all environments under the hood, with all fuels, and packaged with all fittings, brackets and hardware for easy installation.

CAUTION

Read the service precautions at the beginning of this section before attempting repairs or tests.

PRELIMINARY CHECKS

Before assuming the EFI system is at fault, make the standard preliminary checks of ignition system and engine conditions before going any further. Check the battery.

Check the electrical connections; EFI electrical systems won't tolerate corrosion or sloppy connections. Clean connections with one of the electrical cleaners available and be absolutely certain that spaded terminals snap tightly into their plastic connector bodies. Carefully crimp these where necessary with needlenose pliers. Also, check the main harness connector that plugs into the ECU for dirt, corrosion, or bent or spread-out terminals.

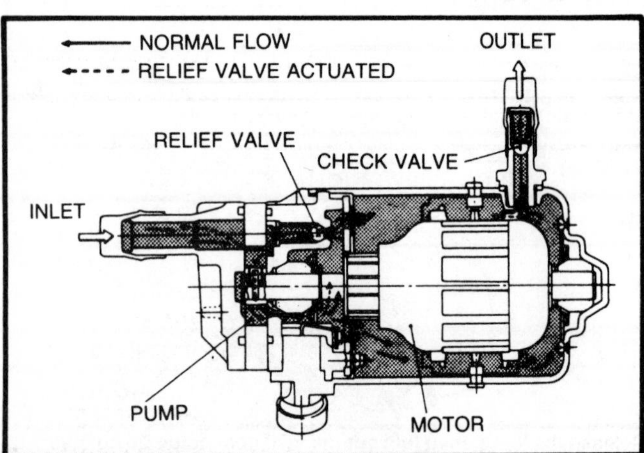

Typical fuel pump used on L-Jetronic systems

This check alone has solved many intermittent EFI performance problems.

NOTE: Sealing electrical connections against moisture is important. Make sure all connectors snap tightly together and no bare wire protrudes from the back of the connector body. Replace any rubber protector caps removed during testing.

A bad set of ignition wires will cause the same symptoms as a bad pressure regulator, so check the ignition system first. Steady vacuum that's within specs is a must on these cars. A vacuum leak unbalances the system throughout its entire operating range. A leaky engine will start, idle, and run poorly.

NOTE: L-Jetronic systems tend to be so "mixture-sensitive" that crankcase air leaks will upset them. If the sealing ring around the oil filter cap is broken or missing, a VW for instance, with L-Jetronic just won't perform.

FUEL PUMP TESTS

Fuel is drawn from the fuel tank into the fuel pump, from which it is discharged under pressure. As it flows through the mechanical fuel damper (if so equipped), pulsation in the fuel flow is damped. Then, the fuel is filtered in the fuel filter, goes through the fuel line, and is injected into the intake port. Surplus fuel is bled through the pressure regulator and is returned to the fuel tank. The pressure regulator controls the injection pressure in such a manner that the pressure difference between the fuel pressure and the intake manifold vacuum is always approximately 36 psi.

The fuel pump is a wet type pump where the vane rollers are directly coupled to a motor which is filled with fuel. The fuel cools and lubricates the pump internal components as it flows through, so the fuel pump should never be allowed to run dry. A relief valve in the pump is designed to open when the pressure in the fuel lines rises over 43–64 psi due to a malfunction in the pressure system. A check valve prevents abrupt drop of pressure in the fuel pipe when stopping the engine.

CAUTION

Operating the fuel pump dry for more than a few seconds can cause pump seizure.

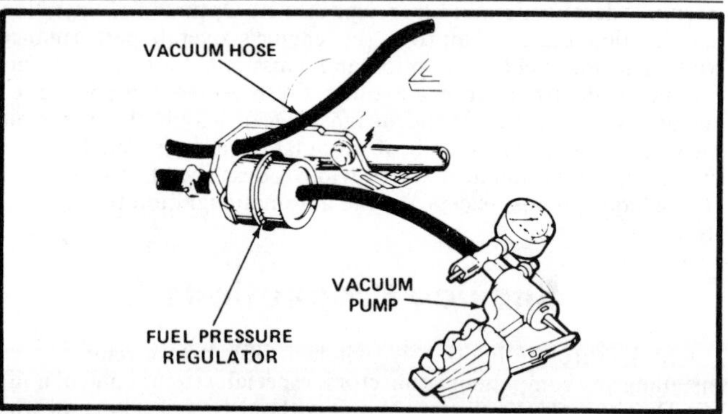

Relieving fuel system pressure on L-Jetronic systems

Relieving Fuel System Pressure

1. Remove the vacuum hose from the fuel pressure regulator.
2. Connect a hand vacuum pump to the regulator and pump vacuum up to 20 in. Hg.
3. Wrap a clean rag around a fuel connection to catch any fuel while carefully loosening connection. Listen for sound of venting into fuel tank.

CAUTION

Fire hazard. Relieve fuel system pressure before attempting to disconnect any fuel injection system components. This procedure allows the fuel system pressure to vent into the tank, it does not however, remove all fuel from the lines.

Alternate Methods for Relieving Fuel Pressure

1. Disconnect the battery ground cable.
2. Disconnect the wiring harness to the cold start valve.

CAUTION

Do not disconnect any components with the engine running on later LH, LH II, LU and Motronic systems. Removing the battery cable will clear the trouble code memory on self-diagnosing injection systems.

3. Using two jumper wires from the battery, energize the cold start valve for two or three seconds to relieve pressure in the fuel system. Be careful not to short the jumpers together.
4. If an early L-Jet system is used, start the engine, then remove the fuel pump relay or unplug the pump connector and allow the engine to stall.
5. Carefully loosen the fuel line as in primary method.

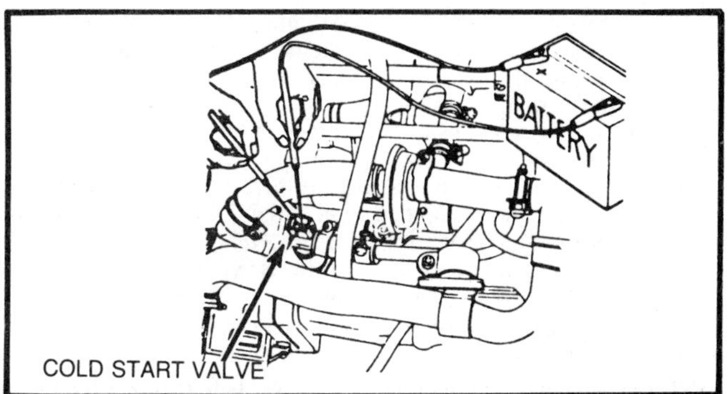

Alternate method for relieving fuel pressure

Fuel Pump Power Check

If an AFC equipped car doesn't start, there's a good chance it might be the fuel pump or its circuit. First, visually check all the pump and relay wiring, as it could be nothing more than a bad connection. Be sure that the pump is grounded. Corrosion can break the ground circuit. Check the pump fuse and look to see if the fusible link in the system is burned. Don't forget to clean the battery cables—the pump calls for 12 volts. If the link or fuse is burned, it could be a seized or worn out pump. Or it could be watered gas, which will freeze the pump in winter. The last item to try is bypassing the relay with a jumper wire. If it works, then the relay's bad. Some models have a convenient fuel pump check connector under the air flow meter or fuel rail. Pry the cap off and, with a jumper wire, you can run the pump without starting the engine. If you can feel fuel pressure at the cold start valve and hear fuel returning through the regulator, the pump's working.

Exercise caution with the dual relays used in L-Jetronic cars. Refer to a wiring diagram of the particular vehicle because the main relay side and the pump relay side of the dual relay are seldom labeled by the manufacturer. Make sure the correct terminals connect the jumper across the *pump relay* side of the dual relay. Remember, too, that all AFC electric fuel pumps use a separate ground wire. So don't neglect to ground check during pump troubleshooting. As is the case with the fuse, these relays could be anywhere. Datsun hides a dual relay inside the kick panel; VW may have it under the dashboard; and others may have their pump relays under the hood. Some AFC systems use a single, main relay on later models.

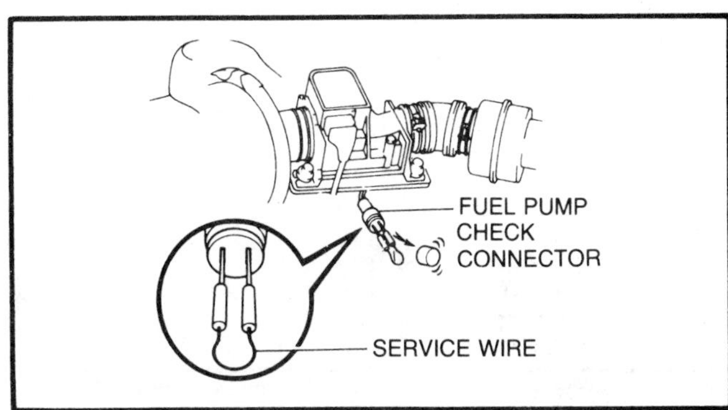

Testing the fuel pump on the L-Jetronic system. Not all models have the test connector

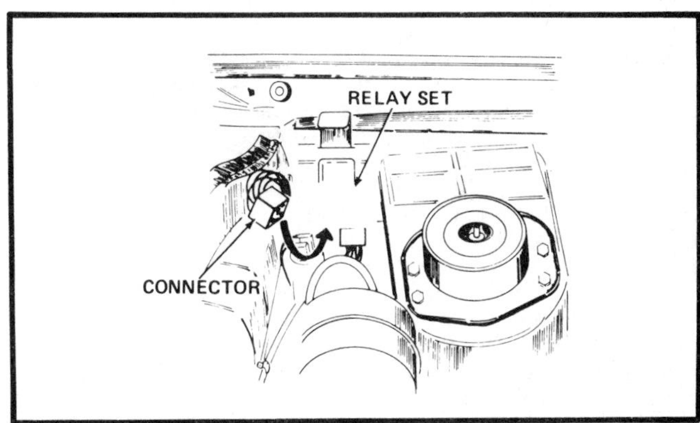

Typical L-Jetronic dual relay set, usually mounted somewhere in the engine compartment

Fuel Pressure Check

NOTE: Use this check to determine if the fuel pump is operating properly and to check for restrictions in fuel lines.

To test the L-Jetronic systems, you must have an extremely accurate pressure gauge that reads 0–50 psi in 0.5-psi increments. Also, remember that high operating pressures mean that connections must be tight. 30–40 psi pressures won't tolerate any haphazard connections. Don't risk a fire—use a reinforced fuel line hose that'll withstand 30–40 psi and use worm gear type clamps to secure the hose. Keep an eye out for leakage in the fuel plumbing and check the braided hose used as OEM equipment.

CAUTION

The fuel system must be depressurized before disconnecting any fuel system components. The cold start valve body is plastic, use care in removing the fuel hose.

Connect the pressure gauge to the cold start injector connection on the fuel rail, or that line between the fuel rail and filter. Banjo fittings are used throughout the Toyota AFC system. An adapter with a banjo fitting at one end will be needed to hook up the gauge. Take a good look at the hoses on these cars. Hoses must be in decent shape and fittings or clamps should be tight.

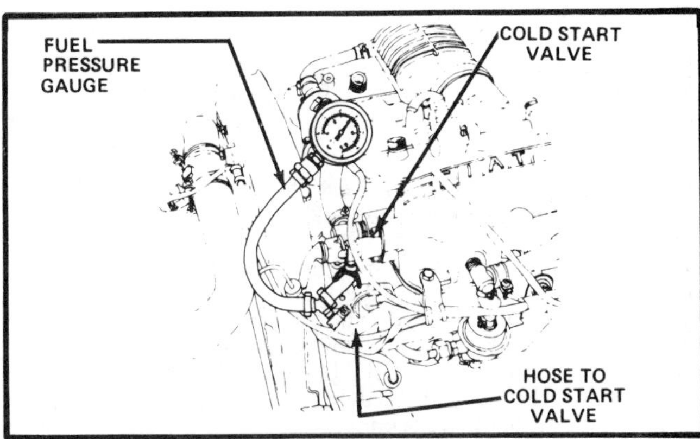

Checking the fuel pressure on L-Jetronic systems-typical

Start the engine and disconnect and plug the vacuum line to the pressure regulator. Pressure at idle should be approximately 33–38 psi (34–45 psi on later models). A bad pressure regulator is indicated by pressure that's too high. Look for kinked fuel lines, restrictive fuel filter, or a faulty pump or pump check valve, if the pressure is too low.

Connect the vacuum hose to the pressure regulator. Watch the gauge. Pressure should drop to 28 psi. A reading over 33 psi means the regulator's bad. Check pressure at full throttle. Pressure should climb to 37 psi as soon as the throttle is wide open. On all L-Jetronic systems, pressure should hold when the engine is shut off. An immediate pressure drop means an internally leaking injector(s), pressure regulator, or fuel pump.

NOTE: The pressure reading may slowly drop through the regulator valve seating or the pump non-return valve. A slow, steady drop is permissable, a rapid fall is not.

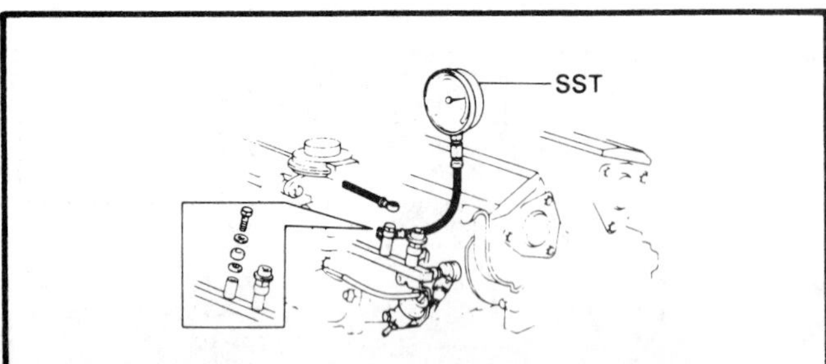

Typical fuel pressure gauge hookup on Japanese L-Jetronic systems—note the banjo fittings used for fuel connections

When taking a fuel pressure reading at idle, normal injector pulsation will cause the gauge to fluctuate. This pulsation is most pronounced on 4 cylinder engines. Watch the needle carefully and take the *average* pressure value. At idle, simply take your pressure reading as being the mid-point between the highest and lowest fluctuations of the pressure gauge needle. If fuel pressure isn't within specs at idle, you could have a clogged fuel filter or either a restricted fuel line or fuel return line. Be sure to change fuel filters at least every 7,500–10,000 miles. Make sure any additives will not damage the oxygen sensor or converter before using them in the fuel system.

After taking the idle pressure reading, shut off the engine and watch the pressure gauge. The system should hold 17–20 psi. If the pressure drops to zero, there's a leak in the system. On a car with a conventional fuel system, such a pressure loss would suggest that either the fuel pump or the carburetor inlet valve were leaking. But assuming that they're not leaking fuel *externally,* there are four places an L-Jetronic system can leak *internally:*

1. The check valve in the output side of the electric fuel pump.
2. The pressure regulator.
3. One or more injectors.
4. The cold start valve.

To begin isolating a leak, crimp the hose connecting the fuel pressure regulator to the fuel tank. Turn the ignition switch off and on to cycle the fuel pump so you build up pressure in the system again.

CAUTION

Do not allow fuel system pressure to exceed 85 psi.

If the system now holds pressure, then the regulator is leaking. If the pressure still drops, keep the locking pliers on the regulator hose and have someone cycle the fuel pump with the ignition switch. Then quickly crimp the hose connecting the pump to the main fuel rail. If the pressure holds steady, the check valve in the output side of the fuel pump is defective. A bad pump check valve usually causes a hard start condition on an EFI equipped engine when hot.

Volume Test

A pressure test alone won't reveal trouble with an EFI fuel pump. It's important to also perform a volume test as well as a pressure test. With 12 volts at the pump and at least half a tank of gas in the car, a typical L-Jetronic pump should deliver about a quart (almost a liter) of fuel in 30 seconds.

CAUTION

Take precautions to avoid fire hazard when performing pressure and volume tests.

Component Removal and Testing

FUEL PRESSURE REGULATOR

Removal and Installation

1. Relieve fuel system pressure.
2. Disengage the vacuum line connecting the regulator to the intake manifold from the pressure regulator.
3. Remove the bolt and washers securing the pressure regulator mounting bracket and carefully pull the regulator and bracket upward. Note the position of the regulator in the bracket.

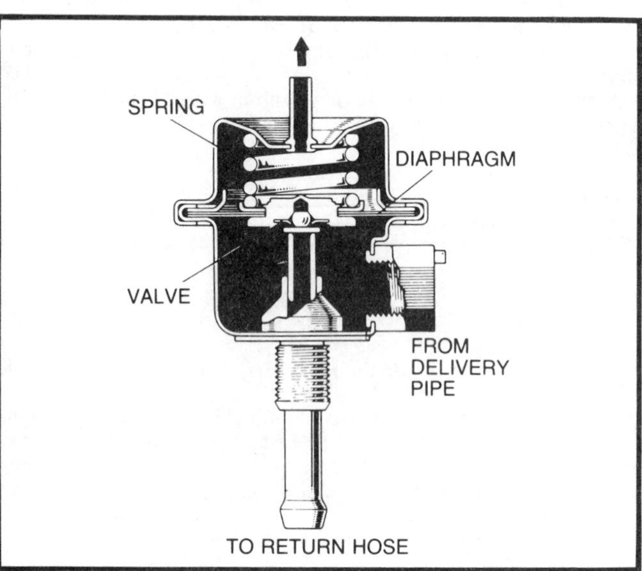

Typical L-Jetronic fuel pressure regulator

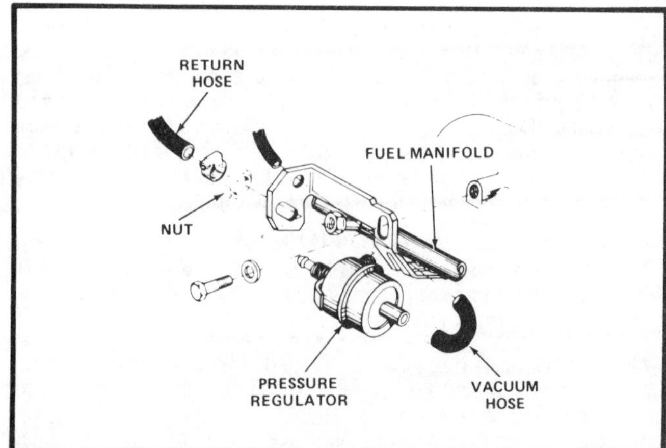

Typical pressure regulator mounting on L-Jetronic systems. Bracket fasteners and fuel connections will vary

4. Unfasten the hose clamps and disconnect the pressure regulator from the fuel hose. Inspect the hose for signs of wear, cracks or fuel leaks.

NOTE: Place a clean rag under the pressure regulator to catch any spilled fuel.

5. Remove the lock nut and remove the pressure regulator.
6. Installation is the reverse of removal.

NOTE: Torque the fuel delivery union bolt to 18–25 ft. lbs. Do not overtighten.

FUEL INJECTORS

Fuel in the L-Jetronic system is not injected directly into the cylinder. Fuel in injected into the intake port, where the air/fuel mixture is drawn into the cylinder when the intake valve opens to start the intake stroke. An electrical signal from each engine sensor is introduced into the control unit for computation. The open valve time period of the injector is controlled by the duration of the pulse computed in the control unit.

The injector operates on the solenoid valve principle. When an electric signal is applied to the coil built into the injector, the plunger is pulled into the solenoid, thereby opening the needle valve for fuel injection. The quantity of injected fuel is in proportion to the duration of the pulse applied from the control unit. The longer the pulse, the more fuel is delivered.

The fuel injectors are electrically connected, in parallel, in the control unit. All injectors receive the injection signal from the control unit at the same time. Therefore, injection is made independently of the engine stroke cycle (intake, combustion, and exhaust).

A bad injection valve can cause a number of problems:

- Hot restart troubles.
- Rough idle.
- Hesitation.
- Poor power.

Hot starting complaints can come from an injector or injectors that are leaking fuel droplets when they're supposed to be completely shut. The next three problems can be caused by a bad spray pattern from one or more of the injectors. Dribble patterns, fire hose shots, and uneven sprays will produce hesitation, stumbling and general lack of power. Replace any injector with a bad spray pattern.

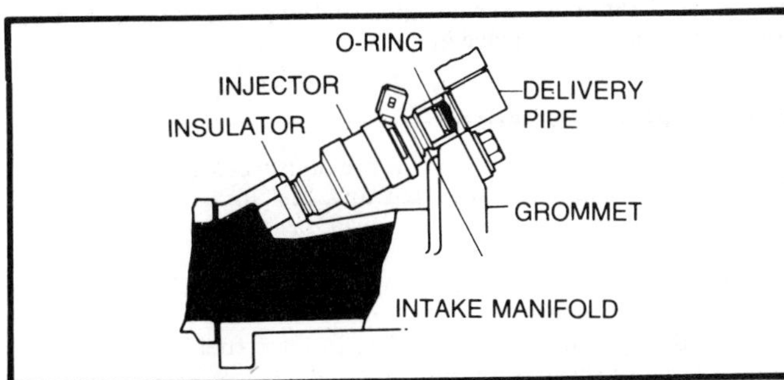

L-Jetronic fuel injector installation showing position of nozzle in the intake port

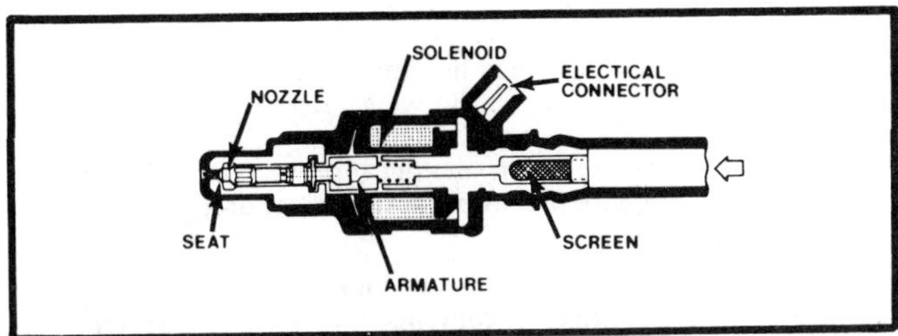

Cross section of a typical L-Jetronic fuel injector

TESTING THE FUEL INJECTOR

With the engine running or cranking, use an engine stethoscope to check that there is normal operating noise in proportion to engine rpm. Using a stethoscope or a screwdriver, you can listen to each injector. A regular clicking means it's working. The interval between clicks should shorten as engine speed is increased. If you don't hear an injector clicking, there may a poor electrical connection at that injector. Try switching injector wires with an adjacent good injector. If swapping wires brings the dead injector to life, look for a corroded or loose connection or a broken injector wire.

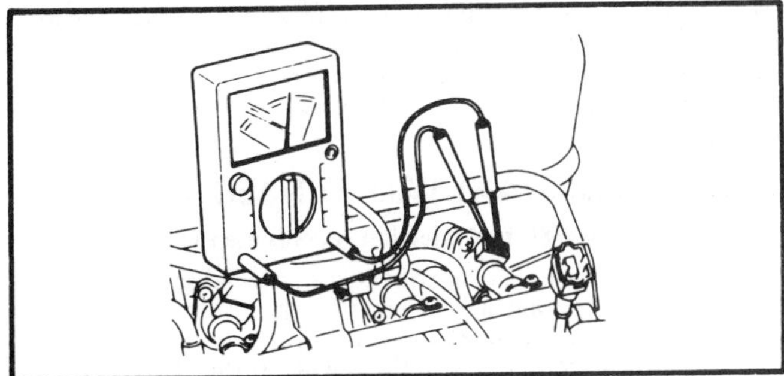

Testing the fuel injector windings with an ohmmeter

Early L-Jetronic systems apply 12 volts to the injectors, but later models usually use only 4 volts to energize the injectors. Check for voltage at the wire harness connector with a voltmeter.

NOTE: Be wary of smelling gasoline when starting a cold engine. Some EFI injectors have a bad habit of allowing fuel to gush out from the seam where the black plastic part of the injector mates with the colored plastic part. But as soon as they warm up, the leak stops.

Injector Winding Test

Use an ohmmeter to measure the resistance value of each injector by connecting the test leads to the terminals on the injector. Resistance should be 1.5–4.0 ohms. On LH II systems, connect ohmmeter between pump relay terminal 87 and control unit connector terminal 13. Resistance should be approximately 16 ohms across the injector terminals.

For continuity tests, remove the ground cable from the battery and disconnect the electrical connectors from the injectors. Check for continuity readings between the two injector terminals. If there is no indication, the injector is faulty.

CAUTION

Applying 12 volts directly to the fuel injector terminals can burn out the winding and ruin the injector. Do not apply battery voltage to fuel injectors under any circumstances.

Dead or Plugged Injectors

You can spot a dead or plugged injector several ways; perform a routine ignition and engine condition test on any EFI equipped

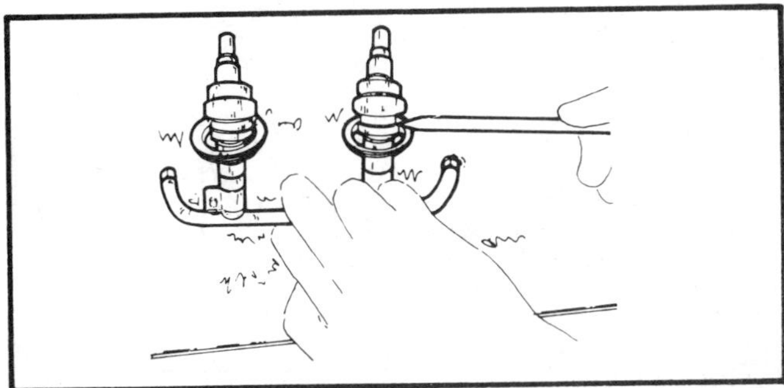

Pointer shows seam where leaks usually develop in fuel injectors

engine before digging into its fuel system. If the ignition is sound and there is little or no rpm drop when you short a cylinder, there could be a vacuum leak at the nozzle seal. Injector O-rings are a potential vacuum leak on any EFI equipment engine.

To determine if there is a leaking O-ring, use the same technique for finding standard intake manifold leaks: spritz the O-ring with some carb and choke cleaner and watch for a response on a vacuum gauge. Needle fluctuations indicate vacuum leaks.

Leaking Injectors

If the system doesn't hold pressure after crimping both the regulator and pump hoses, then an injector is leaking. Remove the entire fuel rail from the engine and switch the pump on and off again, then watch for the injector that's leaking fuel. The injector orifice may become wet, but no more than two drops should form on the valve (injector) per minute.

COLD START VALVE (INJECTOR)

NOTE: Not all AFC fuel injection systems use a cold start valve.

To assist cold starting, a separate cold start valve sprays a fine jet of fuel against the air stream entering the plenum (air intake) chamber before fuel is added to it by the main injectors. The cold start valve is energized from the engine starter motor circuit through a thermo-time switch. The purpose of the thermo-time switch is to ensure that the cold start valve will not be energized when the engine is at normal operating temperature or when the starter motor is used for prolonged periods when the engine is below operating temperature, preventing extra fuel being supplied to the engine when it is not needed. The thermo-time switch will isolate the cold start valve after 8 to 12 seconds.

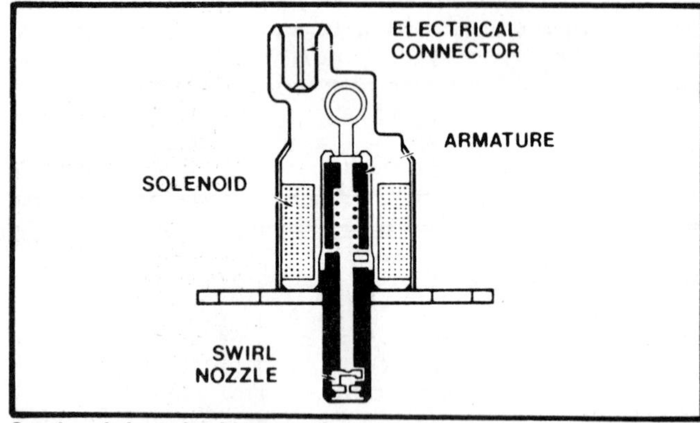

Sectional view of cold start valve

Testing

CAUTION

Fire hazard. Take precautions when performing all tests.

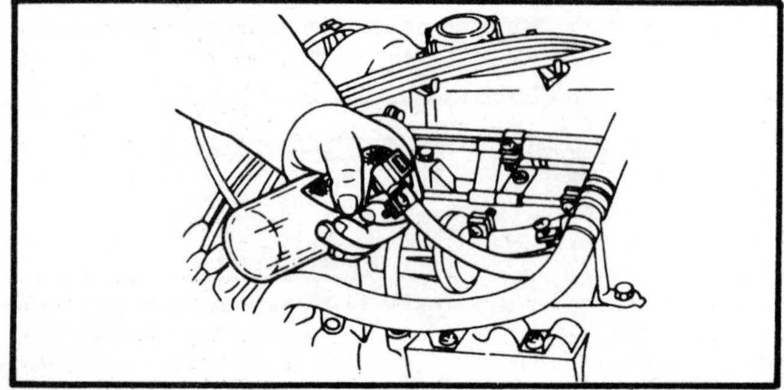
Place the cold start valve nozzle into a clear container when testing

1. Remove the screws holding the valve in the intake manifold. DO NOT disconnect the fuel lines or electrical connector.
2. Place the cold start valve in a container to catch fuel. Wrap a clean rag around the mouth of the container.
3. Operate the starter and note the injection time. Valve should spray fuel for 1–12 seconds if the coolant temperature is lower that approximately 35°C (95°F). Above this temperature, no drip or spray should be noted.
4. If the cold start valve sprays continuously or drips, replace it.
5. If the cold start valve fails to function below 35°C (95°F), replace it.

NOTE: Perform this test as quickly as possible. Avoid energizing the injector for any length of time.

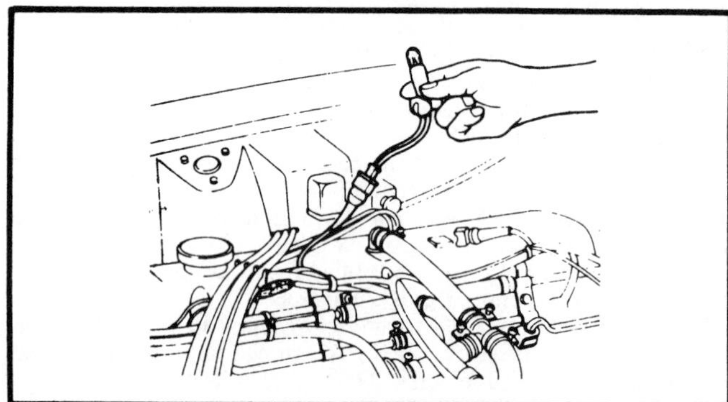
Using a fabricated test light to check the voltage signal from the wiring harness connector. Pulsed injector signals will cause the light to flash

6. Disconnect the cold start valve and hook up a test light across its connector. Ground the No. 1 coil terminal and run the starter. The light should glow for several seconds and then go out. If not, replace the thermo-time switch. Measure the resistance of the cold start valve using an ohmmeter. Correct resistance is 3–5 ohms. Check continuity across the cold start valve terminals.

NOTE: No starts or poor cold starting can be caused by a malfunctioning cold start valve. Cranking a cold engine with the coil wire grounded should produce a cone-shaped spray from the cold start valve. Cranking a warm engine should produce no fuel; if the injector dribbles gas, replace it.

Removal and Installation

1. Relieve fuel system pressure as previously described.
2. Provide a container to catch any fuel, or wrap clean rags around connections. *Use care to prevent dirt from entering the fuel system, and take precautions to avoid the risk of fire.*
3. Disconnect the electrical connector from the cold start valve. Clean any dirt or grease.
4. Remove fuel line from the valve. Pull the fuel hose off of the valve. Inspect the hose condition.

CAUTION

The valve body is plastic. Use care in removing fuel hose.

5. Remove the two fasteners holding the cold start valve in the intake manifold, then remove the valve. Remove and discard O-ring or gasket.
6. Reinstall in reverse order, making sure the system is tight and free from leaks. Replace the rubber sealing ring, or gasket and hose clamp if questionable.

NOTE: Fuel injection systems are highly susceptible to dirt in the system. Take care that all components are clean and free from dirt before reinstalling them.

THERMO-TIME SWITCH

NOTE: Not all AFC fuel injection systems use a thermo-time switch.

The thermo-time switch contains a bimetallic switch and heater coil. The switch completes the circuit for the cold start valve. The harness is connected in series to the cold start valve from the thermo-time switch. The bi-metal contact in the thermo-time switch opens or closes depending on the cooling water temperature, and sends a signal to the cold start valve so that an additional amount of fuel can be injected for starting operation of the engine (mixture enrichment).

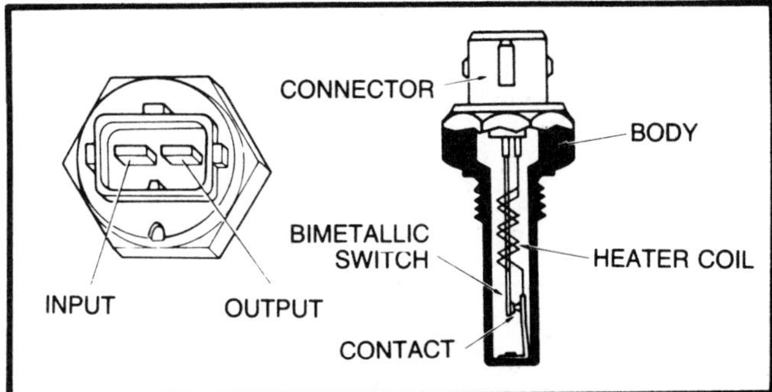

Typical thermo-time switch showing construction and wiring connector pins

During starting, power is supplied to the heater coil which warms the bi-metallic switch. The switch opens within 15 seconds regardless of coolant temperature. The lower the temperature, the longer for the coil to warm up. When the switch opens, the cold start valve is no longer energized and stops enriching the fuel mixture. To remove the thermo-time switch, disconnect the electrical connector and unscrew the switch from its mounting. It may be necessary to drain the cooling system or replace lost coolant.

CAUTION

Do not attempt to remove the thermo-time switch from a hot engine.

Testing

PRIMARY TEST

1. Remove the thermo-time switch and replace it with a plug to prevent loss of coolant liquid. Perform all tests on cold engine.
2. Cool the thermo-time switch by immersing it in cold water. Use a thermometer to check water temperature.
3. Connect the test wiring to the thermo-time switch as follows:
4. Connect one terminal to one of the wires of a test lamp. Connect the other end of the lamp to the positive (+) terminal of the battery.
5. Connect the negative (−) terminal of the battery to the body of thermo-time switch with a jumper wire. The lamp should light.
6. Place the thermo-time switch in a container of water and heat up the water while monitoring its temperature with a thermometer.

NOTE: Do not immerse test connections in water; just the lower end of the thermo-time switch should be submerged.

7. The lamp should go out between 31° and 39°C (88° and 102°F).
8. The nominal values of the thermo-time switch are marked on the six sides of the body (for example: 8 sec./35°C).

SECONDARY TEST

Check the resistance values with an ohmmeter between terminals of the thermo-time switch connector and ground for temperatures lower than 30°C (86°F) and higher than 40°C (104°F). Resistance should be around 50–80 ohms on switches marked 95°F/85; and 25–40 ohms on switches marked 35°C/85.

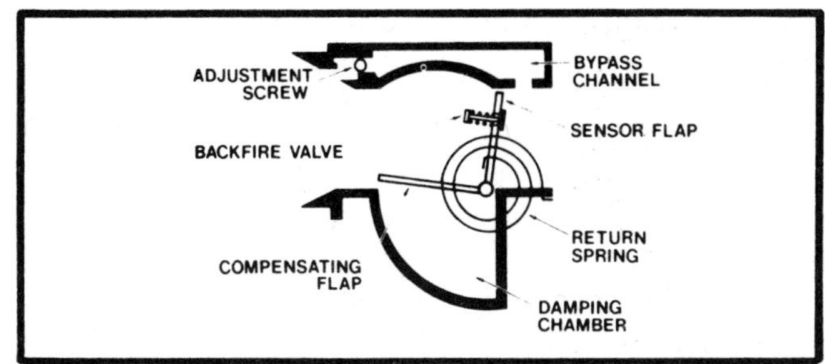

Schematic of an L-Jetronic air flow sensor. Note the sensor flap and idle air adjustment screw

AIR FLOW (OR AIR MASS) SENSOR

On L and LU injection systems, the air flow sensor measures the quantity of intake air, and sends a signal to the control unit so that the base pulse width (voltage signal) can be determined for correct fuel injection by the injector. The air flow sensor is provided with a flap in the air passage. As the air flows through the passage, the flap rotates and its angle of rotation electronically signals the control unit by means of a potentiometer.

The engine will draw in a certain volume of fresh air depending on the throttle valve position and engine speed. This air stream will cause the sensor plate inside the air flow sensor to move against the force of its return spring. The sensor plate is connected to the potentiometer, which sends a voltage signal to the control unit. The temperature sensor in the air flow sensor influences this signal. The

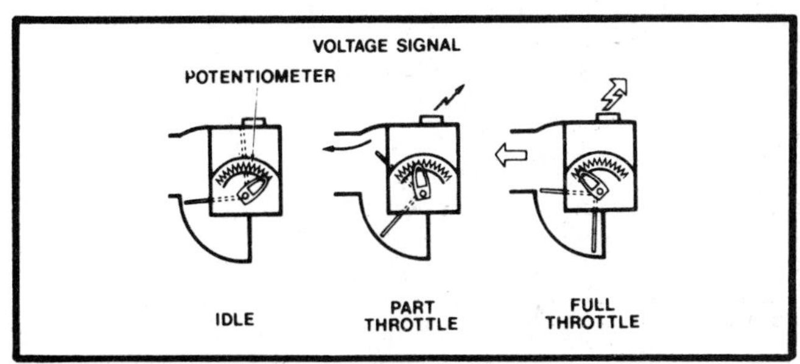

Potentiometer function on the air flow sensor. The voltage signal goes to the ECU

control unit then sends out an opening signal to the fuel injectors to make sure that the volume of injected fuel is exactly right for the volume of intake air. A damping flap in the air flow sensor eliminates unwanted movement of the sensor plate. As the sensor plate moves, the damping flap moves into its damping chamber, acting like a shock absorber for the sensor plate. A small amount of intake air volume moves around the sensor plate via a bypass port. The air/fuel mixture for idling can be adjusted by changing the amount of air flowing through the bypass port with the adjusting screw.

If the sensor plate or its attached damper should become stuck inside the air flow sensor, excessive fuel consumption, marginal performance or a no-start condition could result. LH, LH II and Motronic injection systems incorporate an air mass sensor into the air box. On these systems, the flap valve is replaced with a heated platinum wire that is used to measure air mass without altitude influences.

NOTE: Because of the sensitivity of the air flow (or air mass) meter, there cannot be any air leaks in the ducts. Even the smallest leak could unbalance the system and affect the performance of the engine. During every check, pay attention to hose connections, dipstick and oil filler cap for evidence of air leaks. Should you encounter any, take steps to correct the problem before proceeding with testing.

Removal and Installation

1. Disconnect the air duct hoses from both sides of the air box.
2. Disconnect the electrical connector from the wire harness.
3. Remove all bolts, lockwashers, washers and bushings holding the air flow sensor to the bracket. Remove the unit assembly.
4. Reinstall in reverse order.

Testing Air Flow Sensor

1. Connect an ohmmeter to any terminal on the flow meter. Touch the flow meter body with the other connector. If any continuity is indicated, the unit is defective and must be replaced.
2. Reach into the air flow meter and check the flap operation. If the flap opens and closes smoothly, without binding, the mechanical portion of the unit is working.

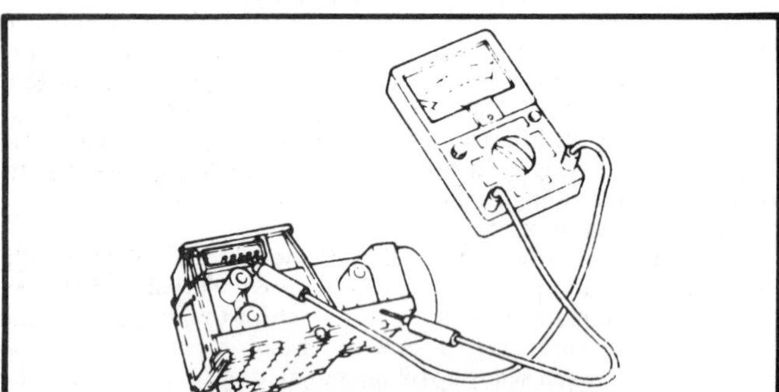

Checking the air flow meter insulation with an ohmmeter. Continuity indicates a problem

NOTE: If the air temperature sensor or potentiometer is malfunctioning, the entire air flow sensor must be replaced.

Testing Air Mass Sensor

NOTE: Replace the air mass sensor as an assembly if the test results indicate a malfunction in any of the following procedures.

1. Ground the main relay terminal No. 21 and make sure the relay switches ON (a slight click should be heard when grounding).
2. Peel back the rubber boot from around the air mass sensor harness connector, but leave it connected to the sensor assembly.
3. Connect a voltmeter between terminal 9 and ground. The voltmeter should read battery voltage (12 volts). Probe all terminals from the rear of the connector.
4. Connect the voltmeter between terminals 9 and 36; the reading should again be 12 volts.
5. If no voltage is present, check for an open circuit or broken wire in the power feed from the battery. If voltage is present, remove the ground connection from relay terminal 21 and proceed with testing. Leave the connector boot peeled back and probe all wire terminals from the rear of the wiring harness connector for all tests.

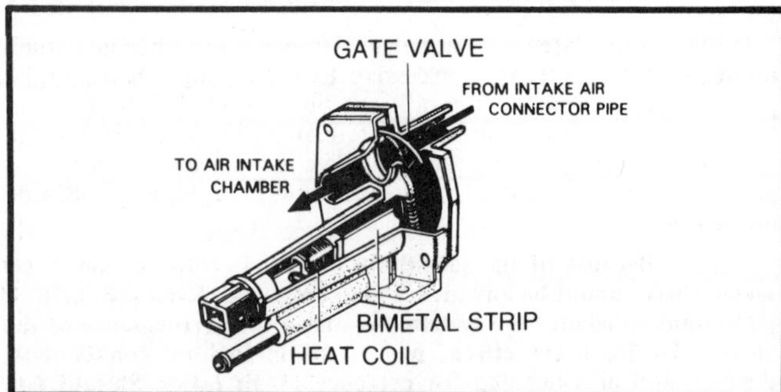

Typical auxiliary air regulator used on Japanese L-Jetronic systems

CAUTION

Harness connectors can be damaged by test leads if probed from the front, causing a poor connection problem that didn't exist before testing. Exercise caution during all electrical test procedures and never use excessive force during probing, removal or installation of wire harness connectors or components.

6. Connect an ohmmeter between terminals 6 and 7 on the air mass meter. Resistance should be about 4 ohms.
7. Connect the ohmmeter between terminals 6 and 14 and read the resistance value. It should be between 0–1000 ohms (resistance varies depending on CO adjustment screw positioning). If values are as described, proceed with testing.
8. Connect a voltmeter between terminal 7 (+) and terminal 6 (−) on the air mass sensor connector.
9. Start the engine and note the voltage reading.
10. The voltage should increase with engine rpm. Slowly increase and decrease engine speed while watching the voltmeter and make sure the voltage changes up and down. Specific values are not as important as the fact that the needle (or digital readout) on the voltmeter swings back and forth as the rpm changes. If the readings change as described, proceed with testing.

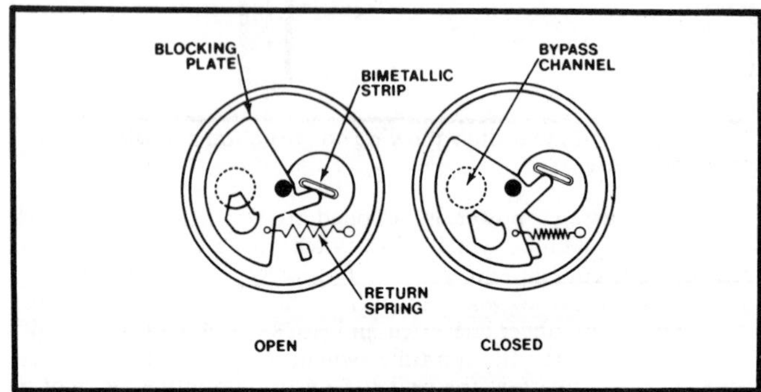

Operation of gate valve in auxiliary air regulator

11. Connect a voltmeter between terminals 8 (+) and 36 (−) to check the air mass sensor wire cleaning (burn-off) operation.
12. Start the engine and increase the engine speed to approximately 2500 rpm for a few seconds.
13. Switch off the engine. After about 5 seconds, the voltmeter should read approximately 1 volt for 1 second. Repeat this test a few times to verify the results.
14. If all tests are as described, remove the voltmeter and reposition the rubber boot cover around the air mass meter connector. Make sure the boot seals correctly to prevent moisture from corroding the electrical connectors.

AUXILIARY AIR REGULATOR

The auxiliary air regulator allows more air to bypass the throttle plate when the engine is cold, triggering a richer fuel mixture for better cold starting.

A rotating metal disc inside the valve contains an opening that lines up with the bypass hoses when the engine is cold. As the engine heats up, a bi-metallic spring gradually closes the opening by rotating the disc. The bi-metallic spring is heated by coolant temperature and a heater coil which is energized through the ignition circuit.

A double relay, usually located near the electronic control unit and grounded by a terminal of the control unit, controls the input to the fuel pump, the cold start valve, the injector positive feed, the control unit, and the auxiliary air regulator.

Removal and Installation

1. Disconnect the air valve hoses.
2. Disconnect the air valve connector.

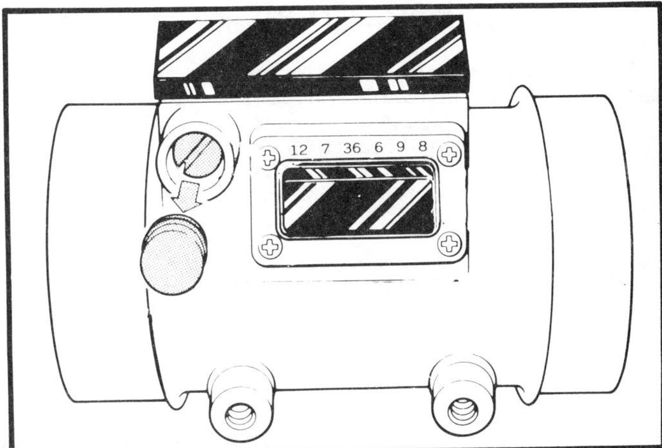

Location of adjustment screw and terminal numbers on air mass sensor

3. Disconnect the water bypass hose.
4. Remove the air valve attaching bolts and air valve.
5. Installation is the reverse of removal. Use a new gasket.

Testing

With the whole assembly at a temperature of about 20°C (about 68°F) and the electrical connectors and air hoses disconnected, visually check through the inside of the air lines to make sure that the diaphragm is partly open.

1. Using an ohmmeter, measure the resistance between the air valve terminals. Normal resistance is 29–49 ohms, depending on manufacturer's specifications.
2. Connect a voltmeter across the terminals of the connector and crank the engine.
3. If there is no voltage in the connector, look for a fault in the electrical system. If voltage is present, reconnect the valve and start the engine.
4. Check the engine rpm while pinching shut the air hose. At low temperature (below 140°F) the engine rpm should drop. After warm up, the engine speed should not drop more than 50 rpm.

AIR CONTROL VALVE

On AFC systems with an air mass sensor, an air control valve is used to regulate the idle speed by bypassing air around the throttle valve. The amount of air bypassed is determined by a signal from the electronic control unit (ECU). Replacement of the air control valve is similar to the auxiliary air valve removal procedure.

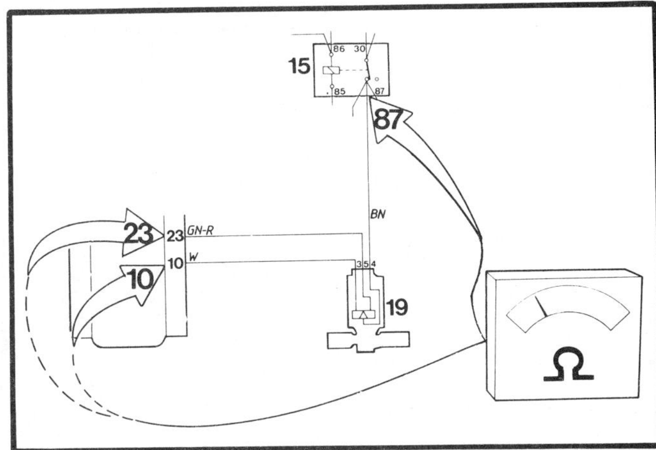

Terminal locations for testing air control valves

Testing

1. Connect an ohmmeter between the fuel pump relay terminal 87 and the control unit connector terminal 10. The reading should be approximately 20 ohms.
2. Connect the ohmmeter between relay terminal 87 and the control unit connector terminal 23. The resistance value should again be about 20 ohms.
3. Remove the ohmmeter and reconnect the harness connectors, if removed. Start and warm-up the engine.
4. Check the idle speed with a tachometer and compare the test results with the idle specification listed on the underhood emission control sticker.
5. Switch on the air conditioner (if equipped) and verify that the idle speed increases when the compressor is energized. If the idle doesn't increase, check for a sticking air control valve or failure of the A/C power circuit (shorted or open wire or microswitch).
6. Switch the engine OFF, then remove the connector from the coolant temperature sensor to simulate a cold engine.
7. Start the engine and verify that the high idle specification is obtained with a tachometer. Turn the ignition OFF and reconnect the coolant temperature sensor if the high idle is correct, then restart the engine and make sure the idle speed is once again at the normal (warm) value.

NOTE: If the air control valve is suspected of malfunction, install a new valve and repeat the test procedures. Adjust idle speed, if necessary.

THROTTLE SWITCH

The throttle switch forms part of the control for the fuel injection system, providing the ECU with information on throttle operating conditions. The switch is grounded at the intake manifold.

The throttle valve switch is attached to the throttle chamber and actuates in response to accelerator pedal movement. This switch has two or three sets of contact points. One set monitors the idle position and the other set monitors full throttle position. Some use additional contacts for mid-range operation.

The idle contact closes when the throttle valve is positioned at idle and opens when it is at any other position. The idle contact compensates for after idle enrichment, and sends the fuel shut-off signal.

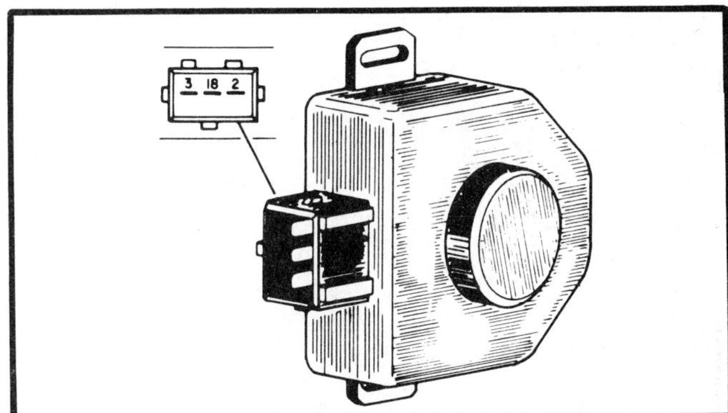

Typical throttle switch showing pin numbers

The full throttle contact closes only when the throttle valve is positioned at full throttle (more than 35 degree opening of the throttle valve). The contact is open while the the throttle valve is at any other position. The full contact compensates for enrichment in full throttle.

On AFC systems with a Constant Idle Speed (CIS) feature incorporated into the injection circuit, a microswitch is used in place of the throttle switch. The microswitch is mounted so as to contact the throttle linkage at idle and signal the ECU as to what throttle position is being used. Some microswitches are set to be open at idle and some are closed. Use a test light to determine which type of microswitch is being used by checking for voltage while moving the throttle lever to open and close the contact lever. The microswitch is usually adjusted when setting the idle speed.

Adjustment

NOTE: A click should be heard when the throttle valve moves and the idle contact opens

1. Make sure that the idle speed is correct. Check the underhood sticker for correct specification.
2. Disconnect the connector from the switch. The center pin is usually the 12 volt input terminal.

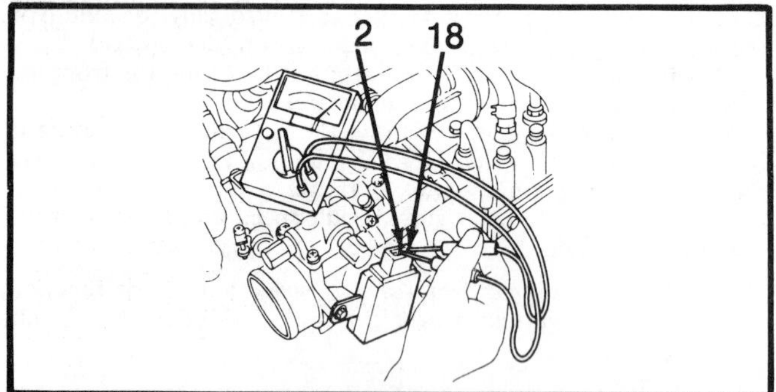

Checking the throttle switch adjustment with an ohmmeter

3. Connect an ohmmeter between terminals 2 and 18 of the switch.
4. Loosen the two screws holding the throttle plate switch.
5. With the engine off, rotate the switch clockwise until the ohmmeter indicates a closed circuit (idle contact closes).
6. At the exact point that the ohmmeter indicates a closed circuit, tighten the two screws holding the switch.
7. Check the adjustment.

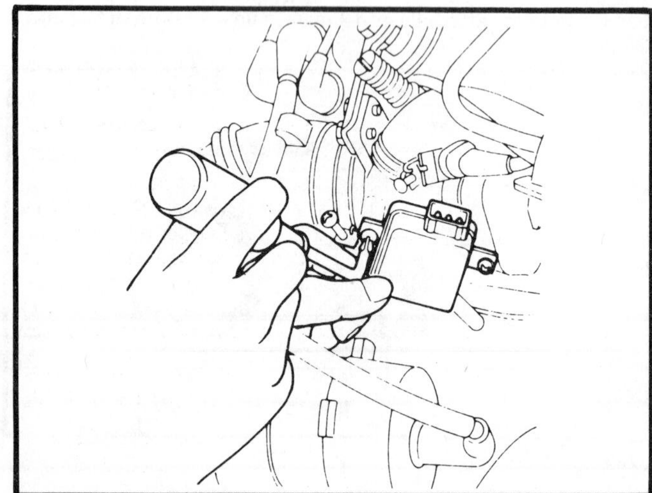
Adjust the throttle switch by loosening the fasteners

Removal and Installation

1. Disconnect the electrical connector.
2. Remove the two screws and washers holding the switch to the intake manifold.
3. Remove the switch by slowly pulling it.
4. Reinstall in the reverse order and adjust the switch.

VACUUM SWITCH—LH-JETRONIC SYSTEM

On some applications, a vacuum switch takes the place of the throttle valve switch to indicate full load and idle (low ported vacuum signal) and part load (maximum vacuum signal) conditions to the electronic control unit. Instead of being mounted directly on the throttle shaft, the vacuum switch is usually located on the bulkhead or fender panel under the hood and is connected to a ported vacuum orifice in the throttle valve assembly by a vacuum hose.

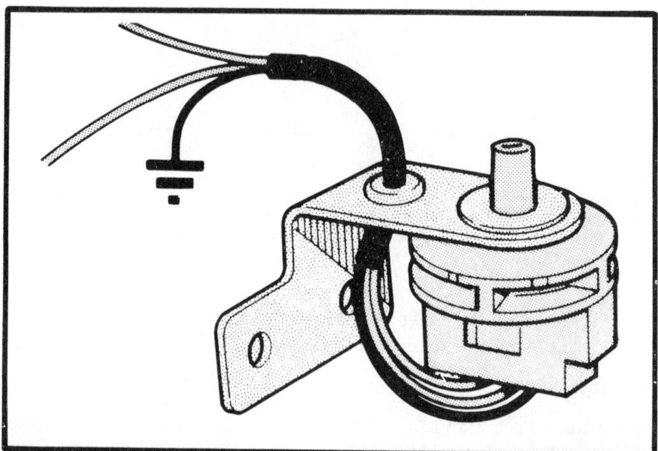
Typical vacuum switch

Testing

To test the vacuum switch, connect a hand vacuum pump to the vacuum fitting and connect an ohmmeter across terminals 3 and 5 at the control unit harness connector. With no vacuum applied to the switch, the resistance should be zero. When approximately 4 in. Hg. is applied with the hand vacuum pump, the resistance should be infinite. A high resistance reading without vacuum indicates an open circuit and zero resistance when vacuum is applied indicates a short circuit. In either case, if incorrect resistance is noted, replace the vacuum switch and repeat the test.

DETONATION SENSOR

On some AFC installations, a detonation sensor is incorporated into the system to detect engine knock conditions. Engine knock (or ping) can be caused by using fuel with too low an octane rating or advanced timing. The detonations within the cylinder set up vibrations within the cylinder block that are detected by a piezoelectric element which converts vibrational pressure into an electrical signal (voltage impulse) which is used as an input signal to the control unit. The ECU uses the signal to measure detonation

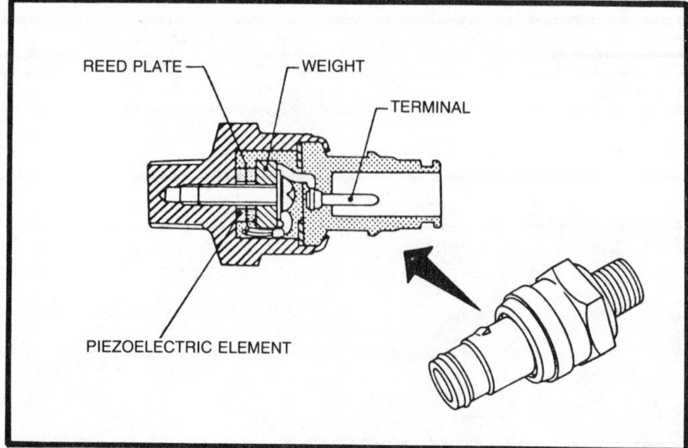

Cross section of typical detonation sensor

and retard the timing to eliminate the problem. A detonation sensor is usually found on turbocharged engines, since serious engine damage can occur if turbocharged engines detonate under boost.

Testing

To test the detonation sensor operation first locate the sensor, remove the harness connector and check for continuity with an ohmmeter. If continuity exists, reconnect the sensor and start the engine after installing a timing light. When the engine bloc is tapped with a small hammer near the detonation sensor, the timing should change. Repeat the test several times and make sure the ignition timing changes (retards) when the block is tapped and returns to normal when the tapping stops. If no change in the timing is noted and the sensor shows continuity, the problem is in the ECU, or the wiring harness.

FUEL TEMPERATURE SENSOR

On some models, a fuel temperature sensor is built into the pressure regulator assembly to monitor fuel temperature. If the temperature rises beyond a preset value, the control unit enriches the mixture to compensate.

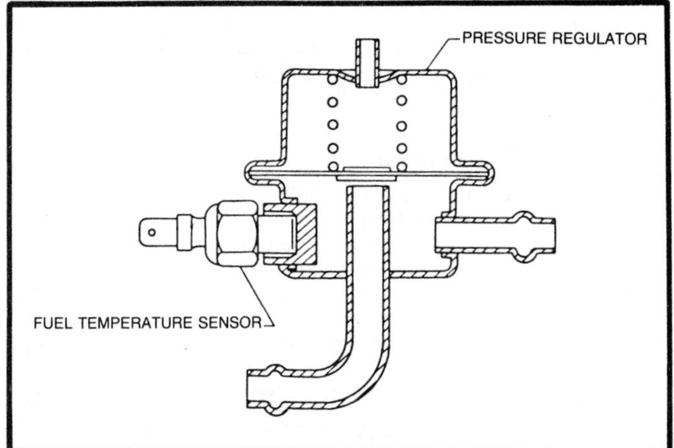

Location of fuel temperature sensor on pressure regulator

Testing

The fuel temperature sensor is tested in the same manner as a coolant sensor. Connect an ohmmeter and check that the resistance value (ohms) changes as the temperature rises. On the Nissan ECCS system, for example, the resistance should fall as the temperature rises. If the temperature sensor checks out, perform a continuity test on the wiring harness connectors at the switch and control unit. If no problem is found, the control unit itself may be malfunctioning.

NOTE: If found to be defective, the fuel temperature sensor and fuel pressure regulator should be replaced as an assembly.

SPEED SENSOR

On some models, a speed sensor is used to provide a signal to the control unit on vehicle road speed. Two general types of sensors are used, depending on whether the instrument cluster is an analog (needle) type or digital (LED) readout type of display. In the analog speedometer, the speed sensor is usually a reed switch which transforms vehicle speed into a pulse signal. On digital type instrument clusters the speed sensor consists of an LED (light emitting diode), photo diode, shutter and wave forming circuit (signal generator) whose operation closely resembles the crank angle sensor described under "ECCS Control Unit" later in the chapter.

Testing

Regardless of instrument cluster type, the easiest method of checking the speed sensor operation is to disconnect the speedometer cable at the transmission and connect an ohmmeter between the control unit harness connector terminal from the speed sensor (e.g. terminal No. 29 on Nissan 300 ZX models) to ground. Check for continuity while slowly rotating the speedometer cable. The ohmmeter should register the pulses from the speed sensor. If no reading is obtained, check the wiring harness to the speed sensor for continuity.

NOTE: Rotate the speedometer slowly and watch for meter changes. Most speed sensors send impulses about 24 times per revolution of the speedometer cable. If found to be defective, the speed sensor is usually replaced with the speedometer assembly as a unit, although some separate sensor assemblies are used in digital instrument cluster displays. Digital cluster speed sensors are checked with the key ON.

BOOST SENSOR SWITCH

TURBOCHARGED MODELS ONLY

On some turbocharged engines, a vacuum-operated boost sensor provides a signal to the ECU for fuel enrichment or electronic wastegate control while under boost. Some manufacturers use vacuum switches to provide protection against a dangerous overboost situation by cutting off the fuel injectors above a certain boost pressure. Boost sensor switches may be located in the engine compartment, or behind the instrument cluster.

Testing

To test vacuum actuated boost pressure switches, connect a hand-pumped vaccum gaugė to the switch vacuum port and an ohmmeter to the terminals of the sensor connector. Pump the vacuum gauge and check for continuity between the terminals when the vacuum is applied. There should be no continuity when the vacuum is absent, and the resistance between the terminals should vary as the vacuum level applied rises and falls. The switch can be inspected in the same manner by using a voltmeter and connecting battery voltage to one side of the switch. Voltage should vary just as resistance did, when the vacuum level changes. Look for changes rather than specific volt or ohm values to determine if the switch is functioning.

COOLANT TEMPERATURE SENSOR

NOTE: Most later models are equipped with cylinder head temperature sensors rather than water temperature sensors. However, the test is the same for both units.

The coolant temperature sensor is located near the cylinder head and provides a signal to the ECU that causes a longer injector open time when the engine is cold. The open time decreases as the engine reaches normal operating temperature. The coolant temperature sensor also completes the circuit for the extra air valve when the engine is cold. Typical resistance values at different temperatures are:

Temperature	Resistance (Ohms)
+80°C (175°F)	270–390
+20°C (68°F)	2100–2900
−10°C (14 °F)	7000–116000

Note: Exact resistance values vary with manufacturer.

The cylinder head temperature sensor, built into the cylinder head, monitors change in cylinder head temperature and transmits a signal to increase the pulse duration (fuel delivery) during the warm-up period. The temperature sensing unit employs a thermistor which is very sensitive in the low temperature range. The elec-

Typical coolant temperature sensor used on L-Jetronic systems

trical resistance of the thermistor decreases in response to the temperature rise.

Removal and Installation

CAUTION

Allow the engine to cool before attempting this procedure.

Partially drain the cooling system. Then, disconnect the electrical terminal plug and unscrew the coolant temperature sensor. Install in reverse order, refilling and bleeding the cooling system.

Testing

In general, coolant and cylinder head temperature sensors have a fairly high mortality rate on EFI systems. VW's, for example, use up cylinder head sensors often. On L-Jetronic systems, the intake air temperature sensor is incorporated into the air flow meter. If it fails, replace the air flow meter. It's impossible to tune or adjust an EFI equipped engine if one of its sensors is bad. Check that the temperature sensor resistance, measured with an ohmmeter, changes when the sensor is checked cold, then after the engine is warmed up. Engine should be off for all tests. Extremely high test values indicates open circuit in temperature sensor or wiring (check temperature sensor ground at intake manifold). Zero resistance indicates short circuit.

Sensor Insulation Check

This check is done on the engine.

1. Disconnect the battery ground cable.
2. Disconnect the sensor harness connector.
3. Connect an ohmmeter to one of the terminals on the sensor and touch the engine block with the other. Any indication of continuity indicates a need to replace the unit.

LAMBDA (OXYGEN) SENSOR

The Lambda or Oxygen sensor is a probe with a ceramic body in a protective housing. It measures the oxygen concentration in the exhaust system and provides a feedback signal to the control unit.

The exhaust gas sensor produces a signal depending on air/fuel mixture ratio. The signal varies directly with the density of oxygen in exhaust gases. The fuel is burned at the best theoretically determined air/fuel ratio of the mixture; the oxygen sensor signal increases when there is a richer mixture, and decreases when there is a lean mixture. The signal from the sensor allows the ECU to fine tune the fuel mixture to allow the catalytic converter to operate at its peak efficiency.

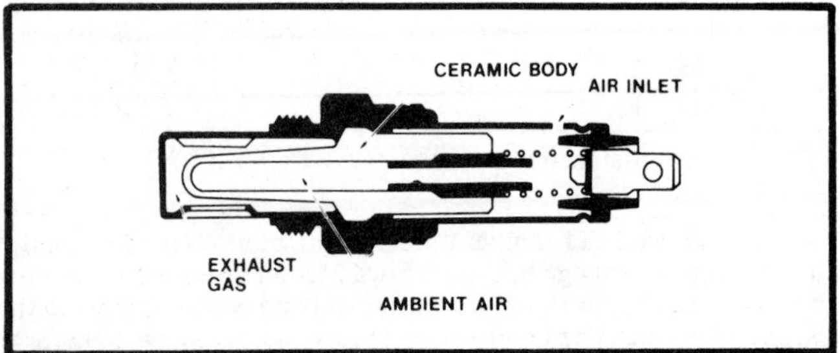

Cross section of a typical oxygen sensor

NOTE: As with the catalytic converter, the use of leaded fuel will contaminate and ruin the Lambda sensor.

The idle switch will turn off the oxygen sensor when coasting due to the excessive oxygen present in the exhaust gas. The idle contact will open when the free play in the accelerator linkage is taken up, even though the throttle valve is still closed. The full throttle contact switches off the oxygen sensor if the throttle is opened more than 30° in order to lower exhaust gas temperature and protect the sensor and the catalytic converter. The control unit will simultaneously switch to a 12% partial and full throttle enrichment.

NOTE: A service interval counter, much like the EGR and Catalytic Converter counters will light a dash warning light when the need for Oxygen Sensor replacement arises (around 30,000 miles). The service indicator must be reset after replacement of the lambda sensor.

Since the oxygen sensor doesn't function until it reaches operating temperature, late models incorporate a heater circuit to warm up the sensor for quicker response and to keep the oxygen sensor at operating temperature during prolonged idle or low speed operation. Oxygen sensors with heaters are recognized by the third wire connector on the sensor body. Do not interchange old and new type sensors.

CAUTION

Testing the oxygen sensor output with a powered voltmeter can damage the sensor by drawing too much current through the sensor. The average oxygen sensor puts out about one volt in operation.

Removal and Installation

1. Allow the exhaust system to cool.
2. Disconnect the cable from the oxygen sensor.
3. Remove the sensor from the exhaust manifold. It may be necessary to use penetrating oil or rust solvent.
4. Coat the threads of the sensor with anti-seize, anti-rust grease. Do not get grease on the sensor body itself.
5. Thread the sensor into the exhaust manifold and torque to 30-36 ft. lb.
6. Connect the electrical cable to the sensor and check the CO.

CAUTION

Do not allow grease to get on the sensor surface. This will contaminate and ruin the Lambda sensor. Exercise care when coating the threads with anti-seize compound. Check CO with suitable meter.

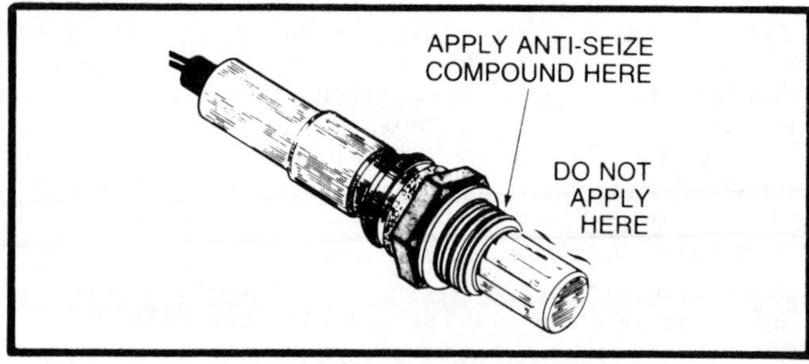

Use anti-seize compound when installing the oxygen sensor

Electronic Control Unit (ECU)

GENERAL INFORMATION

The electronic control unit is the brain of the AFC fuel injection system. It controls various engine operating conditions by constantly measuring specific engine parameters (e.g. temperature, rpm, air mass or volume, etc.) and making adjustments by sending out appropriate voltage signals to components. Each ECU

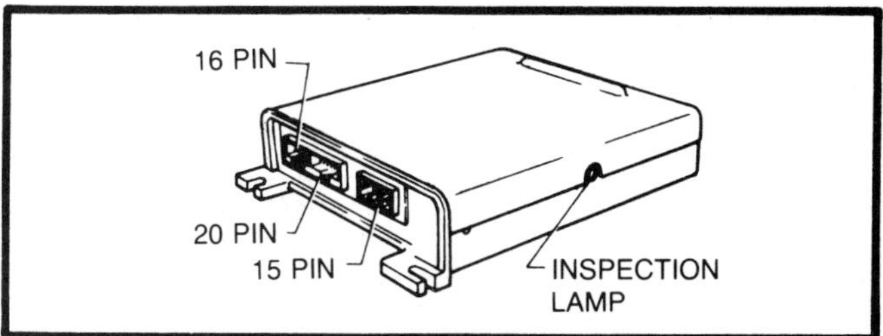

E.C.C.S. control unit used on Nissan models. Except for the pin connectors, all ECU units look pretty much alike

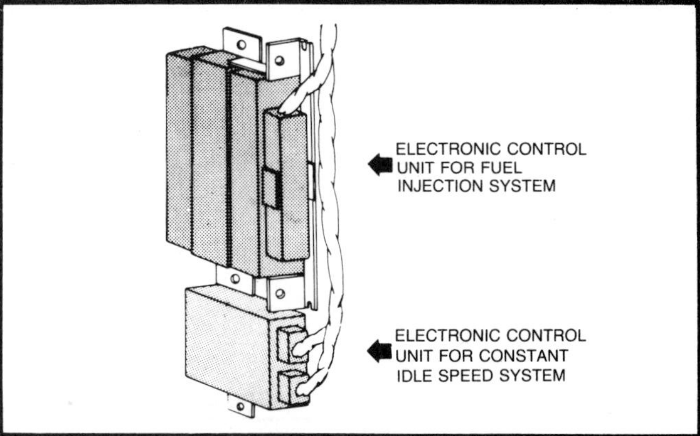

Typical control unit installations

responds to its sensor inputs according to a specific design of its internal integrated circuits, or according to manufacturers instructions programmed into a PROM (Programmed Read Only Memory) microchip computer memory plugged into the solid state circuit board. The number and type of sensors employed by a particular ECU and the manner in which it responds to these sensor input signals is unique to each type of system and different from one manufacturer to another. For this reason, the ECU's are not interchangeable, even if the same type of injection system (L-Jetronic, for example) is used on two different models built by the same manufacturer. When replacing any ECU it is important to correctly identify not only the type of injection system and manufacturer, but also the year and model of car, engine and transmission. Carefully compare the part numbers of both old and new ECU to make sure the correct control unit is being installed or serious damage to the replacement part is likely.

Depending on the type of fuel injection system, the ECU is designed to do different things. The basic L-Jetronic system is designed to control the fuel mixture under various engine load conditions and not much else. Later modifications incorporate other functions such as ignition timing control, EGR and other emission control device operation, idle speed, etc. Some ECU's operate with analog electrical signals while others use digital signals. The manner in which the AFC injection systems measures the intake air can vary from a counterbalanced air flap (L-Jetronic) to a heated platinum wire (LH-Jetronic) and the type and number of engine sensors will vary from one manufacturer to another. Some injection systems (Motronic and the E.C.C.S. system used on the Nissan 300ZX, for example) utilize sensors that measure the crankshaft angle to precisely control ignition timing and injection according to engine dynamometer data programmed into the control unit by the manufacturer. In addition, some injection systems are designed with a self-diagnosis capability that flashes a trouble code to indicate various malfunctions detected during operation. Before attempting any service or diagnosis procedures, familiarize yourself with the specific type of injection system used, as well as the various components, to determine what may be causing a problem and therefore what is in need of testing.

There are certain similarities common to most AFC systems that should be noted when testing. The service precautions outlined at the beginning of this section should be read and carefully followed to avoid causing damage to the control unit or electrical system during service. All ECU's are installed somewhere inside the passenger compartment and all use a large, multi-pin connector to tie the ECU into the wiring harness. This ECU connector is held in place with some form of locking device that must be released before the connector can be removed.

CAUTION

Never attempt to disconnect the ECU with the key ON or the engine running. Serious damage to the ECU can occur if the electrical connectors on any components are allowed to arc or the wrong connections are made when performing tests with a volt/ohmmeter or jumper wire.

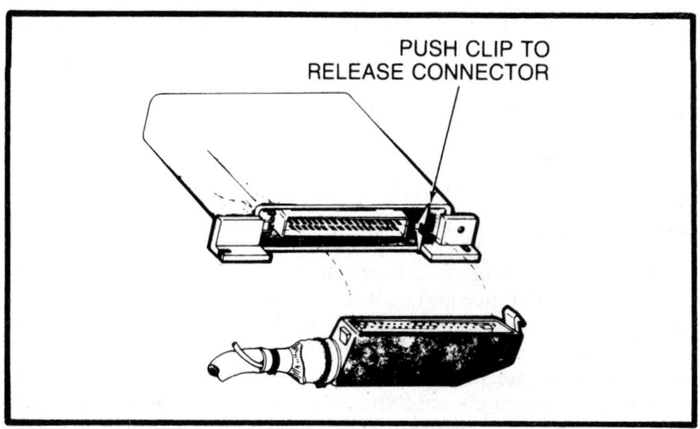

Never disconnect the ECU connector with the key On or the engine running. Exercise care when handling the ECU connector so as not to damage the contact pins

The pin numbers on the ECU and its connector will vary from one model to another, so care must be taken to make sure the correct connections are being made before any test is attempted. These numbers are usually stamped on the side of the main connector. Do not probe the ECU connector terminals from the front, as test probes can damage the connectors. Probe all ECU connections from the rear of the connector, after removing the harness cover, in the test holes provided or with a thin paper clip carefully installed from the wire side to allow the use of small clips. Be careful not to damage any connector during test procedures and never use excessive force when dealing with any electrical component. The control unit should be handled with care when removed and protected from contact with any kind of solvents, water or direct contact with any electrical or magnetic devices. Never apply power directly to a control unit on the bench for any reason.

Most ECU's require special test equipment for pinpoint troubleshooting; however some simple pass/fail procedures are possible using an appropriate tach/dwell meter, volt/ohm meter and fabricating jumper wires and test bulbs with the correct terminal connections. Not all analog (meter) test equipment is compatible with all electronic control systems and a digital multimeter may be necessary for some procedures. Consult the manufacturer's instructions before connecting any powered test equipment to make sure it won't damage the ECU. When testing with an analog volt/ohm or tach/dwell remember that it is usually more important to look for needle movement (swing) than exact test values. Be sure to perform the preliminary checks on fuel, compression and ignition systems before assuming that the ECU is causing a problem. Experience usually shows that simple problems like loose or corroded connections, clogged air, fuel or emission control filters and vacuum or air leaks in the intake system cause most of the problems blamed on a faulty electronic control unit.

Because of their solid state construction and the fact that there are no moving parts, most ECU's are designed to last the life of the car in operation, and are maintenance-free. The ECU should be removed if any AFC-equipped vehicle is to be placed in a paint oven or exposed to any temperature exceeding approximately 176°F. Never attempt to disconnect any control unit with the ignition switch ON or the engine running, and inspect all connectors for damage before reconnecting the wiring harness.

ECU DIAGNOSIS

Quick Test

By fabricating a small test light with an appropriate injector harness connector, the injector signal from the ECU can be ckecked. With the key OFF, connect the test lamp to the harness side of the fuel injector connector. Crank the engine and make sure the light flashes on and off. If the test lamp flashes, it indicates the control unit is supplying a pulse voltage to the injector and chances are it is operating properly.

NOTE: Because two different transistors are used in some AFC electronic control units, it may be necessary to test both the No. 1 and No. 4 cylinder injectors to determine if the control unit is supplying pulses to all of the injectors.

Circuit Tests

Individual control circuits within the ECU can be checked by performing resistance and continuity tests according to the following charts and connector pin locations illustrated. Read the service precautions concerning the ECU before attempting any tests and make sure the test equipment used will not damage the control unit. Make absolutely sure you are probing the correct terminals before making any connections, as serious ECU damage can occur if the wrong two pins are connected in a circuit.

CAUTION

On the Volvo LH-Jetronic system, the test point located behind the battery is NOT to be used in fault tracing the system. Unless special test equipment is used, using this test point can cause serious ECU damage. If equipped with a Constant Idle Speed system, there may be two test points.

L-JETRONIC TEST SPECIFICATIONS

Circuit Tested	Terminal Connection	Correct Value (ohms)
	OHMMETER TESTS	
Wire to coil terminal #1	1-GND	Wire grounded-0 Coil disconnected-∞
Idle circuit through throttle switch	2-18	Throttle closed-0 (Type IV only)
Full throttle enrichment through throttle switch	3-18	0 @ WOT
Ground circuit	5-GND	0
Air sensor circuit	6-9 6-8 8-9 6-7 7-8 6-27	200-400 130-260 70-140 40-300 100-500 2800 @ 68°F
Head sensor	13-GND	2100-2900 @ 68°F 270-390 @ 176°F
Injector wire and resistor	14-10 15-10 32-10 33-10	7 7 7 7
Ground circuit	16-GND 17-GND	0 0
Auxiliary air regulatory and wiring	34 @ ECU- 37 @ relay	30
	VOLTMETER TESTS	
Signal from starter	4-GND	12v-Cranking 0v-Otherwise
Voltage to ECU	10-GND	12v-Key ON 0v-Key OFF
Fuel pump circuit	28-GND	12v-Key ON and sensor flap open

NOTE: Probe all terminals carefully from the rear. Test probes can damage terminals and cause poor connections. Make sure test meters are compatible with injection system

ECU—Electronic Control Unit WOT—Wide Open Throttle
GND—Ground

LH-JETRONIC TEST SPECIFICATIONS

Circuit Tested	Terminal Numbers	Test Results
Battery voltage from ignition switch to ECU	20-GND ①	key ON-12 volts key OFF-0 volts
System Relay test	34-GND ③ 10-GND ①	Voltmeter should indicate 12 volts
Fuel Pump Relay	28-GND ③	Fuel pump runs
Ground Check ④	5-GND ② 16-GND ② 17-GND ②	Resistance should be 0 ohms at all points

NOTE: Test connector located behind the battery is not to be used for fault tracing. Special test equipment must be used or system will be damaged.

GND—Ground
ECU—Electronic Control Unit
① Voltmeter connections
② Ohmmeter connections
③ Jumper wire connections
④ Ground circuits can be checked with 12v test light by connecting one end to the positive battery cable and probing connectors 5, 16 and 17. Test light should come on.

LH II-JETRONIC TEST SPECIFICATIONS

Circuit Tested	Terminal Numbers	Test Results
Battery voltage from ignition switch to ECU	18-GND ①	key ON-12 volts key OFF-0 volts
Fuel Pump relay	17-GND ③	key ON-pump runs
System Relay	9-GND ① 21-GND ③	Voltmeter should read 12 volts. Relay is on
Ground Check	11-GND ② 25-GND ②	0 ohms (no resistance) for both terminals
Ignition coil connection	1-GND ①	Meter should deflect when ignition is switched ON
Air Control Valve circuit	④	20 ohms

NOTE: Terminal numbers should be marked on connectors. Turn ignition OFF when removing or installing connectors. All terminals are at ECU connector.

ECU—Electronic Control Unit GND—Ground
① Voltmeter connections
② Ohmmeter connections
③ Jumper wire connection
④ Connect ohmmeter between pump relay terminal 87 and ECU connector terminal 10; then between pump relay terminal 87 and ECU connector terminal 23. Readings should match.

PORSCHE 928 TEST SPECIFICATIONS

Circuit Tested	Terminal Test Connections	Normal Reading
	OHMMETER TESTS	
Temperature sensor	5-13 ①	2-3 kΩ @ 68 deg. F 1.2-2.4 kΩ @ 83 deg. F
Idle microswitch A	2-18 ①	2-3 kΩ @ 68 deg. F 1.2-2.4 kΩ @ 83 deg. F
	2-18 ②	0Ω
Full throttle microswitch B	3-18 ①	0Ω
	3-18 ③	0 Ω @ 68 deg. F 0-10 Ω @ 83 deg. F
Air flow sensor ④	6-9	400-800 Ω
	6-8	260-520 Ω
	8-9	140-280 Ω
	6-7	80-600 Ω
	7-8	200-1000 Ω
Air flow temperature sensor	27-6	2-3 kΩ @ 68 deg. F 1.2-2.4 kΩ @ 83 deg. F

PORSCHE 928 TEST SPECIFICATIONS

Circuit Tested	Terminal Test Connections	Normal Reading
Fuel injectors ⑤		
1 + 5	10-15	1-1.5 Ω
4 + 8	10-14	1-1.5 Ω
3 + 7	10-32	1-1.5 Ω
2 + 6	10-33	1-1.5 Ω
Ground checks	16-GND	0Ω
	17-GND	0Ω
	35-GND	0Ω
	5-GND	0Ω
	VOLTMETER TESTS	
Test conditions ⑥		
IGN ON	1-GND	1 volt
IGN ON	10-GND	12 volts
IGN ON	29-GND	12 volts
Cranking	4-GND	8 volts (minimum)
Cranking	1-GND	2.5 volts

NOTE: When conducting tests, look for a change rather than specific values. ALl voltage values are approximate.

GND-Ground Ω-Ohms (kΩ-kilo-ohms)
① Accelerator pedal released
② Throttle linkage play eliminated
③ Throttle open more than 1/3 (30 deg.)
④ As of production month 042 or later. For earlier production sensors, divide test values in half.
⑤ 2-3 ohms when each injector is checked separately.
⑥ All voltage values are approximate.

Self-Diagnosing AFC Injection Systems

Some AFC fuel injection systems are equipped with a self-diagnosis capability to allow retrieval of stored trouble codes from the ECU memory. The number of codes stored and the meaning of the code numbers varies from one manufacturer to another. By activating the diagnostic mode and counting the number of flashes on the CHECK ENGINE or ECU lights, it is possible to ask the computer where the problem is (which circuit) and narrow down the number of pin connectors tested when diagnosing an AFC fuel injection problem.

NISSAN E.C.C.S. SYSTEM

On 1984 and later models, the E.C.C.S. control unit used on the 300ZX has a self-diagnosing capability. The E.C.C.S. control unit consists of a microcomputer, connectors for signal input and output and power supply, inspection lamps and a diagnostic mode selector. The control unit calculates basic injector pulse width (fuel delivery) by processing signals from the crank angle sensor and air flow meter. The crank angle sensor monitors engine speed and piston position and sends signals to the ECU for control of fuel injection, ignition timing, idle speed, fuel pump operation and EGR (emission control) operation.

The crank angle sensor consists of a rotor plate and wave forming circuit built into the distributor. The rotor plate has 360 slits for a 1° (engine speed) signal and 6 slits for a 120° signal (crankshaft angle). Light Emitting Diodes (LED's) and Photo Diodes are built

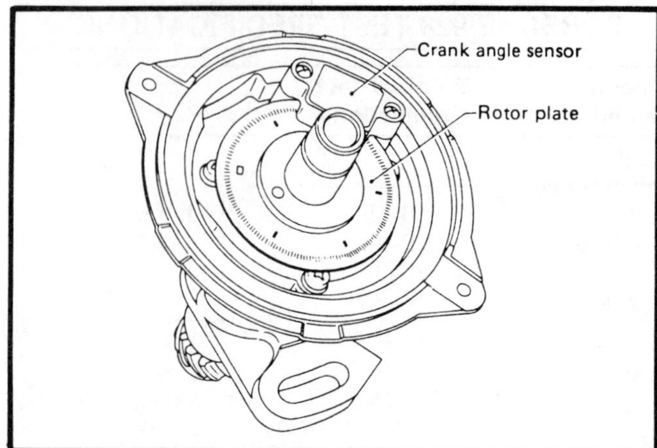

Crank angle sensor used on Nissan E.C.C.S. system

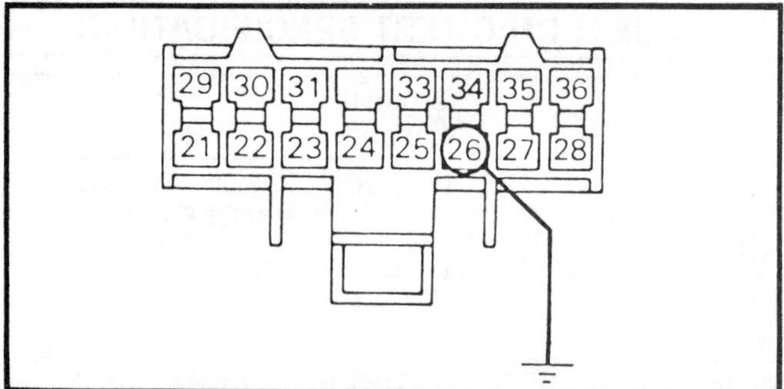

Terminal numbers on 16 pin connector for Nissan E.C.C.S. system

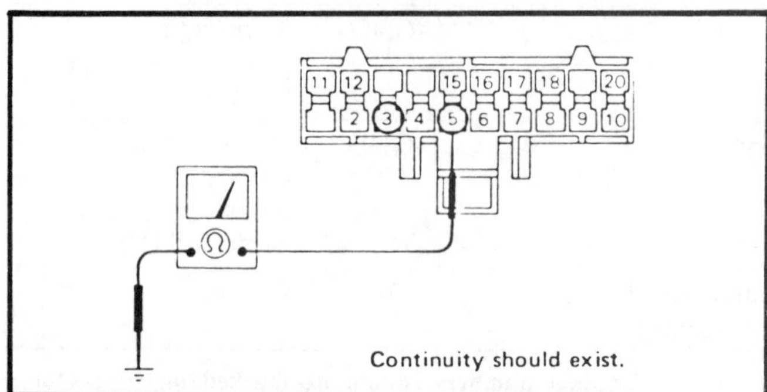

Terminal numbers on 20 pin connector for Nissan E.C.C.S. system. Test ignition signal by checking continuity between terminals 3, 5 and ground

into the wave forming circuit. When the rotor plate cuts the light which is sent to the photo diode from the LED, it causes an alternate voltage which is converted into an on-off pulse by the wave forming circuit and sent to the control unit. Enrichment rates are pre-programmed into the control unit for engine speed and basic injection pulse width.

NOTE: The ECU will shut off the injectors if the engine speed exceeds 6500 rpm, or road speed exceeds 137 mph, whichever comes first. The crank angle sensor is an important component in the E.C.C.S. system and a malfunctioning sensor is sometimes accompanied by a display which shows faults in other systems. Check the crank angle sensor first.

The self-diagnostic system determines the malfunctions of signal systems such as sensors, actuators and wire harness connectors based on the status of the input signals received by the E.C.C.S. control unit. Malfunction codes are displayed by two LED's (red and green) mounted on the side of the control unit. The self-diagnosis results are retained in the memory chip of the ECU and displayed only when the diagnosis mode selector (located on the left side of the ECU) is turned fully clockwise. The self-diagnosis system on the E.C.C.S. control unit is capable of displaying malfunctions being checked, as well as trouble codes stored in the memory. In this manner, an intermittent malfunction can be detected during service procedures.

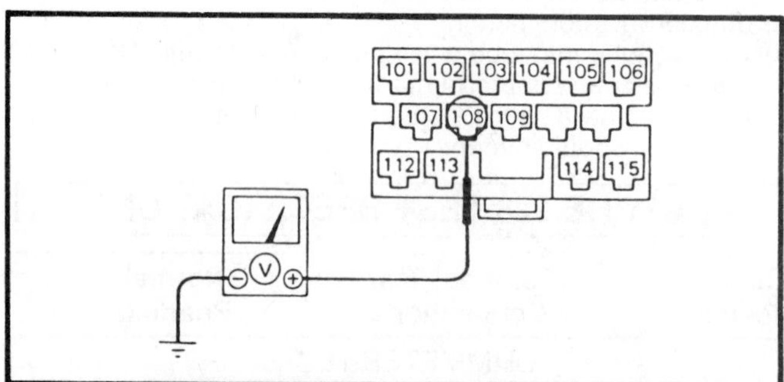

Terminal numbers on 15 pin connector for Nissan E.C.C.S. system. Test the fuel pump circuit by checking for current (with the key ON) between terminal 108 and ground

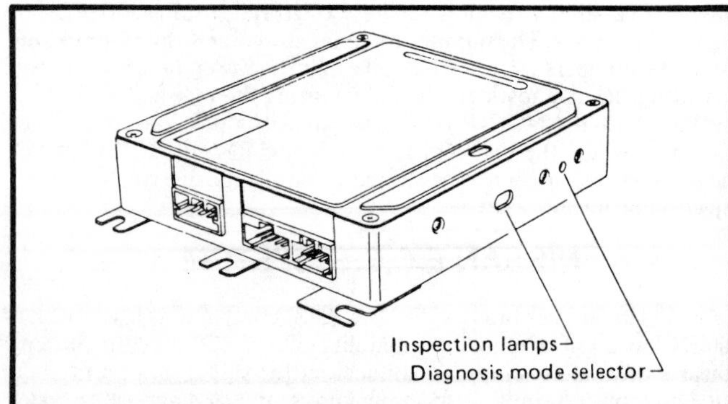

Control unit for Nissan E.C.C.S. system showing location of diagnostic mode selector and readout light

CAUTION

Turn the diagnostic mode selector carefully with a small screwdriver. Do not press hard to turn or the selector may be damaged.

Service codes are displayed as flashes of both the red and green LED. The red LED blinks first, followed by the green LED, and the two together indicate a code number. The red LED is the tenth digit, and the green LED is the unit digit. For example, when the red light blinks three times and the green light blinks twice, the code displayed is 32. All malfunctions are classified by code numbers. When all service procedures are complete, erase the memory by disconnecting the battery cable or the ECU harness connector. *Removing the power to the control unit automatically erases all trouble codes from the memory. Never erase the stored memory before performing self diagnosis tests.*

NISSAN E.C.C.S. TROUBLE CODES

Code Number	ECU Circuit	Test Point Pin Numbers	Normal Test Results
11	Crank Angle Sensor	Check harness for open circuit	Continuity

NISSAN E.C.C.S. TROUBLE CODES

Code Number	ECU Circuit	Test Point Pin Numbers	Normal Test Results
12	Air Flow Meter	Ground terminal 26 ①connect VOM @ 26-31 Apply 12v @ E-D ② connect VOM @ B-D VOM @ 12-GND ③ VOM @ C-F ②	IGN ON—1.5-1.7 volts 1.5-1.7 volts Continuity Continuity
13	Cylinder Head Temperature Sensor	VOM @ 23-26 ①	Above 68 deg F-2.9 kΩ Below 68 deg F-2.1 kΩ
14	Speed Sensor	VOM @ 29-GND ④	Continuity
21	Ignition Signal	VOM @ 3-GND VOM @ 5-GND Check power transistor terminals to base plate	Continuity Continuity Continuity
22	Fuel Pump	VOM @ 108-GND ⑤ Pump connectors Pump relay: VOM @ 1-2 VOM @ 3-4 12v @ 1-2, VOM @ 3-4	IGN ON-12 volts Continuity Continuity ∞
23	Throttle Valve Switch	VOM @ 18-25 ⑥ VOM @ 18-GND VOM @ 25-GND	Continuity ∞ ∞
24	Neutral/Park Switch	VOM @ Switch terminals	Neutral-0Ω Drive-∞Ω
31	Air Conditioner	VOM @ 22-GND ①	IGN ON-12 volts
32	Start Signal	VOM @ 9-GND ③	12 volts with starter S terminal disconnected
34	Detonation Sensor	Disconnect sensor and check timing with engine running	Timing should retard 5 degrees above 2000 rpm
41	Fuel Temperature Sensor	VOM @ 15-GND ③ VOM @ Sensor terminals	Above 68 deg. F-2.9 kΩ Below 68 deg. F-2.1 kΩ Resistance (ohms) should decrease as temperature rises
44	Normal Operation—no further testing required		

NOTE: Make sure test equipment will not damage the control unit before testing

VOM—Volt/ohm meter
GND—Ground
Ω—Ohms (kΩ = kilo-ohms)
∞—Infinite resistance
① 16-pin harness connector
② 6-pin air flow meter connection
③ 20-pin harness connector
④ 16-pin connector at ECU
⑤ Throttle valve switch connector

ISUZU I-TEC SYSTEM

The self diagnosis system is designed to monitor the input and output signals of the sensors and actuators and to store any malfunctions in its memory as a trouble code. When the electronic control unit detects a problem, it activates a CHECK ENGINE light on the dash to alert the driver. The computer will store trouble codes indicating problems in monitored systems including:

- Air flow sensor system
- Coolant temperature sensor
- Speed sensor
- Oxygen sensor
- Detonation sensor
- Injector system
- Electronic Control Unit (ECU)

NOTE: The self diagnosis system is only capable of troubleshooting the circuits in the I-TEC system, not in the components themselves. Malfunctions in a circuit may be caused by a variety of problems, including loose connections, failed components or damage to wiring.

Activating Diagnosis Mode

Locate the diagnosis lead near the control unit and connect the two leads with the ignition ON. The trouble codes stored in the memory are displayed by the CHECK ENGINE light as flashes that indicate numbers (trouble codes). The I-TEC system ECU memory is capable of storing three different trouble codes which are displayed in numerical sequence no matter when the malfunctions occur. Each trouble code is repeated three times, then the next code is displayed. The ECU memory will display all stored trouble codes as long as the diagnosis lead is connected with the key ON. A code 12 indicates that the I-TEC system is functioning normally and that no further testing is necessary. After servicing, clear the trouble codes by disconnecting the No. 4 fuse in the fuse block.

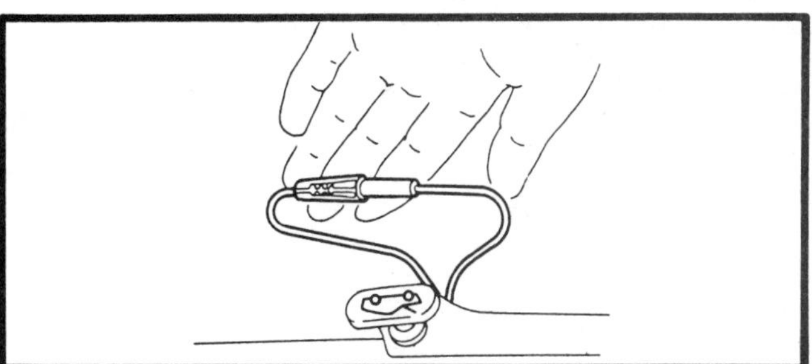

Diagnostic lead for Isuzu I-TEC injection system

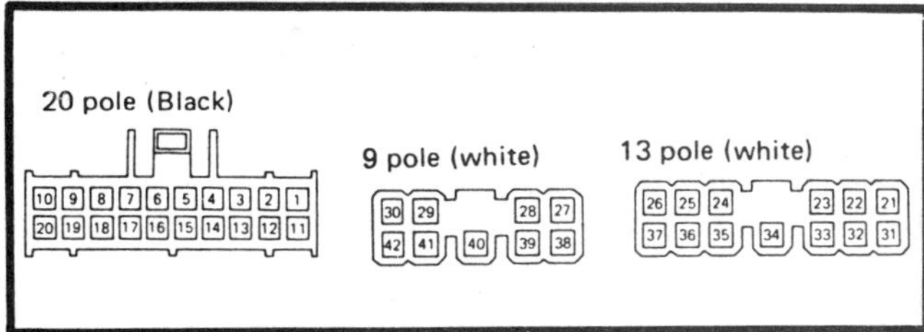

ECU terminal numbers for Isuzu I-TEC injection system

NOTE: All codes stored in the memory will be automatically cleared whenever the main harness connector is removed from the control unit.

Testing

The control unit wiring harness has three types of connectors with specifically numbered terminals. All inspection procedures refer to these numbers when describing terminal connections.

ISUZU I-TEC SYSTEM TROUBLE CODES

Trouble Code	ECU Circuit	Indicated Problem
12	Normal operation	No testing required
13	Oxygen sensor	Open or short circuit, failed sensor
44	Oxygen sensor	Low voltage signal
45	Oxygen sensor	High voltage signal
14	Coolant temperature sensor	Shorted with ground (no signal)
15	Coolant temperature sensor	Incorrect signal (too high or low)
16	Coolant temperature sensor	Excessive signal (harness open)
21	Throttle valve switch	Idle and WOT contacts closed at the same time
43	Throttle valve switch ①	Idle contact shorted
65	Throttle valve switch	Full throttle contact shorted
22	Starter signal	No signal to ECU
41	Crank angle sensor	No signal or wrong signal
61	Air flow sensor	Weak signal (harness shorted or open hot wire)
62	Air flow sensor	Excessive signal (open cold wire)
63	Speed sensor ①	No signal to ECU
66	Detonation sensor	Harness open or shorted to ground
51, 52, 55	ECU malfunction	Incorrect injection pulse or fixed timing
23	Power transistor for ignition	Output terminal shorted to ground
35	Power transistor for ignition	Open harness wire
54	Power transistor for ignition	Faulty transistor or ground
25	Vacuum switching valve	Output terminal shorted to ground or open harness
53	Vacuum switching valve	Faulty transistor or ground
33	Fuel injector	Output terminal shorted to ground or open harness wire
64	Fuel injector	Faulty transistor or ground

NOTE: Make sure test equipment will not damage ECU before making any connections. See manufacturer's instructions to verify compatibility with I-TEC System.

ECU—Electronic Control Unit
① Not diagnosed when the air flow sensor is defective

CAUTION

Make sure of the terminal numbers and location when making test connections, since battery power is supplied to some terminals only when the ignition is ON. Do not probe connector terminals from the front as test probes can damage the pins and connectors. Use a pin to probe connectors from the rear.

TOYOTA EFI SYSTEM

NOTE: The Toyota ECU employs a dash-mounted CHECK ENGINE light that illuminates when the control unit detects a malfunction. The memory will store the trouble codes until the system is cleared by removing the EFI fuse with the engine off.

Activating Diagnosis Mode

To activate the trouble code readout and obtain the diagnostic codes stored in the memory, first check that the battery voltage is at least 11 volts, the throttle valve is fully closed, transmission is in neutral, the engine is at normal operating temperature and all accessories are turned off.

1. Turn the ignition switch ON, but do not start the engine.
2. Locate the Check Engine Connector under the hood, near the ignition coil, and use a short jumper wire to connect the terminals together.
3. Read the diagnostic code as indicated by the number of flashes of the CHECK ENGINE light. If normal system operation is occurring (no malfunctions), the light will blink once every 3 seconds (code 1).
4. The light will blink once every second to indicate a trouble code stored in the memory, with 3 second pauses between each code number. For example, three blinks, a pause, then three blinks indicates a code 3 (air flow meter malfunction) stored in the memory.

NOTE: The diagnostic code series will be repeated as long as the CHECK ENGINE terminals are connected by the jumper wire.

5. After all trouble codes are recorded, remove the jumper wire and replace the rubber cap on the connector.

Diagnostic mode connector used on Toyota models—1984 Starlet shown

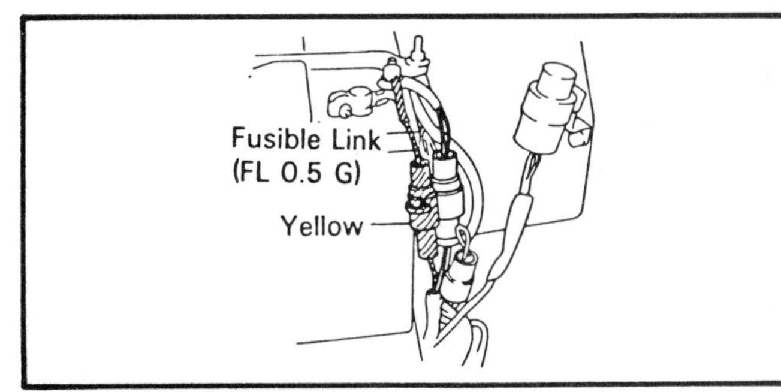

Disconnect the fusible link for 30 seconds to clear the trouble code memory on Toyota Starlet models

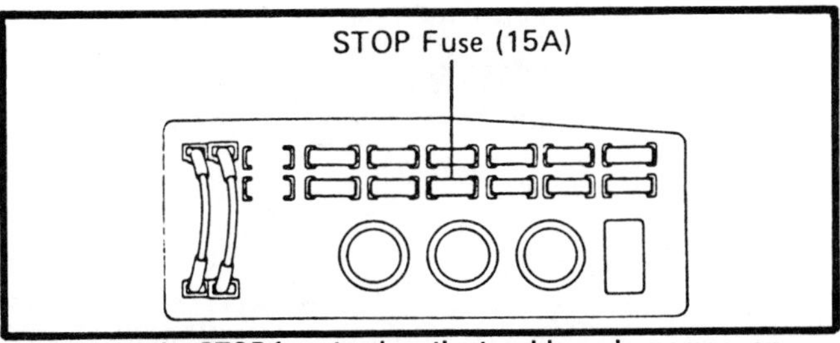

Disconnect the STOP fuse to clear the trouble code memory on Toyota Celica models

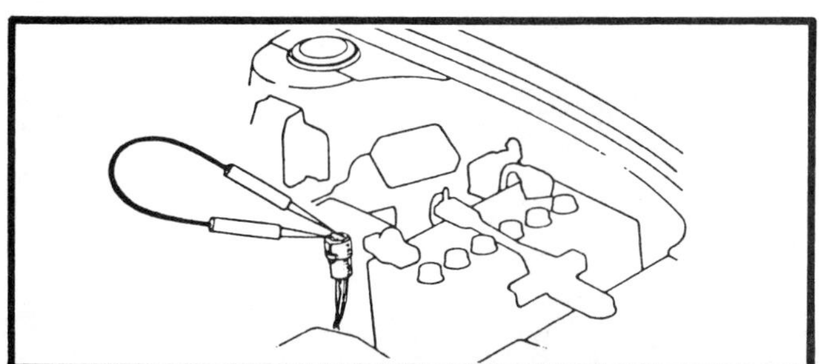
Diagnostic mode connector used on Toyota Celica models

6. Cancel the trouble codes in the memory after repairs by removing the STOP fuse for about 30 seconds (longer at lower ambient temperatures). If the diagnostic codes are not removed from the memory, they will be retained and reported as new problems the next time a malfunction occurs in any other system. Repair success can be verified by clearing the memory and road testing the car to see if any malfunction codes appear. Code 1 should appear when the diagnostic connector is shorted with a jumper wire.

NOTE: Cancellation can also be done by removing the battery negative (−) terminal, but in this case other memory systems (radio ETR, etc.) will also be cancelled out.

TOYOTA EFI TROUBLE CODES

Code No.	ECU Circuit	Possible Cause	Diagnosis Testing
1	Normal operation	This appears when one of the other codes are stored in the memory	EFI system operating normally. No further testing required
2	Air Flow Meter signal (V_C)	Open circuit in V_C or $V_C - V_S$ short circuited. Open circuit in V_B	Air flow meter circuit (V_C, V_S) Air flow meter EFI computer
3	Air flow meter signal (V_S)	Open circuit in V_S, or $V_S - E_2$ short ciruited. Open circuit in V_B	Air flow meter circuit (V_B, V_C, V_S) Air flow meter EFI computer
4	Water thermo sensor signal (THW)	Open circuit in coolant temperature sensor signal.	Coolant temperature sensor circuit Coolant temperature sensor EFI computer
5	O_2 sensor signal	Open or short circuit in O_2 sensor signal (only lean or rich indication)	O_2 sensor circuit O_2 sensor EFI computer
6	Ignition signal	No ignition signal	Ignition system circuit Distributor Ignition coil and igniter EFI computer
7	Throttle position sensor signal	IDL-Psw short circuited	Throttle position sensor circuit Throttle position sensor EFI computer

NOTE: 5-speed transmission models do not use code 7

TOYOTA STARLET EFI TESTING

ECU Terminal Connection	Test Meter Reading	Test Condition(s)
	VOLTMETER TESTS	
$+B$–E_1	10–14 V	Ignition switch ON
BATT–E_1	10–14 V	Ignition switch OFF
IDL–E_1	8–14 V	Throttle valve fully closed
P_{sw}–E_1	8–14 V	Throttle valve fully open
TL–E_1	8–14 V	Ignition switch OFF
IG–E_1	Above 3 V	Cranking and engine running
STA–E_1	6–12 V	Cranking
No. 10–E_1	9–14 V	Ignition switch ON
No. 20–E_1	9–14 V	Ignition switch ON
$+B$–E_2	8–14 V	Ignition switch OFF
V_c–E_2	4–9 V	Ignition switch OFF
V_s–E_2	0.5–2.5 V	Measuring plate fully closed
	5–8 V	Measuring plate fully open
	2.5–5.5 V	Idling
THA–E_2	2–6 V	Intake air temperature 20°C (68°F)
THW–E_2	0.5–2.5 V	Coolant temperature 80°C (176°F)
B/K–E_1	8–14 V	Stop light switch ON
	OHMMETER TESTS	
TL-IDL	0	Throttle valve fully closed
TL-IDL	∞	Throttle valve fully open
TL-P_{sw}	∞	Throttle valve fully closed
TL-P_{sw}	0	Throttle valve fully open
IDL, TL, P_{sw}–Ground	∞	Ignition switch OFF
THW–E_2	200–400 Ω	Coolant temp. 80°C (176°F)
THA–E_2	2–3 kΩ	Intake air temp. 20°C (68°F)
THW, THA–Ground	∞	Ignition switch OFF
$+B$–E_2	200–400 Ω	Ignition switch OFF
V_c–E_2	100–300 Ω	Ignition switch OFF
V_s–E_2	20–100 Ω	Measuring plate fully closed
V_s–E_2	20–1,000 Ω	Measuring plate fully open
$+B$, V_c, V_s–Ground	∞	Ignition switch OFF
E_1, E_2, E_{01}, E_{02}–Ground	0	Ignition switch OFF

NOTE: Make sure test equipment will not damage the control unit before testing

Ω—Ohms
V—Volts
∞—Infinity

E_2	Vs	Vc	BATT	THA	B/K	STA		Ox	THW	IDL	VF	T	#10	E_{01}
IG	E_3	W	+B		SPD				E_1	TL	Psw		#20	E_{02}

ECU connector terminal numbers—1984 Toyota Starlet

TOYOTA CELICA EFI TESTING

ECU Terminal Connection	Test Meter Reading	Test Condition(s)
	OHMMETER TESTS	
TL-IDL	0	Throttle valve fully closed
TL-IDL	∞	Throttle valve fully open
TL-P_{sw}	∞	Throttle valve fully closed
TL-P_{sw}	0	Throttle valve fully open
IDL, TL, P_{sw}-Ground	∞	Ignition switch OFF
THW-E_2	200–400 Ω	Coolant temp. 80°C (176°F)
THA-E_2	2–3 kΩ	Intake air temp. 20°C (68°F)
THW, THA-Ground	∞	Ignition switch OFF
V_B–E_2	200–400 Ω	Ignition switch OFF
V_C–E_2	100–300 Ω	Ignition switch OFF
V_S–E_2	20–400 Ω	Measuring plate fully closed
V_S–E_2	20–1,000 Ω	Measuring plate fully open
V_B, V_C, V_S-Ground	∞	Ignition switch OFF
E_1, E_{01}, E_{02}-Ground	0	Ignition switch OFF
	VOLTMETER TESTS	
+B__E_1	10–14 V	Ignition switch ON
BAT-E_1	10–14 V	Ignition switch OFF
IDL–E_1	8–14 V	Throttle valve fully closed
P_{sw}–E_1	8–14 V	Throttle valve fully open
TL–E_1	8–14 V	Ignition switch OFF
IG–E_1	Above 3 V	Cranking and engine running
STA–E_1	6–12 V	Cranking
No. 10–E_1	9–14 V	Ignition switch ON
No. 20–E_1	9–14 V	Ignition switch ON
W–E_1	8–14 V	No trouble and engine running
MS–E_1	8–14 V	Idling
V_c–E_2	4–9 V	Ignition switch OFF
V_s–E_2	0.5–2.5 V	Measuring plate fully closed
	5–8 V	Measuring plate fully open
	2.5–5.5 V	Idling
THA–E_2	2–6 V	Intake air temperature 20°C (68°F)
THW–E_2	0.5–2.5 V	Coolant temperature 80°C (176°F)
A/C–E_1	8–14 V	Air conditioning ON

NOTE: Make sure test equipment will not damage control unit or components before testing

Ω—Ohms
V—Volts
∞—Infinity

E_2	Vs	Vc	BAT	THA	/	STA	A/C	O_2	THW	IDL	VF	T	#10	E01
IG	E_3	W	+B	MS	/	/	/	/	E_1	TL	Psw	/	#20	E02

ECU connector terminal numbers—1984 Toyota Celica

Idle Speed Control System

IDLE-UP SOLENOID VALVE

On the Nissan E.C.C.S. system, the idle-up solenoid valve is attached to the throttle body and responds to signals from the control unit (ECU) to stabilize the idle speed when the engine is loaded by accessories (A/C, power steering pump, electrical loads, etc.). The operation of the solenoid is part of the ECU's programmed capabilities. As with all AFC fuel injection systems, the solenoid controls an auxiliary air control valve that bypasses air around the throttle plate to raise the engine rpm. By regulating the amount of bypassed air the ECU maintains the idle at a preset rpm value for the engine load sensed.

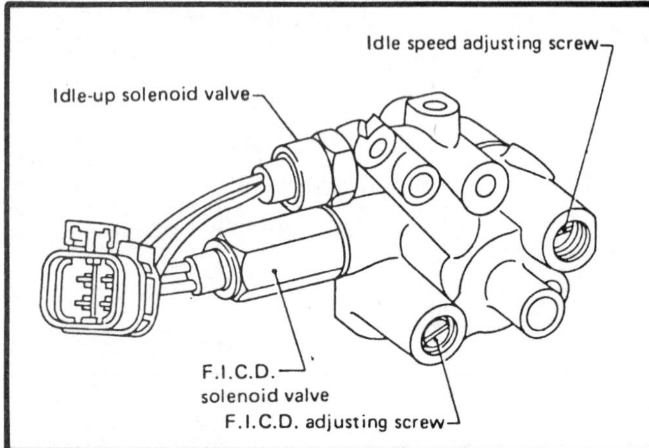

Idle-up solenoid valve used on Nissan E.C.C.S. system

NOTE: Bosch LH and Motronic injection systems also have idle speed control as one of their programmed functions. On the LH system, an idle speed test point is provided that, when grounded, locks the air control valve in its minimum position for base idle setting.

Testing

On the E.C.C.S. system, check the voltage between terminal 2 at the control unit connector and ground when the ignition switch is turned ON. If battery voltage (12v) is not present, check the wiring harness for continuity and check the EFI relay operation. The solenoid valve itself is checked by making sure continuity exists between the connector terminals on the switch.

Bosch Constant Idle Speed (CIS) System

GENERAL INFORMATIOM

Some AFC fuel injection systems are equipped with a separate idle speed control system designed to correct engine speed deviations very quickly, regardless of wear or changes in ignition timing. In the Bosch system, sensors read engine speed, temperature and throttle position and send impulses to an electronic control unit. The control unit continuously compares the sensor inputs with the ideal engine specifications programmed into the system memory by the manufacturer after dynamometer testing and tailored to every type of engine. The system is designed to be maintenance-free for the life of the engine.

The CIS system consists of the following components:

- Coolant Temperature Sensor
- Tachometer Connection (via ignition coil)
- Throttle Switch (position sensor)

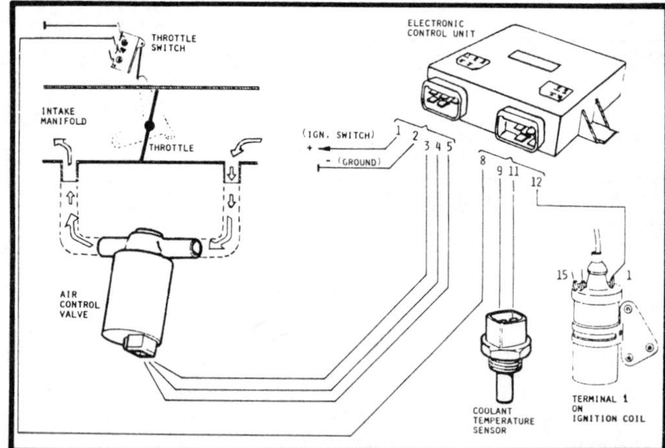

Schematic of Bosch Constant Idle Speed (CIS) system used with LH-Jetronic injection system. Note the separate electronic control unit

- Air Control Valve (bypass air controls rpm)
- Electronic Control Unit (ECU)

NOTE: The ECU for idle speed control is a separate unit from the fuel injection control system. The CIS control unit is usually mounted near the fuel injection ECU. The two are very similar in appearance and care should be taken not to confuse one with the other during testing. The same service precautions apply as previously described for the fuel injection ECU.

The CIS system controls the idle speed by regulating an air control valve to bypass more air around the throttle valve, thereby raising the idle speed in the same manner as described under "Auxiliary Air Valve." A small electric motor rotates clockwise or counterclockwise, depending on the signal from the control unit, opening or closing the valve very quickly and precisely to regulate the air flow. There are three basic air flow modes: the low flow or deceleration mode reduces the air flow when the throttle switch circuit is closed during deceleration. The high or driving mode increases the air bypass flow at normal road speeds with the accelerator depressed. The regulated or idle flow mode maintains a steady idle speed under all temperature conditions. Some newer models incorporate a fourth air flow mode which increases the idle speed when the air conditioner is turned on or other engine accessory loads are applied, providing improved cooling for both the engine and passenger compartment.

TROUBLESHOOTING

The CIS system is diagnosed in the same manner as described for the AFC fuel injection electrical system and control unit. Make all of the preliminary checks for obvious problems such as loose or corroded connectors and broken wires. Bad contacts cause many problems. On Volvo models, there are two connectors on the firewall and at the control unit. This dual wiring harness connection at the control unit differs from the standard single harness connector common to fuel injection ECU assemblies. If there is a problem with the engine idle speed on models equipped with a Bosch CIS system, check the air hoses for obstruction and the air control valve for sticking due to deposits from the PCV (positive crankcase ventilation) system. If there is a problem with periodic buildup of crankcase deposits in the air control valve, more frequent oil change intervals may be all that is necessary to correct the problem.

If the CIS system is suspected of causing a problem, most testing can be done with a good quality volt/ohmmeter and a 12V test light. The test light uses more current than a volt meter and sometimes is better at finding bad connections. The most common faults are poor or corroded contacts at the multipin connectors on the control unit, firewall and sensors.

CAUTION

Do not force the test probes directly into the wire harness connectors from the front. Test probes can spread the terminal contacts and cause a bad connection where none existed. Always test ECU connectors from the rear, using a pin to provide an accessible connection for test clips, if necessary. DO NOT pierce any wires with probes.

CIS System Testing

1. Locate the CIS system electronic control unit and remove any cover panels as necessary to gain access to the wire harness connectors.
2. Make sure the ignition switch is OFF and disconnect both connectors at the electronic control unit.

NOTE: The ignition switch must be OFF whenever disconnecting or reconnecting components to avoid the possibility of arcing which can damage the ECU. Pin numbers are for Volvo models; others may vary.

3. Switch the ignition ON and check for 12 volts at harness connector No. 1 (ECU power input) and a good ground at terminal No. 2. If no battery voltage is present, check the fuse block for a blown CIS system fuse and the ignition switch for continuity.
4. Check the throttle switch (microswitch) with an ohmmeter connected between terminal 8 and ground with the ignition OFF. By operating the throttle and observing the resistance valve of the microswitch, correct operation can be verified. This on/off function can also be checked using a test light connected between harness terminals 1 and 8.

NOTE: It is important to determine how (open or closed) the throttle switch is indicating idle position to the control unit before any adjustment or testing is possible.

5. Check the coolant temperature sensor by connecting an ohmmeter across terminals 9 and 11 and measuring the resistance change at various engine temperatures. Look for a change rather than a specific value; if the resistance changes as the temperature rises, the sensor is operating properly.
6. Connect a tachometer between terminal 12 and ground, then start the engine and note the tach reading. The engine speed as measured by the tachometer should match the actual rpm at idle. If an obviously wrong value is indicated, or no reading at all is obtained, check the firewall connectors. There should be a tach signal from the ignition switch.
7. To check the air control valve, fabricate two appropriate jumper wires and connect one across terminals 1 and 4; and the other across terminals 2 and 5. Start the engine and note the idle speed. It should be 1600-2400 rpm. If not, the idle air control valve is malfunctioning.

VOLVO CIS THROTTLE SWITCH OPERATION

Year/ Model	Throttle Position	Resistance ohms	Test Light
1981 B21F	idle	∞	OFF
	above idle	0	ON
B28F	idle	0	ON
	above idle	∞	OFF
1982 B21F	idle	∞	OFF
	above idle	0	ON
B28F	idle	∞	OFF
	above idle	0	ON

Note: Make sure of correct microswitch operation before attempting any adjustments.

CADILLAC DIGITAL ELECTRONIC FUEL INJECTION

General Information

Digital electronic fuel injection consists of a pair of electrically actuated fuel metering valves which, when actuated, spray a calculated quantity of fuel into the engine intake manifold. These valves or injectors are mounted on the throttle body above the throttle blades with the metering tip pointed into the throttle throats. The injectors are normally actuated alternately.

Gasoline is supplied to the inlet of the injectors through the fuel lines and is maintained at a constant pressure across the injector inlets. When the solenoid-operated valves are energized, the injector ball valve moves to the full open position. Since the pressure differential across the valve is constant, the fuel quantity is changed by varying the time that the injector is held open.

The amount of air entering the engine is measured by monitoring the intake manifold absolute pressure (MAP), the intake manifold air temperature (MAT) and the engine speed (in rpm). This information allows the computer to compute the flow rate of air being inducted into the engine and, consequently, the flow rate of fuel required to achieve the desired air-fuel mixture for the particular engine operating condition.

The following abbreviations are used in this section:

TPS: throttle position sensor
ECM: electronic control module
ISC: idle speed control (includes idle speed motor and throttle switch)
HEI: high energy igition
EST: electric spark timing
MAP: manifold absolute pressure (sensor)
BARO: barometric pressure (sensor)
ECC: electronic climate control
MAT: manifold air temperature (sensor)
CTS: coolant temperature sensor
MPG: miles per gallon (display panel)
EGR: exhaust gas recirculation

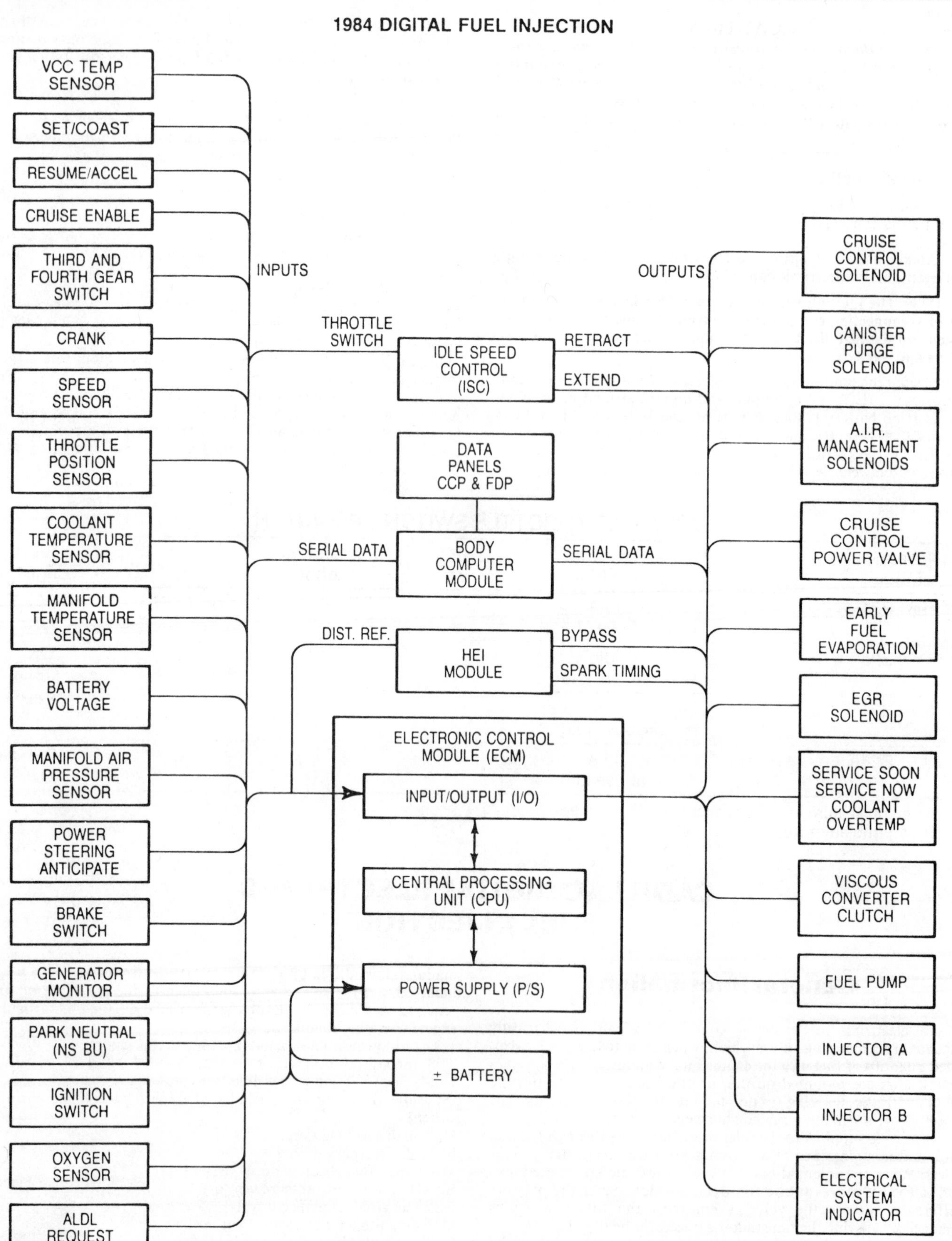

Schematic of Cadillac DFI system operation

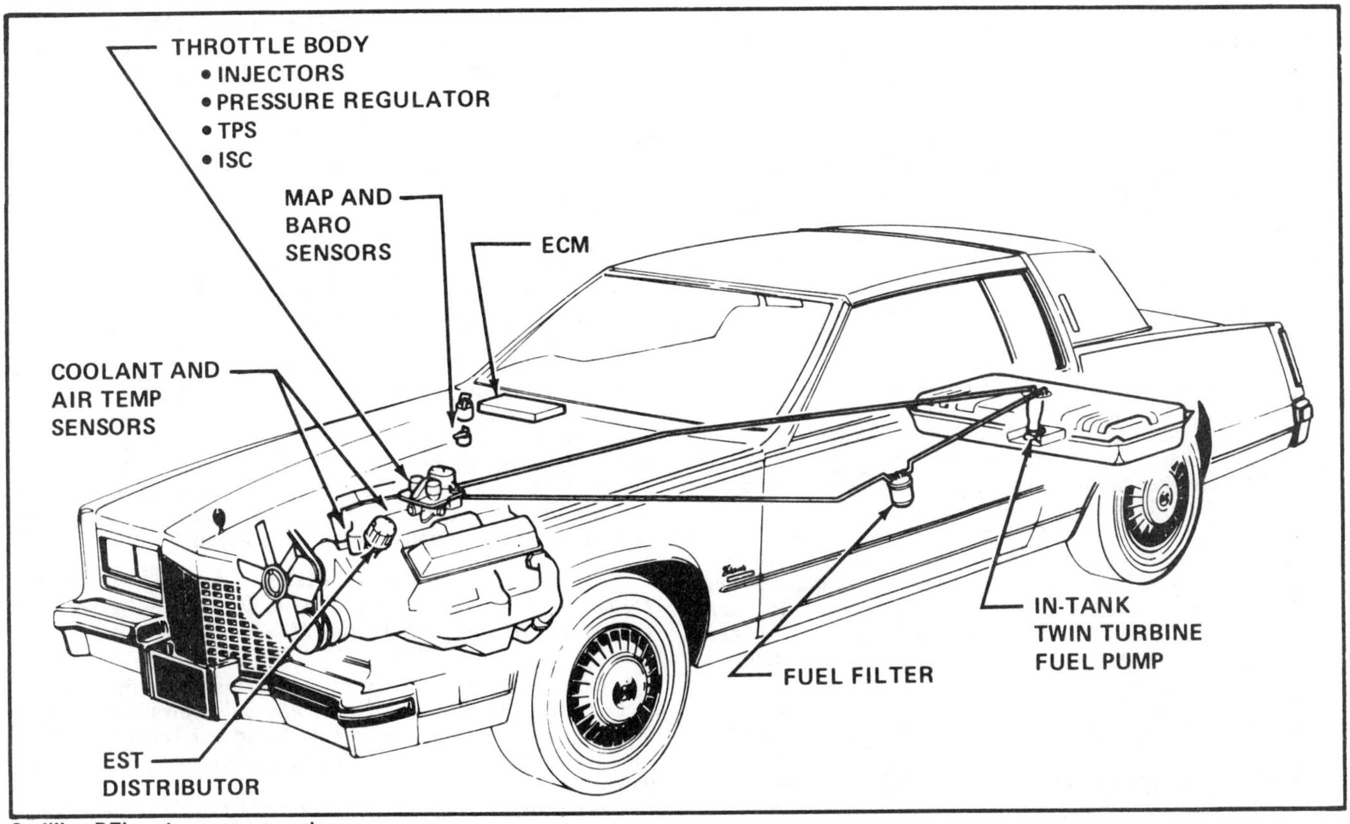

THROTTLE BODY
• INJECTORS
• PRESSURE REGULATOR
• TPS
• ISC
MAP AND BARO SENSORS
ECM
COOLANT AND AIR TEMP SENSORS
IN-TANK TWIN TURBINE FUEL PUMP
FUEL FILTER
EST DISTRIBUTOR

Cadillac DFI system components

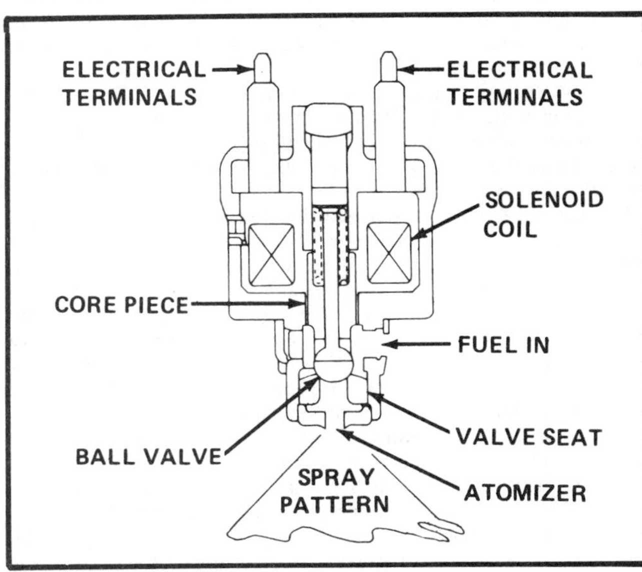

Typical DFI fuel injector

FUEL SUPPLY SYSTEM

The fuel supply system components provide fuel at the correct pressure for metering into the throttle bores by the injectors. The pressure regulator controls fuel pressure to a nominal 10.5 psi across the injectors. The fuel supply system is made up of a fuel tank mounted electric pump, a full-flow fuel filter mounted on the vehicle frame, a fuel pressure regulator integral with the throttle body, fuel supply and fuel return lines and two fuel injectors. The timing and amount of fuel supplied is controlled by the computer.

An electric motor-driven twin turbine-type pump is integral with the fuel tank float unit. It provides fuel at a positive pressure to the throttle body and fuel pressure regulator. The pump is specific for DEFI application and is not repairable. However, the pump may be serviced separately from the fuel gauge unit.

Fuel pump operation is controlled by the fuel pump relay, located in the relay center. Operation of the relay is controlled by a signal from the computer. The fuel pump circuit is protected by a 10 amp fuse, located in the minifuse block. The computer turns the pump on with the ignition "ON" or "START." However, if the engine is not cranked within one second after the ignition is turned on, the computer signal is removed and the pump turns off.

Fuel is pumped from the fuel tank through the supply line and the filter to the throttle body and pressure regulator. The injectors supply fuel to the engine in precisely timed bursts as a result of electrical signals from the computer. Excess fuel is returned to the fuel tank through the fuel return line.

The fuel tank incorporates a reservoir directly below the sending unit-in-tank pump assembly. The "bathtub" shaped reservoir is used to ensure a constant supply of fuel for the in-tank pump even at low fuel level and severe maneuvering conditions.

On-Car Service

FUEL SYSTEM DIAGNOSIS

1. Low or no fuel pressure diagnosis should begin by trying to determine if the fuel pump is operating or not. This is most easily accomplished by turning the ignition on and listening for the one second "run" of the fuel pump and the associated relay clicks. Since this may not be possible in some shops, the best test is to probe both sides of the fuel pump fuse in the mini-fuse block with a voltmeter. Observe the meter as the ignition is turned on. It should go to battery voltage (12 volts) and then, after one second, to zero volts.

2. If the fuel pump circuit is operating properly, the computer signal and the relay are okay. The last connector in the fuel pump circuit is the six-way connector at the tail panel. The voltage actions seen at the fuse should be repeated. If not, there is an open in the circuit. Check the connectors and repair wiring as required. Individual sections of the wiring can be tested with an ohmmeter.

3. If the fuel pump signal is correct at the tail panel connector, a fuel delivery system situation exists. If the pump cannot be heard to run during the one second on period, the pump should be replaced. This observation is more easily made if helper turns the ignition on as the technician listens at the fuel tank or filler neck area. If the fuel pump can be heard to run, disconnect the fuel return line at the throttle body and install a plug in the throttle body opening. This will effectively "dead head" the fuel pump and eliminate the pressure regulator. If the pump is able to produce above 9 psi under these conditions (with the ignition on), replace the fuel pressure regulator as it is controlling at too low pressure.

4. If the fuel pressure remains below 9 psi with the return line plugged, a restriction in the fuel supply line may exist. A blocked fuel line or fuel filter can be determined by visually inspecting the filter element and the fuel line routing for kinks, damage, etc. If lines and filter are okay, replace fuel pump.

5. If the voltage at the fuel pump fuse does not go to 12 volts, first inspect the fuse. If the fuse is okay, check the contacts at terminals 1 and 3 of the fuel pump relay connecting block. The contacts close when the relay coil is energized by the computer. Remove the relay. Probe the relay center socket, which corresponds to relay terminal 4, with a voltmeter.

After ignition has been off for at least 10 seconds, turn the ignition on. If the voltage does not go to 12 volts and back to zero, inspect for opens or shorts to ground. Checking for opens can be done with an ohmmeter connected to both ends of the circuit. If continuity is indicated (0 ohms), check for a short to ground by jumping one end to ground and probing the opposite end with an ohmmeter to ground. An infinite reading indicates the wire is not grounded and the circuit is OK. Replace the computer. If a short to ground in the circuit is found the computer will be damaged. Replace the computer after repair.

6. If voltage at relay terminal 4 goes to 12 volts and then to zero, the computer is performing properly. Probe the relay center socket which corresponds to relay pin 3 with a voltmeter. Since this circuit comes directly from the battery terminal of the starter solenoid, it should be 12 volts at all times. If voltage is zero, inspect the circuit for open.

7. Twelve volts at pin 3 indicates that most of the relay's requirements are met. However, there are still two wiring circuits which could prevent proper relay operation. The computer signal has been proven okay. However, if the coil ground is open, the coil will not be energized. To check, probe the relay center socket which corresponds to relay pin 5, with an ohmmeter connected to ground. Continuity (O ohms) indicates a good circuit. An infinite reading indicates an open. Repair wire as required.

The second circuit in which an open could occur, preventing the proper voltage at the fuse, is between the fuel pump relay terminal 1 and the mini-fuse block. This circuit can be checked by probing both ends with an ohmmeter. If the circuit has continuity as indicated by a zero ohms reading, replace fuel pump relay.

8. High fuel pressure is caused by either a malfunction of the pressure regulator or a restriction in the fuel return line. To determine which of these problems exists, disconnect the fuel return line at the throttle body and connect a suitable fitting to the throttle body to accept a length of flexible rubber fuel hose. Insert the open end of the hose into a suitable fuel container. Observe the fuel pressure as the ignition switch is turned on. If the fuel pressure remains above 12 psi, replace the pressure regulator as it is unable to regulate with no return restriction.

9. If the fuel pressure now falls into the correct pressure range of 9–12 psi, the restriction has been eliminated by bypassing the return system. A restricted fuel return line can be located by visually inspecting the line routing for kinks, damage, etc.

DIAGNOSING POOR PERFORMANCE

1. Unsatisfactory engine performance complaints which are related to the digital electronic fuel injection system are caused either by improper fuel delivery or improper ignition advance (controlled by the computer).

To isolate the problem to one of these systems, remove the air cleaner and observe the injector spray pattern of both injectors at idle. The spray pattern should be compared to a proper injector spray pattern of a known good car.

2. Improper fuel delivery which affects both injectors is most likely a fuel delivery system problem. By switching injector connectors, it can be determined if the problem is the injector assembly or the signal to the injector. If the problem remains with the original injector, it is most likely an injector problem. Replace the injector. If the problem moves with the injector connector, an improper signal circuit is indicated. The injectors are powered through the 3 amp fuses in the mini-fuse block. The fuse block receives battery voltage from the starter solenoid battery terminal when the fuel pump relay is energized by the computer during crank or run. With the relay closed, the circuits apply 12 volts to the injector. The computer provides the ground to energize the solenoid.

CAUTION

Do not apply direct battery voltage to the injector. 12 volts will destroy the fuel injector coil within 1/2 second.

3. Check the injector fuses by visually inspecting the fuse filament. If the fuses are okay there is a harness problem in the voltage feed or ground. The troubled circuit should be investigated. Check for opens with an ohmmeter. If harness checks okay, replace computer.

4. A blown injector fuse should be replaced. If it blows again, a short to ground is indicated. To check for this condition, connect ohmmeter between the red or white wire at the injector connector and ground (ignition off and computer disconnected). A low reading indicates a short circuit. Repair as required. If harness is okay, replace the computer.

5. A proper injector spray pattern indicates that improper ignition timing is the main DEFI component which can cause poor performance. Ground the "set timing" pigtail and check timing. It should be at the base (or initial) value of 10°BTDC (800 rpm or less). If not, reset to 10°.

6. Once it has been established that the ignition timing is using the proper reference signal, the system's ability to advance the spark must also be determined. Since the actual ignition advance produced is the result of other variables besides engine rpm and manifold vacuum, it is not possible to establish checkpoints. However, if the system does advance, the shape of the advance curve can be assumed to be correct, since it is determined by the electronic circuitry which was able to recognize that some advance was required and did respond to this information.

Disconnect the "set timing" jumper and check ignition timing. At normal idle, this should be approximately 20° to 30°BTDC.

7. If the ignition timing does not advance as a result of disconnecting the "set timing" jumper, a problem with the advance system is indicated. Since the computer selects and determines the spark advance curve, it is necessary to determine if the computer is operating properly or not.

During cranking, no spark advance is desired. The computer limits the advance to base timing by turning off the voltage signal. The pick-up coil pulse is used directly to turn the HEI module on. When the engine starts, the computer turns on the voltage in the circuit and the pick-up coil pulse is sent to the computer, modified by the computer and sent back to the module.

Whether the computer advances the timing or not depends upon whether it applied a voltage to the circuit or not (no voltage = base timing; voltage = electric spark timing). To check the circuit, disconnect the four-way connector at the distributor while the engine is idling. This will stop the engine. Probe the harness side pin C

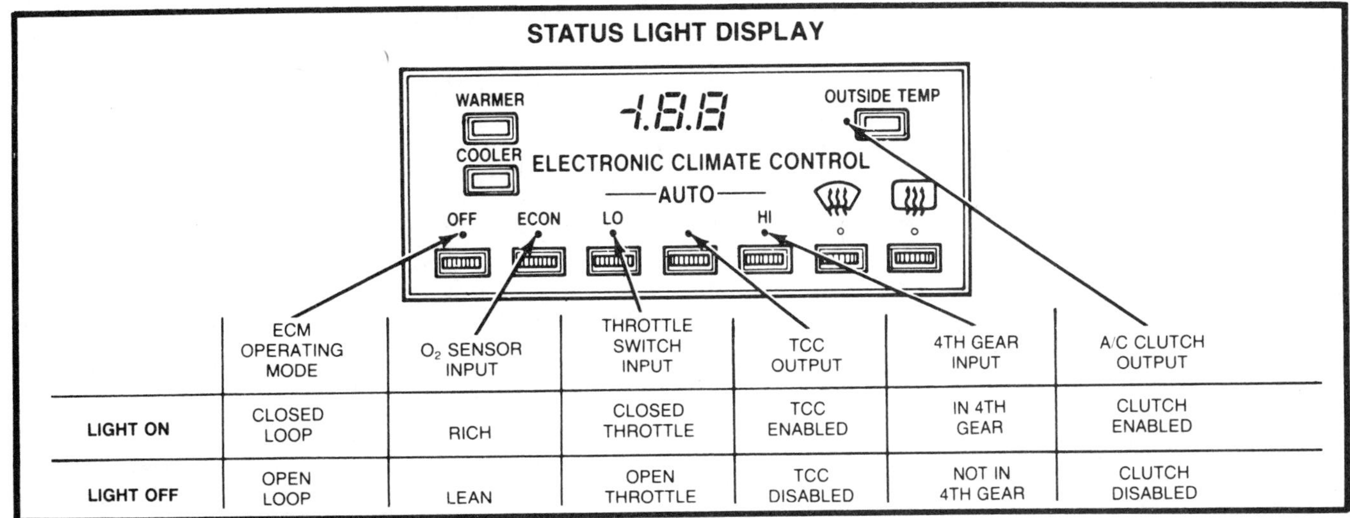

	ECM OPERATING MODE	O_2 SENSOR INPUT	THROTTLE SWITCH INPUT	TCC OUTPUT	4TH GEAR INPUT	A/C CLUTCH OUTPUT
LIGHT ON	CLOSED LOOP	RICH	CLOSED THROTTLE	TCC ENABLED	IN 4TH GEAR	CLUTCH ENABLED
LIGHT OFF	OPEN LOOP	LEAN	OPEN THROTTLE	TCC DISABLED	NOT IN 4TH GEAR	CLUTCH DISABLED

Control head status light display—1982 shown

(not distributor side harness) with a voltmeter while the ignition switch remains on. If voltage is greater than 4 volts, refer to HEI diagnosis because the computer has signaled that EST should be used, but timing did not advance when checked.

8. Voltage less than 1 volt indicates that either the computer signal is not being produced or the circuit is open or shorted to the ground. To check for shorts, disconnect the black/green computer connector and probe harness pin D with an ohmmeter to ground (ignition off and distributor connector disconnected). A zero ohm reading indicates a short. Repair as required.

If the ohmmeter reads infinity, check for opens by jumping distributor connector pin C to ground. If the ohmmeter reading remains infinite, circuit is open. Repair as required. If the ohmmeter reads zero ohms, the circuit is okay.

Check MAT and coolant sensor circuits for an open between the splice and the computer. Attention should be focused on the bulk head and computer connectors. If the circuit is okay, substitute a new computer and observe performance.

9. The EGR system utilizes various controls in order to provide EGR gases only when they are needed for emission control. One of these controls is the EGR solenoid, with power feed from the ignition switch through the 20 amp fuse. Ground for the solenoid is provided by the computer. The computer provides this ground whenever the coolant temperature signal from the coolant sensor says the temperature is below 43°C (110°F). This energizes the solenoid and blocks the flow of vacuum to the EGR transducer thus preventing EGR operation at cold engine temperatures. Above 71°C (160°F), the solenoid ground is removed and the solenoid opens, allowing vacuum to the EGR valve. This vacuum signal is a ported vacuum which exists only off idle. This means that even on a warm engine, there is no EGR vacuum signal and no EGR flow at idle.

10. If EGR operation is okay, check to make sure that throttle valves open to wide open throttle when the accelerator pedal is wide open. If this is okay, the performance problem is not related to the DEFI system. If EGR problems are found, check hoses, etc.

Electronic Control Module (ECM)

GENERAL INFORMATION

The Electronic Control Module, (ECM) or computer provides all computation and controls for the DEFI system. Sensor inputs are fed into the computer from the various sensors. They are processed to produce the appropriate pulse duration for the injectors, the correct idle speed for the particular operating condition and the proper spark advance. Analog inputs from the sensors are converted to digital signals before processing. The computer assembly is mounted under the instrument panel and consists of various printed circuit boards mounted in a protective metal box.

The computer receives power from the vehicle battery. When the ignition is set to the "ON" or "CRANK" position, the following information is received from the sensors:

1. Engine coolant temperature
2. Intake manifold air temperature
3. Intake manifold absolute pressure

Typical electronic climate control head—1982 shown

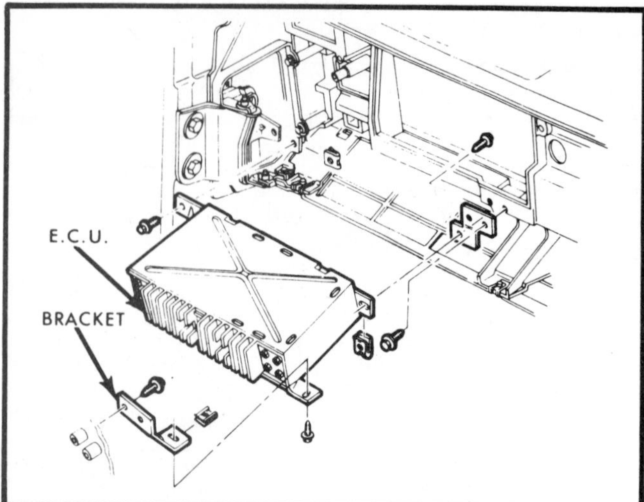

Cadillac DFI electronic control module location

4. Barometric pressure
5. Engine speed
6. Throttle position

The following commands are transmitted by the ECM:

1. Electric fuel pump activation
2. Idle speed control
3. Spark advance control
4. Injection valve activation
5. EGR solenoid activation

The desired air-fuel mixture for various driving and atmospheric conditions are programmed into the computer. As signals are received from the sensors, the computer processes the signals and computes the engine's fuel requirements. The computer issues commands to the injection valves to open for a specific time duration. The duration of the command pulses varies as the operating conditions change.

The digital electronic fuel injection system is activated when the ignition switch is turned to the "ON" position. The following events occur at this moment.

1. The computer receives the ignition "ON" signal.
2. The fuel pump is activated by the ECM. The pump will operate for approximately one second only, unless the engine is cranking or running.
3. All engine sensors are activated and begin transmitting signals to the computer.
4. The EGR solenoid is activated to block the vacuum signal to the EGR valve at coolant temperatures below 110°F.
5. The "CHECK ENGINE" and "COOLANT" lights are illuminated as a functional check of the bulb and circuit.
6. Operation of the fuel economy lamps begins.

The following events occur when the engine is started.

1. The fuel pump is activated for continuous operation.
2. The idle speed control motor will begin controlling idle speed, including fast idle speed, if the throttle switch is closed.
3. The spark advance shifts from base (bypass) timing to the computer programmed spark curve.
4. The fuel pressure regulator maintains the fuel pressure at 10.5 psi by returning excess fuel to the fuel tank.
5. The following sensor signals are continuously received and processed by the computer:
   a. Engine coolant temperature
   b. Intake manifold air temperature
   c. Barometric pressure
   d. Intake manifold absolute air pressure
   e. Engine speed
   f. Throttle position changes
6. The computer alternately grounds each injector, precisely controlling the opening and closing time (pulse width) to deliver fuel to the engine.

Fuel Delivery System

The computer's control of fuel delivery can be considered in three basic modes: cranking, part throttle and wide open throttle.

If the engine is determined to be in the cranking mode by the presence of a voltage in the cranking signal wire from the ignition switch, the starting fuel delivery consists of one long "prime" pulse from both injectors followed by a series of "starting" pulses until the cranking mode signal is no longer present.

In addition, there is a "clear flood" condition in which smaller alternating fuel pulses are delivered if the throttle is held wide open and cranking exceeds five seconds.

Once the engine is running, injector pulse width is then adjusted to account for operating conditions such as idle, part throttle, acceleration, deceleration and altitude.

For wide open throttle conditions, which are sensed by matching the manifold absolute pressure and barometric pressure sensor inputs, additional enrichment is provided.

Electronic Spark Timing (EST)

The EST type HEI distributor receives all spark timing information from the computer when the engine is running. The computer provides spark plug firing pulses based upon the various engine operating parameters. The electronic components for the electronic spark control system are integral with the computer. The two basic operating modes are cranking (or bypass) and normal engine operation.

When the engine is in the cranking/bypass mode, ignition timing occurs at a reference setting (distributor timing set point) regardless of other engine operating parameters. Under all other normal operating conditions, basic engine ignition timing is controlled by the computer and modified or added to, depending on particular conditions such as altitude and/or engine loading.

Idle Speed Control System

The idle speed control system is controlled by the computer. The system acts to control engine idle speed in three ways; as a normal idle (rpm) control, as a fast idle device and as a "dashpot" on decelerations and throttle closing.

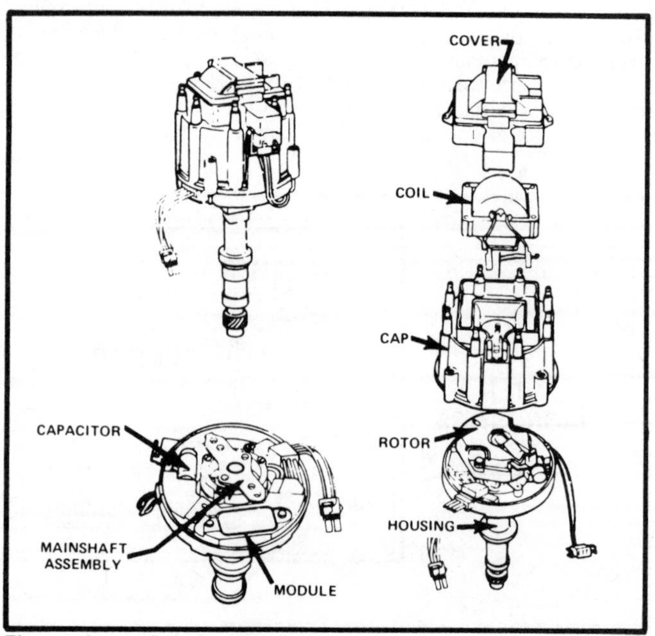

Electronic spark timing distributor

The normal engine idle speed is programmed into the computer and no adjustments are possible. Under normal engine operating conditions, idle speed is maintained by monitoring idle speed in a closed loop fashion. To accomplish this loop, the computer periodically senses the engine idle speed and issues commands to the idle speed control to move the throttle stop to maintain the correct speed.

For engine starting, the throttle is either held open by the idle speed control for a longer (cold) or a shorter (hot) period to provide adequate engine warm-up prior to normal operation. When the engine is shut off, the throttle is opened by fully extending the idle speed control actuator to get ready for the next start.

Signal inputs for transmission gear, air conditioning compressor clutch (engaged or not engaged) and throttle (open or closed) are used to either increase or decrease throttle angle in response to these particular engine loadings.

Vehicle idle speed is controlled by an electrically driven actuator (idle speed control) which changes the throttle angle by acting as a movable idle stop. Inputs to the ISC actuator motor come from the ECM and are determined by the idle speed required for the particular operating condition. The electronic components for the ISC system are integral with the ECM. An integral part of the ISC is the throttle switch. The position of the switch determines whether the ISC should control idle speed or not. When the swtich is closed, as determined by the throttle lever resting upon the end of the ISC actuator, the ECM will issue the appropriate commands to move the idle speed control to provide the programmed idle speed. When the throttle lever moves off the idle speed control actuator from idle, the throttle switch is opened. The computer then extends the actuator and stops sending idle speed commands and the driver controls the engine speed.

Self-Diagnosis System

An amber dash-mounted digital display panel normally used for the Electronic Climate Control (ECC) system, is used to display trouble codes stored in the computer when desired. Any codes that may be stored can be called up and/or cleared by properly exercising the ECC controls.

In the event the computer detects a system malfunction, the "CHECK ENGINE" light will be activated, the corresponding trouble code stored and substitute values to replace missing data may be made available for computations by the computer. This can be thought of as a "Fail-Safe" operation. In this mode, driveability of the car may be poor under certain conditions and the diagnostic procedures should be exercised.

How to Enter Diagnostic Mode

To enter diagnostics, proceed as follows:

1. Turn ignition "ON".
2. depress "OFF" and "WARMER" buttons on the ECC panel simultaneously and hold until ".." appears. "88" will then be displayed, which indicates the beginning of the diagnostic readout.
3. Trouble codes will be displayed on the digital ECC panel beginning with the lowest numbered code. Note that the test panel does not display when the system is in the diagnostic mode.

How to Clear Trouble Codes

Trouble codes stored in the ECM's memory may be cleared (erased) by entering the diagnostic mode and then depressing the "OFF" and "HI" buttons simultaneously. Hold until "00" appears.

How to Exit Diagnostic Mode

To get out of the diagnostic mode, depress any of the ECC function keys (Auto, Econ, etc. except Rear Defog) or turn ignition switch off for 10 seconds. Trouble codes are not erased when this is done.

DIAGNOSIS PROCEDURE

Illumination of the "CHECK ENGINE" light indicates that a malfunction has occurred for which a trouble code has been stored and can be displayed on the ECC control panel. The malfunction may or may not result in abnormal engine operation. To determine which system(s) has malfunctioned, proceed as follows.

1. Turn ignition switch "ON" for 5 seconds.
2. Depress the "OFF" and "WARMER" buttons on the electronic climate control panel simultaneously and hold until ".." appears.
3. Numerals "88" should then appear. The purpose of the "88" display is to check that all segments of the display are working. Diagnosis should not be attempted unless the entire "88" appears, as this could lead to misdiagnosis (Code 31 could be Code 34 with two segments of the display inoperative, etc).
4. Trouble codes will then be displayed for approximately three seconds.
   a. The lowest numbered code will be displayed for approximately three seconds.
   b. Progressively higher codes, if present, will be displayed consecutively for three second intervals until the highest code present has been displayed.
   c. "88" is again displayed.
   d. Displays from steps a., b. and c. will be repeated a second time.
   e. Displays from steps a. and b. will be repeated a third time.
   f. Code 70 will then be displayed, which signals the beginning of the "switch tests" section.
   g. Switch tests require some action on the part of the technician. This action is analyzed by the computer for proper operation.

Switch Test Procedure

When all stored trouble codes have been displayed for the third time, the computer will automatically begin the switch tests. To perform these checks, proceed as follows:

1. Display of trouble Code 70 signals the beginning of this section. This code will continue to be displayed until the proper test action is taken. When ready to begin tests, depress service brake pedal. This begins the test sequence by displaying Code 71.
2. With Code 71 displayed, depress the service brake pedal again to test the brake light circuit. When this check is completed, the test program will automatically sequence to Code 72. If the test action is not performed within 10 seconds, the test program will automatically sequence to "72" and Code 71 will be stored in the computer memory as "not passed".
3. With Code 72 displayed, depress the throttle from idle to wide open throttle and release. This action allows the computer to analyze the operation of the throttle switch. When this check is completed, the test program will automatically sequence to Code 73. Again, if action is not taken within 10 seconds, a Code 72 will be stored as "not passed".
4. With Code 73 displayed, shift the transmission lever to Drive and then to Neutral. When this check is completed, the test program will automatically sequence to code 74. This action must be taken within 10 seconds or a Code 73 will be set.
5. With Code 74 displayed, shift the transmission lever to Reverse and then to Park. Shift transmission within 10 seconds or a Code 74 will be set. When this check is completed, the test program will automatically sequence to Code 78.
6. With code 78 displayed, depress the "AVERAGE" button on the test panel. When this check is completed, the test program will automatically sequence to Code 79. Again, this must be done within 10 seconds or a Code 78 will be set.
7. With Code 79 displayed, depress the "RESET" button on the test panel within 10 seconds to test the function of this switch. This is the end of the switch tests.
8. With the switch tests completed, the computer will now go back and display the switch test code(s) which did not test properly.

DIAGNOSTIC DISPLAY

The dash-mounted digital display panel normally used for the electronic climate control (ECC) system, is used to display trouble codes stored in the computer when desired. Any codes that may be stored can be called up and/or cleared by properly exercising the ECC controls.

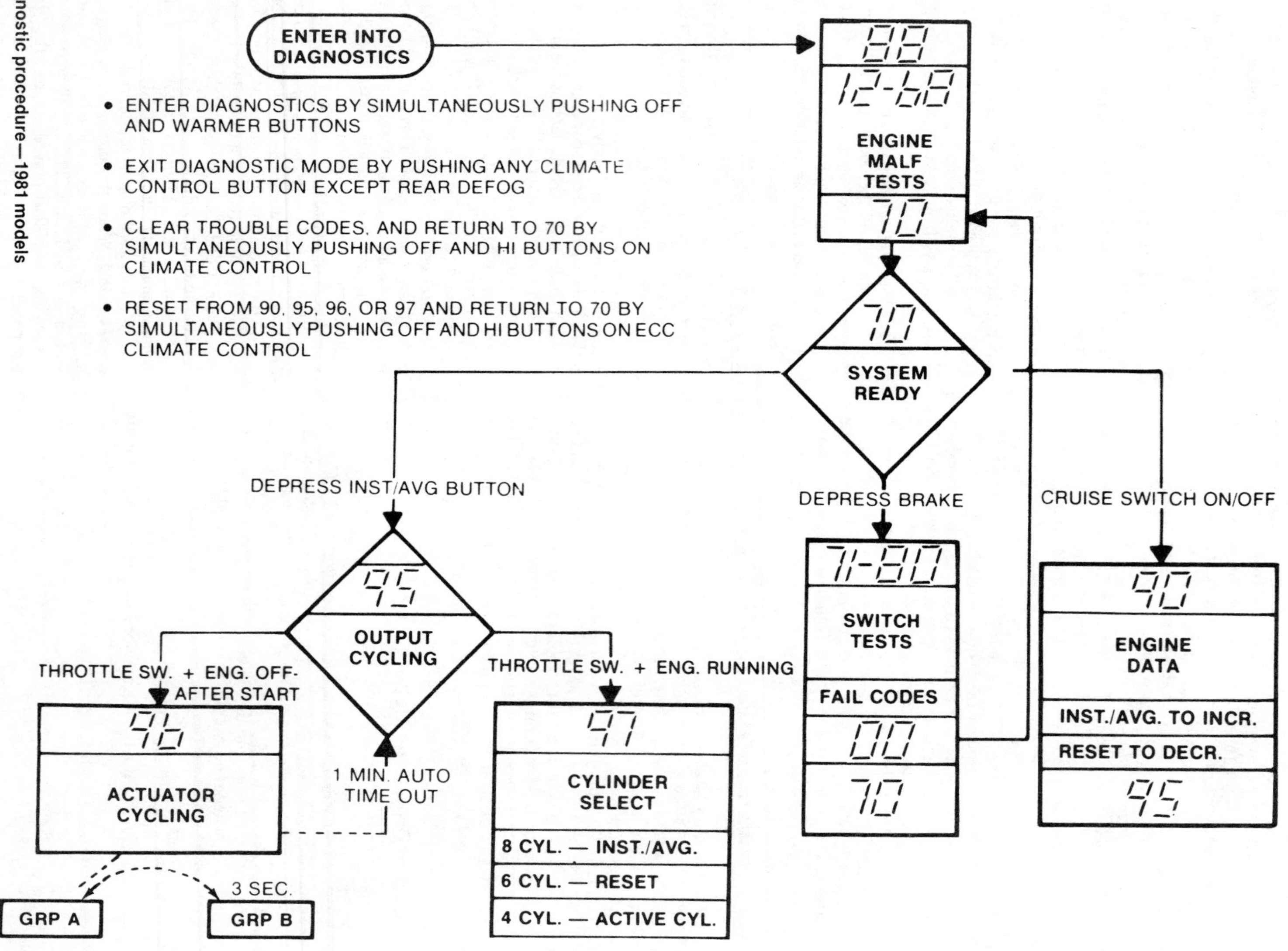

Cadillac DFI diagnostic procedure—1981 models

OPERATION WITH SYSTEM FAILURES

In the event the computer detects a system malfunction, the "CHECK ENGINE" light will be activated, the corresponding trouble code stored and substitute values to replace missing date may be made available for computations by the computer. This can be thought of as a "Fail Safe" operation. In this mode, driveability of the car may be poor under certain conditions and the diagnostic procedures should be exercised.

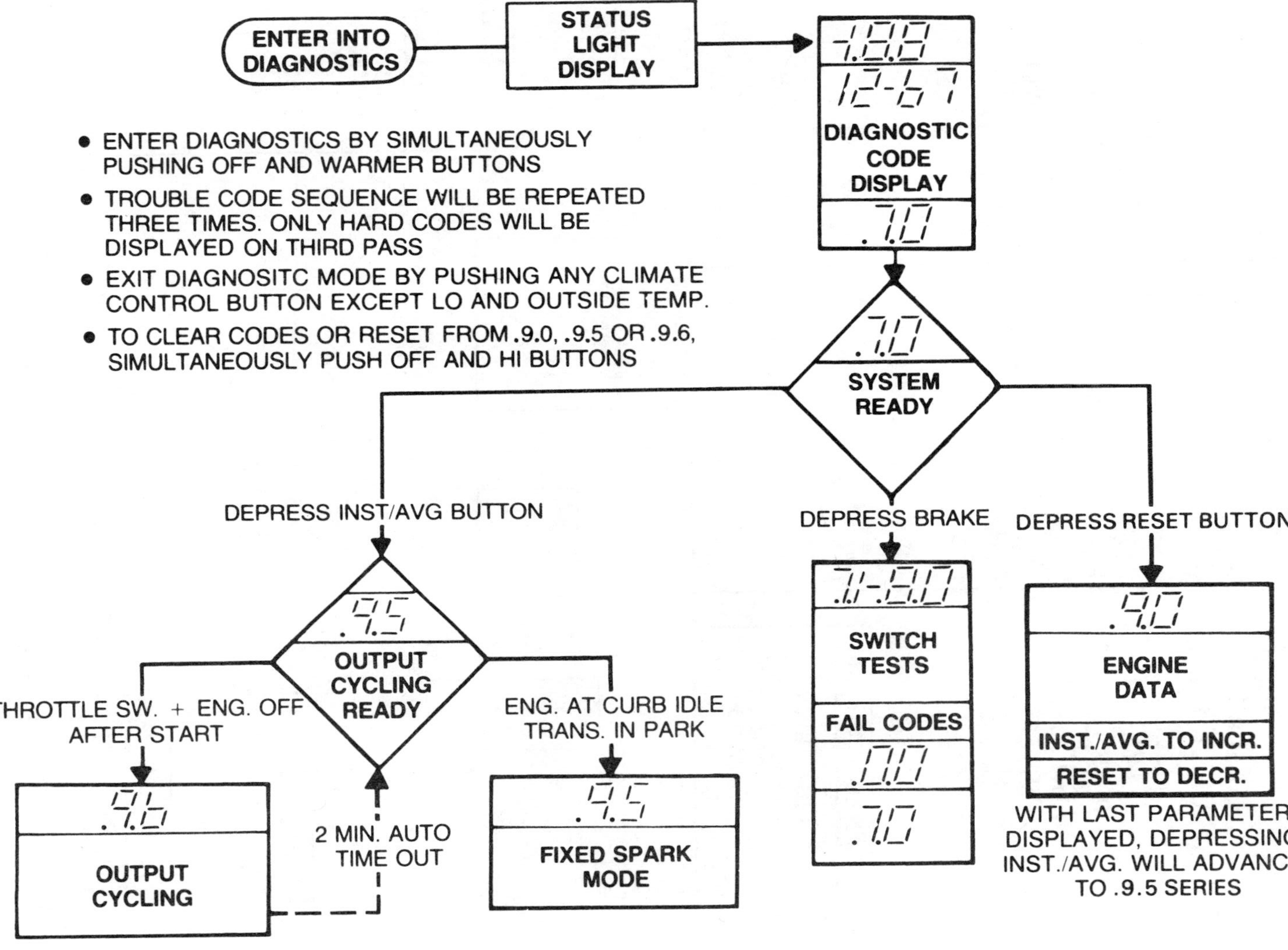
ENTER INTO DIAGNOSTICS
STATUS LIGHT DISPLAY
DIAGNOSTIC CODE DISPLAY
• ENTER DIAGNOSTICS BY SIMULTANEOUSLY PUSHING OFF AND WARMER BUTTONS
• TROUBLE CODE SEQUENCE WILL BE REPEATED THREE TIMES. ONLY HARD CODES WILL BE DISPLAYED ON THIRD PASS
• EXIT DIAGNOSITC MODE BY PUSHING ANY CLIMATE CONTROL BUTTON EXCEPT LO AND OUTSIDE TEMP.
• TO CLEAR CODES OR RESET FROM .9.0, .9.5 OR .9.6, SIMULTANEOUSLY PUSH OFF AND HI BUTTONS
SYSTEM READY
DEPRESS INST/AVG BUTTON
DEPRESS BRAKE
DEPRESS RESET BUTTON
OUTPUT CYCLING READY
THROTTLE SW. + ENG. OFF AFTER START
ENG. AT CURB IDLE TRANS. IN PARK
OUTPUT CYCLING
2 MIN. AUTO TIME OUT
FIXED SPARK MODE
SWITCH TESTS
FAIL CODES
ENGINE DATA
INST./AVG. TO INCR.
RESET TO DECR.
WITH LAST PARAMETER DISPLAYED, DEPRESSING INST./AVG. WILL ADVANCE TO .9.5 SERIES

Cadillac DFI diagnostic procedure—1982 models

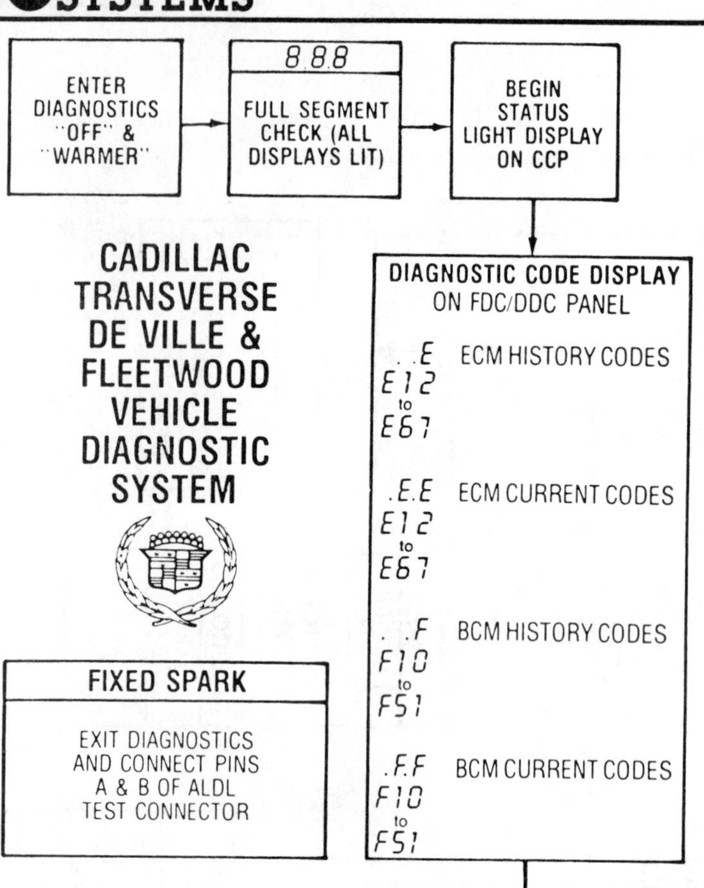

DIAGNOSTICS — BASIC OPERATION

- ENTER DIAGNOSTICS BY SIMULTANEOUSLY PUSHING CCP "OFF" AND "WARMER" BUTTONS UNTIL ALL DISPLAYS ARE LIT.
- MALFUNCTION CODE SEQUENCE BEGINS WITH ECM CODES FOLLOWED BY BCM CODES (DIESEL HAS BCM DIAGNOSTICS ONLY).
- TO PROCEED FROM ".7.0" TO THE OTHER DIAGNOSTIC FEATURES, PRESS AND RELEASE THE INDICATED BUTTON OR PEDAL.
- RETURN TO ".7.0" BY CLEARING EITHER ECM OR BCM CODES.
- EXIT DIAGNOSTICS BY PUSHING "AUTO".

FDC/DDC DISPLAY MODE CODES

DIAGNOSTIC DISPLAY	DESCRIPTION
8.8.8	FDC/DDC Display Check ("-1.8.8" On CCP)
.7.0	System Ready For Further Tests
E.0.0	ECM Malfunction Codes Cleared Or Switch Tests Completed
F.0.0	BCM Malfunction Codes Cleared
E.9.0	System Ready To Display ECM Data
E.9.5	System Ready For Output Cycling
E.9.6	Output Cycling
F.8.0	System Ready To Display BCM Data, ECC Program Number On CCP
F.8.5	Cooling Fans Override

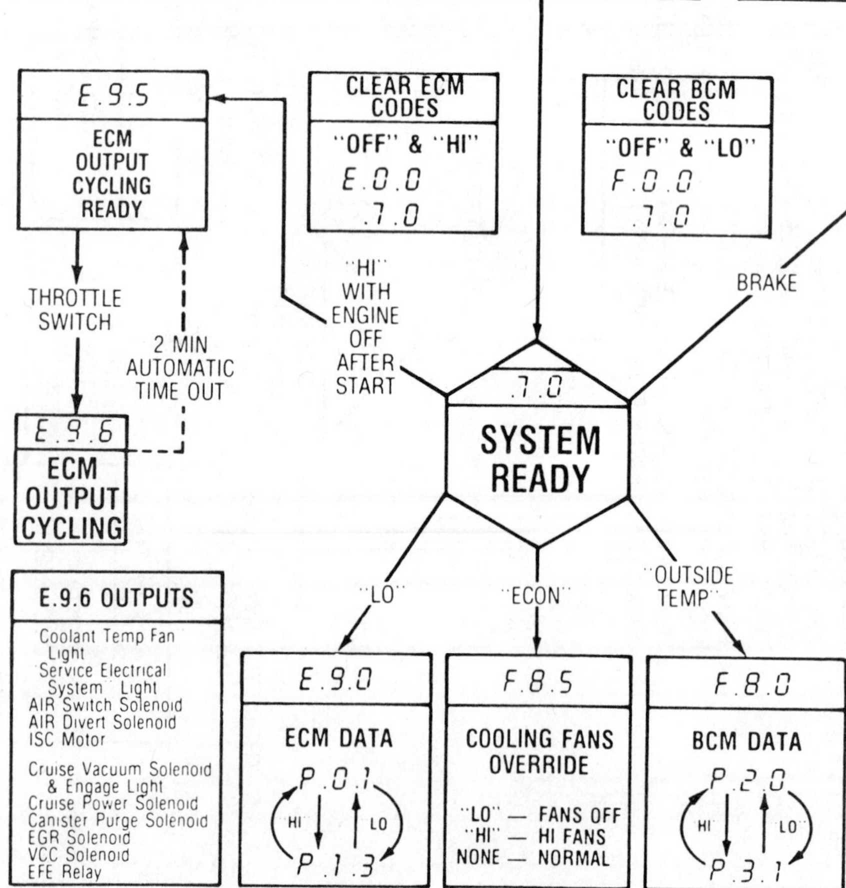

E.7.1 – E.7.8
ECM SWITCH TESTS
FAIL CODES
E.0.0
.7.0

E.7.1 — E.7.8 SWITCH TESTS

DIAGNOSTIC DISPLAY	CIRCUIT TEST
E.7.1	Cruise Control Brake
E.7.2	Throttle Switch
E.7.4	Park/Neutral Switch
E.7.5	Cruise "On-Off" — Return Switch to "On" Before Testing E.7.6 & E.7.7
E.7.6	Cruise "Set/Coast"
E.7.7	Cruise "Resume/Accel"
E.7.8	Power Steering Pressure Switch (Engine Running)

ECC PROGRAM NUMBER OVERRIDE

ECC PROGRAM NUMBER (0-100) IS DISPLAYED ON CCP DURING F.8.0 SEQUENCE.

0 = MAX A/C
100 = MAX HEAT

TO INCREASE PROGRAM NUMBER, PUSH "WARMER".
TO DECREASE PROGRAM NUMBER, PUSH "COOLER".

Cadillac diagnostic procedure—1983 and later models with HT4100 engine

E.9.0 ENGINE DATA DISPLAY

PARAMETER NUMBER	PARAMETER	PARAMETER RANGE	DISPLAY UNITS
P.0.1	Throttle Position	-10 - 90	Degrees
P.0.2	MAP	14 - 109	kPa
P.0.3	Computed BARO	61 - 103	kPa
P.0.4	Coolant Temperature	-40 - 151	°C
P.0.5	MAT	-40 - 151	°C
P.0.6	Injector Pulse Width	0 - 99.9	ms
P.0.7	Oxygen Sensor Voltage	0 - 1.14	Volts
P.0.8	Spark Advance	0 - 52	Degrees
P.0.9	Ignition Cycle Counter	0 - 50	Key Cycles
P.1.0	Battery Voltage	0 - 25.5	Volts
P.1.1	Engine RPM	0 - 6370	RPM ÷ 10
P.1.2	Car Speed	0 - 255	MPH
P.1.3	ECM PROM I.D.	0 - 255	Code

F.8.0 BCM DATA DISPLAY

PARAMETER NUMBER	PARAMETER	PARAMETER RANGE	DISPLAY UNITS
P.2.0	Commanded Blower Voltage	-3.3 - 18.0	Volts
P.2.1	Coolant Temperature	-40 - 215	°C
P.2.2	Commanded Air Mix Door Position	0 - 100	%
P.2.3	Actual Air Mix Door Position	0 - 100	%
P.2.4	Air Delivery Mode 0 = Max A/C 4 = Off 1 = A/C 5 = Normal Purge 2 = Intermediate 6 = Cold Purge 3 = Heater 7 = Front Defog	0 - 7	Code
P.2.5	In-Car Temperature	-40 - 102	°C
P.2.6	Actual Outside Temperature	-40 - 93	°C
P.2.7	High Side Temperature (Condenser Out)	-40 - 215	°C
P.2.8	Low Side Temperature (Evaporator In)	-40 - 93	°C
P.2.9	Actual Fuel Level	0 - 19.0	Gallons
P.3.0	Ignition Cycle Counter	0 - 99	Key Cycles
P.3.1	BCM PROM I.D.	0 - 255	Code

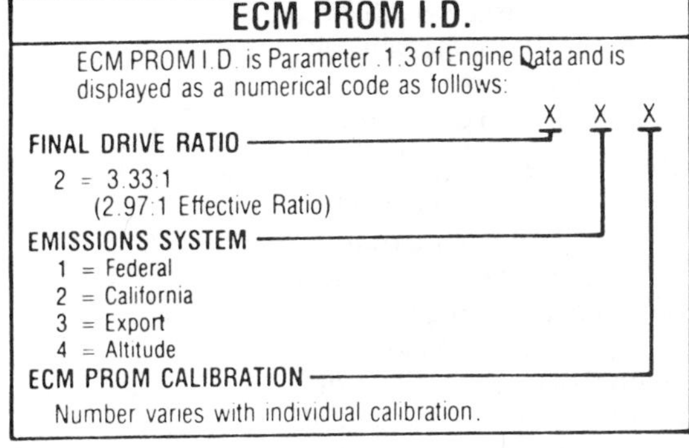

ECM PROM I.D.

ECM PROM I.D. is Parameter .1.3 of Engine Data and is displayed as a numerical code as follows:

X X X

FINAL DRIVE RATIO
2 = 3.33:1
(2.97:1 Effective Ratio)

EMISSIONS SYSTEM
1 = Federal
2 = California
3 = Export
4 = Altitude

ECM PROM CALIBRATION
Number varies with individual calibration.

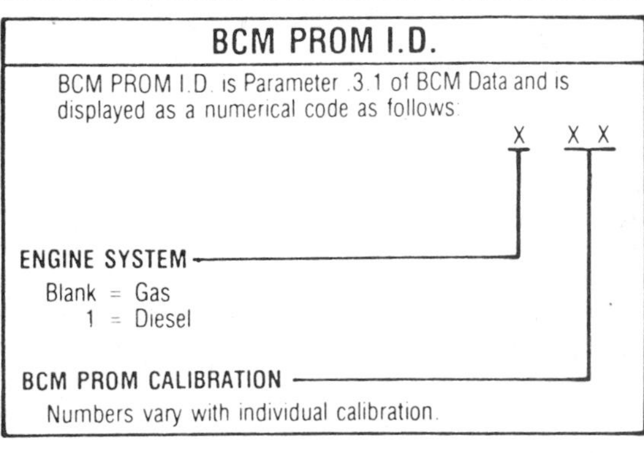

BCM PROM I.D.

BCM PROM I.D. is Parameter .3.1 of BCM Data and is displayed as a numerical code as follows:

X X X

ENGINE SYSTEM
Blank = Gas
1 = Diesel

BCM PROM CALIBRATION
Numbers vary with individual calibration.

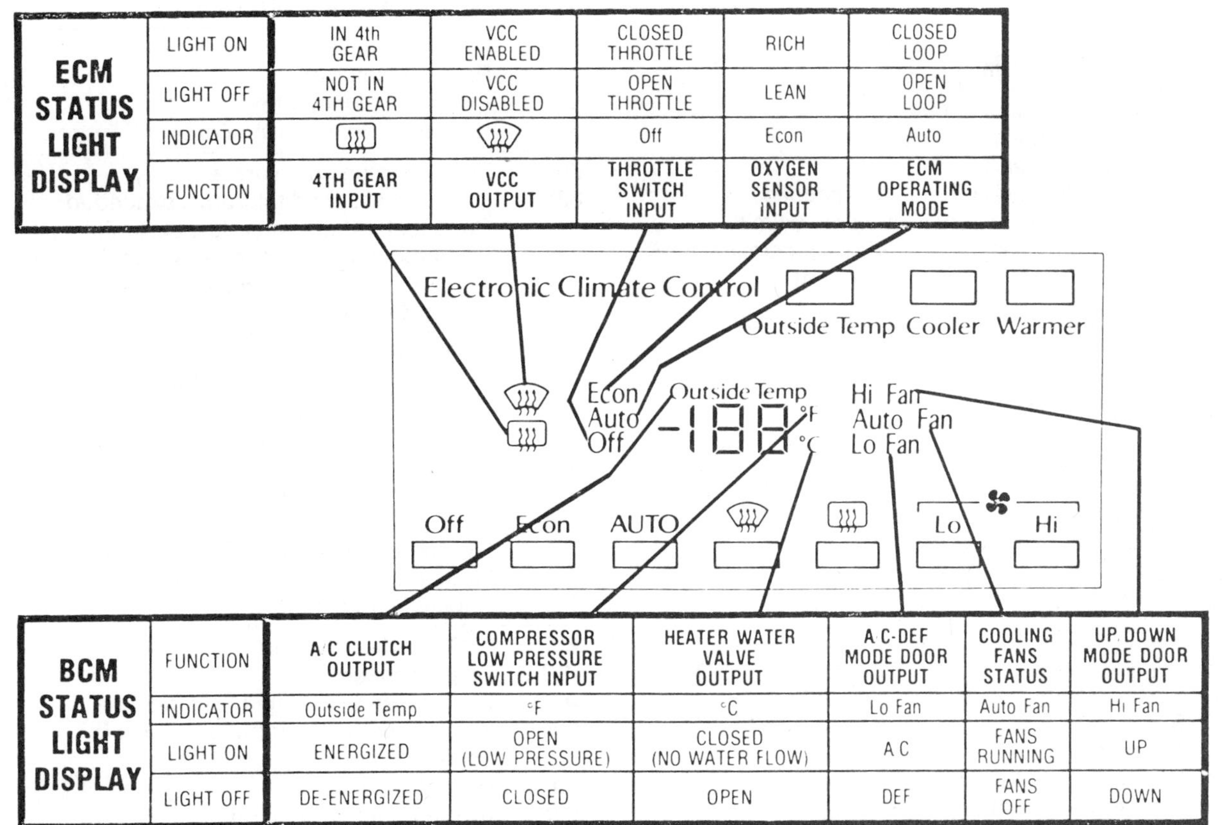

ECM STATUS LIGHT DISPLAY

LIGHT ON	IN 4th GEAR	VCC ENABLED	CLOSED THROTTLE	RICH	CLOSED LOOP
LIGHT OFF	NOT IN 4TH GEAR	VCC DISABLED	OPEN THROTTLE	LEAN	OPEN LOOP
INDICATOR	(rear defog symbol)	(front defog symbol)	Off	Econ	Auto
FUNCTION	**4TH GEAR INPUT**	**VCC OUTPUT**	**THROTTLE SWITCH INPUT**	**OXYGEN SENSOR INPUT**	**ECM OPERATING MODE**

BCM STATUS LIGHT DISPLAY

FUNCTION	A/C CLUTCH OUTPUT	COMPRESSOR LOW PRESSURE SWITCH INPUT	HEATER WATER VALVE OUTPUT	A/C-DEF MODE DOOR OUTPUT	COOLING FANS STATUS	UP/DOWN MODE DOOR OUTPUT
INDICATOR	Outside Temp	°F	°C	Lo Fan	Auto Fan	Hi Fan
LIGHT ON	ENERGIZED	OPEN (LOW PRESSURE)	CLOSED (NO WATER FLOW)	A/C	FANS RUNNING	UP
LIGHT OFF	DE-ENERGIZED	CLOSED	OPEN	DEF	FANS OFF	DOWN

Cadillac diagnostic procedure—1983 and later models with HT4100 engine

1981 DIAGNOSTIC CODES

(The following codes are programmed into the ECM.)

Code	Circuit Affected
12	No tach signal
13	O2 sensor not ready
14	Shorted coolant sensor
15	Open coolant sensor circuit
16	Generator voltage out of range
17	Crank signal circuit high
18	Open crank signal circuit
19	Fuel pump circuit high
20	Open fuel pump circuit
21	Shorted throttle position sensor circuit
22	Open throttle position sensor circuit
23	EST/bypass circuit shorted or open
24	Speed sensor failure
25	Modulated displacement failure
26	Shorted throttle switch circuit
27	Open throttle switch circuit
30	Idle speed control circuit
31	Short MAP sensor circuit
32	Open MAP sensor circuit
33	MAP/BARO sensor correlation
34	MAP hose
35	Shorted BARO sensor circuit
36	Open BARO sensor circuit
37	Shorted MAT sensor circuit
38	Open MAT sensor circuit
44	O2 sensor lean
45	O2 sensor rich
51	PROM insertion faulty
60	Drive (ADL) switch circuit
61	Set and resume switch circuit
62	Car speed exceeds maximum limit
63	Car and set speed tolerance exceeded
64	Car acceleration exceeds maximum limit
65	Coolant temperature exceeds maximum limit
66	Engine rpm exceeds maximum limit
68	Set and resume switch circuit
70	System ready—switch tests
71	Brake light switch
72	ISC throttle switch
73	Drive (ADL) switch
74	Back-up lamp switch
75	Cruise on/off circuit
76	Set/coast circuit
77	Resume/acceleration circuit
78	Instant/average mpg button
79	Reset mpg button
80	A/C clutch circuit
88	Display check
90	System ready to display engine data
95	System ready for actuator cycling
96	Actuator cycling
97	MD cylinder solenoid cycling
00	All diagnostic complete

1982 DIAGNOSTIC CODES

Code	Circuit Affected
12	No distributor (tach) signal
13	O_2 sensor not ready
14	Shorted coolant sensor circuit
15	Open coolant sensor circuit
16	Generator voltage out of range
18	Open crank signal circuit
19	Shorted fuel pump circuit
20	Open fuel pump circuit
21	Shorted throttle position sensor circuit
22	Open throttle position sensor circuit
23	EST circuit problem in run mode
24	Speed sensor circuit problem
25	EST circuit problem in bypass mode
26	Shorted throttle switch circuit
27	Open throttle switch circuit
28	Open fourth gear circuit
29	Shorted fourth gear circuit
30	ISC circuit problem
31	Short MAP sensor circuit
32	Open MAP sensor circuit
33	MAP/BARO sensor correlation
34	MAP signal too high
35	Shorted BARO sensor circuit
36	Open BARO sensor circuit
37	Shorted MAT sensor circuit
38	Open MAT sensor circuit
39	TCC engagement problem
44	Lean exhaust signal
45	Rich exhaust signal
51	PROM error indicator
52	ECM memory reset indicator
60	Transmission not in drive
63	Car and set speed tolerance exceeded
64	Car acceleration exceeds max. limit
65	Coolant temperature exceeds maximum limit
66	Engine rpm exceeds maximum limit
67	Shorted set or resume circuit
.7.0	System ready for further tests
.7.1	Cruise control brake circuit test
.7.2	Throttle switch circuit test
.7.3	Drive (ADL) circuit test
.7.4	Reverse circuit test
.7.5	Cruise on/off circuit test
.7.6	"Set/coast" circuit test
.7.7	"Resume/acceleration" circuit test
.7.8	"Instant/average" circuit test
.7.9	"Reset" circuit test
.8.0	A/C clutch circuit test
-1.8.8	Display check
.9.0	System ready to display engine data
.9.5	System ready for output cycling or in fixed spark mode
.9.6	Output cycling
.0.0	All diagnostics complete

Each code which did not pass the interrogation will be displayed beginning with the lowest number. This time through, the codes will not disappear until the tested component has been repaired and/or tested for proper operation.

9. Upon completion of the trouble code and switch test displays, the ECC panel will remain in the diagnostic mode and display "00" until an ECC mode is selected or the ignition is turned off.

10. Malfunctioning circuits should be analyzed.

"Intermittent" Codes vs. "Hard Failure" Codes

Trouble codes stored in the ECM's memory at any time can be either of the following.

1. A code for malfunctions which are occurring now ("HARD FAILURE"). This malfunction will cause illumination of the "CHECK ENGINE" light.
2. A code for any intermittent malfunctions which have occurred within the last 20 ignition switch cycles. These codes will not cause the "CHECK ENGINE" light to be on now.

Intermittent codes should be diagnosed by inspecting the connectors. During any diagnostic interrogation which displays more than one diagnostic code, it is necessary to determine which code is for the "HARD FAILURE" and which is the "INTERMITTENT." To make this determination, proceed as follows.

1. Enter diagnostics, read and record stored trouble codes.
2. Clear trouble codes.
3. Exit diagnostics by turning the ignition switch off for ten seconds.
4. Turn ignition on and wait 5 seconds, then start engine.
5. Accelerate the engine (to approximately 2000 rpm) for a few seconds.
6. Return to idle.
7. Shift transmission into Drive.
8. Shift to Park.
9. If the "CHECK ENGINE" light comes on, enter diagnostics. Read and record trouble codes. This will reveal only "HARD FAILURE" codes. If the light does not come on, then all stored codes are "INTERMITTENTS."
10. Begin diagnosis with lowest numbered code displayed.

CHART A
MULTIPLE DFI CODES STORED HARD

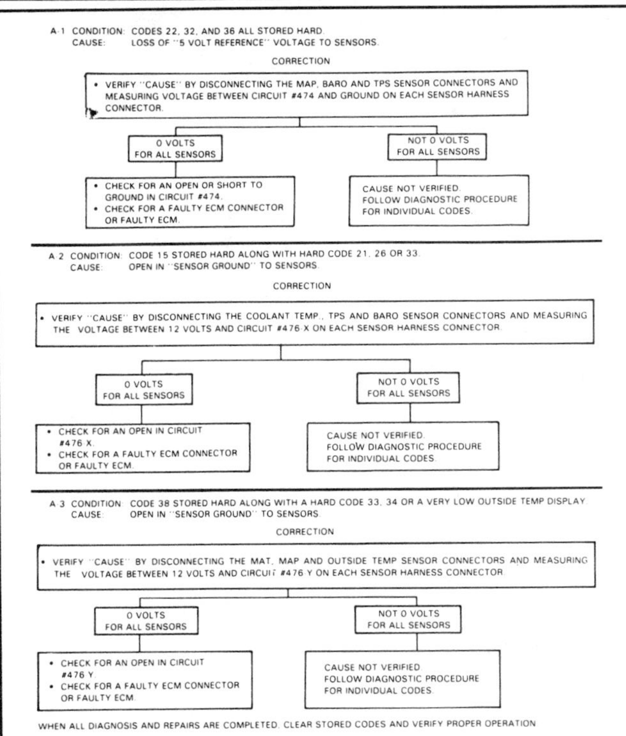

HARD DFI CODE 12
NO DISTRIBUTOR (TACH) SIGNAL

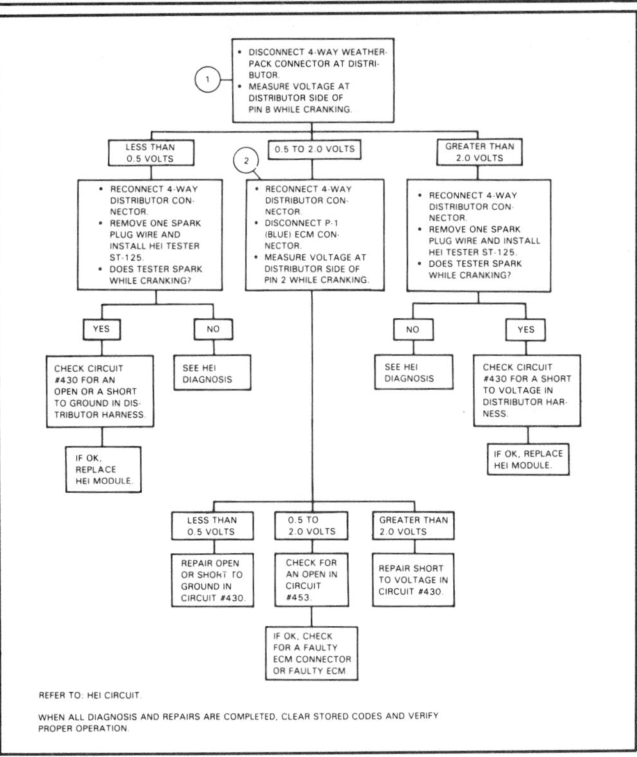

HARD OR INTERMITTENT DFI CODE 13
OXYGEN SENSOR NOT READY

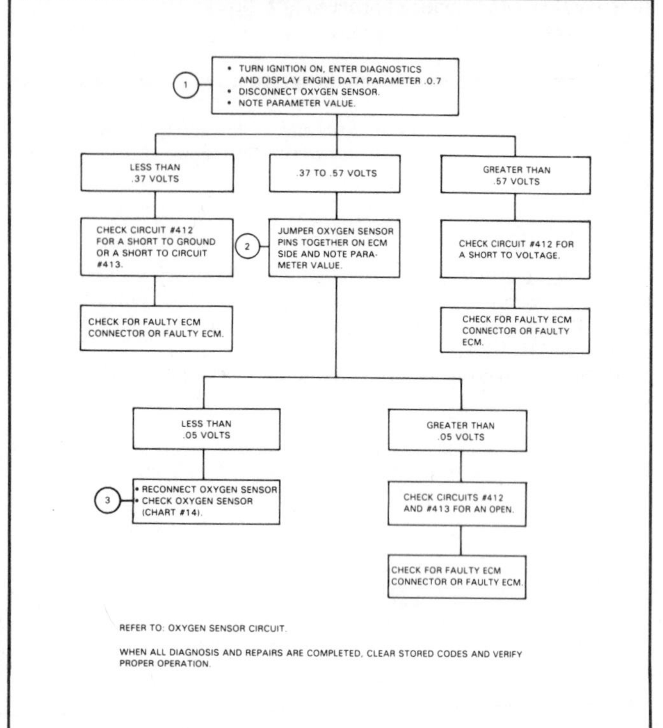
1
• TURN IGNITION ON, ENTER DIAGNOSTICS AND DISPLAY ENGINE DATA PARAMETER .0.7
• DISCONNECT OXYGEN SENSOR.
• NOTE PARAMETER VALUE.
LESS THAN .37 VOLTS
.37 TO .57 VOLTS
GREATER THAN .57 VOLTS
CHECK CIRCUIT #412 FOR A SHORT TO GROUND OR A SHORT TO CIRCUIT #413.
2
JUMPER OXYGEN SENSOR PINS TOGETHER ON ECM SIDE AND NOTE PARAMETER VALUE.
CHECK CIRCUIT #412 FOR A SHORT TO VOLTAGE.
CHECK FOR FAULTY ECM CONNECTOR OR FAULTY ECM.
CHECK FOR FAULTY ECM CONNECTOR OR FAULTY ECM.
LESS THAN .05 VOLTS
GREATER THAN .05 VOLTS
3
• RECONNECT OXYGEN SENSOR
• CHECK OXYGEN SENSOR (CHART #14).
CHECK CIRCUITS #412 AND #413 FOR AN OPEN.
CHECK FOR FAULTY ECM CONNECTOR OR FAULTY ECM.
REFER TO: OXYGEN SENSOR CIRCUIT.
WHEN ALL DIAGNOSIS AND REPAIRS ARE COMPLETED, CLEAR STORED CODES AND VERIFY PROPER OPERATION.

HARD DFI CODE 14
SHORTED COOLANT SENSOR CIRCUIT

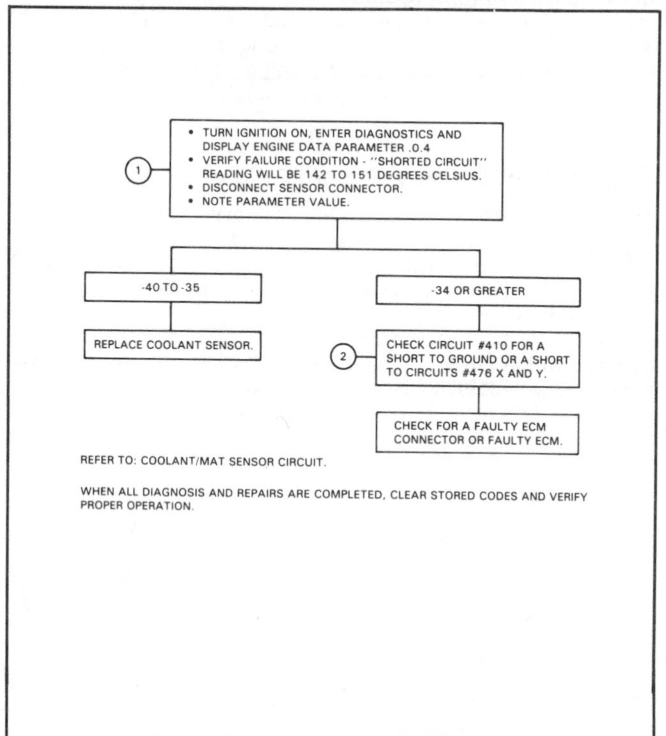
1
• TURN IGNITION ON, ENTER DIAGNOSTICS AND DISPLAY ENGINE DATA PARAMETER .0.4
• VERIFY FAILURE CONDITION - "SHORTED CIRCUIT" READING WILL BE 142 TO 151 DEGREES CELSIUS.
• DISCONNECT SENSOR CONNECTOR.
• NOTE PARAMETER VALUE.
-40 TO -35
-34 OR GREATER
REPLACE COOLANT SENSOR.
2
CHECK CIRCUIT #410 FOR A SHORT TO GROUND OR A SHORT TO CIRCUITS #476 X AND Y.
CHECK FOR A FAULTY ECM CONNECTOR OR FAULTY ECM.
REFER TO: COOLANT/MAT SENSOR CIRCUIT.
WHEN ALL DIAGNOSIS AND REPAIRS ARE COMPLETED, CLEAR STORED CODES AND VERIFY PROPER OPERATION.

HARD DFI CODE 15
OPEN COOLANT SENSOR CIRCUIT

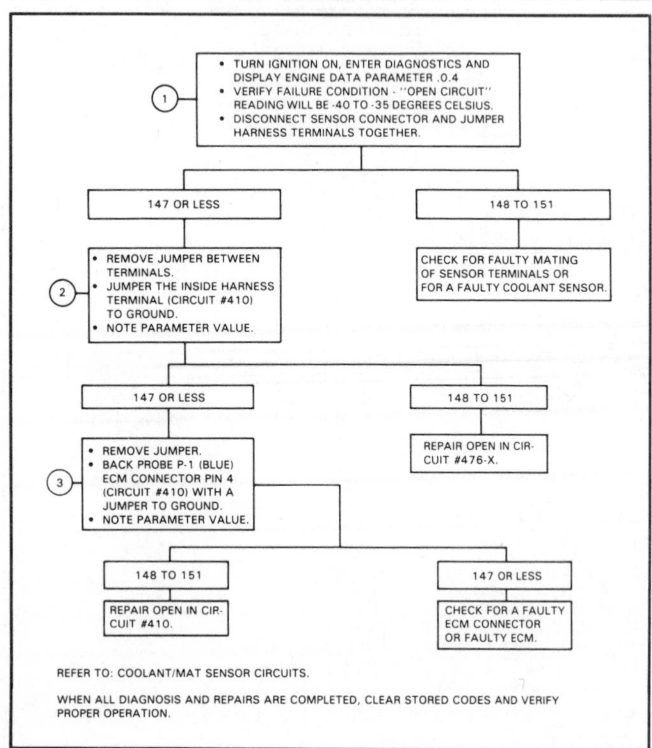
1
• TURN IGNITION ON, ENTER DIAGNOSTICS AND DISPLAY ENGINE DATA PARAMETER .0.4
• VERIFY FAILURE CONDITION - "OPEN CIRCUIT" READING WILL BE -40 TO -35 DEGREES CELSIUS.
• DISCONNECT SENSOR CONNECTOR AND JUMPER HARNESS TERMINALS TOGETHER.
147 OR LESS
148 TO 151
2
• REMOVE JUMPER BETWEEN TERMINALS.
• JUMPER THE INSIDE HARNESS TERMINAL (CIRCUIT #410) TO GROUND.
• NOTE PARAMETER VALUE.
CHECK FOR FAULTY MATING OF SENSOR TERMINALS OR FOR A FAULTY COOLANT SENSOR.
147 OR LESS
148 TO 151
3
• REMOVE JUMPER.
• BACK PROBE P-1 (BLUE) ECM CONNECTOR PIN 4 (CIRCUIT #410) WITH A JUMPER TO GROUND.
• NOTE PARAMETER VALUE.
REPAIR OPEN IN CIRCUIT #476-X.
148 TO 151
147 OR LESS
REPAIR OPEN IN CIRCUIT #410.
CHECK FOR A FAULTY ECM CONNECTOR OR FAULTY ECM.
REFER TO: COOLANT/MAT SENSOR CIRCUITS.
WHEN ALL DIAGNOSIS AND REPAIRS ARE COMPLETED, CLEAR STORED CODES AND VERIFY PROPER OPERATION.

HARD DFI CODE 16
GENERATOR VOLTAGE OUT OF RANGE

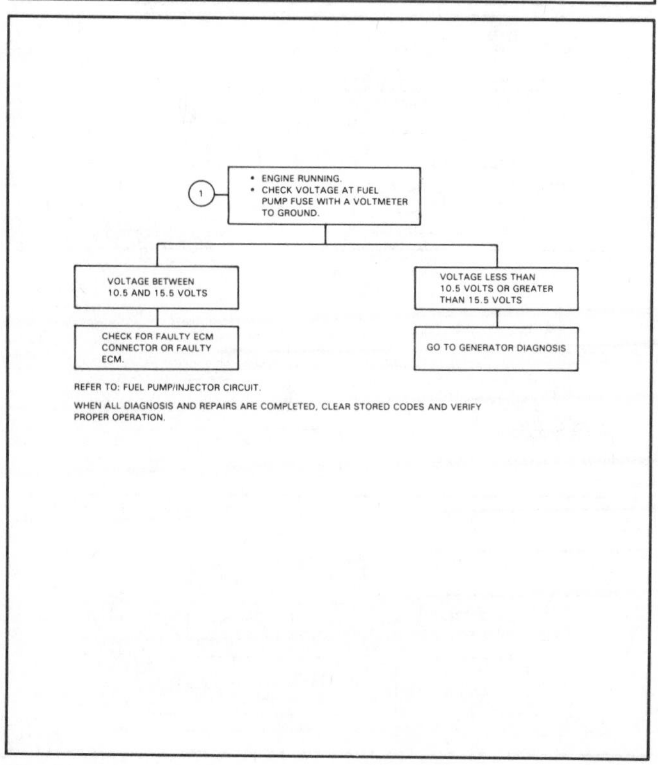
1
• ENGINE RUNNING.
• CHECK VOLTAGE AT FUEL PUMP FUSE WITH A VOLTMETER TO GROUND.
VOLTAGE BETWEEN 10.5 AND 15.5 VOLTS
VOLTAGE LESS THAN 10.5 VOLTS OR GREATER THAN 15.5 VOLTS
CHECK FOR FAULTY ECM CONNECTOR OR FAULTY ECM.
GO TO GENERATOR DIAGNOSIS
REFER TO: FUEL PUMP/INJECTOR CIRCUIT.
WHEN ALL DIAGNOSIS AND REPAIRS ARE COMPLETED, CLEAR STORED CODES AND VERIFY PROPER OPERATION.

HARD DFI CODE 18
OPEN CRANK SIGNAL CIRCUIT

(1) CHECK FOR OPEN 3 AMP CRANK FUSE.

FUSE BLOWN

TEST FOR SHORT TO GROUND IN CIRCUIT #6 BETWEEN FUSE AND ECM CONNECTOR.

SHORT TO GROUND

REPAIR SHORT AND REPLACE FUSE.

CIRCUIT OK

REPLACE FUSE DRIVE CAR RECHECK FUSE.

FUSE OK

STOP

FUSE BLOWN

CHECK FOR A FAULTY ECM CONNECTOR OR FAULTY ECM.

FUSE OK

(2) CHECK VOLTAGE AT CRANK FUSE WHILE CRANKING.

12.0 VOLTS

(3) BACK PROBE P-1 (BLUE) ECM CONNECTOR PIN 1 WHILE CRANKING.

12.0 VOLTS

0 VOLTS

REPAIR OPEN IN CIRCUIT #6 BETWEEN FUSE AND ECM.

0 VOLTS

REPAIR OPEN WIRE TO IGNITION SWITCH IN CIRCUIT #6

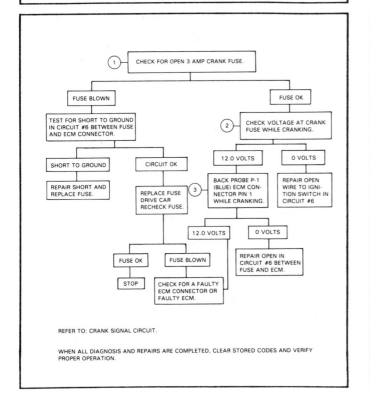

REFER TO: CRANK SIGNAL CIRCUIT.

WHEN ALL DIAGNOSIS AND REPAIRS ARE COMPLETED, CLEAR STORED CODES AND VERIFY PROPER OPERATION.

HARD DFI CODE 19
SHORTED FUEL PUMP CIRCUIT

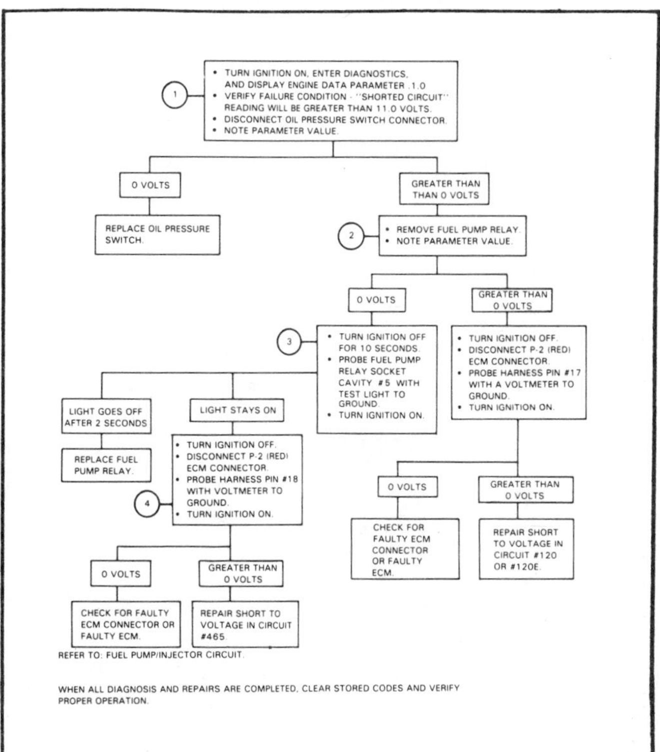

HARD OR INTERMITTENT DFI CODE 20
OPEN FUEL PUMP CIRCUIT

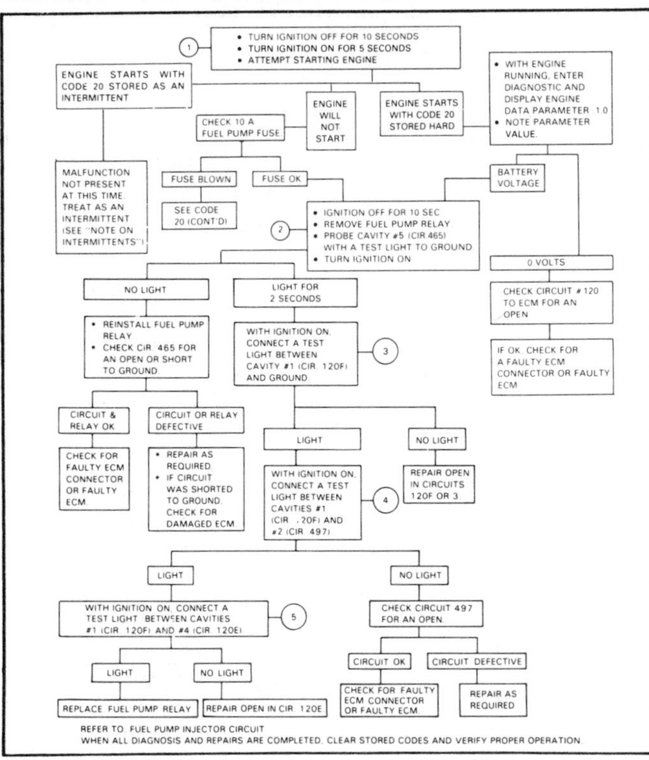

DFI CODE 20 - (CONT'D)
FUSE BLOWN

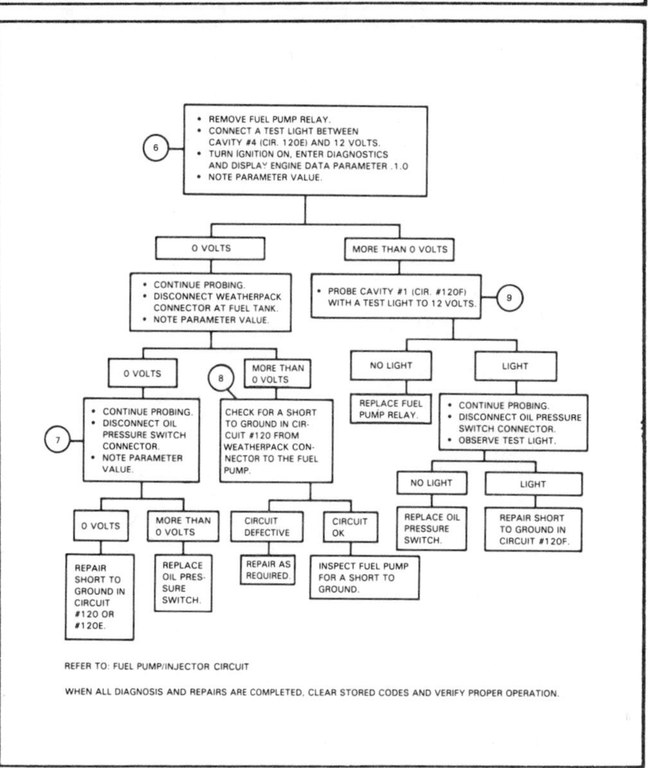

HARD DFI CODE 21
SHORTED THROTTLE POSITION SENSOR CIRCUIT

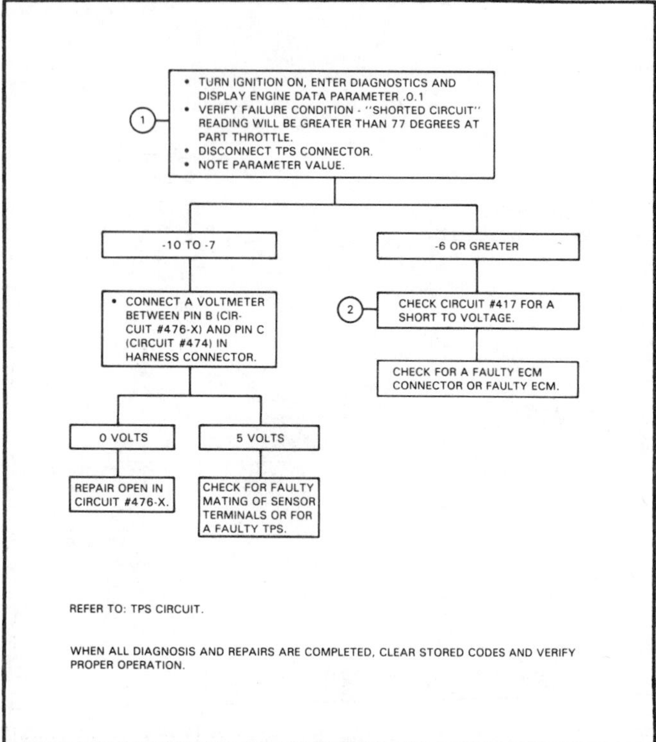

HARD DFI CODE 22
OPEN THROTTLE POSITION SENSOR CIRCUIT

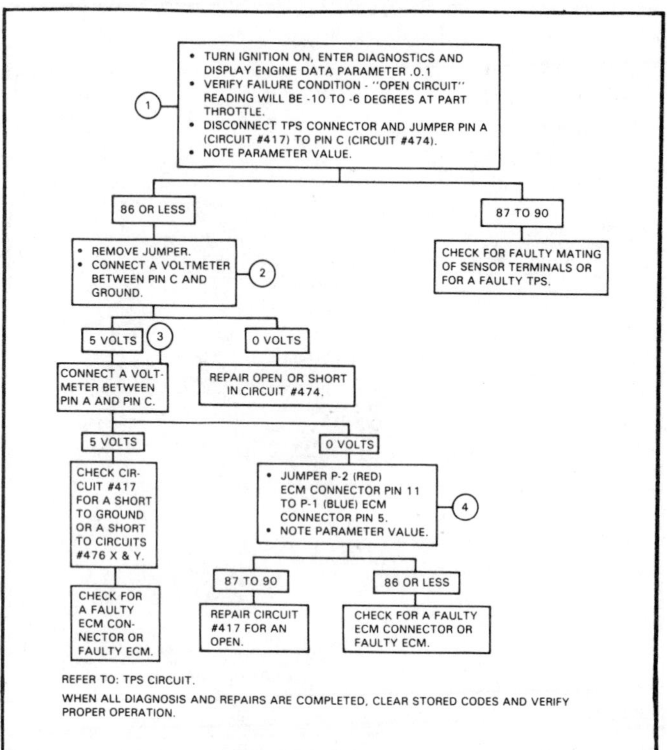

HARD DFI CODE 23
EST CIRCUIT PROBLEM

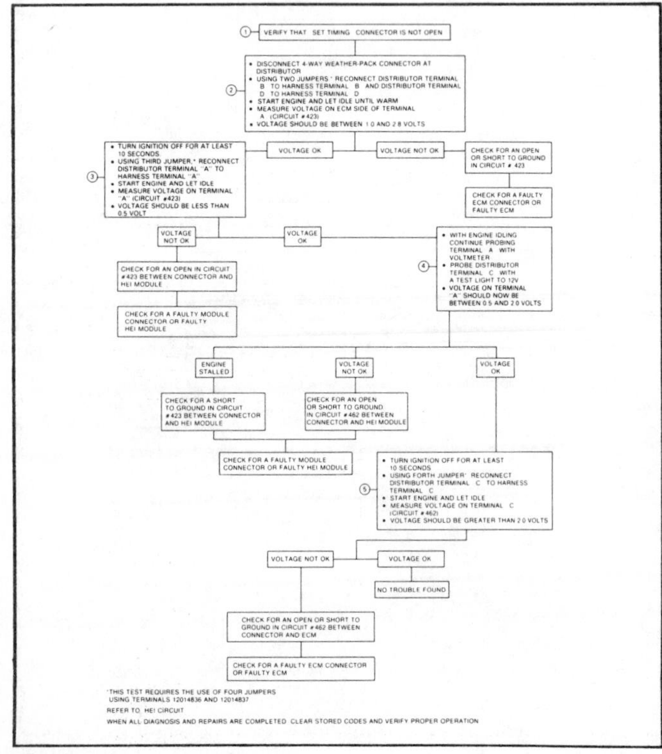

HARD DFI CODE 24(A)
SPEED SENSOR CIRCUIT PROBLEM
(FOR VEHICLES WITHOUT DIGITAL CLUSTER)

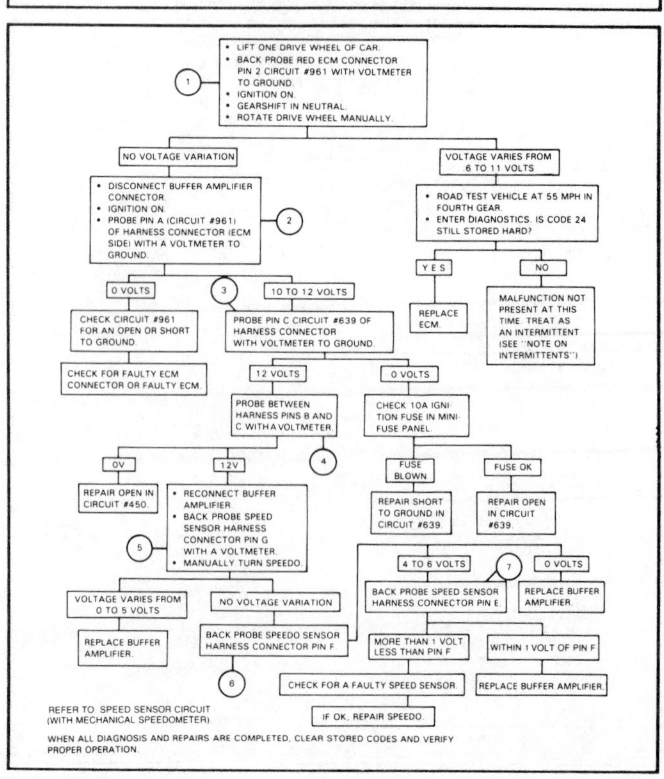

HARD DFI CODE 24(B)
SPEED SENSOR CIRCUIT PROBLEM
(FOR VEHICLES WITH DIGITAL SPEEDOMETER)

(1) DOES DIGITAL SPEEDOMETER WORK PROPERLY?

NO

(3) VERIFY THAT SPEEDO CABLE IS INSTALLED PROPERLY.

- DISCONNECT BUFFER AMPLIFIER CONNECTOR.
- WITH IGNITION ON PROBE PIN D CIRCUIT #39 OF HARNESS CONNECTOR WITH A VOLTMETER.

12 VOLTS

(4) PROBE PIN D TO PIN E WITH VOLTMETER

0 VOLTS — REPAIR OPEN IN CIRCUIT #151

12 VOLTS

- RECONNECT BUFFER AMPLIFIER.
- BACK PROBE SPEED SENSOR PIN G WITH A VOLTMETER. MANUALLY TURN SPEEDO. (5)

VOLTAGE VARIES FROM 0 TO 5V. — REPLACE BUFFER AMPLIFIER.

NO VOLTAGE VARIATION. — (6) BACK PROBE PIN F.

4 TO 6 VOLTS — (7) BACK PROBE PIN E.

WITHIN 1 VOLT OF PIN F — REPLACE BUFFER AMPLIFIER

MORE THAN 1 VOLT LESS THAN PIN F — CHECK FOR A FAULTY SPEED SENSOR — IF OK, REPAIR SPEEDO

0 VOLTS — REPLACE BUFFER AMPLIFIER.

0 VOLTS — CHECK 20A GA-TRANS FUSE

FUSE OK — REPAIR OPEN IN CIRCUIT #39

FUSE BLOWN — REPAIR SHORT IN CIRCUITS FED BY GA-TRANS FUSE.

YES

- LIFT ONE DRIVE WHEEL OF CAR.
- BACK PROBE RED ECM CONNECTOR PIN 2 CIRCUIT #961 WITH VOLTMETER TO GROUND.
- IGNITION ON.
- GEARSHIFT IN NEUTRAL.
- ROTATE DRIVE WHEEL MANUALLY.

VOLTAGE VARIES FROM 6 TO 11 VOLTS

- ROAD TEST VEHICLE AT 55 MPH IN FOURTH GEAR.
- ENTER DIAGNOSTICS. IS CODE 24 STILL STORED HARD?

YES — REPLACE ECM.

NO — MALFUNCTION NOT PRESENT AT THIS TIME. TREAT AS AN INTERMITTENT (SEE "NOTE ON INTERMITTENTS")

NO VOLTAGE VARIATION.

(2)
- DISCONNECT BUFFER AMPLIFIER CONNECTOR.
- IGNITION ON.
- PROBE PIN C CIRCUIT #961 OF HARNESS CONNECTOR WITH A VOLTMETER TO GROUND.

10 TO 12 VOLTS — REPLACE BUFFER

0 VOLTS — CHECK CIRCUIT #961 FOR AN OPEN OR SHORT TO GROUND — CHECK FOR FAULTY ECM CONNECTOR OR FAULTY ECM

REFER TO: SPEED SENSOR CIRCUIT (WITH DIGITAL SPEEDOMETER)
WHEN ALL DIAGNOSIS AND REPAIRS ARE COMPLETED, CLEAR CODES AND VERIFY PROPER OPERATION.

HARD DFI CODE 26
SHORTED THROTTLE SWITCH CIRCUIT

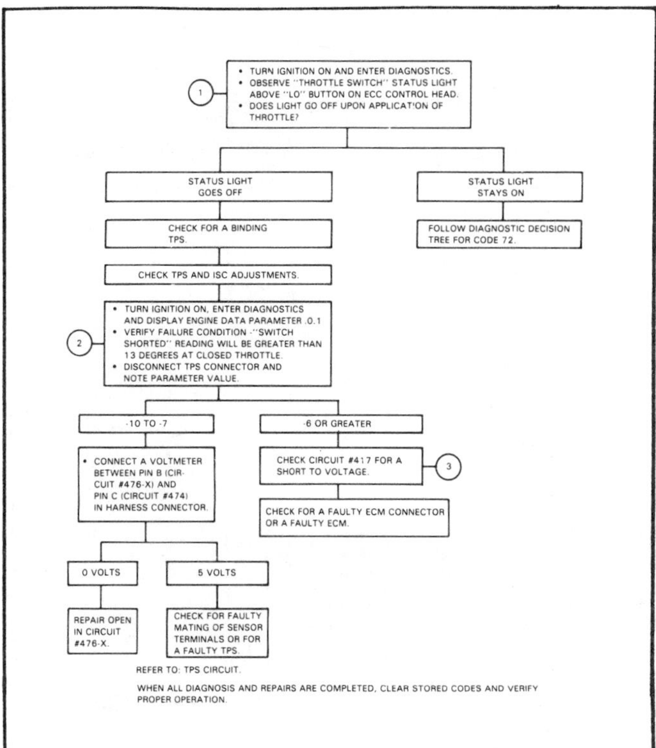

HARD DFI CODE 27
OPEN THROTTLE SWITCH CIRCUIT

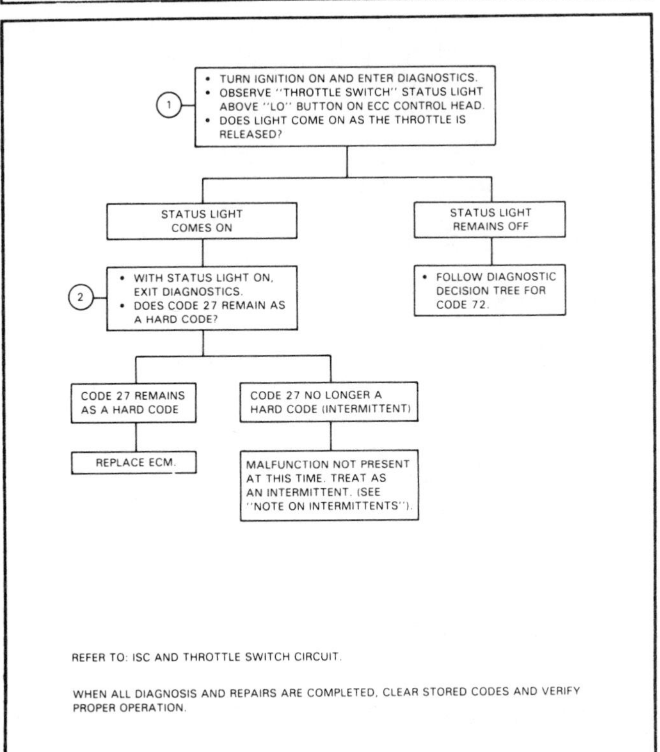

HARD DFI CODE 28
OPEN 4TH GEAR CIRCUIT

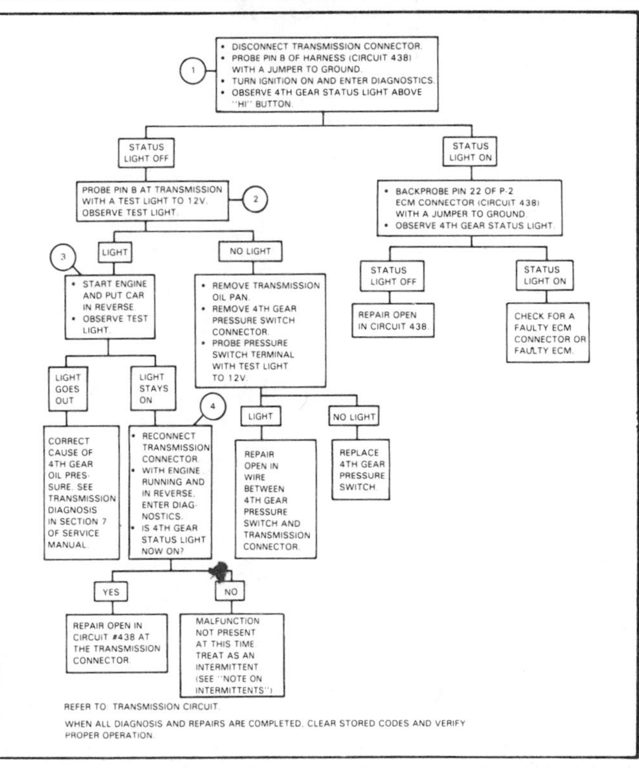

HARD DFI CODE 29
SHORTED 4TH GEAR CIRCUIT

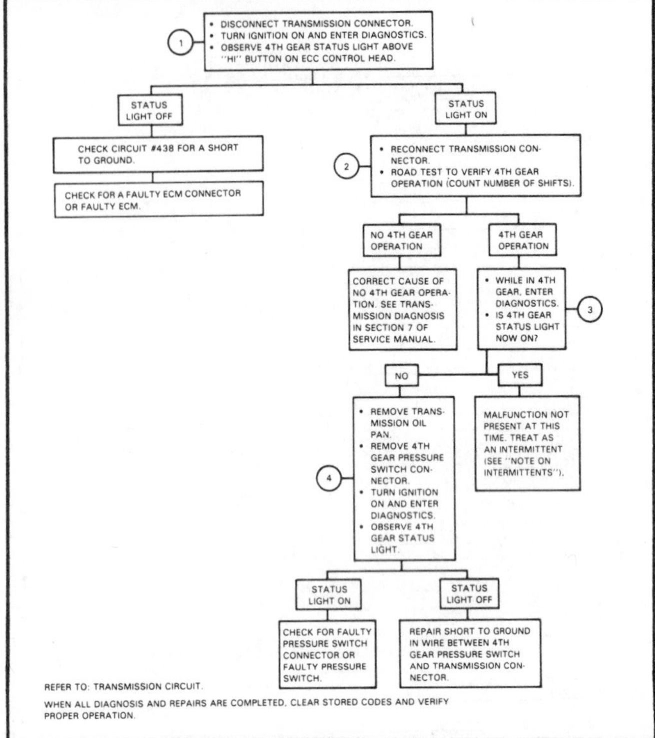
1
• DISCONNECT TRANSMISSION CONNECTOR.
• TURN IGNITION ON AND ENTER DIAGNOSTICS.
• OBSERVE 4TH GEAR STATUS LIGHT ABOVE "HI" BUTTON ON ECC CONTROL HEAD.
STATUS LIGHT OFF
CHECK CIRCUIT #438 FOR A SHORT TO GROUND.
CHECK FOR A FAULTY ECM CONNECTOR OR FAULTY ECM.
STATUS LIGHT ON
2
• RECONNECT TRANSMISSION CONNECTOR.
• ROAD TEST TO VERIFY 4TH GEAR OPERATION (COUNT NUMBER OF SHIFTS).
NO 4TH GEAR OPERATION
CORRECT CAUSE OF NO 4TH GEAR OPERATION. SEE TRANSMISSION DIAGNOSIS IN SECTION 7 OF SERVICE MANUAL.
4TH GEAR OPERATION
3
• WHILE IN 4TH GEAR, ENTER DIAGNOSTICS.
• IS 4TH GEAR STATUS LIGHT NOW ON?
NO
YES
MALFUNCTION NOT PRESENT AT THIS TIME. TREAT AS AN INTERMITTENT (SEE "NOTE ON INTERMITTENTS").
4
• REMOVE TRANSMISSION OIL PAN.
• REMOVE 4TH GEAR PRESSURE SWITCH CONNECTOR.
• TURN IGNITION ON AND ENTER DIAGNOSTICS.
• OBSERVE 4TH GEAR STATUS LIGHT.
STATUS LIGHT ON
CHECK FOR FAULTY PRESSURE SWITCH CONNECTOR OR FAULTY PRESSURE SWITCH.
STATUS LIGHT OFF
REPAIR SHORT TO GROUND IN WIRE BETWEEN 4TH GEAR PRESSURE SWITCH AND TRANSMISSION CONNECTOR.
REFER TO: TRANSMISSION CIRCUIT.
WHEN ALL DIAGNOSIS AND REPAIRS ARE COMPLETED, CLEAR STORED CODES AND VERIFY PROPER OPERATION.

HARD OR INTERMITTENT DFI CODE 30
IDLE SPEED CONTROL CIRCUIT PROBLEM

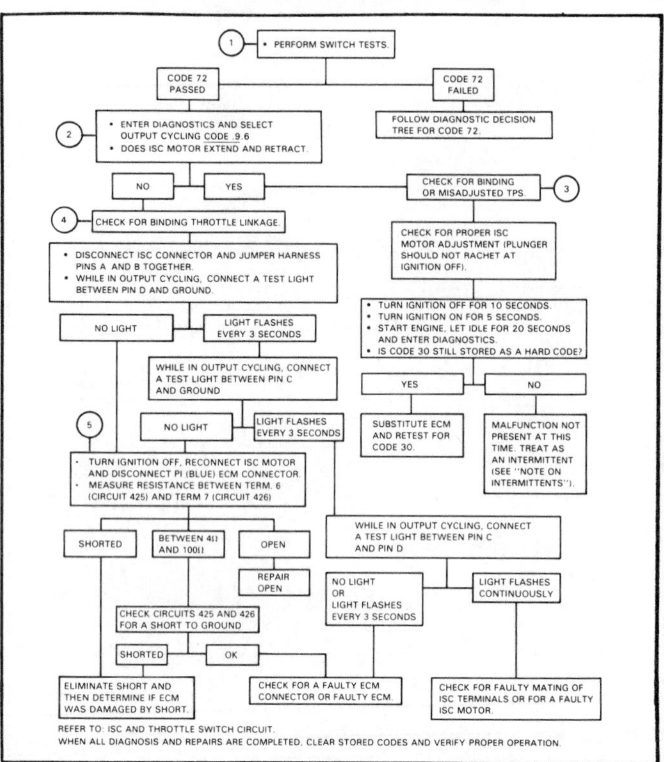
1
• PERFORM SWITCH TESTS.
CODE 72 PASSED
CODE 72 FAILED
FOLLOW DIAGNOSTIC DECISION TREE FOR CODE 72.
2
• ENTER DIAGNOSTICS AND SELECT OUTPUT CYCLING CODE .9.6
• DOES ISC MOTOR EXTEND AND RETRACT.
NO
YES
CHECK FOR BINDING OR MISADJUSTED TPS.
3
CHECK FOR PROPER ISC MOTOR ADJUSTMENT (PLUNGER SHOULD NOT RACHET AT IGNITION OFF).
• TURN IGNITION OFF FOR 10 SECONDS.
• TURN IGNITION ON FOR 5 SECONDS.
• START ENGINE, LET IDLE FOR 20 SECONDS AND ENTER DIAGNOSTICS.
• IS CODE 30 STILL STORED AS A HARD CODE?
YES
SUBSTITUTE ECM AND RETEST FOR CODE 30.
NO
MALFUNCTION NOT PRESENT AT THIS TIME. TREAT AS AN INTERMITTENT (SEE "NOTE ON INTERMITTENTS").
4
CHECK FOR BINDING THROTTLE LINKAGE.
• DISCONNECT ISC CONNECTOR AND JUMPER HARNESS PINS A AND B TOGETHER.
• WHILE IN OUTPUT CYCLING, CONNECT A TEST LIGHT BETWEEN PIN D AND GROUND.
NO LIGHT
LIGHT FLASHES EVERY 3 SECONDS
WHILE IN OUTPUT CYCLING, CONNECT A TEST LIGHT BETWEEN PIN C AND GROUND
NO LIGHT
LIGHT FLASHES EVERY 3 SECONDS
5
• TURN IGNITION OFF, RECONNECT ISC MOTOR AND DISCONNECT PI (BLUE) ECM CONNECTOR.
• MEASURE RESISTANCE BETWEEN TERM. 6 (CIRCUIT 425) AND TERM 7 (CIRCUIT 426)
SHORTED
BETWEEN 4Ω AND 100Ω
OPEN
REPAIR OPEN
CHECK CIRCUITS 425 AND 426 FOR A SHORT TO GROUND
SHORTED
OK
ELIMINATE SHORT AND THEN DETERMINE IF ECM WAS DAMAGED BY SHORT.
CHECK FOR A FAULTY ECM CONNECTOR OR FAULTY ECM.
WHILE IN OUTPUT CYCLING, CONNECT A TEST LIGHT BETWEEN PIN C AND PIN D
NO LIGHT OR LIGHT FLASHES EVERY 3 SECONDS
LIGHT FLASHES CONTINUOUSLY
CHECK FOR FAULTY MATING OF ISC TERMINALS OR FOR A FAULTY ISC MOTOR.
REFER TO: ISC AND THROTTLE SWITCH CIRCUIT.
WHEN ALL DIAGNOSIS AND REPAIRS ARE COMPLETED, CLEAR STORED CODES AND VERIFY PROPER OPERATION.

HARD DFI CODE 31
SHORTED MAP SENSOR CIRCUIT

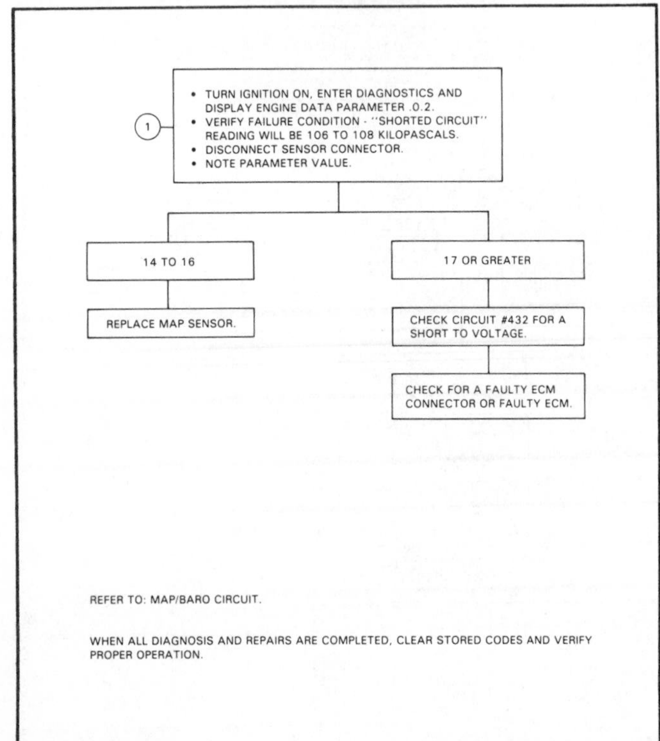
1
• TURN IGNITION ON, ENTER DIAGNOSTICS AND DISPLAY ENGINE DATA PARAMETER .0.2.
• VERIFY FAILURE CONDITION - "SHORTED CIRCUIT" READING WILL BE 106 TO 108 KILOPASCALS.
• DISCONNECT SENSOR CONNECTOR.
• NOTE PARAMETER VALUE.
14 TO 16
REPLACE MAP SENSOR.
17 OR GREATER
CHECK CIRCUIT #432 FOR A SHORT TO VOLTAGE.
CHECK FOR A FAULTY ECM CONNECTOR OR FAULTY ECM.
REFER TO: MAP/BARO CIRCUIT.
WHEN ALL DIAGNOSIS AND REPAIRS ARE COMPLETED, CLEAR STORED CODES AND VERIFY PROPER OPERATION.

HARD DFI CODE 32
OPEN MAP SENSOR CIRCUIT

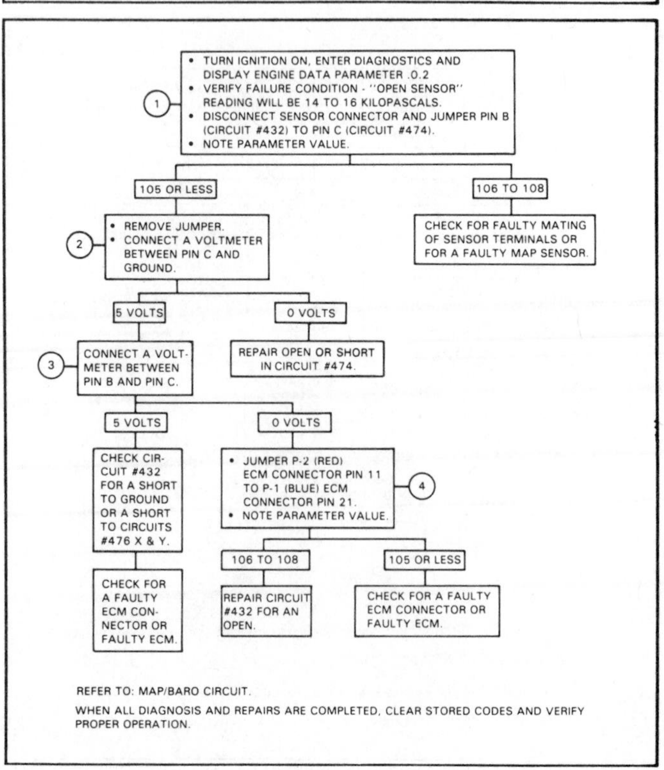
1
• TURN IGNITION ON, ENTER DIAGNOSTICS AND DISPLAY ENGINE DATA PARAMETER .0.2
• VERIFY FAILURE CONDITION - "OPEN SENSOR" READING WILL BE 14 TO 16 KILOPASCALS.
• DISCONNECT SENSOR CONNECTOR AND JUMPER PIN B (CIRCUIT #432) TO PIN C (CIRCUIT #474).
• NOTE PARAMETER VALUE.
105 OR LESS
106 TO 108
CHECK FOR FAULTY MATING OF SENSOR TERMINALS OR FOR A FAULTY MAP SENSOR.
2
• REMOVE JUMPER.
• CONNECT A VOLTMETER BETWEEN PIN C AND GROUND.
5 VOLTS
0 VOLTS
REPAIR OPEN OR SHORT IN CIRCUIT #474.
3
CONNECT A VOLTMETER BETWEEN PIN B AND PIN C.
5 VOLTS
CHECK CIRCUIT #432 FOR A SHORT TO GROUND OR A SHORT TO CIRCUITS #476 X & Y.
CHECK FOR A FAULTY ECM CONNECTOR OR FAULTY ECM.
0 VOLTS
4
• JUMPER P-2 (RED) ECM CONNECTOR PIN 11 TO P-1 (BLUE) ECM CONNECTOR PIN 21.
• NOTE PARAMETER VALUE.
106 TO 108
REPAIR CIRCUIT #432 FOR AN OPEN.
105 OR LESS
CHECK FOR A FAULTY ECM CONNECTOR OR FAULTY ECM.
REFER TO: MAP/BARO CIRCUIT.
WHEN ALL DIAGNOSIS AND REPAIRS ARE COMPLETED, CLEAR STORED CODES AND VERIFY PROPER OPERATION.

HARD OR INTERMITTENT DFI CODE 33
MAP/BARO SENSOR CORRELATION

1
- CLEAR CODES AND TURN IGNITION OFF FOR 10 SECONDS.
- TURN IGNITION ON FOR 10 SECONDS AND ENTER DIAGNOSTICS.
- READ STORED CODES.

CODE 33 STORED AS AN INTERMITTENT

4
- DISCONNECT MAP AND BARO SENSORS.
- OBSERVE ENGINE DATA PARAMETER .0.2 AS THE MAP SENSOR IS RECONNECTED.
- OBSERVE ENGINE DATA PARAMETER .0.3 AS THE BARO SENSOR IS RECONNECTED.

MAP DISPLAYED INTERMEDIATE VALUES BEFORE REACHING MAXIMUM → REPLACE MAP SENSOR.

BARO DISPLAYED INTERMEDIATE VALUES BEFORE REACHING MAXIMUM → REPLACE BARO SENSOR.

BOTH SENSORS WENT IMMEDIATELY TO MAXIMUM VALUE → SUBSTITUTE ECM AND RETEST FOR CODE 33.

CODE 33 NO LONGER STORED → MALFUNCTION NOT PRESENT AT THIS TIME. TREAT AS AN INTERMITTENT. (SEE "NOTE ON INTERMITTENTS".)

CODE 33 STORED AS A HARD FAILURE

2
- DISPLAY ENGINE DATA PARAMETERS .0.2 AND .0.3 AND RECORD VALUES.
- VERIFY FAILURE CONDITION - "CORRELATION" READINGS WILL DIFFER BY 2 OR MORE KPA.
- JUMPER PIN A OF MAP SENSOR TO PIN A OF BARO SENSOR AND NOTE PARAMETER VALUES.

MAP MOVED CLOSER TO BARO → REPAIR OPEN IN CIRCUIT #476-Y.

BARO MOVED CLOSER TO MAP → REPAIR OPEN IN CIRCUIT #476-X.

NO CHANGE

3
- OBTAIN A REPLACEMENT MAP OR BARO SENSOR AND CONNECT TO APPROPRIATE HARNESS.
- COMPARE THIS SENSOR'S PARAMETER VALUE TO THOSE RECORDED PREVIOUSLY.

CLOSER TO MAP → REPLACE BARO SENSOR.

CLOSER TO BARO → REPLACE MAP SENSOR.

WHEN ALL DIAGNOSIS AND REPAIRS ARE COMPLETED, CLEAR STORED CODES AND VERIFY PROPER OPERATION.

HARD DFI CODE 34
MAP SIGNAL TOO HIGH

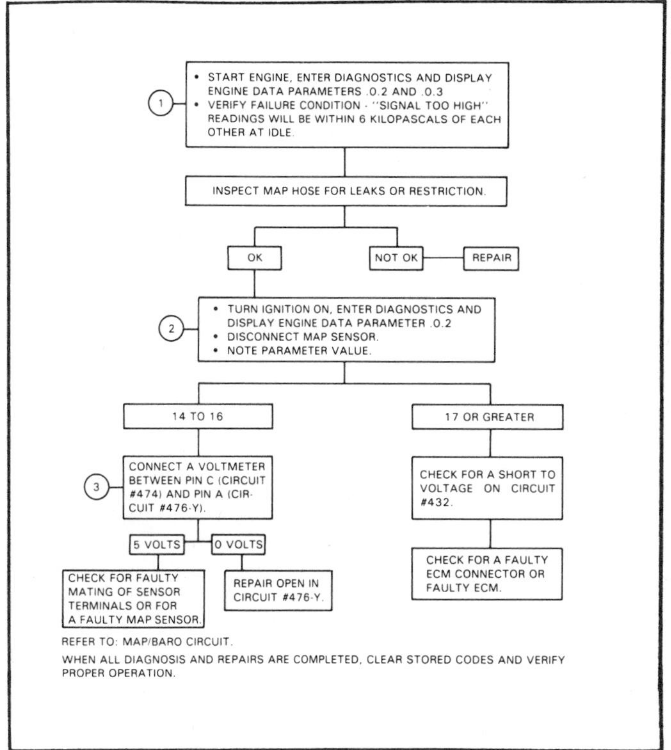

HARD DFI CODE 35
SHORTED BARO SENSOR CIRCUIT

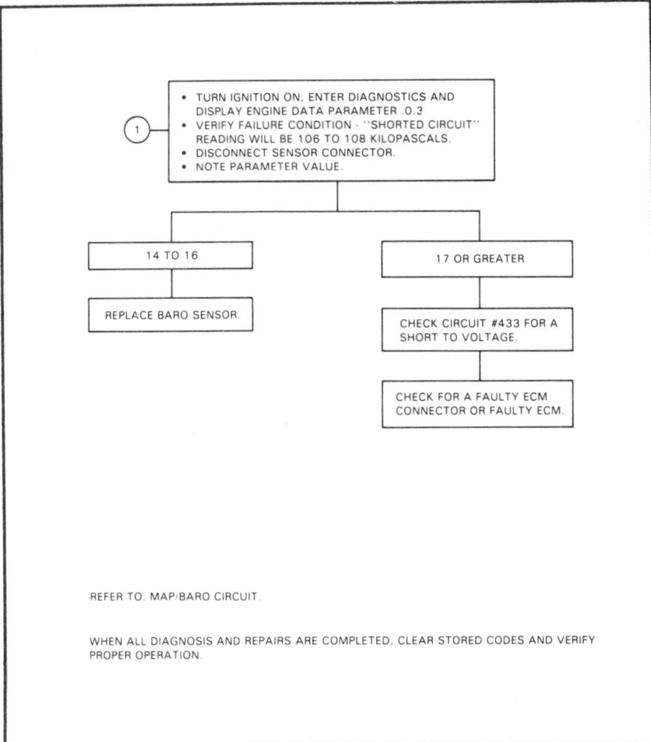

HARD DFI CODE 36
OPEN BARO SENSOR CIRCUIT

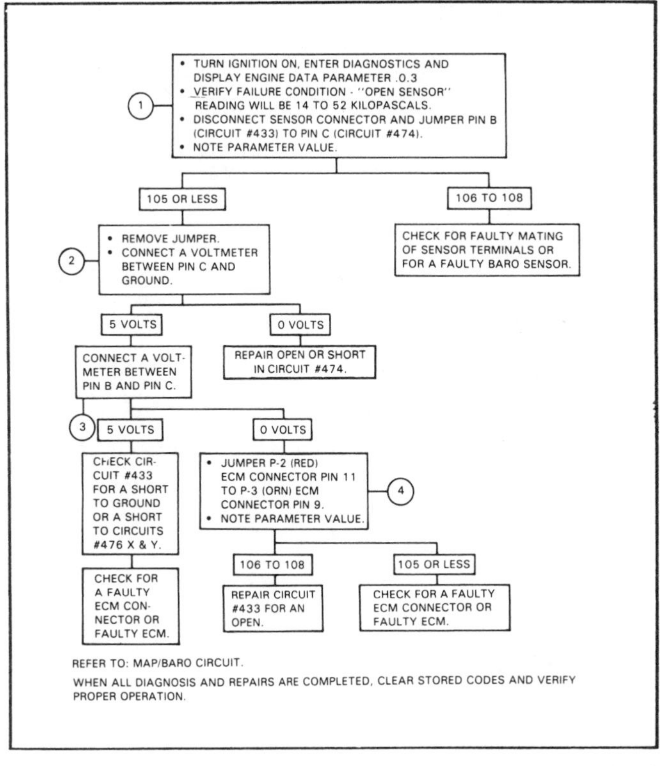

HARD DFI CODE 37
SHORTED MAT SENSOR CIRCUIT

(1)
- TURN IGNITION ON, ENTER DIAGNOSTICS AND DISPLAY ENGINE DATA PARAMETER .0.5
- VERIFY FAILURE CONDITION - "SHORTED CIRCUIT" READING WILL BE 142 TO 151 DEGREES CELSIUS.
- DISCONNECT SENSOR CONNECTOR.
- NOTE PARAMETER VALUE.

-40 TO -35 → REPLACE MAT SENSOR.

-34 OR GREATER → (2) CHECK CIRCUIT #472 FOR A SHORT TO GROUND OR A SHORT TO CIRCUITS #476 X & Y. → CHECK FOR A FAULTY ECM CONNECTOR OR FAULTY ECM.

REFER TO: COOLANT/MAT SENSOR CIRCUIT.

WHEN ALL DIAGNOSIS AND REPAIRS ARE COMPLETED, CLEAR STORED CODES AND VERIFY PROPER OPERATION.

HARD DFI CODE 38
OPEN MAT SENSOR CIRCUIT

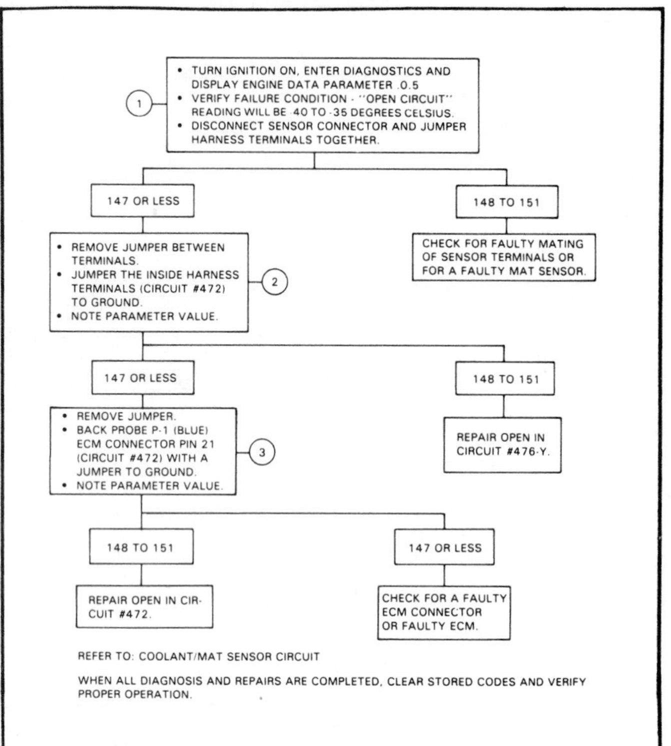

1
• TURN IGNITION ON, ENTER DIAGNOSTICS AND DISPLAY ENGINE DATA PARAMETER .0.5
• VERIFY FAILURE CONDITION - "OPEN CIRCUIT" READING WILL BE -40 TO -35 DEGREES CELSIUS.
• DISCONNECT SENSOR CONNECTOR AND JUMPER HARNESS TERMINALS TOGETHER.
147 OR LESS
148 TO 151
• REMOVE JUMPER BETWEEN TERMINALS.
• JUMPER THE INSIDE HARNESS TERMINALS (CIRCUIT #472) TO GROUND.
• NOTE PARAMETER VALUE.
2
CHECK FOR FAULTY MATING OF SENSOR TERMINALS OR FOR A FAULTY MAT SENSOR.
147 OR LESS
148 TO 151
• REMOVE JUMPER.
• BACK PROBE P-1 (BLUE) ECM CONNECTOR PIN 21 (CIRCUIT #472) WITH A JUMPER TO GROUND.
• NOTE PARAMETER VALUE.
3
REPAIR OPEN IN CIRCUIT #476-Y.
148 TO 151
147 OR LESS
REPAIR OPEN IN CIRCUIT #472.
CHECK FOR A FAULTY ECM CONNECTOR OR FAULTY ECM.
REFER TO: COOLANT/MAT SENSOR CIRCUIT
WHEN ALL DIAGNOSIS AND REPAIRS ARE COMPLETED, CLEAR STORED CODES AND VERIFY PROPER OPERATION.

HARD OR INTERMITTENT DFI
CODE 39 TCC ENGAGEMENT PROBLEM

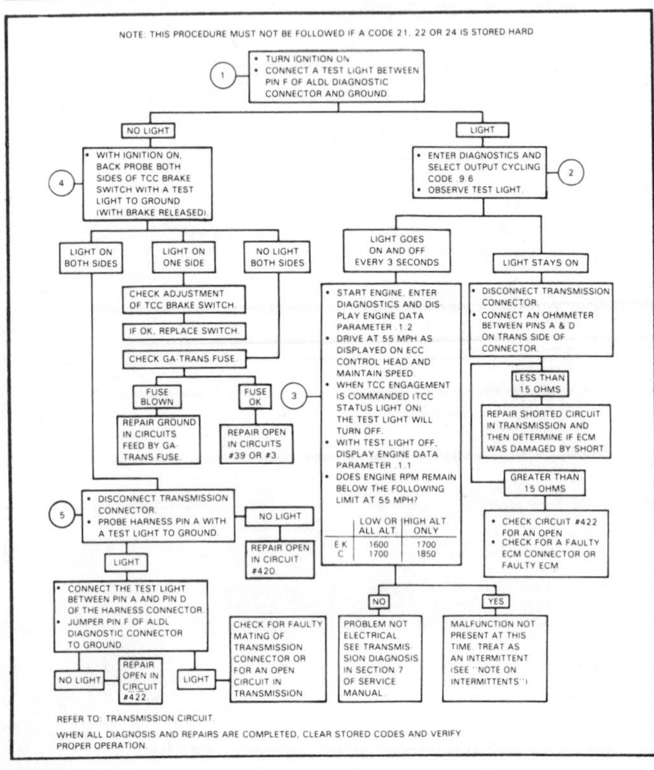

NOTE: THIS PROCEDURE MUST NOT BE FOLLOWED IF A CODE 21, 22 OR 24 IS STORED HARD
1
• TURN IGNITION ON
• CONNECT A TEST LIGHT BETWEEN PIN F OF ALDL DIAGNOSTIC CONNECTOR AND GROUND.
NO LIGHT
LIGHT
• WITH IGNITION ON, BACK PROBE BOTH SIDES OF TCC BRAKE SWITCH WITH A TEST LIGHT TO GROUND (WITH BRAKE RELEASED).
4
• ENTER DIAGNOSTICS AND SELECT OUTPUT CYCLING CODE .9.6
• OBSERVE TEST LIGHT.
2
LIGHT ON BOTH SIDES
LIGHT ON ONE SIDE
NO LIGHT BOTH SIDES
CHECK ADJUSTMENT OF TCC BRAKE SWITCH.
IF OK, REPLACE SWITCH.
CHECK GA TRANS FUSE.
FUSE BLOWN
FUSE OK
REPAIR GROUND IN CIRCUITS FEED BY GA TRANS FUSE.
REPAIR OPEN IN CIRCUITS #39 OR #3.
LIGHT GOES ON AND OFF EVERY 3 SECONDS
LIGHT STAYS ON
• START ENGINE, ENTER DIAGNOSTICS AND DISPLAY ENGINE DATA PARAMETER .1.2
• DRIVE AT 55 MPH AS DISPLAYED ON ECC CONTROL HEAD AND MAINTAIN SPEED.
• WHEN TCC ENGAGEMENT IS COMMANDED (TCC STATUS LIGHT ON) THE TEST LIGHT WILL TURN OFF.
• WITH TEST LIGHT OFF, DISPLAY ENGINE DATA PARAMETER .1.1
• DOES ENGINE RPM REMAIN BELOW THE FOLLOWING LIMIT AT 55 MPH?
3
LOW OR ALL ALT
HIGH ALT ONLY
E K 1600 1700
C 1700 1850
• DISCONNECT TRANSMISSION CONNECTOR.
• CONNECT AN OHMMETER BETWEEN PINS A & D ON TRANS SIDE OF CONNECTOR.
LESS THAN 15 OHMS
REPAIR SHORTED CIRCUIT IN TRANSMISSION AND THEN DETERMINE IF ECM WAS DAMAGED BY SHORT.
GREATER THAN 15 OHMS
• CHECK CIRCUIT #422 FOR AN OPEN.
• CHECK FOR A FAULTY ECM CONNECTOR OR FAULTY ECM.
5
• DISCONNECT TRANSMISSION CONNECTOR.
• PROBE HARNESS PIN A WITH A TEST LIGHT TO GROUND.
NO LIGHT
REPAIR OPEN IN CIRCUIT #420.
LIGHT
• CONNECT THE TEST LIGHT BETWEEN PIN A AND PIN D OF THE HARNESS CONNECTOR.
• JUMPER PIN F OF ALDL DIAGNOSTIC CONNECTOR TO GROUND.
NO LIGHT
REPAIR OPEN IN CIRCUIT #422
LIGHT
CHECK FOR FAULTY MATING OF TRANSMISSION CONNECTOR OR FOR AN OPEN CIRCUIT IN TRANSMISSION
NO
YES
PROBLEM NOT ELECTRICAL. SEE TRANSMISSION DIAGNOSIS IN SECTION 7 OF SERVICE MANUAL
MALFUNCTION NOT PRESENT AT THIS TIME. TREAT AS AN INTERMITTENT (SEE "NOTE ON INTERMITTENTS")
REFER TO: TRANSMISSION CIRCUIT.
WHEN ALL DIAGNOSIS AND REPAIRS ARE COMPLETED, CLEAR STORED CODES AND VERIFY PROPER OPERATION.

HARD OR INTERMITTENT DFI CODE 44
LEAN EXHAUST SIGNAL

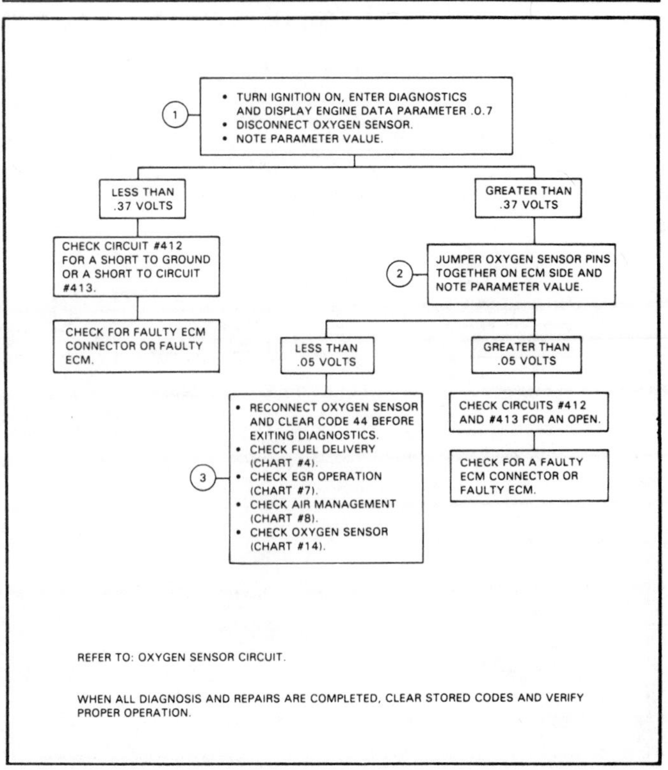

HARD OR INTERMITTENT DFI CODE 45
RICH EXHAUST SIGNAL

(1)
- TURN IGNITION ON, ENTER DIAGNOSTICS AND DISPLAY ENGINE DATA PARAMETER .0.7
- DISCONNECT OXYGEN SENSOR.
- NOTE PARAMETER VALUE.

LESS THAN .57 VOLTS

(2) JUMPER OXYGEN SENSOR PINS TOGETHER ON ECM SIDE AND NOTE PARAMETER VALUE.

LESS THAN .05 VOLTS

- RECONNECT OXYGEN SENSOR
- CHECK FUEL DELIVERY (CHART #4).
- CHECK FOR RESTRICTED AIR CLEANER.
- CHECK OXYGEN SENSOR (CHART #14).

GREATER THAN .05 VOLTS

(3) CHECK CIRCUITS #412 AND #413 FOR AN OPEN.

CHECK FOR A FAULTY ECM CONNECTOR OR FAULTY ECM.

GREATER THAN .57 VOLTS

CHECK CIRCUIT #412 FOR A SHORT TO VOLTAGE.

CHECK FOR A FAULTY ECM CONNECTOR OR FAULTY ECM.

REFER TO: OXYGEN SENSOR CIRCUIT.

WHEN ALL DIAGNOSIS AND REPAIRS ARE COMPLETED, CLEAR STORED CODES AND VERIFY PROPER OPERATION.

DFI CHART #1
NO START OR STALL AFTER START

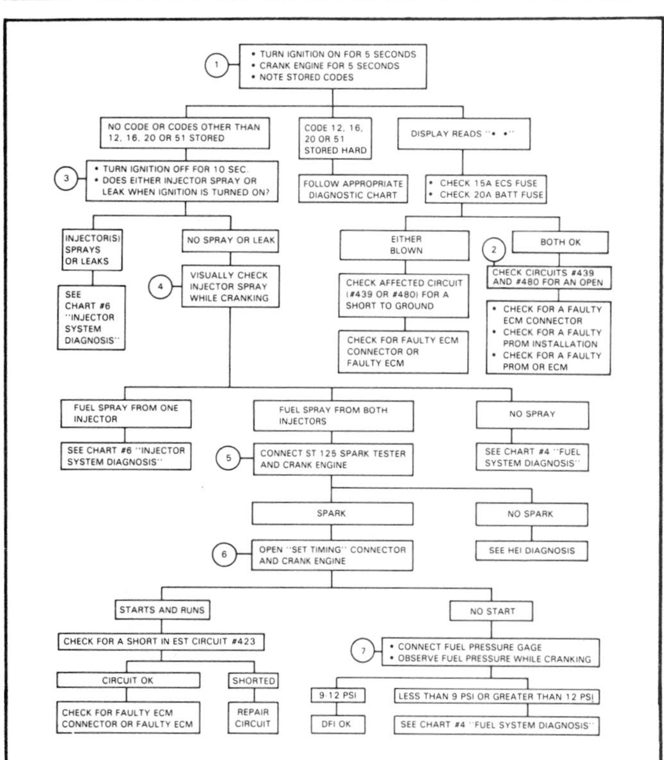

DFI CHART #2
"SERVICE NOW" AND/OR "SERVICE SOON" LIGHTS ON — NO HARD CODES

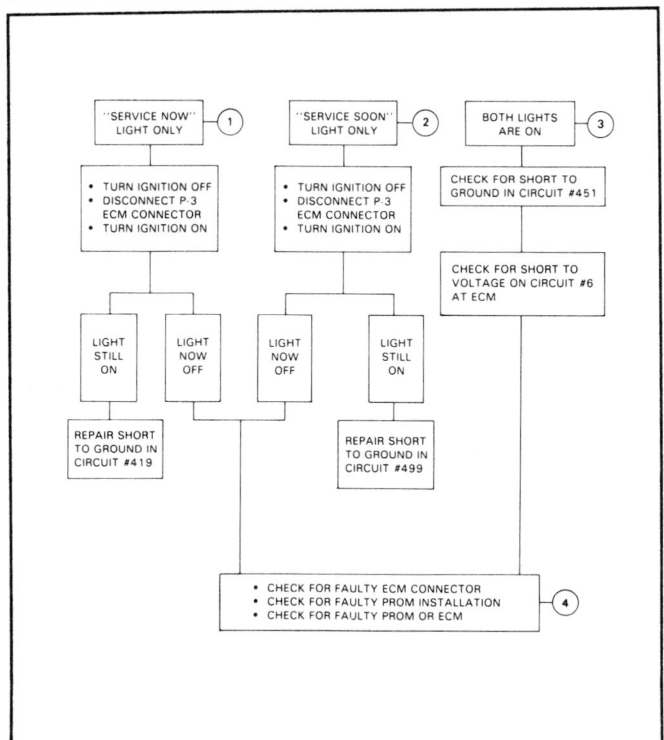

DFI CHART #3A
'SERVICE NOW' LIGHT INOPERATIVE

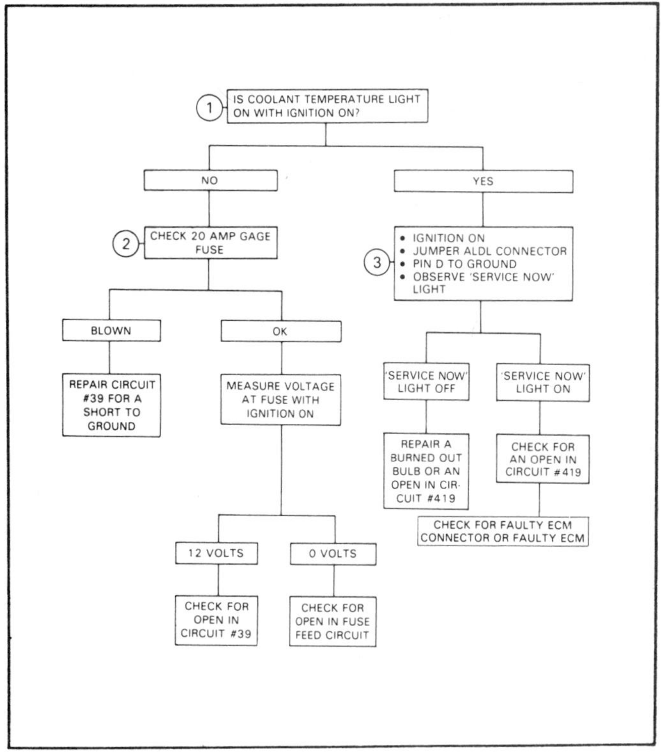

DFI CHART #3B 'SERVICE SOON' LIGHT INOPERATIVE

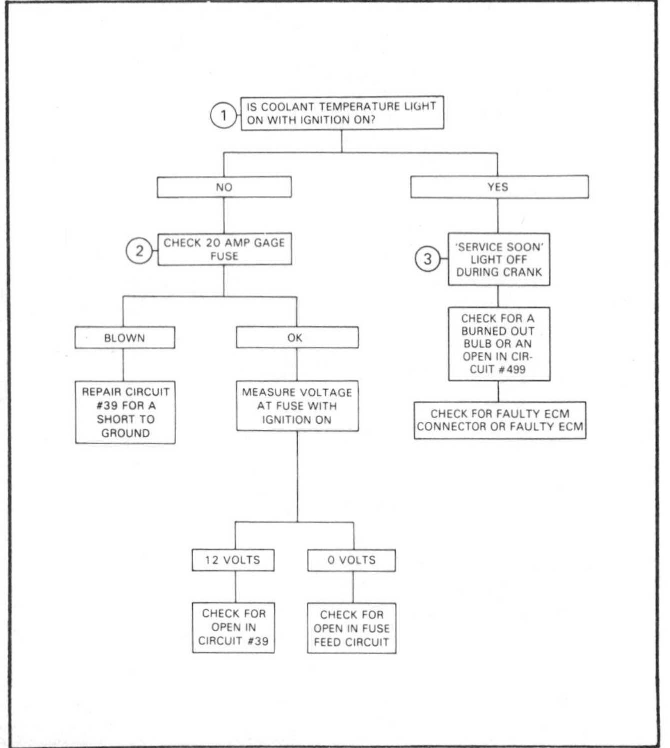

DFI CHART #4 FUEL SYSTEM DIAGNOSIS

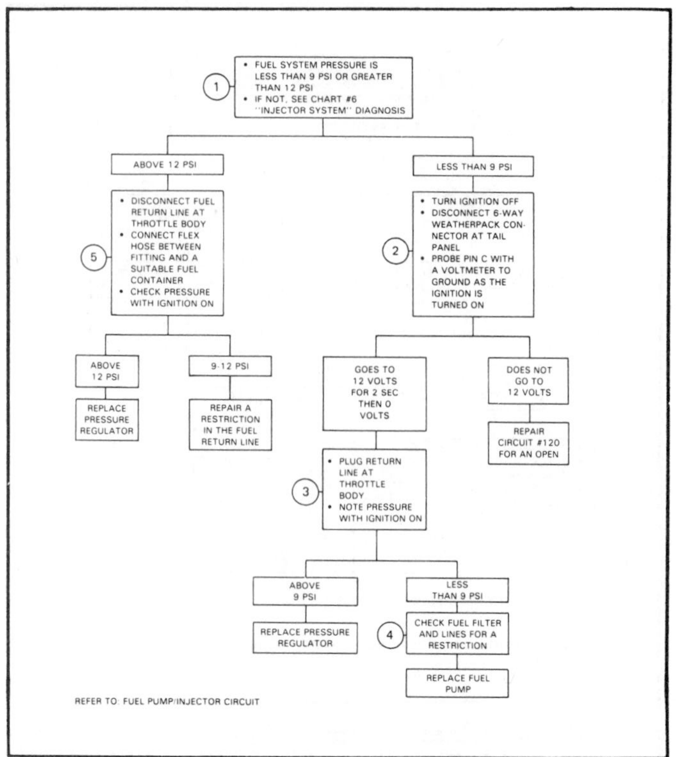

DFI CHART #5 POOR PERFORMANCE AND/OR POOR FUEL ECONOMY

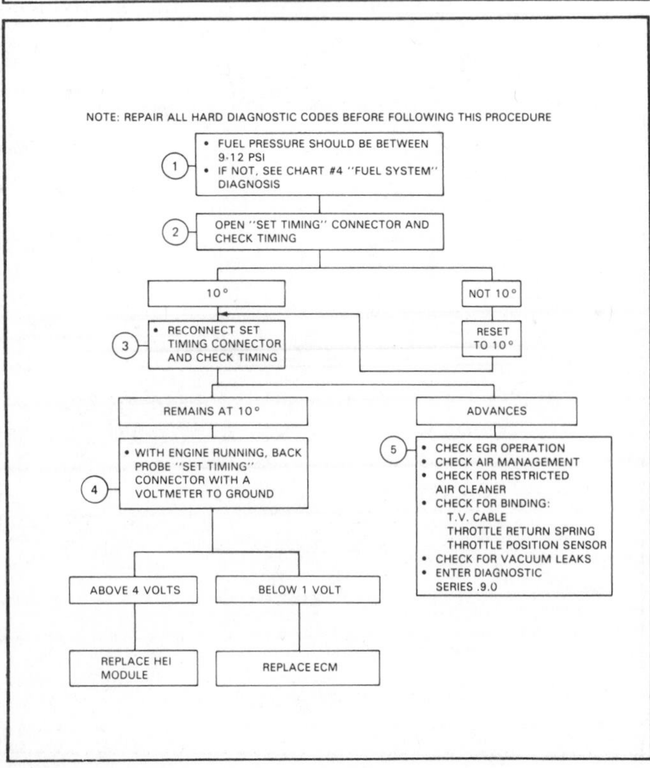

DFI CHART #6 INJECTOR SYSTEM DIAGNOSIS

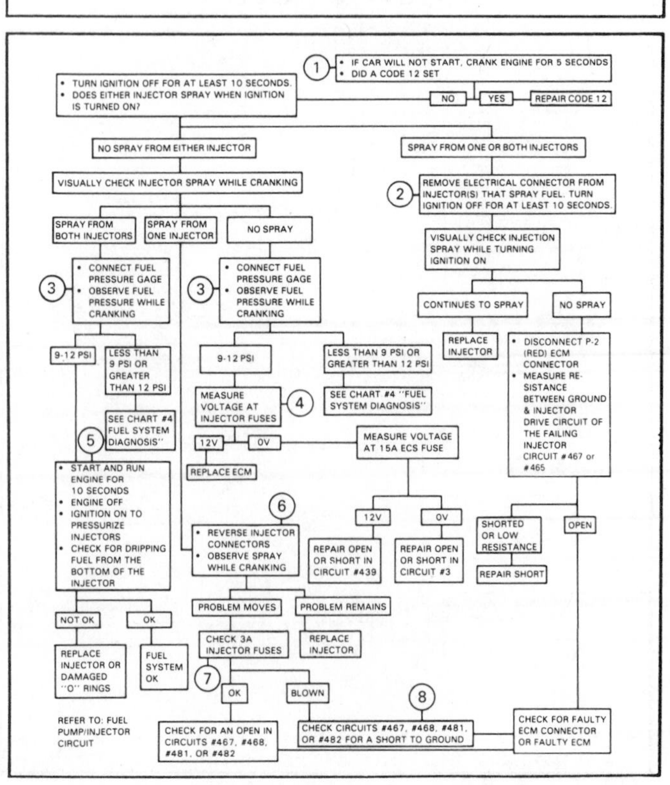

DFI CHART #7
EGR DIAGNOSIS

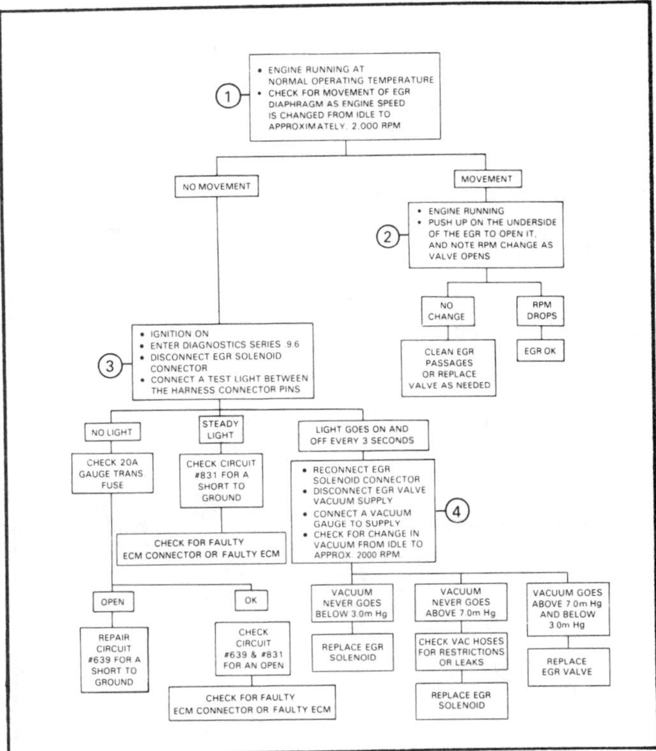

DFI CHART #8
AIR MANAGEMENT DIAGNOSIS

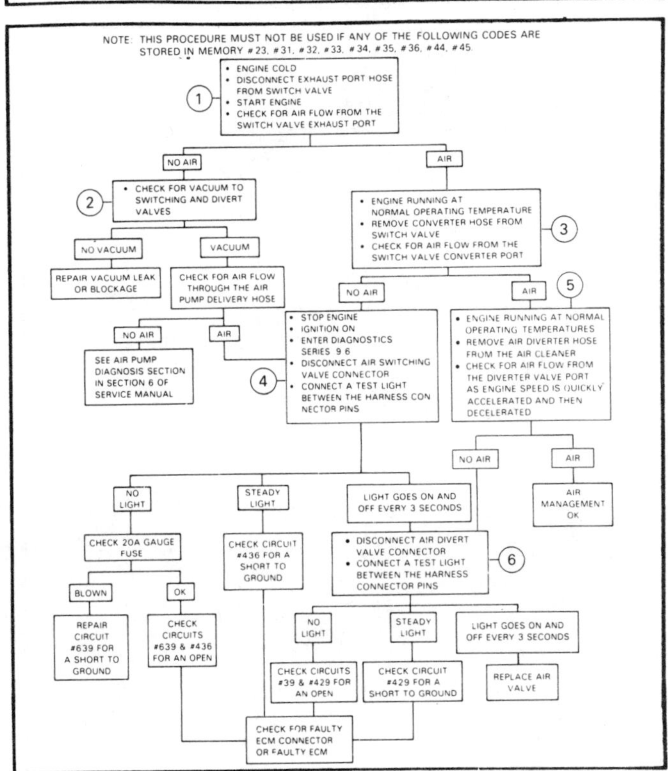

DFI CHART #9
CANISTER PURGE CONTROL DIAGNOSIS

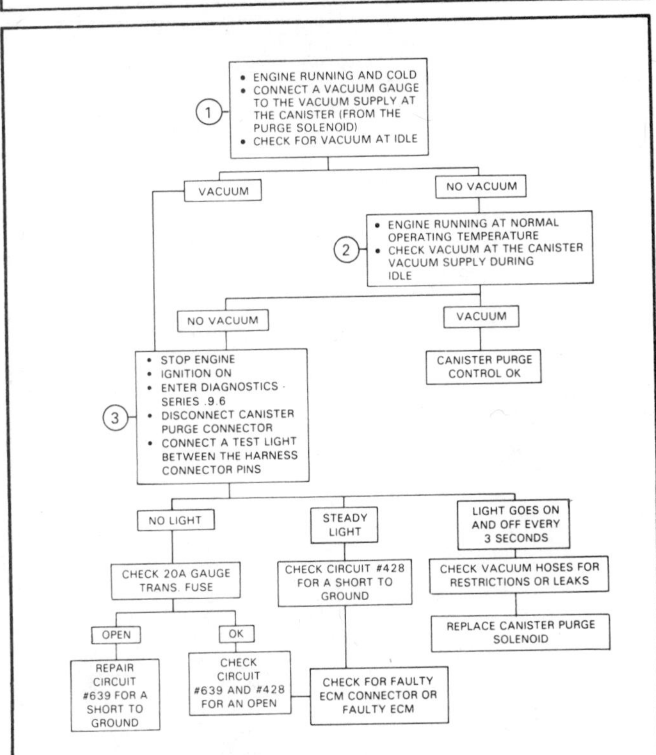

DFI CHART #10
NO CRUISE CONTROL

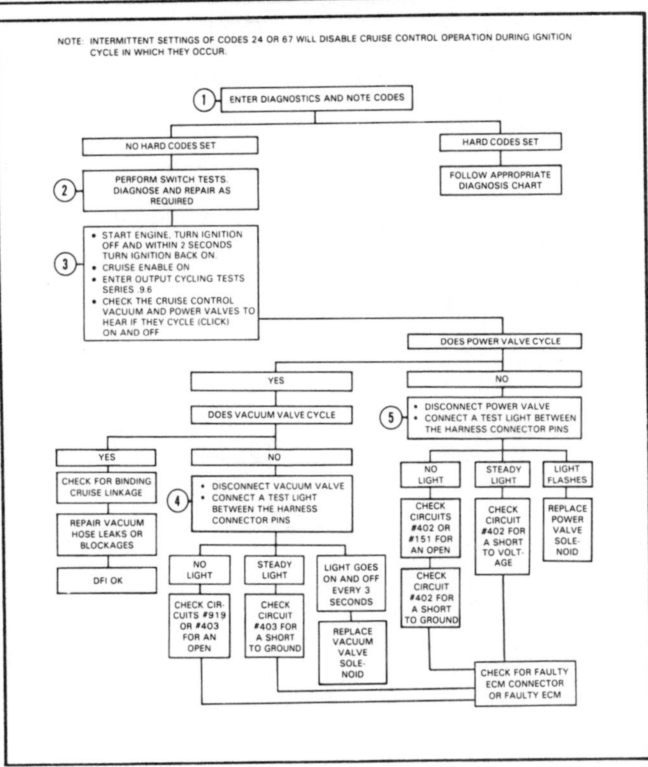

DFI CHART #11A
BLANK FUEL DATA DISPLAY
(IF NOT BLANK, FOLLOW CHART #11B)

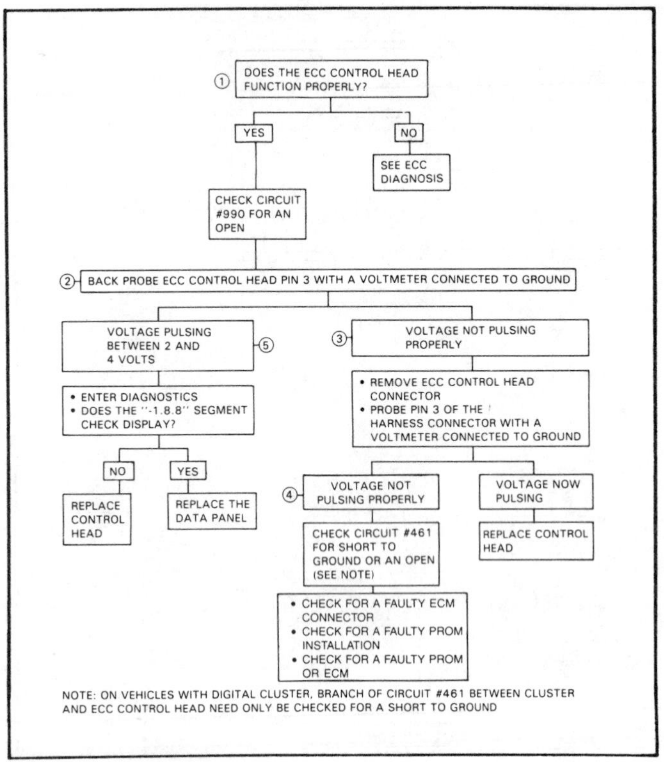

DFI CHART #11B
IMPROPER FUEL DATA DISPLAY
(IF BLANK, FOLLOW CHART #11A)

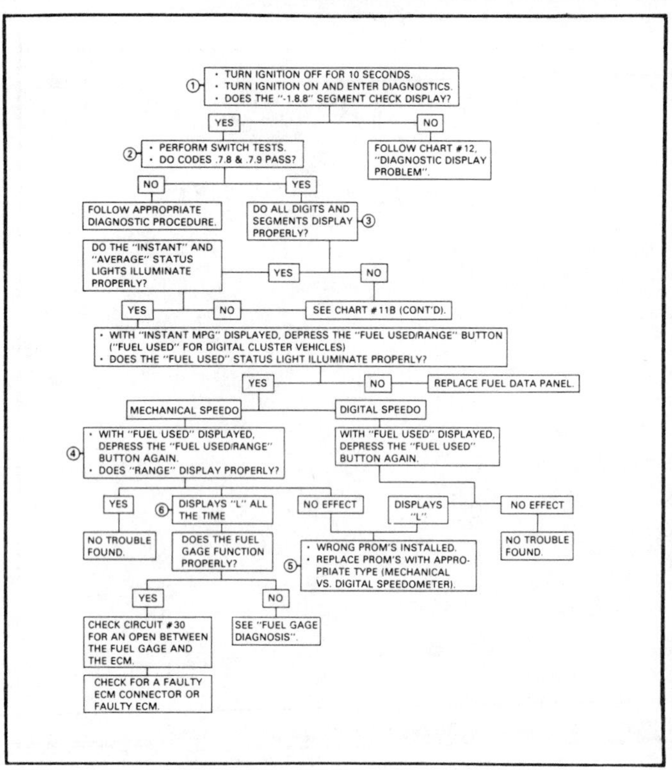

DFI CHART #12
DIAGNOSTIC DISPLAY PROBLEMS

• TURN IGNITION OFF FOR 10 SEC.
• IGNITION ON
• ENTER DIAGNOSTICS

DISPLAYS ". . ."	OUTSIDE TEMPERATURE IS DISPLAYED	DIAGNOSTICS IS ENTERED WITHOUT THE "-1.8.8" DISPLAY CHECK
WILL CAR START	CHECK FOR OPEN CIRCUIT #451	CHECK FOR SHORT TO GROUND CIRCUIT #451
YES: SEE CHART #11A / NO: SEE CHART #1	CHECK FOR FAULTY ECM CONNECTOR OR FAULTY ECM	CHECK FOR FAULTY ECM CONNECTOR OR FAULTY ECM

DFI CHART #13
IMPROPER IDLE SPEED

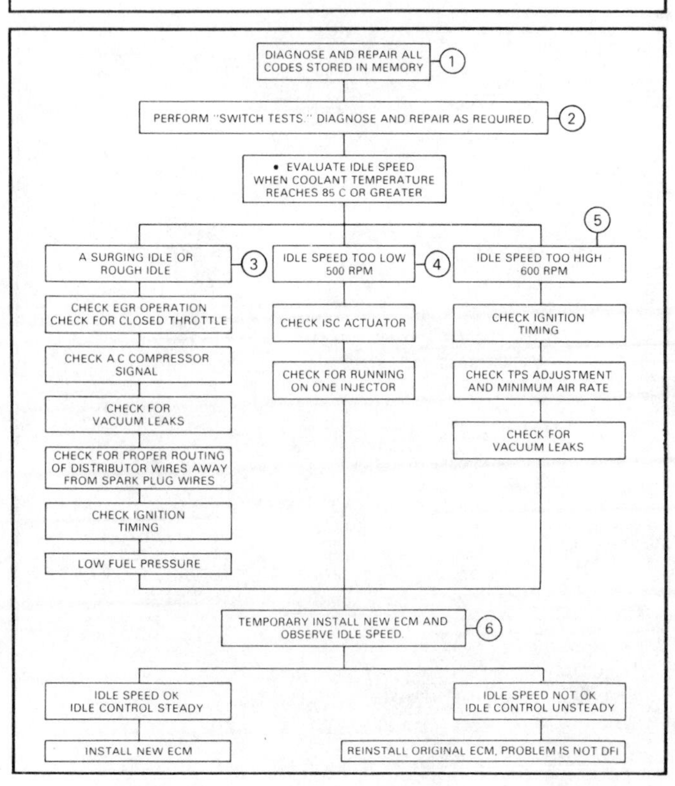

DFI CHART #14
OXYGEN SENSOR TEST

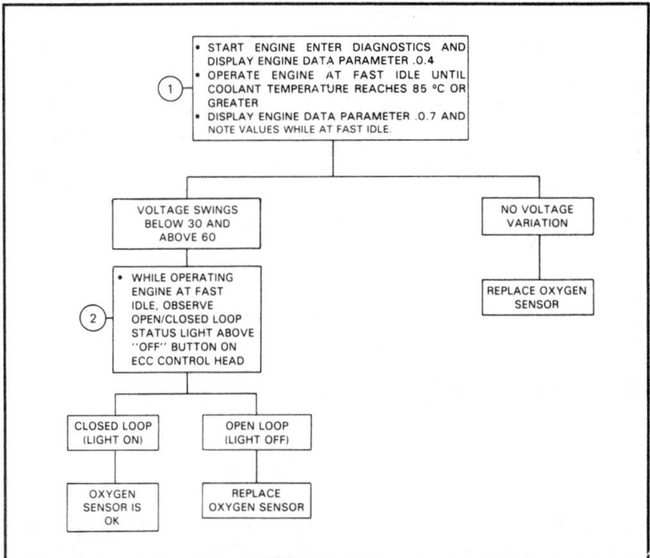
1
• START ENGINE ENTER DIAGNOSTICS AND DISPLAY ENGINE DATA PARAMETER .0.4
• OPERATE ENGINE AT FAST IDLE UNTIL COOLANT TEMPERATURE REACHES 85 °C OR GREATER
• DISPLAY ENGINE DATA PARAMETER .0.7 AND NOTE VALUES WHILE AT FAST IDLE.
VOLTAGE SWINGS BELOW 30 AND ABOVE 60
NO VOLTAGE VARIATION
REPLACE OXYGEN SENSOR
2
• WHILE OPERATING ENGINE AT FAST IDLE, OBSERVE OPEN/CLOSED LOOP STATUS LIGHT ABOVE "OFF" BUTTON ON ECC CONTROL HEAD
CLOSED LOOP (LIGHT ON)
OPEN LOOP (LIGHT OFF)
OXYGEN SENSOR IS OK
REPLACE OXYGEN SENSOR

DFI CHART #15
IMPROPER COOLANT LIGHT OPERATION

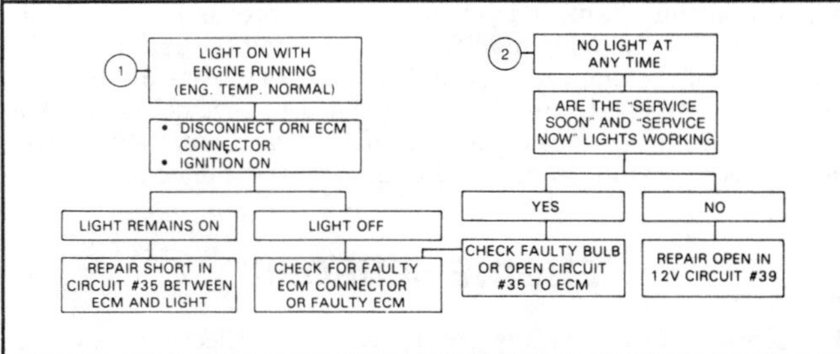
1
LIGHT ON WITH ENGINE RUNNING (ENG. TEMP. NORMAL)
• DISCONNECT ORN ECM CONNECTOR
• IGNITION ON
LIGHT REMAINS ON
LIGHT OFF
REPAIR SHORT IN CIRCUIT #35 BETWEEN ECM AND LIGHT
CHECK FOR FAULTY ECM CONNECTOR OR FAULTY ECM
2
NO LIGHT AT ANY TIME
ARE THE "SERVICE SOON" AND "SERVICE NOW" LIGHTS WORKING
YES
NO
CHECK FAULTY BULB OR OPEN CIRCUIT #35 TO ECM
REPAIR OPEN IN 12V CIRCUIT #39

DFI CHART #16
TCC ELECTRICAL TEST

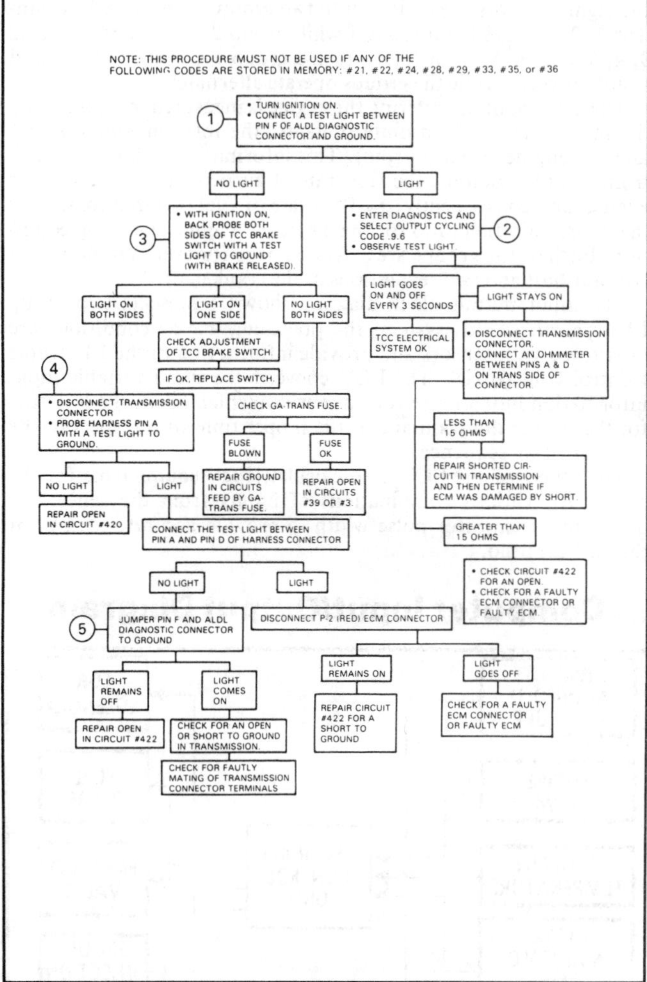
NOTE: THIS PROCEDURE MUST NOT BE USED IF ANY OF THE FOLLOWING CODES ARE STORED IN MEMORY: #21, #22, #24, #28, #29, #33, #35, or #36
1
• TURN IGNITION ON
• CONNECT A TEST LIGHT BETWEEN PIN F OF ALDL DIAGNOSTIC CONNECTOR AND GROUND.
NO LIGHT
LIGHT
3
• WITH IGNITION ON, BACK PROBE BOTH SIDES OF TCC BRAKE SWITCH WITH A TEST LIGHT TO GROUND (WITH BRAKE RELEASED)
2
• ENTER DIAGNOSTICS AND SELECT OUTPUT CYCLING CODE .9.6
• OBSERVE TEST LIGHT
LIGHT ON BOTH SIDES
LIGHT ON ONE SIDE
NO LIGHT BOTH SIDES
LIGHT GOES ON AND OFF EVERY 3 SECONDS
LIGHT STAYS ON
CHECK ADJUSTMENT OF TCC BRAKE SWITCH
IF OK, REPLACE SWITCH
TCC ELECTRICAL SYSTEM OK
• DISCONNECT TRANSMISSION CONNECTOR
• CONNECT AN OHMMETER BETWEEN PINS A & D ON TRANS SIDE OF CONNECTOR
4
• DISCONNECT TRANSMISSION CONNECTOR
• PROBE HARNESS PIN A WITH A TEST LIGHT TO GROUND.
CHECK GA-TRANS FUSE.
FUSE BLOWN
FUSE OK
LESS THAN 15 OHMS
REPAIR SHORTED CIRCUIT IN TRANSMISSION AND THEN DETERMINE IF ECM WAS DAMAGED BY SHORT
NO LIGHT
LIGHT
REPAIR GROUND IN CIRCUITS FEED BY GA-TRANS FUSE.
REPAIR OPEN IN CIRCUITS #39 OR #3.
REPAIR OPEN IN CIRCUIT #420
CONNECT THE TEST LIGHT BETWEEN PIN A AND PIN D OF HARNESS CONNECTOR
GREATER THAN 15 OHMS
• CHECK CIRCUIT #422 FOR AN OPEN.
• CHECK FOR A FAULTY ECM CONNECTOR OR FAULTY ECM.
NO LIGHT
LIGHT
5
JUMPER PIN F AND ALDL DIAGNOSTIC CONNECTOR TO GROUND
DISCONNECT P-2 (RED) ECM CONNECTOR
LIGHT REMAINS ON
LIGHT GOES OFF
LIGHT REMAINS OFF
LIGHT COMES ON
REPAIR CIRCUIT #422 FOR A SHORT TO GROUND
CHECK FOR A FAULTY ECM CONNECTOR OR FAULTY ECM
REPAIR OPEN IN CIRCUIT #422
CHECK FOR AN OPEN OR SHORT TO GROUND IN TRANSMISSION.
CHECK FOR FAUTLY MATING OF TRANSMISSION CONNECTOR TERMINALS

CADILLAC ELECTRONIC FUEL INJECTION SYSTEM

General Information

Electronic Fuel Injection (EFI) precisely controlls the air/fuel mixture for combustion. This is accomplished by monitoring selected engine operating conditions, and electronically metering the fuel requirements to meet those conditions.

EFI basically involves electrically actuated fuel metering valves which, when actuated, spray a predetermined quantity of fuel into the engine.

This arrangement is commonly known as port injection. The injector opening is timed in accordance with the engine frequency so that the fuel charge is in place prior to the intake stroke for the cylinder.

Gasoline is supplied to the inlet of the injectors through the fuel rail at high enough pressure to obtain good fuel atomization and to prevent vapor formation the fuel system. When the solenoid operated valves are energized the injector metering valve (pintle) moves to the full open position and since the pressure differential across the valve is constant, the fuel quantity is changed by varying the time that the injector is held open by the computer.

The injectors could be energized all at one time (called continu-

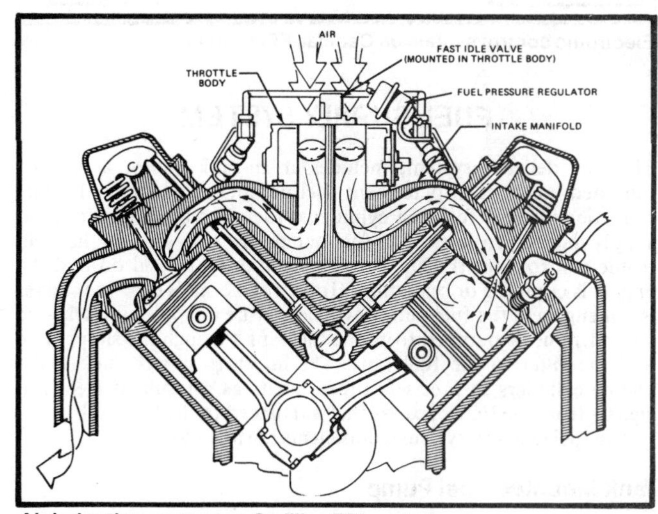
AIR
FAST IDLE VALVE (MOUNTED IN THROTTLE BODY)
THROTTLE BODY
FUEL PRESSURE REGULATOR
INTAKE MANIFOLD
Air induction system on Cadillac EFI system

ous injection) or one after the other in phase with the opening of each intake valve (called sequential injection) or they can be energized in groups.

The EFI System is a two-group system. In a two-group system, the eight injectors are divided into two groups of four each. Cylinders 1, 2, 7 and 8 form group 1 while group 2 consists of cylinders 3, 4, 5, and 6. All four injectors in a group are opened and closed simultaneously while the groups operate alternately.

The amount of air entering the engine is measured by monitoring the intake manifold absolute pressure, the inlet air temperature, and the engine speed (in rpm). This information, allows the electronic unit to compute the flow rate of air being inducted into the engine, and consequently, the flow rate of fuel required to achieve the desired air/fuel ratio for the particular engine operating condition. Each of the groups are activated once for every revolution of the camshaft and two revolutions of the crankshaft.

The input/output block diagram shown represents two group EFI system. In this system, the prevailing engine conditions are monitored with sensors and provide information to the Electronic Control Unit (ECU). The ECU converts the multi-variable input information into an injector pulse width which opens the injectors for the proper duration and at the proper time with respect to the cylinder firing sequence.

The object of the ECU is to calculate fuel requirements for the engine for various combinations of inputs from the sensors to determine an injector pulse width to provide accurate control of the air/fuel ratio.

Computer Input/Output Diagram

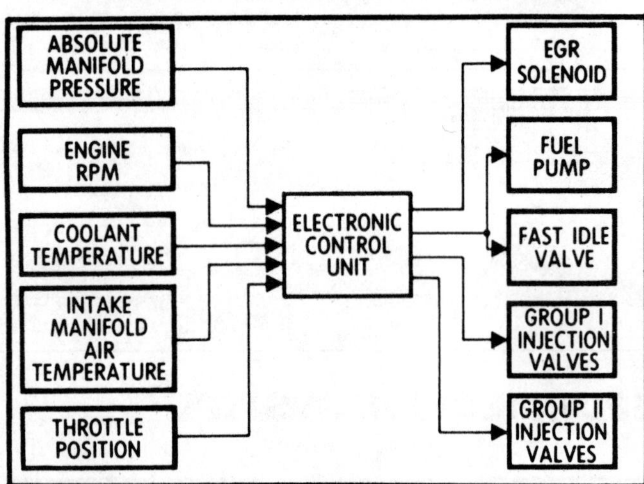

Electronic control system on Cadillac EFI system

FUEL SUPPLY SYSTEM

The fuel delivery system, includes an in-tank pump, a chassis-mounted constant-displacement fuel pump, a fuel filter, the fuel rails, one injector for each cylinder, a fuel pressure regulator, and supply and return lines. The fuel pumps are activated by the electronic control unit (ECU) when the ignition is turned on and the engine is cranking or operating. (If the engine stalls or if the starter is not engaged, the fuel pumps will deactivate in approximately one second.) Fuel is pumped from the fuel tank through the supply line and the filter to the fuel rails. The injectors supply fuel to the engine cylinders in precisely timed bursts as a result of electrical signals from the ECU. Excess fuel is returned to the fuel tank.

The fuel delivery system components are as follows:

Tank-Mounted Fuel Pump

Located inside the fuel tank and an integral part of the fuel gauge tank unit, the in-tank boost pump, used to supply fuel to the chassis mounted pump.

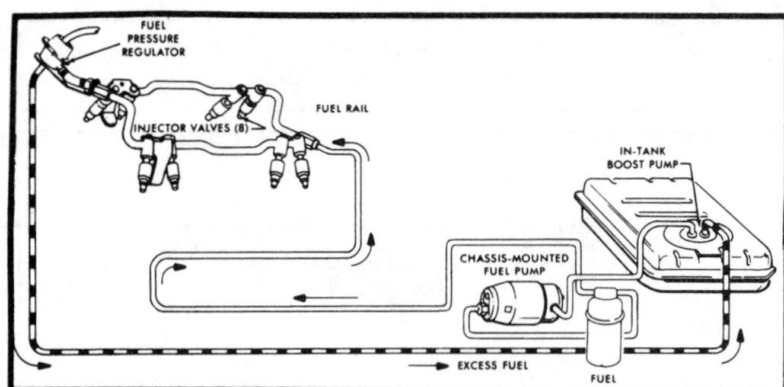

Fuel delivery system on Cadillac EFI system

Chassis-Mounted Fuel Pump

The Chassis-Mounted Fuel Pump is a constant-displacement, roller-vane pump driven by a 12 volt motor. The pump incorporates a check valve to prevent backflow. This maintains fuel pressure when the pump is off. The pump has a flow rate of 33 gallons per hour under normal operating conditions (39 PSI). An internal relief valve provides over-pressure protection by opening at an excessive pressure.

Fuel Pressure Regulator

The Fuel Pressure Regulator contains an air chamber and fuel chamber separated by a spring-loaded diaphragm. The air chamber is connected by a rubber hose to the throttle body assembly. The pressure in this chamber is identical to the pressure of the intake manifold. The changing manifold pressure and the spring, control the action of the diaphragm valve, opening or closing an orifice in the fuel chamber. (Excess fuel is returned to the fuel tank.) This regulator, being connected to the fuel rail and intake manifold, maintains a constant 39 psi differential across the injectors.

ENGINE SENSORS

The sensors are electrically connected to the Electronic Control Unit and all operate independently of each other. Each sensor transmits a signal to the ECU, relating a specific engine operating condition. The ECU analyzes all the signals and transmits the appropriate commands.

Manifold Absolute Pressure Sensor

The Manifold Absolute Pressure (MAP) Sensor monitors the changes in intake manifold pressure which result from engine load, speed and barometric pressure variations. These pressure changes are supplied to the electronic control unit circuitry in the form of electrical signals. The sensor also monitors the changes in the intake manifold pressure due to changes in altitude. As intake manifold pressure increases, additional fuel is required. The MAP sensor sends this information to the ECU so that the pulse width will be increased. Conversely as manifold pressure decreases the pulse width will be shortened.

The sensor is mounted within the electronic control unit. A manifold pressure line, routed with the engine harness, connects it to the front of the throttle body.

Throttle Position switch

The Throttle Position Switch, is mounted to the throttle body and connected to the throttle valve shaft. Movements of the accelerator causes the throttle shaft to rotate (opening or closing the throttle blades). The switch senses the shaft movement and position (closed throttle, wide open throttle, or position changes), and transmits

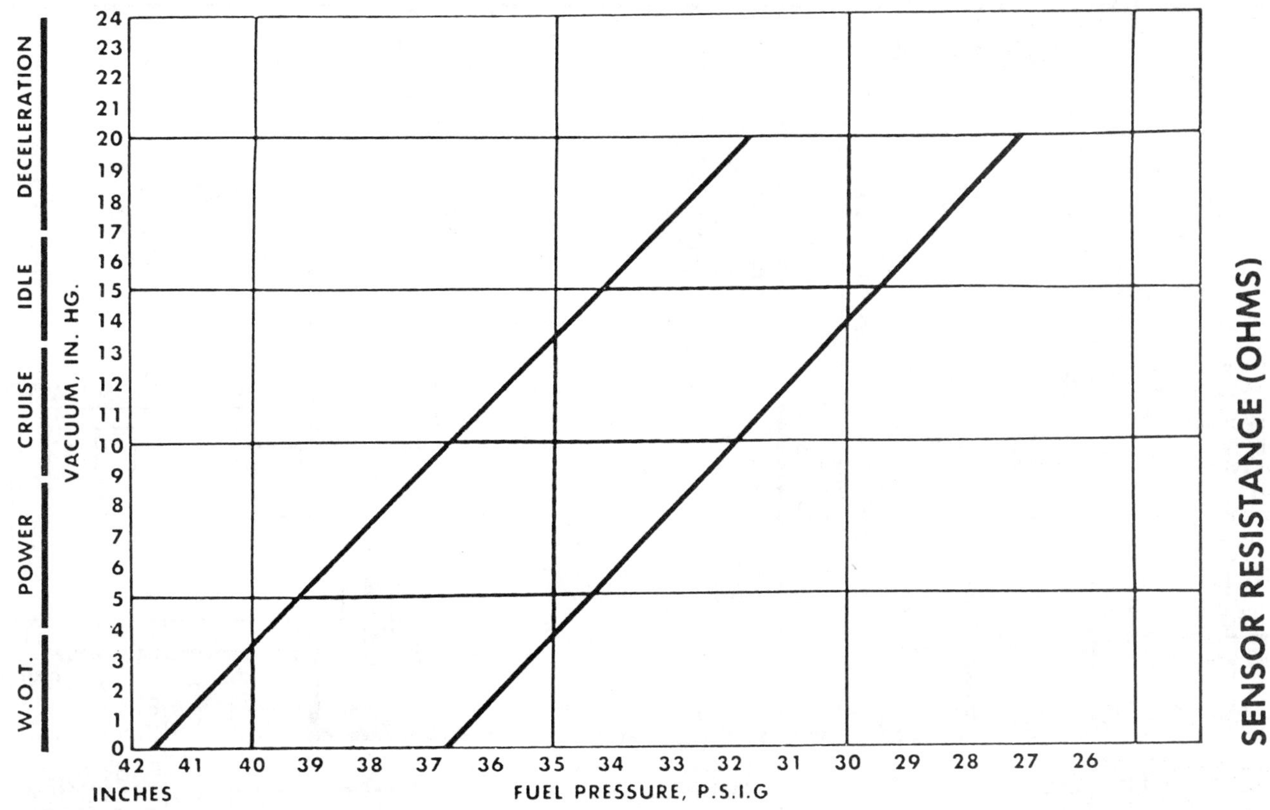

Normal fuel pressure on Cadillac EFI system

appropriate electrical signals to the electronic control unit. The electronic control unit processes these signals to determine the fuel requirements for the particular situation.

Temperature Sensors

The Temperature Sensors, (Coolant and Air) are comprised of a coil of high temperature nickel wire sealed into an epoxy case, and molded into a brass housing with two wires and a connector extending from the body. The resistance of the wire changes as a function of temperature. Low temperature provides low resistance and as temperatures increase, so does resistance. The voltage drop across each sensor is monitored by the ECU.

Speed Sensor

The Speed Sensor, is incorporated within the ignition distributor assembly (HEI). It consists of two components. The first has two reed switches mounted to a plastic housing. The housing is affixed to the distributor shaft housing. The second is a rotor with two magnets, attached to and rotating with the distributor shaft.

The rotor rotates past the reed switches causing them to open and close. This provides two types of information: synchronization of the ECU and the proper injector group with the intake valve timing (phasing); and engine rpm for fuel scheduling.

ELECTRONIC CONTROL UNIT

The Electronic Control Unit, is a pre-programmed analog computer. The ECU is electrically connected to the vehicle power supply and the other EFI components by a harness that is routed through the firewall.

The ECU receives power from the vehicle battery when the ignition is set to the ON or CRANK position. During cranking and engine operation, the following events occur.

The following information is received from the EFI sensors:
1. Engine coolant temperature
2. Intake manifold air temperature

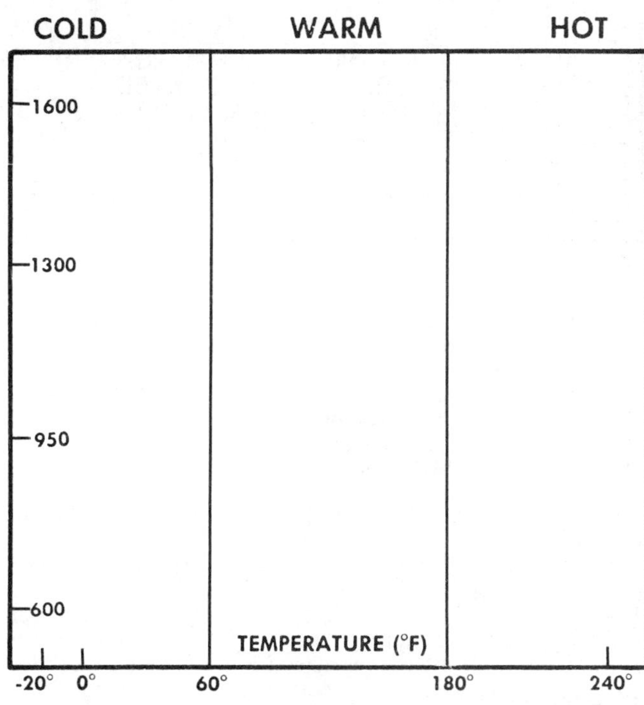

Temperature sensor resistance characteristics on Cadillac EFI system

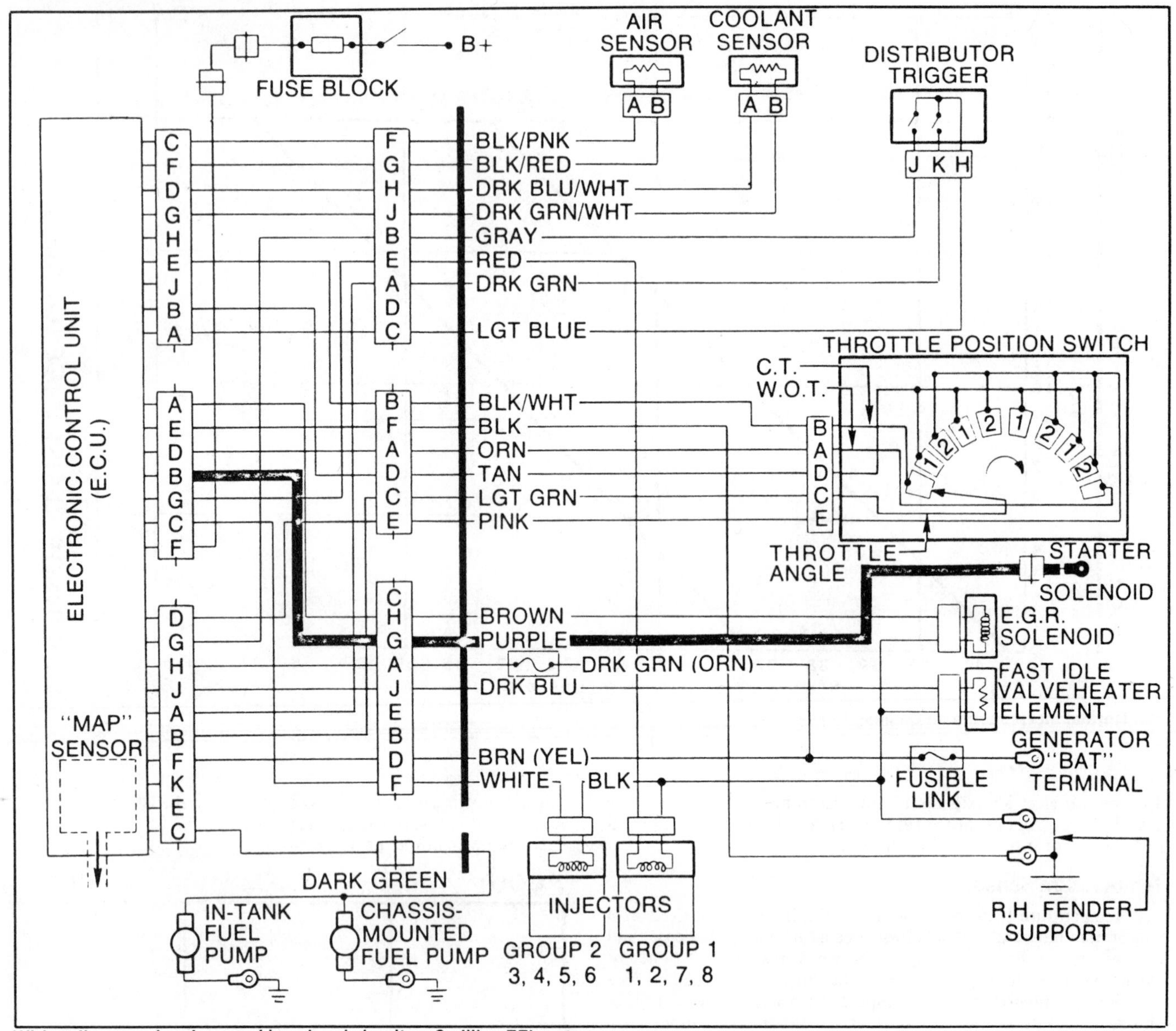

Wiring diagram showing cranking signal circuit on Cadillac EFI system

3. Intake manifold absolute pressure
4. Engine speed and firing position
5. Throttle position and change of position

The following commands are transmitted by the ECU:

1. Electric fuel pump activation
2. Fast idle valve activation
3. Injection valve activation
4. EGR solenoid activation
5. Vacuum retard solenoid activation (California Seville and Eldorado only)

The desired air/fuel ratios for various driving and atmospheric conditions are designed into the ECU. As the above signals are received from the sensors, the ECU processes the signals and computes the engines fuel requirements. The ECU issues commands to the injection valves to open for a specific time duration. The duration of the command pulses varies as the operating conditions change. All injection valves in each group open simultaneously upon command.

Removal

1. Disconnect negative battery cable.
2. Remove three climate control outlet grilles right, left and right center.
3. Working through outlet openings, remove 3 fasteners securing pad to instrument panel support.
4. Remove screws securing pad to instrument panel horizontal support.
5. Pull pad outward and disconnect electrical connector from windshield wiper switch.
6. Remove pad.

NOTE: To facilitate removal or installation, place shift lever in lo range and, on cars equipped with tilt wheel, place wheel in lowest position.

7. Remove MAP sensor hose.
8. Remove three mounting screws holding ECU in position (one in front, two at side bracket, and remove ECU.

CADILLAC ELECTRONIC FUEL INJECTION SYSTEM DIAGNOSIS

THE PROBLEM	THE CAUSE
Engine Cranks But Will Not Staril Not Start	(NOTE: The following problems assume that the rest of the car electrical system is functioning properly.) 1. Blown 10 amp in-line fuel pump fuse located below instrument panel near ECU connectors.* 2. Open circuit in 12 purple wire between starter solenoid and ECU. 3. Open circuit in 18 dark green wire between generator "BAT" terminal and ECU (fusible link).* 4. Poor connection at ECU jumper harness (below instrument panel) or at ECU. 5. Poor connection at fuel pump jumper harness (below instrument panel near ECU), 14 dark green wire.* 6. Poor connection at engine coolant sensor or open circuit in sensor or wiring (cold engine only).** 7. Poor connection at distributor trigger (speed sensor). 8. Distributor trigger (speed sensor) stuck closed. 9. Malfunction in chassis-mounted pump. 10. Malfunction in throttle position switch (W.O.T. section shorted). To check, disconnect switch—engine should start. 11. Fuel flow restriction.
Hard Starting	1. Open engine coolant sensor (cold or partially warm engine only—starts ok hot).** 2. Malfunction in throttle position switch (W.O.T. section shorted). To check, disconnect switch—Engine should start normally. 3. Malfunction in chassis-mounted fuel pump. (Check valves leaking back). 4. Malfunction in pressure regulator.
Poor Fuel Economy	1. Disconnected or leaking MAP sensor hose. 2. Disconnected vacuum hose at fuel pressure regulator or at throttle body. 3. Malfunction of air or coolan sensor.***
Engine Stalls After Start	1. Open circuit in 12 black/yellow ignition signal wire between fuse block and ECU or poor connection at connector (12 black/yellow wire) located below instrument panel near ECU. 2. Poor connection at engine coolant sensor or open circuit in sensor or wiring (cold or warm engine only).**
Rough Idle	1. Disconnected, leaking or pinched MAP sensor hose. If plastic harness line requires replacement, replace entire EFI engine harness. 2. Poor connection at air or coolant sensor or open circuit in sensor or wiring (cold engine only).** 3. Poor connection at injection valve(s). 4. Shorted engine coolant sensor.*** 5. Speed sensor harness located to close to secondary ignition wires.
Prolonged Fast Idle	1. Poor connection at fast idle valve or open circuit in heating element. 2. Throttle position switch misadjusted. 3. Vacuum Leak.
No Fast Idle	1. Bent fast idle valve micro switch causing heater to malfunction and drive valve section down to locked closed position.
Engine Hesitates or Stumbles on Acceleration	1. Disconnected, leaking on pinched MAP sensor hose. If plastic harness line requires replacement, replace entire EFI engine harness. 2. Throttle position switch misadjusted. 3. Malfunction in throttle position switch. 4. Intermittent malfunction in distributor trigger (speed sensor). 5. Poor connection at 6 pin connector of ECU. 6. Poor connection at EGR solenoid or open solenoid (cold engine only).
Lack of High Speed Performance	1. Misadjusted throttle position switch (W.O.T. only). 2. Malfunction in throttle position switch. 3. Malfunction of chassis-mounted fuel pump. 4. Intermittent malfunction in distributor trigger (speed sensor). 5. Fuel filter blocked or restricted. 6. Open circuit in 12 purple wire between starter solenoid and ECU.

*To check, listen for chassis-mounted fuel pump "Whine" (one second only) as key is turned to "ON" position (not to "START" position).

**To check for an "open" circuit in an EFI temperature sensor, connect an ohmmeter to the sensor connector terminals. If the sensor resistance is greater than 1600 ohms, replace the sensor.

***To check for a "closed" (short) circuit in an EFI temperature sensor, connect an ohmmeter to the sensor connector terminals. If the sensor resistance is less than 700 ohms, replace the sensor.

9. Remove electrical connectors ECU.

Installation

1. Connect electrical harness to ECU.
2. Position ECU with electrical connectors on right side of car and install three mounting screws.
3. Install MAP sensor hose.
4. Position pad to instrument panel and connect electrical connector for wiper switch.
5. Install screws securing pad to instrument panel horizontal support.
6. Working through climate control outlet openings, install 3 fasteners securing pad to instrument panel support.
7. Install air outlet grilles as described in Section 1, Note.
8. Connect negative battery cable.

Component Removal

THROTTLE BODY ASSEMBLY

Removal

1. Remove air cleaner.
2. Disconnect throttle return springs (2) from throttle lever.
3. Remove retainer and remove cruise control chain from throttle lever on cars so equipped.
4. Remove "hairpin" clip and disconnect throttle cable from throttle lever.
5. Remove left rear throttle body mounting screw and remove one screw holding throttle bracket to intake manifold.
6. Disengage downshift switch from throttle lever and position bracket, switch and linkage out of way.
7. Disconnect throttle position switch electrical connector and fast idle valve electrical connector. Slide fast idle valve wiring out of notch in throttle body.
8. Remove vacuum hoses from nipples on throttle body. Use back-up wrench when removing power brake vacuum line.
9. Remove remaining throttle body mounting screws and remove throttle body.
10. Remove gasket material from intake manifold and bottom of throttle body. Clean all foreign material from area around intake manifold throttle bores.
11. The following parts are not included in a new throttle body assembly and should be removed as necessary:
    a. Throttle position switch.
    b. Fast idle valve seat.
    c. Fast idle valve spring.
    d. Fast idle valve.
    e. Fast idle valve heater assembly.
    f. Power brake vacuum fitting.

Installation

1. Position a new throttle body gasket to intake manifold with identification tab on left.
2. Install throttle position switch to right side of throttle body, and adjust switch.
3. Position throttle body to intake manifold and loosely install both front and right rear mounting screws.
4. Move throttle linkage bracket into position over left rear mounting screw and install throttle body mounting screw.
5. Install one additional bracket mounting screw and torque screws 4–15 ft. lbs. Adjust transmission downshift.
6. Position throttle cable to throttle lever and secure with "hairpin" clip. Check for proper operation and wide-open-throttle.
7. Position cruise control chain to throttle lever and secure with clip on cars so equipped. Check for proper operation.
8. Install power brake vacuum fitting to rear of throttle body and tighten securely. Use back-up wrench.
9. Install throttle return springs (2) between throttle lever and pressure regulator bracket with open end of spring on outside of throttle lever.
10. Install vacuum hoses to appropriate nipples on throttle body.
11. Install fast idle valve.
12. Install air cleaner.

THROTTLE POSITION SWITCH

Removal

1. Remove throttle body as described earlier in this section.
2. Remove two mounting screws and remove switch from throttle body.

Installation

1. Install throttle position switch to right side of throttle body so that tab on switch engages flat on throttle shaft.
2. Install two mounting screws and tighten screws so that switch will still move but is not loose.
3. Adjust throttle position switch.
4. Install throttle body.

Adjustment

1. Loosen two throttle position switch mounting screws to permit rotation of the switch.
2. Hold the throttle valves in the idle position while performing step 3 and 4.
3. Turn the throttle position switch carefully counterclockwise until the end-stop has been reached.
4. Tighten throttle position switch mounting screws to 11 in. lbs.
5. Check to insure that throttle valves close to the throttle stop. If not, repeat steps 2, 3 and 4.

FAST IDLE VALVE

Removal

1. Remove air cleaner and disconnect electrical connector from fast idle valve heater.
2. Remove air cleaner stud.
3. To remove fast idle valve heater, push down and twist 90° counter clockwise.
4. Remove fast idle valve, spring and seat from position in throttle body.

Installation

1. Install fast idle valve seat, spring and valve in position in throttle body.
2. Position heater on top of fast idle valve and push down to compress spring. Care should be taken to avoid damaging micro-switch contact arm on bottom of heater housing.
3. Align tabs on fast idle valve heater with cut-out portion of throttle body and compress spring further.
4. Rotate heater 90° clockwise to secure in position.
5. Connect electrical connector.
6. Install air cleaner stud and air cleaner.

CAUTION

Do not loosen fittings until all precautions have been taken to relieve pressure. Fuel in system may be under high pressure which could spray out and result in a fire hazard and possible personal injury.

INJECTION VALVE

Removal

1. Remove front and rear fuel rails.
2. Remove electrical conduit from injector brackets 4 places each side.

3. Remove screws holding each injector bracket to intake manifold and remove brackets and grommets.

4. Disconnect electrical lead from all injectors on fuel rail being removed.

5. Remove fuel rail and injectors from engine as a unit. Some injectors will stick in fuel rail while others may remain in manifold.

6. Remove injectors from fuel rail and from intake manifold as required.

7. Injection valves are sealed by O-rings at both fuel rail and intake manifold. Remove and discard all used O-rings.

CAUTION

EFI fuel injectors can be damaged by full battery voltage (12v). Using the battery to apply test voltage will destroy the injector coil within 1/2 second.

Installation

1. Lubricate and install a new O-ring on the fuel rail end of each injector.

2. Install injectors into fuel rail with electrical connector facing inboard.

NOTE: Fuel rails are specific for right and left sides of engine.

3. Lubricate and install a new O-ring into each injector port in the intake manifold.

4. Install fuel rail-injector assembly to intake manifold making sure that each injector is properly installed in manifold O-ring.

5. Install rubber grommets on fuel rail (flange down) and install injector brackets in position.

6. Secure each bracket with two screws.

7. Route electrical harness along bracket and secure to brackets—4 positions.

8. Connect injector leads as follows:
Front cylinder—red/black wires
Front-center cylinder—black/white wires
Rear-center cylinder—black/white wires
Rear cylinder—red/black wires

NOTE: Injectors may be rotated to provide proper harness routing.

9. Repeat steps 1 thru 8 for opposite side if necessary.

10. Install front and rear fuel rails.

11. Turn ignition ON and OFF several times to build up fuel rail pressure and check for fuel leaks.

12. Start engine and check for leaks.

NOTE: A "dry" fuel system on cars equipped with EFI may require a substantial cranking period.

FUEL PRESSURE REGULATOR

Removal

1. Remove vacuum hose from nipple on top of pressure regulator.

2. Remove and discard clamps securing flexible fuel hose connecting regulator to fuel rail. Remove return line.

3. Remove one nut securing pressure regulator to bracket.

NOTE: This nut has metric threads

4. Work regulator off of flexible fuel hose and out of bracket.

Installation

1. Install new hose clamp over flexible fuel hose.

2. Position regulator to bracket and work flexible fuel hose to fuel rail over nipple on side of regulator. Secure to bracket with one nut.

NOTE: This nut has metric threads. Use the nut supplied with each new regulator or the nut removed.

3. Connect return line to fitting on end of regulator.

4. Tighten clamps, securing flexible fuel hose to regulator.

5. Install vacuum hose to remaining nipple on pressure regulator.

6. A "dry" fuel system on cars with EFI may require a substantial cranking period before starting.

FUEL RAIL

Front—Removal

1. Remove and discard hose clamp securing pressure regulator hose to front fuel rail.

2. Using a back-up wrench at side rail fitting, remove flare nut from each end of fuel rail.

3. Disengage front rail from pressure regulator hose and remove from vehicle.

Front—Installation

1. Install new hose clamp over flexible hose.

2. Position front rail to pressure regulator hose and force hose over nipple.

3. Move rail into position and tighten flare nut into fitting on each side rail to *Use a back-up wrench to hold side rail fittings.* Do *not* use teflon tape on flare nuts.

4. Tighten hose clamp securing pressure regulator hose to rail.

5. Turn ignition ON and OFF several times to build up fuel rail pressure and check for fuel leaks.

Rear—Removal

1. Using a back-up wrench on fuel rail remove fuel inlet line from rear rail.

2. Using a back-up wrench on fuel rail, remove nut at each side rail and remove rear fuel rail.

Rear—Installation

1. Position rear fuel rail and thread flare nuts into side rails. Do *not* use teflon tape on flare nuts.

2. Thread fuel inlet line flare nut into rear rail. Do *not* use teflon on flare nut.

3. *Using a back-up wrench on fuel rail,* tighten flare nuts.

4. Turn ignition ON and OFF several times to build up fuel rail pressure and check for fuel leaks.

NOTE: Right or Left side fuel rail removal is the same as injector removal.

FUEL FILTER

The fuel filter element is replaced by unscrewing the bottom cover and removing the filter element. Replace element and gasket with AC type GF 157 or equivalent. Hand tighten bottom cover.

A "Dry" fuel system on cars equipped with EFI may require a substantial cranking period before starting.

CHASSIS-MOUNTED FUEL PUMP

Removal and Installation

The chassis-mounted pump is located forward of the rear wheel along the frame side rail. The mounting nuts have metric threads. Peel back rubber boot and remove two nuts, one from each electrical terminal. Remove electrical leads. Reinstall in reverse order, making sure all connections are tight.

TANK FUEL PUMP

Removal

1. Remove fuel inlet and outlet hoses from nipples on pump.

2. Remove locknuts securing fuel gauge tank unit and fuel pump feed wires to tank unit.

3. Position fuel tank sending unit remover and installer J-24187 on cam locking ring so that tool engages three tabs on ring.

4. Install ratchet and turn counter clockwise to disengage lock ring from fuel tank. Remove tool and lift gauge-pump unit from tank.

Installation

1. Install gage-pump unit in fuel tank, using new gasket.
2. Install fuel tank sending unit remover and installer J-24187 so that it engages three tabs on cam locking ring.
3. Turn clockwise until locking rings is fully engaged in fuel tank.
4. Connect electrical leads to fuel gauge and secure with locknut.
5. Position tank near underbody and attach fuel pump lead wire (14 green) to terminal marked "pump".
6. Move tank to underbody; position fuel tank support straps under tank and loosely install screws securing straps to body.
7. Secure ground wire to rear cross member.
8. Tighten tank strap screws until bottomed. Torque to 25 ft. lbs.
9. Connect fuel line, evaporative loss control system line and return line to fittings at front of tank. Secure with new clamps.
10. Lower car.
11. With fuel filler door open, connect sending unit feed wire (tan).
12. Replace drained fuel in tank.
13. A "dry" fuel system on cars equipped with EFI may require a substantial cranking period before starting.

AIR TEMPERATURE SENSOR

Removal and Installation

1. Locate air temperature sensor at right rear of intake manifold and disconnect sensor from car harness.
2. Remove sensor from intake manifold.
3. Apply a non-hardening sealer to threads of sensor and install sensor in intake manifold.
4. Tighten sensor to 15 ft. lbs.
5. Connect air temperature sensor connector to car harness.

COOLANT TEMPERATURE SENSOR

Removal and Installation

1. Drain radiator until coolant level is below level of cylinder heads.
2. Locate water temperature sensor in heater hose outlet at rear of right hand cylinder head, and disconnect sensor from car harness.
3. Remove sensor from position in heater hose.
4. Apply a non-hardening sealer to threads of sensor and install sensor in heater hose outlet fitting.
5. Tighten sensor to 15 ft. lbs.
6. Connect coolant temperature sensor connector to car harness.
7. Fill cooling system to proper level.

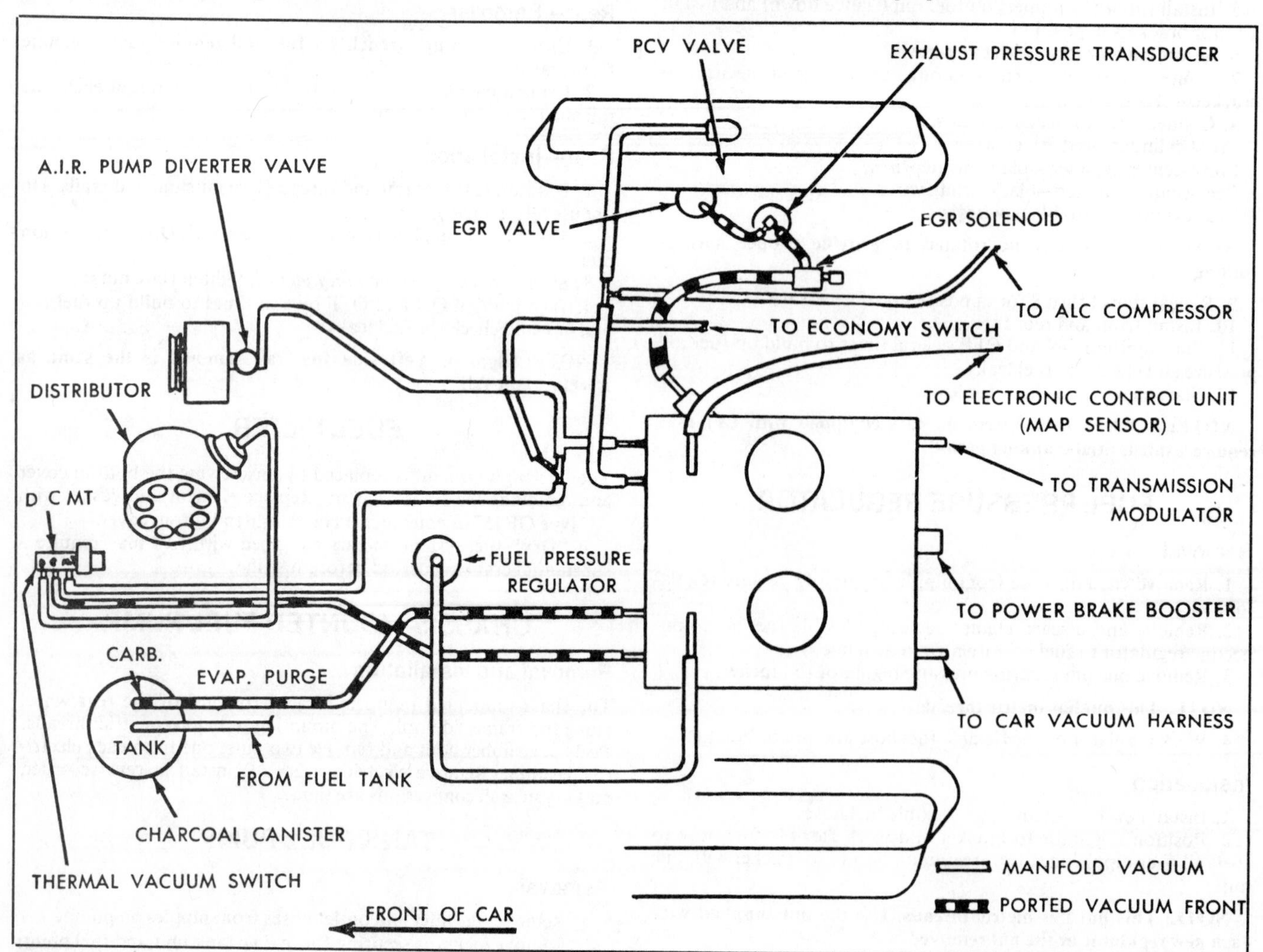

Vacuum circuits on early DeVille, Fleetwood and Eldorado (Federal)

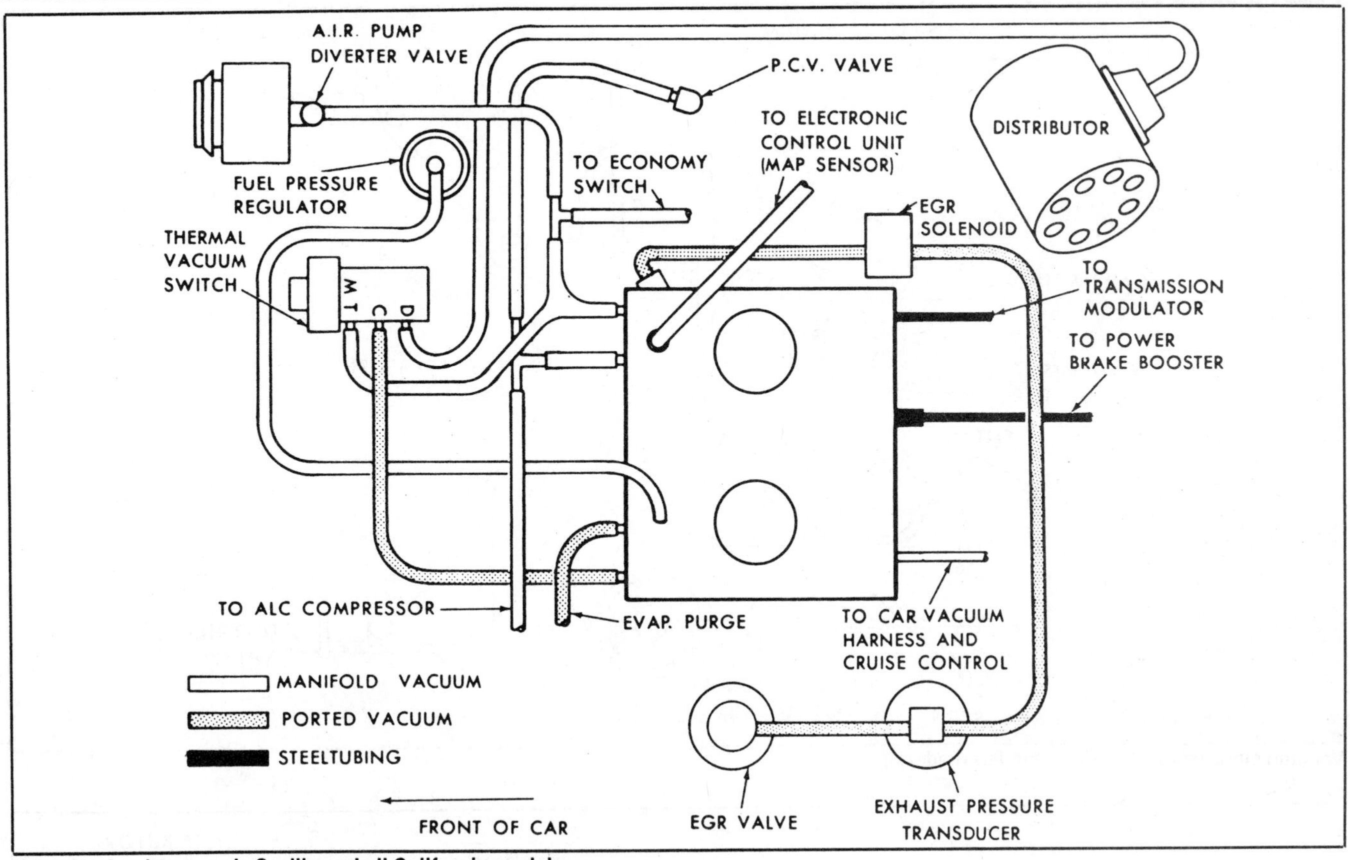

A.I.R. PUMP DIVERTER VALVE
P.C.V. VALVE
TO ELECTRONIC CONTROL UNIT (MAP SENSOR)
DISTRIBUTOR
TO ECONOMY SWITCH
FUEL PRESSURE REGULATOR
THERMAL VACUUM SWITCH
M T C D
EGR SOLENOID
TO TRANSMISSION MODULATOR
TO POWER BRAKE BOOSTER
TO ALC COMPRESSOR
EVAP. PURGE
TO CAR VACUUM HARNESS AND CRUISE CONTROL
MANIFOLD VACUUM
PORTED VACUUM
STEELTUBING
FRONT OF CAR
EGR VALVE
EXHAUST PRESSURE TRANSDUCER

Vacuum circuits on early Seville and all California models

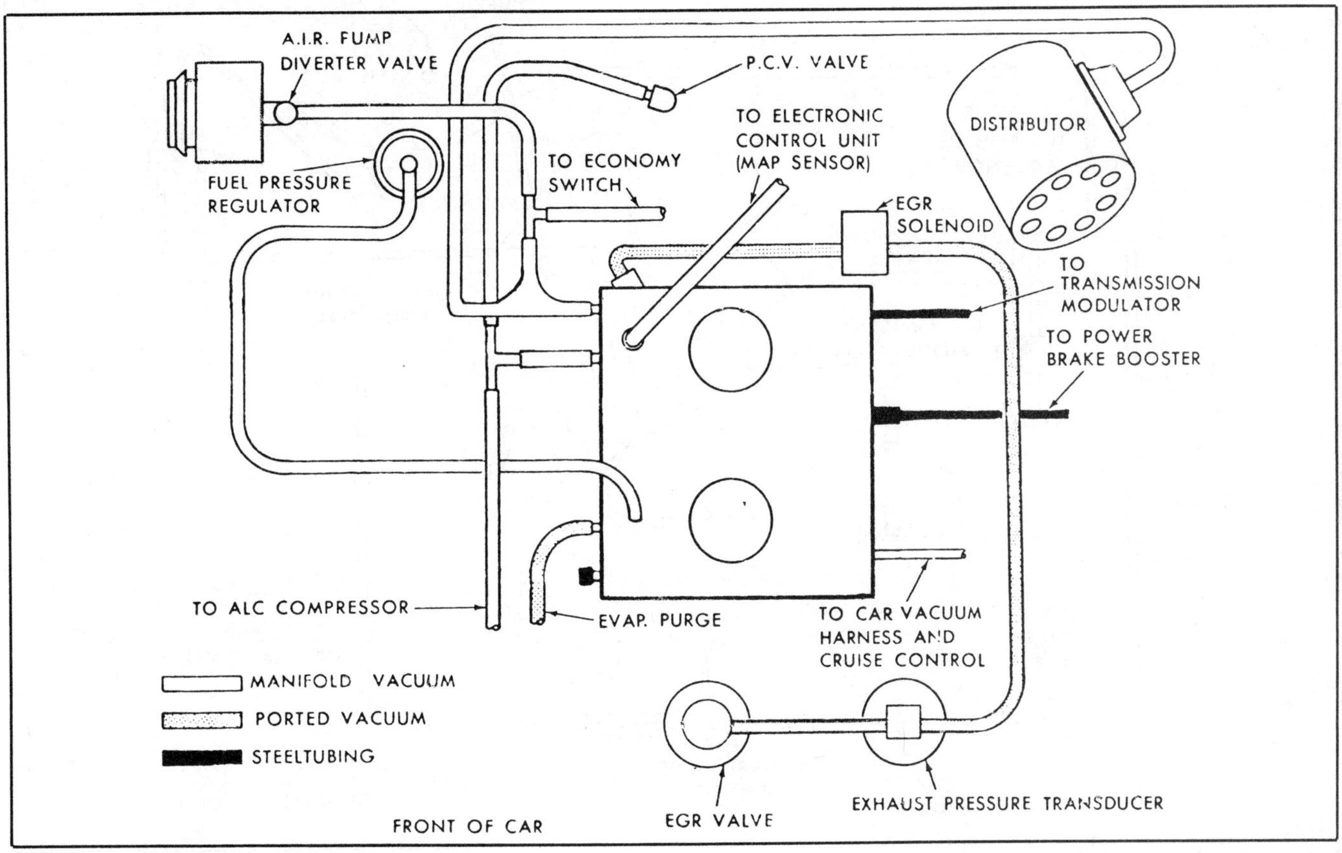

A.I.R. PUMP DIVERTER VALVE
P.C.V. VALVE
TO ELECTRONIC CONTROL UNIT (MAP SENSOR)
DISTRIBUTOR
TO ECONOMY SWITCH
FUEL PRESSURE REGULATOR
EGR SOLENOID
TO TRANSMISSION MODULATOR
TO POWER BRAKE BOOSTER
TO ALC COMPRESSOR
EVAP. PURGE
TO CAR VACUUM HARNESS AND CRUISE CONTROL
MANIFOLD VACUUM
PORTED VACUUM
STEELTUBING
EXHAUST PRESSURE TRANSDUCER
FRONT OF CAR
EGR VALVE

Vacuum circuits on late Seville models

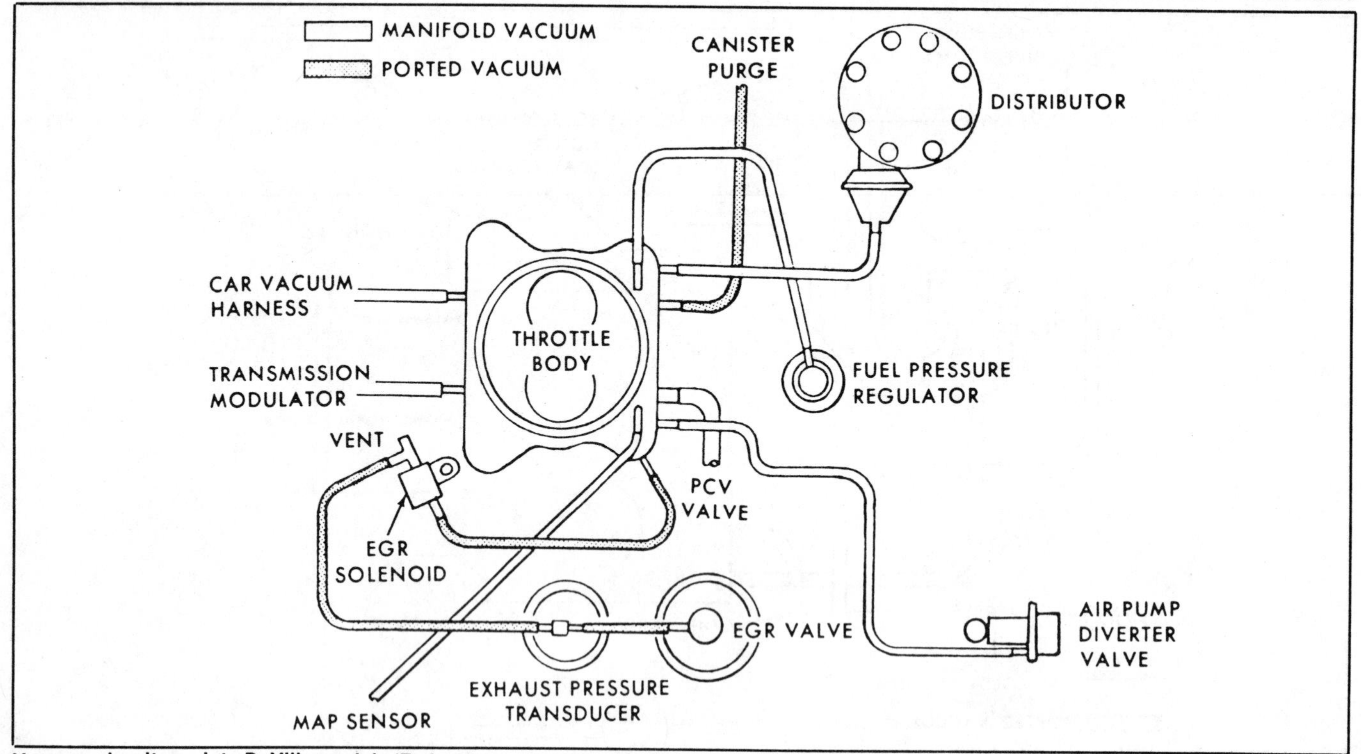

Vacuum circuits on late DeVille models (Federal)

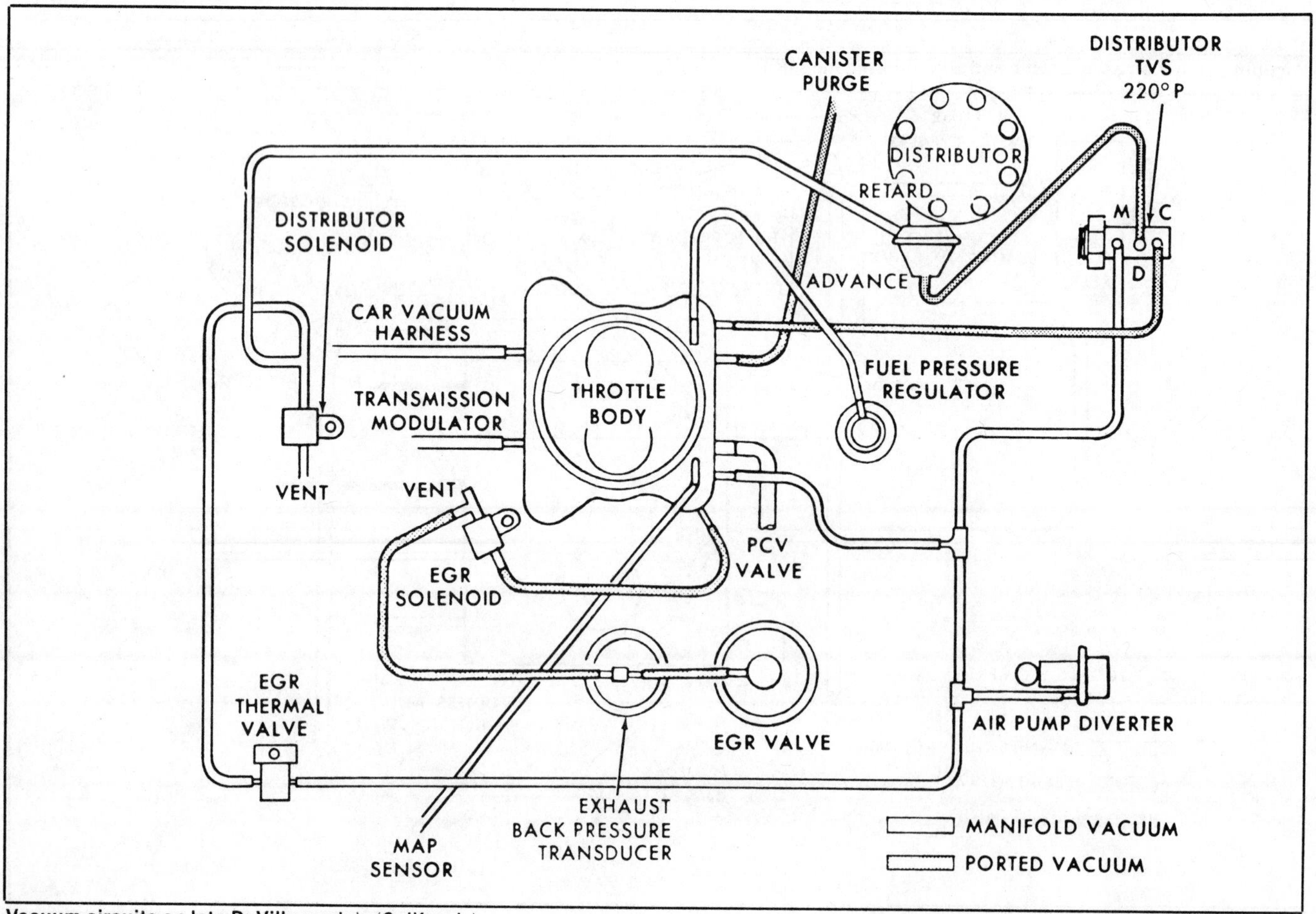

Vacuum circuits on late DeVille models (California)

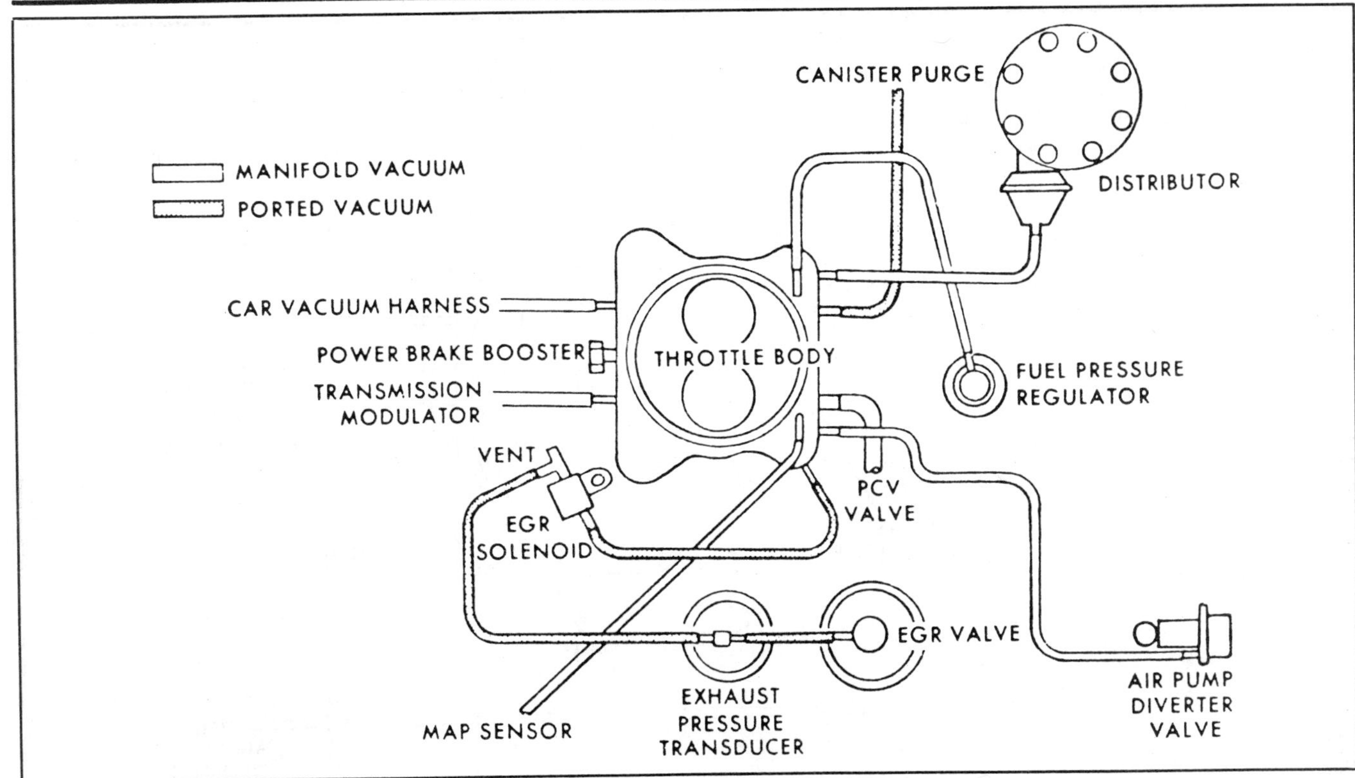

CANISTER PURGE
MANIFOLD VACUUM
PORTED VACUUM
DISTRIBUTOR
CAR VACUUM HARNESS
POWER BRAKE BOOSTER
THROTTLE BODY
FUEL PRESSURE REGULATOR
TRANSMISSION MODULATOR
VENT
EGR SOLENOID
PCV VALVE
EGR VALVE
AIR PUMP DIVERTER VALVE
EXHAUST PRESSURE TRANSDUCER
MAP SENSOR

Vacuum circuits on late Eldorado models (Federal)

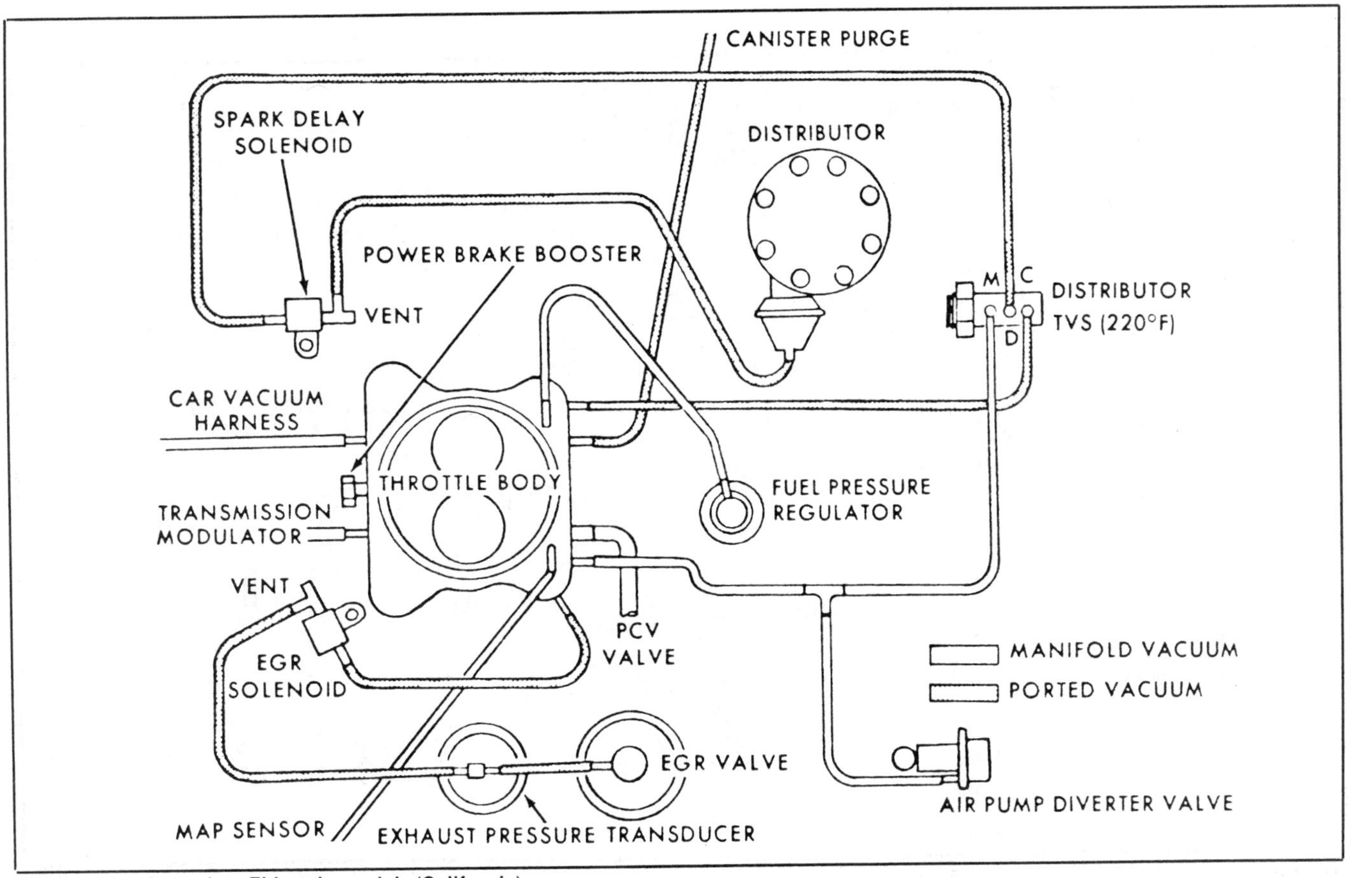

CANISTER PURGE
SPARK DELAY SOLENOID
DISTRIBUTOR
POWER BRAKE BOOSTER
VENT
M
C
DISTRIBUTOR TVS (220°F)
D
CAR VACUUM HARNESS
THROTTLE BODY
FUEL PRESSURE REGULATOR
TRANSMISSION MODULATOR
VENT
PCV VALVE
EGR SOLENOID
MANIFOLD VACUUM
PORTED VACUUM
EGR VALVE
AIR PUMP DIVERTER VALVE
MAP SENSOR
EXHAUST PRESSURE TRANSDUCER

Vacuum circuits on late Eldorado models (California)

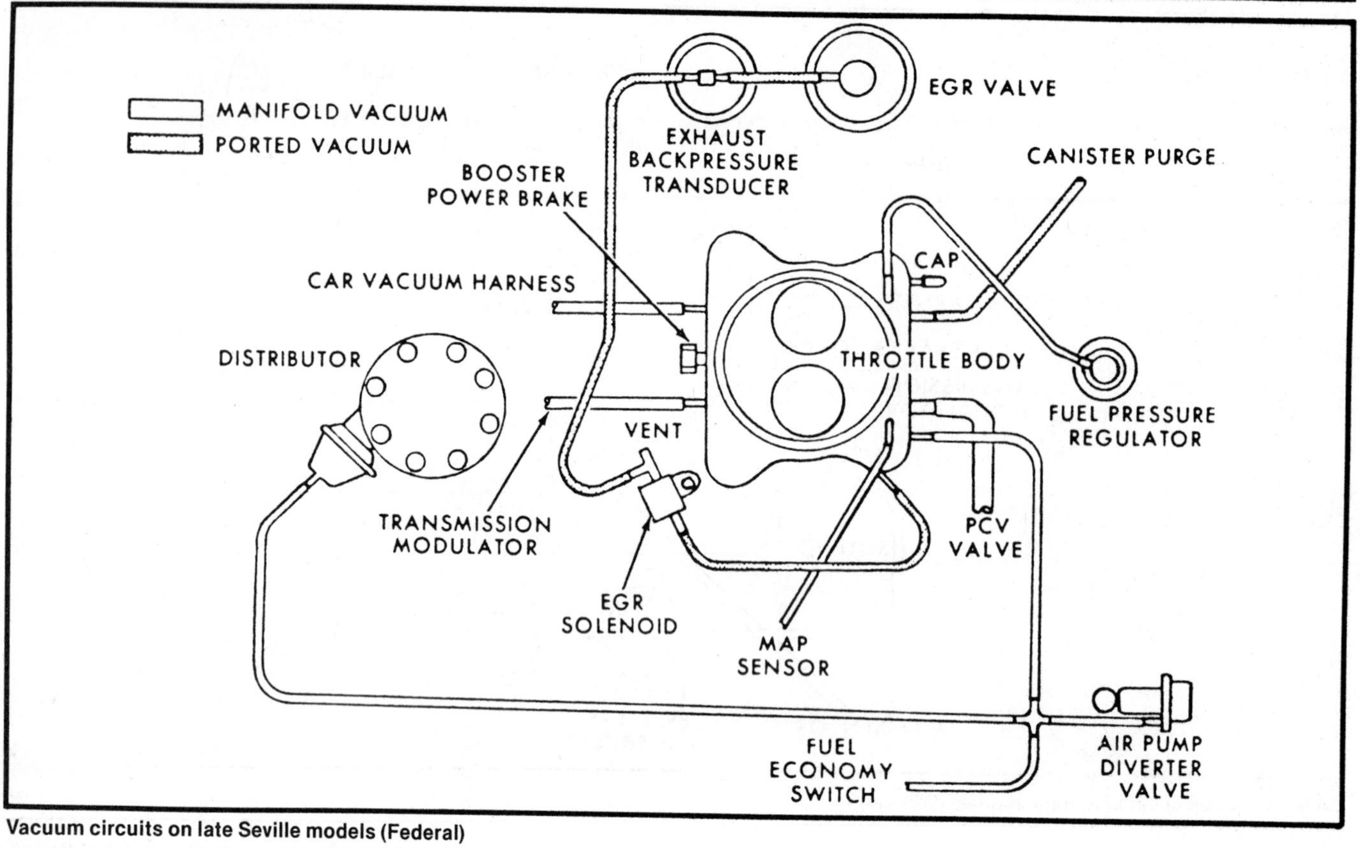

MANIFOLD VACUUM
PORTED VACUUM
EGR VALVE
EXHAUST BACKPRESSURE TRANSDUCER
BOOSTER POWER BRAKE
CANISTER PURGE
CAR VACUUM HARNESS
CAP
DISTRIBUTOR
THROTTLE BODY
FUEL PRESSURE REGULATOR
VENT
TRANSMISSION MODULATOR
PCV VALVE
EGR SOLENOID
MAP SENSOR
FUEL ECONOMY SWITCH
AIR PUMP DIVERTER VALVE

Vacuum circuits on late Seville models (Federal)

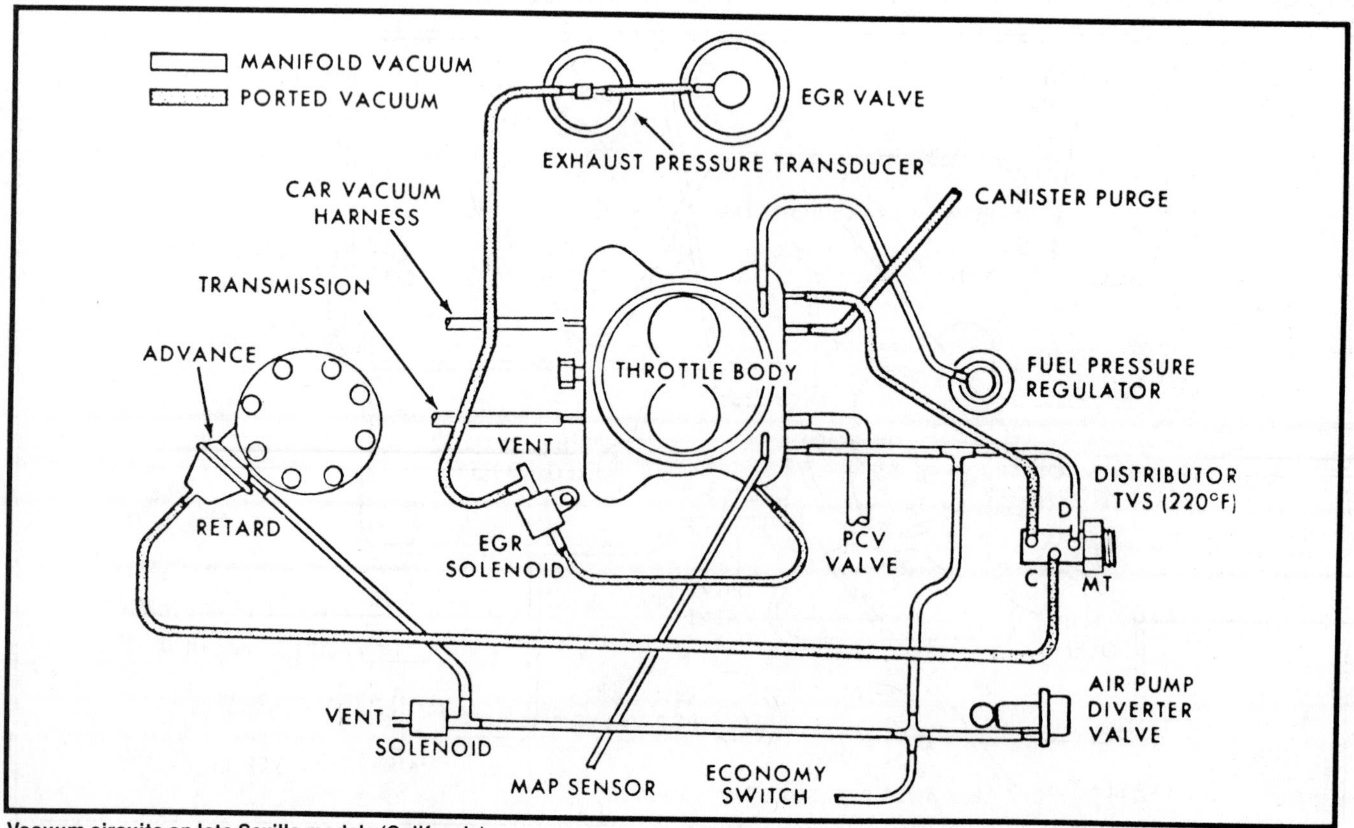

MANIFOLD VACUUM
PORTED VACUUM
EGR VALVE
EXHAUST PRESSURE TRANSDUCER
CAR VACUUM HARNESS
CANISTER PURGE
TRANSMISSION
ADVANCE
THROTTLE BODY
FUEL PRESSURE REGULATOR
VENT
DISTRIBUTOR TVS (220°F)
D
RETARD
EGR SOLENOID
PCV VALVE
C
MT
VENT
SOLENOID
AIR PUMP DIVERTER VALVE
MAP SENSOR
ECONOMY SWITCH

Vacuum circuits on late Seville models (California)

Diagnosis and Testing

The diagnostic tree charts are designed to help solve problems on fuel injected vehicles by leading the user through closely defined conditions and tests so that only the most likely components, vacuum and electrical circuits are checked for proper operation when troubleshooting a particular malfunction. By using the trouble trees to eliminate those systems and components which normally will not cause the condition described, a problem can be isolated within one or more systems or circuits without wasting time on unnecessary testing. Experience has shown that most problems tend to be the result of a fairly simple and obvious cause, such as loose or corroded connectors or air leaks in the intake system; making careful inspection of components during testing essential to quick and accurate troubleshooting. Frequent references to the Electronic Fuel Injection Analyzer J-25400 will be found in the text and in the diagnosis charts. This device or its compatible equivalent is necessary to perform some of the more complicated test procedures listed, but many of the electronic fuel injection components can be functionally tested with the quick checks outlined in the "On-Car Service" procedures. Aftermarket testers are available from a variety of sources, as well as from the manufacturer, but care should be taken that the test equipment being used is designed to diagnose this particular system accurately without damaging the control unit (ECU) or components being tested.

NOTE: Some test procedures incorporate other engine systems in addition to the EFI, since many conditions are the result of something other than the fuel injection system itself. Electrical and vacuum circuits are at the back of this section to help identify small changes incorporated into later models that may affect the diagnosis results. Circuit numbers are given whenever possible.

Because of the specific nature of the conditions listed in the diagnosis procedures, it's important to understand exactly what the definitions of various problems are:

• STALLS—engine stops running at idle or when driving. Determine if the stalling condition is only present when the engine is either hot or cold, or if it happens consistently regardless of operating temperature.

• LOADS UP—engine misses due to excessively rich mixture. This usually occurs during cold engine operation and is characterized by black smoke from the tailpipe.

• ROUGH IDLE—engine runs unevenly at idle. This condition can range from a slight stumble or miss up to a severe shake.

• TIP IN STUMBLE—a delay or hesitation in engine response when accelerating from idle with the car at a standstill. Some slight hesitation conditions are considered normal when they only occur during cold operation and gradually vanish as the engine warms up.

• MISFIRE—rough engine operation due to a lack of combustion in one or more cylinders. Fouled spark plugs or loose ignition wires are the most common cause.

• HESITATION—a delay in engine response when accelerating from cruise or steady throttle operation at road speed. Not to be confused with the tip in stumble described above.

• SAG—engine responds initially, then flattens out or slows down before recovering. Severe sags can cause the engine to stall.

• SURGE—engine power variation under steady throttle or cruise. Engine will speed up or slow down with no change in the throttle position. Can happen at a variety of speeds.

• SLUGGISH—engine delivers limited power under load or at high speeds. Engine loses speed going up hills, doesn't accelerate as fast as normal, or has less top speed than was noted previously.

• CUTS OUT—temporary complete loss of power at sharp, irregular intervals. May occur repeatedly or intermittently, but is usually worse under heavy acceleration.

• POOR FUEL ECONOMY—significantly lower gas mileage than is considered normal for the model and drive-train in question. Always perform a careful mileage test under a variety of road conditions to determine the severity of the problem before attempting corrective measures. Fuel economy is influenced more by external conditions, such as driving habits and terrain, than by a minor malfunction in the fuel injection system that doesn't cause another problem (like rough operation).

Stall After Start—Cold

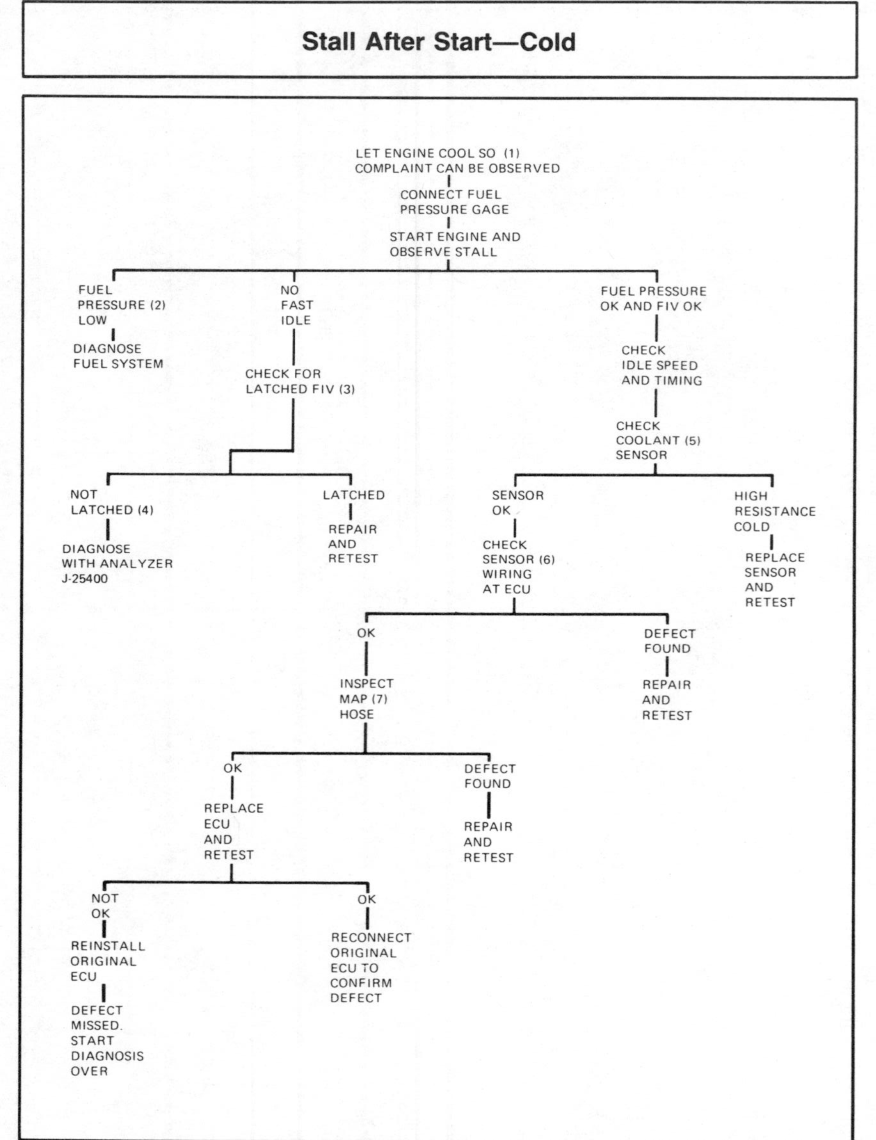

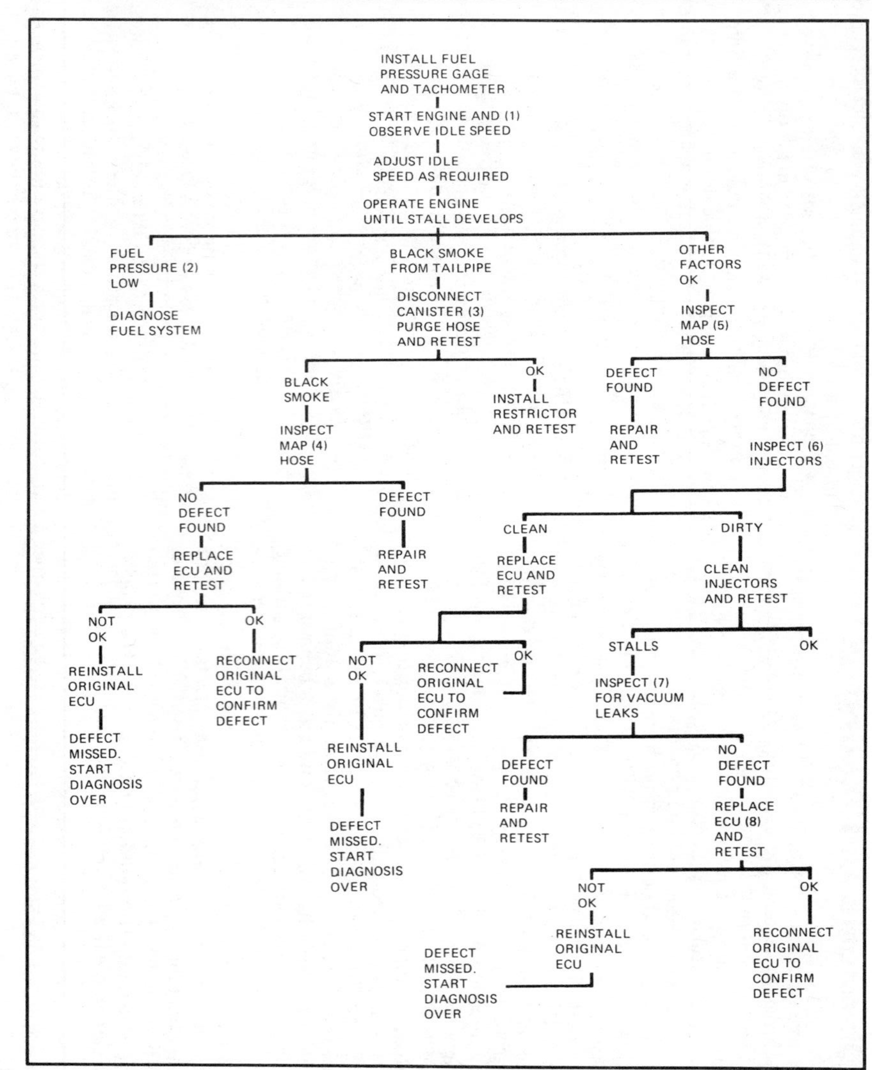

No Start

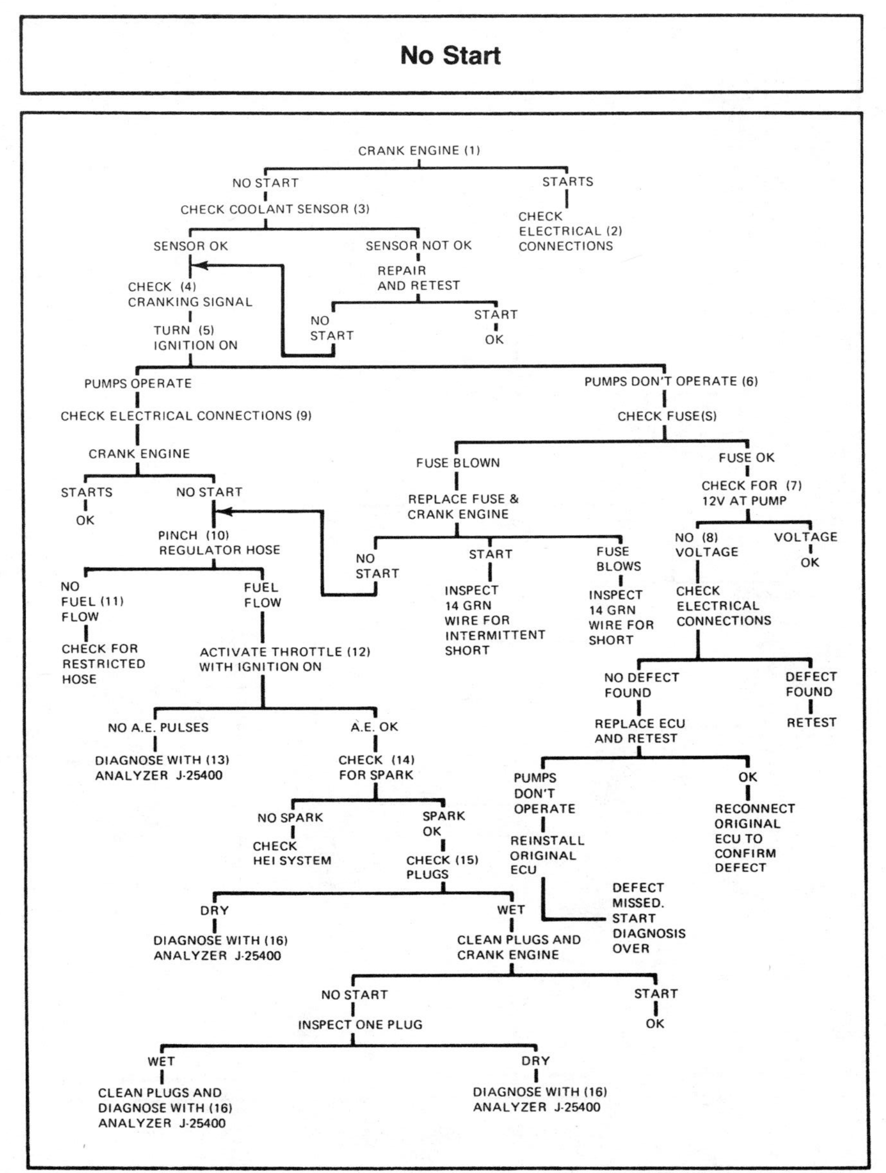

Hard Start—Cold

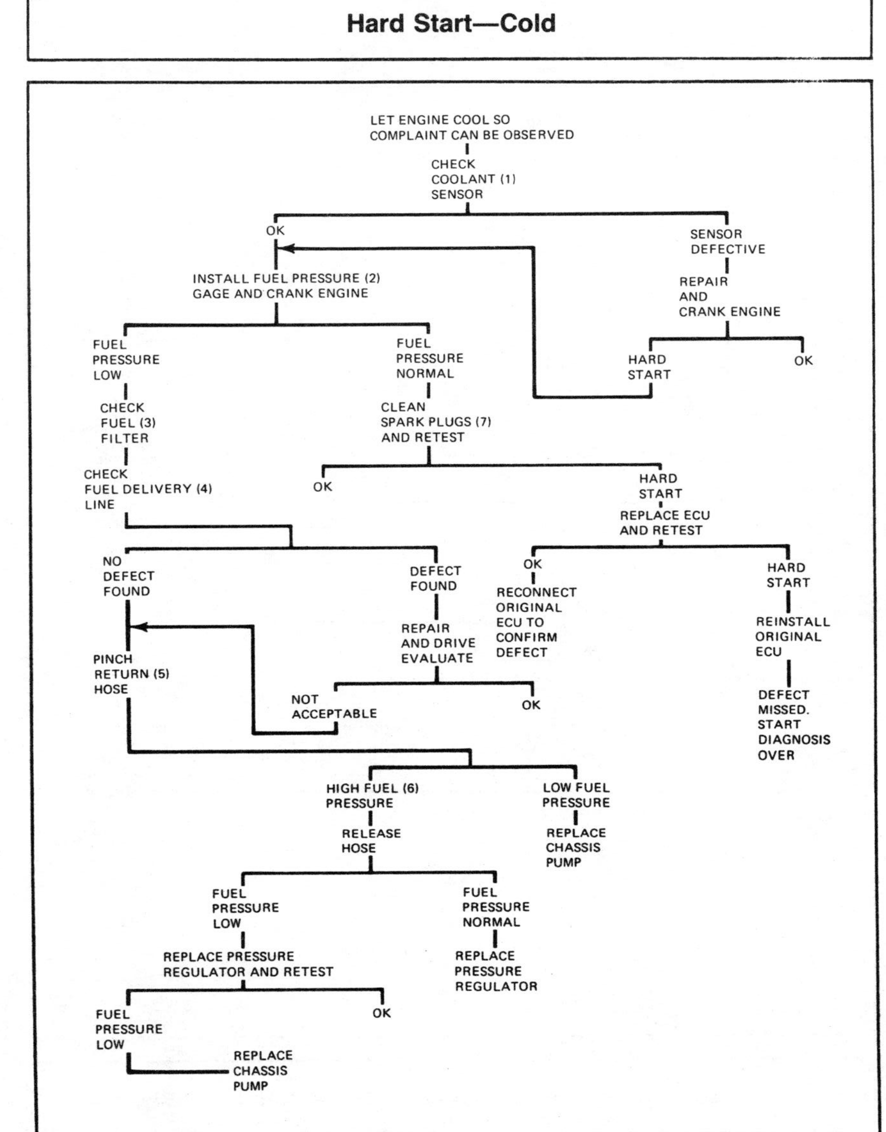

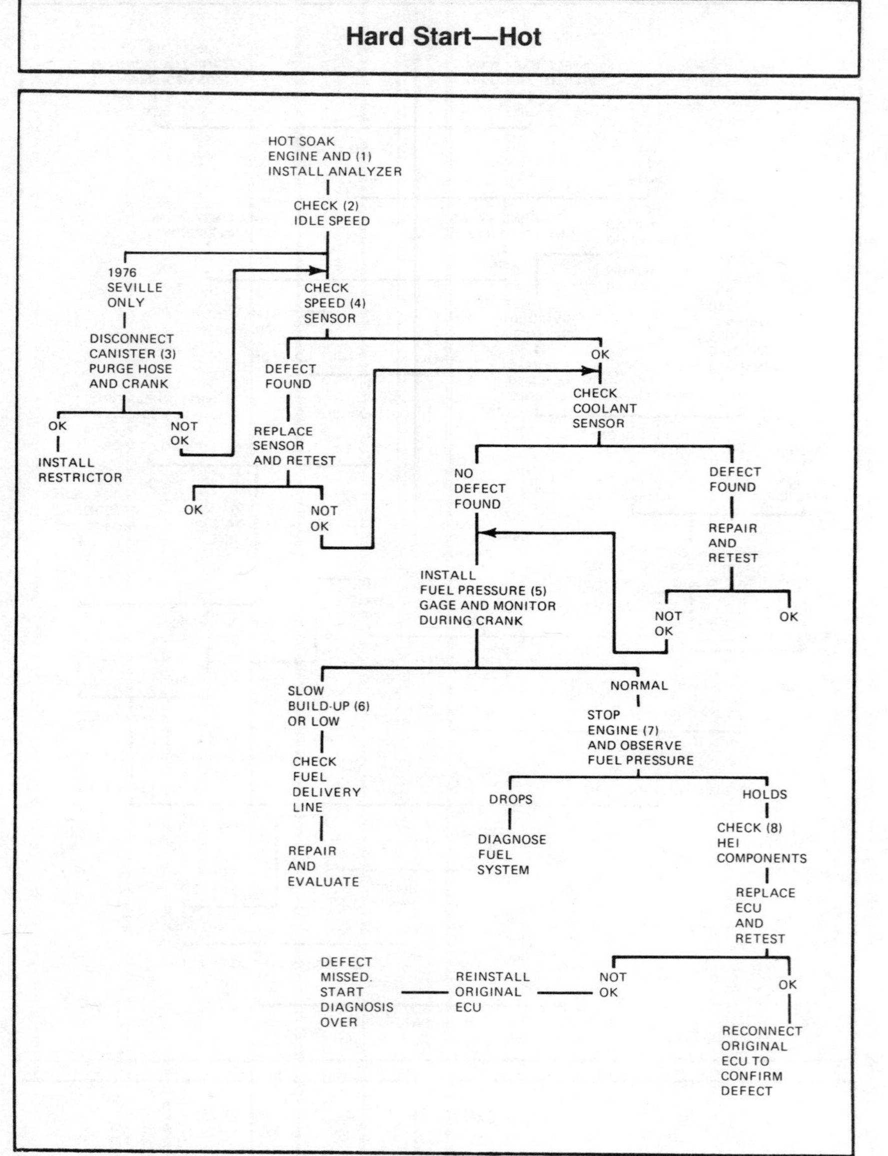
Hard Start—Hot
HOT SOAK ENGINE AND (1) INSTALL ANALYZER
CHECK (2) IDLE SPEED
1976 SEVILLE ONLY
DISCONNECT CANISTER (3) PURGE HOSE AND CRANK
OK
INSTALL RESTRICTOR
NOT OK
CHECK SPEED (4) SENSOR
DEFECT FOUND
REPLACE SENSOR AND RETEST
OK
NOT OK
OK
CHECK COOLANT SENSOR
NO DEFECT FOUND
DEFECT FOUND
REPAIR AND RETEST
NOT OK
OK
INSTALL FUEL PRESSURE (5) GAGE AND MONITOR DURING CRANK
SLOW BUILD-UP (6) OR LOW
CHECK FUEL DELIVERY LINE
REPAIR AND EVALUATE
NORMAL
STOP ENGINE (7) AND OBSERVE FUEL PRESSURE
DROPS
DIAGNOSE FUEL SYSTEM
HOLDS
CHECK (8) HEI COMPONENTS
REPLACE ECU AND RETEST
NOT OK
REINSTALL ORIGINAL ECU
DEFECT MISSED. START DIAGNOSIS OVER
OK
RECONNECT ORIGINAL ECU TO CONFIRM DEFECT

Stalls While Driving—Immediate Restart (Intermit.)

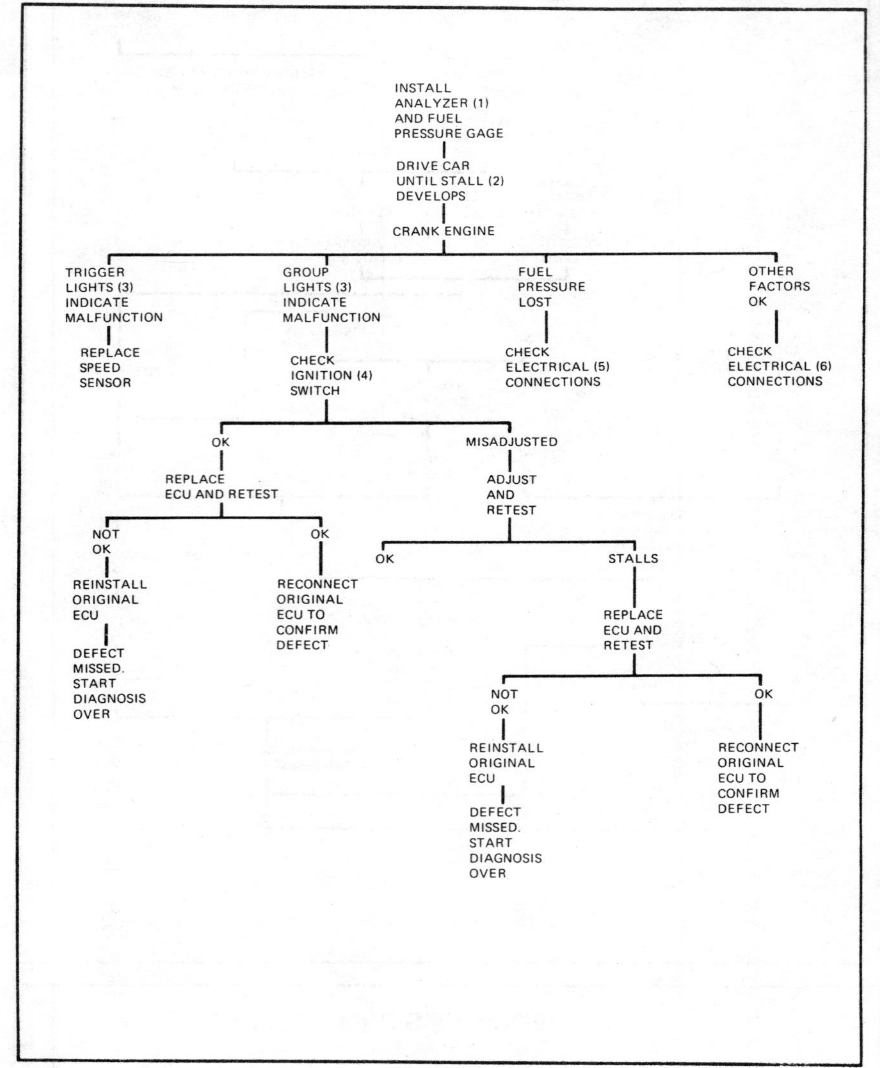
INSTALL ANALYZER (1) AND FUEL PRESSURE GAGE
DRIVE CAR UNTIL STALL (2) DEVELOPS
CRANK ENGINE
TRIGGER LIGHTS (3) INDICATE MALFUNCTION
REPLACE SPEED SENSOR
GROUP LIGHTS (3) INDICATE MALFUNCTION
CHECK IGNITION (4) SWITCH
FUEL PRESSURE LOST
CHECK ELECTRICAL (5) CONNECTIONS
OTHER FACTORS OK
CHECK ELECTRICAL (6) CONNECTIONS
OK
REPLACE ECU AND RETEST
NOT OK
REINSTALL ORIGINAL ECU
DEFECT MISSED. START DIAGNOSIS OVER
OK
RECONNECT ORIGINAL ECU TO CONFIRM DEFECT
MISADJUSTED
ADJUST AND RETEST
OK
STALLS
REPLACE ECU AND RETEST
NOT OK
REINSTALL ORIGINAL ECU
DEFECT MISSED. START DIAGNOSIS OVER
OK
RECONNECT ORIGINAL ECU TO CONFIRM DEFECT

Stalls While Driving—No Immediate Restart (Intermit.)

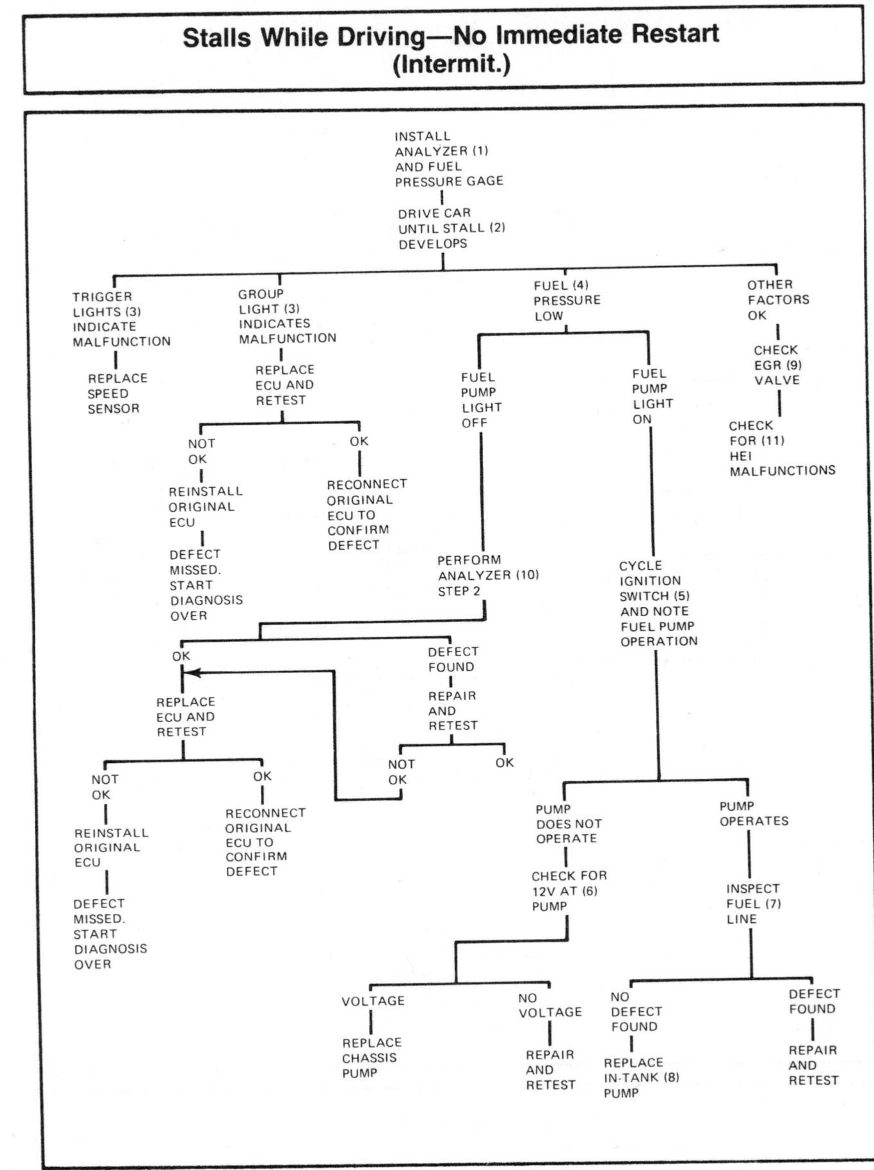

Tip-In Stumble

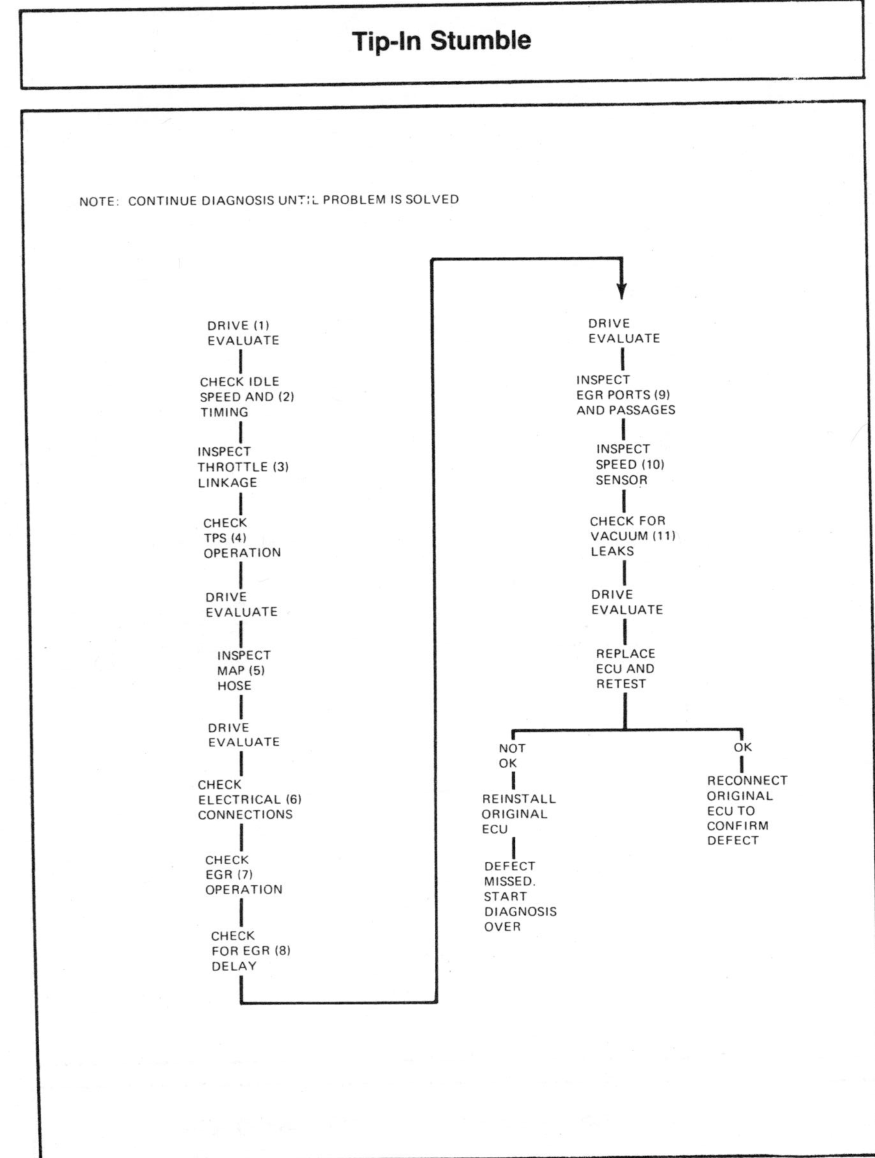

Surge (Lean Operation)

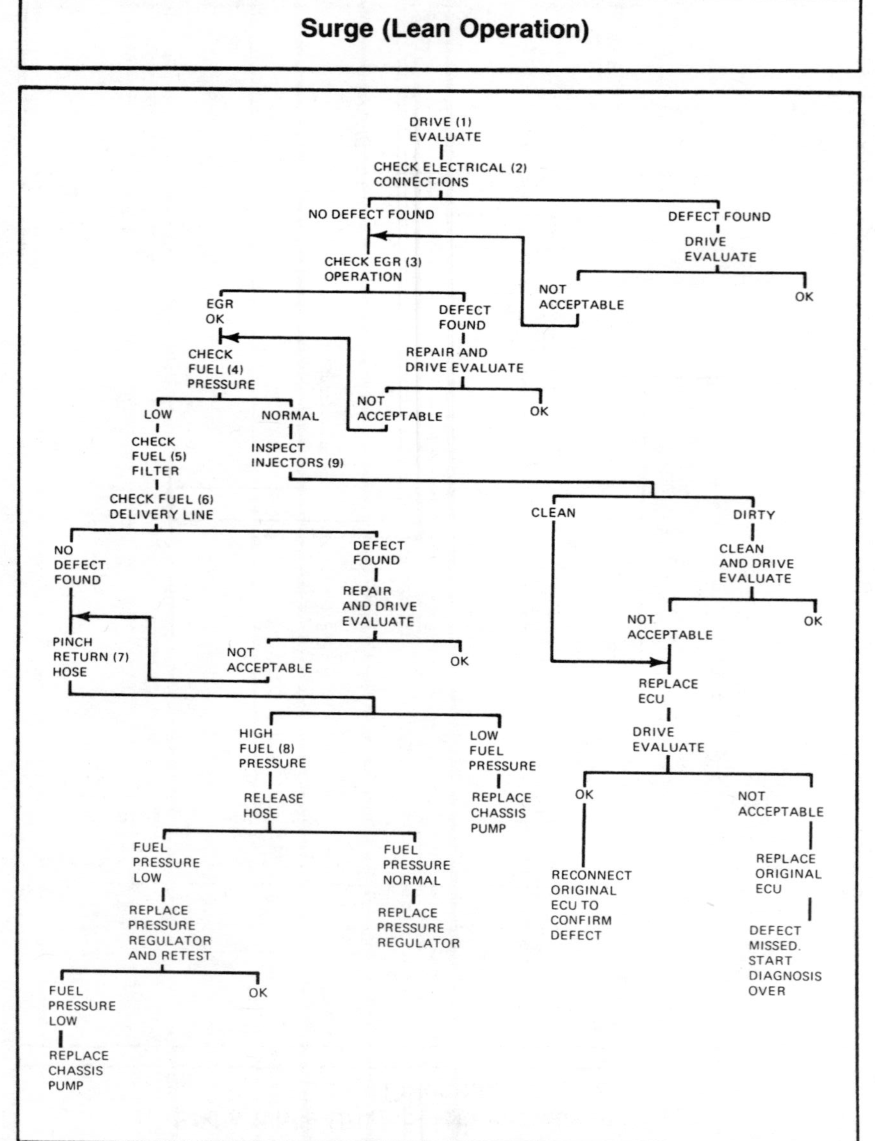

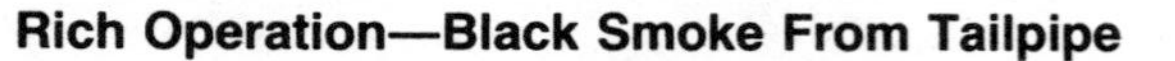

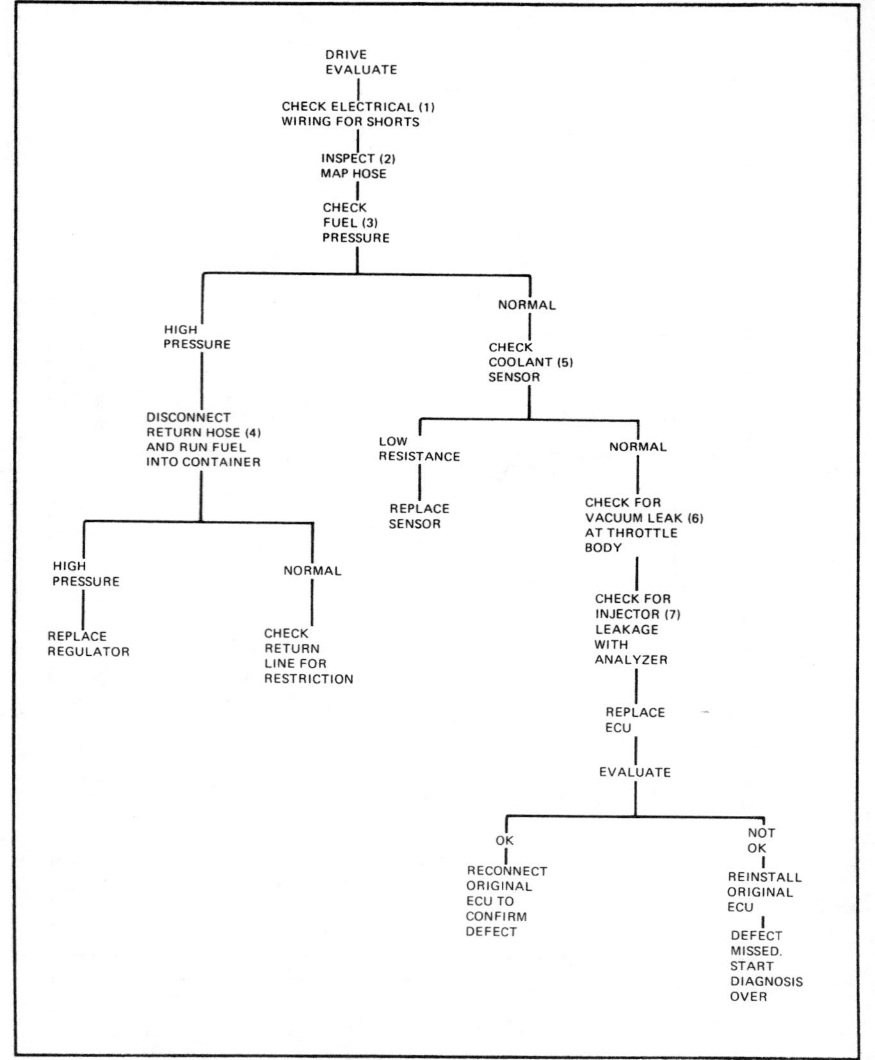

Hesitation

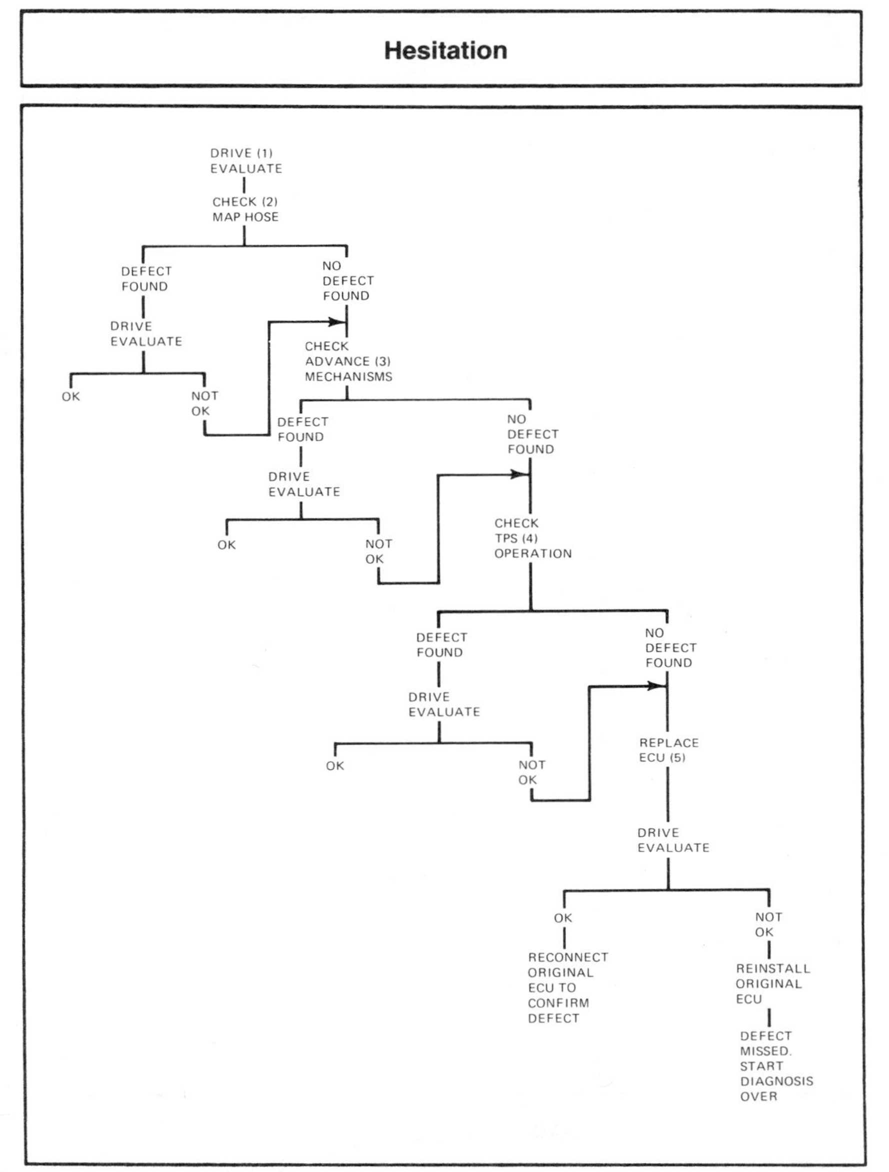

Rough Idle

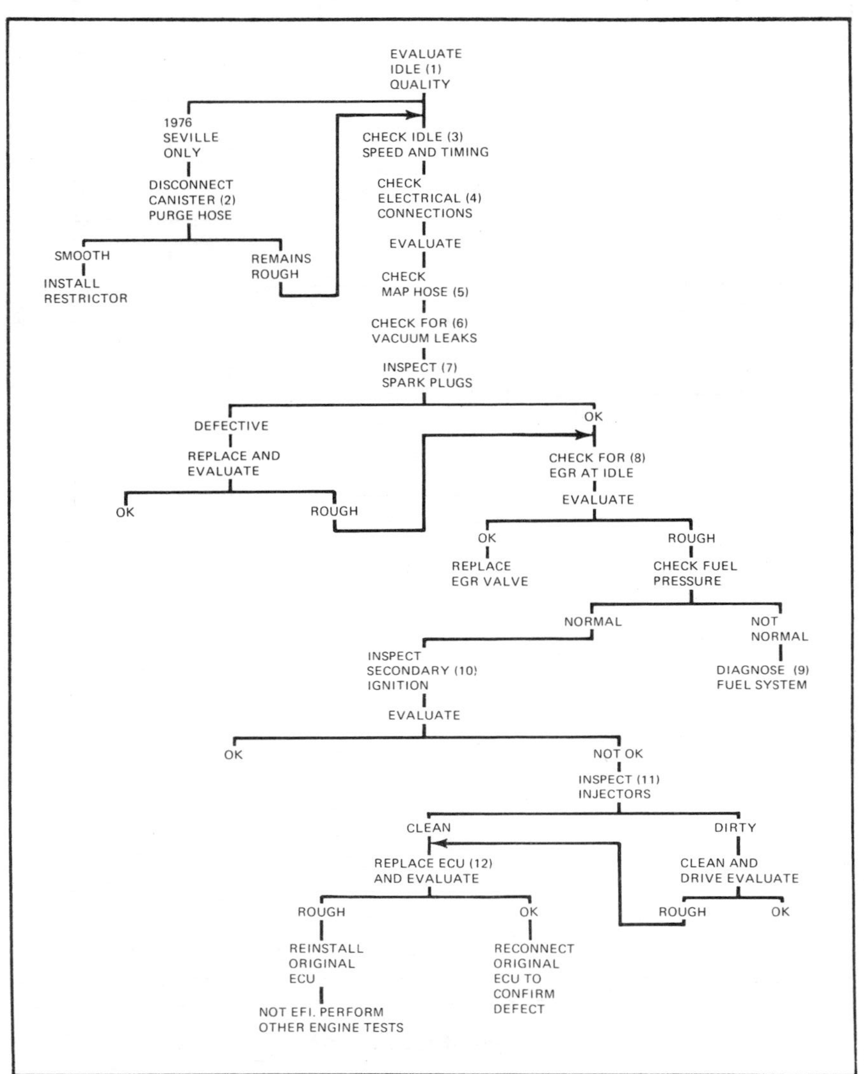

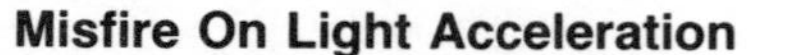

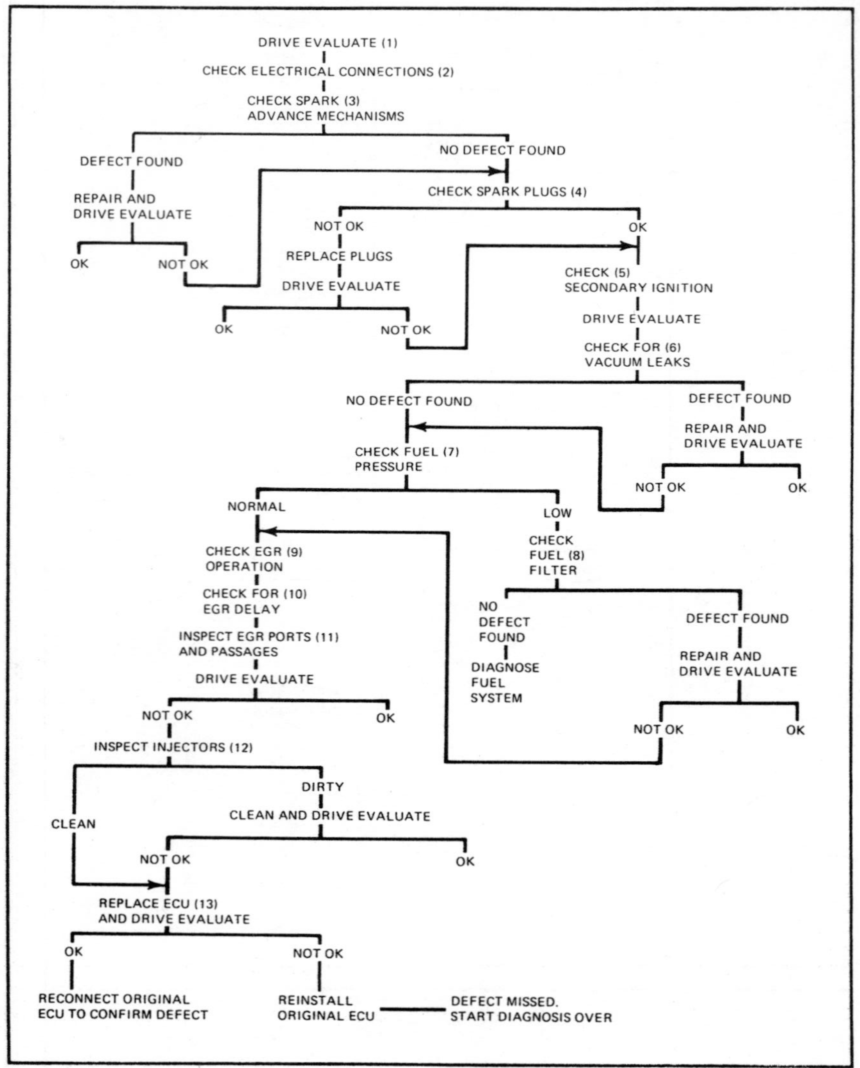

Idle Speed Change

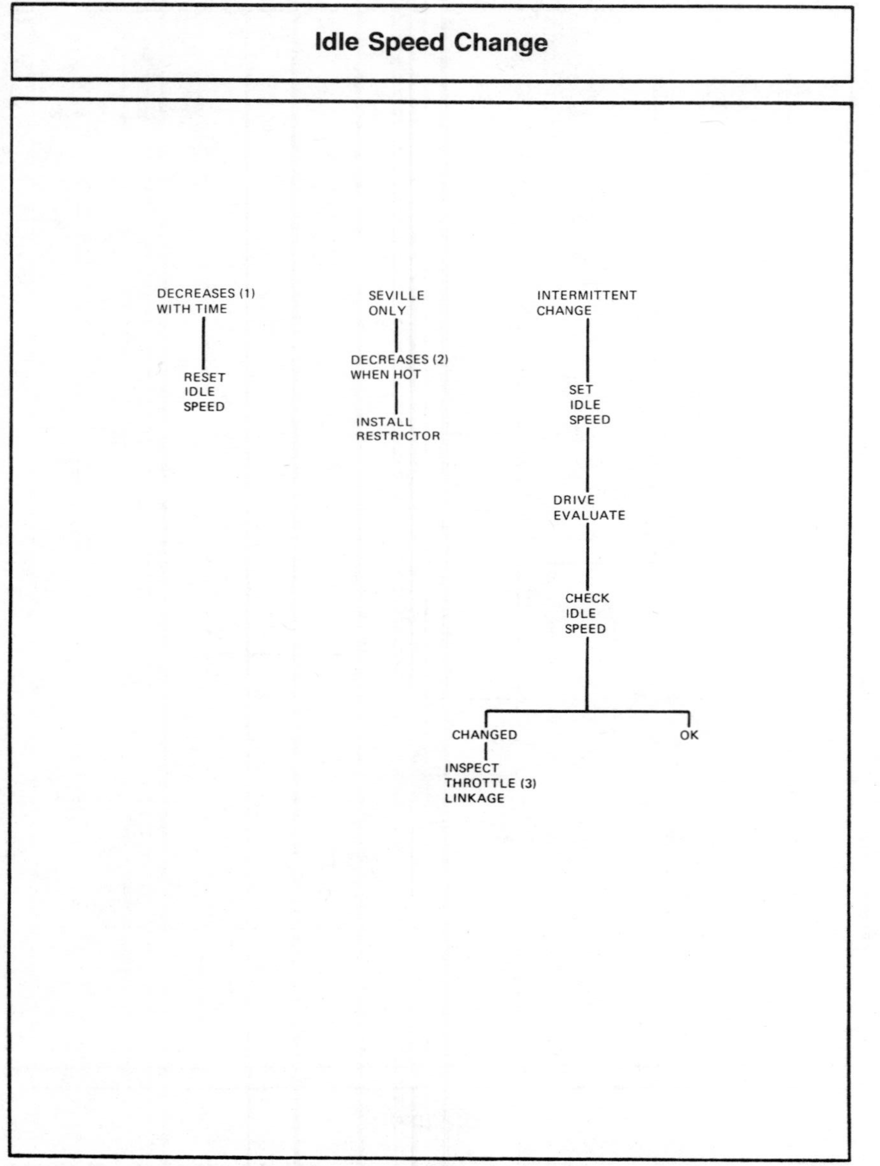

No Fast Idle

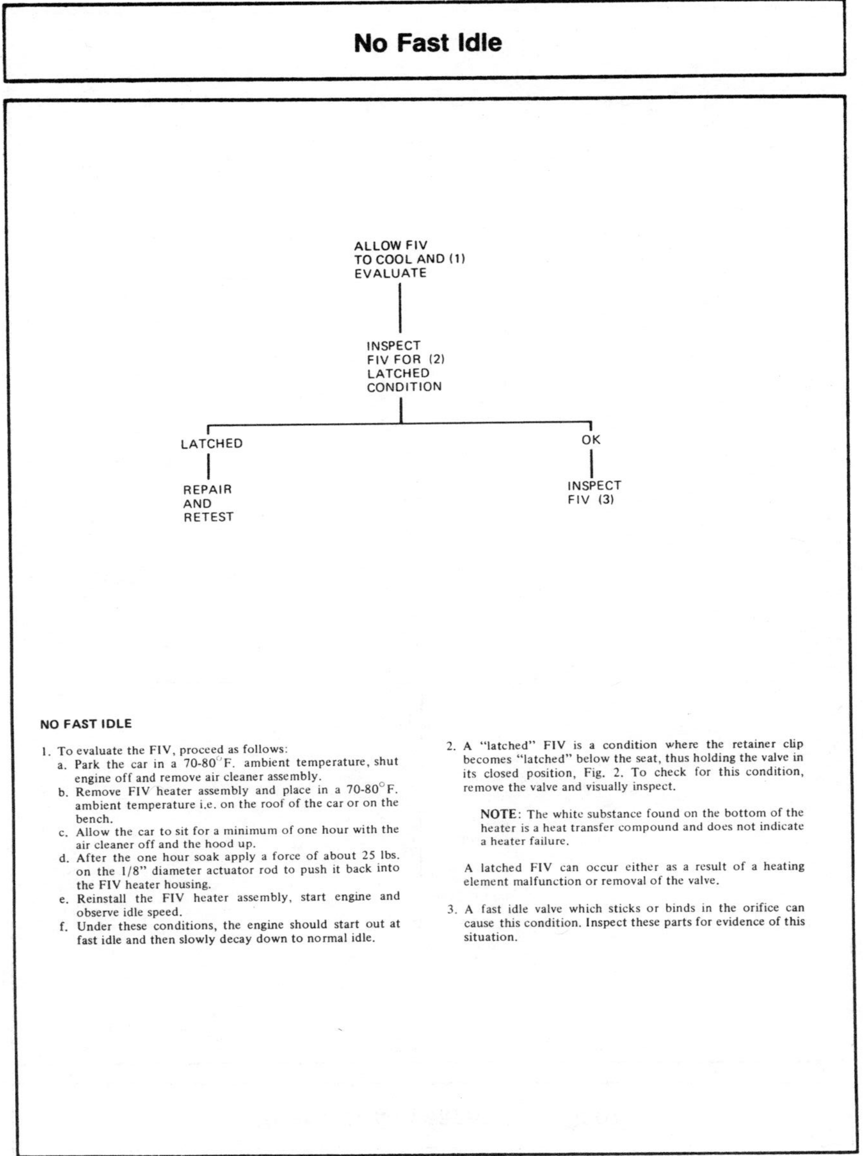

NO FAST IDLE

1. To evaluate the FIV, proceed as follows:
   a. Park the car in a 70-80°F. ambient temperature, shut engine off and remove air cleaner assembly.
   b. Remove FIV heater assembly and place in a 70-80°F. ambient temperature i.e. on the roof of the car or on the bench.
   c. Allow the car to sit for a minimum of one hour with the air cleaner off and the hood up.
   d. After the one hour soak apply a force of about 25 lbs. on the 1/8" diameter actuator rod to push it back into the FIV heater housing.
   e. Reinstall the FIV heater assembly, start engine and observe idle speed.
   f. Under these conditions, the engine should start out at fast idle and then slowly decay down to normal idle.

2. A "latched" FIV is a condition where the retainer clip becomes "latched" below the seat, thus holding the valve in its closed position, Fig. 2. To check for this condition, remove the valve and visually inspect.

   NOTE: The white substance found on the bottom of the heater is a heat transfer compound and does not indicate a heater failure.

   A latched FIV can occur either as a result of a heating element malfunction or removal of the valve.

3. A fast idle valve which sticks or binds in the orifice can cause this condition. Inspect these parts for evidence of this situation.

Poor Cold Driveability

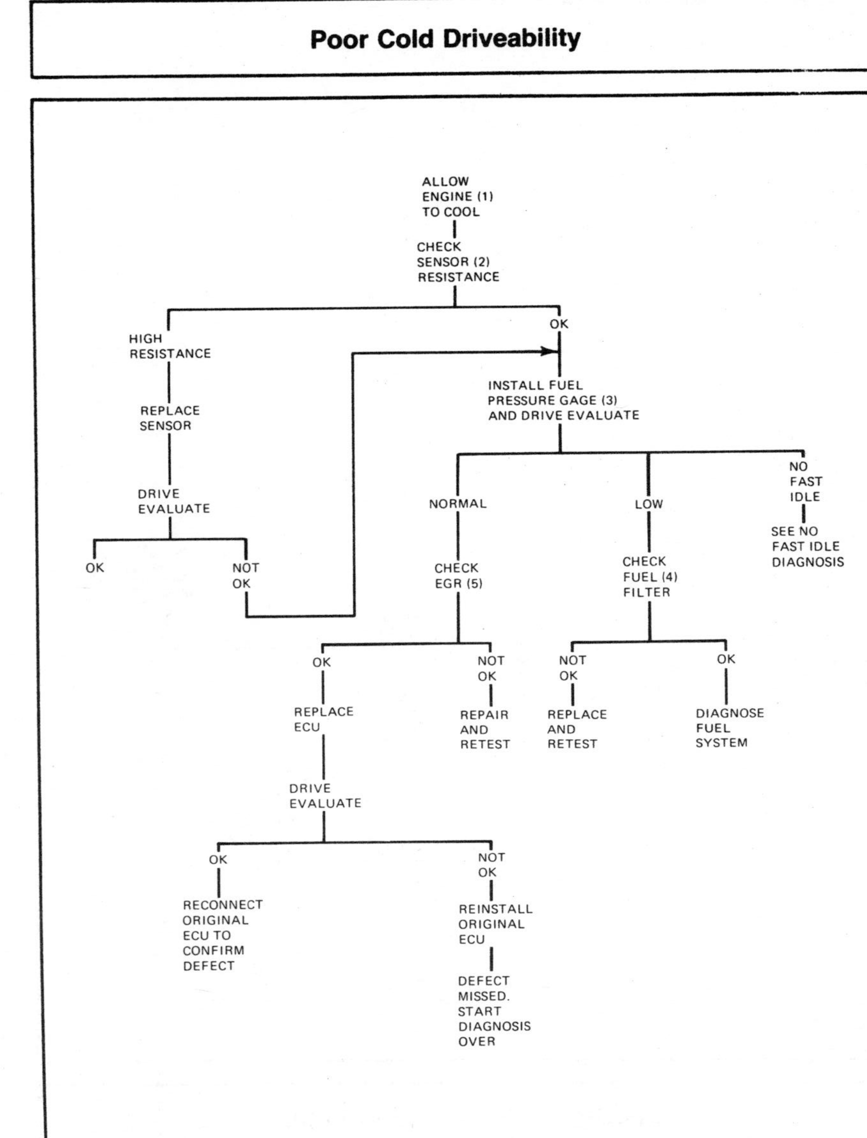

Poor Hot Driveability (or Sag)

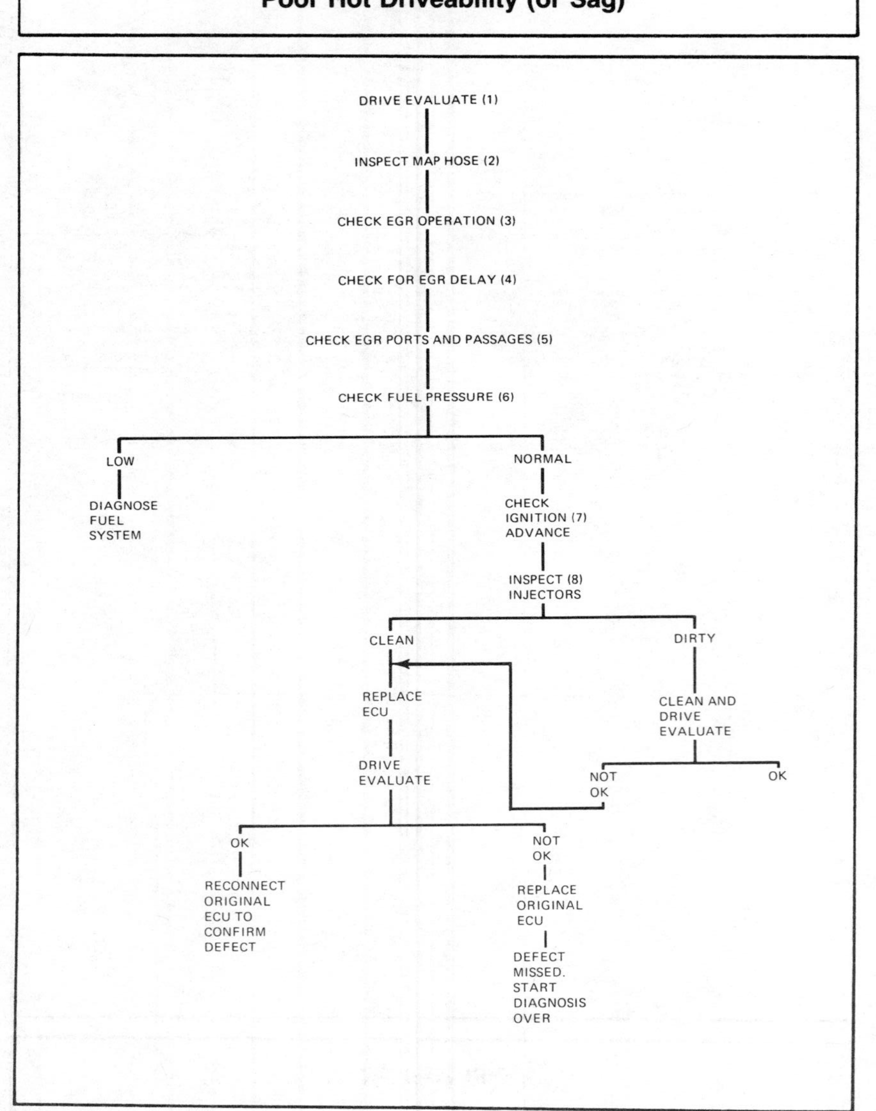

Cuts Out At Wide Open Throttle

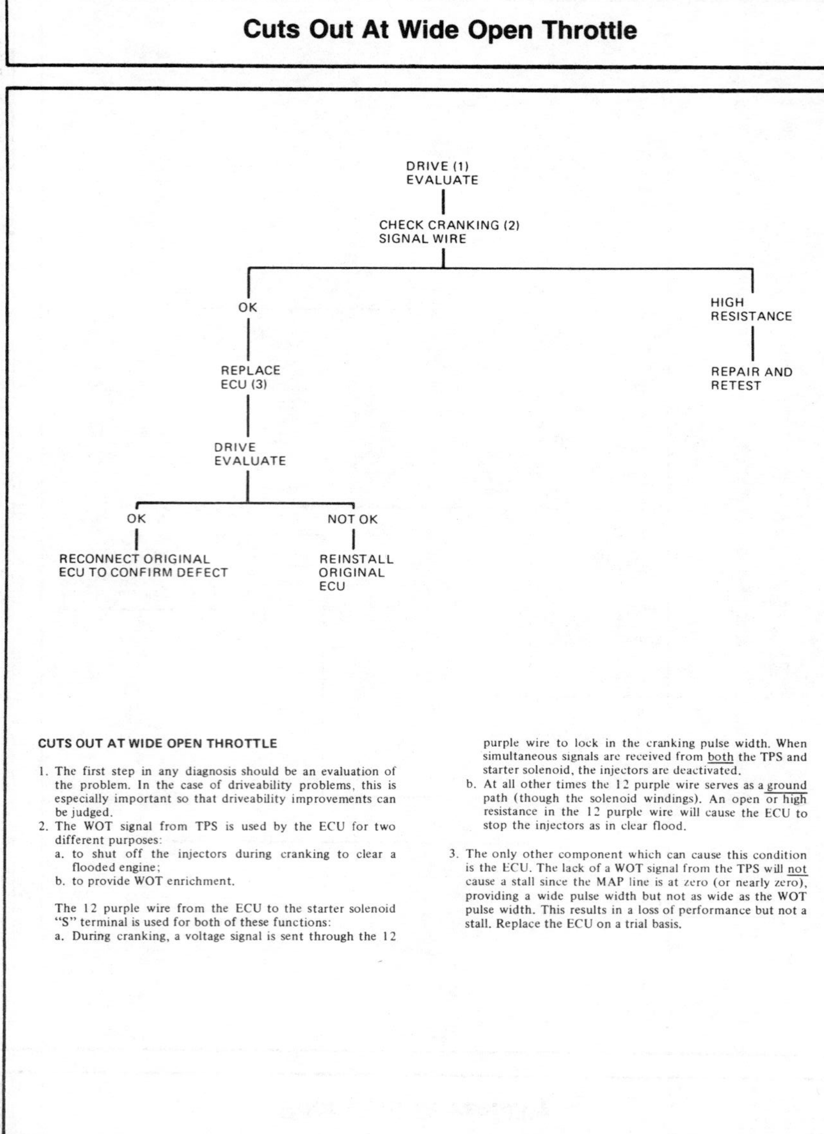

CUTS OUT AT WIDE OPEN THROTTLE

1. The first step in any diagnosis should be an evaluation of the problem. In the case of driveability problems, this is especially important so that driveability improvements can be judged.
2. The WOT signal from TPS is used by the ECU for two different purposes:
   a. to shut off the injectors during cranking to clear a flooded engine;
   b. to provide WOT enrichment.

   The 12 purple wire from the ECU to the starter solenoid "S" terminal is used for both of these functions:
   a. During cranking, a voltage signal is sent through the 12 purple wire to lock in the cranking pulse width. When simultaneous signals are received from both the TPS and starter solenoid, the injectors are deactivated.
   b. At all other times the 12 purple wire serves as a ground path (though the solenoid windings). An open or high resistance in the 12 purple wire will cause the ECU to stop the injectors as in clear flood.
3. The only other component which can cause this condition is the ECU. The lack of a WOT signal from the TPS will not cause a stall since the MAP line is at zero (or nearly zero), providing a wide pulse width but not as wide as the WOT pulse width. This results in a loss of performance but not a stall. Replace the ECU on a trial basis.

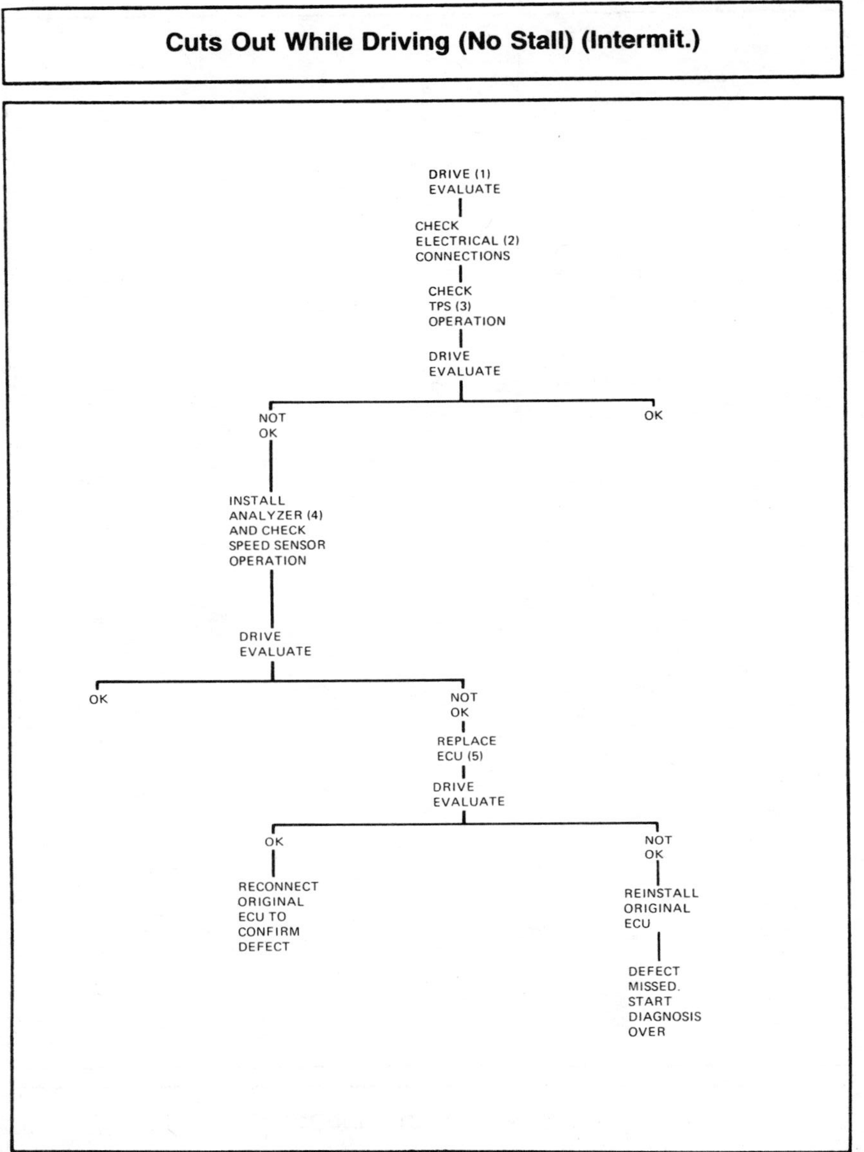
Cuts Out While Driving (No Stall) (Intermit.)
DRIVE (1) EVALUATE
CHECK ELECTRICAL (2) CONNECTIONS
CHECK TPS (3) OPERATION
DRIVE EVALUATE
NOT OK
OK
INSTALL ANALYZER (4) AND CHECK SPEED SENSOR OPERATION
DRIVE EVALUATE
OK
NOT OK
REPLACE ECU (5)
DRIVE EVALUATE
OK
NOT OK
RECONNECT ORIGINAL ECU TO CONFIRM DEFECT
REINSTALL ORIGINAL ECU
DEFECT MISSED. START DIAGNOSIS OVER

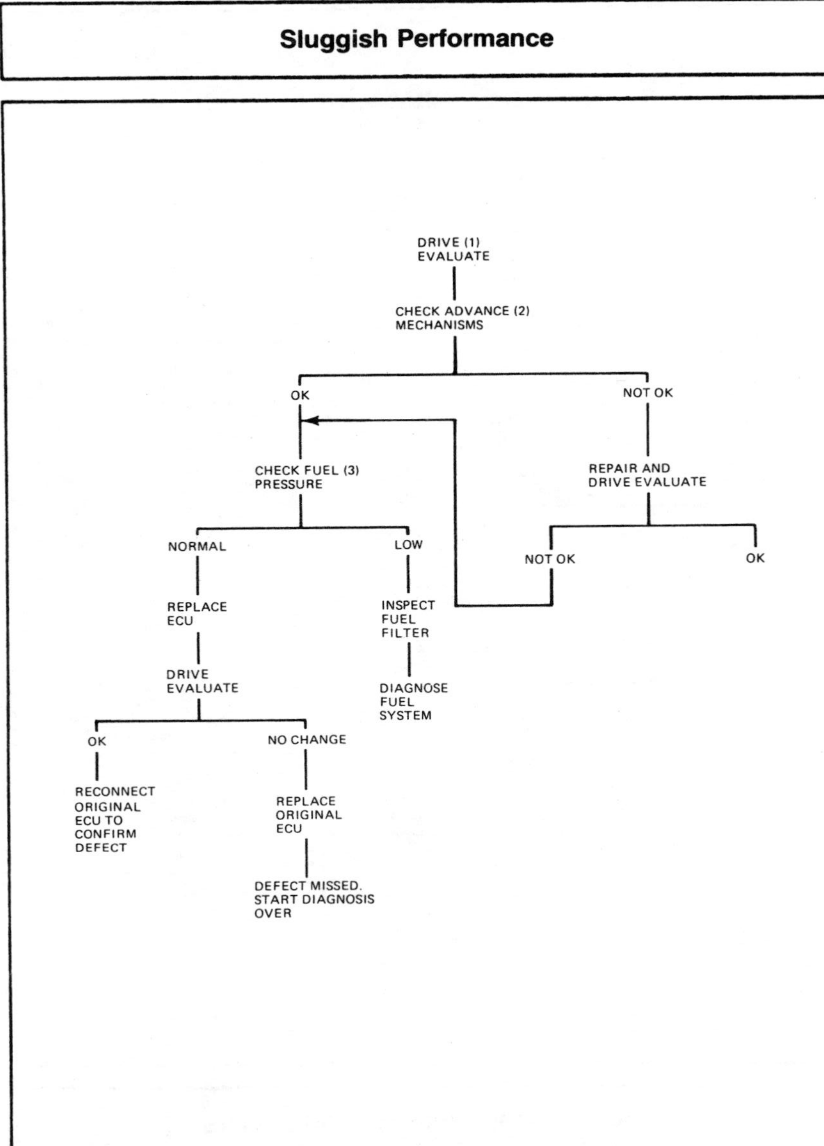
Sluggish Performance
DRIVE (1) EVALUATE
CHECK ADVANCE (2) MECHANISMS
OK
NOT OK
CHECK FUEL (3) PRESSURE
REPAIR AND DRIVE EVALUATE
NORMAL
LOW
NOT OK
OK
REPLACE ECU
INSPECT FUEL FILTER
DRIVE EVALUATE
DIAGNOSE FUEL SYSTEM
OK
NO CHANGE
RECONNECT ORIGINAL ECU TO CONFIRM DEFECT
REPLACE ORIGINAL ECU
DEFECT MISSED. START DIAGNOSIS OVER

Poor Fuel Economy

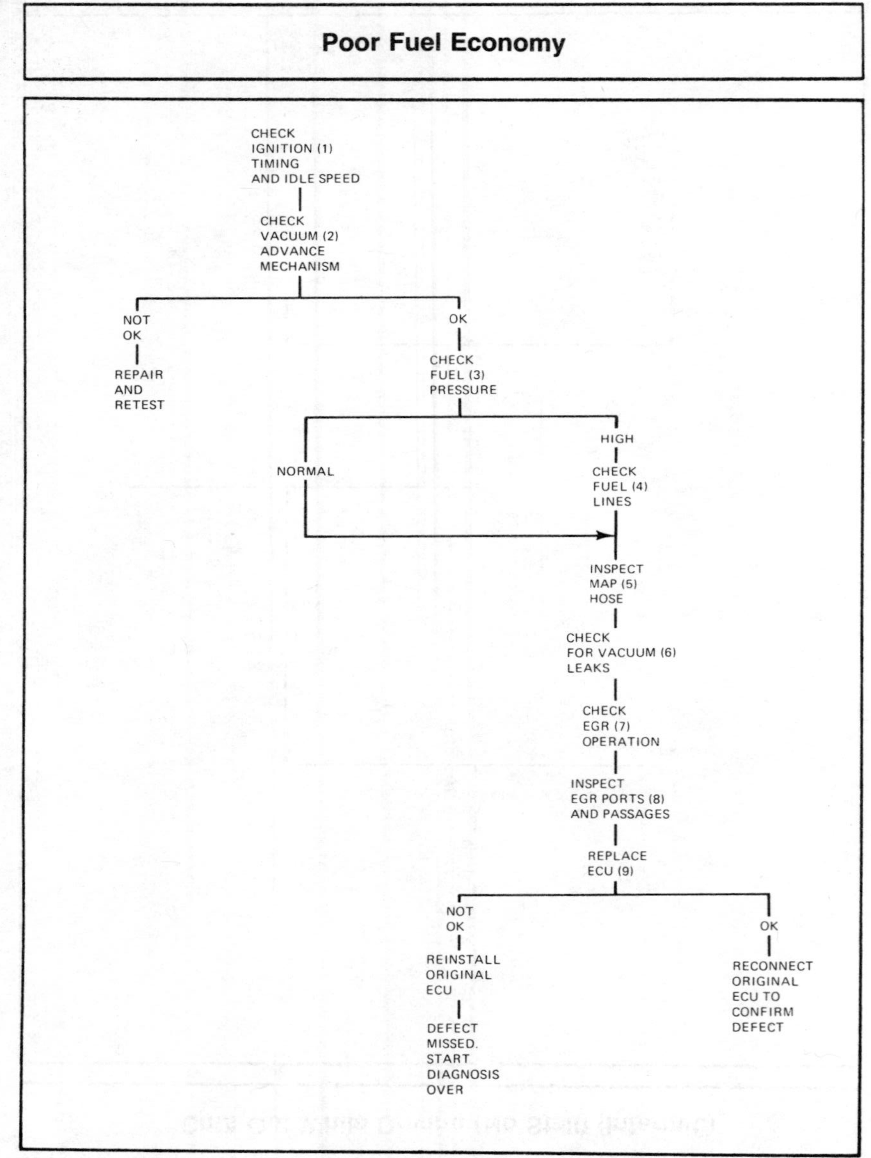

Static Fuel System Diagnosis—Part 1

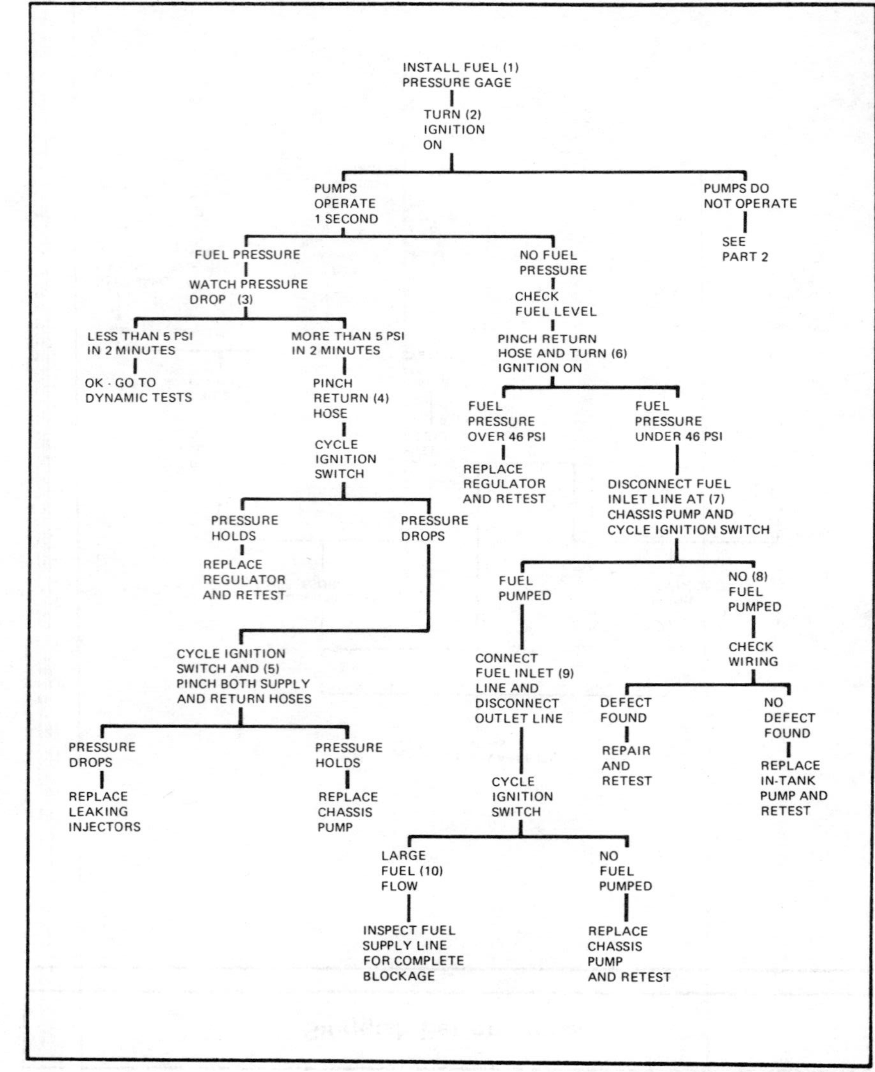

Static Fual System Diagnosis—Part 2

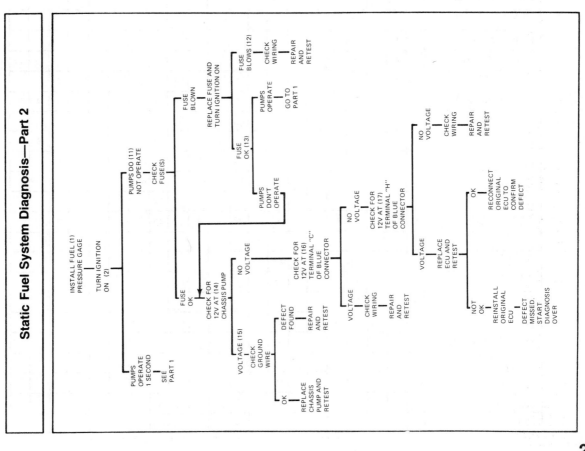

INSTALL FUEL (1) PRESSURE GAGE
TURN IGNITION ON (2)
PUMPS OPERATE 1 SECOND
SEE PART 1
PUMPS DO (11) NOT OPERATE
CHECK FUSE(S)
FUSE OK
FUSE BLOWN
REPLACE FUSE AND TURN IGNITION ON
FUSE OK (13)
FUSE BLOWS (12)
PUMPS DON'T OPERATE
PUMPS OPERATE
GO TO PART 1
CHECK WIRING
REPAIR AND RETEST
CHECK FOR 12V AT (14) CHASSIS PUMP
VOLTAGE (15)
CHECK GROUND WIRE
OK
REPLACE CHASSIS PUMP AND RETEST
DEFECT FOUND
REPAIR AND RETEST
NO VOLTAGE
CHECK FOR 12V AT (16) TERMINAL "C" OF BLUE CONNECTOR
VOLTAGE
CHECK WIRING
REPAIR AND RETEST
NO VOLTAGE
CHECK FOR 12V AT (17) TERMINAL "H" OF BLUE CONNECTOR
VOLTAGE
REPLACE ECU AND RETEST
NOT OK
REINSTALL ORIGINAL ECU
DEFECT MISSED. START DIAGNOSIS OVER
OK
RECONNECT ORIGINAL ECU TO CONFIRM DEFECT
NO VOLTAGE
CHECK WIRING
REPAIR AND RETEST

Dynamic Fuel System Diagnosis

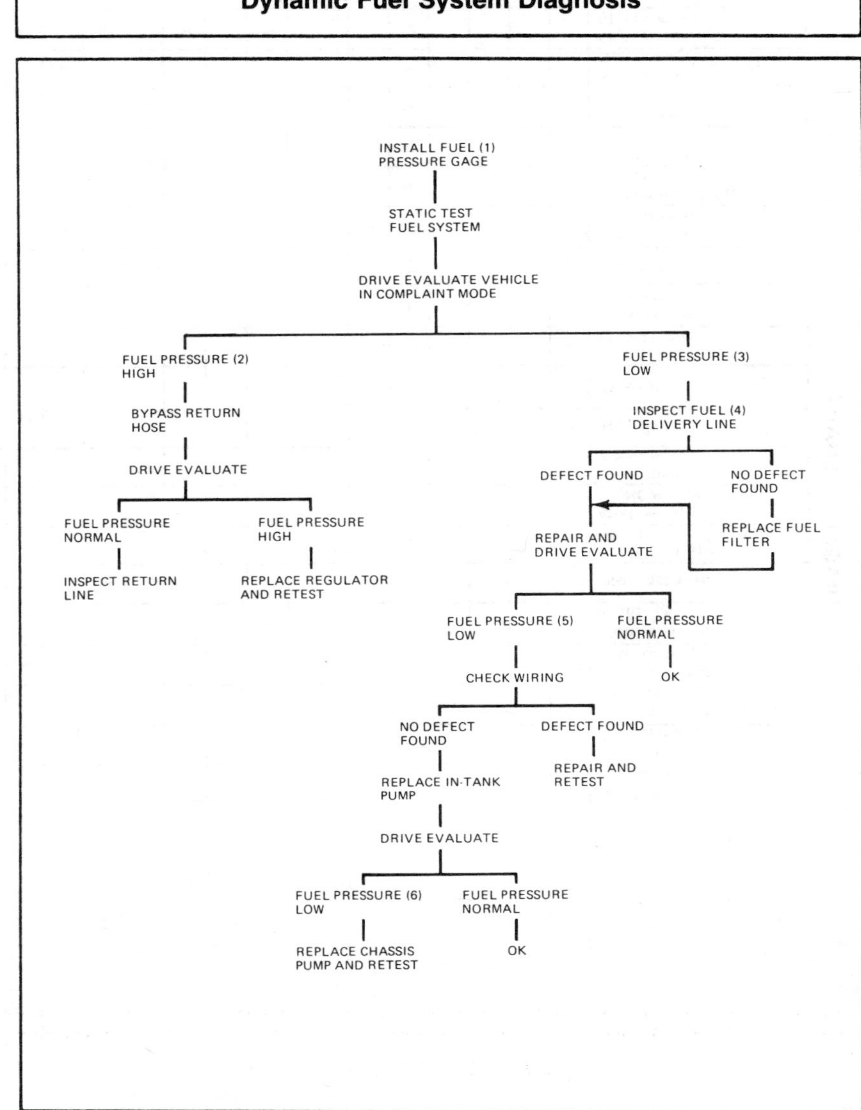

DIAGNOSIS CHART

CAUSES ↓ / COMPLAINTS →	Poor Fuel Economy	Sluggish Performance	Intermittent Cut-Out No Stall	Poor Driveability Hot or Sag	Poor Driveability Cold	No Fast Idle	Cuts Out at WOT	Misfire on Light Acceleration	Idle Speed Change	Rough Idle	Hesitation	Rich Operation	Surge - Lean Operation	Tip-in Stumble	Intermittent Stall - No Restart	Intermittent Stall - Restart	Hard Start Hot	Hard Start Cold	Stall After Start - Hot	Stall After Start - Cold	No Start
Electrical Connections			1					1		3		1	2	4		2					1
Cranking Signal Wire Open							1														7
TPS WOT Circuit																					8
TPS Adjustment and Operation			2						3		3			2							
ECU	9	6	4	11			2	10		15	4	6	6	14	5	3	8	3	7	8	9
Speed Sensor Malfunction			3												2	1	5				6
Speed Sensor Magnet And Sensor Orientation														11							
Coolant Sensor - High Resistance Cold					1													1		1	5
Coolant Sensor - Low Resistance Hot																	4				
Air Temperature Sensor					3															7	
FIV Inoperative						2															
FIV Latched					5	1														3	
Intank Pump															1		6		4		
Fuel Pressure Low Or Drops Off		3		6	2			7		10			4				2	2	3	2	2
Fuel Pressure High	5									11		3									
Fuel Pressure Low At Speed		7																			
Restricted Filter		5											1								
Damaged Fuel Lines		4		7																	
Evaporative Canister Excessive Purge									2	2							3		2		
Low Idle Speed										1				1			1		1	4	
High Idle Speed	3																				
MAP Hose Leak Or Obstruction	4			1						8	1	2		3					5	6	
Dirty or Sticking Injectors				10				9		14		5	5						6		
Injector Tip Damage								8		13											
Sticking Throttle Linkage														12							
Vacuum Leaks at Throttle Body	6							6		5		4		13							
Throttle Body Bore Contamination									1												
Restricted EGR Port In Throttle Body				5				11						10							
Restricted EGR Passage	8			4				12						9							
Wrong EGR Orifice	7			2				13						8							
EGR Solenoid Inoperative					4			14													
EGR Transducer				3				15													
EGR Operating At Idle										9					3						
No EGR								16					3	7							
EGR Delay								17						6							
HEI										12					4		7				3

CHRYSLER ELECTRONIC FUEL INJECTION

General Information

The Chrysler Electronic Fuel Injection (EFI) system is used on the 1981–82 Imperial exclusively. The system is broken down into three parts; the Fuel Hydraulic System, the Air Induction System, and the Fuel, Air and Ignition Command System.

The fuel hydraulic system includes all parts of the EFI system which are in physical contact with the fuel. Together they form the fuel flow path from the fuel tank to the fuel injection assembly and back again. The fuel hydraulic system is divided into two subsystems; the fuel supply subsystem and the fuel control subsystem.

Fuel Supply Subsystem

This system is composed of the in-tank fuel pump, the fuel delivery and return lines, a pair of parallel fuel filters, the control pump housing, the pressure regulator and bypass orifice and a pair of check valves.

In operation, the in-tank fuel pump picks up fuel from the fuel tank and delivers it forward through the fuel delivery line and pair of fuel filters until it reaches the control pump housing. The control pump housing serves as a connecting link between the fuel delivery and return lines. It also serves as a fuel reservoir for the control pump and ensures that the pump is always primed with fuel. A check valve in the control housing prevents the fuel from draining back into the lines when the engine is off (in-tank fuel pump not running).

The pressure regulator maintains the desired fuel pressure within the control pump housing by releasing excess fuel into the fuel return line, which carries it back to the fuel tank. The pressure regulator is equipped with a bypass orifice which purges fuel vapors from the control pump housing when the system is shut down. The fuel return line has a check valve which prevents fuel from running backwards out the return line in the event of an accident in which the vehicle rolls over.

Fuel Control Subsystem

The fuel control subsystem consists of the control pump, located in the control pump housing, the fuel flowmeter and temperature sensor, the fuel pressure switch and the fuel injection assembly.

The control pump is a positive displacement pump driven by a variable-speed electric motor. The control pump delivers fuel at high pressure (24–60 psi) through the fuel flowmeter, temperature sensor and fuel pressure switch to the fuel injection assembly.

The fuel flowmeter consists of a cylindrical cavity containing a free-turning vaned wheel. The fuel flowing through the flowmeter causes the wheel to spin at a rate proportionate to fuel flow. As the wheel spins, the vanes interrupt the light path between a light-emitting diode (LED) and a phototransistor. The frequency of the interruptions (pulses) is interpreted as flow rate by the fuel flowmeter module. A temperature sensor is used in conjunction with the fuel flowmeter to monitor fuel temperature. Together, these two sensors relay part of the information needed for precise control of the control pump motor speed.

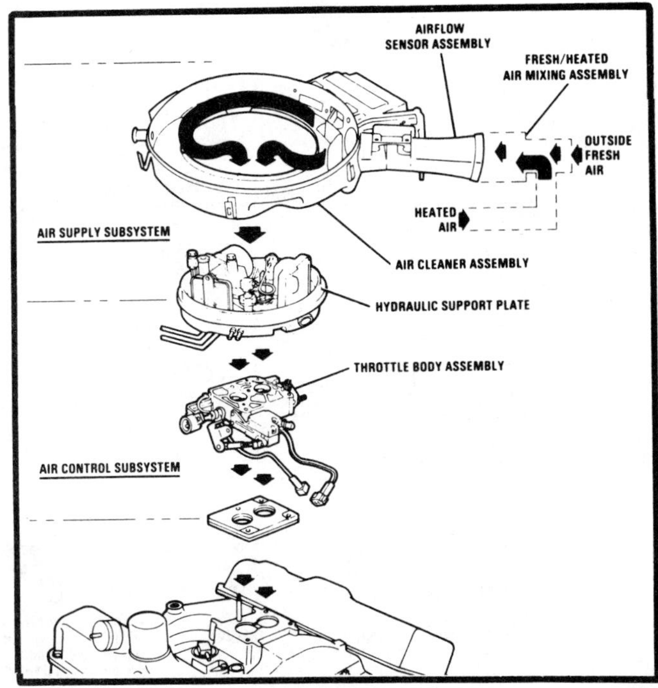

Chrysler EFI air induction components

A fuel pressure switch is located between the fuel flowmeter and the fuel injection assembly which opens when there is sufficient fuel pressure to start or run the engine, and closes when pressure is insufficient. When closed, the fuel pressure switch completes a bypass circuit which drives the control pump at full speed (with the ignition key in the start position). This pressurizes the fuel control circuit and insures quick starts. It also prevents vapor locks in the control pump. This entire pressurization process is completed within the time it takes the engine to revolve once.

Metered fuel entering the fuel injection assembly is directed to two pressure-regulating valves. Each valve feeds into its own U-shaped fuel injection bar, located over the throttle body assembly. The light load regulator valve opens when fuel pressure reaches or exceeds 21 psi and delivers fuel to the light load injector bar. Four tiny holes in the lower surface of the injector bar spray fuel onto crescent-shaped ridges at the edges of the throttle plates, where the actual fuel-air mixing occurs. Airfoil-shaped nozzles around the injector holes help refine fuel spray patterns and promote fuel

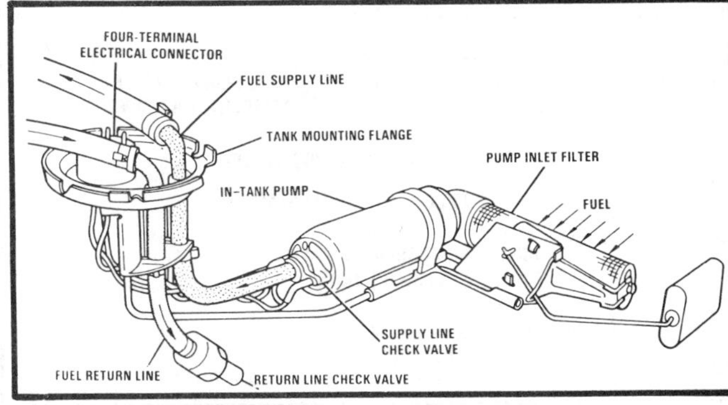

In-tank fuel pump assembly

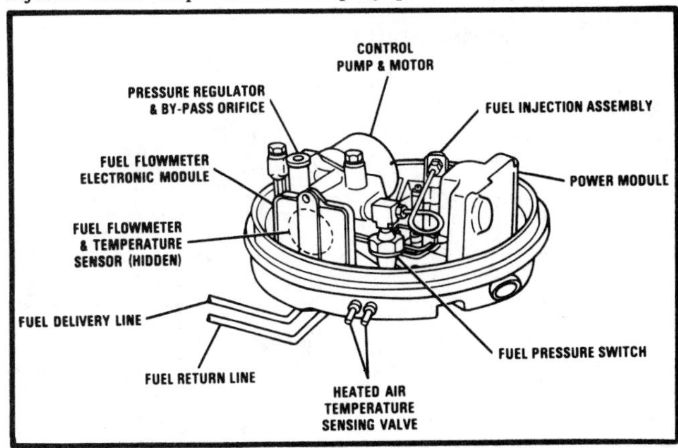

Hydraulic support plate assembly showing components

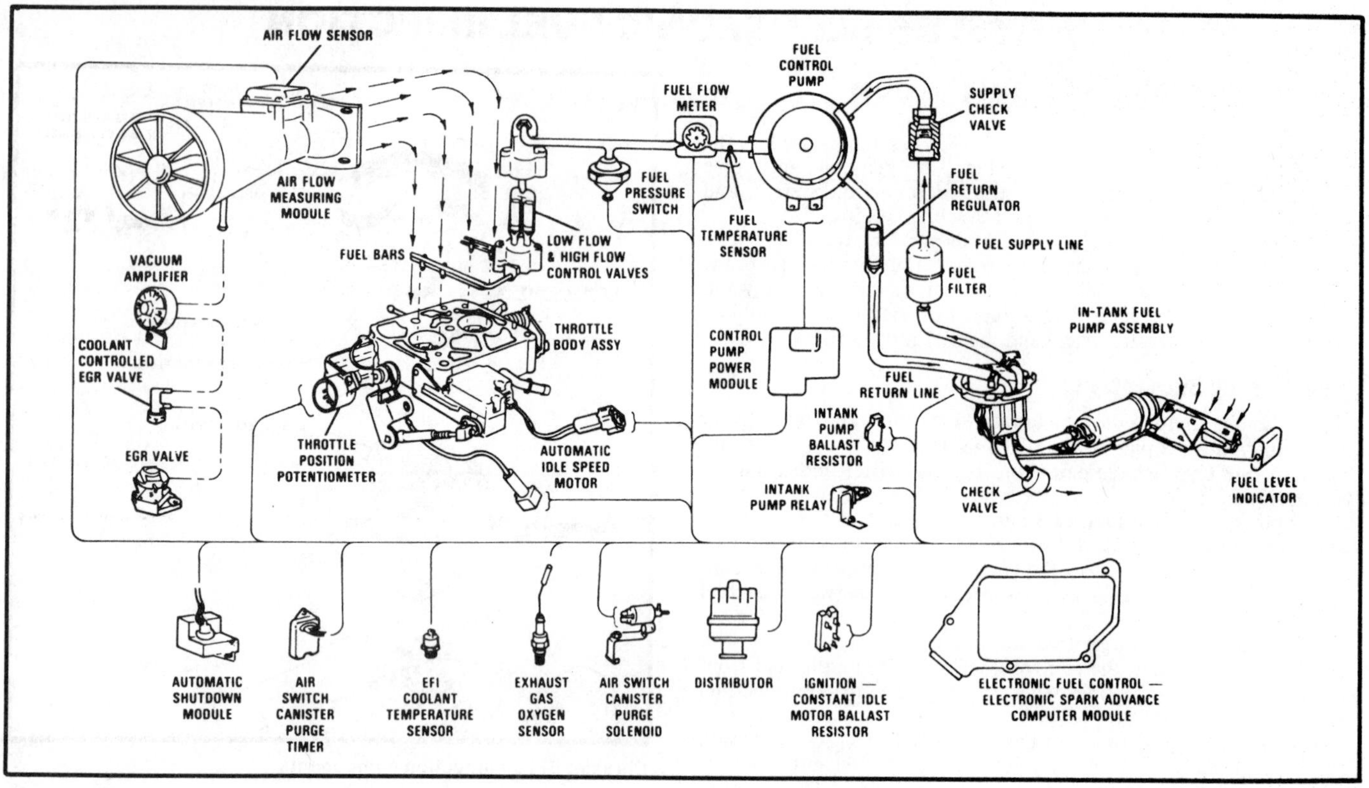

Chrysler EFI functional command components

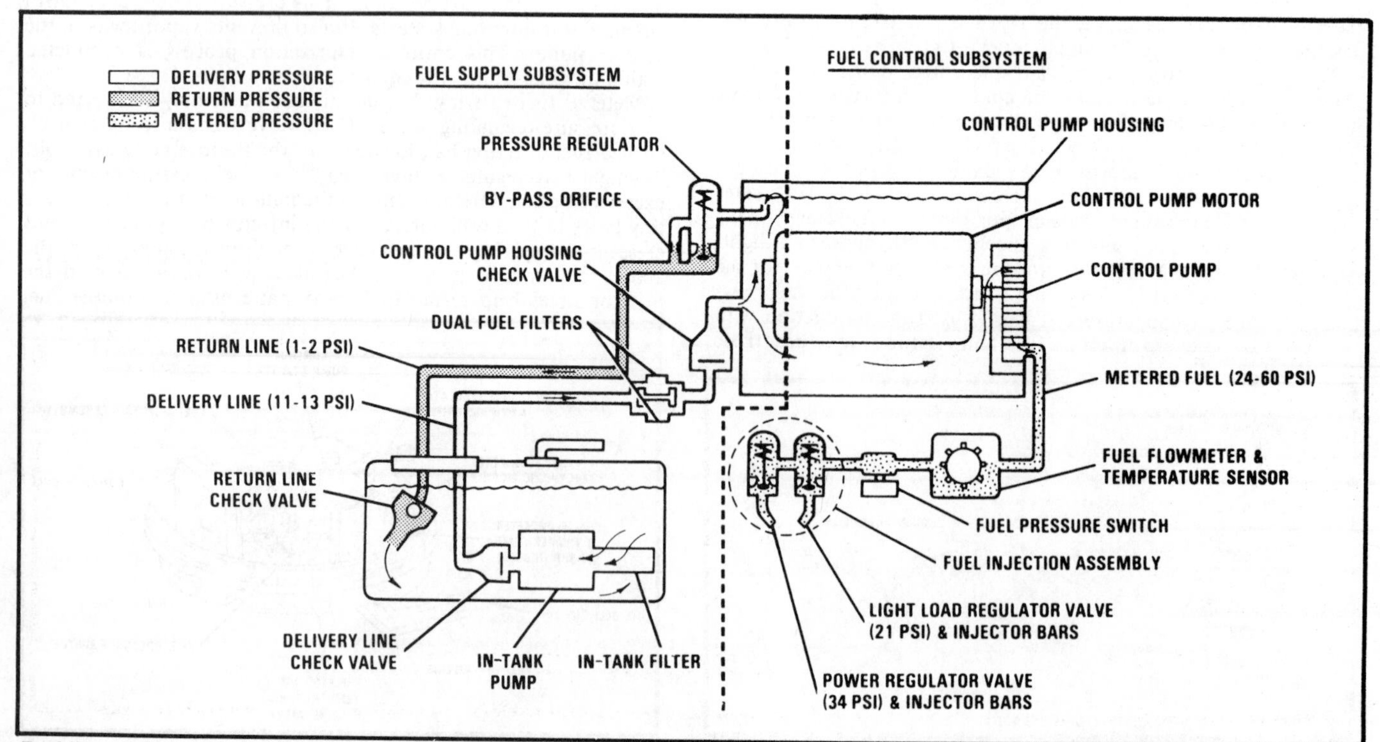

Fuel flow and control circuit for Chrysler EFI system

atomization. The light load circuit supplies all engine fuel when fuel pressure is between 21 and 34 psi, and some of the requirements beyond these pressures. At pressures above 34 psi (heavy engine loads, starting, etc.) the power regulator valve opens and allows the power fuel injection bar to add its spray pattern to the air/fuel mixing process.

Air Supply Subsystem

The air supply subsystem is comprised of the fresh/heated air mixing unit, which provides heated intake air during engine warmup, the air cleaner assembly and the airflow sensor assembly. The airflow sensor assembly is located inside the inlet duct on the air cleaner and measures engine intake airflow volume. This information is compared with fuel flowmeter information electronically and insures precise air/fuel mixture control.

Air Control Subsystem

The air control subsystem is contained in the throttle body assembly. Major parts include the throttle plate and blade subassembly, a throttle position potentiometer, a closed throttle switch and an automatic idle speed motor.

The throttle plates are similar to those used on a carburetor with the exception of a crescent-shaped ridge on the leading edge of each plate which promotes uniform air/fuel mixing.

The throttle position potentiometer senses the angle of throttle blade opening and sends this information to the combustion control computer which then adjusts the air/fuel mixture.

The closed throttle switch activates the automatic idle speed circuit and returns the ignition timing to its basic (minimum) advance timing when the throttle valves are closed (idling). In the event of malfunction, the brake signal circuit acts as a back-up circuit.

Fuel, Air and Ignition Command System

The EFI command system includes the following functions: automatic fuel flow metering to provide optimum air/fuel ratios for every engine operating mode; automatic advance or retardation of ignition timing to optimum points for every engine operating mode; automatic throttle opening adjustment to maintain optimum idling speed for every engine condition when the driver releases the accelerator pedal; automatic fuel flow shut off if certain ignition, engine speed or time requirements are not satisfied.

The heart of the command system is the Combustion Controlled Computer (CCC) which acts in conjunction with two other modules, the power module and the automatic shutdown module. The CCC receives input signals from a wide array of sensors and uses this information to adjust fuel flow and ignition timing to the correct levels.

The CCC also controls the feedback loop oxygen sensor system which has two modes of operation, closed loop and open loop. Under closed loop operation, the CCC receives signals from the oxygen sensor (located in the exhaust gas stream) and adjusts the air/fuel mixture in accordance with that signal. The sensor measures oxygen content in the exhaust system. Under closed loop operation, engine emissions are kept to a minimum. When the system is in open loop (initial start-up, engine cold, etc), the CCC disregards the oxygen signal and substitutes a pre-programmed air/fuel mixture circuit.

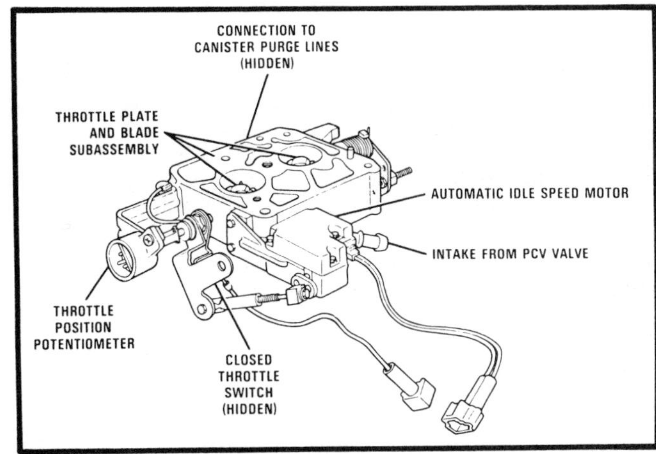

Throttle body assembly

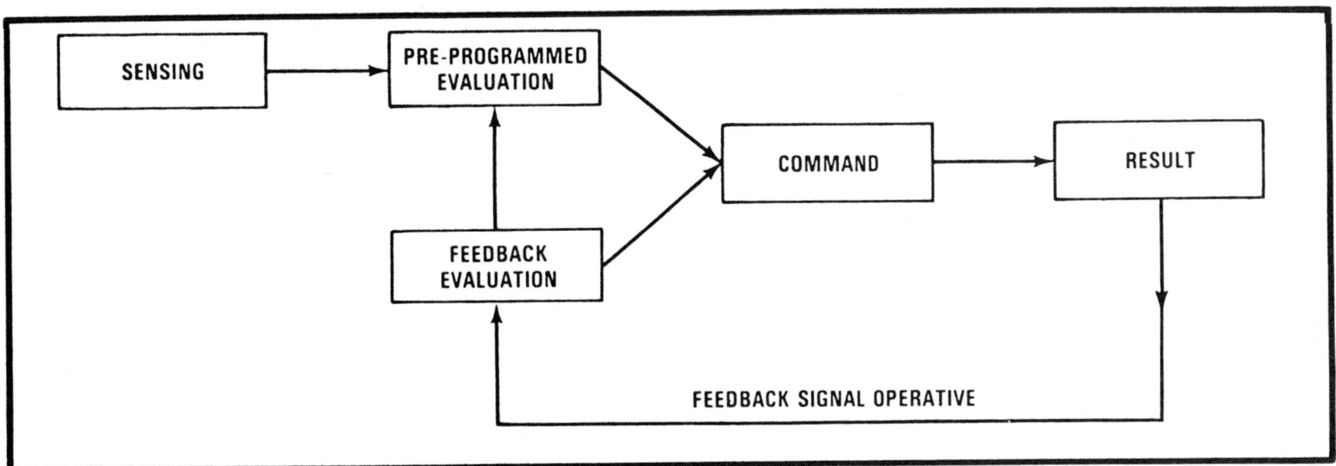

Closed loop command operation

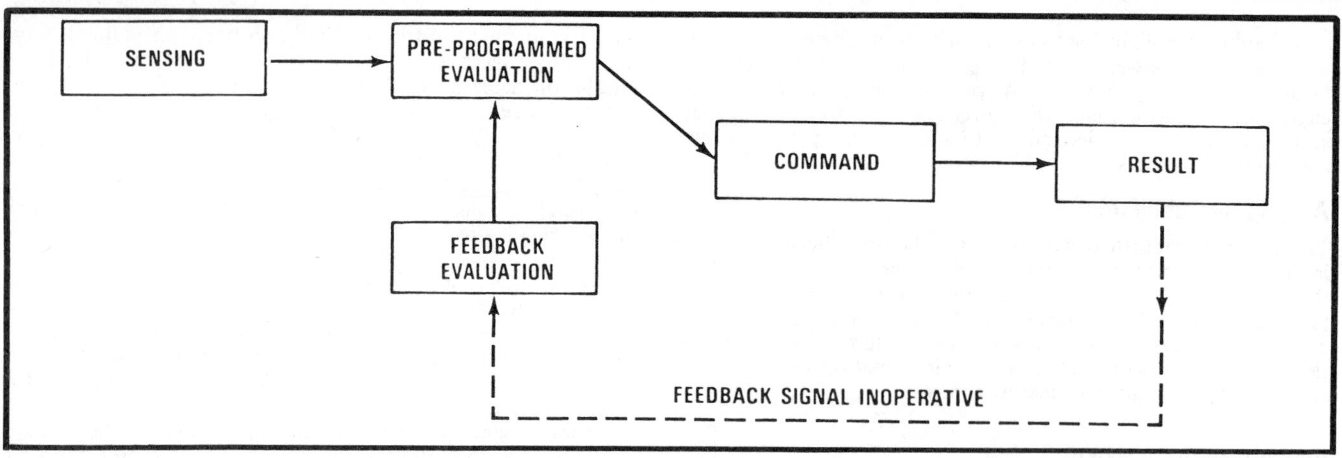

Open loop command operation

On-Car Service

The following is a general diagnostic procedure used to identify which system (ignition, fuel, etc.) is causing the no-start condition. If this procedure fails to produce positive results, the EFI system should be fully tested using a suitable EFI tester.

Take the time to make a preliminary check of all electrical wiring and vacuum hose connections for damage or looseness. Many times, problems occur simply because a connector or hose has become disconnected.

Step 1 (Spark Test)

1. Remove a spark plug wire and insert a well insulated screwdriver into the terminal of the disconnected plug wire. During the spark test, wear a heavy glove on the hand holding the screwdriver and hold the screwdriver by its insulated handle.
2. While holding the screwdriver shaft about 3/16 in. away from a good ground (metal alternator bracket, etc.), have someone crank the engine. Make sure no clothing, etc., is in the way of moving engine parts.
3. If there is a good spark between the screwdriver and ground, perform Step 2 (Fuel Flow Test). If there is no spark, the problem is in the ignition system, not in the fuel injection system.

Step 2 (Fuel Flow Test)

1. Remove the air cleaner.
2. Remove the secondary coil wire from the distributor cap and *ground this wire.*
3. Have someone crank the engine while observing the fuel flow from the fuel injection nozzles on the hydraulic support plate.
4. If there is an adequate fuel flow from the nozzles, the fuel system is probably OK. Perform Step 3 (Spark Plug Test). If there is little or no flow from the injection nozzles, or if there seems to be too much flow from the nozzles (along with other indications of flooding) the fuel system is faulty and should be tested by a qualified technician.

Step 3 (Spark Plug Test)

1. Remove the spark plugs and check for plug fouling. Clean or replace as necessary.
2. Attempt to start the engine. If the engine still won't start, the fault could be with incorrect ignition timeing. Have the timing checked. If it is not 12° ± BTDC, have it reset to 12° BTDC and attempt to restart.

NO START/IGNITION TEST

OPERATION	PROCEDURE	TEST INDICATION	ACTION REQUIRED
1. NO START/IGNITION TEST #1. (Checking for spark at secondary coil wire.)	• Remove secondary coil wire from distributor. • Hold secondary coil wire terminal near (3/16") good ground. • Crank engine.	Good spark between terminal and ground.	• **Fault in distributor rotor, cap or secondary wires.** • Repair or replace as required. • Attempt engine start. • If no start, go back and perform sequence starting with SPARK PLUG TEST (Step 3).
		No spark between terminal and ground.	**Ignition fault not identified.** • Perform NO START/IGNITION TEST #2.

NO START/IGNITION TEST

OPERATION	PROCEDURE	TEST INDICATION	ACTION REQUIRED
2. NO START/IGNITION TEST #2. (Checking pick-up coil resistance at ESA connector.)	• Remove ESA connector from Combustion Control Computer. • Select OHM setting on EFI tester. • Connect test leads to Pins #5 and #9 of ESA connector.	Ohmmeter reads between 150 and 900 ohms.	**No shorts or breaks, but possible ground in circuit.** **Resistance OK.** • Perform NO START/IGNITION TEST #2A.
		Ohmmeter reads at or near zero ohms.	**Short in circuit.** • Perform NO START/IGNITION TEST #2B.
		Ohmmeter reads at or near maximum reading. (Check for good connection and repeat.)	**Break in circuit.** • Perform NO START/IGNITION TEST #2C.
2A. NO START/IGNITION TEST #2A. (Checking for possible ground in ESA harness.)	(Use 200 OHM scale.) • Connect test leads between ESA harness terminals #5 and ground, then #9 and ground.	Ohmmeter reads near or at maximum resistance, with either terminal.	**No Fault in pick-up circuit.** • Perform NO START/IGNITION TEST #3.
		Ohmmeter reads low or no resistance.	**Ground in either ESA harness or pick-up coil.** • Perform NO START/IGNITION TEST #2B.

NO START/IGNITION TEST

OPERATION	PROCEDURE	TEST INDICATION	ACTION REQUIRED
2B. NO START/IGNITION TEST #2B. (Checking whether ground is in ESA harness or in Pick-Up Coil.)	• Disconnect Pick-Up Coil connector at distributor. • Select 200 OHM setting. • Connect Ohmmeter test leads to either **terminal** of pick-up coil connector on distributor and to **ground.** Repeat with other terminal.	Ohmmeter shows continuity at either or both terminals.	**Pick-Up Coil has short.** • Replace pick-up coil. • Reconnect ESA harness and pick-up coil connectors. • Attempt restart. • If no start, go back and perform SPARK PLUG TEST (Step 3).
		Ohmmeter does not show continuity at either terminal.	**Pick-up Coil not grounded.** • Perform NO START/IGNITION TEST #4.
2C. NO START/IGNITION TEST #2C. (Checking for break in Pick-Up Coil.)	• Disconnect Pick-Up Coil connector at distributor. • Connect Ohmmeter test leads to both terminals of Pick-Up Coil connector.	Ohmmeter reads between 150 and 900 ohms.	**ESA harness has fault.** • Repair harness. • Reconnect ESA and Pick-Up Coil connector. • Attempt restart. • If no start, go back and perform sequence starting with SPARK PLUG CHECK (Step 3).
		Ohmmeter reading not to specs.	**Harness OK.** • Replace Pick-Up Coil. • Reconnect connector. • Attempt engine start. • If no start, go back and perform sequence starting with SPARK PLUG TEST (Step 3).

NO START/IGNITION TEST

OPERATION	PROCEDURE	TEST INDICATION	ACTION REQUIRED
3. NO START/IGNITION TEST #3. (Checking Pick-Up Coil Air Gap.)	• Remove distributor cap. • Measure Pick-Up Coil Air Gap with non-magnetic feeler gauge.	Air Gap is .006″.	**Air Gap OK.** • Perform NO START/IGNITION TEST #4.
		Air gap is not .006″.	**Air Gap not at specification.** • Readjust air gap to .006″. • Reconnect Pick-Up Coil connector at distributor. • Attempt engine start. • If no start, go back and perform sequence starting with SPARK PLUG TEST (Step 3).
4. NO START/IGNITION TEST #4. (Checking voltage at coil positive terminal.)	• Set EFI tester to voltmeter position. • Connect test leads to coil **positive terminal** and to **ground.** • Crank engine.	Voltmeter reads 9 volts or more.	**Voltage OK** • Perform NO START/IGNITION #6.
		Voltmeter reads LESS than 9 volts.	**Voltage not up to specification.** • Perform NO START/IGNITION TEST #5.

NO START/IGNITION TEST

OPERATION	PROCEDURE	TEST INDICATION	ACTION REQUIRED
5. NO START/IGNITION TEST #5. (Checking voltage of **BAL** terminal on starter relay.)	• Connect test leads to **BAL terminal** on starter relay and **ground.** • Crank engine. BAL	Voltmeter reads 9 volts or more.	**Fault in harness to coil.** • Repair as required. • Attempt engine start. • If no start, go back and perform sequence starting with SPARK PLUG TEST (Step 3).
		Voltmeter reads less than 9 volts.	**Voltage not up to specification.** • Test starter relay per car service manual procedure and repair or replace if necessary. • Attempt engine start. • If no start, go back and perform sequence starting with SPARK PLUG TEST (Step 3).
6. NO START/IGNITION TEST #6. (Checking voltage at coil negative terminal.)	• Remove ESA connector from Combustion Control Computer. • Connect test leads to ESA Connector **pin #1** and to ground. • Crank engine. 10 9 8 7 6 / 1 2 3 4 5; 1	Voltmeter reads 9 volts or more.	**Voltage OK.** • Perform NO START/IGNITION TEST #7.
		Voltmeter reads less than 9 volts.	**Voltage not to specs.** • Check harness and coil continuity. • Repair as required. • Attempt restart. • If no start, go back and perform sequence starting with SPARK PLUG TEST (Step 3).

NO START/IGNITION TEST

OPERATION	PROCEDURE	TEST INDICATION	ACTION REQUIRED
7. NO START/IGNITION TEST #7. (Checking that ESA Pin #10 is grounded.)	• Set tester in Ohmmeter position. • Connect test leads to ESA connector **Pin #10** and to **ground.** 10 9 8 7 6 / 1 2 3 4 5; 10	Ohmmeter shows continuity.	**GROUND OK.** • Perform NO START/IGNITION TEST #8.
		Ohmmeter shows no continuity.	**Break in ground wire #10 in ESA harness.** • Repair ESA harness. • Reconnect circuit. • Attempt engine start. • If no start, go back and perform sequence starting with SPARK PLUG TEST (Step 3).
8. NO START IGNITION/TEST #8. (Checking for possible fault in ASD circuit.)	• Remove ASD connector. • Repeat Step #1 - SPARK TEST PROCEDURE.	Good spark between screwdriver and ground.	**Faulty ASD.** • Replace ASD module. • Attempt engine start. • If no start, go back and perform sequence starting with SPARK PLUG TEST (Step 3).
		No spark between screwdriver and ground.	**Faulty CCC.** • Replace CCC. • Attempt engine start. • If no start, go back and perform sequence starting with SPARK PLUG TEST (Step 3).

NO START/FUEL TEST

OPERATION	PROCEDURE	TEST INDICATION	ACTION REQUIRED
1. NO START/FUEL TEST #1. (Check for fault In-Tank pump and ASD circuits.)	• Connect EFI Tester to vehicle (do not connect Diagnostic Aid). • Set: - Air By-Pass switch to "OFF" setting. - Rotary switch to Volt setting. - Toggle switch to "Volt - OHM" setting. • Depress ASD By-Pass button momentarily.	Listen for sound of In-Tank Pump running; ASD light, off.	**OK** • Perform NO START/FUEL TEST #2.
		Sound of In-Tank Pump running; ASD light, on.	**Fault in ASD circuit.** • Replace ASD module. • Attempt engine start. • If no start, go back and perform sequence starting with SPARK PLUG TEST (Step 3).
		No sound of In-Tank Pump running.	**Fault in In-Tank Pump circuit.** • Check fuse (cavity #2) 20 amp. • If **OK,** perform NO START FUEL TEST #7. • If **blown,** replace fuse and correct cause (short, defective pump ballast resistor, or low voltage). • Attempt engine start. • If no start, go back and perform sequence starting with SPARK PLUG TEST (Step 3).

NO START/FUEL TEST

OPERATION	PROCEDURE	TEST INDICATION	ACTION REQUIRED
2. NO START/FUEL TEST #2. (Check Power Module drive signal.)	• Repeat NO START/ FUEL TEST #1 procedure, including depressing ASD button momentarily.	Power Module drive signal light OFF.	**Power Module Drive Signal OK.** • Perform NO START/FUEL TEST #3.
		Power Module drive signal test light ON.	**No Power Module Drive Signal.** • Replace CCC. • Attempt engine start. • If no start, go back and perform sequence starting with SPARK PLUG TEST (Step 3).
3. NO START/FUEL TEST #3. (Check Control Pump operation.)	• Remove Air Cleaner Cover. • Remove secondary coil wire from distributor and **ground this wire.** • Have someone observe fuel flow. • Repeat NO START/FUEL TEST #1 procedure.	Short flow of fuel from Injector Nozzles.	**Control Pump operating.** • Perform NO START/FUEL TEST #5. Note: after test replace Air Cleaner Cover.
		No fuel flow.	**Control Pump not operating.** • Perform NO START/FUEL TEST #4.

NO START/FUEL TEST

OPERATION	PROCEDURE	TEST INDICATION	ACTION REQUIRED
4. NO START/FUEL TEST #4. (Checking Control Pump voltage.)	• Disconnect terminal at Control Pump. • Connect voltmeter leads across both terminals of Control Pump harness. • Depress ASD button momentarily. 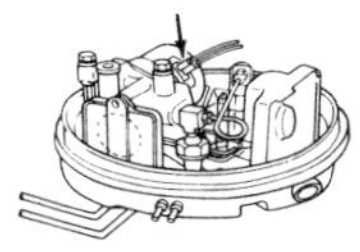	Voltmeter reads 9.5 volts or higher.	**Fault in Hydraulic Support Plate.** • Replace Hydraulic Support Plate. • Attempt engine start. • If no start, reconnect terminals and go back and perform sequence starting with SPARK PLUG TEST (Step 3).
		Voltmeter reads less than 9.5 volts.	**Fault not located.** • Replace terminal. Perform NO START/FUEL TEST #14.
5. NO START/FUEL TEST #5. (Checking Power Module operation.)	• Repeat NO START/FUEL TEST #1 set-up procedure. • Depress ASD By-Pass button.	Power Module light OFF.	**Power Module operation OK.** • Perform NO START/FUEL TEST #13.
		Power Module light ON.	**Fault in Power Module or in Power Module loads.** • Perform NO START/FUEL TEST #6.

NO START/FUEL TEST

OPERATION	PROCEDURE	TEST INDICATION	ACTION REQUIRED
6. NO START/FUEL TEST #6. (Checking for Power Module fault.)	• Repeat NO START/FUEL TEST #1 set-up procedure. • Disconnect - air flowmeter - fuel flowmeter - throttle position potentiometer connectors. - coolant sensor • Depress ASD By-Pass button.	Power Module light ON.	**Fault in Power Module.** • Replace power module. • Attempt engine start. • If no start, reconnect components and go back and perform sequence starting with SPARK PLUG TEST (Step 3).
		Power Module light OFF.	**Fault somewhere in Power Module loads.** • Perform NO START/FUEL TEST #6A.
6A. NO START/FUEL TEST #6A. (Check for fault in Power Module loads.)	Same set-up as TEST #6. • Reconnect Air Flowmeter. • Depress ASD By-Pass button momentarily.	If light comes ON . . .	**Fault in Air Flowmeter.** • If light comes on after circuit is reconnected, locate fault in harness and repair. Otherwise, replace Air Flowmeter. • Attempt engine start. • If no start, reconnect other components and go back and perform sequence starting with SPARK PLUG TEST (Step 3).

NO START/FUEL TEST

OPERATION	PROCEDURE	TEST INDICATION	ACTION REQUIRED
6A. (Continued) NO START/FUEL TEST #6A. (Check for fault in Power Module loads.)	• Reconnect Fuel Flowmeter. • Depress ASD by-pass button momentarily.	If light comes ON . . .	**Fault is in Fuel Flowmeter circuit.** • If light comes on after circuit is reconnected, locate fault in harness and repair. Otherwise, replace Fuel Flowmeter. • Attempt engine start. • If no start, reconnect other components and go back and perform sequence starting with SPARK PLUG TEST (Step 3).
	• Reconnect throttle position potentiometer and depress ASD by-pass button momentarily.	If light comes ON . . .	**Fault in Throttle Position Potentiometer** • Check and correct possible fault in throttle pot harness. • If no fault replace throttle pot. • Recalibrate potentiometer . • Attempt restart.
	• Reconnect Coolant Sensor		**Fault is in Coolant Sensor circuit.** • If light comes on after circuit is reconnected, locate fault in harness and repair. Otherwise, replace Coolant Sensor. • Attempt engine start. • If no start, reconnect other components and go back and perform sequence starting with SPARK PLUG TEST (Step 3).

NO START/FUEL TEST

OPERATION	PROCEDURE	TEST INDICATION	ACTION REQUIRED
7. NO START/FUEL TEST #7. (Checking voltage at In-Tank Ballast Resistor.)	• Repeat NO START/FUEL TEST #1 set-up procedure. • Connect voltmeter leads to green wire on the In-Tank resistor and to ground. • Depress ASD button.	Voltage reads at least 10.5 volts.	**Voltage OK.** • Perform NO START/FUEL TEST #8.
		Voltage reads less than 10.5 volts.	**Voltage not to specs.** • Check for continuity, leak or ground between Point A on Ballast and at ASD terminal connector. (See wiring diagram). • Repair or replace if faulted. • If OK replace ASD. • Attempt engine start. • If no start, go back and perform sequence starting with SPARK PLUG TEST (Step 3).
8. NO START/FUEL TEST #8. (Checking resistance of In-Tank Ballast Resistor.)	• Put rotary switch to OHM position. (use 200 OHM scale.) • Connect Ohmmeter leads to points A & B of Ballast Resistor. A B	Ohmmeter reads .4 OHMS.	**Resistance OK.** • Perform NO START/FUEL TEST #9.
		Ohmmeter does not read .4 OHMS.	**Resistance not to specs.** • Replace In-Tank Ballast Resistor. • Attempt engine start. • If no start, go back and perform sequence starting with SPARK PLUG TEST (Step 3).

NO START/FUEL TEST

OPERATION	PROCEDURE	TEST INDICATION	ACTION REQUIRED
9. NO START/FUEL TEST #9. (Checking voltage by In-Tank Pump Relay.)	• Put rotary switch in Volt position. • Connect leads to **Point A** of In-Tank Relay and to **ground.** B D A C • Crank engine.	Voltage reads 8-10 volts.	**Voltage OK.** • Perform NO START/FUEL TEST #10.
		Voltage reads less than 8-10 volts.	**Voltage not to specs.** • Perform NO START/FUEL TEST #11.
10. NO START/FUEL TEST #10. (Checking for break in harness by-pass to In-Tank Pump.)	• Put rotary switch to Ohmmeter setting. • Connect leads to **point A** of In-Tank Relay at point indicated in circuit (Z1, 12" DG) and to Eight-way connector above right Rocker Cover. B D A C	Ohmmeter shows continuity.	**Harness OK.** • Perform NO START/FUEL TEST #13.
		Ohmmeter shows no continuity.	**Break in harness.** • Repair harness. • Attempt engine start. • If no start, go back and perform sequence starting with SPARK PLUG TEST (Step 3).

NO START/FUEL TEST

OPERATION	PROCEDURE	TEST INDICATION	ACTION REQUIRED
11. NO START/FUEL TEST #11. (Checking for fault in In-Tank Relay or in harness to Starter Relay.)	• Connect voltmeter leads to **Point C** of In-Tank Pump Relay and to ground. B D A C • Crank engine.	Voltmeter reads 9 volts.	**Voltage OK.** • Replace In-Tank Relay. • Attempt engine start. • If no start, go back and perform sequence starting with SPARK PLUG TEST (Step 3).
		Voltmeter reads less than 9 volts.	**Voltage not up to specs.** • Perform NO START FUEL/TEST #12.
12. NO START FUEL/TEST #12. (Checking for break in harness to Starter Relay.)	• Put rotary switch to Ohmmeter position. • Connect test leads to **Point C** and solenoid terminal of Starter Relay. B D A C	Ohmmeter shows continuity.	**Harness OK.** Possible fault in Starter Relay. • Check relay per service manual procedure and replace if necessary. • Attempt engine start. • If no start, go back and perform sequence starting with SPARK PLUG TEST (Step 3).
		Ohmmeter shows no continuity.	**Break in harness.** • Repair harness as required. • Attempt engine start. • If no start, go back and perform sequence starting with SPARK PLUG TEST (Step 3).

NO START/FUEL TEST

OPERATION	PROCEDURE	TEST INDICATION	ACTION REQUIRED
13. NO START/FUEL TEST #13. (Checking fuel supply line pressure.)	• Check battery voltage is at least 12 volts. • Connect pressure gauge to fuel supply line "tee" fitting on hydraulic support plate. (You may have to hook in "tee" fitting if not already on Hydraulic Support Plate.) • Crank engine.	Pressure at least 8 psi.	**Fuel Supply Line Pressure OK.** • Perform NO START/FUEL TEST #14.
		Pressure is much less than 8 psi.	**Fuel Supply Line Pressure Not Up to specs.** Possible leaks, mechanical or vacuum restriction in fuel supply system. Check for the following: • frozen lines. • pinched lines. • leaks in lines or fittings. • restricted fuel tank vent lines. • blocked fuel filter. • If found, correct and retest. • When pressure is at least 8 psi — • Attempt engine start. • If no start, go to Step #14.

NO START/FUEL TEST

OPERATION	PROCEDURE	TEST INDICATION	ACTION REQUIRED
14. NO START/FUEL TEST #14. (Checking for proper EFI crank signals.)	• Remove coil secondary wire from distributor **and ground wire with jumper.** • Connect EFI tester to vehicle. - Toggle switch in "EFI" position. - Air By-Pass "Off" - Rotary switch in "FUEL" position. • Crank engine. • Record signals in each of the following positions: Fuel Air Fuel Temp. Throttle Pot. Vac. Sol. Water Temp.	Engine cold. Crank signals should be: Fuel - 3-25 Air - 2-25 Fuel Temp. - 16-21 Throttle Pot. - Not to exceed 13.8 Vac. Sol. - .1 - .2 Water Temp. - 8-12	Crank signals should all be to specs. If not, with: Fuel - Replace Hydraulic Support Plate. Air - Replace Air Flowmeter. Fuel Temp. - Replace Hydraulic Support Plate. Throttle Pot. - Replace and adjust as required (See recalibration procedure . Vac. Sol. - Verify circuit replace Vac. Solenoid. Water Temp. - Replace water temp. sensor. Note: Before any components are replaced verify that all connectors and harness' are properly installed and have continuity.

NO START/FUEL TEST (Cold Engine)

OPERATION	PROCEDURE	TEST INDICATION	ACTION REQUIRED
15. NO START/FUEL TEST #15. (Checking for fault in CCC or harness to Starter Relay.)	• Disconnect EFI connector from CCC. • Connect voltmeter leads to **pin #8** and to **ground.** • Crank engine.	Voltmeter reads at least 9 volts.	**Voltage OK.** • Replace CCC. • Attempt engine start. • If no start, go back and perform sequence starting with SPARK PLUG TEST (Step 3).
		Voltmeter reads less than 9 volts.	**Voltage not up to specs.** • Check harness to starter relay. (NO START/FUEL TEST #12.)

NO START/FUEL TEST (Excessive Fuel Spray)

OPERATION	PROCEDURE	TEST INDICATION	ACTION REQUIRED
16. NO START/FUEL TEST #16.	• Air Cleaner cover still removed. • Turn ignition key to "RUN" position. • If fuel is continually flowing from injection nozzle, disconnect Control Pump connector.	Fuel continues flowing from Injection Nozzles.	**Fault in Hydraulic Support Plate.** • Replace Hydraulic Support Plate. • Attempt engine start. • If no start, go back and perform sequence starting with SPARK PLUG TEST (Step 3).
	• Recycle ignition key.	No fuel flow from Injection Nozzles, or fuel flow from all Nozzles.	**Fault in CCC.** • Replace CCC. • Attempt engine start. • If no start, go back and perform sequence starting with SPARK PLUG TEST (Step 3).

STARTS BUT STALLS TEST

OPERATION	PROCEDURE	TEST INDICATION	ACTION REQUIRED
1. STARTS BUT STALLS TEST #1. (Checking ignition coil ballast resistor circuit voltage.)	• Remove secondary coil wire from distributor. • Crank engine. • Hold secondary coil wire terminal near (3/16") good ground. (Refer to Spark Test Step #1.) • Connect voltmeter leads to ignition coil ballast resistor **Point A** and to **Ground.**	Voltmeter reads 6 volts or more.	**Voltage OK.** • Verify 6 volts at positive side of ignition coil. If not correct, fault in circuit. • Perform STARTS BUT STALLS TEST #3.
	• Turn ignition key to "RUN" position.	Voltmeter reads less than 6 volts.	**Voltage not to specs.** • Perform STARTS BUT STALLS TEST #2.

STARTS BUT STALLS TEST

OPERATION	PROCEDURE	TEST INDICATION	ACTION REQUIRED
2. STARTS BUT STALLS TEST #2. (Checking voltage to Ballast Resistor.)	• Connect voltmeter to terminal B and to ground. A B	Voltmeter reads at least 10 volts.	**Voltage OK.** • Replace Ballast Resistor. • Restart to verify stalling complaint corrected.
		Voltmeter reads less than 10 volts.	**Voltage not to specs.** • Check circuit to ignition switch and correct fault. • Restart to verify stalling complaint corrected.
3. START BUT STALLS TEST #3. (Checking AIS motor for start position.)	• Turn ignition key to RUN position. • Visually check position of throttle arm at AIS motor. FRONT OF ENGINE MINIMUM IDLE START POSITION NORMAL IDLE	Throttle arm is in START position.	**AIS motor is OK.** • Perform STARTS BUT STALLS TEST #7.
		Throttle arm not in START position.	• Perform STARTS BUT STALLS TEST #4.

STARTS BUT STALLS TEST

OPERATION	PROCEDURE	TEST INDICATION	ACTION REQUIRED
4. STARTS BUT STALLS TEST #4. (Checking Ballast Resistor resistance.)	• Reset rotary switch to Ohmmeter position. • Connect Ohmmeter leads to ballast resistor terminals **C** and **D.** C D	Ohmmeter reading 9-11 ohms.	**Resistance OK.** • Perform STARTS BUT STALLS TEST #5.
		Ohmmeter does not read 9-11 ohms.	**Resistance not to specs.** • Replace Ballast Resistor. • Reconnect harness connector. • Restart to verify stalling complaint corrected.
5. STARTS BUT STALLS TEST #5. (Checking Ballast Resistor resistance.)	• Reset rotary switch to Ohmmeter position. • Remove harness connector from Ballast Resistor. • Connect Ohmmeter leads to ballast resistor terminals **D** and **E.** D E	Ohmmeter reads 4-6 ohms.	**Ballast resistance OK.** • Perform STARTS BUT STALLS TEST #6.
		Ohmmeter does not read 4-6 ohms.	**Ballast resistance not to specs.** • Replace Ballast Resistor. • Reconnect harness connector. • Restart to verify stalling complaint corrected.

STARTS BUT STALLS TEST

OPERATION	PROCEDURE	TEST INDICATION	ACTION REQUIRED
6. STARTS BUT STALLS TEST #6. (Checking circuit from CCC to AIS motor.)	• Move rotary switch to voltmeter position. • Connect voltmeter leads to Pin 6 of ESA connector and to ground. 10 9 8 7 6 / 1 2 3 4 5 / 6	Voltmeter reads 8 volts or more.	**Voltage OK.** • Replace CCC. • Restart to verify stalling complaint corrected.
		Voltmeter reads less than 8 volts.	**Voltage not to specs.** • Check for fault in ESA harness to AIS motor and repair if necessary. —OR— • If harness OK, replace AIS motor. • Restart to verify stalling problem corrected.
7. STARTS BUT STALLS TEST #7. (Checking ASD module.)	• Connect EFI tester to vehicle. • Put rotary switch to FUEL position. • Put air By-Pass switch ON. • Turn ignition key to RUN position while noting Air By-Pass light and ASD light.	ASD light on.	• Perform additional procedure below.
		Air By-Pass light comes on and ASD light goes out momentarily - then - Air By-Pass light goes out and ASD light comes back on.	**Proper test light sequence.** • Perform STARTS BUT STALLS TEST #8.
		Above light sequence does not occur.	• Check ASD ground and retest. • If ground OK replace ASD module. • Restart and verify stalling complaint corrected.

STARTS BUT STALLS TEST

OPERATION	PROCEDURE	TEST INDICATION	ACTION REQUIRED
8. STARTS BUT STALLS TEST #8. (Checking fuel flow.)	• Remove Air Cleaner cover. • Turn ignition key to RUN position. • Have someone observe fuel flow.	Brief fuel flow from injector bars.	**Fuel flow normal.** • Perform STARTS BUT STALLS TEST #17.
		No fuel flow.	• Perform STARTS BUT STALLS TEST #9.
		Fuel flows continually from injector bars.	• Perform STARTS BUT STALLS TEST #8A.
8A. STARTS BUT STALLS TEST #8A. (Checking cause of continuous fuel flow.)	• Disconnect Control Pump connector. • Recycle ignition key OFF-ON.	Fuel flow stops.	• Replace CCC. • Replace Air Cleaner cover and reconnect connector. • Restart to verify stalling complaint corrected.
		Fuel flow continues.	**Defective Hydraulic Support Plate.** • Replace Hydraulic Support Plate. • Replace Air Cleaner cover and reconnect connector. • Restart and verify stalling complaint corrected.

STARTS BUT STALLS TEST

OPERATION	PROCEDURE	TEST INDICATION	ACTION REQUIRED
9. STARTS BUT STALLS TEST #9. (Check for fault in In-Tank Pump and ASD circuits.)	• Connect EFI Tester to vehicle (do not connect Diagnostic Aid). • Set: - Air By-Pass switch to OFF setting. - Rotary switch to VOLT setting. - Toggle switch to VOLT - OHM setting. • Depress ASD By-Pass button momentarily.	Sound of In-Tank Pump running; ASD light, OFF.	**OK.** • Perform NO START/FUEL TEST #2.
		Sound of In-Tank Pump running; ASD light, ON.	**Fault in ASD circuit.** • Replace ASD module. • Attempt engine start. • If no start, go back and perform sequence starting with SPARK PLUG TEST (Step 3.)
		No Sound of In-Tank Pump running.	**Fault in In-Tank Pump circuit.** • Check fuse (cavity #2) 20 amp. • If **OK,** perform STARTS BUT STALLS TEST #10. • If **blown,** replace fuse and correct cause (short, defective pump ballast resistor, or low voltage.) • Attempt engine start. • If no start, go back and perform sequence starting with SPARK PLUG TEST (Step 3).

STARTS BUT STALLS TEST

OPERATION	PROCEDURE	TEST INDICATION	ACTION REQUIRED
10. STARTS BUT STALLS TEST #10. (Checking voltage at In-Tank Ballast Resistor.)	• Repeat NO START/FUEL TEST #1 set-up procedure. • Connect voltmeter leads to green wire on the In-Tank resistor and to ground. • Depress ASD button.	Voltage reads at least 10.5 volts.	**Voltage OK.** • Perform STARTS BUT STALLS TEST #12.
		Voltage reads less than 10.5 volts.	**Voltage not to specs.** • Check for continuity, leak or ground between Point A (Z1, 12, DG) and at ASD terminal connector. • Repair or replace if faulted. • If OK replace ASD. • Attempt engine start. • If no start, go back and perform sequence starting with SPARK PLUG TEST (Step 3).
11. STARTS BUT STALLS TEST #11. (Checking resistance of In-Tank Ballast Resistor.)	• Put rotary switch to OHM position. • Select 200 OHM setting. • Connect Ohmmeter leads to points A & B of ballast resistor. A B	Ohmmeter reads .4 ohms.	**Resistance OK.** • Perform STARTS BUT STALLS TEST #12.
		Ohmmeter does not read .4 ohms.	**Resistance not to specs.** • Replace In-Tank Ballast Resistor. • Attempt engine start. • If no start, go back and perform sequence starting with SPARK PLUG TEST (Step 3).

STARTS BUT STALLS TEST

OPERATION	PROCEDURE	TEST INDICATION	ACTION REQUIRED
12. STARTS BUT STALLS TEST #12. (Checking voltage by In-Tank Pump Relay.)	• Put rotary switch in VOLT position. • Connect leads to **Point A** of In-Tank Relay and to **ground.** 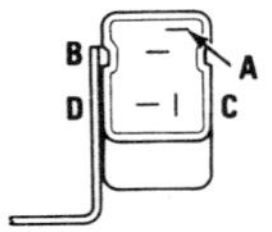 • Crank engine.	Voltage reads 8-10 volts.	**Voltage OK.** • Perform STARTS BUT STALLS TEST #13.
		Voltage reads less than 8-10 volts.	**Voltage not to specs.** • Perform NO START/FUEL TEST #14.
13. STARTS BUT STALLS TEST #13. (Checking for break in harness by-pass to In-Tank Pump.)	• Put rotary switch to Ohmmeter setting. • Connect leads to **Point A** of In-Tank relay at point indicated in circuit (Z1, 12 DG) and to eight-way connector above right Rocker Cover. 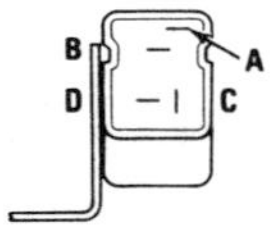	Ohmmeter shows continuity.	**Harness OK.** • Perform STARTS BUT STALLS TEST #14.
		Ohmmeter shows no continuity.	**Break in harness.** • Repair harness. • Attempt engine start. • If no start, go back and perform sequence starting with SPARK PLUG TEST (Step 3).

STARTS BUT STALLS TEST

OPERATION	PROCEDURE	TEST INDICATION	ACTION REQUIRED
14. STARTS BUT STALLS TEST #14. (Checking for fault in In-Tank Relay or in harness to Starter Relay.)	• Connect voltmeter leads to **Point C** of In-Tank Pump Relay and to ground. 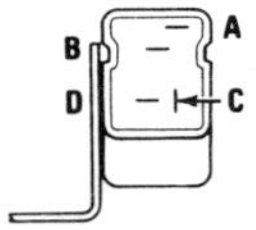 • Crank engine.	Voltmeter reads 9 volts.	**Voltage OK.** • Replace In-Tank Relay. • Attempt engine start. • If no start, go back and perform sequence starting with SPARK PLUG TEST (Step 3).
		Voltmeter reads less than 9 volts.	**Voltage not up to specs.** • Perform STARTS BUT STALLS TEST #15.
15. STARTS BUT STALLS TEST #15. (Checking for break in harness to Starter Relay.)	• Put rotary switch to Ohmmeter position. • Connect test leads to **Point C** and solenoid terminal of starter relay. 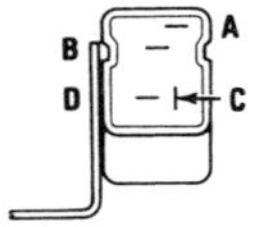	Ohmmeter shows continuity.	**Harness OK.** Possible fault in Starter Relay. • Check relay per service manual procedure and replace if necessary. • Attempt engine start. • If no start, go back and perform sequence starting with SPARK PLUG TEST (Step 3).
		Ohmmeter shows no continuity.	**Break in harness.** • Repair harness as required. • Attempt engine start. • If no start, go back and perform sequence starting with SPARK PLUG TEST (Step 3).

STARTS BUT STALLS TEST

OPERATION	PROCEDURE	TEST INDICATION	ACTION REQUIRED
16. STARTS BUT STALLS TEST #16. (Checking fuel supply line pressure.)	• Check battery voltage is at least 12 volts. • Connect pressure gauge to fuel supply line "tee" fitting on Hydraulic Support Plate. (You may have to hook in "tee" fitting if not already on Hydraulic Support Plate.) • Crank engine.	Pressure at least 8 psi.	**Fuel Supply Line Pressure OK.** • Perform STARTS BUT STALLS TEST #17.
		Pressure is much less than 8 psi.	**Fuel Supply Line Pressure Not Up to specs.** Possible leaks, mechanical or vacuum restriction in fuel supply system. Check for the following: • pinched lines. • leaks in lines or fittings. • restricted fuel tank vent lines. • If found, correct and retest. • When pressure is at least 8 psi — • Attempt engine start. • If no start, go back and perform sequence starting with SPARK PLUG TEST (Step 3).

STARTS BUT STALLS TEST

OPERATION	PROCEDURE	TEST INDICATION	ACTION REQUIRED
17. STARTS BUT STALLS TEST #17. (Checking for proper EFI Crank signals.)	• Remove coil secondary wire from distributor and **ground wire with jumper.** • Connect EFI tester to vehicle: - Toggle switch in EFI position. - Air by-pass OFF - Rotary switch in FUEL position. • Crank engine. • Record signals in each of the following positions: Fuel. Air. Fuel Temp. Throttle Pot. Vac. Sol. Water Temp.	Engine cold. Crank signals should be: Fuel - 3-25. Air - 2-25. Fuel Temp. - 16-21. Throttle Pot. - Not to exceed 13.8 Vac. Sol. - .1 - .2 Water Temp. - 8-12	Crank signals should all be to specs. If not, with: Fuel - Replace Hydraulic Support Plate. Air - Replace Air Flowmeter. Fuel Temp. - Replace Hydraulic Support plate. Throttle Pot. - Replace and adjust as required (see recalibration procedure). Vac. Sol. - Verify circuit replace Vac. Solenoid. Water Temp.-Replace Water Temp. Sensor. Note: Before any components are replaced verify that all connectors and harnesses are properly installed.

STARTS BUT STALLS TEST

OPERATION	PROCEDURE	TEST INDICATION	ACTION REQUIRED
18. STARTS BUT STALLS TEST #18. (Checking for fault in CCC or harness to Starter Relay.)	• Disconnect EFI connector from CCC. • Connect voltmeter leads to **pin #8** and to **ground.** 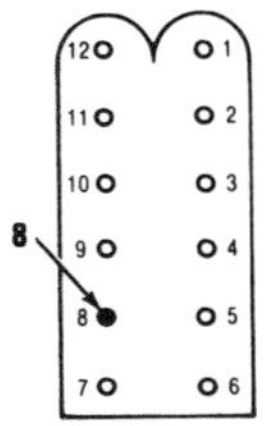 • Crank engine.	Voltmeter reads at least 9 volts.	**Voltage OK.** • Replace CCC. • Attempt engine start. • If no start, go back and perform sequence starting with SPARK PLUG TEST (Step 3).
		Voltmeter reads less than 9 volts.	**Voltage not up to specs.** • Reconnect connector. • Check harness to Starter Relay. (NO START/FUEL TEST #12.)

DRIVEABILITY TEST

OPERATION	PROCEDURE	TEST INDICATION	ACTION REQUIRED
Preliminary Check. (Idle Speed.)	• Connect EFI Tachometer to engine. • Set rotary switch to RPM position. • Turn all vehicle accessories off. • Set parking brake and block front wheels. • Place transmission in "P" position. • Start engine. • Allow engine to reach normal operating temperature. • Place transmission in "D" position.	Tachometer should read 580 RPM (±50) with smooth idle in Drive. (Disregard ASD light ON.)	• **If idle speed not to specs —** • Adjust using automatic idle speed system procedures. • **If rough idle,** check for fouled plugs, cross firing, etc. and correct. • Check basic timing. (Follow procedures.) • Shut down engine and place transmission in "P" position.
Preliminary Check. (Basic Timing.)	• Connect magnetic timing probe to tester and engine. • Ground closed throttle switch with jumper wire. • Rotary switch to TIMING position. • Start engine.	Basic timing should read 12° (±2°). (Disregard ASD light ON).	• Adjust to specified basic timing, if necessary. • Perform DRIVEABILITY TEST #1 (engine still running).

DRIVEABILITY TEST

OPERATION	PROCEDURE	TEST INDICATION	ACTION REQUIRED
1. DRIVEABILITY TEST #1. (Checking spark curve.)	• Increase engine RPM to 1,500 RPM. • Throttle switch still grounded with jumper wire.	Timing unchanged at 12° ±2°.	**Timing OK.** • Perform DRIVEABILITY TEST #2 (engine still running).
		Timing no longer at 12° ±2°.	**Timing not at specs.** • Replace CCC. • Road test to verify driveability complaint corrected.
2. DRIVEABILITY TEST #2. (Further checking of spark curve.)	• Remove ground jumper wire. • Decrease engine speed to 1,000 RPM.	Timing should read: Fed: 15° ± 3° plus Basic Timing Cal: 5° ± 2° plus Basic Timing	**Timing OK.** • Perform DRIVEABILITY TEST #3 (engine still running).
		Timing not within above specifications.	• Shut down engine. • Replace CCC. • Road test to verify driveability complaint corrected.

DRIVEABILITY TEST

OPERATION	PROCEDURE	TEST INDICATION	ACTION REQUIRED
3. DRIVEABILITY TEST #3. (Further checking of spark curve.)	• Increase engine speed to 2,000 RPM.	Timing should read: Fed: 36° ± 2° plus Basic Timing. Cal: 17° ± 2° plus Basic Timing.	**Timing OK.** • Shut down engine. • Perform DRIVEABILITY TEST #4.
		Timing not within above specifications.	• Shut down engine. • Replace CCC. • Road test to verify driveability complaint corrected.
4. DRIVEABILITY TEST #4. (Checking primary fuel flow-normal performance complaint.)	• Connect EFI tester to car: - rotary switch in FUEL position. - Air By-Pass OFF. • Start engine. • Turn Air By-Pass ON. • Remove Air Flowmeter harness connector. • Remove Air Cleaner cover. • Observer fuel spray bars.	Fuel flowing from all four primary nozzles in uniform and steady streams.	**Primary Fuel flow OK.** • Perform DRIVEABILITY TEST #5.
		Fuel not flowing from all four primary nozzles in uniform and steady streams.	• Replace Hydraulic Support Plate. • Remove tester, replace connector and Air Cleaner cover. • Road test to verify normal performance complaint corrected.

DRIVEABILITY TEST

OPERATION	PROCEDURE	TEST INDICATION	ACTION REQUIRED
5. DRIVEABILITY TEST #5. (Checking secondary fuel flow - w.o.t. performance complaint.)	• Same procedure as in #4 except engine off: - toggle switch in EFI position. - remove connector from fuel pressure switch and ground connector. • Depress ASD By-Pass button momentarily. • Observe fuel spray bars.	Fuel flowing from all eight nozzles in uniform and steady streams.	**Secondary fuel flow OK.** • Perform DRIVEABILITY TEST #6.
		Fuel not flowing from all eight nozzles in uniform and steady streams.	• Replace Hydraulic Support Plate. • Remove tester, replace connectors and Air Cleaner cover. • Road test to verify w.o.t. performance complaint corrected.
6. DRIVEABILITY TEST #6. (Checking EFI run signals.)	• Connect EFI tester to vehicle: - toggle switch to EFI position. - Air By-Pass OFF. - AIS Control in normal position; Rotary switch to fuel position. • Start engine. Note: Engine should be at normal operating temp. unless complaint is a cold driveability problem.	Check run signals as follows: (see table below)	If all run signals are to specs., perform DRIVEABILITY TEST #7.

	Idle N	After 70 Sec.	2,000 RPM
Fuel	3.5-6.0	3.5-6.0	10.0-15.0
Air	15-35	15-35	35-55
O_2 Sensor	—	45-70	60-75
Fuel Temp.	16-21	16-21	16-21
Throttle Pos.	3.5-4.5	3.5-4.5	4.6-6.0
Vac. Sol.	.1	13.0	13.0
Water Temp.	8-12	8-12	8-12

DRIVEABILITY TEST

OPERATION	PROCEDURE	TEST INDICATION	ACTION REQUIRED
6. DRIVEABILITY TEST #6. (Continued)			If run signals are not to specs: Fuel - Replace Hydraulic Support Plate. Air - Replace Air Flowmeter. O_2 Sensor - Perform DRIVEABILITY TEST #8. Fuel Temp. -Replace Hydraulic Support Plate. Throttle Position - Adjust or replace as required. Vac. Sol - Perform DRIVEABILITY TEST #10. Water Temp. - Replace water sensor.
7. DRIVEABILITY TEST #7. (Checking ignition primary & secondary with ignition oscilloscope.)	• Connect suitable ignition oscilloscope to engine.	Follow equipment manufacturer's instructions in analyzing ignition system.	• If ignition system fault, repair as required. • If no ignition faults are indicated, replace CCC. • Road test to verify driveability complaint corrected.

DRIVEABILITY TEST

OPERATION	PROCEDURE	TEST INDICATION	ACTION REQUIRED
8. DRIVEABILITY TEST #8. (Checking O_2 Sensor circuit for performance or fuel economy complaint.)	• Connect EFI Tester to engine, including Diagnostic Aid. • Start engine. • Run until engine at normal operating temperature (engine must run at least 70 seconds) AFTER WARM. • Put rotary switch in O_2 Sensor position.	Tester reads in range from 45 to 70.	**O_2 Sensor OK.** • Perform calibration verification. • If no improvement, tune engine and road test.
		Tester reads 99-100.	**Break in O_2 harness to CCC.** • Check harness from O^2 Sensor circuit (N11 18 BK) to terminal #12 of CCC. Repair if necessary. • Retest. If still high perform DRIVEABILITY TEST #9.
		Tester reads in range from 0 to 10.	**Possible bad connection from battery feed.** • Shut down engine. • Disconnect and reconnect battery feed connector or (8 gauge red wire at battery connector). • Perform auto calibration. • Retest. If still low, perform DRIVEABILITY TEST #9.

DRIVEABILITY TEST

OPERATION	PROCEDURE	TEST INDICATION	ACTION REQUIRED
9. DRIVEABILITY TEST #9. (Identifying whether fault in CCC or O_2 Sensor.)	• Disconnect O_2 Sensor wire at the sensor.	Tester reading changes to 90 - 100.	• If readings change as indicated, replace O_2 Sensor and repeat DRIVEABILITY TEST #8.
	• Ground harness end of wire.	Tester reading changes to 0-10.	• If readings do not change as indicated, replace CCC and repeat DRIVEABILITY TEST #8.
		Tester reading does not change to 90 - 100 range when O_2 sensor wire disconnected.	• Check for short in harness from O_2 sensor to CCC. Repair and retest. Still no change, perform DRIVEABILITY TEST #10.
9A. DRIVEABILITY TEST # 9A. Checking fuel flow meter to control pump match.	• Connect EFI Tester to engine including Diagnostic Aid. • Start engine. • Run until engine at normal operating temperature (engine must run at least 70 seconds) after warm. • Put Rotary Switch in O^2 Sensor Position.	Tester reading indicates 0 at idle and 45 to 70 at 2000 RPM.	Replace Support Plate Assembly.

DRIVEABILITY TEST

OPERATION	PROCEDURE	TEST INDICATION	ACTION REQUIRED
10. DRIVEABILITY TEST #10. (Verifying air switching operation during first 70 seconds.)	• Engine off. • Disconnect downstream air hose from Air Switching Valve. • Put tester switch in Vac. Sol. position. • Start engine (note time of start.)	During first 70 seconds: • reading is 0.1 with no air coming from downstream port of Air Switching Valve.	**Reading OK.** • Continue timing sequence into DRIVEABILITY TEST #11.
		During first 70 seconds: • reading goes to and remains at 13.0 with air coming from downstream port of Air Switching Valve.	**Reading not within specs.** • Check harness and connectors at CCC and Timer. Correct any fault and repeat test. • If CCC and Timer harness and connectors OK, fault is in Air Switching Timer. • Replace Timer and repeat test. • Verify complaint corrected.
		During first 70 seconds: • reading is 0.1 with air coming from Air Switching Valve.	• Replace Air Switching Valve and repeat test.

DRIVEABILITY TEST

OPERATION	PROCEDURE	TEST INDICATION	ACTION REQUIRED
11. DRIVEABILITY TEST #11. (Verifying air switching operation after 70 seconds.)	• Note reading on tester after 70 seconds have elapsed from engine start.	After 70 seconds: • reading switches to 13.0 with air flow coming from downstream port of air switching valve.	**Air switching OK.** • Replace CCC and repeat DRIVEABILITY TEST #6.
		After 70 seconds: • reading switches to 13.0 but no air coming from downstream port of Air Switching Valve.	**Fault in air switching vacuum circuit.** • Check vacuum circuit back to tree. If fault identified, repair and retest. • If no fault in vacuum circuit, fault is in Air Switching Valve. • Replace Air Switching Valve and retest Vac. Sol. valve. • Verify complaint corrected.
		After 70 seconds: • reading is not within 1 volt of battery voltage.	**Fault in air switching electrical circuit.** • Check harness and connectors at power module and CCC. • Repair as necessary and retest. • If harness and connectors are OK: - replace timer and retest. - verify complaint corrected.

CHRYSLER MULTI-PORT ELECTRONIC FUEL INJECTION

General Information

The turbocharged multi-point Electronic Fuel Injection system combines an electronic fuel and spark advance control system with a turbocharged intake system. At the center of this system is a digital pre-programmed computer known as a Logic Module that regulates ignition timing, air-fuel ratio, emission control devices and idle speed. This component has the ability to update and revise its programming to meet changing operating conditions.

Various sensors provide the input necessary for the Logic Module to correctly regulate fuel flow at the fuel injectors. These include the Manifold Absolute Pressure, Throttle Position, Oxygen Feedback, Coolant Temperature, Charge Temperature, and Vehicle Speed Sensors. In addition to the sensors, various switches also provide important information. These include the Transmission Neutral-Safety, Heated Backlite, Air Conditioning, and the Air Conditioning Clutch Switches.

Inputs to the Logic Module are converted into signals sent to the Power Module. These signals cause the Power Module to change either the fuel flow at the injector or ignition timing or both. The Logic Module tests many of its own input and output circuits. If a fault is found in a major circuit, this information is stored in the Logic Module. Information on this fault can be displayed to a technician by means of the instrument panel power loss lamp or by connecting a diagnostic readout and observing a numbered display code which directly relates to a general fault.

ELECTRONIC CONTROL SYSTEM

Power Module

The Power Module contains the circuits necessary to power the ignition coil and the fuel injector. These are high current devices and their power supply has been isolated to minimize any "electrical noise" reaching the Logic Module. The Power Module also energizes the Automatic Shut Down (ASD) Relay which activates the fuel pump, ignition coil, and the Power Module itself. The module also receives a signal from the distributor. In the event of no distributor signal, the ASD relay is not activated and power is shut off from the fuel pump and ignition coil. The Power Module contains a voltage converter which reduces battery voltage to a regulated 8.0V output. This 8.0V output powers the distributor and also powers the Logic Module.

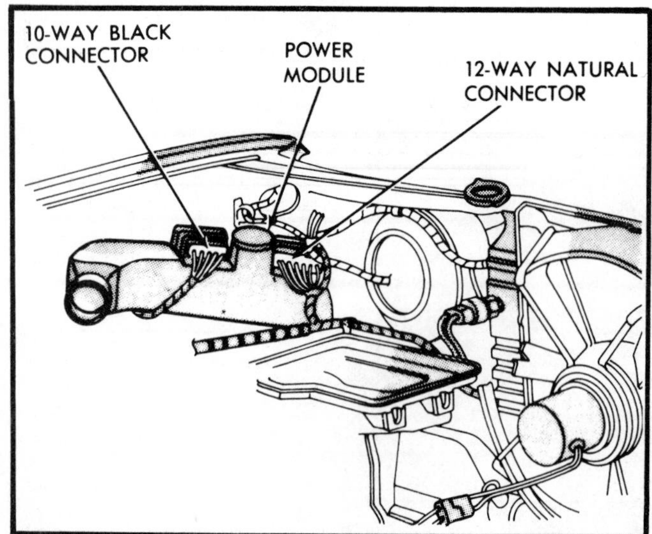

Power module on Chrysler MFI system

Logic Module

The logic module is a digital computer containing a microprocessor. The module receives input signals from various switches, sensors, and components. It then computes the fuel injector pulse width, spark advance, ignition coil dwell, idle speed, and purge and EGR solenoid cycles from this information. The Logic Module tests many of its own input and output circuits. If a fault is found in a major system, this information is stored in the Logic Module.

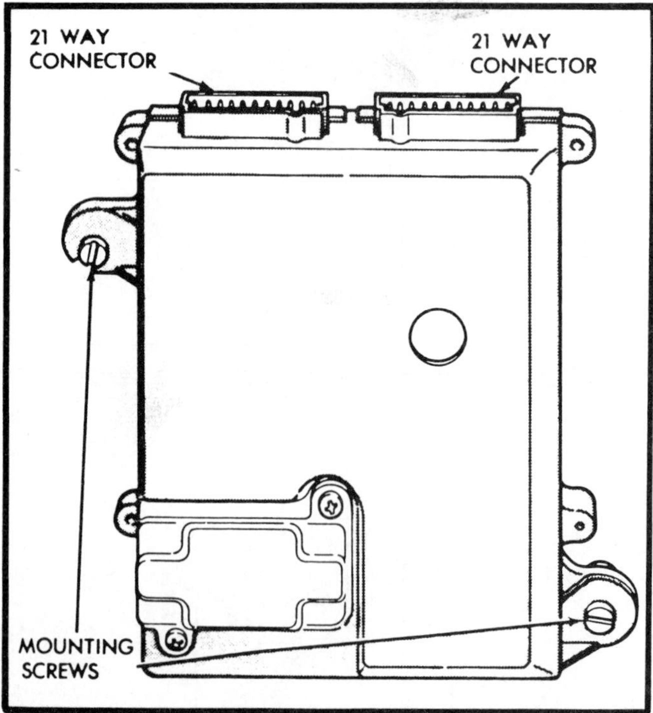

Logic module on Chrysler MFI system showing pin connector locations

Information on this fault can be displayed to a technician by means of flashing lamp on the instrument panel or by connecting a diagnostic readout tool and reading a numbered display code which relates to a general fault.

Automatic Shutdown Relay (ASD)

The Automatic Shutdown Relay (ASD) is powered and controlled through the Power Module. When the Power Module senses a distributor signal during cranking, it grounds the ASD closing its contacts. This completes the circuit for the electric fuel pump, Power Module, and ignition coil. If the distributor signal is lost for any reason the ASD interrupts this circuit in less than one second preventing fuel, spark, and engine operations.

ENGINE SENSORS

Manifold Absolute Pressure (MAP) Sensor

The Manifold Absolute Pressure (MAP) sensor is a device which monitors manifold vacuum. It is mounted in the right side passenger compartment and is connected to a vacuum nipple on the throttle body and, electrically to the Logic Module. The sensor transmits information on manifold vacuum conditions and barometric pressure to the Logic Module. The MAP sensor data on engine load is used with data from other sensors to determine the correct air-fuel mixture.

Oxygen Sensor (O_2 Sensor)

The Oxygen Sensor (O_2 Sensor) is a device which produces an electrical voltage when exposed to the oxygen present in the exhaust gasses. The sensor is mounted in the exhaust manifold and must be heated by the exhaust gasses before producing the voltage.

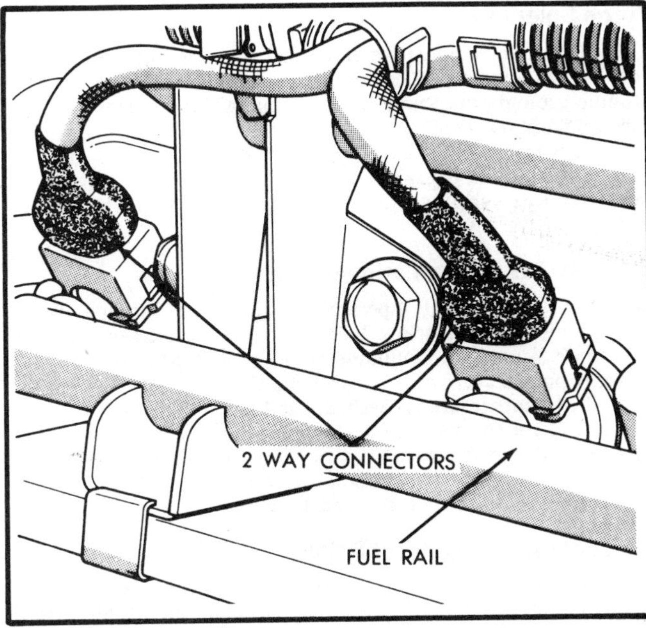

Fuel injection connections

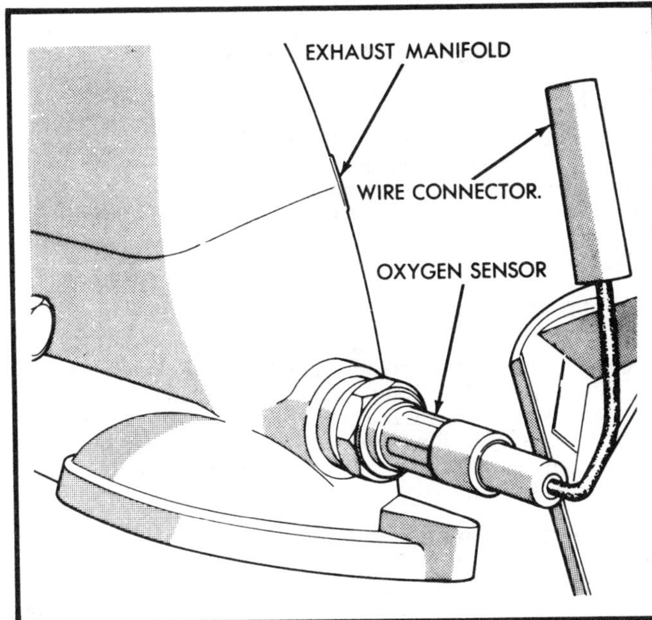

Oxygen sensory connections

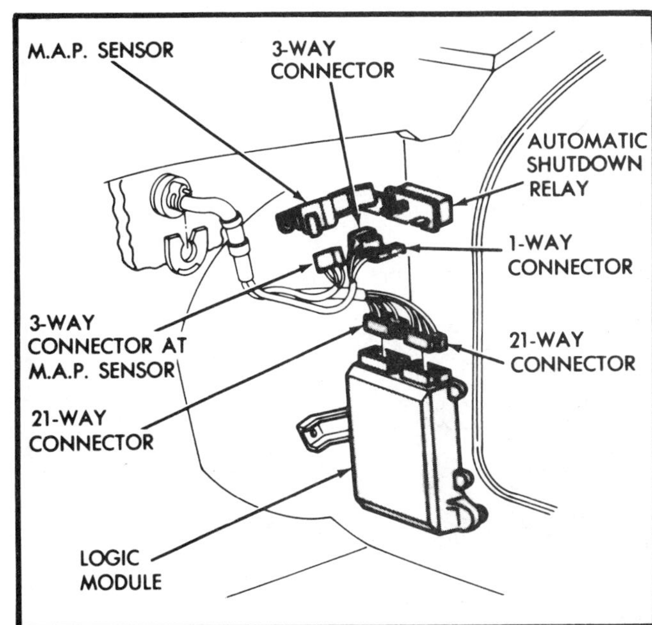

Location of logic module and component connectors on Chrysler MFI system

When there is a large amount of oxygen present (lean mixture), the sensor produces a low voltage. When there is a lesser amount present (rich mixture) it produces a higher voltage. By monitoring the oxygen content and converting it to electrical voltage, the sensor acts as a rich-lean switch. The voltage is transmitted to the Logic Module. The Logic Module signals the Power Module to trigger the fuel injector. The injector changes the mixture.

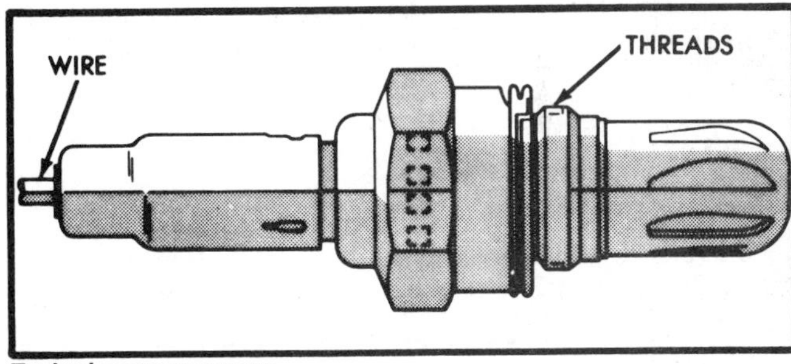

Typical oxygen sensor

Charge Temperature Sensor

The Charge Temperature Sensor is a device mounted in the intake manifold which measures the temperature of the air-fuel mixture. This information is used by the Logic Module to determine engine operating temperature and engine warm-up cycles in the event of a Coolant Temperature Sensor failure.

Coolant Temperature Sensor

The Coolant Termperature Sensor is a device which monitors coolant temperature (which is the same as engine operating temperature). It is mounted in the thermostat housing. This sensor provides data on engine operating temperature to the Logic Module. This data along with data provided by the Charge Temperature Switch allows the Logic Module to demand slightly richer air-fuel mixtures and higher idle speeds until normal operating temperatures are reached. The sensor is a variable resistor with a range of —60°F to 300°F.

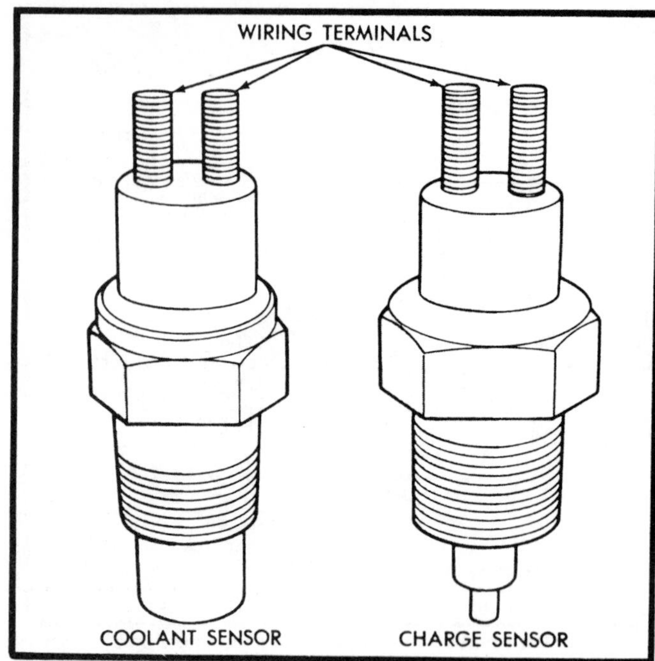

Coolant and charge temperature sensors—typical

SWITCHES AND SOLENOIDS

Various switches provide information to the Logic Module. These include the Neutral Safety, Air Conditioning Clutch, and Brake Light switches. If one or more of these switches is sensed as being in the on position, the Logic Module signals the Automatic Idle Speed Motor to increase idle speed to a scheduled rpm. With the air conditioning on and the throttle blade above a specific angle, the wide open throttle cut-out relay prevents the air conditioning clutch from engaging until the throttle blade is below this angle.

Power Loss Lamp

The Power Loss Lamp comes on each time the ignition key is turned on and stays on for a few seconds as a bulb test. If the Logic Module receives an incorrect signal or no signal from either the Coolant Temperature Sensor, Manifold Absolute Pressure Sensor, or the Throttle Position Sensor, the Power Loss Lamp on the instrument panel is illuminated. This is a warning that the Logic Module has gone into Limp in Mode in an attempt to keep the system operational. It signals an immediate need for service. The Power Loss can also be used to display fault codes. Cycle the ignition switch on, off, on, off, on within five seconds and any fault codes stored in the Logic Module will be displayed.

Limp In Mode is the attempt by the Logic Module to compensate for the failure of certain components by substituting information from other sources. If the Logic Module senses incorrect data or no data at all from the MAP Sensor, Throttle Position Sensor, Charge Temperature Sensor or Coolant Temperature Sensor, the system is placed into Limp In Mode and the Power Loss lamp on the instrument panel is activated.

EGR and canister purge solenoids

Purge Solenoid

The Purge Solenoid works in the same fashion as the EGR solenoid. When engine temperature is below 61°C (145°F) the Logic Module grounds the Purge Solenoid energizing it. This prevents vacuum from reaching the charcoal canister valve. When this temperature is reached the Logic Module de-energizes the solenoid by turning the ground off. Once this occurs vacuum will flow to the canister purge valve and purge fuel vapors through the throttle body.

Exhaust Gas Recirculation Solenoid

The EGR solenoid is operated by the Logic Module. When engine temperature is below 21°C (70°F), the Logic Module energizes the solenoid by grounding it. This closes the solenoid and prevents ported vacuum from reaching the EGR valve. When the prescribed temperature is reached, the logic module will turn off the ground for the solenoid de-energizing it. Once the solenoid is de-energized, ported vacuum from the throttle body will pass through to the EGR valve. At idle and wide open throttle the solenoid is energized which prevents EGR operation.

Air Conditioning Cut Out Relay

The air conditioning cut out relay is electrically in series with the cycling clutch switch and low pressure cut out switch. This relay is in the normally closed (on) position during engine operation. When the Logic Module senses wide open throttle through the Throttle Position Sensor, it will energize the relay, open its contacts, and prevent air conditioning clutch engagement.

FUEL CONTROL SYSTEM

Throttle Body

The throttle body assembly replaces a conventional carburetor air intake system and is connected to both the turbocharger and the intake manifold. The throttle body houses the Throttle Postion Sensor and the Automatic idle Speed Motor. Air flow through the throttle body is controlled by a cable operated throttle blade located in the base of the throttle body.

Fuel Supply Circuit

Fuel is pumped to the fuel rail by an electrical pump which is mounted in the fuel tank. The pump inlet is fitted with a filter to prevent water and other contaminents from entering the fuel supply circuit. Fuel pressure is controlled to a preset level above intake manifold pressure by a pressure regulator which is mounted near the fuel rail. The regulator uses intake manifold pressure at the vacuum tee as a reference.

Fuel Injectors and Fuel Rail Assembly

The four fuel injectors are retained in the fuel rail by lock rings. The rail and injector assembly is then bolted in position with the injectors inserted in the recessed holes in the intake manifold. The Fuel Injector is an electric solenoid powered by the Power Module but, controlled by the Logic Module. The Logic Module, based on ambient, mechanical, and sensor input, determines when and how long the Power Module should operate the injector. When an electric current is supplied to the injector, the armature and pintle move a short distance against a spring, opening a small orifice. Fuel is supplied to the inlet of the injector by the fuel pump, then passes through the injector, around the pintle, and out the orifice. Since the fuel is under high pressure a fine spray is developed in the shape of a hollow cone. The injector, through this spraying action, atomizes the fuel and distributes it into the air entering the combustion chamber.

Fuel Pressure Regulator

The pressure regulator is a mechanical device located downstream of the fuel injector on the throttle body. Its function is to maintain a constant 53 psi (380kPa) across the fuel injector tip. The regulator uses a spring loaded rubber diaphragm to uncover a fuel return port. When the fuel pump becomes operational, fuel flows past the injector into the regulator and is restricted from flowing any further by the blocked return port. When fuel pressure reaches 53 psi (380kPa), it pushes on the diaphragm, compressing the spring, and uncovers the fuel return port. The diaphragm and spring will constantly move from an open to closed position to keep the fuel pressure constant. An assist to the spring loaded diaphragm comes from vacuum in the throttle body above the throttle blade. As venturi vacuum increases less pressure is required to supply the same amount of fuel into the air flow. The vacuum assists in opening the fuel port during high vacuum conditions. This fine tunes the fuel pressure for all operating conditions.

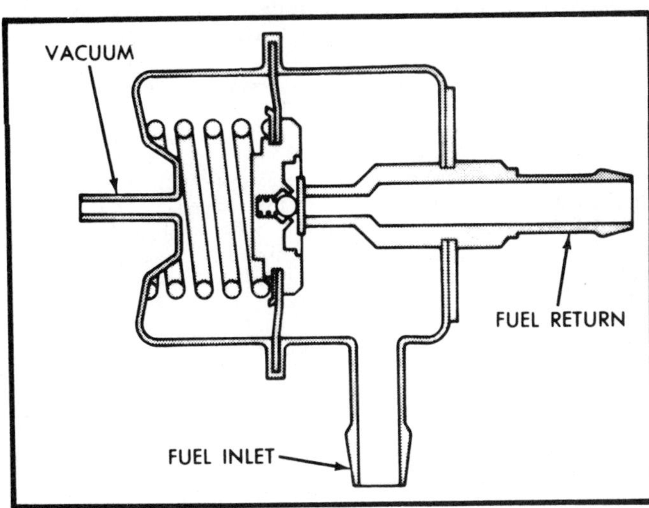

Cross section of fuel pressure regulator

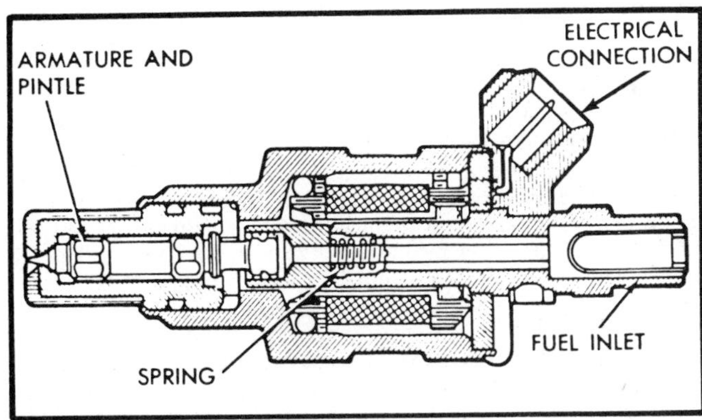

Typical solenoid-type fuel injector

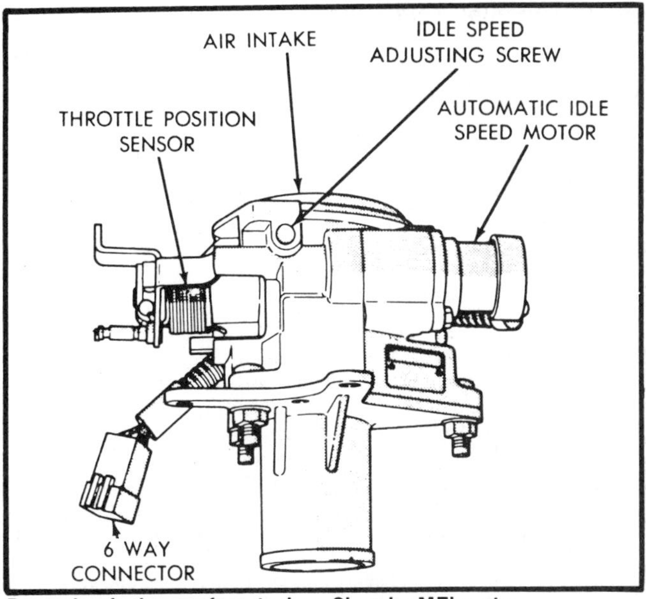

Removing fuel pump from tank on Chrysler MFI system

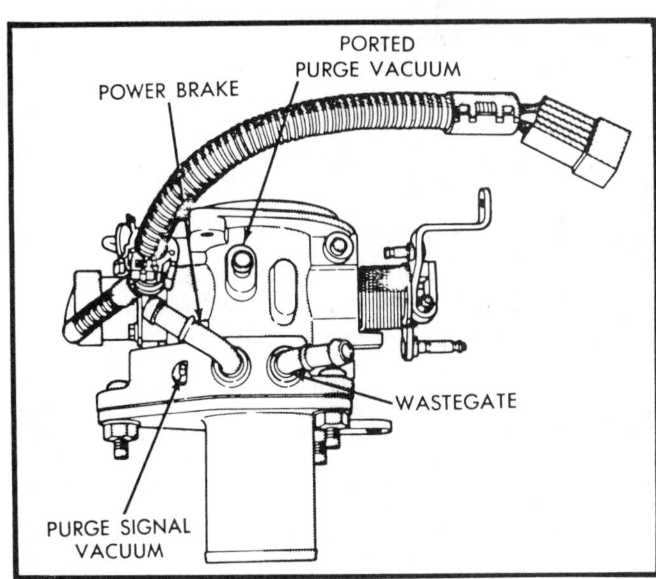

Throttle body assembly on Chrysler MFI system

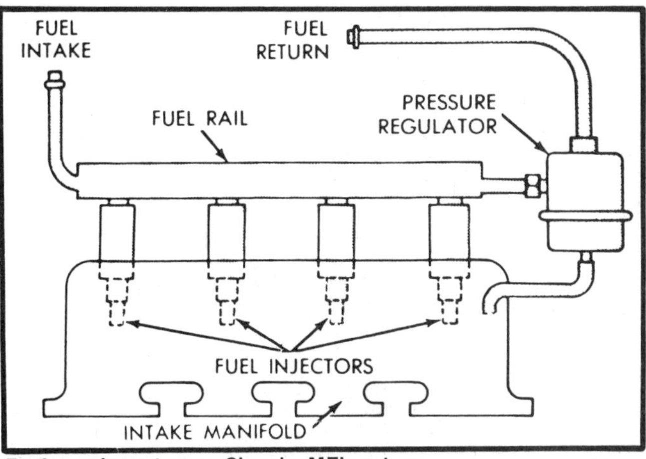

Fuel supply system on Chrysler MFI system

Throttle Position Sensor (TPS)

The throttle Position Sensor (TPS) is an electric resistor which is activated by the movement of the throttle shaft. It is mounted on the throttle body and senses the angle of the throttle body and senses the angle of the throttle blade opening. The voltage that the sensor produces increases or decreases according to the throttle blade opening. This voltage is transmitted to the Logic Module where it is used along with data from other sensors to adjust the air-fuel ratio to varying conditions and during acceleration, deceleration, idle, and wide open throttle operations.

Automatic Idle Speed (AIS) Motor

The Automatic Idle Speed Motor (AIS) is operated by the Logic Module. Data from the Throttle Position Sensor, Speed Sensor, Coolant Temperature Sensor, and various switch operations, (Electric Backlite, Air Conditioning, Safety/Neutral, Brake) are used by the Logic Module to adjust engine idle to an optimum during all idle conditions. The AIS adjusts the air portion of the air-fuel mixture through an air bypass on the back of the throttle body. Basic (no load) idle is determined by the minimum air flow through the throttle body. The AIS opens or closes off the air bypass as an increase or decrease is needed due to engine loads or ambient conditions. The Logic Module senses an air/fuel change and increases or decreases fuel proportionally to change engine idle. Deceleration die out is also prevented by increasing engine idle when the throttle is closed quickly after a driving (speed) condition.

Fuel Pump

The fuel pump used in this system is a positive displacement, roller vane immersible pump with a permanent magnet electric motor. The fuel is drawn in through a filter sock and pushed through the electric motor to the outlet. The pump contains two check valves. One valve is used to relieve internal fuel pump pressure and regulate maximum pump output. The other check valve, located near the pump outlet, restricts fuel movement in either direction when the pump is not operational. Voltage to operate the pump is supplied through the Auto Shutdown Relay.

Fuel Reservoir

The fuel pump is mounted within a fuel reservoir in the fuel tank. The purpose of the reservoir is to provide fuel at the pump intake furing all driving conditions, especially those when low fuel levels are present. The fuel return line directs fuel into a cup on the side of the reservoir. The stream of fuel coming into this cup creates a low pressure area and causes additional fuel from the main tank to flow into the reservoir. This combination of return fuel and fuel from the main tank keeps the reservoir full even when the fuel level is below the reservoir walls.

Exhaust Gas Recirculation (EGR)

The Exhaust Gas Recirculation system is a back pressure type and is controlled two ways. The Logic Module controls vacuum through the EGR solenoid, turning the vacuum circuit on or off. A back pressure transducer measures the amount of exhaust back pressure on the exhaust side of the EGR valve and varies the strength of the vacuum signal applied to the EGR valve. The Logic Module will prevent EGR operation by turning the EGR solenoid off at idle, wide open throttle or when engine temperature falls below 70°F (21°C). The back pressure transducer adjusts the EGR signal to provide programmed amounts of Exhaust Gas Recirculation under all other conditions.

Air Aspirator System

The air aspirator system uses exhaust pressure pulses to draw air into the exhaust system. This reduces carbon monoxide (CO) and hydrocarbon (HC) emissions. It draws fresh air from the clean side of the air cleaner past a one-way diaphragm in the aspirator valve. The diaphragm opens to allow fresh air to mix with exhaust gasses during negative pressure pulses. If pressure pulses are positive, the diaphragm closes, which prevents exhaust gasses from entering the air cleaner. The air aspirator is most effective at idle and slightly off idle where negative pressure pulses are greatest.

On-Car Service

NOTE: Most complaints that may occur with turbocharged multi-point Electronic Fuel Injection can be traced to poor wiring or hose connections. A visual check will help spot these faults and save unnecessary test and diagnosis time.

ON BOARD DIAGNOSTICS

The Logic Module has been programmed to monitor several different circuits of the fuel injection system. This monitoring is called On Board Diagnosis. If a problem is sensed with a monitored circuit, often enough to indicate an actual problem, its Fault Code is stored in the Logic Module for eventual display to the service technician. If the problem is repaired or ceases to exist, the Logic Module cancels the Fault Code after 30 ignition key on/off cycles.

Fault Codes

When a fault code appears (either by flashes of the power loss lamp or by watching the diagnostic readout—Tool C-4805 or equivalent), it indicates that the Logic Module has recognized an abnormal signal in the system. Fault codes indicate the results of a failure do not always identify the failed component.

CODE 11 indicates a problem in the distributor circuit. This code appears if the Logic Module has not sensed a distributor signal since the battery was reconnected.

CODE 12 indicates a problem in the stand-by memory circuit. This code appears if direct memory feed to the Logic Module is interrupted.

CODE 13 indicates a problem in the MAP sensor pneumatic system. This code appears if the MAP sensor vacuum level does not change between start and start/run transfer speed (500-600 rpm).

CODE 14 indicates a problem in the MAP sensor electrical system. This code appears if the map sensor signal is either too low (below .02 volts) or too high (above 4.9 volts).

CODE 15 indicates a problem in the vehicle Speed Sensor circuit. This code appears if engine speed is at idle and speed sensor indicates less than 2 mph. This code is valid only if it is sensed while moving.

CODE 21 indicates a problem in the O_2 feedback circuit. This code appears if engine temperature is above 170°F (77°C), engine speed is above 1500 rpm, and there has been no O_2 signal for more than 5 seconds.

CODE 22 indicates a problem in the coolant temperature circuit. This code appears if the temperature sensor indicates an improbable temperature or a temperature that changes too fast to be real.

CODE 23 indicates a problem in the charge temperature circuit. This code appears if the charge temperature is an improbable temperature or a temperature that changes too fast to be real.

CODE 24 indicates a problem in the Throttle Position Sensor circuit. This code appears if the sensor signal is either below .16 volts or above 4.7 volts.

CODE 25 indicates a problem in the Automatic Idle Speed system. This code appears if the proper voltage from the AIS system is not present. An open motor or harness will not activate this code.

CODE 31 indicates a problem in the Canister Purge Solenoid circuit. This code appears when the proper voltage at the purge solenoid is not present (open or shorted system).

CODE 32 indicates a problem in the Power Loss Lamp circuit. This code appears if proper voltage to the Power Loss Lamp is not present (open or shorted system).

CODE 33 indicates a problem in the Air Conditioning Wide Open Throttle Cut Out Relay circuit. This code appears if the proper voltage at the EGR Solenoid is not present (open or shorted).

CODE 34 indicates a problem in the EGR Solenoid circuit. This code appears if proper voltage at the EGR Solenoid is not present (open or shorted).

CODE 35 indicates a problem in the Fan Relay circuit. This code appears if the radiator fan is either not operating or operating at the wrong time.

CODE 41 indicates a problem in the Charging System. This code appears if battery voltage from the Automatic Shut Down Relay is below 11.75 volts.

CODE 42 indicates a problem in the Automatic Shut Down Relay (ASD) circuit. This code appears if during cranking, battery voltage from the ASD relay is not present for at least 1/3 second after the first distributor pulse or, after engine stall, battery voltage is not off within 3 seconds after last distributor pulse.

CODE 43 indicates a problem in the interface circuit. This code appears if the anti-dwell or injector control signal is not present between the Logic Module and Power Module.

CODE 44 indicates a problem in the Logic Module. This code appears if an incorrect PROM has been installed in the Logic Module.

CODE 45 indicates a problem in the Overboost Shut Off circuit. This code appears if MAP sensor electrical signal rises above 10 psi boost.

CODE 51 indicates a problem in the closed loop fuel system. This code appears if during closed loop conditions, the O_2 signal is either low or high for more than 2 minutes.

CODE 52 indicates a problem in the Logic Module. This code appears if an internal failure exists in the Logic Module.

CODE 53 indicates a problem in the Logic Module. This code appears if an internal failure exists in the Logic Module.

CODE 54 indicates a problem in the Synchronization pick-up circuit. This code appears, if at start/run transfer speed, the reference pick-up signal is present but the synchronization pick-up signal is missing at the logic module.

CODE 55 indicates message complete. This code appears after all fault codes are displayed.

CODE 88 indicates start of message. This code appears at start of fault code messages. This code only appears on Readout Tool C-4805, or equivalent, and may also be used for switch check.

SYSTEM TESTS

Obtaining Fault Codes

1. Connect Diagnostic Readout Box Tool C-4805, or equivalent, to the diagnostic connector located in the engine compartment near the passenger side strut tower.
2. Start the engine if possible, cycle the transmission selector and the A/C switch if applicable. Shut off the engine.
3. Turn the ignition switch on, off, on, off, on. Within 5 seconds record all the diagnostic codes shown on the diagnostic readout box tool, observe the power loss lamp on the instrument panel the lamp should light for 2 seconds then go out (bulb check).

Switch Test

After all codes have been shown and has indicated Code 55 end of message, actuate the following component switches. The digital display must change its numbers when the switch is activated and released:

- Brake Pedal
- Gear Shift Selector park, reverse, park.
- A/C Switch (if applicable).
- Electric Backlite Switch (if applicable).

Actuator Test Mode (ATM)

1. Remove coil wire from cap and place 1/4 in. from a ground.

CAUTION

Coil wire must be 1/4 in. or less from ground or power module damage may result.

2. Remove air cleaner hose from throttle body.
3. Press the ATM button on the diagnostic readout box tool and observe the following:
   - 3 sparks from the coil wire to ground.
   - 2 AILS motor movements (1 open 1 close) you must listen carefully for AIS operation.
   - With the ATM button still depressed, install a jumper wire between pins 2 and 3 of the gray distributor synch. connector. Listen for the click which indicates one set of injectors has been activated. Remove jumper wire and second set of injectors will be activated. Reconnect distributor connector.
4. The ATM capability is cancelled 5 minutes after the ignition switch is turned on. To reinstate this capability cycle the ignition on and off three times ending in the on position.
5. When the ATM button is pressed, fault Code 42 is generated because the ASD relay is by passed. Do not use this code for diagnostics after ATM operation.
6. The ATM test will check 3 categories of operation:
   - When coil fires three times.
     a. Coil operational
     b. Logic Module portion operational
     c. Power Module portion operational
     d. Interface between Power Module and Logic Module is working.
   - AIS is operational
   - Injector fuel pulse into Throttle Body:
     a. Fuel injector operational
     b. Fuel pump operational
     c. Fuel lines intact.
7. The Electronic Fuel Injection system must be evaluated using all the information found in the systems test:
   - Start/No Start
   - Fault Codes
   - Loss of Power Lamp on or off (limp in)
   - ATM Results:
     Spark yes/no
     Fuel yes/no
     AIS movement yes/no

   Once this information is found it will be easier to determine what circuit to look at for further testing.

Relieving Fuel System Pressure

The E.F.I. fuel system is under a constant pressure of approximately 53 psi (380 kPa). Before servicing the fuel tank, fuel pump, fuel lines, fuel filter, or fuel components of the throttle body the fuel pressure must be released as follows:

1. Loosen gas cap to release any in tank pressure.
2. Remove wiring harness connector from any injector.
3. Ground one injector terminal with a jumper.
4. Connect a jumper wire to second terminal and touch battery positive post for no longer than 10 seconds.
5. Remove jumper wires.
6. Continue fuel system service.

Fuel System Pressure Test

CAUTION

Fuel system pressure must be released each time a fuel hose is to be disconnected.

1. Remove fuel intake hose from throttle body and connect fuel system pressure testers C-3292, and C-4749, or equivalent, between fuel filter hose and throttle body.
2. Start engine. If gauge reads 380 kPa = 14, Pa (53 psi = 2 psi) pressure is correct and no further testing is required. Reinstall fuel hose using a new original equipment type clamp and torque to 10 inch lbs. (1 nm).

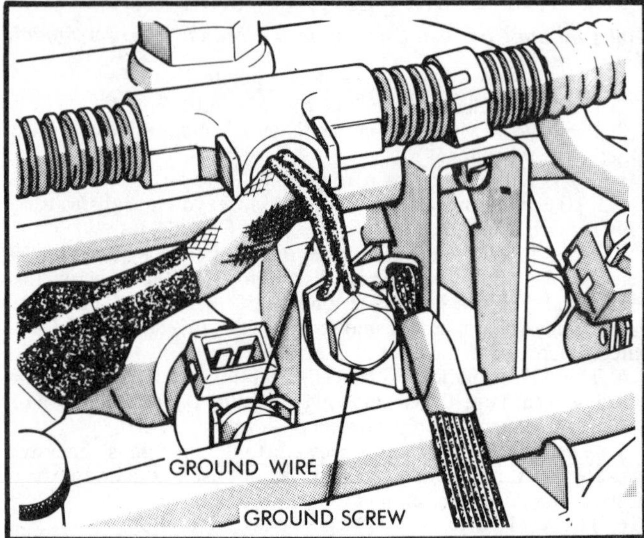

Location of ground wire

3. If fuel pressure is below specifications, install tester between fuel filter hose and fuel line.

4. Start engine. If pressure is now correct, replace fuel filter. If no change is observed, gently squeeze return hose. If pressure increases, replace pressure regulator. If no change is observed, problem is either a plugged pump filter sock or defective fuel pump.

5. If pressure is above specifications, remove fuel return hose from pressure regulator end. Connect a substitute hose and place other end of hose in clean container. Start engine. If pressure is now correct, check for restricted fuel return lilne. If no change is observed, replace fuel regulator.

Component Removal

Mechanical malfunctions are more difficult to diagnose with the EFI system. The Logic Module has been programmed to compensate for some mechanical malfunctions such as incorrect cam timing, vacuum leaks, etc. If engine performance problems are encountered, and no fault codes are displayed, the problem may be mechanical rather than electronic.

THROTTLE BODY

When servicing the fuel portion of the throttle body it will be necessary to bleed fuel pressure before opening any hoses. Always reassemble throttle body components with new O-rings and seals where applicable. Never use lubricants on O-rings or seals, damage may result. If assembly of components is difficult use water to aid assembly. Use care when removing fuel hoses to prevent damage to hose or hose nipple. Always use new hose clamps of the correct type when reassembling and torque hose clamps to 10 inch lbs. (1 Nm). Do not use aviation-style clamps on this system or hose damage may result.

NOTE: It is not necessary to remove the throttle body from the intake manifold to perform component disassembly. If fuel system hoses are to be replaced, only hoses marked EFI/EFM may be used.

Removal and Installation

1. Disconnect negative battery cable.
2. Remove air cleaner to throttle body screws, loosen hose clamp and remove air cleaner adaptor.
3. Remove accelerator, speed control, and transmission kickdown cables and return spring.
4. Remove throttle cable bracket from throttle body.
5. Disconnect 6 way connector.
6. Disconnect vacuum hoses from throttle body.
7. Loosen throttle body to turbocharger hose clamp.
8. Remove throttle body to intake manifold screws.
9. Remove throttle body.
10. Reverse the above procedure for installation.

THROTTLE POSITION SENSOR

Removal and Installation

1. Disconnect negative battery cable and 6-way throttle body connector.
2. Remove 2 screws mounting throttle position sensor to throttle body.
3. Unclip wiring clip from convoluted tube and remove mounting bracket.
4. Lift throttle position sensor off throttle shaft and remove O-ring.
5. Pull the 3 wires of the throttle position sensor from the convoluted tubing.
6. Look inside the 6-way throttle body connector and lift a locking tab with a small screwdriver for each T.P.S. wire blade terminal. Remove each blade from connector. (Not wiring position for reassembly.)
7. Insert each wire blade terminal into throttle body connector. Make sure wires are inserted into correct locations.
8. Insert wires from throttle position sensor into convoluted tube.
9. Install throttle position sensor and new O-ring with mounting bracket to throttle body. Torque screws to 20 inch lbs. (2 Nm).
10. Install wiring clips to convoluted tube.
11. Connect 6 way connector and battery cable.

AUTOMATIC IDLE SPEED MOTOR

Removal and Installation

1. Disconnect negative battery cable and 6-way throttle body connector.
2. Remove 2 screws that mount the A.I.S. to its adaptor. (Do not remove the clamp on the A.I.S. or damage will result.)
3. Remove wiring clips and remove the two A.I.S. wires from the 6-way throttle body connector. Lift each locking tab with a small screwdriver and remove each blade terminal. (Note wiring position for reassembly).
4. Lift A.I.S. from its adaptor.
5. Remove the 2 O-rings on the A.I.S. carefully.
6. Install 2 new O-rings on A.I.S.
7. Carefully work A.I.S. into its adaptor.
8. Install 2 mounting screws and torque to 20 inch lbs. (2 Nm).
9. Route A.I.S. wiring to 6-way connector and install each wire blade terminal into the connector. Make sure wires are inserted in correct locations.
10. Connect wiring clips, 6-way connector, and battery cable.

AUTOMATIC IDLE SPEED MOTOR ASSEMBLY

Removal and Installation

1. Disconnect negative battery cable and 6-way throttle body connector.
2. Remove 2 screws on back of throttle body from A.I.S. adaptor.
3. Remove wiring clips and remove the two A.I.S. wires from the 6-way throttle body connector. Lift each locking tab with a small screwdriver and remove each blade terminal. (Note wiring position for reassembly).
4. Carefully pull the assembly from the rear of the throttle body. The O-ring at the top and seal at the bottom may fall off adaptor.

5. Remove O-ring and seal.
6. Place a new O-ring and seal on adaptor.
7. Carefully position assembly onto back of throttle body (make sure seals stay in place) insert screws and torque to 65 inch lbs. (7 Nm).
8. Route A.I.S. wiring to 6-way connector and install each wire blade terminal into the connector. Make sure wires are inserted in correct locations.
9. Connect wiring clips, 6-way connector, and battery cable.

OXYGEN SENSOR

Removal and Installation

Removing the oxygen sensor from the exhaust manifold may be difficult if the sensor was overtorqued during installation. Use Tool C-4589, or equivalent, to remove the sensor. The threads in the exhaust manifold must be cleaned with a 18mm x 1.5 x 6E tap. If the same sensor is to be reinstalled, the threads must be coated with an anti-seize compound such as Loctite® 771-64 or equivalent. New sensors are packaged with anti-seize compound on the threads and no additional compound is required. Sensors must be torqued to 20 ft. lbs. (27 Nm).

IDLE SPEED

Adjustment

Before adjusting the idle on an electronic fuel injected vehicle the following items must be checked:

a. AIS motor has been checked for operation.
b. Engine has been checked for vacuum or EGR leaks.
c. Engine timing has been checked and set to specifications.
d. Coolant temperature sensor has been checked for operation.

1. Install a tachometer.
2. Warm up engine to normal operating temperature (accessories off).
3. Shut engine off and disconnect radiator fan.
4. Disconnect Throttle Body 6-way connector. Remove the brown with white tracer AIS wire from the connector and reconnect connector.
5. Start engine with transaxle selector in park or neutral.
6. Apply 12 volts to AIS brown with white tracer wire. This will drive the AIS fully closed and the idle should drop.
7. Disconnect then reconnect coolant temperature sensor.
8. With transaxle in neutral, idle speed should be 775 ± 25 (700 ± 25 green engine).
9. If idle is not to specifications adjust idle air bypass screw.
10. If idle will not adjust down, check for vacuum leaks, AIS motor damage, throttle body damage, or speed control cable adjustment.

IGNITION TIMING

Adjustment

1. Connect a power timing light to the number one cylinder, or a magnetic timing unit to the engine. (Use a 10° degree offset when required).
2. Connect a tachometer to the engine and turn selector to the proper cylinder position.
3. Start engine and run until operating temperature is reached.
4. Disconnect and reconnect the water temperature sensor connector on the thermostat housing. The loss of power lamp on the dash must come on and stay on. Engine rpm should be within emission label specifications.
5. Aim power timing light at timing hole in bell housing or read the magnetic timing unit.
6. Loosen distributor and adjust timing to emission label specifications if necessary.
7. Shut engine off, disconnect and reconnect positive battery quick disconnect. Start vehicle, the loss of power lamp should be off.
8. Shut engine off, then turn ignition on, off, on, off, on. Fault codes should be clear with 88-51-55 shown.

CHRYSLER THROTTLE BODY INJECTION SYSTEM

General Information

This electronic fuel injection system is a computer regulated single point fuel injection system that provides precise air/fuel ratio for all driving conditions. At the center of this system is a digital preprogrammed computer known as a logic module that regulates ignition timing, air-fuel ratio, emission control devices and idle speed. This component has the ability to update and revise its programming to meet changing operating conditions.

Various sensors provide the input necessary for the logic module to correctly regulate the fuel flow at the fuel injector. These include the manifold absolute pressure, throttle position, oxygen feedback, coolant temperature, charge temperature and vehicle speed sensors. In addition to the sensors, various switches also provide important information. These include the neutral-safety, heated back lite, air conditioning, air conditioning clutch switches, and an electronic idle switch.

All inputs to the logic module are converted into signals sent to the power module. These signals cause the power module to change either the fuel flow at the injector or ignition timing or both.

The logic module tests many of its own input and output circuits. If a fault is found in a major system this information is stored in the logic module. Information on this fault can be displayed to a technician by means of a flashing light emitting diode (LED) or by connecting a diagnostic read out and reading a numbered display code which directly relates to a specific fault.

ELECTRONIC CONTROL SYSTEM

Power Module

The power module contains the circuits necessary to power the ignition coil and the fuel injector. These are high current devices and their power supply has been isolated to minimize any "electri-

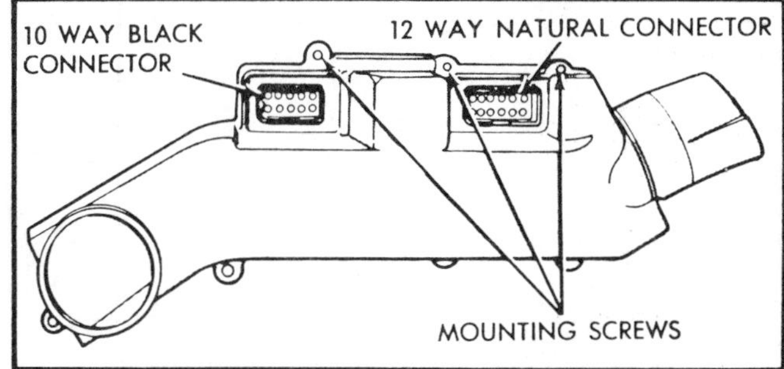

Typical power module showing harness connector terminals

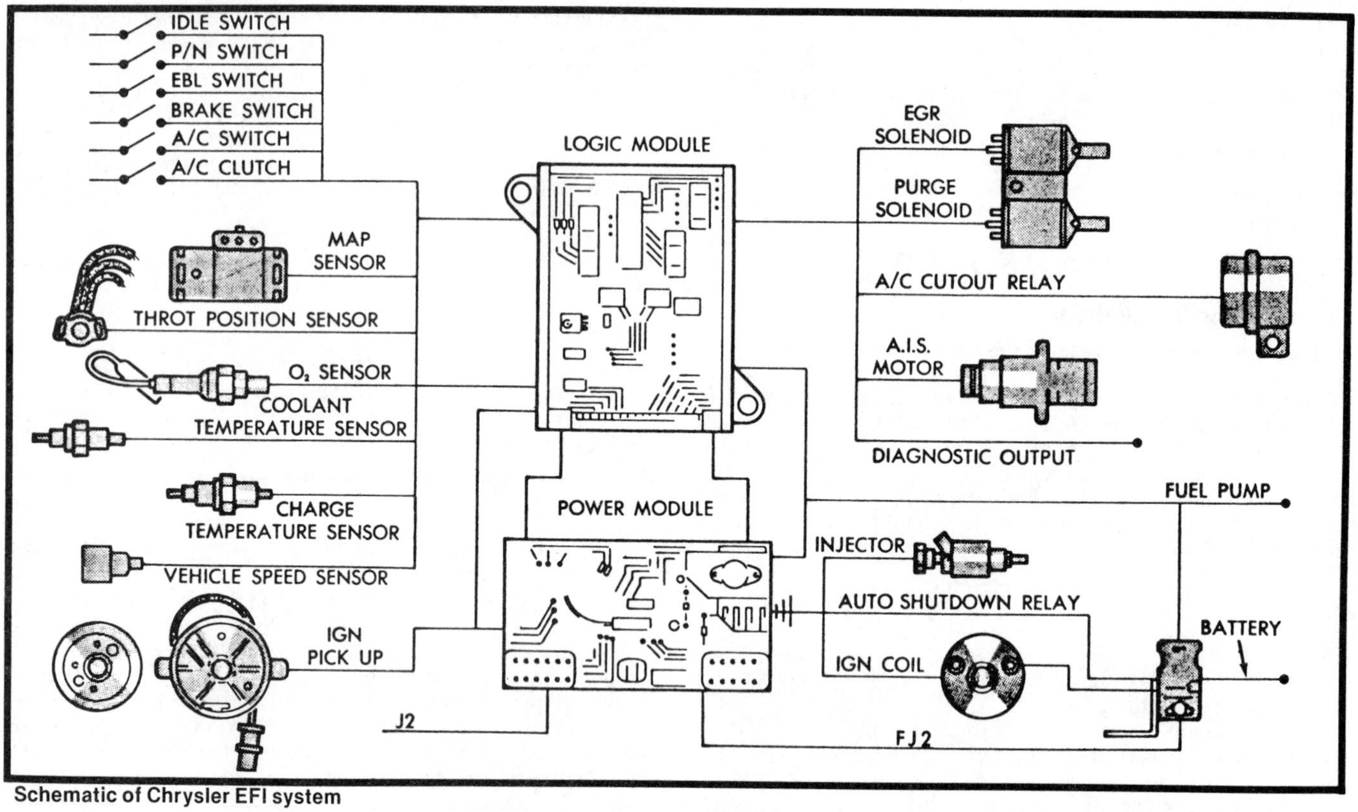

Schematic of Chrysler EFI system

cal noise" reaching the logic module. The power module also energizes the automatic shut down (ASD) relay which activates the fuel pump, ignition coil, and the power module itself. The module also receives a signal from the distributor and sends this signal to the logic module. In the event of no distributor signal, the ASD relay is not activated and power is shut off from the fuel pump and ignition coil. The power module contains a voltage converter which reduces battery voltage to a regulated 8.0V output. This 8.0V output powers the distributor and also powers the logic module.

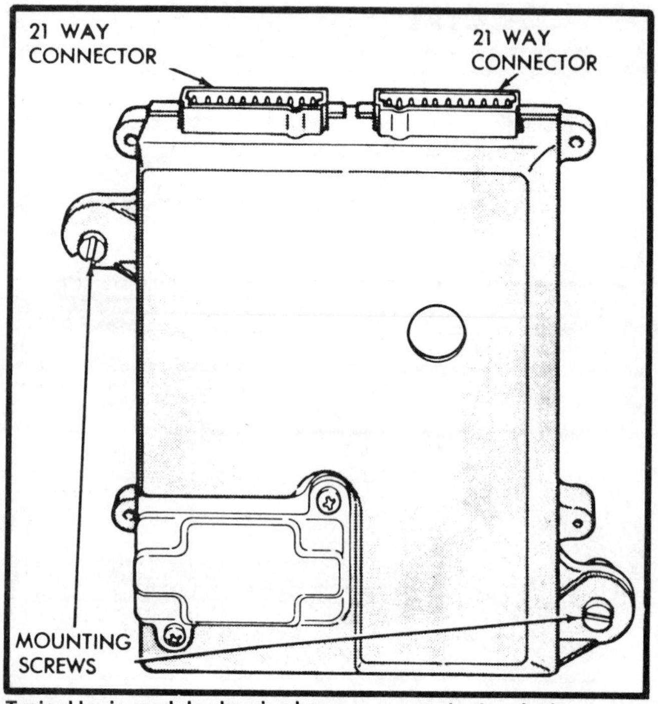

Typical logic module showing harness connector terminals

Logic Module

The logic module is a digital computer containing a microprocessor. The module receives input signals from various switches, sensors, and components. It then computes the fuel injector pulse width, spark advance, ignition coil dwell, automatic idle speed actuation, and purge, and EGR control solenoid cycles.

The logic module tests many of its own input and output circuits.

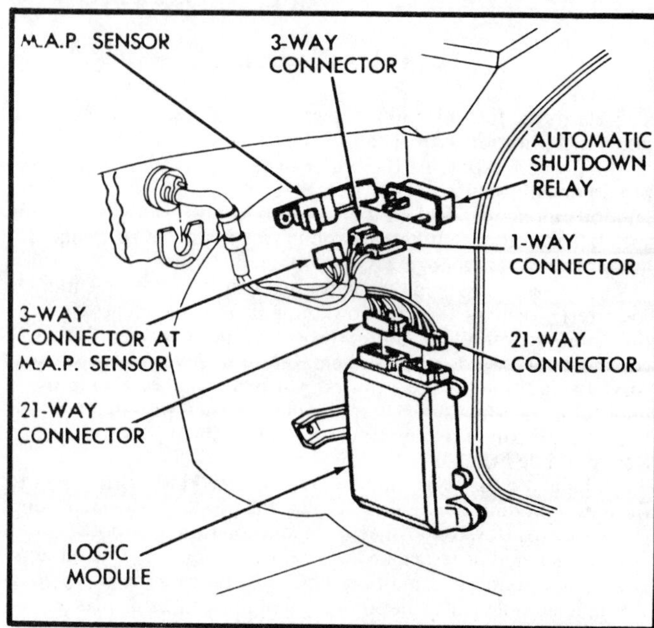

Location of logic module, components and harness connectors

If a fault is found in a major system, this information is stored in the logic module. Information on this fault can be displayed to a technician by means of a flashing, light emitting diode (LED) or by connecting a diagnostic read out and reading a numbered display code which directly relates to a specific fault.

When the power module senses a distributor signal during cranking, it grounds the ASD closing its contacts. This completes the circuit for the electric fuel pump, power module, and ignition coil. If the distributor signal is lost for any reason the ASD interrupts this circuit in less than one second preventing fuel, spark, and engine operations. This fast shutdown serves as a safety feature in the event of an accident.

ENGINE SENSORS

Manifold Absolute Pressure (MAP) Sensor

The manifold absolute pressure (MAP) sensor is a device which monitors manifold vacuum. It is mounted in the right side passenger compartment and is connected to a vacuum nipple on the throttle body and, electrically to the logic module. The sensor transmits information on manifold vacuum conditions and barometric pressure to the logic module. The MAP sensor data on engine load is used with data from other sensors to determine the correct air-fuel mixture.

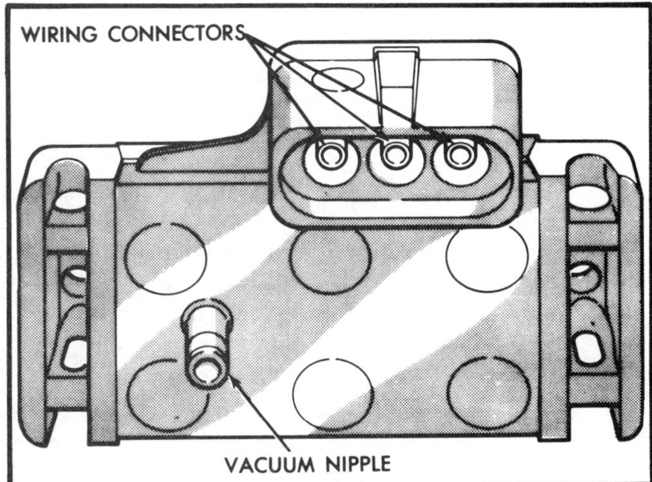

Manifold absolute pressure (MAP) sensor

Oxygen Sensor

The oxygen sensor (O_2 sensor) is a device which produces an electrical voltage when exposed to the oxygen present in the exhaust gases. The sensor is mounted in the exhaust manifold and must be heated by the exhaust gases before producing the voltage. When there is a large amount of oxygen present (lean mixture), the sensor produces a low voltage. When there is a lesser amount present (rich mixture) it produces a higher voltage. By monitoring the oxygen content and converting it to electrical voltage, the sensor acts as a rich-lean switch. The voltage is transmitted to the logic module. The logic module signals the power module to trigger the fuel injector. The injector changes the mixture.

Coolant Temperature Sensor

The coolant temperature sensor is mounted in the thermostat housing. This sensor provides data on engine operating temperature to the logic module. This data along with data provided by the charge temperature switch allows the logic module to demand slightly richer air-fuel mixtures and higher idle speeds until normal operating temperatures are reached. The coolant temperature sensor allows the logic module to act as an automatic choke.

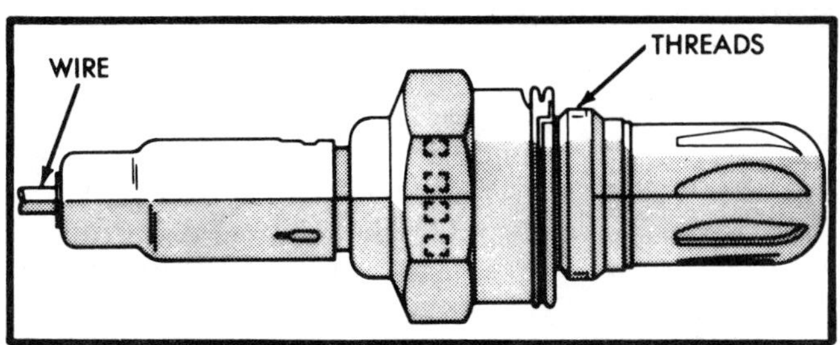

Typical oxygen sensor

Charge Temperature sensor

The charge temperature sensor is a device mounted in the intake manifold which measures the temperature of the air-fuel mixture. This information is used by the logic module to determine engine operating temperature and engine warm-up cycles in the event of a coolant temperature sensor failure.

Switch Input

Various switches provide information to the logic module. These include the idle, neutral safety, electric backlite, air conditioning, air conditioning clutch, and brake light switches. If one or more of these switches is sensed as being in the on position, the logic module signals the automatic idle speed motor to increase idle speed to a scheduled rpm.

With the air conditioning on and the throttle blade above a specific angle, the wide open throttle cut-out relay prevents the air conditioning clutch from engaging until the throttle blade is below this angle.

Power Loss Lamp

The power loss lamp comes on each time the ignition key is turned on and stays on for a few seconds as a bulb test.

If the logic module receives an incorrect signal or no signal from either the coolant temperature sensor, manifold absolute pressure sensor, or the throttle position sensor, the power loss lamp on the instrument panel is illuminated. This is a warning that the logic module has gone into limp in mode in an attempt to keep the system operational.

Limp in mode is the attempt by the logic module to compensate for the failure of certain components by substituting information from other sources. If the logic module senses incorrect data or no data at all from the MAP sensor, throttle position sensor or coolant temperature sensor, the system is placed into Limp in Mode and the power loss lamp on the instrument panel is activated.

FUEL CONTROL SYSTEM

Throttle Body

The throttle body assembly replaces a conventional carburetor and is mounted on top of the intake manifold. The throttle body houses the fuel injector, pressure regulator, throttle position sensor, and automatic idle speed motor. Air flow through the throttle body is

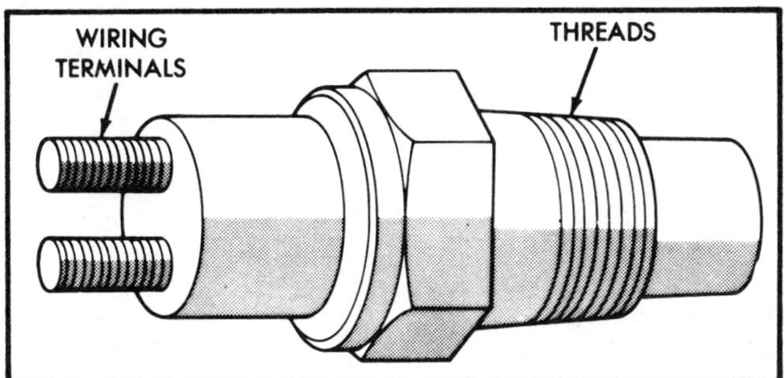

Typical coolant sensor

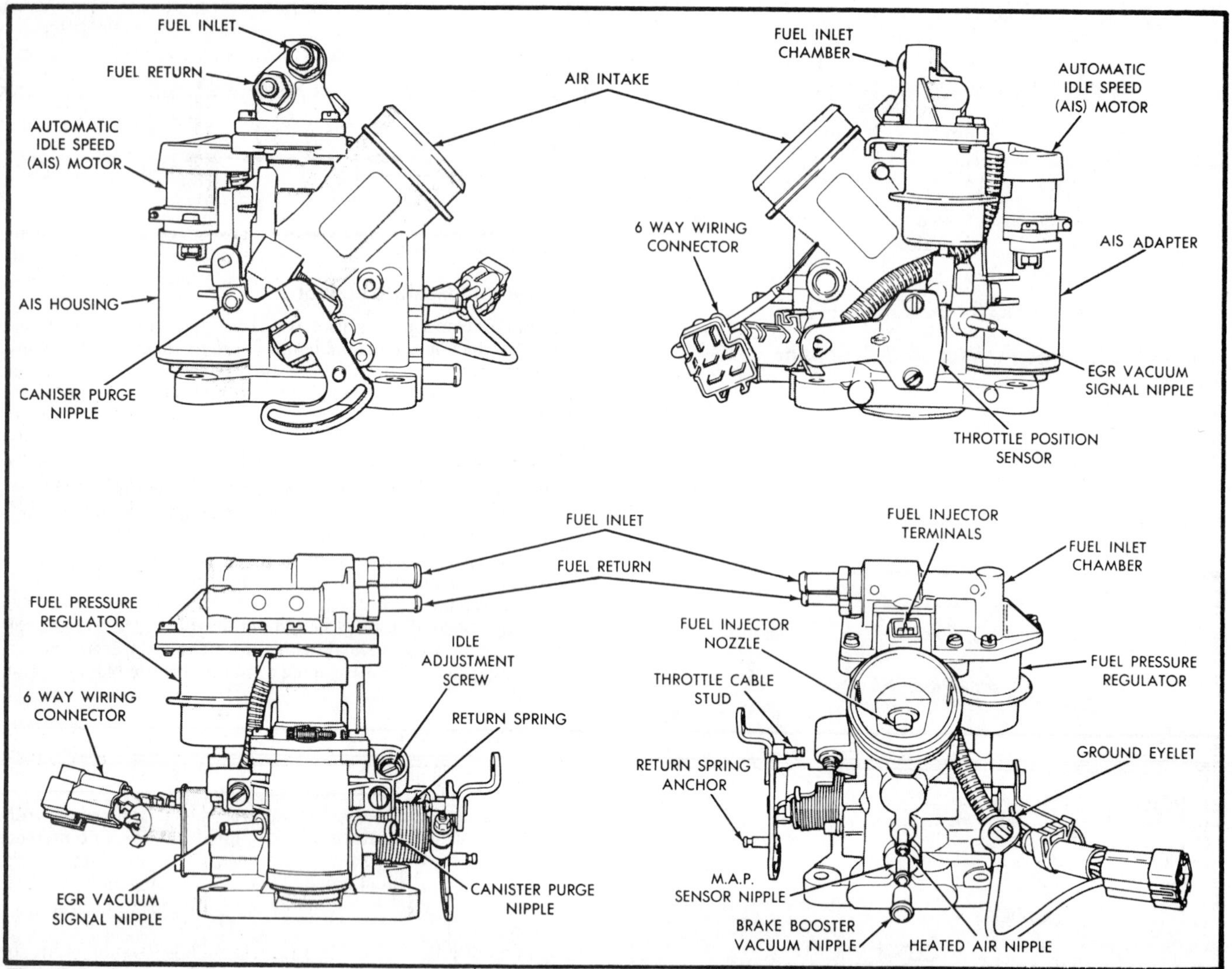

Throttle body assembly

controlled by a cable operated throttle blade located in the base of the throttle body. The throttle body itself provides the chamber for metering atomizing and distributing fuel through out the air entering the engine.

Fuel Injector

The fuel injector is an electric solenoid powered by the power module but, controlled by the logic module. The logic module, based on ambient, mechanical, and sensor input, determines when and how long the power module should operate the injector. When an electric current is supplied to the injector, the armature and pintle move a short distance against a spring, opening a small orifice. Fuel is supplied to the inlet of the injector by the fuel pump, then passes through the injector, around the pintle, and out the orifice. Since the fuel is under high pressure a fine spray is developed in the shape of a hollow cone. The injector, through this spraying action, atomizes the fuel and distributes it into the air entering the throttle body.

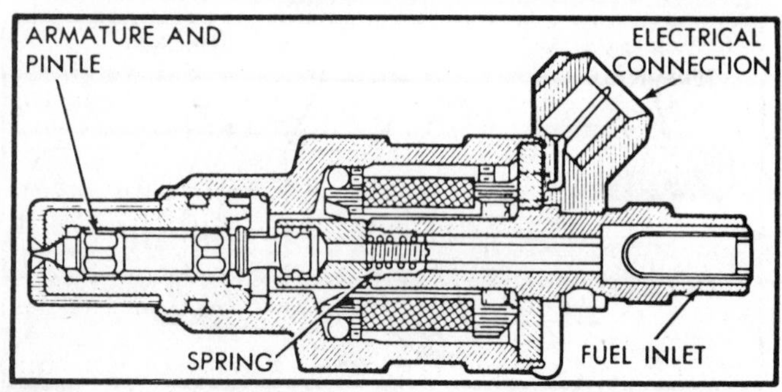

Cross section of typical solenoid-type fuel injector

Fuel Pressure Regulator

The pressure regulator is a mechanical device located downstream of the fuel injector on the throttle body. Its function is to maintain a constant 250kPa (36PSI) across the fuel injector tip. The regulator uses a spring loaded rubber diaphragm to uncover a fuel return port. When the fuel pump becomes operational, fuel flows past the injector into the regulator, and is restricted from flowing any further by the blocked return port. When fuel pressure reaches 250kPa (36PSI) it pushes on the diaphragm, compressing the spring, and uncovers the fuel return port. The diaphragm and spring will constantly move from an open to closed position to keep the fuel pressure constant. An assist to the spring loaded diaphragm comes from vacuum in the throttle body above the throttle blade. As venturi vacuum increases less pressure is required to supply the same amount of fuel into the air flow. The vacuum assists

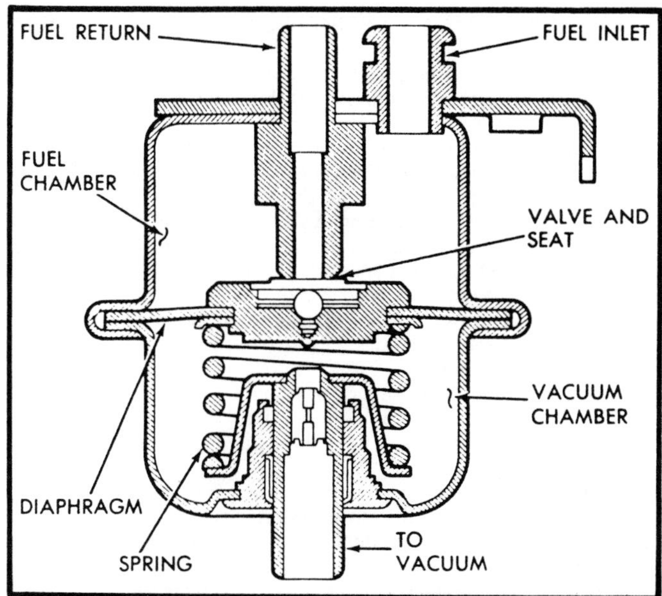

Cross section of typical fuel pressure regulator

in opening the fuel port during high vacuum conditions. This fine tunes the fuel pressure for all operating conditions.

Throttle Position Sensor (TPS)

The throttle position sensor (TPS) is an electric resistor which is activated by the movement of the throttle shaft. It is mounted on the throttle body and senses the angle of the throttle blade opening. The voltage that the sensor produces increases or decreases according to the throttle blade opening. This voltage is transmitted to the logic module where it is used along with data from other sensors to adjust the air-fuel ratio to varying conditions and during acceleration, deceleration, idle, and wide open throttle operations.

Automatic Idle Speed (AIS) Motor

The automatic idle speed motor (AIS) is operated by the logic module. Data from the throttle position sensor, speed sensor, coolant temperature sensor, and various switch operations, (electric backlite, air conditioning, safety/neutral, brake) are used by the logic module to adjust engine idle to an optimum during all idle conditions. The AIS adjusts the air portion of the air/fuel mixture through an air bypass on the back of the throttle body. Basic (no load) idle is determined by the minimum air flow through the throttle body. The AIS opens or closes off the air bypass as an increase or decrease is needed due to engine loads or ambient conditions. The logic module senses an air/fuel change and increases or decreases fuel proportionally to change engine idle. Deceleration die out is also prevented by increasing engine idle when the throttle is closed quickly after a driving (speed) condition.

Fuel Pump

The fuel pump used in this system is a positive displacement, roller vane immersible pump with a permanent magnet electric motor.

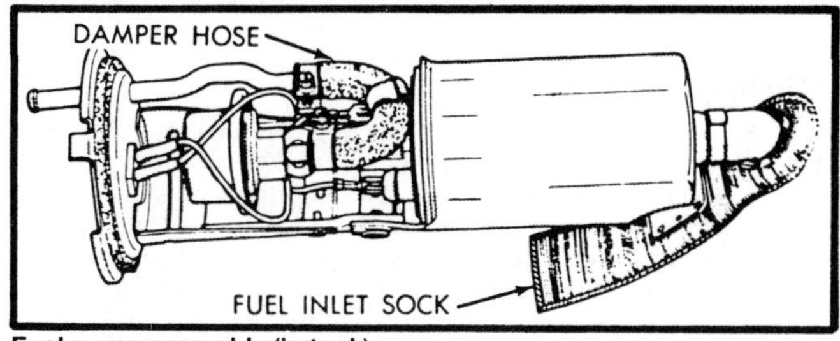

Fuel pump assembly (in-tank)

The fuel is drawn in through a filter sock and pushed through the electric motor to the outlet. The pump contains two check valves. One valve is used to relieve internal fuel pump pressure and regulate maximum pump output. The other check valve, located near the pump outlet, restricts fuel movement in either direction when the pump is not operational. Voltage to operate the pump is supplied through the auto shutdown relay.

On-Car Service

NOTE: Experience has shown that most complaints that may occur with EFI can be traced to poor wiring or hose connections. A visual check will help spot these most common faults and save unnecessary test and diagnosis time.

ON BOARD DIAGNOSTICS

The logic module has been programmed to monitor several different circuits of the fuel injection system. This monitoring is called On Board Diagnosis. If a problem is sensed with a monitored circuit, often enough to indicate an actual problem, its fault code is stored in the logic module for eventual display to the service technician. If the problem is repaired or ceases to exist, the logic module cancels the fault code after 30 ignition key on/off cycles.

Fault Codes

When a fault code appears (either by flashes of the light emitting diode or by watching the diagnostic readout—Tool C-4805 or equivalent, it indicates that the logic module has recognized an abnormal signal in the system. Fault codes indicate the results of a failure but do not always identify the failed component.

CODE 11 means a problem in the distributor circuit. This code appears when the logic module has not seen a distributor signal since the battery was reconnected.

CODE 12 indicates a problem in the stand-by memory circuit. This code appears if direct battery feed to the logic module is interrupted.

CODE 13 means a problem exists in the MAP sensor pneumatic system. This code appears if the MAP sensor vacuum level does not change between start and start/run transfer speed (500–600 rpm).

CODE 14 means a problem exists in the MAP sensor electrical system. This code appears if the MAP sensor signal is either too low (below .02 volts) or too high (above 4.9 volts).

CODE 15 means a problem exists in the vehicle Speed Sensor ciruit. This code appears if engine speed is above 1468 rpm and speed sensor indicates less than 2 mph. This code is valid only if it is sensed while vehicle is moving.

CODE 21 indicates a problem in the O_2 sensor circuit. This code appears if there has been no O_2 signal for more than 5 seconds.

CODE 22 means a problem exists in the coolant temperature sensor circuit. This code appears if the temperature sensor circuit indicates an incorrect temperature or a temperature that changes too fast to be real.

CODE 24 means a problem exists in the throttle position sensor circuit. This code appears if the sensor signal is either below 0.16 vcolts or above 4.7 volts.

CODE 25 means a problem in the automatic idle speed (AIS) control circuit. This code appears if the proper voltage from the AIS system is not present. An open harness or motor will not activate code.

CODE 31 means a problem in the canister purge solenoid circuit. This code appears when the proper voltage at the purge solenoid is not present (open or shorted system).

CODE 32 means a problem in the power loss lamp circuit. This code appears when proper voltage to the power loss lamp circuit is not present (open or shorted system).

CODE 33 means a problem in the air conditioning wide open throttle cut out relay circuit. This code appears if the proper voltage at the air conditioning wide open throttle relay circuit is not present (open or shorted).

CODE 34 means a problem in the EGR solenoid circuit. This code appears if proper voltage at the EGR solenoid circuit is not present (open or shorted system).

CODE 35 indicates a problem in the fan relay circuit. This code appears if the radiator fan is either not operating or operating at wrong time.

CODE 41 means a problem in the charging system. This code appears if battery voltage from the automatic shut down relay is below 11.75 volts.

CODE 42 means a problem in the automatic shut down relay (ASD) circuit. This code appears if during cranking, battery voltage from ASD relay is not present for at least 1/3 second after first distributor pulse or after engine stall, battery voltage is not off within 3 seconds after last distributor pulse.

CODE 43 means a problem in the interface circuit. This code appears if the anti-dwell or injector control signal is not present between the logic module and power module.

CODE 44 means a problem in the logic module. This code appears if an internal failure exists in the logic module.

CODE 51 indicates a problem in the closed loop fuel system. This code appears if during closed loop conditions, the O_2 signal is either low or high for more than 2 minutes.

CODE 52 means a problem in the logic module. This code appears if an internal failure exists in the logic module.

CODE 53 means a problem in the logic module. This code appears if an internal failure exists in the logic module.

CODE 54 means a problem in the logic module. This code appears if an internal failure exists in the logic module.

CODE 55 means "end of message". This code appears as the final code after all other fault codes have been displayed and means "end of message".

CODE 88 means start of message. This code only appears on the diagnostic readout Tool C-4805 or equivalent, and means start of message.

SYSTEMS TEST

Obtaining Fault Codes

1. Connect diagnostic readout box tool C-4805 or equivalent, to the diagnostic connector located in the engine compartment near the passenger side strut tower.
2. Start the engine if possible, cycle the transmission selector and the A/C switch if applicable. Shut off the engine.
3. Turn the ignition switch on, off, on, off, on. Within 5 seconds record all the diagnostic codes shown on the diagnostic readout box tool, observe the power loss lamp on the instrument panel the lamp should light for 2 seconds then go out (bulb check).

Switch Test

After all codes have been shown and has indicated Code 55 end of message, actuate the following component switches. The digital display must change its numbers when the switch is activated and released:

- Brake pedal
- Gear shift selector park, reverse, park.
- A/C switch (if applicable).
- Electric backlite switch (if applicable).

Actuator Test Mode (ATM)

1. Remove coil wire from cap and place 1/4 in. from a ground.

CAUTION

Coil wire must be 1/4 in. or less from ground or power module damage may result.

2. Remove air cleaner hose from throttle body.
3. Press the ATM button on the diagnostic readout box tool and observe the following:
   - 3 sparks from the coil wire to ground.
   - 2 AIS motor movements (1 open, 1 close) you must listen carefully for AIS operation.
   - 1 fuel pulse from the injector into the throttle body.
4. The ATM capability is cancelled 5 minutes after the ignition switch is turned on. To reinstate this capability cycle the ignition ON and OFF three times ending in the ON position.
5. When the ATM button is pressed, fault Code 42 is generated because the ASD relay is bypassed. Do not use this code for diagnostics after ATM operation.
6. The ATM test will check 3 categories of operation:
   - When coil fires three times:
     a. Coil operational
     b. Logic module portion operational
     c. Power module portion operational
     d. Interface between power module and logic module is working.
   - AIS is operational
   - Injector fuel pulse into Throttle Body:
     a. Fuel injector operational
     b. Fuel pump operational
     c. Fuel lines intact
7. The electronic fuel injection system must be evaluated using all the information found in the systems test:
   - Start/no start
   - Fault codes
   - Loss of power lamp on or off (limp in)
   - ATM results:
     a. Spark yes/no
     b. Fuel yes/no
     c. AIS movement yes/no

Once this information is found, it will be easier to determine what circuit to look at for further testing.

Ignition Timing Adjustment

1. Connect a power timing light to the number one cylinder, or a magnetic timing unit to the engine. (Use a 10° degree offset when required).
2. Connect a tachometer to the engine and turn selector to the proper cylinder position.
3. Start engine and run until operating temperature is reached.
4. Disconnect and reconnect the water temperature sensor connector on the thermostat housing. The loss of power lamp on the dash must come on and stay on. Engine rpm should be within emission label specifications.
5. Aim power timing light at timing hole in bell housing or read the magnetic timing unit.
6. Loosen distributor and adjust timing to emission label specifications if necessary.
7. Shut engine off, disconnect and reconnect positive battery quick disconnect. Start vehicle, the loss of power lamp should be off.
8. Shut engine off, then turn ignition on, off, on, off, on. Fault codes should be clear with 88–51–55 shown.
9. Increase engine to 2000 rpm.
10. Read timing it should be approximately 40 degrees.
11. If timing advance does not reach specifications, replace logic module.

Idle Speed Adjustment

1. Before adjusting the idle on an electronic fuel injected vehicle the following items must be checked.
   a. AIS motor has been checked for operation.
   b. Engine has been checked for vacuum or EGR leaks.
   c. Engine timing has been checked and set to specifications.
   d. Coolant temperature sensor has been checked for operation.
2. Connect a tachometer and timing light to engine.

3. Disconnect throttle body 6-way connector. Remove brown with white tracer AIS wire from connector and rejoin connector.
4. Connect one end of a jumper wire to AIS wire and other end to battery positive post for 5 seconds.
5. Connect a jumper to radiator fan so that it will run continuously.
6. Start and run engine for 3 minutes to allow speed to stabilize.
7. Using tool C-4804 or equivalent, turn idle speed adjusting screw to obtain 800 ± 10 rpm (Manual) 725 ± 10 rpm (Automatic) with transaxle in neutral.

NOTE: If idle will not adjust down, check for binding linkage, speed control servo cable adjustments, or throttle shaft binding.

8. Check that timing is 18 ± 2° BTDC (Manual) 12 ± 2° BTDC (Automatic).
9. If timing is not to above specifications turn idle speed adjusting screw until correct idle speed and ignition timing are obtained.
10. Turn off engine, disconnect tachometer and timing light, reinstall AIS wire and remove jumper wire.

Relieving Fuel System Pressure

The E.F.I. fuel system is under a constant pressure of approximately 36 psi (250 kPa). Before servicing the fuel tank, fuel pump, fuel lines, fuel filter, or fuel components of the throttle body the fuel pressure must be released as follows.

1. Loosen gas cap to release any in tank pressure.
2. Remove wiring harness connector from injector.
3. Ground one injector terminal with a jumper.
4. Connect a jumper wire to second terminal and touch battery positive post for no longer than 10 seconds.
5. Remove jumper wires.
6. Continue fuel system service.

Fuel System Pressure Test

CAUTION

Fuel system pressure must be released each time a fuel hose is to be disconnected.

1. Remove fuel intake hose from throttle body and connect fuel system pressure testers C-3292, and C-4749 or equivalent, between fuel filter hose and throttle body.
2. Start engine. If gauge reads 250 kPa ± 14 kPa (36 psi ± 2 psi) pressure is correct and no further testing is required. Reinstall fuel hose using a new original equipment type clamp and torque to 10 inch lbs. (1 Nm).
3. If fuel pressure is below specifications, install tester between fuel filter hose and fuel line.
4. Start engine. If pressure is now correct, replace fuel filter. If no change is observed, gently squeeze return hose. If pressure increases, replace pressure regulator. If no change is observed, problem is either a plugged pump filter sock or defective fuel pump.
5. If pressure is above specifications, remove fuel return hose from throttle body. Connect a substitute hose and place other end of hose in clean container. Start engine. If pressure is now correct, check for restricted fuel return line. If no change is observed, replace fuel regulator.

Component Removal

Mechanical malfunctions are more difficult to diagnose with the EFI system. The logic module has been programmed to compensate for some mechanical malfunctions such as incorrect cam timing, vacuum leaks, etc. If engine performance problems are encountered, and no fault codes are displayed, the problem may be mechanical rather than electronic.

THROTTLE BODY

Removal and Installation

1. Release fuel system pressure.
2. Disconnect negative battery terminal.
3. Disconnect fuel injector wiring connector and throttle body 6-way connector.
4. Remove electrical ground wire from 6-way wiring connector.
5. Remove air cleaner hose.
6. Remove throttle cable and if so equipped, the speed control and transmission kickdown cables.
7. Remove return spring.
8. Remove vacuum hoses.
9. Loosen fuel intake and return hose clamps. Wrap a shop towel around each hose, twist and pull off each hose.
10. Remove throttle body mounting screws and lift throttle body from vehicle.
11. Installation is the reverse of removal. Using a new gasket, with tabs facing forward, install throttle body and torque mounting screws to 17 ft. lbs. (23 Nm).

Disassembly

When servicing the fuel portion of the throttle body it will be necessary to bleed fuel pressure before opening any hoses. Always reassemble throttle body components with new O-rings and seals where applicable. Never use lubricants on O-rings or seals, damage may result. If assembly of components is difficult use water to aid assembly. Use care when removing fuel hoses to prevent damage to hose or hose nipple. Always use new hose clamps of the correct type when reassembling and torque hose clamps to 10 inch lbs. (1 Nm). Do not use aviation-style clamps.

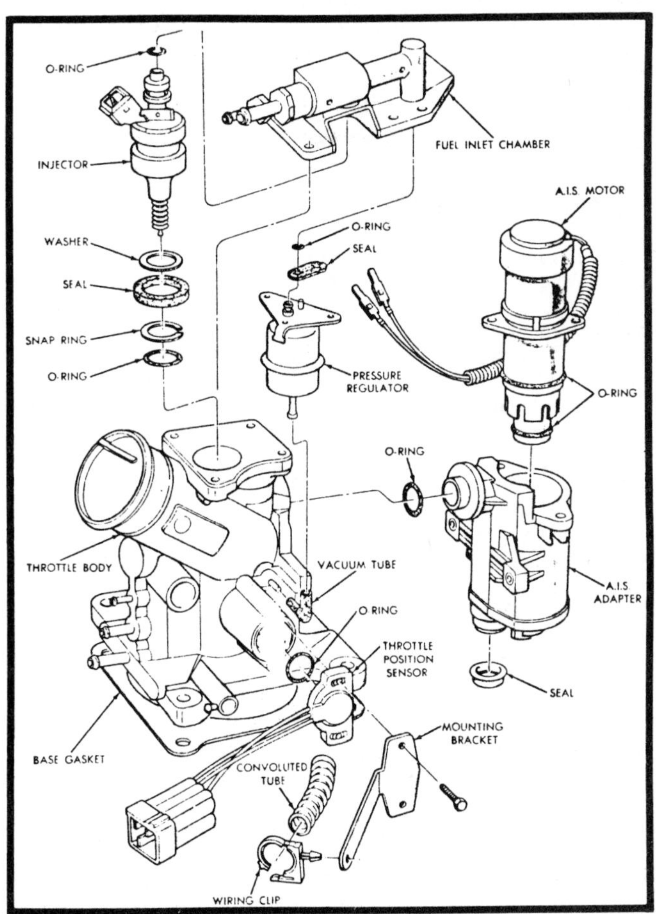

Exploded view of throttle body

NOTE: It is not necessary to remove the throttle body from the intake manifold to perform component disassembly. If fuel system hoses are to be replaced, only hoses marked EFI/EFM may be used.

Injector Removal

1. Perform fuel system pressure release.
2. Disconect negative battery cable.
3. Remove 4 Torx® screws holding fuel inlet chamber to throttle body.
4. Remove vacuum tube from pressure regulator to throttle body.

CAUTION

Place a shop towel around fuel inlet chamber to contain any fuel left in system.

5. Lift fuel inlet chamber and injector off throttle body.
6. Pull injector from fuel inlet chamber.
7. Remove upper and lower O-ring from fuel injector by peeling them off.
8. Remove snap ring that retains seal and washer on injector and remove seal and washer.
9. Installation is the reverse of removal. Place new O-ring washer, and seal on injector and install snap ring.
10. Place assembly into throttle body, install 4 Torx® screws and torque these screws to 35 inch lbs. (4 Nm).

PRESSURE REGULATOR

Removal and Installation

1. Perform fuel system pressure release.
2. Disconnect negative battery cable.
3. Remove 3 Torx® screws mounting pressure regulator to fuel inlet chamber.

CAUTION

Place a shop towel around fuel inlet chamber to contain any fuel left in system.

4. Remove vacuum tube from pressure regulator to throttle body.
5. Pull pressure regulator from throttle body.
6. Carefully peel O-ring off pressure regulator and remove flat seal.
7. Place new seal on pressure regulator and new O-ring.
8. Position pressure regulator on throttle body, press into place, install 3 Torx® screws and torque to 40 inch lbs. (5 Nm).
9. Install vacuum tube from pressure regulator to throttle body.
10. Connect battery, start vehicle, and check for any fuel leaks.

THROTTLE POSITION SENSOR

Removal and Installation

1. Disconnect negative battery cable and 6-way throttle body connector.
2. Remove 2 screws mounting throttle position sensor to throttle body.
3. Unclip wiring clip from convoluted tube and remove mounting bracket.
4. Lift throttle position sensor off throttle shaft and remove O-ring.
5. Pull the 3 wires of the throttle position sensor from the convoluted tubing.
6. Look inside the 6-way throttle body connector and lift a locking tab with a small screwdriver for each T.P.S. wire blade terminal. Removal each blade from connector. (Note wiring position for reassembly.)
7. Insert each wire blade terminal into throttle body connector. Make sure wires are inserted into correct locations.
8. Insert wires from throttle position sensor into convoluted tube.
9. Install throttle position sensor and new O-ring with mounting bracket to throttle body. Torque screws to 20 inch lbs. (2 Nm).
10. Install wiring clips to convoluted tube.
11. Connect 6-way connector and battery cable.

AUTOMATIC IDLE SPEED MOTOR

Removal and Installation

1. Disconnect negative battery cable and 6-way throttle body connector.
2. Remove screws that mount the A.I.S. to its adaptor. (Do not remove the clamp on the A.I.S. or damage will result.)
3. Remove wiring clips and remove the two A.I.S. wires from the 6-way throttle body connector. Lift each locking tab with a small screwdriver and remove each blade terminal. (Note wiring position for reassembly).
4. Lift A.I.S. from its adaptor.
5. Remove the O-rings on the A.I.S. carefully.
6. Install new O-rings on A.I.S.
7. Carefully work A.I.S. into its adaptor.
8. Install 2 mounting screws and torque to 20 inch lbs. (2 Nm).
9. Route A.I.S. wiring to 6-way connector and install each wire blade terminal into the connector. Make sure wires are inserted in correct locations.
10. Connect wiring clip, 6-way connector, and battery cable.

AUTOMATIC IDLE SPEED MOTOR ASSEMBLY

Removal and Installation

1. Disconnect negative battery cable and 6-way throttle body connector.
2. Remove 2 screws on back of throttle body from A.I.S. adaptor.
3. Remove wiring clips and remove the two A.I.S. wires from the 6-way throttle body connector. Lift each locking tab with a small screwdriver and remove each blade terminal. (Note wiring position for reassembly).
4. Carefully pull the assembly from the rear of the throttle body. The O-ring at the top and seal at the bottom may fall off adaptor.
5. Remove O-ring and seal.
6. Place a new O-ring and seal on adaptor.
7. Carefully position assembly onto back of throttle body (make sure seals stay in place) insert screws and torque to 65 inch lbs. (20 Nm).
8. Route A.I.S. wiring to 6-way connector and install each wire blade terminal into the connector. Make sure wires are inserted in correct locations.
9. Connect wiring clips, 6-way connector, and battery cable.

FORD EFI-EEC IV PORT FUEL INJECTION SYSTEM

General Information

The EFI-EEC IV System combines an electronic engine control module with a port fuel injection system to provide a more precise control over the air/fuel ratio, spark timing, deceleration fuel shut-off, EGR, curb and fast idle speed, evaporative emission control, and cold engine enrichment. The EFI-EEC IV system can be divided into three basic subsystems—Fuel, Air and Electronic Engine Control. This section will deal with the Fuel and Air systems only; for all service information on the Electronic Engine

Control system, please see the "Engine Controls" Unit Repair section.

NOTE: For wiring diagrams and diagnosis charts on the EEC IV System, see the "Engine Controls" section.

EFI-EEC IV Fuel System

COMPONENTS AND OPERATION

The fuel subsystem includes a high pressure electric fuel pump, fuel charging manifold, pressure regulator, fuel filter and both solid and flexible fuel lines. The fuel charging manifold includes four electronically controlled fuel injectors, each mounted directly above an intake port in the lower intake manifold. All injectors are energized simultaneously and spray once every crankshaft revolution, delivering a predetermined quantity of fuel into the intake airstream.

The fuel pressure regulator maintains a constant pressure drop across the injector nozzles. The regulator is referenced to intake manifold vacuum and is connected parallel to the fuel injectors and positioned on the far end of the fuel rail. Any excess fuel supplied by the pump passes through the regulator and is returned to the fuel tank via a return line.

NOTE: The pressure regulator reduces fuel pressure to 39–40 psi under normal operating conditions. At idle or high manifold vacuum condition, fuel pressure is reduced to about 30 psi.

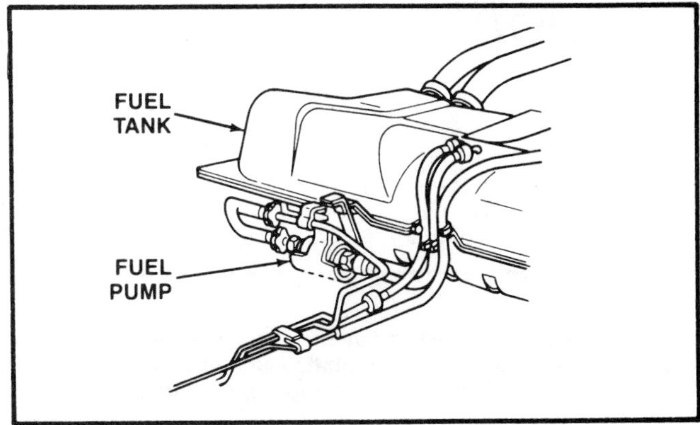

Location of fuel pump on 1.6L engine

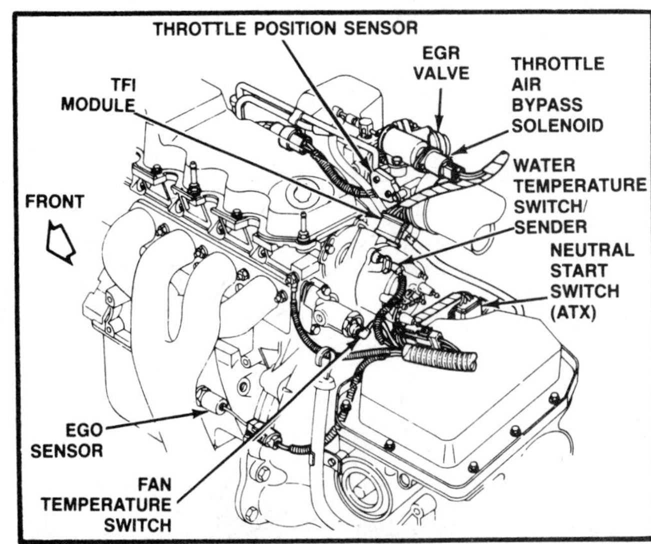

Location of components on 1.6L EFI-EEC IV engine—front view

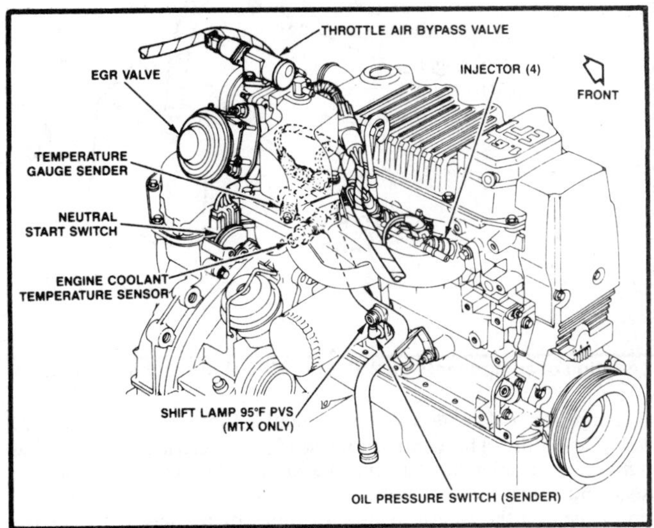

Location of components on 1.6L EFI-EEC IV engine—rear view

The fuel pressure regulator is a diaphragm operated relief valve in which one side of the diaphragm senses fuel pressure and the other side senses manifold vacuum. Normal fuel pressure is established by a spring preload applied to the diaphragm. Control of the fuel system is maintained through the EEC power relay and the EEC IV control unit, although electrical power is routed through the fuel pump relay and an inertia switch. The fuel pump relay is normally located on a bracket somewhere above the Electronic Control Assembly (ECA) and the Inertia Switch is located in the left rear kick panel. The fuel pump is usually mounted on a bracket at the fuel tank.

The inertia switch opens the power circuit to the fuel pump in the event of a collision. Once tripped, the switch must be reset manually by pushing the reset button on the assembly. Check that the inertia switch is reset before diagnosing power supply problems.

FUEL INJECTORS

The fuel injectors used with the EFI-EEC IV system are electromechanical (solenoid) type designed to meter and atomize fuel delivered to the intake ports of the engine. The injectors are mounted in the lower intake manifold and positioned so that their spray nozzles direct the fuel charge in front of the intake valves.

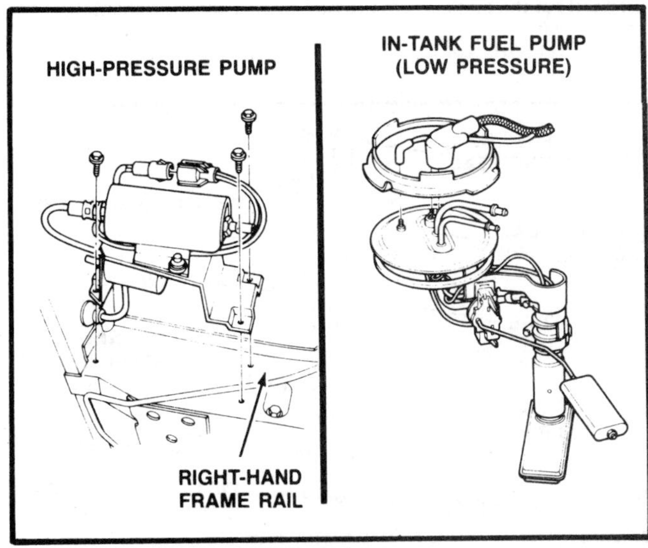

Location of fuel pumps on 2.3L engine

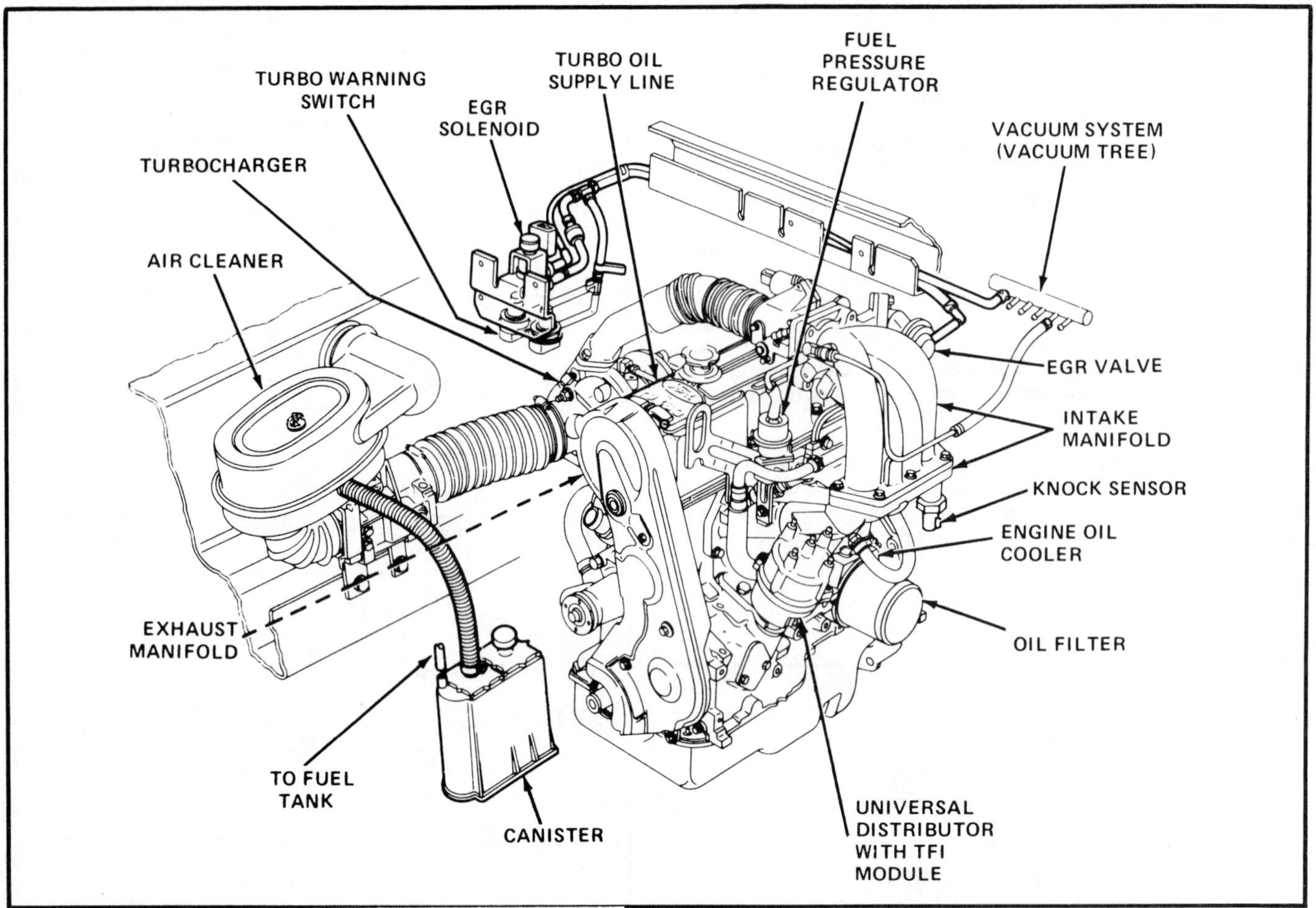

TURBO WARNING SWITCH
TURBO OIL SUPPLY LINE
FUEL PRESSURE REGULATOR
EGR SOLENOID
VACUUM SYSTEM (VACUUM TREE)
TURBOCHARGER
AIR CLEANER
EGR VALVE
INTAKE MANIFOLD
KNOCK SENSOR
ENGINE OIL COOLER
OIL FILTER
EXHAUST MANIFOLD
TO FUEL TANK
CANISTER
UNIVERSAL DISTRIBUTOR WITH TFI MODULE

Location of components on 2.3L EFI—EEC IV turbocharged engine

The injector body consists of a solenoid actuated pintle and needle valve assembly. The control unit sends an electrical impulse that activates the solenoid, causing the pintle to move inward off the seat and allow the fuel to flow. The amount of fuel delivered is controlled by the length of time the injector is energized since the fuel flow orifice is fixed and the fuel pressure drop across the injector tip is constant. Correct atomization is achieved by contouring the pintle at the point where the fuel enters the pintle chamber.

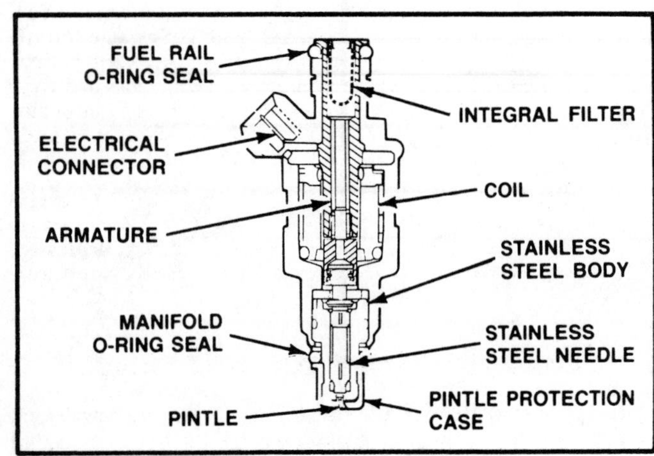

Cross section of typical solenoid—type fuel injector

NOTE: Exercise care when handling fuel injectors during service. Be careful not to lose the pintle cap and replace damaged O-rings to assure a tight seal. Never apply direct battery voltage to test an EFI fuel injector.

The injectors receive high pressure fuel from the fuel manifold (fuel rail) assembly. The complete assembly includes a single, preformed tube with four injector connectors, mounting flange for the pressure regulator, mounting attachments to locate the manifold and provide the fuel injector retainers and Schrader® quick disconnect fitting used to perform fuel pressure tests.

NOTE: The fuel manifold is normally removed with fuel injectors and pressure regulator attached. Fuel injector electrical connectors are plastic and have locking tabs that must be released when disconnecting.

EFI-EEC IV Air System

COMPONENTS AND OPERATION

The EFI-EEC IV air subsystem components include the air cleaner assembly, air flow (vane) meter, throttle air bypass valve and air ducts that connect the air system to the throttle body assembly. The throttle body regulates the air flow to the engine through a single butterfly-type throttle plate controlled by conventional accelerator linkage. The throttle body has an idle adjustment screw (throttle air bypass valve) to set the throttle plate position, a PCV fresh air source upstream of the throttle plate, individual vacuum taps for PCV and control signals and a throttle position sensor that provides a voltage signal for the EEC IV control unit.

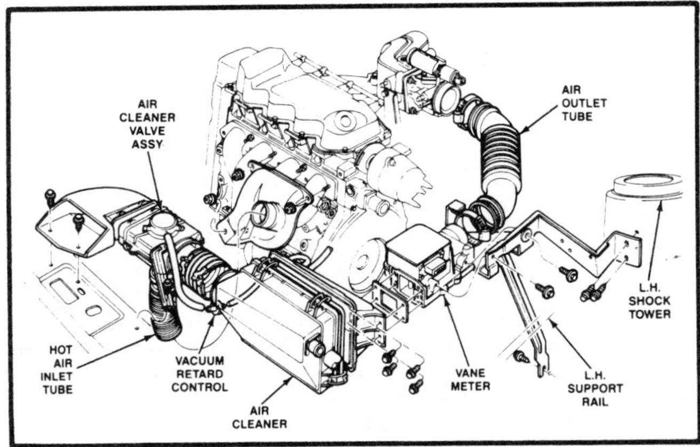

Air flow meter installation on 1.6L engine

NOTE: For information on diagnosis and testing of all EFI-EEC IV system sensors, see the "Engine Controls" section.

The hot air intake system uses a thermostatic flap valve assembly whose components and operation are similar to previous hot air intake systems. Intake air volume and temperature are measured by the vane meter assembly which is mounted between the air cleaner and throttle body. The vane meter consists of two separate devices; the vane airflow sensor (VAF) uses a counterbalanced L-shaped flap valve mounted on a pivot pin and connected to a variable resistor (potentiometer). The control unit measures the amount of deflection of the flap vane by measuring the voltage signal from the potentiometer mounted on top of the meter body; larger air volume moves the vane further and produces a higher voltage signal. The vane air temperature (VAT) sensor is mounted in the middle of the air stream just before the flap valve. Since the mass (weight) of a specific volume of air varies with pressure and temperature, the control unit uses the voltage signal from the air temperature sensor to compensate for these variables and provide a more exact measurement of actual air mass that is necessary to calculate the fuel required to obtain the optimum air/fuel ratio under a wide range of operating conditions. On the EEC IV system, the VAT sensor affects spark timing as a function of air temperature.

NOTE: Make sure all air intake connections are tight before testing. Air leaking into the engine through a loose bellows connection can result in abnormal engine operation and affect the air/fuel mixture ratio.

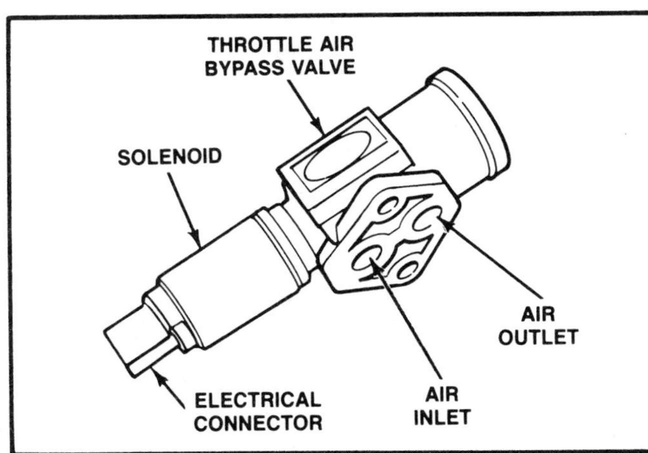

Typical throttle air bypass valve

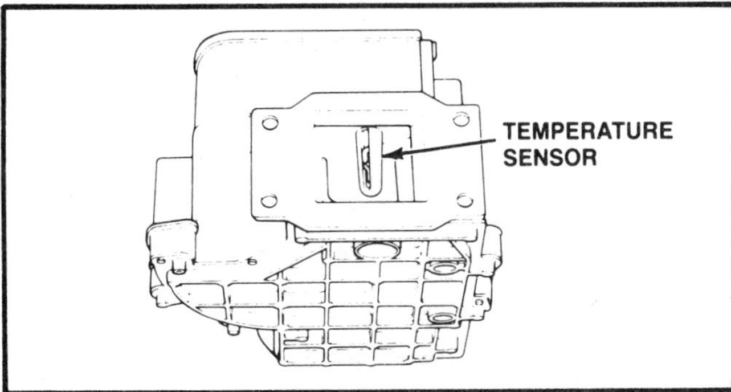

Typical vane air temperature sensor located in air flow meter

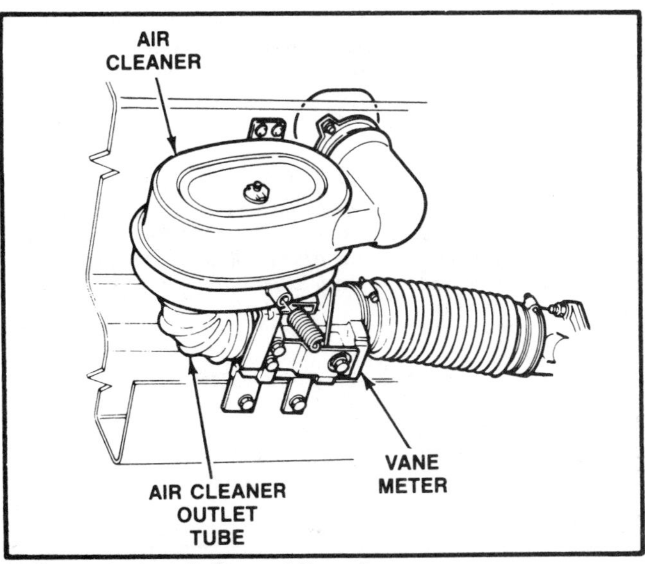

Air flow meter installation on 2.3L engine

THROTTLE AIR BYPASS VALVE

The throttle air bypass valve is an electro-mechanical (solenoid) device whose operation is controlled by the EEC IV control unit. A variable air metering valve controls both cold and warm idle airflow in response to commands from the control unit. The valve operates by bypassing a regulated amount of air around the throttle plate; the higher the voltage signal from the control unit, the more air is bypassed through the valve. In this manner, additional air can be added to the fuel mixture without moving the throttle plate. At curb idle, the valve provides smooth idle for various engine coolant temperatures, compensates for A/C load and compensates for transaxle load and no-load conditions. The valve also provides fast idle for start-up, replacing the fast idle cam, throttle kicker and anti-dieseling solenoid common to previous models.

NOTE: Curb and fast idle speeds are proportional to engine coolant temperature and controlled through the EEC IV control unit. Fast idle kick-down will occur when the throttle is depressed, or after approximately 15–25 seconds after coolant temperature reaches 160°F.

EXHAUST GAS OXYGEN (EGO) SENSOR

The oxygen sensor used on the EFI-EEC IV system is a new design, located between the two downstream tubes in the header in its own mounting boss. The sensor works between zero and one volt output

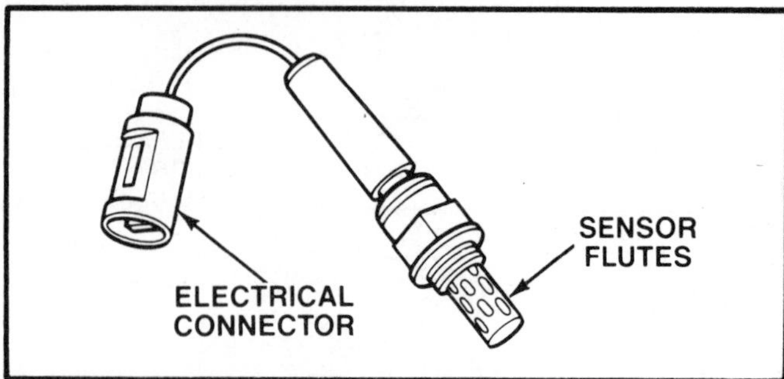

Typical oxygen sensor

depending on the amount of oxygen in the exhaust gas. A voltage reading above 0.6 volt indicates a rich fuel mixture, while a reading below 0.4 volt indicates a lean fuel mixture. Operation of the oxygen sensor is the same as similar systems used on other EEC and MCU models. A new type oxygen sensor can be identified by the metal cap that replaces the rubber protective cover on earlier designs.

CAUTION

Applying direct battery voltage to the oxygen sensor will destroy the sensor's calibration. Even the use of an ohmmeter could cause damage. Before connecting and using a voltmeter, make sure it has a high-input impedence (at least 10 megohms) and is set on the correct voltage range.

On-Car Service

The EFI-EEC IV system has a self-diagnostic capability to aid the technician in locating faults and troubleshooting components. All information on testing procedures is contained in the EEC IV Engine Controls Unity Repair Section. Before removing any fuel lines or fuel system components, first relieve the fuel system pressure by using the same Schrader adapter and fitting that is used to check fuel pressure at the fuel rail.

CAUTION

Exercise care to avoid the chance of fire whenever removing or installing fuel system components.

NOTE: The 2.3L system is similar to the 1.6L, with the addition of a "keep alive" memory in the ECA that retains any intermittent trouble codes stored within the last 20 engine starts. With this system, the memory is not erased when the ignition is switched OFF. In addition, the 2.3L EEC IV system incorporates a knock sensor to detect engine detonation (mounted in the lower intake manifold at the rear of the engine), and a barometric pressure sensor to compensate for altitude variations. The barometric pressure sensor is mounted on the right fender apron.

EEC IV TESTING

As in any service procedure, a routine inspection of the system for loose connections, broken wires or obvious damage is the best way to start. Perform the system Quick Test outlined below before going any further. Check all vacuum connections and secondary ignition wiring before assuming that the problem lies with the EEC IV system. A self-diagnosis capability is built in to the EEC IV system to aid in troubleshooting. The primary tool necessary to read the trouble codes stored in the system is an analog voltmeter or special Self Test Automatic Readout (STAR) tester (Motorcraft No. 007-0M004, or equivalent). While the self-test is not conclusive by itself, when activated it checks the EEC IV system by testing its memory integrity and processing capability. The self-test also verifies that all sensors and actuators are connected and working properly.

When a service code is displayed on an analog voltmeter, each code number is represented by pulses or sweeps of the meter needle. A code 3, for example, will be read as three needle pulses followed by a six-second delay. If a two digit code is stored, there will

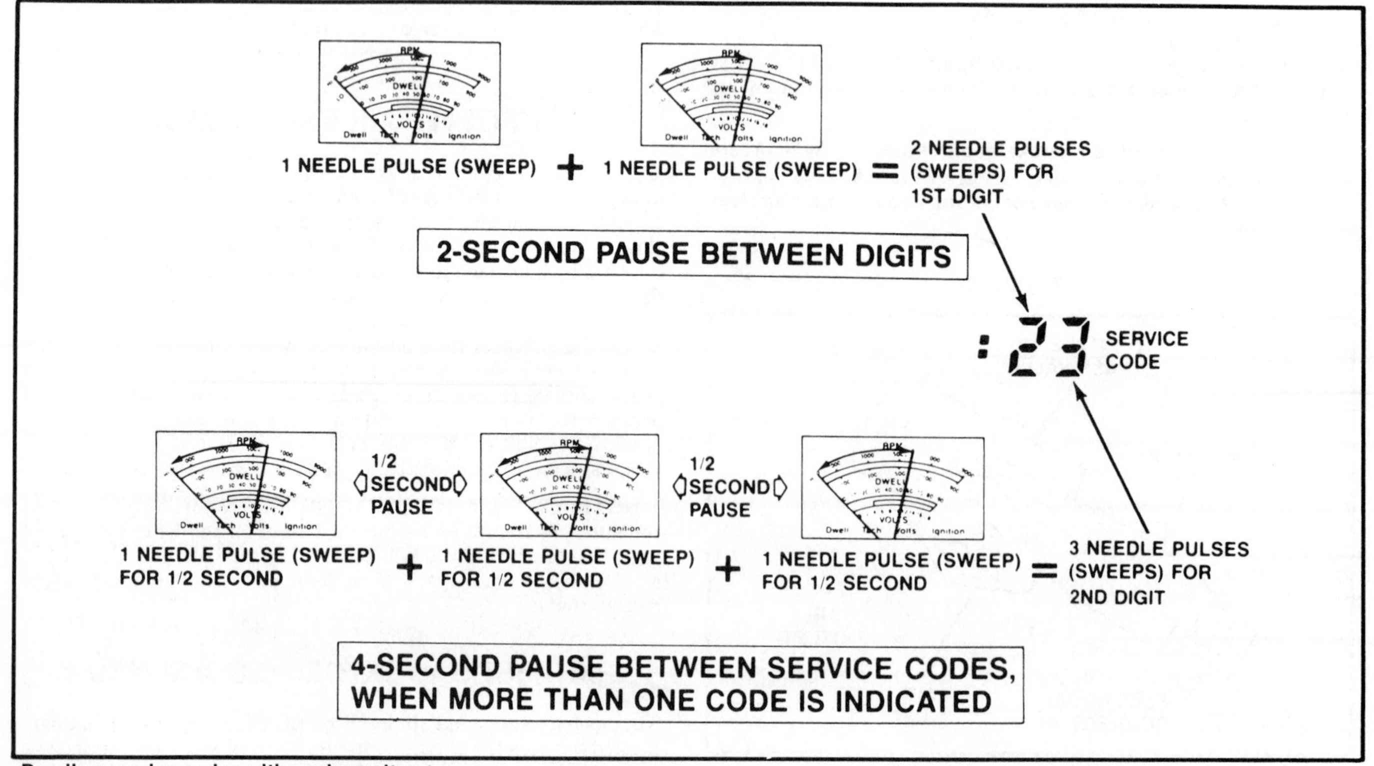

Reading service codes with analog voltmeter

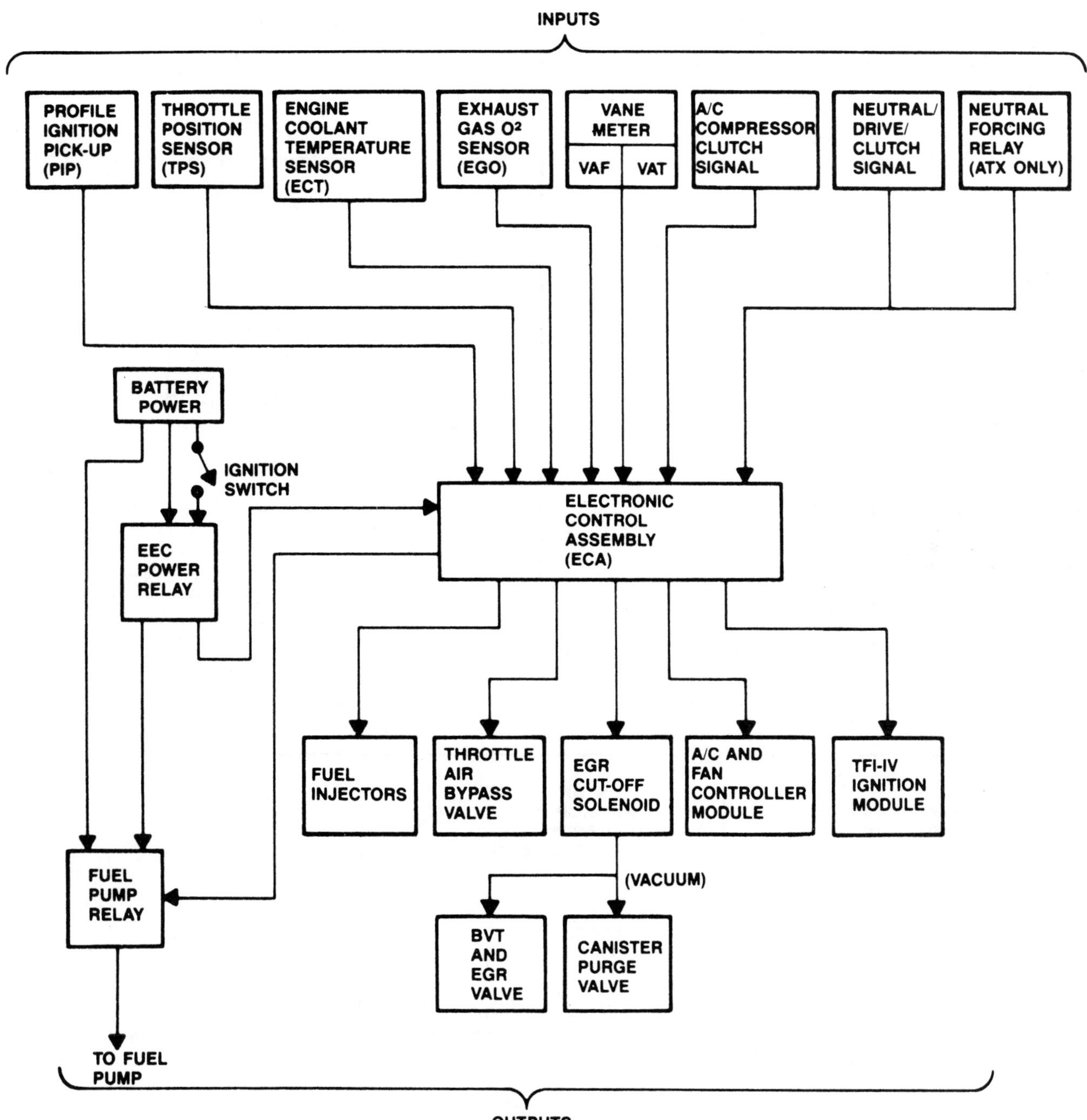

Schematic of EFI-EEC IV system operation—1.6L engine shown

be a two second delay between the pulses for each digit of the number. Code 23, for example, will be displayed as two needle pulses, a two second pause, then three more pulses followed by a four second pause. All testing is complete when the codes have been repeated once. The pulse format in ½ second ON-time for each digit, 2 seconds OFF-time between digits, 4 seconds OFF-time between codes and either 6 seconds (1.6L) or 10 seconds (2.3L) OFF-time before and after the half-second separator pulse.

NOTE: If using the STAR tester, or equivalent, consult the manufacturers instructions included with the unit for correct hookup and trouble code interpretation.

In addition to the service codes, two other types of coded information are outputted during the self-test; engine identification and fast codes. Engine ID codes are one digit numbers equal to one-half the number of engine cylinders (e.g. 4 cylinder is code 2, 8 cylinder is code 4, etc.). Fast codes are simply the service codes transmitted at 100 times the normal rate in a short burst of information. Some meters may detect these codes and register a slight meter deflection just before the trouble codes are flashed. Both the ID and fast codes serve no purpose in the field and this meter deflection should be ignored.

Activating Self-Test Mode on EEC IV

Turn the ignition key OFF. On the 2.3L engine, connect a jumper wire from the self-test input (STI) to pin 2 (signal return) on the self-test connector. On the 1.6L engine, connect a jumper wire from

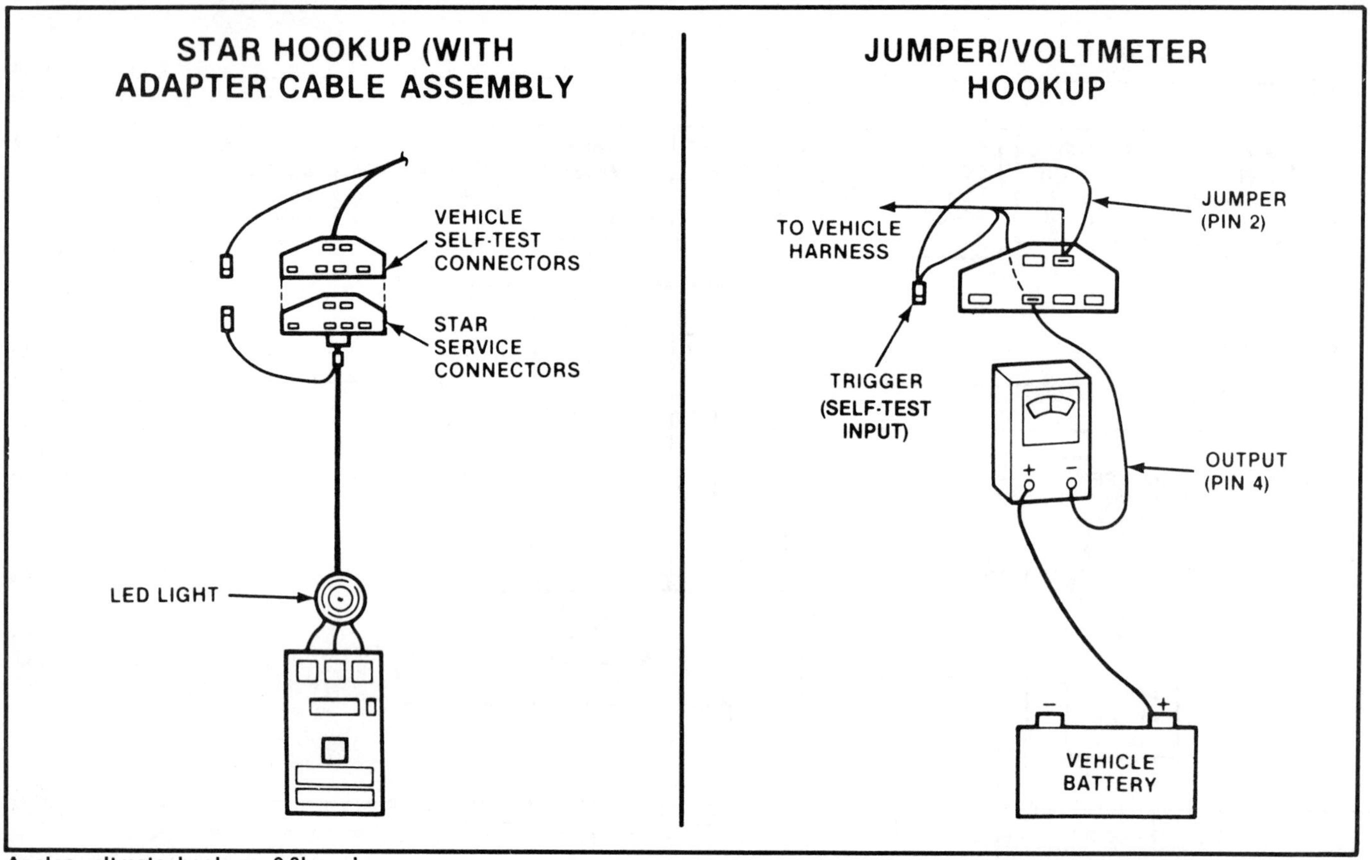

STAR HOOKUP (WITH ADAPTER CABLE ASSEMBLY

VEHICLE SELF-TEST CONNECTORS
STAR SERVICE CONNECTORS
LED LIGHT

JUMPER/VOLTMETER HOOKUP

TO VEHICLE HARNESS
JUMPER (PIN 2)
TRIGGER (SELF-TEST INPUT)
OUTPUT (PIN 4)
VEHICLE BATTERY

Analog voltmeter hookup—2.3L engine

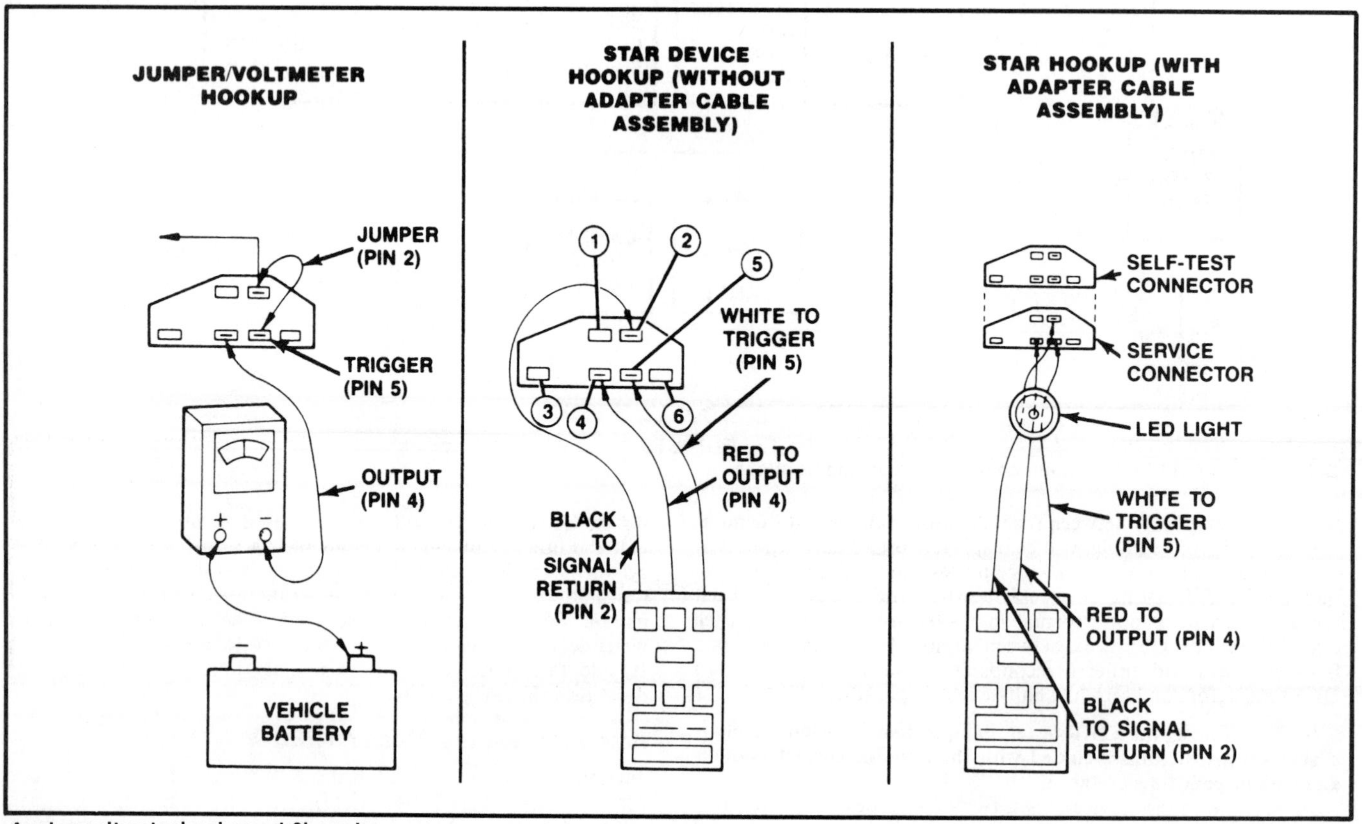

JUMPER/VOLTMETER HOOKUP

JUMPER (PIN 2)
TRIGGER (PIN 5)
OUTPUT (PIN 4)
VEHICLE BATTERY

STAR DEVICE HOOKUP (WITHOUT ADAPTER CABLE ASSEMBLY)

WHITE TO TRIGGER (PIN 5)
RED TO OUTPUT (PIN 4)
BLACK TO SIGNAL RETURN (PIN 2)

STAR HOOKUP (WITH ADAPTER CABLE ASSEMBLY)

SELF-TEST CONNECTOR
SERVICE CONNECTOR
LED LIGHT
WHITE TO TRIGGER (PIN 5)
RED TO OUTPUT (PIN 4)
BLACK TO SIGNAL RETURN (PIN 2)

Analog voltmeter hookup—1.6L engine

pin 5 self-test input to pin 2 (signal return) on the self-test connector. Set the analog voltmeter on a DC voltage range to read from 0–15 volts, then connect the voltmeter from the battery positive (+) terminal to pin 4 self-test output in the self-test connector. Turn the ignition switch ON (engine off) and read the trouble codes on the meter needle as previously described. A code 11 means that the EEC IV system is operating properly and no faults are detected by the computer.

NOTE: This test will only detect "hard" failures that are present when the self-test is activated. For intermittent problems, remove the voltmeter clip from the self-test trigger terminal and wiggle the wiring harness. With the voltmeter still attached to the self-test output, watch for a needle deflection that signals an intermittent condition has occurred. The meter will deflect each time the fault is induced and a trouble code will be stored. Reconnect the self-test trigger terminal to the voltmeter to retrieve the code.

Output Cycling Test

2.3L ENGINE ONLY

This test is performed with the key ON and the engine OFF after the self-test codes have been sent and recorded. Without disconnecting the voltmeter or turning the key OFF, momentarily depress the accelerator pedal to the floor and then release it. All auxiliary EEC IV codes (including the self-test) will be activated and can be read on the voltmeter as before. Another pedal depresssion will turn them off. This cycle may be repeated as necessary, but if activated for more that 10 minutes, the cycle will automatically cancel. This feature forces the processor to activate these outputs for additional diagnosis.

EEC IV System Quick Test

Correct test results for the quick test are dependent on the correct operation of related non-EEC components, such as ignition wires, battery, etc. It may be necessary to correct defects in these areas before the EEC IV system will pass the quick test. Before connecting any test equipment to check the EEC system, make the following checks:

1. Check the air cleaner and intake ducts for leaks or restrictions. Replace the air cleaner if excessive amounts of dust or dirt are found.
2. Check all engine vacuum hoses for proper routing according to the vacuum schematic on the underhood sticker.

Check for proper connections and repair any broken, cracked or pinched hoses or fittings.

3. Check the EEC system wiring harness connectors for tight fit, loose or detached terminals, corrosion, broken or frayed wires, short circuits to metal in the engine compartment or melted insulation exposing bare wire.

NOTE: It may be necessary to disconnect or disassemble the connector to check for terminal damage or corrosion and perform some of the inspections. Note the location of each pin in the connector before disassembly. When doing continuity checks to make sure there are no breaks in the wire, shake or wiggle the harness and connector during testing to check for looseness or intermittent contact.

4. Check the control module, sensors and actuators for obvious physical damage.
5. Turn off all electrical loads when testing and make sure the doors are closed whenever readings are made. DO NOT disconnect any electrical connector with the key ON. Turn the key off to disconnect or reconnect the wiring harness to any sensor or the control unit.
6. Make sure the engine coolant and oil are at the proper level.
7. Check for leaks around the exhaust manifold, oxygen sensor and vacuum hoses connections with the engine idling at normal operating temperature.
8. Only after all the above checks have been performed should the voltmeter be connected to read the trouble codes. If not, the self-diagnosis system may indicate a failed component when all that is wrong is a loose or broken connection.

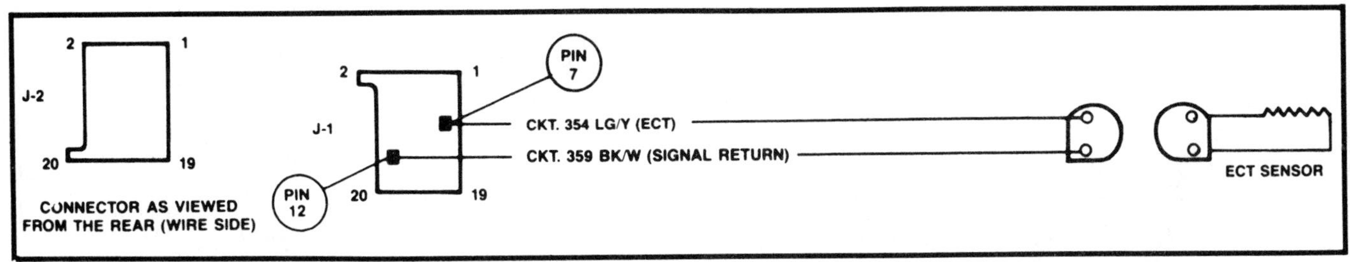

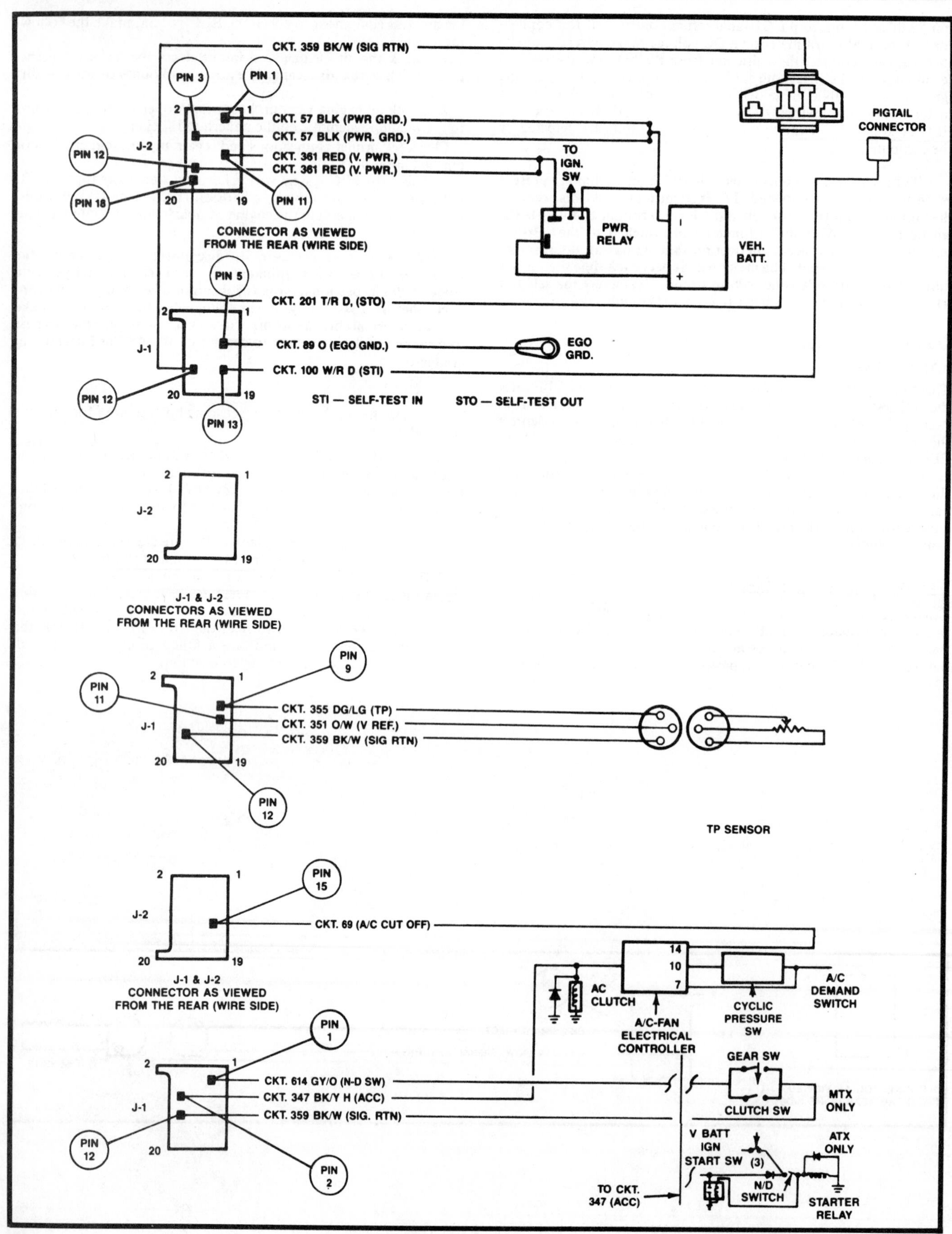
CKT. 359 BK/W (SIG RTN)
PIN 3
PIN 1
J-2
PIN 12
PIN 18
PIN 11
CKT. 57 BLK (PWR GRD.)
CKT. 57 BLK (PWR. GRD.)
CKT. 361 RED (V. PWR.)
CKT. 361 RED (V. PWR.)
TO IGN. SW
PWR RELAY
VEH. BATT.
PIGTAIL CONNECTOR
CONNECTOR AS VIEWED FROM THE REAR (WIRE SIDE)
PIN 5
CKT. 201 T/R D, (STO)
J-1
CKT. 89 O (EGO GND.)
EGO GRD.
CKT. 100 W/R D (STI)
PIN 12
PIN 13
STI — SELF-TEST IN
STO — SELF-TEST OUT
J-1 & J-2 CONNECTORS AS VIEWED FROM THE REAR (WIRE SIDE)
PIN 9
PIN 11
CKT. 355 DG/LG (TP)
CKT. 351 O/W (V REF.)
CKT. 359 BK/W (SIG RTN)
PIN 12
TP SENSOR
PIN 15
CKT. 69 (A/C CUT OFF)
J-1 & J-2 CONNECTOR AS VIEWED FROM THE REAR (WIRE SIDE)
AC CLUTCH
A/C-FAN ELECTRICAL CONTROLLER
CYCLIC PRESSURE SW
A/C DEMAND SWITCH
PIN 1
CKT. 614 GY/O (N-D SW)
CKT. 347 BK/Y H (ACC)
CKT. 359 BK/W (SIG. RTN)
PIN 12
PIN 2
GEAR SW
CLUTCH SW
MTX ONLY
V BATT IGN START SW (3)
ATX ONLY
N/D SWITCH
TO CKT. 347 (ACC)
STARTER RELAY

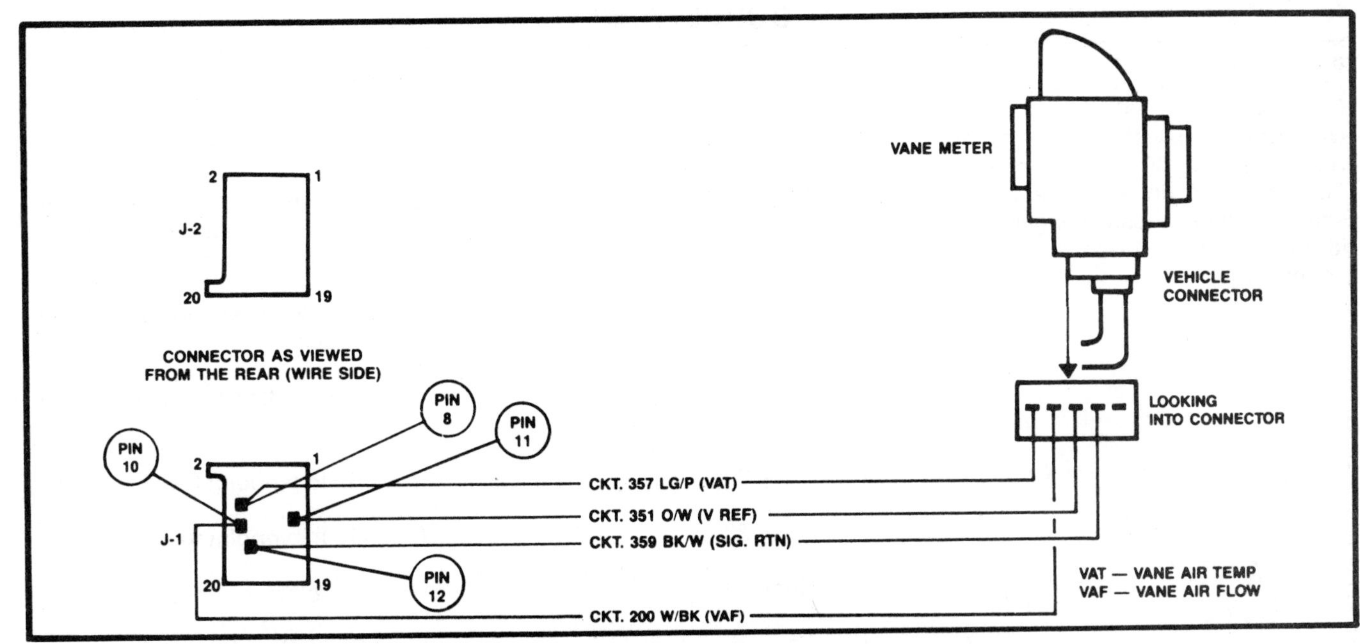

QUICK TEST

Procedure	Result	Action
Voltage Check Turn all accessories off. Check that open circuit battery voltage is greater than 12.4 volts. If engine runs, operate at 2000 rpm for 2 minutes and verify battery voltage of 14.2-14.7 volts with engine running.	a. Battery OK b. Battery not within specs	a. Go to next step b. Recharge battery, check charging system
Key ON/Engine OFF Self-Test Transmission in Neutral or Park, A/C off and key ON. Observe and record trouble codes. When more than one code is present, always start with the first code received.	a. Code 11	a. Go to next step. For no-starts go to pinpoint step A1.
	b. No Codes	b. Go to pinpoint test A
	c. Meter always high	c. Go to pinpoint test step
	d. Code 15	d. Replace processor and retest
	e. Code 21	e. Go to pinpoint test B
	f. Code 23	f. Go to pinpoint test C
	g. Code 24	g. Go to pinpoint test D
	h. Code 26	h. Go to pinpoint test E
	i. Code 67	i. Go to pinpoint test F
Check Timing Verify self-test trigger is installed and check timing while system is in self-test mode	a. Timing is 27-33° BTDC b. Timing out of spec	a. Go to next step b. Reset timing

QUICK TEST

Procedure	Result	Action
Engine Running Test Disconnect self-test trigger, start and run engine at 2000 rpm for 2 minutes. Turn engine off and reconnect self-test trigger. Restart engine and idle. Observe and record trouble codes.	a. No code or meter reads above 10 volts	a. Go to pinpoint test A
	b. Code 11	b. System OK. Go to continuous testing to check intermittent failure
	c. Code 12	c. Check idle speed control (isc) circuit
	d. Code 13	d. Check idle speed control (isc) circuit
	e. Code 21	e. Go to pinpoint test B
	f. Code 23	f. Go to pinpoint test C
	g. Code 24	g. Go to pinpoint test D
	h. Code 26	h. Go to pinpoint test E
	i. Code 41	i. Adjust fuel mixture (CO)
	j. Code 42	j. Adjust fuel mixture (CO)

Continuous Testing
This test is used to detect intermittent failures at the moment the interrupt occurs. It will store the service code until the key is turned off. To begin the test, turn the key off and disconnect the self-trigger. Connect an analog voltmeter negative (−) lead to the self-test out pin and the positive (+) lead to the positive battery post. Turn the key on or start the engine, but do not turn the key off if the engine stalls or the stored service codes will be lost. Monitor the voltmeter reading while wiggling or moving the wiring harness, tap the sensors and connectors and look for any meter deflection that will indicate a short or failure. If any deflection is noted on the voltmeter, initiate the self-test before turning the key off and record the trouble codes stored. The vehicle can also be driven if necessary while in the continuous testing mode to isolate a drive complaint.

PINPOINT TEST

Procedure	Result	Action
No-Start Problem Check for fuel leaks and correct fuel pressure. Check for spark at plug wire and check plugs for fouling or damage.	a. Fuel leaks	a. Correct as necessary
	b. Fouled spark plugs	b. Clean or replace plugs
	c. No spark at plug	c. Go to next step
Ignition Module Signal (IMS) or Spark Output Check. Voltmeter on 20 volt range, probe J2 pin 14 to chassis ground. Crank engine and record reading.	a. Reading is 3-6 volts	a. EEC OK, check cap, rotor and wires
	b. Any other voltage reading	b. Go to next step
Harness Check (IMS Circuit) Key OFF, J2 disconnected. Check circuit 324, J2 pin 14 for shorts to ground.	a. Reading above 2000 ohms	a. Go to next step
	b. Reading below 2000 ohms	b. Repair short in circuit

PINPOINT TEST

Procedure	Result	Action
Distributor Signal Check Voltmeter on 20 volt range, disconnect J1 and J2. Connect voltmeter to J1 pin 14 (circuit 349) and J2 pin 14 (circuit 60). Crank engine and record reading.	a. Reading is 3-6 volts	a. Inspect processor connectors for bent, damaged or corroded terminals. If OK, replace processor
	b. Any other voltage reading	b. Go to next step
Harness Check (Circuit 349) Disconnect vehicle harness from TFI at the distributor and J1 from the processor. Check distributor signal J1 pin 14 (circuit 349) for opens or shorts.	a. No shorts or opens	a. Go to next step
	b. Short or open indicated	b. Repair wire harness as necessary
Harness Check (Circuit 60) Check ignition ground J2 pin 5 (circuit 60) to TFI connector pin 6 for opens.	a. No open indicated	a. Check Thick Film Ignition (TFI) module. Replace if necessary
	b. Open circuit	b. Repair as necessary
Pinpoint Test A Key ON/Engine OFF, processor connected. Voltmeter on 20v range, take voltage measurement at J1 pin 11 to J2 pin 12.	a. Reading 6 volts or more	a. Check for short to battery voltage or processor power. Replace processor if necessary
	b. Reading 4 volts or less	b. Check battery power circuit. Check for shorted throttle position sensor
	c. Reading between 4 and 6 volts	c. Check battery connections
Pinpoint Test B Voltmeter on 20v range, processor connected. Key ON/Engine OFF, take voltage reading from J1 pin 11 to J1 pin 12.	a. Reading 4-6 volts	a. Check coolant temperature sensor and harness for short or open
	b. Any other reading	b. Check battery power cicuit for shorts to processor power or shorted throttle position sensor, vane air meter or harness
Pinpoint Test C Disconnect idle speed control solenoid and verify curb idle per emission decal. Adjust if necessary. Connect voltmeter to J1 pin 9 and J1 pin 12 and rerun self test while monitoring meter.	a. Reading is .5-1.3 volts during test	a. Disconnect J1 and J2 and inspect for damage or corrosion. If none is present, replace processor
	b. Any other reading	b. Check throttle position sensor circuit for shorts or opens. Check for short to battery voltage
Pinpoint Test D Check heated air inlet system. Ambient air temperature must exceed 40°F for this test. Disconnect J1 from the processor and measure resistance from J1 pin 8 to J1 pin 12.	a. Volt-ohmmeter reading between 100-4000 ohms	a. Check vane air temperature sensor for shorts or opens
	b. Reading under 100 or over 4000 ohms	b. Check circuits 357 and 359 for shorts or opens and service as required. If no problem is found, replace vane air flow meter

PINPOINT TEST

Procedure	Result	Action
Pinpoint Test E Connect voltmeter to J1 pin 10 and J1 pin 12. Rerun self-test while monitoring meter.	a. Reading .2-.5 volts with Key ON/Engine OFF or 1.4-2.7 volts with engine running	a. Disconnect J1 and J2 and inspect for damage or corrosion. If no problem is found, replace processor
	b. Any other reading	b. Check circuits 200, 351 and 359 for opens or shorts to ground (sensor and processor must be disconnected). If no problem is found, replace vane air flow meter
Pinpoint Test F Manual Transaxle A/C OFF, shifter in neutral. With key OFF, disconnect J1 and measure resistance from J1 pin 1 and J1 pin 12.	a. Reading less than 5 ohms	a. Take voltage measurement at J1 pin 2 (circuit 347). If reading is above 1 volt, correct short. If reading is below 1 volt, replace processor
	b. Reading above 5 ohms	b. Correct open in circuit or replace gear/clutch switch
Automatic Transaxle Heater controls OFF, shifter in neutral or park. Key ON/Engine OFF. Set voltmeter on 20v range and take measurement from connected J1 pin 1 (circuit 150) to chassis ground.	a. Reading below .5 volts	a. Take voltage measurement at J1 pin 2 (circuit 347). If reading is above 1 volt, correct short. If reading is below 1 volt, replace processor
	b. Reading above .5 volts	b. Correct open in circuit 150 harness or replace N/D switch

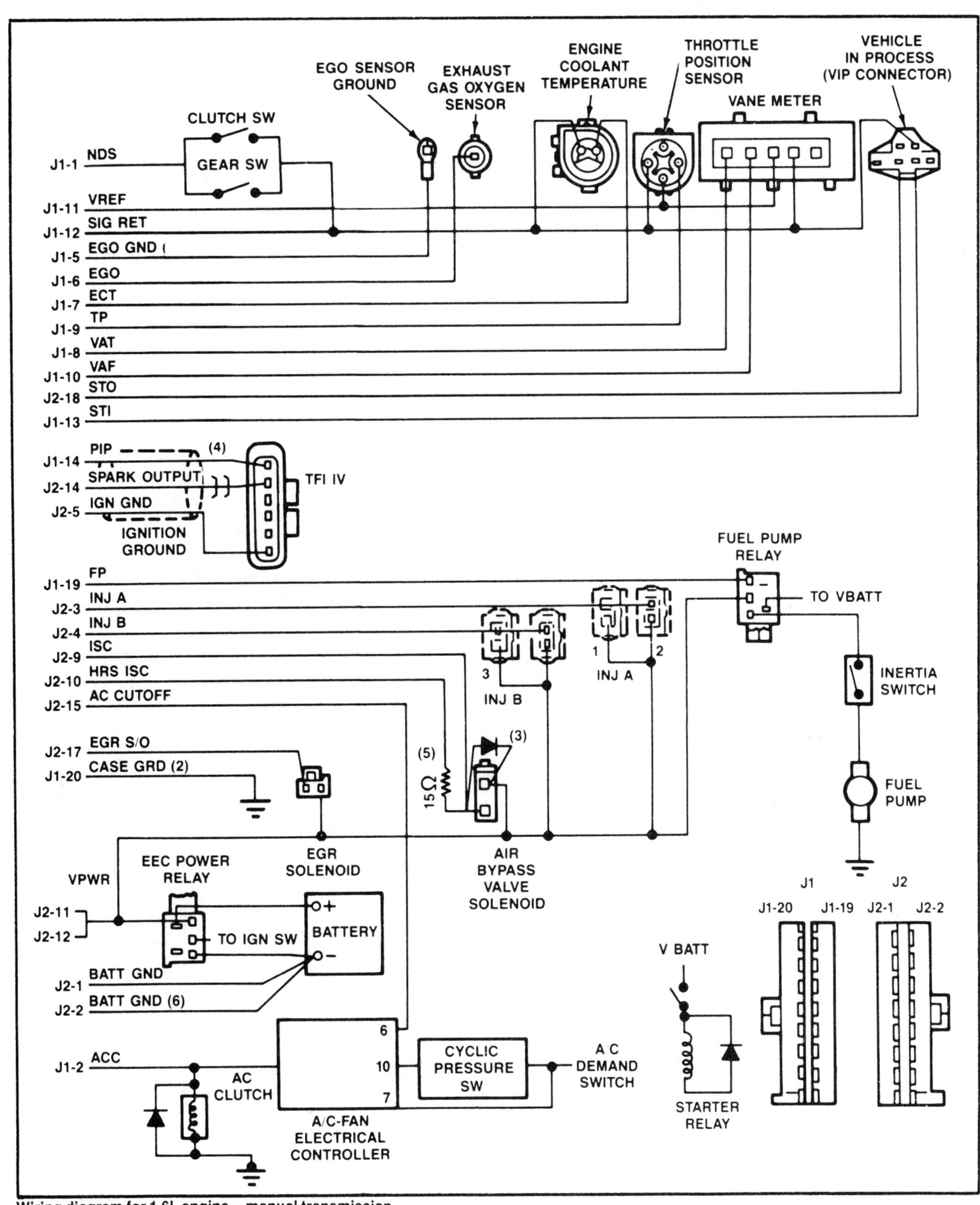

Wiring diagram for 1.6L engine—manual transmission

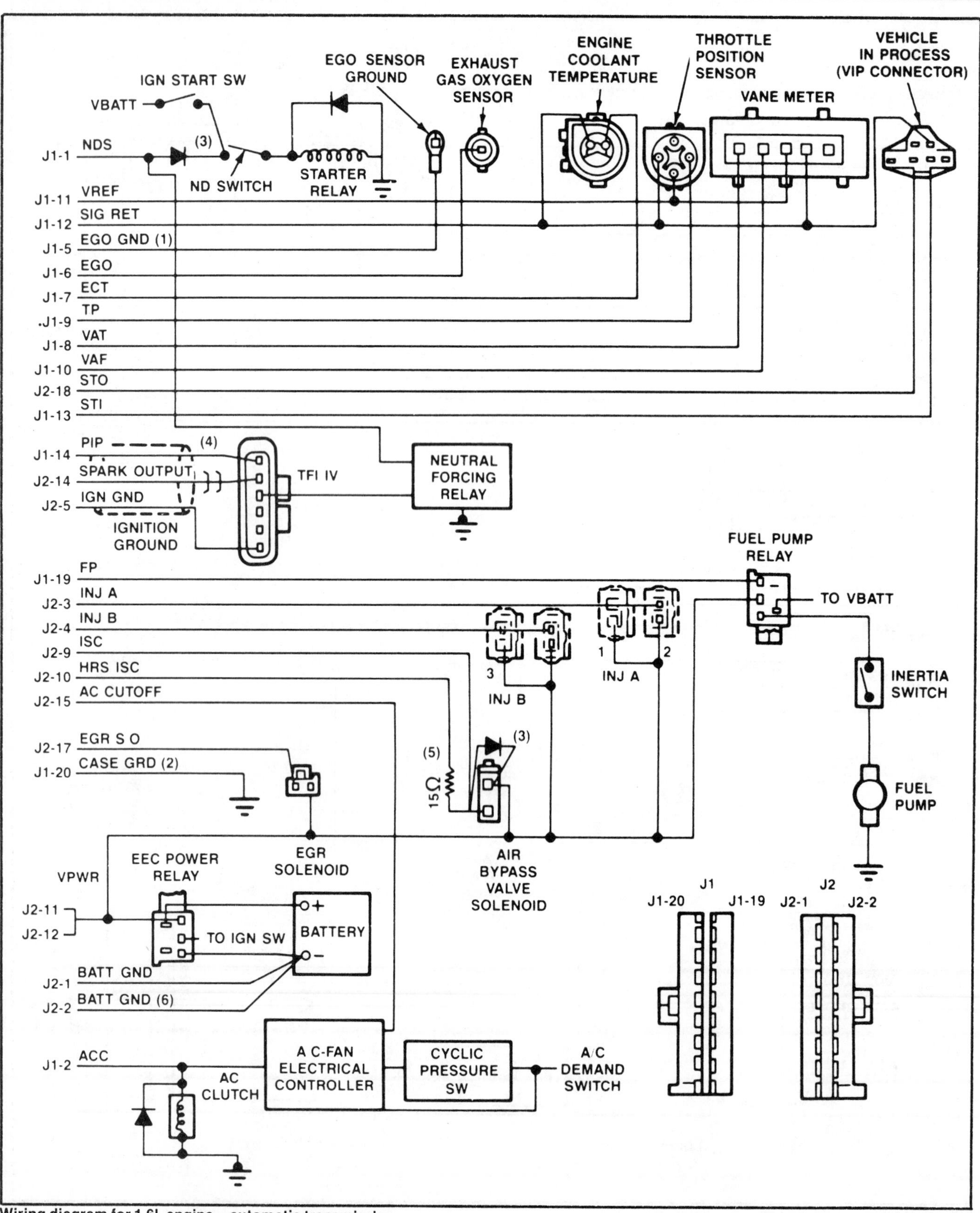

Wiring diagram for 1.6L engine—automatic transmission

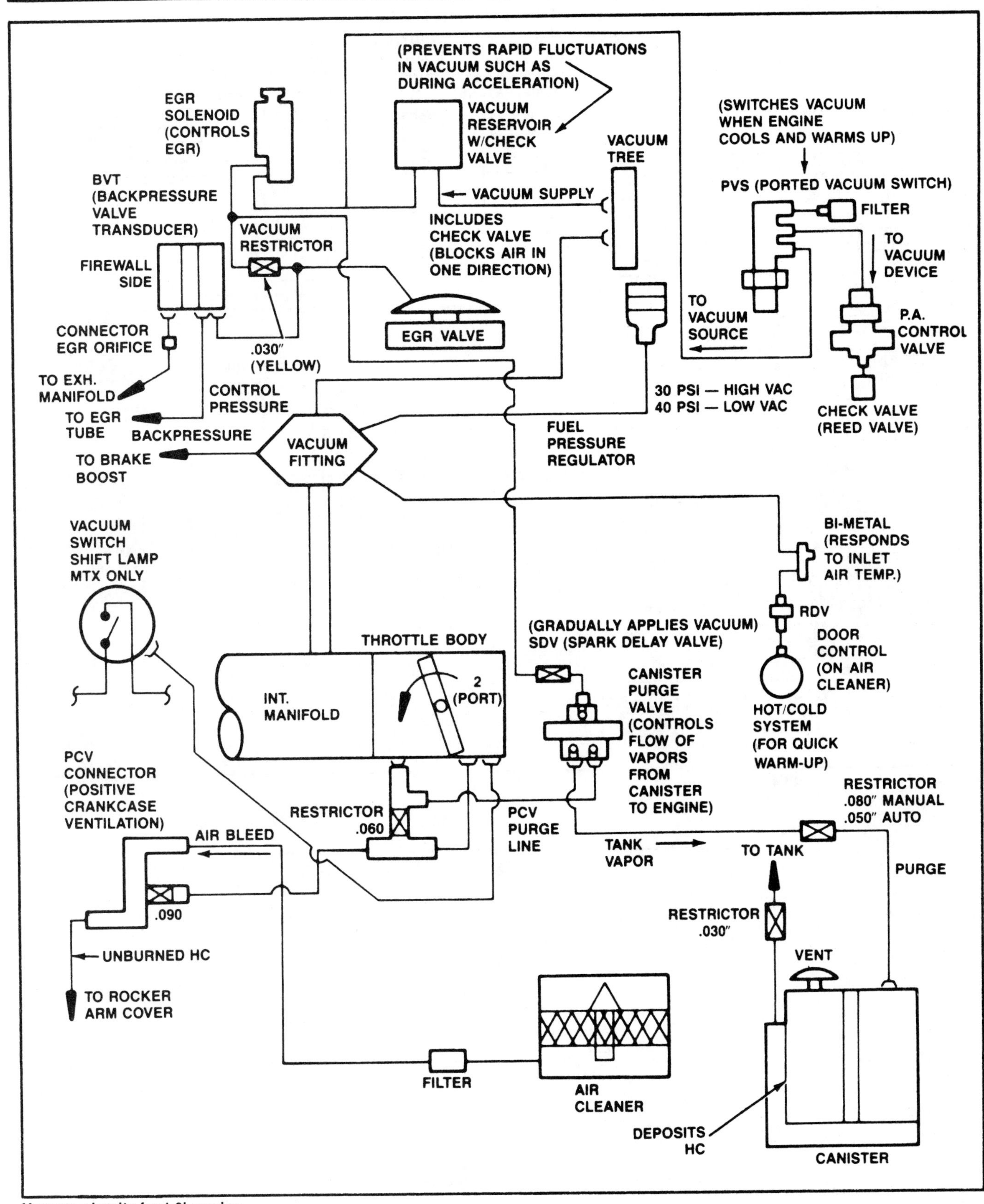

Vacuum circuits for 1.6L engine

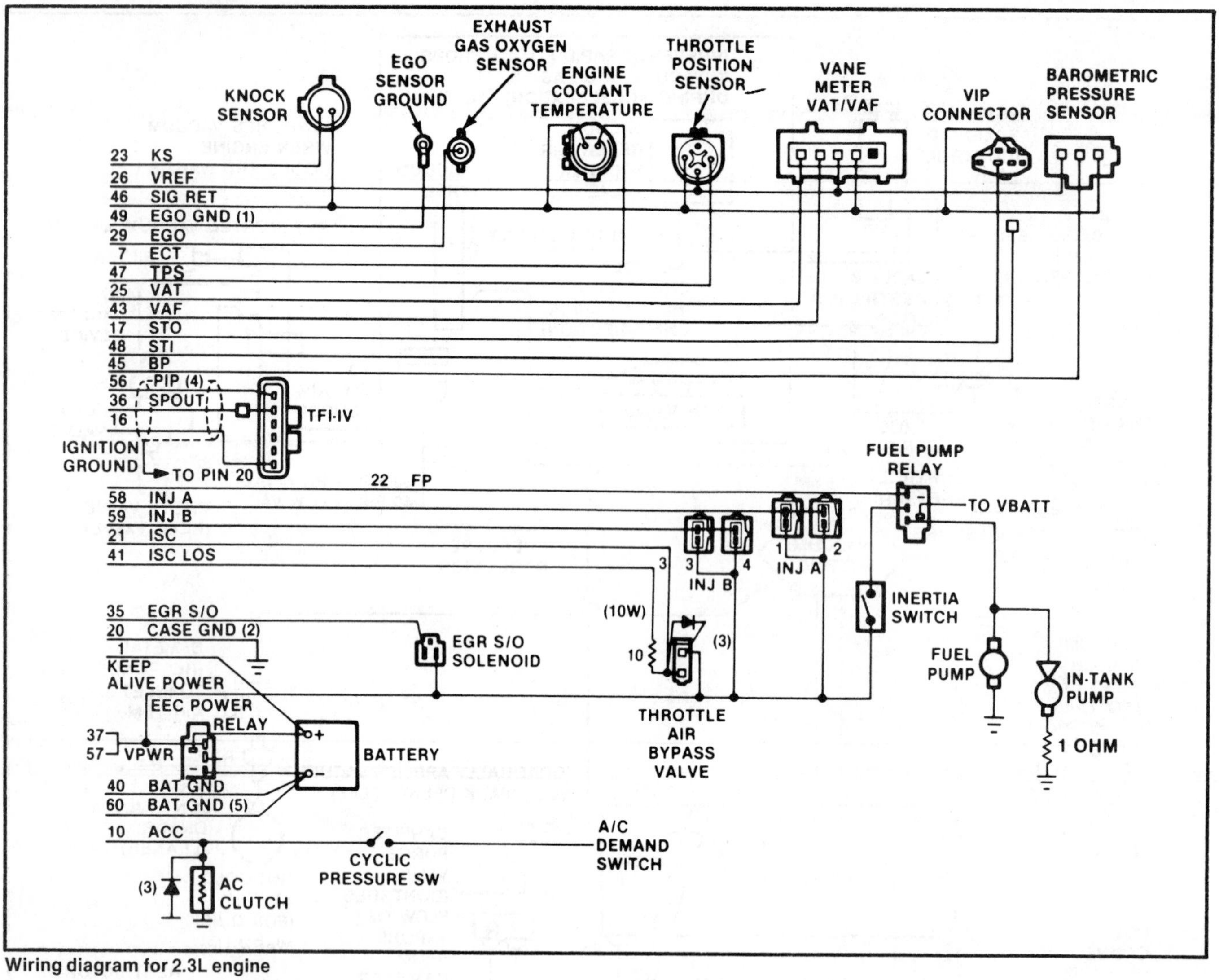

Wiring diagram for 2.3L engine

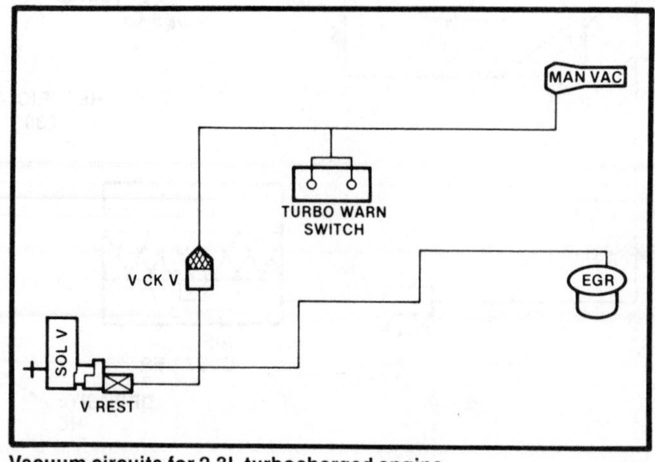

Vacuum circuits for 2.3L turbocharged engine

FORD CENTRAL FUEL INJECTION (CFI) SYSTEM

General Information

The Ford Central Fuel Injection (CFI) System is a single point, pulse time modulated injection system. Fuel is metered into the air intake stream according to engine demands by two solenoid injection valves, mounted in a throttle body on the intake manifold. Fuel is supplied from the fuel tank by a high pressure, electric fuel pump, either by itself or in addition to a low-pressure pump. The fuel is filtered, and sent to the air throttle body where a regulator keeps the fuel delivery pressure at a constant 39 psi (269 kPa). Two injector nozzles are mounted vertically above the throttle plates and connected in parallel with the fuel pressure regulator. Excess fuel supplied by the pump but not needed by the engine, is returned to the fuel tank by a steel fuel return line.

The electronic fuel injection system has several distinct advantages over conventional carburetion. It has improved fuel distribution, capability of fine tuning for altitude and temperature variations, fuel vapor formation and vapor lock largely eliminated due to high pressure in fuel system and reduced evaporative losses due to elimination of the fuel bowl. It also has eliminated engine run-on, since fuel flow immediately stops electrically with engine shut down and elimination of fuel starvation during hard driving maneuvers provided by constant high pressure injection. It provides the precise air-fuel control required for efficient three-way catalyst operation.

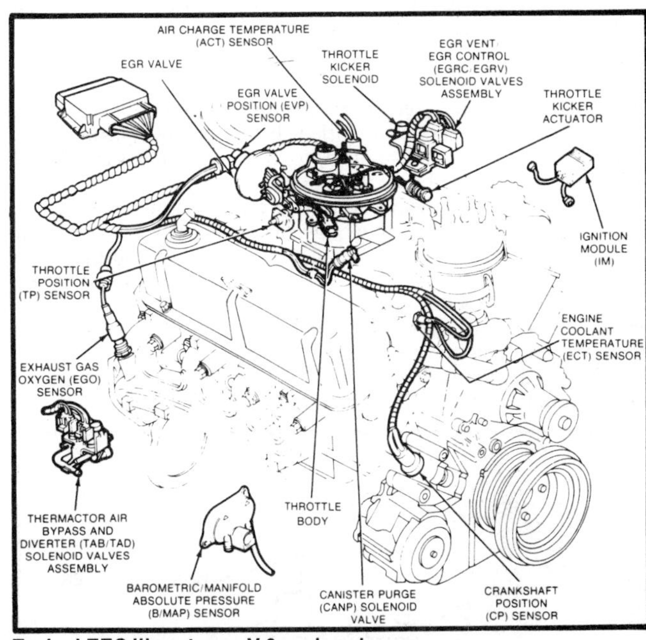

Typical EEC III system—V-8 engine shown

FUEL DELIVERY SYSTEM

Fuel Charging Assembly

The fuel charging assembly controls air-fuel ratio. It consists of a typical carburetor throttle body. It has two bores without venturis. The throttle shaft and valves control engine air flow based on driver demand. The throttle body attaches to the intake manifold mounting pad.

A throttle position sensor is attached to the throttle shaft. It includes a potentiometer (or rheostat) that electrically senses throttle opening. A throttle kicker solenoid fastens opposite the throttle position sensor. During air conditioning operation, the solenoid extends to slightly increase engine idle speed.

Cold engine speed is controlled by an automatic kick-down vacuum motor. There is also an all-electric, bi-metal coil spring which controls cold idle speed. The bi-metal electric coil operates like a conventional carburetor choke coil, but the electronic fuel injection system uses no choke. Fuel enrichment for cold starts is controlled by the computer and injectors.

Fuel Pressure Regulator

The fuel pressure regulator controls critical injector fuel pressure. The regulator receives high pressure fuel from the electric fuel pump. It then adjusts the fuel to the desired pressure for uniform fuel injection. The regulator sets fuel pressure at 39 psi.

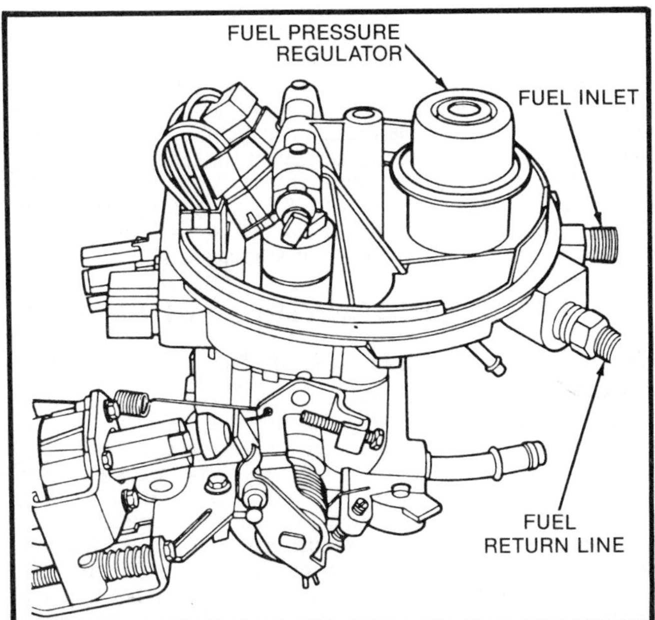

Central Fuel Injection (CFI) fuel charging assembly (left side view)

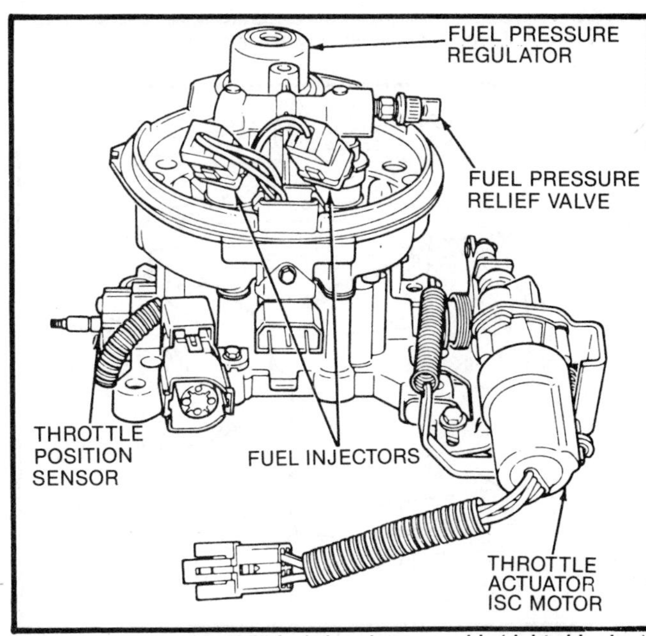

Central Fuel Injection (CFI) fuel charging assembly (right side view)

Fuel Rail

The fuel rail evenly distributes fuel to each injector. Its main purpose is to equalize the fuel flow. One end of the fuel rail contains a relief valve for testing fuel pressure during operation.

Fuel Injectors

The two identical fuel injectors are electro-mechanical devices. The electrical solenoid operates a pintle valve which always travels the same distance from closed to open to closed. Injection is controlled by varying the length of time the pintle valve is open.

The delivery end of the injector is a precisely ground nozzle. The manufacturing and handling of this component is very important for proper operation. When closed, the valve must seal tightly to shut off fuel flow completely. It must seat itself repeatedly with the same precision. Any dirt particles from contaminated fuel can prevent the valve from seating. The shape of the pintle valve and nozzle also determines the fuel spray pattern during injection. Since the injectors are atomizing fuel into droplets, this fuel mist pattern is important to fast vaporization and good combustion.

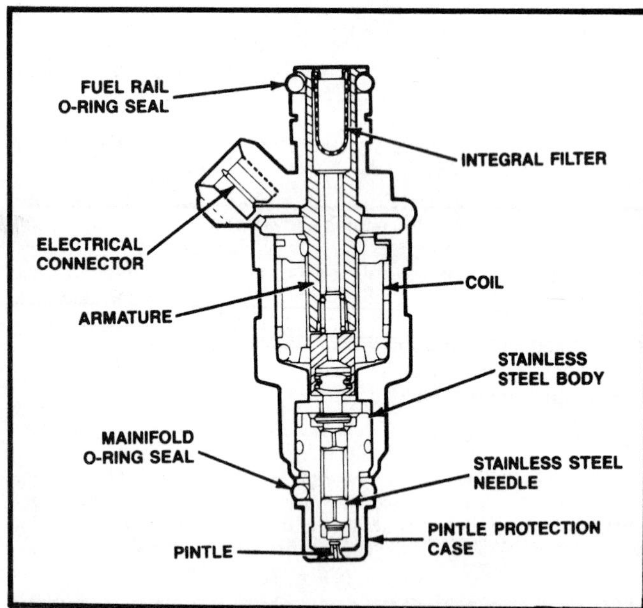

Cross section of typical solenoid-type fuel injector

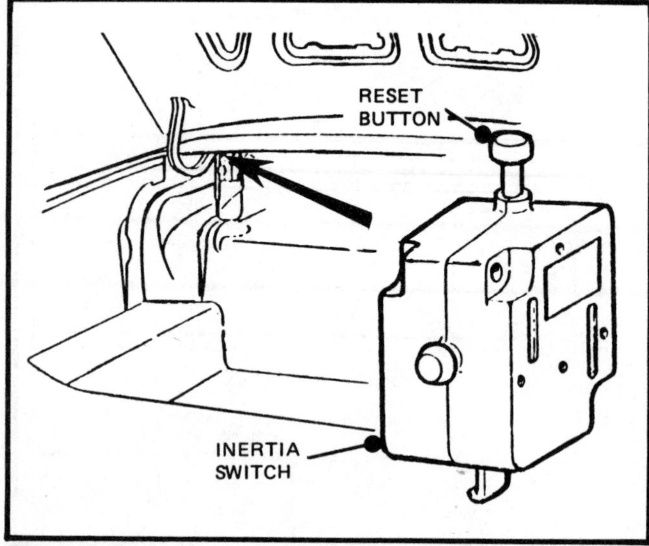

Electric fuel pump inertia switch

The computer, based on voltage inputs from the crank position sensor, operates the injector solenoids four (two per injector) times per engine revolution. When the injector pintle valve unseats, fuel is sprayed in a fine mist into the intake manifold. The computer varies fuel enrichment based on voltage inputs from the exhaust gas oxygen sensor, barometric pressure sensor, manifold absolute pressure sensor, etc., by calculating how long to hold the injectors open. The longer the injectors remain open, the richer the mixture.

Fuel Pump

The fuel delivery system includes an in-line or in-tank high pressure fuel pump, an in-tank low pressure fuel pump (on some 1983–84 models when in-line high pressure pump is used), a primary fuel filter, a secondary fuel filter, fuel supply and return lines, fuel injectors and a fuel pressure regulator. It is a recirculating system that delivers fuel to a pressure regulating valve in the throttle body and returns excess fuel from the throttle body regulator back to the fuel tank. The electrical system has two control relays, one controlled by a vacuum switch and the other controlled by an electronic engine control module. These provide for power to the fuel pump under various operating conditions.

CAUTION

Fuel supply lines on vehicles equipped with fuel injected engine will remain pressurized for long periods of time after engine shutdown. The pressure must be relieved before servicing the fuel system.

An inertia switch is used as a safety device in the fuel system. The inertia switch is located in the trunk, near the left rear wheel well. It is designed to open the fuel pump power circuit in the event of a collision. The switch is reset by pushing each of 2 buttons on the switch simultaneously (some models use switches with only 1 reset button). The inertia switch should not be reset until the fuel system has been inspected for damage or leaks.

With the ignition switch off, the vacuum switch controlled relay is closed and the EEC controlled relay is open. When the ignition switch is first turned to ignition "ON" position, the vacuum switch controlled relay remains closed and the EEC controlled relay also closes. This provides power to the fuel pump to pre-pressurize the fuel system. If the ignition switch is not turned to the "CRANK" position, the EEC module will open its relay after approximately two seconds and shut off power to the pump. When the ignition switch is turned to the "CRANK" position, both the vacuum switch controlled relay and the EEC controlled relay are closed. This provides full battery power to the pump. When the engine starts, manifold vacuum increases and causes the vacuum switch to close and the vacuum controlled relay to open. This provides reduced normal operating voltage to the fuel pump through the resistor which by-passes the vacuum controlled relay. Under heavy engine load conditions, manifold vacuum will reduce, causing the vacuum switch to open. This causes the vacuum controlled relay to close, thus providing the return of full battery power to the pump. The EEC module senses engine speed and shuts off the pump by opening the EEC controlled relay when the engine stops.

ELECTRONIC CONTROL SYSTEM

Electronic Control Assembly (ECA)

The Electronic Control Assembly (ECA) is a solid-state microcomputer consisting of a processor assembly and a calibration assembly. It is located under the instrument panel or passenger's seat and is usually covered by a kick panel. 1981–82 models use an EEC III engine control system, while 1983 and later models use the EEC IV. Although the two systems are similar in appearance and operation, the ECA units are not interchangeable. A multipin connector links the ECA with all system components. The processor assembly is housed in an aluminum case. It contains circuits designed to continuously sample input signals from the engine sensors. It then calculates and sends out proper control signals to adjust air/fuel ratio, spark timing and emission system operation. The processor also provides a continuous reference voltage to the B/MAP, EVP and TPS sensors. EEC III reference voltage is 8–10 volts, while EEC IV systems use a 5 volt reference signal. The calibration assembly is contained in a black plastic housing which plugs into the top of the processor assembly. It contains the memory and programming information used by the processor to determine optimum operating conditions. Different calibration information is used in different vehicle applications, such as California or Federal models. For this reason, careful identification of the engine, year, model and type of electronic control system is essential to insure correct component replacement.

ENGINE SENSORS

Air Charge Temperature Sensor (ACT)

The ACT is threaded into the intake manifold air runner. It is located behind the distributor on V6 engines and directly below the accelerator linkage on V8 engines. The ACT monitors air/fuel charge temperature and sends an appropriate signal to the ECA. This information is used to correct fuel enrichment for variations in intake air density due to temperature changes.

Barometric & Manifold Absolute Pressure Sensors (B/MAP)

The B/MAP sensor on V8 engines is located on the right fender panel in the engine compartment. The MAP sensor used on V6 engines is separate from the barometric sensor and is located on the left fender panel in the engine compartment. The barometric sensor signals the ECA of changes in atmospheric pressure and density to regulate calculated air flow into the engine. The MAP sensor monitors and signals the ECA of changes in intake manifold pressure which result from engine load, speed and atmospheric pressure changes.

Crankshaft Position (CP) Sensor

The CP sensor is mounted on the right front of 1982–83 and some California 1984 5.0L V8 engines. Its purpose is to provide the ECA with an accurate ignition timing reference (when the piston reaches 10° BTDC) and injector operation information (twice each crankshaft revolution). The crankshaft vibration damper is fitted with a 4 lobe "pulse ring". As the crankshaft rotates, the pulse ring lobes interrupt the magnetic field at the tip of the CP sensor.

EGR Valve Position Sensor (EVP)

This sensor, mounted on EGR valve, signals the computer of EGR opening so that it may subtract EGR flow from total air flow into the manifold. In this way, EGR flow is excluded from air flow information used to determine mixture requirements.

Engine Coolant Temperature Sensor (ECT)

The ECT is threaded into the intake manifold water jacket directly above the water pump by-pass hose. The ECT monitors coolant temperature and signals the ECA, which then uses these signals for mixture enrichment (during cool operation), ignition timing and EGR operation. The resistance value of the ECT increases with temperature, causing a voltage signal drop as the engine warms up.

Exhaust Gas Oxygen Sensor (EGO)

The EGO is mounted in the right side exhaust manifold on V8 engines, in the left and right side exhaust manifolds on V6 models. The EGO monitors oxygen content of exhaust gases and sends a constantly changing voltage signal to the ECA. The ECA analyzes this signal and adjusts the air/fuel mixture to obtain the optimum (stoichiometric) ratio.

Knock Sensor (KS)

This sensor is used on 1984 Mustangs and Capris equipped with the 3.8L V6 engine, only. It is attached to the intake manifold in front of the ACT sensor. The KS detects engine vibrations caused by pre-ignition or detonation and provides information to the ECA, which then retards the timing to eliminate detonation.

Thick Film Integrated Module sensor (TFI)

The TFI module sensor plugs into the distributor just below the distributor cap on 3.8L V6 engines and replaces the CP sensor for some 1984 5.0L V8 engines. Its function is to provide the ECA with ignition timing information, similar to what the CP sensor provides.

Throttle Position Sensor (TPS)

The TPS is mounted on the right side of the throttle body, directly connected to the throttle shaft. The TPS senses throttle movement and position and transmits an appropriate electrical signal to the ECA. These signals are used by the ECA to adjust the air/fuel mixture, spark timing and EGR operation according to engine load at idle, part throttle, or full throttle. The TPS is non-adjustable.

On-Car Service

NOTE: For all diagnosis and test procedures on the EEC III and EEC IV electronic control systems, see the "Engine Controls" section.

FUEL PRESSURE TESTS

The diagnostic pressure valve (Schrader type) is located at the top of the Fuel charging main body. This valve provides a covenient point for service personnel to monitor fuel pressure, bleed down the system pressure prior to maintenance, and to bleed out air which may become trapped in the system during filter replacement. A pressure gauge with an adapter is required to perform pressure tests.

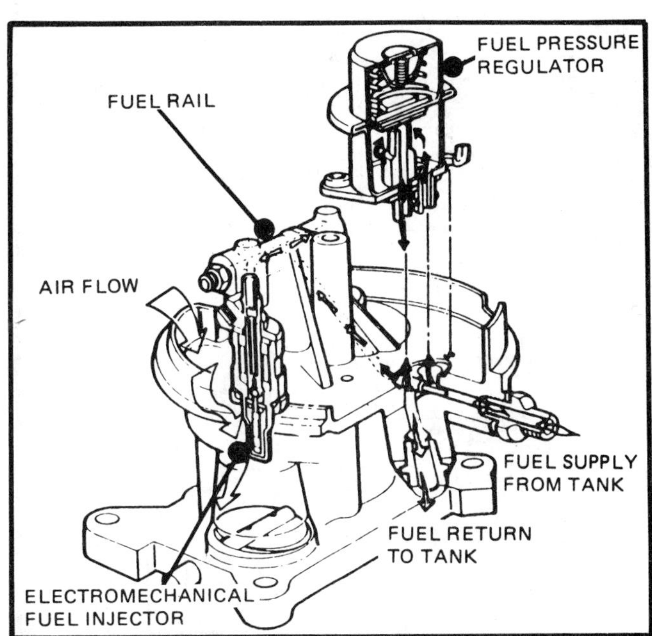

Cross section of CFI fuel charging assembly

CAUTION

Under no circumstances should compressed air be forced into the fuel system using the diagnostic valve. Depressing the pin in the diagnostic valve will relieve system pressure by expelling fuel into the throttle body.

System Pressure Test

Testing fuel pressure requires the use of a special pressure gauge (T80L-9974-A or equivalent) that attaches to the diagnostic pressure tap on the fuel charging assembly. Depressurize the fuel system before disconnecting any lines.

1. Disconnect fuel return line at throttle body (in-tank high pressure pump) or at fuel rail (in-line high pressure and in-tank low pressure pumps) and connect the hose to a one-quart calibrated container. Connect pressure gauge.

2. Disconnect the electrical connector to the fuel pump. The connector is located ahead of fuel tank (in-tank high pressure pump) or just forward of pump outlet (in-line high pressure pump). Connect auxiliary wiring harness to connector of fuel pump. Energize the pump for 10 seconds by applying 12 volts to the auxiliary harness connector, allowing the fuel to drain into the calibrated container. Note the fuel volume and pressure gauge reading.

3. Correct fuel pressure should be 35–45 psi (241–310 kPa). Fuel volume should be 10 ozs. in 10 seconds (minimum) and fuel pressure should maintain minimum 30 psi (206 kPa) immediately after pump cut-off.

If pressure condition is met, but fuel flow is not, check for blocked filter(s) and fuel supply lines. After correcting problem, repeat test procedure. If fuel flow is still inadequate, replace high pressure pump. If flow specification is met but pressure is not, check for worn or damaged pressure regulator valve on throttle body. If both pressure and fuel flow specifications are met, but pressure drops excessively after de-energization, check for leaking injector valve(s) and/or pressure regulator valve. If injector valves and pressure regulator valve are okay, replace high pressure pump. If no pressure or flow is seen in fuel system, check for blocked filters and fuel lines. If no trouble is found, replace in-line fuel pump, in-tank fuel pump and the fuel filter inside the tank.

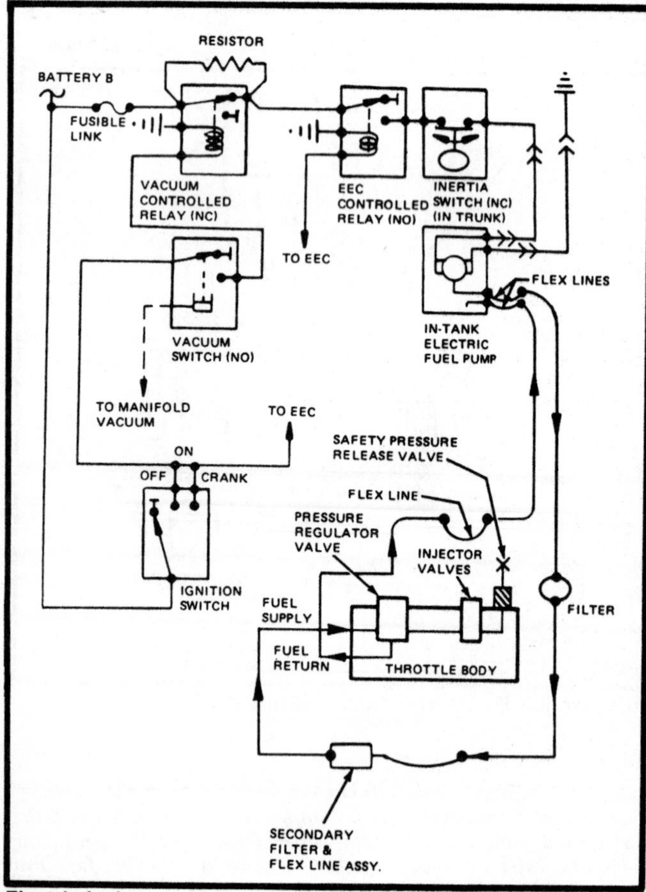

Electric fuel pump wiring and fuel lines routing schematic

Fuel Injector Pressure Test

1. Connect pressure gauge T80L-9974-A, or equivalent, to fuel pressure test fitting. Disconnect coil connector from coil. Disconnect electrical lead from one injector and pressurize fuel system. Disable fuel pump by disconnecting inertia switch or fuel pump relay and observe pressure gauge reading.

2. Crank engine for 2 seconds. Turn ignition OFF and wait 5 seconds, then observe pressure drop. If pressure drop is 2-16 psi (14–110 kPa), the injector is operating properly. Reconnect injector, activate fuel pump, then repeat the procedure for other injector.

3. If pressure drop is less than 2 psi (14 kPa) or more than 16 psi (110 kPa), switch electrical connectors on injectors and repeat test. If pressure drop is still incorrect, replace disconnected injector with one of the same color code, then reconnect both injectors properly and repeat test.

4. Disconnect and plug vacuum hose to EGR valve. It may be necessary to disconnect the idle speed control (3.8L V6) or throttle kicker solenoid (5.0L V8) and use the throttle body stop screw to set engine speed. Start and run the engine at 1800 RPM (2000 rpm on 1984 and later models). Disconnect left injector electrical connector. Note rpm after engine stabilizes (around 1200 rpm). Reconnect injector and allow engine to return to high idle.

5. Perform same procedure for right injector. Note difference between rpm readings of left and right injectors. If difference is 100 rpm or less, check the oxygen sensor. If difference is more than 100 rpm, replace both injectors.

CFI COMPONENT TESTS

NOTE: Note: Complete CFI system diagnosis requires the use of a special tester. See the "Engine Controls" section for details.

Before beginning any component testing, always check the following:

- Check ignition and fuel systems to ensure there is fuel and spark.
- Remove air cleaner assembly and inspect all vacuum and pressure hoses for proper connection to fittings. Check for damaged or pinched hoses.
- Inspect all sub-system wiring harnesses for proper connections to the EGR solenoid valves, injectors, sensors, etc.
- Check for loose or detached connectors and broken or detached wires. Check that all terminals are seated firmly and are not corroded. Look for partially broken or frayed wires or any shorting between wires.
- Inspect sensors for physical damage. Inspect vehicle electrical system. Check battery for full charge and cable connections for tightness.
- Inspect the relay connector and make sure the ECA power relay is securely attached and making a good ground connection.

Fuel Pump Circuit Test

HIGH PRESSURE IN-TANK PUMP

Disconnect electrical connector just forward of the fuel tank. Connect voltmeter to body wiring harness connector. Turn key ON while watching voltmeter. Voltage should rise to battery voltage, then return to zero after about 1 second. Momentarily turn key to START position. Voltage should rise to about 8 volts while cranking. If voltage is not as specified, check electrical system.

HIGH PRESSURE IN-LINE & LOW PRESSURE IN-TANK PUMPS

Disconnect electrical connector at fuel pumps. Connect voltmeter to body wiring harness connector. Turn key ON while watching voltmeter. Voltage should rise to battery voltage, then return to zero after about 1 second. If voltage is not as specified, check inertia switch and electrical system. Connect ohmmeter to in-line pump wiring harness connector. If no continuity is present, check continuity directly at in-line pump terminals. If no continuity at in-line

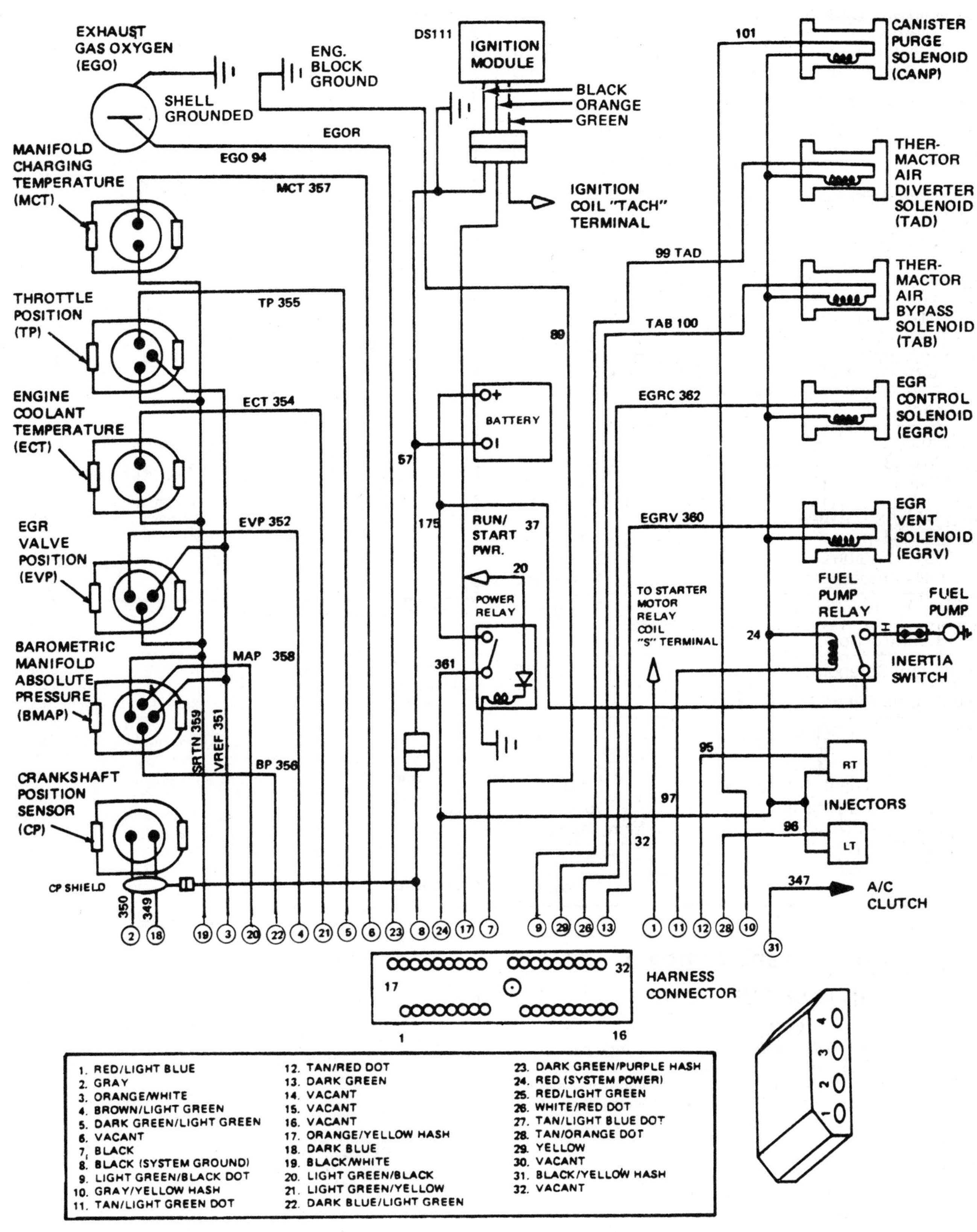
EXHAUST GAS OXYGEN (EGO)
SHELL GROUNDED
ENG. BLOCK GROUND
DS111
IGNITION MODULE
BLACK
ORANGE
GREEN
IGNITION COIL "TACH" TERMINAL
EGOR
EGO 94
MANIFOLD CHARGING TEMPERATURE (MCT)
MCT 357
THROTTLE POSITION (TP)
TP 355
ENGINE COOLANT TEMPERATURE (ECT)
ECT 354
EGR VALVE POSITION (EVP)
EVP 352
BAROMETRIC MANIFOLD ABSOLUTE PRESSURE (BMAP)
MAP 358
SRTN 359
VREF 351
BP 356
CRANKSHAFT POSITION SENSOR (CP)
CP SHIELD
350
349
BATTERY
57
175
RUN/START PWR.
37
20
POWER RELAY
361
89
101
99 TAD
TAB 100
EGRC 362
EGRV 360
TO STARTER MOTOR RELAY COIL "S" TERMINAL
32
95
97
96
24
347
CANISTER PURGE SOLENOID (CANP)
THERMACTOR AIR DIVERTER SOLENOID (TAD)
THERMACTOR AIR BYPASS SOLENOID (TAB)
EGR CONTROL SOLENOID (EGRC)
EGR VENT SOLENOID (EGRV)
FUEL PUMP RELAY
FUEL PUMP
INERTIA SWITCH
RT
LT
INJECTORS
A/C CLUTCH
2 18 19 3 20 22 4 21 5 6 23 8 24 17 7 9 29 26 13 1 11 12 28 10 31
HARNESS CONNECTOR
17
32
1
16
1 2 3 4
1. RED/LIGHT BLUE
2. GRAY
3. ORANGE/WHITE
4. BROWN/LIGHT GREEN
5. DARK GREEN/LIGHT GREEN
6. VACANT
7. BLACK
8. BLACK (SYSTEM GROUND)
9. LIGHT GREEN/BLACK DOT
10. GRAY/YELLOW HASH
11. TAN/LIGHT GREEN DOT
12. TAN/RED DOT
13. DARK GREEN
14. VACANT
15. VACANT
16. VACANT
17. ORANGE/YELLOW HASH
18. DARK BLUE
19. BLACK/WHITE
20. LIGHT GREEN/BLACK
21. LIGHT GREEN/YELLOW
22. DARK BLUE/LIGHT GREEN
23. DARK GREEN/PURPLE HASH
24. RED (SYSTEM POWER)
25. RED/LIGHT GREEN
26. WHITE/RED DOT
27. TAN/LIGHT BLUE DOT
28. TAN/ORANGE DOT
29. YELLOW
30. VACANT
31. BLACK/YELLOW HASH
32. VACANT

Ford CFI wiring diagram

pump terminals, replace in-line pump. If continuity is present, service or replace wiring harness.

Connect ohmmeter across body wiring harness connector. If continuity is present (about 5 ohms), low pressure pump circuit is OK. If no continuity is present, remove fuel tank and check for continuity at in-tank pump flange terminals on top of tank. If continuity is absent at in-tank pump flange terminals, replace assembly. If continuity is present at in-tank pump but not in harness connector, service or replace wiring harness to in-tank pump.

Solenoid and Sensor Resistance Tests

All CFI components must be disconnected from the circuit before testing resistance with a suitable ohmmeter. Repace any component whose measured resistance does not agree with the specifications chart. Shorting the wiring harness across a solenoid valve can burn out the circuitry in the ECA that controls the solenoid valve actuator. Exercise caution when testing solenoid valves to avoid accidental damage to ECA.

CFI RESISTANCE SPECIFICATIONS

Component	Resistance (Ohms)
Air Charge Temp (ACT)	
1981-83	1700-60,000
1984	1100-58,000
Coolant (ECT) Sensor	
1981-83	1100-8000
1984—Engine Off	1300-7700
1984—Engine On	1500-4550
Crank Position Sensor	100-640
FGB Control Solenoid	30-70
EGR Vent Solenoid	30-70
Fuel Pump Relay	50-100
Throttle Kicker Solenoid	50-100
Throttle Position Sensor	
1981-83 Closed Throttle	3000-5000
1984 Closed Throttle	550-1100
Wide Open Throttle	More than 2100
TAB Solenoid	50-100
TAD Solenoid	50-100

Component Removal

IN-TANK FUEL PUMP

CAUTION

The fuel supply lines will remain pressurized for long periods of time after the engine is turned off. A valve is provided on the throttle body for this purpose. Remove the air cleaner and relieve system pressure by depressing the pin in the relief valve cautiously. Fuel will be expelled into the throttle body.

Removal

1. It is necessary to remove the fuel tank.
2. Depressurize the fuel system
3. Remove fuel from the fuel tank by pumping out through the filter tube.
4. Disconnect the supply and return line fittings and the vent line.
5. Disconnect and remove the fuel filler tube.

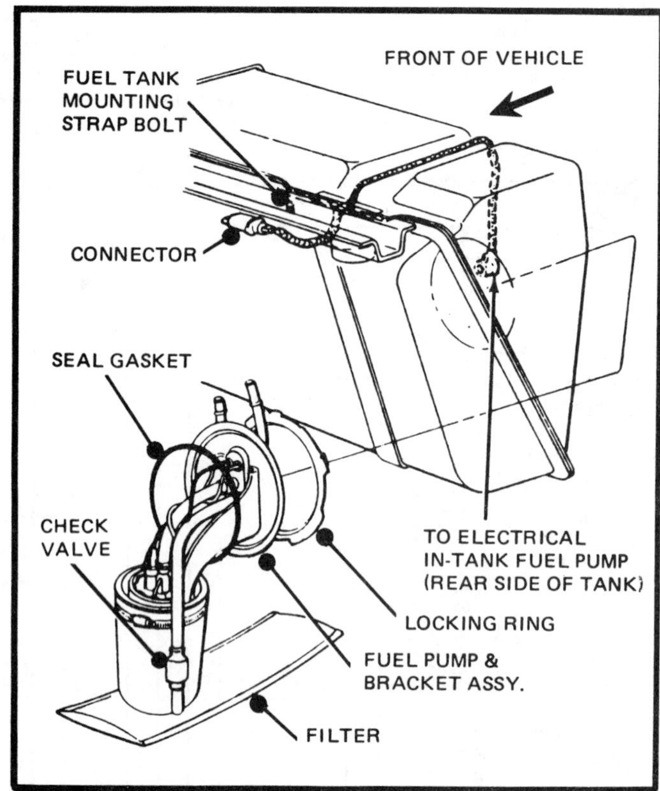

Electric fuel pump installation

6. Disconnect the electrical connections to both the fuel sender and the fuel pump wiring harness.
7. Remove the fuel tank support straps and remove the fuel tank.
8. Turn the fuel pump locking ring counter-clockwise with the necessary tool and remove the locking ring.
9. Remove the fuel pump and bracket assembly.
10. Remove the seal gasket and discard.
11. Remove any dirt that has accumulated around the fuel pump attaching flange, to prevent it from entering the tank during removal and installation.

Installation

1. Put a light coating of heavy grease on a new seal ring to hold it in place during assembly. Install it in fuel tank ring groove.
2. Install the tank in the vehicle.
3. Install the electrical connector.
4. Install the fuel line fittings and tighten to 40-54 N·m (30-40 ft. lbs.).
5. Install a minimum of 10 gallons of fuel and inspect for leaks.
6. Install pressure gauge on valve on throttle body and turn ignition to "ON" position for 3 seconds. Turn ignition key off and back on for 3 seconds repeatedly, 5 to 10 times, until pressure gauge shows at least 35 psi. Reinspect for leaks at fittings.
7. Remove pressure gauge. Start engine and reinspect for leaks.

FUEL CHARGING ASSEMBLY

Removal and Installation

1. Remove the air cleaner.
2. Release pressure from the fuel system at the diagnostic valve on the fuel charging assembly by carefully depressing the pin and discharging fuel into the throttle body.
3. Disconnect the throttle cable and transmission throttle valve lever.

4. Disconnect fuel, vacuum and electrical connections.

NOTE: Either the multi or single ten pin connectors may be used on the system. To disconnect electrical ten pin connectors, push in or squeeze on the right side lower locking tab while pulling up on the connection. Multi connectors disconnect by pulling apart. The ISC connector tab must be moved out while pulling apart.

5. Remove fuel charging assembly retaining nuts, then, remove fuel charging assembly.
6. Remove mounting gasket from intake manifold. Always use a new gasket for installation.
7. Clean gasket mounting surfaces of spacer and fuel charging assembly.
8. Place spacer between two new gaskets and place spacer and gaskets on the intake manifold. Position the charging assembly on the spacer and gasket.
9. Secure fuel charging assembly with attaching nuts. Tighten to 10 ft. lbs (14Nm). To prevent leakage, distortion or damage to the fuel charging assembly body flange, snug the nuts; then, alternately tighten each nut in a criss cross pattern. Tighten to specifications.
10. Connect the fuel line, electrical connectors, throttle cable and all emission lines.
11. Start the engine, check for leaks. Adjust engine idle speed if necessary. Refer to Engine/Emission Control Decal for idle speed specifications.

Disassembly

To prevent damage to the throttle plates, the fuel charging assembly should be placed on a work stand during disassembly and assembly procedures. If a proper stand is not available, use four bolts $^5/_{16} \times 2^1/_2$ inches as legs. Install nuts on the bolts above and below the throttle body. The following is a step-by-step sequence of operations for completely overhauling the fuel charging assembly. Most components may be serviced without a complete disassembly of the fuel charging assembly. To replace individual components follow only the applicable steps.

NOTE: Use a separate container for the component parts of each sub-assembly to insure proper assembly. The automatic transmission throttle valve lever must be adjusted whenever the fuel charging assembly is removed for service or replacement.

1. Remove the air cleaner stud. The air cleaner stud must be removed to separate the upper body from the throttle body.
2. Turn the fuel charging assembly over and remove four screws from the bottom of the throttle body.
3. Separate throttle body from main body. Set throttle body aside.
4. Carefully remove and discard gasket. Note if scraping is necessary, be careful not to damage gasket surfaces of main and throttle screws.
5. Remove three pressure regulator retaining screws.
6. Remove pressure regulator. Inspect condition of gasket and O-ring.
7. Disconnect electrical connectors at each injector. Pull the connectors outward.

NOTE: Pull the connector, not the wire. Tape Identify the connectors. They must be installed on same injector as removed.

8. Loosen, DO NOT REMOVE, wiring harness retaining screw with multi connector; with single ten-pin connector loosen the two retaining screws.
9. Push in on tabs on harness to remove from upper body.
10. Remove fuel injector retainer screw.
11. Remove the injector retainer.
12. One at a time, pull injectors out of upper body. Identify each injector as "choke" or "throttle" side.

NOTE: Each injector has a small O-ring at its top. If the O-ring does not come out with the injector, carefully pick the O-ring out of the cavity in the throttle body.

13. Remove fuel diagnostic valve assembly.
14. Note the position of index mark on choke cap housing.
15. Remove three retaining ring screws.
16. Remove choke cap retaining ring, choke cap, and gasket, if so equipped.
17. Remove thermostat lever screw, and lever, if so equipped.
18. Remove fast idle cam assembly, if so equipped.
19. Remove fast idle control rod positioner, if so equipped.
20. Hold control diaphragm cover tightly in position, while removing two retaining screws, if so equipped.
21. Carefully, remove cover, spring, and pulldown control diaphragm, if so equipped.
22. Remove fast idle retaining nut, if so equipped.
23. Remove fast idle cam adjuster lever, fast idle lever, spring and E-clip, if so equipped.
24. Remove throttle position sensor connector bracket retaining screw.
25. Remove throttle position sensor retaining screws and slide throttle position sensor off the throttle shaft.
26. If CFI assembly is equipped with a throttle positioner, remove the throttle positioner retaining screw, and remove the throttle positioner. If the CFI assembly is equipped with an ISC DC Motor, remove the motor.

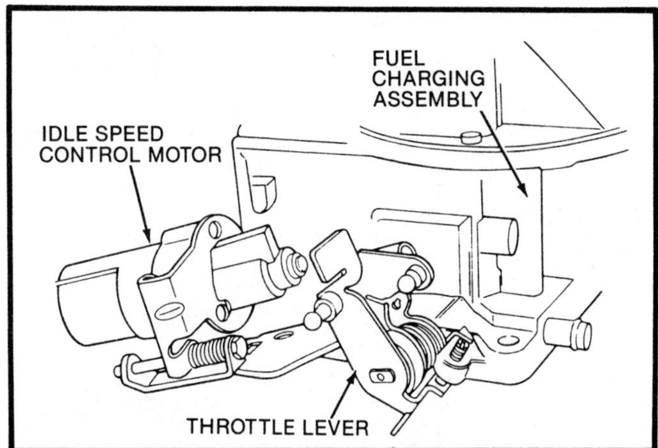

Idle speed control actuator—3.8L engine only

Assembly

1. Install fuel pressure diagnostic valve and cap. Tighten valve to 48–84 inch lbs. (5–9 Nm). Tighten cap to 5–10 inch lbs. (0.6–2 Nm).
2. Lubricate new O-rings and install on each injector (use a light grade oil).
3. Identify injectors and install them in their appropriate locations (choke or throttle side). Use a light twisting, pushing motion to install the injectors.
4. With injectors installed, install injector retainer into position.
5. Install injector retainer screw, and tighten to 36–60 inch lbs. (4–7 Nm).
6. Install injector wiring harness in upper body. Snap harness into position.
7. Tighten injector wiring harness retaining screw, (two screws if equipped with a single ten pin connector), to 8–10 inch lbs. (1 Nm).
8. Snap electrical connectors into position on injectors.
9. Lubricate new fuel pressure regulator O-ring with light oil. Install O-ring and new gasket on regulator.
10. Install pressure regulator in upper body. Tighten retaining screws to 27–40 inch lbs. (3–4 Nm).
11. Depending upon CFI assembly, install either the throttle positioner, or the ISC DC Motor.

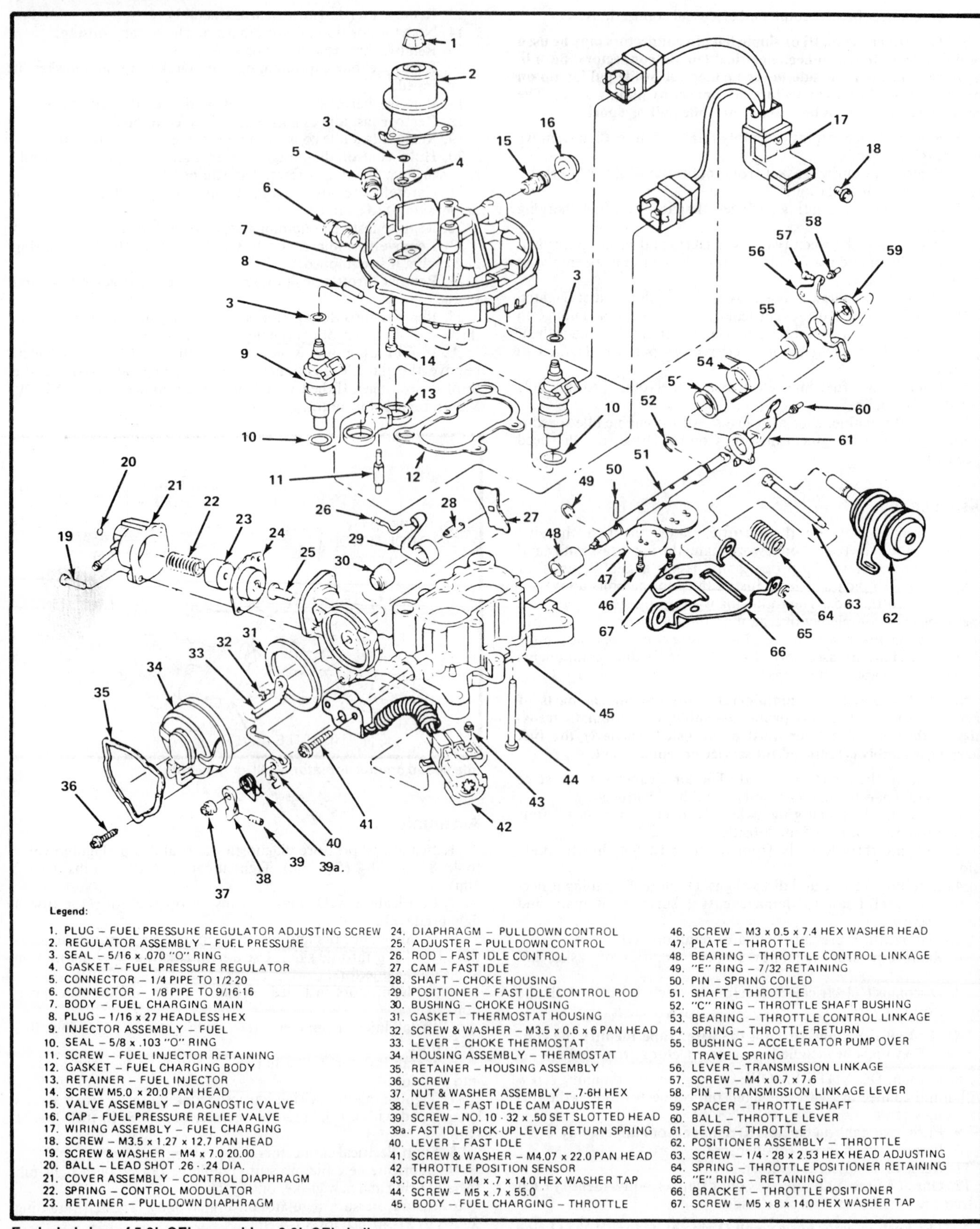
Legend:
1. PLUG – FUEL PRESSURE REGULATOR ADJUSTING SCREW
2. REGULATOR ASSEMBLY – FUEL PRESSURE
3. SEAL – 5/16 x .070 "O" RING
4. GASKET – FUEL PRESSURE REGULATOR
5. CONNECTOR – 1/4 PIPE TO 1/2-20
6. CONNECTOR – 1/8 PIPE TO 9/16-16
7. BODY – FUEL CHARGING MAIN
8. PLUG – 1/16 x 27 HEADLESS HEX
9. INJECTOR ASSEMBLY – FUEL
10. SEAL – 5/8 x .103 "O" RING
11. SCREW – FUEL INJECTOR RETAINING
12. GASKET – FUEL CHARGING BODY
13. RETAINER – FUEL INJECTOR
14. SCREW M5.0 x 20.0 PAN HEAD
15. VALVE ASSEMBLY – DIAGNOSTIC VALVE
16. CAP – FUEL PRESSURE RELIEF VALVE
17. WIRING ASSEMBLY – FUEL CHARGING
18. SCREW – M3.5 x 1.27 x 12.7 PAN HEAD
19. SCREW & WASHER – M4 x 7.0 20.00
20. BALL – LEAD SHOT .26 - .24 DIA.
21. COVER ASSEMBLY – CONTROL DIAPHRAGM
22. SPRING – CONTROL MODULATOR
23. RETAINER – PULLDOWN DIAPHRAGM
24. DIAPHRAGM – PULLDOWN CONTROL
25. ADJUSTER – PULLDOWN CONTROL
26. ROD – FAST IDLE CONTROL
27. CAM – FAST IDLE
28. SHAFT – CHOKE HOUSING
29. POSITIONER – FAST IDLE CONTROL ROD
30. BUSHING – CHOKE HOUSING
31. GASKET – THERMOSTAT HOUSING
32. SCREW & WASHER – M3.5 x 0.6 x 6 PAN HEAD
33. LEVER – CHOKE THERMOSTAT
34. HOUSING ASSEMBLY – THERMOSTAT
35. RETAINER – HOUSING ASSEMBLY
36. SCREW
37. NUT & WASHER ASSEMBLY – .7-6H HEX
38. LEVER – FAST IDLE CAM ADJUSTER
39. SCREW – NO. 10 - 32 x .50 SET SLOTTED HEAD
39a. FAST IDLE PICK-UP LEVER RETURN SPRING
40. LEVER – FAST IDLE
41. SCREW & WASHER – M4.07 x 22.0 PAN HEAD
42. THROTTLE POSITION SENSOR
43. SCREW – M4 x .7 x 14.0 HEX WASHER TAP
44. SCREW – M5 x .7 x 55.0
45. BODY – FUEL CHARGING – THROTTLE
46. SCREW – M3 x 0.5 x 7.4 HEX WASHER HEAD
47. PLATE – THROTTLE
48. BEARING – THROTTLE CONTROL LINKAGE
49. "E" RING – 7/32 RETAINING
50. PIN – SPRING COILED
51. SHAFT – THROTTLE
52. "C" RING – THROTTLE SHAFT BUSHING
53. BEARING – THROTTLE CONTROL LINKAGE
54. SPRING – THROTTLE RETURN
55. BUSHING – ACCELERATOR PUMP OVER TRAVEL SPRING
56. LEVER – TRANSMISSION LINKAGE
57. SCREW – M4 x 0.7 x 7.6
58. PIN – TRANSMISSION LINKAGE LEVER
59. SPACER – THROTTLE SHAFT
60. BALL – THROTTLE LEVER
61. LEVER – THROTTLE
62. POSITIONER ASSEMBLY – THROTTLE
63. SCREW – 1/4 - 28 x 2.53 HEX HEAD ADJUSTING
64. SPRING – THROTTLE POSITIONER RETAINING
65. "E" RING – RETAINING
66. BRACKET – THROTTLE POSITIONER
67. SCREW – M5 x 8 x 14.0 HEX WASHER TAP

Exploded view of 5.0L CFI assembly—3.8L CFI similar

12. Hold throttle position sensor so wire faces up.
13. Slide throttle position sensor on throttle shaft.
14. Rotate throttle position sensor clockwise until aligned with screw holes on throttle body. Install retaining screws and tighten to 11-16 inch lbs. (1-2 Nm).
15. Install throttle position wiring harness bracket retaining screw. Tighten screw to 18-22 inch lbs. (2-3 Nm).
16. Install E-clip, fast idle lever and spring, fast idle adjustment lever and fast idle retaining nut, if so equipped.
17. Tighten fast idle retaining nut to 16-20 inch lbs. (1-2 Nm), if so equipped.
18. Install pull down control diaphragm, control modulator spring and cover, if so equipped. Hold cover in position and install two retaining screws, and tighten to 13-19 inch lbs. (1-2 Nm).
19. Install fast idle control rod positions, if so equipped.
20. Install fast idle cam, if so equipped.
21. Install thermostat lever and retaining screws, if so equipped. Tighten to 13-19 inch lbs. (1-2 Nm).
22. Install choke cap gasket, choke cap, and retaining ring, if so equipped.

NOTE: Be sure the choke cap bimetal spring is properly inserted between the fingers of the thermostat lever and choke cap index mark is properly aligned.

23. Install choke cap retaining screws, tighten to 13-18 inch lbs. (1-2 Nm).
24. Install fuel charging gasket on upper body. Be sure gasket is positioned over bosses. Place throttle body in position on upper body.

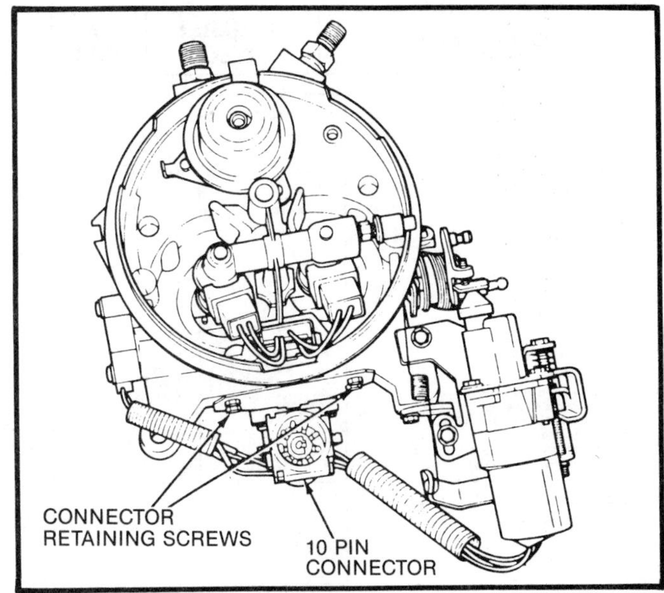

3.8L CFI with 10 pin connector

25. Install four upper body to throttle body retaining screws. Tighten to specifications.
26. Install air cleaner stud. Tighten stud to 70-95 inch lbs. (8-11 Nm).

Wiring and Vacuum Diagrams

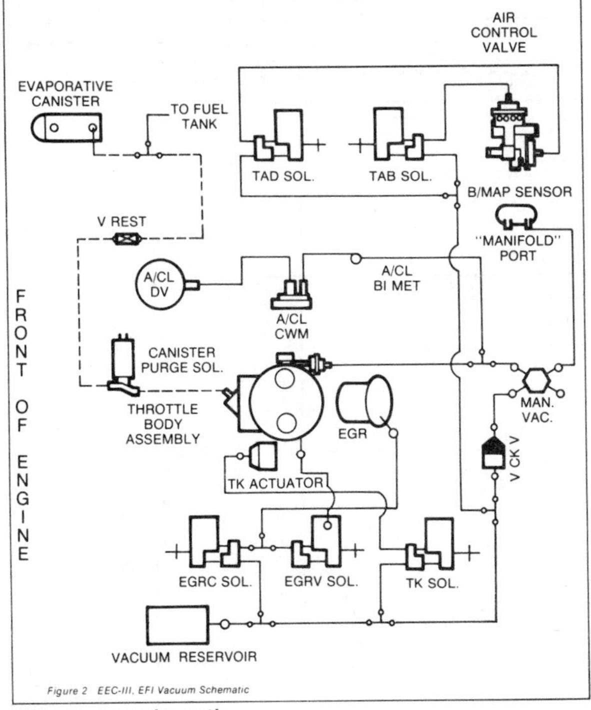

CFI vacuum schematic

Wiring and Vacuum Diagrams

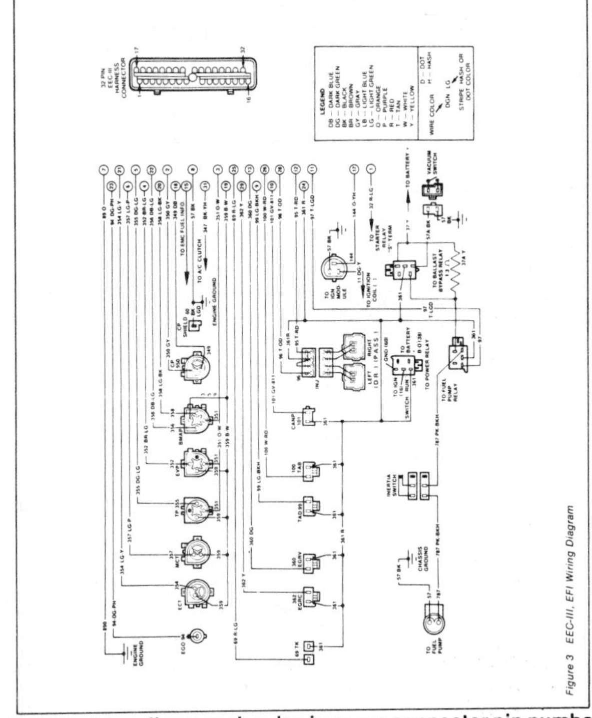

EEC III wiring diagram showing harness connector pin numbers

No Start Problem — Pinpoint Test A (EFI)

TEST STEP	RESULT ▶	ACTION TO TAKE
A1 CHECK FOR FUEL LEAKS		
NO LIGHTED TOBACCO NEARBY • Remove air cleaner • Inspect and service any: — Loose connections — Corrosion — Water or other foreign matter in fuel — Fuel Leaks	Fuel not leaking ▶	GO to A2
	Fuel being injected ▶	CHECK for short to ground in circuits 95 and 96* in harness or processor. If none found, replace injectors.
• Watch for leaks* as you pressurize fuel system (Turn key ON and OFF five times with one second periods ON before switching OFF) • Key OFF *WARNING: If fuel starts leaking or injectors discharge, IMMEDIATELY turn key OFF. No Smoking. NOTE: This step requires two technicians	Fuel is leaking elsewhere than from injectors ▶	REPLACE leaking component* *If no short or leaking found, GO to A34
A2 TRY TO START ENGINE		
	Engine cranks, but does not start or stalls out ▶	GO to A3
	Engine does not crank ▶	REFER to the Shop Manual. If battery were discharged and you suspect the EEC-III system, charge the battery, and perform the QUICK TEST

No Start Problem — Pinpoint Test A (EFI)

TEST STEP	RESULT ▶	ACTION TO TAKE
A7 ENGINE COOLANT TEMPERATURE SENSOR (ECT) — ENGINE AT NORMAL OPERATING TEMPERATURE	DVOM Reading:	
• Tester to A-5	Between 0.83V-7.4V ▶	GO to A8
	Under 0.83V ▶	GO to E3
	More than 7.4V ▶	GO to E5
A8 AIR CHARGE TEMPERATURE SENSOR (ACT)	DVOM Reading:	
• Tester to A-6	Between 0.83V-6.9V ▶	GO to A9
	Under 0.83 or more than 6.9V ▶	GO to S1
A9 MANIFOLD ABSOLUTE PRESSURE SENSOR (MAP)	ELEVATION (FT.)	
• Tester to A-7 [—] • Fill in appropriate limits for your local elevation. (For future reference.)	LIMITS DVOM Reading: 0-1500 7.6-8.4 1501-2500 7.2-8.1 2501-3500 6.9-7.8 3501-4500 6.7-7.3 4501-5500 6.6-7.2 5501-6500 6.3-7.0 6501-7500 6.1-6.7 7501-8500 5.8-6.5 8501-9500 5.6-6.3 9501-10500 5.4-6.0 ▶	GO to A10
	DVOM Reading does not correspond to your local elevation. ▶	GO to F1

No Start Problem — Pinpoint Test A (EFI)

TEST STEP	RESULT ▶	ACTION TO TAKE
A3 CHECK EEC-III SYSTEM	DVOM Reading:	
• Turn key to OFF • Connect Tester • Tester to A1 • Disconnect injectors electrically • DVOM switch to Tester • Turn key to RUN — engine OFF	10.50V or more ▶	GO to A4
	Less than 10.50V ▶	GO to B1
A4 SENSOR REFERENCE VOLTAGE (VREF)	DVOM Reading:	
• Tester to A-3	Between 8-10V ▶	GO to A5
	Under 8V or more than 10V ▶	GO to C1
A5 THROTTLE POSITION SENSOR (TP)	DVOM Reading:	
• Tester to A-4	Between 1.8-2.8V ▶	GO to A6
	Under 1.8 or more than 2.8V ▶	GO to D1
A6 THROTTLE POSITION SENSOR MOVEMENT CHECK	DVOM Reading:	
• Tester to A-4 • Depress accelerator pedal	6V or more ▶	GO to A7
	Under 6V ▶	GO to D2

No Start Problem — Pinpoint Test A (EFI)

TEST STEP	RESULT ▶	ACTION TO TAKE
A10 CHECK MANIFOLD VACUUM ON ENGINE CRANK	DIFFERENCE IN READINGS:	
• Tester to A-7 • Key to RUN, and observe initial DVOM reading • Crank engine, and note reading NOTE: Do not depress gas pedal during test.	0.4V or more ▶	GO to A11
	Under 0.4V ▶	GO to F1
A11 CHECK BMAP SENSOR	DIFFERENCE BETWEEN THE TWO READINGS:	
• Tester to A-8 • Compare reading with MAP reading A19	0.75V or less ▶	GO to A12
	More than 0.75V ▶	GO to F1
A12 CHECK EGR VALVE POSITION SENSOR AND EGR LIGHTS	DVOM Reading:	
• Tester to A-9 [● ON EGRV OFF] [● ON EGRC OFF]	1.4 to 2.45 ▶	GO to A12a
	Under 1.4 or over 2.45 ▶	GO to G1
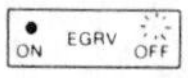 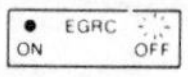	If lights are not as shown ▶	GO to G6

No Start Problem — Pinpoint Test A (EFI)

TEST STEP	RESULT ▶	ACTION TO TAKE
A12a CHECK FOR FULL EGR		
• Disconnect EGR valve vacuum hose at EGR solenoid assembly [DISCONNECT VACUUM LINE] • Attempt to start engine	Engine does not start or stalls when starting ▶	GO to A13
	Engine starts ▶	GO to G12
A13 CRANKING SIGNAL VOLTAGE	DVOM Reading:	
• Tester to A-2	1V or less ▶	GO to A14
	Over 1V ▶	GO to R1
A14 CHECK CP AND IMS SIGNALS AND POWER TO STARTER VOLTAGE	WHILE CRANKING DVOM Reading:	
• Tester connected • Tester to A-2 • Observe DVOM reading and Tester CP and IMS lights while cranking engine for 5 seconds	Under 6.5V on A-2 ▶	GO to R1
	6.5V or more on A-2 AND TESTER LIGHTS	
	IMS ⁘ CP ⁘ ▶	GO to A15
	IMS ● CP ⁘ ▶	GO to A17
	IMS ⁘ CP ● ▶	GO to A20
	IMS ● CP ● ▶	GO to A20

No Start Problem — Pinpoint Test A (EFI)

TEST STEP	RESULT ▶	ACTION TO TAKE
A17 CHECK IGNITION MODULE	TESTER LIGHTS:	
• Disconnect Ignition Module • Crank engine for 5 seconds, and observe CP and IMS light after 3 seconds of cranking	IMS ● CP ⁘ ▶	GO to A18
	IMS ⁘ CP ⁘ ▶	RECONNECT Ignition Module REFER to Ignition System in Part 29-02
A18 CHECK WIRING HARNESS TO IGNITION MODULE	DVOM Reading:	
• Disconnect processor assembly from vehicle harness • Tester to B-11 • Turn key to RUN — engine OFF • Turn key to OFF	0.5V or less ▶	GO to A19
	More than 0.5V ▶	SERVICE short to voltage source in wire 144 (orange/yellow hash) RECONNECT Ignition Module
A19 CHECK HARNESS	DVOM Reading:	
• Turn key to OFF • Tester to 3-11 • Depress and hold Ohms button • Reconnect Ignition Module	1000 ohms or more ▶	REPLACE Processor Assembly
	Under 1000 ohms ▶	SERVICE short-to-ground in wire 144 (orange/yellow hash)
A20 CHECK CP SENSOR	DVOM Reading:	
• Turn key to OFF • Tester to A-10 • Depress and hold Ohms button	Between 100 and 640 ohms ▶	GO to A21
	Under 100 or more than 640 ohms ▶	GO to A24

No Start Problem — Pinpoint Test A (EFI)

TEST STEP	RESULT ▶	ACTION TO TAKE
A15 CHECK FOR SPARK		
• Reconnect coil harness clip • Disconnect the Spark Plug wire to cylinder No. 2 or No. 4 • Connect spark tester between spark plug wire and engine ground • Crank engine • Reconnect the spark plug wire to the spark plug	Spark ▶	CHECK rotor, CP ring, and damper for proper alignment. If OK, go to A29
	No spark ▶	GO to A16
A16 CHECK IGNITION MODULE		
• Disconnect Ignition Module harness. Install Rotunda Test Adapter T79P-12-127A. • Turn key to RUN • Connect the spark tester between the cap end of the ignition coil wire and engine ground	No Spark ▶	REFER to Ignition System
• Repeatedly touch battery terminal with the orange diagnosis test lead, and check for ignition sparks (at coil) when the lead is touched to battery positive • Turn key to OFF • Remove test adapter, and reconnect the ignition module • Reconnect the coil wire to the distributor cap	Spark at coil ▶	CHECK for open in Circuit 144 from EEC harness to Ignition Module. If OK, replace processor

No Start Problem — Pinpoint Test A (EFI)

TEST STEP	RESULT ▶	ACTION TO TAKE
A21 CHECK FOR PHYSICAL DAMAGE		
• Check CP Sensor and pulse ring for damage, such as missing lobes, sensor cracks, damaged wiring, etc. • Check for proper CP Sensor seating	No damage and seating is proper ▶	GO to A22
	Damage or improper seating ▶	SERVICE as required
A22 CHECK PROCESSOR ASSEMBLY	TESTER LIGHT:	
• Disconnect processor assembly from vehicle harness • Crank engine for 5 seconds, and observe CP light after 6 seconds of cranking	CP ● ▶	GO to A23
	CP ⁘ ▶	REPLACE Processor Assembly
A23 CHECK CP HARNESS FOR CONTINUITY	DVOM Reading:	
• Turn key to OFF • Tester to A-10 • Disconnect CP Sensor from vehicle harness • Jumper a wire between CP + (349) and CP − (350) in CP harness [CP+ (349); CP SENSOR HARNESS CONNECTOR; CP− (350)] • Depress and hold Ohms button	5 ohms or more ▶	SERVICE wire 350 (gray) or wire 349 (dark blue) for open circuit or high resistance
	Under 5 ohms ▶	REPLACE CP Sensor

No Start Problem — Pinpoint Test — A (EFI)

TEST STEP	RESULT ▶	ACTION TO TAKE
A24 CHECK CP SENSOR RESISTANCE	DVOM Reading:	
• Disconnect CP sensor from vehicle harness 2000 OHMS	Between 100-640 ohms ▶	GO to [A25]
	Under 100 or more than 640 ohms ▶	REPLACE CP Sensor
A25 CHECK CP+ TO GROUND SHEATH RESISTANCE		
• Measure resistance from CP+ in sensor to both CP shield connectors (sensor and harness) for shorts CP+ CP SENSOR CONNECTOR CP GROUND SHEATH CONNECTOR HARNESS CIRCUIT 60 200 OHMS	No short ▶	GO to [A26]
	Short ▶	REPLACE CP Sensor
A26 CHECK CP SENSOR HARNESS FOR SHORT	DVOM Reading:	
• Turn key to OFF • Tester to A-10 • Disconnect Processor Assembly from Tester • DVOM switch to Tester • Depress and hold Ohms button	+1 (overrange, no shorts) ▶	GO to [A27]
	Under 1999 ohms (Short) ▶	SERVICE short in harness between circuits 350 and 349

No Start Problem — Pinpoint Test — A (EFI)

TEST STEP	RESULT ▶	ACTION TO TAKE
A27 CHECK CP HARNESS FOR CONTINUITY	DVOM Reading:	
• Turn key to OFF • Tester to A-10 • Disconnect CP Sensor from vehicle harness • Jumper a wire between CP+ and CP− in harness • Depress and hold Ohms button CP+ (349) CP (350) CP SENSOR HARNESS CONNECTOR	5 ohms or less ▶	GO to [A28]
	More than 5 ohms ▶	SERVICE circuits 350 (gray) or 349 (dark blue) for open circuit or high resistance
A28 CHECK CP SENSOR HARNESS FOR SHORTS FROM CP+ TO GROUND (CIRCUIT 60)	DVOM Reading:	
CP+ CP HARNESS CONNECTOR CIRCUIT 60 HARNESS CONNECTOR 200 OHMS	5 ohms or more ▶	REPLACE Processor Assembly
	Under 5 ohms ▶	SERVICE short in Harness

No Start Problem — Pinpoint Test — A (EFI)

TEST STEP	RESULT ▶	ACTION TO TAKE
A29 FUEL PUMP CHECK		
NO SMOKING NEARBY • Disconnect ignition coil connector • Disconnect both fuel injector electrical connections at the injectors • Connect pressure gauge to Schraeder valve on Throttle Body injector bar • Note initial pressure reading • Observe pressure gauge as you pressurize fuel system (turn key to RUN for 1 second, then turn key to OFF. Repeat 5 times) • Turn key to OFF WARNING: If fuel starts leaking or injectors discharging, turn key OFF immediately. No smoking. Reconnect ignition coil connector. GO to [A1]	PRESSURE GAUGE READING: Increase ▶	GO to [A31]
	Did not increase ▶	TURN key OFF, reconnect ignition coil connector, and go to [A30]

No Start Problem — Pinpoint Test — A (EFI)

TEST STEP	RESULT ▶	ACTION TO TAKE
A30 CHECK INERTIA SWITCH	PRESSURE GAUGE READING:	
• Locate inertia switch (in trunk) • Push the button of Inertia switch to turn it on • Watch pressure gauge as you attempt to pressurize fuel system NOTE: If switch will not turn "ON", replace it.	Increases ▶	GO to QUICK TEST
	Did not increase ▶	GO to [Q2]
A31 CHECK FUEL DELIVERY	PRESSURE GAUGE READING:	
• Pressurize fuel system • Turn key to OFF • Wait for pressure to become steady • Read pressure gauge	Between 35 and 45 psi ▶	GO to [A32]
	Over 45 psi or under 35 psi ▶	RECONNECT ignition coil. REFER to Part 24-29 of the Shop Manual
A32 CHECK FOR LEAK DOWN	PRESSURE DROP:	
• Wait 2 minutes after pressure gauge reading of test Step [A31] then note drop in gauge reading	4 psi drop or less in 2 min. ▶	GO to [A33]
	More than 4 psi drop in 2 min. ▶	RECONNECT ignition coil.

No Start Problem — Pinpoint Test A (EFI)

TEST STEP	RESULT ▶	ACTION TO TAKE
A33 CHECK FOR FUEL DISCHARGE WITH INJECTORS DISCONNECTED		
• Crank engine for 3 seconds	Pressure does not drop, AND Injectors do not discharge ▶	GO to A34
• Watch pressure gauge and injectors	Pressure drops, AND Injectors discharge ▶	TURN key OFF immediately CHECK for hydraulic lockup or fuel fouled spark plugs RECONNECT ignition coil REPLACE leaking injector(s) with the same color code
A34 CHECK SIGNAL AT DRIVER SIDE INJECTOR	TEST LIGHT:	
• Reconnect injectors	Off ▶	GO to A35
• Connect a 12 V Test Light across 2 pins on the driver side of Throttle Body connector • Turn key to RUN	On ▶	REMOVE Test Light GO to A41

No Start Problem — Pinpoint Test A (EFI)

TEST STEP	RESULT ▶	ACTION TO TAKE
A35 CHECK PROCESSOR OUTPUT TO DRIVER SIDE INJECTOR	TEST LIGHT:	
• Crank engine	Flickers ▶	GO to A36
• Leave test light connected as in A34	Does not flicker ▶	DISCONNECT Test Light GO to A41
A36 CHECK SIGNAL AT PASSENGER SIDE INJECTOR	TEST LIGHT:	
• Connect the 12V Test Light across 2 pins on the passenger side of Throttle Body right connector	Off ▶	GO to A37
• Turn key to RUN	On ▶	DISCONNECT Test Light GO to A41
A37 CHECK PROCESSOR OUTPUT TO PASSENGER SIDE INJECTOR	TEST LIGHT:	
• Crank engine	Flickers ▶	GO to A38
• Leave test light connected as in A36	Does not flicker ▶	DISCONNECT Test Light GO to A41

No Start Problem — Pinpoint Test A (EFI)

TEST STEP	RESULT ▶	ACTION TO TAKE
A38 CHECK ONE INJECTOR'S PRESSURE	PRESSURE DROP:	
• Disconnect one injector electrically (leave other one connected) • Pressurize fuel system • Disable fuel pump by disconnecting Inertia Switch or disconnecting Fuel Pump Relay • Observe pressure gauge reading • Crank engine for 2 seconds • Turn key to OFF • Wait 5 seconds then observe pressure gauge reading drop	Between 2 and 16 psi drop (1 to 8 psi drop per second of cranking) ▶	GO to A39
	Less than 2 or more than 16 psi drop (less than 1 or more than 8 psi drop per second of cranking) ▶	GO to A40
A39 CHECK OTHER INJECTOR'S PRESSURE	PRESSURE DROP:	
• Reactive fuel pump • Reconnect injector, and disconnect the other one • Repeat A38 for this injector	Between 2 and 16 psi drop (1 to 8 psi drop per second of cranking) ▶	RECONNECT ignition coil Spark plug and fuel is present. Check to make sure plugs are not fouled and all plugs are firing and the rotor is turning. Be sure there is no mechanical problem like distributor cross-fire, flooding, or improper fuel.
	Less than 2 or more than 16 psi drop (less than 1 or more than 8 psi drop per second of cranking) ▶	GO to A40

No Start Problem — Pinpoint Test A (EFI)

TEST STEP	RESULT ▶	ACTION TO TAKE
A40 ISOLATE IMPROPER FUEL DELIVERY	PRESSURE DROP:	
• Remove the connected electrical connector from the injector, and connect it to the other injector	Between 2 and 16 psi (1 to 8 psi per second of cranking) ▶	REPLACE the Disconnected injector with same color code RECONNECT both injectors correctly RETEST.
• Reactivate fuel pump • Pressurize fuel system • Disable fuel pump by turning off Inertia Switch or disconnecting Fuel Pump Relay • Observe pressure gauge reading • Crank engine for 2 seconds • Turn key to OFF • Wait 5 seconds then observe pressure gauge reading drop	Less than 2 or more than 16 psi (less than 1 or more than 8 psi per second of cranking) ▶	GO to A41

No Start Problem — Pinpoint Test — A (EFI)

TEST STEP	RESULT ▶	ACTION TO TAKE
A41 CHECK FOR INJECTOR SHORT-TO-GROUND		
• Turn key to OFF • Reconnect ignition coil connector • Reconnect injectors properly • Check each pin on Throttle Body connector for short-to-ground	No short at any pin ▶	GO to A45
	Short at one or more pins ▶	GO to A42
A42 CHECK FOR PROCESSOR SHORT-TO-GROUND		
• Turn key to OFF • Disconnect harness from Processor Assembly • Check each pin of Throttle Body for short-to-ground (See A41)	No short any pin ▶	REPLACE Processor Assembly
	Short at one or more pins ▶	GO to A43
A43 CHECK FOR SHORT-TO-GROUND IN DRIVER SIDE INJECTOR		
(Processor disconnected) • Disconnect driver side injector • Check each pin of Throttle Body connector for short-to-ground (See A41)	No short at any pin ▶	REPLACE driver side injector
	Short at one or more pins ▶	GO to A44

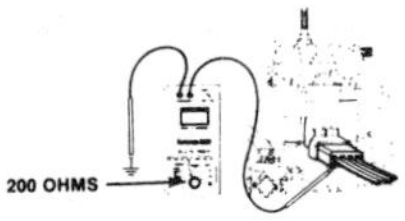

No Start Problem — Pinpoint Test — A (EFI)

TEST STEP	RESULT ▶	ACTION TO TAKE
A47 CHECK FOR SHORT BETWEEN INJECTOR LINES	DVOM Reading:	
• Disconnect harness from Processor Assembly • Using DVOM Test Leads, measure resistance between pins: — 12 and 28 — 24 and 28 (left injector) — 24 and 12 (right injector) on Harness connector	2 or more ohms for each pair of pins ▶	GO to A49
	Under 2 ohms for one or more pair of pins ▶	NOTE which lines, GO to A48
A48 ISOLATE OBSERVED SHORT TO INJECTORS OR HARNESS	DVOM Reading:	
• Disconnect both injectors electrically • Using DVOM Test Leads, measure resistance (200 x 1000 scale) between pins: — 12 and 28 — 24 and 28 (left injector) — 24 and 12 (right injector) on Harness connector	10,000 ohms or more for each pair of pins ▶	REPLACE injector(s) connected to the pin(s) that were shorted in A47 (replace with the same color code)
	Under 10,000 ohms for one or more pair of pins ▶	SERVICE short in harness

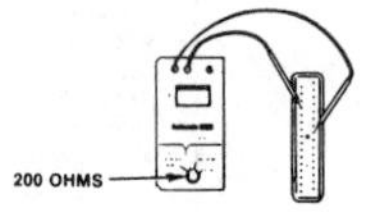

No Start Problem — Pinpoint Test — A (EFI)

TEST STEP	RESULT ▶	ACTION TO TAKE
A44 CHECK FOR SHORT-TO-GROUND IN PASSENGER SIDE INJECTOR		
(Left injector disconnected) • Disconnect passenger side injector • Check each pin on Throttle Body connector again for short-to-ground	No short at any pin ▶	REPLACE passenger side injector
	Short at one or more pins	SERVICE short circuit
A45 CHECK INJECTOR CIRCUITS TO BATTERY VOLTAGE	DVOM Reading:	
• Using DVOM Test Leads, measure voltage between battery negative and each injector signal line	1V or less for both lines ▶	GO to A46
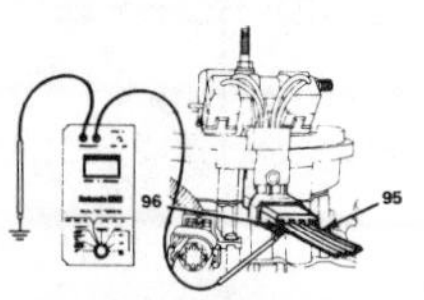	Over 1V for either line ▶	SERVICE short in circuit 95 or 96 between battery positive and circuit(s) out of limits
A46 CHECK RESISTANCE OF INJECTORS HARNESS AND PROCESSOR CONNECTION	DVOM Reading:	
• Using DVOM Test Leads, measure resistance from 361 to 95 and from 361 to 96	Under 2 ohms ▶	GO to A47
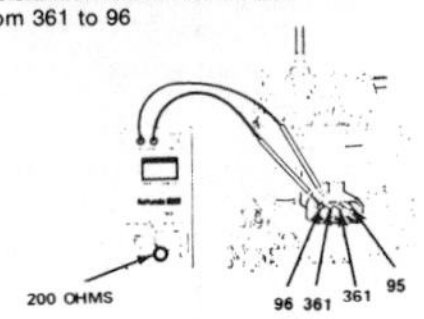	2 or more ohms ▶	GO to A49

No Start Problem — Pinpoint Test — A (EFI)

TEST STEP	RESULT ▶	ACTION TO TAKE
A49 CHECK FOR OPENS IN LINES TO INJECTORS	DVOM Reading:	
• Using DVOM Test Leads, measure resistance (200 Ohms Scale) between pins: — 24 and 28 (driver side injector) — 24 and 12 (passenger side injector) on Harness connector	4 or less ohms for each line ▶	REPLACE Processor Assembly
	Over 4 ohms between 24 and 28 ▶	GO to A50
	Over 4 ohms between 24 and 12 ▶	GO to A51
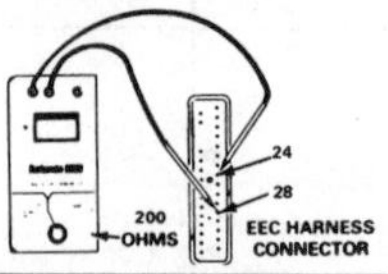	Over 4 ohms on both ▶	SERVICE open in Circuit 361
A50 CHECK FOR OPEN IN DRIVER SIDE INJECTOR HARNESS	DVOM Reading:	
• Disconnect driver side injector electrically • Connect jumper wire across injector harness pins • Using DVOM Test Leads, measure resistance between pins 24 and 28 on Harness connector	5 or less ohms ▶	REPLACE driver side injector
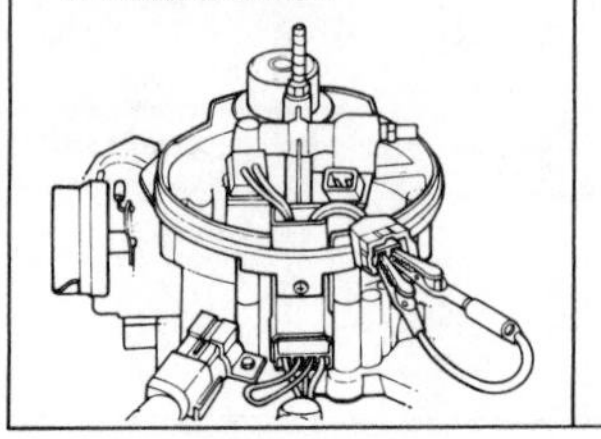	Over 5 ohms ▶	SERVICE open in circuit 96 or 361 to injector

No Start Problem — Pinpoint Test A (EFI)

TEST STEP	RESULT ▶	ACTION TO TAKE
A57 CHECK FOR OPEN IN PASSENGER SIDE INJECTOR HARNESS	DVOM Reading:	
• Disconnect passenger side injector electrically • Connect jumper wire across injector harness pins • Using DVOM Test Leads measure resistance between pins 24 and 12 of Harness connector	5 or less ohms ▶	REPLACE passenger side injector
	Over 5 ohms ▶	SERVICE open in circuit 95 or 361 to injector

Fuel Pump Relays — Pinpoint Test Q (EFI)

TEST STEP	RESULT ▶	ACTION TO TAKE
Q2 CHECK IF FUEL PUMP RUNS		
• Turn key to RUN, but keep engine OFF	Pump runs briefly ▶	GO to Q11
• Listen* for fuel pump to run briefly when key is first turned on *You may want to open gas filler cap to hear the pump. No smoking	Pump does not run ▶	GO to Q3
Q3 RECHECK IF FUEL PUMP RUNS		
• Tester hooked up • Press tester LOS button to "ON" (green flag), and leave LOS button "ON" during entire PINPOINT TEST	Pump runs ▶	REPLACE Calibration Assembly. Retest. If not OK, replace Processor
• Turn key to RUN, but keep engine OFF • Listen closely for fuel pump; it should run constantly	Pump does not run ▶	GO to Q4
Q4 CHECK VOLTAGE ON WIRE TO PROCESSOR	DVOM Reading:	
• Tester to C-4 (the processor grounds this lead to turn on the fuel pump)	Under 2.0V ▶	GO to Q5
• LOS button ON	2.0V or more ▶	GO to Q14

Fuel Pump Relays — Pinpoint Test Q (EFI)

This Test Checks Cause of Low/No Fuel Pressure

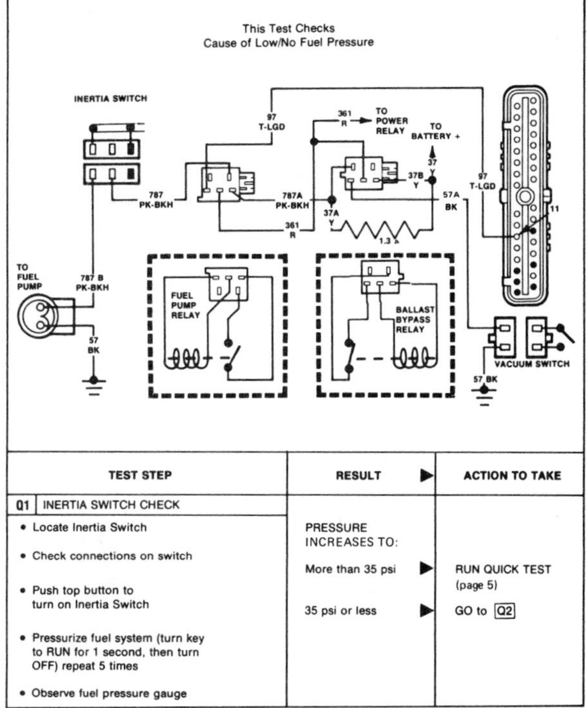

TEST STEP	RESULT ▶	ACTION TO TAKE
Q1 INERTIA SWITCH CHECK	PRESSURE INCREASES TO:	
• Locate Inertia Switch • Check connections on switch	More than 35 psi ▶	RUN QUICK TEST (page 5)
• Push top button to turn on Inertia Switch • Pressurize fuel system (turn key to RUN for 1 second, then turn OFF) repeat 5 times • Observe fuel pressure gauge	35 psi or less ▶	GO to Q2

Fuel Pump Relays — Pinpoint Test Q (EFI)

TEST STEP	RESULT ▶	ACTION TO TAKE
Q5 CHECK FOR POWER TO FUEL PUMP	DVOM Reading:	
• Use DVOM Test Leads to measure voltage on circuit 787 at the Inertia Switch to fuel pump	Under 6V ▶	GO to Q6
• LOS button ON (TO FUEL PUMP RELAY; TO FUEL PUMP)	6V or more ▶	GO to Q17
Q6 CHECK FOR POWER TO INERTIA SWITCH	DVOM Reading:	
• Use DVOM to measure voltage on circuit 787 at Inertia Switch to fuel pump relay	6V or more ▶	REPLACE Inertia Switch
• LOS button ON (TO FUEL PUMP RELAY; TO FUEL PUMP; 20 VOLTS D.C.)	Under 6V ▶	GO to Q7

Fuel Pump Relays — Pinpoint Test — Q (EFI)

TEST STEP	RESULT ▶	ACTION TO TAKE
Q7 CHECK FOR VOLTAGE OUT OF FUEL PUMP RELAY	DVOM Reading:	
• Use DVOM Test Leads to measure voltage on circuit 787 at Fuel Pump Relay	Under 6V ▶	GO to Q8
• LOS button ON	6V or more ▶	SERVICE open in circuit 787 from fuel pump relay to inertia switch
FUEL PUMP RELAY; 20 VOLTS D.C.; CIRCUIT 787		
Q8 CHECK FOR VOLTAGE INTO FUEL PUMP RELAY	DVOM Reading:	
• Use DVOM Test Leads to measure voltage on circuit 787A at the Ballast Bypass Relay	Under 6V ▶	GO to Q9
	6V or more ▶	GO to Q19
FUEL PUMP RELAY; 20 VOLTS D.C.; CIRCUIT 787A		

Fuel Pump Relays — Pinpoint Test — Q (EFI)

TEST STEP	RESULT ▶	ACTION TO TAKE
Q10 CHECK OUTPUT OF BALLAST BYPASS RELAY	DVOM Reading:	
• Use DVOM Test Leads to measure voltage on circuit 787A at the Ballast Bypass Relay	6V or more ▶	SERVICE open in circuit 787A between bypass relay and fuel pump relay, circuit 787A. If reading were between 6V and 10V, GO to Q11. Relay is open or energized or contacts opened
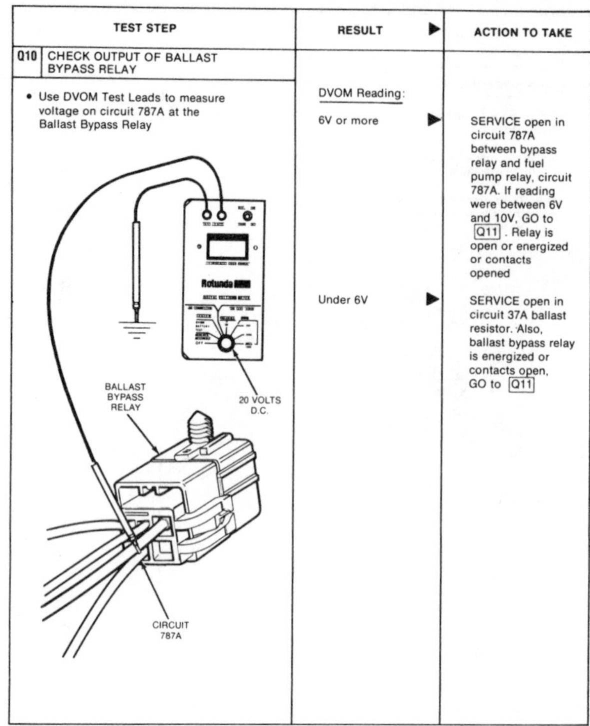	Under 6V ▶	SERVICE open in circuit 37A ballast resistor. Also, ballast bypass relay is energized or contacts open, GO to Q11

Fuel Pump Relays — Pinpoint Test — Q (EFI)

TEST STEP	RESULT ▶	ACTION TO TAKE
Q9 CHECK FOR VOLTAGE INTO BALLAST BYPASS RELAY	DVOM Reading:	
• Use DVOM Test Leads to measure voltage on circuit 37B at the Ballast Bypass Relay	More than 10V ▶	GO to Q10
	Less than 10V ▶	SERVICE open or bad connection in circuit 37B to power. If fuse link is blown, service short to ground in 787, 787A, 787B, 37, or the fuel pump itself; this must be serviced prior to fixing the fuse link
BALLAST BYPASS RELAY; 20 VOLTS D.C.; CIRCUIT 37		

Fuel Pump Relays — Pinpoint Test — Q (EFI)

TEST STEP	RESULT ▶	ACTION TO TAKE
Q11 CHECK FOR VOLTAGE DROP ACROSS FUEL PUMP BALLAST BYPASS RELAY	DVOM Reading:	
• Use DVOM Test Leads to measure voltage from circuit 37 to circuit 787A at Ballast Bypass Relay	More than 1V ▶	GO to Q12
	1V or less ▶	GO to Part 24-29 in the Shop Manual, and check for clogged fuel line, filter, or defective Throttle Body
BALLAST BYPASS RELAY; 20 VOLTS D.C.; CIRCUIT 787A; CIRCUIT 37		
Q12 CHECK FOR RELAY ENERGIZED	DVOM Reading:	
• Use DVOM Test Leads to measure voltage on circuit 57A at Ballast Bypass Relay	Under 10V ▶	GO to Q13
	10V or more ▶	REPLACE Ballast Bypass Relay
BALLAST BYPASS RELAY; 20 VOLTS D.C.; CIRCUIT 57A		

Fuel Pump Relays — Pinpoint Test — Q (EFI)

TEST STEP	RESULT ▶	ACTION TO TAKE
Q13 CHECK VACUUM SWITCH		
• Disconnect Vacuum Swich from vehicle harness	DVOM Reading:	
• Use DVOM Test Leads to check for short-to-ground in circuit 57A	More than 10V ▶	REPLACE Vacuum Switch
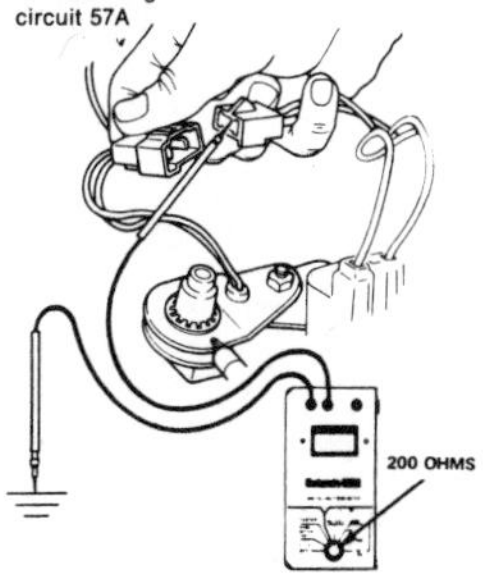	10V or less ▶	SERVICE short in circuit 57A
200 OHMS		
Q14 ISOLATE REASON FOR HIGH VOLTAGE (AT TESTER C-4)		
• Turn key to off • Disconnect fuel pump relay • Disconnect Adapter Harness from processor assembly • Turn key to RUN, but keep engine off • Tester to C-4	DVOM Reading:	
	Under 1V ▶	GO to Q15
	1V or more ▶	SERVICE short to power in circuit 97 or fuel pump relay

Fuel Pump Relays — Pinpoint Test — Q (EFI)

TEST STEP	RESULT ▶	ACTION TO TAKE
Q18 CHECK WIRES TO FUEL PUMP	DVOM Reading:	
• Use DVOM to measure voltage on each of the two pins of the fuel pump. Measure voltage between one of the pins and ground, then between the other pin and ground	Under 6V on both pins ▶	SERVICE open in circuit 787 between Inertia Switch and Fuel Tank
• Do not disconnect connector from fuel tank	6V or more for both ▶	SERVICE open in circuit 57 ground wire
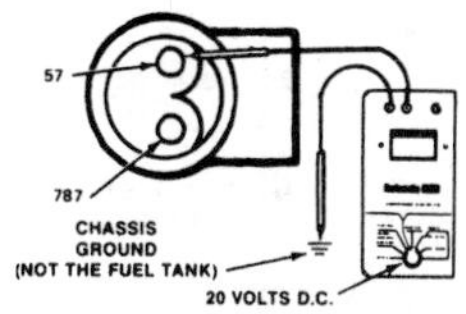 57 787 CHASSIS GROUND (NOT THE FUEL TANK) 20 VOLTS D.C.	One reads more than 7V and other reads under 7V ▶	REPLACE Fuel Pump Assembly
CAUTION: Fuel supply lines will remain pressurized for long periods of time after key is turned OFF. This pressure must be relieved before servicing of the fuel system has begun. A valve is provided on the throttle body for this purpose. Remove air cleaner, and relieve system pressure by depressing pin in relief valve CAUTIOUSLY; fuel will be expelled into throttle body		

Fuel Pump Relays — Pinpoint Test — Q (EFI)

TEST STEP	RESULT ▶	ACTION TO TAKE
Q15 CHECK FOR SHORTED FUEL PUMP RELAY COIL		
• Measure resistance of fuel pump relay coil	More than 40 ohms ▶	GO to Q15
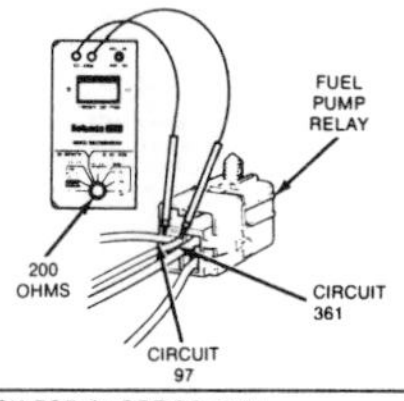 FUEL PUMP RELAY 200 OHMS CIRCUIT 361 CIRCUIT 97	Less than 40 ohms ▶	REPLACE Relay
Q16 CHECK FOR SHORT TO 787A IN RELAY COIL		
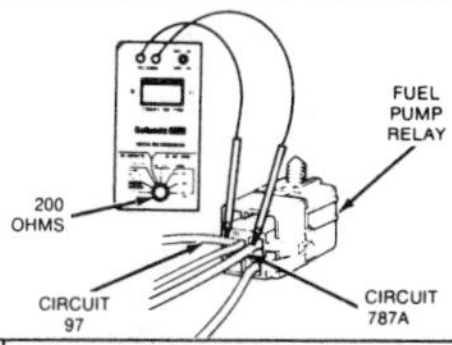 FUEL PUMP RELAY 200 OHMS CIRCUIT 97 CIRCUIT 787A	No short ▶	REPLACE Processor Assembly, and RUN QUICK TEST. If still not OK, replace Calibration Assembly using original Processor Assembly
	Short ▶	REPLACE fuel pump relay
Q17 CHECK WIRES TO FUEL PUMP		
• Visually check wires to fuel Pump for bad connections or opens	Problem identified as bad wiring ▶	SERVICE as needed
	Problem not seen in wiring ▶	GO to Q18

Fuel Pump Relays — Pinpoint Test — Q (EFI)

TEST STEP	RESULT ▶	ACTION TO TAKE
Q19 CHECK FOR VOLTAGE TO RELAY COIL		
• Use DVOM Test Leads to measure voltage on circuit 361 at the Fuel Pump Relay	DVOM Reading:	
	More than 10V ▶	GO to Q20
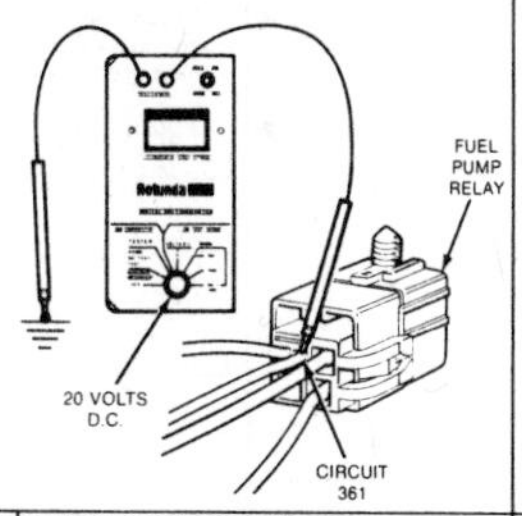 FUEL PUMP RELAY 20 VOLTS D.C. CIRCUIT 361	10V or less ▶	SERVICE open in circuit 361 between Power Relay and Fuel Pump Relay
Q20 CHECK PROCESSOR SIDE OF RELAY COIL		
• Use DVOM test leads to measure voltage on circuit 97 at the fuel pump relay	More than 2V ▶	SERVICE open in circuit 97
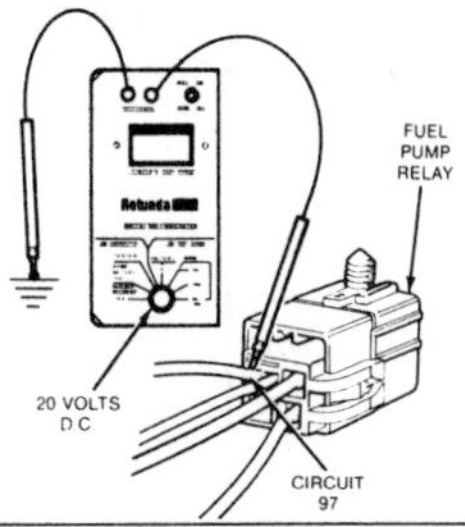 FUEL PUMP RELAY 20 VOLTS D.C. CIRCUIT 97	2V or less ▶	REPLACE Fuel Pump Relay

Fuel Pump Relays — Pinpoint Test — Q (EFI)

TEST STEP	RESULT ▶	ACTION TO TAKE
Q21 CHECK FUEL PUMP RELAY		
NOTE: This check is made if Fuel Pump is always ON • Turn key to OFF • Disconnect Fuel Pump Relay • Listen for Fuel Pump	Fuel Pump OFF ▶ Fuel Pump RUNS ▶	GO to Q22 SERVICE short to power in circuit 787 between Fuel Pump Relay and Fuel Pump
Q22 CHECK CIRCUIT 97 FOR SHORT-TO-GROUND IN HARNESS	DVOM Reading:	
• Using DVOM, check for a short between circuit 97 and ground at fuel pump relay harness connector • Check at the relay connector with relay disconnected FUEL PUMP RELAY; 200 OHMS; CIRCUIT 97	Short ▶ No short ▶	GO to Q23 REPLACE Fuel Pump Relay
Q23 CHECK CIRCUIT 97 FOR SHORT-TO-GROUND WITH ECA DISCONNECTED	DVOM Reading:	
• Disconnect harness from Processor Assembly • Tester to C-4 • Press Ohms button DVOM	150 ohms or less ▶ More than 150 ohms ▶	SERVICE short-to-ground in wire 97 REPLACE Processor Assembly. If still not OK, replace calibration assembly using original processor

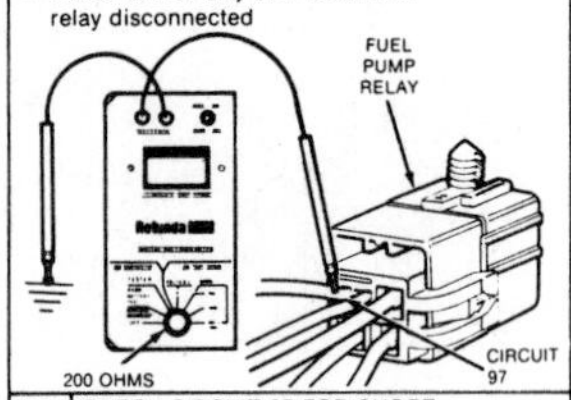

Fuel Pump Relays — Pinpoint Test — Q (EFI)

TEST STEP	RESULT ▶	ACTION TO TAKE
Q26 CHECK IF BALLAST BYPASS RELAY IS ENERGIZED	DVOM Reading:	
• Measure voltage on Circuit 57A at the relay BALLAST BYPASS RELAY; 20 VOLTS D.C.; CIRCUIT 57A	10V or more ▶ Under 10V ▶	GO to Q27 GO to Q29
Q27 CHECK HOSE TO VACUUM SWITCH		
• Check hose from manifold to Vacuum Switch for good connections, leaks, cracks, blockage, etc.	Hose good ▶ Hose faulty ▶	GO to Q28 SERVICE hose
Q28 CHECK VACUUM SWITCH VOLTAGE	DVOM Reading:	
• Use DVOM Test Leads to measure voltage from each Vacuum Switch connector pin to ground (leave switch connected) with vehicle at idle 57A; 57B; 20 VOLTS D.C.	Both readings 10V or more ▶ Both readings under 10V ▶ One reading 10V or more and other reading under 10V ▶	SERVICE open in ground wire 57B SERVICE open or bad connection in circuit 57A between Ballast Bypass Relay and vacuum switch (57A) REPLACE vacuum switch

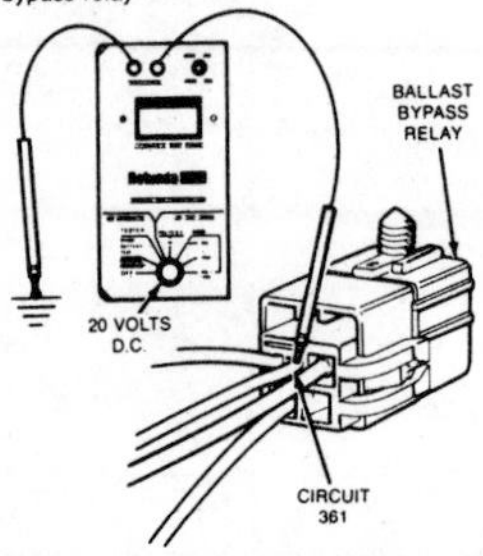

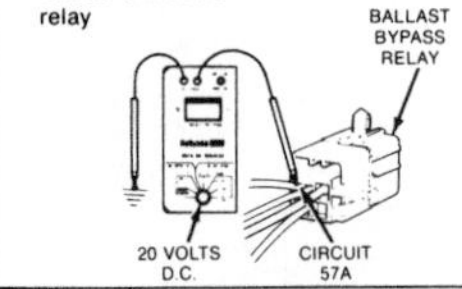

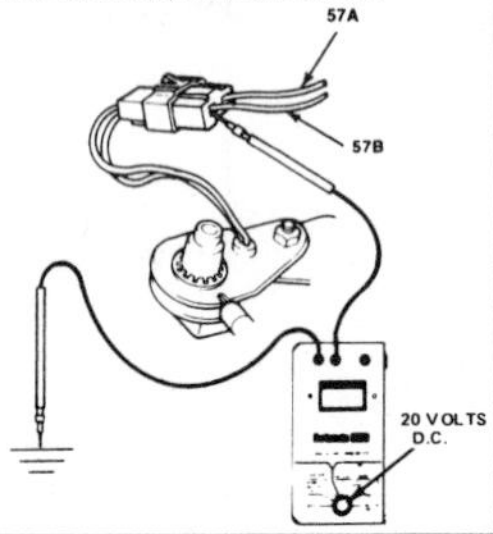

Fuel Pump Relays — Pinpoint Test — Q (EFI)

TEST STEP	RESULT ▶	ACTION TO TAKE
Q24 CHECK FUEL PUMP VOLTAGE	DVOM Reading:	
NOTE: This check is made if Fuel Pump is always noisy • Engine at idle • Measure Fuel Pump voltage at Inertia Switch TO FUEL PUMP RELAY; TO FUEL PUMP; 20 VOLTS D.C.	Under 10V ▶ 10V or more ▶	CHECK Fuel Pump mechanically if noise is excessive GO to Q25
Q25 CHECK VOLTAGE ON CIRCUIT 361 AT THE BALLAST BYPASS RELAY	DVOM Reading:	
• Use DVOM Test Leads to measure voltage on circuit 361 at the ballast bypass relay BALLAST BYPASS RELAY; 20 VOLTS D.C.; CIRCUIT 361	10V or more ▶ Under 10V ▶	GO to Q26 SERVICE open circuit in 361

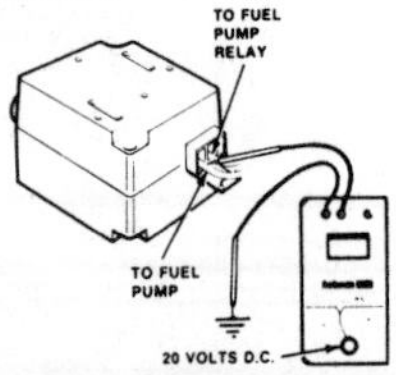

Fuel Pump Relay — Pinpoint Test — Q (EFI)

TEST STEP	RESULT ▶	ACTION TO TAKE
Q29 CHECK "FUEL PUMP BALLAST BYPASS" RELAY	DVOM Reading:	
• Disconnect ballast bypass relay from the harness • Use DVOM Test Lead to measure voltage on circuit 787A at relay connector 787A; 200 VOLTS D.C.	10V or more ▶ Under 10V ▶	SERVICE short to power in circuit 787 between Ballast Bypass Relay and Fuel Pump Relay REPLACE the Ballast Bypass Relay
Q30 FUEL PUMP BALLAST WIRE CHECK	DVOM Reading:	
NOTE: This check is made if vehicle starts but quickly stalls • Turn key to OFF • Disconnect Ballast Bypass Relay from harness • Use DVOM Test Leads to measure resistance of Ballast wire 37; 787A; 200 OHMS	3 ohms or less ▶ Over 3 ohms ▶	GO to A1 SERVICE open circuit or bad connection in Ballast wire circuit 37A

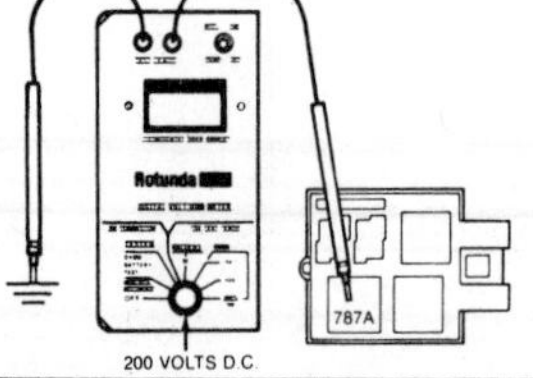

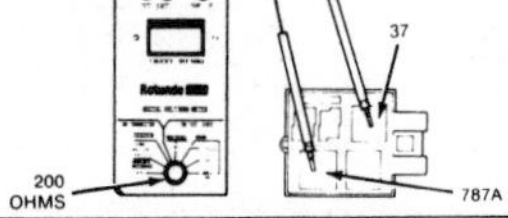

Cranking Signal — Pinpoint Test R (EFI)

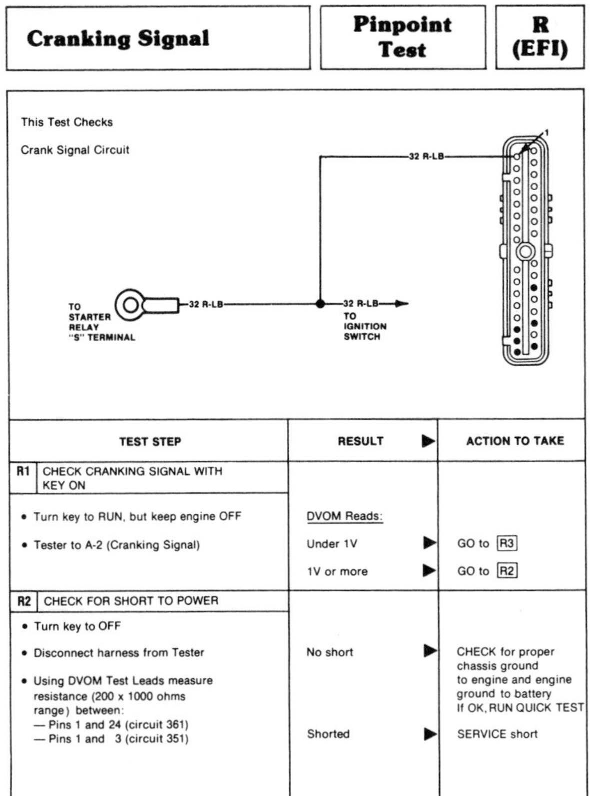

TEST STEP	RESULT ▶	ACTION TO TAKE
R1 CHECK CRANKING SIGNAL WITH KEY ON		
• Turn key to RUN, but keep engine OFF	DVOM Reads:	
• Tester to A-2 (Cranking Signal)	Under 1V ▶	GO to R3
	1V or more ▶	GO to R2
R2 CHECK FOR SHORT TO POWER		
• Turn key to OFF • Disconnect harness from Tester • Using DVOM Test Leads measure resistance (200 x 1000 ohms range) between: — Pins 1 and 24 (circuit 361) — Pins 1 and 3 (circuit 351)	No short ▶	CHECK for proper chassis ground to engine and engine ground to battery If OK, RUN QUICK TEST
	Shorted ▶	SERVICE short

Cranking Signal — Pinpoint Test R (EFI)

TEST STEP	RESULT ▶	ACTION TO TAKE
R3 CHECK FOR OPENS		
• Disconnect wire from Starter Relay "S" terminal • Tester at A-2 • Turn key to START	DVOM Reading:	
	More than 6.5V ▶	RUN QUICK TEST
	6.5V or less ▶	GO to R4
R4 CONTINUE CHECK FOR OPENS		
• Key OFF • Disconnect Tester from EEC harness • Using DVOM Test Leads, measure resistance (200 ohm range) between: — Pin 1 to Starter Relay "S" wire (circuit 32); — Pin 8 to Battery Negative (circuit 57)	DVOM Reading:	
	Under 5 ohms ▶	GO to R5
	5 or more ohms ▶	SERVICE open circuit RECONNECT "S" terminal RUN QUICK TEST
R5 CHECK FOR CRANK SHORT-TO-GROUND		
• Using DVOM Test Leads, measure resistance between: — Pin 1 to Battery Negative (PIN 1; 200X1000 OHMS)	DVOM Reading:	
	More than 10,000 ohms ▶	RECONNECT Tester RECONNECT Starter "S" terminal RUN QUICK TEST
	Under 10,000 ohms ▶	SERVICE short-to-ground in circuit 32 RECONNECT Tester RECONNECT Starter "S" terminal RUN QUICK TEST

Air Charge Temperature — Pinpoint Test S (EFI)

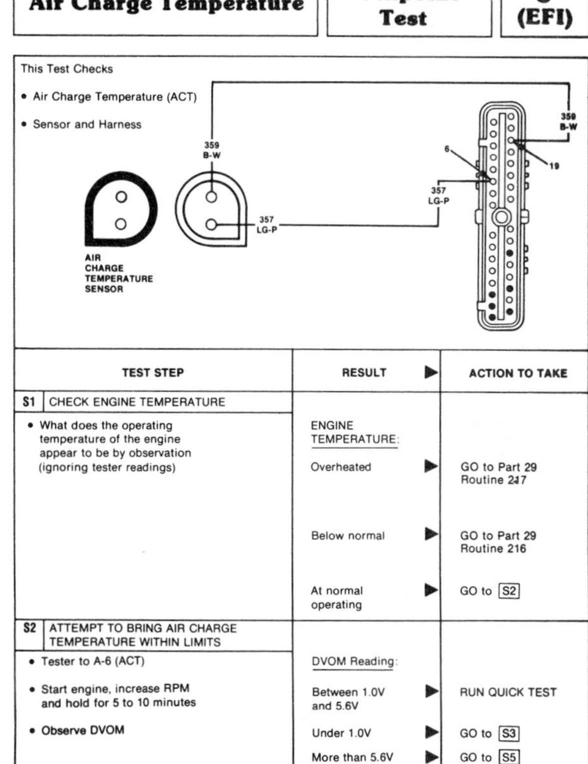

TEST STEP	RESULT ▶	ACTION TO TAKE
S1 CHECK ENGINE TEMPERATURE		
• What does the operating temperature of the engine appear to be by observation (ignoring tester readings)	ENGINE TEMPERATURE:	
	Overheated ▶	GO to Part 29 Routine 247
	Below normal ▶	GO to Part 29 Routine 216
	At normal operating ▶	GO to S2
S2 ATTEMPT TO BRING AIR CHARGE TEMPERATURE WITHIN LIMITS		
• Tester to A-6 (ACT) • Start engine, increase RPM and hold for 5 to 10 minutes • Observe DVOM	DVOM Reading:	
	Between 1.0V and 5.6V ▶	RUN QUICK TEST
	Under 1.0V ▶	GO to S3
	More than 5.6V ▶	GO to S5

Air Charge Temperature — Pinpoint Test S (EFI)

TEST STEP	RESULT ▶	ACTION TO TAKE
S3 CHECK ACT SENSOR RESISTANCE	DVOM Reading:	
• Disconnect harness from ACT sensor • Switch tester to A-6 • Read DVOM (359 B-W; 357 LG-P)	Less than 8.00 ▶	GO to S4
	More than 8.00 ▶	CHECK engine for overtemperate, and let cool down. If this fails to correct problem, replace ACT sensor, and retest
S4 CHECK ACT SENSOR HARNESS FOR SHORTS		
• Turn key to OFF • Use DVOM Test Leads to check for short from pin 6 (ACT) to all remaining 31 connector pins (6; 200 x 1000 OHMS)	DVOM Reading:	
	More than 10,000 ohms ▶	REPLACE ACT SENSOR, and run QUICK TEST. If problem still exists, REPLACE Processor Assembly
	10,000 or less ohms ▶	SERVICE short(s) in EEC harness
S5 CHECK FOR ACT SHORT TO POWER		
• Disconnect EEC Harness from Tester • Turn key to RUN, but keep engine OFF • Use DVOM Test Leads to measure voltage from each Sensor harness pin to chassis ground	DVOM Reading:	
	Under 0.2V on both pins ▶	GO to S6
	0.2V or more on either pin ▶	SERVICE harness short to power

Air Charge Temperature	Pinpoint Test	S (EFI)

TEST STEP	RESULT ▶	ACTION TO TAKE
S6 CHECK ACT SENSOR HARNESS CONTINUITY	DVOM Reading:	
• Turn key to OFF	Under 5.0 ohms ▶	GO to S6a
• Use DVOM Test Leads to check continuity from Sensor connector to the ECA harness connector pins 19 and 6	5.0 or more ohms ▶	SERVICE open or bad harness connection
S6a CHECK ACT SENSOR RESISTANCE	DVOM Reading:	
• Turn key to OFF		
• Disconnect harness from ACT Sensor	1,700-60,000 ohms ▶	REPLACE processor
• Use DVOM Test Leads to measure resistance across sensor pins	Under 1,700 or over 60,000 ohms ▶	REPLACE ACT sensor
S7 LOCATE CAUSE OF CODE 24	DVOM Reading:	
(ECA has interpreted air temperature to be outside the 40-240°F range)	Remains Under 1.0V or more than 5.6V ▶	GO to S1
• Engine at idle	Jumps to under 1.0V or more than 5.6V ▶	SERVICE connection
• Tester to A-6 (ACT)		
• While observing DVOM, wiggle ACT Sensor and harness, checking for bad connections	Remains between 1.0-5.6V ▶	REPLACE Processor Assembly

GENERAL MOTORS MULTI-PORT (MFI) AND SEQUENTIAL (SFI) FUEL INJECTION SYSTEMS

General Information

On 1984 and later non-turbocharged models, a new multi-port fuel injection (MFI) system is available. The MFI system is controlled by an electronic control module (ECM) which monitors engine operations and generates output signals to provide the correct air/fuel mixture, ignition timing and engine idle speed control. Input to the control unit is provided by an oxygen sensor, coolant temperature sensor, detonation sensor, hot film air mass sensor and throttle position sensor. The ECM also receives information concerning engine rpm, road speed, transmission gear position, power steering and air conditioning.

On turbocharged models, a sequential port fuel injection system (SFI) is used for more precise fuel control. With SFI, metered fuel is timed and injected sequentially through six Bosch injectors into individual cylinder ports. Each cylinder receives one injeection per working cycle (every two revolutions), just prior to the opening of the intake valve. The main difference between the two types of fuel injection systems is the manner in which fuel is injected. In the multiport system, all injectors work simultaneously, injecting half the fuel charge each engine revolution. The control units are different for SFI and MFI systems, but most other components are similar. In addition, the SFI system incorporates a new Computer Controlled Coil Ignition system that uses an electronic coil module that replaces the conventional distributor and coil used on most engines. An electronic spark control (ESC) is used to adjust the spark timing.

Both systems use Bosch injectors, one at each intake port, rather than the single injector found on the earlier throttle body system. The injectors are mounted on a fuel rail and are activated by a signal from the electronic control module. The injector is a solenoid-operated valve which remains open depending on the width of the electronic pulses (length of the signal) from the ECM; the longer the open time, the more fuel is injected. In this manner, the air/fuel mixture can be precisely controlled for maximum performance with minimum emissions.

Fuel is pumped from the tank by a high pressure fuel pump, located inside the fuel tank. It is a positive displacement roller vane pump. The impeller serves as a vapor separator and pre-charges the high pressure assembly. A pressure regulator maintains 28-36 psi (28-50 psi on turbocharged engines) in the fuel line to the injectors and the excess fuel is fed back to the tank. On MFI systems, a fuel accumulator is used to dampen the hydraulic line hammer in the system created when all injectors open simultaneously.

The Mass Air Flow Sensor is used to measure the mass of air that is drawn into the engine cylinders. It is located just ahead of the air throttle in the intake system and consists of a heated film which measures the mass of air, rather than just the volume. A resistor is used to measure the temperature of the incoming air and the air mass sensor maintains the temperature of the film at 75 degrees above ambient temperature. As the ambient (outside) air temperature rises, more energy is required to maintain the heated film at the higher temperature and the control unit uses this difference in required energy to calculate the mass of the incoming air. The control unit uses this information to determine the duration of fuel injection pulse, timing and EGR.

The throttle body incorporates an idle air control (IAC) that provides for a bypass channel through which air can flow. It consists

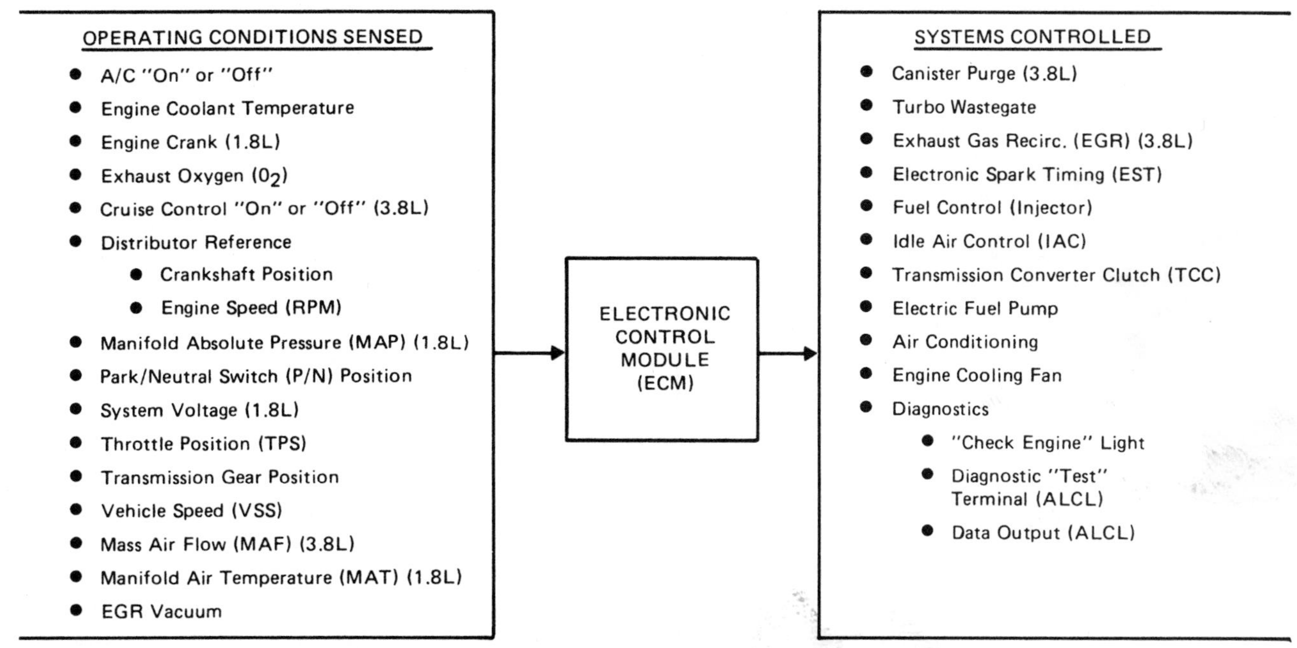

OPERATING CONDITIONS SENSED
A/C "On" or "Off"
Engine Coolant Temperature
Engine Crank (1.8L)
Exhaust Oxygen (O2)
Cruise Control "On" or "Off" (3.8L)
Distributor Reference
Crankshaft Position
Engine Speed (RPM)
Manifold Absolute Pressure (MAP) (1.8L)
Park/Neutral Switch (P/N) Position
System Voltage (1.8L)
Throttle Position (TPS)
Transmission Gear Position
Vehicle Speed (VSS)
Mass Air Flow (MAF) (3.8L)
Manifold Air Temperature (MAT) (1.8L)
EGR Vacuum
ELECTRONIC CONTROL MODULE (ECM)
SYSTEMS CONTROLLED
Canister Purge (3.8L)
Turbo Wastegate
Exhaust Gas Recirc. (EGR) (3.8L)
Electronic Spark Timing (EST)
Fuel Control (Injector)
Idle Air Control (IAC)
Transmission Converter Clutch (TCC)
Electric Fuel Pump
Air Conditioning
Engine Cooling Fan
Diagnostics
"Check Engine" Light
Diagnostic "Test" Terminal (ALCL)
Data Output (ALCL)

Operation schematic of GM MFI and SFI systems. Not all systems are used on all engines

of an orifice and pintle which is controlled by the ECM through a stepper motor. The IAC provides air flow for idle and allows additional air during cold start until the engine reaches operating temperature. As the engine temperature rises, the opening through which air passes is slowly closed.

The throttle position sensor (TPS) provides the control unit with information on throttle position, in order to determine injector pulse width and hence correct mixture. The TPS is connected to the throttle shaft on the throttle body and consists of a potentiometer with one end connected to a 5 volt source from the ECM and the other to ground. A third wire is connected to the ECM to measure the voltage output from the TPS which changes as the throttle valve angle is changed (accelerator pedal moves). At the closed throttle position, the output is low (approximately .4 volts); as the throttle valve opens, the ouput increases to a maximum 5 volts at wide open throttle (WOT). The TPS can be misadjusted open, shorted, or loose and if it is out of adjustment, the idle quality or WOT performance may be poor. A loose TPS can cause intermittent bursts of fuel from the injectors and an unstable idle because the ECM thinks the throttle is moving. This should cause a trouble code to be set. Once a trouble code is set, the ECM will use a preset value for TPS and some vehicle performance may return. A small amount of engine coolant is routed through the throttle assembly to prevent freezing inside the throttle bore during cold operation.

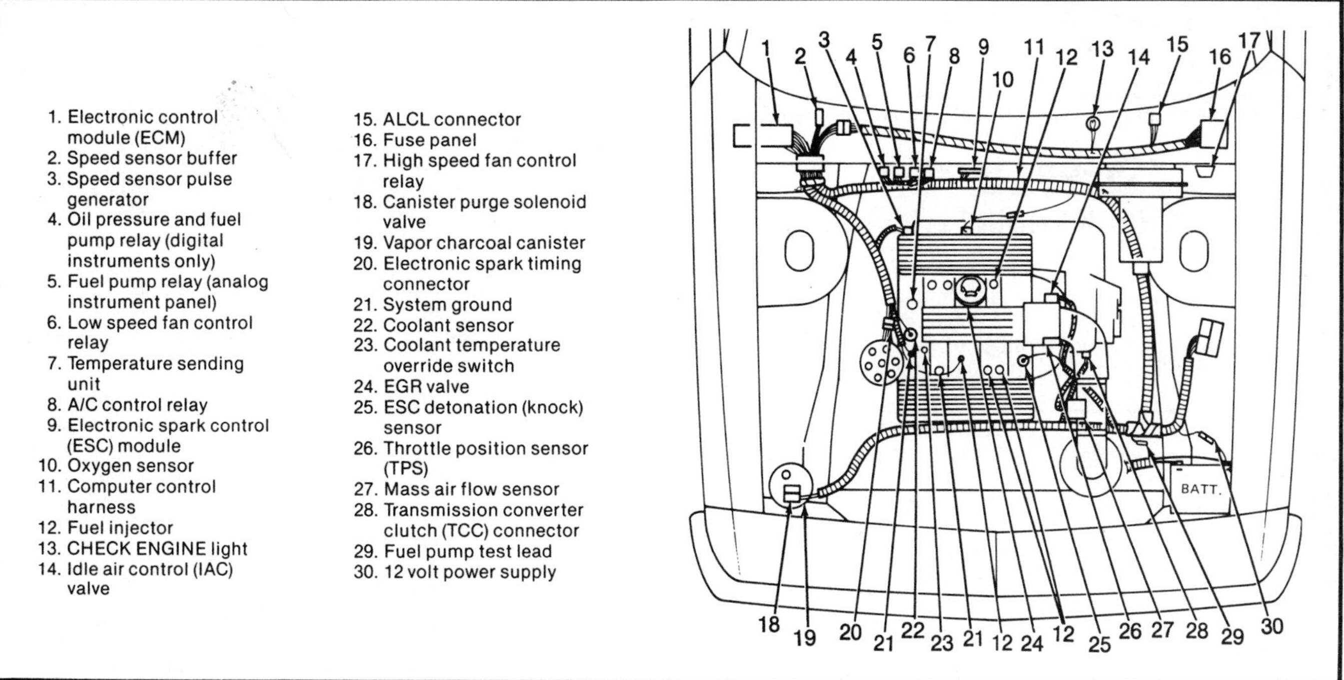

GM MFI component locations on 3.8L V6 engine

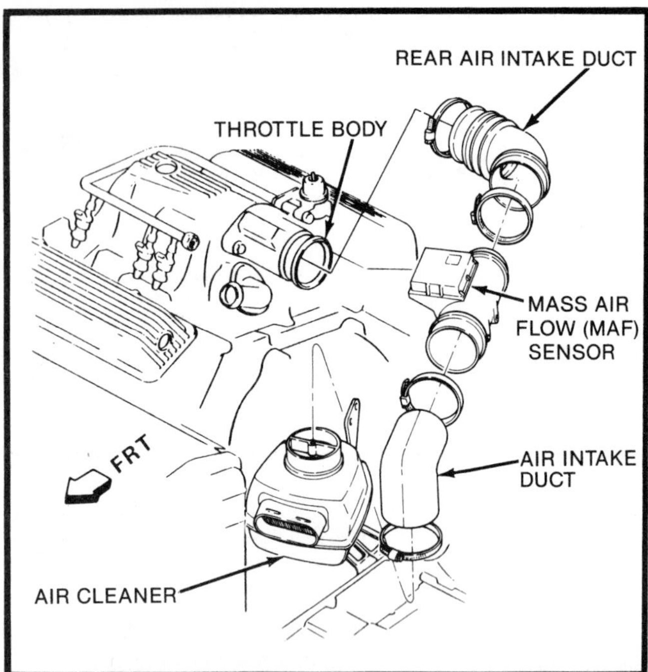

Mass air flow sensor assembly-type

Assembly line communication link (ALCL) connector—typical

On-Car Service

CHECK ENGINE LIGHTS

The "check engine" light on the instrument panel is used as a warning lamp to tell the driver that a problem has occured in the electronic engine control system. When the self-diagnosis mode is activated by grounding the test terminal of the diagnostic connector, the check engine light will flash stored trouble codes to help isolate system problems. The electronic control module (ECM) has a memory that knows what certain engine sensors should be under certain conditions. If a sensor reading is not what the ECM thinks it should be, the control unit will illuminate the check engine light and store a trouble code in its memory. The trouble code indicates what circuit the problem is in, each circuit consisting of a sensor, the wiring harness and connectors to it and the ECM.

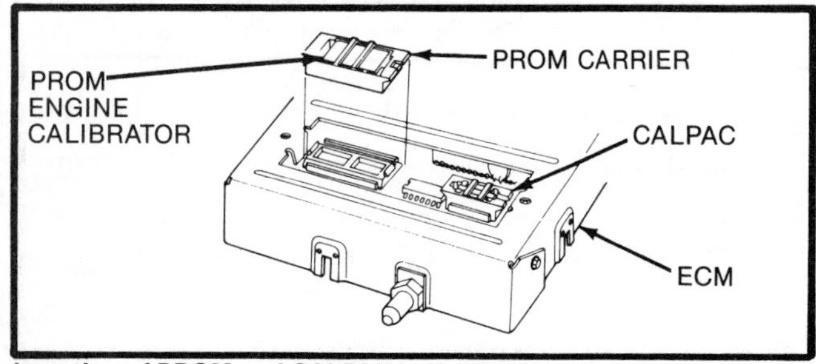

Location of PROM and CALPAK in the ECM

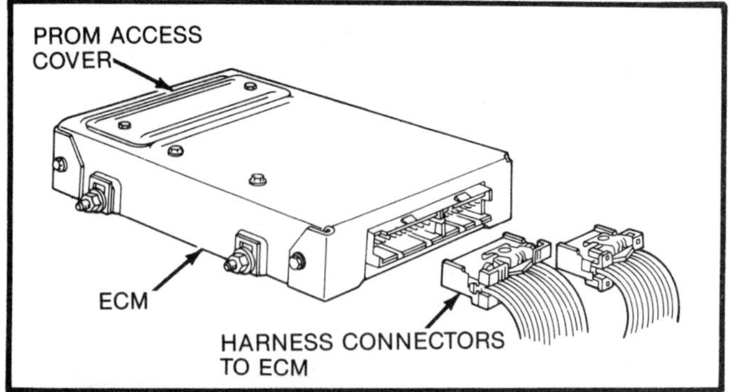

Electronic control module (ECM) used on GM MFI and SFI systems

The Assembly Line Communications Link (ALCL) is a diagnostic connector located in the passenger compartment, usually under the left side of the instrument panel. It has terminals which are used in the assembly plant to check that the engine is operating properly before shipment. Terminal B is the diagnostic test terminal and Terminal A is the ground. By connecting the two terminals together with a jumper wire, the diagnostic mode is activated and the control unit will begin to flash trouble codes using the check engine light.

NOTE: Some models have a "Service Engine Soon" light instead of a "Check Engine" display.

When the test terminal is grounded with the key ON and the engine stopped, the ECM will display code 12 to show that the system is working. The ECM will usually display code 12 three times, then start to display any stored trouble codes. If no trouble codes are stored, the ECM will continue to display code 12 until the test terminal is disconnected. Each trouble code will be flashed three times, then code 12 will display again. The ECM will also energize all controlled relays and solenoids when in the diagnostic mode to check function.

When the test terminal is grounded with the engine running, it will cause the ECM to enter the Field Service Mode. In this mode, the service engine soon light will indicate whether the system is in Open or Closed loop operation. In open loop, the light will flash $2^1/_2$ times per second; in closed loop, the light will flash once per second. In closed loop, the light will stay out most of the time if the system is too lean and will stay on most of the time if the system is too rich.

NOTE: The vehicle may be driven in the Field Service mode and system evaluated at any steady road speed. This mode is useful in diagnosing driveability problems where the system is rich or lean too long.

Trouble codes should be cleared after service is completed. To clear the trouble code memory, disconnect the battery for at least 10 seconds. This may be accomplished by disconnecting the ECM harness from the positive battery pigtail or by removing the ECM fuse.

CAUTION

The ignition switch must be OFF when disconnecting or reconnecting power to the ECM. The vehicle should be driven after the ECM memory is cleared to allow the system to readjust itself. The vehicle should be driven at part throttle under moderate acceleration with the engine at normal operating temperature. A change in performance should be noted initially, but normal performance should return quickly.

GM PORT INJECTION TROUBLE CODES

Trouble Code	Circuit
12	Normal operation
13	Oxygen sensor
14	Coolant sensor (low voltage)
15	Coolant sensor (high voltage)
21	Throttle position sensor (high voltage)
22	Throttle position sensor (low voltage)
24	Speed sensor
32	EGR vacuum control
33	Mass air flow sensor
34	Mass air flow sensor
42	Electronic spark timing
43	Electronic spark control
44	Lean exhaust
45	Rich exhaust
51	PROM failure
52	CALPAK
55	ECM failure

FUEL SYSTEM PRESSURE TEST

When the ignition switch is turned ON, the in-tank fuel pump is energized for as long as the engine is cranking or running and the control unit is receiving signals from the HEI distributor. If there are no reference pulses, the control unit will shut off the fuel pump within two seconds. The pump will deliver fuel to the fuel rail and injectors, then the pressure regulator where the system pressure is controlled to maintain 26-46 psi.

1. Connect pressure gauge J-34370-1, or equivalent, to fuel pressure test point on the fuel rail. Wrap a rag around the pressure tap to absorb any leakage that may occur when installing the gauge.
2. Turn the ignition ON and check that pump pressure is 34-40 psi. This pressure is controlled by spring pressure within the regulator assembly.
3. Start the engine and allow it to idle. The fuel pressure should drop to 28-32 psi due to the lower manifold pressure.

NOTE: The idle pressure will vary somewhat depending on barometric pressure. Check for a drop in pressure indicating regulator control, rather than specific values.

4. On turbocharged models, use a low pressure air pump to apply air pressure to the regulator to simulate turbocharger boost pressure. Boost pressure should increase fuel pressure one pound for every pound of boost. Again, look for changes rather than specific pressures. The maximum fuel pressure should not exceed 46 psi.
5. If the fuel pressure drops, check the operation of the check valve, the pump coupling connection, fuel pressure regulator valve and the injectors. A restricted fuel line or filter may also cause a pressure drop. To check the fuel pump output, restrict the fuel return line and run 12 volts to the pump. The fuel pressure should rise to approximately 75 psi with the return line restricted.

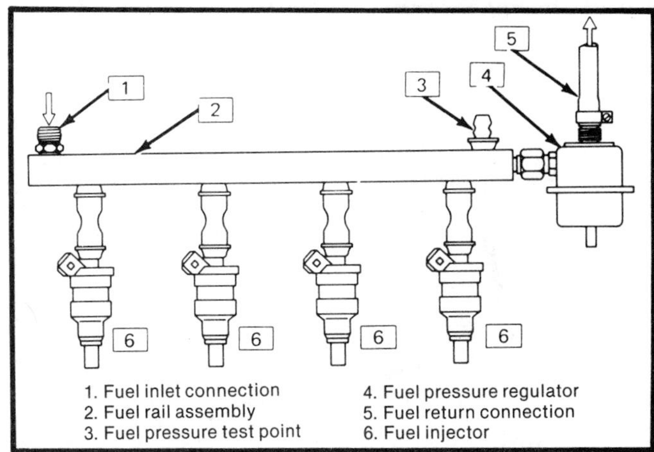

Fuel rail assembly

CAUTION

Before attempting to remove or service any fuel system component, it is necessary to relieve the fuel system pressure.

Relieving Fuel System Pressure

1. Remove the fuel pump fuse from the fuse block.
2. Start the engine. It should run and then stall when the fuel in the lines is exhausted. When the engine stops, crank the starter for about three seconds to make sure all pressure in the fuel lines is released.
3. Replace the fuel pump fuse.

Component Removal

FUEL INJECTORS

Removal and Installation

Use care in removing the fuel injectors to prevent damage to the electrical connector pins on the injector and the nozzle. The fuel injector is serviced as a complete assembly only and should not be immersed in any kind of cleaner.

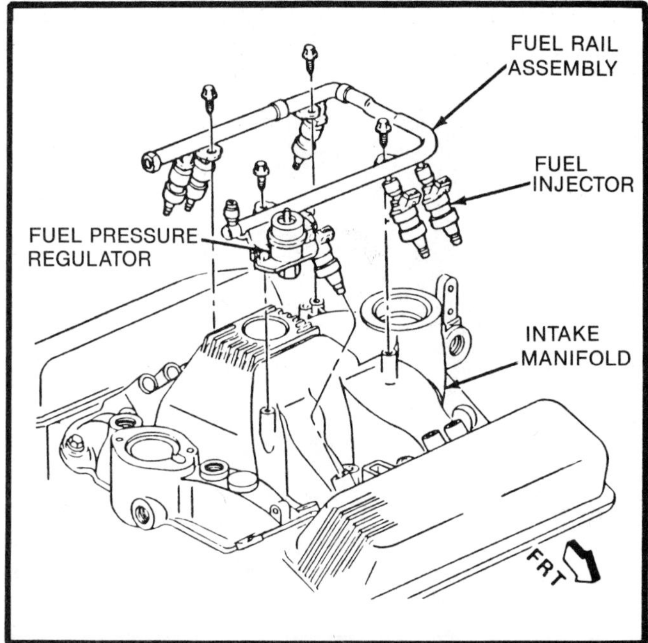

Fuel rail and injectors—3.8L turbocharged engine

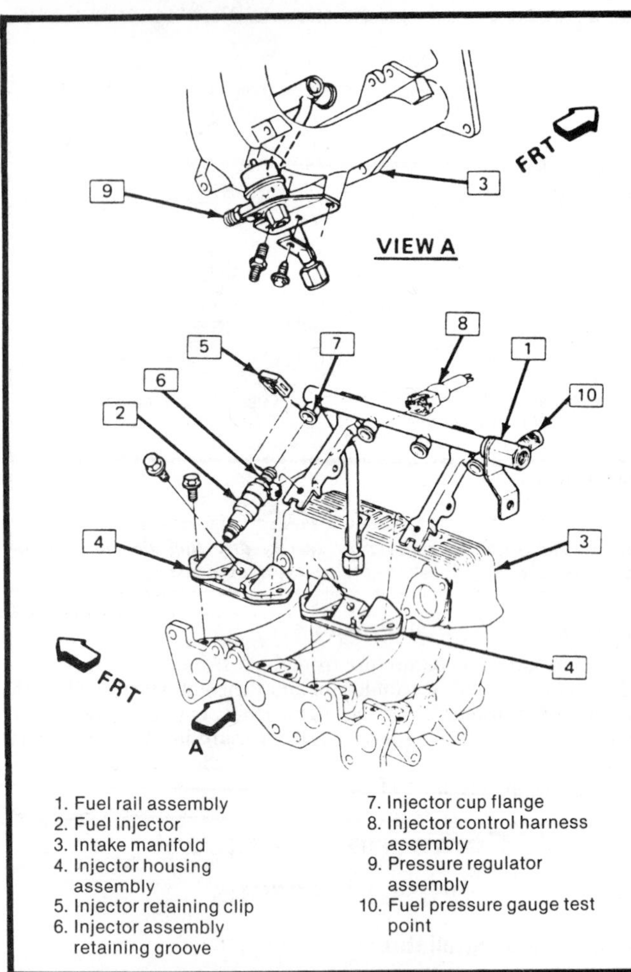

Fuel rail and injector assembly on 4 cyl. engine

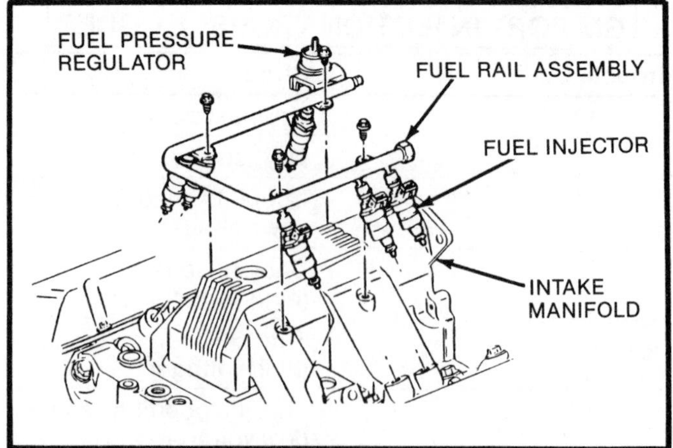

Fuel rail and injectors—3.8L V6 engine

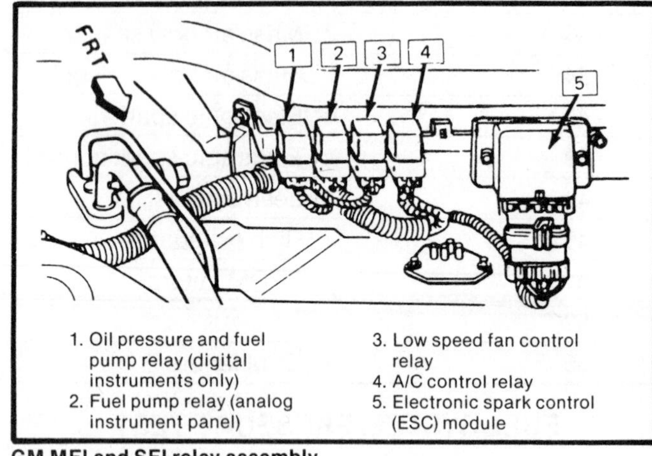

GM MFI and SFI relay assembly

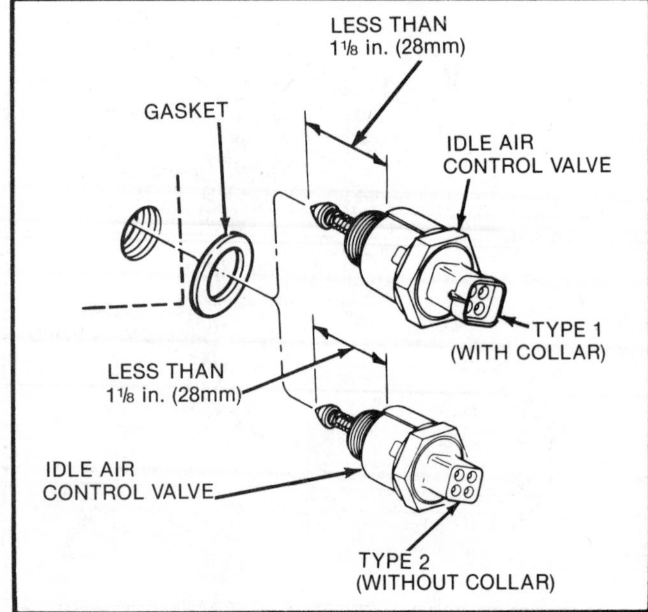

Idle air control valve installation

1. Relieve fuel system pressure.
2. Remove the injector electrical connections.
3. Remove the fuel rail.
4. Separate the injector from the fuel rail.
5. Installation is the reverse of removal. Replace the O-rings when installing injectors into intake manifold.

FUEL PRESSURE REGULATOR

Removal and Installation

1. Relieve fuel system pressure.
2. Remove pressure regulator from fuel rail. Place a rag around the base of the regulator to catch any spilled fuel.
3. Installation is the reverse of removal.

IDLE AIR CONTROL VALVE

Removal and Installation

1. Remove electrical connector from idle air control valve.
2. Remove the idle air control valve using a suitable wrench.
3. Installation is the reverse of removal. Before installing the idle air control valve, measure the distance that the valve is extended. Measurement should be made from the motor housing to the end of the cone. The distance should not exceed 1⅛ inches, or damage to the valve may occur when installed. Use a new gasket and turn the ignition on then off again to allow the ECM to reset the idle air control valve.

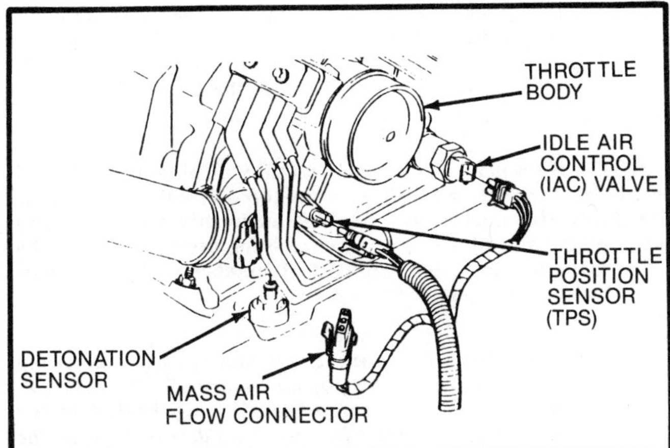

Engine sensors on GM MFI and SFI systems—3.8L engine shown

NOTE: Identify replacement IAC valve as being either Type 1 (with collar at electric terminal end) or Type 2 (without collar). If measuring distance is greater than specified above, proceed as follows:

Type 1: Press on valve firmly to retract it

Type 2: Compress retaining spring from valve while turning valve in with a clockwise motion. Return spring to original position with straight portion of spring end aligned with flat surface of valve.

THROTTLE POSITION SENSOR

Removal and Installation

1. Disconnect the electrical connector from the sensor.
2. Remove the attaching screws, lockwashers and retainers.
3. Remove the throttle position sensor. If necessary, remove the screw holding the actuator to the end of the throttle shaft.
4. With the throttle valve in the normal closed idle position, install the throttle position sensor on the throttle body assembly, making sure the sensor pickup lever is located above the tang on the throttle actuator lever.
5. Install the retainers, screws and lockwashers using a thread locking compound. DO NOT tighten the screws until the throttle position switch is adjusted.

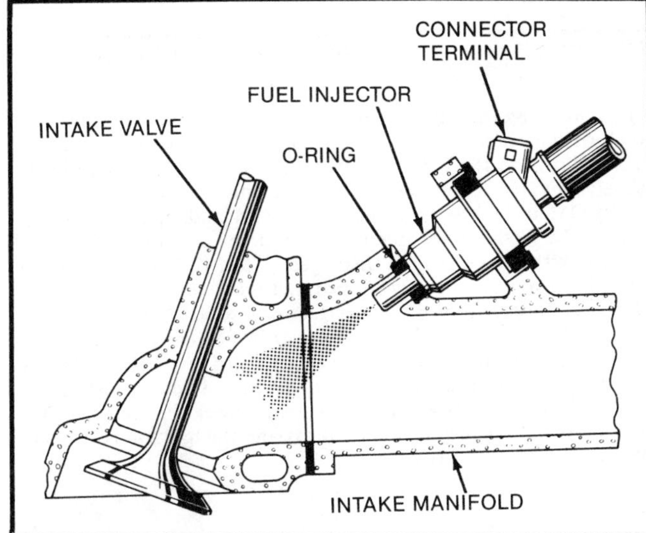

Typical port fuel injector installation

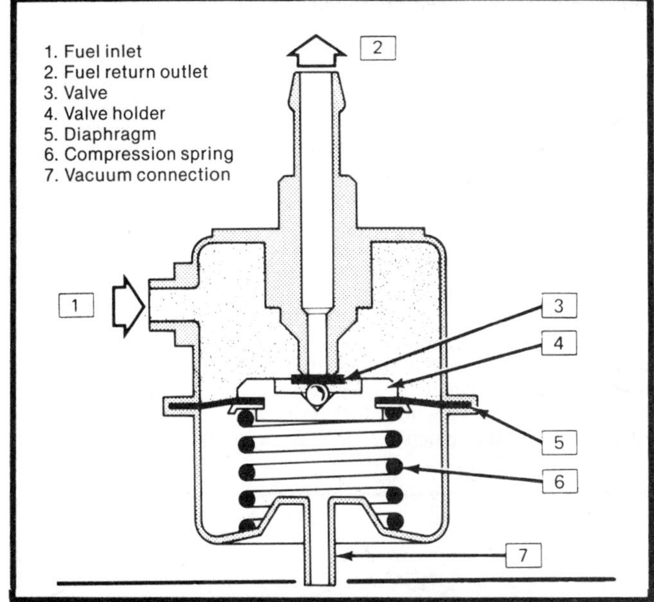

Cross section of typical fuel pressure regulator

6. Install three jumper wires between the throttle position switch and the harness connector.
7. With the ignition switch ON, use a digital voltmeter connected to terminals B and C and adjust the switch to obtain 0.35-0.45 volts.
8. Tighten the mounting screws, then recheck the reading to insure that the adjustment hasn't changed.
9. Turn ignition OFF, remove jumper wires, then reconnect harness to throttle position switch.

OXYGEN SENSOR

Removal and Installation

NOTE: The oxygen sensor uses a permanently attached pigtail and connector. This pigtail should not be removed from the oxygen sensor. Damage or removal of the pigtail or connector could affect proper operation of the oxygen sensor.

The oxygen sensor is installed in the exhaust manifold and is removed in the same manner as a spark plug. The sensor may be

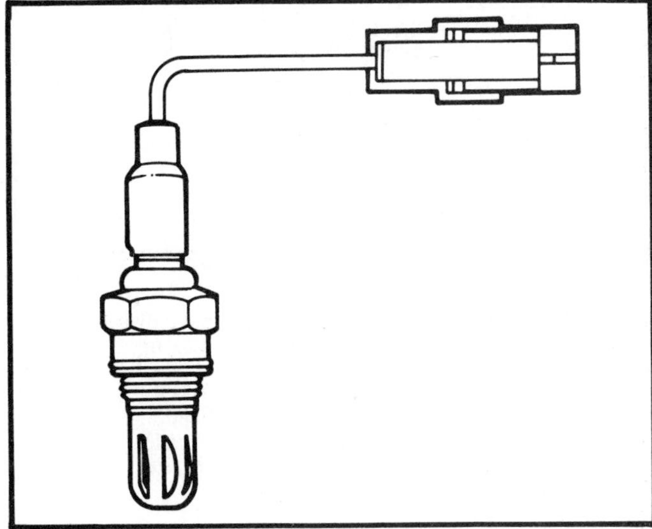
Typical oxygen sensor

difficult to remove when the engine temperature is below 120 deg. F (48 deg. C) and excessive force may damage threads in the exhaust manifold or exhaust pipe. Exercise care when handling the oxygen sensor; the electrical connector and louvered end must be kept free of grease, dirt, or other contaminants. Avoid using cleaning solvents of any kind and don't drop or roughly handle the sensor. A special anti-seize compound is used on the oxygen sensor threads when installing and care should be used NOT to get compound on the sensor itself. Disconnect the negative battery cable when servicing the oxygen sensor and torque to 30 ft. lbs. (41 Nm) when installing.

ELECTRONIC CONTROL MODULE (ECM)

The electronic control module (ECM) is located under the instrument panel. To allow one model of ECM to be used on different models, a device called a calibrator or PRCM (Programmable Read Only Memory) is installed inside the ECM which contains information on the vehicle weight, engine, transmission, axle ratio,

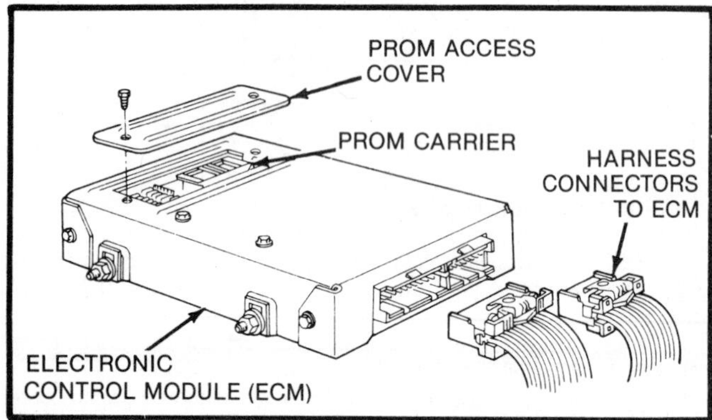

Location of PROM and CALPAK microprocessors in the control unit

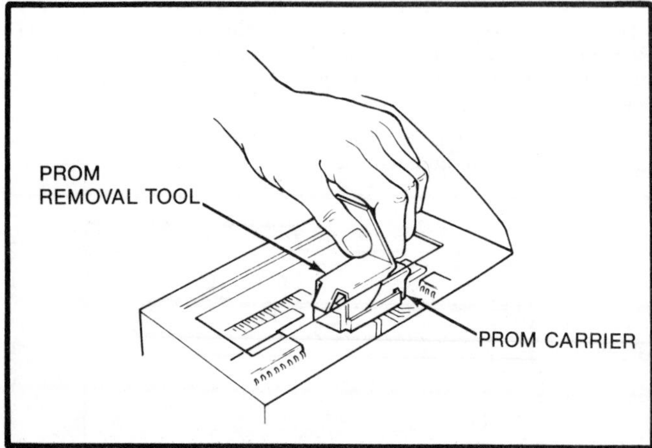

Remove the PROM using the special removal tool

etc. The PROM is specific to the exact model and replacement part numbers must be checked carefully to make sure the correct PROM is being installed during service. Replacement ECM units (called Controllers) are supplied WITHOUT a PROM. The PROM from the old ECM must be carefully removed and installed in the replacement unit during service. Another device called a CALPAK is used to allow fuel delivery if other parts of the ECM are damaged (the "limp home" mode). The CALPAK is similiar in appearance to the PROM and is located in the same place in the ECM, under an access cover. Like the PROM, the CALPAK must be removed and transferred to the new ECM unit being installed.

NOTE: If the diagnosis indicates a faulty ECM unit, the PROM should be checked to see if they are the correct parts. Trouble code 51 indicates that the PROM is installed incorrectly. When replacing the production ECM with a new part, it is important to transfer the Broadcast code and production ECM number to the new part label. Do not record on the ECM cover.

CAUTION

The ignition must be OFF whenever disconnecting or connecting the ECM electrical harness. It is possible to install a PROM backwards during service. Exercise care when replacing the PROM that it is installed correctly, or the PROM will be destroyed when the ignition is switched ON.

Removal and Installation

To remove the ECM, first disconnect the battery. Remove the wiring harness and mounting hardware, then remove the ECM from the passenger compartment. The PROM and CALPAK are located under the access cover on the top of the control unit. Using the rocker type PROM removal tool, or equivalent, engage one end of the PROM carrier with the hook end of the tool. Press on the vertical bar end of the tool and rock the engaged end of the PROM carrier up as far as possible. Engage the opposite end of the PROM carrier in the same manner and rock this end up as far as possible. Repeat this process until the PROM carrier and PROM are free of the socket. The PROM carrier should only be removed with the removal tool or damage to the PROM or PROM socket may occur.

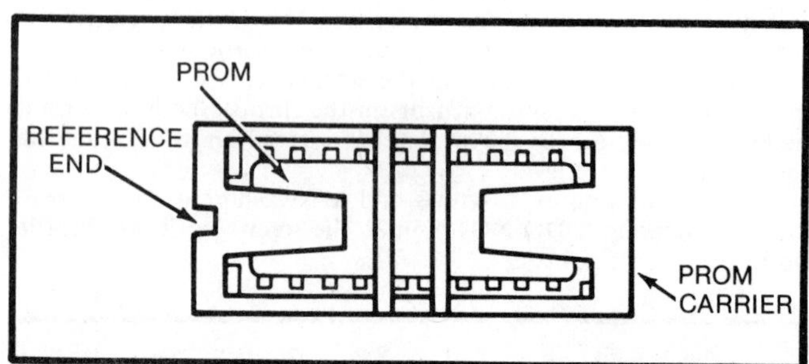

Note the position of the reference notch when installing the PROM

When installing the PROM carrier in the PROM socket, the small notch of the carrier should be aligned with the small notch in the socket. Press on the PROM carrier until it is firmly seated in the socket. DO NOT press on the PROM; only the carrier. To check the PROM installation, reinstall the ECM and turn the ignition switch ON. Activate the diagnostic mode as previously described and check that a code 12 is displayed. Code 12 indicates that the PROM is installed correctly and is functioning normally. If trouble code 51 is displayed, or the "service engine soon" light is on steadily with no codes, the PROM is not fully seated, installed backwards, has bent pins, or is defective. Bent pins may be straightened and the PROM can be seated properly with a gentle push, but a PROM that has been installed backwards should be replaced. Any time the PROM is installed backwards and the ignition is switched ON, the PROM is destroyed.

Schematic

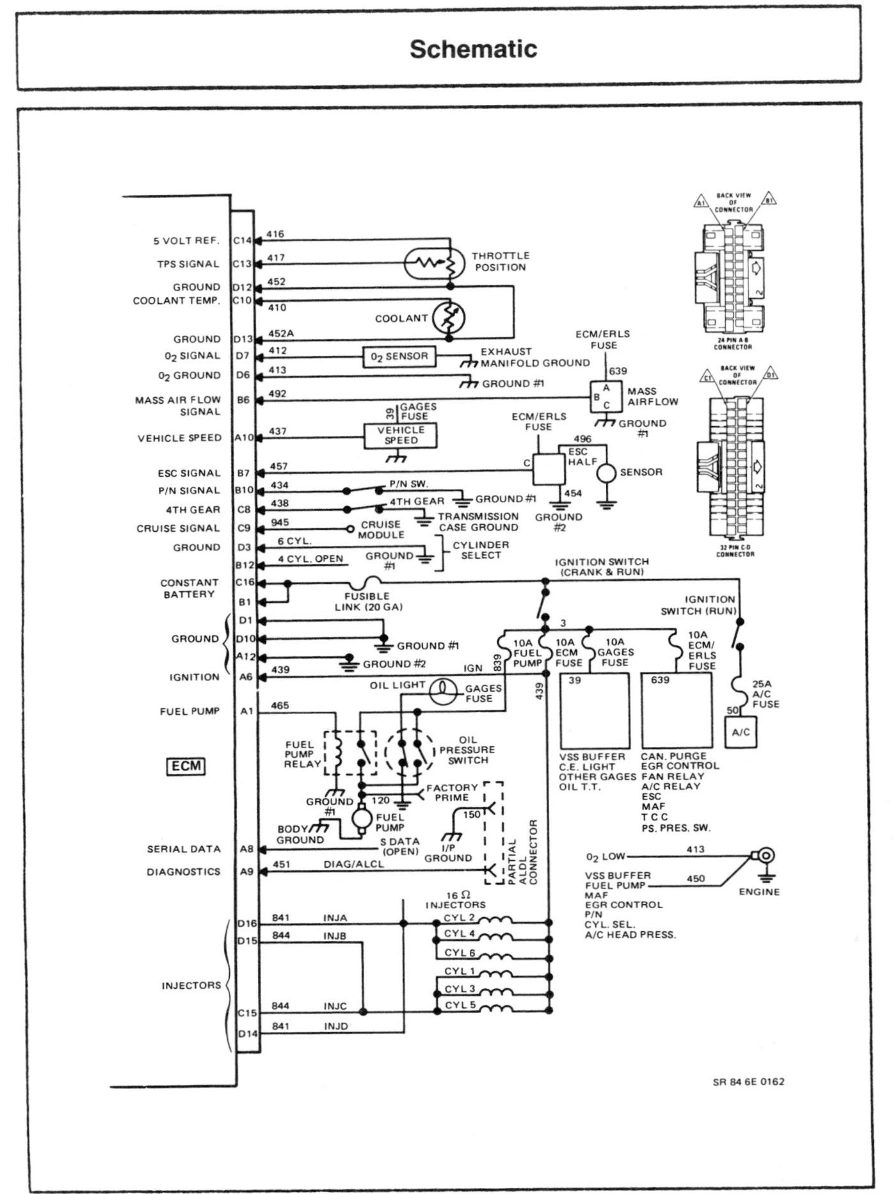

ECM Connector ID

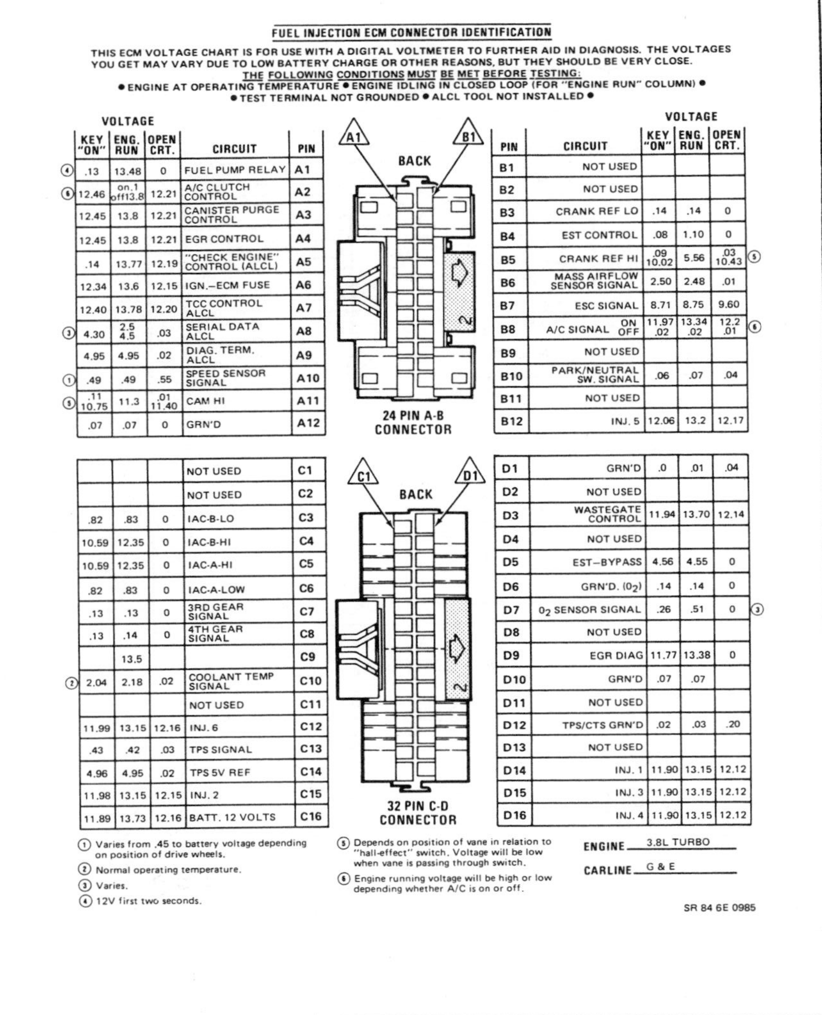

FUEL INJECTION ECM CONNECTOR IDENTIFICATION

THIS ECM VOLTAGE CHART IS FOR USE WITH A DIGITAL VOLTMETER TO FURTHER AID IN DIAGNOSIS. THE VOLTAGES YOU GET MAY VARY DUE TO LOW BATTERY CHARGE OR OTHER REASONS, BUT THEY SHOULD BE VERY CLOSE.

THE FOLLOWING CONDITIONS MUST BE MET BEFORE TESTING:

● ENGINE AT OPERATING TEMPERATURE ● ENGINE IDLING IN CLOSED LOOP (FOR "ENGINE RUN" COLUMN) ● ● TEST TERMINAL NOT GROUNDED ● ALCL TOOL NOT INSTALLED ●

Note	VOLTAGE KEY "ON"	ENG. RUN	OPEN CRT.	CIRCUIT	PIN
④	.13	13.48	0	FUEL PUMP RELAY	A1
⑥	12.46	on 1 off 13.8	12.21	A/C CLUTCH CONTROL	A2
	12.45	13.8	12.21	CANISTER PURGE CONTROL	A3
	12.45	13.8	12.21	EGR CONTROL	A4
	.14	13.77	12.19	"CHECK ENGINE" CONTROL (ALCL)	A5
	12.34	13.6	12.15	IGN.–ECM FUSE	A6
	12.40	13.78	12.20	TCC CONTROL ALCL	A7
③	4.30	2.5 4.5	.03	SERIAL DATA ALCL	A8
	4.95	4.95	.02	DIAG. TERM. ALCL	A9
①	.49	.49	.55	SPEED SENSOR SIGNAL	A10
⑤	.11 10.75	11.3	.01 11.40	CAM HI	A11
	.07	.07	0	GRN'D	A12

24 PIN A-B CONNECTOR (BACK)

PIN	CIRCUIT	VOLTAGE KEY "ON"	ENG. RUN	OPEN CRT.	Note
B1	NOT USED				
B2	NOT USED				
B3	CRANK REF LO	.14	.14	0	
B4	EST CONTROL	.08	1.10	0	
B5	CRANK REF HI	.09 10.02	5.56	.03 10.43	⑤
B6	MASS AIRFLOW SENSOR SIGNAL	2.50	2.48	.01	
B7	ESC SIGNAL	8.71	8.75	9.60	
B8	A/C SIGNAL ON / OFF	11.97 .02	13.34 .02	12.2 .01	⑥
B9	NOT USED				
B10	PARK/NEUTRAL SW. SIGNAL	.06	.07	.04	
B11	NOT USED				
B12	INJ. 5	12.06	13.2	12.17	

Note	KEY "ON"	ENG. RUN	OPEN CRT.	CIRCUIT	PIN
				NOT USED	C1
				NOT USED	C2
	.82	.83	0	IAC-B-LO	C3
	10.59	12.35	0	IAC-B-HI	C4
	10.59	12.35	0	IAC-A-HI	C5
	.82	.83	0	IAC-A-LOW	C6
	.13	.13	0	3RD GEAR SIGNAL	C7
	.13	.14	0	4TH GEAR SIGNAL	C8
		13.5			C9
②	2.04	2.18	.02	COOLANT TEMP SIGNAL	C10
				NOT USED	C11
	11.99	13.15	12.16	INJ. 6	C12
	.43	.42	.03	TPS SIGNAL	C13
	4.96	4.95	.02	TPS 5V REF	C14
	11.98	13.15	12.15	INJ. 2	C15
	11.89	13.73	12.16	BATT. 12 VOLTS	C16

32 PIN C-D CONNECTOR (BACK)

PIN	CIRCUIT	KEY "ON"	ENG. RUN	OPEN CRT.	Note
D1	GRN'D	.0	.01	.04	
D2	NOT USED				
D3	WASTEGATE CONTROL	11.94	13.70	12.14	
D4	NOT USED				
D5	EST–BYPASS	4.56	4.55	0	
D6	GRN'D. (O_2)	.14	.14	0	
D7	O_2 SENSOR SIGNAL	.26	.51	0	③
D8	NOT USED				
D9	EGR DIAG	11.77	13.38	0	
D10	GRN'D	.07	.07		
D11	NOT USED				
D12	TPS/CTS GRN'D	.02	.03	.20	
D13	NOT USED				
D14	INJ. 1	11.90	13.15	12.12	
D15	INJ. 3	11.90	13.15	12.12	
D16	INJ. 4	11.90	13.15	12.12	

① Varies from .45 to battery voltage depending on position of drive wheels.
② Normal operating temperature.
③ Varies.
④ 12V first two seconds.
⑤ Depends on position of vane in relation to "hall-effect" switch. Voltage will be low when vane is passing through switch.
⑥ Engine running voltage will be high or low depending whether A/C is on or off.

ENGINE 3.8L TURBO
CARLINE G & E

SR 84 6E 0985

Diagnostic Circuit Check—3.8L

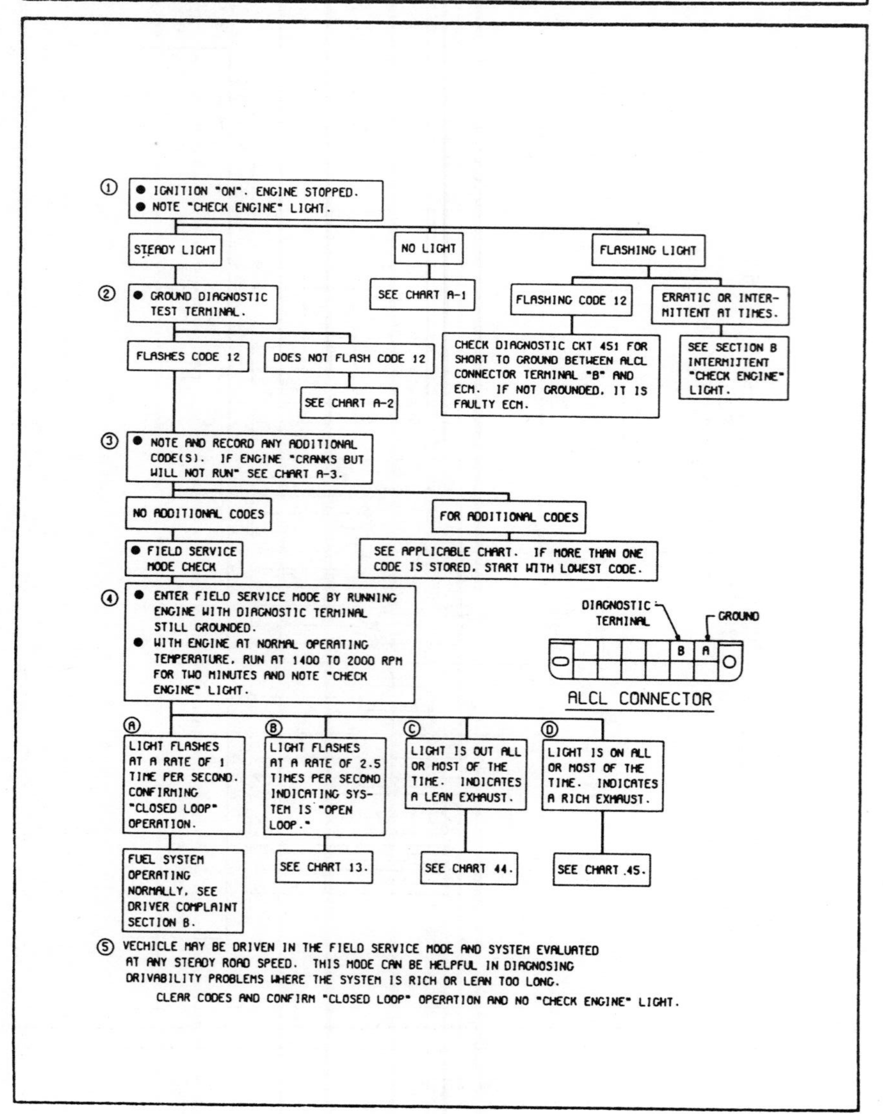

Schematic

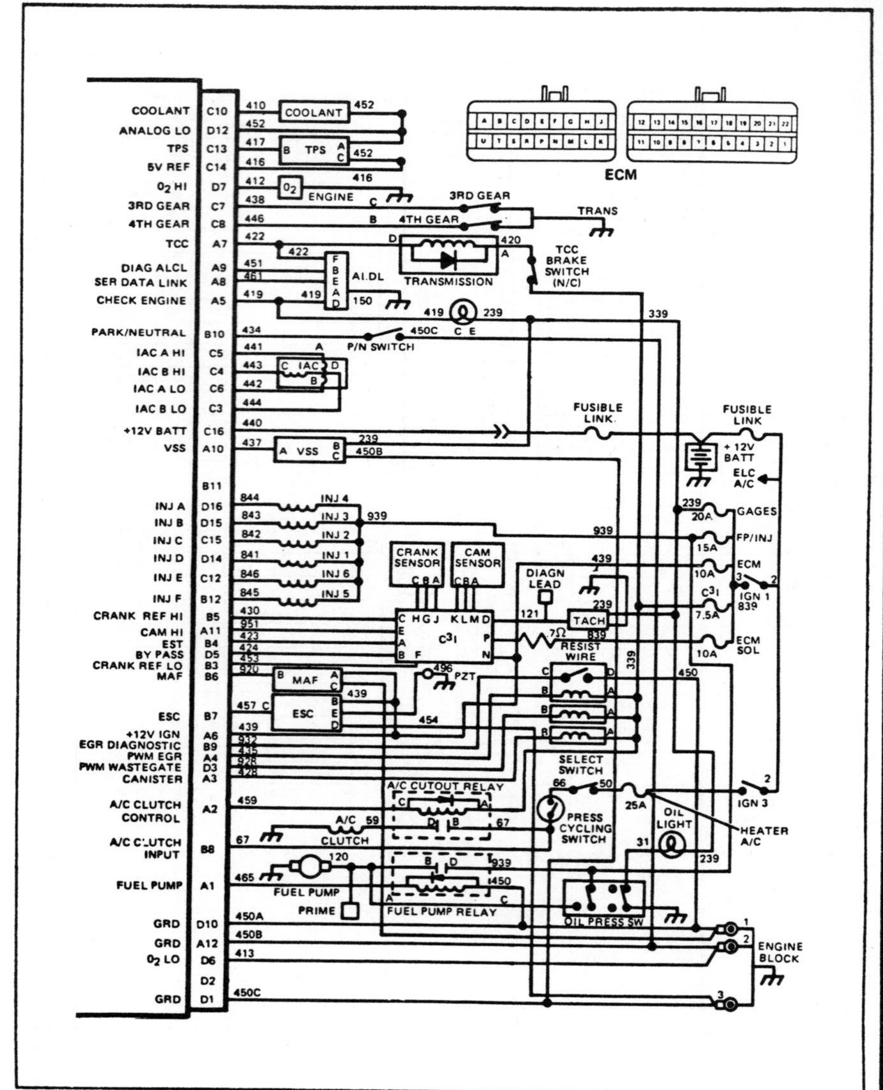

CHART A-1
No "Check Engine" Light—3.8L

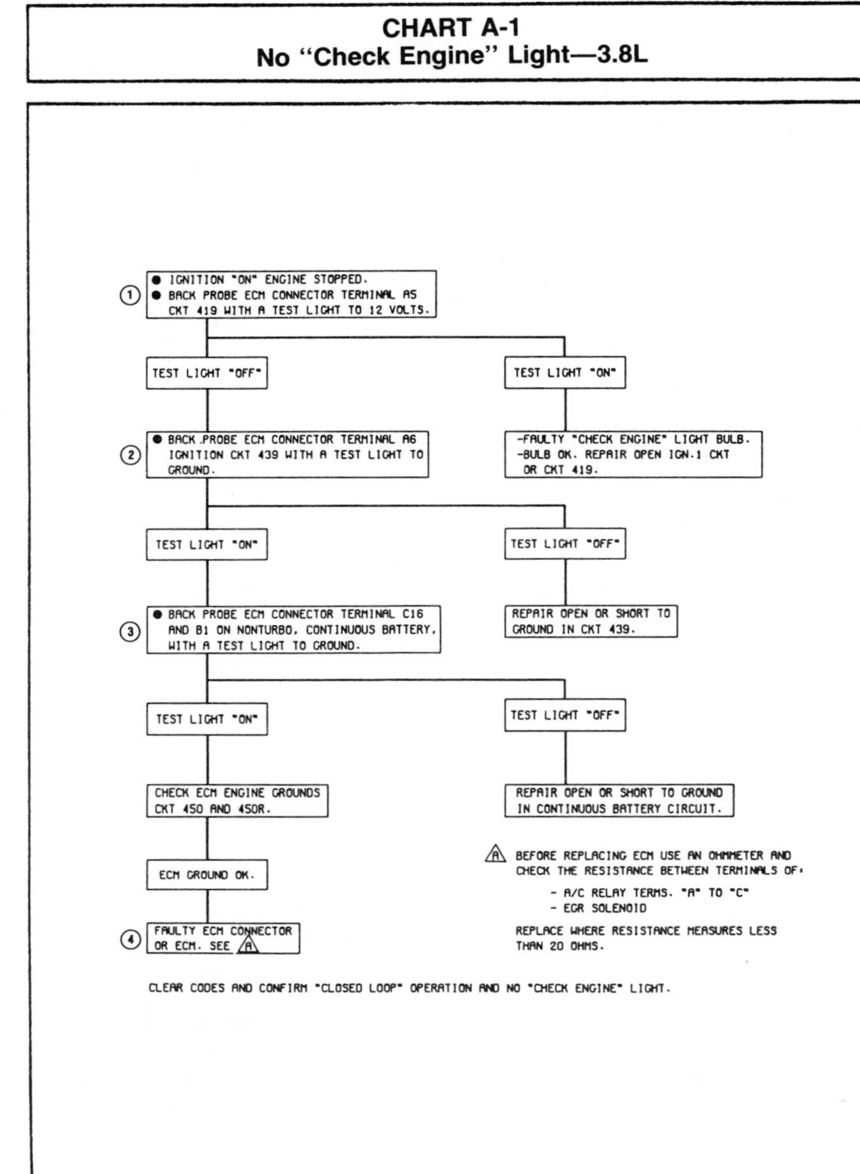
① ● IGNITION "ON" ENGINE STOPPED.
● BACK PROBE ECM CONNECTOR TERMINAL A5 CKT 419 WITH A TEST LIGHT TO 12 VOLTS.
TEST LIGHT "OFF"
TEST LIGHT "ON"
② ● BACK PROBE ECM CONNECTOR TERMINAL A6 IGNITION CKT 439 WITH A TEST LIGHT TO GROUND.
-FAULTY "CHECK ENGINE" LIGHT BULB.
-BULB OK. REPAIR OPEN IGN.1 CKT OR CKT 419.
TEST LIGHT "ON"
TEST LIGHT "OFF"
③ ● BACK PROBE ECM CONNECTOR TERMINAL C16 AND B1 ON NONTURBO, CONTINUOUS BATTERY, WITH A TEST LIGHT TO GROUND.
REPAIR OPEN OR SHORT TO GROUND IN CKT 439.
TEST LIGHT "ON"
TEST LIGHT "OFF"
CHECK ECM ENGINE GROUNDS CKT 450 AND 450R.
REPAIR OPEN OR SHORT TO GROUND IN CONTINUOUS BATTERY CIRCUIT.
ECM GROUND OK.
④ FAULTY ECM CONNECTOR OR ECM. SEE A
A BEFORE REPLACING ECM USE AN OHMMETER AND CHECK THE RESISTANCE BETWEEN TERMINALS OF:
- A/C RELAY TERMS. "A" TO "C"
- EGR SOLENOID
REPLACE WHERE RESISTANCE MEASURES LESS THAN 20 OHMS.
CLEAR CODES AND CONFIRM "CLOSED LOOP" OPERATION AND NO "CHECK ENGINE" LIGHT.

CHART A-2
Won't Flash Code 12—3.8L

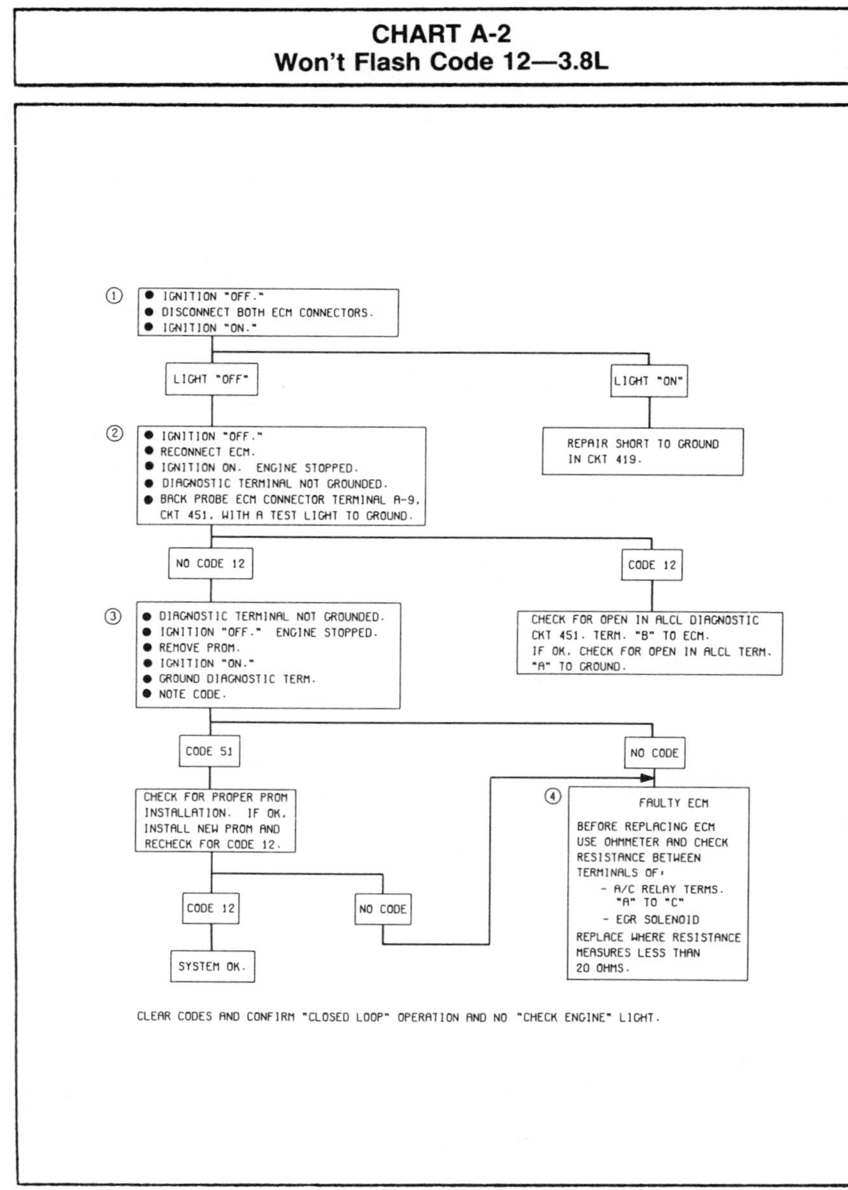
① ● IGNITION "OFF."
● DISCONNECT BOTH ECM CONNECTORS.
● IGNITION "ON."
LIGHT "OFF"
LIGHT "ON"
② ● IGNITION "OFF."
● RECONNECT ECM.
● IGNITION ON. ENGINE STOPPED.
● DIAGNOSTIC TERMINAL NOT GROUNDED.
● BACK PROBE ECM CONNECTOR TERMINAL A-9, CKT 451, WITH A TEST LIGHT TO GROUND.
REPAIR SHORT TO GROUND IN CKT 419.
NO CODE 12
CODE 12
③ ● DIAGNOSTIC TERMINAL NOT GROUNDED.
● IGNITION "OFF." ENGINE STOPPED.
● REMOVE PROM.
● IGNITION "ON."
● GROUND DIAGNOSTIC TERM.
● NOTE CODE.
CHECK FOR OPEN IN ALCL DIAGNOSTIC CKT 451. TERM. "B" TO ECM.
IF OK, CHECK FOR OPEN IN ALCL TERM. "A" TO GROUND.
CODE 51
NO CODE
CHECK FOR PROPER PROM INSTALLATION. IF OK, INSTALL NEW PROM AND RECHECK FOR CODE 12.
④ FAULTY ECM
BEFORE REPLACING ECM USE OHMMETER AND CHECK RESISTANCE BETWEEN TERMINALS OF:
- A/C RELAY TERMS. "A" TO "C"
- EGR SOLENOID
REPLACE WHERE RESISTANCE MEASURES LESS THAN 20 OHMS.
CODE 12
NO CODE
SYSTEM OK.
CLEAR CODES AND CONFIRM "CLOSED LOOP" OPERATION AND NO "CHECK ENGINE" LIGHT.

CHART A-3
Engine Cranks, But Will Not Run—1984 3.8L

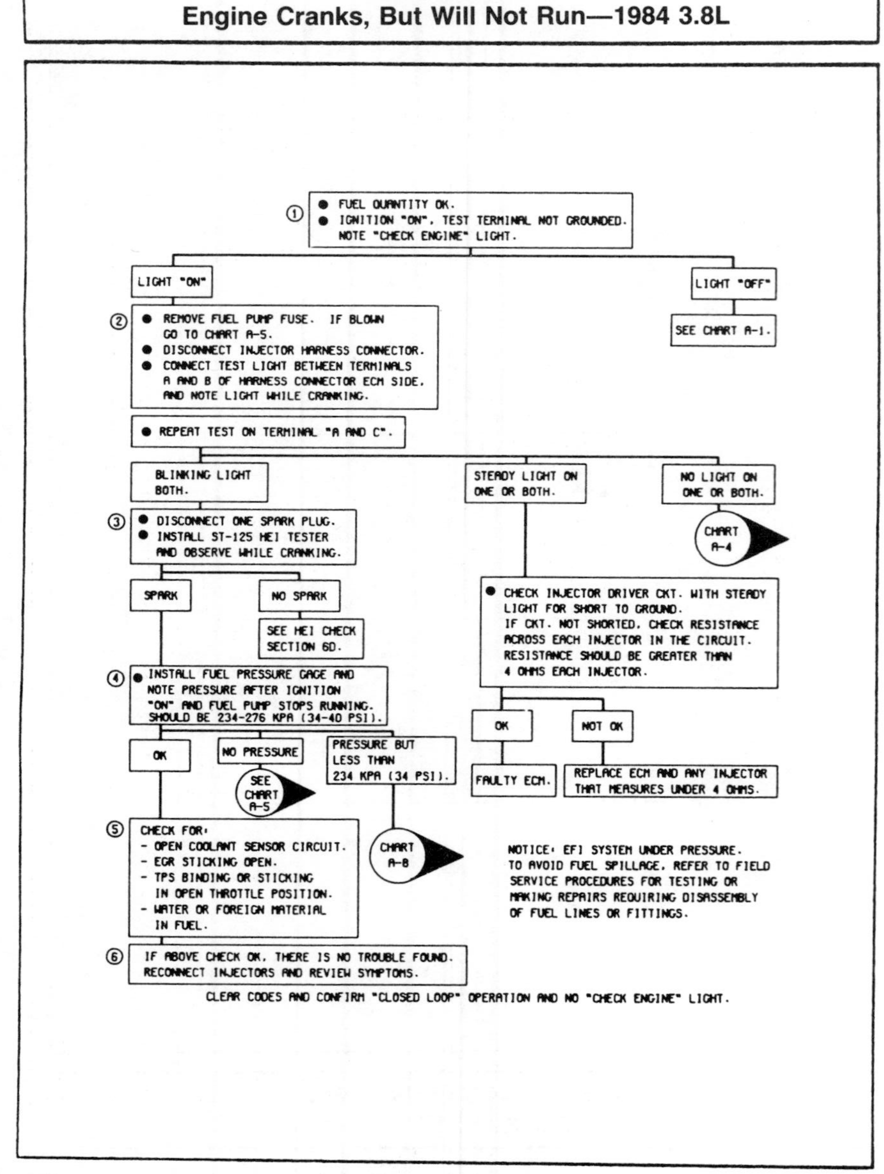

CHART A-4
Engine Cranks, But Will Not Run—3.8L

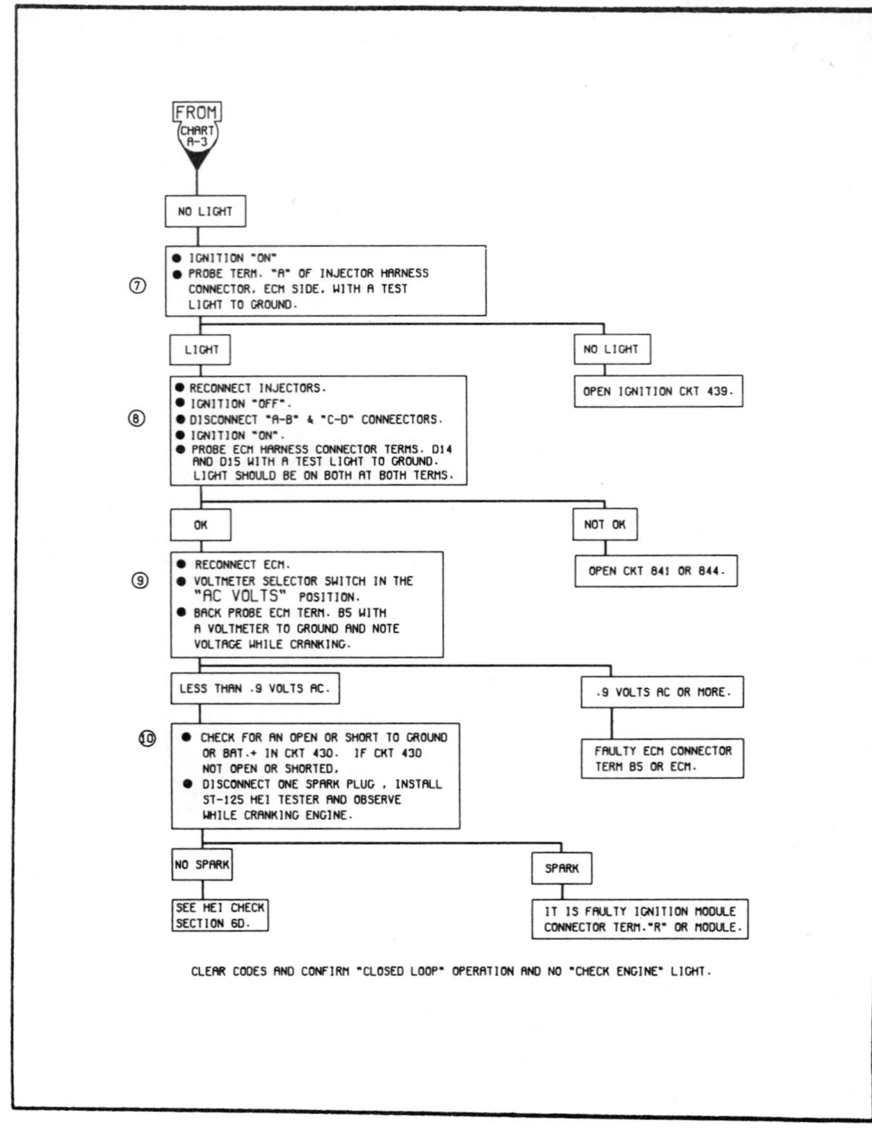

CHART A-5
Engine Cranks, But Will Not Run—3.8L

FROM CHART A-3

NO PRESSURE

- CHECK FUEL PUMP FUSE.

NOT OK

①
- DISCONNECT FUEL PUMP AT REAR BODY CONNECTOR.
- IGNITION "OFF."
- PROBE FUEL PUMP TEST CONNECTOR CKT 120. WITH A TEST LIGHT TO 12 VOLTS.

LIGHT "OFF"

- RECONNECT FUEL PUMP.
- REPLACE FUSE.
- IGNITION "ON."
- RECHECK FUSE.

OK

DEFECTIVE FUSE OR INTERMITTENT SHORT TO GROUND IN CKT 120.

NOT OK

IN TANK FUEL PUMP OR PUMP HARNESS SHORTED TO GROUND.

LIGHT "ON"

REPAIR SHORT TO GROUND IN CKT 120.

OK

TEST POINT ONE FROM SECTION B

②
- PROBE FUEL PUMP TEST CONNECTOR WITH A TEST LIGHT TO GROUND.
- IGNITION "OFF" FOR 10 SECONDS.
- NOTE LIGHT WITHIN 2 SECONDS AFTER IGNITION "ON."

LIGHT "ON"

③
- APPLY BATTERY VOLTAGE TO PUMP TEST CONNECTOR.
- LISTEN FOR PUMP RUNNING AT FUEL TANK.

PUMP RUNS

CHECK FOR:
- PLUGGED IN-LINE FILTER.
- PLUGGED PUMP INLET FILTER.
- RESTRICTED FUEL LINE.

IF OK

PUMP NOT RUNNING

CHECK FOR:
- OPEN WIRE IN CKT 120.
- OPEN PUMP GROUND WIRE CKT 150.

IF OK

REPLACE IN-TANK FUEL PUMP

LIGHT "OFF"

④
- DISCONNECT PUMP RELAY.
- IGNITION "ON." ENGINE STOPPED.
- PROBE RELAY HARNESS CONNECTOR TERMINAL "D" WITH A TEST LIGHT TO GROUND.

LIGHT "ON"

CHART A-6

LIGHT "OFF"

REPAIR OPEN CKT 839 OR 939.

CLEAR CODES AND CONFIRM "CLOSED LOOP" OPERATION AND NO "CHECK ENGINE" LIGHT.

CHART A-6
Engine Cranks, But Will Not Run—3.8L

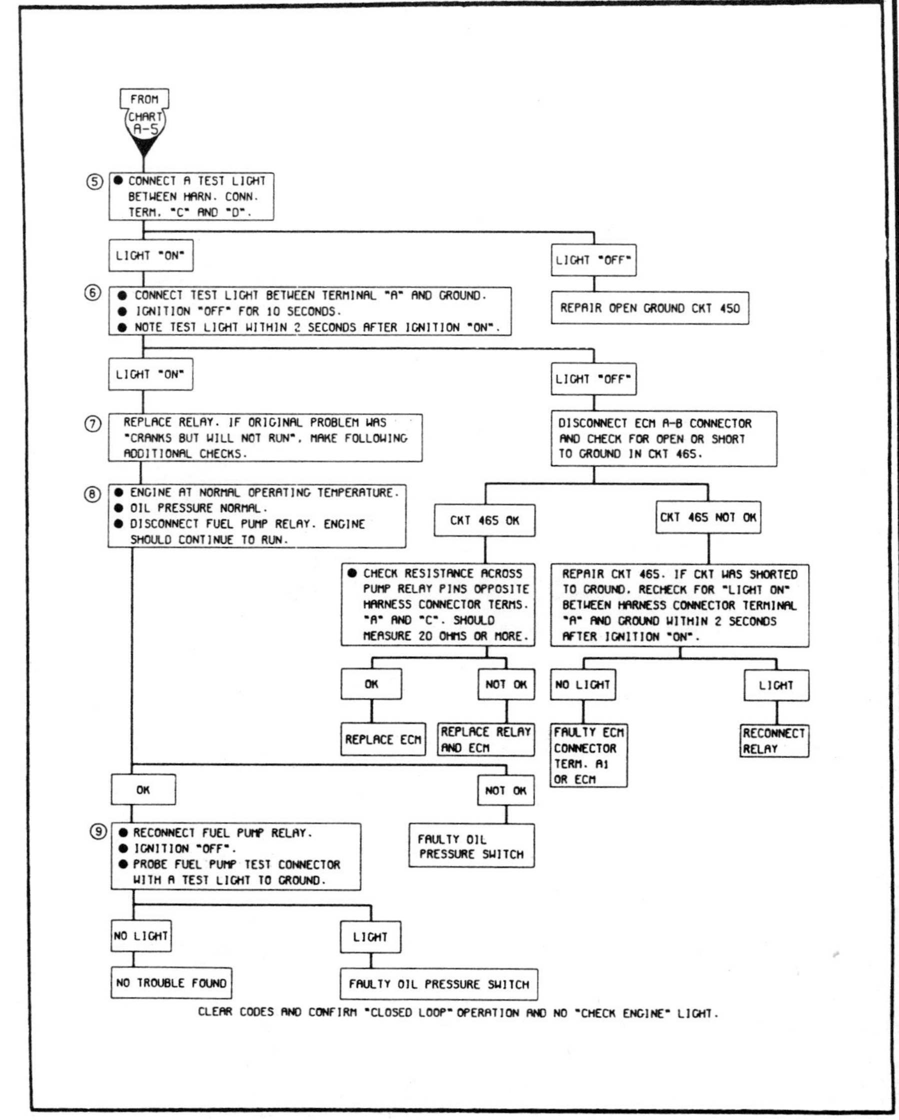

FROM CHART A-5
⑤ ● CONNECT A TEST LIGHT BETWEEN HARN. CONN. TERM. "C" AND "D".
LIGHT "ON"
LIGHT "OFF"
REPAIR OPEN GROUND CKT 450
⑥ ● CONNECT TEST LIGHT BETWEEN TERMINAL "A" AND GROUND.
● IGNITION "OFF" FOR 10 SECONDS.
● NOTE TEST LIGHT WITHIN 2 SECONDS AFTER IGNITION "ON".
LIGHT "ON"
LIGHT "OFF"
DISCONNECT ECM A-B CONNECTOR AND CHECK FOR OPEN OR SHORT TO GROUND IN CKT 465.
CKT 465 OK
CKT 465 NOT OK
● CHECK RESISTANCE ACROSS PUMP RELAY PINS OPPOSITE HARNESS CONNECTOR TERMS. "A" AND "C". SHOULD MEASURE 20 OHMS OR MORE.
OK
REPLACE ECM
NOT OK
REPLACE RELAY AND ECM
REPAIR CKT 465. IF CKT WAS SHORTED TO GROUND, RECHECK FOR "LIGHT ON" BETWEEN HARNESS CONNECTOR TERMINAL "A" AND GROUND WITHIN 2 SECONDS AFTER IGNITION "ON".
NO LIGHT
FAULTY ECM CONNECTOR TERM. A1 OR ECM
LIGHT
RECONNECT RELAY
⑦ REPLACE RELAY. IF ORIGINAL PROBLEM WAS "CRANKS BUT WILL NOT RUN", MAKE FOLLOWING ADDITIONAL CHECKS.
⑧ ● ENGINE AT NORMAL OPERATING TEMPERATURE.
● OIL PRESSURE NORMAL.
● DISCONNECT FUEL PUMP RELAY. ENGINE SHOULD CONTINUE TO RUN.
OK
NOT OK
FAULTY OIL PRESSURE SWITCH
⑨ ● RECONNECT FUEL PUMP RELAY.
● IGNITION "OFF".
● PROBE FUEL PUMP TEST CONNECTOR WITH A TEST LIGHT TO GROUND.
NO LIGHT
NO TROUBLE FOUND
LIGHT
FAULTY OIL PRESSURE SWITCH
CLEAR CODES AND CONFIRM "CLOSED LOOP" OPERATION AND NO "CHECK ENGINE" LIGHT.

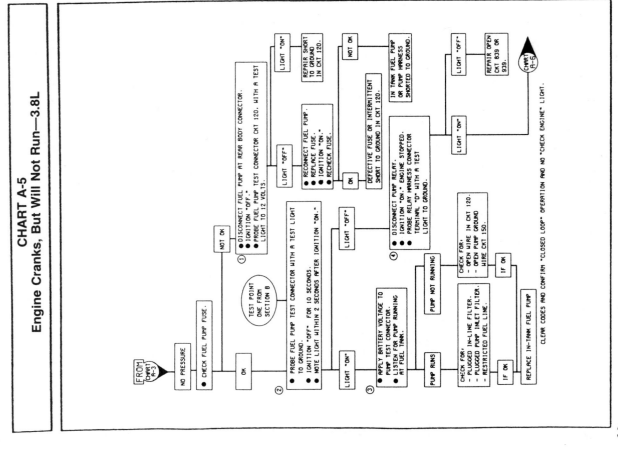

CHART A-7
Fuel System Diagnosis (1 of 2)

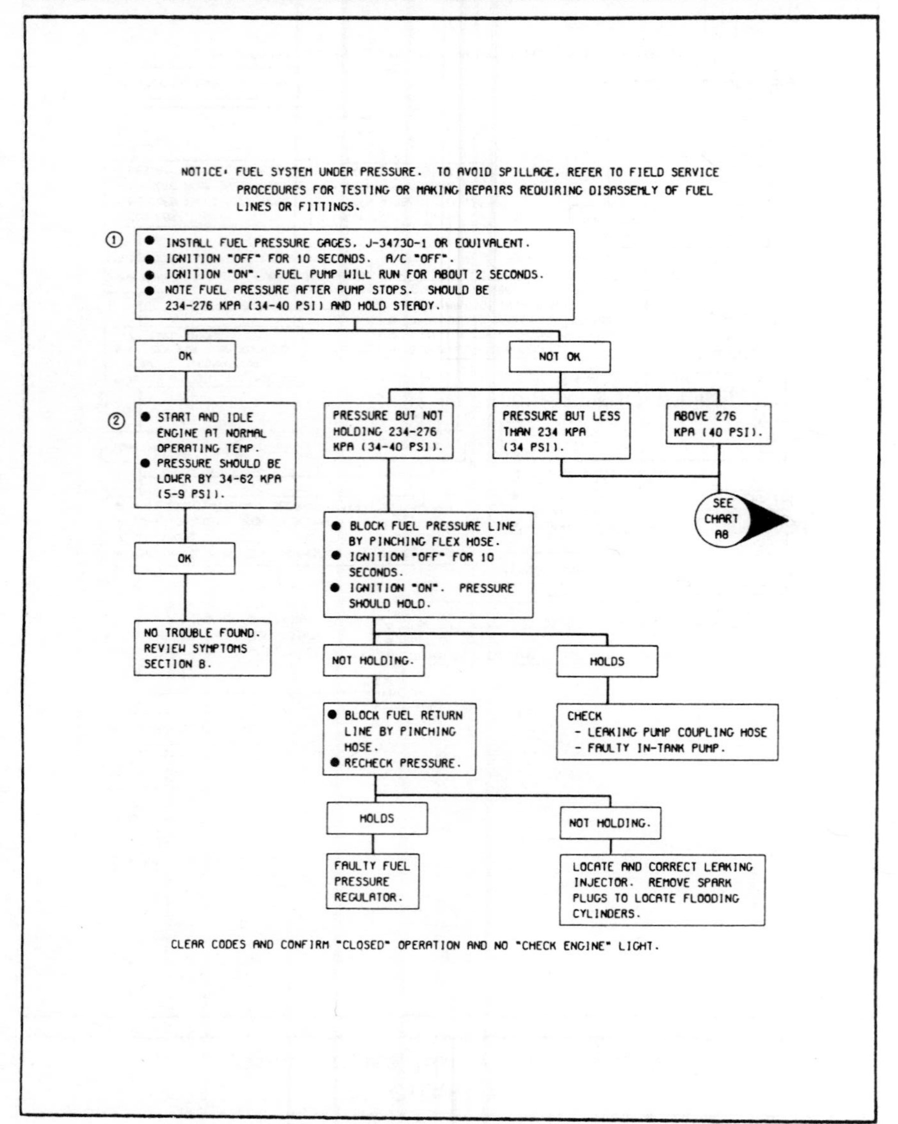

NOTICE: FUEL SYSTEM UNDER PRESSURE. TO AVOID SPILLAGE, REFER TO FIELD SERVICE PROCEDURES FOR TESTING OR MAKING REPAIRS REQUIRING DISASSEMLY OF FUEL LINES OR FITTINGS.
①
● INSTALL FUEL PRESSURE GAGES. J-34730-1 OR EQUIVALENT.
● IGNITION "OFF" FOR 10 SECONDS. A/C "OFF".
● IGNITION "ON". FUEL PUMP WILL RUN FOR ABOUT 2 SECONDS.
● NOTE FUEL PRESSURE AFTER PUMP STOPS. SHOULD BE 234-276 KPA (34-40 PSI) AND HOLD STEADY.
OK
NOT OK
②
● START AND IDLE ENGINE AT NORMAL OPERATING TEMP.
● PRESSURE SHOULD BE LOWER BY 34-62 KPA (5-9 PSI).
OK
NO TROUBLE FOUND. REVIEW SYMPTOMS SECTION B.
PRESSURE BUT NOT HOLDING 234-276 KPA (34-40 PSI).
PRESSURE BUT LESS THAN 234 KPA (34 PSI).
ABOVE 276 KPA (40 PSI).
SEE CHART A8
● BLOCK FUEL PRESSURE LINE BY PINCHING FLEX HOSE.
● IGNITION "OFF" FOR 10 SECONDS.
● IGNITION "ON". PRESSURE SHOULD HOLD.
NOT HOLDING.
HOLDS
● BLOCK FUEL RETURN LINE BY PINCHING HOSE.
● RECHECK PRESSURE.
CHECK
- LEAKING PUMP COUPLING HOSE
- FAULTY IN-TANK PUMP.
HOLDS
NOT HOLDING.
FAULTY FUEL PRESSURE REGULATOR.
LOCATE AND CORRECT LEAKING INJECTOR. REMOVE SPARK PLUGS TO LOCATE FLOODING CYLINDERS.
CLEAR CODES AND CONFIRM "CLOSED" OPERATION AND NO "CHECK ENGINE" LIGHT.

CHART A-8
Fuel System Diagnosis (2 of 2)

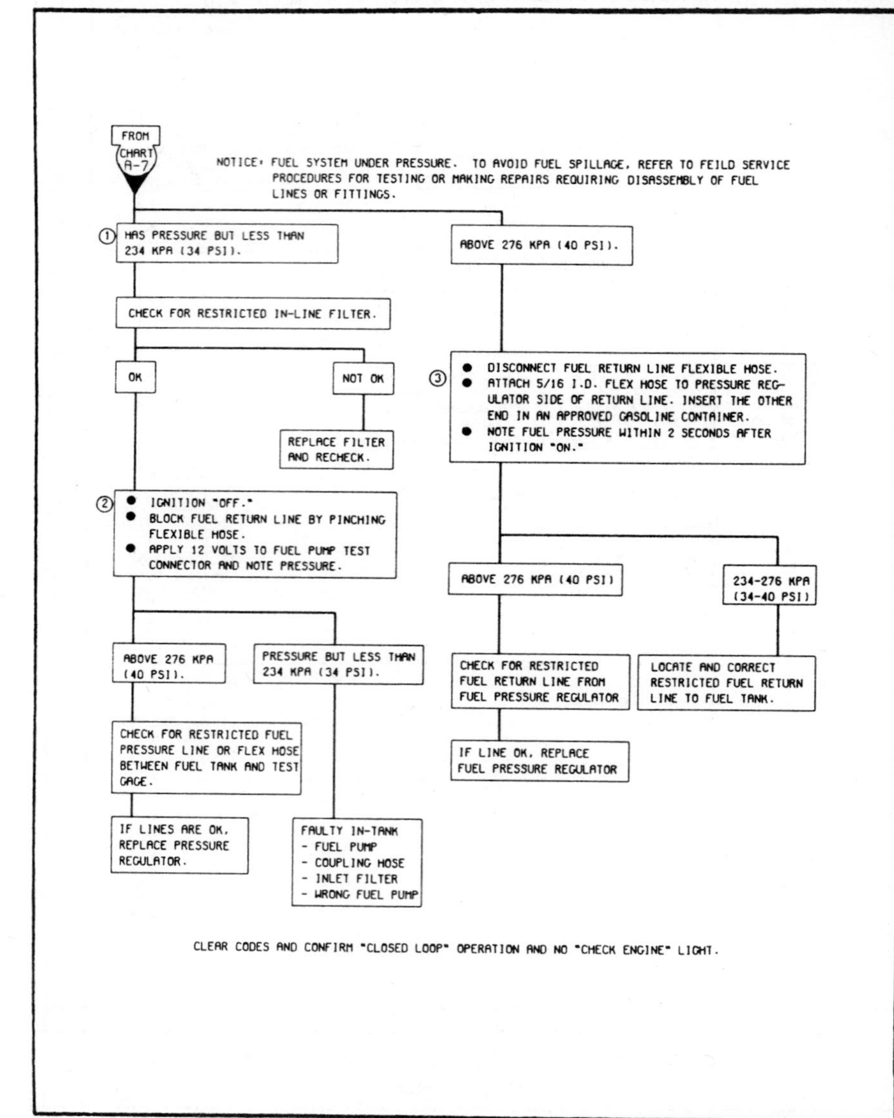

CODE 13
Open Oxygen Sensor Circuit—3.8L

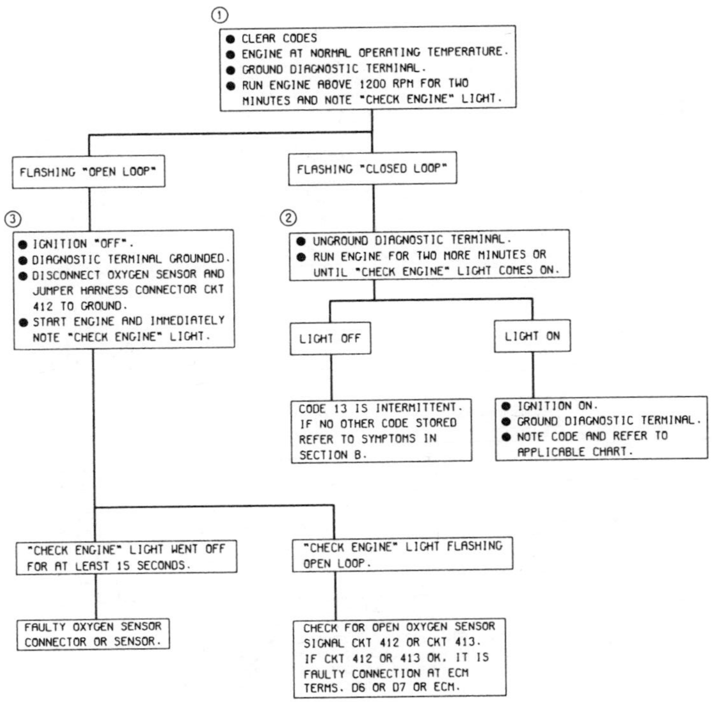

①
● CLEAR CODES
● ENGINE AT NORMAL OPERATING TEMPERATURE.
● GROUND DIAGNOSTIC TERMINAL.
● RUN ENGINE ABOVE 1200 RPM FOR TWO MINUTES AND NOTE "CHECK ENGINE" LIGHT.
FLASHING "OPEN LOOP"
FLASHING "CLOSED LOOP"
③
● IGNITION "OFF".
● DIAGNOSTIC TERMINAL GROUNDED.
● DISCONNECT OXYGEN SENSOR AND JUMPER HARNESS CONNECTOR CKT 412 TO GROUND.
● START ENGINE AND IMMEDIATELY NOTE "CHECK ENGINE" LIGHT.
②
● UNGROUND DIAGNOSTIC TERMINAL.
● RUN ENGINE FOR TWO MORE MINUTES OR UNTIL "CHECK ENGINE" LIGHT COMES ON.
LIGHT OFF
LIGHT ON
CODE 13 IS INTERMITTENT. IF NO OTHER CODE STORED REFER TO SYMPTOMS IN SECTION B.
● IGNITION ON.
● GROUND DIAGNOSTIC TERMINAL.
● NOTE CODE AND REFER TO APPLICABLE CHART.
"CHECK ENGINE" LIGHT WENT OFF FOR AT LEAST 15 SECONDS.
"CHECK ENGINE" LIGHT FLASHING OPEN LOOP.
FAULTY OXYGEN SENSOR CONNECTOR OR SENSOR.
CHECK FOR OPEN OXYGEN SENSOR SIGNAL CKT 412 OR CKT 413. IF CKT 412 OR 413 OK, IT IS FAULTY CONNECTION AT ECM TERMS. D6 OR D7 OR ECM.

CLEAR CODES AND CONFIRM "CLOSED LOOP" OPERATION AND NO "CHECK ENGINE" LIGHT.

CODE 14
Coolant Sensor Circuit—Signal Voltage Low

①
● IGNITION "OFF", CLEAR CODES.
● DIAGNOSTIC TERMINAL NOT GROUNDED.
● START WARM ENGINE AND RUN FOR 1 MINUTE OR UNTIL "CHECK ENGINE" LIGHT COMES ON.
● IGNITION "ON", ENGINE STOPPED.
● GROUND DIAGNOSTIC TERMINAL AND NOTE CODE.
CODE 14
NO CODE STORED. PROBLEM IS INTERMITTENT. IF NO OTHER CODES WERE STORED SEE SYMPTOMS SECTION B.
②
● DISCONNECT COOLANT SENSOR.
● IGNITION "ON",ENGINE STOPPED.
● CHECK VOLTAGE BETWEEN HARNESS CONNECTOR TERMINALS.
● CKT 410 AND 452.
OVER 4 VOLTS
BELOW 4 VOLTS
CHECK RESISTANCE ACROSS COOLANT SENSOR TERMINALS. SHOULD BE MORE THAN 100 OHMS.
● DISCONNECT ECM C-D CONNECTOR.
● CHECK SIGNAL CKT 410 FOR SHORT TO CKT 452 OR CHASSIS GROUND.
OK
NOT OK
INTERMITTENT FAULT IN SENSOR CIRCUIT OR CONN. IF ADDITIONAL CODES WERE STORED, SEE APPLI-CABLE CHARTS. IF NO CODE STORED SEE SYMPTOMS, SECTION B.
REPLACE SENSOR.
IF CKT 410 NOT SHORTED, IT IS A FAULTY ECM.

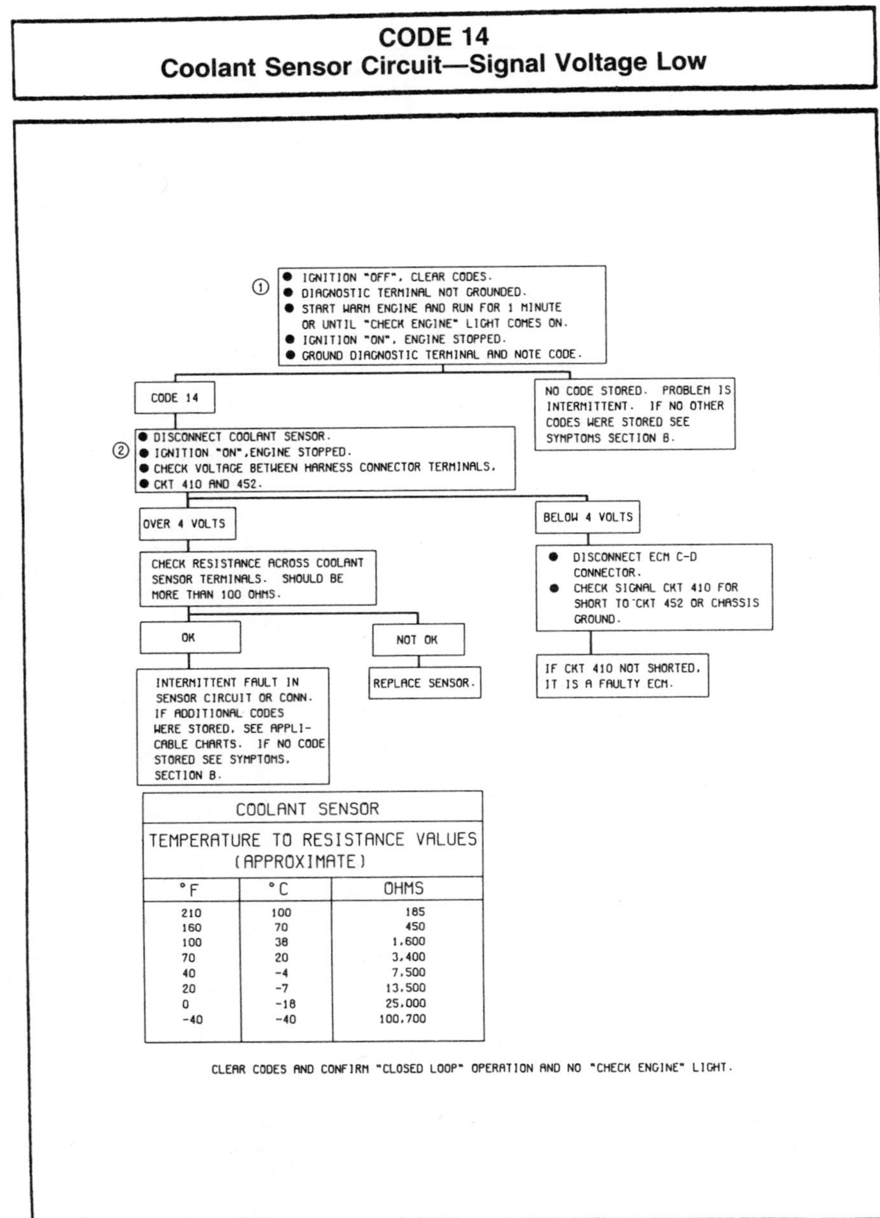

COOLANT SENSOR

TEMPERATURE TO RESISTANCE VALUES (APPROXIMATE)

°F	°C	OHMS
210	100	185
160	70	450
100	38	1,600
70	20	3,400
40	4	7,500
20	-7	13,500
0	-18	25,000
-40	-40	100,700

CLEAR CODES AND CONFIRM "CLOSED LOOP" OPERATION AND NO "CHECK ENGINE" LIGHT.

CODE 15
Coolant Sensor Circuit—Signal Voltage High

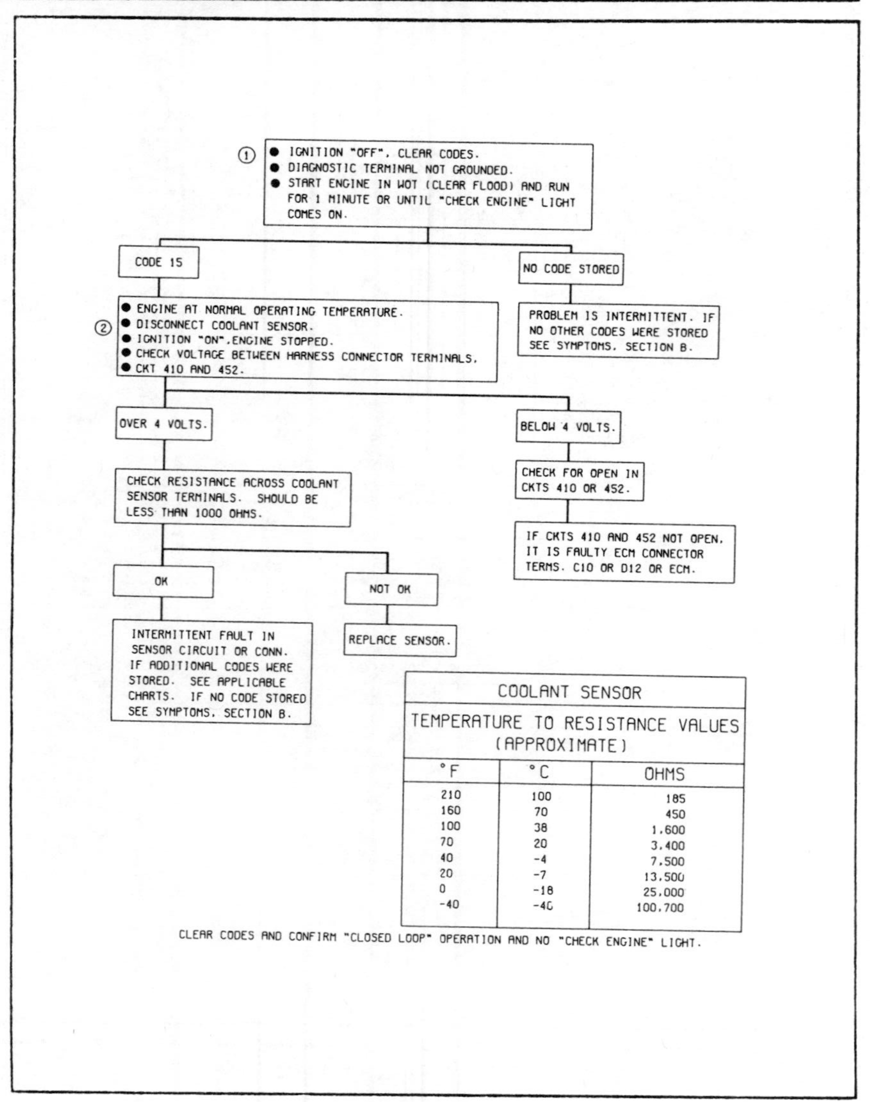
① ● IGNITION "OFF", CLEAR CODES.
● DIAGNOSTIC TERMINAL NOT GROUNDED.
● START ENGINE IN WOT (CLEAR FLOOD) AND RUN FOR 1 MINUTE OR UNTIL "CHECK ENGINE" LIGHT COMES ON.
CODE 15
NO CODE STORED
② ● ENGINE AT NORMAL OPERATING TEMPERATURE.
● DISCONNECT COOLANT SENSOR.
● IGNITION "ON", ENGINE STOPPED.
● CHECK VOLTAGE BETWEEN HARNESS CONNECTOR TERMINALS.
● CKT 410 AND 452.
PROBLEM IS INTERMITTENT. IF NO OTHER CODES WERE STORED SEE SYMPTOMS, SECTION B.
OVER 4 VOLTS.
BELOW 4 VOLTS.
CHECK RESISTANCE ACROSS COOLANT SENSOR TERMINALS. SHOULD BE LESS THAN 1000 OHMS.
CHECK FOR OPEN IN CKTS 410 OR 452.
IF CKTS 410 AND 452 NOT OPEN, IT IS FAULTY ECM CONNECTOR TERMS. C10 OR D12 OR ECM.
OK
NOT OK
INTERMITTENT FAULT IN SENSOR CIRCUIT OR CONN. IF ADDITIONAL CODES WERE STORED. SEE APPLICABLE CHARTS. IF NO CODE STORED SEE SYMPTOMS, SECTION B.
REPLACE SENSOR.
CLEAR CODES AND CONFIRM "CLOSED LOOP" OPERATION AND NO "CHECK ENGINE" LIGHT.

COOLANT SENSOR
TEMPERATURE TO RESISTANCE VALUES (APPROXIMATE)

°F	°C	OHMS
210	100	185
160	70	450
100	38	1,600
70	20	3,400
40	-4	7,500
20	-7	13,500
0	-18	25,000
-40	-40	100,700

CODE 21
Throttle Position Sensor—Signal Voltage High

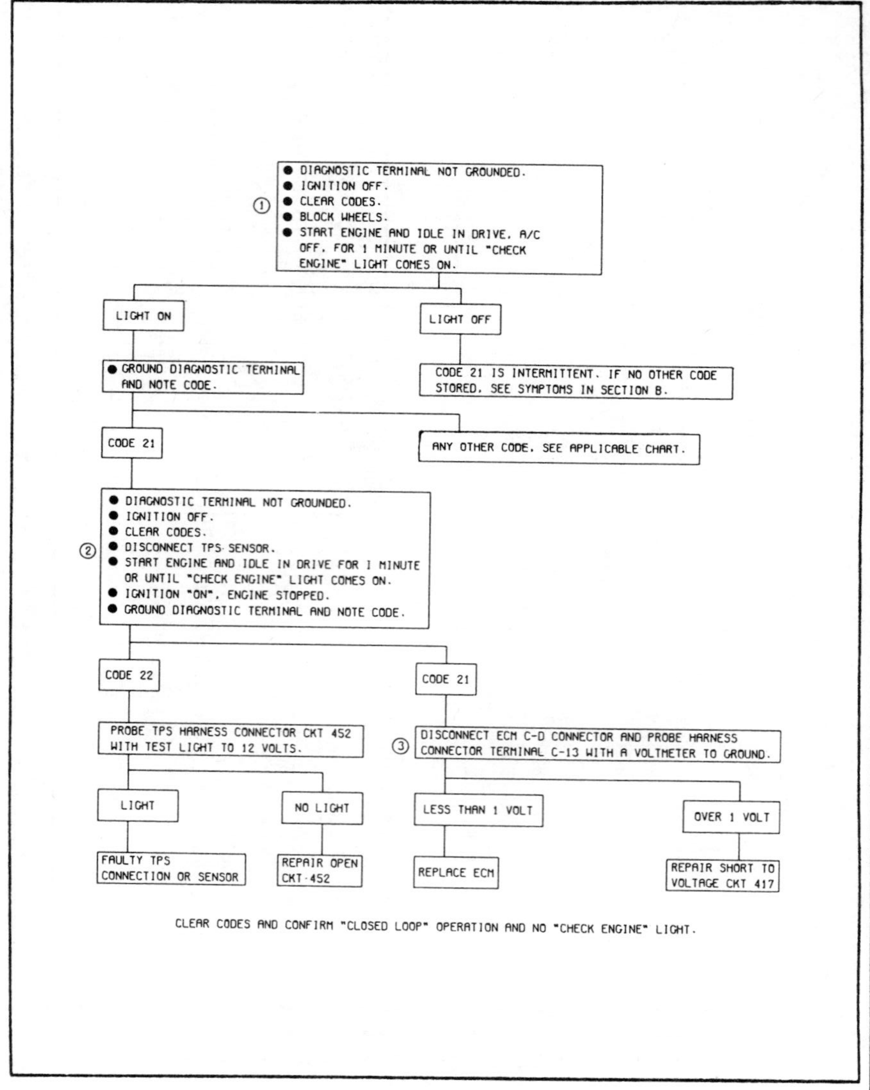
① ● DIAGNOSTIC TERMINAL NOT GROUNDED.
● IGNITION OFF.
● CLEAR CODES.
● BLOCK WHEELS.
● START ENGINE AND IDLE IN DRIVE, A/C OFF, FOR 1 MINUTE OR UNTIL "CHECK ENGINE" LIGHT COMES ON.
LIGHT ON
LIGHT OFF
● GROUND DIAGNOSTIC TERMINAL AND NOTE CODE.
CODE 21 IS INTERMITTENT. IF NO OTHER CODE STORED, SEE SYMPTOMS IN SECTION B.
CODE 21
ANY OTHER CODE, SEE APPLICABLE CHART.
② ● DIAGNOSTIC TERMINAL NOT GROUNDED.
● IGNITION OFF.
● CLEAR CODES.
● DISCONNECT TPS SENSOR.
● START ENGINE AND IDLE IN DRIVE FOR 1 MINUTE OR UNTIL "CHECK ENGINE" LIGHT COMES ON.
● IGNITION "ON", ENGINE STOPPED.
● GROUND DIAGNOSTIC TERMINAL AND NOTE CODE.
CODE 22
CODE 21
PROBE TPS HARNESS CONNECTOR CKT 452 WITH TEST LIGHT TO 12 VOLTS.
③ DISCONNECT ECM C-D CONNECTOR AND PROBE HARNESS CONNECTOR TERMINAL C-13 WITH A VOLTMETER TO GROUND.
LIGHT
NO LIGHT
LESS THAN 1 VOLT
OVER 1 VOLT
FAULTY TPS CONNECTION OR SENSOR
REPAIR OPEN CKT 452
REPLACE ECM
REPAIR SHORT TO VOLTAGE CKT 417
CLEAR CODES AND CONFIRM "CLOSED LOOP" OPERATION AND NO "CHECK ENGINE" LIGHT.

CODE 22
Throttle Position Sensor—Signal Voltage Low

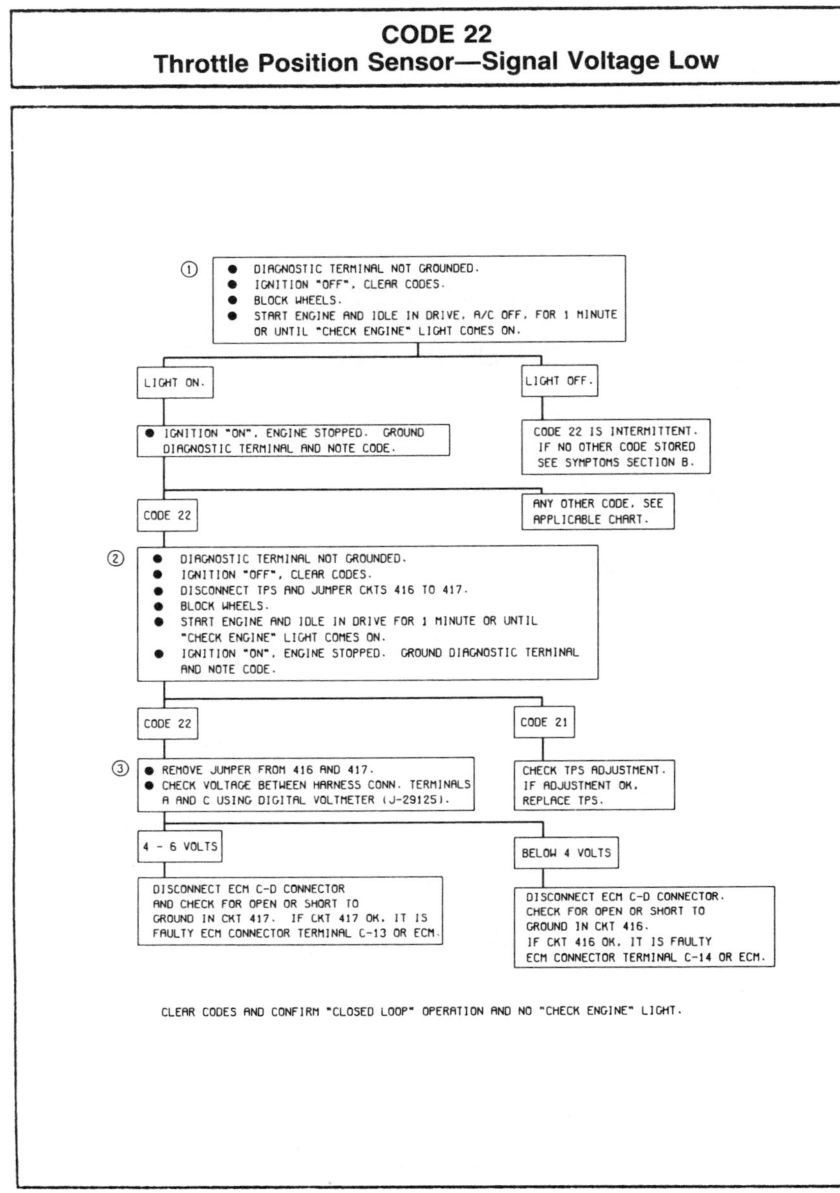

1
DIAGNOSTIC TERMINAL NOT GROUNDED.
IGNITION "OFF", CLEAR CODES.
BLOCK WHEELS.
START ENGINE AND IDLE IN DRIVE, A/C OFF, FOR 1 MINUTE OR UNTIL "CHECK ENGINE" LIGHT COMES ON.
LIGHT ON.
LIGHT OFF.
IGNITION "ON", ENGINE STOPPED. GROUND DIAGNOSTIC TERMINAL AND NOTE CODE.
CODE 22 IS INTERMITTENT. IF NO OTHER CODE STORED SEE SYMPTOMS SECTION B.
CODE 22
ANY OTHER CODE, SEE APPLICABLE CHART.
2
DIAGNOSTIC TERMINAL NOT GROUNDED.
IGNITION "OFF", CLEAR CODES.
DISCONNECT TPS AND JUMPER CKTS 416 TO 417.
BLOCK WHEELS.
START ENGINE AND IDLE IN DRIVE FOR 1 MINUTE OR UNTIL "CHECK ENGINE" LIGHT COMES ON.
IGNITION "ON", ENGINE STOPPED. GROUND DIAGNOSTIC TERMINAL AND NOTE CODE.
CODE 22
CODE 21
3
REMOVE JUMPER FROM 416 AND 417.
CHECK VOLTAGE BETWEEN HARNESS CONN. TERMINALS A AND C USING DIGITAL VOLTMETER (J-29125).
CHECK TPS ADJUSTMENT. IF ADJUSTMENT OK, REPLACE TPS.
4 - 6 VOLTS
BELOW 4 VOLTS
DISCONNECT ECM C-D CONNECTOR AND CHECK FOR OPEN OR SHORT TO GROUND IN CKT 417. IF CKT 417 OK, IT IS FAULTY ECM CONNECTOR TERMINAL C-13 OR ECM.
DISCONNECT ECM C-D CONNECTOR. CHECK FOR OPEN OR SHORT TO GROUND IN CKT 416. IF CKT 416 OK, IT IS FAULTY ECM CONNECTOR TERMINAL C-14 OR ECM.
CLEAR CODES AND CONFIRM "CLOSED LOOP" OPERATION AND NO "CHECK ENGINE" LIGHT.

CODE 24
Vehicle Speed Sensor (VSS)—3.8L 'C' Series

NOTICE: TO PREVENT MISDIAGNOSIS, DISREGARD CODE 24 IF SET WHEN DRIVE WHEELS ARE NOT TURNING.

1.
- SPEEDOMETER WORKING OK. IF NOT, REPAIR.
- LIFT DRIVE WHEELS.
- ENGINE IDLING, IN DRIVE.
- BACK PROBE ECM CONNECTOR TERMINAL A-10, CKT 437 WITH A VOLTMETER TO GROUND, AND OBSERVE VOLTAGE.

NO VOLTAGE VARIATION.

2.
- DISCONNECT VSS BUFFER.
- IGNITION "ON", ENGINE STOPPED.
- CONNECT VOLTMETER BETWEEN BUFFER HARNESS CONNECTOR TERMINALS "A" AND "F". SHOULD BE BATTERY VOLTAGE.

OK → FAULTY BUFFER CONNECTION OR BUFFER.

NOT OK → CONNECT VOLTMETER BETWEEN TERM. "F" AND GROUND.

READS BATTERY VOLTAGE. → REPAIR OPEN CKT 450.

LESS THAN 1 VOLT. → CHECK CKT 437 FOR OPEN OR SHORT TO GROUND. IF CKT NOT OPEN OR SHORTED, IT IS FAULTY ECM CONNECTOR TERMINAL A10 OR ECM.

VOLTAGE VARIES.

CHECK PARK/NEUTRAL SWITCH, CHART C-2A.

PARK/NEUTRAL SWITCH OK.

3.
- IGNITION "OFF", CLEAR CODES.
- DIAGNOSTIC TERMINAL NOT GROUNDED.
- START ENGINE, TRANSMISSION IN LOW RANGE.
- INCREASE VEHICLE SPEED TO A STEADY 20 MPH FOR 20 SEC., THEN SLOWLY DECELERATE TO 10 MPH.
- NOTE "CHECK ENGINE" LIGHT.

NO LIGHT. → VSS SYSTEM OK. NO TROUBLE FOUND.

LIGHT "ON".
- STOP ENGINE.
- IGNITION "ON".
- GROUND DIAGNOSTIC TERMINAL.

NO CODE 24. → SEE CHART FOR CODE INDICATED.

CODE 24. → FAULTY ECM CONNECTOR OR ECM.

CLEAR CODES AND CONFIRM "CLOSED LOOP" OPERATION AND NO "CHECK ENGINE" LIGHT.

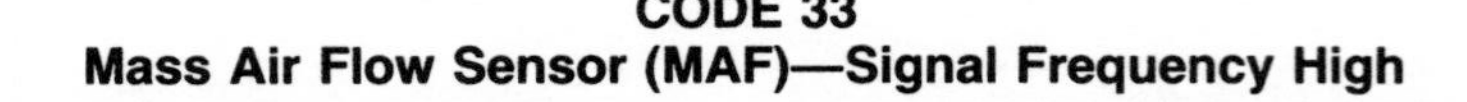

CODE 32
EGR Vacuum Control—3.8L

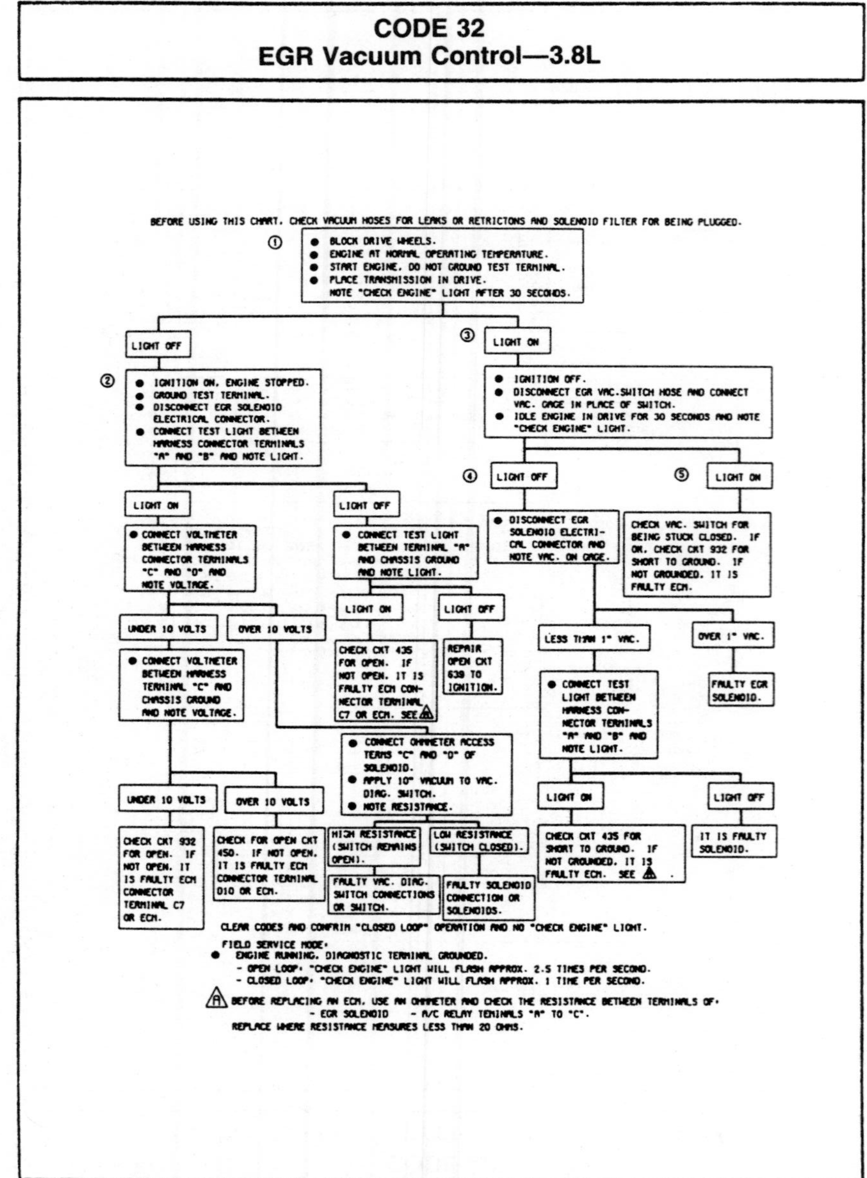

BEFORE USING THIS CHART, CHECK VACUUM HOSES FOR LEAKS OR RETRICTONS AND SOLENOID FILTER FOR BEING PLUGGED.
① ● BLOCK DRIVE WHEELS. ● ENGINE AT NORMAL OPERATING TEMPERATURE. ● START ENGINE. DO NOT GROUND TEST TERMINAL. ● PLACE TRANSMISSION IN DRIVE. NOTE "CHECK ENGINE" LIGHT AFTER 30 SECONDS.
LIGHT OFF
② ● IGNITION ON, ENGINE STOPPED. ● GROUND TEST TERMINAL. ● DISCONNECT EGR SOLENOID ELECTRICAL CONNECTOR. ● CONNECT TEST LIGHT BETWEEN HARNESS CONNECTOR TERMINALS "A" AND "B" AND NOTE LIGHT.
LIGHT ON
● CONNECT VOLTMETER BETWEEN HARNESS CONNECTOR TERMINALS "C" AND "D" AND NOTE VOLTAGE.
UNDER 10 VOLTS
OVER 10 VOLTS
● CONNECT VOLTMETER BETWEEN HARNESS TERMINAL "C" AND CHASSIS GROUND AND NOTE VOLTAGE.
UNDER 10 VOLTS
OVER 10 VOLTS
CHECK CKT 932 FOR OPEN. IF NOT OPEN, IT IS FAULTY ECM CONNECTOR TERMINAL C7 OR ECM.
CHECK FOR OPEN CKT 450. IF NOT OPEN, IT IS FAULTY ECM CONNECTOR TERMINAL D10 OR ECM.
LIGHT OFF
● CONNECT TEST LIGHT BETWEEN TERMINAL "A" AND CHASSIS GROUND AND NOTE LIGHT.
LIGHT ON
LIGHT OFF
CHECK CKT 435 FOR OPEN. IF NOT OPEN, IT IS FAULTY ECM CONNECTOR TERMINAL C7 OR ECM. SEE A
REPAIR OPEN CKT 639 TO IGNITION.
● CONNECT OHMMETER ACCESS TERMS "C" AND "D" OF SOLENOID. ● APPLY 10" VACUUM TO VAC. DIAG. SWITCH. ● NOTE RESISTANCE.
HIGH RESISTANCE (SWITCH REMAINS OPEN).
LOW RESISTANCE (SWITCH CLOSED).
FAULTY VAC. DIAG. SWITCH CONNECTIONS OR SWITCH.
FAULTY SOLENOID CONNECTION OR SOLENOIDS.
③ LIGHT ON
● IGNITION OFF. ● DISCONNECT EGR VAC. SWITCH HOSE AND CONNECT VAC. GAGE IN PLACE OF SWITCH. ● IDLE ENGINE IN DRIVE FOR 30 SECONDS AND NOTE "CHECK ENGINE" LIGHT.
④ LIGHT OFF
● DISCONNECT EGR SOLENOID ELECTRICAL CONNECTOR AND NOTE VAC. ON GAGE.
LESS THAN 1" VAC.
● CONNECT TEST LIGHT BETWEEN HARNESS CONNECTOR TERMINALS "A" AND "B" AND NOTE LIGHT.
LIGHT ON
LIGHT OFF
CHECK CKT 435 FOR SHORT TO GROUND. IF NOT GROUNDED, IT IS FAULTY ECM. SEE A.
IT IS FAULTY SOLENOID.
OVER 1" VAC.
FAULTY EGR SOLENOID.
⑤ LIGHT ON
CHECK VAC. SWITCH FOR BEING STUCK CLOSED. IF OK, CHECK CKT 932 FOR SHORT TO GROUND. IF NOT GROUNDED, IT IS FAULTY ECM.
CLEAR CODES AND CONFIRM "CLOSED LOOP" OPERATION AND NO "CHECK ENGINE" LIGHT.
FIELD SERVICE MODE:
● ENGINE RUNNING, DIAGNOSTIC TERMINAL GROUNDED.
- OPEN LOOP: "CHECK ENGINE" LIGHT WILL FLASH APPROX. 2.5 TIMES PER SECOND.
- CLOSED LOOP: "CHECK ENGINE" LIGHT WILL FLASH APPROX. 1 TIME PER SECOND.
A BEFORE REPLACING AN ECM, USE AN OHMMETER AND CHECK THE RESISTANCE BETWEEN TERMINALS OF:
- EGR SOLENOID - A/C RELAY TERMINALS "A" TO "C".
REPLACE WHERE RESISTANCE MEASURES LESS THAN 20 OHMS.

CODE 33
Mass Air Flow Sensor (MAF)—Signal Frequency High

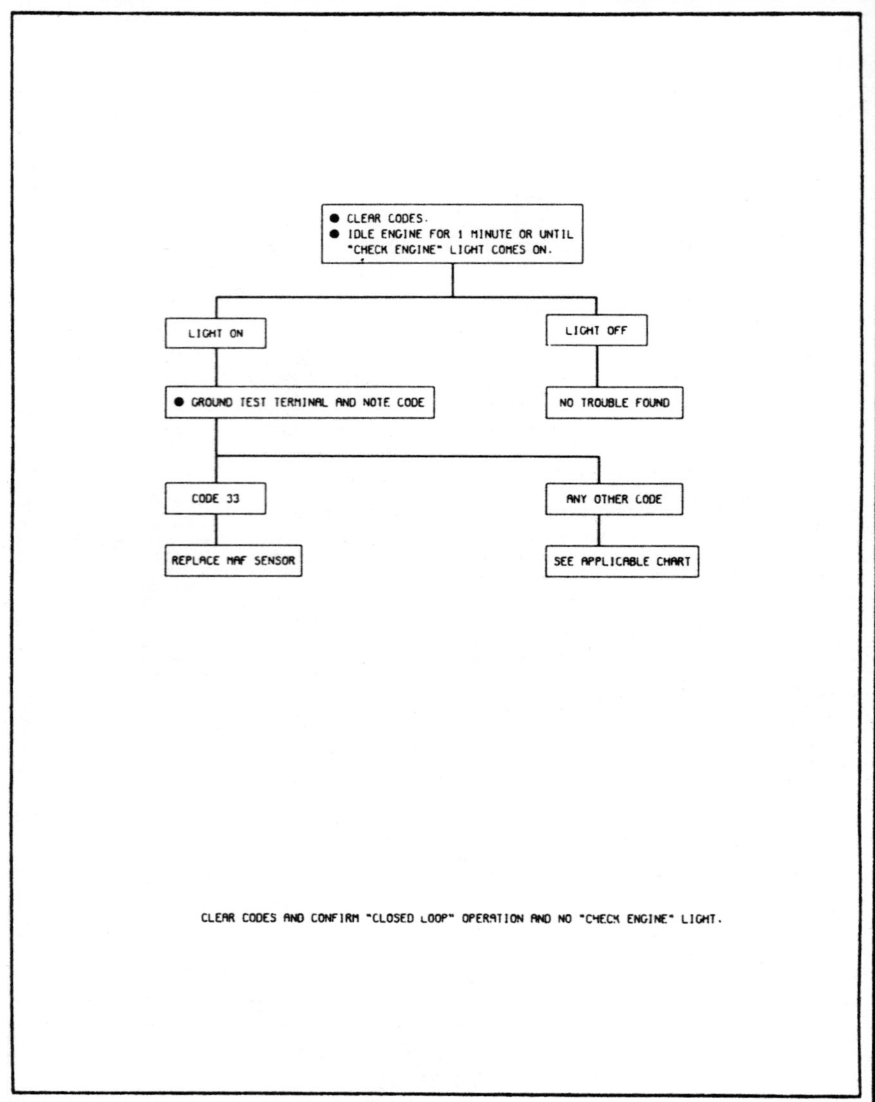

CODE 34
Mass Air Flow Sensor (MAF)—Signal Frequency Low

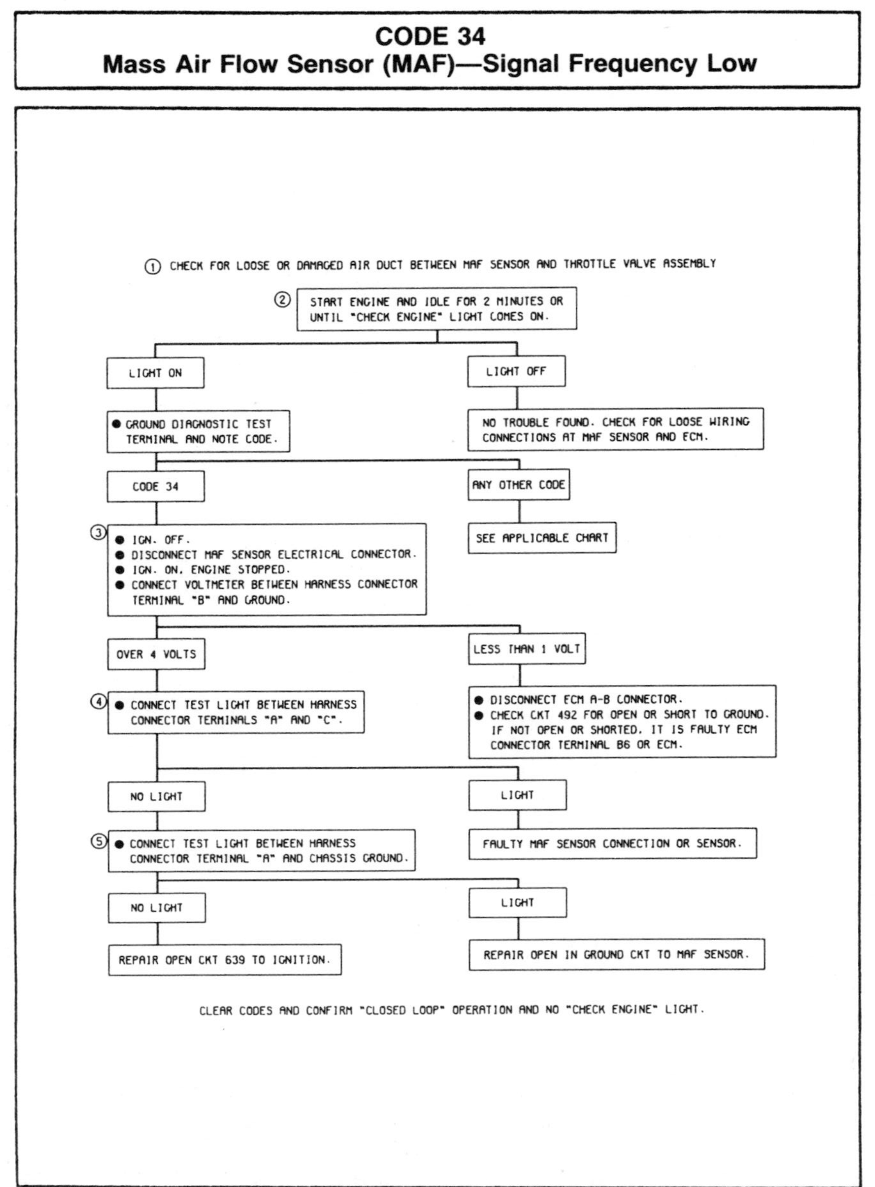

① CHECK FOR LOOSE OR DAMAGED AIR DUCT BETWEEN MAF SENSOR AND THROTTLE VALVE ASSEMBLY
② START ENGINE AND IDLE FOR 2 MINUTES OR UNTIL "CHECK ENGINE" LIGHT COMES ON.
LIGHT ON
● GROUND DIAGNOSTIC TEST TERMINAL AND NOTE CODE.
LIGHT OFF
NO TROUBLE FOUND. CHECK FOR LOOSE WIRING CONNECTIONS AT MAF SENSOR AND ECM.
CODE 34
ANY OTHER CODE
SEE APPLICABLE CHART
③ ● IGN. OFF.
● DISCONNECT MAF SENSOR ELECTRICAL CONNECTOR.
● IGN. ON, ENGINE STOPPED.
● CONNECT VOLTMETER BETWEEN HARNESS CONNECTOR TERMINAL "B" AND GROUND.
OVER 4 VOLTS
LESS THAN 1 VOLT
④ ● CONNECT TEST LIGHT BETWEEN HARNESS CONNECTOR TERMINALS "A" AND "C".
● DISCONNECT ECM A-B CONNECTOR.
● CHECK CKT 492 FOR OPEN OR SHORT TO GROUND. IF NOT OPEN OR SHORTED, IT IS FAULTY ECM CONNECTOR TERMINAL B6 OR ECM.
NO LIGHT
LIGHT
FAULTY MAF SENSOR CONNECTION OR SENSOR.
⑤ ● CONNECT TEST LIGHT BETWEEN HARNESS CONNECTOR TERMINAL "A" AND CHASSIS GROUND.
NO LIGHT
LIGHT
REPAIR OPEN CKT 639 TO IGNITION.
REPAIR OPEN IN GROUND CKT TO MAF SENSOR.
CLEAR CODES AND CONFIRM "CLOSED LOOP" OPERATION AND NO "CHECK ENGINE" LIGHT.

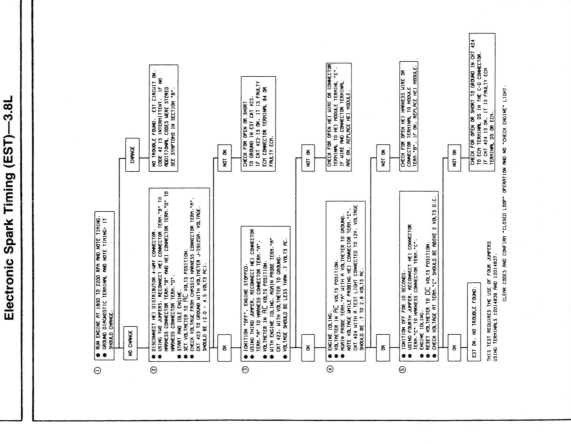

CODE 42
Electronic Spark Timing (EST)—3.8L

① ● RUN ENGINE AT 1800 TO 2200 RPM AND NOTE TIMING.
● GROUND DIAGNOSTIC TERMINAL AND NOTE TIMING; IT SHOULD CHANGE.

CHANGE: NO TROUBLE FOUND. EST CIRCUIT OK. CODE 42 IS INTERMITTENT. IF NO ADDITIONAL CODES WERE STORED SEE SYMPTOMS IN SECTION "B".

NO CHANGE:

② ● DISCONNECT HEI DISTRIBUTOR 4-WAY CONNECTOR.
● USING TWO JUMPERS, RECONNECT HEI CONNECTOR TERM."B" TO HARNESS CONNECTOR TERM."B" AND HEI CONNECTOR TERM."D" TO HARNESS CONNECTOR TERM."D".
● START AND IDLE ENGINE.
● SET VOLTMETER TO AC VOLTS POSITION.
● CHECK VOLTAGE FROM CHASSIS HARNESS CONNECTOR TERM."A". CKT 423 TO GROUND WITH VOLTMETER J-29125A. VOLTAGE SHOULD BE (2.0 - 4.5 VOLTS AC).

NOT OK: CHECK FOR OPEN OR SHORT TO GROUND IN EST CKT 423. IF CKT 423 IS OK, IT IS FAULTY ECM CONNECTOR TERMINAL B4 OR FAULTY ECM.

OK:

③ ● IGNITION "OFF", ENGINE STOPPED.
● USING THIRD JUMPER, RECONNECT HEI CONNECTOR TERM."A" TO HARNESS CONNECTOR TERM."A".
● VOLTMETER IN AC VOLTS POSITION.
● WITH ENGINE IDLING, AGAIN PROBE TERM."A" CKT 423, WITH VOLTMETER TO GROUND.
● VOLTAGE SHOULD BE LESS THAN .7 VOLTS AC.

NOT OK: CHECK FOR OPEN HEI WIRE OR CONNECTOR TERMINAL TO HEI MODULE TERMINAL "E". IF WIRE AND CONNECTOR TERMINAL ARE OK, REPLACE HEI MODULE.

OK:

④ ● ENGINE IDLING.
● VOLTMETER IN AC VOLTS POSITION.
● AGAIN PROBE TERM."A" WITH A VOLTMETER TO GROUND. NOTE VOLTAGE WHILE PROBING HEI CONNECTOR TERM."C". CKT 424 WITH A TEST LIGHT CONNECTED TO 12V. VOLTAGE SHOULD BE .7 TO 2.8 VOLTS AC.

NOT OK: CHECK FOR OPEN HEI HARNESS WIRE OR CONNECTOR TERMINAL TO MODULE TERM."B". IF OK, REPLACE HEI MODULE.

OK:

⑤ ● IGNITION OFF FOR 10 SECONDS.
● USING FOURTH JUMPER, RECONNECT HEI CONNECTOR TERM."C" TO HARNESS CONNECTOR TERM."C".
● ENGINE IDLING.
● RESET VOLTMETER TO DC VOLTS POSITION.
● CHECK VOLTAGE AT TERM."C" SHOULD BE ABOVE 2 VOLTS D.C.

OK: EST OK. NO TROUBLE FOUND.

NOT OK: CHECK FOR OPEN OR SHORT TO GROUND IN CKT 424 TO ECM TERMINAL D5 IN THE C-D CONNECTOR. IF CKT 424 IS OK, IT IS FAULTY ECM TERMINAL D5 OR ECM.

THIS TEST REQUIRES THE USE OF FOUR JUMPERS USING TERMINALS 12014836 AND 12014837.

CLEAR CODES AND CONFIRM "CLOSED LOOP" OPERATION AND NO "CHECK ENGINE" LIGHT.

CODE 43
Electronic Spark Control (ESC)—3.8L

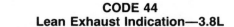

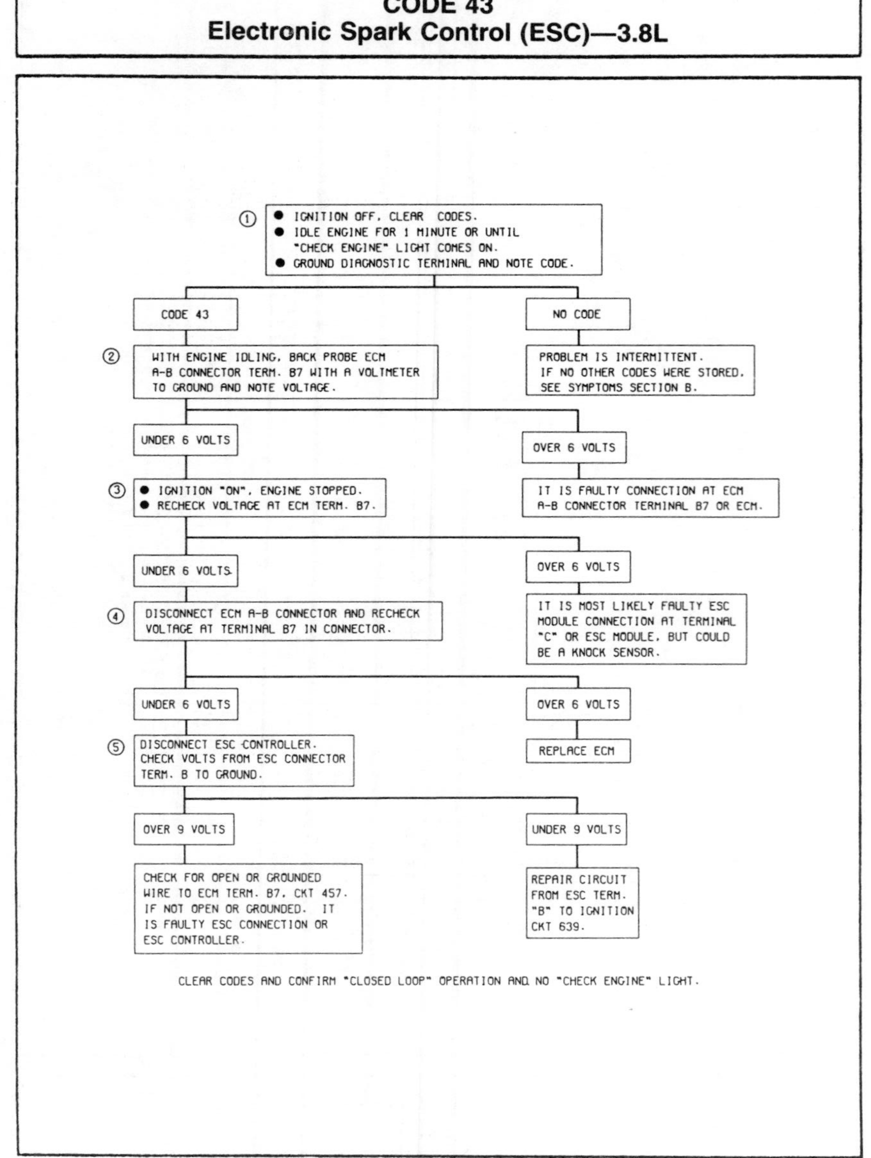

① ● IGNITION OFF, CLEAR CODES.
● IDLE ENGINE FOR 1 MINUTE OR UNTIL "CHECK ENGINE" LIGHT COMES ON.
● GROUND DIAGNOSTIC TERMINAL AND NOTE CODE.
CODE 43
NO CODE
② WITH ENGINE IDLING, BACK PROBE ECM A-B CONNECTOR TERM. B7 WITH A VOLTMETER TO GROUND AND NOTE VOLTAGE.
PROBLEM IS INTERMITTENT. IF NO OTHER CODES WERE STORED, SEE SYMPTOMS SECTION B.
UNDER 6 VOLTS
OVER 6 VOLTS
③ ● IGNITION "ON", ENGINE STOPPED.
● RECHECK VOLTAGE AT ECM TERM. B7.
IT IS FAULTY CONNECTION AT ECM A-B CONNECTOR TERMINAL B7 OR ECM.
UNDER 6 VOLTS
OVER 6 VOLTS
④ DISCONNECT ECM A-B CONNECTOR AND RECHECK VOLTAGE AT TERMINAL B7 IN CONNECTOR.
IT IS MOST LIKELY FAULTY ESC MODULE CONNECTION AT TERMINAL "C" OR ESC MODULE, BUT COULD BE A KNOCK SENSOR.
UNDER 6 VOLTS
OVER 6 VOLTS
⑤ DISCONNECT ESC CONTROLLER. CHECK VOLTS FROM ESC CONNECTOR TERM. B TO GROUND.
REPLACE ECM
OVER 9 VOLTS
UNDER 9 VOLTS
CHECK FOR OPEN OR GROUNDED WIRE TO ECM TERM. B7, CKT 457. IF NOT OPEN OR GROUNDED. IT IS FAULTY ESC CONNECTION OR ESC CONTROLLER.
REPAIR CIRCUIT FROM ESC TERM. "B" TO IGNITION CKT 639.
CLEAR CODES AND CONFIRM "CLOSED LOOP" OPERATION AND NO "CHECK ENGINE" LIGHT.

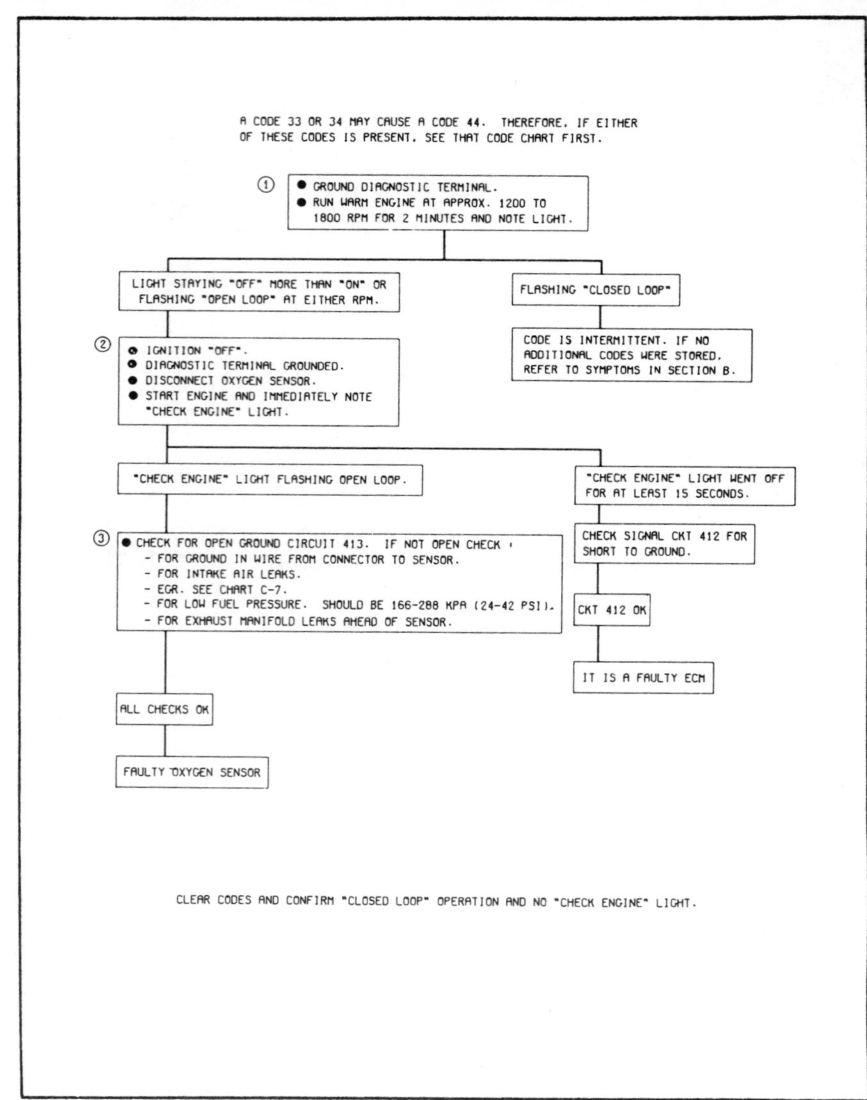

A CODE 33 OR 34 MAY CAUSE A CODE 44. THEREFORE, IF EITHER OF THESE CODES IS PRESENT, SEE THAT CODE CHART FIRST.
① ● GROUND DIAGNOSTIC TERMINAL.
● RUN WARM ENGINE AT APPROX. 1200 TO 1800 RPM FOR 2 MINUTES AND NOTE LIGHT.
LIGHT STAYING "OFF" MORE THAN "ON" OR FLASHING "OPEN LOOP" AT EITHER RPM.
FLASHING "CLOSED LOOP"
② ● IGNITION "OFF".
● DIAGNOSTIC TERMINAL GROUNDED.
● DISCONNECT OXYGEN SENSOR.
● START ENGINE AND IMMEDIATELY NOTE "CHECK ENGINE" LIGHT.
CODE IS INTERMITTENT. IF NO ADDITIONAL CODES WERE STORED, REFER TO SYMPTOMS IN SECTION B.
"CHECK ENGINE" LIGHT FLASHING OPEN LOOP.
"CHECK ENGINE" LIGHT WENT OFF FOR AT LEAST 15 SECONDS.
③ ● CHECK FOR OPEN GROUND CIRCUIT 413. IF NOT OPEN CHECK ,
- FOR GROUND IN WIRE FROM CONNECTOR TO SENSOR.
- FOR INTAKE AIR LEAKS.
- EGR. SEE CHART C-7.
- FOR LOW FUEL PRESSURE. SHOULD BE 166-288 KPA (24-42 PSI).
- FOR EXHAUST MANIFOLD LEAKS AHEAD OF SENSOR.
CHECK SIGNAL CKT 412 FOR SHORT TO GROUND.
CKT 412 OK
IT IS A FAULTY ECM
ALL CHECKS OK
FAULTY OXYGEN SENSOR
CLEAR CODES AND CONFIRM "CLOSED LOOP" OPERATION AND NO "CHECK ENGINE" LIGHT.

CODE 45
Rich Exhaust Indication—3.8L

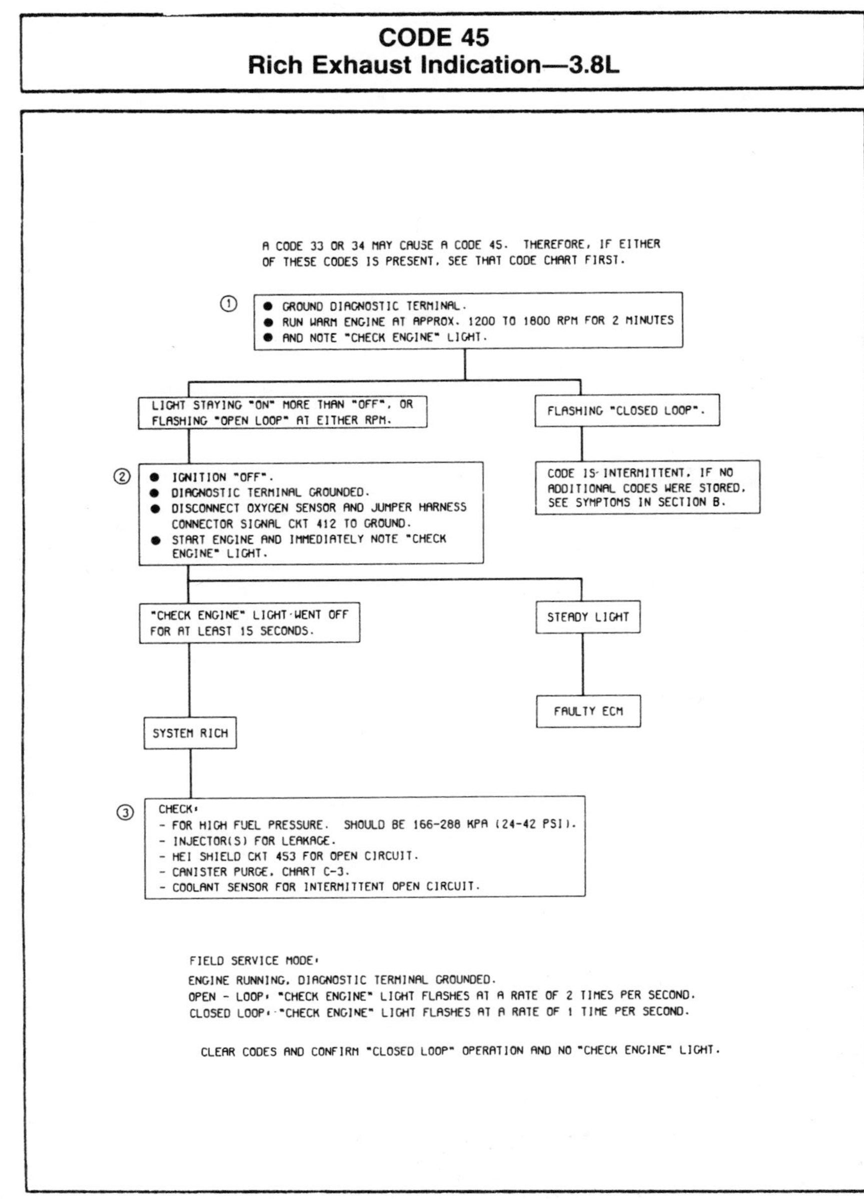

Port Fuel Injection—3.8L

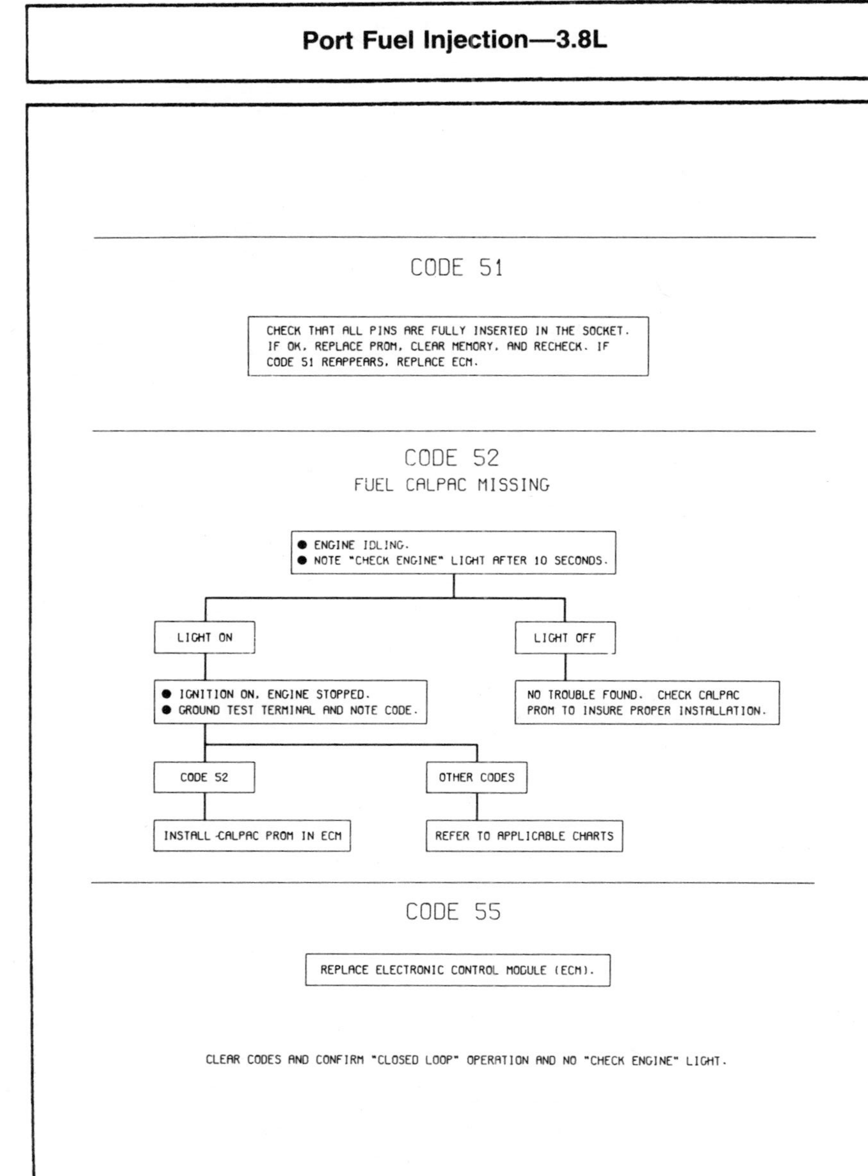

CHART C-1A
Park/Neutral Switch—3.8L

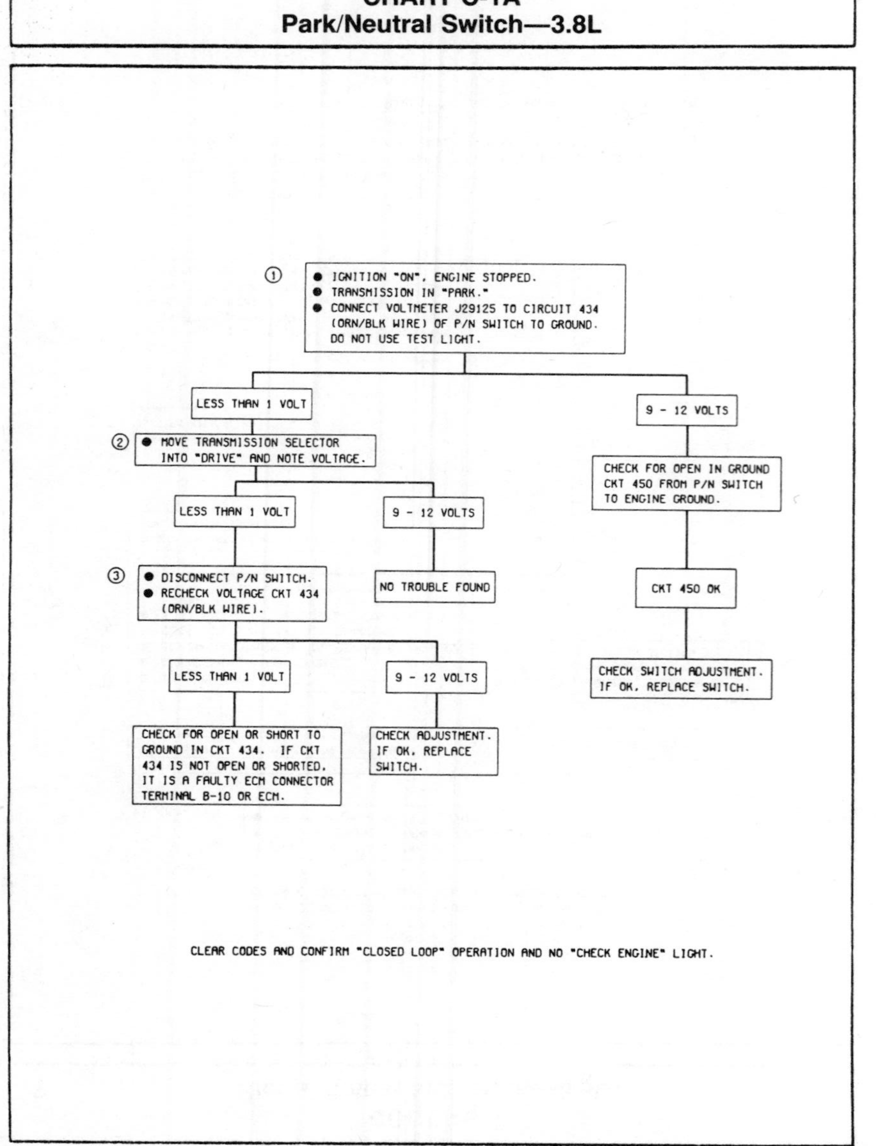
① ● IGNITION "ON", ENGINE STOPPED.
● TRANSMISSION IN "PARK."
● CONNECT VOLTMETER J29125 TO CIRCUIT 434 (ORN/BLK WIRE) OF P/N SWITCH TO GROUND. DO NOT USE TEST LIGHT.
LESS THAN 1 VOLT
9 - 12 VOLTS
② ● MOVE TRANSMISSION SELECTOR INTO "DRIVE" AND NOTE VOLTAGE.
LESS THAN 1 VOLT
9 - 12 VOLTS
NO TROUBLE FOUND
③ ● DISCONNECT P/N SWITCH.
● RECHECK VOLTAGE CKT 434 (ORN/BLK WIRE).
LESS THAN 1 VOLT
9 - 12 VOLTS
CHECK FOR OPEN OR SHORT TO GROUND IN CKT 434. IF CKT 434 IS NOT OPEN OR SHORTED, IT IS A FAULTY ECM CONNECTOR TERMINAL B-10 OR ECM.
CHECK ADJUSTMENT. IF OK, REPLACE SWITCH.
CHECK FOR OPEN IN GROUND CKT 450 FROM P/N SWITCH TO ENGINE GROUND.
CKT 450 OK
CHECK SWITCH ADJUSTMENT. IF OK, REPLACE SWITCH.
CLEAR CODES AND CONFIRM "CLOSED LOOP" OPERATION AND NO "CHECK ENGINE" LIGHT.

CHART C-1E
Power Steering Pressure Switch Check—3.8L

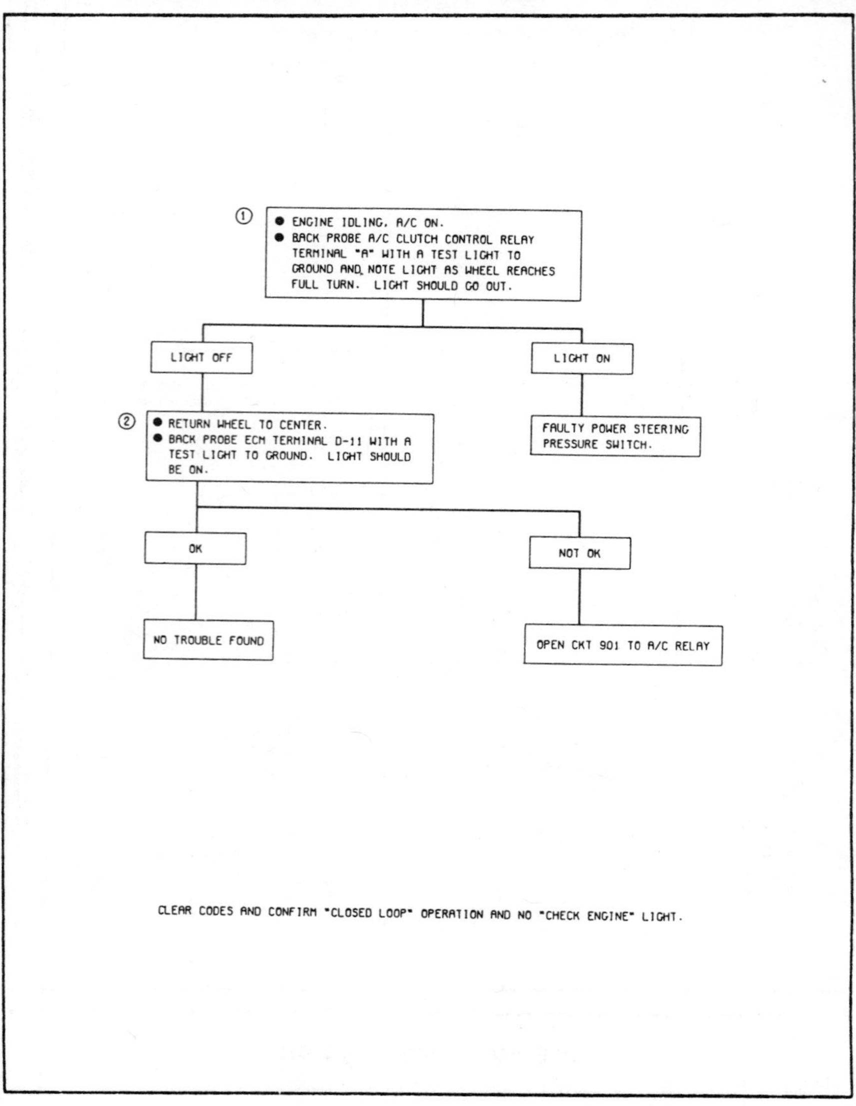
① ● ENGINE IDLING, A/C ON.
● BACK PROBE A/C CLUTCH CONTROL RELAY TERMINAL "A" WITH A TEST LIGHT TO GROUND AND, NOTE LIGHT AS WHEEL REACHES FULL TURN. LIGHT SHOULD GO OUT.
LIGHT OFF
LIGHT ON
② ● RETURN WHEEL TO CENTER.
● BACK PROBE ECM TERMINAL D-11 WITH A TEST LIGHT TO GROUND. LIGHT SHOULD BE ON.
FAULTY POWER STEERING PRESSURE SWITCH.
OK
NOT OK
NO TROUBLE FOUND
OPEN CKT 901 TO A/C RELAY
CLEAR CODES AND CONFIRM "CLOSED LOOP" OPERATION AND NO "CHECK ENGINE" LIGHT.

CHART C-2A
Injector Balance Test

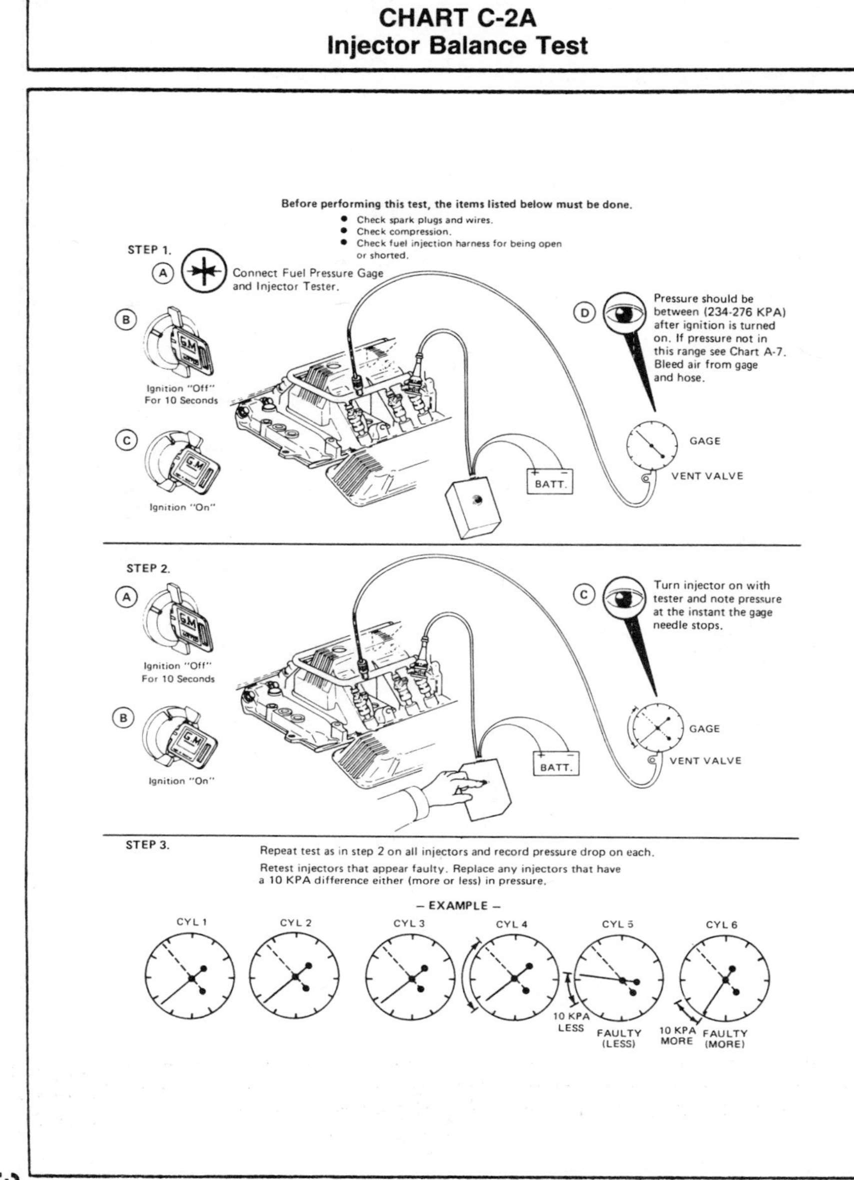

Before performing this test, the items listed below must be done.
● Check spark plugs and wires.
● Check compression.
● Check fuel injection harness for being open or shorted.
STEP 1.
A
Connect Fuel Pressure Gage and Injector Tester.
B
Ignition "Off" For 10 Seconds
C
Ignition "On"
D
Pressure should be between (234-276 KPA) after ignition is turned on. If pressure not in this range see Chart A-7. Bleed air from gage and hose.
GAGE
VENT VALVE
BATT.
STEP 2.
A
Ignition "Off" For 10 Seconds
B
Ignition "On"
C
Turn injector on with tester and note pressure at the instant the gage needle stops.
GAGE
VENT VALVE
BATT.
STEP 3.
Repeat test as in step 2 on all injectors and record pressure drop on each.
Retest injectors that appear faulty. Replace any injectors that have a 10 KPA difference either (more or less) in pressure.
– EXAMPLE –
CYL 1
CYL 2
CYL 3
CYL 4
CYL 5
CYL 6
10 KPA LESS
FAULTY (LESS)
10 KPA MORE
FAULTY (MORE)

CHART C-2C
Idle Air Control—3.8L

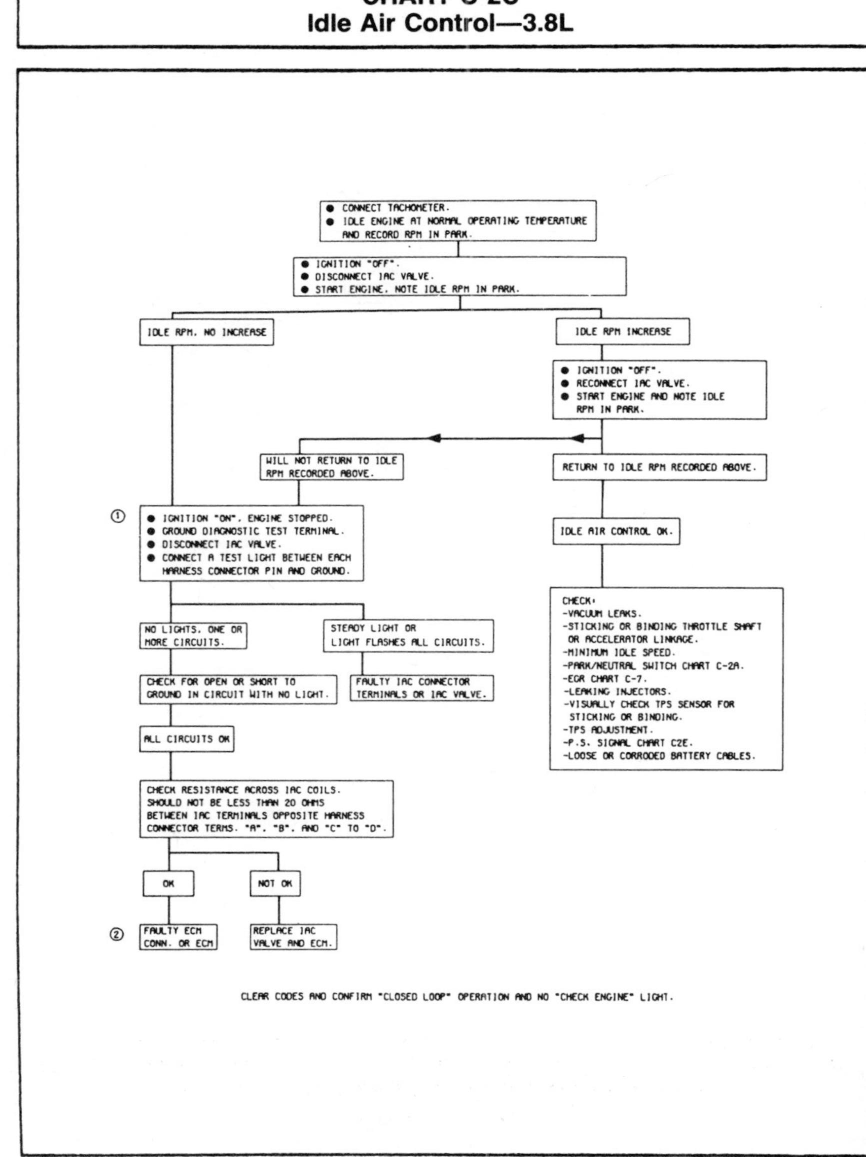

● CONNECT TACHOMETER.
● IDLE ENGINE AT NORMAL OPERATING TEMPERATURE AND RECORD RPM IN PARK.
● IGNITION "OFF".
● DISCONNECT IAC VALVE.
● START ENGINE. NOTE IDLE RPM IN PARK.
IDLE RPM. NO INCREASE
IDLE RPM INCREASE
● IGNITION "OFF".
● RECONNECT IAC VALVE.
● START ENGINE AND NOTE IDLE RPM IN PARK.
WILL NOT RETURN TO IDLE RPM RECORDED ABOVE.
RETURN TO IDLE RPM RECORDED ABOVE.
IDLE AIR CONTROL OK.
①
● IGNITION "ON", ENGINE STOPPED.
● GROUND DIAGNOSTIC TEST TERMINAL.
● DISCONNECT IAC VALVE.
● CONNECT A TEST LIGHT BETWEEN EACH HARNESS CONNECTOR PIN AND GROUND.
NO LIGHTS, ONE OR MORE CIRCUITS.
STEADY LIGHT OR LIGHT FLASHES ALL CIRCUITS.
CHECK FOR OPEN OR SHORT TO GROUND IN CIRCUIT WITH NO LIGHT.
FAULTY IAC CONNECTOR TERMINALS OR IAC VALVE.
ALL CIRCUITS OK
CHECK:
-VACUUM LEAKS.
-STICKING OR BINDING THROTTLE SHAFT OR ACCELERATOR LINKAGE.
-MINIMUM IDLE SPEED.
-PARK/NEUTRAL SWITCH CHART C-2A.
-EGR CHART C-7.
-LEAKING INJECTORS.
-VISUALLY CHECK TPS SENSOR FOR STICKING OR BINDING.
-TPS ADJUSTMENT.
-P.S. SIGNAL CHART C2E.
-LOOSE OR CORRODED BATTERY CABLES.
CHECK RESISTANCE ACROSS IAC COILS. SHOULD NOT BE LESS THAN 20 OHMS BETWEEN IAC TERMINALS OPPOSITE HARNESS CONNECTOR TERMS. "A", "B", AND "C" TO "D".
OK
NOT OK
②
FAULTY ECM CONN. OR ECM
REPLACE IAC VALVE AND ECM.
CLEAR CODES AND CONFIRM "CLOSED LOOP" OPERATION AND NO "CHECK ENGINE" LIGHT.

CHART C-3
Canister Purge Valve—3.8L

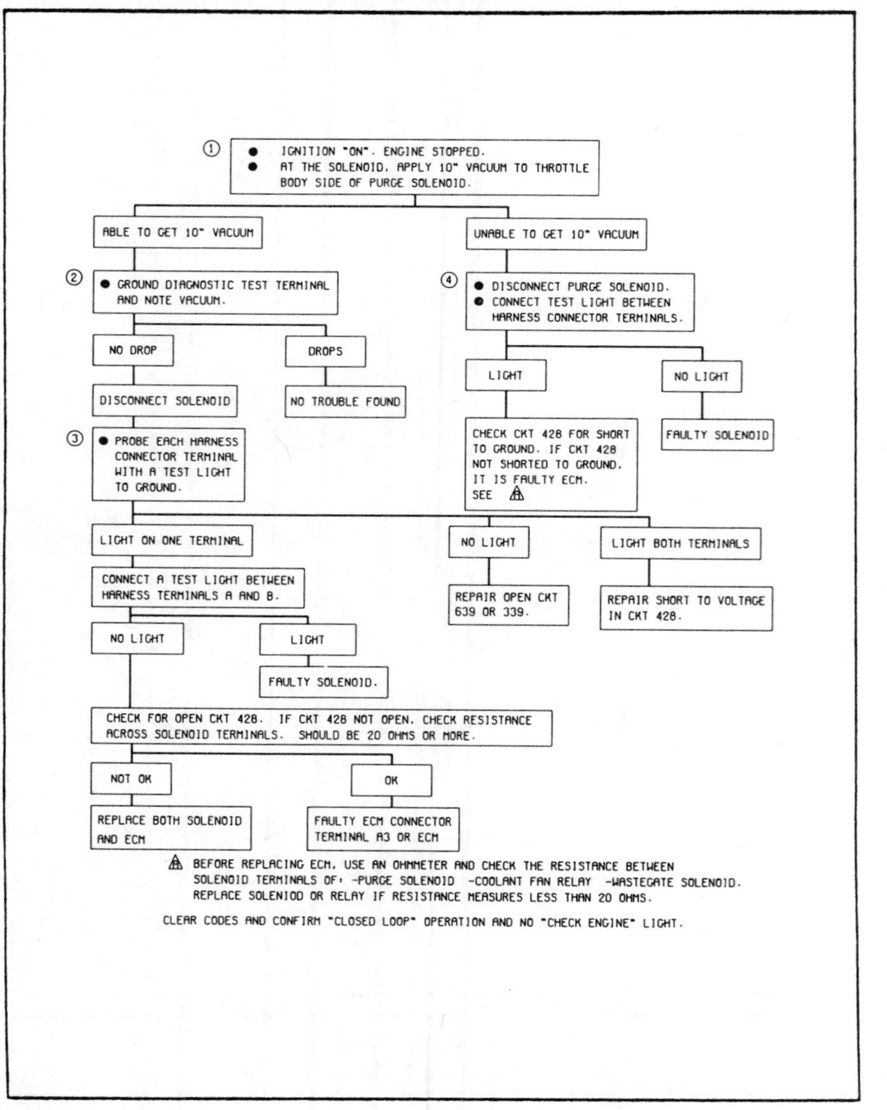

CHART C-4A
Ignition System Check—3.8L Turbo

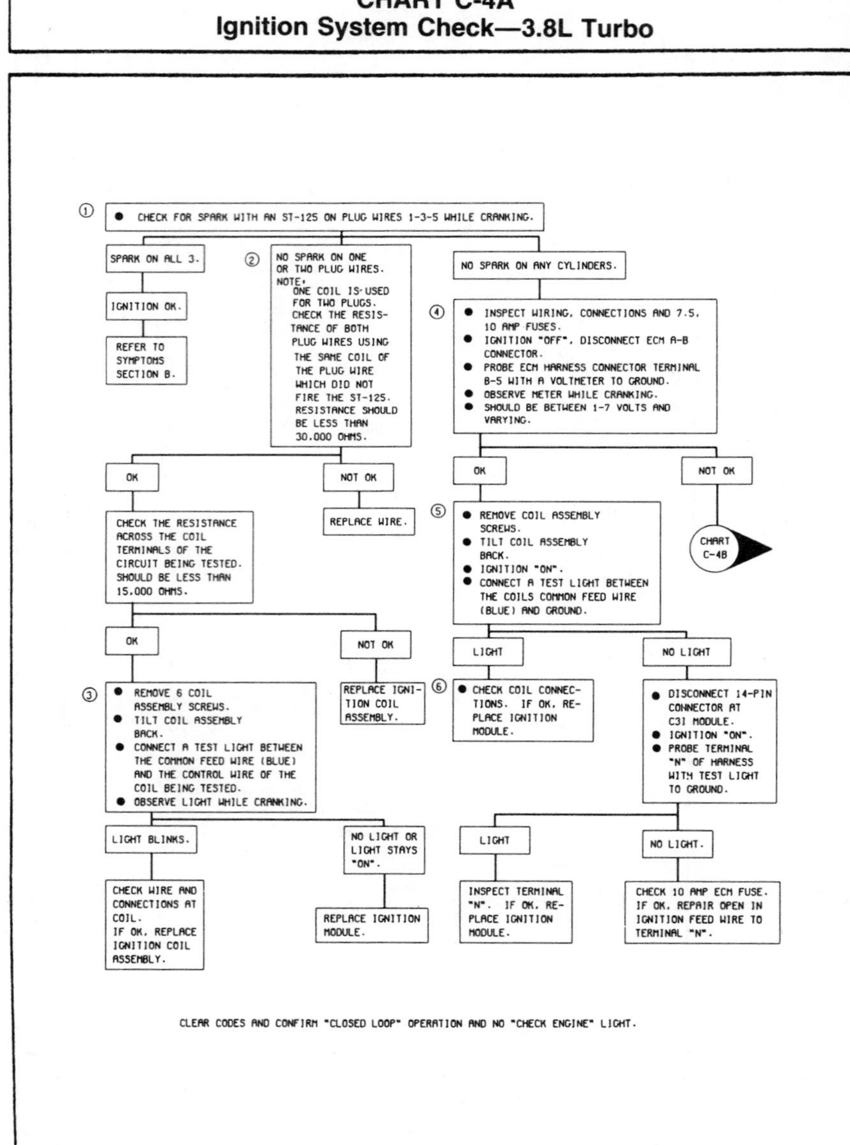

CHART C-4B
Ignition System Check—3.8L Turbo

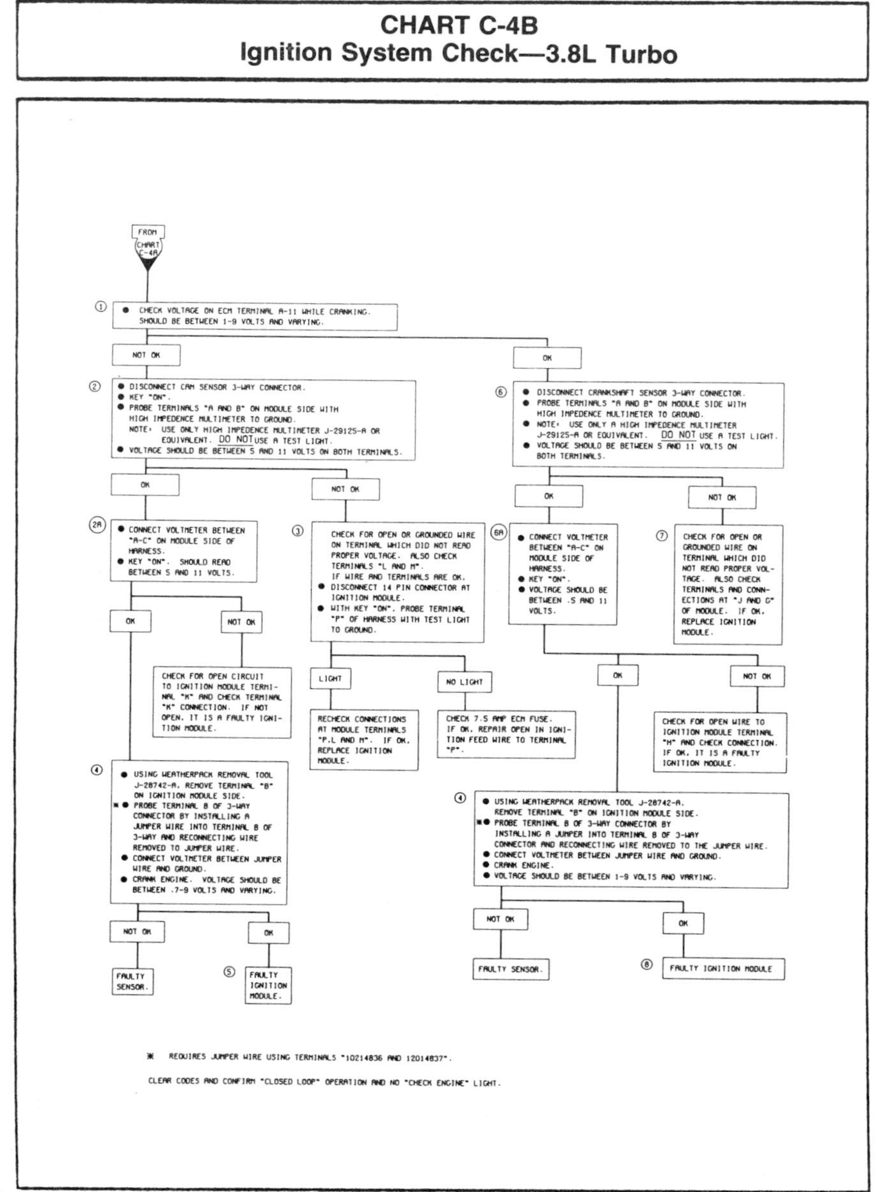

CHART C-5
Electronic Spark Control (ESC)—3.8L

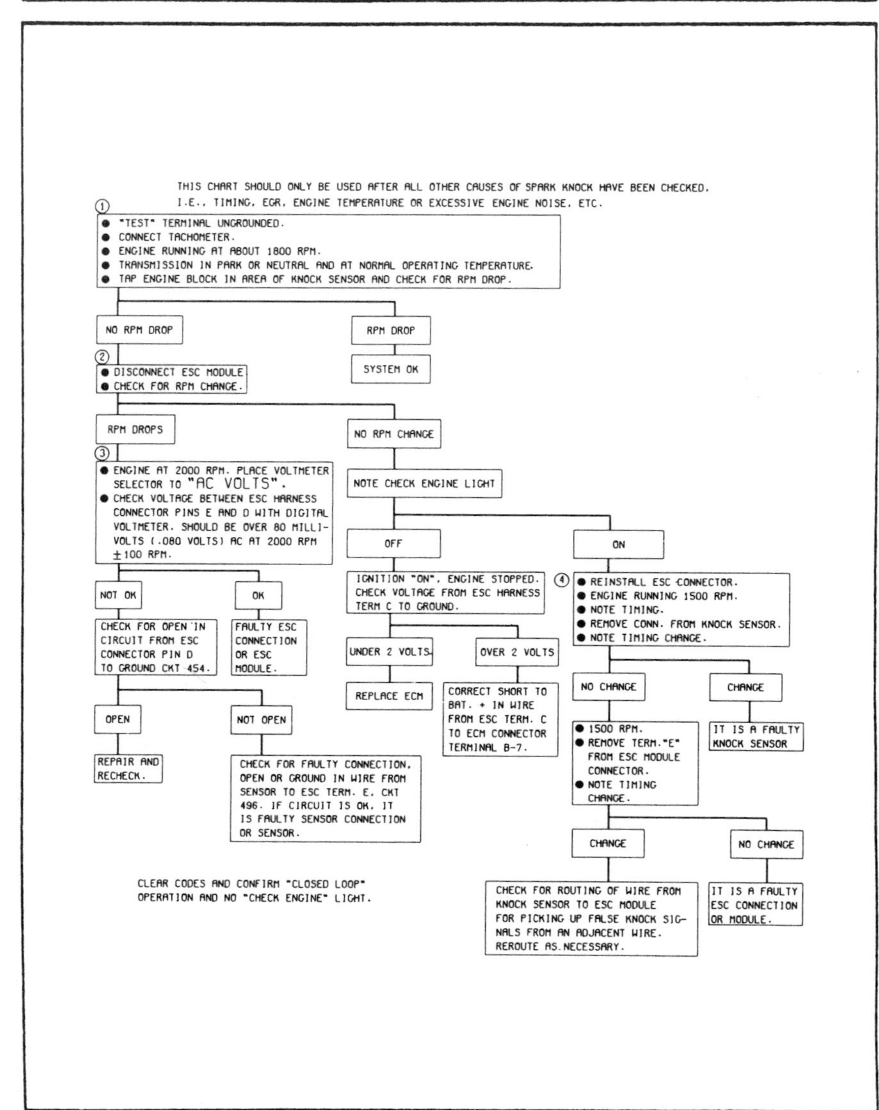

CHART C-7
EGR Check—3.8L

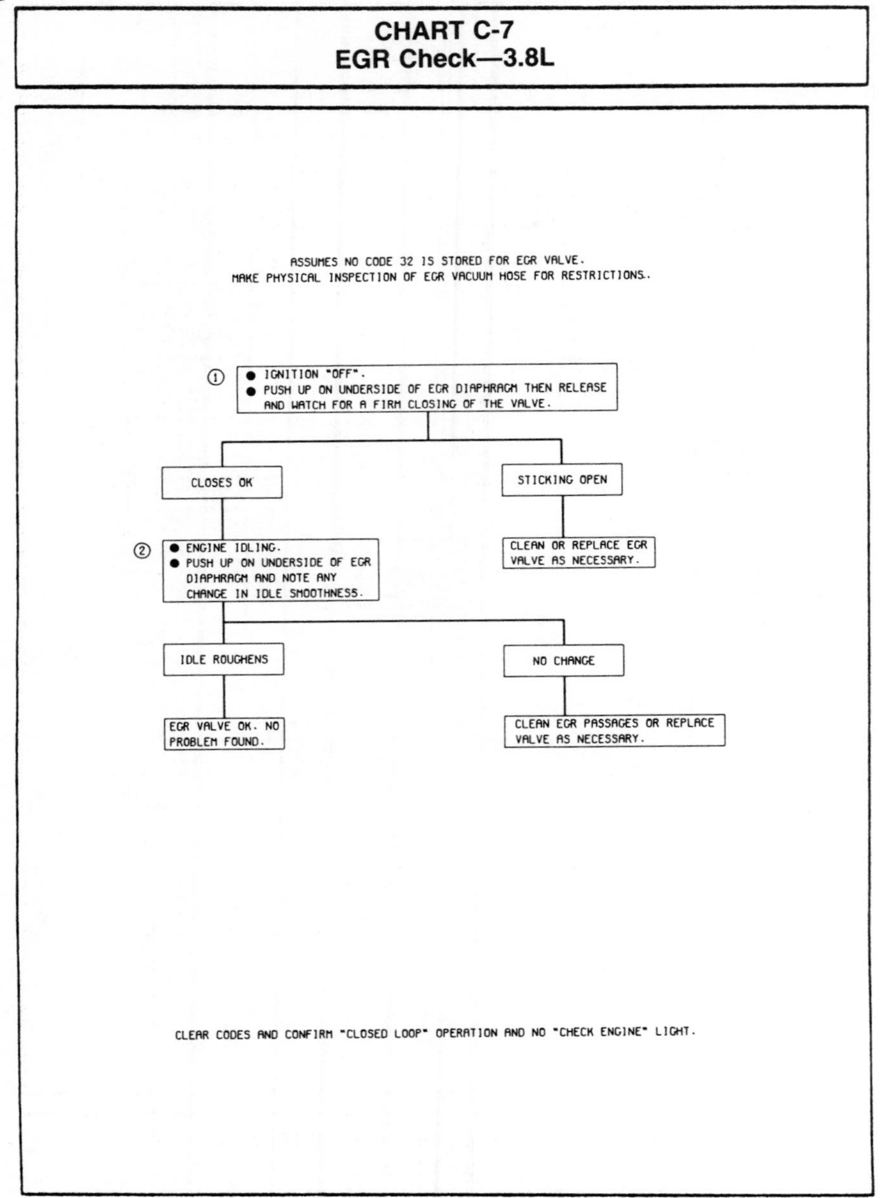

CHART C-8A
Transmission Converter Clutch (TCC)—3.8L W/440- T4

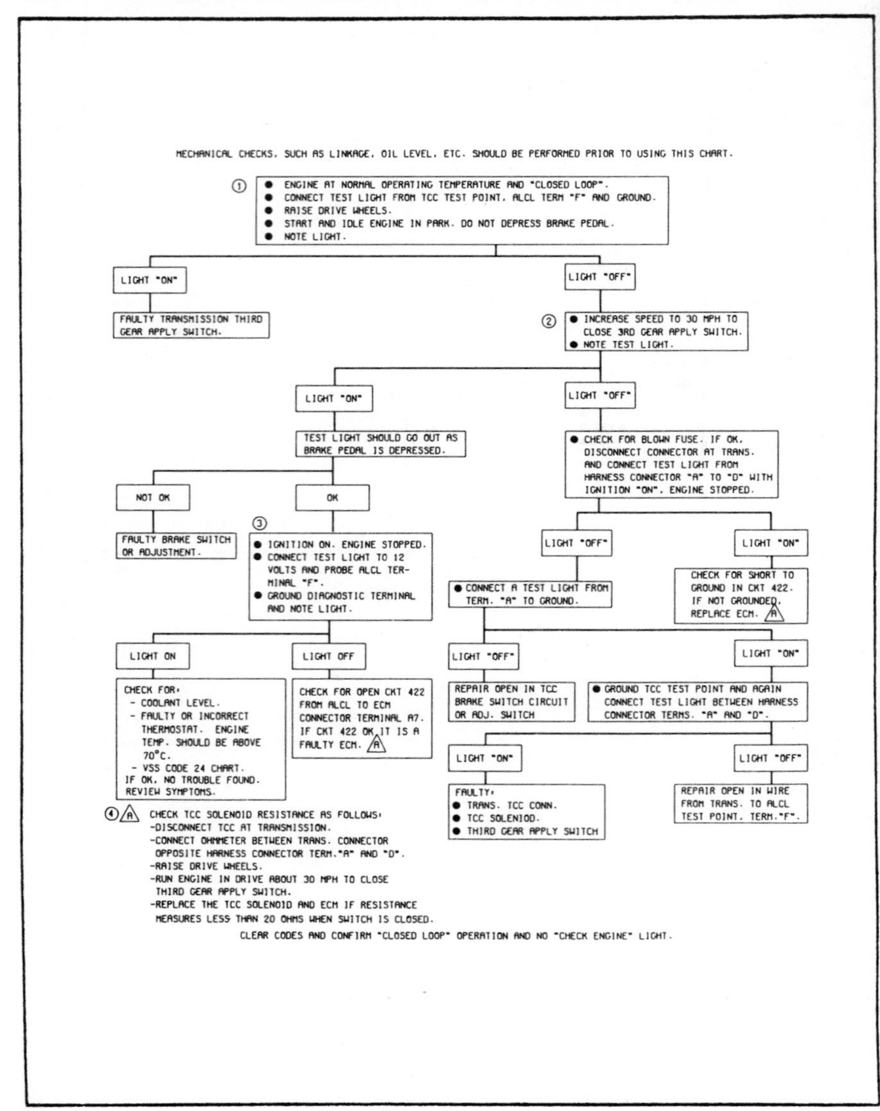

CHART C-8B
Transmission Converter Clutch (TCC)—3.8L W/400- T4

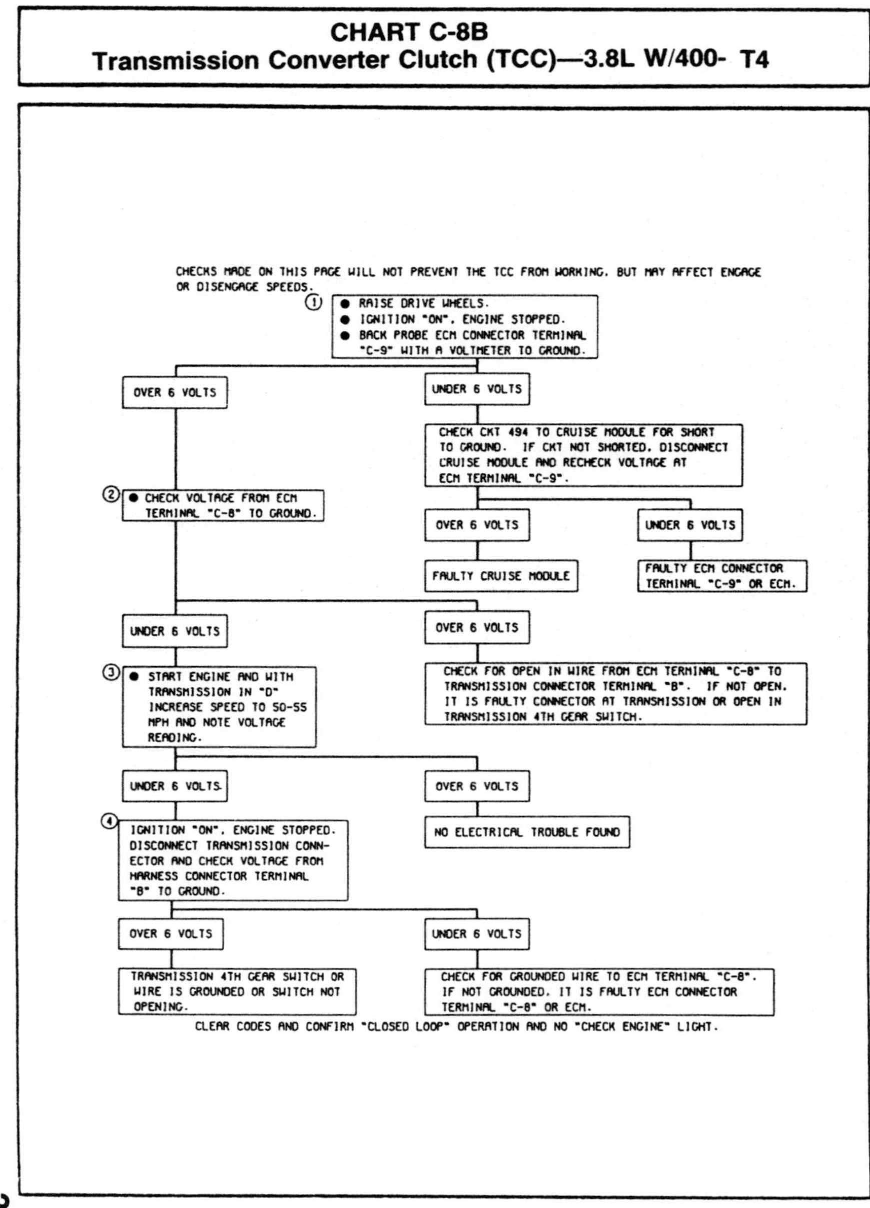

CHART C-10A
A/C Clutch Control (1 of 2)—3.8L

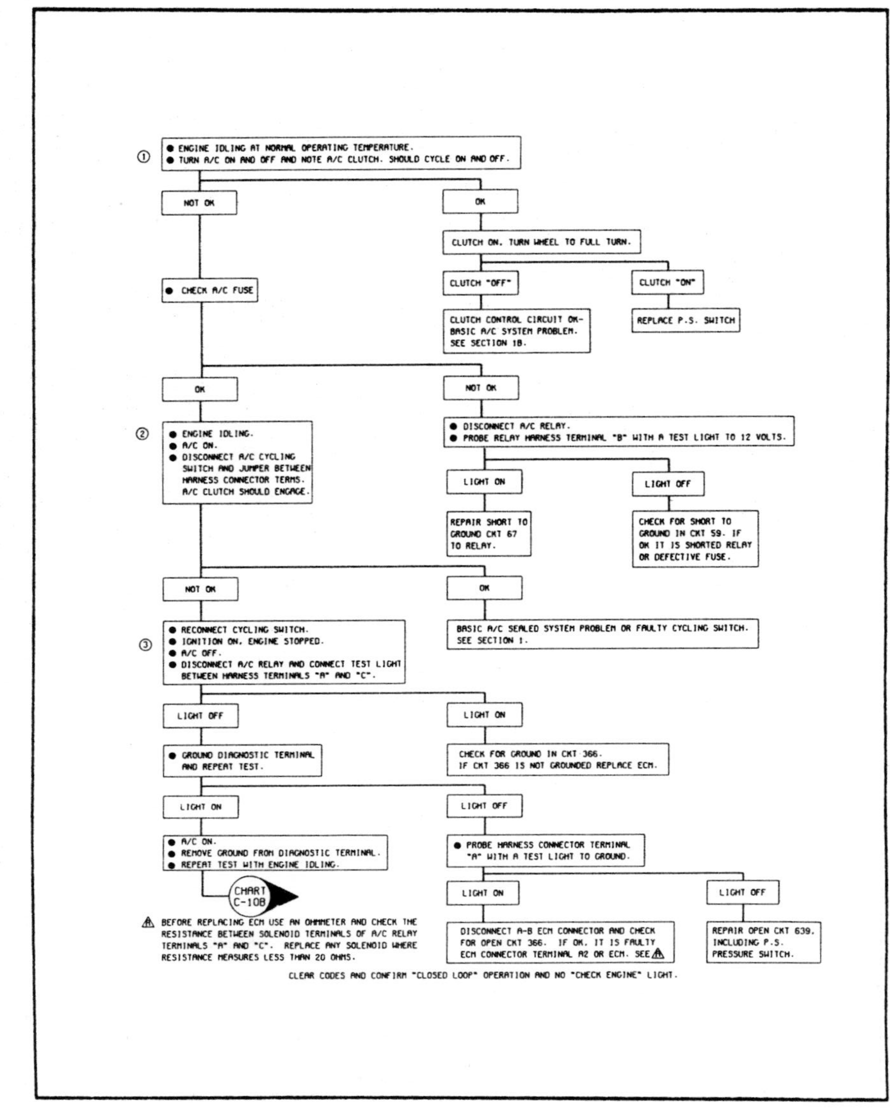

CHART C-10B
A/C Clutch Control (2 of 2)—3.8L

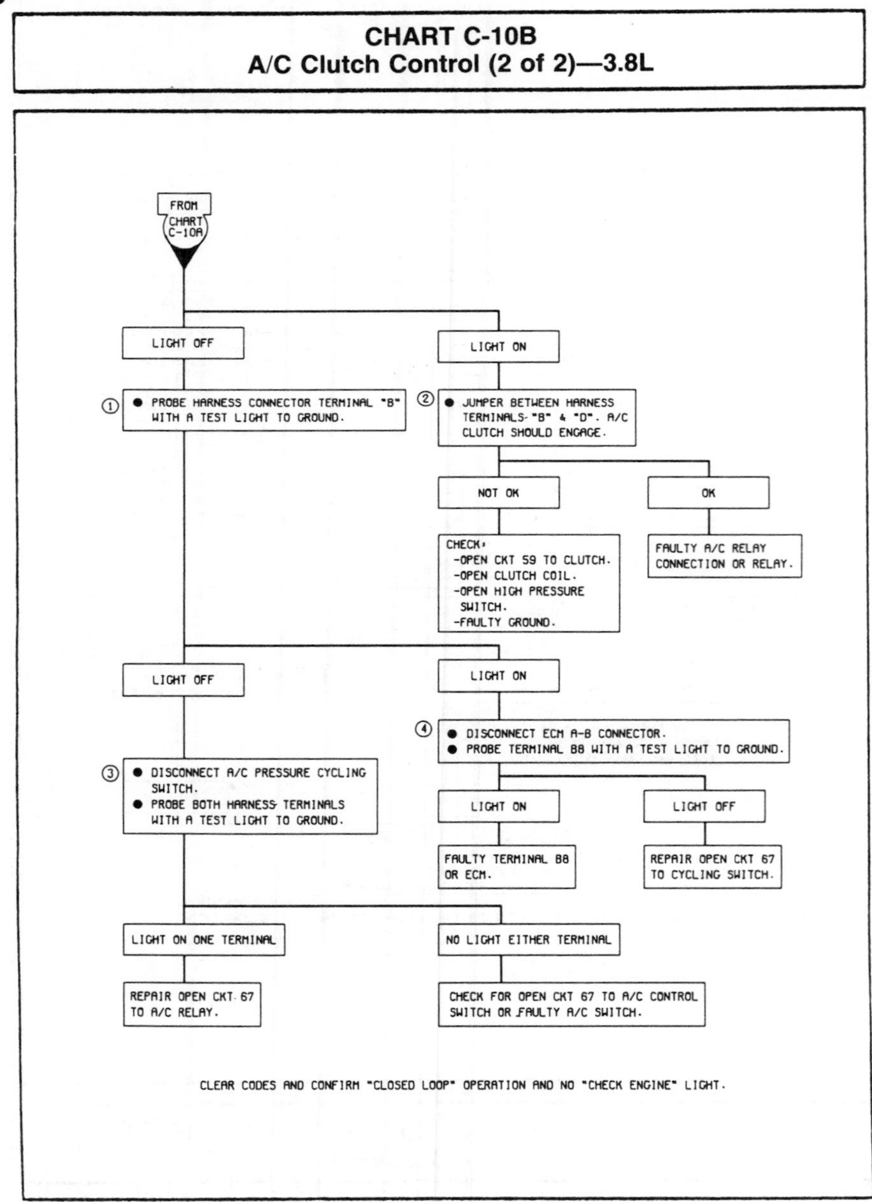

CHART C-12A
Coolant Fan Functional Check—3.8L

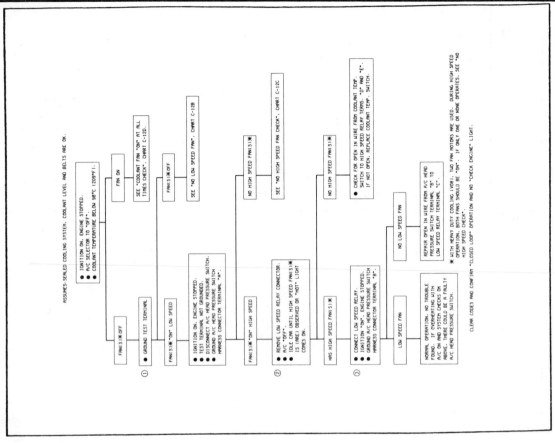

ASSUMES- SEALED COOLING SYSTEM, COOLANT LEVEL AND BELTS ARE OK.

- IGNITION ON, ENGINE STOPPED.
- A/C SELECTOR TO "OFF".
- COOLANT TEMPERATURE BELOW 98°C (209°F).

FAN(S)✱ OFF

FAN ON

SEE "COOLANT FAN "ON" AT ALL TIMES CHECK", CHART C-12D.

① ● GROUND TEST TERMINAL

FAN(S)✱ "ON" LOW SPEED

FAN(S)✱ OFF

SEE "NO LOW SPEED FAN", CHART C-12B

- IGNITION ON, ENGINE STOPPED.
- TEST TERMINAL NOT GROUNDED.
- DISCONNECT A/C HEAD PRESSURE SWITCH.
- GROUND A/C HEAD PRESSURE SWITCH HARNESS CONNECTOR TERMINAL "A".

FAN(S)✱ "ON" HIGH SPEED

NO HIGH SPEED FAN(S)✱

SEE "NO HIGH SPEED FAN CHECK", CHART C-12C

② ● REMOVE LOW SPEED RELAY CONNECTOR.
● A/C "OFF".
● IDLE CAR UNTIL HIGH SPEED FAN(S)✱ IS (ARE) OBSERVED OR "HOT" LIGHT COMES ON.

HAS HIGH SPEED FAN(S)✱

NO HIGH SPEED FAN(S)✱

● CHECK FOR OPEN IN WIRE FROM COOLANT TEMP. SWITCH TO HIGH SPEED RELAY TERMS. "D" AND "E". IF NOT OPEN, REPLACE COOLANT TEMP. SWITCH.

③ ● CONNECT LOW SPEED RELAY.
● IGNITION "ON", ENGINE STOPPED.
● GROUND A/C HEAD PRESSURE SWITCH HARNESS CONNECTOR TERMINAL "B".

LOW SPEED FAN

NO LOW SPEED FAN

NORMAL OPERATION, NO TROUBLE FOUND. IF OVERHEATING WITH A/C ON AND SYSTEM CHECKS OK ABOVE, THERE COULD BE A FAULTY A/C HEAD PRESSURE SWITCH.

REPAIR OPEN IN WIRE FROM A/C HEAD PRESSURE SWITCH TERMINAL "B" TO LOW SPEED RELAY TERMINAL "C".

✱ WITH HEAVY DUTY COOLING (V08), TWO FAN MOTORS ARE USED. DURING HIGH SPEED OPERATION, BOTH FANS SHOULD BE "ON". IF ONLY ONE OR NONE OPERATES, SEE "NO HIGH SPEED CHECK".

CLEAR CODES AND CONFIRM "CLOSED LOOP" OPERATION AND NO "CHECK ENGINE" LIGHT.

CHART C-12B
No Low Speed Fan—3.8L

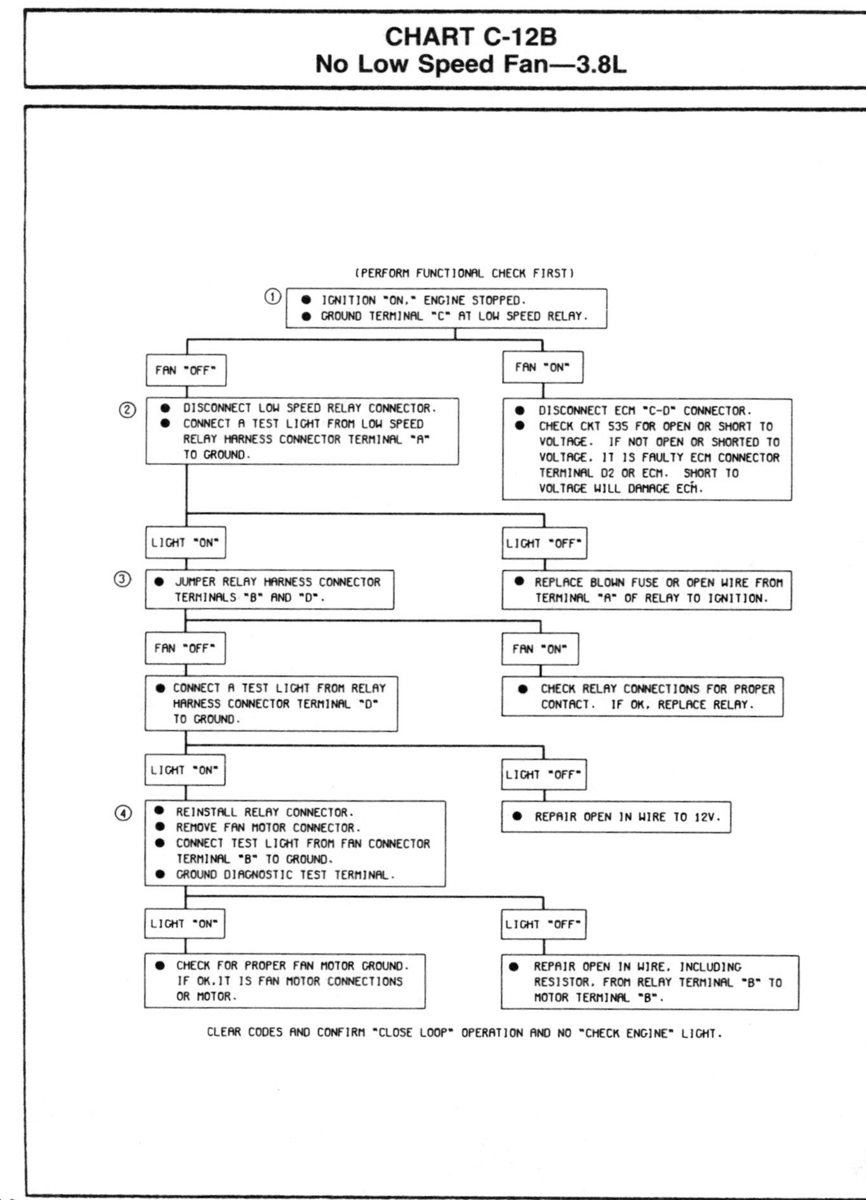

CHART C-12C
No High Speed Fan—3.8L

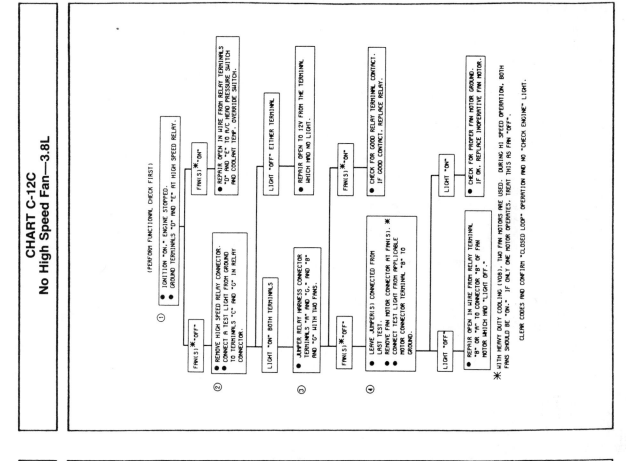

(PERFORM FUNCTIONAL CHECK FIRST)

① • IGNITION "ON," ENGINE STOPPED.
• GROUND TERMINALS "D" AND "E" AT HIGH SPEED RELAY.

FAN(S)* "OFF"

② • REMOVE HIGH SPEED RELAY CONNECTOR.
• CONNECT A TEST LIGHT FROM GROUND TO TERMINALS "C" AND "G" IN RELAY CONNECTOR.

FAN(S)* "ON"

• REPAIR OPEN IN WIRE FROM RELAY TERMINALS "D" AND "E" TO A/C HEAD PRESSURE SWITCH AND COOLANT TEMP. OVERRIDE SWITCH.

LIGHT "ON" BOTH TERMINALS

③ • JUMPER RELAY HARNESS CONNECTOR TERMINALS "A" AND "G," AND "B" AND "G" WITH TWO FANS.

LIGHT "OFF" EITHER TERMINAL

• REPAIR OPEN TO 12V FROM THE TERMINAL WHICH HAD NO LIGHT.

FAN(S)* "OFF"

④ • LEAVE JUMPER(S) CONNECTED FROM LAST TEST.
• REMOVE FAN MOTOR CONNECTOR AT FAN(S). *
• CONNECT TEST LIGHT FROM APPLICABLE MOTOR CONNECTOR TERMINAL "B" TO GROUND.

FAN(S)* "ON"

• CHECK FOR GOOD RELAY TERMINAL CONTACT. IF GOOD CONTACT, REPLACE RELAY.

LIGHT "OFF"

• REPAIR OPEN IN WIRE FROM RELAY TERMINAL "B" OR "A" TO CONNECTOR "B" OF FAN MOTOR WHICH HAD "LIGHT OFF."

LIGHT "ON"

• CHECK FOR PROPER FAN MOTOR GROUND. IF OK, REPLACE INOPERATIVE FAN MOTOR.

* WITH HEAVY DUTY COOLING (V08), TWO FAN MOTORS ARE USED. DURING HI SPEED OPERATION, BOTH FANS SHOULD BE "ON." IF ONLY ONE MOTOR OPERATES, TREAT THIS AS FAN "OFF".

CLEAR CODES AND CONFIRM "CLOSED LOOP" OPERATION AND NO "CHECK ENGINE" LIGHT.

ECM Connector ID—1.8L

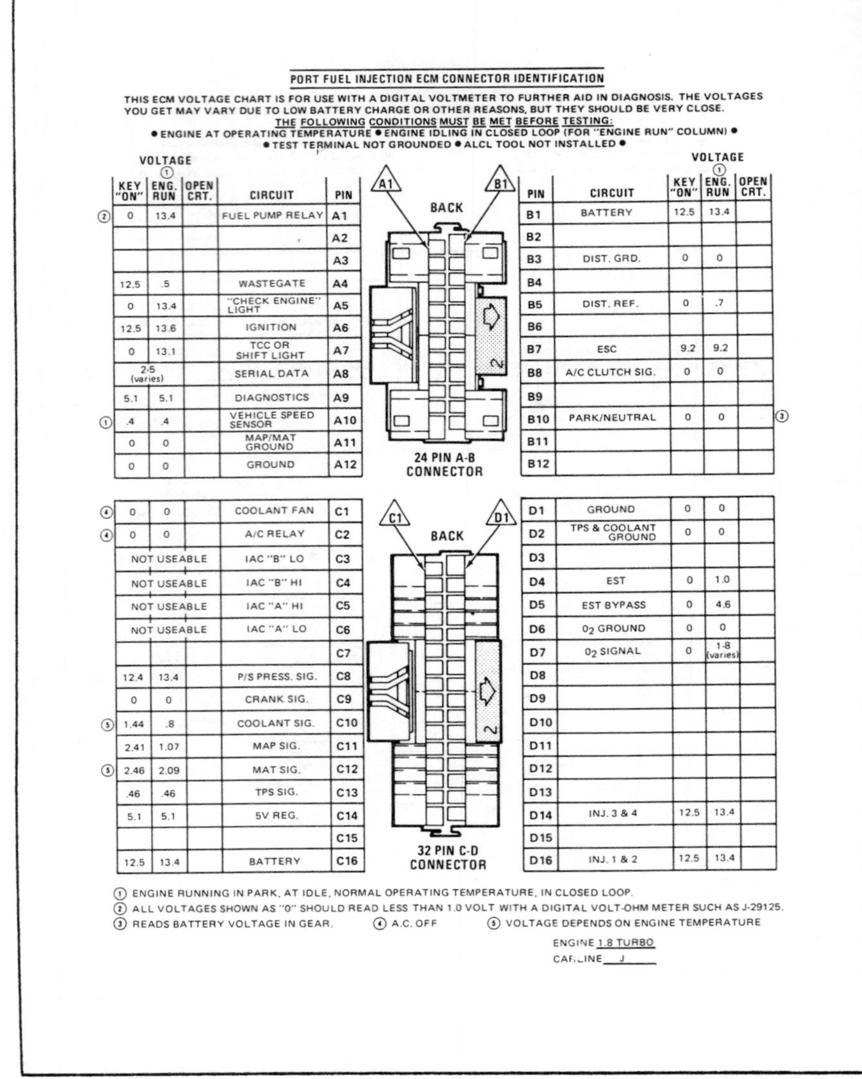

PORT FUEL INJECTION ECM CONNECTOR IDENTIFICATION

THIS ECM VOLTAGE CHART IS FOR USE WITH A DIGITAL VOLTMETER TO FURTHER AID IN DIAGNOSIS. THE VOLTAGES YOU GET MAY VARY DUE TO LOW BATTERY CHARGE OR OTHER REASONS, BUT THEY SHOULD BE VERY CLOSE.
THE FOLLOWING CONDITIONS MUST BE MET BEFORE TESTING:
• ENGINE AT OPERATING TEMPERATURE • ENGINE IDLING IN CLOSED LOOP (FOR "ENGINE RUN" COLUMN) •
• TEST TERMINAL NOT GROUNDED • ALCL TOOL NOT INSTALLED •

24 PIN A-B CONNECTOR

Note	VOLTAGE KEY "ON"	VOLTAGE ENG. RUN ①	OPEN CRT.	CIRCUIT	PIN
②	0	13.4		FUEL PUMP RELAY	A1
					A2
					A3
	12.5	.5		WASTEGATE	A4
	0	13.4		"CHECK ENGINE" LIGHT	A5
	12.5	13.6		IGNITION	A6
	0	13.1		TCC OR SHIFT LIGHT	A7
	2-5 (varies)			SERIAL DATA	A8
	5.1	5.1		DIAGNOSTICS	A9
①	.4	.4		VEHICLE SPEED SENSOR	A10
	0	0		MAP/MAT GROUND	A11
	0	0		GROUND	A12

PIN	CIRCUIT	VOLTAGE KEY "ON"	VOLTAGE ENG. RUN ①	OPEN CRT.	Note
B1	BATTERY	12.5	13.4		
B2					
B3	DIST. GRD.	0	0		
B4					
B5	DIST. REF.	0	.7		
B6					
B7	ESC	9.2	9.2		
B8	A/C CLUTCH SIG.	0	0		
B9					
B10	PARK/NEUTRAL	0	0		③
B11					
B12					

32 PIN C-D CONNECTOR

Note	KEY "ON"	ENG. RUN	OPEN CRT.	CIRCUIT	PIN
④	0	0		COOLANT FAN	C1
④	0	0		A/C RELAY	C2
	NOT USEABLE			IAC "B" LO	C3
	NOT USEABLE			IAC "B" HI	C4
	NOT USEABLE			IAC "A" HI	C5
	NOT USEABLE			IAC "A" LO	C6
					C7
	12.4	13.4		P/S PRESS. SIG.	C8
	0	0		CRANK SIG.	C9
⑤	1.44	.8		COOLANT SIG.	C10
	2.41	1.07		MAP SIG.	C11
⑤	2.46	2.09		MAT SIG.	C12
	.46	.46		TPS SIG.	C13
	5.1	5.1		5V REG.	C14
					C15
	12.5	13.4		BATTERY	C16

PIN	CIRCUIT	KEY "ON"	ENG. RUN	OPEN CRT.
D1	GROUND	0	0	
D2	TPS & COOLANT GROUND	0	0	
D3				
D4	EST	0	1.0	
D5	EST BYPASS	0	4.6	
D6	O_2 GROUND	0	0	
D7	O_2 SIGNAL	0	1-8 (varies)	
D8				
D9				
D10				
D11				
D12				
D13				
D14	INJ. 3 & 4	12.5	13.4	
D15				
D16	INJ. 1 & 2	12.5	13.4	

① ENGINE RUNNING IN PARK, AT IDLE, NORMAL OPERATING TEMPERATURE, IN CLOSED LOOP.
② ALL VOLTAGES SHOWN AS "0" SHOULD READ LESS THAN 1.0 VOLT WITH A DIGITAL VOLT-OHM METER SUCH AS J-29125.
③ READS BATTERY VOLTAGE IN GEAR. ④ A.C. OFF ⑤ VOLTAGE DEPENDS ON ENGINE TEMPERATURE

ENGINE 1.8 TURBO
CARLINE J

CHART C-12D
Fan 'On' At All Times—3.8L

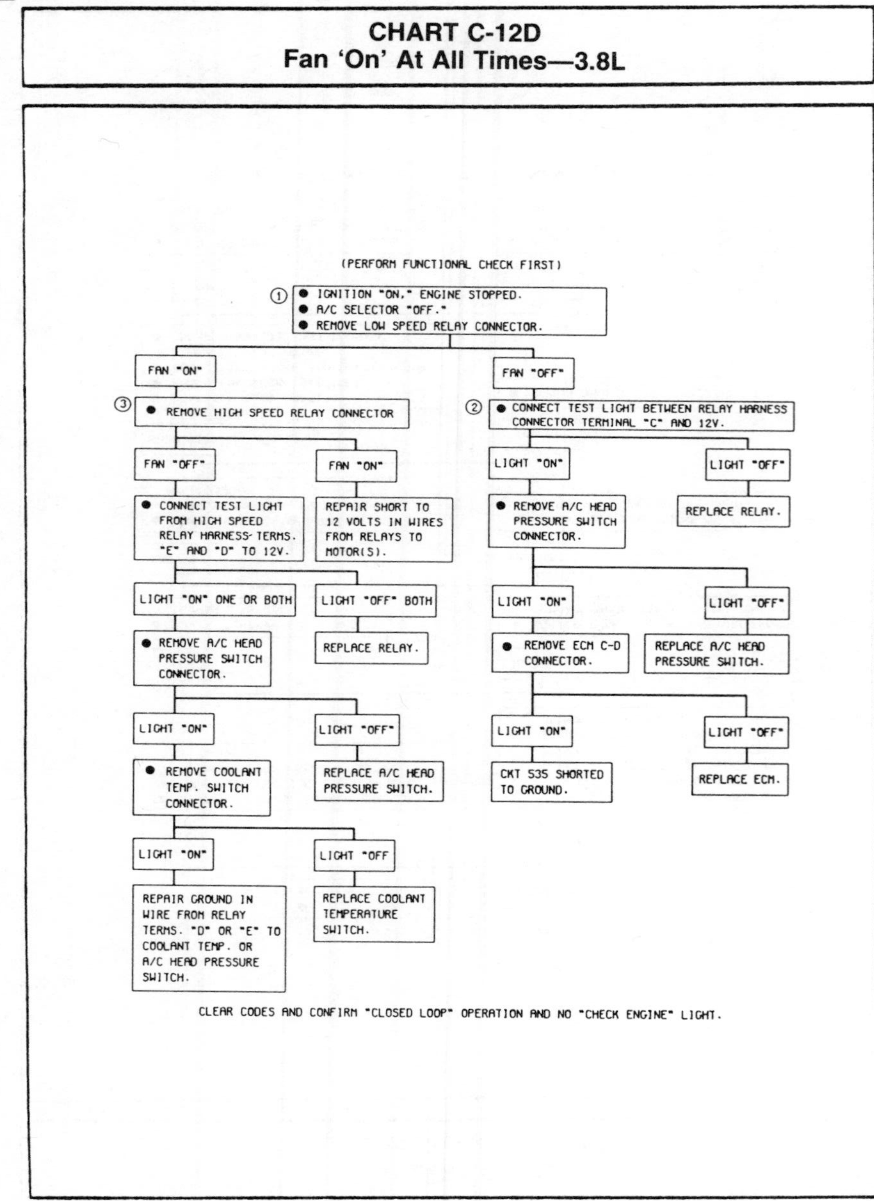

ECM Schematic—1.8L

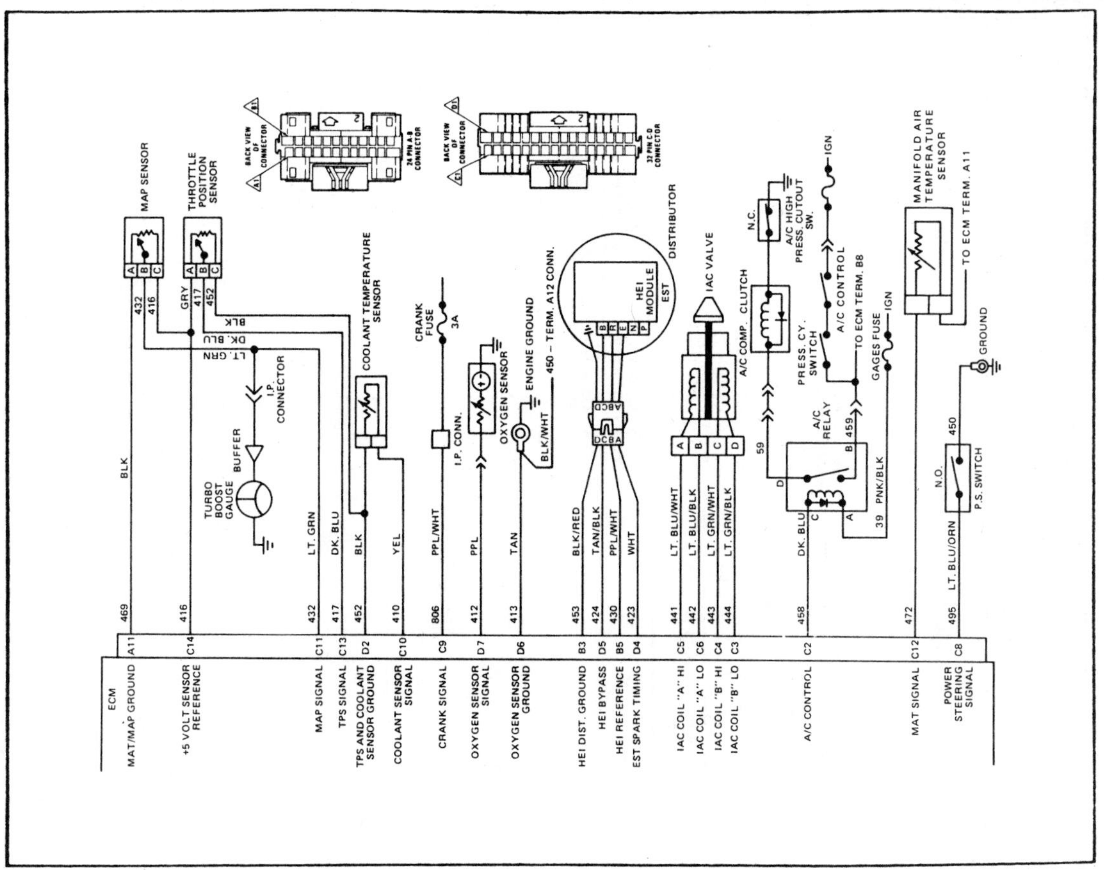

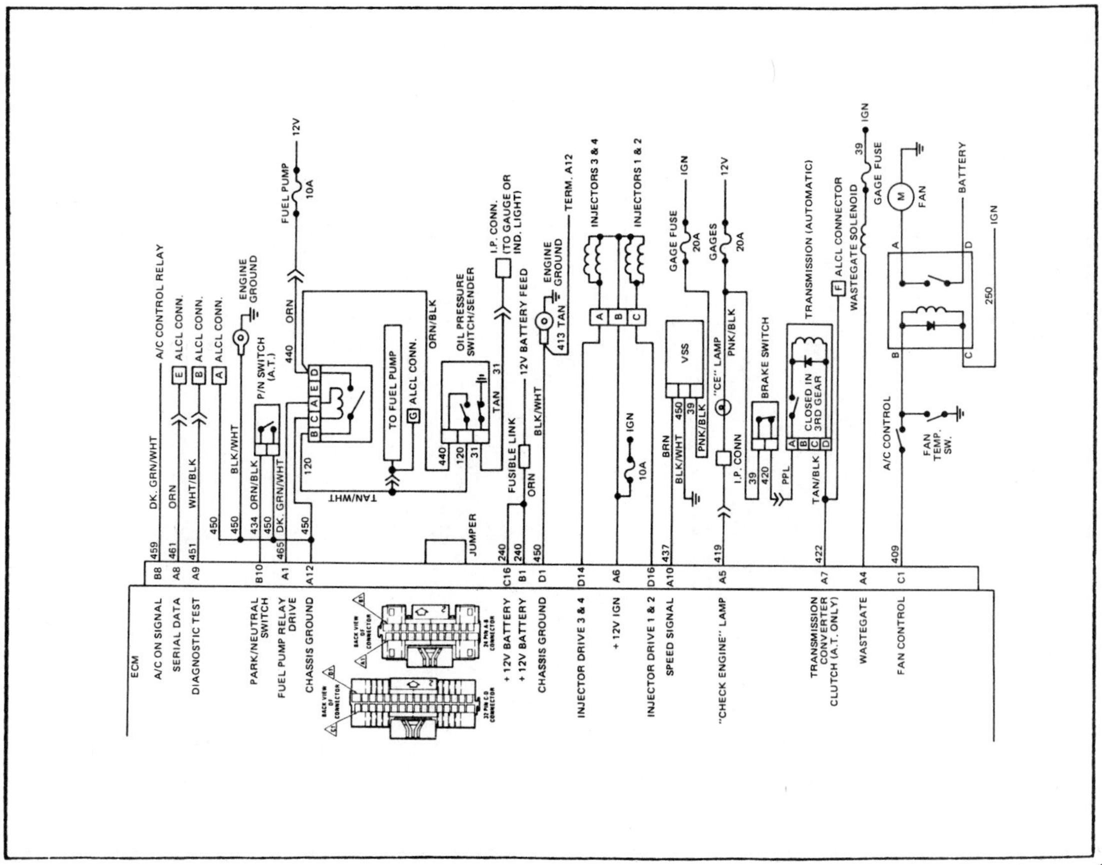

ECM Schematic—1.8L

ECM
A/C ON SIGNAL
SERIAL DATA
DIAGNOSTIC TEST
PARK/NEUTRAL SWITCH
FUEL PUMP RELAY DRIVE
CHASSIS GROUND
+12V BATTERY
+12V BATTERY
CHASSIS GROUND
INJECTOR DRIVE 3 & 4
+12V IGN
INJECTOR DRIVE 1 & 2
SPEED SIGNAL
"CHECK ENGINE" LAMP
TRANSMISSION CONVERTER CLUTCH (A.T. ONLY)
WASTEGATE
FAN CONTROL
A/C CONTROL RELAY
ALCL CONN.
ENGINE GROUND
P/N SWITCH (A.T.)
FUEL PUMP 10A
12V
TO FUEL PUMP
ALCL CONN.
OIL PRESSURE SWITCH/SENDER
I.P. CONN. (TO GAUGE OR IND. LIGHT)
12V BATTERY FEED
FUSIBLE LINK
JUMPER
TERM. A12
INJECTORS 3 & 4
INJECTORS 1 & 2
GAGE FUSE 20A
GAGES 20A
VSS
"CE" LAMP
BRAKE SWITCH
TRANSMISSION (AUTOMATIC)
CLOSED IN 3RD GEAR
ALCL CONNECTOR
WASTEGATE SOLENOID
GAGE FUSE
FAN
BATTERY
A/C CONTROL
FAN TEMP. SW.
IGN

CHART A-7
Fuel System Diagnosis (1 of 2)—1.8L

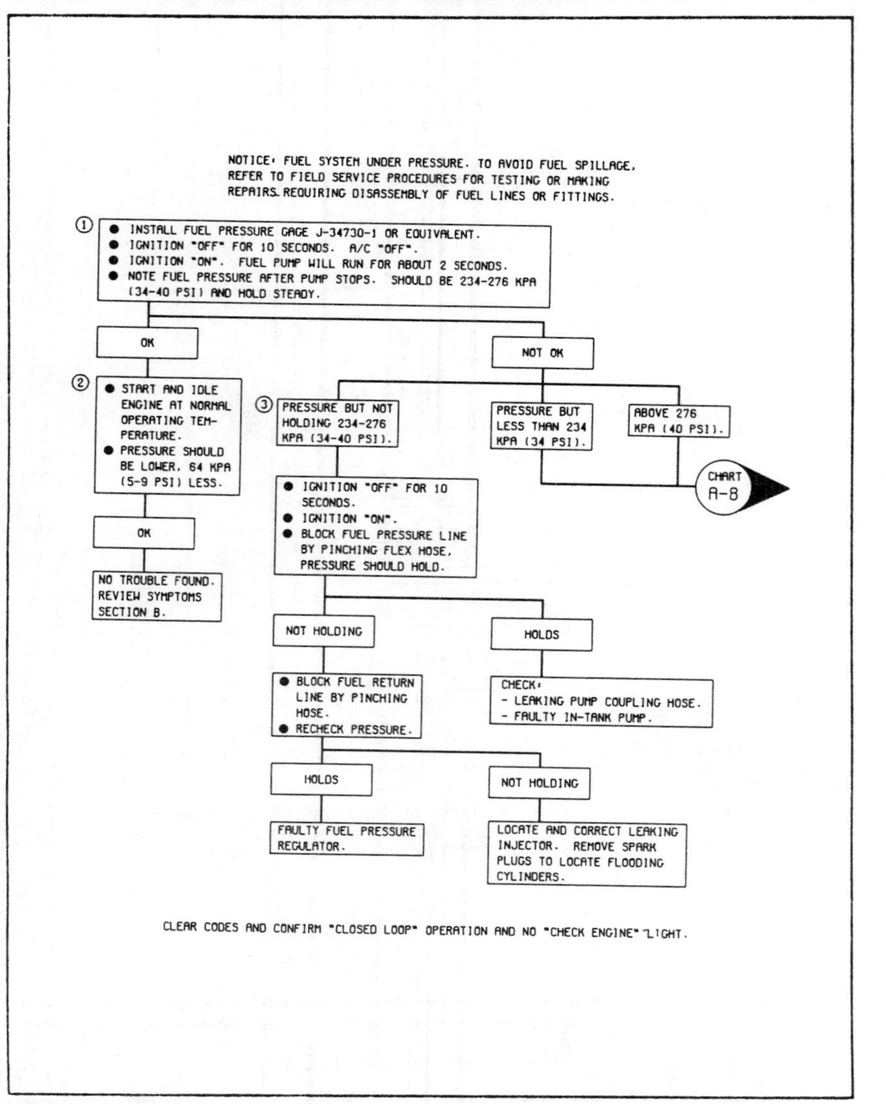
NOTICE: FUEL SYSTEM UNDER PRESSURE. TO AVOID FUEL SPILLAGE, REFER TO FIELD SERVICE PROCEDURES FOR TESTING OR MAKING REPAIRS REQUIRING DISASSEMBLY OF FUEL LINES OR FITTINGS.
① ● INSTALL FUEL PRESSURE GAGE J-34730-1 OR EQUIVALENT. ● IGNITION "OFF" FOR 10 SECONDS. A/C "OFF". ● IGNITION "ON". FUEL PUMP WILL RUN FOR ABOUT 2 SECONDS. ● NOTE FUEL PRESSURE AFTER PUMP STOPS. SHOULD BE 234-276 KPA (34-40 PSI) AND HOLD STEADY.
OK
NOT OK
② ● START AND IDLE ENGINE AT NORMAL OPERATING TEMPERATURE. ● PRESSURE SHOULD BE LOWER, 64 KPA (5-9 PSI) LESS.
OK
NO TROUBLE FOUND. REVIEW SYMPTOMS SECTION B.
③ PRESSURE BUT NOT HOLDING 234-276 KPA (34-40 PSI).
PRESSURE BUT LESS THAN 234 KPA (34 PSI).
ABOVE 276 KPA (40 PSI).
CHART A-8
● IGNITION "OFF" FOR 10 SECONDS. ● IGNITION "ON". ● BLOCK FUEL PRESSURE LINE BY PINCHING FLEX HOSE. PRESSURE SHOULD HOLD.
NOT HOLDING
HOLDS
● BLOCK FUEL RETURN LINE BY PINCHING HOSE. ● RECHECK PRESSURE.
CHECK: - LEAKING PUMP COUPLING HOSE. - FAULTY IN-TANK PUMP.
HOLDS
NOT HOLDING
FAULTY FUEL PRESSURE REGULATOR.
LOCATE AND CORRECT LEAKING INJECTOR. REMOVE SPARK PLUGS TO LOCATE FLOODING CYLINDERS.
CLEAR CODES AND CONFIRM "CLOSED LOOP" OPERATION AND NO "CHECK ENGINE" LIGHT.

CHART A-8
Fuel System Diagnosis (2 of 2)—1.8L

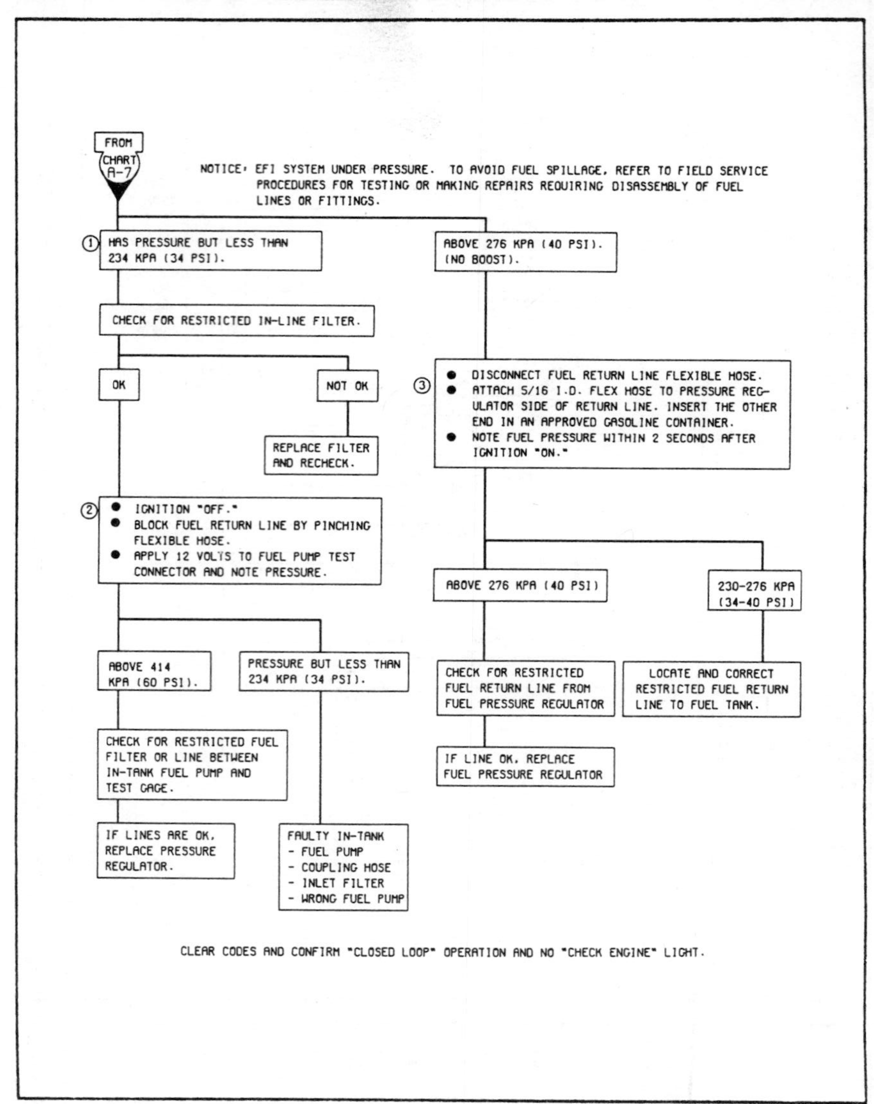
FROM CHART A-7
NOTICE: EFI SYSTEM UNDER PRESSURE. TO AVOID FUEL SPILLAGE, REFER TO FIELD SERVICE PROCEDURES FOR TESTING OR MAKING REPAIRS REQUIRING DISASSEMBLY OF FUEL LINES OR FITTINGS.
① HAS PRESSURE BUT LESS THAN 234 KPA (34 PSI).
ABOVE 276 KPA (40 PSI). (NO BOOST).
CHECK FOR RESTRICTED IN-LINE FILTER.
OK
NOT OK
REPLACE FILTER AND RECHECK.
③ ● DISCONNECT FUEL RETURN LINE FLEXIBLE HOSE. ● ATTACH 5/16 I.D. FLEX HOSE TO PRESSURE REGULATOR SIDE OF RETURN LINE. INSERT THE OTHER END IN AN APPROVED GASOLINE CONTAINER. ● NOTE FUEL PRESSURE WITHIN 2 SECONDS AFTER IGNITION "ON."
② ● IGNITION "OFF." ● BLOCK FUEL RETURN LINE BY PINCHING FLEXIBLE HOSE. ● APPLY 12 VOLTS TO FUEL PUMP TEST CONNECTOR AND NOTE PRESSURE.
ABOVE 276 KPA (40 PSI)
230-276 KPA (34-40 PSI)
ABOVE 414 KPA (60 PSI).
PRESSURE BUT LESS THAN 234 KPA (34 PSI).
CHECK FOR RESTRICTED FUEL RETURN LINE FROM FUEL PRESSURE REGULATOR
LOCATE AND CORRECT RESTRICTED FUEL RETURN LINE TO FUEL TANK.
CHECK FOR RESTRICTED FUEL FILTER OR LINE BETWEEN IN-TANK FUEL PUMP AND TEST GAGE.
IF LINE OK, REPLACE FUEL PRESSURE REGULATOR
IF LINES ARE OK, REPLACE PRESSURE REGULATOR.
FAULTY IN-TANK - FUEL PUMP - COUPLING HOSE - INLET FILTER - WRONG FUEL PUMP
CLEAR CODES AND CONFIRM "CLOSED LOOP" OPERATION AND NO "CHECK ENGINE" LIGHT.

GM THROTTLE BODY INJECTION

General Information

All 1982 and later USA cars equipped with the Pontiac 4-151, the 1.8L OHC or the 2.0L OHV engine use a single bore, throttle body fuel injection unit. All Canadian 4-151s retain the 2 bbl carburetors. The 1982 and later Corvette equipped with the V8 350 and the 1982 and later Camaro and Firebird equipped with the V8 305 engine use a pair of single bore throttle body injection units.

In this throttle body system, a single fuel injector mounted at the top of the throttle body sprays fuel down through the throttle valve and into the intake manifold. The throttle body resembles a carburetor in appearance but does away with much of the carburetor's complexity (choke system and linkage, power valves, accelerator pump, jets, fuel circuits, etc.), replacing these with the electrically operated fuel injector.

The injector is actually a solenoid which when activated lifts a pintle valve off its seat, allowing the pressurized (10 psi) fuel behind the valve to spray out. The nozzle of the injector is designed to atomize the fuel for complete air/fuel mixture.

The activating signal for the injector originates with the Electronic Control Module (ECM), which monitors engine temperature, throttle position, vehicle speed and several other engine-related conditions then continuously updates injector opening times in relation to the information given by these sensors.

The throttle body is also equipped with an idle air control motor. The idle air control motor operates a pintle valve at the side of the throttle body. When the valve opens it allows air to bypass the throttle, which provides the additional air required to idle at elevated speed when the engine is cold. The idle air control motor also compensates for accessory loads and changing engine friction during break-in. The idle speed control motor is controlled by the ECM.

Fuel pressure for the system is provided by an in-tank fuel pump. The pump is a two-stage turbine design powered by a DC motor. It is designed for smooth, quiet operation, high flow and fast priming. The design of the fuel inlet reduces the possibility of vapor lock under hot fuel conditions. The pump sends fuel forward through the fuel line to a stainless steel high-flow fuel filter mounted on the engine. From the filter the fuel moves to the throttle body. The fuel pump inlet is located in a reservoir in the fuel tank which insures a constant supply of fuel to the pump during hard cornering and on steep inclines. The fuel pump is controlled by a fuel pump relay, which in turn receives its signal from the ECM. A fuel pressure regulator inside the throttle body maintains fuel pressure at 10 psi and routes unused fuel back to the fuel tank through a fuel return line. On the dual throttle body system, a fuel pressure compensator is used on the second throttle body assembly to compensate for a momentary fuel pressure drop between the two units. This constant circulation of fuel through the throttle body prevents component overheating and vapor lock.

The electronic control module (ECM), also called a microcomputer, is the brain of the fuel injection system. After receiving inputs from various sensing elements in the system. The ECM commands the fuel injector, idle air control motor, EST distributor, torque converter clutch and other engine actuators to operate in a pre-programmed manner to improve driveability and fuel economy while controlling emissions. The sensing elements update the computer every tenth of a second for general information and every 12.5 milli-seconds for critical emissions and driveability information.

The ECM has limited system diagnostic capability. If certain system malfunctions occur, the diagnostic "check engine" light in the instrument panel will light, alerting the driver to the need for service.

Since both idle speed and mixture are controlled by the ECM on this system, no adjustments are possible or necessary.

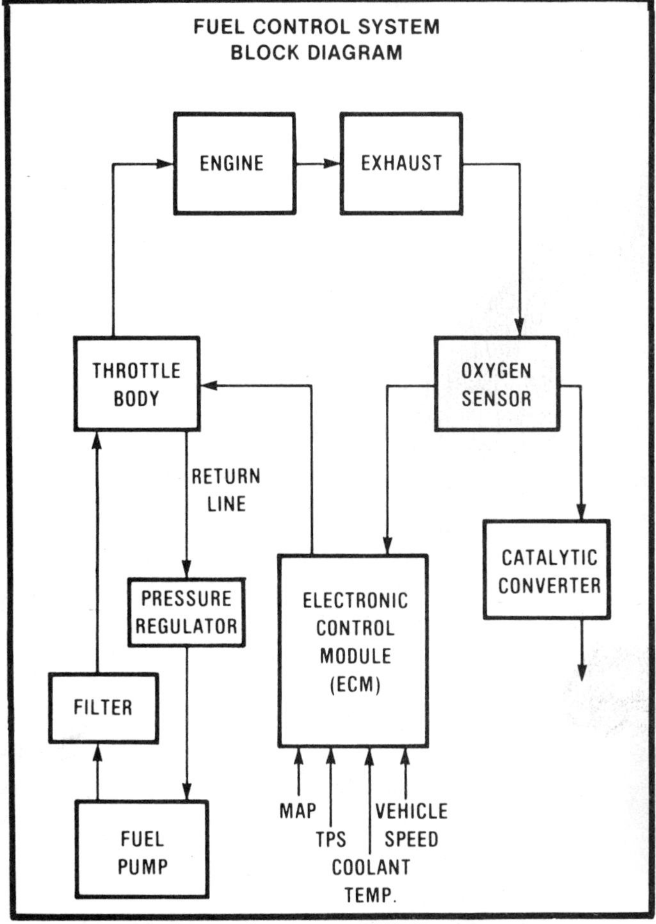

Schematic of fuel control system on GM TBI system

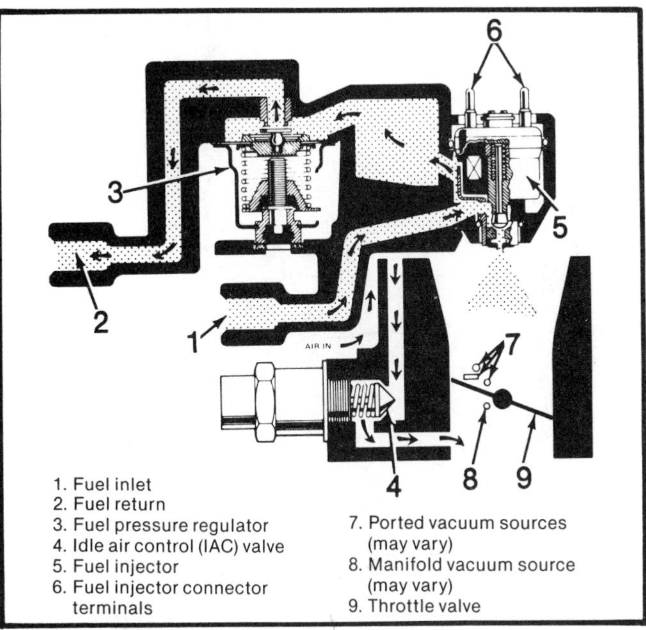

Throttle body unit operation

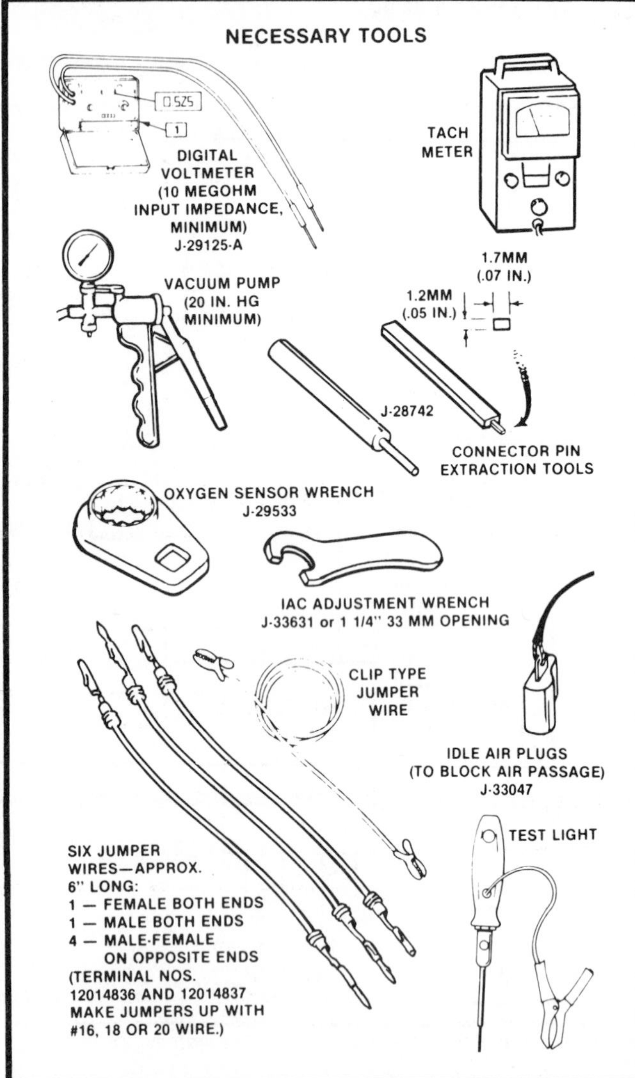

Special tools used to service the GM TBI system

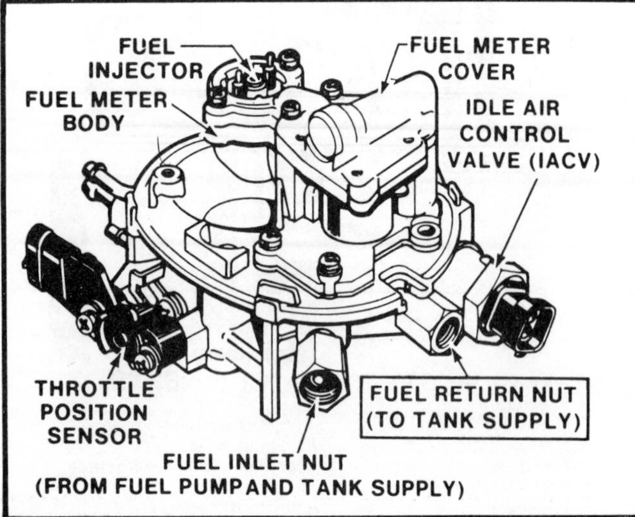

Typical GM throttle body unit assembly

On-Car Service

The system does not require special tools for diagnosis. A tachometer, test light, ohmmeter, digital voltmeter with 10 megohms impedance, vacuum pump, vacuum gauge and jumper wires are required for diagnosis. A test light or voltmeter must be used when specified in the procedures.

FUEL SYSTEM PRESSURE TEST

CAUTION

To reduce the risk of fire and personal injury, it is necessary to relieve the fuel system pressure before servicing fuel system components.

1. Remove the fuel pump fuse from the fuse block.
2. Crank the engine. The engine will run until it runs out of fuel. Crank the engine again for 3 seconds making sure it is out of fuel.
3. Turn the ignition off and replace the fuse.
4. Remove the air cleaner. Plug the thermal vacuum port on the throttle body.
5. Remove the fuel line between the throttle body and filter.
6. Install a fuel pressure gauge between the throttle body and fuel filter. The gauge should be able to register at least 15 psi.
7. Start the car. Observe the fuel pressure reading. It should be 9–13 psi.

NOTE: Before removing the fuel pressure gauge the fuel system must be depressurized.

8. Reinstall the parts in the reverse order of removal.

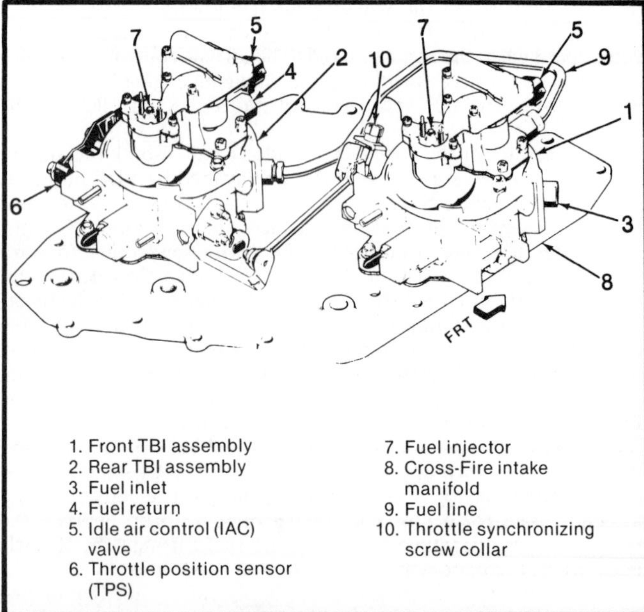

5.7L V8 Cross Fire throttle body units

MINIMUM AIR RATE ADJUSTMENT (2.5L)

This adjustment should be performed only when the throttle body parts have been replaced or required to do so by the T.P.S. adjustment. Engine should be at normal operating temperature before making adjustment.

1. Remove air cleaner and air cleaner to TBI gasket. Plug vacuum port on TBI unit for THERMAC.

NOTE: On vehicles equipped with a tamper resistant plug covering the minimum air adjustment screw, the throttle body unit must be removed from the engine to remove the plug.

2. Remove T.V. cable from throttle control bracket to allow access to minimum air adjustment screw.
3. Connect a tachometer to engine.
4. Start engine, transmission in Park (Neutral on manual transmission) and allow engine rpm to stabilize.
5. Install tool J-33047, or its equal, and idle air passage of throttle body. Be certain that tool seats fully in passage and no air leaks exist.
6. Using appropriate screwdriver, turn minimum air screw until engine rpm is 500 ± 25 in neutral with automatic transaxle, and 775 ± 25 in neutral with manual transaxle.
7. Stop engine and remove tool J-33047 from throttle body.
8. Reinstall T.V. cable into throttle control bracket.
9. Use silicone sealant or equivalent to cover minimum air adjustment screw.
10. Install air cleaner gasket and air cleaner to engine.

CURB IDLE AIR RATE (5.0L CROSSFIRE INJECTION SYSTEM)

The throttle position of each throttle body must be balanced so that the throttle plates are synchronized to open simultaneously. This is a checking and adjustment procedure; adjustment should be performed only when a throttle body has been replaced or when checking procedure indicates an adjustment is required.

1. Remove air cleaner and air cleaner to TBI unit for THERMAC.
2. Start engine and allow engine rpm to stabilize.
3. Plug idle air passages of each throttle body with plugs J-33047, or their equal. Be certain plugs are fully seated in passages and no air leaks exist. Engine rpm should decrease to curb idle air rate. If engine rpm does not decrease, check for vacuum leak.
4. Remove cap from ported tube on rear TBI unit and connect the vacuum gauge.
5. Observe the gauge, reading should be approximately 0.45 in. Hg. If adjustment is required proceed as follows:
   a. Remove tamper resistant screw covering the minimum air adjustment screw if required.
   b. Adjust minimum air adjustment screw to obtain approximately 0.45 in. Hg.
   c. After adjustment, proceed to front TBI unit.
6. Remove gauge from rear TBI unit and re-install cap on ported tube.
7. Remove cap from ported tube on front TBI unit and connect vacuum gauge. Reading should also be approximately 0.45 in. Hg. If adjustment is required proceed as follows:
   a. Locate split lever screw on throttle linkage. If screw is welded for tamper resistance, break weld and install new screw with thread locking compound applied.
   b. Adjust split lever screw to obtain approximately 0.45 in. Hg.

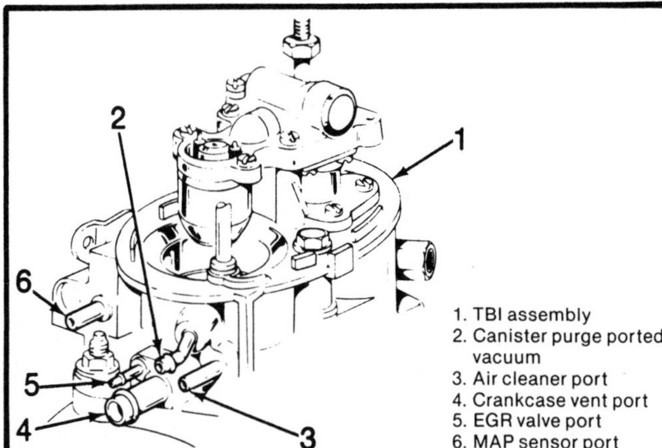

TBI port and manifold vacuum sources

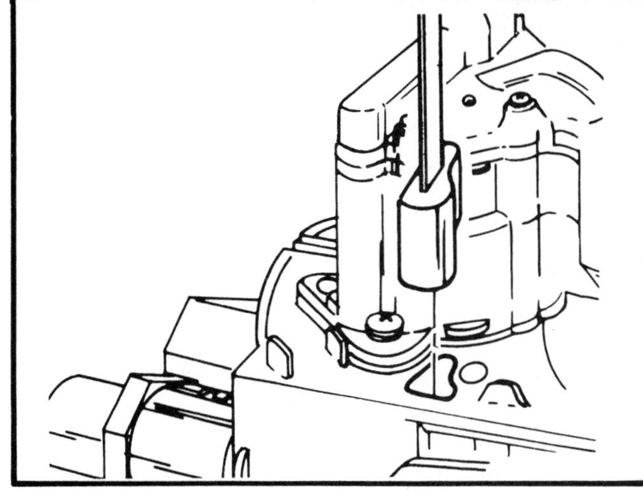
Installing idle air passage blocking tool

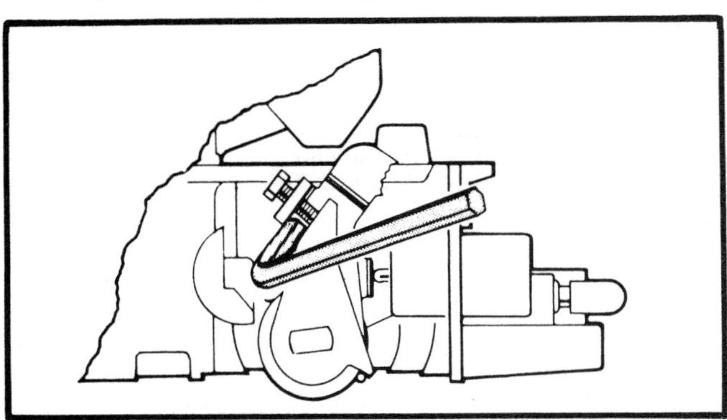
Blocking throttle lever

8. Remove gauge from front TBI unit and re-install cap on ported tube.
9. If both readings are approximately 0.45 in. Hg., no adjustment is required; throttle plates are synchronized.
10. Stop engine and remove idle air passage plugs.
11. Check T.P.S. voltage and adjust if required.
12. Install air cleaner gaskets, connect vacuum line to TBI unit and install air cleaner.

THROTTLE POSITION SENSOR

Testing

Throttle position sensor adjustment should be checked after minimum air adjustment is completed.

1. Remove air cleaner.

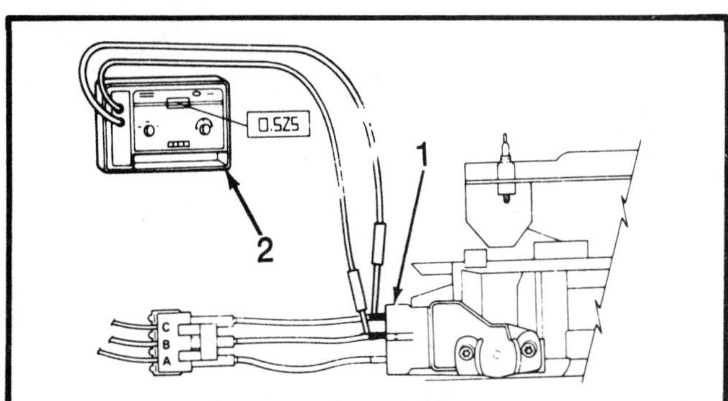

Adjusting throttle position switch (1) using digital volt-ohmmeter (2) with 10 mega-ohm impedance

2. Disconnect T.P.S. harness from T.P.S.
3. Using three jumper wires connect T.P.S. harness to T.P.S.
4. With ignition "ON," engine stopped, used a digital voltmeter to measure voltage between terminals B and C.
5. Voltage should read .525 ± .075 volts. 5.0L engine and .820 + .250 volts 2.5L engine.
6. Adjust T.P.S. if required.
7. With ignition "OFF," remove jumpers and connect T.P.S. harness to T.P.S.
8. Install air cleaner.

Adjustment

1. After installing TPS to throttle body, install throttle body unit to engine.
2. Remove EGR valve and heat shield from engine.
3. Using three six inch jumpers, connect TPS harness to TPS
4. With ignition "ON," engine stopped, use a digital voltmeter to measure voltage between TPS terminals B and C.
5. Loosen two TPS attaching screws and rotate throttle position sensor to obtain a voltage reading of .525 ± .075 volts 5.OL engine and .820 ± .250 volts 2.5L engine.
6. With ignition "OFF," remove jumpers and reconnect TPS harness to TPS
7. Install EGR valve and heat shield to engine, using new gasket as necessary.
8. Install air cleaner gasket and air cleaner to throttle body unit.

NOTE: Throttle position sensors are not adjustable on 1983–85 2.5L and 1982–85 1.8L engines with TBI.

Removal

The throttle position sensor (TPS) is an electrical unit and must not be immersed in any type of liquid solvent or cleaner. The TPS is factory adjusted and the retaining screws are spot welded in place to retain the critical setting. With these considerations, it is possible to clean the throttle body assembly without removing the TPS if care is used. Should TPS replacement be required however, proceed using the following steps:

1. Invert throttle body and place on a clean, flat surface.

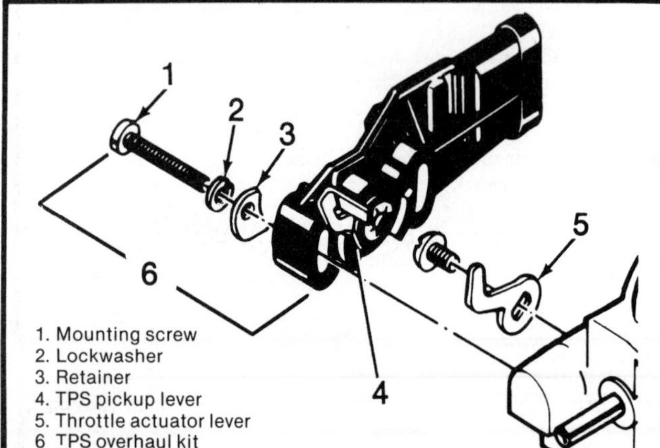

Throttle position switch (TPS) attaching screws. Note lever position when removing the TPS

2. Using a 5/16 in. drill bit, drill completely through two (2) TPS screw access holes in base of throttle body to be sure of removing the spot welds holding TPS screws in place.
3. Remove the two TPS attaching screws, lockwashers, and retainers. Then, remove TPS sensor from throttle body. DISCARD SCREWS. New screws are supplied in service kits.
4. If necessary, remove screw holding throttle position sensor actuator lever to end of throttle shaft.
5. Remove the idle air control assembly and gasket from the throttle body.

NOTE: DO NOT immerse the idle air control motor in any type of cleaner and it should always be removed before throttle body cleaning. Immersion in cleaner will damage the IAC assembly. It is replaced only as a complete assembly.

Further disassembly of the throttle body is not required for cleaning purposes. The throttle valve screws are permanently staked in place and should not be removed. The throttle body is serviced as a complete assembly.

Installation

1. Place throttle body assembly on holding fixture to avoid damaging throttle valve.
2. Using a new sealing gasket, install idle air control motor in throttle body. Tighten motor securely.

NOTE: DO NOT overtighten to prevent damage to valve.

3. If removed, install throttle position sensor actuator lever by aligning flats on lever with flats on end of shaft. Install retaining screw and tighten securely.

NOTE: Install throttle position sensor after completion of assembly of the throttle body unit. Use thread locking compound supplied in service kit on attaching screws.

Component Removal

FUEL INJECTOR

Removal and Installation

1. Remove the air cleaner.
2. Disconnect injector electrical connector by squeezing two tabs together and pulling straight up.

NOTE: Use care in removing to prevent damage to the electrical connector pins on top of the injector, injector fuel filter and nozzle. The fuel injector is only serviced as a complete assembly. Do not immerse it in any type of cleaner.

3. Remove the fuel meter cover.
4. Using a small awl, gently pry up on the injector evenly and carefully remove it.

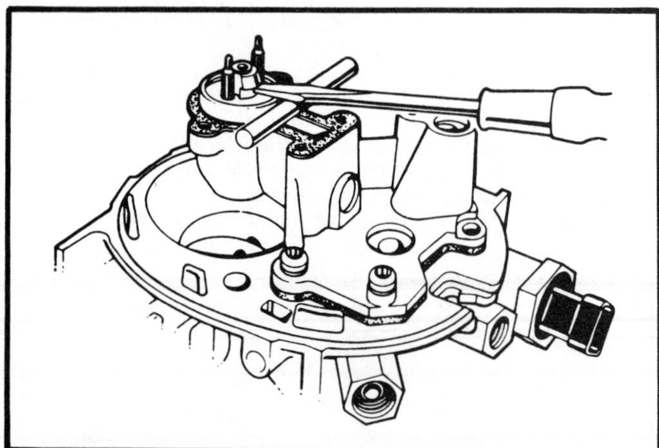

Removing fuel injector

5. Installation is the reverse of removal, with the following recommendations.

Use Dextron®II transmission fluid to lubricate all O-rings. Install the steel backup washer in the recess of the fuel meter body. Then, install the O-ring directly above backup washer, pressing the O-ring into the recess.

NOTE: Do not attempt to reverse this procedure and install backup washer and O-ring after injector is located in the cavity. To do so will prevent seating of the O-ring in the recess.

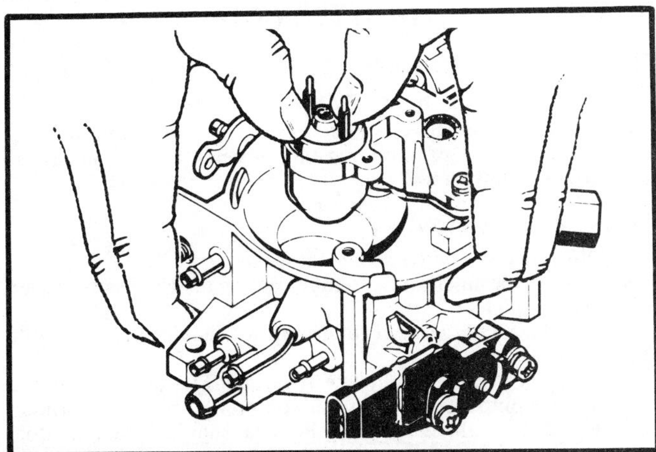

Installing fuel injector

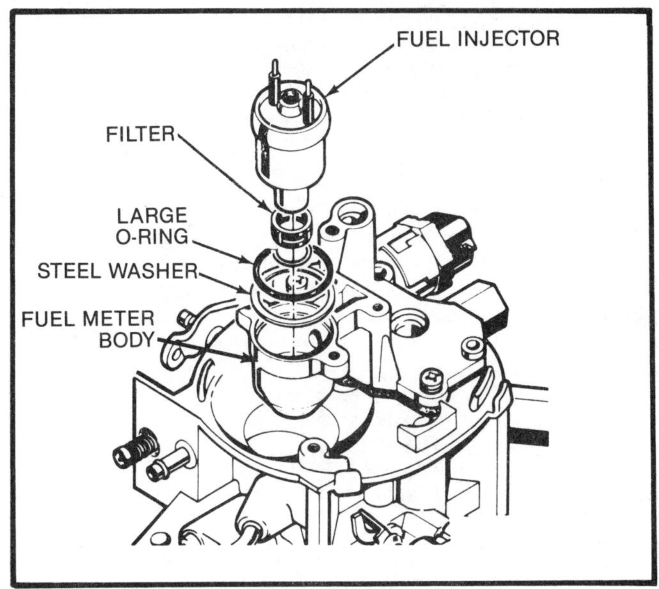

Fuel injector components

IDLE AIR CONTROL ASSEMBLY

Removal and Installation

1. Remove the air cleaner.
2. Disconnect the electrical connection from the idle air control assembly.
3. Using a 1¼ in. wrench, remove the idle air control assembly from the throttle body.

NOTE: Before installing a new assembly, measure the distance that the conical valve is extended. This measurement should be made from motor housing to end of cone. It should be greater than 1.259 in. If the cone is extended too far damage to the motor may result.

On cars with manual transmissions, idle speed will be controlled when operating temperature is reached. For automatic transmission cars, engage the transmission in drive after operating temperature is reached. This will allow the ECM to control idle speed.

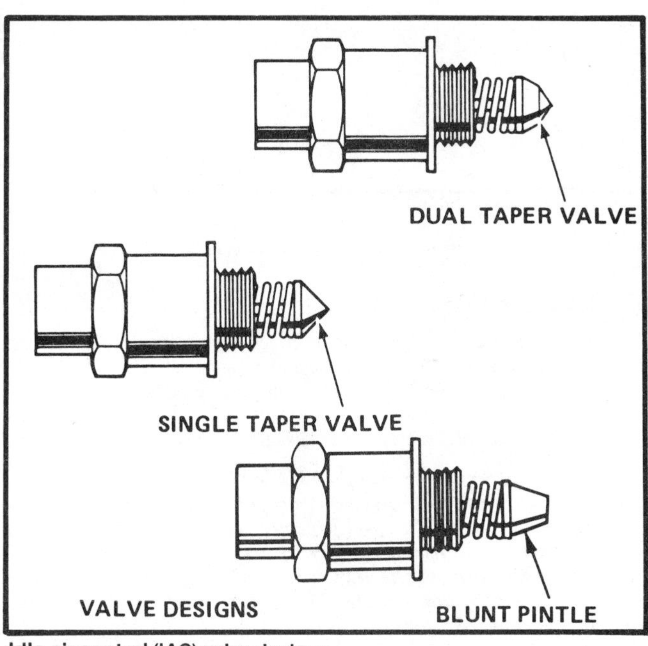

Idle air control (IAC) valve designs

FUEL PRESSURE REGULATOR/ COMPENSATOR

Removal and Installation

1. Remove air cleaner.
2. Disconnect electrical connector to injector by squeezing on two tabs and pulling straight up.
3. Remove five screws securing fuel meter cover to fuel meter body. Notice location of two short screws during removal.

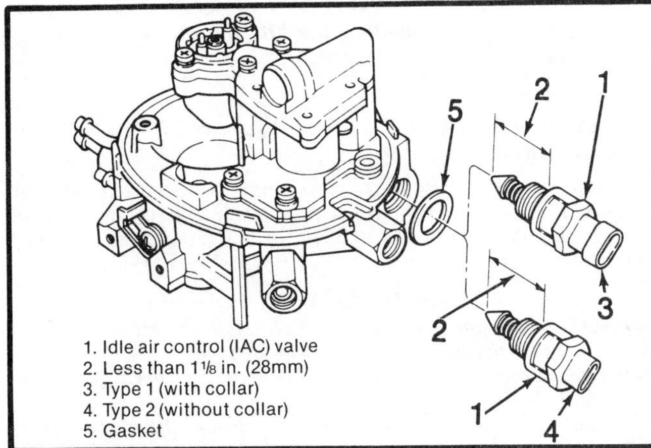

Idle air control valve installation

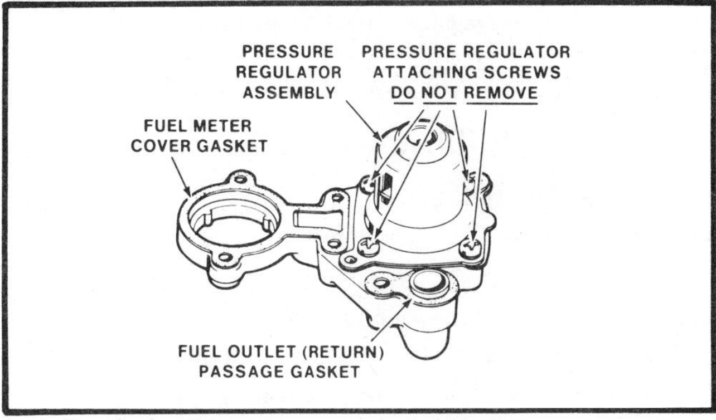

Installing fuel meter cover

CAUTION

Do not remove the four screws securing the pressure regulator to the fuel meter cover. The fuel pressure regulator includes a large spring under heavy tension which, if accidentally released, could cause personal injury. The fuel meter cover is only serviced as a complete assembly and includes the fuel pressure regulator preset and plugged at the factory.

NOTE: DO NOT immerse the fuel meter cover (with pressure regulator) in any type of cleaner. Immersion in cleaner will damage the internal fuel pressure regulator diaphragms and gaskets.

4. Installation is the reverse of removal.

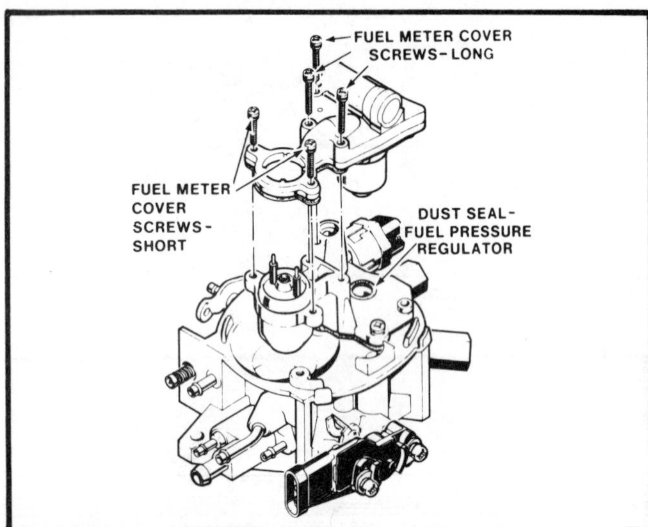

Fuel meter cover removal

FUEL METER COVER

Removal and Installation

1. Remove the five fuel meter cover screws and lockwashers holding the cover on the fuel meter body.
2. Lift off fuel meter cover (with fuel pressure regulator assembly).
3. Remove the fuel meter cover gaskets.

CAUTION

Do not remove the four screws securing the pressure regulator to the fuel meter cover. The fuel pressure regulator includes a large spring under heavy tension which, if accidentally released, could cause personal injury. The fuel meter cover is only serviced as a complete assembly and includes the fuel pressure regulator preset and plugged at the factory.

NOTE: Do not immerse the fuel meter cover (with pressure regulator) in any type of cleaner. Immersion in cleaner will damage the internal fuel pressure regulator diaphragms and gaskets.

4. Remove the sealing ring (dust seal from the fuel meter body).
5. Installation is the reverse of removal.

FUEL METER BODY

Removal and Installation

1. Remove the fuel inlet and outlet nuts and gaskets from fuel meter body.
2. Remove three screws and lockwashers. Remove fuel meter body from throttle body assembly.

NOTE: The air cleaner stud must have been removed previously.

3. Remove fuel meter body insulator gasket.
4. Installation is the reverse of removal.

THROTTLE BODY

Removal and Installation

SINGLE UNIT—FOUR CYLINDER ENGINE

Refer to the "Rear Unit-V8 Engine" procedure which follows. Disregard Steps 6 and 7 of that procedure, instead, just disconnect the fuel feed and return lines.

FRONT UNIT—V8 ENGINE

1. Disconnect the battery cables at the battery.
2. Remove the air cleaner assembly, noting the connection points of the vacuum lines.
3. disconnect the electrical connectors at the injector and the idle air control motor.
4. Disconnect the vacuum line from the TBI unit, noting the connection points. During installation, refer to the underhood emission control information decal for vacuum line routing information.
5. Disconnect the transmission detent cable from the TBI unit.
6. Disconnect the fuel inlet (feed) and fuel balance line connections at the front TBI unit.
7. Disconnect the throttle control rod between the two TBI units.
8. Unbolt and remove the TBI unit.
9. Installation is the reverse of the previous steps. Torque the TBI bolts to 120-168 inch lbs. during installation.

REAR UNIT—V8 ENGINE

1. Disconnect the battery cables at the battery.
2. Remove the air cleaner assembly, noting the connection points of the vacuum lines.
3. Disconnect the electrical connectors at the injector, idle air control motor, and throttle position sensor.
4. Disconnect the vacuum lines from the TBI unit, noting the connection points. During installation, refer to the underhood emission control information decal for vacuum line routing information.
5. Disconnect the throttle and cruise control (if so equipped) cables at the TBI unit.
6. Disconnect the fuel return and balance line connections from the rear TBI unit.
7. Disconnect the throttle control rod between the two units.
8. Unbolt and remove the TBI unit.
9. Installation is the reverse of the previous steps. Torque the TBI bolts to 120-168 inch lbs. during installation.

Throttle Body Disassembly

When servicing the single TBI unit on four cylinder engines, follow all steps except those specified "front unit." "Rear Unit" steps DO apply.

CAUTION

Use extreme care when handling the TBI unit to avoid damage to the swirl plates located beneath the throttle valve.

NOTE: If both TBI units are to be disassembled, DO NOT mix parts between either unit.

1. Remove the fuel meter cover assembly (five screws). Remove the gaskets after the cover has been removed. The fuel meter cover assembly is serviced only as a unit. If necessary, the entire unit must be replaced.

CAUTION

DO NOT remove the four screws which retain the pressure regulator (rear unit) or pressure compensator (front unit). There is a spring beneath the cover which is under great pressure. If the cover is accidentally released, personal injury could result.

Do not immerse the fuel meter cover in any type of cleaning solvent.

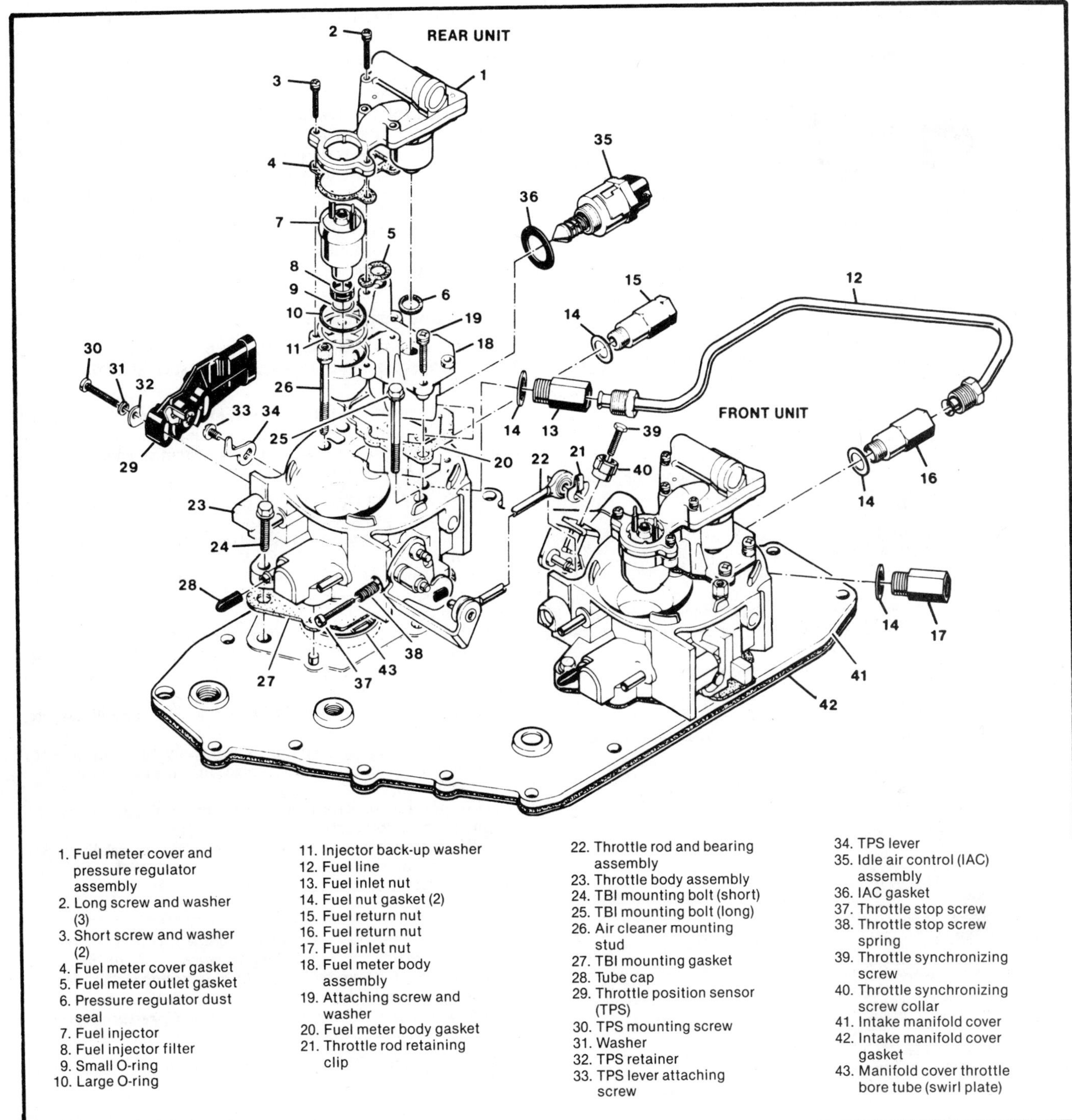

Exploded view of Cross Fire TBI units

2. Remove the foam dust seal from the meter body of the rear unit.

3. Remove the fuel injector using a pair of small pliers as follows:

  a. Grasp the injector collar, between the electrical terminals.

  b. Carefully pull the injector upward, in a twisting motion.

  c. If the injectors are to be removed from both TBI units, mark them so that they may be installed in their original units.

4. Remove the filter from the base of the injector by rotating it back and forth.

5. Remove the O-ring and the steel washer from the top of the fuel meter body, then remove the small O-ring from the bottom of the injector cavity.

6. Remove the fuel inlet and outlet nuts (and gaskets) from the fuel meter body.

7. Remove the fuel meter body assembly and gasket from the throttle body assembly (three screws).

8. For the rear TBI unit only: Remove the throttle position sensor (TPS) from the throttle body (two screws). If necessary, remove the screw which holds the TPS actuator lever to the end of the throttle shaft.

9. Remove the idle air control motor from the throttle body.

CAUTION

Because the TPS and idle air control motors are electrical units, they must not be immersed in any type of cleaning solvent.

Throttle Body Assembly

NOTE: During assembly, replace the gaskets, injector washer, O-rings, and pressure regulator dust seal with new parts.

1. Install the idle air control motor in the throttle body, using a new gasket. Torque the retaining screws to 13 ft. lbs.

NOTE: DO NOT overtighten the screws.

2. For the rear TBI unit only: If removed, install the TPS actuator lever by aligning the flats of the lever and the shaft. Install and tighten the retaining screw.
3. Install the fuel meter body on the throttle body, using a new gasket. Also, apply thread locking compound to the three fuel meter body screws according to the chemical manufacturers instructions. Torque the screws to 35 inch lbs.
4. Install the fuel inlet and outlet nuts, using new gaskets. Torque the nuts to 260 inch lbs.
5. Carefully twist the fuel filter onto the injector base.
6. Lubricate the new O-rings with Dexron®II transmission fluid.
7. Install the small O-ring onto the injector, pressing it up against the fuel filter.
8. Install the steel washer into the injector cavity recess of the fuel meter body. Install the large O-ring above the steel washer, in the cavity recess. The O-ring must be flush with the fuel meter body surface.
9. Using a pushing/twisting motion, carefully install the injector. Center the nozzle O-ring in the bottom of the injector cavity and align the raised lug on the injector base with the notch in the fuel meter body cavity. Make sure the injector is seated fully in the cavity. The electrical connections should be parallel to the throttle shaft of the throttle body.

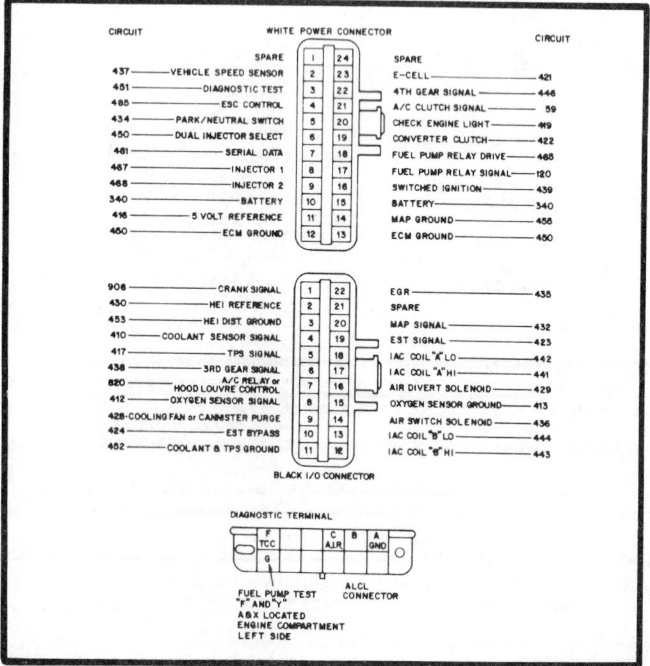

10. For the rear TBI unit only: Install the new pressure regulator dust seal into the fuel meter body recess.
11. Install the new fuel meter cover and fuel outlet passage gaskets on the fuel meter cover.
12. Install the fuel meter cover assembly, using thread locking compound on the five retaining screws. Torque the screws to 28 inch lbs. Note that the two short screws must be installed alongside the fuel injector (one screw each side).
13. For the rear TBI unit only: With the throttle valve in the closed (idle) position, install the TPS but do not tighten the attaching screws. The TPS lever must be located ABOVE the tang on the throttle acutator lever.
14. Install the TBI unit(s) as previously outlined and adjust the throttle position sensor.

Cleaning and Inspection

The throttle body injection parts, except as noted below, should be cleaned in a cold immersion-type cleaner such as Carbon X (X-55) or its equivalent.

NOTE: The throttle position sensor, idle air control motor, fuel meter cover (with pressure regulator), fuel injector, fuel filter, rubber parts, diaphragms, etc., should NOT be immersed in cleaner as they will swell, harden or distort.

1. Thoroughly clean all metal parts and blow dry with shop air. Make sure all fuel passages are free of burrs and dirt.
2. Inspect casting mating surfaces for damage that could affect gasket sealing.
3. Check, repair or replace parts as required, if the following problems are encountered:

a. Flooding

(1) Inspect large and small fuel injector O-rings for damage such as cuts, distortion, etc. check that the steel backup washer is located beneath the large (upper) O-ring. Use new O-rings when reinstalling injector.

(2) Inspect fuel injector fuel filter for damage, cleanliness, etc. Clean or replace as necessary.

(3) If the fuel injector continues to supply fuel with injector electrical connections removed, replace injector as required.

b. Hesitation

(1) Inspect fuel injector fuel filter for being plugged, dirty, etc. Clean or replace as necessary.

(2) If improper fuel inlet and outlet pressure readings, are noted check for restricted passages or inoperative fuel pressure regulator. Repair or replace as required.

CAUTION

DO NOT remove the four screws securing the fuel pressure regulator to the fuel meter cover. The fuel pressure regulator includes a large spring under heavy tension, which if accidentally released, could cause personal injury. The fuel meter cover is only serviced as a complete assembly and includes the fuel pressure regulator preset and plugged at the factory.

c. Hard Starting - Poor Cold Operation (See items listed under "Hesitation," above.)

d. Rough Idle

(1) Inspect large and small fuel injector O-rings for damage such as cuts, distortion, etc. check that the steel backup washer is located beneath the large (upper) O-ring. Use new O-rings when reinstalling injector.

(2) Inspect fuel injector fuel filter for damage, cleanliness, etc. Clean or replace as necessary.

(3) If the fuel injector continues to supply fuel with injector electrical connections removed, replace injector as required.

Diagnostic Circuit Check

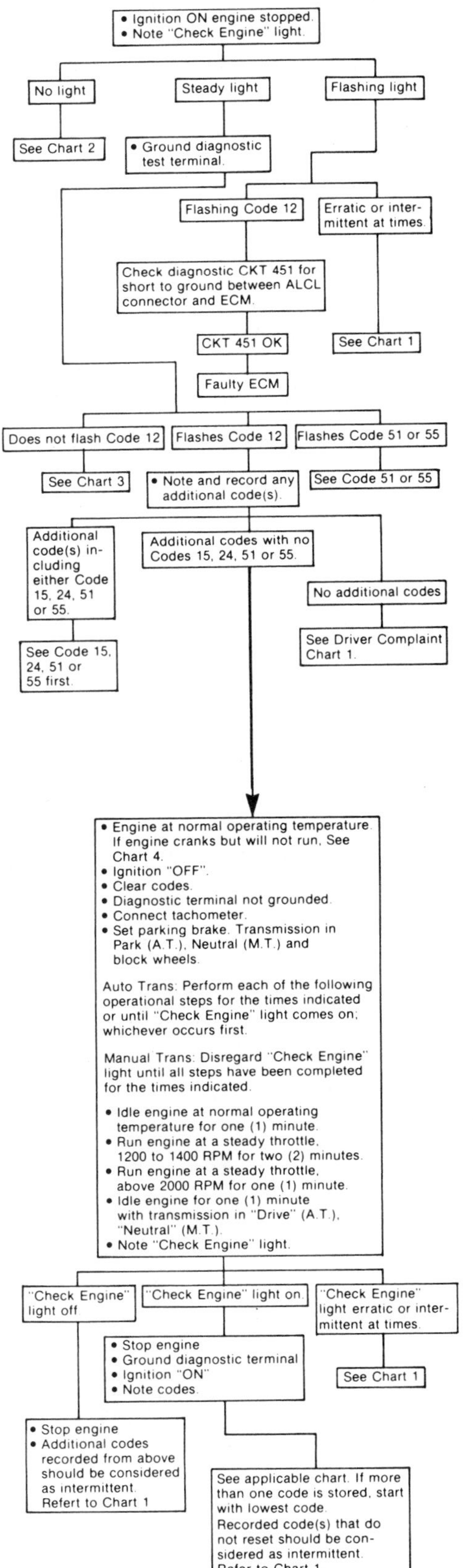
• Ignition ON engine stopped.
• Note "Check Engine" light.
No light
Steady light
Flashing light
See Chart 2
• Ground diagnostic test terminal.
Flashing Code 12
Erratic or intermittent at times.
Check diagnostic CKT 451 for short to ground between ALCL connector and ECM.
CKT 451 OK
See Chart 1
Faulty ECM
Does not flash Code 12
Flashes Code 12
Flashes Code 51 or 55
See Chart 3
• Note and record any additional code(s).
See Code 51 or 55
Additional code(s) including either Code 15, 24, 51 or 55.
Additional codes with no Codes 15, 24, 51 or 55.
No additional codes
See Code 15, 24, 51 or 55 first.
See Driver Complaint Chart 1.
• Engine at normal operating temperature. If engine cranks but will not run, See Chart 4.
• Ignition "OFF".
• Clear codes.
• Diagnostic terminal not grounded.
• Connect tachometer.
• Set parking brake. Transmission in Park (A.T.), Neutral (M.T.) and block wheels.
Auto Trans: Perform each of the following operational steps for the times indicated or until "Check Engine" light comes on; whichever occurs first.
Manual Trans: Disregard "Check Engine" light until all steps have been completed for the times indicated.
• Idle engine at normal operating temperature for one (1) minute.
• Run engine at a steady throttle, 1200 to 1400 RPM for two (2) minutes.
• Run engine at a steady throttle, above 2000 RPM for one (1) minute.
• Idle engine for one (1) minute with transmission in "Drive" (A.T.), "Neutral" (M.T.).
• Note "Check Engine" light.
"Check Engine" light off
"Check Engine" light on
"Check Engine" light erratic or intermittent at times.
• Stop engine
• Ground diagnostic terminal
• Ignition "ON"
• Note codes.
See Chart 1
• Stop engine
• Additional codes recorded from above should be considered as intermittent. Refert to Chart 1
See applicable chart. If more than one code is stored, start with lowest code. Recorded code(s) that do not reset should be considered as intermittent. Refer to Chart 1.

Chart 2

No "Check Engine" Light

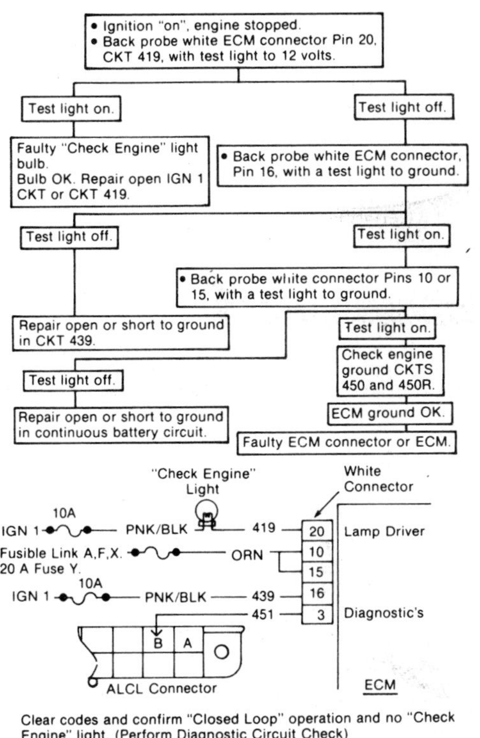
• Ignition "on", engine stopped.
• Back probe white ECM connector Pin 20, CKT 419, with test light to 12 volts.
Test light on.
Test light off.
Faulty "Check Engine" light bulb. Bulb OK. Repair open IGN 1 CKT or CKT 419.
• Back probe white ECM connector, Pin 16, with a test light to ground.
Test light off.
Test light on.
• Back probe white connector Pins 10 or 15, with a test light to ground.
Repair open or short to ground in CKT 439.
Test light on.
Check engine ground CKTS 450 and 450R.
Test light off.
Repair open or short to ground in continuous battery circuit.
ECM ground OK.
Faulty ECM connector or ECM.
"Check Engine" Light
White Connector
10A
IGN 1
PNK/BLK
419
20
Lamp Driver
Fusible Link A,F,X. 20 A Fuse Y.
ORN
10
15
10A
IGN 1
PNK/BLK
439
16
451
3
Diagnostic's
B
A
ALCL Connector
ECM

Clear codes and confirm "Closed Loop" operation and no "Check Engine" light. (Perform Diagnostic Circuit Check)

Chart 3

Won't Flash Code 12

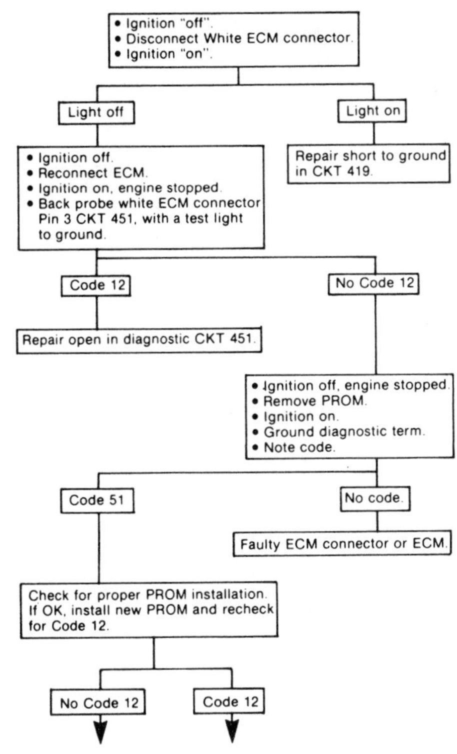
• Ignition "off".
• Disconnect White ECM connector.
• Ignition "on".
Light off
Light on
• Ignition off.
• Reconnect ECM.
• Ignition on, engine stopped.
• Back probe white ECM connector Pin 3 CKT 451, with a test light to ground.
Repair short to ground in CKT 419.
Code 12
No Code 12
Repair open in diagnostic CKT 451.
• Ignition off, engine stopped.
• Remove PROM.
• Ignition on.
• Ground diagnostic term.
• Note code.
Code 51
No code.
Faulty ECM connector or ECM.
Check for proper PROM installation. If OK, install new PROM and recheck for Code 12.
No Code 12
Code 12

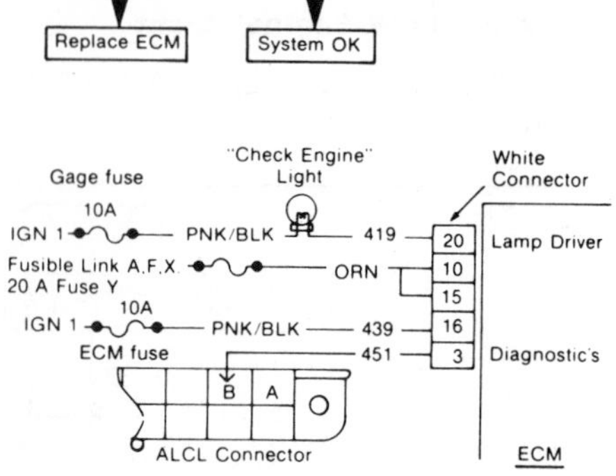

Replace ECM
System OK
"Check Engine" Light
Gage fuse
White Connector
10A
IGN 1
PNK/BLK
419
20
Lamp Driver
Fusible Link A,F,X
20 A Fuse Y
ORN
10
15
10A
IGN 1
PNK/BLK
439
16
ECM fuse
451
3
Diagnostic's
B
A
ALCL Connector
ECM

Clear codes and confirm "Closed Loop" operation and no "Check Engine" light. (Perform Diagnostic Circuit Check)

Chart 4

2.5 ℓ

— Engine cranks but will not run.

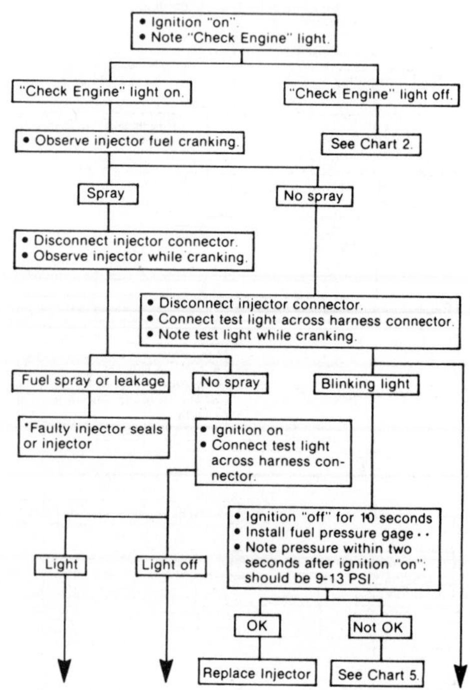

• Ignition "on".
• Note "Check Engine" light.
"Check Engine" light on.
"Check Engine" light off.
• Observe injector fuel cranking.
See Chart 2.
Spray
No spray
• Disconnect injector connector.
• Observe injector while cranking.
• Disconnect injector connector.
• Connect test light across harness connector.
• Note test light while cranking.
Fuel spray or leakage
No spray
Blinking light
*Faulty injector seals or injector
• Ignition on
• Connect test light across harness connector.
• Ignition "off" for 10 seconds
• Install fuel pressure gage**
• Note pressure within two seconds after ignition "on"; should be 9-13 PSI.
Light
Light off
OK
Not OK
Replace Injector
See Chart 5.

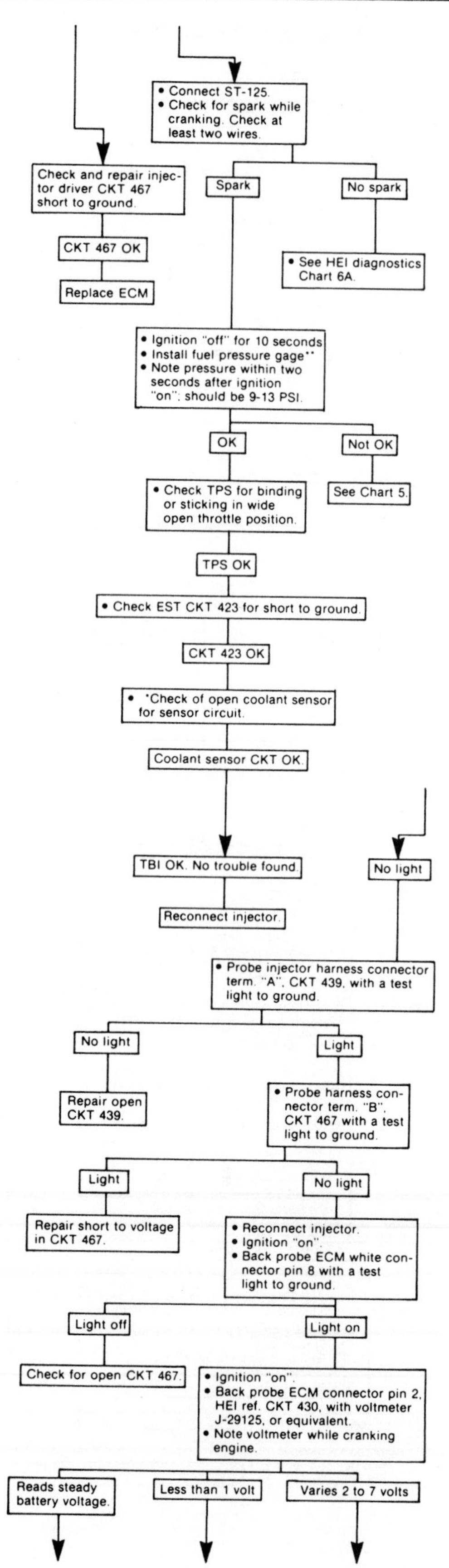

• Connect ST-125.
• Check for spark while cranking. Check at least two wires.
Check and repair injector driver CKT 467 short to ground.
Spark
No spark
CKT 467 OK
Replace ECM
• See HEI diagnostics Chart 6A.
• Ignition "off" for 10 seconds
• Install fuel pressure gage**
• Note pressure within two seconds after ignition "on"; should be 9-13 PSI.
OK
Not OK
• Check TPS for binding or sticking in wide open throttle position.
See Chart 5.
TPS OK
• Check EST CKT 423 for short to ground.
CKT 423 OK
• *Check of open coolant sensor for sensor circuit.
Coolant sensor CKT OK.
TBI OK. No trouble found.
No light
Reconnect injector.
• Probe injector harness connector term. "A", CKT 439, with a test light to ground.
No light
Light
Repair open CKT 439.
• Probe harness connector term. "B", CKT 467 with a test light to ground.
Light
No light
Repair short to voltage in CKT 467.
• Reconnect injector.
• Ignition "on".
• Back probe ECM white connector pin 8 with a test light to ground.
Light off
Light on
Check for open CKT 467.
• Ignition "on".
• Back probe ECM connector pin 2, HEI ref. CKT 430, with voltmeter J-29125, or equivalent.
• Note voltmeter while cranking engine.
Reads steady battery voltage.
Less than 1 volt
Varies 2 to 7 volts

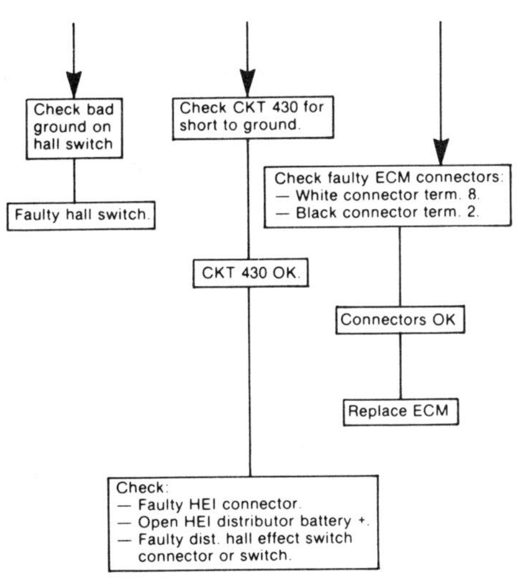

*May require spark plug cleaning to start engine.

**EFI fuel system under pressure. To avoid fuel spillage, refer to field service procedure for testing or making repairs requiring the disassembly of fuel line or fittings.

SCHEMATIC CHARTS 4 and 5 (2.5L)

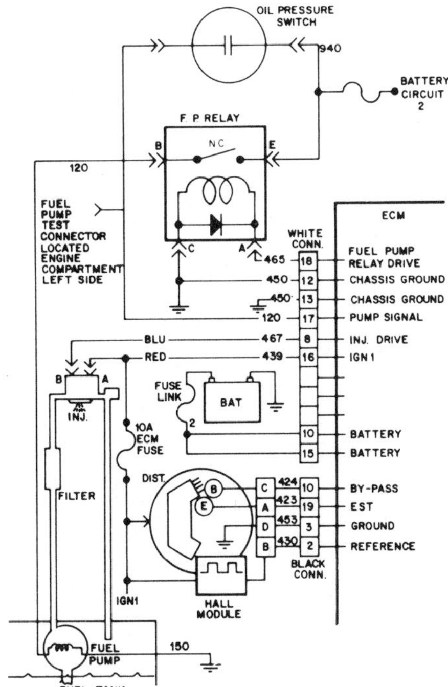

Chart 4A

5.0L and 5.7L

— Engine cranks but will not run.

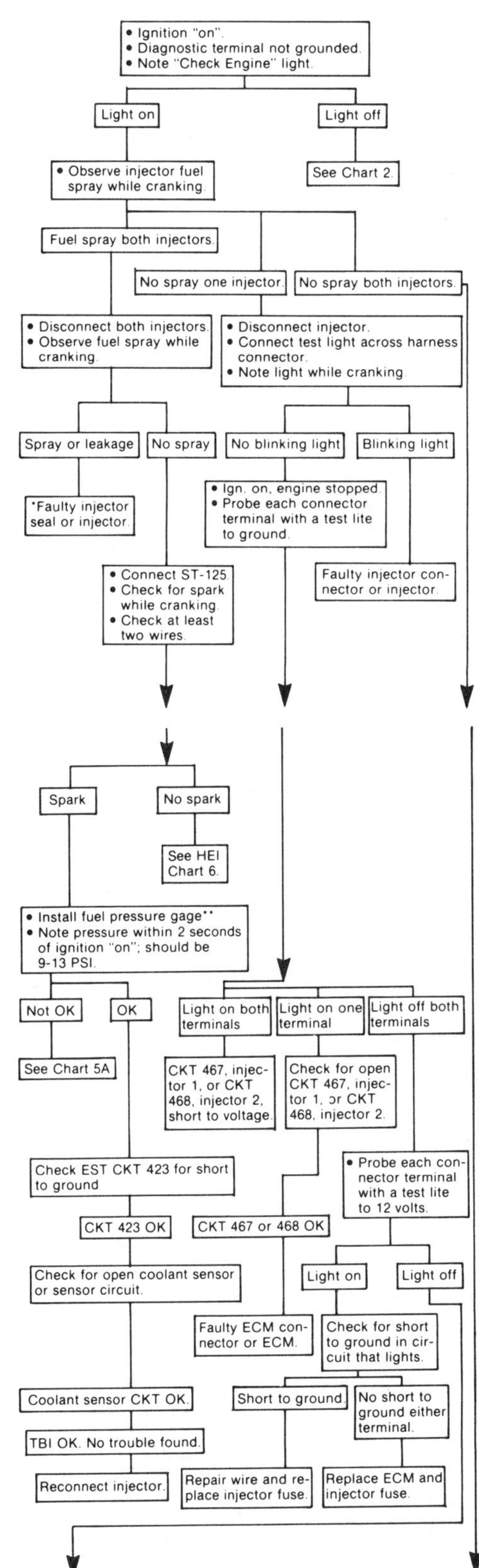

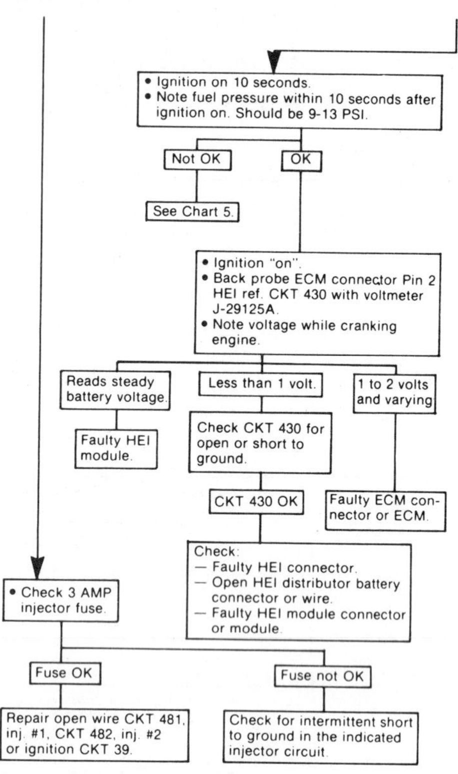

• Ignition on 10 seconds.
• Note fuel pressure within 10 seconds after ignition on. Should be 9-13 PSI.
Not OK
OK
See Chart 5.
• Ignition "on".
• Back probe ECM connector Pin 2 HEI ref. CKT 430 with voltmeter J-29125A.
• Note voltage while cranking engine.
Reads steady battery voltage.
Less than 1 volt.
1 to 2 volts and varying
Faulty HEI module.
Check CKT 430 for open or short to ground.
CKT 430 OK
Faulty ECM connector or ECM.
Check:
— Faulty HEI connector.
— Open HEI distributor battery connector or wire.
— Faulty HEI module connector or module.
• Check 3 AMP injector fuse.
Fuse OK
Fuse not OK
Repair open wire CKT 481, inj. #1, CKT 482, inj. #2 or ignition CKT 39.
Check for intermittent short to ground in the indicated injector circuit.

*May require spark plug cleaning to start engine.

**EFI fuel system under pressure. To avoid fuel spillage, refer to field service procedure for testing or making repairs requiring the disassembly of fuel line or fittings.

SCHEMATIC CHARTS
4A and 5A (5.0L and 5.7L)

OIL PRESSURE SWITCH
FUSE 20A 5.7ℓ FUSIBLE LINK 5.0ℓ
RED 5.0ℓ
BLU 5.7ℓ
ORN
340
BAT
FUEL PUMP TEST CONNECTOR "F" AND "Y" ALCL TERM. G
N.C.
TO ECM POWER CONNECTOR PINS 10 & 15
458
10A
120
465 — 18 — FUEL PUMP RELAY DRIVE
450 — 12 — CHASSIS GROUND
450R — 13 — CHASSIS GROUND
120 — 17 — F. P. SIGNAL
467 — 8 — INJ. 1 DRIVE
481 3A 10A 439 — 16 — IGNITION 1
482 3A IGN1
468 — 9 — INJ. 2 DRIVE
10 — BATTERY
15 — BATTERY
WHITE CONN.
INJ. 1 FRONT
INJ. 2 REAR
ECM
IN-LINE FILTER
C 424 10 HEI BYPASS
A 423 19 EST
D 453 3 GROUND
B 430 2 REFERENCE
BLACK CONN.
IGN1
HEI DISTRIBUTOR
FUEL PUMP
150

Chart 5

2.5 ℓ Fuel System Diagnosis

Schematic See Page 33

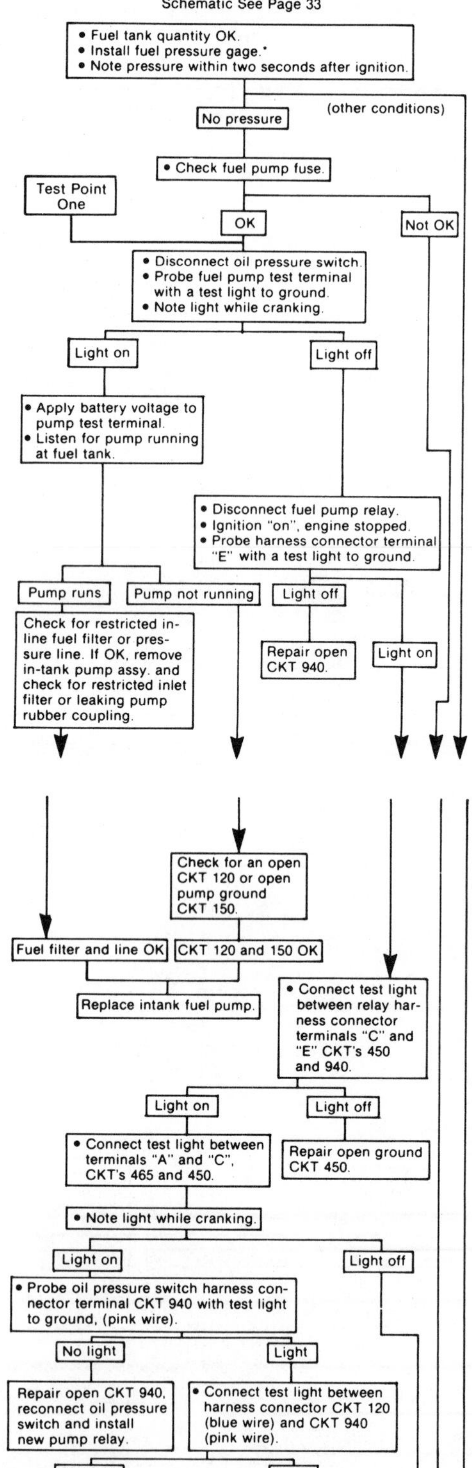

• Fuel tank quantity OK.
• Install fuel pressure gage.*
• Note pressure within two seconds after ignition.
No pressure
(other conditions)
• Check fuel pump fuse.
Test Point One
OK
Not OK
• Disconnect oil pressure switch.
• Probe fuel pump test terminal with a test light to ground.
• Note light while cranking.
Light on
Light off
• Apply battery voltage to pump test terminal.
• Listen for pump running at fuel tank.
• Disconnect fuel pump relay.
• Ignition "on", engine stopped.
• Probe harness connector terminal "E" with a test light to ground.
Pump runs
Pump not running
Light off
Check for restricted in-line fuel filter or pressure line. If OK, remove in-tank pump assy. and check for restricted inlet filter or leaking pump rubber coupling.
Repair open CKT 940.
Light on
Check for an open CKT 120 or open pump ground CKT 150.
Fuel filter and line OK
CKT 120 and 150 OK
Replace intank fuel pump.
• Connect test light between relay harness connector terminals "C" and "E" CKT's 450 and 940.
Light on
Light off
• Connect test light between terminals "A" and "C", CKT's 465 and 450.
Repair open ground CKT 450.
• Note light while cranking.
Light on
Light off
• Probe oil pressure switch harness connector terminal CKT 940 with test light to ground, (pink wire).
No light
Light
Repair open CKT 940, reconnect oil pressure switch and install new pump relay.
• Connect test light between harness connector CKT 120 (blue wire) and CKT 940 (pink wire).
No light
Light
Repair open CKT 120. Reconnect oil pressure switch and install new relay.
Install new pump relay.
C
B
A

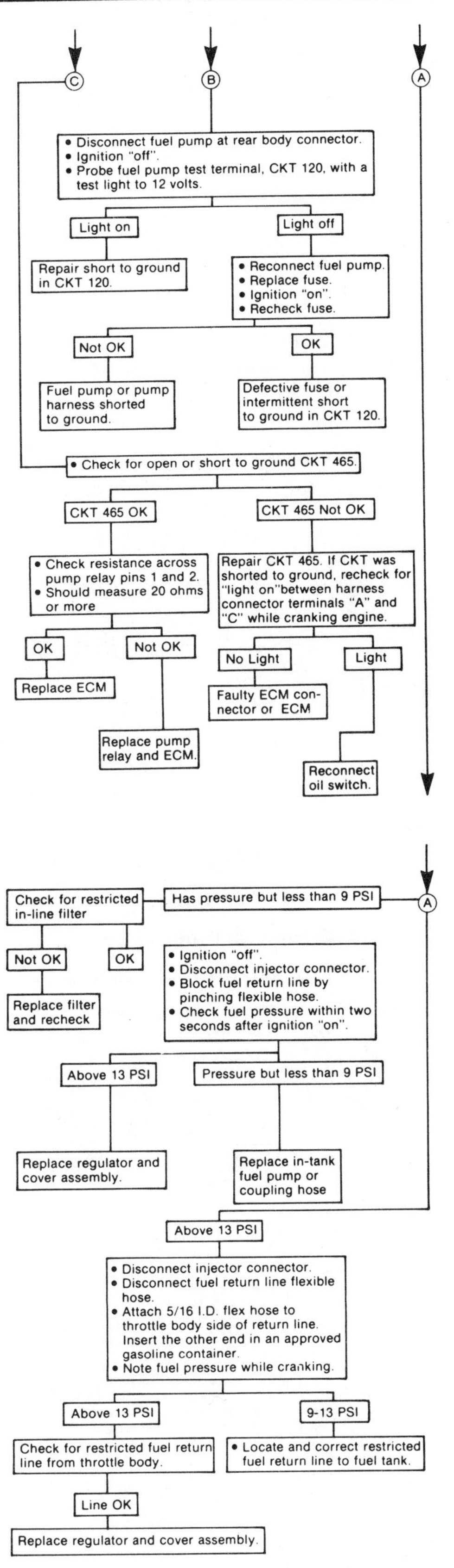
C
B
A
• Disconnect fuel pump at rear body connector.
• Ignition "off".
• Probe fuel pump test terminal, CKT 120, with a test light to 12 volts.
Light on
Light off
Repair short to ground in CKT 120.
• Reconnect fuel pump.
• Replace fuse.
• Ignition "on".
• Recheck fuse.
Not OK
OK
Fuel pump or pump harness shorted to ground.
Defective fuse or intermittent short to ground in CKT 120.
• Check for open or short to ground CKT 465.
CKT 465 OK
CKT 465 Not OK
• Check resistance across pump relay pins 1 and 2.
• Should measure 20 ohms or more
Repair CKT 465. If CKT was shorted to ground, recheck for "light on" between harness connector terminals "A" and "C" while cranking engine.
OK
Not OK
No Light
Light
Replace ECM
Faulty ECM connector or ECM
Replace pump relay and ECM.
Reconnect oil switch.
Check for restricted in-line filter
Has pressure but less than 9 PSI
A
Not OK
OK
Replace filter and recheck
• Ignition "off".
• Disconnect injector connector.
• Block fuel return line by pinching flexible hose.
• Check fuel pressure within two seconds after ignition "on".
Above 13 PSI
Pressure but less than 9 PSI
Replace regulator and cover assembly.
Replace in-tank fuel pump or coupling hose
Above 13 PSI
• Disconnect injector connector.
• Disconnect fuel return line flexible hose.
• Attach 5/16 I.D. flex hose to throttle body side of return line. Insert the other end in an approved gasoline container.
• Note fuel pressure while cranking.
Above 13 PSI
9-13 PSI
Check for restricted fuel return line from throttle body.
• Locate and correct restricted fuel return line to fuel tank.
Line OK
Replace regulator and cover assembly.

*EFI fuel system under pressure. To avoid fuel spillage, refer to field service procedure for testing or making repairs requiring the disassembly of fuel line or fittings.

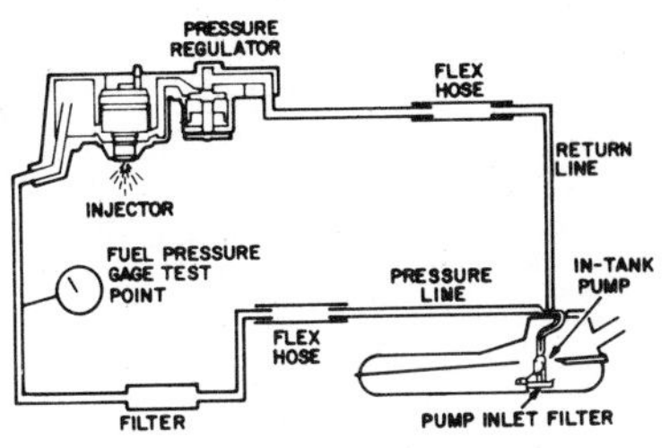
PRESSURE REGULATOR
FLEX HOSE
RETURN LINE
INJECTOR
FUEL PRESSURE GAGE TEST POINT
PRESSURE LINE
IN-TANK PUMP
FLEX HOSE
FILTER
PUMP INLET FILTER

Chart 5A

Fuel System Diagnosis 5.0ℓ - 5.7ℓ

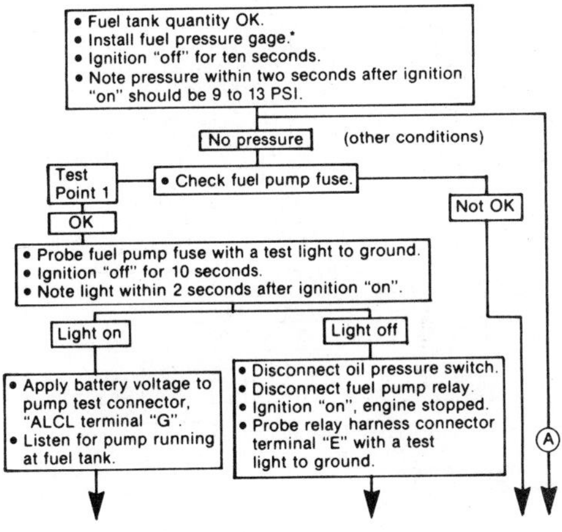
• Fuel tank quantity OK.
• Install fuel pressure gage.*
• Ignition "off" for ten seconds.
• Note pressure within two seconds after ignition "on" should be 9 to 13 PSI.
No pressure
(other conditions)
Test Point 1
• Check fuel pump fuse.
Not OK
OK
• Probe fuel pump fuse with a test light to ground.
• Ignition "off" for 10 seconds.
• Note light within 2 seconds after ignition "on".
Light on
Light off
• Apply battery voltage to pump test connector, "ALCL terminal "G".
• Listen for pump running at fuel tank.
• Disconnect oil pressure switch.
• Disconnect fuel pump relay.
• Ignition "on", engine stopped.
• Probe relay harness connector terminal "E" with a test light to ground.
A

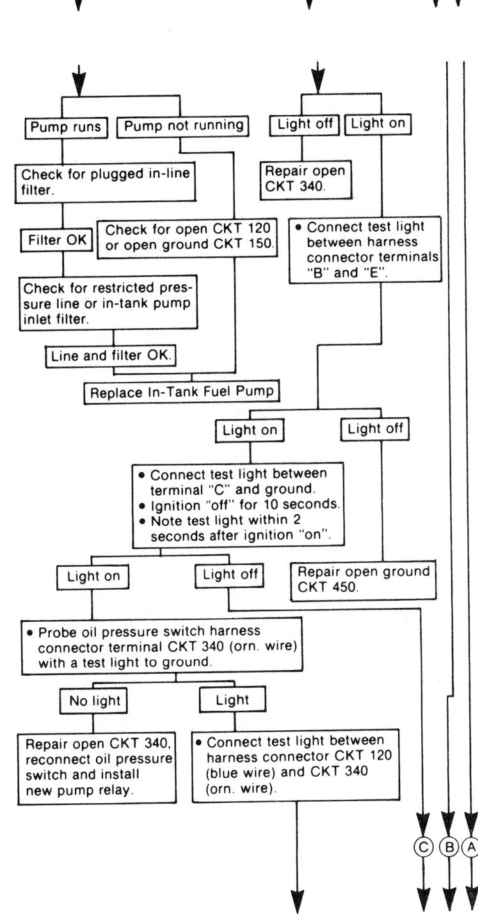
Pump runs
Pump not running
Light off
Light on
Check for plugged in-line filter.
Repair open CKT 340.
Filter OK
Check for open CKT 120 or open ground CKT 150.
• Connect test light between harness connector terminals "B" and "E".
Check for restricted pressure line or in-tank pump inlet filter.
Line and filter OK.
Replace In-Tank Fuel Pump
Light on
Light off
• Connect test light between terminal "C" and ground.
• Ignition "off" for 10 seconds.
• Note test light within 2 seconds after ignition "on".
Light on
Light off
Repair open ground CKT 450.
• Probe oil pressure switch harness connector terminal CKT 340 (orn. wire) with a test light to ground.
No light
Light
Repair open CKT 340, reconnect oil pressure switch and install new pump relay.
• Connect test light between harness connector CKT 120 (blue wire) and CKT 340 (orn. wire).
C
B
A

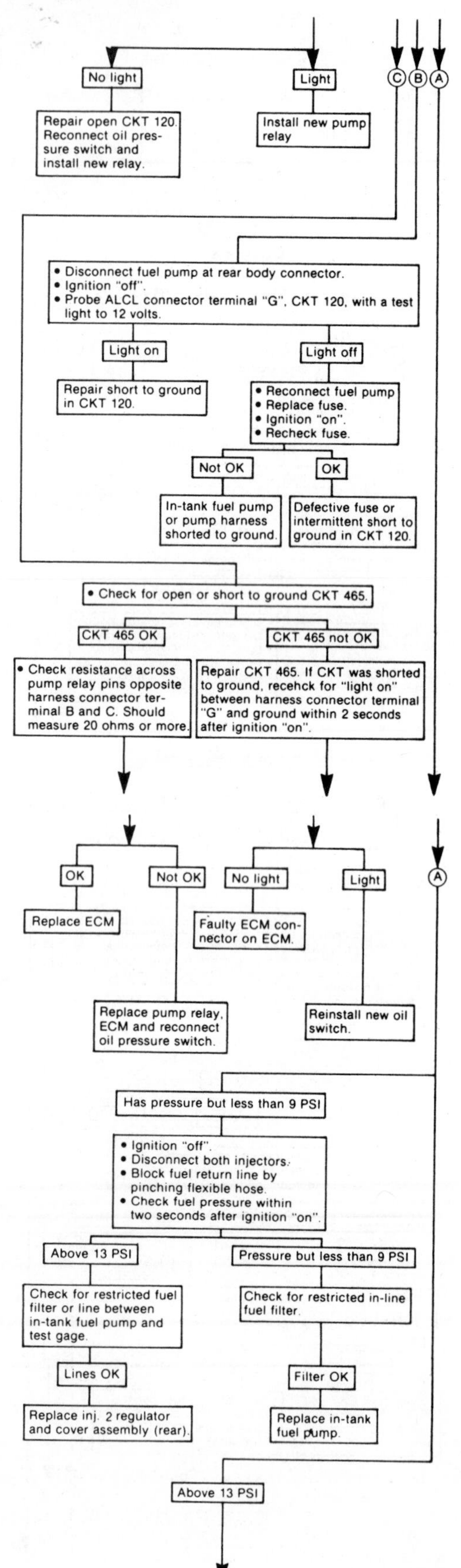
No light
Light
C B A
Repair open CKT 120. Reconnect oil pressure switch and install new relay.
Install new pump relay
• Disconnect fuel pump at rear body connector.
• Ignition "off".
• Probe ALCL connector terminal "G", CKT 120, with a test light to 12 volts.
Light on
Light off
Repair short to ground in CKT 120.
• Reconnect fuel pump
• Replace fuse.
• Ignition "on".
• Recheck fuse.
Not OK
OK
In-tank fuel pump or pump harness shorted to ground.
Defective fuse or intermittent short to ground in CKT 120.
• Check for open or short to ground CKT 465.
CKT 465 OK
CKT 465 not OK
• Check resistance across pump relay pins opposite harness connector terminal B and C. Should measure 20 ohms or more.
Repair CKT 465. If CKT was shorted to ground, recehck for "light on" between harness connector terminal "G" and ground within 2 seconds after ignition "on".
OK
Not OK
No light
Light
A
Replace ECM
Faulty ECM connector on ECM.
Replace pump relay, ECM and reconnect oil pressure switch.
Reinstall new oil switch.
Has pressure but less than 9 PSI
• Ignition "off".
• Disconnect both injectors.
• Block fuel return line by pinching flexible hose.
• Check fuel pressure within two seconds after ignition "on".
Above 13 PSI
Pressure but less than 9 PSI
Check for restricted fuel filter or line between in-tank fuel pump and test gage.
Check for restricted in-line fuel filter.
Lines OK
Filter OK
Replace inj. 2 regulator and cover assembly (rear).
Replace in-tank fuel pump.
Above 13 PSI

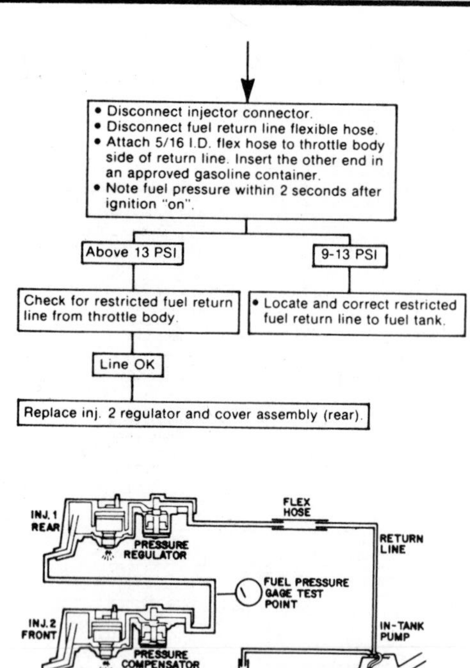
• Disconnect injector connector.
• Disconnect fuel return line flexible hose.
• Attach 5/16 I.D. flex hose to throttle body side of return line. Insert the other end in an approved gasoline container.
• Note fuel pressure within 2 seconds after ignition "on".
Above 13 PSI
9-13 PSI
Check for restricted fuel return line from throttle body.
• Locate and correct restricted fuel return line to fuel tank.
Line OK
Replace inj. 2 regulator and cover assembly (rear).
INJ. 1 REAR
PRESSURE REGULATOR
FLEX HOSE
RETURN LINE
FUEL PRESSURE GAGE TEST POINT
INJ. 2 FRONT
PRESSURE COMPENSATOR
IN-TANK PUMP
FILTER
PUMP INLET FILTER

*EFI fuel system under pressure. To avoid fuel spillage, refer to field service procedure for testing or making repairs requiring the disassembly of fuel line or fittings.

Chart 6

HEI Distributor

(With Integral Ignition Coil)

NOTE: If a tachometer is connected to the tachometer terminal, disconnect it before proceeding with the test.

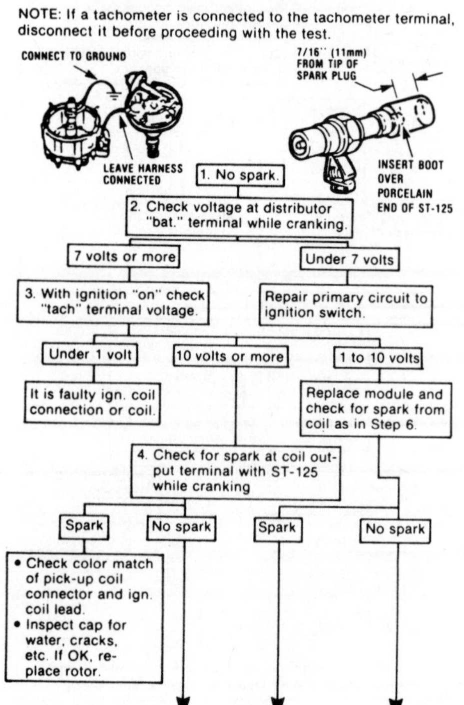
CONNECT TO GROUND
7/16" (11mm) FROM TIP OF SPARK PLUG
LEAVE HARNESS CONNECTED
INSERT BOOT OVER PORCELAIN END OF ST-125
1. No spark.
2. Check voltage at distributor "bat." terminal while cranking.
7 volts or more
Under 7 volts
3. With ignition "on" check "tach" terminal voltage.
Repair primary circuit to ignition switch.
Under 1 volt
10 volts or more
1 to 10 volts
It is faulty ign. coil connection or coil.
Replace module and check for spark from coil as in Step 6.
4. Check for spark at coil output terminal with ST-125 while cranking
Spark
No spark
Spark
No spark
• Check color match of pick-up coil connector and ign. coil lead.
• Inspect cap for water, cracks, etc. If OK, replace rotor.

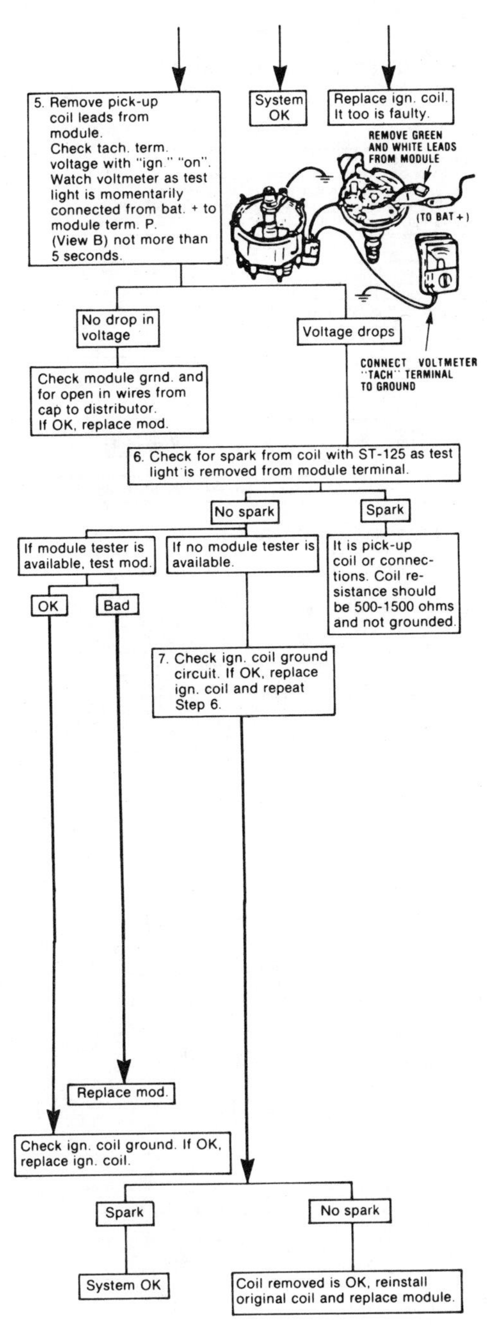

Chart 6A

HEI Distributor

(With Remote Coil)

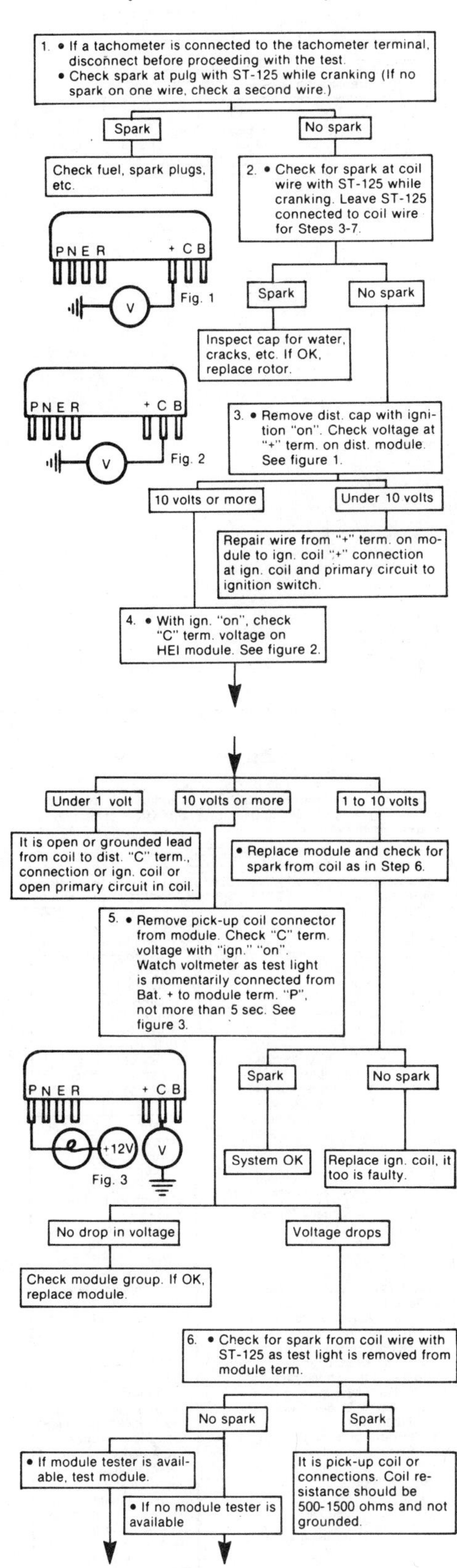

1. • If a tachometer is connected to the tachometer terminal, disconnect before proceeding with the test.
• Check spark at pulg with ST-125 while cranking (If no spark on one wire, check a second wire.)
Spark
No spark
Check fuel, spark plugs, etc.
2. • Check for spark at coil wire with ST-125 while cranking. Leave ST-125 connected to coil wire for Steps 3-7.
P N E R
+ C B
V
Fig. 1
Spark
No spark
Inspect cap for water, cracks, etc. If OK, replace rotor.
Fig. 2
3. • Remove dist. cap with ignition "on". Check voltage at "+" term. on dist. module. See figure 1.
10 volts or more
Under 10 volts
Repair wire from "+" term. on module to ign. coil "+" connection at ign. coil and primary circuit to ignition switch.
4. • With ign. "on", check "C" term. voltage on HEI module. See figure 2.
Under 1 volt
10 volts or more
1 to 10 volts
It is open or grounded lead from coil to dist. "C" term., connection or ign. coil or open primary circuit in coil.
• Replace module and check for spark from coil as in Step 6.
5. • Remove pick-up coil connector from module. Check "C" term. voltage with "ign." "on". Watch voltmeter as test light is momentarily connected from Bat. + to module term. "P", not more than 5 sec. See figure 3.
+12V
Fig. 3
Spark
No spark
System OK
Replace ign. coil, it too is faulty.
No drop in voltage
Voltage drops
Check module group. If OK, replace module.
6. • Check for spark from coil wire with ST-125 as test light is removed from module term.
No spark
Spark
• If module tester is available, test module.
• If no module tester is available
It is pick-up coil or connections. Coil resistance should be 500-1500 ohms and not grounded.

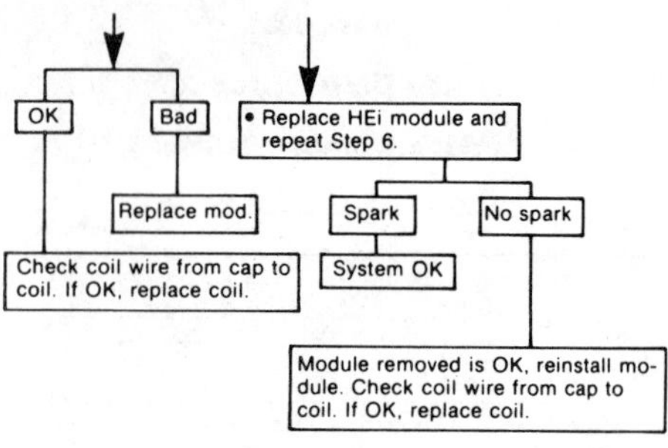

Chart 7

(No Stored Code)

These conditions, related to EFI, are in addition to those listed under "Engine Performance Diagnosis" at the front of Section 6 of Chassis Repair Manual.

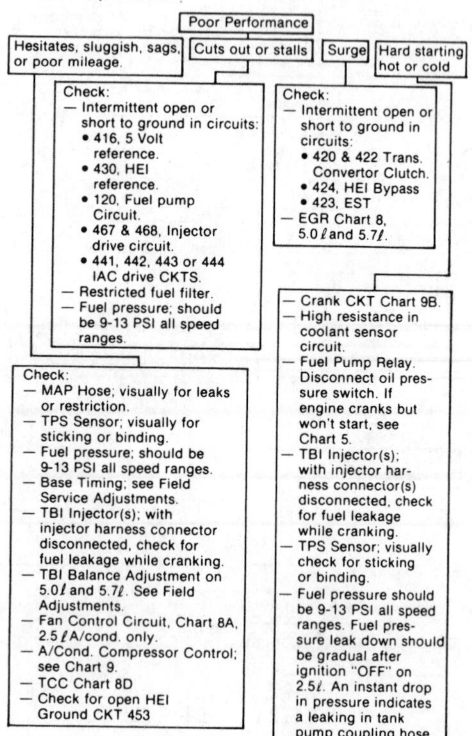

Poor Performance
Hesitates, sluggish, sags, or poor mileage
Cuts out or stalls
Surge
Hard starting hot or cold
Check:
— Intermittent open or short to ground in circuits:
• 416, 5 Volt reference.
• 430, HEI reference.
• 120, Fuel pump Circuit.
• 467 & 468, Injector drive circuit.
• 441, 442, 443 or 444 IAC drive CKTS.
— Restricted fuel filter.
— Fuel pressure; should be 9-13 PSI all speed ranges.
Check:
— Intermittent open or short to ground in circuits:
• 420 & 422 Trans. Convertor Clutch.
• 424, HEI Bypass
• 423, EST
— EGR Chart 8, 5.0ℓ and 5.7ℓ.
Check:
— MAP Hose; visually for leaks or restriction.
— TPS Sensor; visually for sticking or binding.
— Fuel pressure; should be 9-13 PSI all speed ranges.
— Base Timing; see Field Service Adjustments.
— TBI Injector(s); with injector harness connector disconnected, check for fuel leakage while cranking.
— TBI Balance Adjustment on 5.0ℓ and 5.7ℓ. See Field Adjustments.
— Fan Control Circuit, Chart 8A, 2.5ℓ A/cond. only.
— A/Cond. Compressor Control; see Chart 9.
— TCC Chart 8D
— Check for open HEI Ground CKT 453
— Crank CKT Chart 9B.
— High resistance in coolant sensor circuit.
— Fuel Pump Relay. Disconnect oil pressure switch. If engine cranks but won't start, see Chart 5.
— TBI Injector(s); with injector harness connecior(s) disconnected, check for fuel leakage while cranking.
— TPS Sensor; visually check for sticking or binding.
— Fuel pressure should be 9-13 PSI all speed ranges. Fuel pressure leak down should be gradual after ignition "OFF" on 2.5ℓ. An instant drop in pressure indicates a leaking in tank pump coupling hose or check valve.

Chart 8

EGR Check 5.0ℓ and 5.7ℓ

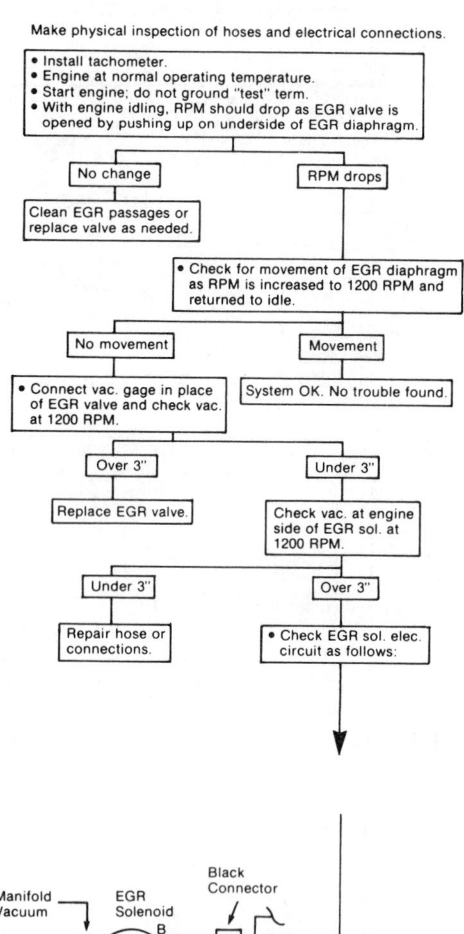

Make physical inspection of hoses and electrical connections.
• Install tachometer.
• Engine at normal operating temperature.
• Start engine; do not ground "test" term.
• With engine idling, RPM should drop as EGR valve is opened by pushing up on underside of EGR diaphragm.
No change
RPM drops
Clean EGR passages or replace valve as needed.
• Check for movement of EGR diaphragm as RPM is increased to 1200 RPM and returned to idle.
No movement
Movement
• Connect vac. gage in place of EGR valve and check vac. at 1200 RPM.
System OK. No trouble found.
Over 3"
Under 3"
Replace EGR valve.
Check vac. at engine side of EGR sol. at 1200 RPM.
Under 3"
Over 3"
Repair hose or connections.
• Check EGR sol. elec. circuit as follows:

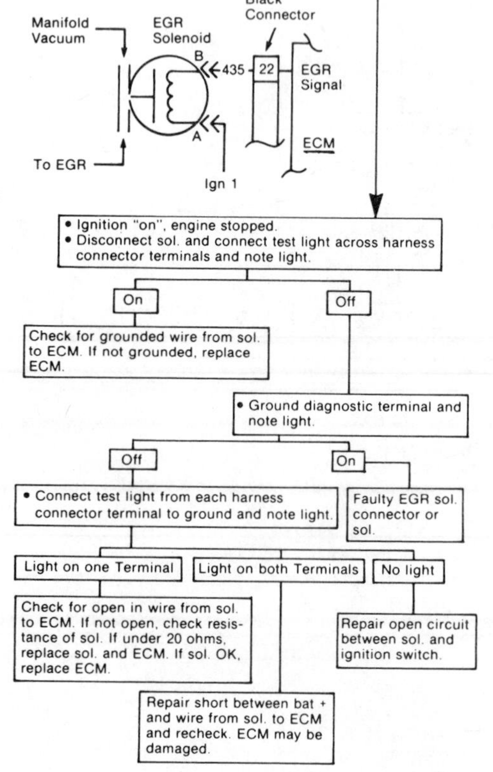

Chart 8A

Coolant Fan Control Circuit 2.5ℓ With Air Conditioning

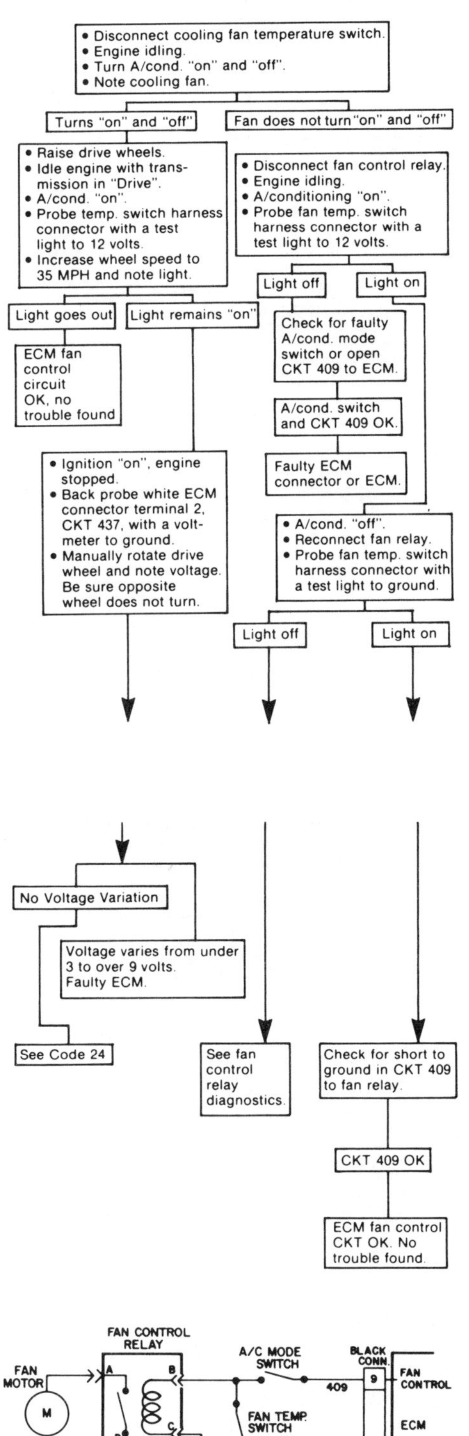

• Disconnect cooling fan temperature switch.
• Engine idling.
• Turn A/cond. "on" and "off".
• Note cooling fan.
Turns "on" and "off"
Fan does not turn "on" and "off"
• Raise drive wheels.
• Idle engine with transmission in "Drive".
• A/cond. "on".
• Probe temp. switch harness connector with a test light to 12 volts.
• Increase wheel speed to 35 MPH and note light.
• Disconnect fan control relay.
• Engine idling.
• A/conditioning "on".
• Probe fan temp. switch harness connector with a test light to 12 volts.
Light off
Light on
Light goes out
Light remains "on"
Check for faulty A/cond. mode switch or open CKT 409 to ECM.
ECM fan control circuit OK, no trouble found
A/cond. switch and CKT 409 OK.
Faulty ECM connector or ECM.
• Ignition "on", engine stopped.
• Back probe white ECM connector terminal 2, CKT 437, with a voltmeter to ground.
• Manually rotate drive wheel and note voltage. Be sure opposite wheel does not turn.
• A/cond. "off".
• Reconnect fan relay.
• Probe fan temp. switch harness connector with a test light to ground.
Light off
Light on
No Voltage Variation
Voltage varies from under 3 to over 9 volts. Faulty ECM.
See Code 24
See fan control relay diagnostics.
Check for short to ground in CKT 409 to fan relay.
CKT 409 OK
ECM fan control CKT OK. No trouble found.
FAN CONTROL RELAY
FAN MOTOR
M
A
B
C
D
BATTERY
IGN 1
A/C MODE SWITCH
FAN TEMP. SWITCH
409
BLACK CONN.
9
FAN CONTROL
ECM

Chart 8B

Air Management Check 5.0ℓ 5.7ℓ With Electric Divert And Electric Switching Valve (ED/ES)

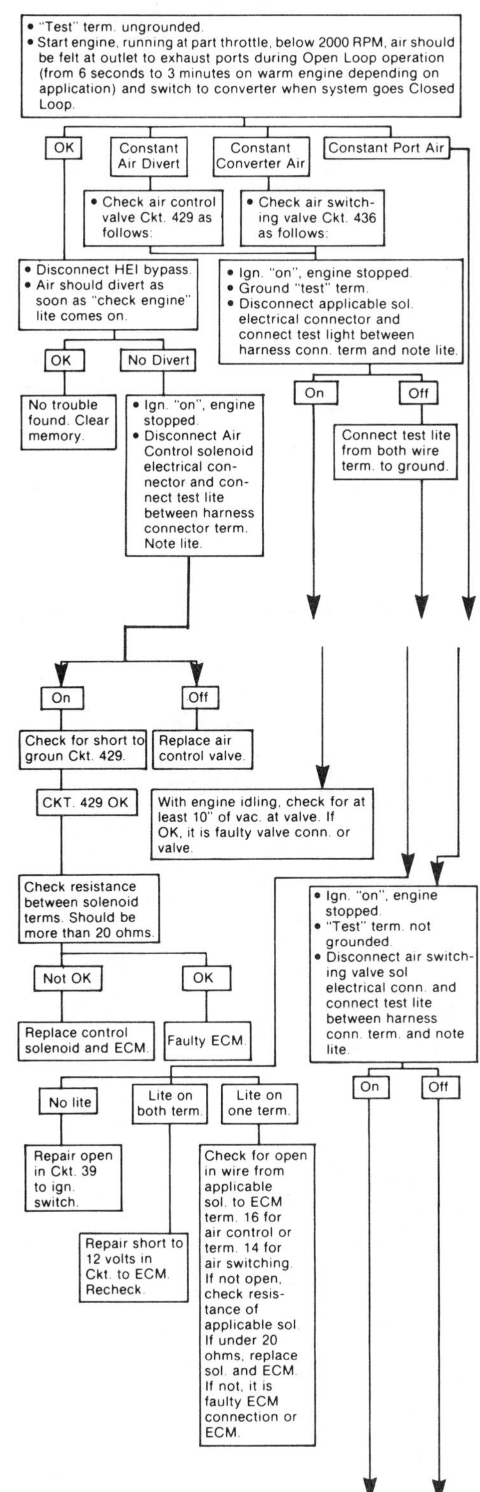

Check vacuum hose condition and connection. Repair as necessary. On some applications, some air may divert above 2000 RPM.
• "Test" term. ungrounded.
• Start engine, running at part throttle, below 2000 RPM, air should be felt at outlet to exhaust ports during Open Loop operation (from 6 seconds to 3 minutes on warm engine depending on application) and switch to converter when system goes Closed Loop.
OK
Constant Air Divert
Constant Converter Air
Constant Port Air
• Check air control valve Ckt. 429 as follows:
• Check air switching valve Ckt. 436 as follows:
• Disconnect HEI bypass.
• Air should divert as soon as "check engine" lite comes on.
• Ign. "on", engine stopped.
• Ground "test" term.
• Disconnect applicable sol. electrical connector and connect test light between harness conn. term and note lite.
OK
No Divert
On
Off
No trouble found. Clear memory.
• Ign. "on", engine stopped.
• Disconnect Air Control solenoid electrical connector and connect test lite between harness connector term. Note lite.
Connect test lite from both wire term. to ground.
On
Off
Check for short to groun Ckt. 429.
Replace air control valve.
CKT. 429 OK
With engine idling, check for at least 10" of vac. at valve. If OK, it is faulty valve conn. or valve.
Check resistance between solenoid terms. Should be more than 20 ohms.
• Ign. "on", engine stopped.
• "Test" term. not grounded.
• Disconnect air switching valve sol electrical conn. and connect test lite between harness conn. term. and note lite.
Not OK
OK
Replace control solenoid and ECM.
Faulty ECM.
On
Off
No lite
Lite on both term.
Lite on one term.
Repair open in Ckt. 39 to ign. switch.
Repair short to 12 volts in Ckt. to ECM. Recheck.
Check for open in wire from applicable sol. to ECM term. 16 for air control or term. 14 for air switching. If not open, check resistance of applicable sol. If under 20 ohms, replace sol. and ECM. If not, it is faulty ECM connection or ECM.

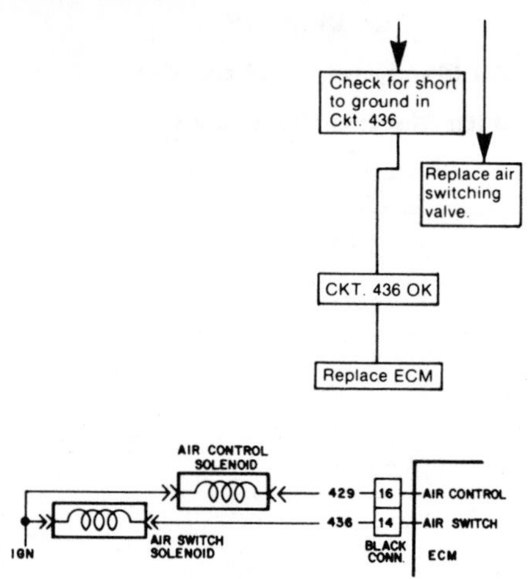

Check for short to ground in Ckt. 436
Replace air switching valve.
CKT. 436 OK
Replace ECM
AIR CONTROL SOLENOID
429
16
AIR CONTROL
436
14
AIR SWITCH
IGN
AIR SWITCH SOLENOID
BLACK CONN.
ECM

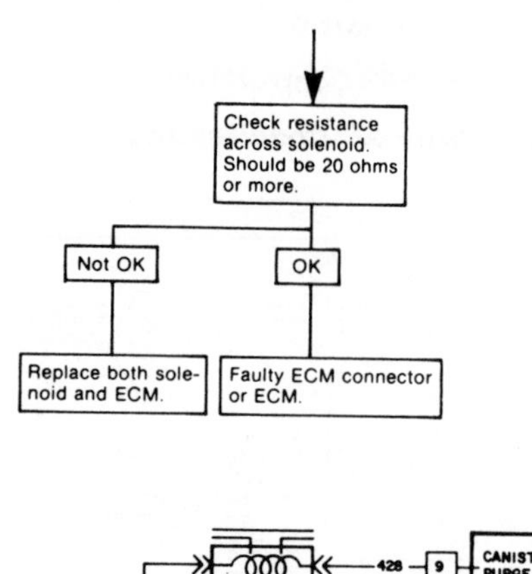

Check resistance across solenoid. Should be 20 ohms or more.
Not OK
OK
Replace both solenoid and ECM.
Faulty ECM connector or ECM.
IGN CKT39
CANISTER PURGE SOLENOID
428
9
BLACK CONN.
CANISTER PURGE
ECM

Chart 8C

Canister Purge Valve

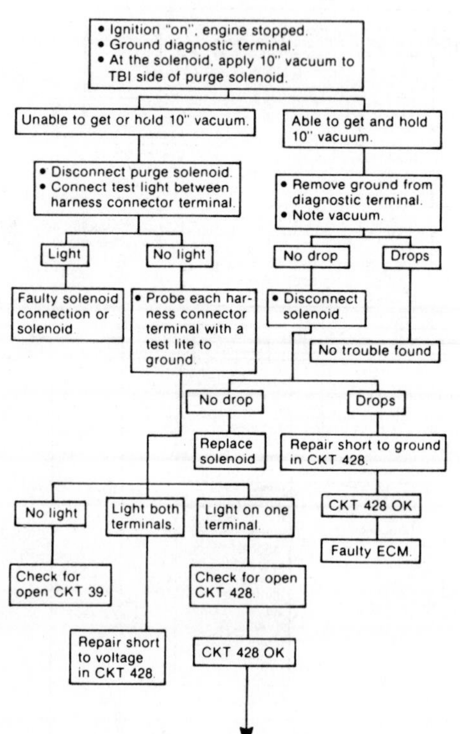

• Ignition "on", engine stopped.
• Ground diagnostic terminal.
• At the solenoid, apply 10" vacuum to TBI side of purge solenoid.
Unable to get or hold 10" vacuum.
Able to get and hold 10" vacuum.
• Disconnect purge solenoid.
• Connect test light between harness connector terminal.
• Remove ground from diagnostic terminal.
• Note vacuum.
Light
No light
No drop
Drops
Faulty solenoid connection or solenoid.
• Probe each harness connector terminal with a test lite to ground.
• Disconnect solenoid.
No trouble found
No drop
Drops
Replace solenoid
Repair short to ground in CKT 428.
No light
Light both terminals.
Light on one terminal.
CKT 428 OK
Faulty ECM.
Check for open CKT 39.
Check for open CKT 428.
Repair short to voltage in CKT 428.
CKT 428 OK

Chart 8D

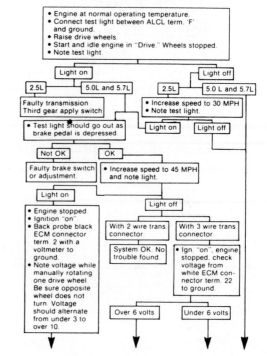

Torque Converter Clutch (TCC)

Electrical Diagnosis

Mechanical checks such as linkage, oil level, etc., should be performed prior to using this chart.

- Engine at normal operating temperature.
- Connect test light between ALCL term. 'F' and ground.
- Raise drive wheels.
- Start and idle engine in "Drive." Wheels stopped.
- Note test light.

Light on

- 2.5L
  - Faulty transmission Third gear apply switch.
- 5.0L and 5.7L
  - Test light should go out as brake pedal is depressed.
    - Not OK
      - Faulty brake switch or adjustment.
    - OK
      - Increase speed to 45 MPH and note light.
        - Light on
          - Engine stopped.
          - Ignition "on".
          - Back probe black ECM connector term. 2 with a voltmeter to ground.
          - Note voltage while manually rotating one drive wheel. Be sure opposite wheel does not turn. Voltage should alternate from under 3 to over 10.
        - Light off
          - With 2 wire trans. connector.
            - System OK. No trouble found.
          - With 3 wire trans. connector.
            - Ign. "on", engine stopped, check voltage from white ECM connector term. 22 to ground.
              - Over 6 volts
              - Under 6 volts

Light off

- 2.5L
  - Increase speed to 30 MPH
  - Note test light.
    - Light on
    - Light off
- 5.0 L and 5.7L

Check for open in wire from ECM term. 22 to trans. connector term. "B", CKT 446. If not open, it is faulty connector at trans. or open in trans. 4th gear switch.

OK

Not OK

See Chart 24

Check:
— Low coolant level
— Higher than normal resistance in coolant sensor circuit . . . if OK, it is faulty ECM white connector term. 19 or ECM.

• Start engine and with trans. in "D" increase speed to 50-55 MPH and note voltage reading.

Over 6 volts

No elc. trouble found.

Under 6 volts

• Ign. "on", engine stopped. Disconnect trans. connector and check volt from harness connector term. "B" to ground.

Over 6 volts

Trans. 4th gear switch or wire is grounded or switch not opening.

Under 6 volts

Check for ground wire to ECM term. 22, CKT 446. If not grounded, it is faulty ECM connector or ECM.

• Check for blown fuse. If OK, disconnect connector at trans. and connect test light from harness connector "A" to "D". With ign. "on", engine stopped, note test light.

Light off

• Connect test light from term. "A" to ground.

Light off

Repair open in TCC brake switch circuit or adj. switch.

Light on

Ground TCC test point and again connect test light between harness connector term's. 'A' & 'D'.

Light on

Faulty:
— Trans. TCC connection.
— TCC solenoid
— Third gear apply switch
— 4-3 downshift switch

Light off

Repair open in wire from trans. to ALCL test point, term. "F".

Light on

Check for ground in wire to ECM, CKT 422. If not grounded, replace ECM.

NC B D A TRANSMISSION 420 PPL
446 BLU 22
422 TAN/BLK 19
437 BRN 2
4TH GEAR
TCC
ECM
VSS
WHITE CONN.
TCC TEST POINT ALCL TERM. "F"
VSS
IGN 1 PNK/BLK
NC BRAKE SWITCH

Chart 8E

Hood Louvre Solenoid — 5.0ℓ and 5.7ℓ

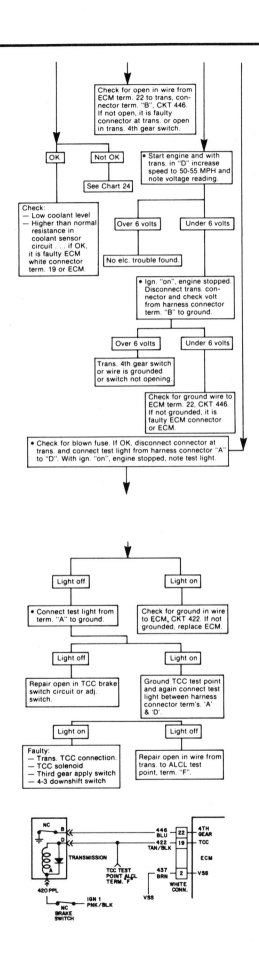

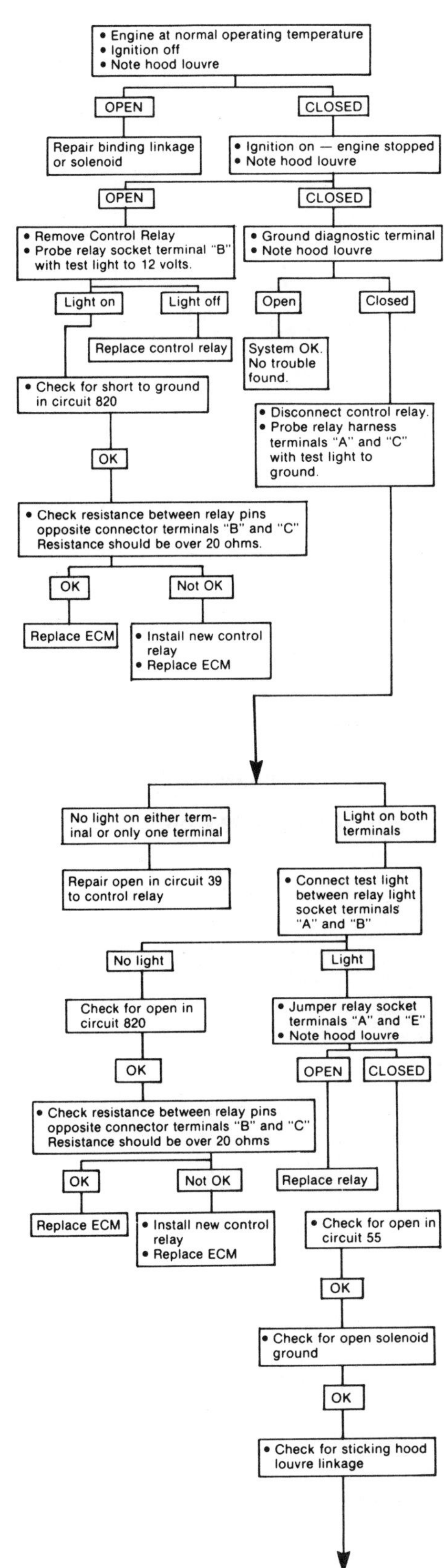
• Engine at normal operating temperature
• Ignition off
• Note hood louvre
OPEN
Repair binding linkage or solenoid
CLOSED
• Ignition on — engine stopped
• Note hood louvre
OPEN
• Remove Control Relay
• Probe relay socket terminal "B" with test light to 12 volts.
Light on
• Check for short to ground in circuit 820
OK
• Check resistance between relay pins opposite connector terminals "B" and "C" Resistance should be over 20 ohms.
OK
Replace ECM
Not OK
• Install new control relay
• Replace ECM
Light off
Replace control relay
CLOSED
• Ground diagnostic terminal
• Note hood louvre
Open
System OK. No trouble found.
Closed
• Disconnect control relay.
• Probe relay harness terminals "A" and "C" with test light to ground.
No light on either terminal or only one terminal
Repair open in circuit 39 to control relay
Light on both terminals
• Connect test light between relay light socket terminals "A" and "B"
No light
Check for open in circuit 820
OK
• Check resistance between relay pins opposite connector terminals "B" and "C" Resistance should be over 20 ohms
OK
Replace ECM
Not OK
• Install new control relay
• Replace ECM
Light
• Jumper relay socket terminals "A" and "E"
• Note hood louvre
OPEN
Replace relay
CLOSED
• Check for open in circuit 55
OK
• Check for open solenoid ground
OK
• Check for sticking hood louvre linkage

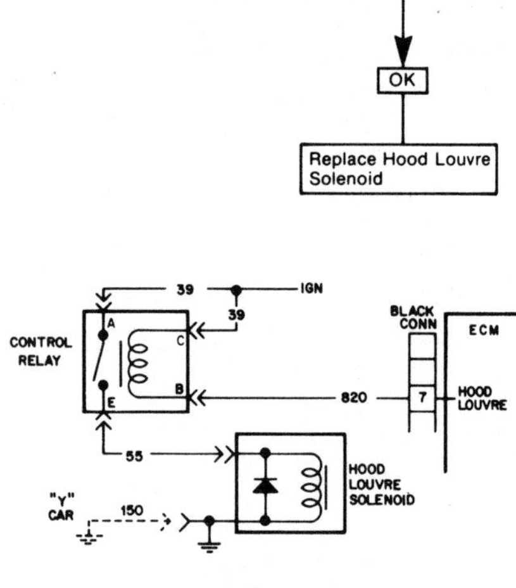

Chart 9

Air Conditioning Control 2.5 ℓ

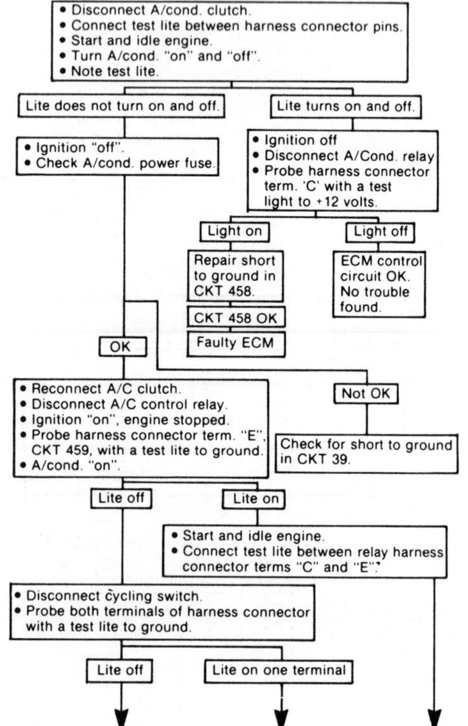

• Disconnect A/cond. clutch.
• Connect test lite between harness connector pins.
• Start and idle engine.
• Turn A/cond. "on" and "off".
• Note test lite.
Lite does not turn on and off.
Lite turns on and off.
• Ignition "off".
• Check A/cond. power fuse.
• Ignition off
• Disconnect A/Cond. relay
• Probe harness connector term. 'C' with a test light to +12 volts.
Light on
Light off
Repair short to ground in CKT 458.
ECM control circuit OK. No trouble found.
CKT 458 OK
Faulty ECM
OK
Not OK
• Reconnect A/C clutch.
• Disconnect A/C control relay.
• Ignition "on", engine stopped.
• Probe harness connector term. "E", CKT 459, with a test lite to ground.
• A/cond. "on".
Check for short to ground in CKT 39.
Lite off
Lite on
• Start and idle engine.
• Connect test lite between relay harness connector terms "C" and "E".
• Disconnect cycling switch.
• Probe both terminals of harness connector with a test lite to ground.
Lite off
Lite on one terminal

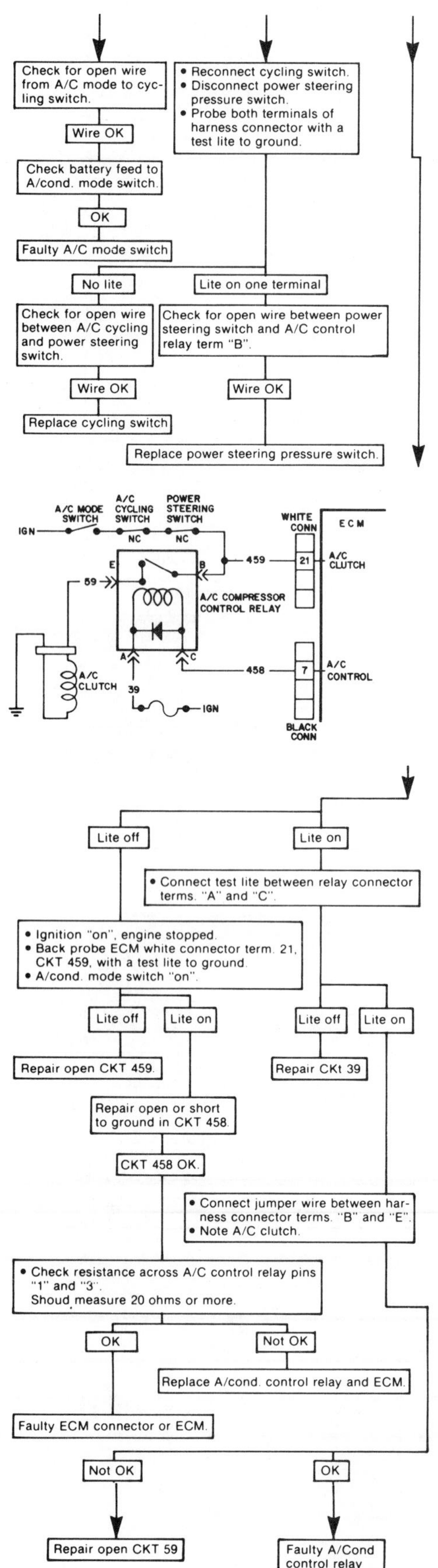

Chart 9A

Park/Neutral Switch

2.5ℓ and 5.0ℓ

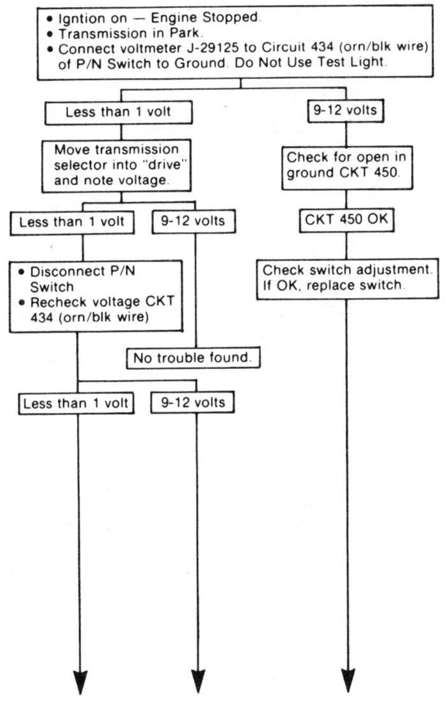
• Igntion on — Engine Stopped.
• Transmission in Park.
• Connect voltmeter J-29125 to Circuit 434 (orn/blk wire) of P/N Switch to Ground. Do Not Use Test Light.
Less than 1 volt
9-12 volts
Move transmission selector into "drive" and note voltage.
Check for open in ground CKT 450.
Less than 1 volt
9-12 volts
CKT 450 OK
• Disconnect P/N Switch
• Recheck voltage CKT 434 (orn/blk wire)
Check switch adjustment. If OK, replace switch.
No trouble found.
Less than 1 volt
9-12 volts

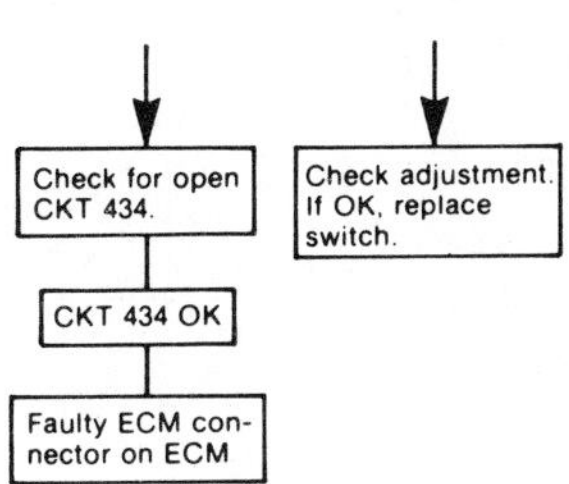
Check for open CKT 434.
Check adjustment. If OK, replace switch.
CKT 434 OK
Faulty ECM connector on ECM

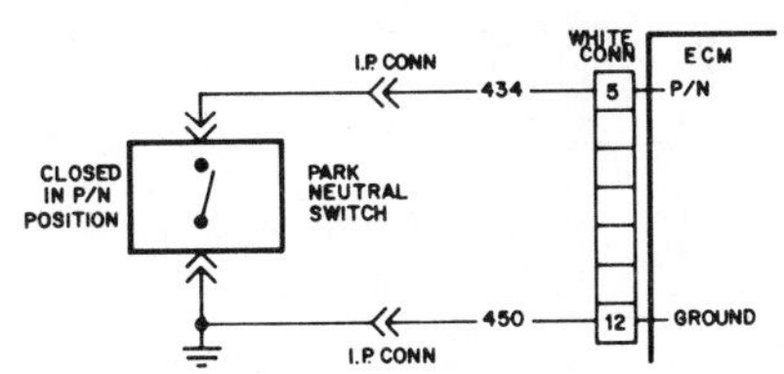
I.P. CONN
434
WHITE CONN
ECM
5
P/N
CLOSED IN P/N POSITION
PARK NEUTRAL SWITCH
450
12
GROUND
I.P. CONN

Chart 9B

Crank Signal

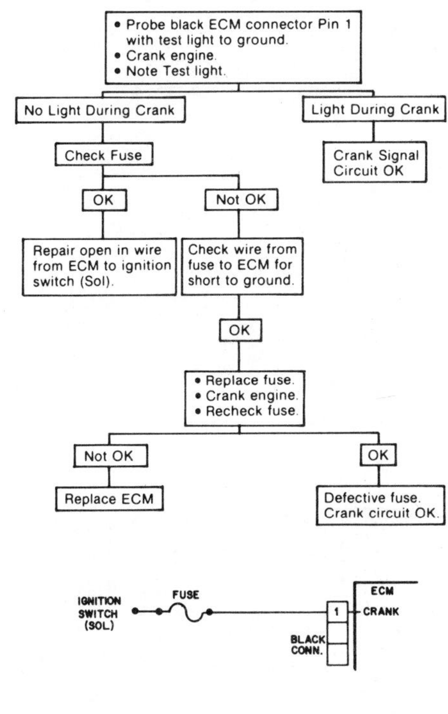
• Probe black ECM connector Pin 1 with test light to ground.
• Crank engine.
• Note Test light.
No Light During Crank
Light During Crank
Check Fuse
Crank Signal Circuit OK
OK
Not OK
Repair open in wire from ECM to ignition switch (Sol).
Check wire from fuse to ECM for short to ground.
OK
• Replace fuse.
• Crank engine.
• Recheck fuse.
Not OK
OK
Replace ECM
Defective fuse. Crank circuit OK.
IGNITION SWITCH (SOL)
FUSE
ECM
1
CRANK
BLACK CONN.

Chart 9C

Park/Neutral Relay

1982 Corvette

Be sure park/neutral switch is OK and properly adjusted before using chart.

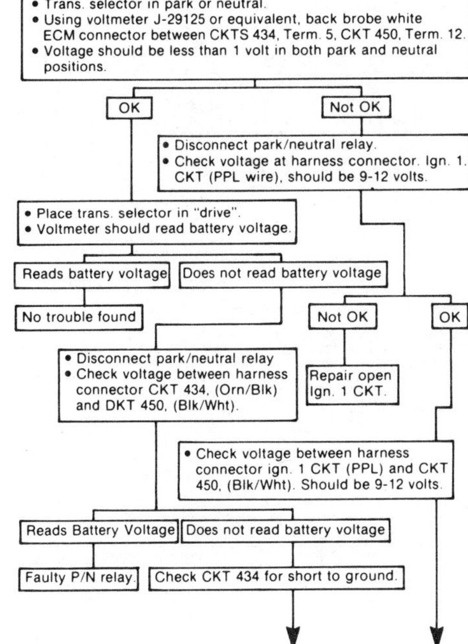
• Ignition on. Engine stopped.
• Trans. selector in park or neutral.
• Using voltmeter J-29125 or equivalent, back brobe white ECM connector between CKTS 434, Term. 5, CKT 450, Term. 12.
• Voltage should be less than 1 volt in both park and neutral positions.
OK
Not OK
• Disconnect park/neutral relay.
• Check voltage at harness connector. Ign. 1. CKT (PPL wire), should be 9-12 volts.
• Place trans. selector in "drive".
• Voltmeter should read battery voltage.
Reads battery voltage
Does not read battery voltage
No trouble found
Not OK
OK
• Disconnect park/neutral relay
• Check voltage between harness connector CKT 434, (Orn/Blk) and DKT 450, (Blk/Wht).
Repair open Ign. 1 CKT.
• Check voltage between harness connector ign. 1 CKT (PPL) and CKT 450, (Blk/Wht). Should be 9-12 volts.
Reads Battery Voltage
Does not read battery voltage
Faulty P/N relay.
Check CKT 434 for short to ground.

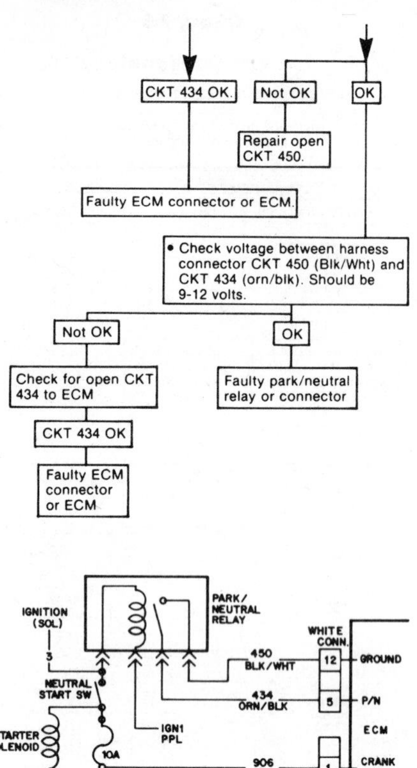

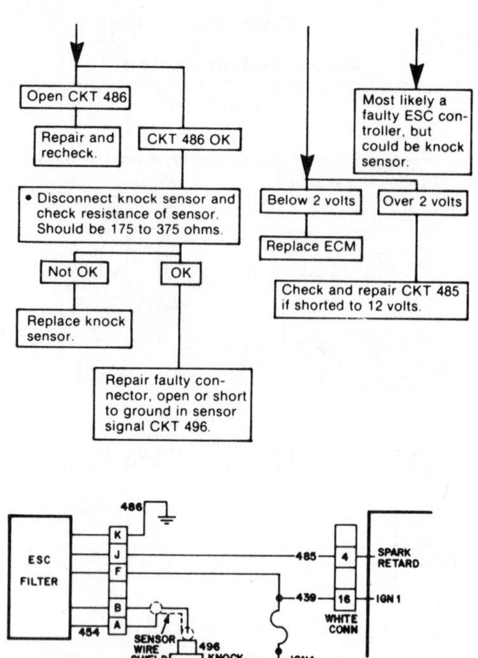

Clear codes and confirm "Closed Loop" operation and no "Check Engine" light. (Perform Diagnostic Circuit Check)

Chart 10

Electronic Spark Control (ESC)

System Check

Engine Knock — No Code 43

This chart should only be used after all other causes of Spark Knock have been checked, i.e., Base Timing, EGR, Engine Temperature, or Excessive Engine Noise.

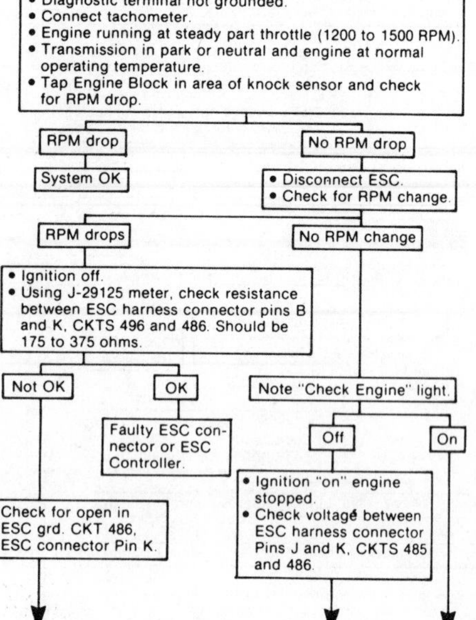

Chart 11

Idle Air Control

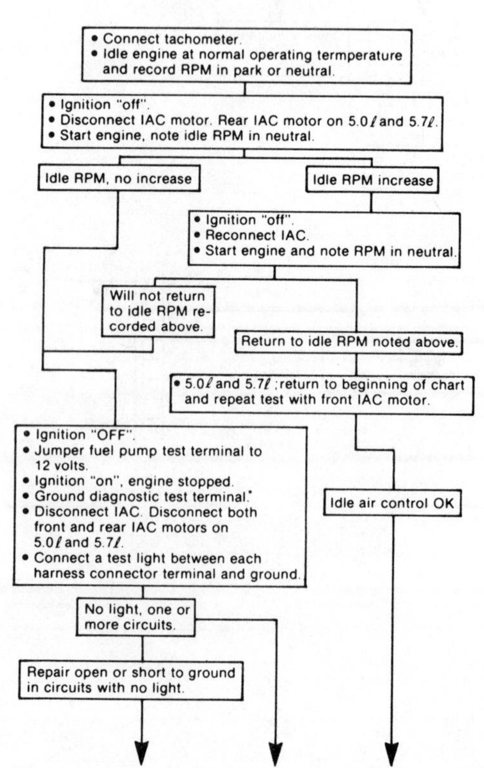

• Connect tachometer.
• Idle engine at normal operating termperature and record RPM in park or neutral.
• Ignition "off".
• Disconnect IAC motor. Rear IAC motor on 5.0ℓ and 5.7ℓ.
• Start engine, note idle RPM in neutral.
Idle RPM, no increase
Idle RPM increase
• Ignition "off".
• Reconnect IAC.
• Start engine and note RPM in neutral.
Will not return to idle RPM recorded above.
Return to idle RPM noted above.
• 5.0ℓ and 5.7ℓ; return to beginning of chart and repeat test with front IAC motor.
• Ignition "OFF".
• Jumper fuel pump test terminal to 12 volts.
• Ignition "on", engine stopped.
• Ground diagnostic test terminal.
• Disconnect IAC. Disconnect both front and rear IAC motors on 5.0ℓ and 5.7ℓ.
• Connect a test light between each harness connector terminal and ground.
Idle air control OK
No light, one or more circuits.
Repair open or short to ground in circuits with no light.

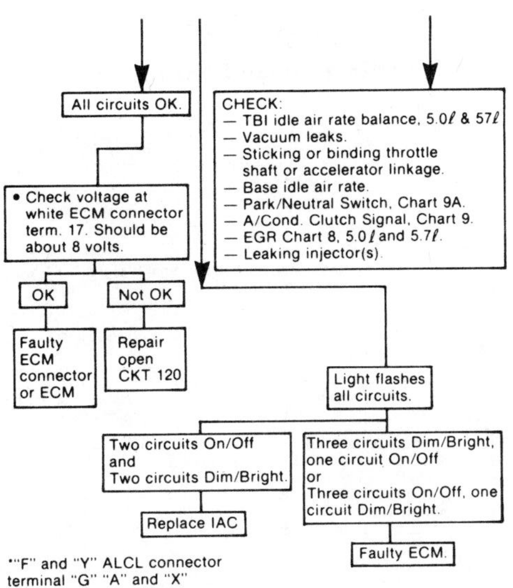

*"F" and "Y" ALCL connector terminal "G" "A" and "X" engine compartment left side.

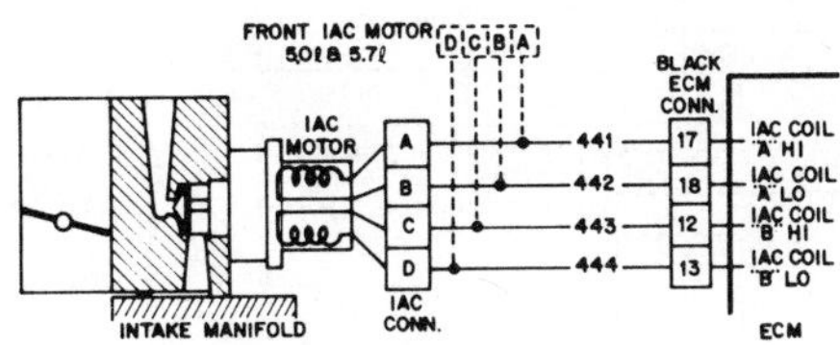

Clear codes and confirm "Closed Loop" operation and no "Check Engine" light. (Perform Diagnostic Circuit Check)

Chart 14

Coolant Sensor Circuit (Signal Voltage Low)

If the "Engine Hot" light is "on", check for overheating condition before making following test:

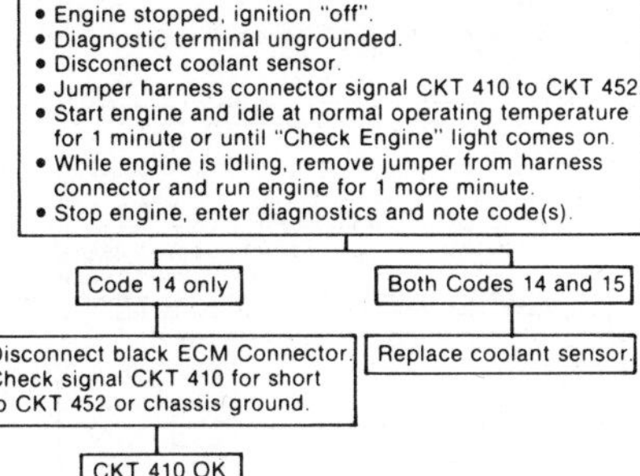

• Engine stopped, ignition "off".
• Diagnostic terminal ungrounded.
• Disconnect coolant sensor.
• Jumper harness connector signal CKT 410 to CKT 452.
• Start engine and idle at normal operating temperature for 1 minute or until "Check Engine" light comes on.
• While engine is idling, remove jumper from harness connector and run engine for 1 more minute.
• Stop engine, enter diagnostics and note code(s).
Code 14 only
Both Codes 14 and 15.
Disconnect black ECM Connector. Check signal CKT 410 for short to CKT 452 or chassis ground.
Replace coolant sensor.
CKT 410 OK
Replace ECM.

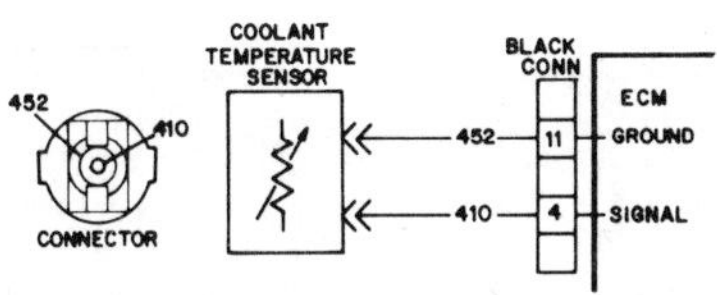

Clear codes and confirm "Closed Loop" operation and no "Check Engine" light. (Perform Diagnostic Circuit Check)

Chart 13

Oxygen Sensor Circuit

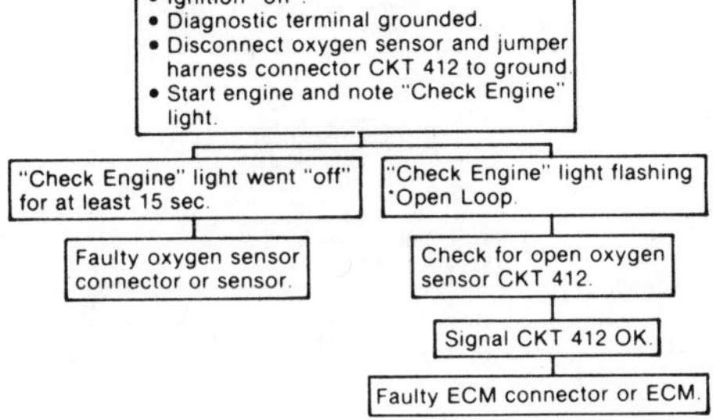

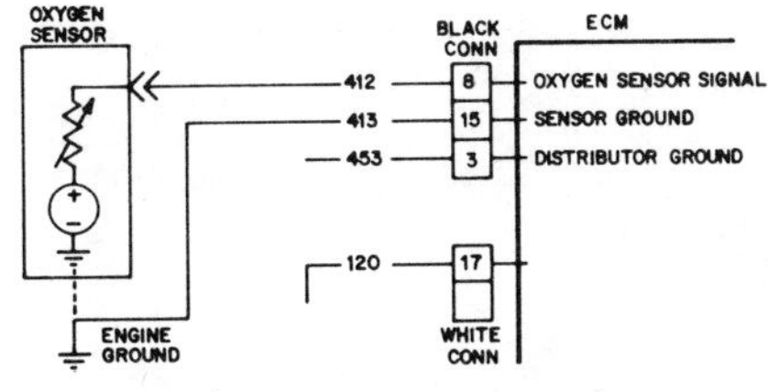

Field Service Mode:

- Engine running — diagnostic terminal grounded.
  - Open Loop; "Check Engine" light flashes at the rate of 2 times per second.
  - Closed Loop; "Check Engine" light flashes at a rate of 1 time per second.

Clear codes and confirm "Closed Loop" operation and no "Check Engine" light. (Perform Diagnostic Circuit Check)

Chart 15

Coolant Sensor Circuit (Signal Voltage High)

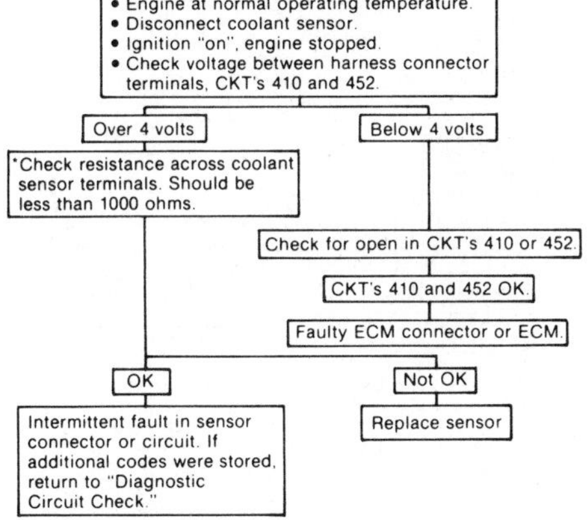

*Sensor check may require use of wire and connector assembly, Part No. 12026621.

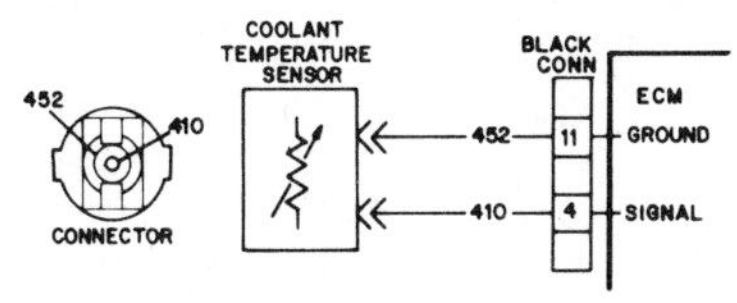

Clear codes and confirm "Closed Loop" operation and no "Check Engine" light. (Perform Diagnostic Circuit Check)

Code 21

Throttle Position Sensor

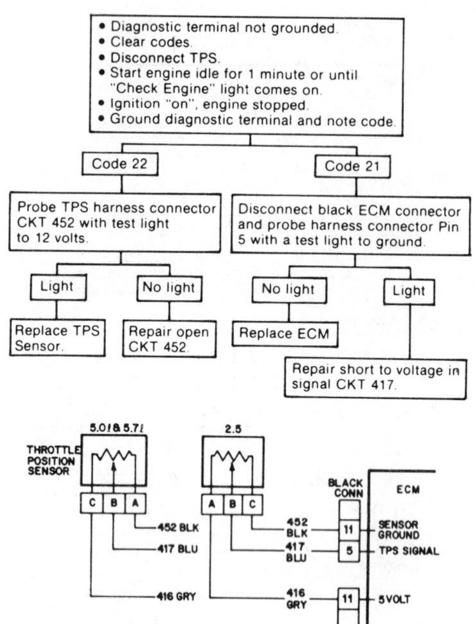

• Diagnostic terminal not grounded.
• Clear codes.
• Disconnect TPS.
• Start engine idle for 1 minute or until "Check Engine" light comes on.
• Ignition "on", engine stopped.
• Ground diagnostic terminal and note code.
Code 22
Code 21
Probe TPS harness connector CKT 452 with test light to 12 volts.
Disconnect black ECM connector and probe harness connector Pin 5 with a test light to ground.
Light
No light
No light
Light
Replace TPS Sensor.
Repair open CKT 452.
Replace ECM
Repair short to voltage in signal CKT 417.
5.0 & 5.7 l
2.5
THROTTLE POSITION SENSOR
C B A
A B C
452 BLK
417 BLU
416 GRY
BLACK CONN
ECM
11
5
SENSOR GROUND
TPS SIGNAL
5 VOLT
WHITE CONN.

Clear codes and confirm "Closed Loop" operation and no "Check Engine" light. (Perform Diagnostic Circuit Check)

Code 22

Throttle Position Sensor (Signal Voltage Low)

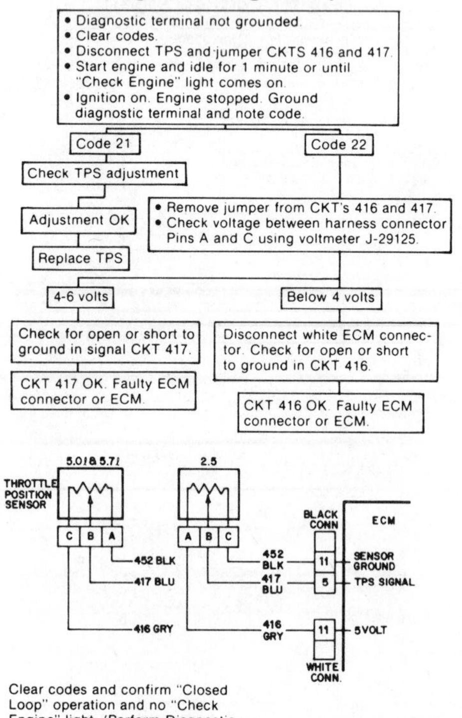

• Diagnostic terminal not grounded.
• Clear codes.
• Disconnect TPS and jumper CKTS 416 and 417.
• Start engine and idle for 1 minute or until "Check Engine" light comes on.
• Ignition on. Engine stopped. Ground diagnostic terminal and note code.
Code 21
Code 22
Check TPS adjustment
Adjustment OK
Replace TPS
• Remove jumper from CKT's 416 and 417.
• Check voltage between harness connector Pins A and C using voltmeter J-29125.
4-6 volts
Below 4 volts
Check for open or short to ground in signal CKT 417.
CKT 417 OK. Faulty ECM connector or ECM.
Disconnect white ECM connector. Check for open or short to ground in CKT 416.
CKT 416 OK. Faulty ECM connector or ECM.
5.0 & 5.7 l
2.5
THROTTLE POSITION SENSOR
C B A
A B C
452 BLK
417 BLU
416 GRY
BLACK CONN
ECM
11
5
SENSOR GROUND
TPS SIGNAL
5 VOLT
WHITE CONN.

Clear codes and confirm "Closed Loop" operation and no "Check Engine" light. (Perform Diagnostic Circuit Check)

Codes 24

Vehicle Speed Sensor (VSS)

NOTICE: A false Code 24 and "Check Engine" light may be set if engine is run above 1200 RPM in neutral on vehicles equipped with manual transmissions. To prevent misdiagnosis, disregard Code 24 if set under these circumstances.

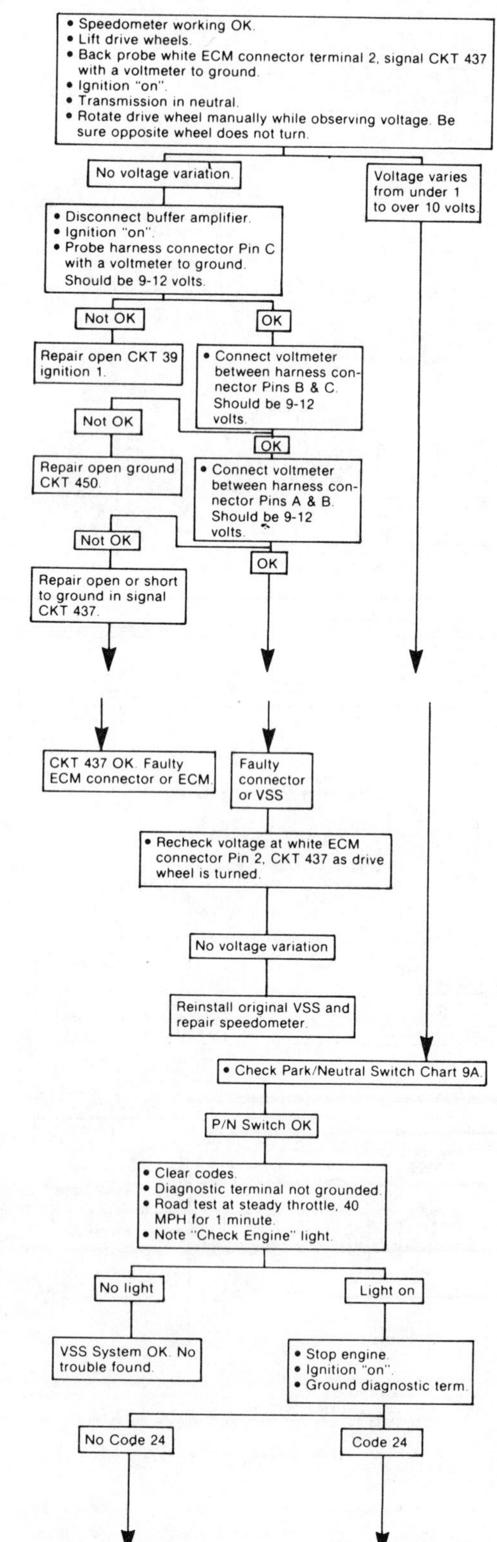

• Speedometer working OK.
• Lift drive wheels.
• Back probe white ECM connector terminal 2, signal CKT 437 with a voltmeter to ground.
• Ignition "on".
• Transmission in neutral.
• Rotate drive wheel manually while observing voltage. Be sure opposite wheel does not turn.
No voltage variation.
Voltage varies from under 1 to over 10 volts.
• Disconnect buffer amplifier.
• Ignition "on".
• Probe harness connector Pin C with a voltmeter to ground. Should be 9-12 volts.
Not OK
OK
Repair open CKT 39 ignition 1.
• Connect voltmeter between harness connector Pins B & C. Should be 9-12 volts.
Not OK
OK
Repair open ground CKT 450.
• Connect voltmeter between harness connector Pins A & B. Should be 9-12 volts.
Not OK
OK
Repair open or short to ground in signal CKT 437.
CKT 437 OK. Faulty ECM connector or ECM.
Faulty connector or VSS
• Recheck voltage at white ECM connector Pin 2, CKT 437 as drive wheel is turned.
No voltage variation
Reinstall original VSS and repair speedometer.
• Check Park/Neutral Switch Chart 9A.
P/N Switch OK
• Clear codes.
• Diagnostic terminal not grounded.
• Road test at steady throttle, 40 MPH for 1 minute.
• Note "Check Engine" light.
No light
Light on
VSS System OK. No trouble found.
• Stop engine.
• Ignition "on".
• Ground diagnostic term.
No Code 24
Code 24

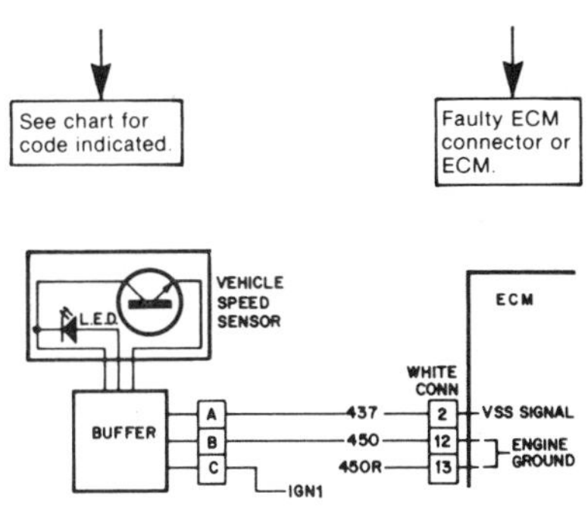

Clear codes and confirm "Closed Loop" operation and no "Check Engine" light. (Perform Diagnostic Circuit Check)

Code 34

MAP Sensor

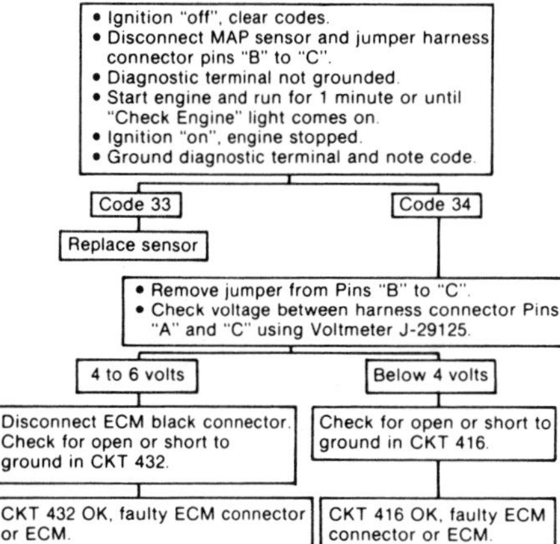

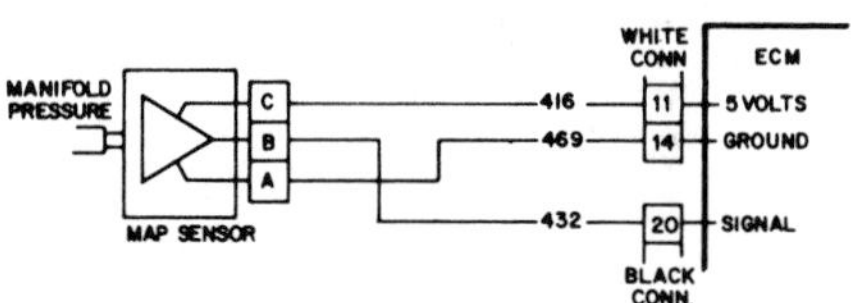

Clear codes and confirm "Closed Loop" operation and no "Check Engine" light. (Perform Diagnostic Circuit Check)

Code 33

MAP Sensor

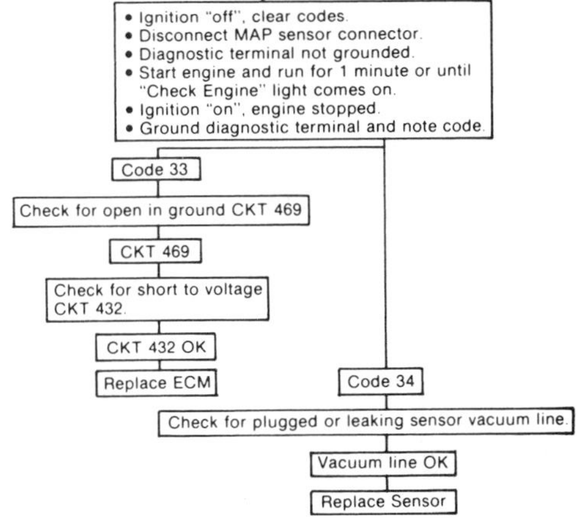

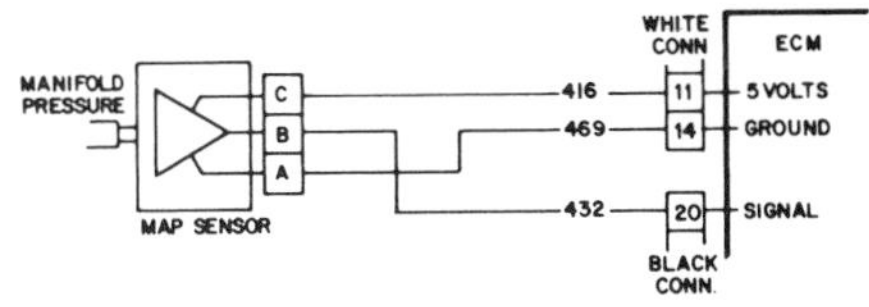

Clear codes and confirm "Closed Loop" operation and no "Check Engine" light. (Perform Diagnostic Circuit Check)

Code 42

Electronic Spark Timing (EST)

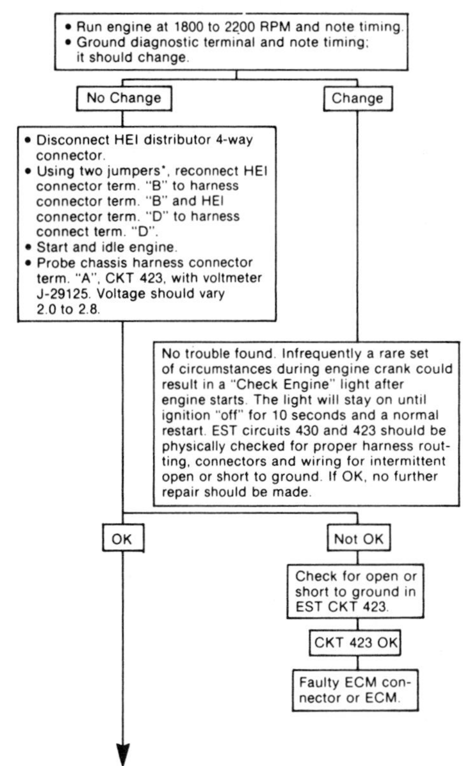

• Run engine at 1800 to 2200 RPM and note timing.
• Ground diagnostic terminal and note timing; it should change.
No Change
Change
• Disconnect HEI distributor 4-way connector.
• Using two jumpers*, reconnect HEI connector term. "B" to harness connector term. "B" and HEI connector term. "D" to harness connect term. "D".
• Start and idle engine.
• Probe chassis harness connector term. "A", CKT 423, with voltmeter J-29125. Voltage should vary 2.0 to 2.8.
No trouble found. Infrequently a rare set of circumstances during engine crank could result in a "Check Engine" light after engine starts. The light will stay on until ignition "off" for 10 seconds and a normal restart. EST circuits 430 and 423 should be physically checked for proper harness routing, connectors and wiring for intermittent open or short to ground. If OK, no further repair should be made.
OK
Not OK
Check for open or short to ground in EST CKT 423.
CKT 423 OK
Faulty ECM connector or ECM.

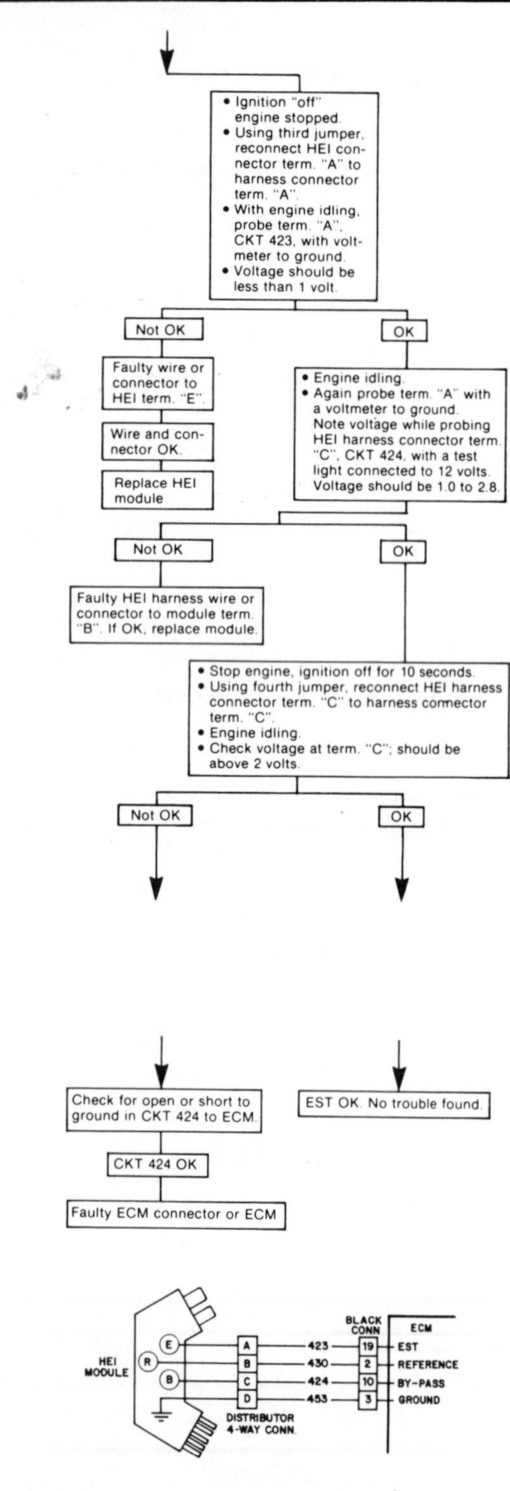

• Ignition "off" engine stopped.
• Using third jumper, reconnect HEI connector term. "A" to harness connector term. "A".
• With engine idling, probe term. "A", CKT 423, with voltmeter to ground.
• Voltage should be less than 1 volt.
Not OK
OK
Faulty wire or connector to HEI term. "E".
Wire and connector OK.
Replace HEI module
• Engine idling.
• Again probe term. "A" with a voltmeter to ground.
Note voltage while probing HEI harness connector term. "C", CKT 424, with a test light connected to 12 volts.
Voltage should be 1.0 to 2.8.
Not OK
OK
Faulty HEI harness wire or connector to module term. "B". If OK, replace module.
• Stop engine, ignition off for 10 seconds.
• Using fourth jumper, reconnect HEI harness connector term. "C" to harness connector term. "C".
• Engine idling.
• Check voltage at term. "C"; should be above 2 volts.
Not OK
OK
Check for open or short to ground in CKT 424 to ECM.
CKT 424 OK
Faulty ECM connector or ECM
EST OK. No trouble found.
HEI MODULE
E
R
B
A
B
C
D
DISTRIBUTOR 4-WAY CONN.
423
430
424
453
BLACK CONN
19
2
10
3
ECM
EST
REFERENCE
BY-PASS
GROUND

Clear codes and confirm "Closed Loop" operation and no "Check Engine" light. (Perform Diagnostic Circuit Check)

*This test requires the use of four jumpers using terminals 12014836 and 12014837.

Code 43

Electronic Spark Control (ESC)

5.0ℓ and 5.7ℓ

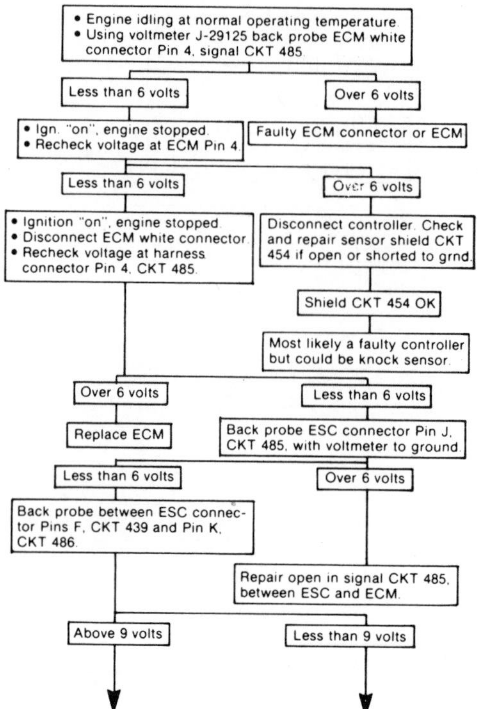

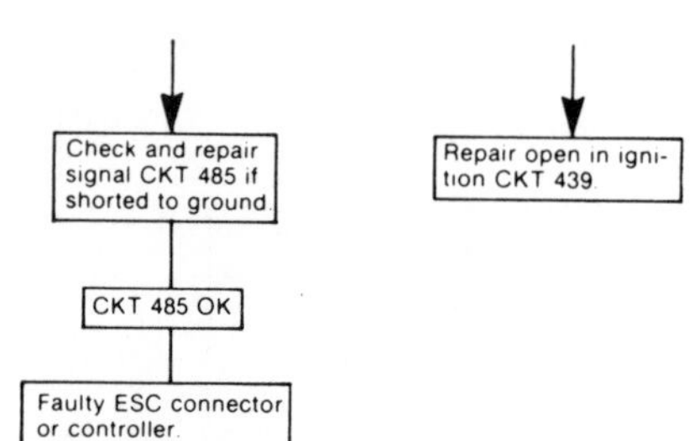

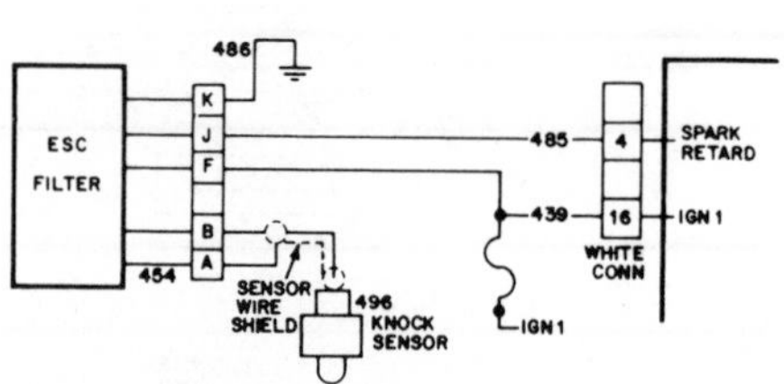

Clear codes and confirm "Closed Loop" operation and no "Check Engine" light. (Perform Diagnostic Circuit Check)

Code 44

Lean Exhaust Indication

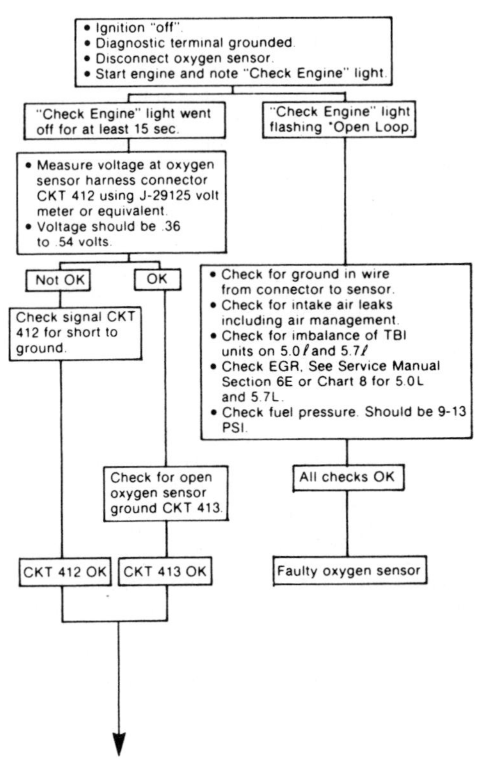

• Ignition "off".
• Diagnostic terminal grounded.
• Disconnect oxygen sensor.
• Start engine and note "Check Engine" light.
"Check Engine" light went off for at least 15 sec.
"Check Engine" light flashing *Open Loop.
• Measure voltage at oxygen sensor harness connector CKT 412 using J-29125 volt meter or equivalent.
• Voltage should be .36 to .54 volts.
Not OK
OK
Check signal CKT 412 for short to ground.
• Check for ground in wire from connector to sensor.
• Check for intake air leaks including air management.
• Check for imbalance of TBI units on 5.0ℓ and 5.7ℓ
• Check EGR. See Service Manual Section 6E or Chart 8 for 5.0L and 5.7L.
• Check fuel pressure. Should be 9-13 PSI.
Check for open oxygen sensor ground CKT 413.
All checks OK
CKT 412 OK
CKT 413 OK
Faulty oxygen sensor

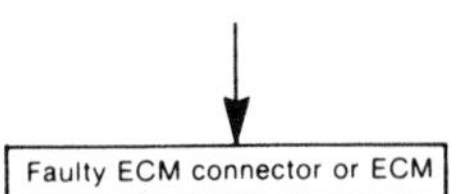

Field Service Mode:
Engine running — diagnostic terminal grounded.
*— Open Loop: "Check Engine" light flashes at the rate of 2 times per second.
— Closed Loop: "Check Engine" light flashes at a rate of 1 time per second.

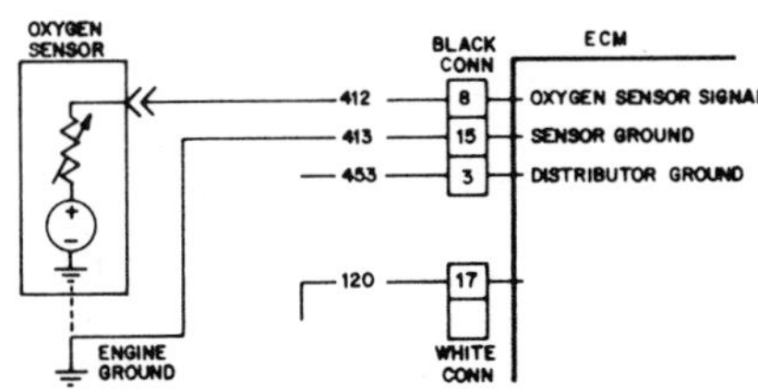

Clear codes and confirm "Closed Loop" operation and no "Check Engine" light. (Perform Diagnostic Circuit Check)

Code 45

Rich Exhaust Indication

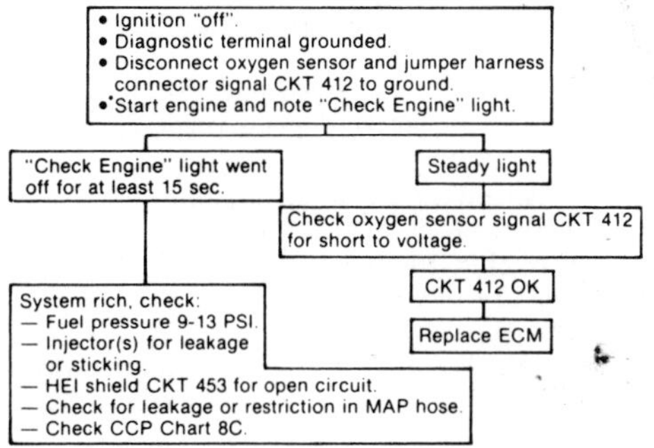

• Ignition "off".
• Diagnostic terminal grounded.
• Disconnect oxygen sensor and jumper harness connector signal CKT 412 to ground.
•*Start engine and note "Check Engine" light.
"Check Engine" light went off for at least 15 sec.
Steady light
Check oxygen sensor signal CKT 412 for short to voltage.
CKT 412 OK
Replace ECM
System rich, check:
— Fuel pressure 9-13 PSI.
— Injector(s) for leakage or sticking.
— HEI shield CKT 453 for open circuit.
— Check for leakage or restriction in MAP hose.
— Check CCP Chart 8C.

*Field Service Mode:
Engine running — diagnostic terminal grounded.
— Open Loop: "Check Engine" light flashes at the rate of 2 times per second.
— Closed Loop: "Check Engine" light flashes at a rate of 1 time per second.

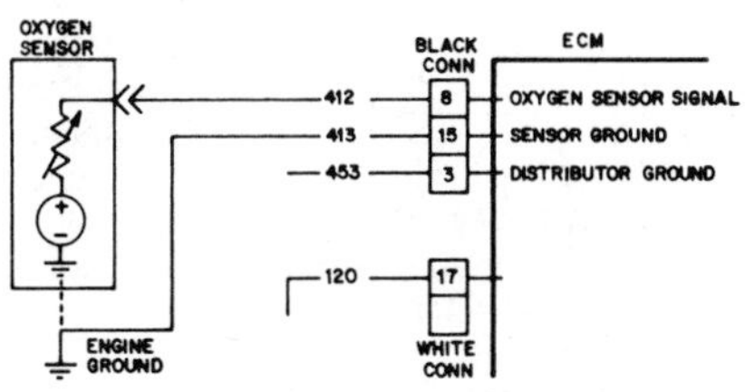

Clear codes and confirm "Closed Loop" operation and no "Check Engine" light. (Perform Diagnostic Circuit Check)

Code 51

Check that all pins are fully inserted in the socket. If OK, replace PROM, clear memory, and recheck. If code 51 reappears, replace ECM.

Code 55

Replace Electronic Control Module (ECM)

Clear codes and confirm "Closed Loop" operation and no "Check Engine" light. (Perform Diagnostic Circuit Check)

MAZDA ELECTRONIC GASOLINE INJECTION (EGI) SYSTEM

General Information

The Mazda Electronic Gasoline Injection (EGI) system is a computer controlled, port-type fuel injection system used on 1984 RX-7 models equipped with the 13B rotary engine. Fuel is metered to the engine by two solenoid-type injectors that open and close in response to signals from the elctronic control unit (ECU). By varying the amount of time the injectors remain open, the ECU precisely controls the fuel mixture according to data on engine operating conditions provided by various engine sensors. The EGI fuel injection system can be divided into three main subsystems; the electronic control system, the air induction system and the fuel supply system.

ELECTRONIC CONTROL SYSTEM

The heart of the EGI system is a micro-processor called the electronic control unit (ECU). The ECU receives data from the engine sensors and, based on the information received and the instructions programmed into the ECU memory, generates output signals to control fuel delivery, idle speed and various solenoid valves to control engine emissions and allow the most efficient operation of the three-way catalytic converter by controlling the secondary air supply system and air switching valve to divert air from the air pump to either the air cleaner or catalytic converter. The ECU receives data on engine speed (rpm), engine and coolant temperature, throttle position, manifold vacuum, intake air volume, intake air temperature, atmospheric pressure, and the amount of oxygen in the exhaust gasses. In addition, the ECU is provided with signals for starter operation (cranking), A/C operation and in-gear operation (neutral/clutch switch). The ECU is located under the floor pan on the right (passenger) side of the vehicle.

AIR INDUCTION SYSTEM

The EGI air induction system consists of the air cleaner assembly and ducting, an air flow meter with a sensor flap that measures the amount of intake air entering the system, a throttle chamber and dynamic chamber or air plenum which is attached to the intake manifold. A throttle position sensor is mounted on the throttle valve assembly to provide the ECU with information on the throttle opening angle. A By-Pass Air Control (BAC) valve controls idle speed under various engine conditions and loads by allowing a regulated amount of air to bypass the throttle plates. The BAC valve is controlled by the vent solenoid valve and the vacuum solenoid valve in response to signals from the ECU. The air temperature sensor (thermistor) is mounted in the dynamic chamber and provides a signal to the ECU to allow "fine tuning" of the mixture to compensate for air density changes due to temperature. The ECU also uses the air temperature signal to control the BAC valve operation and to shut the vacuum passage between the dynamic chamber and the fuel pressure regulator control valve when the intake air temperature is above 122 degrees F (50 degrees C).

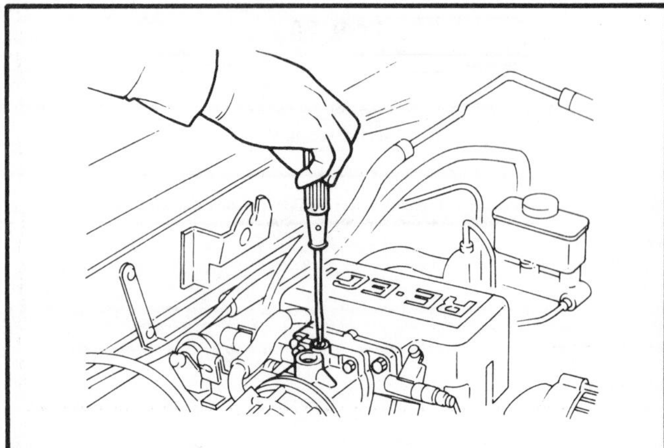

Adjusting idle speed

FUEL SUPPLY SYSTEM

The EGI fuel supply system consists of the fuel tank and lines, fuel filter, electric fuel pump, pulsation damper, fuel pressure regulator, fuel delivery manifold and two solenoid-type fuel injectors. A check and cut valve releases excessive pressure or vacuum in the fuel tank to the atmosphere and prevents the loss of fuel if the vehicle overturns. The fuel system is constantly under pressure, which is regulated to 37 psi.

The fuel pump provides more fuel than the injection system needs for normal operation, with excess fuel being returned to the tank via the fuel return line. The fuel pump is a roller cell type which is capable of providing 49–71 psi of pressure as measured at the pump outlet. The pump is located near the fuel tank, under a protective shield on the underbody of the vehicle. The pump connector is located under the storage compartment located behind the driver's seat. The pressure regulator and pulsation damper are mounted on the fuel delivery manifold, located under the dynamic chamber just above the injectors. The operation of the fuel delivery system is controlled through the main relay mounted in the engine compartment next to the master cylinder, with power feeding through a fusible link located near the battery.

On-Car Service

IDLE SPEED

Adjustment

Before checking or adjusting the idle speed, switch off all accessories, remove the fuel filler cap and connect a suitable tachometer. The throttle sensor should also be checked for proper adjustment.

1. Start the engine and allow it to reach normal operating temperature.
2. Disconnect the vent and vacuum solenoid valve connector.
3. Remove the blind cap and adjust the idle speed to 800 rpm by turning the air adjust screw. Raise the engine speed, then allow it to return to idle and recheck rpm with the tachometer.
4. After idle speed adjustment is complete, reinstall the blind cap onto the air adjust screw.

IDLE MIXTURE

Adjustment

The idle mixture is preset at the factor, and should not normally require adjustment unless the variable resistor is replaced. Disconnect the variable resistor connector and connect an ohmmeter to the resistor terminals. If continuity does not exist, or if the resistance value is not within 0.5–4.5 kilo-ohms, replace the variable resistor. Before adjusting the idle mixture, switch off all accessories, remove the fuel filler cap and connect a suitable tachometer to the engine. The throttle sensor adjustment should also be checked.

1. Start the engine and allow it to reach normal operating temperature.
2. Disconnect the vent and vacuum solenoid valve connector.
3. Remove the blind cap and adjust the idle speed as previously described.
4. Turn the variable resistor adjusting screw until the highest idle

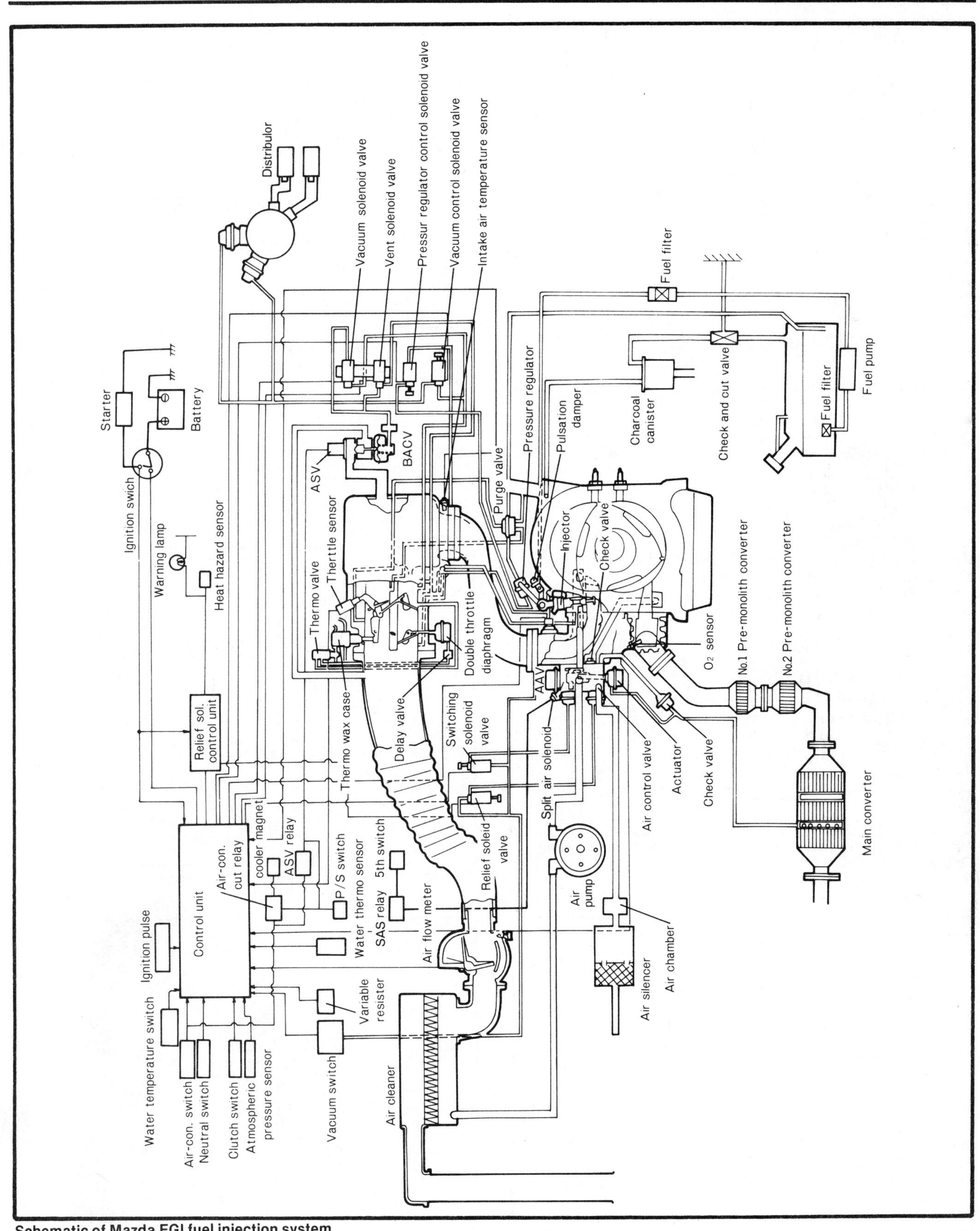

Schematic of Mazda EGI fuel injection system

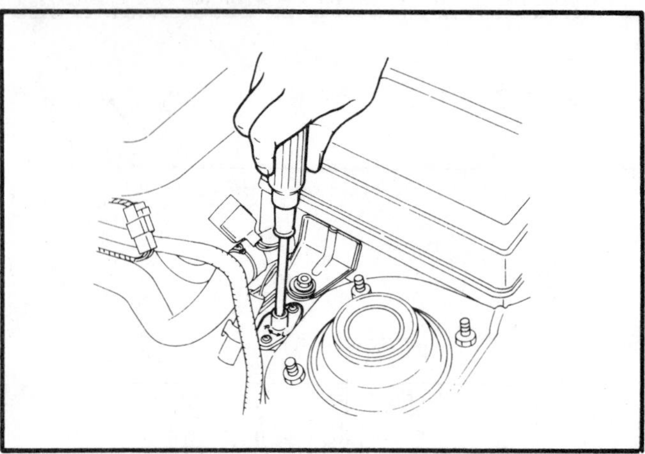

Adjusting idle mixture with variable resistor

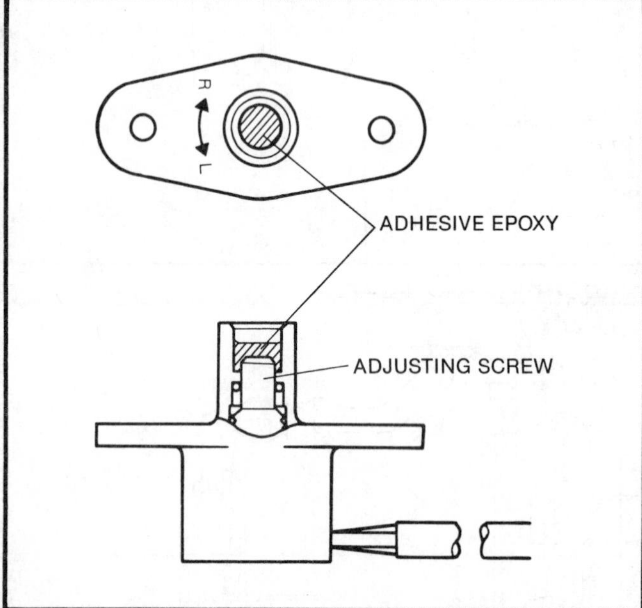

Variable resistor used to adjust idle speed on Mazda EGI system. See text for details

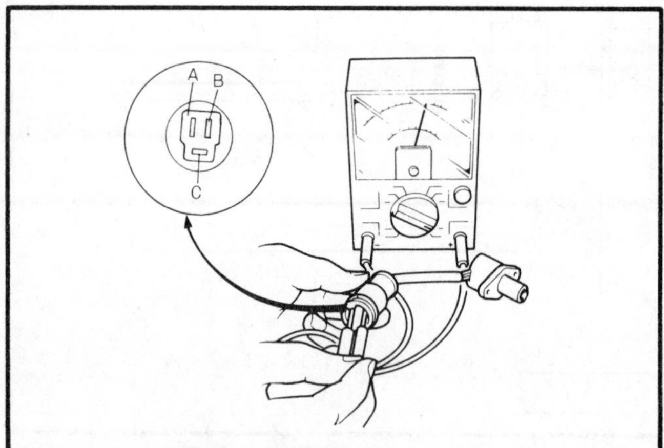

Checking resistance for idle mixture adjustment. Note the terminal identification

is obtained, then reset the idle speed to 800 rpm by turning the idle adjustment screw.

5. Turn the variable resistor adjusting screw counterclockwise until the idle speed reaches 780 rpm, then turn clockwise until the idle is again 800 rpm.

6. Connect the vent and vacuum solenoid connector.

7. Install the blind cap onto the idle adjustment screw and fill the head of the variable resistor adjustment screw with epoxy cement.

PRESSURE TESTS

Fuel Pump Pressure Test

1. Disconnect the negative battery terminal.

2. Disconnect the main fuel hose from the fuel pipe and connect a suitable pressure gauge.

CAUTION

Fuel system is under pressure. Wrap a clean cloth around the fuel connection when loosening to catch the fuel spray and take precautions to avoid the risk of fire.

3. Reconnect the negative battery terminal.

4. Turn the ignition switch ON and use a jumper to energize the fuel pump through the shortcircuit terminal.

5. Read the fuel pump pressure on the pressure gauge. It should be between 49–71 psi. If not, replace the fuel pump.

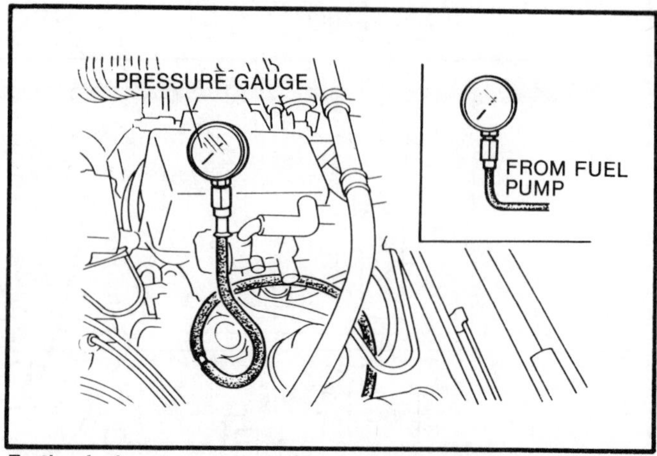

Testing fuel pump pressure with gauge connected to fuel feed line

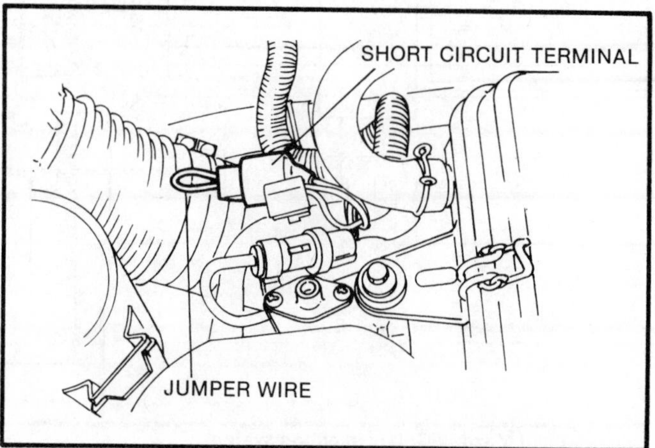

Install a jumper wire into the fuel pump short circuit terminal to energize the fuel pump during test procedures

System Pressure Test

1. Disconnect the negative battery terminal.
2. Disconnect the fuel main hose from the fuel pipe and connect pressure gauge using a three-way joint.

CAUTION

Fuel system is under pressure. Wrap a clean cloth around the fuel connection when loosening to catch any fuel spray and take precautions to avoid the risk of fire.

3. Connect the negative battery cable and start the engine.
4. Disconnect the vacuum hose to the pressure regulator control valve.
5. Measure the fuel system pressure at idle. It should be about 37 psi.
6. Connect the vacuum hose to the pressure regulator control valve and again measure the fuel system pressure. It should drop to about 28 psi.
7. Replace the fuel pressure regulator if any other results are obtained.

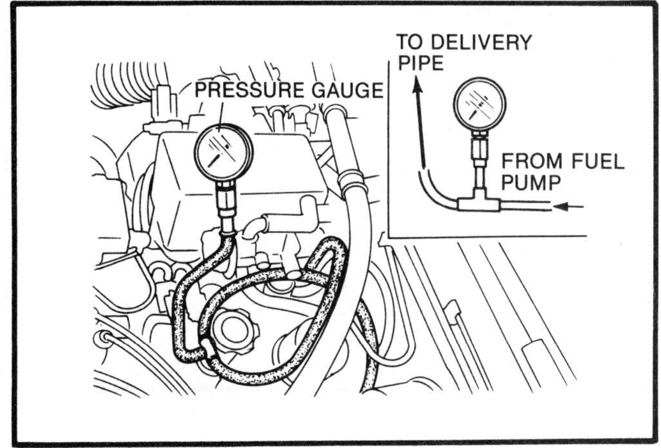

Testing fuel pressure regulator with tee adapter. Make sure all connections are tight

Testing Fuel Injector Operation

Use a mechanic's stethescope to check for operation noise from the injectors at idle and under acceleration. The injectors should buzz as the pintle opens and closes rapidly. If the injectors are not operating, check that continuity is present in the wire from the trailing coil negative (–) coil to terminal U in the control unit connector. Check the EGI main fusible link and turn the ignition on and off and make sure the relays click. If clicking is not heard in the main relay when the ignition is switched on, check for 12 volts at main relay connector No. 2 terminal.

BYPASS AIR CONTROL (BAC) SYSTEM

Testing

Start the engine and allow it to reach normal operating temperature. With the engine idling, turn the headlight switch ON and disconnect the vent and vacuum solenoid valve connector. The engine speed should decrease. Reconnect the vent and vacuum solenoid valve connector and make sure the engine speed increases back up to 800 rpm. If the test results are not as described, perform the following tests to isolate which solenoid is malfunctioning.

VENT SOLENOID VALVE

Testing

1. Disconnect the vacuum line from the control solenoid to the vent solenoid valve.
2. Disconnect the vent solenoid connector.
3. Blow through the vent solenoid valve vacuum line and make sure that air does not pass.
4. Apply 12 volts to the vent solenoid valve connector and again blow through the vacuum line. Air should now pass through.
5. If the test results are not as described, replace the vent solenoid valve and retest.

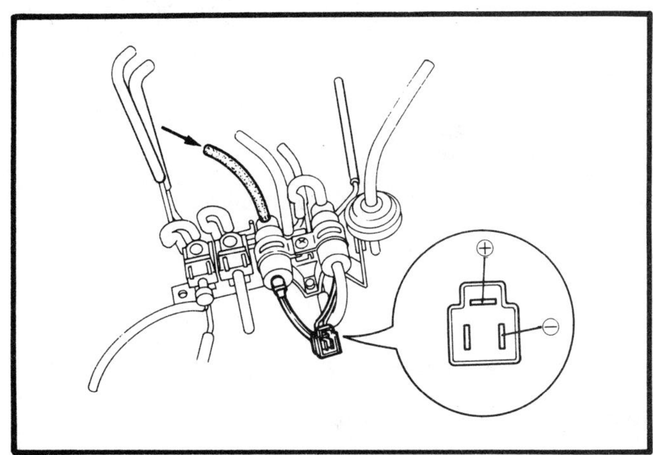

Testing vent solenoid valve. See text for details

VACUUM SOLENOID VALVE

Testing

1. Disconnect the vacuum line from the control solenoid to the vacuum solenoid valve.
2. Disconnect the vacuum solenoid connector.
3. Blow through the vacuum solenoid line and make sure that air passes.
4. Apply 12 volts to the vacuum solenoid connector and again blow through the disconnected line. Air should not pass through.
5. If the test results are not as described, replace the vacuum solenoid valve and retest.

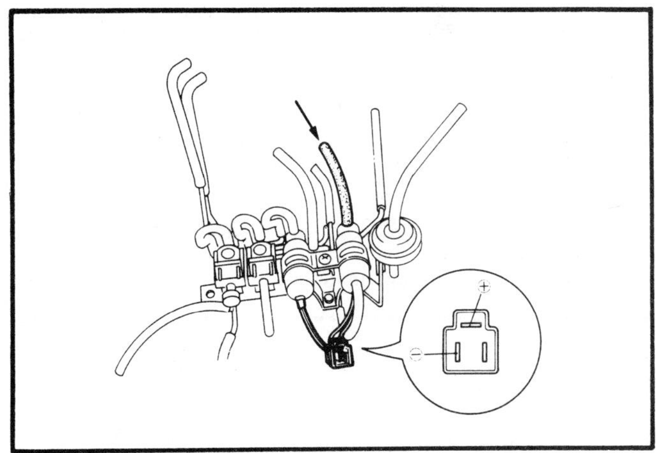

Test connections for checking vacuum solenoid operation.

AIR SUPPLY VALVE

Testing

1. Start the engine and allow it to idle at normal operating temperature.
2. Turn the A/C switch ON and make sure the engine speed does not decrease.
3. Disconnect the air supply valve connector. The engine speed should drop.

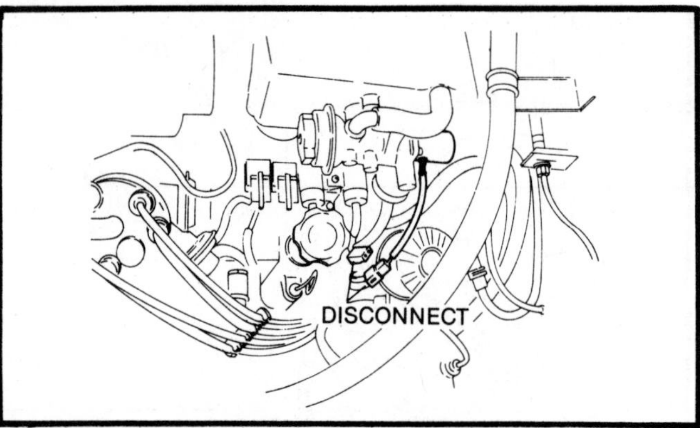

Disconnect the air supply valve when testing

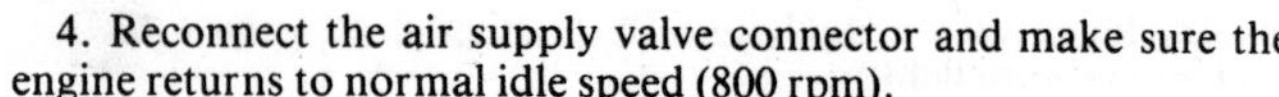

4. Reconnect the air supply valve connector and make sure the engine returns to normal idle speed (800 rpm).
5. Turn the steering wheel to the right and left and make sure the magnetic clutch of the compressor turns off.
6. If the test results are not as described, connect a tachometer to the engine.
7. Disconnect the vent and and vacuum solenoid valve connector.
8. Check and adjust the idle speed, if necessary.
9. Disconnect the air supply valve connector and apply 12 volts to the connector. The engine speed should be within 1000–1070 rpm.
10. If the engine speed is not as specified, remove the blind cap and adjust the engine speed with the adjusting screw on the air supply valve.
11. Install the blind cap after adjustment and repeat the test.

THROTTLE SENSOR

Testing

1. Start the engine and allow it to reach normal operating temperature, then turn the ignition OFF.
2. Connect two voltmeters to the test connector.
3. Turn the ignition ON and check that current flows to only one voltmeter.
4. If both voltmeters show current flow, remove the cap and turn the adjusting screw counterclockwise until only one voltmeter shows current.
5. If no current is registered by either voltmeter, turn the adjusting screw clockwise until only one voltmeter shows current.
6. Install the cap on the adjusting screw and disconnect the throttle sensor connector. Connect an ohmmeter to the throttle sensor terminals.
7. Open the throttle valve and check the resistance reading from the throttle sensor. At idle, the throttle sensor should show about 1 kilo-ohm resistance. At wide open throttle, the sensor should show about 5 kilo-ohms resistance.
8. Any results other than these, replace the throttle sensor and retest.

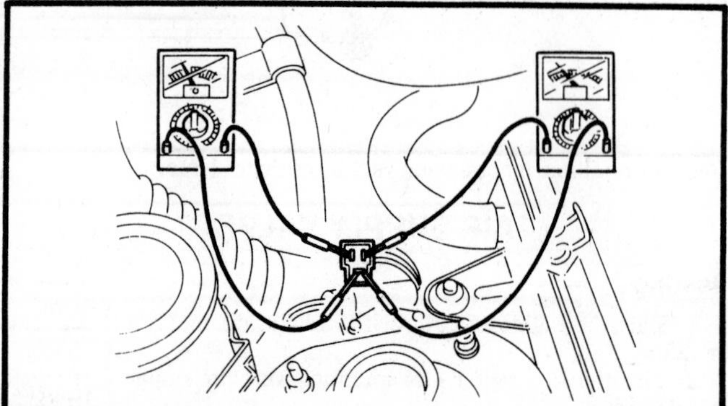

Adjusting the throttle sensor on Mazda EGI system requires the use of two ohmmeters

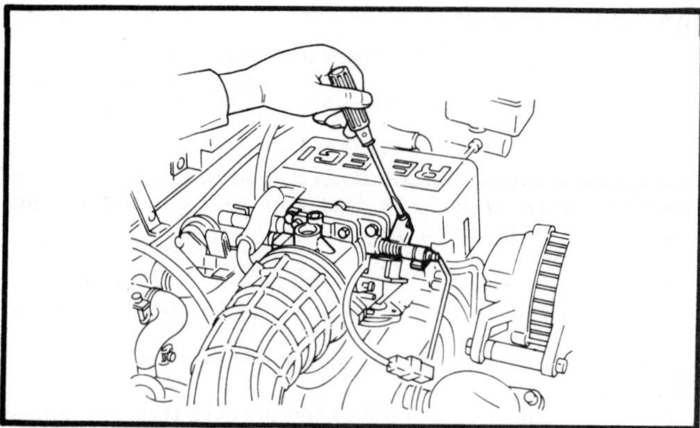

Location of throttle sensor adjusting screw.

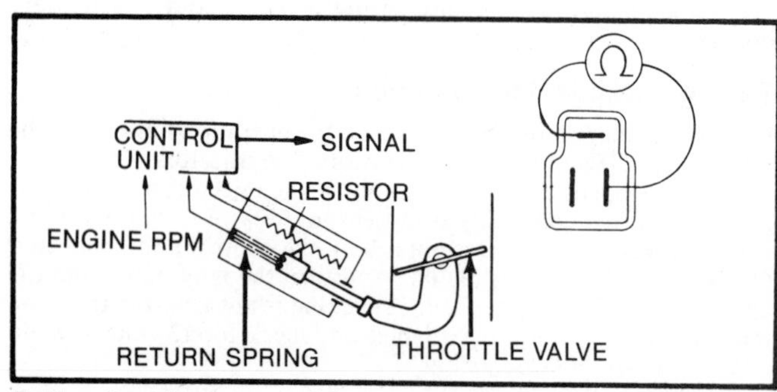

Testing throttle sensor resistance. It should be about 1kΩ @ idle and 5kΩ @ full throttle

NOTE: During deceleration above a certain engine speed, fuel is not injected to reduce emissions. Check this fuel cut operation by holding the engine at 2000 rpm in neutral and making sure the engine speed varies when the throttle sensor is pushed in with a finger.

DASH POT

Testing

1. Make sure the dash pot rod does not prevent the throttle lever from returning to the idle stop.
2. Open the throttle lever fully and make sure the dash pot lever extends quickly. Release the throttle lever and make sure it returns slowly to the idle position after touching the dash pot rod.
3. Connect a suitable tachometer to the engine.

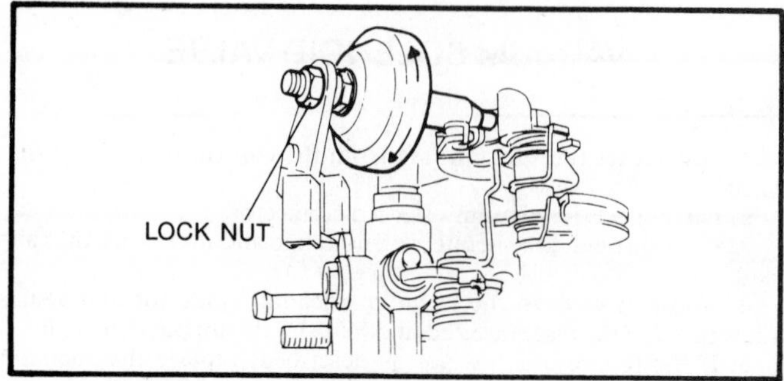

Loosen the locknut to adjust the dash pot

4. Start the engine and allow it to reach normal operating temperature. Check the idle speed and adjust if necessary.

5. Operate the throttle lever until it is away from the dash pot rod, then slowly decrease the engine speed and note the rpm at which the throttle lever just touches the dash pot rod. It should contact the rod at between 2350–2650 rpm.

6. If the engine speed is not within the range specified when the throttle lever contacts the dash pot rod, loosen the lock nut and adjust the contact point by turning the dash pot diaphragm. Tighten the lock nut and retest after all adjustments are complete.

ANTI-BACKFIRE VALVE

Testing

Start the engine and allow it to reach normal operating temperature. With the engine at idle, disconnect the hose from the air control valve to the air pump at the air pump. Place a finger over the air hose and make sure no vacuum is present at idle. Increase the engine speed to 3000 rpm, then allow the engine to return to idle rapidly. Vacuum should be present at the air hose for a few seconds while decelerating. If no air is being drawn into the air hose during deceleration, or if air is drawn in at idle, replace the air control valve and retest.

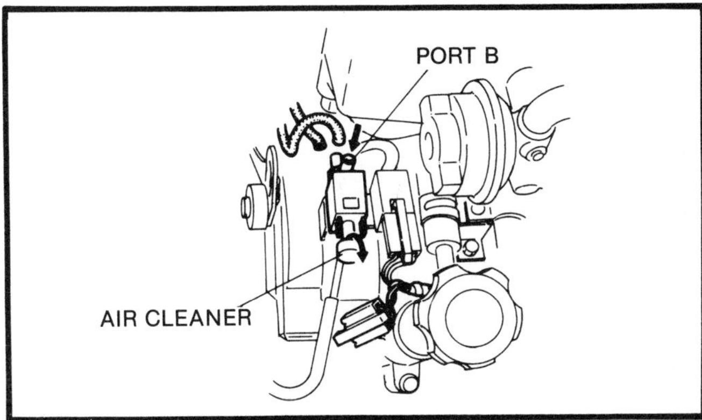

Test the vacuum control solenoid valve by disconnecting the hoses and making sure air passes from port B to the air cleaner.

VACUUM CONTROL SOLENOID VALVE

Testing

1. Start the engine and allow it to reach normal operating temperature.

2. Connect a suitable tachometer to the engine.

3. Disconnect the vacuum line at the solenoid and place a finger over the solenoid connection. Make sure no vacuum is present through the solenoid valve at idle.

4. Gradually increase the engine speed to about 1200 rpm and make sure vacuum is now present at the solenoid valve connection.

5. Decrease engine speed rapidly from 4000 rpm and make sure no vacuum is present at the solenoid connection while decelerating.

6. Turn the A/C switch ON and make sure vacuum is present at the solenoid valve connection at idle.

7. Disconnect the vacuum sensing lines from the solenoid valve and blow through the valve from port B. Make sure air passes through the valve and comes out the filter (C).

8. Disconnect the connector from the vacuum control solenoid valve and apply 12 volts to the connector terminals at the valve.

9. Blow through the valve at port B and make sure air passes through the valve and comes out port A.

10. If any other test results are obtained, replace the vacuum control solenoid valve and retest.

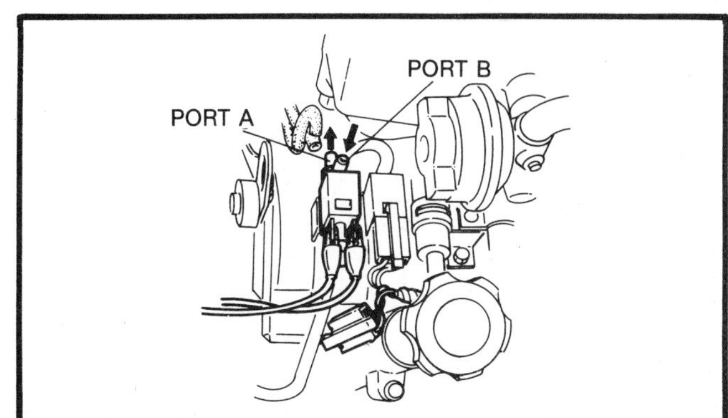

When testing, air should pass from port B to port A with battery voltage applied to terminals

PRESSURE REGULATOR CONTROL SOLENOID

Valve Test

Disconnect the vacuum lines from the solenoid valve and blow through the solenoid valve at port B. Air should pass through the valve and come out port A at the air filter. Disconnect the solenoid valve connector and apply 12 volts to the terminals of the valve. Blow through the valve from port B and make sure air passes and comes out port C. If any other results are obtained, replace the pressure regulator control solenoid valve and retest.

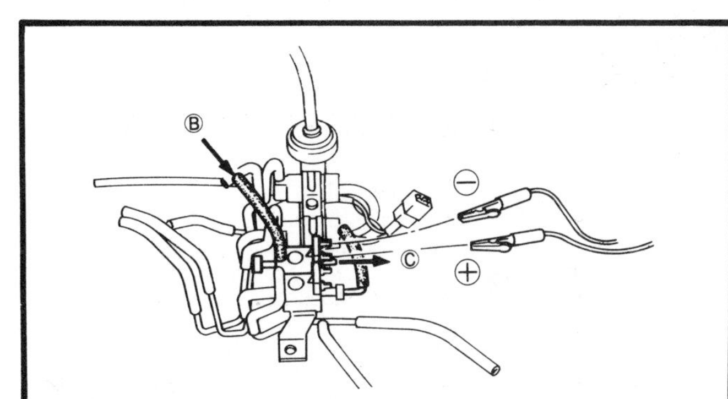

Testing pressure regulator control solenoid valve. Air should pass from port B to port C with battery voltage applied to terminals

INTAKE AIR TEMPERATURE SENSOR

Testing

Remove the intake air temperature sensor from the dynamic air chamber and connect an ohmmeter to the sensor terminals. Check the sensor calibration by reading the resistance values at various temperatures. If the resistance is out of range or fails to vary at different temperatures, replace the sensor.

AIR TEMPERATURE SENSOR TEST

Temperature	Resistance
20°C (68°F)	41.5 ± 4.15 kΩ
50°C (122°F)	11.85 ± 1.19 kΩ
85°C (185°F)	3.5 ± 0.35 kΩ

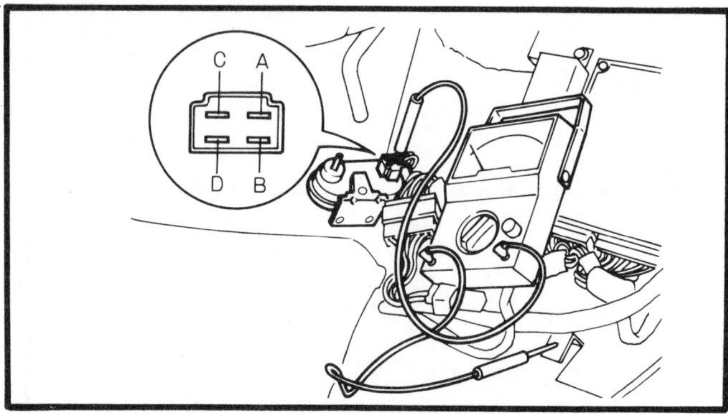

Testing atmospheric pressure sensor. See text for details

ATMOSPHERIC PRESSURE SENSOR

Testing

Connect the voltmeter to atmospheric pressure sensor terminal D, turn the ignition switch ON and take a voltage reading. It should read 3.5–4.5 volts at sea level and 2.5–3.5 volts at high altitude (areas above 8500 ft. in elevation.)

ELECTRONIC CONTROL UNIT

Testing

Connect a voltmeter to the control unit by probing the main harness connector carefully from the rear and grounding the other voltmeter probe. All control unit testing is limited to simple voltage checks with the ignition ON and the engine OFF at normal operating temperature. If the voltage is not as specified in the test specifications chart, first check all wiring, connections and components before assuming that the control unit is defective. The Mazda System Checker 83 (40 G040 920) is necessary to troubleshoot the electronic control system with the OEM test connector located next to the ECU harness connectors at the control unit. With the tester the on-board diagnosis system will read out trouble codes (1 through 6) to indicate problems in different circuits within the fuel injection system. Follow the manufacturer's instructions included with the tester for all wiring and sensor checks using the special tester.

CAUTION

Do not attempt to disconnect or reconnect the control unit main harness connector with the ignition switch ON, or the ECU can be damaged or destroyed.

MAZDA ECU TEST SPECIFICATIONS

Terminal	Connection to	Voltage with ignition ON (when functioning properly)
A	Main relay	approx. 12V
B	Ground	0V
C	Water thermo sensor	1 ~ 2V (warm engine)
D	Ground	0V
E	Air flow meter	4 ~ 6V . . . at 20°C 1.5 ~ 3.5V . . . at 50°C
F	Injector (#20)	approx. 12V
G	Throttle sensor & Atmospheric pressure sensor	4.5 ~ 5.5V
H	Injector (#10)	approx. 12V
I	Throttle sensor	approx. 1V
J	Vacuum switch	approx. 12V
L	Variable resistor (V/R)	0 ~ 12V (Varies according to the V/R adjustment)
M	Ignition switch "START" terminal	below 1.5V
N	O_2 sensor	0V
O	Air flow meter	approx. 12V
P	Atmospheric pressure sensor	approx. 4V
Q	Air flow meter	approx. 2V
R	Air flow meter	approx. 7.5V
S	Ground	0V
T	Ground	0V
U	Ignition coil (T) - terminal	approx. 12V
V	Main relay	approx. 12V
a	Switching solenoid valve	approx. 12V
b	Relief solenoid valve control unit	approx. 12V
c	Checking connector	0V
d	Vacuum control solenoid valve (T/L)	approx. 12V
e	Pressure regulator control valve	below 1.5V
f	Checking connector	0V
h	Vent solenoid valve	below 1.5V (throttle sensor is adjusted properly)
i	Clutch switch	below 1.5V . . . pedal released approx. 12V . . . pedal depressed
j	Neutral switch	below 1.5V . . . in neutral approx. 12V . . . in gear
k	Water temperature switch	below 1.5V . . . above 15°C
l	Intake air temperature sensor	8.5 ~ 10.5V . . . at 20°C 5 ~ 7V . . . at 50°C
m	Air-con. switch	below 1.5V . . . air-con. switch OFF
n	Vacuum control valve	approx. 12V (throttle sensor is adjusted properly)

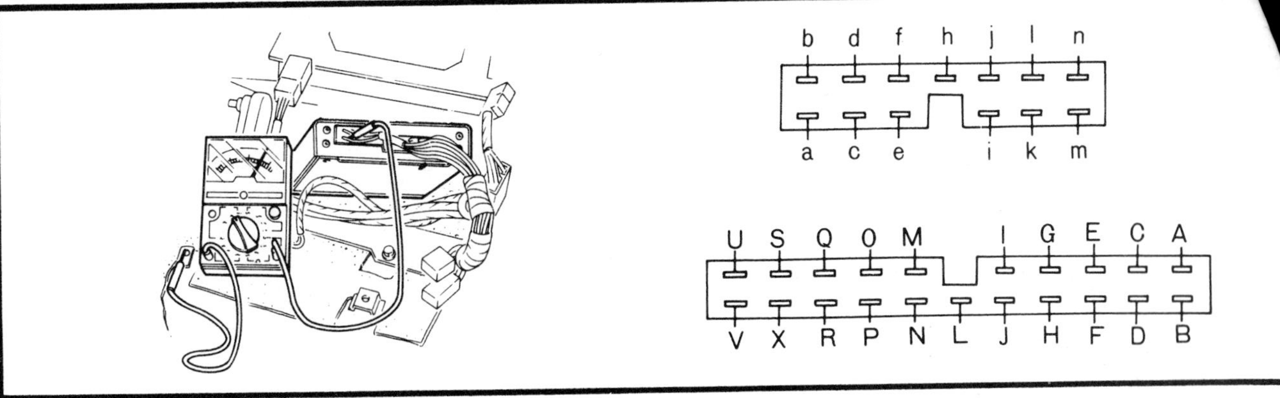

Component Removal and Testing

FUEL PUMP

Removal and Installation

CAUTION

Fuel system is under pressure. Wrap a clean cloth around the fuel connection when loosening to catch any fuel spray and take precautions to avoid the risk of fire.

1. Remove the storage compartment located behind the driver's seat and disconnect the fuel pump connector.
2. Raise the vehicle and support it safely.
3. Remove the pump bracket clamp bolt.
4. Disconnect the inlet and outlet hoses from the fuel pump and plug the lines.
5. Remove the fuel pump from the bracket.
6. Installation is the reverse of removal.

FUEL PRESSURE REGULATOR

Removal and Installation

1. Remove the dynamic chamber as detailed later in this section.
2. Disconnect the vacuum hose.
3. Disconnect the fuel return hose after covering the starter to prevent fuel from splashing on it. Wrap a clean cloth around all fuel connections when disconnecting to catch any fuel spray from the pressurized system.
4. Remove the pressure regulator mounting nut and remove the pressure regulator.
5. Installation is the reverse of removal. Check for fuel leaks with fuel pressure applied before installing the dynamic chamber.

FUEL INJECTOR

Removal and Installation

1. Remove the dynamic chamber as detailed later in this section.
2. Remove the fuel feed and return lines from the fuel manifold and damper. Wrap a clean cloth around the fuel connections to catch any fuel spray from the pressurized system.
3. Remove the vacuum line from the fuel pressure regulator.
4. Remove the fuel injector harness connector.
5. Remove the fuel manifold by attaching bolts and remove the fuel manifold with the damper and pressure regulator attached.
6. Remove the injector(s) from the intermediate casing, making sure the insulator comes out with the injector.

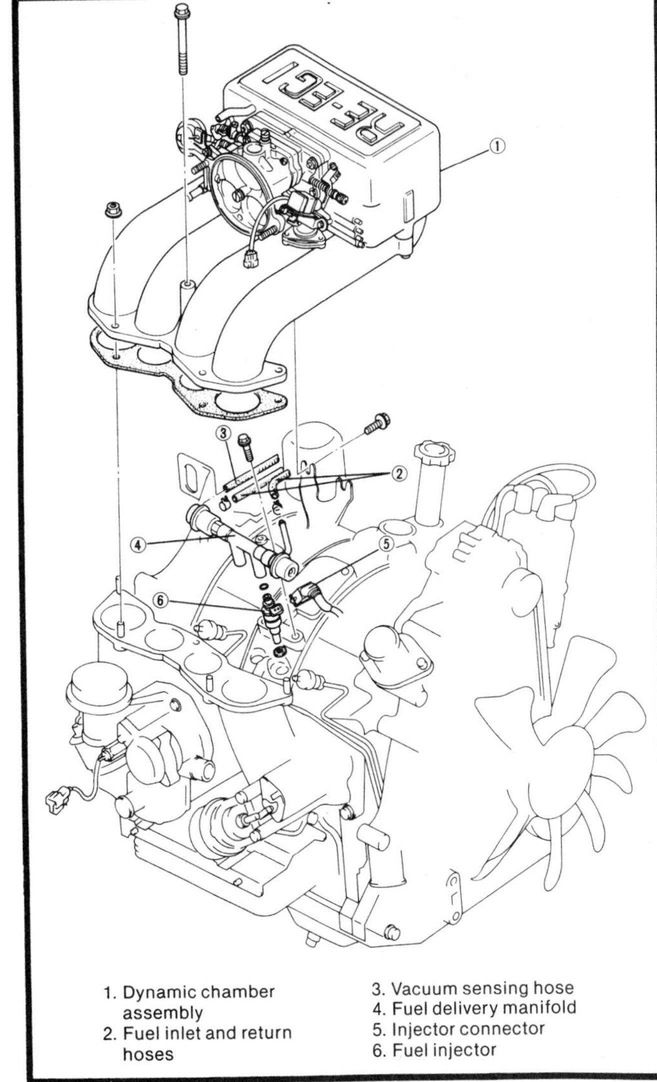

Exploded view of injector mounting and dynamic chamber assembly

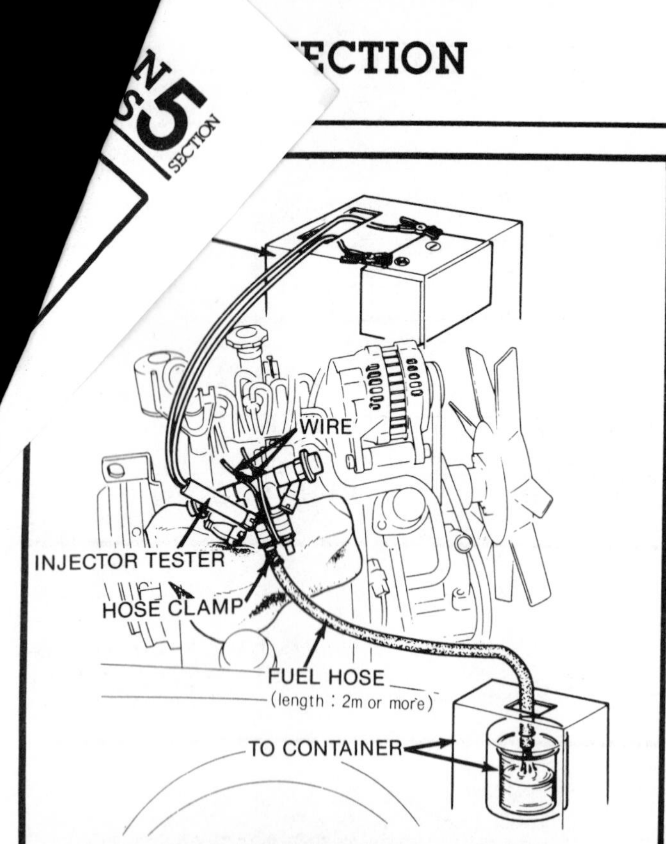

Injector fuel leak test. Wire the injectors to the fuel rail as shown to prevent fuel pressure from blowing off hose connections.

7. Installation is the reverse of removal. Replace the O-ring and insulator. Lubricate the O-ring with clean gasoline when installing on the injector tip. Check for fuel leaks with fuel pressure applied before installing dynamic chamber.

Injector Winding Test

Measure the resistance of the injector with an ohmmeter across the injector terminals. It should be 1.5–3 ohms. If no resistance is measured or if the resistance exceeds 3 ohms, replace the fuel injector.

Injector Fuel Leak Test

Install the fuel injectors onto the fuel manifold and connect all fuel lines, vacuum lines and harness connectors. Use mechanic's wire to secure the injectors to the fuel manifold so that no movement is possible. If the injectors are not securely attached to the fuel manifold, they may blow off when fuel pressure is applied. Install a jumper wire on the fuel pump short circuit terminal and turn the ignition ON to pressurize the fuel system. Check for leaks from the fuel injector nozzles. After 5 minutes, a very slight amount of fuel leakage is acceptable. If the injectors drip continuously or more than a very slight amount of fuel leaks, replace the fuel injectors and retest.

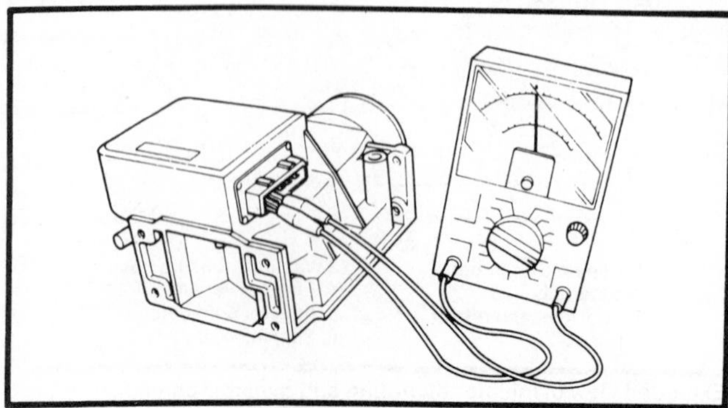

Testing the air flow meter with an ohmmeter

NOTE: Injectors may be replaced individually if only one is found to be defective. Use a clean cloth to catch any fuel during testing.

Injector Volume Test

With the injectors still attached to the fuel manifold as described above, connect Injector Tester 49 9200 040, or equivalent, to the injector to be tested. Install a suitable fuel hose from the end of the test injector to a graduated container. Short the fuel pump test terminal and apply 12 volts to the injector tester. The fuel injector is good if 149–197 cc of fuel is delivered within 15 seconds.

CAUTION

Do not apply 12 volts directly to the injector or the windings will be destroyed.

AIR FLOW METER

Removal and Installation

Remove the air cleaner element, then loosen the air flow meter attaching bolts and remove the air cleaner. Loosen the air funnel (duct) band and remove the air flow meter assembly. Installation is the reverse of removal.

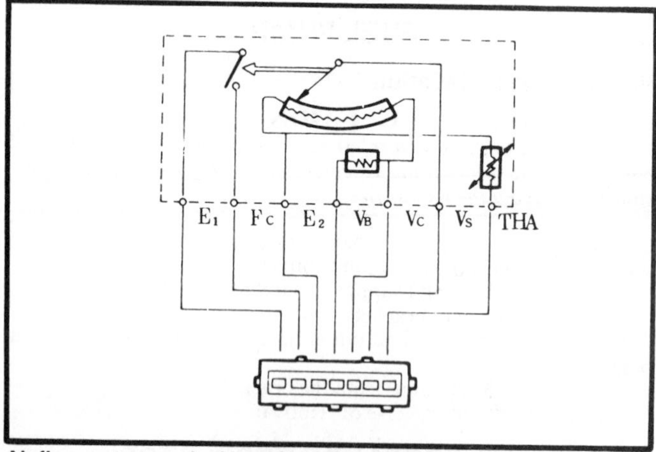

Air flow meter terminal locations. See the specifications chart for test connections and resistance values

Testing

Check the air flow meter body for cracks and check the resistance of each terminal using an ohmmeter. Compare the resistance measurements with the specification chart and replace the air flow meter if there is no reading or the reading is out of range. Press on the measuring plate and measure the resistance between the terminals as detailed in the chart to check vane meter operation. Replace the air flow meter if no change is noted when the sensor plate is moved.

AIR FLOW METER SPECIFICATIONS

Terminal	Resistance (ohms)
E_2-V_s	20–400
E_2-V_c	100–300
E_2-V_b	200–400
E_2-THA (Intake air temperature sensor)	−4 F—10k-20k 32 F—4k-7k 68 F—2k-3k 104 F—900-1300 140 F—400-700

AIR FLOW METER SPECIFICATIONS

Terminal	Resistance (ohms)
E_1-F_c	Infinity (plate closed) Zero (plate open)
E_2-V_s	20-400 (plate closed) 20-1000 (plate open)

k—one thousand

THROTTLE CHAMBER

Removal and Installation

1. Remove the air funnel (duct) from the throttle chamber.
2. Disconnect the accelerator cable.
3. Disconnect the throttle sensor harness connector.
4. Disconnect the metering oil pump connecting rod.
5. Disconnect the water hoses.
6. Loosen and remove the mounting bolts and remove the throttle chamber.
7. Installation is the reverse of removal.

DYNAMIC CHAMBER

Removal and Installation

1. Disconnect the air funnel (duct) to the throttle chamber.
2. Disconnect the accelerator cable.
3. Disconnect the throttle sensor harness connector.
4. Disconnect the metering oil pump connecting rod.
5. Disconnect the water hoses.
6. Disconnect the negative battery cable.
7. Remove the terminal cover.
8. Disconnect the vacuum sensing lines.
9. Disconnect the air supply valve harness connector.
10. Disconnect the intake air temperature sensor connector.
11. Loosen the mounting bolts holding the dynamic chamber to the intake manifold ports and lift off the dynamic chamber.
12. Installation is the reverse of removal. Use a new gasket when installing the dynamic chamber.

NOTE: Cover the intake manifold ports with a clean cloth when the dynamic chamber is removed to prevent debris from entering the engine during service and testing procedures.

MITSUBISHI ELECTRONICALLY CONTROLLED INJECTION (ECI) SYSTEM

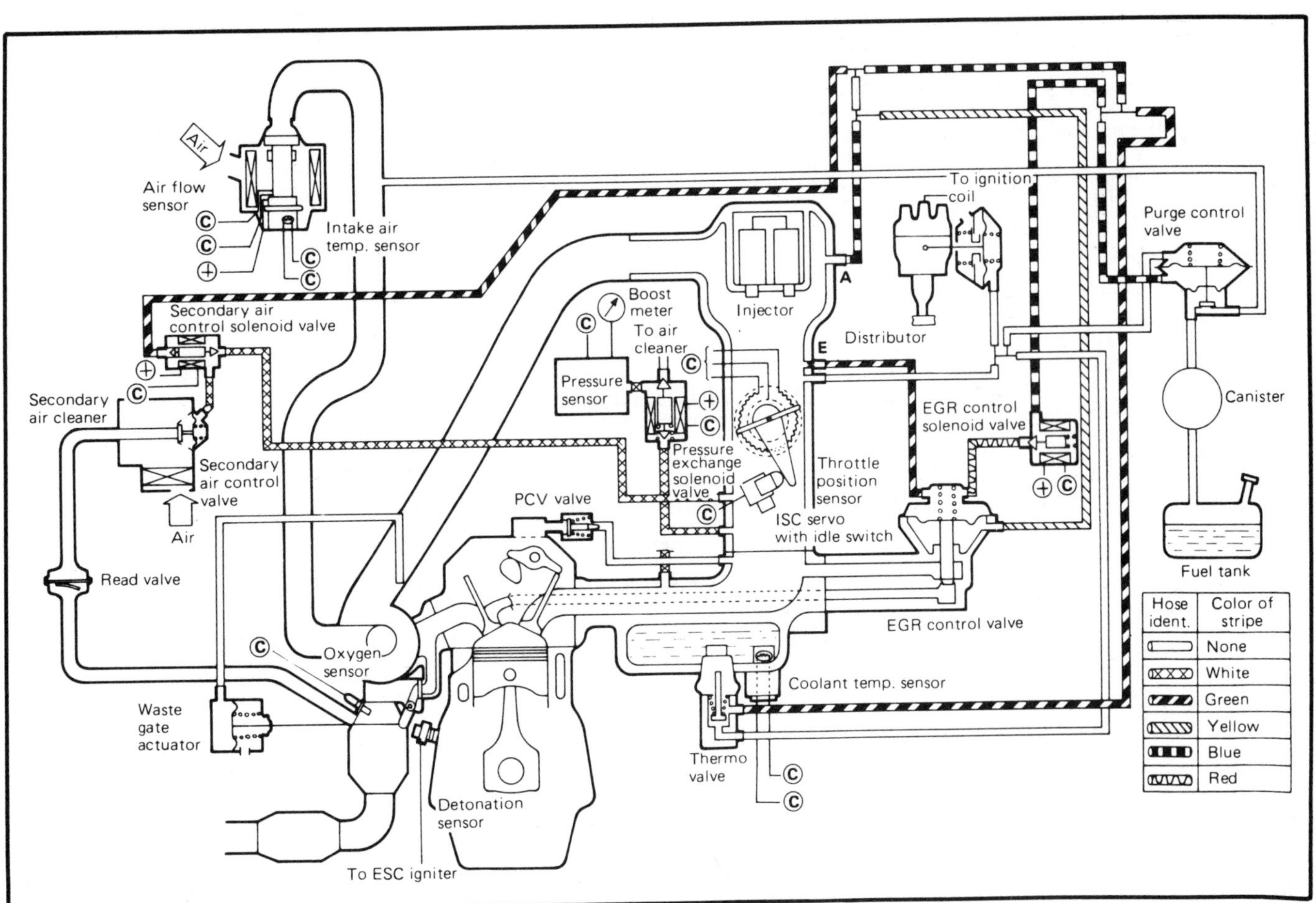

Mitsubishi throttle body system schematic

...CTION

...nformation

...ly Controlled Injection (ECI) system is ...ottle body system with some variations in ...nents. The fuel control system consists of an ...nit (ECU), two solenoid-type fuel injectors, an ...d several engine sensors. The ECU receives volt... ...n the engine sensors on operating conditions, then ...pulses to the injectors to constantly adjust the fuel ...n addition, the ECU controls starting enrichment, warm... ...chment, fast idle, deceleration fuel cut-off and overboost ...ut-off on turbocharged models.

...ne of the primary components is the air flow sensor with its ...evice for generating Karman vortexes. Ultrasonic waves are transmitted across the air flow containing the Karman vortexes, which are generated in proportion to the air flow rate. The greater the number of vortexes, the more the frequency of the ultrasonic waves are changed (modulated). These modulated ultrasonic waves are picked up by the receiver in the air flow sensor and converted into a voltage signal for the ECU. The ECU uses this signal to measure air flow and control fuel delivery and secondary air management. An intake air temperature sensor is used to provide a signal so that air density changes due to temperature can be calculated.

Other components in the system are common to all throttle body systems. During closed loop operation, the ECU monitors the oxygen sensor to determine the correct fuel mixture according to the oxygen content of the exhaust gases. In the open loop mode, fuel mixture is preprogrammed into the control unit memory.

When the ECI system is activated by the ignition switch, the fuel pump is energized by the ECU. The pump will only operate for about one second unless the engine is running or the starter is cranking. When the engine starts, the fuel pump relay switches to continuous operation and all engine sensors are activated and begin providing input for the ECU. The ISC motor will control idle speed (including fast idle) if the throttle position switch is in the idle position, and the ignition advance shifts from base timing to the preprogrammed ignition advance curve. The fuel pressure regulator maintains system pressure at approximately 14.5 psi (1 bar) by returning excess fuel provided by the fuel pump to the tank.

The ECU provides a ground for the injectors to precisely control the open and closing time (pulse width) to deliver exact amounts of fuel to the engine, continuously adjusting the air/fuel mixture while monitoring signals from the various engine sensors including:

- Engine coolant temperature
- Intake manifold air temperature and volume
- Barometric pressure
- Intake manifold absolute pressure
- Engine speed (rpm)
- Idle speed
- Detonation
- Boost pressure (turbo models only)
- Throttle position
- Exhaust gas content (oxygen level)

On-Car Service

SELF-DIAGNOSIS SYSTEM

The Mitsubishi self-diagnosis system monitors the various input signals from the engine sensors and enters a trouble code in the onboard computer memory if a problem is detected. There are nine monitored items, including the "normal operation" code which can be read by using a special ECI tester and adapter. The adapter connects the ECI tester to the diagnosis connector located on the right cowl, next to the control unit. Because the computer memory draws its power directly from the battery, the trouble codes are not erased when the ignition is switched OFF. The memory can only, be cleared (trouble codes erased) if a battery cable is disconnected or the main ECU wiring harness connector is disconnected from the computer module.

NOTE: The trouble codes will not be erased if the battery cable or harness connector is reconnected within 10 seconds.

If two or more trouble codes are stored in the memory, the computer will read out the codes in order beginning with the lowest number. The needle of the ECI tester will sing back and forth between 0 and 12 volts to indicate the trouble code stored. There is no memory for code No. 1 (oxygen sensor) once the ignition is switched OFF, so it is necessary to perform this diagnosis with the

MITSUBISHI ECI TROUBLE CODES

Trouble Code	ECU Circuit	Indicated Problem
1	Oxygen sensor	Open circuit in wire harness, faulty oxygen sensor or connector
2	Ignition signal	Open or shorted wire harness, faulty igniter
3	Air flow sensor	Open or shorted wire harness, loose connector, defective air flow sensor
4	Boost pressure sensor	Defective boost sensor, open or shorted wire harness or connector
5	Throttle position sensor	Sensor contacts shorted, open or shorted wire harness or connector
6	ISC motor position sensor	Defective throttle sensor open or shorted wire harness or connector defective ISC servo
7	Coolant temperature sensor	Defective sensor, open or shorted wire harness or connector
8	Speed sensor	Malfunction in speed sensor circuit, open or shorted wire harness or connector

ISC—Idle Speed Control

engine running. The oxygen sensor should be allowed to warm up for testing (engine at normal operating temperature) and the trouble code should be read before the ignition is switched OFF. All other codes will be read out with the engine ON or OFF. If there are no trouble codes stored in the computer (system is operating normally), the ECI tester will indicate a constant 12 volts on the meter. Consult the instructions supplied with the test equipment to insure proper connections for diagnosis and testing of all components.

If there is a problem stored, the meter needle will swing back and forth every 0.4 seconds. Trouble codes are read by counting the pulses, with a two second pause between different codes. If the battery voltage is low, the self-diagnosis system will not operate properly, so the battery condition and state of charge should be checked before attempting any self-diagnosis inspection procedures. After completing service procedures, the computer trouble code memory should be erased by disconnecting the battery cable or main harness connector to the control unit for at least 10 seconds.

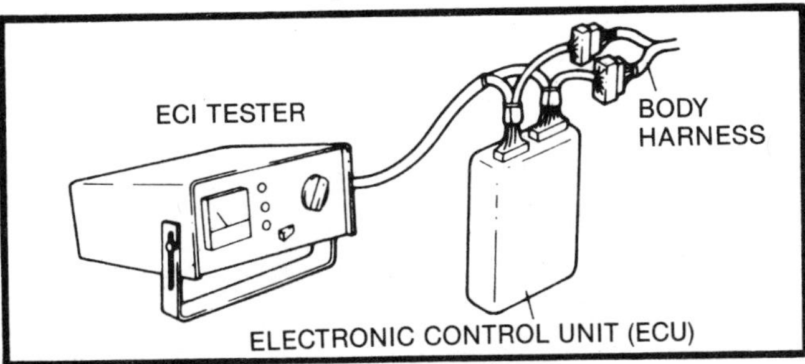

Connecting ECI tester to control unit harness

NOTE: Installation of CB or other two-way radio equipment may affect the operation of the electronic control unit. Antennas and other radio equipment should be installed as far away from the ECU as possible.

IDLE SPEED CONTROL (ISC) SERVO AND THROTTLE POSITION SENSOR

Adjustment

If the ISC Servo, throttle position sensor, throttle body or injectors have been removed or replaced, the following adjustments should be made. These adjustments are important to driveability.

1. Start the engine and allow it to reach operating temperature.
2. Stop the engine.
3. Loosen the throttle position sensor mounting screws the throttle position sensor clockwise as far as it will go, t porarily tighten the screws.
4. Set the ISC Servo position by turning the ignition ON seconds, then turn the ignition OFF.
5. Disconnect the ISC servo harness connector.
6. Start the engine and check the idle speed. Adjust to specifi tions as listed on the underhood emission control sticker, if nec sary, by turning the adjusting screw clockwise to increase rpm counterclockwise to decrease rpm.

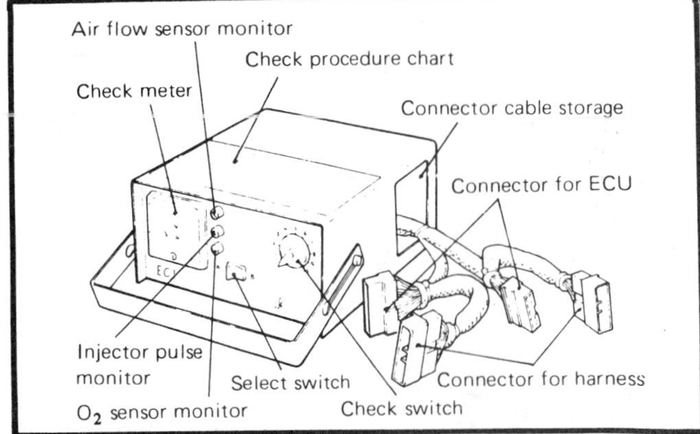

Mitsubishi fuel injection tester

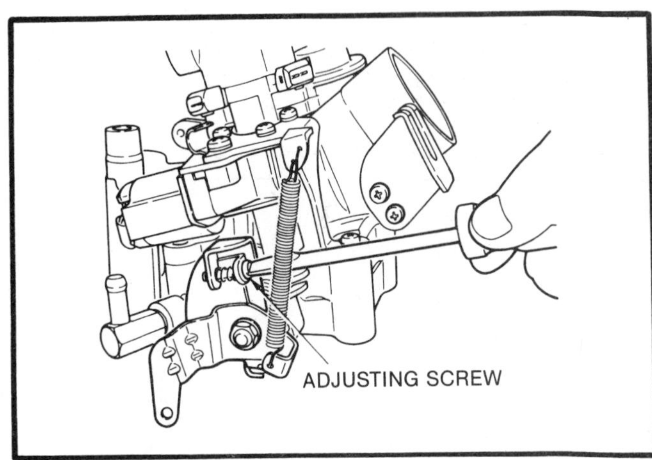

Idle speed adjustment

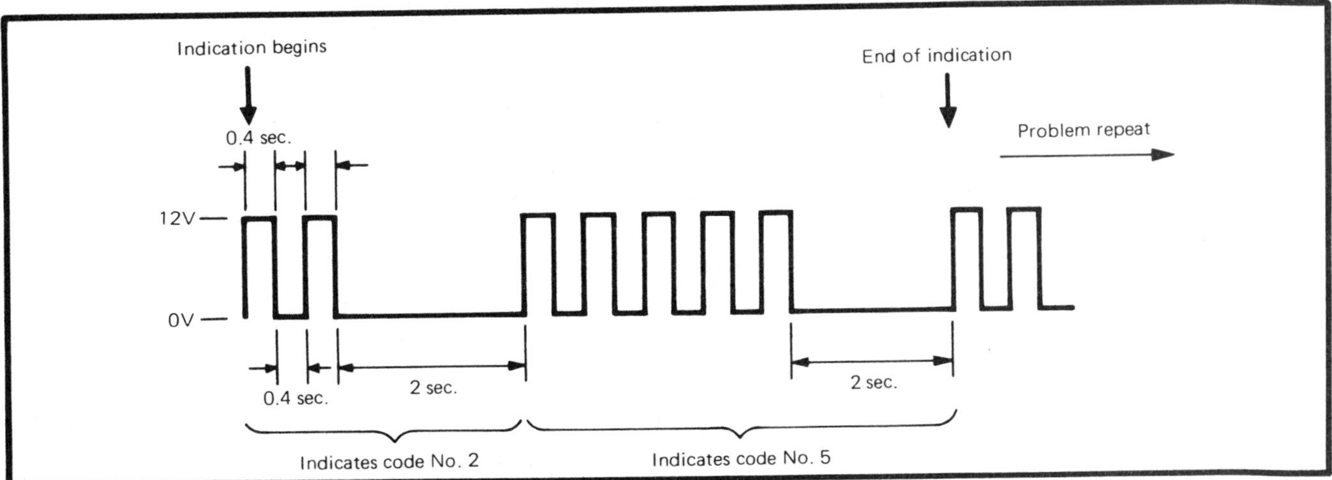

Reading Mitsubishi trouble codes. Note the time between pulses

ECTION

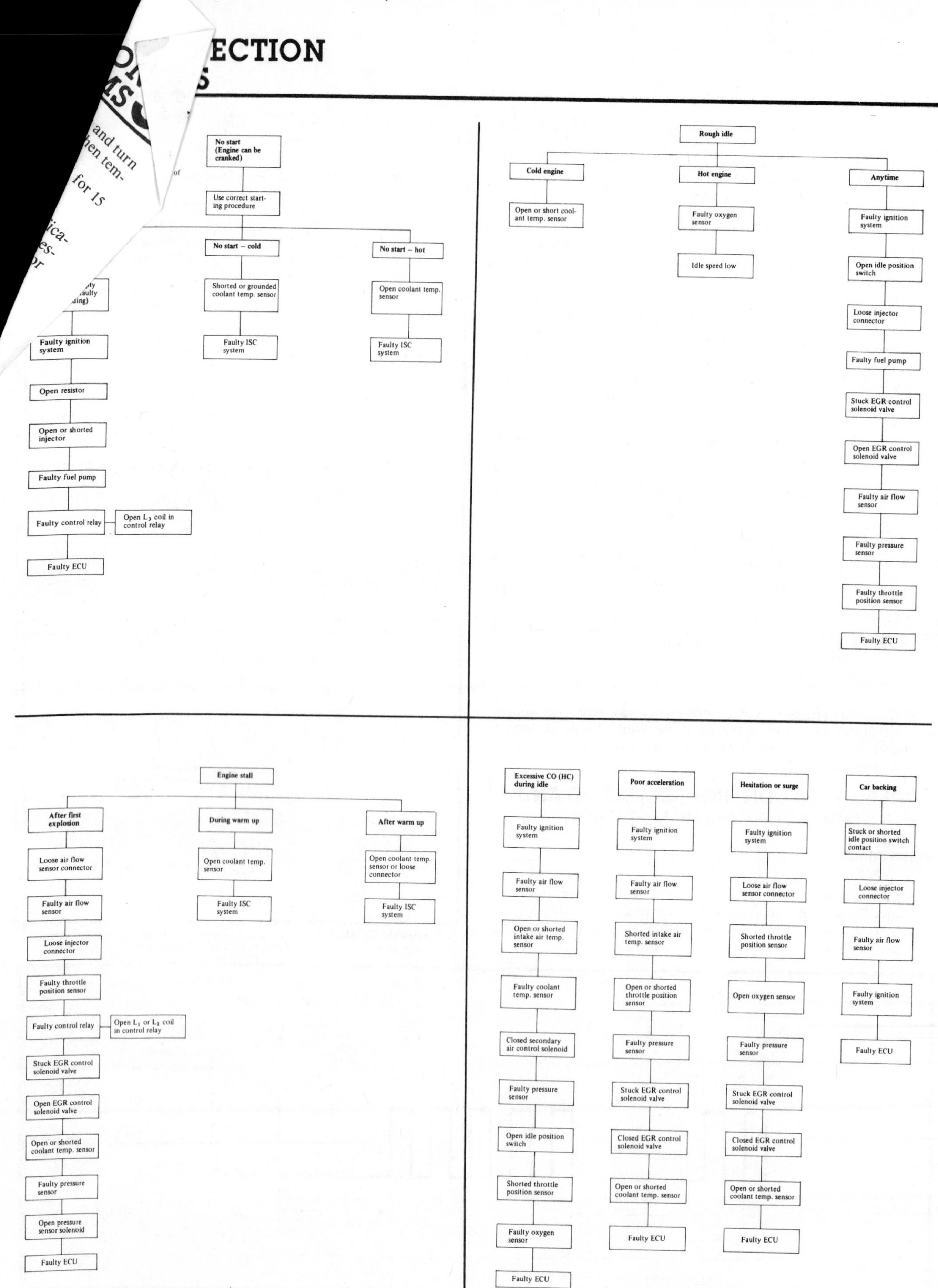

No start (Engine can be cranked)
Use correct starting procedure
No start – cold
No start – hot
Shorted or grounded coolant temp. sensor
Faulty ISC system
Open coolant temp. sensor
Faulty ISC system
Faulty ignition system
Open resistor
Open or shorted injector
Faulty fuel pump
Faulty control relay
Open L_3 coil in control relay
Faulty ECU
Rough idle
Cold engine
Hot engine
Anytime
Open or short coolant temp. sensor
Faulty oxygen sensor
Idle speed low
Faulty ignition system
Open idle position switch
Loose injector connector
Faulty fuel pump
Stuck EGR control solenoid valve
Open EGR control solenoid valve
Faulty air flow sensor
Faulty pressure sensor
Faulty throttle position sensor
Faulty ECU
Engine stall
After first explosion
During warm up
After warm up
Loose air flow sensor connector
Faulty air flow sensor
Loose injector connector
Faulty throttle position sensor
Faulty control relay
Open L_1 or L_2 coil in control relay
Stuck EGR control solenoid valve
Open EGR control solenoid valve
Open or shorted coolant temp. sensor
Faulty pressure sensor
Open pressure sensor solenoid
Faulty ECU
Open coolant temp. sensor
Faulty ISC system
Open coolant temp. sensor or loose connector
Faulty ISC system
Excessive CO (HC) during idle
Faulty ignition system
Faulty air flow sensor
Open or shorted intake air temp. sensor
Faulty coolant temp. sensor
Closed secondary air control solenoid
Faulty pressure sensor
Open idle position switch
Shorted throttle position sensor
Faulty oxygen sensor
Faulty ECU
Poor acceleration
Faulty ignition system
Faulty air flow sensor
Shorted intake air temp. sensor
Open or shorted throttle position sensor
Faulty pressure sensor
Stuck EGR control solenoid valve
Closed EGR control solenoid valve
Open or shorted coolant temp. sensor
Faulty ECU
Hesitation or surge
Faulty ignition system
Loose air flow sensor connector
Shorted throttle position sensor
Open oxygen sensor
Faulty pressure sensor
Stuck EGR control solenoid valve
Closed EGR control solenoid valve
Open or shorted coolant temp. sensor
Faulty ECU
Car backing
Stuck or shorted idle position switch contact
Loose injector connector
Faulty air flow sensor
Faulty ignition system
Faulty ECU

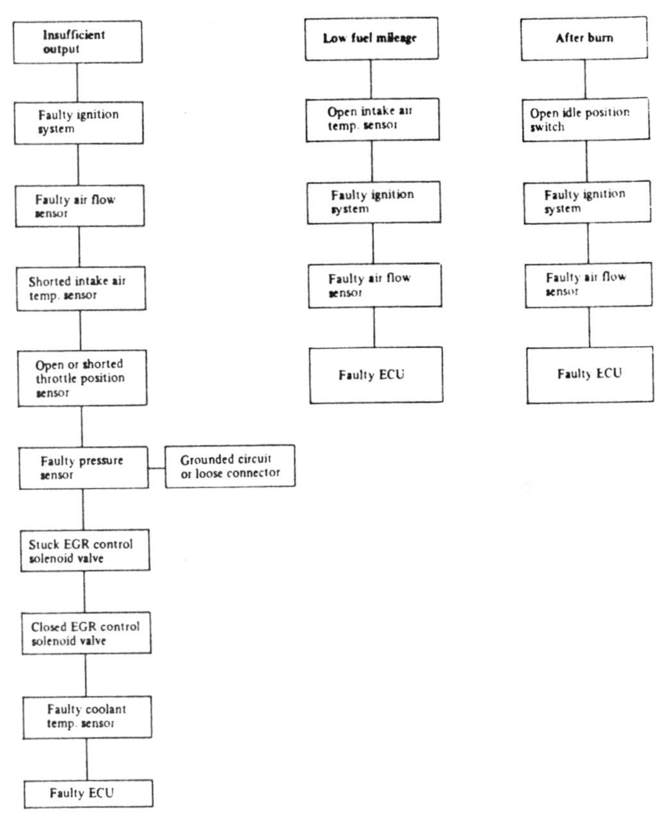

7. Stop the engine and disconnect the throttle position sensor harness connector.
8. Connect adapter and digital voltmeter between the throttle position sensor connector.
9. Turn the ignition switch ON but do not start the engine.
10. Read the throttle position sensor output voltage on the digital voltmeter. If the voltage is not 0.45–0.51 volts, loosen the throttle position sensor mounting screws and turn the sensor until the reading is within the specified range, then tighten the mounting screws.
11. Open the throttle fully and confirm that the output voltage is correct when the throttle valve is returned to the idle position.
12. Remove the digital voltmeter connections and recheck idle speed.

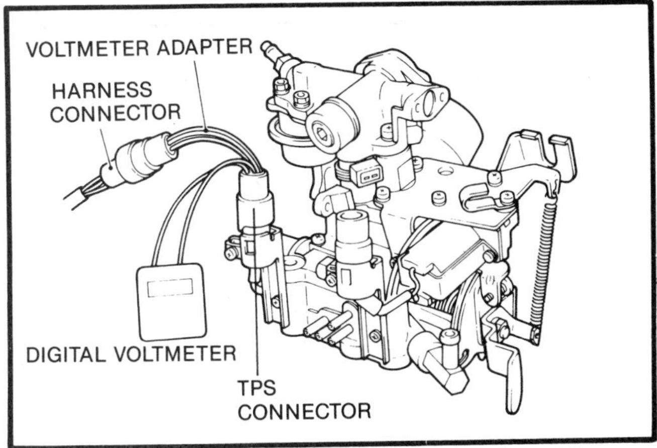

Using adapter to connect digital voltmeter to throttle position sensor

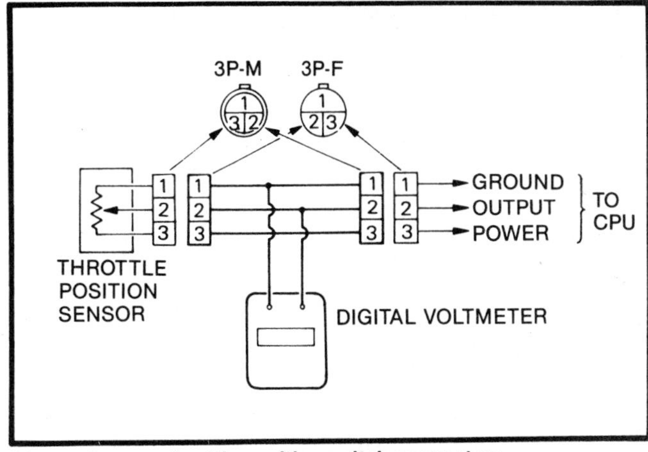

Pin numbers on throttle position switch connectors

Component Removal and Testing

AIR FLOW SENSOR

Testing

Check the air cleaner element for contamination and damage and replace if necessary. Check the air cleaner case and cover for cracks and damage that could allow air leaks in the system. Check the

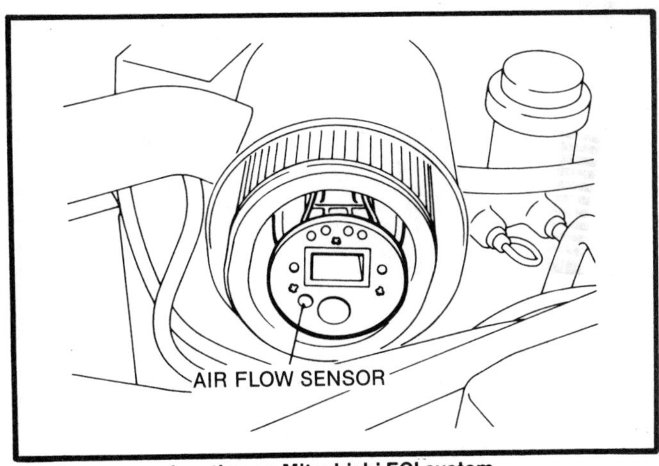

Air flow sensor location on Mitsubishi ECI system

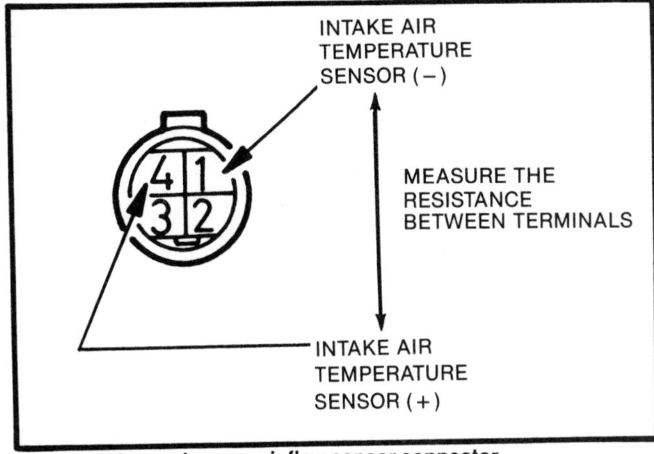

Terminal pin numbers on air flow sensor connector

resistance of the air intake temperature sensor between terminals 1 and 4 of the connector. Resistance should be approximately 2.5 ohms at 68 degrees F (20 degrees C).

Removal and Installation

1. Disconnect the air flow sensor connector. The connector must be disconnected before removing the air cleaner cover and air cleaner body.
2. Unsnap the finger clip and remove the air cleaner cover.
3. Remove the filter element from the air cleaner body.
4. Remove the air flow sensor.
5. Installation is the reverse of removal. Make sure the connector is correctly fastened.

FUEL INJECTORS AND THROTTLE BODY

Removal and Installation

1. Disconnect the battery ground cable.
2. Drain the engine coolant down to the intake manifold level or below.
3. Disconnect the air intake hose from the throttle body assembly.

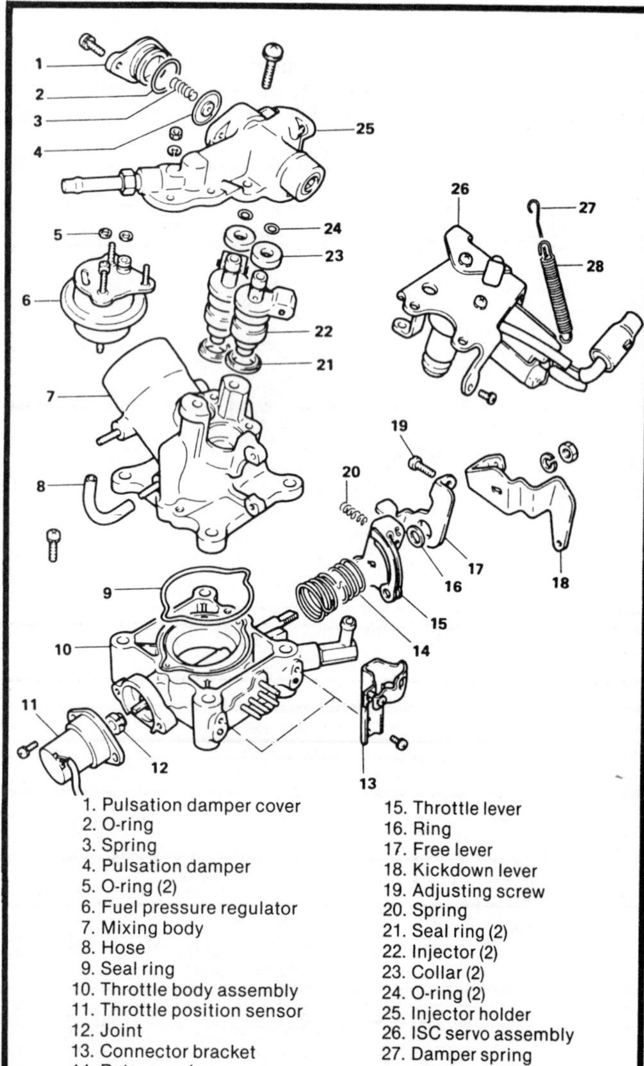

Exploded view of throttle body assembly for Mitsubishi ECI system

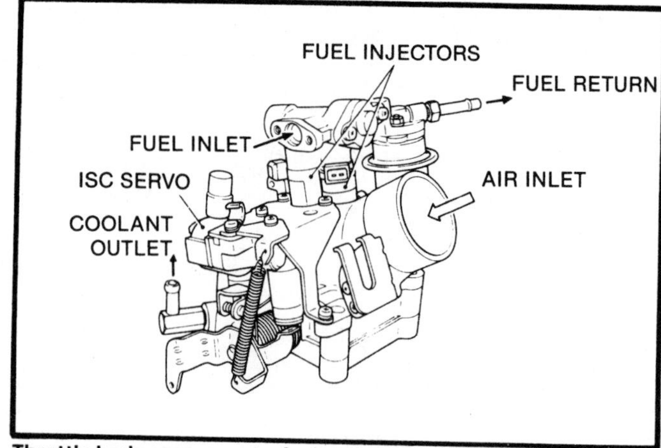

Throttle body components, fuel and coolant connections

4. Disconnect the throttle cable from the throttle lever.
5. Disconnect the fuel inlet pipe and fuel return hose from the assembly.
6. Disconnect the harness connectors from the injectors.
7. Disconnect the ISC servo and throttle position sensor connectors.
8. Disconnect the vacuum hose from the throttle body.
9. Remove the four mounting bolts and remove the throttle body and injectors as an assembly.
10. Installation is the reverse of removal. Adjust the ISC servo and throttle position sensor as previously described. Check for fuel leaks.

Overhaul

1. Clamp the throttle body assembly in a soft-jawed vise, being careful not to damage the housing.
2. Disconnect the rubber hose from the fuel pressure regulator and throttle body.
3. Remove the injector retainer tightening screws and remove the retainer.
4. Remove the pressure regulator from the retainer.
5. Remove the pulsation damper cover from the retainer and take out the spring and diaphragm.
6. Pull the injectors from the throttle body. Do not use pliers to remove the injectors. Remove the gaskets from the throttle body.
7. Remove the throttle return spring and the damper spring.
8. Remove the connector bracket.
9. Remove the ISC servo mounting bracket with the servo attached. Do not remove the ISC servo from the bracket unless it is being replaced.

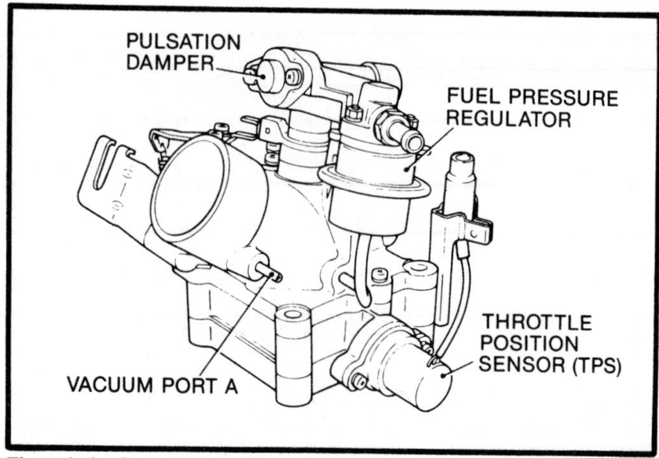

Throttle body component locations

10. Remove the two screws, then remove the mixing body and seal ring from the throttle body.

11. Remove the throttle position sensor only if it is being replaced.

CAUTION

Do not immerse parts in cleaning solvent. Immersing the ISC servo, throttle position sensor and injector will damage the insulation. Wipe all parts clean with a dry cloth. Clean vacuum and fuel ports with compressed air.

12. Reassemble throttle body and injectors in reverse order of removal. Adjust the throttle position sensor, if removed, and make sure the resistance value changes when the throttle lever is moved. Install new seal rings in the throttle body and mixing body and replace O-rings on injectors. Press the injectors into the throttle body cavities firmly using finger pressure only. Install the pulsation damper diaphragm as illustrated and replace the O-rings on the fuel pressure regulator. Check the filters in the retainer for clogging or damage and replace if necessary. Tighten the retainer screws alternately, but do not overtighten.

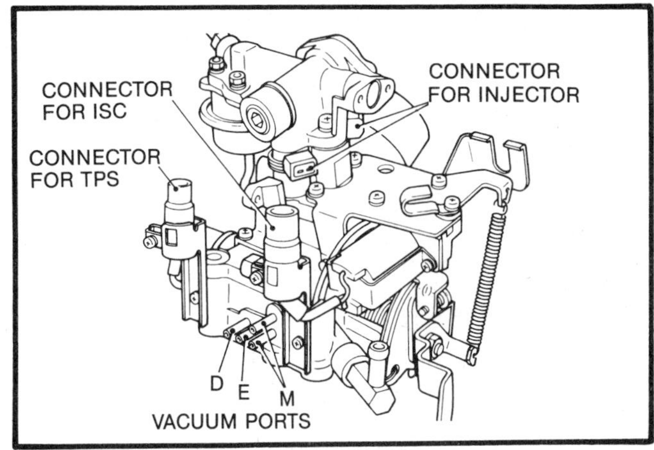

Connector and vacuum port locations

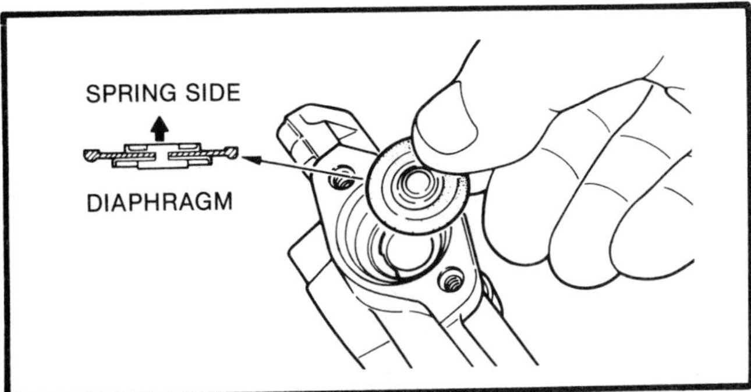

Installing pulsation damper diaphragm

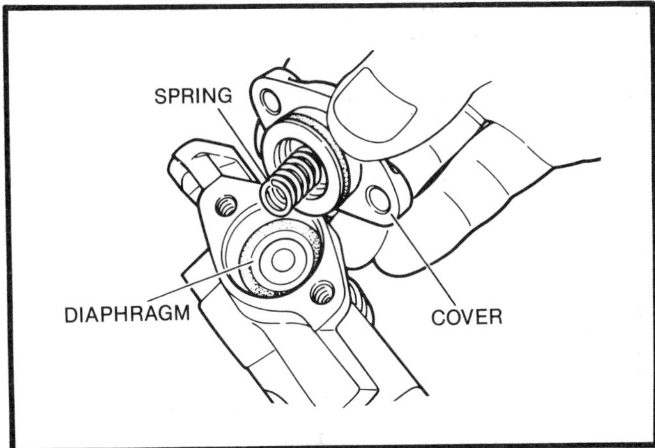

Removing pulsation damper cover

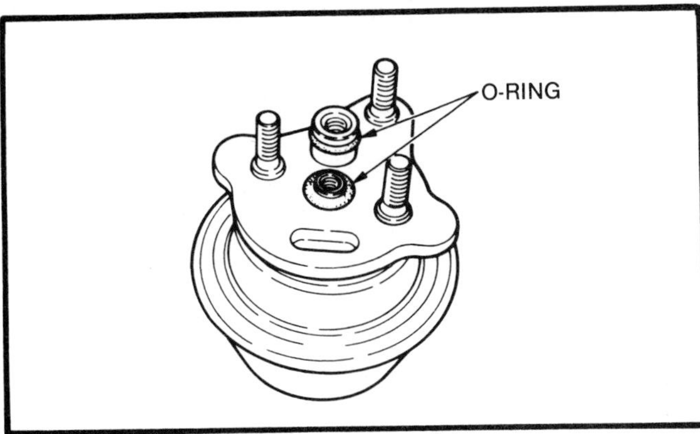

Location of O-rings on fuel pressure regulator

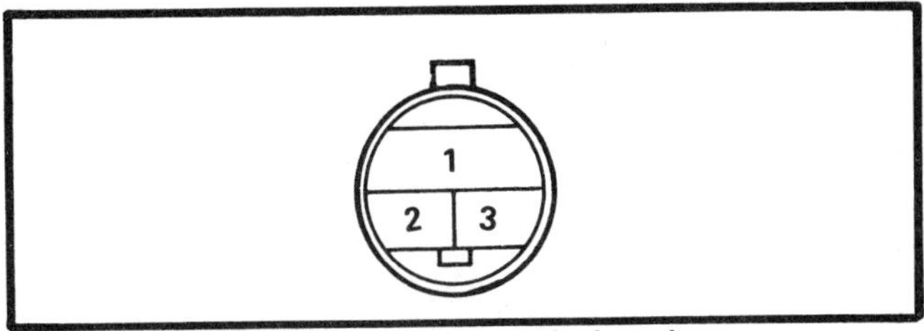

Throttle position sensor connector terminal numbers

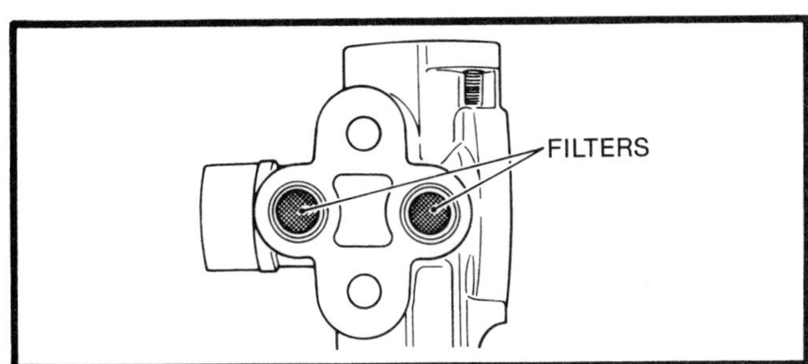

Location of filters in injector retainer

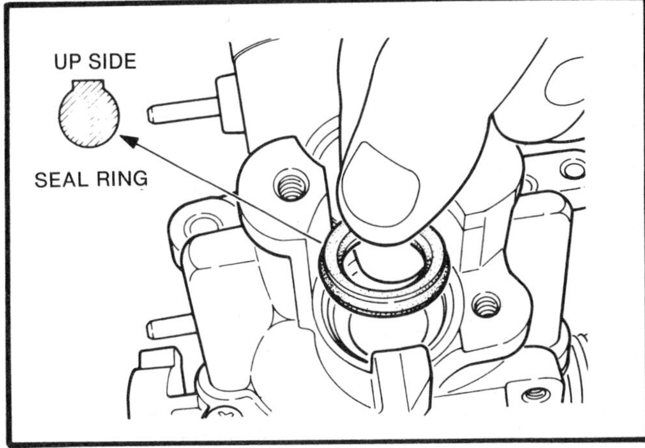

Install sealing ring with the flat side up

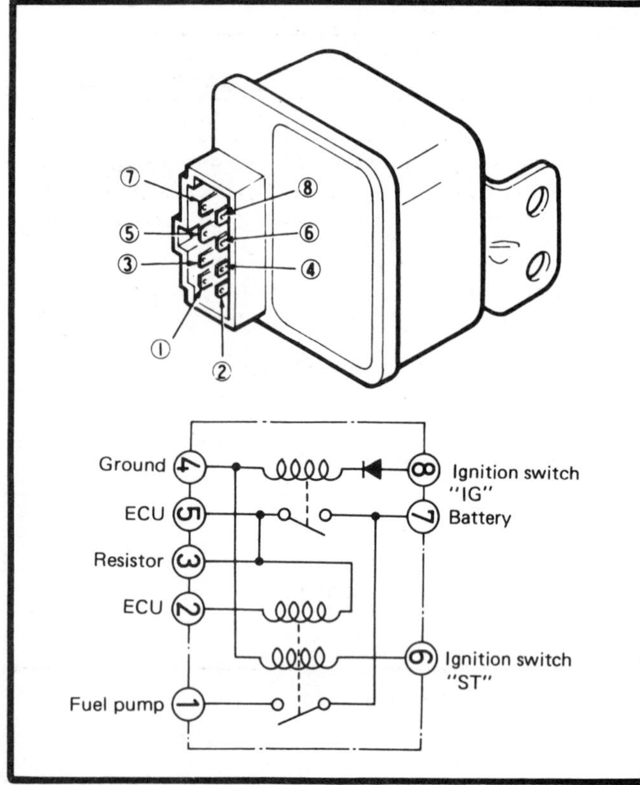

Control relay on Mitsubishi ECI system showing pin numbers

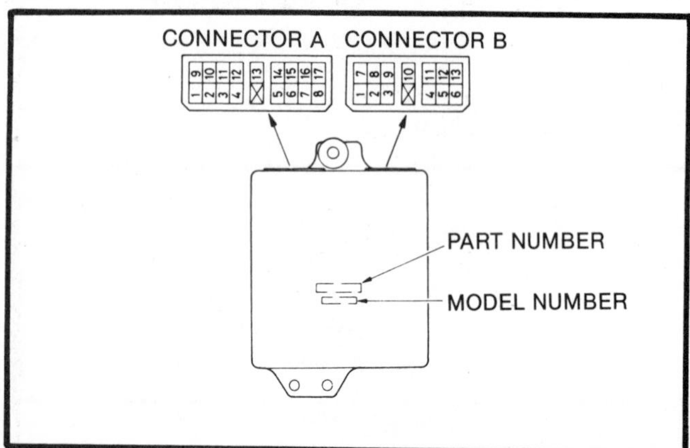

ECU used on Mitsubishi ECI system showing harness connector pin numbers

ECI RESISTOR

Testing

The resistor is mounted on the left side of the firewall. To test, disconnect the resistor harness connector and measure the resistance across terminals 1 and 2, then across terminals 1 and 3. Resistance should be about 6 ohms. If the resistance measures zero or infinity, replace the resistor. If resistance is as specified, make sure the connector terminals are tight and free from corrosion.

COOLANT TEMPERATURE SENSOR

Testing

Remove the coolant temperature sensor from the intake manifold and immerse the sensor probe in water. Do not immerse the electrical connections, just the sensor itself. Heat the water and check

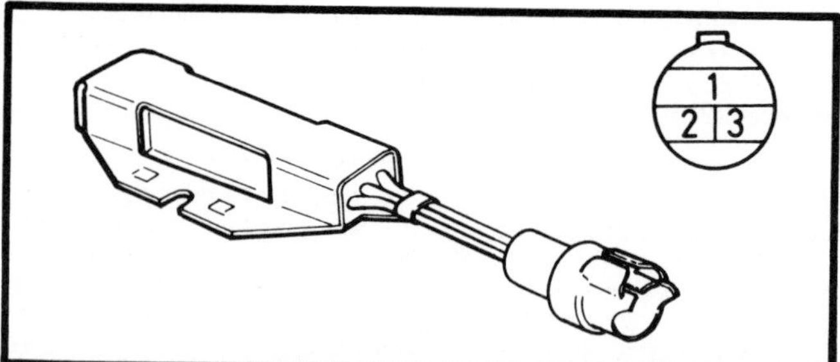

Resistor used on Mitsubishi ECI system showing terminal numbers

that the resistance changes from approximately 16 kilo-ohms at −4 degrees F (−20 degrees C) to 296 ohms at 176 degrees F (80 degrees C). At room temperature, the sensor should read about 2.45 kilo-ohms. If the resistance is as described, apply sealant to the threaded portion of the sensor and install into the intake manifold. If not, replace the sensor. Make sure all harness connections are fastened securely. Torque the sensor to 21 ft. lbs. (30 Nm).

ECI CONTROL RELAY

Testing

The control relay is mounted on top of the ECU, under the air duct of the instrument panel. Test for continuity between terminals 1 and 7, then between terminals 3 and 7. If there is no continuity, the relay is good; if continuity exists, replace the relay. Apply 12 volts across terminals 8 and 4 while testing continuity between terminals 3 and 7; if continuity exists, the relay is good. Apply 12 volts across terminals 6 and 4 while testing continuity between terminals 1 and 7; again, if continuity exists, the relay is good.

MITSUBISHI ECI TEST SPECIFICATIONS

Circuit Tested	ECU Terminal Location	Normal Reading
Power supply	B-1	Key OFF-0 volts Key ON- 11–13 volts
Secondary air control valve	A-10	Reading should change from 0.2–15 volts within 30 seconds at warm idle
Throttle position switch	A-1	Key ON- 0.4–1.5 volts (4.5–5.0 volts @ WOT)
Coolant temperature sensor	A-3	3.5 volts @ 32 deg. F 0.5 volts @ 176 deg. F
Air temperature sensor	A-4	3.5 volts @ 32 deg. F 0.6 volts @ 176 deg. F
Idle position switch	A-5	Key ON- 0–0.4 volts @ idle 11–13 volts @ WOT
ISC motor position	A-14	Key ON- 11–13 volts ①
EGR control valve solenoid	B-4	0–0.5 volts above 3000 rpm 13–15 volts @ idle ②
Speed sensor	A-15	0.2–5 volts with transmission in gear and slowly accelerating
Cranking signal	A-13	Over 8 volts
Control relay	B-5	0–1 volts @ idle

MITSUBISHI ECI TEST SPECIFICATIONS

Circuit Tested	ECU Terminal Location	Normal Reading
Ignition pulse signal	A-8	12–15 volts @ idle 11–13 volts @ 3000 rpm
Air flow sensor signal	A-7	2.7–3.2 volts between idle and 3000 rpm
Injector No. 1	B-9	13–15 volts @ idle 12–13 volts @ 3000 rpm
Injector No. 2	B-10	13–15 volts @ idle 12–13 volts @ 3000 rpm
Oxygen sensor	A-6	0–2.7 volts ③
Pressure sensor	A-17	1.5–2.6 volts

NOTE: Turn ignition OFF between tests when making connections. All testing done with ignition ON unless noted otherwise. Make sure test equipment is compatible with injection system before making any voltage checks.

WOT = Wide Open Throttle

① If ignition switch is turned on for 15 seconds or more, the reading drops to 1 volt or less momentarily, then returns to 6–13 volts

② Engine warmed up to operating temperature

③ Do not use a power voltmeter to test oxygen sensor. The sensor can be damaged by the voltage draw

RENAULT ELECTRONIC FUEL INJECTION

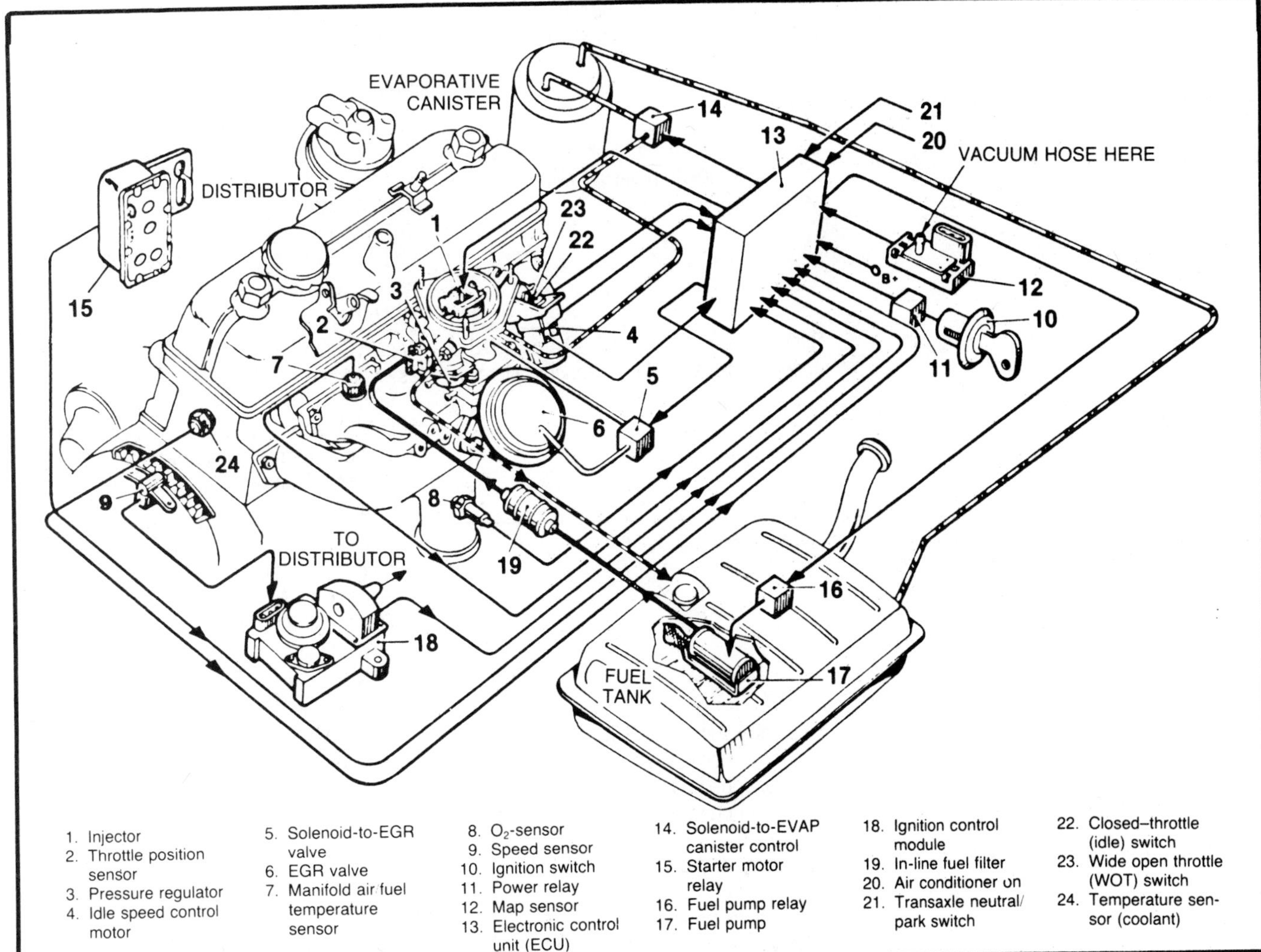

Throttle body fuel injection system used on Renault Alliance

General Information

The Throttle Body Fuel Injection (TBI) system is a "pulse time" system that injects fuel into the throttle body above the throttle blade. Fuel is metered to the engine by one or more electronically controlled fuel injector(s). The Electronic Control Unit (ECU) controls injection according to input provided from sensors that detect exhaust gas oxygen content, coolant temperature, manifold absolute pressure, crankshaft position and throttle position. The sensors provide an electronic signal usually modulated by varying resistance within the sensor itself.

TBI fuel injection has two main sub-systems. The fuel system consists of an electric fuel pump (in tank), a fuel filter, a pressure regulator and a fuel injector. The control system consists of a manifold air/fuel mixture temperature (MAT) sensor, a coolant temperature sensor (CTS), a manifold absolute pressure (MAP) sensor, a wide open throttle (WOT) switch, a closed throttle (idle) switch, an exhaust oxygen (O_2) sensor, an electronic control unit (ECU), a gear position indicator (automatic transmission models only), a throttle position sensor and an idle speed control (ISC) motor. There may be more than one fuel injector used.

TBI TROUBLE DIAGNOSIS

Condition or Trouble Code	Possible Cause	Correction
CODE 1 (poor low air temp. engine performance).	Manifold air/fuel temperature (MAT) sensor resistance is not less than 1000 ohms (HOT) or more than 100 kohms (VERY COLD).	Replace MAT sensor if not within specifications. Refer to MAT sensor test procedure.
CODE 2 (poor warm temp. engine performance-engine lacks power).	Coolant temperature sensor resistance is less than 300 ohms or more than 300 kohms (10 kohms at room temp.).	Replace coolant temperature sensor. Test MAT sensor. Refer to coolant temp. sensor test and MAT sensor test procedures.
CODE 3 (poor fuel economy, hard cold engine starting, stalling, and rough idle).	Defective wide open throttle (WOT) switch or closed (idle) throttle switch or both, and/or associated wire harness.	Test WOT switch operation and associated circuit. Refer to WOT switch test procedure. Test closed throttle switch operation and associated circuit. Refer to closed throttle switch test procedure.
CODE 4 (poor engine acceleration, sluggish performance, poor fuel economy).	Simultaneous closed throttle switch and manifold absolute pressure (MAP) sensor failure.	Test closed throttle switch and repair/replace as necessary. Refer to closed throttle switch test procedure. Test MAP sensor and associated hoses and wire harness. Repair or replace as necessary. Refer to MAP sensor test procedure.
CODE 5 (poor acceleration, sluggish performance).	Simultaneous WOT switch and manifold absolute pressure (MAP) sensor failure.	Test WOT switch and repair or replace as necessary. Refer to WOT switch test procedure. Test MAP sensor and associated hoses and wire harness. Repair or replace as necessary. Refer to MAP sensor test procedure.
CODE 6 (poor fuel economy, bad driveability, poor idle, black smoke from tailpipe).	Inoperative oxygen sensor.	Test oxygen sensor operation and replace if necessary. Test the fuel system for correct pressure. Test the EGR solenoid control. Test canister purge. Test secondary ignition circuit. Test PCV circuit. Refer to individual component test procedure.
No test bulb flash.	No battery voltage at ECU (J1-A with key on). No ground at ECU (J1-F). Simultaneous WOT and CTS switch contact (Ground at both D2 Pin 6 and D2 Pin 13). No battery voltage at test bulb (D2 Pin 4). Defective test bulb. Battery voltage low (less than 11.5V).	Repair or replace wire harness, connectors or relays. Repair or replace WOT switch, CTS switch, harness or connectors. Repair wire harness or connector. Replace test bulb. Charge or replace battery, repair vehicle wire harness.

On-Car Service

PRELIMINARY CHECKS

The Throttle Body Fuel Injection (TBI) System should be considered as a possible source of trouble for engine performance, fuel economy and exhaust emission complaints only after normal tests and inspections of other engine components have been performed. An integral self-diagnostic system within the ECU detects common malfunctions that are most likely to occur.

On Renault models the self-diagnostic system will illuminate a test bulb if a malfunction exists. When the trouble code terminal (at the diagnostic connector in the engine compartment) is connected to a test bulb the system will flash a trouble code if a malfunction has been detected.

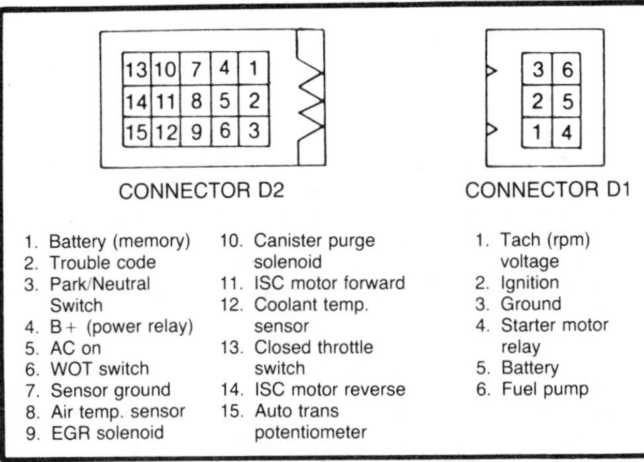

Diagnostic connector terminal identification. Make sure the correct terminals are connected during testing

Before performing any test, make sure that the malfunction is not caused by a component other than the fuel injection system (e.g., spark plugs, distributor, advance timing, etc.); that no air is entering the intake and exhaust system above the catalytic converter and, finally, that fuel is actually reaching the injector (test the pressure in the circuit).

TBI SERVICE PRECAUTIONS

1. Never connect or disconnect any electrical component without turning off the ignition switch.
2. With the engine stopped and the ignition on, the fuel pump should not be operative. The fuel pump is controlled by the electronic control unit (ECU).
3. Disconnect the battery cables before charging.
4. If the temperature is likely to exceed 80°C (176°F), as in a paint shop bake oven, remove the electronic control unit (ECU).
5. Misconnections should be carefully avoided because even momentary contact may cause serious ECU or system damage.

1983–84 Renault TBI

CAUTION

Be extremely careful when making any test connections. Never apply more than 12 volts to any point or component in the TBI system.

DIAGNOSIS PROCEDURE

The self-diagnostic feature of the electronic control unit (ECU) provides support for diagnosing system problems by recording six possible failures should they be encountered during normal engine operation. Additional tests should allow specific tracing of a failure to a single-component source. Multiple disassociated failures must be diagnosed separately. It is possible that the test procedures can cause false interpretations of certain irregularities and consider them as ECU failures.

NOTE: In the following procedures, no specialized service equipment is necessary. It is necessary, however, to have available a volt/ohmmeter, a 12V test lamp, and an assortment of jumper wires and probes.

Trouble Code Test Lamp

Poor fuel economy, erratic idle speed, power surging and excessive engine stalling are typical symptoms when the fuel system has a component failure. If the ECU is functional, service diagnostic codes can be obtained by connecting a No. 158 test bulb (or equivalent) to pins D2-2 and D2-4 of the large diagnostic connector. With the test bulb installed, push the WOT switch lever on the throttle body and with the ISC motor plunger (closed throttle switch) also closed, have a helper turn on the ignition switch while observing the test bulb.

NOTE: If the ECU if functioning normally, the test bulb should light for a moment then go out. This will always occur regardless of the failure condition, and serves as an indication that the ECU is functional.

After the initial illumination the ECU will cycle through and flash a single digit code if any system malfunctions have been detected by the ECU during engine operation.

The ECU is capable of storing various trouble codes in memory. The initial trouble detected will be flashed first and then, followed by a short pause, the second trouble detected will be flashed. There will be a somewhat longer pause between the second code and the repeat cycle of the first code again. This provides distinction between codes 3 - 6 and 6 - 3. While both codes indicate the same two failures, the first digit indicates the initial trouble while the last digit indicates the most recent failure.

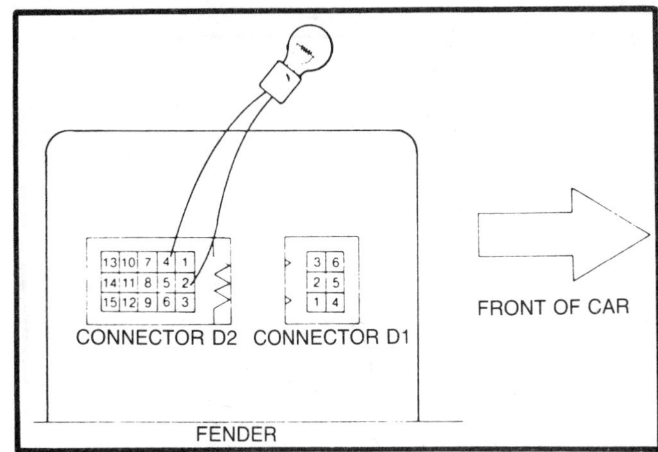

Trouble code test lamp connections

RENAULT TBI TROUBLE CODES

In a given situation where further testing indicates no apparent cause for the failure indicated by the ECU self-diagnostic system, an intermittent failure, except that marginal components must be more closely examined. If the trouble code is erased and quickly returns with no other symptoms, the ECU should be suspected. However, in the absence of other negative symptoms, replacement of the ECU can be suspected. Again, because of the cost involved and the relative reliability of the ECU, the ECU should never be replaced without testing with dealer test equipment. If the ECU is determined to be defective, it must be replaced. There are no dealer serviceable parts in the ECU. No repairs should be attempted.

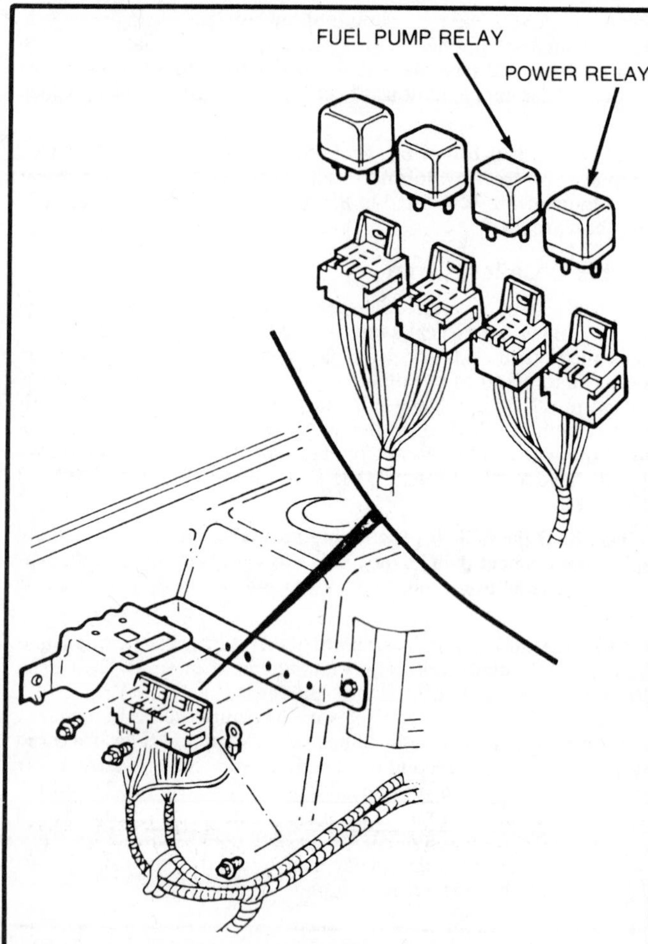

Relay block connector and bracket

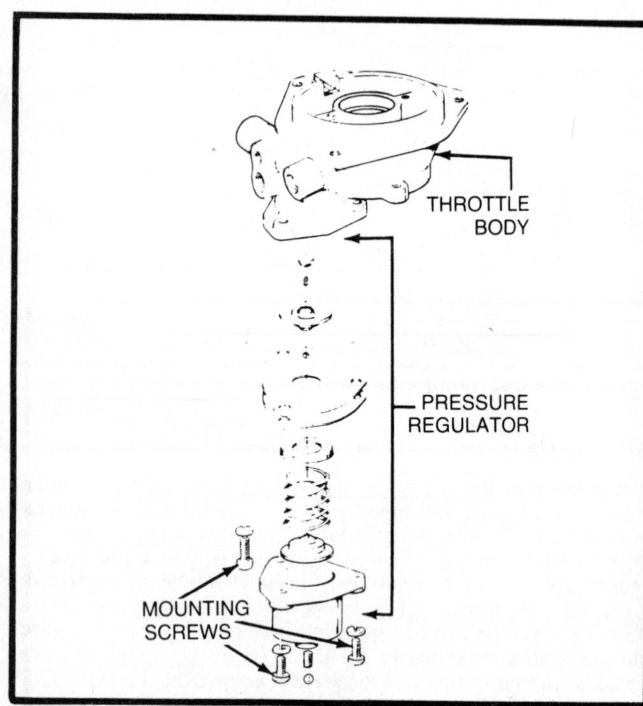

Exploded view of pressure regulator assembly

It is important to note that the trouble memory is erased if the ECU power is interrupted by disconnecting either battery cable terminal, or allowing the engine to remain unstarted in excess of five days.

It is equally important to erase the trouble memory when a defective component is replaced. This can be done by any one of the actions listed above.

Component Removal and Testing

The fuel pump is a roller type with a permanent magnet electric motor that is immersed in fuel. The check valve is designed to maintain the fuel pressure after the engine has stopped. The pump has two electric wire terminals marked + and – to ensure that it rotates in the correct direction.

The ECU controls the fuel pump, which is usually located inside the fuel tank and is attached to the fuel gauge sending unit assembly. Fuel pump power is supplied through two relays, located in the engine compartment.

PRESSURE REGULATOR

The fuel pressure regulator (overflow-type) is integral with the throttle body. The valve is a diaphragm-operated relief valve in which one side of the valve is exposed to fuel pressure and the other side is exposed to air horn pressure. Nominal pressure is established by a calibrated spring.

The pressure regulator has a spring chamber that is vented to the same pressure as the tip of the injector. Because the differential pressure between the injector nozzle and the spring chamber is the same, the volume of fuel injected is dependent only on the length of time the injector is energized. The pump delivers fuel in excess of the maximum required by the engine and the excess fuel flows back to the fuel tank from the pressure regulator via the fuel return hose.

Adjustment

Adjustment of the fuel pressure regulator is necessary only to establish the correct pressure after a replacement has been installed.

1. Remove the air filter.
2. Connect a tachometer to the diagnostic connector terminals D1-1 and D1-3.
3. Connect an accurate fuel pressure gauge to the fuel body pressure test fitting. Exercise caution when removing fuel lines.
4. Start the engine and accelerate to approximately 2000 rpm.
5. Turn the allen head adjustment screw on the bottom of the regulator to obtain 1 bar (14.5 psi) of pressure.

NOTE: Turning the screw inward increases the pressure and turning the screw outward decreases the pressure.

6. Install a lead seal ball to cover the regulator adjustment screw after adjusting the pressure to specification.
7. Turn the ignition off, then disconnect the tachometer.
8. Disconnect the fuel pressure gauge and install the cap on the test fitting. Install the air filter.

Removal and Installation

1. Remove the three retaining screws that secure the pressure regulator to the throttle body.
2. Remove the pressure regulator assembly. Note the location of the components for assembly reference.
3. Discard the gasket.
4. Position the pressure regulator assembly with a replacement gasket.
5. Install the three retaining screws to secure the pressure regulator to the throttle body.
6. Start the engine and inspect for leaks.

FUEL INJECTOR

The injector body contains a solenoid with a plunger or core piece that is pulled upward by the solenoid armature allowing the spring loaded ball valve to come off the valve seat. This allows fuel to pass through to the atomizer/spray nozzle. During engine operation, the fuel injector is energized by the ECU and serves to meter and direct atomized fuel into the throttle bore above the throttle blade. During engine start-up the injector is energized to deliver a predetermined volume of fuel to aid starting.

NOTE: More than one injector may be used.

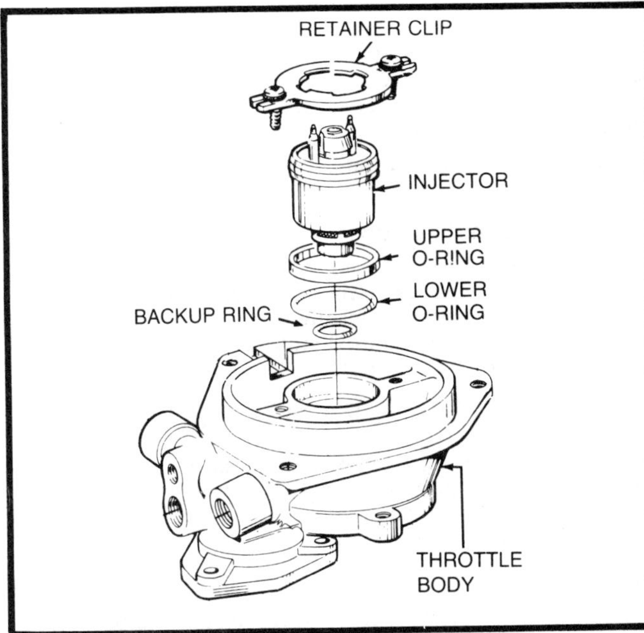

TBI fuel injector installation

Removal and Installation

1. Remove the air filter.
2. Remove the injector wire connector.
3. Remove the injector retainer clip screws.
4. Remove the injector retainer clip.
5. Using a small pair of pliers, gently grasp the center collar of the injector (between electrical terminals) and carefully remove the injector with a lifting/twisting motion.
6. Discard the upper and lower O-rings. Note that the back-up ring fits over the upper O-ring.
7. To install the injector, proceed as follows: Lubricate with light oil and install a replacement lower O-ring in the housing bore.
8. Lubricate with light oil and install a replacement upper O-ring into housing bore.
9. Install the back-up ring over the upper O-ring.
10. Position the replacement injector in the throttle body and center the nozzle in the lower housing bore. Seat the injector using a pushing/twisting motion.
11. Align the wire terminals.
12. Install the retainer clip and screws.
13. Install the injector wire connector.
14. Install the air filter.

THROTTLE BODY

Removal

1. Remove the throttle cable and return spring.
2. Disconnect the wire harness connector from the injector.

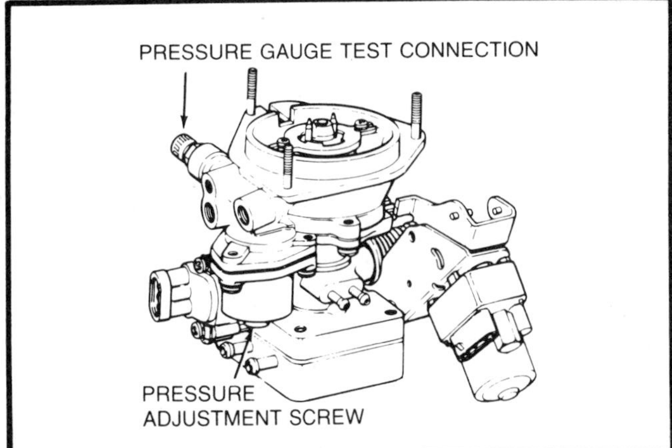

Throttle body assembly. Note testing and adjustment points for fuel pressure

3. Disconnect the wire harness connector from the wide open throttle (WOT) switch.
4. Disconnect the wire harness connector from the ISC motor.
5. Disconnect the fuel supply pipe from the throttle body.
6. Disconnect the fuel return pipe from the throttle body.
7. Disconnect the vacuum hoses from the throttle body assembly. Identify and tag the hoses for installation reference.
8. Remove the throttle body-to-manifold retaining nuts from the studs.
9. Remove the throttle body assembly from the intake manifold.

NOTE: If the throttle body assembly is being replaced, transfer the following components to the replacement throttle body; idle speed control (ISC) motor/WOT switch and bracket assembly.

Installation

1. Install the replacement throttle body assembly on the intake manifold. Use a replacement gasket between the components.
2. Install the throttle body-to-manifold retaining nuts on the studs.
3. Connect the vacuum hoses.
4. Connect the fuel return pipe to the throttle body.
5. Connect the fuel supply pipe to the throttle body.
6. Connect the wire harness connector to the injector.
7. Connect the wire harness connector to the WOT switch.
8. Connect the wire harness connector to the ISC motor.
9. Install the throttle cable and return spring.

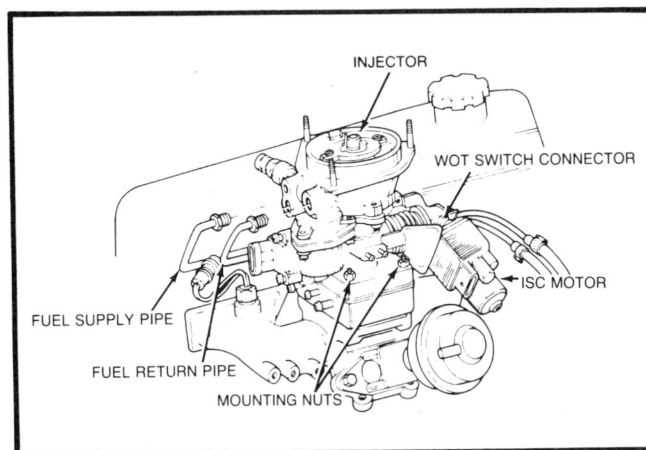

Removing the throttle body assembly

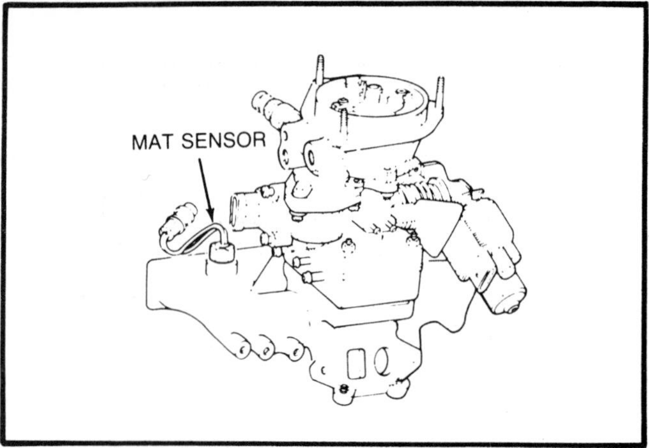

Location of the manifold air/fuel temperature (MAT) sensor

FUEL BODY ASSEMBLY

Removal and Installation

1. Remove the throttle body assembly from the intake manifold.
2. Remove the three Torx® screws that retain the fuel body to the throttle body.
3. Remove the original gasket and discard it.
4. Install the replacement fuel body on the throttle body using a replacement gasket.
5. Install the three fuel body-to-throttle body retaining Torx® head screws and tighten securely.
6. Install the throttle body assembly on the intake manifold.

MANIFOLD AIR/FUEL TEMPERATURE (MAT) SENSOR

The manifold air/fuel temperature (MAT) sensor is installed in the intake manifold in front of an intake port. This sensor reacts to the temperature of the air/fuel mixture (charge) in the intake manifold and provides an analog voltage for the ECU.

Testing

1. Disconnect the wire harness connector from the MAT sensor.
2. Test the resistance of the sensor with a high input impedance (digital) ohmmeter. Resistance ranges from 300 ohms to 300K ohms (10K ohms at room temperature).
3. Replace the sensor if it is not within the range of resistance specified above.

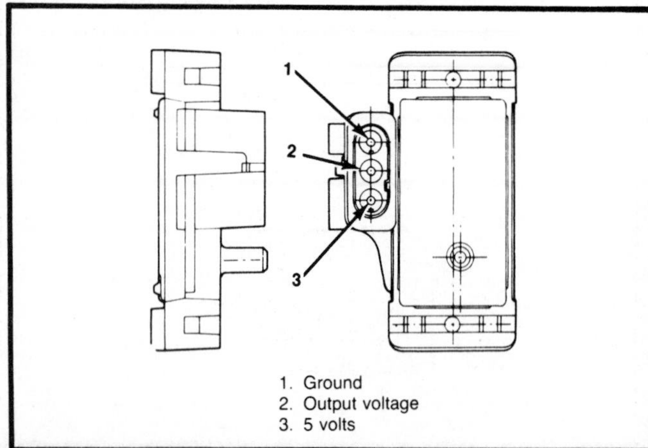

Manifold absolute pressure (MAP) sensor

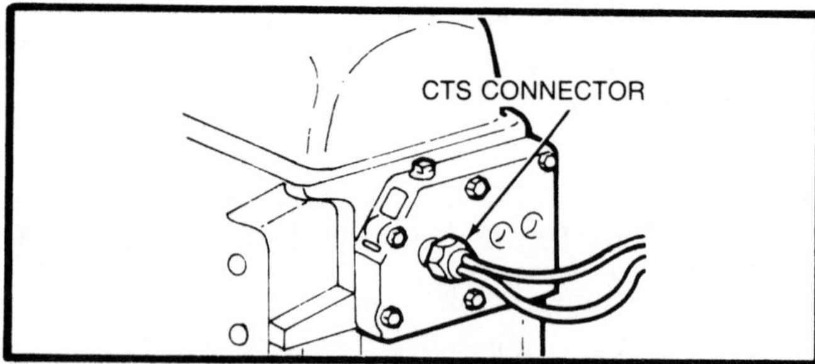

Coolant temperature sensor (CTS) installation

4. Test the resistance of the wire harness between the ECU harness connector J2-13 and the sensor connector, and J2-11 to sensor connector. Repair as necessary if the resistance is greater than 1 ohm.

NOTE: J2-13 and J2-11 are at ECU harness conductor.

Removal and Installation

1. Disconnect the wire harness connector from the MAT sensor.
2. Remove the MAT sensor from the intake manifold.
3. Clean the threads in the manifold.
4. Install the replacement MAT sensor in the intake manifold.
5. Connect the wire harness connector to the MAT sensor.

COOLANT TEMPERATURE SENSOR (CTS)

The coolant temperature sensor is installed in the engine water jacket. This sensor provides an analog voltage for the ECU. The ECU will enrich the fuel mixture delivered by the injector as long as the engine is cold.

Testing

1. Disconnect the harness from the coolant temperature sensor.
2. Disconnect the wire harness connector.
3. Test the resistance of the sensor with a high input impedance (digital) ohmmeter. The resistance ranges from 300 ohms to more than 300K ohms (10K ohms at room temperature).
4. Replace the sensor if not within the range of resistance specified above.
5. Test the resistance of the wire harness between J2-14 and the sensor connector, and J2-11 to the sensor connector. Find and repair open wire in harness, if indicated.

NOTE: J2-14 and J2-11 are at the ECU connector.

Removal and Installation

1. Remove the wire harness connector from the coolant temperature sensor (CTS).
2. Remove the CTS from the cylinder head and rapidly plug the hole to prevent loss of coolant. Installation is the reverse of removal.

CAUTION

DO NOT remove the CTS with the cooling system hot and under pressure. Serious burns from coolant can result.

MANIFOLD ABSOLUTE PRESSURE (MAP) SENSOR

The MAP sensor reacts to absolute pressure in the intake manifold and provides an analog voltage for the ECU. Manifold pressure is used to supply mixture density information and ambient barometric pressure information to the ECU. The sensor is remote mounted and a tube from the throttle body provides its input pressure.

Testing

1. Inspect the MAP sensor vacuum hose connections at the throttle body and sensor. Repair as necessary.
2. Test the MAP sensor output voltage at the MAP sensor connector pin B (as marked on the sensor body) with the ignition switch ON and the engine OFF. Output voltage should be 4–5 volts.

NOTE: The voltage should drop 0.5–1.5 volts with a hot neutral idle speed condition.

3. Test ECU pin J2-12 for the same voltage described above to verify the wire harness condition. Repair as necessary.
4. Test the MAP sensor supply voltage at the sensor connector pin C with the ignition ON. The voltage should be 5 volts (± 0.5V). The voltage should also be at pin J2-2 of the ECU wire harness connector. Repair or replace the wire harness as necessary.
5. Test the MAP sensor ground at sensor connector pin A and ECU connector pin J2-13. Repair the wire harness if necessary.
6. Test the sensor ground at the ECU connector between pin J2-13 and J1-F with an ohmmeter. If the ohmmeter indicates an open circuit, inspect for a good sensor ground on the flywheel housing near the starter motor. If the ground is good, replace the ECU. If J2-13 is shorted to 12 volts, correct this condition before replacing the ECU.

Removal and Installation

1. Disconnect the electrical connector.
2. Disconnect the vacuum hose.
3. Remove the retaining nuts.
4. Remove the MAP sensor from the cowl panel.
5. Installation is the reverse of removal.

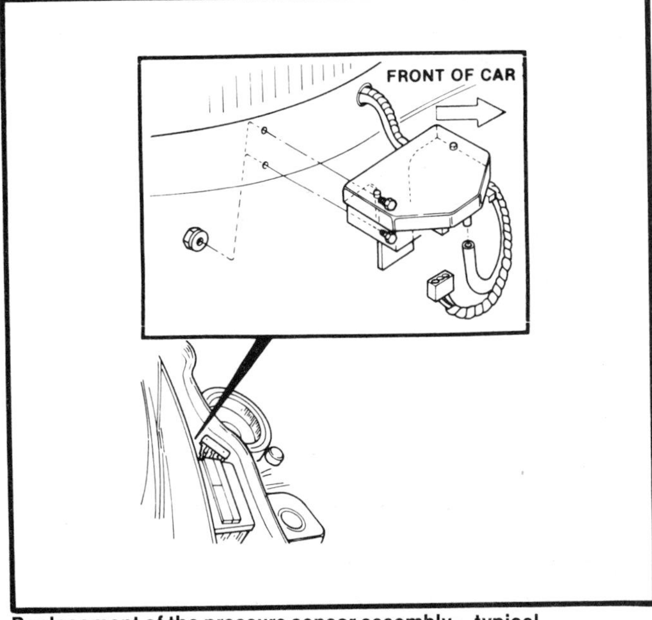

Replacement of the pressure sensor assembly—typical

WIDE OPEN THROTTLE (WOT) SWITCH

The WOT switch is mounted on the side of the throttle body. This switch provides a digital voltage for the ECU, which in turn enriches the fuel mixture delivered by the injector that is necessary for the increased air flow.

Testing

1. Disconnect the wire harness from the WOT switch.
2. Test the operation of the WOT switch with a high input impedance ohmmeter. Open and close the switch manually.
3. Resistance should be infinite when the throttle is closed. A low resistance should be indicated at the WOT position. Test the switch operation several times.
4. Replace the WOT switch if defective. Connect the wire harness connector.
5. With the ignition switch ON, test the WOT switch voltage at the diagnostic connector (D2-6 to ground, D2-7). Voltage should be zero (0) at the WOT position; greater than 2 volts if not at WOT position.
6. If the voltage is always zero, test for a short circuit to ground in the wire harness or switch. Check for an open circuit between J2-19 (ECU) and the switch connector. Repair or replace the wire harness as necessary.
7. If the voltage is always greater than 2 volts, test for an open wire or connector between the switch and ground. Repair as necessary.

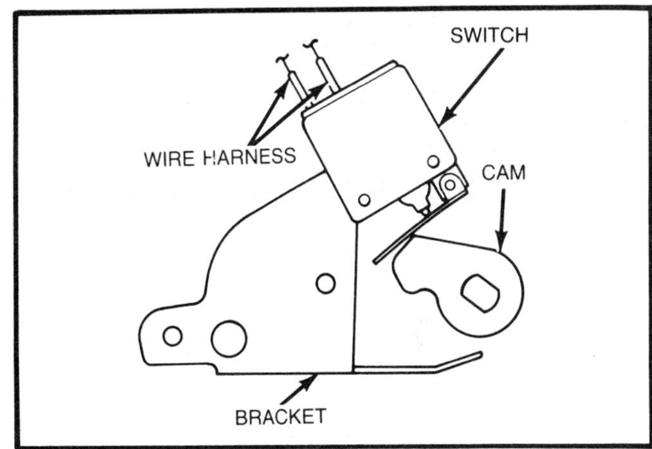

Wide open throttle (WOT) switch mounted on throttle body

Adjustment

NOTE: Adjustment of the WOT switch is necessary only to establish the initial position of the switch after it has been replaced.

1. With the throttle body assembly removed from the engine, loosen the two retaining screws that attach the WOT switch to the bracket.
2. Open the throttle to wide open position.
3. Attach the alignment gauge J-26701 or equivalent on the flat surface of the throttle lever.
4. Rotate the degree scale of the gauge until the 15 degree mark is aligned with the pointer.

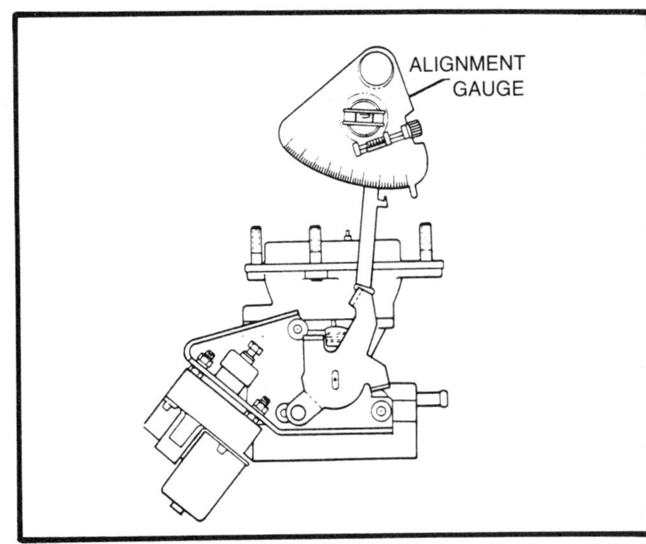

Wide open throttle switch adjustment

5. Adjust the bubble until it is centered.
6. Rotate the degree scale to align zero degrees with the pointer.
7. Close the throttle slightly to center the bubble on the gauge. The throttle is now at 15 degrees before WOT position.
8. Position the WOT switch lever on the throttle cam so that the switch plunger is just closed at 15 degrees before WOT position.
9. Tighten the WOT switch retaining screws.
10. Remove the alignment bubble gauge.

Removal and Installation

1. Remove the air filter.
2. Disconnect the throttle return spring.
3. Disconnect the throttle cable.
4. Remove the ISC motor bracket-to-throttle body screws.
5. Disconnect the wire connectors from the WOT switch and ISC motor.
6. Remove the bracket, ISC motor and WOT switch assembly from the throttle body.
7. Remove the ISC motor.
8. Remove the two WOT switch-to-bracket screws.
9. Remove the WOT switch.
10. Installation is the reverse of removal.

IDLE SPEED CONTROL (ISC) MOTOR

Engine idle speed and engine deceleration throttle stop angle are controlled by an electric motor driven actuator that changes the throttle stop angle by being a movable idle stop. The ECU controls the ISC motor actuator by providing the appropriate voltage outputs to produce the idle speed or throttle stop angle required for the particular engine operating condition. The electronic components that control the ISC motor are integral with the ECU.

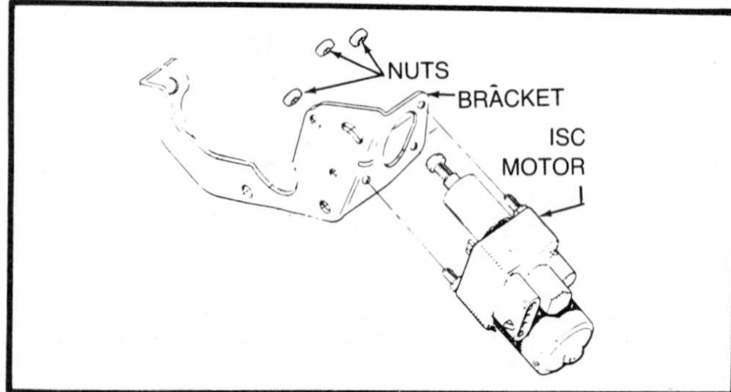

Idle speed control (ISC) motor replacement

The system controls engine idle speed or throttle stop angle as would a dashpot for deceleration, as would a fast idle device for cold engine operation, and as would a normal idle speed device for warm engine operation. The inputs from the air conditioner compressor (ON/OFF), transaxle (PARK, NEUTRAL), and throttle extremities (OPEN/CLOSED) are used to increase or decrease the throttle stop angle in response to particular engine operating conditions.

For engine starting, the throttle is either held open for a longer period (COLD) or a short time (HOT) to provide adequate engine warm-up prior to normal engine operation. With normal engine idle speed operation, the idle speed is maintained at a programmed rpm and varies slightly according to engine operating conditions. Additionally, with certain engine deceleration conditions, the throttle is held slightly open.

Adjustment

NOTE: Adjustment is necessary only to establish the initial position of the plunger after the ISC motor has been replaced.

1. Remove the air cleaner.
2. Start the engine and allow it to reach normal operating temperature. Make sure that the air conditioning is turned off.
3. Connect a tachometer to terminal D1-1 and D1-3 of the diagnostic connector.
4. Turn the ignition off. The ISC motor plunger should move to the fully extended position.
5. Disconnect the ISC motor wire and start the engine. The engine idle should be 3300–3700 rpm. If not, turn the hex screw on the end of the plunger to achieve a 3500 rpm reading.
6. Fully retract the plunger by holding the closed throttle switch plunger in while the throttle is open. If the closed throttle switch plunger touches the throttle lever when the throttle is closed, check the throttle linkage for binding or damage and correct as required.
7. Connect the ISC motor wire.
8. Turn the ignition off for 10 seconds. The ISC plunger should fully extend.
9. Start the engine. The idle should be 3500 rpm for a short period, then gradually reduce to the specified idle speed.
10. Turn off the ignition and disconnect the tachometer.

NOTE: Holding the closed throttle switch plunger as described above may activate an intermittant trouble code in the ECU memory. To erase the code, disconnect the battery for at least 10 seconds.

Closed Throttle Switch Test

NOTE: It is important that all testing be done with the idle speed control motor plunger in the fully extended position, as it would be after a normal engine shut down. If it is necessary to extend the motor plunger to test the switch an ISC motor failure can be suspected. Refer to ISC motor test if necessary.

1. With the ignition switch ON, test the switch voltage at the diagnostic connector (D2-13 and D2-7, ground). Voltage should be close to zero at closed throttle and greater than 2 volts off the closed throttle position.
2. If the voltage is always zero, test for a short circuit to ground in the wire harness or switch. Test for an open circuit between J2-20 (ECU connector) and the switch.
3. If voltage is always more than 2 volts, test for an open circuit in the wire harness between the ECU and the switch connector, and between the switch connector, and ground. Repair or replace the wire harness as necessary.

Removal and Installation

NOTE: The closed throttle (idle) switch is integral with the motor.

1. Disconnect the throttle return spring.
2. Disconnect the wire connector from the motor.
3. Remove the motor-to-bracket retaining nuts.
4. Remove the motor from its bracket.
5. Installation is the reverse of removal.

OXYGEN (O_2) SENSOR

The oxygen sensor is located in the exhaust pipe adaptor. The analog voltage output from this sensor, which varies with the oxygen content of the exhaust gas, is supplied to the ECU. The ECU utilizes it as a reference voltage.

Testing

1. Test the continuity of the harness between the O_2 sensor connector and J2-9 on the ECU wire harness connector with an ohmmeter. Ensure that the wire harness is not shorted to ground. Repair or replace the wire harness as necessary.
2. Test the continuity between the sensor ground (exhaust manifold) and Pin J2-13 on the ECU wire harness connector. Repair the wire harness if necessary.
3. Test the fuel system for the correct pressure. Pressure should be approximately 1 bar (14–15 psi) with the engine at idle speed.

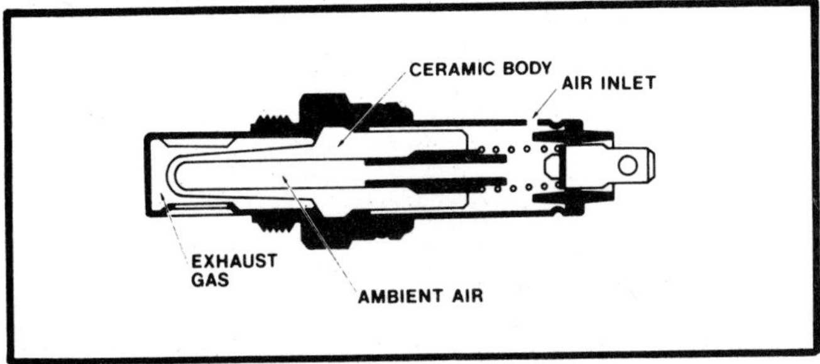

Schematic of typical oxygen sensor

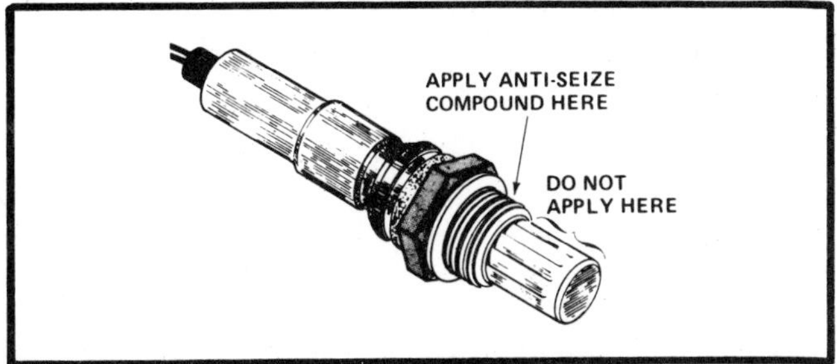

Use anti-seize compound when installing the oxygen sensor

Refer to the fuel system pressure test. Repair the fuel system if necessary.

4. Check the sensor operation by driving the vehicle with a test lamp (No. 158 bulb) connected between the diagnostic connector D2-2 and D2-4.

5. Bulb lighted at start is normal operation for test circuit. If the bulb does not light after warm up, the O_2 sensor is functioning normally. If the bulb stays lit or lights after the engine warms up, replace the O_2 sensor.

NOTE: Additional testing may be required to locate the cause of an oxygen sensor failure. Other system failures that could cause an O_2 failure include EGR solenoid control, canister purge control, PCV system, secondary ignition circuit and fuel delivery system.

Removal and Installation

1. Disconnect the wire connector from the O_2 sensor.
2. Remove the O_2 sensor from the exhaust pipe adaptor.
3. Clean the threads in the adaptor.
4. Apply anti-seize sealer to the threads on the O_2 sensor.

CAUTION

Apply anti-seize sealer only to the threads and not to any other part of the sensor.

5. Hand thread the sensor into the exhaust pipe adaptor.
6. Tighten the sensor with 20–25 ft. lbs. (27–34 Nm) torque.
7. Connect the wire connector.

CAUTION

Make sure that the wire terminal ends are properly seated in the connector prior to joining the connectors. Do not push the rubber boot down on the sensor body beyond 13 mm (0.5 in.) above the base. Also, the Oxygen Sensor pigtail wires cannot be spliced or soldered. If broken, replace the sensor.

ELECTRONIC CONTROL UNIT (ECU)

The Electronic Control Unit (ECU) is located below the glove box adjacent to the fuse panel. The ECU controls the injector fuel delivery time and changes the injected flow according to inputs received from sensors that react to exhaust gas oxygen, air temperature, coolant temperature, manifold absolute pressure, and crankshaft and throttle positions. The ECU is powered by the vehicle battery and, when the ignition is turned to the ON or START position, the voltage inputs are received from the sensors and switches. The desired air/fuel mixtures for various driving and atmospheric conditions are programmed into the ECU. As inputs are received from the sensors and switches, the ECU processes the inputs and computes the engine fuel requirements. The ECU energizes the injector for a specific time duration. The duration of pulse varies as engine operating conditions change.

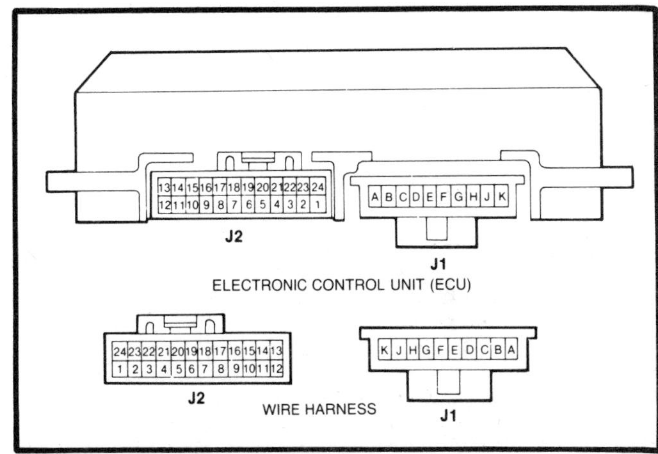

Location of test pins on the ECU and wiring harness on Renault models

Testing

The only accurate and safe way to test the ECU is by using the self-diagnosis feature described earlier.

Removal and Installation

1. Remove the retaining screws and bracket that support the ECU below the glove box.
2. Remove the ECU.
3. Disconnect the wire harness connectors from the ECU.
4. Installation is the reverse of removal.

NOTE: It should be understood that the ECU is extremely reliable and must be the final component to be replaced if a doubt exists concerning the cause of an injection system failure.

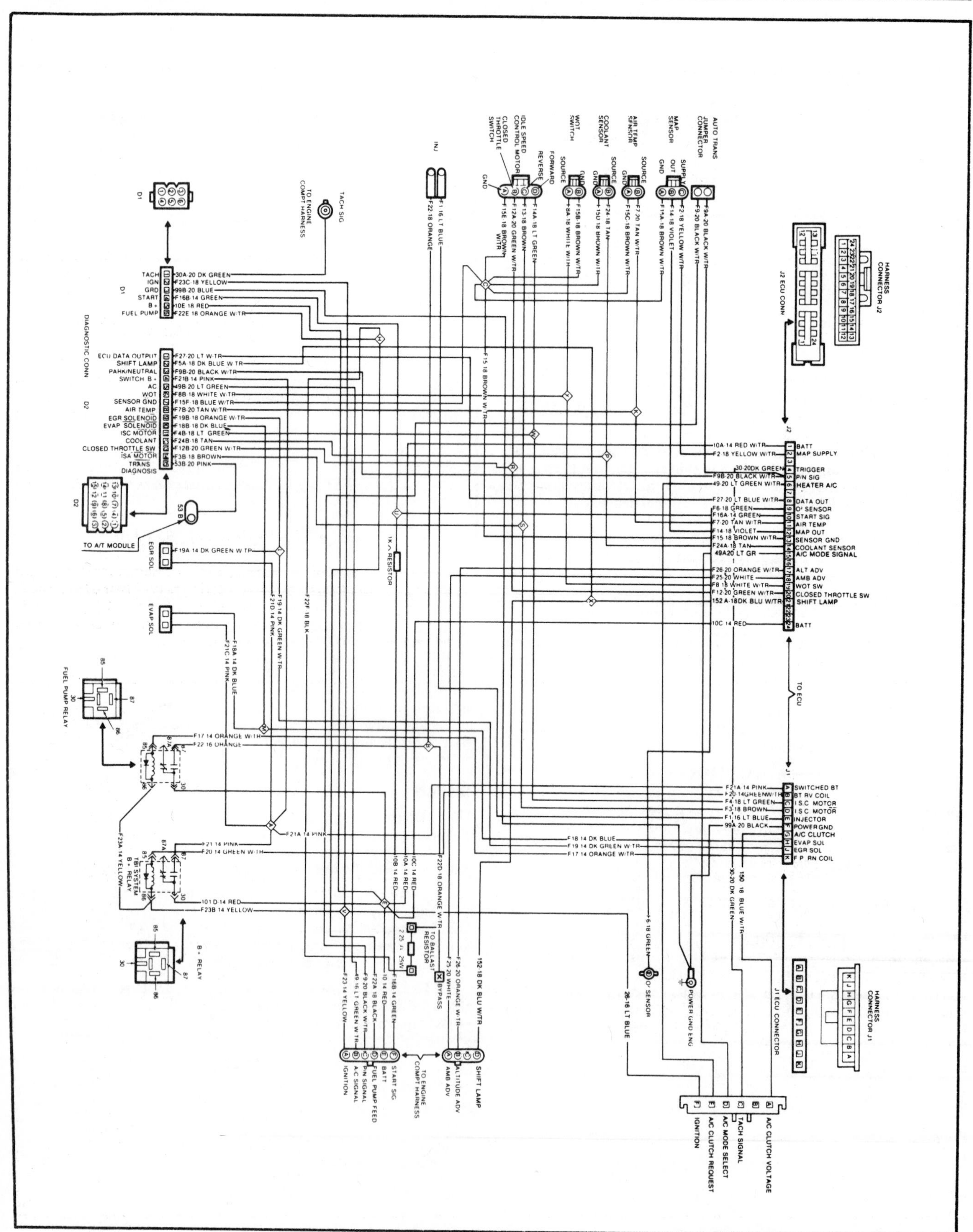

GM TURBOCHARGERS

General Description

A turbocharger is used to increase power on a demand basis, thus allowing a smaller, more economical engine to perform the job of a larger engine. As load on the engine is increased and the throttle is opened, more air-fuel mixture flows into the combustion chambers. As this increased flow is burned, a larger volume of higher energy exhaust gas enters the engine exhaust system and is directed through the turbocharger turbine housing. Some of this energy is used to increase the speed of the turbine wheel. The turbine wheel is connected by a shaft to the compressor wheel. The increased speed of the compressor wheel allows it to compress the air-fuel mixture it receives from the carburetor and delivers it to the intake manifold. The resulting higher pressure in the intake manifold allows a denser charge to enter the combustion chambers. The denser charge can develop more power during the combustion cycle.

The intake manifold pressure (boost) is controlled to a correct maximum value by an exhaust bypass valve (wastegate). The valve allows a portion of the exhaust gas to bypass the turbine wheel, thus not increasing turbine speed. The wastegate is operated by a spring loaded diaphragm device (actuator assembly) that senses the pressure differential across the compressor. When boost reaches a set value about ambient pressure, the wastegate beings to bypass exhaust gas.

NOTE: Any alteration to the air intake or exhaust system which upsets the air flow balance may result in serious damage to the turbocharged engine.

Operation

ELECTRONIC SPARK CONTROL

Turbocharged engines use a modified HEI system which is called Electronic Spark Control (ESC). The ESC system is used to control engine detonation by automatically retarding ignition timing during periods of engine operation when detonation occurs.

The intake manifold transmits the vibrations caused by detonation to the sensor mounting location. The sensor detects the presence and intensity of detonation and feeds this information to the controller. The controller, which is mounted on the fan shroud, evaluates the sensor signal and sends a command signal to the distributor to adjust timing.

The HEI distributor has a modified electronic module which responds to signals from the controller. Electronic spark control is continually monitoring engine operation for detonation and retarding ignition timing up to 18 to 20° (V-6) or 13 to 17° (V-8) to minimize detonation levels, as necessary.

ENRICHMENT VACUUM REGULATOR

The Power Enrichment Vacuum Regulator (PEVR) is designed to control vacuum flow to the carburetor power piston on turbocharged engines. The PEVR regulates vacuum to the remote power enrichment port on the carburetor based on the manifold vacuum/pressure signal.

The PEVR has an input port and an output port. The vacuum input port is in the center of the PEVR. The vacuum output port is located on the perimeter of the PEVR. The manifold signal port extends into the intake manifold.

TURBOCHARGER OIL SUPPLY

An adequate supply of clean engine oil is essential to the proper operation of the turbocharger. The rotating assembly (turbine wheel, connecting shaft and compressor wheel) can attain speeds of 130,000 to 140,000 rpm during boost. Interruption or contamination of the oil supply to the support bearings in the center housing rotating assembly

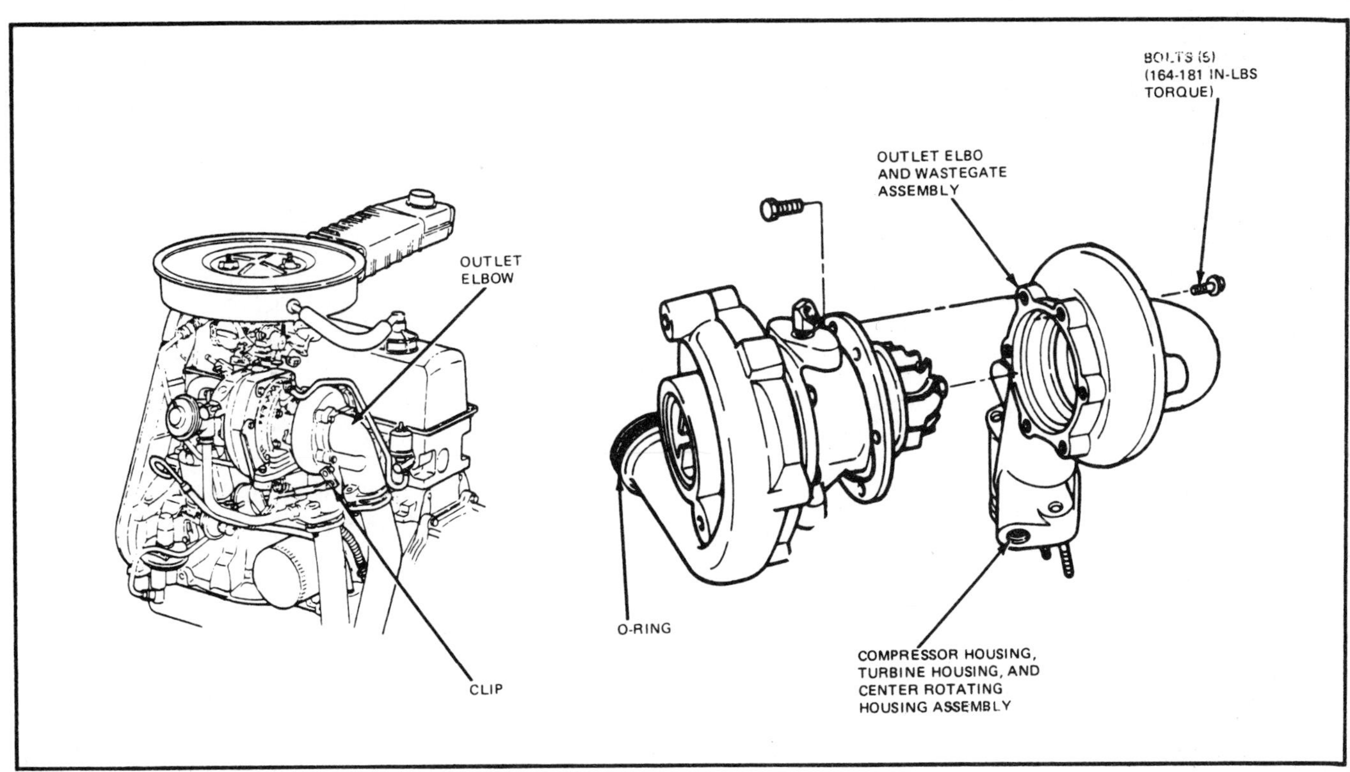

Turbine housing removal

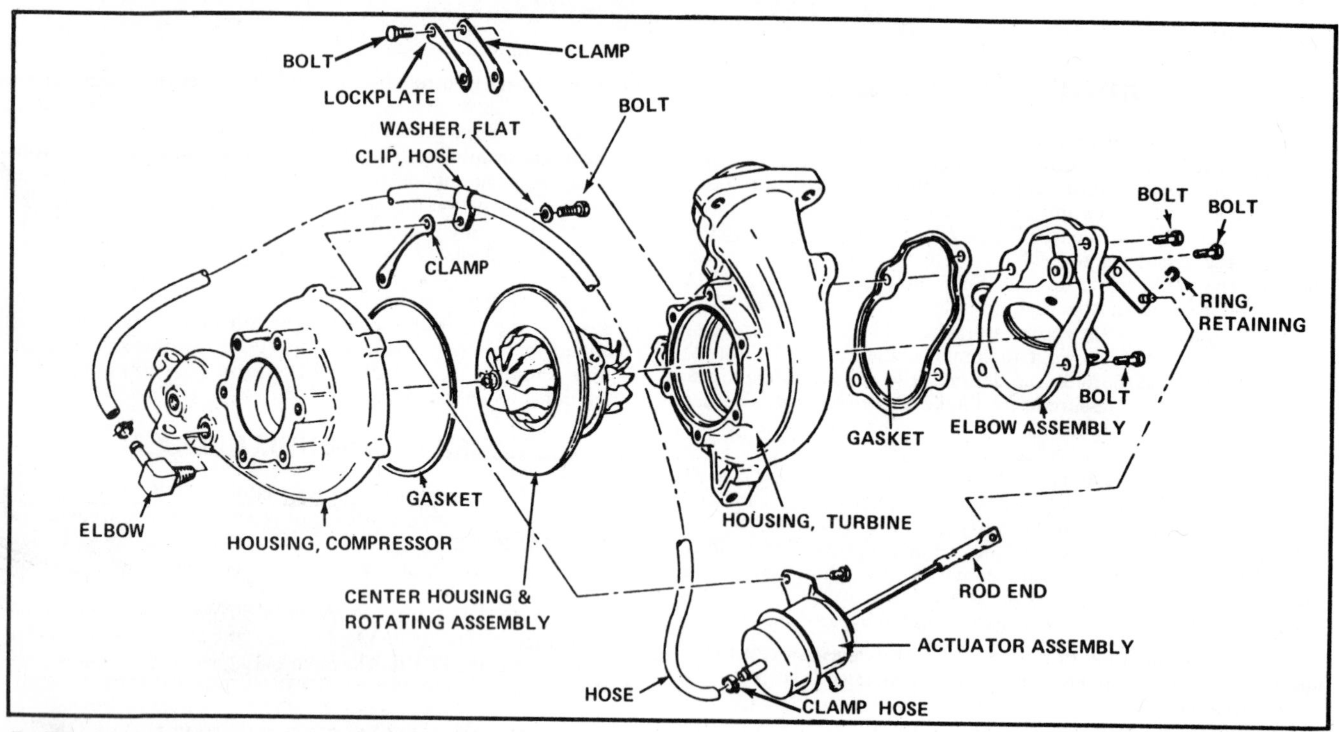

Turbocharger components

(CHRA) can result in major turbocharger damage.

When changing the oil and oil filter on a turbocharged engine or performing any operation which results in oil drainage or loss, use the following procedures before starting the engine:

1. Disconnect ignition switch connector (pink wire) from the HEI distributor.
2. Crank engine several times (not to exceed 30 seconds for each cranking interval) until oil light goes out.
3. Reconnect pink wire to the distributor. This procedure will aid the filling of the oil system.

NOTE: Any time a basic engine bearing (main bearing, connecting rod bearing and camshaft bearing) has been damaged in a turbocharged engine, the oil and oil filter should be changed as a part of the repair procedure. In addition, the turbocharger should be flushed with clean engine oil to reduce the possibility of contamination.

Any time a center housing rotating assembly, or any part of a turbocharger which includes the center housing rotating assembly, is being replaced, the oil and oil filter should be changed as a part of the repair procedure.

Test Procedures

GENERAL PRECAUTIONS

Before starting any turbocharger unit repair procedure, several general precautions should be considered.

1. Clean area around turbocharger with non-caustic solution before removal of assembly.
2. When removing turbocharger assembly, take special care not to bend, nick or in any way damage compressor or turbine wheel blades. Any damage may result in rotating assembly imbalance, failure of center housing rotating assembly (CHRA) and failure of compressor and/or turbine housings.
3. Before disconnecting center housing rotating assembly from either compressor housing or turbine housing, scribe the components in order that they may be reassembled in the same relative position.
4. If silastic sealer, or equivalent, is found at any point in turbocharger disassembly (such as between center housing rotating assembly backplate and compressor housing), the area should be cleaned and sealed with an equivalent sealer during reassembly.

WASTEGATE/BOOST PRESSURE

1. Visually inspect wastegate-actuator mechanical linkage for damage.
2. Check hose from compressor housing to actuator assembly and return tubing from actuator to PCV tee.
3. Attach hand operated vacuum/pressure pump J-23738 in series with compound gauge J-28474 to actuator assembly. Replace compressor housing to actuator assembly hose.
4. Apply pressure to actuator assembly. At approximately 9 psi (8.5 to 9.5 psi), the actuator rod end should move .015 inch, actuating the wastegate linkage. If not, replace the actuator assembly and check that opening calibration pressure is 9 psi. Crimp threads on actuator rod to maintain correct calibration.
5. Remove test equipment and reconnect compressor housing to actuator assembly hose.
6. An alternative method of checking wastegate operation is to perform a road test which measures boost pressure.

POWER ENRICHMENT VACUUM REGULATOR (PEVR)

1. Visually check the PEVR and attaching hoses for deterioration, cracking or other damage.
2. Tee one hose from manometer J-23951 between the yellow-striped input hose and the input port. Connect the other manometer hose directly to the output port of the PEVR.
3. Start the engine and let it idle. There should be no more than a 1.0 in. Hg difference. If there is, replace the PEVR.
4. If the PEVR passes the preceding test and is still considered to be a possible problem source, remove the PEVR from the intake manifold.

5. Plug the intake manifold and connect the input and output hoses to the PEVR.

6. Tee compound gauge J-28474 into the output hose of the PEVR.

7. Start the engine and let it idle. The compound gauge reading from the output port should be 7.0 to 9.0 in. Hg. (V-6), or 8.0 to 10.0 in. Hg. (V-8).

8. Apply 3 psi to the manifold signal port of the PEVR. The vacuum reading from the output port should be 1.4 to 2.6 in. Hg. If there is difficulty in measuring this low level of vacuum output, an additional requirement can be used. Apply a minimum of 5 psi to the manifold signal port of the PEVR. There should be no vacuum output from the PEVR.

9. If the PEVR does not meet requirements 7 and 8, replace it.

TURBOCHARGER INTERNAL INSPECTION

1. Remove turbocharger exhaust outlet pipe from the elbow assembly. Using a mirror, observe movement of wastegate while manually operating actuator linkage. Replace elbow assembly if wastegate fails to open or close. Inspect wastegate poppet valve for deterioration and warpage. Replace elbow assembly if poppet valve is damaged.

2. Remove turbocharger assembly from engine.

3. Check for loose backplate to CHRA bolts and missing gasket or O-ring. Tighten or replace as necessary.

4. Gently spin compressor wheel. If rotating assembly binds, replace CHRA.

5. Remove oil drain from CHRA. Check CHRA for sludging in oil drain area. Clean, if minor. Replace CHRA if severely sludged or coked.

6. Inspect compressor wheel area for oil leakage from CHRA. If leakage is present, replace CHRA.

7. If compressor wheel is damaged or severely cocked, replace CHRA.

8. If CHRA is being replaced, pre-lubricate with clean engine oil and proceed to step 9. If CHRA is not being replaced, proceed to step 10.

9. Inspect compressor housing (still attached to engine) and turbine housing. Replace either housing if gouged, nicked or distorted.

10. If CHRA is not being replaced, remove turbine housing from CHRA.

11. Check the journal bearings for radial clearance as follows.

a. Attach a dial indicator with a two inch long, 3/4 to 1 inch offset extension rod to the center housing such that the indicator

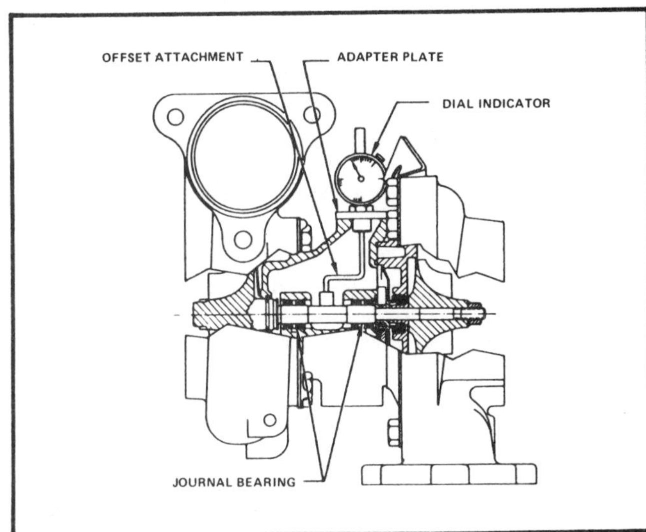

Journal bearing clearance measurement

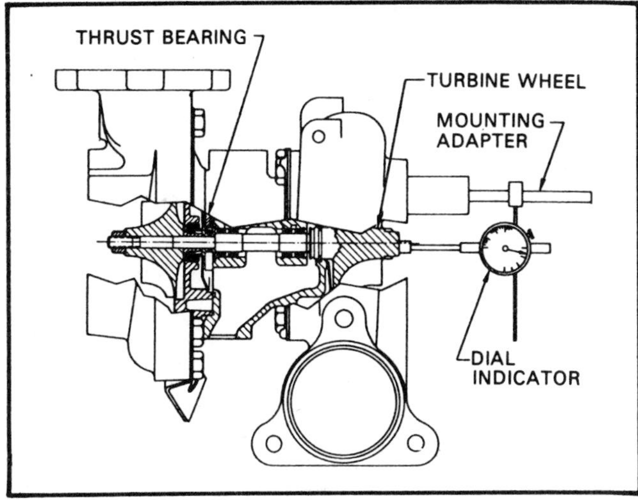

Thrust bearing clearance measurement

plunger extends through the oil outlet port and contacts the shaft of the rotating assembly.

b. Manually apply pressure equally and at the same time to both the compressor and turbine wheels as required to move the shaft away from the dial indicator plunger as far as it will go.

c. Set the dial indicator to zero.

d. Manually apply pressure equally and at the same time to both the compressor and turbine wheels to move the shaft toward the dial indicator plunger as far as it will go. Move the maximum value on the indicator dial.

NOTE: Make sure that the dial indicator reading noted is the maximum reading obtainable, which can be verified by rolling the wheels slightly in both directions while applying pressure.

e. Manually apply pressure equally and at the same time to the compressor and turbine wheels as required to move the shaft away from the dial indicator plunger as far as it will go. Note that the indicator pointer returns exactly to zero.

f. Repeat steps a. through f. as required to make sure that the maximum clearance between the center housing bores and the shaft bearing diameters, as indicated by the maximum shaft travel, has been obtained.

g. If the maximum bearing radial clearance is less than 0.003 inch or greater than 0.006 inch, replace CHRA and inspect housings as indicated in step 9.

NOTE: Continued operation of a turbocharger having improper bearing radial clearance will result in severe damage to the compressor wheel and housing or to the turbine wheel and housing.

12. Check for thrust bearing axial clearance as follows.

a. Mount a dial indicator at the turbine end of the turbocharger such that the dial indicator tip rests on the end of the turbine wheel.

b. Manually move the compressor wheel and turbine wheel assembly alternately toward and away from the dial indicator plunger. Note the travel of the shaft in each direction, as shown on the dial indicator.

c. Repeat step b. as required to make sure that the maximum clearance between the thrust bearing components has been obtained.

d. If the maximum thrust bearing axial clearance is less than 0.001 inch or greater than 0.003 inch, replace CHRA and inspect housings as indicated in step 9.

NOTE: Continued operation of a turbocharger having an improper amount of thrust bearing axial clearance will result in severe damage to the compressor wheel and housing or to the turbine wheel and housing.

13. Install oil drain adapter and tube on CHRA.

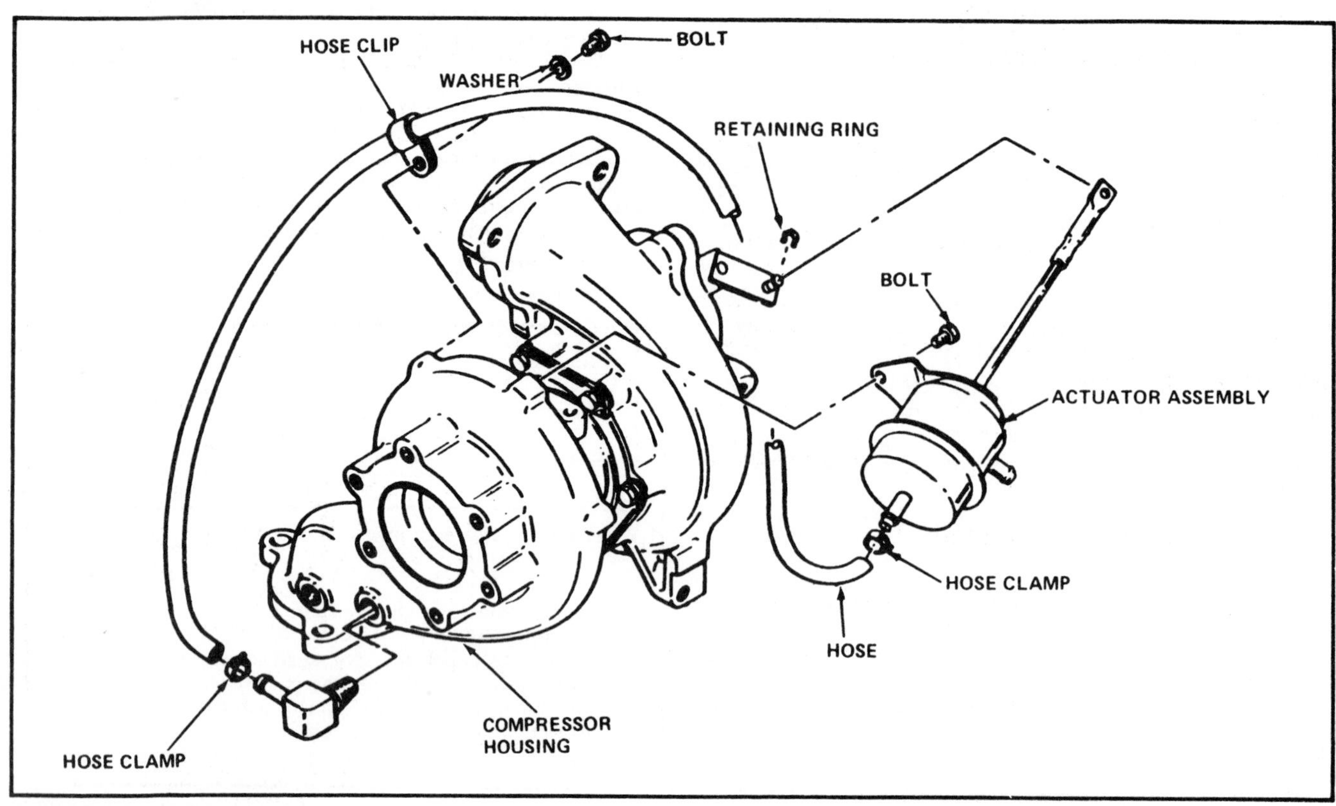

Wastegate actuator attachment

14. Install turbocharger assembly to engine.

NOTE: Before connecting turbocharger exhaust outlet pipe to elbow assembly, gently spin the turbine wheel to be certain that the rotating assembly (turbine wheel, connecting shaft and compressor wheel) does not bind.

ROAD TEST

V-6

1. Tee compound gauge J-28474 into tubing between compressor housing and boost gauge switches with sufficient length of hose to place gauge in passenger compartment.

CAUTION

Determine that hose and compound gauge are in proper operating condition to avoid possible leakage of air-fuel mixture into passenger compartment during road test.

2. Conditions and speed limits permitting, perform a zero to 40 to 50 mph wide open throttle acceleration. Boost pressure, as measured by the compound gauge during road testing, should reach 9-10 psi. If not, replace actuator assembly and check for proper calibration. Actuator rod end should move .015 in. at approximately 9 psi.

V-8

1. Remove 1/4 in. pipe plug or vacuum switch located in the power enrichment adapter. Install a straight vacuum fitting and compound gauge tubing J-28474 into the power enrichment adapter, with sufficient length of hose to place gauge in passenger compartment.

CAUTION

Determine that the hose and compound gauge are in proper operating condition to avoid possible leakage of air-fuel mixture into the passenger compartment during road test.

2. Conditions and speed limits permitting, perform a zero to 40 or 50 mph wide open throttle acceleration. Boost pressure, as measured by the compound gauge during road testing, should reach 9-10 psi. If not, replace actuator assembly and check for proper calibration. Actuator rod end should move .015 in. at approximately 9 psi, as detailed in the wastegate test procedure.

Component Service

WASTEGATE ACTUATOR

V-6

1. Disconnect hoses. Remove the clip attaching wastegate linkage to actuator rod and remove mounting bolts.
2. Installation is the reverse of removal.

V-8

1. Disconnect hose clamp and two hoses from actuator assembly.
2. Remove retainer attaching wastegate linkage to actuator rod.
3. Remove two bolts attaching actuator assembly to compressor housing.
4. Installation is the reverse of removal.

ELECTRONIC SPARK CONTROL (ESC) DETONATION SENSOR

V-6 and V-8

1. Squeeze sides of metal connector crosswise to wire from controller and gently pull straight up to remove connector. Do not pull up on the wire.
2. Remove detonation sensor.
3. When installing, torque detonation sensor to 19 N•m (14 ft. lb.).

NOTE: Do not over-torque. Proper sensor torque is critical to the sensor performance. Do not use impact tool. Do not apply a side load to the detonation sensor. Do not attempt to repair tapped hole in intake manifold for detonation sensor.

4. Squeeze sides of metal connector crosswise to wire from controller and start straight down over detonation sensor terminal. Release sides of metal connector and push down until connector snaps into place.

ELBOW ASSEMBLY

V-6

1. Loosen turbocharger exhaust outlet pipe at the catalytic converter and disconnect it from the elbow assembly.
2. Remove the clip attaching wastegate linkage to actuator rod. Remove the bolts which mount the elbow to the turbine housing.
3. Installation is the reverse of removal.

V-8

NOTE: Whenever the elbow assembly is replaced, the wastegate actuator assembly must also be replaced with the adjustable type. Adjust as described in wastegate test procedure.

1. Detach turbocharger exhaust outlet pipe and catalytic converter at the intermediate pipe.
2. Disconnect turbocharger exhaust outlet pipe from elbow assembly.
3. Disconnect turbocharger inlet pipe at elbow assembly. Loosen at exhaust manifold and swing out of the way.
4. Remove retainer attaching wastegate linkage to actuator rod.
5. Remove elbow assembly support bracket bolts at elbow. Loosen bracket bolts at intake manifold and swing out of the way.
6. Remove six bolts attaching elbow assembly to turbine housing.
7. Installation is the reverse of removal.

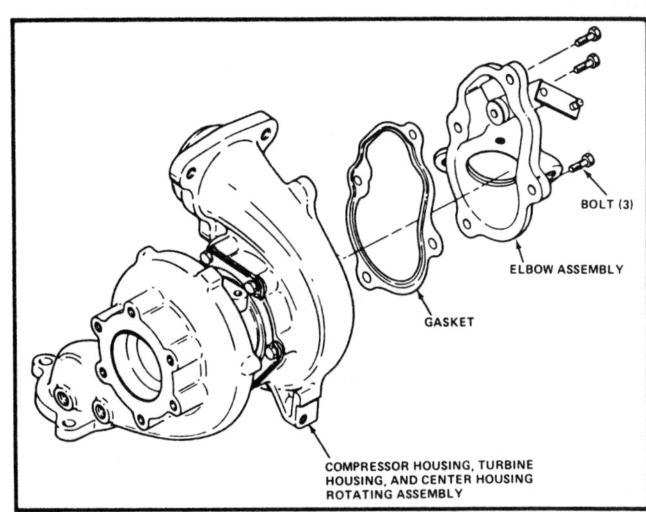

Elbow assembly removal

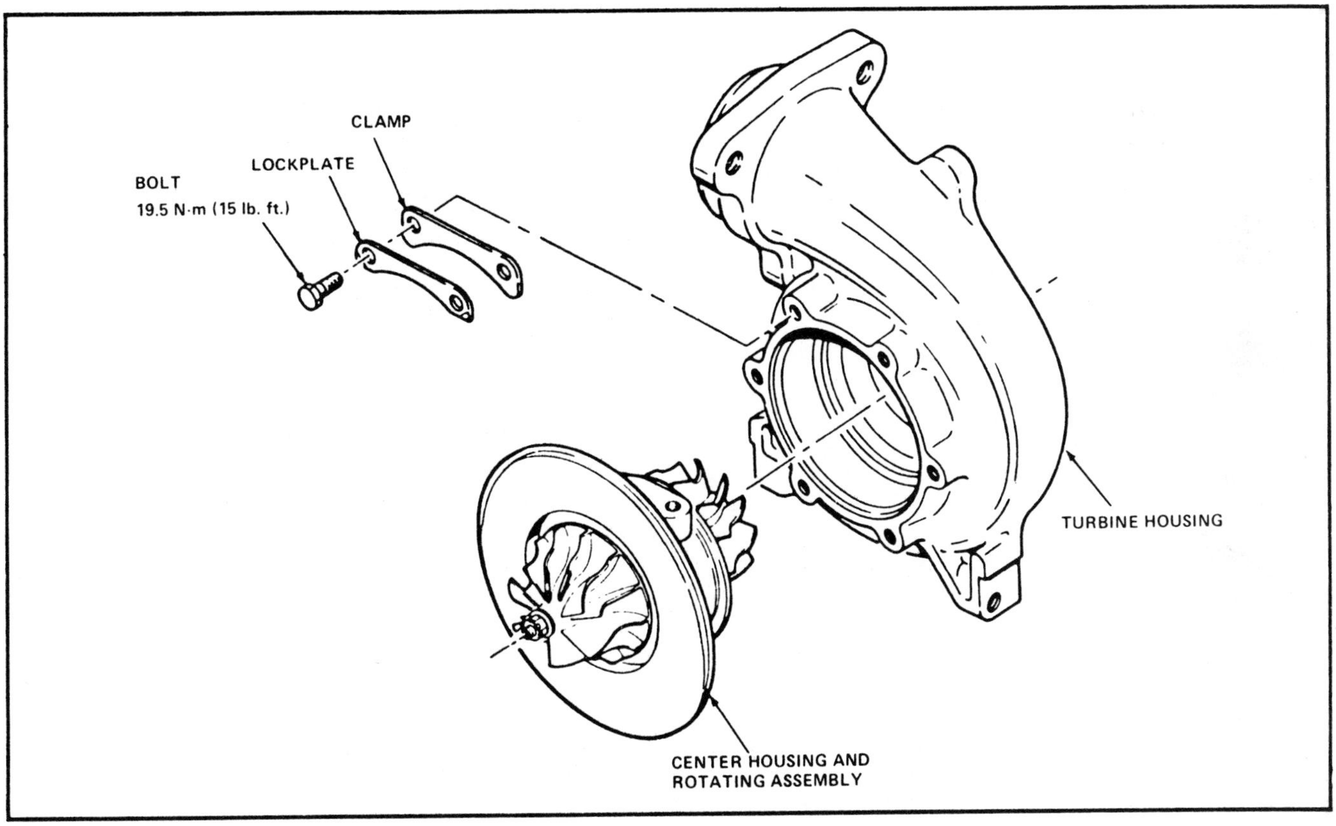

CLAMP
LOCKPLATE
BOLT
19.5 N·m (15 lb. ft.)
TURBINE HOUSING
CENTER HOUSING AND ROTATING ASSEMBLY

Center housing rotating assembly

TURBINE HOUSING AND ELBOW ASSEMBLY/CENTER HOUSING ROTATING ASSEMBLY

V-6

1. Disconnect turbocharger exhaust outlet pipe from the catalytic converter and the elbow assemply.
2. Disconnect turbocharger exhaust inlet pipe from the turbine housing and the exhaust manifold.
3. Remove bolts attaching turbine housing to bracket on intake manifold.
4. Disconnect oil feed pipe from center housing rotating assembly and remove oil drain hose from oil drain pipe.
5. Remove clip attaching wastegate linkage to actuator rod.
6. Remove bolts and clamps attaching CHRA backplate to compressor housing.
7. Remove bolts, clamps, etc. attaching turbine housing to CHRA.
8. Installation is the reverse of removal.

CENTER HOUSING ROTATING ASSEMBLY

V-8

1. Remove elbow assembly.
2. Remove oil feed and return lines from housing.
3. Remove six bolts and three lockplates attaching turbine housing to CHRA.
4. Installation is the reverse of removal.

COMPRESSOR HOUSING

V-6

1. Disconnect turbocharger exhaust outlet pipe from the catalytic converter and the elbow assembly.
2. Disconnect turbocharger exhaust inlet pipe from the turbine housing and the exhaust manifold.
3. Remove bolts attaching turbine housing to bracket on intake manifold.
4. Disconnect oil feed pipe from center housing rotating assembly and remove oil drain hose from oil drain pipe.
5. Remove clip attaching wastegate linkage to actuator rod.
6. Remove bolts and clamps attaching CHRA backplate to compressor housing.
7. Remove bolts attaching compressor housing to plenum.
8. Disconnect boost gauge hose from housing connector.
9. Remove bolts attaching compressor housing to intake manifold.
10. Installation is the reverse of removal.

V-8

1. Remove turbo elbow assembly and CHRA.
2. Remove air cleaner.
3. Remove EGR valve and heat shield.
4. Remove six bolts attaching compressor housing to plenum.
5. Remove three bolts attaching compressor housing to intake manifold.
6. Installation is the reverse of removal.

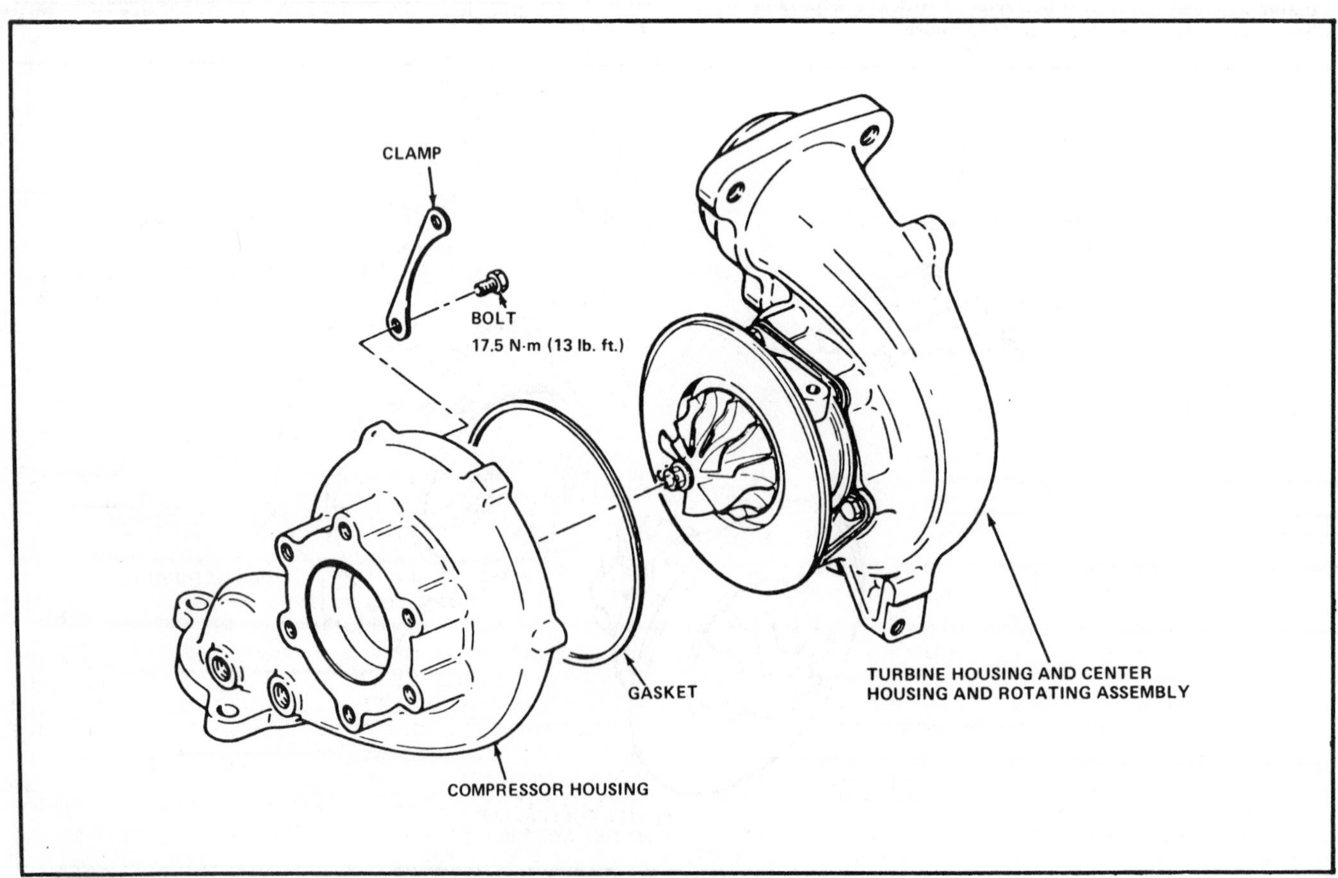

Compressor housing

Glossary

ACT Air Charge Temperature sensor.

AIR GAP the distance or space between the reluctor tooth and pick-up coil.

AFC Air Flow Controlled fuel injection.

AMMETER an electrical meter used to measure current flow (amperes) in an electrical circuit. Ammeter should be connected in series and current flowing in the circuit to be checked.

AMPERE (AMP) the unit current flow is measured in. Amperage equals the voltage divided by the resistance.

ARMATURE another name for reluctor used by Ford. See Reluctor for definition.

BALLAST RESISTOR is a resistor used in the ignition primary circuit between the ignition switch and coil to limit the current flow to the coil when the primary circuit is closed. Can also be used in the form of a resistance wire.

BID Breakerless Inductive Discharge ignition system.

BP Barometric Pressure sensor.

BYPASS system used to bypass ballast resistor during engine cranking to increase voltage supplied to the coil.

CALIBRATION ASSEMBLY memory module that plugs into an on-board computer that contains instructions for engine operation.

CANP Canister Purge solenoid that opens the fuel vapor canister line to the intake manifold when energized.

CAPACITOR a device which stores an electrical charge.

CCC Computer Command Control system used on GM models.

C3I Computer Controlled Coil Ignition used on GM models.

C4 Computer Controlled Catalytic Converter system used on GM and some AMC models.

CFI Central Fuel Injection system used on Ford models.

CIS Constant Injection System or Constant Idle Speed system manufactured by Bosch.

CP Crankshaft Position sensor.

CONDUCTOR any material through which an electrical current can be transmitted easily.

CONTINUITY continuous or complete circuit. Can be checked with an ohmmeter.

DIELECTRIC SILICONE COMPOUND non-conductive silicone grease applied to spark plug wire boots, rotors and connectors to prevent arcing and moisture from entering a connector.

DIODE an electrical device that will allow current to flow in one direction only.

DURA SPARK SYSTEM Ford electronic ignition system that followed the SSI system.

EEC Ford electronic engine control system uses an electronic control assembly (ECA) micro computer to control engine timing, air/fuel ratio and emission control devices.

EFI Electronic Fuel Injection.

EGI Electronic Gasoline Injection system used on Mazda models.

EGO Exhaust Gas Oxygen sensor.

EGR Exhaust Gas Recirculation.

EIS Electronic Ignition System which uses a reluctor and a pick up coil along with a module to replace the ignition points and condenser.

ELECTRONIC CONTROL UNIT (ECU) ignition module, module, amplifier or igniter. See Module for definition.

ESA Chrysler electronic spark advance system uses a spark control computer (SCC) to control ignition timing based on sensor inputs. Also called Electronic Lean Burn (ELB) and Electronic Spark Control (ESC).

EVP EGR valve position sensor.

FBC Feedback Carburetor.

GND Ground or negative (–).

HALL EFFECT PICK-UP ASSEMBLY used to input a signal to the electronic control unit. The system operates on the Hall Effect principle whereby a magnetic field is blocked from the pick-up by a rotating shutter assembly. Used by Chrysler, Bosch and General Motors.

IAT Intake Air Temperature sensor.

IGNITER term used by the Japanese automotive and ignition manufacturers for the electronic control unit or module.

IGNITION COIL a step-up transformer consisting of a primary and a secondary winding with an iron core. As the current flow in the primary winding stops, the magnetic field collapses across the secondary winding inducing the high secondary voltage. The coil may be oil filled or an epoxy design.

INDUCTION a means of transferring electrical energy in the form of a magnetic field. Principle used in the ignition coil to increase voltage.

INFINITY an ohmmeter reading which indicates an open circuit in which no current will flow.

INJECTOR a solenoid or pressure-operated fuel delivery valve used of fuel injection systems.

INTEGRATED CIRCUIT (IC) electronic micro-circuit consisting of a semi-conductor components or elements made using thick-film or thin-film technology. Elements are located on a small chip made of a semi-conducting material, greatly reducing the size of the electronic control unit and allowing it to be incorporated within the distributor.

ISC Idle speed control device.

MAP Manifold absolute pressure sensor.

MCT Manifold charge temperature sensor.

MCU Microprocessor Control Unit used on Ford models.

MFI Multiport Fuel Injection used on GM models.

MICROPROCESSORS a miniature computer on a silicone chip.

MODULE Electronic control unit, amplifier or igniter of solid state or integrated design which controls the current flow in the ignition primary circuit based on input from the pick-up coil. When the module opens the primary circuit, the high secondary voltage is induced in the coil.

OHM the electrical unit of resistance to current flow.

OHMMETER the electrical meter used to measure the resistance in ohms. Self-powered and must be connected to an electrically open circuit or damage to the ohmmeter will result.

OXYGEN SENSOR used with the feedback system to sense the presence of oxygen in the exhaust gas and signal the computer which can reference the voltage signal to an air/fuel ratio.

PICK-UP COIL inputs signal to the electronic control unit to open the primary circuit. Consists of a fine wire coil mounted around a permanent magnet. As the reluctor's ferrous tooth passes through the magnetic field an alternating current is produced, signalling the electronic control unit. Can operate on the principle of metal detecting, magnetic induction or Hall Effect. Is also referred to as a stator or sensor.

POTENTIOMETER a variable resistor used to change a voltage signal.

PRIMARY CIRCUIT is the low voltage side of the ignition system which consists of the ignition switch, ballast resistor or resistance wire, bypass, coil, electronic control unit and pick-up coil as well as the connecting wires and harnesses.

PULSE GENERATOR also called a pulse signal generator. Term used by Japanese and German automotive and ignition manufacturers to describe the pick-up and reluctor assembly. Generates an electrical pulse which triggers the electronic control unit or igniter.

RELUCTOR also called an armature or trigger wheel. Ferrous metal piece attached to the distributor shaft. Made up of teeth of which the number are the same as the number of engine cylinders. As the reluctor teeth pass through the pick-up magnetic field an alternating current is generated in the pick-up coil. The reluctor in effect references the position of the pistons on their compression strokes to the pick-up coil.

RESISTANCE the opposition to the flow of current through a circuit or electrical device, and is measured in ohms. Resistance is equal to the voltage divided by the amperage.

SECONDARY the high voltage side of the ignition system, usually above 20,000 volts. The secondary includes the ignition coil, coil wire, distributor cap and rotor, spark plug wires and spark plugs.

SENSOR also called the pick-up coil or stator. See pick-up coil for definition.

SFI Sequential Fuel Injection system used on GM models.

SHUTTER also called the vane. Used in a Hall Effect distributor to block the magnetic field from the Hall Effect pick-up. It is attached to the rotor and is grounded to the distributor shaft.

SPARK DURATION the length of time measured in milliseconds (1/1000th second) the spark is established across the spark plug gap.

SSI Solid State Ignition used on some Ford and American Motors vehicles.

STATOR another name for a pick-up coil. See pick-up coil for definition.

SWITCHING TRANSISTOR used in some electronic ignition systems, it acts as a switch for high current in response to a low voltage signal applied to the base terminal.

TAB Thermactor air bypass solenoid.

TAD Thermactor air diverter solenoid.

THERMISTOR a device that changes its resistance with temperature.

THICK FILM INTEGRATED (TFI) used by Ford to describe their integrated ignition module electronic ignition system.

TK or TKS throttle kicker solenoid. An actuator moves the throttle linkage to increase idle rpm.

TPS Throttle Position Sensor.

TRANSDUCER a device used to change a force into an electrical signal. Used in the Chrysler electronic spark advance system as the vacuum transducer, throttle position transducer to input a voltage to the spark command computer relating the engine vacuum and throttle position.

TRANSISTOR a semi-conductor component which can be actuated by a small voltage to perform an electrical switching function.

TRIGGER WHEEL see Reluctor for definition.

VOLT the unit of electrical pressure or electromotive force.

VOLTAGE DROP the difference in voltage between one point in a circuit and another, usually across a resistance. Voltage drop is measured in parallel with current flowing in the circuit.

VOLTMETER electrical meter used to measure voltage in a circuit. Voltmeters must be connected in parallel across the load or circuit.

VREF The reference voltage or power supplied by the computer control unit to some sensors regulated at a specific voltage.